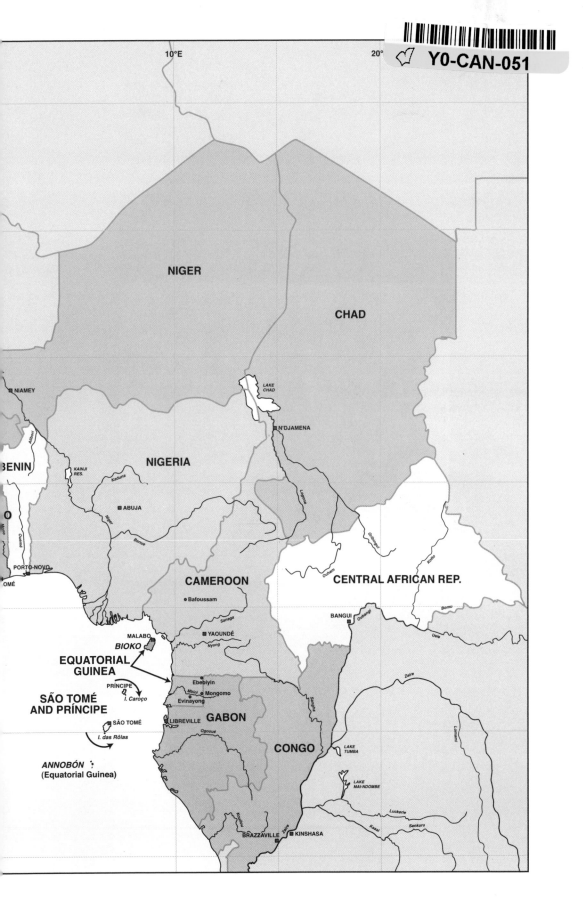

10°E
20°

NIGER

CHAD

LAKE CHAD

NIAMEY

N'DJAMENA

BENIN

NIGERIA

KAINJI RES.

Kaduna

Niger

ABUJA

Benue

Logone

PORTO-NOVO

OMÉ

CAMEROON

CENTRAL AFRICAN REP.

Bafoussam

Oubam

Ouangi

Bomu

Sanaga

BANGUI

Uele

MALABO

BIOKO

YAOUNDÉ

Nyong

EQUATORIAL
GUINEA

Ebebiyin

Mbini

Mongomo

Zaire

Sangha

PRÍNCIPE

I. Caroço

SÃO TOMÉ
AND PRÍNCIPE

Evinayong

GABON

LIBREVILLE

SÃO TOMÉ

Ogooué

CONGO

LAKE
TUMBA

Lomami

I. das Rôlas

ANNOBÓN
(Equatorial Guinea)

LAKE
MAI-NDOMBE

Luckerie

Sankuru

Kasai

Koulilou

Zaire

BRAZZAVILLE

KINSHASA

A Guide to the

BIRDS OF WESTERN
AFRICA

A Guide to the

BIRDS OF WESTERN
AFRICA

NIK BORROW

AND

RON DEMEY

Princeton University Press
Princeton and Oxford

To my parents

Nik

To Rita

Ron

Published in the United States, Canada, and the Philippine Islands by
Princeton University Press, 41 William Street, Princeton, New Jersey 08540

In the United Kingdom and European Union, published by
Christopher Helm, an imprint of A & C Black (Publishers) Ltd.,
37 Soho Square, London W1D 3QZ

Copyright © 2001 Nik Borrow and Ron Demey

Library of Congress Control Number 2001094366

ISBN 0-691-09520-5

This book has been composed in New Baskerville

www.birds.princeton.edu

Printed in Singapore

10 9 8 7 6 5 4 3 2 1

CONTENTS

ACKNOWLEDGEMENTS

A book of this scope would not have been possible without the research of many ornithologists past and present. We have, in particular, drawn extensively upon the works of Bannerman, Chapin, and Mackworth-Praed and Grant and are happy to acknowledge the debt. More recently, *The Birds of Africa* has, with every volume published, gained in importance as the major work of reference. Serle and Morel's field guide served to put us on our way in our early ventures in West Africa, soon to be replaced by Mackworth-Praed and Grant's comprehensive handbooks, which remain battered but highly trusted companions.

The preparation of this book has, in addition, benefited greatly from the encouragement, advice and constructive help of many kinds received at different times from a large number of individuals, including Mark Beaman, Pamela Beresford, Chris Bowden, Robert Cheke, Patrice Christy, William S. Clark, Nigel Cleere, Geoffrey Field, Lincoln Fishpool, Yves Garrino, Wulf Gatter, Tony Harris, Cornelis Hazevoet, Paul Herroelen, James Jobling, Michel Louette, Jaime Pérez del Val, Alison Stattersfield, Lieve Verstraeten, Jan Van de Voorde and Matthias Waltert. Iain Robertson is thanked for his crucial role in getting the project started and for his input at its early stages.

Peter Colston and Robert Prys-Jones, at the Natural History Museum in Tring, UK, and Michel Louette, at the Royal Museum for Central Africa, Tervuren, Belgium, are thanked for facilitating access to skins. Michel Louette is also thanked for stimulating discussions and generous help in various ways.

For help with vocalisations we are grateful to Claude Chappuis, who also generously made available copies of recordings from his invaluable collection, and to Richard Ranft, at the National Sound Archive, London, UK.

A special word of appreciation goes to Robert Dowsett and Françoise Dowsett-Lemaire for their selfless assistance and invaluable help over the years; they willingly reviewed the major part of the species accounts and generously let us benefit from their considerable knowledge and field experience as well as from information in the Tauraco databases.

BirdLife International is thanked for allowing the use of some of its maps.

Particular thanks are due to Robert Kirk and Nigel Redman, the successive managing editors, for their patience and understanding for a project that took many more years to complete than initially planned.

NB would like to thank the many Birdquest clients, friends and companions in the field, including, in particular, Mark Andrews, Ombrou Antoine, Bamenda Highlands Forest Project, Lawrence Bangura, Sering Bojang, Joachim Dibakou, Brian Finch, Germain Gagné, Alan Greensmith, Ndong Bass Innocent, Solomon Jallow, Ekpe Kennedy, Nomo Guirobo Luc, Mount Kupe Forest Project, Gerhard Radl, Ian Sinclair, Sio, Zo Beugre Sylvain, Wandifa Touray and Mark Van Beirs.

RD thanks his companions in the field over the past twenty years, in particular Jane and Peter Chandley, Jean-Michel Borie, Thierry Bara, Ian Davidson, Marc Languy, Michel Nicole, Alain Rousseau, Jan Van de Voorde and the members of the Important Bird Area teams in Cameroon, Ethiopia and Nigeria. Lincoln Fishpool, now at BirdLife International, is thanked for his stimulating companionship and help in numerous ways since the early days together in Ivory Coast.

RD is particularly grateful to Rita Swinnen for her enthusiastic and unstinting support throughout the long period that this work has been in progress, and her always cheerful company in the field. Without her help and assistance in so many ways this book would never have been written.

INTRODUCTION

AREA COVERED

This work describes and illustrates all 1269 species definitely recorded from western Africa, as well as some that have been claimed but whose occurrence requires proof. In total, 1285 species are treated.

Western Africa, as defined here and subsequently referred to as 'the region', comprises the 23 countries, south of the Sahara, from Mauritania in the northwest, to Chad and the Central African Republic in the east, and Congo-Brazzaville in the southeast, including the Cape Verde and Gulf of Guinea islands. The term 'region' is used in a general sense and does not indicate an avifaunal/biogeographical region or subregion. For ease of reference, all range states considered to comprise western Africa are covered in their entirety; thus parts of northern Mauritania, Niger and Chad that are generally considered within the Western Palearctic are included. The four principal Gulf of Guinea islands comprise Bioko (formerly Fernando Po), Príncipe, São Tomé and Annobón (formerly Pagalú). Bioko and Annobón form part of Equatorial Guinea.

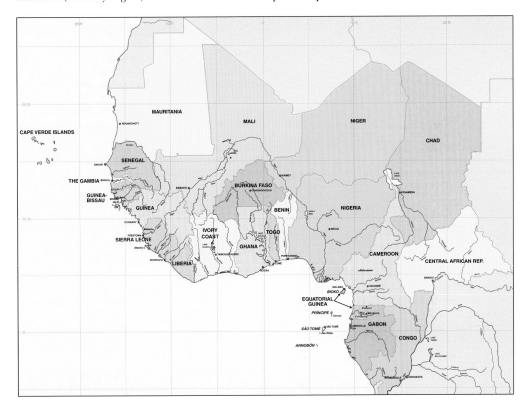

Figure 1. Western Africa.

NOMENCLATURE

Taxonomy, sequence and scientific names

In general, we have followed *The Birds of Africa*, vols 1–6 (Brown *et al.* 1982; Urban *et al.* 1986, 1997; Keith *et al.* 1992; Fry *et al.* 1988, 2000; hereafter abbreviated as *BoA*), although in some cases we have preferred Dowsett & Forbes-Watson (1993) or other recent authors, where these adopt what we consider a more advanced or consistent view. Where our taxonomy differs from *BoA*, rationale is briefly presented in a note at the foot of the relevant species account.

In some cases, where a taxon has been variably treated as a subspecies or a species by different authors, and there is strong evidence to suggest that the latter option may be preferable, the taxon has been treated under a separate heading and English name; however, to indicate that this does not imply an undisputed taxonomic decision, the specific name under which it is also often treated is placed within parentheses, e.g. *Bostrychia (olivacea) bocagei* and *Lagonosticta (rhodopareia) umbrinodorsalis*.

9

English names

With the aim of establishing a standardised world list of English names, many novel bird names have been coined in recent years, which has often resulted in confusion and frustration for users of ornithological works. In order to avoid further complication, we have principally followed *BoA*, supplemented by Dowsett & Forbes-Watson (1993), even if we do not personally favour the name chosen. Where these sources offer alternatives, we have chosen the name most commonly used in western Africa. Where Palearctic bird names, as listed by Beaman (1994), differ from these, based on convincing rationale, we have followed the latter. If none of these three sources proposes a name (e.g. in certain cases arising from taxonomic uncertainty), or if none of the proposed names is in frequent usage in western African ornithological literature, we have used other sources. Some alternative English names are given.

We have not, however, followed *BoA*'s controversial spelling of some names. Debate principally centres on the use of hyphenation and capitalisation. We agree with Beaman (1994) that the current fashion of introducing hyphens into bird names is both inconsistent and ugly, and have therefore used hyphens only when we considered it essential. A clear and consistent explanation of the use of hyphens is given by Inskipp *et al.* (1996), and their reasoning, which is briefly reproduced hereafter, has been largely followed here. They state that hyphens in group names should be used in two circumstances:

1. to link two nouns in apposition
2. between an adjective and a noun, but only in cases to avoid a misleading impression of the species' relationships.

Truly apposed nouns appear to be rare in bird names and most qualifying words within a group name that seem to be nouns are actually adjectival nouns, i.e. nouns functioning as adjectives that, for construction purposes, should be treated as ordinary adjectives and, therefore, should not take a hyphen. Thus, a snake eagle is a 'snake-eating eagle', not a 'snake-eagle' (the latter implies an unlikely intermediate between a snake and an eagle!). Adjectival nouns are abbreviations and can be identified as such if it makes sense to add a suffix such as -like, -sized, -billed, -eating, -nesting, -loving, -dwelling, or -driven. Examples: Tiger Heron, Night Heron, Cuckoo Hawk, Bat Hawk, Fish Eagle, Stone Partridge, Water Rail, Wood Dove, Grass Owl, Bush Lark, Hoopoe Lark, Sand Martin, Cliff Swallow, Ant Thrush, Scrub Robin, Robin Chat, Ground Thrush, Reed Warbler, Swamp Warbler, Wren Warbler, Woodland Warbler, Forest Flycatcher, Hill Babbler, Thrush Babbler, Sparrow Weaver, etc.

Names such as Scops Owl or Turtle Dove are more obscure but their qualifiers will probably also be demonstrated to be adjectival. Thus, they function in the same way as a combination of an overt adjective with a noun, as in Crested Tern, Green Pigeon, Crested Flycatcher, Penduline Tit and Glossy Starling.

Hyphens are placed between nouns and adjectives (or adjectival nouns) only in the interests of matching English with scientific classification. Painted Snipes and Cuckoo Shrikes are not, respectively, snipes and shrikes. The two components of the name are therefore better hyphenated, with the initial letter of the second word being in lower case: Painted-snipe and Cuckoo-shrike. (Hyphenated constructions with the initial letter of the second word being capitalised, e.g. 'Cuckoo-Shrike', are not used in this book.) It is, however, impossible to apply this rule consistently within a family and in some cases it appears preferable to leave well-established names unhyphenated, e.g. European Honey Buzzard (Beaman 1994, *contra* Inskipp *et al.* 1996).

The use of a terminal 's' in possessives is retained in all cases, including those of personal names ending in an 's' sound, following Dowsett & Forbes-Watson (1993) and other recent authors; thus Ayres's and Bates's.

French names

The names proposed in *Noms français des oiseaux du monde* (1993) have been used throughout. Some alternative names, used e.g. by *BoA*, Dowsett & Forbes-Watson (1993) or Serle & Morel (1979), are given.

THE SPECIES ACCOUNTS

Family Introduction

Species are grouped according to family and introduced by a brief family account. Each family is presented by its English and scientific names followed by the total number of species it has in, successively, the world, Africa and western Africa (abbreviated as WA). Family accounts follow a relatively standard format. Priority is given to information helpful in the field. Size and type of bird are followed by a succinct description of the main family characteristics, physical appearance, and general and breeding behaviour. Any feature common to all members of the family, or all those occurring in our region, is mentioned here rather than being repeated for each species. Also mentioned are particular identification problems and whether the family is endemic to Africa, or to the region covered by this book. If useful, different genera are briefly presented. In some large families, brief genus introductions have been inserted prior to the relevant species accounts rather than within the family account.

Species Heading

The first line of each species heading contains the English name followed by its scientific (binominal) name and the number of the plate(s) on which the species is illustrated, followed, within parentheses, by the species number on the plate. The French name is mentioned on the following line. Alternative English and French names (if any) are given as the first line of the species accounts.

Measurements

Each species text begins with length (abbreviated to L) given in cm and (between parentheses) inches (1 inch = 2.54 cm). Wingspan (WS) and tail-streamer length are mentioned where relevant. Measurements are taken from authoritative sources, complemented by our own mensural data obtained from specimens. Total-length measurements represent the length of museum skins stretched out on their backs and measured from bill tip to tail tip. It should be noted that direct comparison of these measurements can be quite misleading because they do not take into account other aspects of physiognomy, such as relative bill, neck and tail length, bulk, etc. It is therefore most useful only when comparing related species.

Subspecies

If the species has a single subspecies occurring in our region, its name is placed immediately prior to the species description (after Length). If more than one subspecies occurs in our region, these are described in the species account. It should be borne in mind that differences between subspecies are often subtle (e.g. colour tones), or based on measurements, and are thus often very hard or impossible to discern in the field.

In a few cases, where the description of the subspecies occurring in our region is, for some reason, not included in the main text of the species account, but mentioned in the NOTE at the end, the symbol ¶ has been placed after Length.

Description

The descriptions usually start with characteristic features common to all, or at least to all adult, plumages. Short descriptions of the different plumages follow, in the order adult male, adult female, juvenile and, if relevant, immature. Descriptions are kept brief, with diagnostic or most important identification features in *italics*. If more than one subspecies occurs, the description starts with the most widespread or characteristic subspecies; if there is no such taxon, the descriptions of the various subspecies generally follow their distribution, from north to south and from west to east, with island forms last.

The terms 'small', 'medium-sized' and 'large' indicate only the relative size of a species compared with its close relatives. For the sake of brevity, and when not indicated otherwise (or obviously different), the term 'top of head' includes forehead and crown, 'throat' includes the chin, and 'legs' includes legs and feet. 'Above' generally encompasses head and upperparts, and sometimes also tail; 'below' the entire underparts; in some cases, certain parts are excluded, but this should be obvious from the context. 'Flight feathers' include primaries and secondaries (but not tail feathers). 'Wing feathers' include flight feathers and wing-coverts. 'Upperparts' often includes scapulars and wing-coverts (if concolorous). When throat and undertail-coverts are described separately, the rest of the underparts (breast, belly and flanks) is sometimes indicated as 'underparts' (and not 'rest of underparts'). Study of the illustrations should preclude any confusion.

The term 'immature' (used generally for a non-adult bird) has normally not been used to indicate juvenile plumage (the first plumage of true, non-downy, feathers). For young passerines of Palearctic origin wintering in our area, 'immature' indicates 'first-winter' plumage. In some cases 'immature' is used when the available information prevents greater precision. The terms 'winter' and 'summer' refer to northern hemisphere seasons.

Abbreviations Subspecies (=races) are indicated by their subspecific name only, if preceded by 'adult', 'male', 'female', 'juvenile' or an indication of their range, the term 'race' being generally omitted, e.g. 'Adult male *gabonensis*' or 'Western *extrema*' and 'Nominate' (not 'Nominate race').

Colours Colour names have deliberately been kept as simple as possible and most will be readily understood. Where two colours are combined (compound colours), the last named is dominant; in other words, they should be interpreted as the second colour tinged with the first; e.g. reddish-brown is brown tinged red. Where the suffix -ish is added to a colour, this indicates a weaker or less distinct shade of that colour.

Frequently used colours include buff (pale or dull yellowish with a brownish or beige tinge), chestnut (dark reddish-brown), olive (dull yellowish-green, like the fruit), and horn (pale brownish-yellow; used solely for bills). Dusky (dirty greyish or brownish) is often used for rather indistinct darkish markings. The term dark (opposite pale) is used for dark plumage markings that do not possess any obvious colour.

Voice

Only the most characteristic vocalisations are given. Transcriptions of calls and songs are placed in *italics*. References to Claude Chappuis' (2000) outstanding collection of 15 CDs of *African Bird Sounds* – an invaluable tool – are presented within square brackets, with CD and track number.

As is widely acknowledged, transcribing bird sounds in such a way that they can be unambiguously interpreted by others is almost impossible. Phonetic renditions and verbal descriptions are necessarily highly subjective and open to misinterpretation, but there appears to be no convenient alternative. The user of this book is therefore advised to listen to sound recordings and compare them with the transcriptions, in order to understand the authors' interpretation.

In the transcriptions of vocalisations the following conventions apply:

single vowels are pronounced short (thus *a* as in 'apple', *e* as in 'extra', *i* as in 'it', *u* as in 'full')
ee as in 'see', 'be'
iiiii is higher pitched than *eeee*
k as in 'cat' (*c* is not used for the hard 'k' sound)
ch as in 'check'
sh as in 'sheep'

CAPITAL LETTERS indicate that the component in question is considerably louder than the others.

Pauses between notes or syllables are denoted as follows (after Lewington *et al.* 1991):

see-see very short pause
see see normal pause, as in ordinary conversation
see, see longer pause (at least *c.* 1 second)
see...see pause of more than 2 seconds

Habits

Includes notes on behaviour relevant to identification (if known) and preferred habitat. Details on nesting are given in only a few cases, where these represent important identification clues. If certain aspects of the species' biology (e.g. nesting) remain unknown, this is usually mentioned to draw attention to the importance of further research.

Similar Species

Indicates the distinguishing features of similar species or refers to their accounts. This section may be omitted if a species does not normally present any identification problems.

Status and Distribution

Presents the species' status, its abundance within its principal habitats and its geographic distribution in our area. Status and distribution on the African mainland is usually mentioned first. When a species is stated to occur 'throughout' (the region), this indicates that it occurs throughout the mainland and available habitat. If a species occurs on Cape Verde or Gulf of Guinea islands, this is specifically mentioned. In this context, 'São Tomé and Príncipe' means the two islands, not the political entity: when a species occurs on only one island, that is specifically stated.

'Claimed from ...' indicates that the species has been reported from the country (or area) mentioned, but that the record is open to doubt or has never been adequately documented and requires confirmation.

The description of the distribution generally proceeds from north to south and from west to east. Some isolated records may be indicated on the maps only, without mention in the text.

Almost all information in this section is based on published data up to March 2001. A few unpublished records have been included, when considered reliable. In the compilation of this section R. J. Dowsett's country lists (Dowsett & Dowsett-Lemaire 1993) proved invaluable.

It should be appreciated that the distribution of many species remains incompletely known and that the large-scale degradation of original habitats in much of the region has had a profound effect on bird distributions. Rainforest inhabitants may become threatened with extinction, while others that favour open habitats such as farmland, may have expanded their range.

Abundance categories Our aim has been to be as user-friendly as possible. Therefore we use only five, easily understood, abundance categories:

common	invariably encountered singly or in significant numbers within its normal habitat
not uncommon	usually, but not invariably, encountered within its normal habitat
uncommon	relatively frequently, but not regularly, encountered within its normal habitat
scarce	only irregularly and infrequently encountered within its normal habitat
rare	rarely encountered, often implying fewer than *c.* 10 records.

Their use normally starts with the category of most frequently occurring abundance; e.g. 'uncommon to rare' means that the species is more often (=in more countries) uncommon than rare; if vice versa it would probably be 'rare to *locally* uncommon'. Where the generally applicable abundance category is deviated from, this is indicated within parentheses, e.g. 'common to uncommon resident in the forest zone, in Gambia (rare) and from Sierra Leone to Togo'.

'Widespread' is not an abundance category but indicates that a species has a wide range. Within the vast area covered, however, the species may occur only in certain places.

Status categories We use the following seven status categories:

resident	a species that resides within its range throughout the year and breeds; the opposite of a migrant.
passage migrant	non-breeding visitor present only during migration.
partial migrant	resident species throughout most of its range, occurring regularly towards the edges of its range only at given periods.
intra-African migrant	a species that breeds in one part of Africa and spends the post-breeding season in a different area, or appears only seasonally in another part.
Palearctic migrant	a species that breeds in the Palearctic region (Europe, N Africa and part of Asia) and spends the northern winter in sub-Saharan Africa.
visitor	non-breeding visitor; for Palearctic migrants, we refer for convenience to the northern hemisphere seasons.
vagrant	a species outside its normal range.

It should be appreciated that it is sometimes difficult to distinguish between a genuine vagrant and a species with only a few records. Some 'vagrants' may prove to be more or less regular, if rare or scarce, with increased observer coverage.

Threatened species If a species is included in *Threatened Birds of the World* (BirdLife International 2000) this is indicated at the end of the 'Status and Distribution' section by '*TBW* category:' followed by one of the following designations:

THREATENED (Critical)	species facing an extremely high risk of extinction in the wild in the immediate future.
THREATENED (Endangered)	species facing a very high risk of extinction in the wild in the near future.
THREATENED (Vulnerable)	species facing a high risk of extinction in the wild in the medium-term future.
Near Threatened	species coming very close to qualifying as threatened.
Data Deficient	species for which there is inadequate information to make an assessment of its risk of extinction.

For a comprehensive discussion of the above categories, see BirdLife International (2000).

Note

Where appropriate, a note, usually on the taxonomic status of the species, is presented.

References

References are only given for single-source data that we were unable to verify, recent discoveries or some distributional data.

THE MAPS

Distribution maps are provided for all species except vagrants, some pelagics and a few species of very restricted range. The maps reflect the known or inferred distribution of a species in areas of suitable habitat within a broadly defined range. As locality data remain scant for many species in our region, the maps should not necessarily be taken as providing a true reflection of actual distributions and must therefore be used with caution and common sense.

Key to the maps:

resident	mainly resident but partially migratory or erratic within range	breeding visitor (intra-African migrant)
non-breeding visitor (main range)	non-breeding visitor (sparse occurrence)	X vagrant or isolated record

THE PLATES

All species recorded or claimed from the region are illustrated in colour. Our aim has been to illustrate as many distinct plumages as space would permit. Thus, distinctive male, female and immature plumages are depicted, subject to the availability of representative specimens or personal field notes. This is also the case with races that are sufficiently distinctive to be separated in the field. Only in a few cases, however, has the juvenile plumage of passerines been illustrated, as this appears unnecessary and potentially confusing. Indeed, this plumage is usually acquired for only a short period and identification is normally facilitated by the presence of the parents.

Wherever possible or desirable, care has been taken to respect family groupings – this has resulted in some plates containing more illustrations than others. A few plates depict groups of birds painted to different scales; in such cases the various groups are divided by a narrow line, both on the plate and on the caption page. On most plates, the species are arranged in order of resemblance. The order of the plates generally follows taxonomic sequence. Species and distinctive subspecies that occur exclusively on the Cape Verde and Gulf of Guinea islands are grouped on the three plates 143–145. The last two plates (146–147) show vagrants and a few species peripheral to the region.

THE PLATE CAPTIONS

These follow a standardised format. All species illustrated are indicated by a number. A few that are very similar to an illustrated species have not been depicted but receive a separate, unnumbered caption, adjacent to that of the similar species. Rare vagrants or localised species that are illustrated on plates 146–147 are, nevertheless, given an unnumbered caption opposite the most appropriate plate (of the same or closest family), to draw attention to their occurrence within our region.

To give an indication of the bird's relative size, length (in cm only) is mentioned after the species name.

The status of each species is denoted by the following letters (in bold):

R   Resident
M   Intra-African migrant
P   Palearctic migrant (including a few species of Nearctic origin)
V   Vagrant

If more than one category is applicable, the commoner is placed first.

Threatened Birds of the World categories are indicated by the following red letters (in bold):

CR   Threatened (Critical)
EN   Threatened (Endangered)
VU   Threatened (Vulnerable)
NT   Near Threatened
DD   Data Deficient

The following line succinctly indicates the habitat or vegetation zone in which the species normally occurs, its range within our region and its abundance. It should be appreciated that space permitted only a rather simplified extract of the main species account. Occurrence as a vagrant has been included wherever possible; in a few cases, however, when there was insufficient space to do so, such information was omitted. If a species is endemic to the region, this has been mentioned.

Subsequently, the most important distinguishing identification features are given. We have attempted to include as much information as space permitted. This inevitably has resulted in some information found in the main text being repeated here, but we considered this preferable to leaving much of the page blank.

For the sake of brevity, subspecies are indicated by their subspecific name only, when preceded by 'Adult' or 'Juvenile' (e.g. 'Adult *gabonensis*').

AUTHORS' NOTE

Many interesting observations remain buried in personal notebooks or unpublished reports. It is therefore recommended that relevant data be documented and submitted to a refereed journal for publication, e.g. *Malimbus* or *Bulletin of the African Bird Club* (see page 22).

In a work of this scope some errors and omissions appear inevitable, despite the care with which museum specimens were examined and data collected. Future fieldwork will also certainly add to our knowledge of the region's avifauna. The authors (c/o the publishers, Christopher Helm) would therefore be pleased to receive any information which updates and corrects that presented in this book.

CLIMATE, TOPOGRAPHY AND MAIN HABITATS

CLIMATE

The principal feature of the region's climate is the alternate wet and dry seasons, which are governed by the movement of the Intertropical Convergence Zone (ITCZ), a zone of low pressure towards which blow winds from the northern and southern high pressure belts. Southwesterly winds from the Atlantic Ocean are warm and moist, while northeasterly winds from the Sahara, known as Harmattan, are hot, dry and dusty. The ITCZ annually oscillates north and south following, with a lag of 1–2 months, the position of the sun. At the northern summer solstice it lies near the Tropic of Cancer and wet maritime winds produce a rainy season north of the equator. When the ITCZ moves south and the sun is over the Tropic of Capricorn, most of the region comes under the influence of the continental, hot, dry Harmattan. Overall, the dry season lengthens and rainfall decreases with increasing latitude. In the south, the dry season generally extends from December to March, while in the north it lengthens to at least seven months, from October to April. This pattern can, however, exhibit considerable annual variation, especially in the north where the wet season is short (July–September) and rainfall is erratic in timing, quantity and distribution. In the south, from Ivory Coast to Nigeria, the rains decrease in August, resulting in a 'short dry season'.

In the part of our region south of the equator, namely Gabon and Congo, the weather pattern is the reverse of that to the north, with a dry season from mid-May through September and a long rainy season from October to mid-May. Rains decrease during a short dry spell in December–January. Here also, monthly rainfall varies greatly from year to year.

Maximum temperatures and temperature ranges increase with latitude. In the forest zone temperatures vary little, averaging around 27°C throughout the year, while in the desert they may range from around 0°C to more than 45°C.

TOPOGRAPHY

Most of the western African mainland is, unlike eastern and southern Africa, largely flat and low lying, with altitudes rarely exceeding 400 m except granite inselbergs that can reach 700 m. Notable montane or hilly areas include the Fouta Djalon (Guinea), the Loma Mts (Sierra Leone), Mt Nimba (1752 m, Liberia/Guinea/Ivory Coast), the Jos Plateau (C Nigeria), the Aïr Mts (Niger), and the Tibesti and Ennedi Mts (N Chad). The interior of Gabon and Congo is occupied by a vast plateau ranging at 300–1000 m.

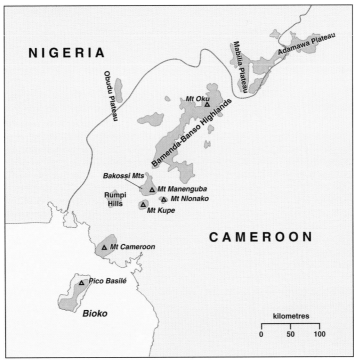

Figure 2. The Cameroon highlands.

The most important, however, is the Cameroon highlands, a chain of mountains in western Cameroon running southwest and extending across the border into SE Nigeria (Mambilla and Obudu Plateaux) (see fig. 2). It contains the highest peak in W Africa, Mt Cameroon (4095 m), which is volcanic and still active. Other mountains include Mt Rata (1768 m) in the Rumpi Hills, Mt Kupe (2064 m), Mt Nlonako (1825 m), Mt Manenguba (2411 m) in the Bakossi Mts, and Mt Oku (3011 m) in the Bamenda Highlands.

The Gulf of Guinea islands are part of the line of volcanoes, extending from the Cameroon highlands southwest into the Atlantic Ocean. Bioko lies on the continental shelf just *c.* 30 km off the coast of Cameroon and was probably linked to the African mainland in the past. Pico Basilé (3011 m) is still active. Príncipe, São Tomé and Annobón, however, are true volcanic islands surrounded by deep seas. This isolation has led to a species-poor avifauna with a high degree of endemism.

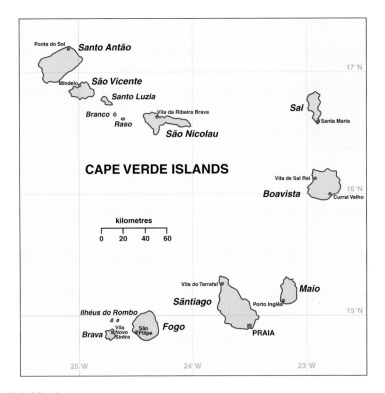

Figure 3. Cape Verde Islands.

The oceanic Cape Verde Islands are situated 460–830 km off the coast of Senegal and consist of ten islands and several islets of volcanic origin (see fig. 3). They have never been connected to the African mainland and are usually divided into two groups: the Windward Islands in the north and Leeward Islands to the south. The former are mountainous, with peaks reaching over 1000 m (and up to 2800 m on Fogo, the only active volcano), while the latter are of low relief and receive only little and irregular rain (Hazevoet 1995).

The region has five or six main river systems. The 4030 km-long Niger River, with its main tributary the Bénoué (or Benue), drains the majority of the region. It arises in the high-rainfall Fouta Djalon hills of Guinea, only a few hundred kilometres from the ocean, flows through every climatic and vegetation zone of the region and forms a huge inland delta, which considerably delays flooding downstream such that maximum water levels here occur well into the dry season, coinciding with the northern hemisphere winter and important for migrant waterbirds. The other main river systems are the Senegal River, the Ogooué, the Congo River and its tributary the Ubangi, and the Chari (sole affluent of Lake Chad).

There are few large natural lakes in the region, the largest being Lake Chad, which exhibits considerable fluctuations in water levels.

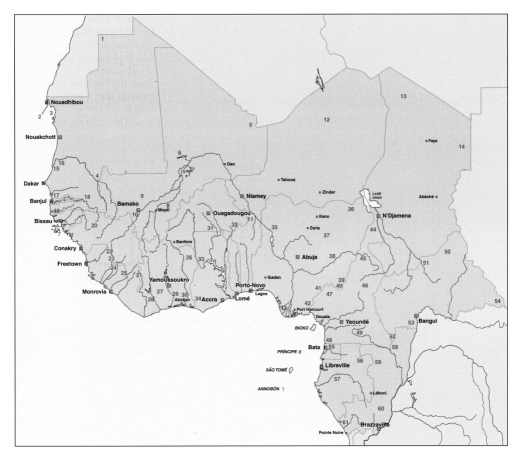

Figure 4. Some localities mentioned in the text.

Mauritania
1   Zemmour
2   Baie de l'Etoile
3   Banc d'Arguin NP
4   Guidimaka

Mali
5   Adrar des Ifôghas (=Iforhas)
6   Lac Faguibine
7   Central Niger delta
8   Bandiagara escarpment
9   Boucle du Baoulé NP
10 Mandingues Mts

Niger
11 'W' NP
12 Aïr Mts

Chad
13 Tibesti
14 Ennedi

Senegal
15 Djoudj NP
16 Lac de Guier
17 Delta du Saloum NP
18 Niokolo-Koba NP
19 Basse Casamance

Guinea
20 Fouta Djalon
21 Mt Nimba

Sierra Leone
22 Loma Mts
23 Tingi Mts
24 Gola Forest

Liberia
21 Mt Nimba
25 Wonegizi Mts

Ivory Coast
21 Mt Nimba
26 Comoé NP
27 Marahoué NP
28 Taï NP
29 Lamto
30 Yapo Forest

Burkina Faso
11 'W' NP
31 Po NP
32 Arli NP

Ghana
33 Mole NP
34 Bia NP

Benin
11 'W' NP

Nigeria
35 Kainji NP
36 Hadejia-Nguru wetlands
37 Jos Plateau
38 Wase Rock

39 Gashaka-Gumti NP
40 Mambilla Plateau; Gotel Mts
41 Obudu Plateau
42 Cross River NP; Oban Hills
43 Lower Niger delta

Cameroon
44 Waza NP
45 Bénoué NP
46 Adamawa Plateau
47 Bamenda Highlands
48 Campo-Ma'an NP
49 Dja Game Reserve

Central African Republic
50 Manovo-Gounda-St Floris NP
51 Bamingui-Bangoran NP
52 Dzanga-Sanga Forest
53 Lobaye Préfecture
54 Ouossi R. (Baroua and Zémio area)

Equatorial Guinea
55 Mt Alen

Gabon
56 Makokou
57 Lopé NP

Congo
58 Nouabalé-Ndoki NP
59 Odzala NP
60 Léfini Reserve (Bateke Plateau)
61 Conkouati Faunal Reserve

MAIN HABITATS

W Africa contains a broad range of habitats, from rainforest to desert. They are arranged in a series of parallel latitudinal bands orientated west–east, reflecting the decreasing northward rainfall gradient. Except locally, the lack of relief means there is little disturbance to the zonal arrangement of both climatic and vegetation belts.

The sea coast is generally flat and sandy, with some intertidal mud and sand flats. Africa's largest intertidal mudflats occur in the region, in N Mauritania (Banc d'Arguin area, 46,000 ha) and along the coast of Guinea-Bissau, Guinea and Sierra Leone (284,300 ha). Smaller mudflats, which seasonally hold significant aggregations of waders, are the Senegal delta and Sine-Saloum, both in Senegal.

Brackish creeks and lagoons may occur behind the narrow coastal belt; these may be very large locally (e.g. in Ivory Coast). Mudflats and lagoons typically are bordered by mangroves. Open swamps and swamp forest may be found further inland.

Africa's two major lowland rainforest blocks occur in the region, often referred to as the Upper Guinea and Lower Guinea forests, separated, from E Ghana to Benin, by the Dahomey Gap, where savanna reaches the coast. These forests, except within the Dahomey Gap, originally covered most of the area between Sierra Leone and SE Guinea to SW CAR and Congo. Much of this area is now deforested and replaced by 'derived savanna', a mosaic of cultivation, farmbush and secondary forest, which permits savanna species to penetrate the forest zone. Of the remaining forest, very little is true primary: most has been modified by man. As the degree of secondary modification may be hard to gauge, the term 'high forest' has been proposed. This forest zone may be subdivided into the moister evergreen forests in the south and the drier semi-deciduous forests further north, but the change is progressive.

The savanna zone, immediately north of the forest, is usually subdivided, on the basis of the density of the woodland and species composition, into two: the Guinea savanna in the south and the drier Sudan savanna further north. The zone is characterised by various types of wooded grassland, in which there is a gradual decrease of tree cover and tree height, reflecting the decline in rainfall, northward. The other dominant feature of the vegetation here is the grasses that grow under and between the trees. In Guinea savanna they are tall (2–3 m), dense perennial species which, with increasing latitude, give way to smaller species in which

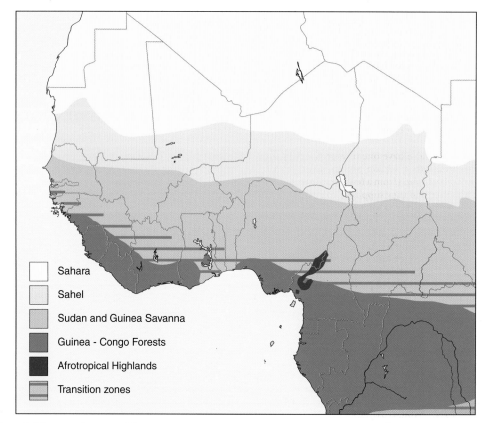

Figure 5. The vegetation zones of Western Africa.

annuals become increasingly common. Watercourses are bordered by gallery forest of various width. Some dense forest patches may remain; these permit forest species to penetrate deep into the savanna zone. Bare, granitic inselbergs are a typical feature of the landscape, especially in the northern Guinea savanna. A feature of these savannas, especially in the south, are the fires that regularly rage through them in the late dry season (December–February). Although a natural phenomenon to which taxa of the region are well adapted, these fires now are almost all deliberately started by man.

In the southeast of our region, in C Congo and Gabon, a northward extension of the vast woodlands of southern Africa penetrates, as a wedge, into the forest zone. This gives rise to the presence of bird species and subspecies of southern affinities.

The Sahel zone, north of the savanna zone, is characterised by thorn scrub (with *Acacia* spp. and *Ziziphus* spp.), sparse, mostly annual, grasses, and very low and often erratic rainfall (mean 100–600 mm). The most important wetlands of the region stretch across this semi-arid belt and include the Senegal delta (with Djoudj National Park), the central Niger delta in Mali, the Hadejia–Nguru wetlands in N Nigeria, and the Logone floodplain in N Cameroon.

The northern border of the region is formed by the Sahara desert, which consists of arid landscapes with sandy, stony or rocky substrates and sparse plant cover, except in depressions, wadis and oases, where water is retained.

Afrotropical highlands with montane forest and montane grassland occur only in W Cameroon and Bioko. Montane forest differs from lowland forest in tree species composition and is also relatively rich in endemic bird species. Local variations in rainfall result in differences in the altitude at which it is found. A combination of high rainfall and reduced temperature causes montane forest to occur at relatively low elevations on Mt Cameroon and Bioko (generally above 800 m). Further inland, montane forest is only found higher, e.g. above 1200 m on Mt Kupe and at 2000–2950 m in the Bamenda Highlands (Stattersfield *et al.* 1998).

RESTRICTED-RANGE SPECIES AND ENDEMIC BIRD AREAS

BirdLife International has analysed the distribution patterns of birds with restricted ranges, defined as landbird species which have or have had a total global breeding range of less than 50,000 km^2 throughout historical times (i.e. post-1800) (Stattersfield *et al.* 1998). The results demonstrate that these restricted-range species tend to cluster, often on islands or in isolated patches of a particular habitat. Regions where two or more species of restricted range share completely or partially overlapping distributions are termed Endemic Bird Areas (EBAs). Where the distributions of such species only partially overlap, the total area of the EBA may be considerably larger than 50,000 km^2. Areas where only one restricted-range species occurs are termed Secondary Areas (SAs).

Most EBAs also have one or more globally threatened bird species and are important for restricted-range species from other wildlife groups. They are therefore clearly priorities for conservation action. EBAs vary considerably in size (from a few square kilometres to more than 100,000 km^2) and in the numbers of restricted-range species that they support (from two to 80), and thus their relative importance also varies.

In western Africa, a total of 87 restricted-range species occur in seven EBAs and four SAs. The region's EBAs comprise the Cape Verde Islands (4 restricted-range species), Annobón (3), São Tomé (21), Príncipe (11), the Upper Guinea forests (15), the Cameroon and Gabon lowlands (6), and the Cameroon mountains (29). The region's Secondary Areas (all with one restricted-range species) are the Upper Niger valley, South-west Nigeria, Lower Niger valley and Gabon–Cabinda coast. Six restricted-range species are shared between EBAs.

Tropical lowland and montane forests are the predominant habitats, and in the Upper Guinea forests and Cameroon mountains, in particular, it is estimated that there has been major (>50%) loss of these key habitat types. The Cameroon mountains are particularly important for the absolute numbers of restricted-range species occurring (the third largest total in the African region) and for the proportion of these that are threatened (12 species). São Tomé and the Upper Guinea forests also rate very highly in terms of their biological importance and threat levels.

Examples of highly threatened (according to BirdLife International 2000) restricted-range species include: Raso Lark *Alauda razae* (confined to one minute island in the Cape Verdes and at risk from introduced predators); Dwarf Olive Ibis *Bostrychia* (*olivacea*) *bocagei*, São Tomé Fiscal *Lanius newtoni* and São Tomé Grosbeak *Neospiza concolor* (all from São Tomé and assumed to have tiny populations in a small area of primary forest); Rufous Fishing Owl *Scotopelia ussheri* (a rarely seen species with a small, fragmented population seriously threatened by habitat loss); Liberian Greenbul *Phyllastrephus leucolepis* (a poorly known species that appears to have an extremely small, severely fragmented range and is inferred to be declining owing to habitat destruction);

Bannerman's Turaco *Tauraco bannermani* (which is only likely to survive if the Kilum-Ijim forest, the largest remaining montane forest area in the Cameroon mountains, is preserved) and Mount Kupe Bush-shrike *Malaconotus kupeensis* (which has a very small population with a small range and suffers from habitat loss).

The BirdLife International African Partnership has identified potentially conservable sites termed Important Bird Areas throughout the continent, including suites of sites within EBAs that seek to protect the unique species within these areas.

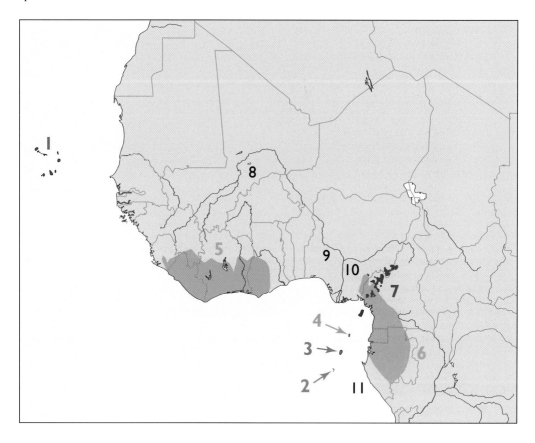

Figure 6. Endemic Bird Areas of western Africa.

Endemic Bird Areas
1 Cape Verde Islands
2 Annobón
3 São Tomé
4 Príncipe
5 Upper Guinea forests
6 Cameroon and Gabon lowlands
7 Cameroon mountains

Secondary Areas
8   Upper Niger valley
9   South-west Nigeria
10 Lower Niger valley
11 Gabon-Cabinda coast

Source: BirdLife International

TAXONOMY: SOME DEFINITIONS

Taxonomy is the science of classification and naming of living organisms. The classification of organisms attempts to reflect relationships between them and works on a hierarchical system by which an organism is placed in categories of decreasing level, from 'Kingdom' to 'Species' and 'Subspecies'. These categories are known as taxa (singular: taxon; hence 'taxonomy'). The main categories useful to the field ornithologist are the following.

Vertebrates are divided into **classes**: mammals, birds, reptiles, amphibians and fish. Birds are vertebrates characterised by the possession of feathers and belong to the class Aves, which is the scientific name for 'birds'.

The class Aves contains approximately 29 large groups called **orders**, with names ending in '-iformes'. Birds of the order Passeriformes (passerines, or songbirds) are placed together because they share certain morphological characters. Other orders include, for example, Ciconiiformes (storks, herons and ibises), Strigiformes (owls) and Piciformes (woodpeckers, honeyguides and barbets).

Orders are divided into *c.* 180 **families**, with names ending in '-idae', following the same logic of shared characters. Examples of families within the order of Passeriformes are: Hirundinidae (swallows and martins), Pycnonotidae (bulbuls) and Ploceidae (weavers).

Within families, birds are clustered into more than 2000 **genera** (singular: genus). Examples of different genera within the family Sylviidae are: *Sylvia, Acrocephalus, Prinia, Apalis, Phylloscopus*, etc.

A genus comprises one or several **species**. For example, four species belonging to the genus *Prinia* occur in western Africa: *Prinia subflava, Prinia fluviatilis, Prinia bairdii* and *Prinia molleri*. According to the biological species concept, which is followed in this book, a species can be defined as a population, or group of populations, of actually or potentially interbreeding individuals, reproductively isolated from all other such populations. Members of a species should be able to interbreed freely and produce fertile offspring.

The scientific name of a species is based on an internationally accepted, binomial system and consists of a two-part, latinised name, which is conventionally written in italics. The first part is the generic name (with first letter capitalised), the second the specific name (all lower case).

Finally, a species may be divided into different **subspecies** (also called races). Subspecies are groups of similar-looking individuals, slightly different from other groups, but belonging to the same species. Members of subspecies can or could still interbreed with other members of the same subspecies. A subspecies' name is added as a third part to the species' scientific name, e.g. *Prinia subflava pallescens*. The first population to be described becomes the nominate subspecies and carries the same subspecific and specific names, e.g. *Prinia subflava subflava*. If there is no doubt as to which species or genus is involved, the name can be shortened to *Prinia s. subflava* or *P. s. subflava*.

ORGANISATIONS

INTERNATIONAL

BirdLife International
Wellbrook Court, Girton Road, Cambridge CB3 0NA, UK.
Tel. +44 (0)1223 277318   Fax. +44 (0)1223 277200
E-mail birdlife@birdlife.org.uk

BirdLife International (formerly the International Council for Bird Preservation) strives to conserve birds, their habitats and global biodiversity, and is the leading authority on the status of the world's birds and the urgent problems that confront them. BirdLife aims to:

Prevent the extinction of any bird species.

Maintain and where possible improve the conservation status of all bird species.

Conserve and where appropriate improve and enlarge sites and habitats important for birds.

Help, through birds, to conserve biodiversity and to improve the quality of people's lives.

Integrate bird conservation into sustaining people's livelihoods.

Publications: *World Birdwatch* magazine (quarterly) and, with Cambridge University Press, *Bird Conservation International* (quarterly).

African Bird Club
c/o BirdLife International, Wellbrook Court, Girton Road, Cambridge CB3 0NA, UK.
ABC Web site http://www.africanbirdclub.org

The African Bird Club aims to:

Provide a worldwide focus for African ornithology.

Encourage an interest in the conservation of the birds of the region.

Liaise with and promote the work of existing regional societies.

Encourage observers to visit lesser known areas of the region.

Encourage observers to actively search for globally threatened and near-threatened species.

Run the ABC Conservation Programme.

Publication: *Bulletin of the African Bird Club* (bi-annual).

West African Ornithological Society / Société d'Ornithologie de l'Ouest Africain
Secretary: R. E. Sharland, 1 Fisher's Heron, East Mills, Fordingbridge, Hampshire, SP6 2JR, UK.

The Society aims to promote scientific interest in the birds of West Africa and to further the region's ornithology, principally through publication of its journal.

Publication: *Malimbus* (bi-annual).

NATIONAL

Burkina Faso: Fondation des Amis de la Nature (Naturama)
01 BP 6133, Ouagadougou 01, Burkina Faso
E-mail: naturama@fasonet.bf

Cameroon: Cameroon Ornithological Club
P O Box 3055, Messa, Yaoundé, Cameroon
E-mail: coc@iccnet.cm

Ghana: Ghana Wildlife Society
P O Box 13252, Accra, Ghana
E-mail: wildsoc@ighmail.com

Nigeria: Nigerian Conservation Society
P O Box 74638, Victoria Island, Lagos, Nigeria
E-mail: ncf@hyperia.com

Sierra Leone: Conservation Society of Sierra Leone
P O Box 1292, Freetown, Sierra Leone
E-mail: cssl@sierratel.sl

BIRD TOPOGRAPHY

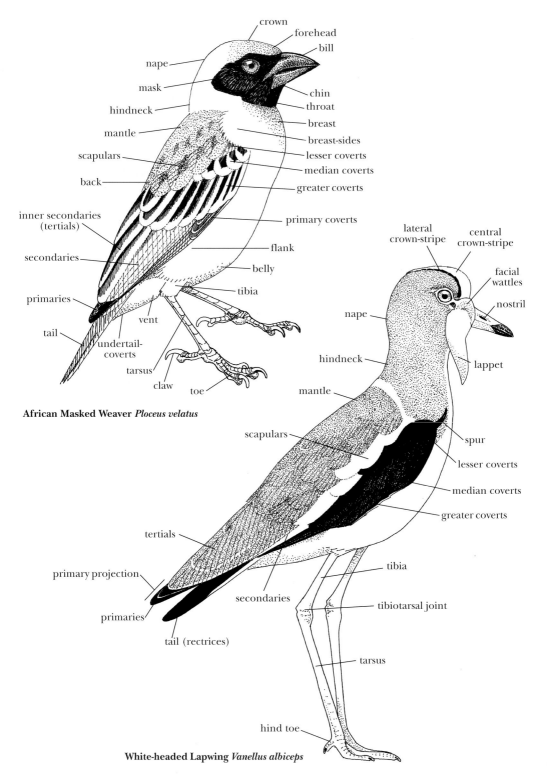

crown
forehead
bill
nape
mask
chin
hindneck
throat
breast
mantle
breast-sides
scapulars
lesser coverts
median coverts
back
greater coverts
inner secondaries (tertials)
primary coverts
flank
secondaries
belly
tibia
primaries
vent
tail
undertail-coverts
tarsus
claw
toe

African Masked Weaver *Ploceus velatus*

lateral crown-stripe
central crown-stripe
facial wattles
nostril
nape
hindneck
lappet
mantle
scapulars
spur
lesser coverts
median coverts
greater coverts
tertials
tibia
primary projection
tibiotarsal joint
primaries
secondaries
tail (rectrices)
tarsus
hind toe

White-headed Lapwing *Vanellus albiceps*

23

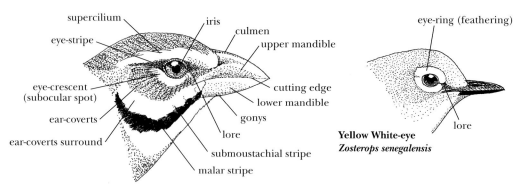

Chestnut-crowned Sparrow Weaver
Plocepasser superciliosus

supercilium
eye-stripe
iris
culmen
upper mandible
eye-crescent (subocular spot)
cutting edge
lower mandible
ear-coverts
gonys
ear-coverts surround
lore
submoustachial stripe
malar stripe

Yellow White-eye
Zosterops senegalensis

eye-ring (feathering)
lore

Congo Serpent Eagle
Dryotriorchis spectabilis

crest
orbital ring (bare skin)
cere
moustachial stripe
gular stripe

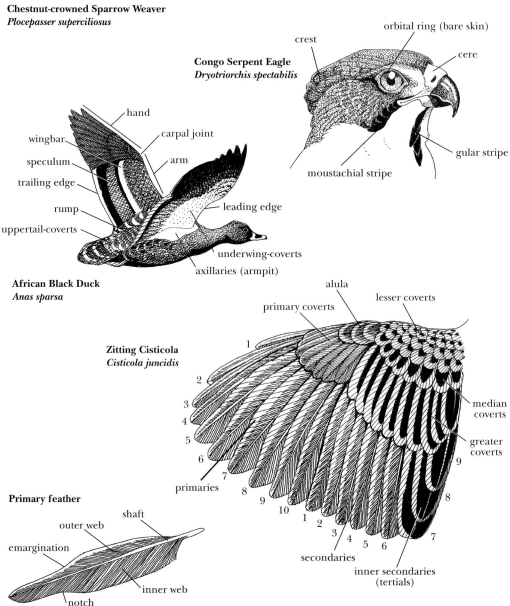

hand
carpal joint
wingbar
speculum
arm
trailing edge
rump
leading edge
uppertail-coverts
underwing-coverts
axillaries (armpit)

African Black Duck
Anas sparsa

alula
lesser coverts
primary coverts

Zitting Cisticola
Cisticola juncidis

median coverts
greater coverts

primaries
secondaries
inner secondaries (tertials)

Primary feather

shaft
outer web
emargination
inner web
notch

GLOSSARY

adult: a bird in final plumage (plumage no longer changing with age).

aerial: air-frequenting (e.g. an aerial feeder).

allopatric: mutually exclusive geographically. Applied to (taxonomically) closely related forms (populations) whose ranges do not overlap. Opposite of sympatric. See also parapatric.

allospecies: one of the species constituting a superspecies.

altitudinal migrant: a species that breeds at high altitudes and moves to lower levels and valleys in the non-breeding season.

antiphonal: referring to precisely timed alternating singing or calling by two birds, usually a mated pair.

aquatic: living in water.

arboreal: living in trees.

arm: inner part of wing, between the body and the carpal joint (wing-coverts and secondaries).

axillaries: the feathers at the junction of the underwing and the body; the 'armpit' of a bird.

bare parts: those parts not covered by feathers (including bill, cere, eyes, wattles, orbital ring and feet).

bib: a contrasting, usually dark area, on the throat and/or upper breast.

brood parasite: a species that lays its eggs in the nest of another (the 'host') and plays no parental role in raising its young, e.g. cuckoos, honeyguides and indigobirds.

call: brief vocalisation, used mainly to maintain contact with conspecifics or to alert to danger. Often consists of a single, simple note (cf. song).

cap: a contrasting patch on the top of the head.

carpal: the bend of the wing between 'hand' and 'arm', or carpal joint.

carpal bar: a contrasting, dark bar on the upperwing, running diagonally from the carpal joint towards the body. Exhibited in certain plumages by some gull species.

carpal joint: the joint at the bend of the wing.

carpal patch: a contrasting mark or area at or near the carpal joint.

casque: an enlargement on the upper surface of the bill, in front of the head, as on hornbills.

cere: bare and often brightly coloured skin at the base of the upper mandible, containing the nostrils.

cheek: loosely applied to the area on the side of the head.

collar: a band of contrasting colour on the neck.

colony: an assemblage of birds within a discrete area to nest (hence colonial species).

congeneric: belonging to the same genus (hence congeners).

conspecific: belonging to the same species.

crepuscular: active at dawn and dusk.

crest: a tuft of elongate feathers on the head.

cryptic: colours or markings aiding concealment (as in nightjars).

culmen: the ridge of the upper mandible.

dambo: seasonally wet grassland on acid soils.

deciduous: a tree that loses its leaves during set periods of the year, remaining leafless for some time.

dimorphic: having two distinct morphs or forms.

dissonant: not in harmony, harsh toned.

distal: furthest from centre of body or point of attachment; opposite of proximal.

disyllabic: consisting of two syllables.

diurnal: active during the day or in daylight.

duetting: male and female of a pair singing simultaneously or antiphonally in response to each other.

eclipse plumage: a female-like plumage acquired by males of some species (e.g. ducks and some sunbirds) during or following breeding.

edgings/edges: in relation to feather patterns, indicates outer feather margins. Edgings can result in distinct paler or darker panels of colour on wings or tail.

emergents: trees whose crowns are conspicuously taller than the surrounding canopy.

endemic: native or confined to a particular area (e.g. turacos are endemic to Africa, Bannerman's Turaco is endemic to montane forests of W Cameroon).

epiphyte: a plant that uses another for support, not for nutrients.

eye-crescent: a crescent above or below the eye; often occurring both above and below (forming a broken eye-ring).

eye-ring: a ring of tiny feathers surrounding the orbital ring.

eye-stripe: a usually dark stripe that extends back from the bill over the lores and through the eye (post-ocular stripe).

face: anthropomorphic term, encompassing the forehead, lores, supercilium, ear-coverts and upper throat, or any combination of these.

feral: a free-flying bird derived from domesticated or captive stock.

filoplume: a thin, hair-like feather.

first-winter: an immature plumage stage acquired after juvenile plumage and recognised by the presence of retained, and more worn, juvenile feathers in wings and tail (refers to winter season in northern hemisphere and is therefore used for Palearctic migrants).

flight feathers: main wing feathers (primaries and secondaries), but excluding tail feathers.

forest: a continuous stand of trees at least 10 m tall, their crowns interlocking.

forest outliers: patches of forest away from main forested areas.

form: a general term for distinguishable entities, including species, subspecies, morphs, etc.

fringes: in relation to feather tracts, indicates complete feather margins (compare edgings). Contrasting fringes can result in a scaly appearance to body feathers or wing-coverts.

frugivorous: fruit-eating (hence frugivore, fruit-eater).

gallery forest: forest along rivers.

gape: the fleshy interior and corners of the bill.

genus (plural genera): a taxonomic category between family and species, representing one or more species with a common ancestor (phylogenetic origin); the genus name forms the first part of the two-part scientific name.

gliding: flight on a direct course without, or between, wingbeats.

gorget: a distinctively coloured or streaked band across the throat or upper breast ('necklace').

graduated: referring to a tail in which the central rectrices are longest, the others becoming progressively shorter toward the sides.

granivorous: feeding on grains or seeds (hence granivore, seed-eater).

grassland: land covered by grasses and other herbs, either with woody plants, or the latter not covering more than 10% of the ground.

gregarious: commonly assembling in groups.

gular: related to the throat. A gular pouch is a loose area of skin extending from the throat (e.g. on pelicans). A gular stripe is a stripe extending on the centre of the throat (usually narrow and dark).

hand: the outer part of the wing, between carpal joint and tip.

herbaceous: of or similar to herbs; referring to fleshy plants that wither after the growing season (as opposed to woody plants).

Holarctic: biogeographical region that includes the Palearctic and Nearctic regions, i.e. most of the temperate zone of the northern hemisphere.

hood: a contrasting area covering all or most of the head and neck.

hybrid: the product of a cross between individuals of unlike genetic constitution; usually used for a cross between individuals of different species.

immature: a general term for a non-adult bird.

inner wing: inner part of the wing; also called arm.

insectivorous: insect eating (hence insectivore, insect-eater).

inselberg: an isolated hill or mountain rising abruptly from its surroundings.

iridescent: the glossy or 'metallic' effect of changing colours caused by reflected light from specially structured feathers, e.g. on glossy starlings and sunbirds.

iris (plural irides): the round, coloured membrane surrounding the pupil of the eye.

juvenile: a bird in its first feathered, non-downy, plumage.

jizz: an overall impression of the appearance a bird in the field, based on a combination of characters.

leading edge: the front edge of the wing. Generally referred to when it is marked with a contrasting (dark or pale) band.

littoral: situated near a (sea) shore.

local: occurring within a small or restricted area.

mangroves: open or closed stands of salt-adapted evergreen trees or bushes occurring on shores between high- and low-water mark.

mask: a dark area of plumage surrounding the eye, usually extending from the base of the bill to the ear-coverts.

melanistic: a blackish morph.

midwing panel: a contrasting pale area in the middle of the inner wing on some gull species.

migratory: making regular geographic movements.

mirror: a subterminal white spot on the wingtip of a gull. Not to be confused with the white primary tips (these, if present, are always visible on a gull at rest; mirrors are usually not visible on the closed wing).

monotypic: a biological group having a single representative. E.g. 'monotypic genus' (a genus with a single species), 'monotypic species' (a species with no subspecies). See polytypic.

montane: growing or living in mountainous areas.

morph: a normal but distinct plumage variant which is not related to sex, age or season.

morphological: pertaining to form and structure.

mottled: plumage marked with coarse spots or irregular blotches.

moult: the process of replacement of old feathers by new. Moult may be complete (all head, body, wing and tail feathers replaced during the same period) or partial (involving the renewal of all or most contour feathers, except those of wings and tail). In smaller species moult is usually repeated during a set period each year; in larger species (e.g. raptors) it may be spread over several years with overlapping moults.

Nearctic: biogeographical region comprising N America south to the tropics.

nocturnal: active at night.

nomadic: referring to a wandering or erratically occurring species with no fixed territory when not breeding.

nominate (race or **subspecies)**: the first described and named form of a species, typified by its subspecific name being the same as the specific e.g. *Sylvietta virens virens*.

non-passerines: all orders of birds except the passerines.

notched: referring to a tail in which the central feathers are slightly shorter than the outer ones, forming a very shallow fork or notch.

nuchal: relating to the hindneck or nape (used with reference to a crest, patch or collar).

orbital ring: ring of bare skin immediately surrounding the eye (not to be confused with eye-ring).

outer wing: outer part of the wing; also called hand.

Palearctic: biogeographical region that includes Europe, N Africa, the Middle East and N Asia south to the Himalayas and the Yangtze River in China.

parapatric: occupying different but contiguous geographical areas. Applied to taxonomically related populations whose ranges are closely contiguous but do not overlap. See also allopatric and sympatric.

partial migration: migration by part of a population.

passage migrant: a migrant that occurs regularly but only briefly at a locality during migration to and from its breeding and wintering grounds.

passerines: members of the large Order Passeriformes, often referred to as 'perching birds' or 'songbirds', characterised by perching with three toes pointing forward and one toe back. Includes all species from broadbills onward.

pectoral tufts: coloured tufts at each side of the breast (frequently invisible at rest).

pelagic: of the open sea. A pelagic species spends most of its life at sea, far from land.

pied: patterned black and white.

polyandrous: a female animal that mates with more than one male.

polygamous: an animal that has two or more mates of the opposite sex at the same time.

polygynous: a male animal that mates with more than one female during the breeding season.

polymorphic: having more than two distinct plumage morphs.

polytypic: a biological group having more than one representative. E.g. 'polytypic genus' (a genus with two or more species), 'polytypic species' (a species with two or more subspecies). See monotypic.

post-ocular stripe: a short, usually pale, stripe, which extends back from the eye. Sometimes reduced to a post-ocular spot.

primary forest: forest in a virgin or undisturbed state.

primary projection: the distance that the tips of the primaries project beyond the tertials on a closed wing.

proximal: nearest to centre of body or to point of attachment; opposite of distal.

race: synonymous with subspecies.

rainforest: closed-canopy forest in areas of high rainfall.

range: geographical area in which a taxon is distributed.

raptor: bird of prey (generally refers to diurnal species, not owls).

rectrices (singular rectrix): main tail feathers.

rictal bristles: sparse, though often prominent, bristles at the base of the bill.

remiges (singular remex): flight feathers (primaries and secondaries).

resident: a species or population that occurs year-round in the same area and breeds there, even though some individuals may not remain in the same area throughout the year.

riparian: bordering water.

riverine: living or growing on a river bank.

roost: a sleeping or resting place.

saddle: generally used to indicate a part of the upperparts contrasting in colour with the rest of the upper surface (e.g. in juvenile marsh terns)

Sahel: semi-arid zone between savanna and desert, characterised by scattered, thorny vegetation.

savanna: habitat dominated by grasses with a varying proportion of trees and shrubs.

secondary forest: forest regenerating after a greater or lesser degree of disturbance, often by selective logging or agriculture. It is characterised by a lack of large trees and a significant proportion of coloniser species.

secondary bar: a contrasting dark bar on the secondaries, as in immature plumages of some gull species.

sedentary: a species that remains in the same site throughout the year, individuals wandering no more than a few kilometres at the most.

serrated: with a fine, saw-like edge (as in the outer edge of the outer primary in saw-wings).

shorebird: another term for wader.

shoulder patch: an area of contrastingly coloured wing-coverts.

soaring: circling flight (often in thermal of warm air).

song: a more complex pattern of vocalisations, mainly uttered in the breeding season (cf. call).

speculum: a usually iridescent panel on the secondaries in dabbling ducks.

stratum (plural **strata**): level (used to indicate levels within forest, e.g. middle strata = mid-levels).

streamer: an exceptionally long, slender tail feather, as in some bee-eaters, paradise flycatchers and sunbirds.

square: referring to a tail in which all feathers are of equal length, forming a straight border.

subadult: an imprecise term indicating a bird in nearly adult plumage; often used when precise age is difficult to establish, e.g. in large raptors.

subspecies: a population of a given species that differs more or less obviously in appearance from one or more other populations of the same species. Subspecies are assumed to be (at least theoretically) capable of interbreeding but to be relatively isolated from each other, with interbreeding limited to areas of contact. The border between species and subspecies is often arbitrary. The nominate subspecies is the one first named: its scientific name has the second and the third terms identical, e.g. *Trochocercus nitens nitens*.

subterminal: near the end. A subterminal band is a contrasting (dark or pale) band, usually broad, situated near the tip of a feather or feather tract (used particularly in reference to the tail).

supercilium: a usually pale stripe, which extends from the base of the upper mandible, above the lores and the eye, and extending behind the eye; compare supraloral stripe.

superspecies: a group of closely related and largely or entirely allopatric species.

supraloral stripe: a short, usually pale, stripe, which extends from the base of the upper mandible, over the lores, to just above or in front of the eye (not extending behind the eye; if it does it becomes a supercilium).

sympatric: occurring in the same geographical area. Opposite of allopatric.

taxon (plural **taxa**): a named form (this unit of biological classification can refer to a group of organisms of any taxonomic rank, e.g. family, genus, species, subspecies, etc.)

taxonomy: the science of classification and naming of all life forms.

terminal: at the end. A terminal band is a contrasting (dark or pale) band, usually broad, situated at the tip of a feather or feather tract (used particularly in reference to the tail).

terrestrial: living or occurring mainly on the ground.

thermal: rising current of heated air (used by e.g. raptors to gain height).

thicket: a closed stand of bushes and climbers usually between 3 and 7 m tall.

top of head: includes forehead and crown.

trailing edge: the rear edge of the wing, often contrasting with the rest of the wing (cf. leading edge).

understorey: the lowest stratum in forest or woodland.

vent: the area around the cloaca (anal opening), just behind the legs (not to be confused with the undertail-coverts).

vermiculations: narrow, often wavy, bars that generally create an overall effect and are visible only at close range.

vestigial: a feature that is very much reduced (almost absent).

wader: general term used, in the plural, for members of the families Rostratulidae (painted-snipes) to Scolopacidae (sandpipers and allies); another term for shorebird.

wattle: naked, fleshy, usually brightly coloured, skin on the head (e.g. around eye), base of bill or throat (e.g. as in wattle-eyes, African Wattled Plover and Black-casqued Hornbill).

window: a contrasting pale panel in the outer wing of some raptors and gulls.

wingbar: generally a narrow and well-defined dark or pale bar on the upperwing, and often referring to a band formed by pale tips to the greater or median coverts (or both, as in 'double wingbar').

wing feathers: includes flight feathers and wing-coverts.

wing panel: a contrasting, usually pale band on the wing; broader and generally more diffuse than a wingbar (often formed by pale edges to the remiges or coverts). At rest, usually refers to a panel on the secondary coverts, secondaries and/or tertials; in flight, usually refers to a panel on the primary bases.

wooded grassland: land covered by grasses and other herbs, with woody plants covering 10–40% of the ground.

woodland: an open stand of trees at least 8 m tall with a canopy cover of 40% or more. The field layer is usually dominated by grasses.

zygodactyl: form of foot with the outer and inner toes pointing backward, the middle toes forward, as in turacos, cuckoos, parrots, owls and woodpeckers.

ABBREVIATIONS

BoA	*The Birds of Africa*
BWP	*Birds of the Western Palearctic* (Cramp & Simmons 1977–1983, Cramp 1985–1992, Cramp & Perrins 1993–1994)
BWPC	*Birds of the Western Palearctic* Concise Edition (Snow & Perrins 1998)
c.	circa (approximately)
C	central
CAR	Central African Republic
esp.	especially
Eq. Guinea	Equatorial Guinea
excl.	excluding
FR	Forest Reserve
HBW	*Handbook of the Birds of the World* (del Hoyo *et al.* 1992–2001)
id.	idem
incl.	including
I.	Island (e.g. Príncipe I.)
Is	Islands (e.g. Cape Verde Is)
L	length
L.	Lake
Mt	Mount, Mountain
Mts	Mountains
N, NE, E, SE, S, SW, W, NW	compass directions (northern, northeastern, eastern, etc.)
	Thus S Africa means southern Africa, not the country South Africa; if the latter is meant, it is written in full 'South Africa'.
NP	National Park
q.v.	*quod vide*, which see
R.	River
resp.	respectively
sp.	species (singular)
spp.	species (plural)
ssp.	subspecies (=race)
TBW	*Threatened Birds of the World*
WA	western Africa
WS	wingspan

COLOUR PLATES
1–147

PLATE 1: OSTRICH, GREBES, DARTER, CORMORANTS AND PELICANS

1 **GREAT CRESTED GREBE** *Podiceps cristatus* 46–51 cm **V**   **Page 333**
Various aquatic habitats. Palearctic vagrant, Senegal, Mali, Niger.
   1a   **Adult non-breeding** Long, mainly white neck; blackish crown; long, sharp-pointed bill.
   1b   **Adult breeding** Double-horned crest; chestnut and black tippets.

2 **BLACK-NECKED GREBE** *Podiceps nigricollis* 28–34 cm **V**   **Page 334**
Various aquatic habitats. Palearctic vagrant, N Senegal, Mauritania, N Nigeria, SW Cameroon.
   2a   **Adult non-breeding** Steep forehead; black cap to below eye; slightly upturned bill.
   2b   **Adult breeding** Black neck; loose golden ear-tufts.

3 **LITTLE GREBE** *Tachybaptus ruficollis* 25–29 cm **R/P?**   **Page 333**
Various aquatic habitats. Throughout. Locally common.
   3a   **Adult breeding** Dumpy; bright chestnut cheeks and foreneck; yellowish gape patch.
   3b   **Adult non-breeding** Duller; pale brown cheeks and foreneck; contrasting dark cap.

4 **OSTRICH** *Struthio camelus* **R**   **Page 327**
Sahel. Mauritania to Chad–NE CAR. Rare/uncommon.
   4a   **Adult male** Huge (height 210–275 cm); black-and-white plumage.
   4b   **Adult female** Smaller (height 175–190 cm); dull grey-brown plumage.
   4c   **Chick at *c.* 4 weeks** Buffish, spiky down tipped black; black spots on neck.

5 **AFRICAN DARTER** *Anhinga rufa* *c.* 80 cm; WS 120 cm **R**   **Page 337**
Various aquatic habitats. Throughout. Scarce/uncommon.
Slender neck; pointed, dagger-shaped bill. Swims with only head and neck above water.
   5a   **Adult male** Mainly black; chestnut foreneck; white stripe on head- and neck-sides.
   5b   **Adult female** Duller and browner than male.
   5c   **Juvenile** As female, but even browner and paler; no white neck-stripe.

6 **LONG-TAILED CORMORANT** *Phalacrocorax africanus* 51–56 cm; WS 85 cm **R**   **Page 337**
Various aquatic habitats. Throughout; also São Tomé. Common.
   6a   **Adult breeding** Black; long tail; short crest on forehead.
   6b   **Adult non-breeding** Duller; no crest. Some brownish like immature but eye red.
   6c   **Juvenile/immature** Mainly dull brownish above, pale brownish to buffish-white below.

7 **GREAT CORMORANT** *Phalacrocorax carbo* 80–100 cm; WS 130–160 cm **R/V**   **Page 336**
Aquatic habitats. Mauritania–Guinea; locally common. L. Chad area; uncommon. Rare elsewhere.
   7a   **Adult *lucidus* breeding** Large; white throat and breast; white thigh patch.
   7b   **Juvenile** Brownish above, whitish below.
   7c   **Adult *maroccanus* breeding** White throat, black breast. Breeds Morocco; possibly moves south
       outside breeding season.

CAPE CORMORANT *Phalacrocorax capensis* 61–64 cm; WS 109 cm **V**   **Page 337**
Vagrant from southern African coasts, north to Gabon. **See illustration on plate 146.**
Adult breeding Medium-sized; black; yellow-orange gular area.
Adult non-breeding/immature Dull brown, underparts paler; yellowish-brown gular area.

8 **GREAT WHITE PELICAN** *Pelecanus onocrotalus* 140–175 cm; WS 270–360 cm **R/M Page 338**
Large aquatic habitats. Breeds Mauritania, Senegal and Nigeria. Scarce/uncommon to locally common.
   8a   **Adult** Huge; mainly white; massive bill with yellow pouch.
   8b   **Immature** Mottled greyish-brown above; dirty white below.
   8c–d Adult in flight Strongly contrasting black-and-white wings.
   8e–f Immature in flight Note underwing pattern (brown on coverts progressively lost with age);
       contrasting white rump.

9 **PINK-BACKED PELICAN** *Pelecanus rufescens* 125–132 cm; WS 265–290 cm **R**   **Page 338**
Large aquatic habitats. Throughout. Rare/uncommon to locally common.
   9a   **Adult** Whitish with pale grey cast; pale yellowish bill and pouch.
   9b   **Immature** Mottled brownish above; bare parts dull coloured.
   9c–d Adult in flight Less contrasting wing pattern than 8c–d.
   9e–f Immature in flight Underwing pattern as adult; contrasting white rump.

PLATE 2: PETRELS AND SHEARWATERS

Medium-sized to large pelagic seabirds with long narrow wings. Occurrence in W Africa inadequately known.

1 FEA'S PETREL *Pterodroma feae* 35 cm; WS 84–91 cm **R/V?** NT   **Page 328**
Breeds Cape Verde Is. Non-breeding visitor offshore south to 9°S; rare.
1a   Dark upperwings contrasting with greyer mantle; pale grey tail; dark mark behind eye.
1b   Dark underwings; white underparts with dusky 'shawl' extending on neck-sides.

ZINO'S PETREL *Pterodroma madeira* 33 cm; WS 78–83 cm **V?** CR   **Page 328**
Possible non-breeding visitor (breeds Madeira). Not illustrated.
Extremely similar to 1; slightly smaller, with slimmer bill and broader, more rounded wingtips.

2 BULWER'S PETREL *Bulweria bulwerii* 27 cm; WS 68–73 cm **R/M**   **Page 329**
Breeds Cape Verde Is. Non-breeding visitor offshore Mauritania to Gulf of Guinea; scarce.
Entirely blackish with pale brown band on upperwing-coverts; long pointed tail.

3 CAPE PETREL *Daption capense* 39 cm; WS 81–91cm **V**   **Page 328**
Vagrant from Southern Ocean, recorded off Togo (Gabon?).
3a–b Striking black-and-white plumage above and below.

4 CORY'S SHEARWATER *Calonectris diomedea* 45–56 cm; WS 100–125 cm **P**   **Page 329**
Non-breeding visitor, Mauritania to Gulf of Guinea. Rare/not uncommon.
Large. Dull brownish upperparts contrast with white underparts; bowed wings; pale bill.

5 CAPE VERDE SHEARWATER *Calonectris edwardsii* 40–41 cm; WS 90–110 cm **R/M**   **Page 329**
Breeds Cape Verde Is. Non-breeding visitor, Mauritania to Gulf of Guinea; rare/not uncommon.
As 4 but slimmer; head smaller, more angular; tail relatively longer; head and upperparts slightly
darker and greyer brown; bill slimmer, grey with dark subterminal band.

6 GREAT SHEARWATER *Puffinus gravis* 43–51 cm; WS 105–122 cm **M/V**   **Page 329**
Non-breeding visitor offshore Mauritania–Gabon; rare. Also Cape Verde waters; not uncommon.
6a–b Black cap and brown upperparts contrast with white collar and white uppertail-coverts. White
        underwing with dark band on axillaries and dark belly patch.

7 SOOTY SHEARWATER *Puffinus griseus* 40–51 cm; WS 94–109 cm **M/V**   **Page 330**
Non-breeding visitor offshore, throughout. Uncommon/rare.
7a–b Entirely dark; silvery underwing-coverts. Long, slim body; long swept-back wings; soaring flight.

8 LITTLE SHEARWATER *Puffinus assimilis* 25–30 cm; WS 58–67 cm **R/M**   **Page 330**
Probably scarce but regular offshore to at least 9°S.
8a   *P. a. baroli* Small; black above; white below; face mainly white. Breeds Atlantic islands.
8b   *P. a. boydi* As 8a but face-sides and undertail-coverts dark. Breeds Cape Verde Is.

9 MANX SHEARWATER *Puffinus puffinus* 30–35 cm; WS 76–82 cm **P**   **Page 330**
Non-breeding visitor offshore, mostly Mauritania–Liberia. Scarce.
9a–b Black above; white below; head-sides black. Banks from side to side in flight.

BALEARIC SHEARWATER *Puffinus mauretanicus* 30–40 cm; WS 76–93 cm **V**   **Page 330**
Vagrant, recorded off Mauritania. **See illustration on plate 146.**
Resembles 9 but upperparts brownish; underparts dirty greyish-buff; axillaries and vent dark.

SOUTHERN GIANT PETREL *Macronectes giganteus* 86–99 cm; WS 185–210 cm **V** NT **Page 328**
Vagrant, offshore Congo. **See illustration on plate 146.**
Very large (like small albatross). Massive, pale bill. Adult dark morph greyish-brown with whitish
head. Adult white morph all white with some black specks.

NORTHERN FULMAR *Fulmarus glacialis* 43–52 cm; WS 101–117 cm **V**   **Page 780**
Palearctic vagrant, offshore Senegal. Not illustrated.
Stocky; head and underparts typically white; upperparts pale grey. Flies on stiff, straight wings.

PLATE 3: STORM-PETRELS AND TROPICBIRDS

Storm-petrels. Small, swallow-like pelagic seabirds. Identification at sea often difficult; flight action is an important clue. Occurrence in W Africa inadequately known.

1 **EUROPEAN STORM-PETREL** *Hydrobates pelagicus* 15–16 cm; WS 38–42 cm **P**   **Page 332**
Regular offshore Mauritania to Ivory Coast. Rare, Gulf of Guinea.
1a–b Small. Wings and tail rounded; conspicuous white band on underwing; white rump patch. Fluttering flight.

2 **WILSON'S STORM-PETREL** *Oceanites oceanicus* 15–19 cm, WS 38–42 cm **M/V**   **Page 331**
Probably uncommon/scarce offshore.
2a–b Small. Distinct pale diagonal band on upperwing; underwing all dark. Feet project beyond square tail. Wings typically held rather straight when gliding. Flies above waves with legs dangling when foraging.

3 **LEACH'S STORM-PETREL** *Oceanodroma leucorhoa* 19–22 cm; WS 45–48 cm **P/V**   **Page 332**
Probably regular offshore.
3a–b Relatively large. Prominent pale diagonal band on upperwing; rump divided in centre or smudged grey. Tail forked; feet do not project. Buoyant, bounding flight. Wings held angled when gliding.

4 **MADEIRAN STORM-PETREL** *Oceanodroma castro* 19–21 cm; WS 43–46 cm **R/M/V** **Page 332**
Breeds Cape Verde Is (locally not uncommon) and possibly São Tomé; probably frequent offshore.
4a–b Relatively large. Tail less forked than 3; more white on rump. Compare also 2. Buoyant and zigzagging flight.

5 **WHITE-FACED STORM-PETREL** *Pelagodroma marina* 20–21 cm; WS 41–43 cm **R/M Page 331**
Breeds Cape Verde Is; probably regular offshore Mauritania to at least Liberia.
5a–b Grey above; white below; distinctive head pattern. Very long legs project beyond slightly forked tail. Flies close to waves with legs dangling.

BLACK-BELLIED STORM-PETREL *Fregetta tropica* 20 cm; WS 46 cm **V**   **Page 331**
Vagrant; recorded off Liberia (Sierra Leone?) and Gulf of Guinea. **See illustration on plate 146**
White rump; white belly with black stripe on centre; white underwing-coverts.

WHITE-BELLIED STORM-PETREL *Fregetta grallaria* 20 cm; WS 46 cm **V?**   **Page 332**
Potential vagrant (single record north of Cape Verde Is). **See illustration on plate 146**
Similar to Black-bellied Storm-petrel; belly all white; underwing with more white, dark leading edge narrower.

Tropicbirds. Medium-sized, graceful, highly aerial seabirds. Local. Breed colonially on islands.

6 **WHITE-TAILED TROPICBIRD** *Phaethon lepturus* 40 cm ; WS 92 cm **R/V**   **Page 334**
Breeds São Tomé (Príncipe?), Annobón; locally common. Offshore vagrant elsewhere.
6a **Adult** Pure white upperparts; diagnostic black diagonal bar on upperwing; long tail streamers (33–40 cm); yellowish-orange bill.
6b **Juvenile/immature** Note coarse barring on upperparts and extensive white on upperwing; pale yellow, black-tipped bill.

7 **RED-BILLED TROPICBIRD** *Phaethon aethereus* 48 cm; WS 103 cm **R/V**   **Page 334**
Breeds Senegal and Cape Verde Is; scarce and local. Offshore vagrant elsewhere.
7a **Adult** Barred upperparts; long tail streamers (46–56 cm); red bill.
7b **Juvenile/immature** Closely barred upperparts appear grey at distance; inner secondaries and upperwing-coverts barred; black nuchal collar; black-tipped yellowish bill.

PLATE 4: GANNETS, BOOBIES, FRIGATEBIRDS AND PENGUINS

Gannets and boobies. Large, conspicuous seabirds with stout, tapering bills, cigar-shaped bodies, long, narrow wings and wedge-shaped tails.

1 NORTHERN GANNET *Sula bassana* 87–100 cm; WS 165–180 cm **P**   **Page 335**
Uncommon non-breeding visitor to at least 10°N. Vagrant, Cape Verde Is.
- **1a Adult** All-white secondaries and tail; yellowish crown and nape.
- **1b Adult (head)** Short gular stripe.
- **1c Second-year** Some white feathers appearing on secondaries and tail.
- **1d Juvenile** Dark brown; uppertail-coverts white. Indistinguishable from juvenile Cape Gannet.

2 CAPE GANNET *Sula capensis* 84–94 cm; WS 165–180 cm **V VU**   **Page 335**
Vagrant or scarce non-breeding visitor to Gulf of Guinea.
- **2a Adult** Black secondaries and tail; yellowish crown and nape. Compare 1.
- **2b Adult (head)** Long gular stripe.
- **2c Second-year** Secondaries and tail all black.

3 MASKED BOOBY *Sula dactylatra* 81–92 cm; WS 152 cm **V**   **Page 335**
Vagrant, Gulf of Guinea and offshore Liberia and Ivory Coast.
- **3a Adult** All-black secondaries and tail; white crown; small black mask.
- **3b Juvenile** All dark above with white collar.
- **3c Immature** Dark head contrasts with white breast and belly; white collar broader and more conspicuous than in juvenile.

4 RED-FOOTED BOOBY *Sula sula* 66–77 cm; WS 91–101 cm **V**   **Page 336**
Vagrant, Cape Verde Is and Gulf of Guinea.
- **4a Adult white morph** Black secondaries; all-white tail; red feet.
- **4b Adult brown morph** Entirely dull brown above and below; red feet.
- **4c Juvenile (all morphs)** As 4b, but with dark bill and yellowish-grey feet.

5 BROWN BOOBY *Sula leucogaster* 64–74 cm; WS 132–150 cm **R/M/V**   **Page 336**
Breeds Cape Verde Is, Alcatraz and Gulf of Guinea Is; locally common. Uncommon elsewhere.
- **5a Adult (perched)** All brown above; white breast and belly.
- **5b Adult** Distinctive underwing pattern; brown of head extending onto breast.
- **5c Juvenile** White parts of adult dusky-brown.

Frigatebirds. Large seabirds with extremely long, angular wings and very long, deeply forked tails. Very buoyant flight; soar effortlessly for long periods. Identification often difficult; note in particular amount of white on underparts and occurrence of 'spurs' on underwing.

6 MAGNIFICENT FRIGATEBIRD *Fregata magnificens* 89–114 cm; WS 217–244 cm **R/V Page 339**
Breeds Cape Verde Is; rare. Offshore Mauritania–Senegambia; rare.
- **6a Adult male** Entirely black; red gular pouch.
- **6b Adult male** Gular pouch inflated during display.
- **6c Adult female** All-dark underwings; white breast.
- **6d Immature** All-dark underwing; broken breast-band.

7 ASCENSION FRIGATEBIRD *Fregata aquila* 89–96 cm; WS 196–201 cm **V VU**   **Page 339**
Vagrant, Gulf of Guinea.
- **7a Adult male** As 6a, but range different. Typical female very similar.
- **7b Adult female pale morph** Note narrow white 'spur' on axillaries.
- **7c Immature** White on underwing-coverts; complete brown breast-band.

Penguins. Stocky, flightless seabirds, with flipper-like wings, and short legs and tails. Float with head held high and most of body submerged. On land, walk upright with waddling gait.

JACKASS PENGUIN *Spheniscus demersus* *c.* 60 cm **V VU**   **Page 333**
S African vagrant, Gabon and Congo. **See illustration on plate 146.**

PLATE 5: HERONS I

Slender, medium-sized to large wading birds with long necks and legs and long, straight, pointed bills. Fly with neck retracted, forming an S-shape. Most species usually found in or near water.

1 GREY HERON *Ardea cinerea* 90–100 cm; WS 175–195 cm **R/P**   **Page 346**
Aquatic habitats. Throughout; common/not uncommon. Also Cape Verde and Gulf of Guinea Is; uncommon/rare.
  1a Adult White crown; narrow black nape plumes; yellow bill.
  1b Juvenile Grey foreneck and crown.
  1c In flight Underwing wholly dark grey.

2 BLACK-HEADED HERON *Ardea melanocephala* 92–96 cm **R/M**   **Page 346**
Grassland, farmland, aquatic habitats. Throughout. Uncommon/common.
  2a Adult Black crown, head-sides and hindneck. White throat contrasts with black-and-white streaked foreneck.
  2b Juvenile Dark crown and head-sides contrast with white foreneck.
  2c In flight White underwing-coverts contrast with dark grey flight feathers.

3 PURPLE HERON *Ardea purpurea* 78–90 cm; WS 120–150 cm **R/P**   **Page 346**
Aquatic habitats. Throughout; uncommon/common. Cape Verde Is; rare.
  3a Adult Rufous-buff head-sides and neck; dark grey back; purple shoulders.
  3b Juvenile Sandy-brown upperparts; snake-like neck; yellow legs.
  3c Adult in flight Distinctive, deeply pouched appearance of neck; large feet; brownish underwing-coverts.
  3d Juvenile/immature in flight Paler than adult.

4 GOLIATH HERON *Ardea goliath* 135–150 cm; WS 210–230 cm **R**   **Page 347**
Aquatic habitats. Throughout; scarce/uncommon. Vagrant, Liberia.
  4a Adult Massive size; rufous-brown head and hindneck; dark grey back.
  4b Juvenile Paler and browner than adult; more white on head-sides and neck; black legs.
  4c In flight Brown underwing-coverts contrast with dark grey flight feathers.

PLATE 6: HERONS II

1 BLACK-CROWNED NIGHT HERON *Nycticorax nycticorax* 56–65 cm **R/P** **Page 341**
Aquatic habitats. Throughout. Uncommon/locally common. Vagrant, Cape Verde Is. WS 90–100 cm
1a **Adult breeding** Stocky; black crown and back; white cheeks and neck; grey wings.
1b **Juvenile** Dark brown streaked buff with conspicuous buff spots on wings.
1c **Adult in flight** Feet project clearly beyond tail.
1d **Juvenile in flight** Note absence of white on back and distinct feet projection.

2 WHITE-BACKED NIGHT HERON *Gorsachius leuconotus* 50–55 cm **R** **Page 341**
Aquatic habitats in forest zone and wooded savanna. Senegambia to Congo. Rare/scarce to locally common.
2a **Adult** Black head; dark upperparts; rufous neck and breast; white throat.
2b **Juvenile** Darker than 1b with fewer wing spots and blackish forehead; huge eye.
2c **Adult in flight** Inconspicuous white patch on back; feet project slightly beyond tail.
2d **Juvenile in flight** Note presence of white scapulars and slight feet projection.

3 SQUACCO HERON *Ardeola ralloides* 42–47 cm; WS 80–92 cm **R/P** **Page 342**
Aquatic habitats. Throughout. Not uncommon. Vagrant, Cape Verde Is, Bioko, Príncipe.
3a **Adult breeding** Wholly warm buff; bright blue bill tipped black.
3b **Juvenile** Dull greyish-brown above; streaked below; white wings.
3c **Adult in flight** White wings contrast with buff upperparts.

4 WESTERN REEF EGRET *Egretta gularis* 55–65 cm; WS 86–104 cm **R/M** **Page 344**
Aquatic habitats, mainly coastal. Mauritania–Gabon and Gulf of Guinea Is. Not uncommon/locally common. Also Cape Verde Is (rare).
4a **Adult dark morph** Slate-black; white throat; greenish-black legs; yellowish feet.
4b **Adult white morph** Very similar to 7; note rather heavier bill with slightly drooped tip and pale base; yellowish feet and lower legs.
4c **Adult dark morph in flight** White patch on primary-coverts.

5 BLACK HERON *Egretta ardesiaca* 48–50 cm **R/M** **Page 343**
Aquatic habitats. S Mauritania–S Nigeria; C Mali–Chad. Uncommon/locally common. Vagrant, Burkina Faso, Gabon, Cape Verde Is, São Tomé.
Entirely black; orange-yellow feet; shaggy plumes on hindneck, mantle and breast.

6 CATTLE EGRET *Bubulcus ibis* 48–56 cm; WS 90–96 cm **R/M** **Page 342**
Various open habitats. Throughout; also Cape Verde and Gulf of Guinea Is. Common.
Small; mainly white; distinct jowl. Tame.
6a **Adult breeding** Crown, mantle and breast buff; bill and legs reddish.
6b **Adult non-breeding** All white; bill yellowish; legs and feet greenish-yellow to greyish.
6c **In flight** Stocky; shortish bill. Rapid, shallow wingbeats.

7 LITTLE EGRET *Egretta garzetta* 55–65 cm; WS 88–95 cm **R/M/P** **Page 345**
Aquatic habitats. Throughout. Common. Also Cape Verde Is (not uncommon), Bioko, São Tomé and Príncipe (rare).
All white; slender, black bill; black legs; yellow feet.
7a **Adult breeding** Plumes on nape, mantle and breast; yellow lores.
7b **Juvenile** Bare parts dull coloured; no plumes.

8 INTERMEDIATE EGRET *Egretta intermedia* 65–72 cm; WS 105–115 cm **R/M** **Page 345**
Aquatic habitats. Throughout. Uncommon/common. Vagrant, Cape Verde Is.
All white; bill yellow; gape line ends below eye; legs and feet black with yellowish tibia.
Adult breeding Extensive plumes hang over back and tail.

9 GREAT EGRET *Egretta alba* 85–95 cm; WS 140–170 cm **R/M/P** **Page 345**
Aquatic habitats. Throughout. Common/not uncommon. Vagrant, Cape Verde Is and São Tomé.
Large; all white; long, kinked neck; gape line ends behind eye; legs and feet black.
9a **Adult *melanorhynchos* breeding** Bill black; lores emerald-green; long plumes hang over back and tail.
9b **Adult *melanorhynchos* non-breeding** Bill all yellow; no long plumes.

PLATE 7: HAMERKOP, HERONS AND BITTERNS

Hamerkop. Unique waterbird. Builds huge domed nest in fork of large trees.

1 HAMERKOP *Scopus umbretta* 50–56 cm **R**   **Page 347**
Aquatic habitats in open woodland. Throughout. Uncommon/common.
1a   Entirely dark brown; characteristic head shape.
1b   **In flight** Broad wings.

Herons (continued)

2 LITTLE BITTERN *Ixobrychus minutus* 27–38 cm; WS 40–58 cm **R/P**   **Page 340**
Aquatic habitats. Throughout; also Príncipe. Uncommon. Vagrant, Cape Verde Is.
2a   **Adult male** *payesii* Small; black crown and back contrast with rich chestnut head-sides and neck.
2b   **Adult male** *minutus* Much paler head-sides and neck than 2a.
2c   **Adult female** Duller than male; upperwing-coverts browner, less contrasting.
2d   **Juvenile** Streaked, pale buffy upperparts.
2e   **Adult male in flight** Conspicuous pale wing panel (upperwing-coverts).

3 DWARF BITTERN *Ixobrychus sturmii* 25–30 cm **M**   **Page 340**
Aquatic habitats. Throughout. Uncommon/scarce.
3a   **Adult** Very small; wholly dark grey above; boldly streaked below.
3b   **Juvenile** Buff edges to feathers of upperparts.
3c   **In flight** Small; dark.

4 GREEN-BACKED HERON *Butorides striatus* 40 cm; WS 54 cm **R**   **Page 343**
Aquatic habitats. Throughout; also Gulf of Guinea Is. Common.
4a   **Adult** Small; dark crown and upperparts; grey head-sides and underparts.
4b   **Juvenile** Brown upperparts spotted buffish.
4c   **Adult in flight** Dark; conspicuous yellow or orange legs.

5 RUFOUS-BELLIED HERON *Ardeola rufiventris* 38 cm **V?**   **Page 342**
Aquatic habitats. S & E African species claimed from S Nigeria.
5a   **Adult male** Black, with maroon wings, belly and tail.
5b   **Juvenile** Streaked buffish-brown on head-sides, neck and upper breast.
5c   **Adult in flight** Dark, with bright yellow legs and feet.

6 WHITE-CRESTED TIGER HERON *Tigriornis leucolophus* 66–80 cm **R**   DD   **Page 341**
Forest streams; also mangroves. Senegambia to SW CAR–Congo. Uncommon/rare.
6a   **Adult** Cryptically patterned; white crest (half concealed by nape feathers).
6b   **Juvenile** More strongly barred than adult.
6c   **In flight** Dark; primaries white tipped.

7 GREAT BITTERN *Botaurus stellaris* 70–80 cm; WS 125–135 cm **P/V**   **Page 340**
Aquatic habitats. Recorded Mauritania, Senegambia, Ghana, Niger, Nigeria, Cameroon, CAR and Gabon. Rare.
7a   **Adult** Black crown and malar stripe. Secretive.
7b   **In flight** Broad, rounded wings strongly bowed downwards.

PLATE 8: STORKS

Large to very large wading and terrestrial birds with long bills, necks and legs. Fly with neck outstretched, except Marabou Stork, and legs trailing. Normal flight action consists of soaring and gliding, often at great heights, alternated with slow wingbeats. Gregarious or solitary.

1 YELLOW-BILLED STORK *Mycteria ibis* 95–105 cm; WS 155–165 cm **M/R**   **Page 348**
Aquatic habitats. Throughout; mainly in savanna belt. Uncommon/rare. Vagrant, São Tomé.
1a Adult Mainly white; yellow bill; red facial skin and legs.
1b Juvenile Mainly greyish-brown; dull-coloured bare parts.
1c Adult in flight Black flight feathers and tail. Compare 3c.

2 AFRICAN OPENBILL STORK *Anastomus lamelligerus* 80–94 cm **M/R**   **Page 348**
Aquatic habitats, mainly fresh water. Throughout. Uncommon/rare to locally not uncommon.
2a Adult All dark; distinctly shaped bill with gap between mandibles.
2b Juvenile Dull brown; buff-edged feathers on upperparts; bill almost straight.

3 WHITE STORK *Ciconia ciconia* 100–120 cm; WS 155–165 cm **P**   **Page 349**
Grassland, open savanna, wetlands. Mainly Sahel; uncommon/locally common. Rare/scarce further south. Vagrant, N Congo, São Tomé.
3a Adult Mainly white with black on wings; red bill and legs.
3b Juvenile Bill dusky.
3c Adult in flight Black flight feathers; white tail. Compare 1c.

4 BLACK STORK *Ciconia nigra* 95–100 cm; WS 145–155 cm **P/V**   **Page 348**
Aquatic and dry habitats. Mauritania–Guinea-Bissau to Chad–CAR. Rare/scarce.
4a Adult Black, with white breast and belly; red bill and legs.
4b Juvenile Sooty-brown; dull-coloured bare parts.
4c–d Adult in flight All-black upperparts. Compare 5c–d.

5 ABDIM'S STORK *Ciconia abdimii* 75–81 cm **M**   **Page 349**
Grassland, cultivated areas. Mainly Sahel/savanna throughout. Locally common/rare.
5a Adult Mainly black, with white breast and belly; greenish bill and legs.
5b Juvenile Duller than adult.
5c–d Adult in flight White lower back and rump. Compare 4c–d.

6 WOOLLY-NECKED STORK *Ciconia episcopus* *c.* 90 cm **R/M**   **Page 349**
Various habitats, usually near water. North to Senegambia to C Chad. Uncommon/rare.
6a Adult Dark; white head and neck; bill black tipped reddish; legs black.
6b Juvenile Lacks white forehead.
6c Adult in flight White head, neck, belly and undertail-coverts.

7 SADDLE-BILLED STORK *Ephippiorhynchus senegalensis* 145–150 cm **R/M**   **Page 350**
Aquatic habitats. Senegambia–Guinea-Bissau to Chad–CAR; also Gabon, Congo. Rare/uncommon. WS 240–270 cm.
7a Adult male Very large; black and white; red, black and yellow bill; dark eyes; yellow wattles.
7b Adult female Yellow eyes; typically no wattles.
7c Juvenile Greyish; bill dusky.
7d Adult in flight White flight feathers.

8 MARABOU STORK *Leptoptilos crumeniferus* *c.* 150 cm; WS 230–285 cm **R/M**   **Page 350**
Open habitats, terrestrial and aquatic. Mainly Sahel/northern savanna; also S Gabon, Congo. Rare/locally not uncommon.
8a Adult Huge; bare pinkish head and neck; massive bill.
8b Juvenile Upperparts dark brown.
8c Adult in flight Black wings contrast with white underparts; head tucked into shoulders.

PLATE 9: SPOONBILLS, SHOEBILL AND FLAMINGOS

Spoonbills. Large, white wading birds with characteristic spoon-shaped bill and long necks and legs. Fly with neck outstretched; sometimes soar. Gregarious or solitary. Feed by sweeping bill from side to side in water.

1   **EURASIAN SPOONBILL** *Platalea leucorodia*   80–90 cm; WS 115–130 cm   **R/P**   **Page 353**
Aquatic habitats.
    1a   **Adult** *leucorodia* Bill black with yellow tip; legs black; yellowish breast-band. Palearctic winter visitor, Mauritania–Guinea-Bissau and patchily to Chad. Uncommon/rare.
    1b   **Adult** *balsaci* Bill all black. Breeds Mauritania (Banc d'Arguin); locally common.
    1c   **Juvenile** Bill dull pinkish, legs yellowish-brown.
    1d   **Juvenile in flight** Outstretched neck; primaries tipped black (all white in adult).

2   **AFRICAN SPOONBILL** *Platalea alba*   *c.* 90 cm   **R**   **Page 354**
Aquatic habitats. Widespread but patchily distributed. Uncommon/scarce.
    2a   **Adult** Bare face red; bill grey with red margins; legs red.
    2b   **Juvenile** Bill yellowish; legs black; no red on face.
    2c   **Juvenile in flight** Note amount of black on wingtips and primary-coverts.

Shoebill. Very large, long-legged wading bird with unique, slightly shoe-shaped bill. Normally silent.

3   **SHOEBILL** *Balaeniceps rex*   110–140 cm   **M**   NT   **Page 351**
Extensive freshwater swamps. N CAR. Rare.
    3a   **Adult** All grey; distinctively shaped, huge pale bill.
    3b   **In flight** Neck usually tucked in.

Flamingos. Large wading birds with extremely long necks and legs, pink plumage and unique down-curved bills adapted for filter-feeding. Fly in lines or V formations with neck outstretched. Highly gregarious and quite noisy. Occur on shallow brackish, alkaline or saline lakes and lagoons.

4   **GREATER FLAMINGO** *Phoenicopterus ruber*   127–140 cm; WS 140–165 cm   **R/P/M**   **Page 354**
Coastal aquatic habitats. Mauritania–Sierra Leone; locally common to uncommon/rare. Vagrant, S Niger, N Cameroon, S Congo.
    4a   **Adult** Pale pink plumage; pink bill with contrasting black tip.
    4b   **Juvenile** Dull greyish plumage; note rather large bill.
    4c–d Adult in flight Scarlet lesser coverts contrast strongly with black flight feathers.5

5   **LESSER FLAMINGO** *Phoeniconaias minor*   80–90 cm; WS 95–100 cm   **M/V**   NT   **Page 355**
Aquatic habitats, mainly coastal. Mainly SW Mauritania (breeding)/Sierra Leone; rare. Scattered records Ghana, Niger, L. Chad area, Nigeria, Cameroon, Gabon, São Tomé and Príncipe.
    5a   **Adult** Rose-pink plumage; dark crimson bill; smaller than 4a.
    5b   **Juvenile** Brownish-grey; bill smaller than 4b.
    5c–d Adult in flight Lesser coverts contrast less than in 4c–d.

PLATE 10: IBISES

Medium-sized terrestrial and wading birds with rather long necks and legs and slender decurved bills. Fly with neck outstretched. Feed in dry habitats, on forest floor or in shallow wet areas by probing in soft mud.

1 GLOSSY IBIS *Plegadis falcinellus* 55–65 cm; WS 80–95 cm **R/P/V** **Page 351**
Aquatic habitats. Mainly Mauritania–Senegambia to Chad–N CAR. Not uncommon/rare. Vagrant, Cape Verde Is.
1a Adult breeding Dark, glossy plumage; slender build.
1b Adult non-breeding Duller; head and neck finely streaked white.
1c Juvenile Sooty-brown; head and throat mottled white.
1d In flight Legs project well beyond tail. Mostly silent.

2 HADADA IBIS *Bostrychia hagedash* 76–89 cm **R** **Page 351**
Rivers, marshes, moist savannas, mangroves. Throughout; also Bioko. Not uncommon/uncommon.
2a Adult Larger and heavier than 1; white malar stripe.
2b In flight Legs do not project beyond tail. Characteristic, loud, far-carrying call.

3 OLIVE IBIS *Bostrychia olivacea* 65–75 cm **R** **Page 352**
Forest. Sierra Leone–Ghana; Cameroon–Congo; Príncipe. Rare.
3a Adult All dark; coral-red bill; larger than 4.
3b In flight Legs do not project beyond tail; red bill conspicuous. Loud, disyllabic call with stress on first syllable. Compare 2b and 4b.

4 SPOT-BREASTED IBIS *Bostrychia rara* 47–55 cm **R** **Page 352**
Forest. Liberia–Ghana; Cameroon to SW CAR–Congo. Rare/scarce to locally not uncommon/common.
4a Adult Small; dark; red bill; crest (hard to see); bare turquoise patches on head.
4b In flight Legs do not project beyond tail. Loud, disyllabic call with stress on second syllable. Compare 3b.

5 NORTHERN BALD IBIS *Geronticus eremita* 70–80 cm; WS *c.* 130 cm **P/V** CR **Page 353**
Dry rocky areas; also open fields, lagoons. Mauritania, Senegal, N Mali. Very rare.
5a Adult Unmistakable. All dark; bare reddish head; reddish bill and legs; shaggy crest.
5b Juvenile Head feathered; no crest; duller plumage.
5c In flight Legs do not project beyond tail.

6 SACRED IBIS *Threskiornis aethiopica* 65–82 cm; WS *c.* 120 cm **R/M** **Page 353**
Grassland, cultivation, floodplains, rivers, etc. Mainly in Sahel and savanna zones. Uncommon/scarce.
6a Adult White with bare black head and neck, and black ornamental plumes at rear.
6b Juvenile Head feathered, blackish mottled white.
6c In flight White wings with narrow black edge.

PLATE 11: DUCKS AND GEESE

1 WHITE-FACED WHISTLING DUCK *Dendrocygna viduata* 43–48 cm **R/M**   **Page 356**
Freshwater wetlands. Throughout. Locally common.
  1a   Adult Dark; contrasting white face. Upright stance, with long neck and legs.
  1b   Juvenile Duller; face buffish.
  1c   Adult in flight Wholly dark; chestnut forewing.

2 FULVOUS WHISTLING DUCK *Dendrocygna bicolor* 45–53 cm **R/M**   **Page 355**
Freshwater wetlands. Mauritania–Guinea to N Cameroon–Chad; also W CAR and SW Gabon. Locally
common.
  2a   Adult Mainly fulvous-brown; bold white flank streaks. Shape and stance as 1.
  2b   Adult in flight Blackish wings; fulvous body; white, U-shaped rump.

3 COMMON SHELDUCK *Tadorna tadorna* 58–71 cm **P/V**   **Page 358**
Coastal and freshwater wetlands. Mauritania; rare. Vagrant, Senegal, Ghana, Niger.
  3a   Adult male Mainly black and white with rufous breast-band; red bill with knob.
  3b   Adult female Duller; often some white at bill base; no knob on bill.
  3c   In flight Heavy; white with black head and black flight feathers.

4 HARTLAUB'S DUCK *Pteronetta hartlaubi* 56–58 cm **R** NT   **Page 359**
Forested streams. Guinea to Ghana; Nigeria to CAR–Congo. Uncommon/locally not uncommon.
  4a   Adult male Chestnut body; black head; some white on forehead (extent variable).
  4b   Adult female Duller; usually no white on head; some pink on bill. Juvenile duller still.
  4c   In flight Blue-grey forewing.

5 AFRICAN PYGMY GOOSE *Nettapus auritus* 30–33 cm **R**   **Page 359**
Freshwater wetlands. Throughout; not uncommon/scarce. Vagrant, São Tomé.
  5a   Adult male Small; white face; green neck-sides; yellow bill; rusty-orange underparts.
  5b   Adult female Duller; head-sides mottled greyish.
  5c   In flight Dark above; white patch on inner secondaries.

6 KNOB-BILLED DUCK *Sarkidiornis melanotos* 56–76 cm **M**   **Page 359**
Freshwater wetlands. Throughout. Uncommon/locally common.
  6a   Adult male Head and underparts mainly white; upperparts black; knob on bill.
  6b   Adult female Smaller than male; no knob on bill.
  6c   Juvenile Dark brown above; buffish-brown below.
  6d   Adult male in flight Black wings and upperparts; white underparts.

7 SPUR-WINGED GOOSE *Plectropterus gambensis* 75–100 cm **R**   **Page 358**
Freshwater wetlands. Throughout. Uncommon/locally common.
  7a   Adult male Huge; black-and-white plumage; pinkish bill and legs; red frontal knob.
  7b   Adult female Smaller and duller; frontal knob reduced or absent.
  7c   Adult male in flight White forewing above and below.

8 EGYPTIAN GOOSE *Alopochen aegyptiacus* 63–73 cm **R/M/V**   **Page 357**
Wetlands. S Mauritania–Guinea to Chad–CAR and Congo. Locally not uncommon.
  8a   Adult Large; pink bill and legs; dark eye patch.
  8b   Juvenile Duller; bill and legs greyish.
  8c   Adult in flight White wing-coverts above and below.

PLATE 12: DUCKS II

1 AFRICAN BLACK DUCK *Anas sparsa* 48–57 cm **R**   **Page 362**
Rocky, forested streams. SE Guinea, E Nigeria, Cameroon, Eq. Guinea, Gabon. Rare. Vagrant, Mali and Congo.
1a   **Adult** Blackish-brown; buffish spots on upperparts; pinkish bill with dusky saddle.
1b   **Juvenile** Duller; buff spots reduced; bill greyish.
1c   **In flight** Dark blue speculum bordered white; barred tail; white underwing-coverts.

2 YELLOW-BILLED DUCK *Anas undulata* 51–58 cm **R?**   **Page 362**
Freshwater wetlands. Cameroon and E Nigeria. Scarce and local.
2a   **Adult** Dark plumage; yellow bill.
2b   **In flight** Dark blue speculum bordered white; white underwing-coverts.

3 GADWALL *Anas strepera* 46–55 cm **P/V**   **Page 360**
Freshwater wetlands. N Senegal and N Nigeria; rare. Vagrant, Mali, Niger, Cameroon.
3a   **Adult male breeding** Mainly greyish; black rear end; blackish bill.
3b   **Adult female** Bill with orange sides and dark culmen. Compare 4b.
3c   **Adult male in flight** Small white speculum; white belly; white underwing.
3d   **Adult female in flight** Well-defined white belly.

4 MALLARD *Anas platyrhynchos* 50–65 cm **P/V**   **Page 361**
Freshwater wetlands. Mauritania; rare. Vagrant, N Senegal, Mali, S Niger, N Nigeria.
4a   **Adult male breeding** Green head; narrow white collar; purplish-brown breast; yellow bill.
4b   **Adult female** Pale brown marked darker; bill orange with irregular dark area on culmen.
4c   **Adult male in flight** Dark blue speculum bordered white.
4d   **Adult female in flight** Brown belly. Compare 3d.

5 NORTHERN SHOVELER *Anas clypeata* 44–52 cm **P**   **Page 364**
Freshwater and coastal wetlands. Widespread; local concentrations. Common/scarce.
5a   **Adult male breeding** Large, spatulate bill; dark green head; white breast; chestnut flanks.
5b   **Adult female** Bill shape diagnostic.
5c   **Adult male in flight** Bluish forewing; front-heavy appearance.
5d   **Adult female in flight** Greyish-blue forewing.

6 NORTHERN PINTAIL *Anas acuta* 51–56 cm **P**   **Page 362**
Freshwater and coastal wetlands. Mauritania–Sierra Leone to Chad–CAR. Common/scarce.
6a   **Adult male breeding** Brown head; white stripe on side of long neck; pointed tail.
6b   **Adult female** Rather plain buff-brown head; slim grey bill.
6c   **Adult male in flight** Long neck and tail; dark green speculum bordered white at rear.
6d   **Adult female in flight** Brownish speculum bordered white; greyish underwing.

7 EURASIAN WIGEON *Anas penelope* 45–51 cm **P**   **Page 360**
Freshwater and coastal wetlands. Mauritania–N Senegal to Chad. Scarce/uncommon. Vagrant, Ghana, SW Nigeria and Cameroon.
7a   **Adult male breeding** Chestnut head; yellow forehead and crown; pinkish breast.
7b   **Adult female** Rather uniform rusty-brown to brownish-grey.
7c   **Adult male in flight** White forewing.
7d   **Adult female in flight** White belly; short pointed tail.

PLATE 13: DUCKS III

1 CAPE TEAL *Anas capensis* 44–48 cm **R/M/V** **Page 361**
Freshwater wetlands. SE Niger, NE Nigeria, Chad. Scarce/uncommon. Vagrant, Ghana.
1a Adult Very pale; pink bill.
1b In flight Green speculum broadly bordered white.

2 MARBLED DUCK *Marmaronetta angustirostris* 39–42 cm **P/V VU** **Page 364**
Freshwater and coastal wetlands. Senegal; not uncommon. Vagrant, Mali, NE Nigeria, N Cameroon, Chad and Cape Verde Is.
2a Adult Pale grey-brown; dark eye patch; large buffish spots on upperparts.
2b In flight Pale secondaries; no speculum nor wingbars; white underwing.

3 BLUE-WINGED TEAL *Anas discors* 37–41 cm **V** **Page 363**
Freshwater and coastal wetlands. N American vagrant, N Senegal.
3a Adult male breeding Dull violet-blue head with large white crescent.
3b Adult female Head pattern as 5b but more indistinct; prominent pale loral spot.
3c Adult male in flight Blue forewing; no white trailing edge.
3d Adult female in flight As male but duller.

4 HOTTENTOT TEAL *Anas hottentota* 30–35 cm **R** **Page 363**
Freshwater wetlands. S Niger, N Nigeria, W Chad, N Cameroon. Uncommon.
4a Adult Very small; dark crown; buff head-sides; blue bill.
4b In flight Green speculum; white trailing edge.

5 GARGANEY *Anas querquedula* 37–41 cm **P** **Page 363**
Freshwater and coastal wetlands. Throughout. Common/uncommon.
5a Adult male breeding White stripe from eye to nape; dark brown breast; pale grey flanks.
5b Adult female Pale supercilium; dark crown and eye-stripe; whitish loral spot.
5c Adult male in flight Pale grey-blue forewing.
5d Adult female in flight Pale greyish forewing; white trailing edge.

6 COMMON TEAL *Anas crecca* 34–38 cm **P** **Page 361**
Freshwater wetlands. Mauritania–Senegambia to Chad–CAR. Uncommon/scarce to locally common. Rare/vagrant, S Ivory Coast, S Ghana, SW Nigeria and Cape Verde Is.
6a Adult male breeding Chestnut head; green band from eye to nape; yellow on rear end.
6b Adult female Rather plain head; white stripe at sides of undertail-coverts. Compare 5b.
6c Adult male in flight Green speculum bordered white (broader in front).
6d Adult female in flight Wing pattern similar to 6c; forewing brown.

7 COMMON POCHARD *Aythya ferina* 42–49 cm **P/V** **Page 364**
Freshwater and coastal wetlands. Mauritania, Senegambia, Mali, N Nigeria. Uncommon/rare. Vagrant elsewhere.
7a Adult male breeding Chestnut head; black breast and rear end; pale grey body.
7b Adult female Brownish; peaked crown; sloping forehead; longish bill.
7c Adult male in flight Pale grey band on flight feathers.
7d Adult female in flight Grey wing band; greyish underwing and belly.

8 WHITE-BACKED DUCK *Thalassornis leuconotus* 38–40 cm **R/V** **Page 356**
Freshwater wetlands. Patchily distributed throughout. Uncommon/rare and local.
8a Adult Buff-and-brown plumage; darkish top of head; white loral patch. Note shape.
8b In flight White lower back.

9 FERRUGINOUS DUCK *Aythya nyroca* 38–42 cm **P/V NT** **Page 365**
Freshwater and coastal wetlands. Mauritania–Senegambia to Chad–CAR. Uncommon/rare. Vagrant, Sierra Leone, Ghana, CAR and Cape Verde Is.
9a Adult male breeding Mainly dark chestnut; white undertail-coverts; white eye.
9b Adult female Duller; eye dark. Compare 10b.
9c Adult male in flight Broad white band on flight feathers; white belly.

10 TUFTED DUCK *Aythya fuligula* 40–47 cm **P/V** **Page 365**
Freshwater and coastal wetlands. Mauritania–Senegal to N Nigeria–Chad. Scarce/locally common. Rare/vagrant, Gambia, Sierra Leone, Ivory Coast, Nigeria, Cameroon and Cape Verde Is.
10a Adult male breeding Black; white flanks; drooping crest; yellow eye.
10b Adult female Mainly dark brown; flanks paler; short tuft; yellow eye.
10c Adult male in flight White band on flight feathers.
10d Adult female in flight Pattern similar to male.

PLATE 14: VAGRANT DUCKS AND GEESE

1 BLACK-BELLIED WHISTLING DUCK *Dendrocygna autumnalis* 48–53 cm **V**   **Page 356**
Neotropical vagrant, Gambia.
Adult Pinkish-red bill; dark cap; whitish eye-ring.

2 BEAN GOOSE *Anser fabalis* 66–88 cm **V**   **Page 357**
Palearctic vagrant, Mali.
Adult Large; brownish-grey; head and neck darker; legs bright orange.

3 GREATER WHITE-FRONTED GOOSE *Anser albifrons* 65–78 cm **V**   **Page 357**
Palearctic vagrant, N Mauritania, Niger and N Nigeria.
Adult Brownish-grey; white blaze on forehead; blackish bars on belly; pink bill.

4 BRENT GOOSE *Branta bernicla* 55–62 cm **V**   **Page 357**
Palearctic vagrant, N Mauritania and S Senegal.
Adult Dark; head, neck and bill black; white vent.

5 RUDDY SHELDUCK *Tadorna ferruginea* 61–67 cm **V?**   **Page 358**
Palearctic vagrant claimed from N Mauritania.
    5a Adult male breeding Plain rusty-orange; narrow blackish neck collar. In flight, white wing-coverts,
        black flight feathers.
    5b Adult female No collar.

6 AMERICAN WIGEON *Anas americana* 45–56 cm **V**   **Page 360**
N American vagrant, N Senegal.
    6a Adult male breeding Crown cream-white; dark metallic green band from eye to nape.
    6b Adult female As Eurasian Wigeon (Plate 12: 7b) but head, neck and upperparts greyer.
    6c Adult male in flight White wing panel; white belly. Compare Eurasian Wigeon (Plate 12: 7c).
    6d Adult female in flight Greyish underwing; white axillaries.

7 LESSER SCAUP *Aythya affinis* 38–45 cm **V**   **Page 366**
N American vagrant, Cape Verde Is.
    7a Adult male breeding Peaked hindcrown; finely vermiculated grey upperparts.
    7b Adult female White ring at base of bill. Compare Tufted Duck (Plate 13: 10b).
    7c Adult male in flight White band on upperside of secondaries; pale grey inner primaries; whitish
        underwing.
    7d Adult female in flight Pattern similar to male.

8 COMMON SCOTER *Melanitta nigra* 44–45 cm **P**   **Page 366**
Open sea and estuaries. Mauritania. Rare/scarce.
    8a Adult male Compact; black; yellow patch on knobbed black bill.
    8b Adult female Dark sooty-brown; pale cheeks and foreneck.
    8c Adult male in flight Uniformly dark; paler flight feathers.

PLATE 15: FISH EAGLE, OSPREY, PALM-NUT VULTURE, HARRIER HAWK, BATELEUR AND SECRETARY BIRD

1 AFRICAN FISH EAGLE *Haliaeetus vocifer* 63–73 cm; WS 190–240 cm **R** Page 370
Large rivers, lagoons, lakes, reservoirs. Throughout. Uncommon/rare to locally common.
  1a **Adult** White head, mantle and breast; chestnut lower underparts.
  1b **Juvenile** Mainly mottled brown; whitish breast streaked brown.
  1c **Adult in flight** White head and breast; white tail; chestnut underwing-coverts and belly.
  1d **Juvenile in flight** Pale 'window' in primaries; whitish tail with black terminal band.

2 OSPREY *Pandion haliaetus* 52–61 cm; WS 145–173 cm **P/R** Page 366
Large rivers, lagoons, lakes, etc. Throughout. Uncommon/locally common. Breeds Cape Verde Is.
  2a **Adult** Broad dark stripe through eye; short nuchal crest.
  2b **In flight** Long, narrow wings with long 'hand'; black carpal patches.

3 PALM-NUT VULTURE *Gypohierax angolensis* c. 60 cm; WS 140–150 cm **R** Page 370
Forest and wooded savanna belt. Throughout; also Bioko. Not uncommon/locally common.
  3a **Adult** Black-and-white plumage; bare reddish-pink face; heavy yellowish-horn bill.
  3b **Juvenile** Drab brown; dull yellowish face; darkish bill; whitish legs.
  3c **Adult in flight from above** Black secondaries, greater coverts, wingtips and tail base.
  3d **Adult in flight from below** Black secondaries, primary tips and tail base.
  3e **Juvenile in flight** Darkish; broad rounded wings; short rounded tail.

4 AFRICAN HARRIER HAWK *Polyboroides typus* 60–68 cm; WS c. 160 cm **R** Page 376
Forest zone and adjacent savanna. Throughout. Common.
  4a **Adult** Mainly grey; small head with bare, yellow to orange face; finely barred underparts; long bare legs.
  4b **Juvenile/immature** Dark brown to tawny, variably mottled, streaked and barred.
  4c **Juvenile** A dark brown individual.
  4d **Adult in flight from above** Broad wings with broad black trailing edge and tips; black tail with white bar.
  4e **Adult in flight from below** Narrow white line along black trailing edge and tips.
  4f **Juvenile in flight** Remiges and rectrices more or less distinctly barred; greater coverts tipped dark.

5 BATELEUR *Terathopius ecaudatus* 55–70 cm; WS 170–187 cm **R** Page 375
Savanna. S Mauritania–Guinea to Chad–CAR. Uncommon/locally common.
  5a **Adult male** Jet-black head and body; chestnut back; grey shoulders; red face and feet.
  5b **Adult male** Cream-backed form (uncommon).
  5c **Adult female** As male but with grey panel on secondaries.
  5d **Juvenile** Entirely brown. Note large head, long wings, invisible tail, dark eye.
  5e **Adult male in flight from above** Black, bow-shaped wings; grey shoulders; chestnut back.
  5f **Adult male in flight from below** Black secondaries contrast with white coverts.
  5g **Adult female in flight** Black trailing edge to largely white underwing.
  5h **Juvenile in flight** All brown; unique silhouette similar to adult, but tail slightly longer.

6 SECRETARY BIRD *Sagittarius serpentarius* 125–150 cm; WS 212 cm **R/M** Page 393
Sahel and northern savannas. S Mauritania–Guinea-Bissau to Chad–CAR. Rare/uncommon.
  6a **Adult** Very long legs; plume-like feathers on nape; bare reddish face. Terrestrial.
  6b **In flight** Black remiges contrast with pale grey wing-coverts; very long central rectrices project beyond long legs.

PLATE 16: VULTURES

Medium-sized to huge with long, broad, strongly 'fingered' wings, and usually unfeathered head and neck. Larger species have long neck with ruff at base, and short tail. Carrion eaters. Given to soaring, often at great height. In flight head is tucked between shoulders and appears relatively small.

1 HOODED VULTURE *Necrosyrtes monachus* 65–75 cm; WS 170–182 cm **R**   **Page 371**
Sahel and savanna belts; patchily in forest zone. Mauritania–Guinea to Chad–CAR. Locally common.
1a **Adult** Dark brown; bare face and neck pink (red in excitement); long, slender bill.
1b **Juvenile** Bare face and neck pale bluish.
1c **Adult in flight** Uniformly dark; broad wings; short, rounded tail; slim bill.

2 EGYPTIAN VULTURE *Neophron percnopterus* 55–75 cm; WS 155–175 cm **R/P**   **Page 371**
Sahel. Mauritania–Senegambia to Chad–N Cameroon; uncommon. Cape Verde Is; not uncommon.
2a **Adult** Mainly white; bare yellow face; long, narrow bill; long feathers at back of head.
2b **Juvenile** Dark brown; greyish face and legs.
2c **Adult in flight** White, wedge-shaped tail; black remiges contrast with white coverts.
2d **Juvenile in flight** Distinctive silhouette similar to adult.

3 WHITE-HEADED VULTURE *Trigonoceps occipitalis* 78–85 cm; WS 202–230 cm **R**   **Page 373**
Sahel and savanna. S Mauritania–Guinea to Chad–CAR. Uncommon. Also reported from Gabon.
3a **Adult female** White to pink face and neck; red bill; blue cere; peaked hindcrown; white inner secondaries (dark in male).
3b **Juvenile** Mainly dark brown; bare parts duller.
3c **Adult male/juvenile in flight** White line on greater coverts; white belly.
3d **Adult female in flight** As 3c but inner secondaries white.

4 AFRICAN WHITE-BACKED VULTURE *Gyps africanus* 80–98 cm; WS 212–218 cm **R Page 371**
Savanna. S Mauritania–Guinea to Chad–CAR. Uncommon. Vagrant, Sierra Leone, Liberia, Gabon.
4a **Adult** Very large; mainly brown; bare blackish face; black bill; white back and rump.
4b **Juvenile** Darker and streaky; brownish ruff; no white on back and rump.
4c **Adult in flight** Black remiges contrast with whitish coverts. From above: white back.
4d **Juvenile in flight** Dark brown; white line along leading edge of wing; no white on back.

5 RÜPPELL'S GRIFFON VULTURE *Gyps rueppellii* 85–107 cm; WS 220–250 cm **R/M Page 372**
Sahel. S Mauritania–Guinea to Chad–CAR. Also montane areas, Cameroon. Not uncommon.
5a **Adult** Grey-brown feathers broadly tipped buffish; yellowish-horn bill; pale yellow eye.
5b **Juvenile** Plain above, streaky below; blackish bill; dark eye.
5c **Adult in flight** Dark; white bar along leading edge of wing; two white lines on coverts.
5d **Juvenile in flight** Very similar to 4d when very young.

6 EURASIAN GRIFFON VULTURE *Gyps fulvus* 95–110 cm; WS 230–280 cm **P**   **Page 372**
Sahel and desert. Mauritania, Senegambia, Mali. Uncommon/rare. Vagrant, Niger.
6a **Adult** Huge; mainly sandy-brown; whitish ruff; yellowish bill and eye.
6b **Juvenile** Slightly darker and streaky; pale brownish ruff; dark bill and eye.
6c **Adult in flight** Blackish remiges; pale brownish underwing-coverts with 1–2 pale lines.
6d **Juvenile in flight** Underwing-coverts paler and plainer than adult.

7 LAPPET-FACED VULTURE *Torgos tracheliotus* 98–115 cm; WS 255–290 cm **R VU**   **Page 373**
Sahel. Mauritania–Senegambia to Chad–CAR. Uncommon.
7a **Adult** Huge; mainly blackish; underparts streaked white; bare pink head and neck; massive yellowish-horn bill.
7b **Juvenile** Duller and more uniformly darkish.
7c **Adult in flight** Distinct white bar along leading edge of wing; white thighs.
7d **Juvenile in flight** Wholly dark. Compare distant birds with 1c.

PLATE 17: KITES AND HARRIERS

Kites. Medium-sized raptors with graceful flight and, in most species, forked tail.

1 BLACK KITE *Milvus migrans* 50–60 cm; WS 130–155 cm **M/P/R**   Page 369
Various habitats. Throughout. Commonest African raptor. Rare resident, Cape Verde Is.
   1a **Adult** *parasitus* Entirely chocolate-brown; yellow bill; long, slightly forked tail. Resident and intra-African migrant.
   1b **Adult** *migrans* Pale greyish head streaked black; black bill. Palearctic migrant.
   1c **Juvenile** *parasitus* Dark bill.
   1d **Adult** *parasitus* **in flight** Long, narrow wings; constantly twisting tail. Nominate has more contrasting wing.

2 RED KITE *Milvus milvus* 55–72 cm; WS 140–180 cm **V/R**   Page 369
Palearctic vagrant, Mauritania and Gambia. Very rare resident, Cape Verde Is.
   2a **Adult** *milvus* Mainly rufous streaked black; pale greyish head; long, deeply forked tail.
   Adult *fasciicauda* (Cape Verde Is endemic; separate species?) has less forked, more barred tail; greyish (not whitish) on inner webs of primaries.
   2b **Adult** *milvus* **in flight** Large whitish 'window' on primaries below.

Harriers. Medium-sized and slender with long wings and tails. Sexually dimorphic; females mainly brownish, some posing identification problems. Flight buoyant with wings characteristically held in shallow V. Hunt low, quartering the ground and dropping on prey. Frequent open country, cultivated areas and marshes.

3 AFRICAN MARSH HARRIER *Circus ranivorus* 44–50 cm; WS 110 cm **R**   Page 378
Open habitats. Congo. Uncommon.
   3a **Adult** Mainly brown variably streaked darker; lower underparts more rufous.
   3b **Juvenile** Darker and plainer; variable whitish areas on nape, throat and breast.
   3c **Adult in flight** Whitish leading edge to wing; barred tail; no white on uppertail-coverts.
   3d **Juvenile in flight** Dark brown with irregular whitish breast-band.

4 EURASIAN MARSH HARRIER *Circus aeruginosus* 42–56 cm; WS 110–140 cm **P**   Page 378
Open habitats. Throughout; not uncommon/locally common. Cape Verde Is; rare.
   4a **Adult male** Mainly brown, with buff-brown head and breast streaked darker.
   4b **Adult female** Mainly dark brown; creamy crown, nape and throat.
   4c **Adult male in flight** Tricoloured wings with large black tips; plain, blue-grey tail.
   4d **Adult female in flight** Leading edge of wing usually creamy; often pale patch on breast.

5 HEN HARRIER *Circus cyaneus* 43–55 cm; WS 97–121 cm **P?**   Page 377
Open habitats. Claimed from Mauritania (rare).
   5a **Adult male** Mainly blue-grey; lower underparts white.
   5b **Adult female** Buffish underparts heavily streaked brown.
   5c **Juvenile** As 5b; underparts usually darker and washed rufous.
   5d **Adult male in flight** Large black wingtips; white uppertail-coverts; dusky trailing edge to underwing.
   5e **Adult female in flight** White uppertail-coverts; three well-defined dark bands on underwing.

6 PALLID HARRIER *Circus macrourus* 40–50 cm; WS 95–120 cm **P** NT   Page 377
Open habitats. Mauritania–Liberia to Chad–CAR. Uncommon/scarce.
   6a **Adult male** Pale grey above, mainly white below.
   6b **Adult female** Narrow collar; less white above eye than 7b.
   6c **Juvenile** Plain rufous below; black cheek patch reaching base of bill; pale collar.
   6d **Adult male in flight** Very pale; black wedge at wingtips.
   6e **Adult female in flight** White uppertail-coverts; upper- and inner underwing darker than 7e.
   6f **Juvenile in flight** Lacks dark tips to primaries below (present in 7f).

7 MONTAGU'S HARRIER *Circus pygargus* 38–50 cm; WS 96–120 cm **P**   Page 377
Open habitats. Throughout; uncommon/scarce to locally not uncommon. Vagrant, Congo and Cape Verde Is.
   7a **Adult male** Mainly blue-grey; belly white variably streaked chestnut.
   7b **Adult female** Facial pattern less distinct than 6b; more white around eye.
   7c **Juvenile** Usually deeper rufous below than 6c; cheek patch smaller; no pale collar.
   7d **Adult male in flight** Wings with black tips and one black bar above, two below.
   7e **Adult female in flight** Dark bar on base of secondaries above; underwing more uniform than 6e, with three well-defined dark bands and broad pale band along dark trailing edge.
   7f **Juvenile in flight** Dark tips to primaries.

PLATE 18: SNAKE EAGLES

Rather large and eagle-like, with relatively large heads and large yellow eyes (producing somewhat owl-like appearance), long, broad wings, bare tarsi and short toes. Given to soaring and perching for long periods. Drop on their prey (mainly snakes and other reptiles) and kill it on ground with their powerful feet. Occur in open country and woodland.

1 SHORT-TOED EAGLE *Circaetus gallicus* 59–69 cm; WS 162–195 cm **P**   **Page 373**
Sahel and northern savanna. S Mauritania–Guinea to Chad. Uncommon.
- **1a**   **Typical adult male** As 1b but breast-band broken.
- **1b**   **Typical adult female** Grey-brown above; upper breast mottled brown; white underparts irregularly blotched and barred brown.
- **1c**   **Juvenile** Head and breast more rufous-brown.
- **1d**   **Typical adult female in flight** White underwing-coverts with dark lines.
- **1e**   **Pale adult/juvenile in flight** Underwing pattern more indistinct.

2 BEAUDOUIN'S SNAKE EAGLE *Circaetus beaudouini* 62–69 cm; WS *c.* 170 cm **R**   **Page 374**
Sahel and northern savanna. S Mauritania–Guinea to Chad–CAR. Uncommon.
- **2a**   **Adult male** As 2b but more white on breast.
- **2b**   **Adult female** Darker brown than 1b; white underparts with narrower bars.
- **2c**   **Juvenile** Dark brown above, more rufous-brown below.
- **2d**   **Adult female in flight** Underwing-coverts white.
- **2e**   **Juvenile in flight** Brownish underwing-coverts; indistinct dusky bars on remiges.
- **2f**   **Adult male hovering.**

3 BLACK-BREASTED SNAKE EAGLE *Circaetus pectoralis* 63–68 cm; WS 178 cm **M**   **Page 374**
Savanna. SE Gabon and S Congo. Rare.
- **3a**   **Adult** Blackish above and on breast, pure white below.
- **3b**   **Immature** Underparts with bars and blotches.
- **3c**   **Juvenile** Entirely rufous-brown, darker on wings.
- **3d**   **Adult in flight** White underwing; dark bars on remiges.
- **3e**   **Juvenile in flight** Rufous-brown underwing-coverts; indistinct dusky bars on remiges.

4 BROWN SNAKE EAGLE *Circaetus cinereus* 66–75 cm, WS 200 cm **R**   **Page 375**
Sahel and savanna. S Mauritania–Liberia to Chad–CAR. Uncommon.
- **4a**   **Adult** Entirely dark brown. Compare 2c and 3c.
- **4b**   **Immature** As adult or streaked white on head; may have white feathers below.
- **4c**   **Adult in flight** Whitish remiges; dark brown body and underwing-coverts.

5 WESTERN BANDED SNAKE EAGLE *Circaetus cinerascens* 55–60 cm; WS 114 cm **R**   **Page 375**
Savanna. Senegambia–Sierra Leone to Chad–CAR. Uncommon.
- **5a**   **Adult male** Grey-brown; broad white band on tail.
- **5b**   **Adult female** As male or all brown as illustrated.
- **5c**   **Juvenile** Whitish head and underparts.
- **5d**   **Adult in flight** Underwing white; four dark bars on remiges; distinctive tail pattern.
- **5e**   **Juvenile in flight** Body whitish; wing and tail pattern as adult but duller.

PLATE 19: SMALL KITES, BAT HAWK, CUCKOO HAWK, LONG-TAILED HAWK AND SERPENT EAGLE

1 BLACK-SHOULDERED KITE *Elanus caeruleus* 31–35 cm; WS 75–85 cm **R**   **Page 368**
Various open habitats. Throughout. Common/uncommon.
- **1a** **Adult** Smallish; blue-grey above; dark eye patch; black shoulders; white below.
- **1b** **Juvenile** Upperparts tinged brownish, feathers edged white; rufous wash on breast.
- **1c** **Adult in flight from above** Long, pointed, broad-based wings; short tail; black shoulders.
- **1d** **Adult in flight from below** All white with black primaries.

2 AFRICAN SWALLOW-TAILED KITE *Chelictinia riocourii* 35–38 cm; WS 90 cm **M**   **Page 369**
Sahel. Mauritania–Senegal to Chad. Uncommon/locally common. Irregular further south.
- **2a** **Adult** Smallish; grey and white; long, deeply forked tail.
- **2b** **Juvenile** Short tail; feathers of upperparts edged rufous; pale rufous wash on breast.
- **2c** **Adult in flight** Slender, tern-like; long pointed wings with black carpal patches.
- **2d** **Juvenile in flight** Almost square tail.

3 AFRICAN CUCKOO HAWK *Aviceda cuculoides* 40 cm; WS 91 cm **R**   **Page 367**
Forest and southern savanna belts. Throughout. Uncommon/not uncommon.
- **3a** **Adult** Dark grey above and on breast; broad chestnut bars on white underparts.
- **3b** **Juvenile** Dark brown above; whitish supercilium; white below with brown spots.
- **3c** **Adult in flight** Chestnut underwing-coverts; longish wings and tail.
- **3d** **Juvenile in flight** Buffish underwing-coverts; remiges barred like adult.

4 BAT HAWK *Macheiramphus alcinus* 45 cm; WS 110 cm **R**   **Page 368**
Forest and savanna zones. Throughout. Uncommon and local.
- **4a** **Adult** Blackish-brown; dark stripe down white throat; short occipital crest.
- **4b** **Juvenile** Underparts with variable amount of white.
- **4c** **Adult in flight** Long, broad and pointed wings; longish, square tail.

5 CONGO SERPENT EAGLE *Dryotriorchis spectabilis* 51 cm; WS 85–96 cm **R**   **Page 376**
Forest. Sierra Leone–SE Guinea to Ghana; W Nigeria to CAR and Congo. Scarce/locally not uncommon.
- **5a** **Adult** *spectabilis* Rather large head; long tail; bare legs; dark throat-stripe, bold spots on breast, more bar-like markings on belly. Upper Guinea forest.
- **5b** **Adult** *batesi* Markings on underparts restricted to flanks. Lower Guinea forest.
- **5c** **Juvenile** Head mottled whitish; underparts white boldly spotted blackish.
- **5d** **Adult** *spectabilis* **in flight** Short wings with barred remiges; long, rounded tail.
- **5e** **Adult** *batesi* **in flight** Whiter below.
- **5f** **Juvenile in flight** Pale head.

6 LONG-TAILED HAWK *Urotriorchis macrourus* 57–73 cm; WS 90 cm **R**   **Page 383**
Forest. E Sierra Leone–SE Guinea to W Togo; W Nigeria to SW CAR–Congo. Uncommon/scarce to locally not uncommon.
- **6a** **Adult typical form** Dark grey above, mainly chestnut below, long black tail.
- **6b** **Adult grey form** Underparts grey. Rare.
- **6c** **Juvenile** Brown above; white variably marked dark brown below.
- **6d** **Adult in flight from above** Very long tail tipped and barred white; white uppertail-coverts.
- **6e** **Adult in flight from below** Chestnut underwing-coverts.
- **6f** **Juvenile in flight** Long tail distinctive.

PLATE 20: SPARROWHAWKS, LIZARD BUZZARD, GABAR GOSHAWK AND CHANTING GOSHAWK

1   SHIKRA *Accipiter badius* 28–30 cm; WS 58–60 cm **R/M**   Page 380
Savanna and Sahel belts. Throughout. Southward movement in dry season. Common.
  1a **Adult** Small; blue-grey above, white below with fine rufous barring; red/orange eye.
  1b **Juvenile** Brown above; dark throat-stripe; blotched and barred below; tail barred.
  1c **Adult in flight from above** Plain grey (no white on rump or tail).
  1d **Adult in flight from below** Very pale; wingtips dusky.
  1e **Juvenile in flight from below** Indistinctly barred on wings and tail.

2   GABAR GOSHAWK *Micronisus gabar* 28–36 cm; WS 60 cm **R**   Page 378
Sahel and northern savanna. S Mauritania–Senegambia to Chad–CAR. Not uncommon/scarce.
  2a **Adult** Smallish; grey breast; cere and legs pinkish-red. 7a is much larger.
  2b **Adult melanistic form** All black; tail faintly barred white.
  2c **Juvenile** Pale, streaky head; cere and legs yellowish to orange.
  2d **Adult in flight from above** White uppertail-coverts.
  2e **Adult in flight from below** Narrowly barred underwing; four black bars on tail.
  2f **Adult melanistic form in flight** White barring on primaries above.
  2g **Juvenile in flight** Underwing-coverts narrowly barred rufous-brown.

3   OVAMBO SPARROWHAWK *Accipiter ovampensis* 33–40 cm; WS 67 cm **M**   Page 382
Savanna zone. Senegal, Mali, and Liberia to Chad–CAR. Uncommon/rare.
  3a **Adult** Underparts, incl. throat, narrowly barred grey; central tail feathers with white shafts between dark bands. Compare 2a.
  3b **Adult melanistic form** All black; paler tail bars; remiges barred dark below.
  3c **Juvenile rufous-breasted form** Whitish supercilium; dark ear-coverts; variably streaked underparts.
  3d **Juvenile pale-breasted form** As 3c but rufous areas whitish.
  3e **Adult in flight from above** Three faint white tail spots.
  3f **Adult in flight from below** Finely barred underwing.

4   LIZARD BUZZARD *Kaupifalco monogrammicus* 35–37 cm; WS 79 cm **R**   Page 384
Savanna and forest zones. Throughout. Not uncommon.
  4a **Adult** Stocky; black throat-stripe; grey breast; pinkish-red cere and legs.
  4b **Adult in flight from above** Black tail with broad white band; white uppertail-coverts.
  4c **Adult in flight from below** Underwing appears white from distance.

5   EURASIAN SPARROWHAWK *Accipiter nisus* 28–41 cm; WS 58–80 cm **P/V**   Page 382
Open country, woodland. Scarce Palearctic migrant, Mauritania. Vagrant, Gambia, Mali and Chad.
  5a **Adult male** Narrowly barred rufous below; rufous ear-coverts.
  5b **Adult female** Much larger, browner; white supercilium.
  5c **Juvenile** Browner than 5b; barring more irregular.
  5d **Adult male in flight** No white on uppertail-coverts; tail square with sharp corners.
  5e **Adult female in flight** As 5d but larger; flight steadier.

6   LEVANT SPARROWHAWK *Accipiter brevipes* 29–39 cm; WS 63–80 cm **V**   Page 381
Palearctic vagrant claimed from N Cameroon.
  6a **Adult male** Grey cheeks; dark eye; tail unbarred above, narrowly barred below.
  6b **Adult female** Only slightly larger; browner; dark throat-stripe.
  6c **Juvenile** Boldly spotted below.
  6d **Adult male in flight** Very pale below; wings tipped black above and below. Compare 1d.
  6e **Adult female in flight** Body, wings and tail narrowly barred below.

7   DARK CHANTING GOSHAWK *Melierax metabates* 38–48 cm; WS 95–110 cm **R**   Page 379
Sahel and northern savanna. Throughout; also SE Gabon and Congo. Not uncommon.
  7a **Adult** Long pinkish-red legs; pinkish-red cere; long, rounded tail. Compare 2a.
  7b **Juvenile** Brown above, variably marked brown and white below; long bare legs.
  7c **Adult in flight from above** Uppertail-coverts finely barred grey; black central rectrices.
  7d **Adult in flight from below** Black tips to long broad wings.
  7e **Juvenile in flight from above** Uppertail-coverts narrowly barred grey-brown.
  7f **Juvenile in flight from below** Note shape of wings and tail.

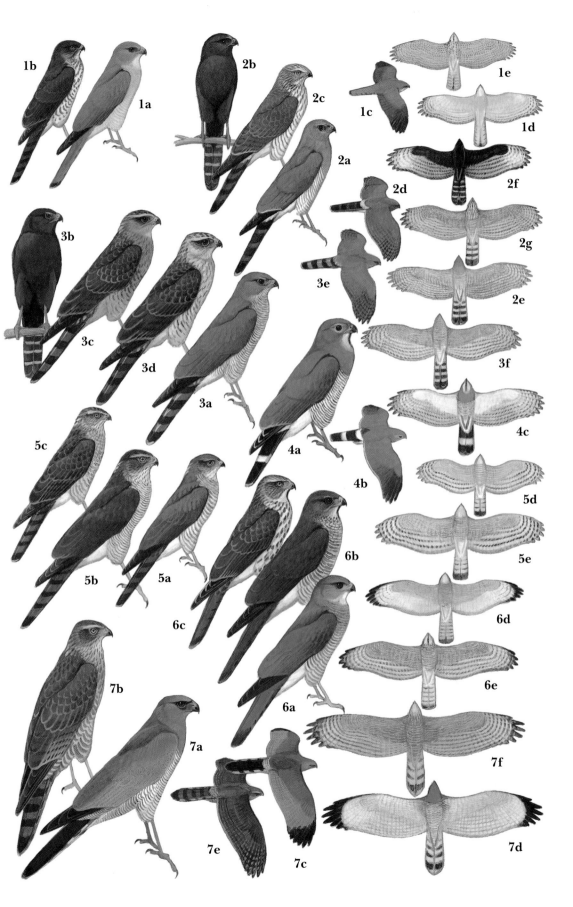

PLATE 21: SPARROWHAWKS AND GOSHAWK

Small to medium-sized, with rather short, rounded wings and long tails. Tarsi bare and long. Large size difference between sexes; females much larger. Swift and agile. Hunt by dashing on their prey (mainly birds and small mammals), following a stealthy approach behind cover.

1   LITTLE SPARROWHAWK *Accipiter minullus* 23–27 cm; WS 39 cm   **Page 381**
Woodland. No certain records from our region. Resident in E Africa and south of equator.
- **1a**   **Adult** Very small; dark grey above; narrowly barred chestnut below.
- **1b**   **Juvenile** Brown above; heavily spotted and barred brown below.
- **1c**   **Adult in flight from above** White band on uppertail-coverts; two white tail spots.
- **1d**   **Adult in flight from below** Finely barred remiges; dark bars on white tail.

2   RED-THIGHED SPARROWHAWK *Accipiter erythropus* 23–28 cm; WS 40 cm   **R**   **Page 381**
Forest. Gambia to S CAR and Congo. Uncommon/scarce.
- **2a**   **Adult** *erythropus* Very small; blackish above; grey-white tinged pinkish below. Upper Guinea forest.
- **2b**   **Adult** *zenkeri* Breast and belly deep rufous. Lower Guinea forest.
- **2c**   **Juvenile** Brown above; sparsely spotted and barred brown below.
- **2d**   **Adult** *erythropus* **in flight from above** White band on uppertail-coverts; three white tail spots partially hidden by dark central tail feathers.
- **2e**   **Adult** *erythropus* **in flight from below** Underwings and tail barred blackish.
- **2f**   **Adult** *zenkeri* **in flight from above** Two unbroken white tail spots.
- **2g**   **Adult** *zenkeri* **in flight from below** As 2e but breast and belly deep rufous.

3   CHESTNUT-FLANKED SPARROWHAWK *Accipiter castanilius* 28–36 cm; WS 60 cm   **R Page 380**
Forest. W Nigeria to S CAR–Congo. Uncommon/scarce.
- **3a**   **Adult** Blackish-grey above (including rump); barred dark below; breast-sides, flanks and thighs bright rufous.
- **3b**   **Juvenile** Probably indistinguishable from 4b.
- **3c**   **Adult in flight from above** Three white spots in centre of blackish tail.
- **3d**   **Adult in flight from below** Underwing whitish with dark barring on remiges.

4   AFRICAN GOSHAWK *Accipiter tachiro* 36–48 cm; WS 70 cm   **R**   **Page 379**
Forest and outliers in savanna. Gambia–Liberia to S Chad–CAR and Congo. Common.
- **4a**   **Adult** *macroscelides* Narrowly barred orange-chestnut below; three white spots in centre of blackish tail. From W Cameroon westwards.
- **4b**   **Juvenile** *macroscelides* Dark throat-stripe; spotted breast; barred flanks and thighs.
- **4c**   **Adult** *toussenelii* Paler and plainer below than 4a. From SE Cameroon east and south.
- **4d**   **Juvenile** *toussenelii* Whiter and less marked below than 4b.
- **4e**   **Adult** *macroscelides* **in flight from above** Three white spots in centre of blackish tail.
- **4f**   **Adult** *macroscelides* **in flight from below** Underwing-coverts barred orange.
- **4g**   **Adult** *toussenelii* **in flight from below** Underwing-coverts white.

5   BLACK SPARROWHAWK *Accipiter melanoleucus* 46–55 cm; WS 102 cm   **R**   **Page 382**
Forest. Gambia–Liberia to S CAR–Congo; not uncommon/rare. Vagrant, S Mali and SW Niger.
- **5a**   **Adult (typical form)** Large; black above; white below; black patches on flanks and thighs.
- **5b**   **Juvenile rufous-breasted form** Compare juvenile Ayres's Hawk Eagle (Plate 25: 3b).
- **5c**   **Juvenile white-breasted form** As 5b but white ground colour to underparts.
- **5d**   **Adult in flight from above** Black.
- **5e**   **Adult in flight from below** Mainly white; irregular blackish patches on flanks.
- **5f**   **Adult melanistic form in flight from below** Black breast and underwing-coverts, throat white. Rare.
- **5g**   **Juvenile rufous-breasted form in flight from below** Rufous underwing-coverts.

PLATE 22: HONEY BUZZARD, GRASSHOPPER BUZZARD AND *BUTEO* BUZZARDS

1   EUROPEAN HONEY BUZZARD *Pernis apivorus* 51–60 cm; WS 113–150 cm   **P**   Page 368
Forest, wooded savanna. Throughout. Not uncommon/scarce. Vagrant, Bioko.
- **1a   Adult (typical)** Small-looking grey head; slender neck; yellow eye.
- **1b   Adult (dark)**
- **1c   Juvenile (pale)**
- **1d   Adult (typical) in flight** Tail longish with rounded corners, dark terminal band and two narrower bars near base; dark carpal patch.
- **1e   Adult (dark) in flight**
- **1f   Juvenile (pale) in flight** Flight and tail feathers with 4–5 more evenly spread bars.

2   GRASSHOPPER BUZZARD *Butastur rufipennis* 41–44 cm; WS 102 cm   **M**   Page 383
Sahel; south to northern edge of forest in dry season. Throughout. Common.
- **2a   Adult** Slender; wings, tail and legs long; plumage mainly brown and rufous.
- **2b   Juvenile** More rufous than adult; feathers of upperparts edged rufous.
- **2c   Adult in flight from above** Conspicuous rufous wing patches diagnostic.
- **2d   Adult in flight from below** Well-defined blackish trailing edge; pale rufous remiges.

3   RED-NECKED BUZZARD *Buteo auguralis* 40-50 cm; WS 95 cm   **R/M**   Page 385
Forest and savanna belts. Throughout. Moves north with rains. Not uncommon/scarce.
- **3a   Adult** Chestnut head-sides and nape; dark brown throat and breast.
- **3b   Juvenile** White below with black blotches on breast and flanks.
- **3c   Adult in flight** Rufous tail with black subterminal band.
- **3d   Juvenile in flight** Brown tail barred black.

4   COMMON BUZZARD *Buteo buteo* 45–58 cm; WS 110–132 cm   **P/R**   Page 384
Wooded habitats. Irregularly distributed Palearctic migrant throughout. Rare/scarce. Rare resident, Cape Verde Is.
- **4a   Adult *vulpinus* (pale)** Pale rufous-brown underparts and tail.
- **4b   Adult *vulpinus* (foxy)** Uniform dark chestnut underparts and tail.
- **4c   Adult *vulpinus* (dark)** Mainly blackish.
- **4d   Adult *vulpinus* (pale) in flight** Broad rounded wings and tail without distinct terminal band.
- **4e   Adult *vulpinus* (foxy) in flight** White remiges with narrow barring; black trailing edge and fingers; small carpal patch; rufous tail with subterminal band.
- **4f   Adult *vulpinus* (dark) in flight**

5   LONG-LEGGED BUZZARD *Buteo rufinus* 50–65 cm; WS 115–163 cm   **P**   Page 385
Sahel. Mauritania–Senegal to Chad. Scarce/uncommon. Vagrant, Gambia, Ghana, Togo, Nigeria, Cameroon.
- **5a   Adult (typical)** Pale head and breast; dark rufous-brown belly and thighs. Compare 4a.
- **5b   Adult (dark)** Mainly blackish; tail broadly barred.
- **5c   Juvenile** Tail narrowly barred.
- **5d   Adult (typical) in flight** Large black carpal patch and trailing edge; plain orange tail paler at base.
- **5e   Adult (dark) in flight** Barred tail with broad subterminal band.
- **5f   Juvenile in flight** No distinct trailing edge to wing; faintly barred buff-grey tail.

6   AUGUR BUZZARD *Buteo augur* 55–60 cm; WS 132 cm   **V?**   Page 385
E African species. One (doubtful) claim from N Cameroon.
- **6a   Adult (typical)** Black above, white below; short rufous tail.
- **6b   Adult (dark)** Entirely black with rufous tail.
- **6c   Juvenile** Brown above; dark-streaked throat; brown tail barred dark.
- **6d   Adult (typical) in flight** Very broad wings; very short rufous tail.
- **6e   Adult (dark) in flight** Black underwing-coverts; distinctive silhouette.
- **6f   Juvenile in flight** Duller than adult; silhouette distinctive.

PLATE 23: *AQUILA* EAGLES

Medium-sized to very large with broad 'fingered' wings. Tarsi feathered. Given to soaring. Prey on large and small mammals, birds, and reptiles; some eat carrion and insects. In general fiercer and more active than buzzards. Occur in open country, woodland and forest.

1 WAHLBERG'S EAGLE *Aquila wahlbergi* 55–61 cm; WS 141 cm **R/M** **Page 388**
Savanna. Mauritania–Liberia to Chad–CAR; not uncommon. Vagrant, Gabon and N Congo.
1a **Adult (dark)** Dark brown; slight occipital crest; dark eye.
1b **Adult (pale)** White to buffy head and underparts. Uncommon.
1c–d **Adult (dark) in flight** Long, parallel wings held flat or slightly arched; longish, square tail held closed.
1e–f **Adult (pale) in flight** White underwing-coverts. Note characteristic silhouette.

2 TAWNY EAGLE *Aquila rapax* 62–75 cm; WS 165–185 cm **R/M** **Page 387**
Savanna. Throughout. Moves south in dry season. Not uncommon/scarce.
2a **Adult (dark)** Yellow gape extending to below centre of eye.
2b **Adult (pale)** Tawny to pale buffish. Appearance often rather ragged.
2c **Juvenile** Typically paler than adult; white-tipped greater coverts and trailing edge.
2d **Adult (typical) in flight from above** Creamy rump patch.
2e **Adult (typical) in flight from below** Usually pale 'window' on inner primaries.
2f **Adult (dark) in flight from below**
2g **Juvenile in flight from above** Narrow pale line on greater coverts.
2h **Juvenile in flight from below** Narrow white trailing edge to wing and tail.

3 STEPPE EAGLE *Aquila nipalensis* 62–81 cm; WS 163–200 cm **V** **Page 387**
Palearctic vagrant, Mali, Niger, NE Nigeria,. Cameroon, Chad and N Congo.
3a **Adult** Long yellow gape extending to below rear of eye; oval nostrils; pale nape patch.
3b **Juvenile** Typically paler than adult; white-tipped greater coverts, trailing edge and tail.
3c **Adult in flight from above** Whitish primary patch; small white patch on back. Compare 4 and 5.
3d **Adult in flight from below** Dark trailing edge to wings and tail.
3e **Juvenile in flight from above** Pale line on greater coverts; pale U on uppertail-coverts.
3f **Juvenile in flight from below** Pale band on central underwing; white trailing edge to wings and tail.

4 LESSER SPOTTED EAGLE *Aquila pomarina* 57–64 cm; WS 145–170 cm **P** **Page 386**
Rare Palearctic migrant, NE Nigeria, Chad, Cameroon, SW CAR and N Congo.
4a **Adult** Dark brown; narrow 'trousers'; yellowish eye. Compare 5a.
4b **Juvenile** Pale nape patch; pale tips to greater coverts and trailing edge of wing.
4c **Adult in flight from above** Small, distinct primary patch; pale U on uppertail-coverts.
4d **Adult in flight from below** Wing-coverts typically paler than flight feathers; two whitish carpal crescents.
4e **Juvenile in flight from above** Primary patch and U on tail-coverts more distinct than 4c.
4f **Juvenile in flight from below** Pale tips to greater coverts and trailing edge of wing (soon abraded).

5 GREATER SPOTTED EAGLE *Aquila clanga* 60–70 cm; WS 155–180 cm **V** **VU** **Page 386**
Palearctic vagrant, N Cameroon and Chad.
5a **Adult** Darker, more uniform and heavier than 4a; 'trousers' more bushy; eye dark.
5b **Juvenile** Darker and more spotted than 4b.
5c **Juvenile '*fulvescens*'** Pale buffish or yellowish-brown to rufous. Rare.
5d **Adult in flight from above** Diffuse primary patch.
5e **Adult in flight from below** Wing-coverts typically as dark or slightly darker than flight feathers; single carpal crescent; tail relatively shorter than in 4.
5f **Juvenile in flight from above** Spotted upperwing-coverts; white trailing edge.
5g **Juvenile in flight from below** Pale vent.
5h **Juvenile '*fulvescens*' in flight from below** Pale carpal crescent; note silhouette.

PLATE 24: LARGE EAGLES

1   EASTERN IMPERIAL EAGLE *Aquila heliaca* 72–85 cm; WS 180–220 cm **V   VU**   **Page 388**
Palearctic vagrant, N Cameroon.
  1a   **Adult** Blackish-brown; pale tawny crown and nape; white 'braces'.
  1b   **Juvenile** Pale and streaky.
  1c   **Adult in flight from above** Pale top of head; broad black terminal band on grey tail.
  1d   **Adult in flight from below** Long parallel wings held flat when soaring; tail held closed.
  1e   **Juvenile in flight from above** Two pale lines on wing-coverts.
  1f   **Juvenile in flight from below** Pale wedge on innermost primaries.

2   GOLDEN EAGLE *Aquila chrysaetos* 75–90 cm; WS 190–225 cm **P/R**   **Page 389**
Scarce Palearctic migrant and winter visitor, Mauritania. Rare resident, NE Mali.
  2a   **Adult** Rufous-yellow crown and nape.
  2b   **Juvenile** Darker than adult; no 'golden' nape; white tail with broad black terminal band.
  2c–d Adult in flight Longish tail; long wings held in shallow V; wings with slight S-curved trailing edge;
      diffuse pale panel on upperwing-coverts.
  2e–f Juvenile in flight Broad black terminal tail band; variably distinct wing patches.

3   VERREAUX'S EAGLE *Aquila verreauxii* 80–90 cm; WS 190–210 cm **R**   **Page 389**
Crags, inselbergs, gorges. Niger (N Aïr) and E Chad (Ennedi). Rare. Vagrant, E Mali.
  3a   **Adult** Jet-black with narrow white V on back.
  3b   **Juvenile** Tawny-rufous crown and mantle.
  3c   **Adult in flight from above** White Y on back; white windows in primaries.
  3d   **Adult in flight from below** Leaf-shaped wings; longish, often partially spread tail.
  3e   **Juvenile in flight from above** Pale windows in primaries; pale crescent at tail base.
  3f   **Juvenile in flight from below** Pale windows in primaries

4   CROWNED EAGLE *Stephanoaetus coronatus* 80–99 cm; WS 163–209 cm **R**   **Page 392**
Forest and outliers in savanna. Gambia–Liberia to SW CAR–Congo. Uncommon/rare.
  4a   **Adult** Very large and powerful; boldly barred and blotched black, rufous and buff below; short
      occipital crest.
  4b   **Juvenile** Head and underparts mainly white.
  4c–d Adult in flight Broad rounded wings; long tail; rufous underwing-coverts; broadly barred tail.
  4e–f Juvenile in flight Pale rufous-buff underwing-coverts; 3–4 bars on remiges.

5   MARTIAL EAGLE *Polemaetus bellicosus* 78–86 cm; WS 195–260 cm **R**   **Page 392**
Savanna. S Mauritania–Sierra Leone to Chad–CAR; also SE Gabon. Uncommon.
  5a   **Adult** Very large and powerful; dark brown above and on breast; white below spotted dark brown;
      short occipital crest.
  5b   **Juvenile** Grey above; white below.
  5c–d Adult in flight Long broad dark wings; short tail; contrastingly white lower underparts.
  5e–f Juvenile in flight Whitish body and underwing-coverts; finely grey-barred remiges.

PLATE 25: HAWK EAGLES

1 LONG-CRESTED EAGLE *Lophaetus occipitalis* 53–58 cm; WS 115 cm **R** Page 391
Savanna and forest zones. Throughout. Uncommon/not uncommon.
1a Adult Entirely blackish; long, loose crest.
1b–c Adult in flight Large white 'windows' on primaries.

2 BOOTED EAGLE *Hieraaetus pennatus* 45–55 cm; WS 110–132 cm **P** Page 390
Sahel and savanna zones. Throughout; also N Congo. Uncommon.
2a Adult pale morph Brown above; creamy-white below.
2b Adult dark morph Entirely dark brown.
2c Adult in flight from above White 'landing lights'; whitish crescent on uppertail-coverts; broad pale band on upperwing-coverts.
2d Adult pale morph in flight from below Blackish remiges; whitish underwing-coverts and body.
2e Adult dark morph in flight from below Dark brown underwing-coverts and body.

3 AYRES'S HAWK EAGLE *Hieraaetus ayresii* 45–60 cm; WS 124 cm **R** Page 391
Forest and its outliers. Guinea–Liberia to S Chad–Congo. Uncommon/scarce. Rare, Gambia.
3a Adult Blackish-brown above; white below boldly streaked and blotched black.
3b Juvenile Pale rufous to whitish head and underparts.
3c White-headed form (age uncertain)
3d Adult in flight from above White 'landing lights'.
3e Adult in flight from below Heavily mottled black-and-white wing-coverts; heavily barred flight and tail feathers; black terminal tail band.
3f Juvenile in flight from below Pale rufous wing-coverts; blackish-barred remiges.

4 CASSIN'S HAWK EAGLE *Spizaetus africanus* 50–61 cm; WS 120 cm **R** Page 392
Forest. Sierra Leone–SE Guinea to SW CAR–Congo. Scarce/locally not uncommon.
4a Adult Blackish above; white below marked black on breast-sides, flanks and thighs.
4b Juvenile Pale rufous head; mainly white underparts streaked blackish.
4c Adult in flight from above Mainly dark; blackish hood contrasts with white neck-sides.
4d Adult in flight from below Mainly black wing-coverts; barred remiges.
4e Juvenile in flight from below Rufous-streaked underwing-coverts.

5 AFRICAN HAWK EAGLE *Hieraaetus spilogaster* 60–74 cm; WS 142 cm **R** Page 390
Savanna. S Mauritania–Sierra Leone to Chad–CAR. Uncommon/scarce.
5a Adult Blackish above; white streaked black below; thighs unstreaked.
5b Juvenile Rufous below.
5c Adult in flight from above White 'windows' on primaries.
5d Adult in flight from below Black-mottled underwing-coverts; black trailing edge to wing; black terminal band to tail.
5e Juvenile in flight from above Dull version of adult.
5f Juvenile in flight from below Rufous underwing-coverts bordered by black line; narrowly barred, grey tail.

6 BONELLI'S EAGLE *Hieraaetus fasciatus* 55–72 cm; WS 145–180 cm **P** Page 389
Palearctic passage migrant and winter visitor, Mauritania–N Senegal. Scarce/rare.
6a Adult Dark brown above; white, lightly streaked dark below.
6b Juvenile Rusty-brown below.
6c Adult in flight from above Small white patch on mantle; no white 'windows' on primaries.
6d Adult in flight from below White body contrasts with dark wings; longish tail with black terminal band.
6e Juvenile in flight from above Resembles 5e, but lacks obvious white on primaries.
6f Juvenile in flight from below Resembles 5f, but has dusky secondaries.

PLATE 26: FALCONS I

1 COMMON KESTREL *Falco tinnunculus* 30–38 cm; WS 65–80 cm **R/P**   **Page 394**
Open habitats. Almost throughout; also Cape Verde Is. Not uncommon/common.
1a Adult male *rufescens* Grey head; rufous-brown upperparts heavily marked black; grey, barred tail.
1b Adult female *rufescens* Brownish head streaked black; tail brown barred black.
1c Adult male *tinnunculus* Paler and less marked than 1a; tail unbarred.
1d–e Adult male *rufescens* **in flight** Dark flight feathers; chestnut upperwing-coverts and upperparts; pale underwing densely barred and spotted dark.
1f–g Adult female *rufescens* **in flight** Underwing more heavily barred than male.

2 LESSER KESTREL *Falco naumanni* 25–33 cm; WS 58–74 cm **P** VU   **Page 393**
Open habitats. Mauritania–Senegambia to Chad–N CAR; also NE Gabon and N Congo. Uncommon/not uncommon.
2a Adult male Plain grey head; plain rufous upperparts; grey wing panel; lightly marked underparts.
2b Adult female As 1b but paler, with less-marked head and underparts.
2c–d Adult male in flight Tricoloured upperwing; very pale underwing tipped dusky; central rectrices often slightly longer.
2e–f Adult female in flight Underwing whiter than 1f–g; dusky tip more distinct.

3 FOX KESTREL *Falco alopex* 35–42 cm; WS *c.* 90 cm **R**   **Page 394**
Inselbergs in savanna. S Mauritania–Sierra Leone to Chad–CAR. Uncommon/not uncommon.
3a Adult Very rufous, finely streaked black; tail narrowly barred black.
3b–c In flight Long wings and tail; underwing-coverts pale rufous; underwing tipped black.

4 MERLIN *Falco columbarius* 24–33 cm; WS 50–69 cm **V**   **Page 397**
Open habitats. Palearctic vagrant, N Senegal.
4a Adult male Very small, compact; blue-grey upperparts.
4b Adult female Brown upperparts fringed rufous or buff; tail barred.
4c Adult male in flight Blue-grey above; tail with broad black subterminal band.
4d–e Adult female in flight Dark brown above; coarsely barred underwing; barred tail.

5 BARBARY FALCON *Falco pelegrinoides* 32–45 cm; WS 76–100 cm **P/V**   **Page 399**
Open habitats. Mauritania, Senegambia, N Burkina Faso, Niger, Chad, N Cameroon. Rare Palearctic migrant/vagrant.
5a Adult As 6a, but rufous on nape; moustachial stripe narrower; white cheek patch larger; underparts pale pinkish-buff narrowly and indistinctly barred brown.
5b Juvenile As 6b, but head with buff or rufous; underparts sandy-buff with finer streaking.
5c–d Adult in flight Similar to 6d–e, but see 5a.
5e Juvenile in flight Note uniformly marked, dark-tipped underwing with sandy-buff ground colour.

6 PEREGRINE FALCON *Falco peregrinus* 33–50 cm; WS 80–115 cm **R/P**   **Page 398**
Cliffs, open habitats. Patchily distributed throughout. Uncommon/rare.
6a Adult *minor* Blackish top of head; broad moustachial stripe; densely spotted and barred black below. Resident.
6b Adult *peregrinus* Paler than 6a; underparts less heavily marked. Palearctic migrant.
6c Juvenile Brown above; heavily streaked below.
6d–e Adult in flight Stocky; broad-based, pointed wings; relatively short tail.
6f Juvenile in flight Streaked underparts; narrowly and uniformly barred underwing.

7 LANNER FALCON *Falco biarmicus* 38–52 cm; WS 90–115 cm **R/P**   **Page 398**
Open habitats. Savanna and Sahel zones. Uncommon/not uncommon.
7a Adult *abyssinicus* Rufous crown and nape; narrow blackish moustachial stripe.
7b Juvenile Browner; crown and nape paler; boldly streaked below.
7c–d Adult in flight Long, rather blunt wings; pale underparts and underwing-coverts.
7e Juvenile in flight Dark underwing-coverts contrast with paler flight feathers.

8 SAKER FALCON *Falco cherrug* 43–60 cm; WS 102–135 cm **P/V**   **Page 398**
Open habitats. Mauritania; rare. Vagrant, N Senegal, Mali, Chad and Cameroon.
8a Adult Pale head; streaked below, especially on flanks; 'trousers' usually dark.
8b Juvenile Head pattern usually more pronounced; underparts more boldly streaked.
8c–d Adult in flight Underwing more contrasting than 7e.

PLATE 27: FALCONS II

1 GREY KESTREL *Falco ardosiaceus* 30–39 cm; WS *c.* 70 cm **R** Page 395
Open habitats. Almost throughout; uncommon/scarce. Rare, SE Gabon and Congo.
1a **Adult** Wholly slate-grey; cere, orbital ring and legs bright yellow.
1b–c **In flight** Slate-grey above and below; primaries blackish.

2 RED-NECKED FALCON *Falco chicquera* 29–36 cm; WS 65–80 cm **R** Page 395
Savanna, typically with palm trees. S Mauritania–Guinea to Chad–CAR. Uncommon.
2a **Adult** Bright chestnut top of head; white cheeks; densely barred above and below.
2b **Juvenile** Duller than adult.
2c–d **Adult in flight** Dark above, primaries blackish; barring below appears grey at distance.
2e **Juvenile in flight** Pale rufous-buff underparts and underwing-coverts.

3 AFRICAN HOBBY *Falco cuvierii* 28–31 cm; WS *c.* 70 cm **R** Page 397
Savanna and extensive clearings in forest zone. Almost throughout. Scarce/locally not uncommon.
3a **Adult** Orange-rufous below; blackish-slate above; short moustachial stripe.
3b **Juvenile** Duller; underparts heavily streaked black.
3c–d **Adult in flight** Swift and slender; mainly orange-rufous below.
3e **Juvenile in flight** Duller, streaky version of adult.

4 EURASIAN HOBBY *Falco subbuteo* 29–36 cm; WS 70–92 cm **P** Page 397
Open habitats. Almost throughout. Scarce/uncommon.
4a **Adult** Narrow black moustachial stripe contrasting with white cheeks and throat; boldly streaked
    underparts; rufous vent and thighs.
4b **Juvenile** Browner; pale forehead; no rufous on vent and thighs.
4c–d **Adult in flight** Long, pointed wings; medium-long tail.

5 RED-FOOTED FALCON *Falco vespertinus* 28–31 cm; WS 60–78 cm **P** Page 395
Open habitats. Almost throughout. Rare/uncommon. Vagrant, São Tomé.
5a **Adult male** Dark blue-grey; rufous vent and thighs; reddish cere, orbital ring and legs.
5b **Adult female** Orangey crown, nape and underparts; small black mask.
5c **Juvenile** Pale forehead; short moustachial streak; pale neck collar. Compare 4b.
5d–e **Adult male in flight** Silvery-grey flight feathers contrast with rest of plumage.
5f–g **Adult female in flight** Plain underwing-coverts; barred flight feathers.
5h **Juvenile in flight** Less boldly streaked below than 4b; paler head pattern.

6 AMUR FALCON *Falco amurensis* 26–32 cm; WS 58–75 cm **V?** Page 396
Open habitats. Passage migrant to E Africa, wintering in S Africa. Vagrant, Gabon.
6a **Adult male** As 5a.
6b **Adult female** Slate-grey crown; short moustachial streak; whitish below barred and blotched dark.
6c **Juvenile** Browner above than 6b; more streaked below.
6d–e **Adult male in flight** As 5d–e, but has white underwing-coverts.
6f–g **Adult female in flight** White underwing strongly barred black.
6h **Juvenile in flight** Similar to adult female.

7 SOOTY FALCON *Falco concolor* 32–38 cm; WS 73–90 cm **V** Page 396
Open habitats. Palearctic vagrant, E Chad.
7a **Adult** All blue-grey; much more slender, with longer wings and tail, than 1.
7b **Juvenile** Head pattern as 4b but cheeks and throat buffy; underparts brownish-buff.
7c **Adult in flight** Slender; tail wedge shaped; underwings and tail blue-grey tipped dark.
7d **Juvenile in flight** Note broad subterminal tail band and slender jizz.

8 ELEONORA'S FALCON *Falco eleonorae* 36–42 cm; WS 84–105 cm **V** Page 396
Coastal. Palearctic vagrant, N Mauritania.
8a **Adult pale morph** Well-defined black moustachial stripe; white cheeks and throat; orangey
    underparts narrowly streaked blackish.
8b **Adult dark morph** Wholly brownish-black.
8c **Juvenile** Browner than 8a; underparts buffish.
8d **Adult pale morph in flight** Slender; long wings and tail; dark underwing-coverts.
8e **Adult dark morph in flight** Underwing-coverts darker than flight feathers.
8f **Juvenile in flight** Dark-looking underwing-coverts contrast with paler flight feathers.

PLATE 28: FRANCOLINS

Terrestrial birds with thickset body, small head, short and robust bill, short wings and tail, and strong feet. Usually in pairs or small groups. Only reluctantly take flight; prefer to run. Identification features include coloration of bare parts, vocalisations, habitat and range. Calls often loud and raucous.

1 MOUNT CAMEROON FRANCOLIN *Francolinus camerunensis* *c.* 33 cm **R** EN **Page 405**
Montane forest. Endemic to Mt Cameroon (850–2100 m). Uncommon.
1a Adult male Very dark; bright red bill, eye patch and legs.
1b Adult female Streaked, whitish throat; whitish marked underparts.

2 FINSCH'S FRANCOLIN *Francolinus finschi* *c.* 35 cm **R** **Page 403**
Open grassland. SE Gabon–Congo. Locally not uncommon/uncommon.
2a Adult Mainly rufous head.
2b In flight Rufous in wings.

3 RING-NECKED FRANCOLIN *Francolinus streptophorus* 30–33 cm **R** **Page 403**
Rocky, grassy slopes. W Cameroon (Foumban area). Scarce and local.
3a Adult male Black-and-white barring around neck and on upper breast.
3b Adult female As male, but finely barred rusty-buff above.

4 HEUGLIN'S FRANCOLIN *Francolinus icterorhynchus* *c.* 32 cm **R** **Page 405**
Open and lightly wooded grassland, cultivation. CAR. Common.
Adult The only francolin with yellowish bill, eye patch and legs.

5 WHITE-THROATED FRANCOLIN *Francolinus albogularis* *c.* 23 cm **R** **Page 402**
Wooded grassland. Senegambia to N Cameroon. Uncommon/rare and local.
5a Adult male *albigularis* Small; whitish supercilium; white throat. Chestnut wings conspicuous in flight. Senegambia to Ivory Coast.
5b Adult female *albigularis* Variable amount of fine dark barring on breast and flanks.
5c Adult male *buckleyi* Head buff to rusty-buff. E Ivory Coast to N Cameroon.
5d Adult female *buckleyi* Wavy barring below extending onto belly.

6 AHANTA FRANCOLIN *Francolinus ahantensis* *c.* 33 cm **R** **Page 404**
Forest edge, second growth and thickets. Senegambia–Liberia to SW Nigeria. Not uncommon/locally scarce.
Adult Dark brown streaked white on hindneck and mantle; orange-red bill and feet.

7 SCALY FRANCOLIN *Francolinus squamatus* 30–33 cm **R** **Page 404**
Forest edge, second growth and thickets. Nigeria to S CAR–Congo. Common/not uncommon.
Adult Dark grey-brown overall; scarlet legs.

8 CLAPPERTON'S FRANCOLIN *Francolinus clappertoni* 30–33 cm **R** **Page 405**
Arid grassland. E Mali to S Chad–N CAR; locally common. Also S Mauritania and C Mali.
Adult Large; red eye patch; white supercilium; black bill; dusky-red legs.

9 RED-NECKED FRANCOLIN *Francolinus afer* 35–41 cm **R** **Page 406**
Grassland. S Gabon–Congo. Locally not uncommon/common.
9a Adult *cranchii* Bare red throat.
9b Adult *afer* Pale brown breast streaked black; rest of underparts blackish, feathers fringed white. Introduced São Tomé; common.

10 DOUBLE-SPURRED FRANCOLIN *Francolinus bicalcaratus* 30–35 cm **R** **Page 404**
Grassland, farmbush, scrub. S Mauritania–Liberia to SW Chad–Cameroon. Common.
Adult Large; mainly brownish; white supercilium; greenish legs.

BARBARY PARTRIDGE *Alectoris barbara* 32–35 cm **R** **Page 401**
Rocky slopes. N Mauritania. Rare and local. **See illustration on plate 146.**
Adult Rufous-brown collar with white spots; boldly barred flanks; bright red bill and legs.

PLATE 29: BUTTONQUAILS, QUAILS, SMALL FRANCOLINS AND STONE PARTRIDGE

Buttonquails. Very small, quail-like terrestrial birds with short, rounded body and wings. Females larger and brighter than males. Secretive and very difficult to flush; usually do not fly far.

1 LITTLE BUTTONQUAIL *Turnix sylvatica* 14–16 cm **R/M**   **Page 408**
Grassland, cultivation, other open habitats almost throughout (not desert). Scarce/locally common.
1a   **Adult** Very small; orangey breast spotted black on sides; pale yellow eye.
1b   **In flight** Brown remiges contrast with pale coverts.

2 BLACK-RUMPED BUTTONQUAIL *Turnix hottentotta* 14–16 cm **R/M/V**   **Page 409**
Grassland. Recorded Sierra Leone, S Ivory Coast, Ghana, Nigeria to SW CAR–Congo. Rare and local.
2a   **Adult** Resembles 1a, but darker; orange-rufous breast-sides barred black and white; eye dark.
2b   **In flight** Black rump.

3 QUAIL-PLOVER *Ortyxelos meiffrenii* 11–13 cm **R/M**   **Page 408**
Arid and semi-arid grassland. S Mauritania–Senegal to N Cameroon and Chad; also coastal Ghana. Uncommon and local.
3a   **Adult** Pale sandy-rufous above; mainly whitish below.
3b   **In flight** Blackish remiges contrast with diagonal band of white on greater coverts. Flight fluttering.

Quails. Very small and rounded. Sexually dimorphic. Secretive. Usually seen when flushed.

4 HARLEQUIN QUAIL *Coturnix delegorguei* 15–16 cm **M/V/R**   **Page 400**
Grassland, cultivation. Nomadic, with sparse records throughout; also Bioko. Common resident, São Tomé; vagrant, Príncipe.
4a   **Adult male** Black-and-white head pattern; black-and-chestnut underparts.
4b   **Adult female** Darker than 6b; pale tawny underparts.

5 BLUE QUAIL *Coturnix chinensis* 13–14 cm **R/M**   **Page 400**
Grassland, edges of cultivation. Sierra Leone–SE Guinea to CAR–N Congo; coastal Gabon. Uncommon/rare and local.
5a   **Adult male** Dark slate-blue; chestnut on wings, flanks and uppertail-coverts.
5b   **Adult female** Heavily barred breast and flanks.

6 COMMON QUAIL *Coturnix coturnix* 16–18 cm **P/R**   **Page 400**
Open grassy habitats. Palearctic migrant, Mauritania–Sierra Leone to Chad; uncommon/rare. Resident, Cape Verde Is; common.
6a   **Adult male** Strong head pattern with blackish anchor on throat; breast-sides and flanks streaked buff, black and chestnut.
6b   **Adult female** Duller than male; no anchor on throat.
6c   **In flight** Plain wings; streaked upperparts.

7 LATHAM'S FOREST FRANCOLIN *Francolinus lathami* 20–25 cm **R**   **Page 401**
Forest interior. Sierra Leone–SE Guinea to SW CAR–Congo. Uncommon/locally not uncommon.
7a   **Adult male** Bold head pattern; white-spotted underparts; yellow legs. Very secretive.
7b   **Adult female** Browner than male; belly white.
7c   **Juvenile** Brownish head; white throat

8 SCHLEGEL'S FRANCOLIN *Francolinus schlegelii* 21–24 cm **R**   **Page 403**
Grassland. WC Cameroon to S Chad–N CAR. Uncommon/rare and local.
8a   **Adult male** Rusty-orange head; dusky eye-stripe; vinous-chestnut upperparts.
8b   **Adult female** Duller; narrow barring on underparts more irregular.

9 COQUI FRANCOLIN *Francolinus coqui* 20–25 cm **R**   **Page 402**
Grassland. Disjunct distribution.
9a   **Adult male** *spinetorum* Rufous-orange head; barred breast and flanks. S Mauritania, Mali, W Niger, N Nigeria. Rare/uncommon.
9b   **Adult female** *spinetorum* Note facial pattern.
9c   **Adult male** *coqui* Bars extending onto belly. SW Gabon–Congo; locally common.
9d   **Adult female** *coqui* Belly heavily barred.

10 STONE PARTRIDGE *Ptilopachus petrosus* c. 25 cm **R**   **Page 401**
Various savanna habitats, often with rocks. S Mauritania–Sierra Leone to Chad–CAR. Locally common/uncommon.
Adult Dark brown; cocked tail.

PLATE 30: GUINEAFOWL

Medium-large terrestrial birds with small, bare head, relatively long neck and thickset body. Bare skin of head and neck often brightly coloured. Gregarious when not breeding. Often vocal.

1   **BLACK GUINEAFOWL** *Agelastes niger*   40–43 cm   **R**   **Page 406**
Lowland forest. SE Nigeria to Congo and SW CAR. Scarce and local.
   1a   **Adult** Entirely black with bare reddish head and throat.
   1b   **Juvenile** Dull black with white belly.

2   **WHITE-BREASTED GUINEAFOWL** *Agelastes meleagrides*   40–45 cm   **R**   VU   **Page 406**
Lowland forest. E Sierra Leone, Liberia, Ivory Coast, W Ghana. Rare to locally not uncommon. Endemic.
   2a   **Adult** Pinkish-red head and upper neck contrast with broad white collar.
   2b   **Juvenile** Dull black; white belly; dark brownish head with two tawny crown-stripes.

3   **PLUMED GUINEAFOWL** *Guttera plumifera*   45–51 cm   **R**   **Page 407**
Lowland forest. S Cameroon to Congo and CAR. Locally not uncommon to scarce.
   3a   **Adult** Bare grey-blue head; stiff feathery crest; plumage finely speckled bluish.
   3b   **Juvenile** Dull upperparts; lightly scaled underparts.

4   **CRESTED GUINEAFOWL** *Guttera pucherani*   46–56 cm   **R**   **Page 407**
Lowland and gallery forest. Guinea-Bissau to SW Cameroon; Congo; CAR. Locally not uncommon. Rare and local, S Mali.
   4a   **Adult** *verreauxi* Similar to 3 but has bare blue head, curly crest and black collar.
   4b   **Adult** *sclateri* As 4a but crest shorter in front. SW Cameroon.
   4c   **Juvenile** Heavily scaled plumage with pale blue on mantle and tertials.

5   **HELMETED GUINEAFOWL** *Numida meleagris*   53–63 cm   **R**   **Page 407**
Savanna; various open habitats. S Mauritania–Sierra Leone to S Chad–CAR; SW Gabon to Congo. Locally common to scarce. Introduced, Cape Verde Is, São Tomé and Annobón.
   5a   **Adult** *galeata* Bare bluish head; small casque; red gape wattles.
   5b   **Adult** *meleagris* Larger casque; blue gape wattles. E Cameroon to CAR.
   5c   **Juvenile** Tawny-buff head with two blackish lateral crown-stripes; no casque; dull grey-brown plumage speckled buffish-white and irregularly barred rusty-buff.

PLATE 31: FLUFFTAILS AND RAILS

Flufftails. Small, secretive rails. Vocalisations highly characteristic. Note, in males, amount of chestnut on head, neck and breast, markings on body (spots or streaks) and colour of tail.

1 WHITE-SPOTTED FLUFFTAIL *Sarothrura pulchra* 15 cm **R**   **Page 410**
Lowland forest and cultivation. Senegambia to S CAR–Congo. Common/scarce.
  1a Adult male Head, breast and tail chestnut; rest of plumage black densely spotted white.
  1b Adult female Black areas barred rufous-buff; tail chestnut barred black.
  1c Juvenile Unique among flufftails in being as respective adult but duller.

2 BUFF-SPOTTED FLUFFTAIL *Sarothrura elegans* 15 cm **R/M**   **Page 410**
Lowland forest and overgrown cultivation. E Sierra Leone–SE Guinea to Ivory Coast; S Nigeria to Congo; also Bioko. Uncommon/rare to locally not uncommon.
  2a Adult male Upperparts spotted buff; tail rufous barred black.
  2b Adult female Brown above finely spotted buff; coarsely barred tawny-and-brown belly.
  2c Juvenile Sooty-brown; belly paler.

3 CHESTNUT-HEADED FLUFFTAIL *Sarothrura lugens* 15 cm **R**   **Page 411**
Swampy and grassy habitats. Cameroon and NE Gabon. Rare/uncommon and local.
  3a Adult male Chestnut confined to head; rest of plumage black finely streaked white.
  3b Adult female Head streaked buff-brown; underparts spotted and streaked.
  3c Juvenile Mainly sooty-black.

4 RED-CHESTED FLUFFTAIL *Sarothrura rufa* 15 cm **R**   **Page 411**
Swampy and grassy habitats. Locally common, Gabon. Rare, Sierra Leone, Liberia, Togo, Nigeria, Cameroon, Congo and SW CAR.
  4a Adult male Chestnut head and breast; rest of plumage black finely streaked white.
  4b Adult female Dark brown to buffish, densely barred.
  4c Juvenile Mainly sooty-black.

5 STREAKY-BREASTED FLUFFTAIL *Sarothrura boehmi* 15 cm **R/V**   **Page 412**
Grassy habitats. NE Gabon, Congo; locally common. Rare/vagrant, Mali, Nigeria, S Cameroon.
  5a Adult male Rufous paler and more extensive than in 3a, underparts whiter with coarser streaking.
  5b Adult female Brownish-black with scalloped and barred appearance.
  5c Juvenile Mainly sooty-black.

Rails. Medium-sized to large terrestrial and aquatic birds. Secretive. Sexes similar.

6 GREY-THROATED RAIL *Canirallus oculeus* 30 cm **R**   **Page 410**
Lowland forest. E Sierra Leone–SE Guinea to SW Ghana; S Nigeria to Congo. Scarce/uncommon to locally not uncommon.
  6a Adult White spots on edge of dark brown wing; grey face and throat; rufous-chestnut neck to upper belly and tail.
  6b Juvenile Dark brown head and underparts.

7 NKULENGU RAIL *Himantornis haematopus* *c.* 43 cm **R**   **Page 409**
Lowland forest. Guinea to Togo; S Nigeria to CAR–Congo. Uncommon/locally common.
  Adult Large. Dark brownish scaled paler; long red legs.

8 AFRICAN WATER RAIL *Rallus caerulescens* 28 cm **R**   **Page 413**
Aquatic habitats. Gambia, Sierra Leone, Cameroon, NE Gabon, NE CAR and S Congo; rare and very local. Vagrant, São Tomé.
  8a Adult Long red bill; red legs; head-sides and underparts grey.
  8b Juvenile As adult but browner.

PLATE 32: CRAKES

Small to medium-sized terrestrial and aquatic birds. Flick short tail while walking. Legs and toes long; wings rounded. Flight over short distances fluttering and seemingly weak. Most species skulking and difficult to observe.

1 BLACK CRAKE *Amaurornis flavirostris* 19–23 cm **R** Page 414
Freshwater wetlands. Throughout. Common.
  1a **Adult** All black; bright yellow bill; red legs. Less shy than other crakes; often in the open.
  1b **Juvenile** Duller; bill and legs dusky.

2 SPOTTED CRAKE *Porzana porzana* 22–24 cm **P** Page 413
Marshes and similar habitats. Patchily recorded, Mauritania–Ivory Coast to Chad. Rare.
  2a **Adult male** Yellow bill with orange-red base; plain buff undertail-coverts.
  2b **Adult female** Usually browner and more speckled than male.

3 LITTLE CRAKE *Porzana parva* 18–20 cm **P** Page 413
Marshes and similar habitats. Mostly Senegal R. delta; also Gambia, Liberia, Ivory Coast, Burkina Faso, Niger and Nigeria. Rare.
  3a **Adult male** Lime-green bill with red base; long primary projection; blue-grey face and underparts; faint barring on flanks.
  3b **Adult female** Whitish throat; buffish underparts.

4 BAILLON'S CRAKE *Porzana pusilla* 17–19 cm **P** Page 413
Marshes and similar habitats. Senegal R. delta. Not uncommon.
  4a **Adult male** Green bill; short primary projection; densely speckled upperparts; obvious barring on flanks.
  4b **Adult female** Throat and underparts often paler, ear-coverts brownish.

5 STRIPED CRAKE *Aenigmatolimnas marginalis* 18–21 cm **M** Page 414
Wet grasslands and similar habitats. Patchily recorded, Ivory Coast to Congo. Scarce/rare.
  5a **Adult male** Greenish bill; dark brown upperparts finely streaked white; russet undertail-coverts. Highly secretive.
  5b **Adult female** Head and breast to flanks grey.
  5c **Juvenile** As male but duller and browner; no white streaking.

6 AFRICAN CRAKE *Crex egregia* 20–23 cm **M/R** Page 412
Grassy habitats. Throughout; also São Tomé and Príncipe. Common/rare. Vagrant, Bioko.
  6a **Adult** Face to breast grey; lower underparts barred black and white.
  6b **Juvenile** Browner below; barring less distinct.
  6c **In flight** Upperparts dark brown mottled black; flight feathers blackish; dangling legs.

7 CORN CRAKE *Crex crex* 27–30 cm **P** VU Page 412
Grassy habitats. Patchily recorded throughout. Rare.
  7a **Adult** Mainly brownish-buff; broad brownish-grey supercilium.
  7b **In flight** Conspicuous chestnut wings.

PLATE 33: GALLINULES, JACANAS AND FINFOOT

Gallinules, moorhens and coots. Less secretive than rails and crakes. Frequent open water or clamber about in dense waterside vegetation.

1 COMMON MOORHEN *Gallinula chloropus* 30–36 cm **R/P**   **Page 415**
Various freshwater wetlands. Throughout but distribution patchy, also São Tomé and Príncipe; locally not uncommon/rare. Vagrant, Cape Verde Is.
1a   **Adult** White line on flanks; red bill with yellow tip; greenish legs.
1b   **Juvenile** Mainly greyish-brown; throat and belly whitish; bill greenish-brown.

2 LESSER MOORHEN *Gallinula angulata* 23 cm **R/M**   **Page 416**
Various freshwater wetlands. Throughout; locally common/uncommon. Also São Tomé and Príncipe; rare. Vagrant, Bioko.
2a   **Adult** As 1a, but bill yellow with red culmen; also smaller.
2b   **Juvenile** Dark brown above; head-sides, neck and breast brownish-buff.

3 ALLEN'S GALLINULE *Porphyrio alleni* 25 cm **M/R**   **Page 415**
Various freshwater wetlands. Throughout; uncommon/locally common. Vagrant, Bioko, São Tomé and Annobón.
3a   **Adult** Glossy purplish and dark green; red bill and legs; bluish frontal shield.
3b   **Adult female at start of breeding** Apple-green frontal shield.
3c   **Juvenile** Mainly buffish-brown; scaly upperparts.

AMERICAN PURPLE GALLINULE *Porphyrio martinica* 33 cm **V**   **Page 415**
American vagrant recorded off Liberian coast. **See illustration on plate 146.**
Adult Yellow-tipped red bill; yellow legs.

4 PURPLE SWAMPHEN *Porphyrio porphyrio* 45–50 cm **R**   **Page 415**
Various freshwater wetlands. Throughout but distribution patchy. Common/rare and local.
4a   **Adult** Large; glossy purple and blue-green; red bill, frontal shield and legs.
4b   **Juvenile** Paler and greyer; bare parts duskier.

5 EURASIAN COOT *Fulica atra* 37 cm **P**   **Page 416**
Open waters. Mauritania–Senegal to Chad. Rare/locally common.
Adult All black; white bill and frontal shield; lobed toes.

Jacanas. Long-legged waterbirds with extremely long toes and claws, enabling them to walk on floating vegetation.

6 LESSER JACANA *Microparra capensis* 16 cm **R/M**   **Page 422**
Various freshwater wetlands. Patchily distributed, C Mali–Ivory Coast to Chad–CAR. Locally common/rare. Vagrant, Sierra Leone, S Mauritania.
6a   **Adult** Small size; chestnut crown; white supercilium; white underparts. Compare 7b.
6b   **In flight** Black flight feathers; white trailing edge; pale midwing panel; trailing feet.

7 AFRICAN JACANA *Actophilornis africana* 30 cm **R**   **Page 421**
Various freshwater wetlands. Throughout. Common.
7a   **Adult** Bright chestnut; white foreneck; blue bill and frontal shield; extremely long toes.
7b   **Juvenile** Underparts mainly white; white supercilium.

Finfoot. Unobtrusive grebe-like waterbird with long neck and small head. Inhabits forested rivers, lakes and lagoons with overhanging vegetation.

8 AFRICAN FINFOOT *Podica senegalensis* 50 cm **R**   **Page 418**
Throughout forest zone; also in wooded savanna. Uncommon/locally common.
8a   **Adult male** *senegalensis* **breeding** Red bill and feet; white line on neck-sides; finely spotted upperparts.
8b   **Adult male** *camerunensis* Much darker than 8a; few or no spots above; no white line on neck-sides. Restricted to S Cameroon, Gabon and Bioko.
8c   **Adult female** Browner; white throat and foreneck.
8d   **Juvenile** As 8c but paler, less spotted; lower neck to flanks tawny-buff.

PLATE 34: CRANES AND BUSTARDS

1 **DEMOISELLE CRANE** *Anthropoides virgo* 90–100 cm; WS 155–180 cm **P**   **Page 417**
Grasslands. NE Nigeria and Chad. Rare/scarce.
  1a **Adult** White head plumes; black foreneck; long, narrow feathers over tail.
  1b **Immature** Washed-out version of 1a.
  1c **In flight** Much as 2c; black on neck reaching breast; greyish tinge to inner primaries.

2 **COMMON CRANE** *Grus grus* 100–120 cm; WS 180–220 cm **V**   **Page 417**
Open habitats. Vagrant, N Mauritania and N Nigeria.
  2a **Adult** Large; grey; white stripe on neck-sides; bushy feathers over tail.
  2b **Immature** Browner than 2a.
  2c **In flight** Grey with black flight feathers; outstretched neck. Compare 1c.

WATTLED CRANE *Bugeranus carunculatus* c. 175 cm; WS 230–260 cm **V?** VU **Page 417**
Marshes and grasslands. Single old record, Guinea-Bissau (vagrant?). **See illustration on plate 146.**

3 **BLACK CROWNED CRANE** *Balearica pavonina* 100 cm; WS 180–200 cm **R/M** NT **Page 417**
Open habitats. Mauritania–Senegambia to Chad; distribution patchy. Rare/locally not uncommon.
  3a **Adult** Dark; large white wing panel; straw-coloured crest.
  3b **Immature** Washed rusty; shorter crest; no bare area on head-sides.
  3c **In flight** Contrasting white wing-coverts.

4 **SAVILE'S BUSTARD** *Eupodotis savilei* 41 cm **R**   **Page 420**
Arid and semi-arid habitats. S Mauritania–Senegambia to Chad; distribution patchy. Locally not uncommon/rare.
  4a **Adult male** Small; short neck and legs; grey foreneck; black underparts. Compare 5.
  4b **Adult female** White throat; deep buff neck; broad white breast-band.
  4c **In flight** White band separating blackish remiges from tawny upperwing-coverts.

5 **BLACK-BELLIED BUSTARD** *Eupodotis melanogaster* 60 cm **R/M**   **Page 421**
Grassland, derived savanna. S Mauritania–Sierra Leone to Chad–CAR; also S Gabon–Congo. Not uncommon/rare.
  5a **Adult male** Long neck and legs; black throat and underparts.
  5b **Adult female** Mainly brownish-buff; slender.
  5c **Adult male in flight** Upperwing with largest area of white of any bustard.
  5d **Adult female in flight** Much less white in wing; flight feathers mainly blackish.

6 **WHITE-BELLIED BUSTARD** *Eupodotis senegalensis* 50 cm **R**   **Page 420**
Open savanna, grassland, thorn scrub. Locally not uncommon/rare.
  6a **Adult male** *senegalensis* Black throat patch; greyish-blue neck; tawny breast; white belly.
    S Mauritania–Senegal to N Cameroon–Chad.
  6b **Adult male** *mackenziei* Tawny-buff hindneck. SE Gabon and Congo.
  6c **Adult female** Throat white; neck mainly tawny-buff.
  6d **In flight** Tawny-buff above with mainly black remiges.

7 **DENHAM'S BUSTARD** *Neotis denhami* 80–100 cm; WS 170–250 cm **R/M** NT   **Page 418**
Wooded savanna. S Mauritania–Sierra Leone to Chad–CAR; also SE Gabon–Congo. Scarce/locally not uncommon.
  7a **Adult male** Large; rufous hindneck; grey foreneck; bold white wing markings.
  7b **Adult female** Smaller; head-sides, throat and foreneck tinged buffish.
  7c **In flight** Mainly dark above, with conspicuous white markings on wings.

8 **ARABIAN BUSTARD** *Ardeotis arabs* 80–100 cm; WS 205–250 cm **R/M**   **Page 419**
Arid grassy plains, Acacia woodland. S Mauritania–N Senegal to N Cameroon–Chad, CAR. Uncommon/rare. Vagrant, N Ivory Coast and N Ghana.
  8a **Adult male** Large; grey, thick-looking neck.
  8b **In flight** Mainly dark above; outer primaries and coverts mainly black. Compare 7c.

9 **NUBIAN BUSTARD** *Neotis nuba* 50–70 cm; WS 140–180 cm **R/M** NT   **Page 419**
Dry thorn scrub. Mauritania to Chad. Uncommon/rare.
  9a **Adult male** Fairly large and pale; rufous, black-edged crown; black throat.
  9b **Adult female** Head markings duller.
  9c **In flight** Black remiges; large white area on primaries; mainly white underwing.

HOUBARA BUSTARD *Chlamydotis undulata* 55–65 cm; WS 135–170 cm **R** NT   **Page 419**
Semi-desert. Mauritania. Uncommon. Black frills on neck-sides. **See illustration on plate 146.**

PLATE 35: THICK-KNEES, OYSTERCATCHER, STILT AND AVOCET

Thick-knees. Have cryptically patterned plumage and large head and eyes. Crepuscular and nocturnal, often encountered on roads at night. Calls melodious and far carrying. Note wing pattern.

1 SENEGAL THICK-KNEE *Burhinus senegalensis* 32–38 cm **R/M** Page 425
River banks, lake shores, etc. S Mauritania–Sierra Leone to Chad–CAR. Common/not uncommon.
1a Adult Broad, pale greyish wing panel bordered above by narrow black bar.
1b In flight Prominent white patches in black primaries; pale wing panel.

2 WATER THICK-KNEE *Burhinus vermiculatus* 38–41 cm **R/M** Page 425
River banks, lake shores, lagoons, etc. S Senegal, Liberia–Ghana, Burkina Faso, SW Niger, Nigeria to CAR–Congo. Locally common/rare.
2a Adult Greyish wing panel bordered above by narrow white bar highlighted by black above.
2b In flight White primary patches; pale wing panel; feet project slightly beyond tail.

3 STONE-CURLEW *Burhinus oedicnemus* 40–45 cm **P/R** Page 424
Arid, stony areas. Palearctic visitor, Mauritania–Senegal and Mali; uncommon/rare. Rare resident, N Mauritania. Vagrant, Guinea, Sierra Leone, Niger and CAR.
3a Adult White wingbar bordered above and below by black. Bill yellow tipped black.
3b In flight White primary patches; pale wing panel bordered by black-and-white bar.

4 SPOTTED THICK-KNEE *Burhinus capensis* *c.* 43 cm **R** Page 425
Dry, open woodland, arid areas. S Mauritania–Senegambia to Chad–NE CAR; uncommon. Vagrant, Guinea and Gabon.
4a Adult Upperparts densely spotted; no bars or panel on wing.
4b Juvenile Slightly duller; upperparts streaked.
4c In flight Small white primary patches.

5 EURASIAN OYSTERCATCHER *Haematopus ostralegus* 40–45 cm **P/V** Page 423
Coast. Mauritania–Senegambia, locally common; uncommon/rare further south and east. Vagrant, Cape Verde Is.
5a Adult non-breeding Bulky; black and white; red bill; white collar.
5b Adult breeding As 5a but lacks white collar.
5c In flight Broad white wingbar.

6 AFRICAN BLACK OYSTERCATCHER *Haematopus moquini* 42–45 cm **V? NT** Page 423
S African species; occurrence in W Africa uncertain. As 5 but plumage entirely black.

7 BLACK-WINGED STILT *Himantopus himantopus* 35–40 cm **R/P** Page 423
Various aquatic habitats. Throughout; also Cape Verde Is. Locally common/uncommon. Vagrant, Congo.
7a Adult Very long pinkish legs; needle-like bill; head white or with variable amount of black.
7b Adult Example of a dark-headed individual.
7c Juvenile Duller; upperparts brownish with narrow buff fringes.
7d Adult in flight Black upperwing; legs extend far beyond tail.

8 PIED AVOCET *Recurvirostra avosetta* 42–45 cm **P** Page 424
Mudflats, estuaries, lake margins. Mainly Mauritania–Guinea and L. Chad area, also Mali. Scarce/rare elsewhere. Vagrant, Cape Verde Is.
8a Adult Slender, upturned black bill.
8b In flight Distinctive black-and-white pattern.

6

PLATE 36: PHALAROPES, SNIPES, PRATINCOLES AND COURSERS

1   RED-NECKED PHALAROPE *Phalaropus lobatus* 18–19 cm **V**   **Page 450**
Pelagic; wetlands. Palearctic/N American vagrant, Mauritania, Senegambia, Sierra Leone, Cape Verde Is.
   1a   **Adult non-breeding** More dainty than 3; needle-like bill.
   1b   **Juvenile** Rusty-yellow V on dark brown upperparts.

2   WILSON'S PHALAROPE *Phalaropus tricolor* 22–24 cm **V**   **Page 449**
Various wetlands. N American vagrant, Ivory Coast.
   2a   **Adult non-breeding** Needle-like bill; very pale grey above; yellowish legs.
   2b   **In flight** Plain upperparts and wings; white rump and uppertail-coverts.

3   RED PHALAROPE *Phalaropus fulicaria* 20–22 cm **P**   **Page 450**
Pelagic; various wetlands. Mauritania–Cameroon, rare. Cape Verde Is, uncommon.
   3a   **Adult non-breeding** Black eye patch; bill thicker than 1. **3b In flight** White wingbar.

4   GREAT SNIPE *Gallinago media* 27–29 cm **P**   **NT**   **Page 443**
Various wetlands. Widespread. Uncommon/rare.
   4a   **Adult** Bulkier than 6; bill shorter; white wingbars; barred underparts.
   4b   **In flight** White lines on upper wing; broad white tail corners. Low, direct flight.

5   JACK SNIPE *Lymnocryptes minimus* 17–19 cm **P/V**   **Page 443**
Various wetlands. Scattered records, almost throughout; also Cape Verde Is. Rare/locally uncommon.
   5a   **Adult** Small; shortish bill; two yellowish stripes on upperparts. Skulking.
   5b   **In flight** Note overall size and bill length. Flushes silently from underfoot.

6   COMMON SNIPE *Gallinago gallinago* 25–27 cm **P**   **Page 443**
Various wetlands. Throughout; uncommon/common. Also Cape Verde Is (rare); Bioko (vagrant).
   6a   **Adult** Cryptically patterned; boldly striped head; very long, straight bill.
   6b   **In flight** Narrow white trailing edge to wing; white edges to tail. Fast zigzagging flight.

7   EGYPTIAN PLOVER *Pluvianus aegyptius* 20 cm **R/M**   **Page 426**
Sandbars in large rivers. S Mauritania–Liberia to Chad–Congo; also Congo. Locally not uncommon/rare.
   7a   **Adult** Distinctive pattern of black, white, grey and creamy-buff.
   7b   **Juvenile** Some rusty on crown, mantle and wing-coverts.
   7c   **In flight** Striking wing pattern with diagonal black band on white.

8   GREATER PAINTED-SNIPE *Rostratula benghalensis* 23–26 cm **R/M**   **Page 422**
Various wetlands. S Mauritania–Liberia to Chad, CAR, Gabon. Uncommon/locally common.
   8a   **Adult male** Rather rail-like; golden-buff 'spectacles' and V on mantle.
   8b   **Adult female** Chestnut head and neck; white 'spectacles'; dark green upperparts.
   8c   **Adult male in flight** Broad, rounded wings; dangling legs; slow wingbeats.
   8d   **Adult female in flight**

9   COLLARED PRATINCOLE *Glareola pratincola* 24–28 cm **R/M/P**   **Page 427**
Various dry and wet, open habitats. Widespread. Uncommon; local. Vagrant, Cape Verde Is.
   9a   **Adult breeding** Creamy-yellow throat bordered by black line; forked tail.
   9b   **Adult non-breeding** Duller; throat outline indistinct; breast mottled.
   9c   **Juvenile** Scaly upperparts; throat pattern indistinct; breast blotched.
   9d   **In flight** Long, pointed wings; chestnut underwing-coverts; white trailing edge to inner wing.

10   BLACK-WINGED PRATINCOLE *Glareola nordmanni* 24–28 cm **P/V**   **DD**   **Page 428**
Various dry and wet, open habitats. Patchily recorded; rare/vagrant to locally uncommon.
   10a–b Adult breeding/non-breeding As 9a–b, but usually has shorter tail.
   10c   **Juvenile** As 9c; differences as adult.
   10d   **In flight** Darker above than 9d; underwing all dark; no white trailing edge to wing.

11   GREY PRATINCOLE *Glareola cinerea* 18–20 cm **M**   **Page 428**
Sandbars in large rivers. Mali; Niger; Benin to S Chad–Congo. Locally common/uncommon.
   11a   **Adult** Pale grey above; pale chestnut nuchal collar. Associated with large sand banks.
   11b   **Juvenile** Duller; head plainer.
   11c   **In flight** Striking pattern; white patch on black outer wing.

12   ROCK PRATINCOLE *Glareola nuchalis* 18–20 cm **R/M**   **Page 428**
Rocky rivers. Sierra Leone–SE Guinea to CAR–Congo; also Bioko. Common/uncommon.
   12a   **Adult** *liberiae* Chestnut nuchal collar; red, black-tipped bill; red legs.
   12b   **Adult** *nuchalis* White nuchal collar.
   12c   **Juvenile** Duller; scaly upperparts; no nuchal collar.
   12d   **In flight** White uppertail-coverts; shallowly forked, mainly black tail.

Continued on page 104

PLATE 37: LAPWINGS

Mostly well marked and easily identified. Several have wattles at base of bill or carpal spurs. Distinctive black-and-white wing pattern in flight. Flight rather slow and heavy.

1 LESSER BLACK-WINGED LAPWING *Vanellus lugubris* 22–26 cm **M**   **Page 436**
Savanna. Patchily distributed, Guinea to Congo; uncommon/not uncommon. Rare, Senegambia.
1a   **Adult** Rather plain looking; white forehead.
1b   **Juvenile** Head pattern duller; wing feathers and scapulars fringed buff.
1c   **In flight** Broad white trailing edge to inner wing.

2 BROWN-CHESTED LAPWING *Vanellus superciliosus* *c.* 23 cm **M**   **Page 435**
Short grassland. Breeds Nigeria–Cameroon, probably CAR. Uncommon and local.
2a   **Adult** Dark chestnut breast-band; yellow wattle in front of eye.
2b   **Juvenile** Duller; upperparts scaled rusty.
2c   **In flight** Diagonal white band on wing.

3 WHITE-TAILED LAPWING *Vanellus leucurus* 26–29 cm **V**   **Page 436**
Damp grassland, marshes, lakes. Palearctic vagrant, S Niger, NE Nigeria and Chad.
3a   **Adult** Rather plain; slender; tail entirely white.
3b   **In flight** Broad diagonal white band on wing; white tail.

4 BLACK-HEADED LAPWING *Vanellus tectus* *c.* 25 cm **R**   **Page 435**
Dry grassland. Senegambia–Guinea Chad–N CAR. Common to rare.
4a   **Adult** Wispy black crest; black-tipped red bill; reddish legs.
4b   **In flight** Large white area from forewing across inner wing.

5 NORTHERN LAPWING *Vanellus vanellus* 28–31 cm **P**   **Page 436**
Open habitats. Mauritania; scarce. Vagrant, N Senegal, Gambia and Cape Verde Is.
5a   **Adult** Wispy crest; dark glossy green above; black breast-band.
5b   **In flight** Strikingly broad, rounded wings; mainly dark above.

6 SPUR-WINGED LAPWING *Vanellus spinosus* 25–28 cm **R**   **Page 435**
Various wetlands. S Mauritania–Sierra Leone to Chad–N CAR. Common/locally uncommon.
6a   **Adult** Top of head black; head-sides and neck white; underparts mainly black.
6b   **In flight** Diagonal white band from carpal joint to inner secondaries.

7 LONG-TOED LAPWING *Vanellus crassirostris* *c.* 31 cm **R**   **Page 436**
Aquatic habitats. L. Chad and CE Nigeria. Not uncommon but local.
7a   **Adult** White face and foreneck; black hindcrown to breast; red legs.
7b   **In flight** Mainly white wing-coverts.

8 WHITE-HEADED LAPWING *Vanellus albiceps* 28–32 cm **R/M**   **Page 434**
Rivers. Throughout except arid north. Common/not uncommon.
8a   **Adult** White band on wing; large yellow wattles; greenish-yellow legs.
8b   **In flight** Wings mainly white with black on coverts and outer primaries.

9 AFRICAN WATTLED LAPWING *Vanellus senegallus* *c.* 34 cm **R/M**   **Page 434**
Grassland. S Mauritania–Liberia to Chad–CAR; common/uncommon. Vagrant, Congo.
9a   **Adult** Mainly pale brown; yellow wattles; long yellow legs.
9b   **In flight** Broad white diagonal band on inner wing; black flight feathers.

Continued from page 102

13 BRONZE-WINGED COURSER *Cursorius chalcopterus* 25–29 cm **R/M?**   **Page 427**
Wooded savanna. S Mauritania–Guinea to Chad–CAR; Gabon, Congo. Uncommon/scarce.
13a   **Adult** Distinctive head pattern; black breast-band; long reddish legs. Nocturnal.
13b   **In flight** Black flight feathers; white uppertail-coverts; black tail band.

14 TEMMINCK'S COURSER *Cursorius temminckii* 19–21 cm **R**   **Page 427**
Savanna. S Mauritania–Guinea to Chad–CAR; SE Gabon–Congo. Uncommon/locally not uncommon.
14a   **Adult** Rufous crown; blackish patch on centre of chestnut belly. **14b Juvenile** Duller.
14c   **In flight** Black outer wing and secondaries; black underwing.

15 CREAM-COLOURED COURSER *Cursorius cursor* 21–24 cm **P/R**   **Page 426**
Desert and semi-desert. Mauritania–Senegal to Chad; uncommon. Cape Verde Is; not uncommon.
15a   **Adult** Sandy-cream; grey hindcrown; black V from eye to nape. **15b Juvenile** Duller.
15c   **In flight** Black outer wing; black underwing.

PLATE 38: TURNSTONE AND PLOVERS

1 RUDDY TURNSTONE *Arenaria interpres* 21–25 cm **P/V**   **Page 449**
Coast. Mauritania to Congo; common/uncommon. Vagrant inland. Also Cape Verde Is, São Tomé and Príncipe.
1a Adult breeding Stocky; rufous-chestnut and black above; short, orange legs.
1b Adult non-breeding Mottled blackish-brown.
1c Juvenile Scaly upperparts.
1d In flight Striking pied pattern.

2 THREE-BANDED PLOVER *Charadrius tricollaris* 17–18 cm **R**   **Page 430**
Near water. Nigeria to W Chad, Gabon, Congo; uncommon/rare, local. Vagrant, Ivory Coast, Ghana.
2a Adult As 3 but forehead white; legs shorter.
2b In flight Narrow white line on wings.

3 FORBES'S PLOVER *Charadrius forbesi* c. 20 cm **R/M**   **Page 431**
Dry and wet habitats. Throughout, north to Senegambia. Locally not uncommon/rare.
3a Adult Two blackish breast-bands; red orbital ring; white band from eye across nape.
3b Juvenile Duller; band across nape buffish; upperparts scaled buff.
3c In flight Appears dark; no white wingbar.

4 KITTLITZ'S PLOVER *Charadrius pecuarius* 14–16 cm **R/M**   **Page 430**
Various open habitats. Widespread; uncommon/locally common. Rare, Liberia, Ivory Coast, Togo.
4a Adult breeding Distinctive head pattern; warm-buff breast.
4b Adult non-breeding Duller; supercilium orange-buff. Juvenile similar but upperparts scaly.
4c In flight White patch on outer wing extends as narrow line on inner wing.

5 COMMON RINGED PLOVER *Charadrius hiaticula* 18–20 cm **P**   **Page 429**
Mudflats, various open habitats. Throughout; also Cape Verde Is and Bioko. Common/not uncommon.
5a Adult breeding More compact than 7; stubby, black-tipped, orange bill; orange legs.
5b Adult non-breeding White forehead and supercilium.
5c In flight Prominent white wingbar.

6 KENTISH PLOVER *Charadrius alexandrinus* 15–17 cm **R/P**   **Page 431**
Coastal and inland shores. Throughout, also Cape Verde Is. Common/rare.
6a Adult male breeding Rufous crown; narrow black patches on breast-sides. Compare 10.
6b Adult female/adult non-breeding Duller; little or no rufous on crown.
6c In flight White wingbar.

7 LITTLE RINGED PLOVER *Charadrius dubius* 15–17 cm **P**   **Page 429**
Various open habitats. Throughout; not uncommon/locally common. Rare, Cape Verde Is. Vagrant, Bioko and São Tomé.
7a Adult breeding Slimmer than 5; thinner, black bill; distinct orbital ring; dull-coloured legs.
7b Adult non-breeding Duller than 7a; plainer head than 5b.
7c In flight No wingbar. Compare 5c.

8 GREATER SAND PLOVER *Charadrius leschenaultii* 20–23 cm **V**   **Page 432**
Mudflats, grassland. Palearctic vagrant, Senegal, Liberia, Ivory Coast, NE Nigeria, Gabon.
8a Adult non-breeding Long legs; relatively long and heavy bill; no white nuchal collar.
8b In flight White wingbar; white tail-sides; feet project beyond tail.

LESSER SAND PLOVER *Charadrius mongolus* 18–20 cm **V**   **Page 780**
Palearctic vagrant, claimed from Gabon. Not illustrated.
As Greater Sand Plover but has smaller bill and shorter legs.

9 CASPIAN PLOVER *Charadrius asiaticus* 19–21 cm **V**   **Page 432**
Grassland. Palearctic vagrant, Mauritania, Mali, Sierra Leone, Nigeria, Cameroon, Gabon, Congo.
9a Adult non-breeding Long, greyish to yellowish legs; broad grey-brown breast-band.
9b In flight Faint wingbar; no white tail-sides; feet project beyond tail.

10 WHITE-FRONTED PLOVER *Charadrius marginatus* c. 18 cm **R**   **Page 431**
Sandy coasts and inland shores. Throughout; locally common/rare. Vagrant, Bioko and São Tomé.
10a Adult male breeding Large white forehead and supercilium; orange-tinged breast.
10b Adult female/adult non-breeding Duller.
10c In flight White wingbar; broad white tail-sides.

PLATE 39: PLOVERS, GODWITS AND CURLEWS

1 EUROPEAN GOLDEN PLOVER *Pluvialis apricaria* 26–29 cm **P/V**   **Page 433**
Tidal mudflats. Mauritania, rare. Vagrant, Senegambia.
1a **Adult breeding** Golden-yellow tones above; white band bordering black below.
1b **Adult non-breeding** Yellowish tones above; yellow-buff below. Compare 2 and 3.
1c **In flight from above** Indistinct wingbar.
1d **In flight from below** White underwing. Compare 2 and 3.

2 PACIFIC GOLDEN PLOVER *Pluvialis fulva* 23–26 cm **P/V**   **Page 433**
Various coastal wetlands and farmland. Gabon, rare. Vagrant, Ivory Coast.
2a **Adult breeding** White 'shawl' from forehead to flanks. Slimmer than 1.
2b **Adult non-breeding** Warm-buff plumage tones; primaries shorter and tertials longer than 3b; wings project slightly beyond tail.
2c **In flight** Grey underwing; indistinct bar on upperwing.

3 AMERICAN GOLDEN PLOVER *Pluvialis dominica* 24–28 cm **V**   **Page 433**
Various wetlands and farmland. N American vagrant, Senegambia–Nigeria; Cape Verde Is.
3a **Adult breeding** White 'shawl' from forehead to breast-sides.
3b **Adult non-breeding** Cold grey plumage tones; 4–5 primary tips project beyond relatively short tertials; wings clearly project beyond tail. In flight: wing pattern as 2c.

4 GREY PLOVER *Pluvialis squatarola* 27–30 cm **P**   **Page 434**
Tidal mudflats. Mauritania–Congo, also Cape Verde Is and São Tomé; not uncommon/common. Rare/vagrant inland and Bioko.
4a **Adult breeding** Spangled black and white above; head-sides to belly black bordered white.
4b **Adult non-breeding** Mainly dull greyish; diffuse supercilium.
4c **Immature** As 4b but neatly chequered above; finely streaked below.
4d **In flight from above** Bold white wingbar; square white rump.
4e **In flight from below** Black axillaries ('armpits'). Compare 1–3.

5 EURASIAN DOTTEREL *Charadrius morinellus* 20–22 cm **V**   **Page 432**
Open habitats. Palearctic vagrant, Mauritania, Gambia.
5a **Adult non-breeding** Broad supercilia form V on nape; narrow pale breast-band; yellowish legs.
5b **In flight** Plain upperwing.

6 BLACK-TAILED GODWIT *Limosa limosa* 37–44 cm **P**   **Page 444**
Various wetlands. Throughout; common to uncommon/rare. Vagrant, Cape Verde Is.
6a **Adult breeding** Long straight bill; long dark legs; rufous-orange head and neck.
6b **Adult non-breeding** Plain grey above. Longer necked and legged than 7.
6c **Juvenile** Rusty-orange fringes to feathers above.
6d **In flight** Broad white wingbar; square white uppertail; black terminal tail band.

7 BAR-TAILED GODWIT *Limosa lapponica* 37–41 cm **P**   **Page 444**
Various wetlands. Entire coast and Cape Verde Is.; common/scarce. A few inland records, Ghana, Nigeria and Chad. Vagrant, Príncipe.
7a **Adult breeding** Slightly upturned bill; legs shorter than 6; underparts dark rufous.
7b **Adult non-breeding** Appears streaked above.
7c **Juvenile** Upperpart feathers notched buff.
7d **In flight** Plain upperwing; white wedge on back; barred tail. Compare 8b.

8 WHIMBREL *Numenius phaeopus* 40–46 cm **P**   **Page 445**
Various wetlands; also dry ground. Entire coast, also Cape Verde and Gulf of Guinea Is; common. Rare inland, but in Mali locally uncommon.
8a **Adult** Long decurved bill; boldly streaked head.
8b **Nominate** *phaeopus* **in flight** White wedge on back; wings more uniform than 7d.
  N. p. hudsonicus (N American vagrant, Sierra Leone and Cape Verde Is) lacks white wedge.

9 EURASIAN CURLEW *Numenius arquata* 50–60 cm **P**   **Page 445**
Various wetlands. Along entire coast; common/scarce. Inland rare/locally uncommon. Cape Verde Is, São Tomé and Príncipe, rare.
9a **Adult** Larger and paler than 8; bill longer; no bold streaks on head.
9b **In flight** Dark outer wing contrasts with paler inner wing.

PLATE 40: SMALL SANDPIPERS

1   WHITE-RUMPED SANDPIPER *Calidris fuscicollis*   15–18 cm   **V**   **Page 440**
Various wetlands. N American vagrant, Ivory Coast and Ghana.
1a   Adult non-breeding Long primary projection; wingtips clearly extend beyond tail.
1b   In flight White uppertail-coverts; dark tail.

2   BAIRD'S SANDPIPER *Calidris bairdii*   14–17 cm   **V?**   **Page 440**
Various wetlands. Possible N American vagrant. Claimed from Mauritania, N Senegal and Gambia.
2a   Adult non-breeding Less prominent supercilium than 1a; slightly shorter, all-black bill.
2b   In flight Resembles 1b but has dark-centred rump and uppertail-coverts.

3   LITTLE STINT *Calidris minuta*   12–14 cm   **P**   **Page 438**
Various wetlands. Throughout, also Cape Verde Is. Common/not uncommon.
3a   Adult breeding Short, straight bill; black legs; rusty on head, upperparts and breast.
3b   Adult non-breeding Grey above with some dark streaking; grey breast-sides.
3c   Juvenile Whitish Vs on mantle and scapulars.
3d   In flight Narrow white wingbar; grey outer tail.

4   BUFF-BREASTED SANDPIPER *Tryngites subruficollis*   18–20 cm   **V**   **Page 442**
Wetlands, grassy areas. N American vagrant, Senegambia, Sierra Leone, Ghana, Gabon.
4a   Adult non-breeding Sandy-buff below; small head; short bill; yellowish legs.
4b   In flight White underwing with dusky half crescent on greater primary-coverts.

5   PECTORAL SANDPIPER *Calidris melanotos*   19–23 cm   **V**   **Page 440**
Various wetlands. N American and E Palearctic vagrant, Mauritania to Gabon; also Príncipe.
5a   Adult non-breeding Sharply demarcated streaked neck and upper breast; pale legs.
5b   In flight Indistinct narrow wingbar; white sides to rump and uppertail.

6   TEMMINCK'S STINT *Calidris temminckii*   13–15 cm   **P**   **Page 439**
Various wetlands. Mauritania–Liberia to Chad–CAR; rare/locally not uncommon. Vagrant, Cape Verde Is.
6a   Adult breeding Brownish to olive-yellow legs; more elongated than 3.
6b   Adult non-breeding Complete grey breast-band. Recalls miniature Common Sandpiper.
6c   Juvenile Buffish fringes above.
6d   In flight As 3d but outer tail white. When flushed, flies high and erratically.

7   SANDERLING *Calidris alba*   20–21 cm   **P**   **Page 438**
Beaches. Entire coast and Cape Verde Is; common/not uncommon. Some inland records.
7a   Adult breeding Black bill and legs; head and breast rufous marked black.
7b   Adult non-breeding Very pale grey above.
7c   Juvenile Chequered black and white above.
7d   In flight Bold white wingbar.

8   BROAD-BILLED SANDPIPER *Limicola falcinellus*   16–18 cm   **V**   **Page 442**
Mudflats and muddy edges. Palearctic vagrant, Mauritania, N Mali, NE Nigeria, C Chad, Cameroon, Gabon.
8a   Adult breeding 'Split' supercilium; boldly streaked breast.
8b   Adult non-breeding Stockier than 9; faint head pattern; dusky streaking above.
8c   In flight Narrow white wingbar; white sides to rump and uppertail.

9   DUNLIN *Calidris alpina*   16–22 cm   **P**   **Page 441**
Various wetlands. Mauritania–Guinea; common/not uncommon. Cape Verde Is; scarce. Rare elsewhere.
9a   Adult breeding Black belly patch.
9b   Adult non-breeding Less elegant than 11b; less evenly decurved bill; legs shorter.
9c   Juvenile Scaly upperparts; streaked breast; variably spotted belly-sides.
9d   In flight Narrow white wingbar; white sides to rump and uppertail.

10   RED KNOT *Calidris canutus*   23–25 cm   **P**   **Page 437**
Mudflats. Along entire coast; common/scarce. Rare, Mali. Vagrant, Cape Verde Is.
10a   Adult breeding Stocky; stout, straight bill; deep rufous face and underparts.
10b   Adult non-breeding Plain, pale grey above.
10c   Juvenile Scaly upperparts.
10d   In flight Narrow white wingbar; whitish-grey rump; grey tail.

11   CURLEW SANDPIPER *Calidris ferruginea*   18–23 cm   **P**   **Page 441**
Various wetlands. Entire coast and Cape Verde Is; common/not uncommon. Small numbers inland.
11a   Adult breeding Evenly decurved bill; longish legs; rusty-red head and underparts.
11b   Adult non-breeding Grey above; prominent supercilium. Compare 9.
11c   Juvenile Scaly upperparts.
11d   In flight Broad white band on uppertail-coverts.

PLATE 41: RUFF, *TRINGA* SANDPIPERS AND ALLIES

1 GREEN SANDPIPER *Tringa ochropus* 21–24 cm **P**   **Page 447**
Various, mainly freshwater wetlands. Throughout; common/not uncommon. Rare, Cape Verde Is.
Vagrant, Bioko and São Tomé.
1a **Adult breeding** Darker above than 2; distinct white eye-ring; greyish-green legs.
1b **Adult non-breeding** Plainer than 1a.
1c **In flight** Blackish upper- and underwings; pure white rump and belly.

2 WOOD SANDPIPER *Tringa glareola* 19–21 cm **P**   **Page 448**
Various wetlands. Throughout; common. Cape Verde Is (uncommon); Bioko, São Tomé (rare).
2a **Adult breeding** Slimmer than 1; densely speckled upperparts; distinct supercilium.
2b **Adult non-breeding** As 2a, but less distinctly marked.
2c **In flight** Pale grey underwing; plain upperwing; white rump. Compare 1c.

3 COMMON SANDPIPER *Actitis hypoleucos* 19–21 cm **P**   **Page 448**
Various wetlands. Throughout, also Cape Verde and Gulf of Guinea Is. Common.
3a **Adult breeding** As 3b but with dark markings above. Constantly bobs rear body.
3b **Adult non-breeding** Plain brown above; white wedge on sides of brown breast.
3c **In flight** White wingbar. Low, fluttering flight on stiff, bowed wings.

4 RUFF *Philomachus pugnax* Male 26–32 cm, female 20–25 cm **P**   **Page 442**
Various wetlands. Throughout, also Cape Verde Is. Common to rare/uncommon.
4a **Adult male** Plumage variable (moulting male illustrated); head relatively small; legs orange-red to
       yellowish; mantle feathers appear loose and are often fluffed.
4b **Adult male non-breeding** Grey-brown and scaly above. Example of white-headed bird.
4c **Adult female non-breeding** Considerably smaller than male.
4d **Juvenile** Neatly patterned upperparts.
4e **In flight** Narrow white wingbar; white oval areas on tail-sides.

5 TEREK SANDPIPER *Xenus cinereus* 22–25 cm **V**   **Page 448**
Various wetlands. Palearctic vagrant, Mauritania to Chad and Gabon.
5a **Adult non-breeding** Gently upturned bill; shortish, pinkish to orange-yellow legs.
5b **In flight** Dark leading edge and whitish trailing edge to wing; grey rump and tail.

6 COMMON REDSHANK *Tringa totanus* 27–29 cm **P**   **Page 446**
Various wetlands. Throughout, also Cape Verde Is; common/scarce. Vagrant, Bioko.
6a **Adult breeding** Orange-red legs; red-based bill; densely streaked brown below.
6b **Adult non-breeding** Plain grey-brown above; white eye-ring; greyish breast.
6c **Juvenile** Brown above with buff spots and notches.
6d **In flight** Broad white trailing edge to wing (distinctive); white wedge on back.

7 SPOTTED REDSHANK *Tringa erythropus* 29–32 cm **P**   **Page 446**
Various wetlands. Mauritania–Liberia to Chad; not uncommon/scarce. Cape Verde Is; rare.
7a **Adult breeding** Black; white spotting on upperparts.
7b **Adult non-breeding** Pale grey above; grey wash to breast; legs and bill longer than 6.
7c **Juvenile** Closely streaked and barred grey below.
7d **Adult non-breeding in flight** Plain wings; white wedge on back; barred uppertail.

8 COMMON GREENSHANK *Tringa nebularia* 30–34 cm **P**   **Page 447**
Various wetlands. Throughout; common. Also Cape Verde and Gulf of Guinea Is.
8a **Adult breeding** Stout, slightly upturned bill; dull greenish legs;
8b **Adult non-breeding** Plainer, greyer than 8a above; less streaked head and breast.
8c **Juvenile** Darker than adult; streaky head and breast; white-fringed upperparts.
8d **In flight** As 7d but wings darker, more contrasting; rump and tail whiter.

9 LESSER YELLOWLEGS *Tringa flavipes* 23–25 cm **V**   **Page 447**
Various wetlands. N American vagrant, Gambia, Ghana, Nigeria, Cape Verde Is.
9a **Adult non-breeding** Slimmer than 6; bill finer and black; legs longer and yellow.
9b **In flight** Square white rump.

10 MARSH SANDPIPER *Tringa stagnatilis* 22–25 cm **P**   **Page 446**
Various wetlands. Throughout; uncommon/locally not uncommon. Rare, Bioko. Vagrant, Cape Verde Is.
10a **Adult breeding** Straight, needle-like bill; long greenish to yellowish legs; spotted breast.
10b **Adult non-breeding** White forehead and supercilium; plain grey above; white below.
10c **Juvenile** Finely streaked crown and hindneck; white-fringed upperparts.
10d **In flight** Long white wedge on back.

PLATE 42: VAGRANT WADERS

1 SEMIPALMATED PLOVER *Charadrius semipalmatus* 17–19 cm **V**   **Page 430**
N American vagrant, Cape Verde Is.
  1a Adult breeding As Common Ringed Plover (Plate 38: 5) but has small webs between all front toes;
  breast-band usually narrower; bill stubbier; supercilium behind eye indistinct or lacking..
  1b Adult non-breeding Duller; white post-ocular streak sometimes joins white forehead.

2 LEAST SANDPIPER *Calidris minutilla* 13–14 cm **V**   **Page 439**
N American vagrant, Cape Verde Is.
  2a Adult non-breeding Usually complete streaked breast-band; greenish to yellowish legs.
  2b Juvenile Recalls Little Stint (Plate 40: 3c) but whitish V on mantle narrower and less distinct,
  primary projection shorter, legs greenish.

3 LONG-TOED STINT *Calidris subminuta* 14–15 cm **V**   **Page 439**
E Palearctic vagrant, coastal Congo.
  3a Adult non-breeding As 2a but slightly larger, with longer neck and longer legs and toes.
  3b Juvenile As 2b but brown-grey coverts fringed whitish (not rufous).

4 SEMIPALMATED SANDPIPER *Calidris pusilla* 13–15 cm **V**   **Page 438**
N American vagrant, NW Mauritania and Cape Verde Is.
  4a Adult non-breeding As Little Stint (Plate 40: 3b) but has half-webbed toes and different call. Also
  slightly paler; bill rather heavier, thick-tipped.
  4b Juvenile Plumage more uniform than Little Stint (Plate 40: 3c); no distinct white Vs on mantle.

5 UPLAND SANDPIPER *Bartramia longicauda* 28–32 cm **V**   **Page 445**
N American vagrant, NW Mauritania.
  Adult Distinctive shape; small head; long, thin neck; shortish, straight bill; long tail.

6 PURPLE SANDPIPER *Calidris maritima* 19–22 cm **V**   **Page 441**
Palearctic vagrant, NW Mauritania.
  Adult non-breeding Stocky; darkish; short, yellowish legs.

7 SHORT-BILLED DOWITCHER *Limnodromus griseus* 25–29 cm **V?**   **Page 443**
N American vagrant, claimed from Ghana.
  7a Adult non-breeding Long, straight, blunt bill; prominent supercilium; greenish legs.
  7b Juvenile Breast and upperparts tinged rusty-buff; upperpart feathers neatly fringed.
  7c Tail Black bars narrower than white bars.
  7d Juvenile tertial Centre irregularly striped and barred buff.

  For comparison:

  LONG-BILLED DOWITCHER *Limnodromus scolopaceus* 27–30 cm
  N American species; potential vagrant. Extremely similar to Short-billed Dowitcher. Differences include:
  7e Tail Black bars broader than white bars.
  7f Juvenile tertial Centre plain dark.

8 SPOTTED SANDPIPER *Actitis macularia* 19–21 cm **V**   **Page 449**
N American vagrant, coastal Cameroon and Cape Verde Is.
  8a Adult breeding Underparts boldly spotted black; black-tipped pinkish bill; pinkish legs.
  8b Adult non-breeding As Common Sandpiper (Plate 41: 3) but tail shorter, legs paler, wingbar
  shorter.

9 SLENDER-BILLED CURLEW *Numenius tenuirostris* 36–41 cm **V?** CR   **Page 445**
Extremely rare Palearctic species. Uncertain record, L. Chad.
  Adult All-black bill shorter, straighter and thinner than Eurasian Curlew (Plate 39: 9) and Whimbrel
  (Plate 39: 8); rounded spots on breast and flanks.

10 SOLITARY SANDPIPER *Tringa solitaria* 18–21 cm **V**   **Page 447**
N American vagrant, Gambia and Cape Verde Is.
  10a Adult non-breeding As Green Sandpiper (Plate 41: 1) but tail dark centred.
  10b In flight Dark band on centre of rump and tail.

PLATE 43: SKUAS

Gull-like seabirds with powerful, steady and fast flight. Predatory and piratical. Feed mainly by harassing other seabirds, especially gulls and terns, to force them to disgorge or drop their catch. Plumages variable and identification often problematic; important features include structure, proportions and flight action; also note presence or absence of tail projection and of prominent pale areas in plumage.

1 LONG-TAILED SKUA *Stercorarius longicaudus* 48–53 cm; WS 105–115 cm **P/V** **Page 451**
Winters mainly off Namibia and W South Africa. Rare/scarce or vagrant, Mauritania–Nigeria. Also Cape Verde and Gulf of Guinea seas.
- **1a Adult breeding** Very long tail streamers, yellowish breast and face, grey belly.
- **1b Juvenile pale morph** Cold tones to plumage; very pale fringes to upperparts.
- **1c Adult breeding in flight** Black flight feathers, grey upperparts; tern-like flight.
- **1d Adult non-breeding in flight** Head duskier; vent barred; tail streamers shorter.
- **1e Juvenile pale morph in flight** Blunt tips of central tail feathers clearly project.
- **1f Juvenile dark morph in flight** Uppertail, underwing and vent barred.

2 ARCTIC SKUA *Stercorarius parasiticus* 41–46 cm; WS 110–120 **P/V** **Page 451**
Winters mainly off Namibia and W South Africa. Not uncommon, Mauritania–Ghana. Rare/vagrant elsewhere. Also Cape Verde and Gulf of Guinea seas.
- **2a Adult breeding pale morph** Pointed tail streamers; grey breast-band.
- **2b Juvenile pale morph** Warm rufous-brown; often paler nuchal band.
- **2c Adult breeding pale morph in flight** Wings more slender than 3; flight falcon-like.
- **2d Adult breeding dark morph in flight** Blackish-brown; pale primary patches.
- **2e Adult non-breeding in flight** Head pattern less distinct; uppertail and vent barred.
- **2f Juvenile pale morph in flight** Points of central tail feathers project slightly.
- **2g Juvenile dark morph in flight**

3 POMARINE SKUA *Stercorarius pomarinus* 46–51 cm; WS 125–135 cm **P/V** **Page 450**
Winters mainly at sea between 20° and 8°N. Not uncommon, Mauritania–Ghana. Rare/vagrant elsewhere. Also Cape Verde and Gulf of Guinea seas.
- **3a Adult breeding pale morph** Spoon-shaped central tail feathers; mottled breast-band.
- **3b Juvenile** Variable; lacks warm tones of typical juvenile Arctic; head unstreaked.
- **3c Adult breeding pale morph in flight** Heavier than 2; wings broad based; breast rounded.
- **3d Adult breeding dark morph in flight** Blackish-brown; white primary patches. Rare.
- **3e Adult non-breeding in flight** Uppertail and vent barred; tail streamers often lacking.
- **3f Juvenile pale morph in flight** Pale crescent at base of primaries creates double wing patch; barred areas on underwings, uppertail-coverts and vent contrastingly pale.
- **3g Juvenile dark morph in flight**

4 GREAT SKUA *Catharacta skua* 51–58 cm; WS 135–145 cm **P/V** **Page 452**
Main wintering area in Atlantic south to W Africa. Uncommon, Mauritania. Rare/vagrant Senegal–Nigeria, also Cape Verde seas and Annobón.
- **4a Adult** Dark; coarsely streaked; heavy. Juvenile similar but less streaky.
- **4b Adult in flight** Conspicuous white patches at base of primaries above and below.

5 SOUTH POLAR SKUA *Catharacta maccormicki* 51–54 cm; WS 125–135 cm **V** **Page 452**
Breeds Antarctica. Recorded off Senegal. Presumably regular passage migrant.
- **5a Adult dark morph in flight** As 4, but slightly slimmer; cold greyish-brown; unstreaked; nape often pale.
- **5b Adult pale morph in flight** (drawn to smaller scale) Head and underparts pale buff-brown.

PLATE 44: GULLS IN FLIGHT

1 LITTLE GULL *Larus minutus* WS 75–80 cm **P/V**   **Page 454**
   1a **Adult non-breeding from above** No black on wings; dusky cap and ear-spot.
   1b **Adult non-breeding from below** Blackish underwing with white trailing edge.
   1c **First-winter** Small size; blackish W on upperwings; black tail band. Compare 4b.

2 SABINE'S GULL *Larus sabini* WS 90–100 cm **P**   **Page 454**
   2a **Adult non-breeding** Triangular-patterned, tricoloured wings; slightly forked tail.
   2b **Juvenile** Same pattern, but grey areas browner; black tail band.

3 COMMON GULL *Larus canus* WS 110–130 cm **V**   **Page 456**
   3a **Adult non-breeding** Black wingtips with rather large white 'mirrors'.
   3b **First-winter** Blackish leading edge to outer wing, secondary bar and tail band.
   3c **Second-winter** Black leading edge to outer wing; small white 'mirrors'.

4 BLACK-LEGGED KITTIWAKE *Rissa tridactyla* WS 95–120 cm **P/V**   **Page 458**
   4a **Adult non-breeding** Clear-cut black wingtips.
   4b **First-winter** Neat black W on upperwings; blackish collar on nape; black tail band. Compare 1c.

5 LAUGHING GULL *Larus atricilla* WS 100–125 cm **V**   **Page 453**
   5a **Adult non-breeding** Slate-grey upperparts; black wingtips without 'mirrors'.
   5b **First-winter** Blackish primaries, secondary bar and tail band; dusky rear of head, breast and flanks.

6 FRANKLIN'S GULL *Larus pipixcan* WS 85–95 cm **V**   **Page 454**
   6a **Adult non-breeding** Black on wingtips surrounded by white; dark half-hood.
   6b **First-winter** Differs from 5b in dark half-hood, paler inner primaries, and narrower tail band, which does not extend to outermost tail feathers.

7 YELLOW-LEGGED GULL *Larus cachinnans* WS 138–155 cm **P**   **Page 457**
   7a **Adult non-breeding** Ashy-grey upperparts; black wingtips with small white 'mirrors'.
   7b **First-winter** Brownish upperparts; outer wing, secondaries and broad tail band blackish.
   7c **Second-winter** Acquires ashy-grey mantle and white head and body.
   7d **Third-winter** As adult, but traces of immaturity on wings and tail.

8 LESSER BLACK-BACKED GULL *Larus fuscus* WS 135–155 cm **P**   **Page 457**
   8a **Adult breeding** Blackish upperparts.
   8b **First-winter** Outer wing and secondaries blackish; greater coverts dark.
   8c **Second-winter** Acquires adult mantle colour and white head and body.

9 GREY-HEADED GULL *Larus cirrocephalus* WS 100–115 cm **R/V**   **Page 455**
   9a **Adult breeding** Black wingtip bordered by white diagonal bar; grey hood.
   9b **First-winter** Black outer primaries, secondary bar and tail band; brown carpal bar.
   9c **Second-winter** Black on wingtip more extensive; dusky trailing edge.

10 KELP GULL *Larus dominicanus* WS 128–142 cm **V**   **Page 457**
   10a **Adult** Black upperparts; dark eye; powerful bill.
   10b **First-winter** Very dark looking, with blackish outer wing and secondaries.
   10c **Second-winter** Largely white head; dark brown saddle; white tail base.

11 BLACK-HEADED GULL *Larus ridibundus* WS 100–110 cm **P/V**   **Page 455**
   11a **Adult non-breeding** Rather pointed wings; white leading edge to outer wing.
   11b **First-winter** Brown carpal bar; blackish secondary bar and tail band.

12 MEDITERRANEAN GULL *Larus melanocephalus* WS 92–100 cm **P/V**   **Page 453**
   12a **Adult non-breeding** Unmarked wings; dusky patch on ear-coverts
   12b **First-winter** Pale saddle; contrasting upperwing pattern; dark patch behind eye.

13 SLENDER-BILLED GULL *Larus genei* WS 100–110 cm **P/M/V**   **Page 456**
   13a **Adult non-breeding** White leading edge to outer wing; elongated head and neck.
   13b **First-winter** Pale brown carpal bar; blackish secondary bar and tail band.

14 AUDOUIN'S GULL *Larus audouinii* WS 115–140 cm **P** **NT**   **Page 456**
   14a **Adult** Black wingtips sharply contrasting with rest of plumage; dark eye and bill.
   14b **First-winter** Blackish primaries and secondary bar; dark greater coverts; black tail; U-shaped band across rump.

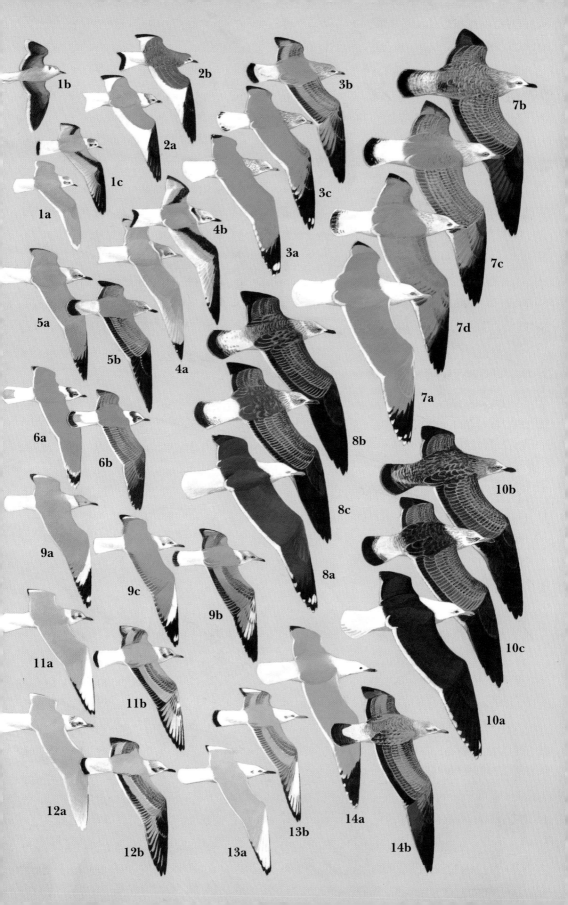

PLATE 45: LARGE GULLS

1 **COMMON GULL** *Larus canus* 40–42 cm **V**   **Page 456**
  Coastal. Palearctic vagrant, Mauritania–Senegambia.
  1a **Adult breeding** Rounded head; dark eye; dark blue-grey upperparts; yellow-green bill.
  1b **Adult non-breeding** Head has dusky markings; pale bill with dark band; greenish legs.
  1c **First-winter** Dusky head and breast; dusky-tipped pinkish bill; pink legs.

2 **AUDOUIN'S GULL** *Larus audouinii* 48–52 cm **P NT**   **Page 456**
  Coastal. Mauritania–Senegambia. Rare.
  2a **Adult** Pale grey upperparts; yellow-tipped dark red bill with dark subterminal band; olive-grey legs.
  2b **First-winter** Whitish head contrasts with grey-brown upperparts; dark grey legs.

3 **LESSER BLACK-BACKED GULL** *Larus fuscus* 52–67 cm **P**   **Page 457**
  Coast, major rivers, lakes. Mauritania–Congo and locally inland; also Cape Verde Is. Common/uncommon.
  3a **Adult** *graellsi/fuscus* **breeding** Similar to 3b–c, but head and neck white.
  3b **Adult** *graellsi* **non-breeding** Slate-grey upperparts; grey-brown streaking on head and breast-sides; yellow legs.
  3c **Adult** *fuscus* **non-breeding** Blackish upperparts; much whiter head and breast-sides than 3b.
  3d **First-winter** Dark grey-brown; bill black; legs pink.

4 **KELP GULL** *Larus dominicanus* 58 cm **M**   **Page 457**
  Coastal. Migrant from southern hemisphere, Gabon, Senegambia and Mauritania. Scarce.
  4a **Adult** More robust than 3; bill heavier; eye dark; legs olive.
  4b **First-winter** Mainly dark brown; bill black; pinkish-brown legs.

5 **YELLOW-LEGGED GULL** *Larus cachinnans* 55–67 cm **P**   **Page 457**
  Coastal. Mauritania–Senegambia; uncommon. Rare further south and Cape Verde Is.
  5a **Adult non-breeding** Ashy-grey upperparts; yellow bill with orange spot; yellow legs.
  5b **First-winter** Mainly brownish, with pale head and underparts; bill black; legs brownish-pink.

PLATE 46: SMALL GULLS AND TERNS

1   COMMON TERN *Sterna hirundo*   31–35 cm; WS 77–98 cm   **R/P/M**   **Page 462**
Coastal. Throughout; common visitor. A few inland records. Rare, Cape Verde Is, Bioko, São Tomé and Príncipe. Breeds Mauritania and Gabon; occasionally elsewhere.
  1a   **Adult breeding** Black-tipped red bill; contrasting darker outer primaries; tail not projecting beyond wingtips. White line between cap and gape broader than in 2a.
  1b   **Adult non-breeding** Dark bill with reddish base; white forehead; no tail streamers.
  1c   **Juvenile** Gingery wash to forehead and scaly upperparts; blackish carpal bar.
  1d   **Adult breeding in flight** Dusky wedge on outer wing.
  1e   **First-winter in flight** Dusky carpal bar and secondary bar; dark outer primaries.

2   ARCTIC TERN *Sterna paradisaea*   33–35 cm; WS 78–85 cm   **P**   **Page 462**
Coastal. Throughout; also Cape Verde seas. Uncommon/rare.
  2a   **Adult breeding** All-red bill; pale grey underparts; tail projecting beyond wingtips.
  2b   **Adult non-breeding** Blackish bill (shorter than 1b); white forehead; no tail streamers.
  2c   **Adult breeding in flight** Uniform upperwings; white underwings, tip bordered by neat black line.
  2d   **First-winter in flight** Dusky carpal bar (less obvious than in 1e); very pale hindwing with broad white trailing edge.

3   ROSEATE TERN *Sterna dougallii*   33–38 cm; WS 72–80 cm   **P**   **Page 461**
Coastal/pelagic. Mauritania–Nigeria; uncommon/rare. Vagrant, Cape Verde Is.
  3a   **Adult breeding** Mainly black bill; very long tail projecting beyond wingtips.
  3b   **Adult non-breeding** Much paler overall than 1b and 2b.
  3c   **Adult breeding in flight** Very pale upperparts; dusky outer primaries.
  3d   **First-winter in flight** Dusky carpal bar; whitish secondaries.

4   BLACK-LEGGED KITTIWAKE *Rissa tridactyla*   38–40 cm   **P/V**   **Page 458**
Pelagic. Mauritania; probably regular offshore. Vagrant, Senegambia. Scarce, Cape Verde Is.
  4a   **Adult breeding** Head white; unmarked greenish-yellow bill.
  4b   **Adult non-breeding** Back of head and neck smudged grey; blackish ear-spot.
  4c   **First-winter** Blackish ear-spot and hindneck collar.

5   FRANKLIN'S GULL *Larus pipixcan*   32–36 cm   **V**   **Page 454**
Coastal. N American vagrant, Senegambia.
  5a   **Adult breeding** Differs from 6a in more rounded head, shorter bill, broader eye-crescents.
  5b   **Adult non-breeding** Dark half-hood; broad eye-crescents.
  5c   **First-winter** Differs from 6c in half-hood and white neck and underparts.

6   LAUGHING GULL *Larus atricilla*   38–41 cm   **V**   **Page 453**
Coastal. American vagrant, Senegambia. Also claimed, Mauritania.
  6a   **Adult breeding** Black hood; white eye-crescents; dull red bill and legs.
  6b   **Adult non-breeding** Dusky-marked head; dark grey upperparts; longish, blackish bill.
  6c   **First-winter** Rear of head to breast and flanks dusky brownish-grey.

7   SABINE'S GULL *Larus sabini*   27–32 cm   **P**   **Page 454**
Pelagic. Arctic passage migrant throughout W African seas.
  7a   **Adult breeding** Dark grey hood; yellow-tipped black bill.
  7b   **Adult non-breeding** White head with blackish nape patch.
  7c   **First-winter** Brownish wing-coverts; all-black bill.

8   LITTLE GULL *Larus minutus*   25–27 cm   **P/V**   **Page 454**
Coastal. Mauritania; scarce. Vagrant, Senegambia–Gabon.
  8a   **Adult breeding** Small size; black hood; no white eye-crescents.
  8b   **Adult non-breeding** Dusky cap and ear-spot; pale grey upperparts.
  8c   **First-winter** Dusky cap and ear-spot; black primaries and wingbar.

9   MEDITERRANEAN GULL *Larus melanocephalus*   36–38 cm   **P**   **Page 453**
Coastal. Mauritania–Senegambia. Scarce/rare.
  9a   **Adult breeding** Slightly larger than 12a; bill stouter; hood extends on nape; legs longer.
  9b   **Adult non-breeding** White wingtips; dark wedge behind eye.
  9c   **First-winter** Dusky patch on ear-coverts; blunt bill.

Continued on page 124

PLATE 47: VAGRANT GULLS AND AUKS

1 COMMON GUILLEMOT *Uria aalge* 38–43 cm **V**   **Page 466**
Palearctic vagrant, NW Mauritania.
Adult non-breeding Slender, pointed, black bill; dark line across ear-coverts; flanks variably streaked.

2 RAZORBILL *Alca torda* 37–39 cm **V**   **Page 467**
Palearctic vagrant, NW Mauritania.
Adult non-breeding Heavy, blunt bill; pointed, often raised, tail.

3 BONAPARTE'S GULL *Larus philadelphia* 28–30 cm; WS 90–100 cm **V**   **Page 455**
N American vagrant, N Senegal.
  3a **Adult non-breeding** As Black-headed Gull (Plate 46: 12b) but smaller; ear-spot more distinct; bill black.
  3b **First-winter** Bill all dark; legs pinkish.
  3c **Adult non-breeding in flight** Compare Black-headed Gull (Plate 44: 11a); underwing white; neat black   edge to primaries.
  3d **First-winter in flight** Compare Black-headed Gull (Plate 44: 11b); primary coverts differently patterned; black trailing edge more distinct.

4 RING-BILLED GULL *Larus delawarensis* 43–47 cm; WS 120–155 cm **V**   **Page 456**
N American vagrant, Senegal.
  4a **Adult non-breeding** Yellow bill with black subterminal band. Compare Common Gull (Plate 45: 1).
  4b **First-winter** Typically heavily spotted on head, neck and breast-sides; pinkish, dark-tipped bill.
  4c **Adult non-breeding in flight** White mirrors smaller than in Common Gull (Plate 44: 3a).
  4d **First-winter in flight** Grey on upperparts paler than in Common Gull; tail band not clear-cut.

5 GREAT BLACK-BACKED GULL *Larus marinus* 64–78 cm; WS 150–165 cm **V**   **Page 458**
Palearctic vagrant, Mauritania.
  5a **Adult non-breeding** Very large and bulky; blackish upperparts; massive bill; pink legs.
  5b **First-winter** Whitish head and breast; deep, heavy black bill.
  5c **Adult non-breeding in flight** Large mirrors on outer two primaries.
  5d **First-winter in flight** Dark flight feathers contrast with paler coverts.

Continued from page 122

10 SLENDER-BILLED GULL *Larus genei* 42–44 cm **P/M/V**   **Page 456**
Coastal. Breeds Mauritania, Senegal. Visitor, Mauritania–Guinea. Locally not uncommon. Vagrant elsewhere, also Cape Verde Is.
  10a **Adult breeding** White head; distinctive, elongated jizz.
  10b **Adult non-breeding** Faint dusky ear-spot.
  10c **First-winter** Brownish wing-coverts; dark tail band; pale bill and legs.

11 GREY-HEADED GULL *Larus cirrocephalus* 39–42 cm **R/V**   **Page 455**
Coast (Mauritania–Benin) and major rivers (Mali–L. Chad; also Cameroon and CAR). Common/scarce.
  11a **Adult breeding** Pale grey hood; red bill and legs.
  11b **Adult non-breeding** White head, often with some greyish; faint dusky ear-spot.
  11c **First-winter** White head with dusky markings; brownish wing-coverts.

12 BLACK-HEADED GULL *Larus ridibundus* 34–37 cm **P/V**   **Page 455**
Coast (mainly Mauritania–Ghana) and inland aquatic habitats; also Cape Verde Is. Uncommon/rare.
  12a **Adult breeding** Dark-brown hood; dark red bill, legs and feet.
  12b **Adult non-breeding** White head with dusky ear-spot.
  12c **First-winter** Brownish wing-coverts; dusky head markings.

PLATE 48: SMALL TERNS AND SKIMMER

1   LITTLE TERN *Sterna albifrons*   22–24 cm; WS 48–55 cm   **R/M/P**   **Page 463**
Coastal; throughout. Inland along major rivers. Common/uncommon.
  1a   **Adult** *guineae* **breeding** Small size; yellow bill and legs; white forehead; white rump.
  1b   **Adult** *albifrons* **breeding** Yellow bill with clear black tip (in 1a little or no black); grey rump.
  1c   **Adult non-breeding** Black bill; black 'shawl'; streaked crown; brownish legs.
  1d   **Juvenile** Upperparts with pale fringes.
  1e   **Adult** *guineae* **breeding in flight** Narrow dusky wedge on upperwing. Flight fluttering.
  1f   **First-winter** *albifrons* **in flight** Dark carpal bar and leading edge.

2   DAMARA TERN *Sterna balaenarum*   23 cm; WS 51 cm   **M**   NT   **Page 463**
Coastal. S African migrant, Congo–Liberia. Rare/scarce.
  2a   **Adult breeding** Small size; black cap; slightly decurved, black bill.
  2b   **Adult non-breeding** White forehead.
  2c   **Juvenile** Upperparts with pale fringes.
  2d   **Adult breeding in flight** Dumpy; dusky leading edge to outer wing.
  2e   **Adult non-breeding in flight** Dusky carpal bar and leading edge to outer wing.

3   WHISKERED TERN *Chlidonias hybridus*   23–25 cm; WS 74–78 cm   **P**   **Page 464**
Wetlands. Widespread; also Bioko. Common/scarce.
  3a   **Adult breeding** Black cap; grey underparts; contrasting white streak on head-sides.
  3b   **Adult non-breeding** White underparts; black band on nape; streaked crown.
  3c   **Adult breeding in flight** *Sterna* jizz; sooty-black belly.
  3d   **Adult non-breeding in flight** Uniformly pale grey upperparts, rump and tail. Compare 4d.
  3e   **Juvenile in flight** Dark brown saddle; dark bill and legs.
  3f   **First-winter in flight** As 3e, but upperparts plain grey.

4   BLACK TERN *Chlidonias niger*   22–24 cm; WS 64–68 cm   **P**   **Page 464**
Coastal. Throughout, also Gulf of Guinea waters; common. Some inland records.
  4a   **Adult breeding** Black head and underparts; slate-grey above; black bill; darkish red legs.
  4b   **Adult non-breeding** White head and underparts; black cap with large 'headphones'.
  4c   **Adult breeding in flight** Contrasting pale grey wings and white undertail-coverts.
  4d   **Adult non-breeding in flight** Dark patch on breast-sides; grey rump and tail.
  4e   **Juvenile in flight** Darkish saddle only faintly contrasting; dark carpal bar.
  4f   **First-winter in flight** As 4e, but upperparts mostly plain grey.

5   WHITE-WINGED TERN *Chlidonias leucopterus*   20–23 cm; WS 63–67 cm   **P**   **Page 465**
Wetlands; also far from water. Throughout. Common/uncommon.
  5a   **Adult breeding** Back; contrasting pale grey wings and white rump, vent and tail.
  5b   **Adult non-breeding** White head and underparts; small black 'headphones'; streaking on rear crown.
  5c   **Adult breeding in flight** Contrasting black underwing-coverts and pale grey remiges.
  5d   **Adult non-breeding in flight** No breast patches. Compare 4d.
  5e   **Juvenile in flight** Dark brown saddle; pale grey wings; white rump.
  5f   **First-winter in flight** As 5e, but upperparts pale grey.

6   AFRICAN SKIMMER *Rynchops flavirostris*   36–42 cm; WS 125–135 cm   **P/M**   NT   **Page 466**
Large rivers, lakes, lagoons. Throughout. Locally not uncommon/rare.
  6a   **Adult breeding** Black above, white below; characteristic, long, orange-red bill.
  6b   **Adult non-breeding** Browner above; whitish collar on lower hindneck.
  6c   **Juvenile** Upperparts browner, fringed buffish; bill has dusky tip.
  6d   **Adult breeding in flight** Long wings; short, forked tail; diagnostic feeding action.

7   SOOTY TERN *Sterna fuscata*   33–36 cm; WS 82–94 cm   **R/M**   **Page 463**
Pelagic. Breeds mainly Príncipe. Disperses from Cameroon to Mauritania.
  7a   **Adult** Broad white forehead extending to just above eye. Compare 8a.
  7b   **Juvenile** Blackish; whitish lower underparts; white fringes to upperparts.
  7c   **Adult in flight** All-black upperparts without neck collar. Compare 8c.
  7d   **Immature in flight** Whitish underwing-coverts and lower underparts.

8   BRIDLED TERN *Sterna anaethetus*   30–32 cm; WS 77–81 cm   **R/M**   **Page 462**
Pelagic. Breeds Mauritania, Senegal, São Tomé and Annobón. Disperses widely. Uncommon.
  8a   **Adult** Narrow white forehead extending over and beyond eye. Compare 7a.
  8b   **Juvenile** Paler; scalloped upperparts; dusky forehead.
  8c   **Adult in flight** Grey collar usually separates black crown from brown upperparts.
  8d   **Immature in flight** Mostly as adult, with white underwing-coverts and dark remiges.

Continued on page 128

PLATE 49: LARGE AND MEDIUM-SIZED TERNS

1 GULL-BILLED TERN *Gelochelidon nilotica* 35–43 cm; WS 86–103 cm **R/P**   **Page 459**
Inland and coastal waters. Visitor, Mauritania–Liberia to Chad–Cameroon; locally common/rare.
Resident Mauritania and Senegal; locally common/uncommon. Vagrant, Cape Verde Is.
  1a Adult breeding Black cap without crest; stout black bill. Compare 2.
  1b Adult non-breeding White head with black streak behind eye.
  1c Juvenile Faint brownish tinge to crown and upperparts; indistinctly marked above.
  1d Adult breeding in flight Upperparts, rump and tail pale grey; short tail.
  1e First-winter in flight Primaries worn and darker than in adult.

2 SANDWICH TERN *Sterna sandvicensis* 36–41cm; WS 95–105 cm **P**   **Page 461**
Coastal. Throughout; common. Rare, Cape Verde Is, São Tomé and Príncipe.
  2a Adult breeding Black cap with shaggy crest; yellow-tipped black bill.
  2b Adult non-breeding Head white with black 'shawl'.
  2c Adult non-breeding in flight Very white looking; dark outer primaries; short tail.
  2d First-winter in flight Darkish carpal and secondary bars; darkish outer tail feathers.

3 LESSER CRESTED TERN *Sterna bengalensis* 35–37 cm; WS 92–105 cm **P**   **Page 460**
Coastal. Mauritania to Guinea; scarce/locally not uncommon. Vagrant, Ghana.
  3a Adult breeding Black cap with shaggy crest, orange-red bill.
  3b Adult non-breeding Head white with black 'shawl'; orange bill.
  3c Adult non-breeding in flight Underwing with narrow dark trailing edge to outer primaries.
  3d First-winter in flight Dark outer primaries, dusky secondary bar.

4 GREATER CRESTED TERN *Sterna bergii* 46–49 cm; WS 125–130 cm **V?**   **Page 460**
Not recorded with certainty in W Africa. One claim from Nigeria.
  4a Adult breeding Black cap with shaggy crest; white forehead; lemon-yellow bill.
  4b Adult non-breeding Black cap receding and speckled white; crest reduced.
  4c Adult non-breeding in flight Upperparts darker grey than in other large terns.
  4d First-winter in flight Upperwing with dark leading edge and dark secondary bar.

5 ROYAL TERN *Sterna maxima* 45–50 cm; WS 125–135 cm **R/M**   **Page 460**
Coastal. Throughout; common. Breeds Mauritania and Senegambia. Vagrant, Cape Verde Is.
  5a Adult breeding Black cap with short crest; deep orange bill. Second largest tern.
  5b Adult non-breeding Black 'shawl' on hindcrown; pale orange bill.
  5c Juvenile Prominent carpal and secondary bars; dull yellowish legs.
  5d Adult non-breeding in flight Dark trailing edge to outer primaries.
  5e First-winter in flight Dark outer wing; three dark bars on innerwing; dark-tipped tail.

6 CASPIAN TERN *Sterna caspia* 47–54 cm; WS 130–145 cm **R/P**   **Page 459**
Coastal; throughout. Locally inland. Breeds Mauritania–Guinea, Eq. Guinea–Gabon. Common/
uncommon. Vagrant, Cape Verde Is.
  6a Adult breeding Large black cap with short rough crest; massive red bill. Largest tern.
  6b Adult non-breeding Dark cap densely streaked white.
  6c Juvenile Scaly pattern on upperparts; bill dull reddish tipped dusky.
  6d Adult non-breeding in flight Dark wedge on tip of underwing.
  6e First-winter in flight Dusky leading edge and secondary bar.

Continued from page 126

9 BROWN NODDY *Anous stolidus* 38–40 cm; WS 77–85 cm **R/V**   **Page 465**
Pelagic. Breeds Gulf of Guinea; locally common. Vagrant elsewhere.
  9a Adult Dark brown with contrasting greyish-white cap. Compare 10a.
  9b Juvenile Almost entirely dark; buffish fringes to upperparts.
  9c Immature Grey restricted to forehead.
  9d Adult in flight Two-toned wing (especially underwing). Compare 10c.

10 BLACK NODDY *Anous minutus* 35–39 cm; WS 66–72 cm **R/V**   **Page 465**
Pelagic. Breeds Gulf of Guinea; locally common. Vagrant elsewhere.
  10a Adult Blacker than 9a, bill longer and more slender, cap whiter and more extensive.
  10b Juvenile White restricted to forehead and sharply demarcated.
  10c Adult in flight Uniformly blackish-brown. More slender, wingbeats faster than 9.

PLATE 50: SANDGROUSE

Medium-sized terrestrial species of arid habitats. Plumage cryptic, making them difficult to see on ground, wings and tail pointed, bill and legs short. Flight swift and direct, reminiscent of pigeons. Can cover considerable distances daily to visit favourite water holes, where they may gather in large flocks, usually at dawn or dusk. Calls are a good identification clue.

1 CHESTNUT-BELLIED SANDGROUSE *Pterocles exustus*   29–33 cm **R/M**   **Page 467**
Semi-desert. Mauritania–Senegal to Niger and Chad. Common/not uncommon. Southward movements in dry season.
- **1a Adult male** Long, finely pointed tail; plain head and throat; narrow black breast-band; dark chestnut belly.
- **1b Adult female** Densely barred dark brown upperparts.
- **1c Adult male in flight** Entirely dark underwing; dark belly.
- **1d Adult female in flight**

2 SPOTTED SANDGROUSE *Pterocles senegallus*   29–35 cm **R**   **Page 467**
Desert. Mauritania, N Mali, Niger and Chad; uncommon. Vagrant, Burkina Faso.
- **2a Adult male** Long, finely pointed tail; dark streak on centre of pale belly; yellow-orange face.
- **2b Adult female** Spotted black upperparts and breast.
- **2c Adult male in flight** Dark flight feathers contrast with pale underwing-coverts.
- **2d Adult female in flight** Mainly pale upperwing with dusky trailing edge.

3 CROWNED SANDGROUSE *Pterocles coronatus*   25–29 cm **R**   **Page 468**
Desert. N Mauritania, N Niger and N Chad; uncommon. Vagrant, Mali.
- **3a Adult male** Black facial mark; well-defined sandy spots on upperparts.
- **3b Adult female** Densely spotted and barred above and below.
- **3c Adult male in flight** Blackish flight feathers; pale upper- and underwing-coverts; short tail.
- **3d Adult female in flight**

4 LICHTENSTEIN'S SANDGROUSE *Pterocles lichtensteinii*   22–26 cm **R/M**   **Page 468**
Desert. Mauritania, Mali, Niger and Chad; uncommon. Vagrant, N Senegal.
- **4a Adult male** Densely barred; black and white forehead; two narrow black bands on plain breast; orange bill.
- **4b Adult female** Finely and densely barred.
- **4c Adult male in flight** Blackish flight feathers; darkish underwings; short tail.
- **4d Adult female in flight**

5 FOUR-BANDED SANDGROUSE *Pterocles quadricinctus*   25–28 cm **R/M**   **Page 468**
Dry savanna. S Mauritania–Guinea-Bissau to Chad–N CAR. Locally common/not uncommon.
- **5a Adult male** Black-and-white forehead; chestnut, white and black breast-bands.
- **5b Adult female** Plain face, throat and upper breast; upperparts barred chestnut and black.
- **5c Adult male in flight** Blackish flight feathers; grey underwing-coverts; short tail.
- **5d Adult female in flight** Blackish flight feathers; yellowish-buff upperwing-coverts.

PLATE 51: PIGEONS AND DOVES I

Arboreal and terrestrial species with compact body, small head and strong, fast flight. Often produce loud and characteristic clapping sound on taking wing or in display. Feed on fruit or seeds. Calls distinctive and an excellent identification clue.

1 WESTERN BRONZE-NAPED PIGEON *Columba iriditorques*   28 cm   **R**   **Page 472**
Forest. Sierra Leone–SE Guinea to Congo. Uncommon/scarce.
1a   Adult male Dark blue-grey head; dark vinous breast and belly. Broad pale terminal tail band evident on landing.
1b   Adult female Cinnamon-rufous crown; greyish-chestnut underparts.

2 BRUCE'S GREEN PIGEON *Treron waalia*   28–30 cm   **R**   **Page 470**
Dry savanna, thorn scrub. S Mauritania–Guinea to Chad–CAR. Not uncommon.
Adult Pale grey head and breast; yellow belly; whitish bill with bluish cere.

3 AFRICAN GREEN PIGEON *Treron calva*   25–28 cm   **R**   **Page 469**
Forest and wooded savanna. Throughout; also Bioko and Príncipe. Common.
Adult Green head, breast and belly; pale bill with large red base.

4 CAMEROON OLIVE PIGEON *Columba sjostedti*   36–40 cm   **R**   **Page 473**
Montane forest and thickets. SE Nigeria–W Cameroon; also Bioko. Common. Endemic.
4a   Adult Very dark; speckled mantle and underparts; yellow bill and eye.
4b   Juvenile Much duller; scaly plumage.

5 WHITE-NAPED PIGEON *Columba albinucha*   36 cm   **R**   **NT**   **Page 473**
Montane and transitional forest. W Cameroon. Rare and very local.
5a   Adult Resembles 4a, but has large white or whitish nuchal patch; broad grey terminal tail band; red feet.
5b   Juvenile Much duller; white-fringed feathers on lower breast.

6 AFEP PIGEON *Columba unicincta*   36–40 cm   **R**   **Page 474**
Forest. Sierra Leone–SE Guinea to Ghana; SE Nigeria to SW CAR–Congo. Not uncommon/scarce.
Adult Large; pale grey; red eye. In flight, contrasting pale grey band on middle of tail.

7 TAMBOURINE DOVE *Turtur tympanistria*   20–22 cm   **R**   **Page 470**
Forest. Guinea to Congo; also Bioko. Common. Rare, S Senegal.
7a   Adult male White face and underparts; dark brown upperparts. In flight, rufous remiges.
7b   Adult female Greyish face and breast.
7c   Juvenile Duller, browner; scaly upperparts.

8 LEMON DOVE *Aplopelia larvata*   24–25 cm   **R**   **Page 475**
Forest. Sierra Leone–SE Guinea to W Ivory Coast; SE Nigeria to Congo. Also Gulf of Guinea Is (see Plate 143: 13). Rare/locally uncommon.
8a   Adult male *inornata* Whitish face; green-and-violet glossed hindcrown and mantle.
8b   Adult female *inornata* Cinnamon underparts.

9 BLUE-HEADED WOOD DOVE *Turtur brehmeri*   25 cm   **R**   **Page 470**
Forest. Sierra Leone–SE Guinea to SW CAR–Congo. Not uncommon.
9a   Adult Rufous-chestnut with blue-grey head.
9b   Juvenile Duller; indistinctly barred upperparts; no metallic wing spots.

10 SPECKLED PIGEON *Columba guinea*   35–40 cm   **R**   **Page 474**
Variety of open country, also towns. S Mauritania–Guinea to Chad–CAR. Reaches coast in Dahomey Gap. Common.
10a   Adult Chestnut upperparts; spotted wing-coverts; pale grey rump; bare red eye patch.
10b   Juvenile Duller.

11 ROCK DOVE *Columba livia*   31–34 cm   **R**   **Page 475**
Rocky cliffs and mountains. Mauritania–Guinea to Burkina Faso–Togo and C Sahara. Also Cape Verdes. Locally common/not uncommon.
Adult *gymnocyclus* Dark slate-grey; white rump; two black wingbars. *C. l. targia* (C Sahara) much paler, with grey rump.

WOOD PIGEON *Columba palumbus*   40–45 cm   **V**   **Page 474**
Palearctic vagrant, Mauritania. **See illustration on plate 146.**

PLATE 52: PIGEONS AND DOVES II

1 EUROPEAN TURTLE DOVE *Streptopelia turtur* 26–28 cm **P/R** **Page 477**
Dry savannas, farmland. Palearctic visitor, S Mauritania–Senegambia to Chad; common. Vagrant elsewhere. Rare, Cape Verde Is. Resident, Saharan mountains.
1a **Adult** Black-and-white striped neck patch; chequered upperparts.
1b **In flight** Blue-grey wing panel; rufous shoulders; white terminal tail band.

2 ADAMAWA TURTLE DOVE *Streptopelia hypopyrrha* 30–31 cm **R** **Page 477**
Open woodland in rocky, hilly areas. NC & E Nigeria to SW Chad. Locally common/uncommon. Also Togo and Senegambia (status unclear).
2a **Adult** Blue-grey head to upper breast; black neck patch.
2b **In flight** Dark overall; pale grey terminal tail band.

3 AFRICAN COLLARED DOVE *Streptopelia roseogrisea* 27–29 cm **R/M** **Page 477**
Thornbush, arid farmland. Mauritania–Senegambia to N Cameroon–Chad; common. Vagrant, N Ghana.
3a **Adult** Pale grey-pink head and underparts; dark eye.
3b **In flight from above** Dark primaries contrast with pale upperwing-coverts.
3c **In flight from below** White underwing.

4 AFRICAN MOURNING DOVE *Streptopelia decipiens* 28–30 cm **R/M** **Page 476**
Riparian Acacia woodland, arid wooded grassland. Mauritania–Senegambia to Chad /N CAR. Common.
4a **Adult** Grey face and crown; yellowish eye; narrow red orbital ring.
4b **In flight** White tail corners.

5 RED-EYED DOVE *Streptopelia semitorquata* 30–34 cm **R** **Page 475**
Various habitats in forest and savanna zones. Throughout; also Bioko. Common.
5a **Adult** Grey crown; dark grey-brown upperparts; dark vinous-pink underparts.
5b **Juvenile** Rufous-buff fringes to upperpart feathers; indistinct half-collar.
5c **In flight** Black band on tail.

6 RING-NECKED DOVE *Streptopelia capicola* 25 cm **R** **Page 476**
Open wooded grassland. SE Gabon–S Congo. Locally scarce/not uncommon.
6a **Adult** Blue-grey head washed pink; pale pinkish-grey underparts.
6b **In flight from above** White tail corners. **6c In flight from below** Grey underwing.

7 VINACEOUS DOVE *Streptopelia vinacea* 25 cm **R/M** **Page 476**
Various savanna habitats (also coastal). Mauritania–Sierra Leone to Chad–CAR. Common.
7a **Adult** Pale vinous-pink head and underparts; dark eye.
7b **In flight from above** White tail corners. **7c In flight from below** Pale grey underwing.

8 LAUGHING DOVE *Streptopelia senegalensis* 23–25 cm **R** **Page 478**
Open woodland, farmland, villages and towns. Widespread. Common. Also São Tomé and Príncipe.
8a **Adult** Rufous-brown above; blue-grey outer wing-coverts; black-speckled necklace.
8b **In flight** Blue-grey wing panel; blue-grey back to uppertail-coverts.

9 NAMAQUA DOVE *Oena capensis* 22–26 cm **R/M** **Page 472**
Thornbush, dry grassland, cultivation. Mauritania–Guinea to Chad–CAR; common/uncommon. Vagrant, Sierra Leone and Cape Verde Is. Scarce/rare, coastal Gabon and Congo.
9a **Adult male** Black mask; long, pointed tail.
9b **Adult female** Greyish face; dusky bill.
9c **Juvenile** Scalloped upperparts.
9d **Adult male in flight** Rufous primaries; long tail.

10 EMERALD-SPOTTED WOOD DOVE *Turtur chalcospilos* 20 cm **R** **Page 471**
Wooded habitats. Coastal Gabon and Congo. Uncommon.
10a **Adult** As 11a, but with dusky bill and metallic green wing spots.
10b **Juvenile** As 11b, but with whitish tips to greater and median coverts.

11 BLUE-SPOTTED WOOD DOVE *Turtur afer* 20 cm **R** **Page 471**
Various wooded habitats. Throughout. Common.
11a **Adult** Red bill with yellow tip; metallic blue wing spots.
11b **Juvenile** Duller, browner; scaly upperparts; dusky bill.
11c **In flight** Rufous primaries; two blackish bands across lower back.

12 BLACK-BILLED WOOD DOVE *Turtur abyssinicus* 20 cm **R** **Page 471**
Dry savanna, thickets. S Mauritania–Guinea to Chad–N CAR; reaches coast in Dahomey Gap. Common.
12a **Adult** As 11a, but with black bill. **12b Juvenile** As 11b, but slightly paler and greyer.

PLATE 53: PARROTS AND LOVEBIRDS

Mostly arboreal species with compact body and stout, strongly hooked and powerful bills. Flight strong, fast and direct. Noisy, especially in flight. Calls include high-pitched screeches (parrots) and shrill twitterings or chirrupings (lovebirds). Somewhat gregarious. Feed mostly on fruits and seeds.

1 BLACK-COLLARED LOVEBIRD *Agapornis swindernianus* 13 cm **R**   **Page 480**
Lowland forest. Scarce/rare to locally not uncommon.
   1a   **Adult** *swindernianus* Small; narrow black collar on nape; yellowish band on mantle; black bill.
        Liberia to Ghana.
   1b   **Adult** *zenkeri* Band on mantle orange-red. S Cameroon to SW CAR–E Gabon–N Congo.
   1c   **Juvenile** All-green head; horn-coloured bill.

2 RED-HEADED LOVEBIRD *Agapornis pullarius* 15 cm **R**   **Page 480**
Wooded grassland. Guinea–Sierra Leone; N Ivory Coast; Ghana to S Chad–CAR and Congo. Also São Tomé. Locally common/uncommon.
   2a   **Adult male** Red face and bill.
   2b   **Adult female** Yellowish-orange face.
   2c   **In flight** Triangular wings; short tail.

3 ROSE-RINGED PARAKEET *Psittacula krameri* 38–42 cm **R**   **Page 481**
Savanna, farmland, mangrove. S Mauritania–Guinea to Chad–N CAR. Not uncommon/common. Introduced, Cape Verde Is.
   3a   **Adult male** Long, pointed tail; red bill; head encircled by narrow ring.
   3b   **Adult female** All-green head; tail shorter.

4 SENEGAL PARROT *Poicephalus senegalus* 23 cm **R**   **Page 480**
Savanna woodland. Senegambia–Guinea to SW Chad. Common/uncommon. Endemic.
   4a   **Adult** *senegalus* Underparts deep yellow becoming orange in centre. Senegambia–Guinea to NW Nigeria.
   4b   **Adult** *versteri* Deep orange to scarlet underparts. W Ivory Coast to SW Nigeria.
        P. s. mesotypus (NE Nigeria to SW Chad) intermediate.
   4c   **Juvenile** Mostly greenish underparts.

5 MEYER'S PARROT *Poicephalus meyeri* 21–23 cm **R**   **Page 479**
Savanna woodland. S Chad–N CAR. Common.
   5a   **Adult** Ash-brown above; yellow patch on crown and on edge of wing.
   5b   **Juvenile** No yellow on crown or wing.

6 NIAM-NIAM PARROT *Poicephalus crassus* 25 cm **R**   **Page 480**
Savanna woodland. SW Chad–CAR. Local. Status inadequately known.
Adult Mainly grass-green with greyish-brown head and breast.

7 RED-FRONTED PARROT *Poicephalus gulielmi* 26–30 cm **R**   **Page 479**
Lowland forest. Liberia to Ghana; SE Nigeria to Congo; S CAR. Rare/uncommon.
   7a   **Adult** Green head; orange or orange-red on forehead; yellowish-green rump.
   7b   **Juvenile** Dusky head; no orange in plumage.

8 BROWN-NECKED PARROT *Poicephalus robustus* 28–33 cm **R**   **Page 479**
Savanna woodland; also mangroves. Patchily distributed, Senegambia to Ghana; EC Nigeria. Rare/locally uncommon.
   8a   **Adult male** Large; greyish head; large bill.
   8b   **Adult female** As 8a but with red forehead. Compare 7a.

9 GREY PARROT *Psittacus erithacus* 28–39 cm **R**   **Page 478**
Lowland and gallery forest, wooded savanna, mangroves. Locally common/scarce.
   9a   **Adult** *timneh* Slate-grey; contrasting paler rump; dark maroon tail. Guinea to Ivory Coast (west of Comoé R.); also Guinea-Bissau, S Mali.
   9b   **Adult** *erithacus* Very large; grey; bright scarlet tail. Ivory Coast (east of Comoé R.) to S CAR–Congo; also Bioko, São Tomé and Príncipe.

PLATE 54: TURACOS

Arboreal species with crested heads and long tails. Flight weak, over short distances only and consisting of several flaps followed by a glide. Agile in trees, characteristically running and bounding along branches (except *Crinifer*). Feed principally on fruit. Vocal. *Turaco* have loud, raucous calls that are overall similar but may, with practice, be distinguished in most cases. The two *Musophaga* utter pleasant, rolling sounds. *Crinifer* have a variety of cackling and yapping calls. *Corythaeola* has a series of unmistakable, impressive calls.

1 WHITE-CRESTED TURACO *Tauraco leucolophus* *c.* 40 cm **R**　　　　　**Page 483**
Wooded savanna, gallery forest. E Nigeria to S Chad–CAR. Locally common.
Conspicuous white head, unique among turacos.

2 BANNERMAN'S TURACO *Tauraco bannermani* *c.* 43 cm **R** EN　　　**Page 483**
Montane forest. W Cameroon. Locally common. Endemic.
Red crest; greyish face.

3 RED-CRESTED TURACO *Tauraco erythrolophus* *c.* 43 cm　　　　**Page 483**
Evergreen and gallery forest (Angola). No certain records. Claimed from S Cameroon.
Red crest; whitish face.

4 BLACK-BILLED TURACO *Tauraco schuetti* *c.* 40 cm **R**　　　　**Page 482**
Lowland forest, savanna woodland. S CAR. Common.
White-tipped crest; black bill.

5 YELLOW-BILLED TURACO *Tauraco macrorhynchus* 40–43 cm **R**　　**Page 482**
Lowland forest. Uncommon/locally common.
5a **Adult *macrorhynchus*** Yellow bill; black-and-white-tipped crest. Sierra Leone–SE Guinea to Ghana.
5b **Adult *verreauxii*** Red-tipped crest. Nigeria to Congo; Bioko.
5c **Juvenile, both races** All-green crest; dull-coloured bill.

6 GREEN TURACO *Tauraco persa* 40–43 cm **R**　　　　　　**Page 482**
Lowland and gallery forest, lower montane forest. S Senegambia–Ivory Coast to S CAR–Congo. Common.
6a **Adult *persa*** All-green crest; orange-red bill, white spot in front of eye. From Ivory Coast east.
6b **Adult *buffoni*** No white line below eye. From Ivory Coast west.
6c **In flight** Mainly bright crimson flight feathers.

7 ROSS'S TURACO *Musophaga rossae* 51–54 cm **R**　　　　　**Page 483**
Gallery forest, forest edge. Cameroon and CAR. Locally not uncommon.
Glossy blackish-violet; yellow bill and orbital skin; crimson crest.

8 WESTERN GREY PLANTAIN-EATER *Crinifer piscator* *c.* 50 cm **R**　**Page 484**
Gallery forest, savanna woodland, cultivated areas. S Mauritania–Sierra Leone to Chad–CAR; also along Congo R. Common.
Mainly grey and white; white wing patch conspicuous in flight; pale yellow bill.

9 VIOLET TURACO *Musophaga violacea* 45–50 cm **R**　　　　**Page 483**
Gallery forest, forest edge. S Senegambia–Guinea to N Cameroon and N CAR. Locally not uncommon. Endemic.
Glossy purple and black; yellow bill; scarlet orbital skin; crimson crown.

10 EASTERN GREY PLANTAIN-EATER *Crinifer zonurus* *c.* 50 cm **R**　**Page 484**
Open wooded savanna. SE Chad and SE CAR. Common.
Outer tail feathers partially white. Compare 8.

11 GREAT BLUE TURACO *Corythaeola cristata* 70–75 cm **R**　　**Page 481**
Forest. Guinea-Bissau; Guinea to Togo; Nigeria to CAR–Congo. Common/uncommon.
Very large; long tail with broad black bar; blackish crest; yellow bill tipped reddish.

PLATE 55: CUCKOOS I

Typical cuckoos. Arboreal with long, pointed wings and long tails, reminiscent of a small raptor in flight. Feed on insects and hairy caterpillars. Brood parasites. Many rather inconspicuous and some quite difficult to observe, but all have distinctive calls.

1   KLAAS'S CUCKOO *Chrysococcyx klaas* 18 cm **M/R**   **Page 489**
Various wooded habitats. Throughout; also Bioko and São Tomé. Generally common.
  1a   **Adult male** Small white streak behind eye; green patch on breast-sides; outer tail feathers mainly white.
  1b   **Adult female** Variable; upperparts more bronzy; flanks barred; post-ocular streak buffy.
  1c   **Juvenile** Barred green and russet above; fine olive-brown barring below; slight post-ocular patch. Compare 3b–c.

2   DIDRIC CUCKOO *Chrysococcyx caprius* 19 cm **M/R**   **Page 489**
Various open and wooded habitats (not closed forest). Throughout; also Bioko. Common.
  2a   **Adult male** White eye-stripe; white spots on wings and outer tail feathers; short malar stripe; red orbital ring.
  2b   **Adult female** Variable; upperparts more bronzy than male; red orbital ring.
  2c   **Juvenile** Highly variable (rufous morph illustrated); coral-red bill.

3   AFRICAN EMERALD CUCKOO *Chrysococcyx cupreus* 23 cm **R/M**   **Page 488**
Lowland forest, gallery forest, dense woodland. S Senegambia–Liberia to CAR–Congo; also Gulf of Guinea Is. Common/scarce.
  3a   **Adult male** Upperparts bright glossy green; belly golden-yellow.
  3b   **Adult female** Barred russet and green above; boldly barred bronzy-green below.
  3c   **Juvenile** As female, but crown and nape feathers green fringed white.

4   YELLOW-THROATED CUCKOO *Chrysococcyx flavigularis* 18 cm **R**   **Page 488**
Forest. Sierra Leone to Togo and Nigeria; rare and patchy. S Cameroon to SW CAR–Congo; uncommon.
  4a   **Adult male** Darkish, with conspicuous golden-yellow stripe on throat; outer tail feathers white with subterminal dark bar.
  4b   **Adult female** Face and underparts buffish narrowly barred darkish.
  4c   **Juvenile** Russet fringes to feathers of upperparts; tawnier head and underparts.

5   GREAT SPOTTED CUCKOO *Clamator glandarius* 35–40 cm **M/R/P**   **Page 485**
Semi-arid open woodland. S Mauritania–Liberia to Chad–CAR. Scarce/locally common. Moves south in dry season. Vagrant, SE Cameroon, Gabon, Congo and Annobón.
  5a   **Adult** Spotted upperparts, long white-tipped tail, grey crest, creamy underparts.
  5b   **Juvenile** Darker; rudimentary crest, rufous primaries.

6   JACOBIN CUCKOO *Oxylophus jacobinus* 33 cm **M**   **Page 485**
Dry savanna woodland. Senegambia to S Chad–N CAR. Locally common. Moves south in dry season, crossing equator. Vagrant, São Tomé.
  6a   **Adult** Black above, white below, crest, white wing patch, long tail.
  6b   **Juvenile** Duskier; smaller crest.

7   LEVAILLANT'S CUCKOO *Oxylophus levaillantii* 38–40 cm **M**   **Page 485**
Dense savanna woodland, forest edge, adjacent cultivated areas. Throughout. Common/uncommon. Moves south in dry season. Vagrant, Bioko.
  7a   **Adult** As 6a, but with heavily streaked throat and breast.
  7b   **Juvenile** Duskier; streaks finer, crest smaller.

140

1   AFRICAN CUCKOO *Cuculus gularis* 31–33 cm **M**   **Page 487**
Wooded savanna. Throughout. Moves north with rains. Uncommon/common.
  1a **Adult male** Grey upperparts and breast; barred underparts; bill yellow tipped black.
  1b **Adult female** May have more buffish underparts, especially throat and breast.
  1c **Juvenile** Grey upperparts finely barred white; dirty buff throat and breast.
  1d **In flight** Long pointed wings; long tail.

2   COMMON CUCKOO *Cuculus canorus* 32–36 cm **P**   **Page 486**
Wooded and derived savanna, open areas in forest. Widespread; scarce. Vagrant, Cape Verde Is.
  2a **Adult male** As 1a, but bill mainly dusky with yellow base. Adult female similar but with rusty-brown tinge.
  2b **Adult female hepatic form** (rare) Rufous upperparts; barred above and below.

3   DUSKY LONG-TAILED CUCKOO *Cercococcyx mechowi* 33 cm **R**   **Page 487**
Forest. Sierra Leone–SE Guinea to S CAR–Congo. Uncommon/rare to locally common.
  3a **Adult** Slender, small bodied, with very long tail. Blackish above; boldly barred below; rich buff undertail-coverts.
  3b **Juvenile** Dark brownish-grey upperparts and throat, with dark rufous fringes to feathers.

4   OLIVE LONG-TAILED CUCKOO *Cercococcyx olivinus* 33 cm **R**   **Page 488**
Forest. Sierra Leone–SE Guinea to Ghana; Nigeria to SW CAR–Congo. Rare/locally common.
  4a **Adult** Jizz as 3. Greyish-brown above; boldly barred below; pale buff undertail-coverts.
  4b **Juvenile** Underparts initially streaked, progressively replaced by barring.

5   RED-CHESTED CUCKOO *Cuculus solitarius* 28–31 cm **M**   **Page 486**
Forest, wooded savanna. Senegambia–Liberia to CAR–Congo; also Bioko. Not uncommon/uncommon.
  5a **Adult** Slate-grey upperparts and throat; chestnut-red breast; barred belly and undertail-coverts.
  5b **Juvenile** Very dark; narrow white crescents; white nape patch; belly and undertail-coverts boldly barred on buffish-white.

6   THICK-BILLED CUCKOO *Pachycoccyx audeberti* 36 cm **R**   **Page 485**
Wooded savanna, gallery forest, forest edge. Distribution patchy, Sierra Leone–Guinea to CAR and Congo. Rare.
  6a **Adult** Slate-grey or slate-brown above, white below.
  6b **Juvenile** Boldly spotted above, white below.

7   BLACK CUCKOO *Cuculus clamosus* 28–31 cm **M/R**   **Page 486**
Forest and woodland. Widespread. Uncommon/locally common.
  7a **Adult** *gabonensis* Variable, but always very dark; dark chestnut throat and breast; boldly barred below. Forest. E Sierra Leone–SE Guinea to CAR–Congo.
  7b **Adult** *gabonensis* A less strongly coloured bird.
  7c **Adult** *gabonensis* A bird with unbarred throat.
  7d **Adult** *clamosus* All black; white-tipped tail. Underparts may be barred. Woodland. S Senegambia to Chad–CAR and Congo.
  7e **Juvenile both races** Wholly dull black.

PLATE 57: YELLOWBILL AND COUCALS

Non-parasitic cuckoos. **Yellowbill** is arboreal and mainly insectivorous. **Coucals** are semi-terrestrial, stoutly built birds with rounded wings and long, broad tails. Flight clumsy and not sustained. Mainly insectivorous, but also take small vertebrates. Utter rapid series of deep, hollow hoots and liquid bubbling notes (like water poured from a bottle).

1 YELLOWBILL *Ceuthmochares aereus* 33 cm **R** **Page 489**
Forest (even where heavily degraded). S Senegambia; rare/uncommon. Guinea to CAR–Congo; common.
Wholly slate-grey with yellow bill.
1a **Adult** *aereus* Bluish gloss to upperparts and tail. East of lower Niger R.
1b **Adult** *flavirostris* Purplish gloss. West of lower Niger R.
1c **Juvenile** Upper mandible brownish; plumage duller.

2 BLACK COUCAL *Centropus grillii* 30–35 cm **M/R** **Page 490**
Moist grassland, marshy areas, shrub. Throughout, but irregularly distributed. Uncommon/locally common.
2a **Adult breeding** Black, with rufous wings.
2b **Juvenile** Pale; head and mantle streaky; rest of upperparts barred. Adult non-breeding similar.

3 SENEGAL COUCAL *Centropus senegalensis* 36–40 cm **R** **Page 491**
Variety of open, grassy and bushy habitats. Throughout. Common.
3a **Adult** Head and tail glossy black; upperparts rufous-brown; underparts whitish.
3b **Juvenile** Strongly barred upperparts; head dark and streaky; underparts tawny.
3c **Adult 'epomidis' morph** Head and throat black; breast and belly chestnut.

4 BLUE-HEADED COUCAL *Centropus monachus* 45 cm **R** **Page 491**
Swampy places, forest edge, savanna woodland. SE Guinea to CAR–Congo. Uncommon/common.
4a **Adult** As 3a but with strong blue gloss to head to mantle.
4b **Juvenile** Similar to 3b, but with black tail and faint buff barring on uppertail-coverts.

5 GABON COUCAL *Centropus anselli* 48–58 cm **R** **Page 490**
Forest. S Cameroon–SW CAR–Congo. Common.
5a **Adult** Dark; underparts tawny.
5b **Juvenile** Barred throat.

6 BLACK-THROATED COUCAL *Centropus leucogaster* 48–58 cm **R** **Page 490**
Forest and forest edges. Guinea to Gabon; also S Senegal and Guinea-Bissau. Not uncommon.
6a **Adult** Head to breast black; rest of underparts white.
6b **Juvenile** Duskier, with barred upperparts and tail.

WHITE-BROWED COUCAL *Centropus superciliosus* 40 cm **R?** **Page 490**
Tall grass and shrubbery. SW Congo. Rare. **See illustration on plate 147**.
The only coucal with a pale supercilium.

PLATE 58: OWLS I

Family **Tytonidae (Barn owls)**. Medium-sized nocturnal birds of prey with a distinctive heart-shaped, pale facial disc. Flight buoyant and silent. Perch upright.

1 BARN OWL *Tyto alba* 33–36 cm **R**   Page 491
Various habitats; not in closed forest; often associated with man. Throughout; also Cape Verde Is and São Tomé. Uncommon/not uncommon.
1a **Adult** *affinis* Pale grey mixed with golden-buff above; white below, variably washed buff. Mainland and Bioko. Nominate *alba* (Aïr, Niger) paler overall.
1b **Adult** *thomensis* Darker; face pale brownish; underparts golden-brown. São Tomé.
  T. a. detorta (Cape Verde Is) similar; face cinnamon.
1c **Adult** *affinis* **in flight** Very pale overall.

2 AFRICAN GRASS OWL *Tyto capensis* 36–40 cm **R**   Page 492
Dambos, montane grassland. W Cameroon and Congo. Rare/locally not uncommon.
2a **Adult** Dark brown above; white face; whitish to pale rufous below; speckled breast.
2b **Juvenile** Dusky-russet face; rufous-buff below.
2c **In flight** Pale patch at base of primaries.

Family **Strigidae (True owls)**. Small to large, mainly nocturnal birds of prey with characteristic large, rounded heads and large, forward-facing eyes within flat facial disc. Plumage mostly cryptic, many species with ear-tufts. Calls distinctive; usually most vocal just after dark and before dawn.

3 SHORT-EARED OWL *Asio flammeus* 35–40 cm **P/V**   Page 499
Various open habitats. Mauritania, Senegambia, Mali, N Guinea; rare/uncommon. Vagrant, Liberia, Niger, Chad, Cape Verde Is.
3a **Adult** Sandy-buff marked dark brown and tawny above; whitish face; yellow eyes.
3b **In flight** Long wings; dark carpal patch; buff patch at base of primaries; black wingtips.

4 MARSH OWL *Asio capensis* 30–35 cm **R/M**   Page 499
Open grassland. S Mauritania–Guinea to W Chad and S Congo. Uncommon/rare and local. Locally not uncommon, Mali (central Niger delta), Nigeria (Jos and Mambilla Plateaux), Cameroon.
4a **Adult** Plain dark brown; rounded head; large dark eyes.
4b **In flight** Large pale patch at base of primaries.

5 RUFOUS FISHING OWL *Scotopelia ussheri* 43–51 cm **R** EN   Page 496
Riverine forest, mangrove. Sierra Leone–SE Guinea to Ghana. Rare/scarce. Endemic.
Adult Pale rufous below with fine streaks; darker above; upperparts plain.

6 PEL'S FISHING OWL *Scotopelia peli* 55–63 cm **R**   Page 496
Large forested rivers, swamps. Throughout savanna and forest zones. Rare/locally not uncommon.
Adult Very large. Orange-rufous; barred above; plain face; large dark eyes; no ear-tufts.

7 VERMICULATED FISHING OWL *Scotopelia bouvieri* 43–51 cm **R**   Page 497
Forested rivers, swamp forest. S Nigeria to CAR–Congo. Rare to locally not uncommon/common.
Adult Rufous above streaked and vermiculated dusky-brown; whitish below boldly streaked dark.

1 CHESTNUT-BACKED OWLET *Glaucidium sjostedti* 24–28 cm **R**   Page 498
Forest interior. SE Nigeria to S Congo and SW CAR. Rare/common.
Adult Large rounded head; finely barred head and mantle; finely barred cinnamon-rufous
underparts.

2 AFRICAN BARRED OWLET *Glaucidium capense* 20–21 cm **R**   Page 498
Forest. Liberia and Ivory Coast; S Cameroon to SW CAR–Congo. Rare/locally not uncommon.
Adult Dumpy; dark brown upperparts barred tawny; white underparts barred and spotted dark
brown.

3 RED-CHESTED OWLET *Glaucidium tephronotum* 20–24 cm **R**   Page 497
Forest. Sierra Leone to Ghana; S Cameroon to SW CAR–Congo. Rare/uncommon.
3a   **Adult** *tephronotum* Rather long tail with three white spots on centre; white underparts spotted
brown; rufous-chestnut breast-sides and flanks. Upper Guinea forest.
3b   **Adult** *pycrafti* Darker above; less rufous below; spots blackish. Lower Guinea forest.

4 PEARL-SPOTTED OWLET *Glaucidium perlatum* 19–21 cm **R**   Page 497
Wooded savanna. S Mauritania–Guinea to Chad–CAR. Uncommon/locally common.
4a   **Adult** Brown spotted white above; white heavily streaked rufous-brown below; tail rather long.
4b   **Back of head** Two dark 'eye-spots' forming false face.

5 EUROPEAN SCOPS OWL *Otus scops* 18–20 cm **P**   Page 493
Various bushy and wooded habitats. Mauritania–Guinea to Chad–Cameroon. Rare/not uncommon.
Adult As 6. In hand: outermost (10th) primary longer than 5th.

6 AFRICAN SCOPS OWL *Otus senegalensis* 16–18 cm **R**   Page 493
Wooded grassland. Throughout; also Annobón. Uncommon/locally common.
Adult Cryptic plumage; prominent ear-tufts; short tail.

7 SANDY SCOPS OWL *Otus icterorhynchus* 18–22 cm **R**   Page 492
Forest. Liberia to Ghana; Cameroon to SW CAR–Congo. Uncommon/rare.
7a   **Adult** *icterorhynchus* Cinnamon-brown speckled white; short ear-tufts. Upper Guinea forest.
7b   **Adult** *holerythrus* More rufous; no white spots on wing-coverts. Lower Guinea forest.

8 WHITE-FACED OWL *Ptilopsis leucotis* 23–28 cm **R**   Page 493
Various wooded habitats, mainly wooded grassland. Widespread. Not uncommon.
8a   **Adult** *leucotis* White face boldly rimmed with black; large orange eyes; prominent ear-tufts.
S Mauritania– Liberia to S Chad–CAR.
8b   **Adult** *granti* Darker; greyer; black markings more pronounced. Gabon and Congo.

9 LITTLE OWL *Athene noctua* 21–25 cm **R**   Page 498
Stony desert. Mauritania, N Mali, N Niger, N Chad. Uncommon and local.
Adult Sandy-brown above dappled white; flat head; 'frowning' face.

PLATE 60: OWLS III

1 AFRICAN WOOD OWL *Strix woodfordii* 33–35 cm **R** **Page 499**
Forest, dense woodland, plantations. Senegambia–Liberia to CAR–Congo; Bioko. Generally common.
Adult Round head; barred underparts; large dark eyes.

2 MANED OWL *Jubula lettii* 34–37 cm **R** **Page 494**
Forest. Liberia to Ghana; S Cameroon to Congo. Rare.
Adult Rufous-brown; bushy ear-tufts; rufous face rimmed with black; deep yellow to crimson eyes.

3 FRASER'S EAGLE OWL *Bubo poensis* 38–45 cm **R** **Page 495**
Forest. Sierra Leone to Ghana; Nigeria to SW CAR–Congo; Bioko. Uncommon/not uncommon.
Adult Dark rufous; narrowly barred above and below; ear-tufts; dark eyes.

4 AKUN EAGLE OWL *Bubo leucostictus* 40–46 cm **R** **Page 496**
Forest. Sierra Leone to Ghana; Nigeria to SW CAR–Congo. Rare/locally not uncommon.
Adult Dark brown; underparts barred and blotched dark; two-toned ear-tufts; yellow eyes.

5 SHELLEY'S EAGLE OWL *Bubo shelleyi* 54–61 cm **R** **Page 495**
Forest. SE Sierra Leone to W Ivory Coast; Ghana, S Cameroon, Gabon and S Congo. Rare.
Adult Very large; dark brown above; whitish below heavily barred dark brown; prominent ear-tufts.

6 SPOTTED EAGLE OWL *Bubo africanus* 43–48 cm **R** **Page 494**
From rocky desert outcrops to wooded grassland and forest edge. Throughout. Uncommon/not uncommon.
6a **Adult** *cinerascens* Large; grey-brown; two-toned ear-tufts; finely barred underparts; dark blotches on breast; dark eyes.
6b **Adult** *africanus* Greyer overall; yellow eyes. SE Gabon and Congo.

7 VERREAUX'S EAGLE OWL *Bubo lacteus* 58–65 cm **R** **Page 495**
Wooded grassland, riverine woodland. S Mauritania–N Liberia to Chad–CAR. Uncommon/rare.
Adult Very large; greyish; pink eyelids; dark eyes; ear-tufts seldom raised.

8 EURASIAN EAGLE OWL *Bubo bubo* 46–50 cm **R** **Page 499**
B. b. ascalaphus (Desert Eagle Owl). Desert and sub-desert. Mauritania to Chad; uncommon. Vagrant, Senegal.
Adult Tawny-rufous streaked and blotched dark and pale; prominent ear-tufts; orange eyes.

PLATE 61: NIGHTJARS I

Crepuscular and nocturnal with very cryptic plumage, long wings, moderately long tail, tiny bill and huge gape for catching insects on the wing. Normally not seen during the day, except when flushed from underfoot. Identification of many species problematic; important features include overall coloration, markings on scapulars and wing-coverts, presence of nuchal collar, and amount of white (male) or buff (female) in wings and tail, but best identified by voice. Typically sing at or just after dusk and just before dawn; often also throughout moonlit nights. Often encountered on dirt roads at night; orange reflection of eyes in headlights of car visible at considerable distance.

1 BROWN NIGHTJAR *Caprimulgus binotatus* 21–23 cm **R**   **Page 500**
Forest and forest edge. Liberia to Ghana; W Cameroon–Gabon; SW CAR–N Congo. Rare.
1a Adult Small; very dark.
1b In flight No white in wings and tail.

2 BATES'S NIGHTJAR *Caprimulgus batesi* 29–31 cm **R**   **Page 500**
Forest and forest edge. S Cameroon, SW CAR, Gabon, S Congo. Uncommon/locally not uncommon.
2a Adult Large; very dark; relatively long tail.
2b Adult male in flight Small white wing patch ; white tail corners.
2c Adult female in flight No wing patch; buff tail corners.

PRIGOGINE'S NIGHTJAR *Caprimulgus prigoginei* 19 cm **R?** **CR**   **Page 500**
Forest, SE Cameroon–N Congo? **See illustration on plate 147.**

3 BLACK-SHOULDERED NIGHTJAR *Caprimulgus nigriscapularis* 23–25 cm **R**   **Page 502**
Woodland, gallery forest, forest edge. S Senegal to CAR. Not uncommon/rare.
3a Adult Rufous-brown; blackish shoulder; rusty-buff line on scapulars.
3b Adult male in flight White wing patch; white tail corners.
3c Adult female in flight Wing patch and tail corners slightly smaller.

4 FIERY-NECKED NIGHTJAR *Caprimulgus pectoralis* 23–25 cm **R**   **Page 502**
Woodland, gallery forest, forest edge. SE Gabon and Congo. Locally common.
4a Adult As 3a but paler, greyer brown; wing-coverts more uniform.
4b Adult male in flight White wing patch (slightly larger than 3b); white tail corners.
4c Adult female in flight Buffish wing patch and tail corners.

5 RUFOUS-CHEEKED NIGHTJAR *Caprimulgus rufigena* 23–24 cm **M**   **Page 505**
Wooded grasslands. Mainly Cameroon. Rare, Congo, Nigeria, Chad.
5a Adult Grey-brown; pale line on scapulars; buff-tipped wing-coverts.
5b Adult male in flight White wing patch; white tail corners.
5c Adult female in flight Wing patch smaller, usually buffy; no white in tail.

6 EUROPEAN NIGHTJAR *Caprimulgus europaeus* 24–28 cm **P**   **Page 504**
Various open and lightly wooded habitats. Throughout, but mainly Nigeria to Gabon. Uncommon/rare.
6a Adult Dark greyish; pale wingbar; pale line on scapulars; whitish moustachial stripe.
6b Adult male in flight White wing patch; white tail corners.
6c Adult female in flight No white in wings and tail.

7 RED-NECKED NIGHTJAR *Caprimulgus ruficollis* 30–34 cm **P**   **Page 504**
Various open and wooded habitats. Mauritania–Liberia to N Ghana. Rare/uncommon.
7a Adult Large; rufous nuchal collar; whitish moustachial stripe; large white throat patch.
7b Adult male in flight White wing patch; white tail corners.
7c Adult female in flight Buffish wing patch and tail corners.

8 LONG-TAILED NIGHTJAR *Caprimulgus climacurus* 28–43 cm **R/M**   **Page 501**
Semi-desert to forest zone. Throughout. Common.
8a Adult male Small; long, graduated tail; white wingbar.
8b Adult male in flight White wing patches; white trailing edge to wing.
8c Adult female in flight Tail shorter than in male; buffish wing patch.

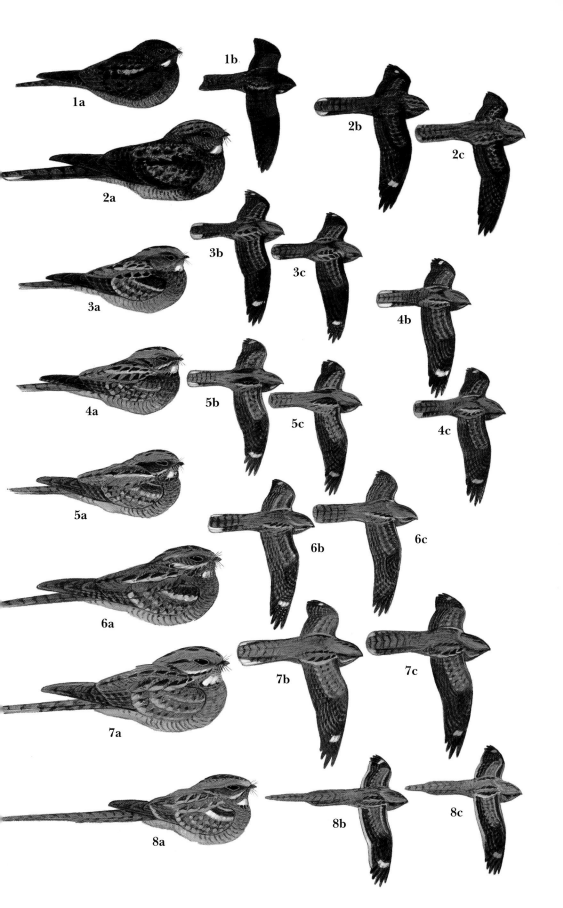

PLATE 62: NIGHTJARS II

1 GOLDEN NIGHTJAR *Caprimulgus eximius*   23–25 cm   **R**   **Page 503**
Sahel. S Mauritania–N Senegal to Chad. Uncommon/rare.
1a Adult Golden-buff with irregular white spots, each bordered and speckled black.
1b Adult male in flight Large white wing patch; white corners to tail.
1c Adult female in flight Smaller and buffish wing patch and tail corners.

2 EGYPTIAN NIGHTJAR *Caprimulgus aegyptius*   24–27 cm   **P**   **Page 504**
Desert and semi-desert. Mauritania–N Senegal to N Togo, S Niger, N Nigeria, Chad. Uncommon/locally
not uncommon.
2a Adult Rather uniform, pale greyish plumage.
2b In flight No white in wings and tail; dark brownish outer wing.

3 FRECKLED NIGHTJAR *Caprimulgus tristigma*   26–28 cm   **R**   **Page 503**
Rocky outcrops. Mali–N Sierra Leone to C CAR and Gabon. Uncommon/locally not uncommon.
3a Adult Dark greyish or blackish-brown finely speckled white or buffish.
3b Adult male in flight Small white wing patch; white tail corners.
3c Adult female in flight Wing patch smaller; no white in tail.

4 PLAIN NIGHTJAR *Caprimulgus inornatus*   22–23 cm   **M/R**   **Page 503**
Various open and wooded habitats. Sahel to forest zone. Common/not uncommon.
4a Adult Variably coloured but uniform plumage; row of small black spots on scapulars.
4b Adult Rufous-coloured individual.
4c Adult male in flight White wing patch and tail corners. Cinnamon-coloured individual illustrated.
4d Adult female in flight Buff wing patch and tail corners. Grey-coloured individual.

5 SQUARE-TAILED NIGHTJAR *Caprimulgus fossii*   23–24 cm   **R**   **Page 502**
Grasslands, open woodland, farmland. Eq. Guinea, Gabon, S Congo. Common.
5a Adult White wingbar; buffish line on scapulars.
5b Adult male in flight White wing patch; white trailing edge to wing; white-edged tail.
5c Adult female in flight Wingbar, wing patch, trailing edge and tail edge buffish.

6 SWAMP NIGHTJAR *Caprimulgus natalensis*   20–24 cm   **R**   **Page 501**
Damp and dry grasslands. Patchily distributed, E Gambia to SW CAR–Congo; Liberia to W Cameroon.
Locally rare/not uncommon.
6a Adult Long pale supercilium and moustachial stripe; dark head-sides; large blackish markings on
scapulars and wing-coverts; round buff spots on breast.
6b Adult male in flight White wing patch; white-edged tail.
6c Adult female in flight Buffish wing patch; buff-edged tail.

7 STANDARD-WINGED NIGHTJAR *Macrodipteryx longipennis*   21–22 cm   **M**   **Page 505**
Various open, wooded habitats. Throughout. Not uncommon/common.
7a Adult male non-breeding/adult female Rufous nuchal collar; buffish line on scapulars; buffish
throat.
7b Adult male in flight Extremely elongated primary with bare shaft and blackish vane at tip.

8 PENNANT-WINGED NIGHTJAR *Macrodipteryx vexillarius*   24–28 cm   **M**   **Page 506**
Wooded grassland. Nigeria to S Chad–CAR and Congo. Locally common/scarce. Vagrant, SW Niger, SE
Burkina Faso, Togo, Bioko.
8a Adult female As 7a, but larger, slightly darker above, white throat patch, paler and more barred
lower underparts.
8b Adult male in flight Broad white band on black remiges; extremely elongated whitish primary.
8c Adult female in flight No white in wings and tail.

PLATE 63: SWIFTS

Highly aerial, anchor-shaped birds with mostly dark plumage. Flight swift and effortless on long, stiff wings. Some difficult to identify; useful features include size, silhouette, rump pattern, tail shape, habitat and locality.

1 MOTTLED SPINETAIL *Telacanthura ussheri* 13–14 cm **R**   **Page 507**
Wooded savanna, open areas in forest zone. Senegambia–Liberia to SW CAR–Congo. Scarce/common.
1a   **Above** Broad white rump. Compare 9.
1b   **Below** Pale, mottled throat; narrow whitish area on vent.

2 SABINE'S SPINETAIL *Rhaphidura sabini* 10.5–11.5 cm **R**   **Page 507**
Rainforest. Sierra Leone–SE Guinea to CAR–Congo; locally common/uncommon. Bioko, scarce.
2a   **Above** White rump and uppertail-coverts. Small.
2b   **Below** White belly and undertail-coverts; tail almost completely hidden.

3 CASSIN'S SPINETAIL *Neafrapus cassini* 12–13 cm **R**   **Page 508**
Rainforest. Sierra Leone to SW CAR–Congo; locally not uncommon. Bioko, rare.
3a   **Above** Narrow white rump band; very short tail.
3b   **Below** White belly and undertail-coverts; distinctly notched wings.

4 BÖHM'S SPINETAIL *Neafrapus boehmi* 9–10 cm **M**   **Page 508**
Rainforest. NE Gabon.
4a   **Above** As 3 but much smaller, rump band broader; tail extremely short.
4b   **Below** Throat paler than 3.

5 BLACK SPINETAIL *Telacanthura melanopygia* 15–17 cm **R**   **Page 507**
Rainforest. SE Sierra Leone to SW CAR–N Congo. Rare/locally uncommon.
Wholly blackish; throat mottled dusky.

6 PALLID SWIFT *Apus pallidus* 16 cm; WS 42–46 cm **P/R**   **Page 510**
Palearctic migrant, Mauritania–Liberia to Chad and N Congo; locally common/rare. Vagrant, Cape Verde Is. Breeds NW Mauritania, Niger, NE Chad, probably E Mali; locally common/uncommon. Paler than 8, with larger throat patch; inner wing contrasting with darker outer wing.

7 NYANZA SWIFT *Apus niansae* 15 cm **V**   **Page 510**
Vagrant, N Congo. Dark brownish; contrasting panel on inner wing. Compare 8 and 11.

8 COMMON SWIFT *Apus apus* 16–17 cm; WS 42–48 cm **P**   **Page 510**
Over all habitats. Throughout. Common. Also Cape Verde Is (uncommon), São Tomé (rare). Uniformly blackish-brown; forked tail; ill-defined throat patch. Compare 6 and 11.

9 LITTLE SWIFT *Apus affinis* 12–13 cm; WS 33–35 cm **R**   **Page 512**
Mainly in towns. Throughout; also Bioko, São Tomé and Príncipe. Common.
9a   **Above** Broad white rump patch; square tail. Compare 1 and 10.
9b   **Below** White throat.

10 HORUS SWIFT *Apus horus* 15 cm **R**   **Page 511**
Wooded savanna. Nigeria to Congo. Rare/scarce to locally uncommon.
10a   **Above** Broad white rump patch; forked tail. Compare 9 and 14.
10b   **Below** White throat.
10c   **Dark morph ('toulsoni')** Dark brown rump; greyer throat. Gabon and Congo.

11 AFRICAN BLACK SWIFT *Apus barbatus* 16 cm **R**   **Page 509**
Various habitats. Patchily distributed, Sierra Leone to Cameroon. Locally uncommon/rare.
11a   **Above** As 8 but darker; contrasting paler inner wing panel.
11b   **Below** As 8 but shallower fork to tail. **11c** **Head** Compare 12.

12 FERNANDO PO SWIFT *Apus (barbatus) sladeniae* 16 cm R **DD**   **Page 509**
Mainly highlands. SE Nigeria–W Cameroon; Bioko. Rare and local.
Blacker than 11; Little or no greyish-white on throat.

13 PLAIN SWIFT *Apus unicolor* 14–15 cm; WS 38–39 cm **V**   **Page 510**
Palearctic vagrant, Mauritania. Smaller and slimmer than 8 and 6; throat darker; flight feathers semi-translucent from below; flight more erratic.

14 WHITE-RUMPED SWIFT *Apus caffer* 14–15 cm; WS 34–36 cm **R/M**   **Page 511**
Savanna and Sahel. Throughout. Rare/locally uncommon.
14a   **Above** Long, deeply forked tail (often held closed); U-shaped rump patch.
14b   **Below** White throat.

Continued on page 158

PLATE 64: MOUSEBIRDS AND TROGONS

Mousebirds Buffish-brown or greyish with long, stiff, graduated tails and short, erectile crests. Always in small parties. Flight on short, whirring wings, fast and direct with long glides. Climb well. Feed mainly on fruit and leaves.

1 BLUE-NAPED MOUSEBIRD *Urocolius macrourus* 33–38 cm **R**   Page 513
Arid open country, thorn scrub. SW Mauritania– Senegambia to Chad. Common/uncommon.
1a Adult Very long, slender tail; blue nape patch; red on bill and around eye.
1b Juvenile No blue on nape; shorter crest; dull-coloured bare parts.

2 SPECKLED MOUSEBIRD *Colius striatus* 30–36 cm **R**   Page 513
Various open and wooded habitats. E Nigeria to CAR–Congo; also N Ghana. Common/locally not uncommon.
2a Adult Very long tail; bushy crest. In small, noisy parties.
2b Juvenile Shorter tail and crest; pale stripe on back.

3 RED-BACKED MOUSEBIRD *Colius castanotus* *c.* 33 cm **?**   Page 513
No certain records: claim from S Gabon not accepted. Locally common in coastal strip, Angola.
Chestnut lower back and rump; no bare whitish spot behind eye.

Trogons. Arboreal forest species with brightly coloured plumage and long, broad tails. Easily overlooked despite their brilliant coloration as they perch motionlessly for long periods at mid-level in forest shade.

4 NARINA'S TROGON *Apaloderma narina* 29–34 cm **R**   Page 514
Forest. Sierra Leone–SE Guinea to Congo. Uncommon/scarce to locally common.
4a Adult male *brachyurum* Bare patches on head-sides green; outer tail feathers white. From Nigeria east.
4b Adult female *brachyurum* Forehead to breast greyish; rest of underparts pink.
4c Adult male *constantia* Bare patch at gape yellow. From Nigeria west.
4d Adult female *constantia* As 4b but lacking fine vermiculations on breast.
4e Juvenile *constantia* White and buff spots on wing.

5 BAR-TAILED TROGON *Apaloderma vittatum* 28 cm **R**   Page 514
Montane forest. SE Nigeria–W Cameroon; Bioko. Uncommon/locally not uncommon.
5a Adult male Darker than 4 and 6; bare patches at gape and below eye orange-yellow; outer tail feathers closely barred black and white.
5b Adult female Head and throat dark brownish.
5c Juvenile White or buff spots on wing.

6 BARE-CHEEKED TROGON *Apaloderma aequatoriale* 28 cm **R**   Page 514
Lowland forest. SE Nigeria to SW CAR–Congo. Uncommon/not uncommon.
6a Adult male Bare patches on head-sides yellow and more conspicuous than in 4. No overlap with 4c.
6b Adult female Throat and breast brownish.
6c Juvenile White or buff spots on wing.

Continued from page 156

15 AFRICAN PALM SWIFT *Cypsiurus parvus* 16 cm, incl. tail of up to 9 cm **R**   Page 509
Near palm trees, frequently in towns. Throughout; also Bioko, São Tomé and Príncipe. Common.
Wholly greyish; long wings and tail.

16 BATES'S SWIFT *Apus batesi* 14 cm **R**   Page 511
Forest. Liberia, Ghana, Nigeria to N Gabon and N Congo. Rare/locally uncommon.
Wholly glossy black; slender tail. Compare 8, 11 and 19.

17 MOTTLED SWIFT *Apus aequatorialis* 23 cm **R/M**   Page 512
Crags, escarpments; disperses widely. Patchily distributed. Guinea to E Chad and NE Gabon. Rare/uncommon; local.
Very large. All dark except whitish throat and greyish-looking belly.

18 ALPINE SWIFT *Apus melba* 20–22 cm; WS 54–60 cm **P/M**   Page 512
Variety of habitats. Palearctic migrant, Mauritania–Liberia to Cameroon–Gabon, Cape Verde Is; not uncommon/rare. Breeds C Mali; common.
Very large. White underparts with dark brown breast-band.

19 SCARCE SWIFT *Schoutedenapus myoptilus* 16.5 cm **R**   Page 508
Montane forest; also lower. W Cameroon and Bioko. Rare/scarce and local.
Dark brown; deeply forked tail (often held closed); pale throat; rest of underparts slightly paler than upperparts. Diagnostic clicking call.

1b

1a

2a

2b

3

4b

4a

4d

5a

5b

5c

4e

4c

6a

6b

6c

PLATE 65: KINGFISHERS I

Characterised by compact silhouette with large head, long dagger-shaped bill, short body and very short legs. Most species brightly coloured. Occur in a variety of habitats. Aquatic species feed mainly on fish captured by plunge-diving; terrestrial species subsist on insects and reptiles on which they swoop from a perch. Flight fast and direct.

1 AFRICAN DWARF KINGFISHER *Ceyx lecontei* 10 cm **R** **Page 517**
Dense rainforest. Sierra Leone–SE Guinea to Ghana; SW Nigeria to CAR–Congo. Rare/uncommon.
  1a **Adult** Tiny. Rufous crown; black forehead.
  1b **Juvenile** Blackish crown and upperparts with blue-tipped feathers; bill blackish tipped white.

2 AFRICAN PYGMY KINGFISHER *Ceyx pictus* 12 cm **R/M** **Page 518**
Various habitats; not usually near water. Throughout. Common/uncommon.
  2a **Adult** Very small. Blue-and-black crown; rufous supercilium; violet wash to ear-coverts.
  2b **Juvenile** Duller; mottled upperparts; bill black tipped white.

3 WHITE-BELLIED KINGFISHER *Alcedo leucogaster* 13 cm **R** **Page 518**
Small forest streams. Guinea-Bissau to CAR–Congo; Bioko. Uncommon/scarce.
  3a **Adult** *leucogaster* Underparts mainly white; head-sides, breast and flanks rufous-chestnut. Nigeria to Gabon and CAR; Bioko.
      A. l. leopoldi has blue superciliary area; intergrades with 3a in Congo R. basin.
  3b **Juvenile** *leucogaster* Duller; black, white-tipped bill.
  3c **Adult** *bowdleri* Rufous on lores more extensive than in 3a. Guinea-Bissau to Togo.

4 MALACHITE KINGFISHER *Alcedo cristata* 13 cm **R/M** **Page 518**
Aquatic habitats. Throughout. Generally common.
  4a **Adult** Brilliant violet-blue above; blue-and-black crest.
  4b **Juvenile** Duller; bill black.

5 SHINING-BLUE KINGFISHER *Alcedo quadribrachys* 16 cm **R** **Page 519**
Forested aquatic habitats. Senegambia to CAR–Congo. Uncommon/scarce to common.
  5a **Adult male** *quadribrachys* Dark blue upperparts; all-black bill; mainly rufous underparts. Senegambia to WC Nigeria; also SW Mali and Burkina Faso.
  5b **Adult female** *quadribrachys* Some dark red on base of lower mandible.
  5c **Juvenile** *quadribrachys* Bill tipped whitish; underparts pale and mottled.
  5d **Adult male** *guentheri* Mantle to uppertail-coverts paler, more contrasting with rest of upperparts. SW Nigeria to CAR–Congo.

6 PIED KINGFISHER *Ceryle rudis* 25 cm **R** **Page 520**
Aquatic habitats. Throughout; also Bioko. Common.
  6a **Adult male** Black and white; double breast-band.
  6b **Adult female** Single incomplete breast-band.
  6c **Adult male hovering**

7 COMMON KINGFISHER *Alcedo atthis* 15–16 cm **V?** **Page 519**
Palearctic vagrant claimed from N Mauritania. Aquatic habitats.
Brilliant blue above; broad blue malar stripe.

PLATE 66: KINGFISHERS II

1 CHOCOLATE-BACKED KINGFISHER *Halcyon badia*   21 cm   **R**   **Page 515**
Rainforest. Sierra Leone–Guinea to Ghana; SW Nigeria to CAR–Congo. Uncommon/locally not
uncommon.
  1a   **Adult** Dark brown upperparts; clean white underparts; azure rump and tail.
  1b   **Juvenile** Black bill, tipped whitish.

2 BROWN-HOODED KINGFISHER *Halcyon albiventris*   22 cm   **R**   **Page 515**
Woodland. S Gabon and Congo. Not uncommon.
  2a   **Adult male** Streaked brownish head; lightly streaked, dirty buff underparts; heavy bill.
  2b   **Adult female** Back browner.
  2c   **Juvenile** Duller; bill dusky, tipped whitish.

3 WOODLAND KINGFISHER *Halcyon senegalensis*   22 cm   **M/R**   **Page 516**
Woodlands, forest clearings, farmland, mangroves. Throughout; also Bioko. Common.
  3a   **Adult** *senegalensis* Bright azure-blue upperparts; black wing-coverts; grey head and mantle; greyish-
white underparts.
  3b   **Adult** *fuscopilea* Darker crown; greyer breast and mantle. Sedentary; forest zone.
  3c   **Juvenile** Duller; bill dusky.

4 GREY-HEADED KINGFISHER *Halcyon leucocephala*   22 cm   **M**   **Page 516**
Woodland and cultivation. Throughout; also Cape Verde Is. Fairly common.
  4a   **Adult** Grey head; bright chestnut belly.
  4b   **Juvenile** Dusky bill; buff or pale chestnut belly.

5 BLUE-BREASTED KINGFISHER *Halcyon malimbica*   25 cm   **R**   **Page 516**
Forest and densely wooded savanna. Throughout; also Príncipe. Uncommon/common.
  5a   **Adult** Black back, scapulars and wing-coverts; black streak through eye. Compare 3.
  5b   **Juvenile** Head and underparts tinged buff.

6 STRIPED KINGFISHER *Halcyon chelicuti*   17 cm   **R**   **Page 517**
Dry savanna woodland and bush. Throughout. Generally not uncommon.
Rather drab coloured. Streaked crown and underparts; black streak through eye.

7 GIANT KINGFISHER *Megaceryle maxima*   42–46 cm   **R**   **Page 520**
Aquatic habitats in forest and savanna zones. Throughout; also Bioko. Not uncommon.
Largest kingfisher. Massive, black bill; white-speckled, slate-grey upperparts.
  7a   **Adult male** Chestnut breast.
  7b   **Adult female** Chestnut belly.

PLATE 67: BEE-EATERS I

Slender, mostly brightly coloured species with long, pointed, slightly decurved bills and triangular wings. Feed mainly on wasps and bees, either caught in short sallies from a perch or in lengthy hawking flights.

1 BLACK BEE-EATER *Merops gularis* 20 cm **R**   Page 521
Forest zone. Sierra Leone–SE Guinea to SW CAR–Congo. Scarce/locally common.
Black upperparts; azure-blue rump; scarlet throat; blue-streaked breast and belly.
  1a **Adult** *gularis* Blue supercilium. Sierra Leone to Nigeria.
  1b **Juvenile** *gularis* Duskier than adult; no red throat.
  1c **Adult** *australis* No supercilium. Cameroon to CAR and Congo.
  1d **In flight** Black with conspicuous azure-blue rump.

2 BLUE-HEADED BEE-EATER *Merops muelleri* 19 cm **R**   Page 521
Forest. Sierra Leone–SE Guinea to SW CAR–Congo; Bioko. Scarce/rare.
Deep purple-blue; dark chestnut back and wings; small red throat patch.
  2a **Adult** *mentalis* Short, blunt tail streamers. West of Douala, Cameroon.
  2b **Juvenile** *mentalis* Duskier than adult; no streamers.
  2c **Adult** *muelleri* Whitish forehead; no streamers. South and east of Douala, Cameroon.
  2d **In flight** Appears mainly dark.

3 BLACK-HEADED BEE-EATER *Merops breweri* 25–28 cm **R**   Page 520
Forest edge and open woodland. Gabon–Congo; also CAR, SE Nigeria, C Ivory Coast. Scarce/rare.
  3a **Adult** Large. Black head; green upperparts; cinnamon breast and belly; long central tail feathers.
  3b **Juvenile** Duskier than adult; no streamers.
  3c **In flight** Green, with black head; tail has some cinnamon; wingtips rather broad.

4 EUROPEAN BEE-EATER *Merops apiaster* 23–25 cm **P**   Page 524
Various open habitats; avoids forest zone. Throughout. Locally common/scarce. Vagrant, Cape Verde Is.
  4a **Adult male** Chestnut crown and mantle; golden scapulars; yellow throat; bluish underparts; tail
      streamers.
  4b **Adult female** As male, but less chestnut on wing; scapulars suffused with green.
  4c **In flight** Chestnut mantle and coverts; golden scapulars.

5 NORTHERN CARMINE BEE-EATER *Merops nubicus* 24–27 cm **M**   Page 525
Dry and open savanna. Throughout. Not uncommon/locally common.
  5a **Adult** Carmine-pink; dark bluish head; very long tail streamers.
  5b **Juvenile** Duskier than adult; short tail streamers.
  5c **In flight** Blue head, rump and undertail-coverts contrast with striking carmine plumage.

6 ROSY BEE-EATER *Merops malimbicus* 22–25 cm **M**   Page 524
Savanna woodland; also over forest. Very local breeder Nigeria, S Gabon and S Congo; disperses as far west as W Ivory Coast.
  6a **Adult** Grey above; reddish-pink below; white streak below black mask; tail streamers.
  6b **Juvenile** Duskier; no streamers.
  6c **In flight** Contrasting under- and upperparts; streamers.

PLATE 68: BEE-EATERS II

1   BLUE-BREASTED BEE-EATER *Merops variegatus*   17 cm   **R**   **Page 522**
Various open habitats. Local and uncommon.
- **1a   Adult** *variegatus* Yellow throat; purple-blue gorget; white spot on neck-sides. S Cameroon to Congo–SW CAR. Compare 2.
- **1b   Juvenile** *variegatus* No gorget; pale greenish breast.
- **1c   Adult** *loringi* Blue supercilium. Highlands of E Nigeria–W Cameroon.
- **1d   In flight** Cinnamon fight feathers tipped black; primaries tinged green.

2   LITTLE BEE-EATER *Merops pusillus*   15–17 cm   **R**   **Page 521**
Various open habitats. S Mauritania–Liberia to CAR–Congo. Common to uncommon/scarce.
- **2a   Adult** *pusillus* Small. Yellow throat; black gorget; nearly square tail. Compare 1.
- **2b   Juvenile** *pusillus* No black gorget; pale greenish breast.
- **2c   Adult** *meridionalis* Short, narrow blue supercilium. S Gabon–Congo.
- **2d   In flight** Cinnamon flight feathers; black trailing edge to wing.

3   RED-THROATED BEE-EATER *Merops bulocki*   20–22 cm   **R**   **Page 522**
Savanna woodland. S Mauritania–Guinea to S Chad–N CAR. Locally common. Vagrant, Sierra Leone.
- **3a   Adult** Bright red throat; deep blue undertail-coverts. Gregarious.
- **3b   Adult** Rare yellow-throated variant.
- **3c   Juvenile** As adult but duller.
- **3d   In flight** Black trailing edge to wing; tail green with some buff, square.

4   WHITE-FRONTED BEE-EATER *Merops bullockoides*   22–24 cm   **R**   **Page 523**
Wooded grassland. Coastal strip, Gabon and Congo. Locally common.
- **4a   Adult** White forehead; white line below black mask; red throat. Note nasal call.
- **4b   In flight** Black trailing edge to wing; deep blue vent; square bluish-green tail.

5   SWALLOW-TAILED BEE-EATER *Merops hirundineus*   20–22 cm   **R/M**   **Page 522**
Savanna woodland. Senegambia–Sierra Leone to S Chad and SW CAR. Rather scarce and local. Vagrant, S Mauritania, Liberia.
- **5a   Adult** Strongly forked, pale blue tail; yellow throat; blue gorget.
- **5b   Juvenile** Throat greenish-white; no gorget.
- **5c   In flight** Mainly green with forked, white-tipped tail.

6   WHITE-THROATED BEE-EATER *Merops albicollis*   19–21 cm   **M**   **Page 523**
From Sahel (during rains) to forest zone (dry season). Throughout. Seasonally common/not uncommon.
- **6a   Adult** Very long tail streamers (up to 12 cm); distinctive black-and-white head pattern.
- **6b   Juvenile** Duller; no streamers; throat tinged yellowish.
- **6c   In flight** Long streamers conspicuous; wings green and pale rufous with black trailing edge.

7   BLUE-CHEEKED BEE-EATER *Merops persicus*   24–26 cm   **M/P**   **Page 524**
Very local breeder in Sahel zone. Intra-African and Palearctic migrant in savanna zones throughout; uncommon/rare.
- **7a   Adult** Large. Bright green; very long tail streamers (up to 11 cm). Gregarious.
- **7b   Juvenile** Duller; no streamers.
- **7c   In flight** All green above; underwing rufous with narrow dusky trailing edge.

8   LITTLE GREEN BEE-EATER *Merops orientalis*   16–18 cm   **R**   **Page 523**
Sahel and northern savannas. S Mauritania–Senegambia to Chad. Common/uncommon.
- **8a   Adult** Small. All green; very long tail streamers; narrow black gorget.
- **8b   Juvenile** Duller; no streamers; no gorget.
- **8c   In flight** Note small size and long tail streamers.

PLATE 69: ROLLERS

Robust, conspicuous birds with colourful plumage, large heads and stout bills, slightly hooked at tip. Hunt from perch, swooping to ground to catch large insects or small vertebrates (*Coracias*), or pursuing insects in the air (*Eurystomus*). Vocal and pugnacious. Voice harsh.

1 BROAD-BILLED ROLLER *Eurystomus glaucurus* 29–30 cm **M**   **Page 527**
Various wooded habitats, from forest to Sahel. Throughout; common. Vagrant, Cape Verde Is, São Tomé and Príncipe.
   1a Adult Chestnut above; deep lilac below; short yellow bill; mainly azure-blue tail and undertail. Compare 2.
   1b Juvenile Duskier; underparts mottled brown and bluish; bill partly dusky.
   1c Adult in flight Pointed wings; dark blue greater coverts.

2 BLUE-THROATED ROLLER *Eurystomus gularis* 25 cm **R**   **Page 527**
Forest. Guinea-Bissau to CAR–Congo; also Bioko. Uncommon/not uncommon.
   2a Adult Dark chestnut; blue throat patch; chestnut undertail. Compare 1.
   2b Juvenile Duskier; lower breast and belly greyish-blue mottled dusky.
   2c Adult in flight Silhouette as 1; dark chestnut greater coverts.

3 RUFOUS-CROWNED ROLLER *Coracias naevius* 35–40 cm **R/M**   **Page 525**
Open woodlands. S Mauritania–Guinea to S Chad and CAR. Generally uncommon. Rare, Sierra Leone and Liberia.
   3a Adult Large; conspicuous white supercilium; brownish-pink underparts streaked white; square tail.
   3b Juvenile Duller, more olive.
   3c Adult in flight Purplish fight feathers; lilac upperwing-coverts.

4 BLUE-BELLIED ROLLER *Coracias cyanogaster* 28–30 cm **R/M**   **Page 526**
Wooded savanna. Senegambia–Sierra Leone to CAR; locally common/scarce. Also S Mauritania; rare.
   4a Adult Pale buffish head and breast; purplish-blue belly.
   4b In flight Pale blue bar on purplish-blue wings; pale blue tail with long tail streamers.

5 EUROPEAN ROLLER *Coracias garrulus* 32 cm **P**   **Page 526**
Open wooded habitats. Throughout; also São Tomé and Príncipe. Uncommon/rare. Vagrant, Cape Verde Is.
   5a Adult Mainly bright pale blue and chestnut; square tail.
   5b In flight Blackish flight feathers.

6 LILAC-BREASTED ROLLER *Coracias caudatus* 28–30 cm **R**   **Page 526**
Wooded grassland. SE Gabon–SC Congo. Not uncommon.
   6a Adult Lilac throat and breast streaked white; long tail streamers (up to 8 cm).
   6b In flight Purplish flight feathers.

7 ABYSSINIAN ROLLER *Coracias abyssinicus* 28–30 cm **M**   **Page 526**
Dry wooded habitats. S Mauritania–Liberia to Chad–CAR. Common.
   7a Adult Mainly turquoise and chestnut; very long tail streamers (up to 12 cm).
   7b Juvenile Duller; no streamers. Compare 5.
   7c Adult in flight Flight feathers, forewing, rump and uppertail-coverts deep violet-blue.

PLATE 70: WOOD-HOOPOES AND HOOPOE

Wood-hoopoes. Slender with mainly dark glossy plumage, slender, decurved bills and long graduated tails. Arboreal and agile, often hanging upside-down while probing bark for insects. Larger species conspicuous, noisy and usually gregarious; smaller ones quieter, mostly solitary or in pairs.

1   GREEN WOOD-HOOPOE *Phoeniculus purpureus* 35–40 cm **R**   **Page 528**
Open wooded habitats. S Mauritania–Sierra Leone to Chad–CAR. Uncommon/locally common.
- **1a   Adult** Large; long graduated tail; red or red-tipped bill; red feet.
- **1b   Juvenile** Smaller and duller; bill and feet black.
- **1c   Adult in flight** Double white wingbar.

2   WHITE-HEADED WOOD-HOOPOE *Phoeniculus bollei* 30–35 cm **R**   **Page 528**
Forest. SE Guinea–Liberia to S Ghana; S Nigeria to S CAR. Uncommon.
- **2a   Adult** *bollei* Buffish-white head; red bill and feet; no white on wings and tail.
- **2b   Juvenile** *bollei* Smaller; variable: a dusky-headed individual.
- **2c   Juvenile** *bollei* A pale-headed individual.
- **2d   Adult** *okuensis* Less buffish-white on head. Mt Oku (W Cameroon) only.

3   BLACK WOOD-HOOPOE *Rhinopomastus aterrimus* 23 cm **R**   **Page 529**
Open wooded habitats, Sahel and savanna zones. Uncommon/locally not uncommon.
- **3a   Adult** *aterrimus* Violet-black; bill and feet black. S Mauritania–Sierra Leone to Chad–CAR.
- **3b   Adult** *aterrimus* **in flight** Double white wingbar.
- **3c   Adult** *anchietae* No white on primary-coverts; subterminal white spots on outer two tail feathers. SE Gabon and SC Congo.

4   FOREST WOOD-HOOPOE *Phoeniculus castaneiceps* 28 cm **R**   **Page 528**
Forest. Liberia–SE Guinea to SW CAR–Congo. Scarce/uncommon and local.
- **4a   Adult** *castaneiceps* Chestnut head; pale bill; black feet. Liberia–SE Guinea to S Ghana, SW Nigeria.
- **4b   Adult male** *brunneiceps* An all-dark, glossy bottle-green individual. S Cameroon to SW CAR–N Congo.
- **4c   Adult female** *brunneiceps* Example of an individual with a whitish head.
- **4d   Juvenile** *brunneiceps* Duller; buffish-white head; feathers narrowly fringed darker.

Hoopoes. Very distinctive, with slender decurved bills, broad rounded wings and erectile, fan-shaped crests. Feed on ground.

5   HOOPOE *Upupa epops* 25–28 cm **R/M/P**   **Page 529**
Wooded savanna, dry scrub. Throughout. Uncommon/not uncommon. Sandy or cinnamon; black-and-white wings and tail.
- **5a   Adult** *epops* Palest race. Palearctic passage migrant and winter visitor; Mauritania–Sierra Leone to Chad; also Gabon. Vagrant, Cape Verde Is.
- **5b   Adult** *epops* **in flight** Broad white band on primaries and four white bands on secondaries and coverts. *U. e. major* slightly duller than 5a–b; less white in secondaries; tail band narrower; belly more streaked. NE Chad.
- **5c   Adult** *senegalensis* Slightly darker than 5a. Resident; Mauritania–Sierra Leone to Chad. Also Liberia (rare).
- **5d   Adult** *senegalensis* **in flight** More white in secondaries than 5b, forming patches.
- **5e   Adult male** *africana* **in flight** No white on primaries; body rich cinnamon. Gabon and Congo. *U. e. waibeli* intermediate between 5c–d and 5e; primaries as 5c–d. South of *senegalensis* in Cameroon, S Chad, CAR.

Medium-sized to very large, with large decurved bills surmounted by horny casque of variable size. Plumage of most species black and white, with variable amount of bare skin around eye and on malar area or throat; tail long. Sexes separable by shape and pattern of casque, coloration of bare parts and/or plumage. Mostly arboreal. Fly with neck outstretched. Most species vocal, with distinctive calls. Those on this plate are small to medium-sized with small casque, limited to low ridge in most.

1 RED-BILLED DWARF HORNBILL *Tockus camurus* 34–39 cm **R** Page 532
Forest. S Sierra Leone–SE Guinea to Ghana; Benin to S CAR–Congo. Not uncommon.
1a Adult male Small; brown above; red bill.
1b Adult female Red bill tipped black.

2 BLACK DWARF HORNBILL *Tockus hartlaubi* 35–39 cm **R** Page 532
Forest. S Sierra Leone–SE Guinea to Togo; S Nigeria to SW CAR–Congo. Uncommon.
2a Adult male *hartlaubi* Small; black above; white supercilium; black bill tipped dark red.
2b Adult female *hartlaubi* Entirely black bill.
2c Adult male *granti* Wing-coverts and inner secondaries tipped white; more red on bill. SW CAR–Congo.

3 AFRICAN PIED HORNBILL *Tockus fasciatus* 48–55 cm **R** Page 533
Forest and adjacent wooded habitats. Senegambia to CAR–Congo; common. Also S Mauritania; rare.
3a Adult male *semifasciatus* Broad black wings; outer tail feathers 3–4 broadly tipped white; cream-yellow bill tipped black. From Nigeria (Owerri area) west.
3b Adult male *fasciatus* Outer tail feathers 3–4 mostly white; bill tipped dark red below. From extreme E Nigeria east.
3c Adult female *fasciatus* Bill tipped black.
3d Juvenile Bill entirely creamy; in both races white in tail restricted to broad tips to outer tail feathers 3–4.

4 PIPING HORNBILL *Bycanistes fistulator* 50–60 cm **R** Page 534
Lowland and gallery forest. S Senegal–Liberia to CAR–Congo. Not uncommon/common.
4a Adult male *fistulator* Wings black with broad white tips to secondaries; outer tail feathers broadly tipped white; grooved bill dusky with pale base and tip. From W Nigeria west.
4b Adult female *fistulator* Bill smaller.
4c Juvenile *fistulator* Bill smaller and entirely dusky.
4d Adult male *sharpii* White in wings extending to inner primaries; outer tail feathers entirely white. E Nigeria to Congo.
4e Adult male *duboisi* Outer primaries also white; well-developed casque. SC Cameroon to CAR.

5 CROWNED HORNBILL *Tockus alboterminatus* 48–55 cm Page 533
Not recorded in region; reaches coastal forest patches in Congo-Kinshasa and Cabinda.
Adult male Dusky-brown above; red bill with yellow band at base. Female has smaller bill.

6 AFRICAN GREY HORNBILL *Tockus nasutus* 45–51 cm **M/R** Page 533
Savanna woodland. S Mauritania–Sierra Leone to Chad–CAR. Common. Vagrant, Liberia.
6a Adult male Dull greyish; white supercilium; black bill with creamy patch at base.
6b Adult female Smaller bill tipped dark red; rest of upper mandible pale yellow.
6c Juvenile Smaller. entirely dusky-grey bill.

7 RED-BILLED HORNBILL *Tockus erythrorhynchus* 40–48 cm **R** Page 532
Savanna woodland, thorn scrub. S Mauritania–Guinea to Chad–N CAR. Common/not uncommon.
7a Adult male *erythrorhynchus* Mainly white head; white spots on wing-coverts; slender red bill; yellowish to pink orbital ring.
7b Adult female *erythrorhynchus* Bill smaller.
7c Juvenile *erythrorhynchus* Bill smaller and duller.
7d Adult male *kempi* As 7a but orbital ring black. From W Mali west.

7d

not to scale

PLATE 72: HORNBILLS II

All species on this plate, except White-crested Hornbill, are very large, black-and-white birds, with large decurved bills surmounted by high horny casques. Their wingbeats make a distinctive loud swishing noise.

1 BROWN-CHEEKED HORNBILL *Bycanistes cylindricus* 65–77 cm **R** NT **Page 535**
Forest. Sierra Leone–SE Guinea to Togo. Uncommon/rare to locally not uncommon. Endemic.
- **1a Adult male** Broad black band on tail; upper part of thighs black; pale bill and casque.
- **1b Adult female** Bill and casque smaller.
- **1c Juvenile** Bill smaller, without casque.

2 WHITE-THIGHED HORNBILL *Bycanistes albotibialis* 65–77 cm **R** **Page 535**
Forest. From S Nigeria (Benin?) to SW CAR–Congo. Uncommon/locally common.
- **2a Adult male** Black band on tail narrower than in 1a; thighs entirely white; bill mainly dusky; casque pale and differently shaped to 1a.
- **2b Adult female** Bill and casque smaller and dusky.

3 YELLOW-CASQUED HORNBILL *Ceratogymna elata* 70–90 cm **R** NT **Page 536**
Forest. SW Senegal to Togo; SW Nigeria to W Cameroon. Rare/uncommon and local. Endemic.
- **3a Adult male** Black with white outer tail feathers; bill blackish; upper part of casque cream.
- **3b Adult female** As 4b but with pale bill, paler rufous head and tail pattern as 3a.

4 BLACK-CASQUED HORNBILL *Ceratogymna atrata* 70–90 cm **R** **Page 535**
Forest. Sierra Leone–SE Guinea to Ghana; E Nigeria to S CAR–Congo; Bioko. Uncommon/rare to locally common. Endemic.
- **4a Adult male** Black with broad white tips to outer tail feathers; bill and casque entirely black. Compare 3a.
- **4b Adult female** Rufous-brown head and neck; smaller bill and casque. Compare 3b.

5 BLACK-AND-WHITE-CASQUED HORNBILL *Bycanistes subcylindricus* 65–78 cm **R** **Page 534**
Forest. Sierra Leone to CAR–N Congo. Locally common/rare.
- **5a Adult male** *subcylindricus* Central tail feathers black; blackish bill with two-toned casque. West of Niger R.
- **5b Adult female** *subcylindricus* Smaller bill and casque.
- **5c Juvenile** *subcylindricus* Smaller, entirely dusky bill without casque.
- **5d Adult male** *subquadratus* Casque with larger pale area. East of Niger R.

6 WHITE-CRESTED HORNBILL *Tropicranus albocristatus* 70–80 cm **R** **Page 530**
Forest. Sierra Leone–SE Guinea to S CAR–Congo. Not uncommon/locally common.
- **6a Adult male** *albocristatus* White crown and head-sides; very long, graduated, white-tipped tail. From W Ivory Coast west.
- **6b Adult male** *cassini* Head-sides black; flight feathers, greater coverts and scapulars tipped white. From Nigeria east.
- **6c Adult male** *macrourus* As 6a but white of head extending to throat and neck. From E Ivory Coast to Benin.
- **6d Juvenile** Smaller; bill entirely darkish without casque.

7 ABYSSINIAN GROUND HORNBILL *Bucorvus abyssinicus* 90–110 cm **R** **Page 530**
Savanna. S Mauritania–Sierra Leone to Chad–CAR. Uncommon/rare to locally not uncommon.
- **7a Adult male** Huge; black with white primaries (conspicuous in flight); bare red skin on neck and throat.
- **7b Adult female** Smaller; bare skin entirely blue.
- **7c Juvenile** Dark sooty-brown; small bill; undeveloped casque.

PLATE 73: BARBETS I

Stocky, with large heads and heavy bills. Mostly arboreal; nest in self-excavated holes in trees. Calls characteristic, typically consisting of a monotonously repeated single note.

1   YELLOW-FRONTED TINKERBIRD *Pogoniulus chrysoconus* 11.5 cm **R**   Page 539
Wooded savanna. SW Mauritania–Guinea to Chad–CAR. Common.
Adult Golden-yellow forecrown; boldly black-and-white streaked face and upperparts.

2   SPECKLED TINKERBIRD *Pogoniulus scolopaceus* 13 cm **R**   Page 537
Forest, forest edge, clearings, bush. Sierra Leone–SE Guinea to SE CAR and Congo; Bioko. Common.
Adult Dark brown; scaly upperparts; pale eye.

3   YELLOW-RUMPED TINKERBIRD *Pogoniulus bilineatus* 10 cm **R**   Page 539
Forest and woodland. Senegambia to CAR and Congo; Bioko. Common.
Adult Black upperparts; bold white facial stripes; yellow rump; white throat.

4   WESTERN GREEN TINKERBIRD *Pogoniulus coryphaeus* 9 cm **R**   Page 538
Montane forest. SE Nigeria–W Cameroon. Locally common.
Adult Black upperparts; yellow stripe from crown to rump.

5   YELLOW-THROATED TINKERBIRD *Pogoniulus subsulphureus* 10 cm **R**   Page 538
Forest. Sierra Leone to SE CAR and Congo; Bioko. Common.
5a   **Adult** *chrysopyga* As 3; distinguished by faster song. From Ghana west.
5b   **Adult** *flavimentum* Yellow throat; yellow facial stripes. S Nigeria (Togo?) to Congo and SE CAR.
          Nominate (Bioko) similar.

6   YELLOW-SPOTTED BARBET *Buccanodon duchaillui* 15 cm **R**   Page 539
Forest. Sierra Leone–SE Guinea to Togo; SW Nigeria to SW CAR–Congo; also N CAR. Common/
uncommon.
Adult Red forehead to crown; yellow stripe on head-sides.

7   RED-RUMPED TINKERBIRD *Pogoniulus atroflavus* 13 cm **R**   Page 538
Forest. S Senegal; SW Guinea to SE CAR and Congo. Uncommon/not uncommon.
Adult Red rump; three yellow facial stripes.

8   HAIRY-BREASTED BARBET *Tricholaema hirsuta* 16.5 cm **R**   Page 539
Forest and forest edge. Locally common.
8a   **Adult** *hirsuta* Head and throat black; two bold facial stripes. S Senegal; Sierra Leone–SE Guinea to
          Ghana. Intergrading with 8c in Togo and W Nigeria.
8b   **Adult** *ansorgii* Throat whitish streaked black; nape spotted yellow. E Cameroon–CAR.
8c   **Adult** *flavipunctata* Head freckled without facial stripes; plumage browner. S Nigeria to N & C
          Gabon.
          T. h. angolensis similar but browner, especially on belly. S Gabon, W & C Congo.

PLATE 74: BARBETS II

1 NAKED-FACED BARBET *Gymnobucco calvus* 17–18 cm **R**   Page 537
Forest and forest edge. SW Guinea to Congo. Locally common.
1a **Adult** Dusky-brown; bare blackish head; sparse short bristles at gape and nostrils.
1b **Juvenile** Head more feathered; darkish-tipped bill.

2 BRISTLE-NOSED BARBET *Gymnobucco peli* 17–18 cm **R**   Page 537
Forest and forest edge. Sierra Leone to Congo. Uncommon/locally common; less common than 1.
2a **Adult** As 1a, but with conspicuous nasal tufts and paler bill.
2b **Juvenile** Head feathered; nasal tufts small or absent.

3 SLADEN'S BARBET *Gymnobucco sladeni* 17–18 cm **R**   Page 536
Lowland forest. S CAR (1 record).
Adult Grey throat and neck-sides; buff to brownish-buff nasal tufts; black bill; reddish eye.

4 GREY-THROATED BARBET *Gymnobucco bonapartei* 16.5–17.5 cm **R**   Page 536
Forest and forest edge. Cameroon to S CAR and Congo. Locally common.
4a **Adult** Greyish head and throat; brownish nasal tufts; blackish bill; brown to red eye.
4b **Juvenile** Bill yellowish with dusky tip; no nasal tufts.

5 WHITE-HEADED BARBET *Lybius leucocephalus* 15.0–16.5 cm **R**   Page 540
Woodland and cultivation. NC Nigeria to S Chad–CAR. Locally common/scarce.
Adult *adamauae* All-white head and breast; white rump.

6 BLACK-BACKED BARBET *Lybius minor* 16.5 cm **R**   Page 541
Woodland and gallery forest. S Gabon–Congo. Scarce/locally common.
6a **Adult** *minor* Dark above, white below; red forehead and crown; pinkish belly.
6b **Adult** *macclounii* Head-sides white; white scapulars form V. Intermediates occur.

7 BLACK-BILLED BARBET *Lybius guifsobalito* 15 cm **V**   Page 540
Open savanna woodland. One record, N Cameroon.
Adult Black with red on face to upper breast; yellow-edged flight feathers.

8 VIEILLOT'S BARBET *Lybius vieilloti* 15 cm **R**   Page 540
Savanna woodland, thorn scrub. S Mauritania–Sierra Leone and Liberia to Chad–CAR. Common/
uncommon.
Adult Red head; pale yellow underparts; red spots from throat to central belly.

9 YELLOW-BREASTED BARBET *Trachyphonus margaritatus* 20 cm **R**   Page 542
Dry wooded grassland, thorn bush, desert edge. S Mauritania to C Chad. Locally common.
Pink bill; yellow head with black cap; yellow breast; white-spotted upperparts.
9a **Adult male** Back patch on centre of breast.
9b **Adult female** No black on breast.

10 YELLOW-BILLED BARBET *Trachylaemus purpuratus* 24 cm **R**   Page 542
Forest and forest edge. Sierra Leone–SE Guinea to S CAR–Congo. Not uncommon.
Yellow bill; yellow orbital patch; blackish upperparts; long tail.
10a **Adult** *purpuratus* Lower breast and belly blotched black. From SE Nigeria east.
10b **Adult** *goffinii* No black on underparts. From Ghana west.
  T. p. togoensis (E Ghana to SW Nigeria) intermediate.

11 BLACK-BREASTED BARBET *Lybius rolleti* 26.5 cm **R**   Page 541
Wooded grassland and cultivation. S Chad–N CAR. Uncommon/locally common.
Adult Massive ivory bill; mainly black plumage; red belly; white flanks.

12 BEARDED BARBET *Lybius dubius* 25 cm **R**   Page 541
Woodland. Senegambia–Guinea-Bissau to SE Chad–C CAR. Locally not uncommon. Endemic.
Adult Massive yellowish bill; black breast-band; no wingbar. Compare 13.

13 DOUBLE-TOOTHED BARBET *Lybius bidentatus* 23 cm **R**   Page 541
Woodland. Mali; Guinea-Bissau–Sierra Leone to Congo; CAR. Not uncommon.
Adult Heavy whitish bill; no breast-band; red (or pinkish to whitish) wingbar; grey legs and feet.

PLATE 75: HONEYGUIDES

Drab coloured and generally inconspicuous. Outer tail feathers white and generally dark tipped, often conspicuous in flight.

1 GREATER HONEYGUIDE *Indicator indicator* 18–19.5 cm **R**   **Page 545**
Savanna woodland. S Mauritania–Sierra Leone to Chad–CAR; also SE Gabon and SW Congo.
Uncommon/rare to locally common.
1a   **Adult male** Black throat; whitish ear patch; pink bill.
1b   **Adult female** Duller than 1a; no distinctive head pattern; bill dark.
1c   **Juvenile** Pale to deep yellow throat and breast; bill dark.

2 SPOTTED HONEYGUIDE *Indicator maculatus* 16.5–17.5 cm **R**   **Page 544**
Evergreen and gallery forest. Senegambia to SE CAR and Congo. Rare/locally not uncommon.
2a   **Adult** *maculatus* Olive underparts spotted creamy-green. From SW Nigeria west.
     I. m. stictothorax has spots paler, more distinct; head-sides finely streaked. From SE Nigeria east and south.
2b   **Juvenile** Speckled forehead; spotting below more extensive and distinct.

3 LESSER HONEYGUIDE *Indicator minor* 14–15 cm **R**   **Page 545**
Wooded savanna. Senegambia–Sierra Leone to Chad–CAR. Uncommon.
Stubby bill; grey head; indistinctly streaked above. Compare 4, 6 and 7.

4 THICK-BILLED HONEYGUIDE *Indicator conirostris* 14–15 cm **R**   **Page 546**
Forest. SE Sierra Leone to Ghana; S Nigeria to S CAR–Congo. Uncommon.
As 3; typically more distinctly streaked above, darker grey below.

5 LYRE-TAILED HONEYGUIDE *Melichneutes robustus* 16.5–17.5 cm **R**   **Page 544**
Lowland, lower montane and gallery forest. Sierra Leone–SE Guinea to Ivory Coast; E Nigeria to SE CAR and Congo. Rare/uncommon.
Distinctly shaped tail. Rarely seen, but sound diagnostic.

6 WILLCOCKS'S HONEYGUIDE *Indicator willcocksi* 11–13 cm **R**   **Page 546**
Forest, gallery forest, dense woodland. Guinea-Bissau; Sierra Leone to Ghana; Nigeria to S Chad–CAR.
Rare/uncommon.
Olive wash on grey breast; no submoustachial stripe; dusky streaks on flanks. Compare 7, 3 and 4.

7 LEAST HONEYGUIDE *Indicator exilis* 11–13 cm **R**   **Page 546**
Forest. Guinea-Bissau to Togo (patchy); S Nigeria to CAR–Congo; Bioko. Uncommon/rare.
Distinctly streaked above and on flanks; dark submoustachial stripe; narrow white line on lores.
Compare 6 and 4.

8 YELLOW-FOOTED HONEYGUIDE *Melignomon eisentrauti* 14–15 cm **R DD**   **Page 544**
Forest. SE Sierra Leone, Liberia, Ivory Coast, SW Ghana, W Cameroon. Rare. Endemic.
Bill and legs yellowish; off-white undertail-coverts; tail mainly white from below.

9 ZENKER'S HONEYGUIDE *Melignomon zenkeri* 14–15 cm **R**   **Page 543**
Forest. SW & S Cameroon, Eq. Guinea, NE Gabon, N Congo, SE CAR. Rare/locally not uncommon.
Darkish, bulbul-like. Wholly greyish-olive below; graduated tail mainly dark; bicoloured bill (dark above, yellow below); yellowish legs. Compare 8.

10 WAHLBERG'S HONEYBIRD *Prodotiscus regulus* 12–13 cm **R**   **Page 543**
Woodland and bush. Patchily distributed. Liberia, N Ivory Coast, Togo, SE Nigeria, Cameroon, W CAR.
Rare.
Resembles 11 but browner; outer tail feathers tipped dark (all white in juvenile).

11 CASSIN'S HONEYBIRD *Prodotiscus insignis* 11–12 cm **R**   **Page 543**
Evergreen and gallery forest. Guinea to Togo; S Nigeria to CAR–Congo. Scarce/uncommon.
Flycatcher-like; outer tail feathers all white; fine bill.

PLATE 76: WOODPECKERS I

Small to large birds with strong, straight, pointed bills. Characteristically cling to trees, using stiff tail as prop. Often in pairs. Sexes differ in head pattern, males usually with (more) red on head. Nest in self-excavated holes in trees or termitaria.

1 GOLDEN-TAILED WOODPECKER *Campethera abingoni* 20 cm **R**   Page 549
Gallery forest and woodland. Patchily distributed throughout. Rare/uncommon.
  1a Adult male Dull green upperparts; barred back; streaked underparts; red crown and moustachial stripe.
  1b Adult female Forecrown and moustachial stripe black speckled white.

2 FINE-SPOTTED WOODPECKER *Campethera punctuligera* 21 cm **R**   Page 548
Woodland. S Mauritania–Sierra Leone to Chad–CAR. Locally common/uncommon.
  2a Adult male Pale yellow underparts; breast finely spotted blackish; red crown and moustachial stripe.
  2b Adult female Forecrown black speckled white; indistinct, speckled moustachial stripe.
  2c Juvenile Forecrown plain black; moustachial stripe blacker than in adult female.

  NUBIAN WOODPECKER *Campethera nubica* 20 cm **R**   Page 548
  Open woodland. NE CAR. Uncommon. **See illustration on plate 147**.
  Adult male Olive-brown upperparts boldly barred and spotted whitish; breast and flanks boldly spotted; red crown and moustachial stripe.
  Adult female Forecrown and moustachial stripe black speckled white.

3 TULLBERG'S WOODPECKER *Campethera tullbergi* 18–20 cm **R**   Page 550
Moist montane forest. SE Nigeria–W Cameroon; Bioko. Uncommon.
  3a Adult male Plain green above; diagnostic red carpal patch; finely spotted face and underparts; red forecrown.
  3b Adult female Forecrown black speckled white.

4 GABON WOODPECKER *Dendropicos gabonensis* c. 15 cm **R**   Page 551
Forest edge, second growth. Sierra Leone to SW CAR–Congo. Uncommon/locally common.
  4a Adult male *lugubris* Small; plain upperparts; heavily streaked underparts; head-sides and malar stripe olive-brown; hindcrown and nape red. Sierra Leone to Togo.
  4b Adult female *lugubris* Forehead and crown dark olive.
  4c Adult male *gabonensis* Boldly spotted underparts; head-sides streaked; red crown. Gabon to Congo. *C. g. reichenowi* intermediate. S Nigeria to SW Cameroon.
  4d Adult female *gabonensis* Forehead and crown black.

5 GREEN-BACKED WOODPECKER *Campethera cailliautii* c. 16 cm **R**   Page 549
Forest, forest edge and adjacent woodland. E Ghana to S CAR–Congo. Uncommon/common.
  5a Adult male Plain green upperparts; densely barred underparts; greenish tail; red forehead and crown.
  5b Adult female Forecrown speckled black.

6 LITTLE GREEN WOODPECKER *Campethera maculosa* c. 16 cm **R**   Page 549
Forest and forest edge. Senegal–Liberia to Ghana; also S Mauritania and SW Mali. Uncommon/locally common. Endemic.
  6a Adult male Golden-olive upperparts; boldly barred underparts; black tail; red forehead and crown.
  6b Adult female Forehead and crown barred black and white.

7 BROWN-EARED WOODPECKER *Campethera caroli* 18–19 cm **R**   Page 550
Forest. Sierra Leone–SE Guinea to Ghana; Cameroon to S CAR–Congo. Locally common/uncommon.
  7a Adult male Dark olive; distinctive brown ear patch; spotted buffish below; some red feathers on nape.
  7b Adult female Entire crown and nape olive.

8 BUFF-SPOTTED WOODPECKER *Campethera nivosa* 14–16 cm **R**   Page 550
Forest. SW Senegambia–Liberia to S CAR–Congo; also S Mauritania, SW Mali. Locally common/rare.
  8a Adult male Dark olive; densely spotted buffish below; red patch on nape.
  8b Adult female Nape olive.

PLATE 77: WOODPECKERS II

1   **LITTLE GREY WOODPECKER** *Dendropicos elachus* 11–13 cm **R**   **Page 551**
Arid country. SW Mauritania–N Senegal to N Cameroon–Chad. Uncommon/rare.
  1a   **Adult male** Very small and pale; red rump; red hindcrown and nape.
  1b   **Adult female** Crown and nape dull brown.

2   **BROWN-BACKED WOODPECKER** *Dendropicos obsoletus* 13–14 cm **R**   **Page 553**
Woodland. S Mauritania–Sierra Leone to Chad–CAR. Uncommon/not uncommon.
  2a   **Adult male** Small; greyish-brown above; plain back; wings and tail heavily spotted white;
      underparts whitish and only faintly streaked; red hindcrown.
  2b   **Adult female** Entire crown dark brown.

3   **ELLIOT'S WOODPECKER** *Dendropicos elliotii* 19–21 cm **R**   **Page 553**
Forest. Uncommon/locally common.
  3a   **Adult male** *elliotii* Plain green upperparts; black forehead and forecrown; red hindcrown and
      nape; plain pale greenish head-sides; heavily streaked underparts. Montane areas, SE Nigeria–W
      Cameroon, Bioko.
  3b   **Adult male** *johnstoni* Underparts mainly plain; red hindcrown and nape. Lowland Cameroon to
      Gabon–NW Congo.
  3c   **Adult female** *johnstoni* Crown and nape all black.

4   **GREY WOODPECKER** *Dendropicos goertae* *c.* 20 cm **R**   **Page 553**
Woodland, forest edge. S Mauritania–Liberia to Chad–CAR; also S Gabon–S Congo. Not uncommon/
common.
  4a   **Adult male** Grey head and underparts; golden-olive upperparts; red rump; orangey streak on belly;
      red crown and nape.
  4b   **Adult female** Entire head grey.

5   **SPECKLE-BREASTED WOODPECKER** *Dendropicos poecilolaemus* 14–15 cm **R**   **Page 551**
Forest edges, light woodland. SE Nigeria to SW CAR; also SE CAR. Rare/locally common.
  5a   **Adult male** Small; distinctive fine spotting on breast; red hindcrown.
  5b   **Adult female** Entire crown black.

6   **CARDINAL WOODPECKER** *Dendropicos fuscescens* 14–15 cm **R**   **Page 552**
Woodland. S Mauritania–Sierra Leone to CAR–Congo. Scarce/uncommon to locally common.
  6a   **Adult male** *lafresnayi* Small; faintly barred back; lightly streaked below; red hindcrown and nape.
      From Nigeria west.
      D. f. sharpii has more heavily streaked underparts. From Cameroon east and south.
  6b   **Adult female** Hindcrown and nape black.

7   **YELLOW-CRESTED WOODPECKER** *Dendropicos xantholophus* 20–23 cm **R**   **Page 552**
Forest and dense woodland. SE Nigeria to SW CAR–Congo. Rare/locally common.
  7a   **Adult male** Plain dark olive above; densely spotted and barred below; bold head markings; yellow
      hindcrown diagnostic.
  7b   **Adult female** Forehead and crown all black.

8   **BEARDED WOODPECKER** *Dendropicos namaquus* 23–25 cm **R**   **Page 552**
Woodland. CAR. Not uncommon.
  8a   **Adult male** Large and dark; bold head markings; golden rump and tail tips; red hindcrown.
  8b   **Adult female** Forehead speckled, hindcrown black.

9   **FIRE-BELLIED WOODPECKER** *Dendropicos pyrrhogaster* *c.* 24 cm **R**   **Page 552**
Forest and forest edge. Guinea to W Cameroon; also SW Mali. Not uncommon/rare. Endemic.
  9a   **Adult male** Large; red rump and centre of underparts; bold head markings; red crown.
  9b   **Adult female** Forehead and crown all black.

PLATE 78: WRYNECKS, PICULET, BROADBILLS AND PITTAS

Wrynecks have cryptic plumage and soft tails, which are not used as brace when clinging to trees. Feed mainly on ants. Nest in natural cavities or old woodpecker or barbet holes.

1 EURASIAN WRYNECK *Jynx torquilla* 16–18 cm **P**   **Page 547**
Savanna. Mauritania–N Liberia to Chad–CAR. Uncommon/rare.
Adult Cryptically patterned, bark-like plumage.

2 RED-THROATED WRYNECK *Jynx ruficollis* 18–19 cm **R**   **Page 547**
Open woodland. Uncommon/locally common.
Brown mottled and barred above; distinctive chestnut breast patch.
2a **Adult** *pulchricollis* Chestnut patch extends onto lower throat. SE Nigeria to CAR.
2b **Adult** *ruficollis* Chestnut patch larger, extending onto chin. SE Gabon and Congo.

Piculets. Tiny woodpecker-like birds with short bills and very short tails, which are not used as brace when clinging to bark. Nest in self-excavated holes in trees and vines.

3 AFRICAN PICULET *Sasia africana* 8–9 cm **R**   **Page 548**
Forest. Liberia to Ghana; S Cameroon to SW CAR–Congo. Rare to uncommon/locally common.
Tiny, appearing tail-less. Sharp pointed bill; bare red skin around eye; red feet.
3a **Adult male** Orange-red forehead.
3b **Adult female** Olive-green forehead.

Broadbills. Small, flycatcher-like forest birds with broad gapes, large heads and short tails. Possess unique short elliptical display flight accompanied by distinctive rattling. Otherwise very inconspicuous.

4 AFRICAN BROADBILL *Smithornis capensis* 13 cm **R**   **Page 555**
Secondary and gallery forest, thickets. Patchily distributed. Scarce/rare.
4a **Adult male** *camarunensis* Streaked underparts; black crown. White patch on back conspicuous in display flight. E Nigeria to SE CAR and Congo.
    S. c. delacouri is slightly paler; nape greyer. Sierra Leone to Togo.
4b **Adult female** *camarunensis* Crown greyish-brown.

5 RUFOUS-SIDED BROADBILL *Smithornis rufolateralis* 11.5 cm **R**   **Page 554**
Forest interior. Sierra Leone to Togo; S Nigeria to SW CAR–Congo. Uncommon/locally not uncommon.
5a **Adult male** Bright rufous-orange patch on breast-sides; black head; white spots on wing-coverts. White patch on back conspicuous in display flight.
5b **Adult female** Breast patches duller; head brown.

6 GREY-HEADED BROADBILL *Smithornis sharpei* 15 cm **R**   **Page 554**
Lowland and montane forest. E Nigeria to Gabon and N Congo; Bioko. Locally not uncommon/rare.
6a **Adult male** Bright rufous breast; dark grey head with rufous loral spot and malar stripe; undertail-coverts rufous.
6b **Adult female** Pale grey head and washed-out rufous on breast.

Pittas. Brilliantly coloured but very secretive terrestrial forest species with long, strong legs and short tails.

7 AFRICAN PITTA *Pitta angolensis* 17–22 cm **M/R**   **Page 556**
Evergreen and semi-deciduous forest. Patchily distributed, Sierra Leone to CAR and Congo. Rare.
7a **Adult** Unmistakable. Robust and brightly coloured. Cinnamon-buff breast; azure-blue rump.
7b **Juvenile** Duller and darker version of 7a; brownish underparts; no wing markings.
7c **Adult in flight** White wing patch.

8 GREEN-BREASTED PITTA *Pitta reichenowii* 19–21 cm **R**   **Page 555**
Old secondary forest, thickets. S Cameroon and NE Gabon. Rare.
Adult Similar to 7a, but with green breast.

PLATE 79: LARKS I

Terrestrial birds with inconspicuous, cryptic plumage pattern and coloration, often varying within a species to match dominant shade of soil. Fresh and worn plumage often strikingly different: with wear, pale feather tips and fringes abrade, leaving only dark centres (which fade from blackish to brownish), resulting in more uniform plumage.

1 SINGING BUSH LARK *Mirafra cantillans*   13 cm   **R/M**   Page 557
Dry, open grassland, semi-arid thornbush. S Mauritania–Senegambia to Niger, N Cameroon, Chad. Locally not uncommon.
Sandy grey-brown above, streaked dusky; whitish below; pale bill.

2 WHITE-TAILED BUSH LARK *Mirafra albicauda*   13 cm   **R**   Page 557
Open grassland. Chad (L. Chad area). Rare.
Blackish, scaly-looking above; short tail with white outer feathers; rufous wing panel.

3 KORDOFAN LARK *Mirafra cordofanica*   14 cm   **R/M**   Page 557
Sub-desert with red sandy soil. Mauritania–N Senegal to W Niger; also E Chad. Uncommon/rare.
Pale sandy-rufous above; tricoloured tail (rufous, black and white); stout whitish bill.

4 RUSTY BUSH LARK *Mirafra rufa*   13–14 cm   **R**   Page 558
Dry, open, rocky bush. NE Mali–W Niger; Chad. Locally not uncommon. Vagrant, Togo.
Rufous-brown above, variably streaked; no white in comparatively long tail; bicoloured bill.

5 FLAPPET LARK *Mirafra rufocinnamomea*   13–14 cm   **R**   Page 558
Various grassy habitats. S Mauritania–Guinea to W Chad, CAR and Congo. Uncommon/common; often local.
Dry rattling sound in display flight diagnostic. In flight, rufous wing patch. Plumage variable.
5a Adult *tigrina* Rufous above and below. E Cameroon–E CAR.
    M. r. serlei (SE Nigeria) slightly deeper rufous above.
5b Adult *buckleyi* Cinnamon-brown streaked blackish above. From Nigeria west.
    M. r. schoutedeni (Gabon–Congo, S CAR) pale brown, lightly marked above.

6 RUFOUS-NAPED LARK *Mirafra africana*   16–18 cm   **R**   Page 558
Various open habitats. Patchily distributed (see below). Uncommon/common and local.
Short crest. In flight, rufous wing patch, relatively short tail. Plumage variable.
6a Adult *henrici* Upperpart feathers blackish fringed sandy-rufous; boldly streaked breast. Sierra Leone; Mt Nimba.
    M. a. batesi (E Mali, Nigeria, SE Niger) slightly paler.
6b Adult *stresemanni* Rufous-brown lightly streaked black above; rich cinnamon below. N Cameroon.
    M. a. bamendae (W Cameroon) similar below; more like *henrici* above.
6c Adult *malbranti* Dull brown above ; almost unstreaked, buffish, below. SE Gabon–Congo.

7 SUN LARK *Galerida modesta*   14 cm   **R**   Page 562
Various open grassy habitats. S Senegambia–Sierra Leone to S Chad–CAR. Locally not uncommon/common.
Prominent supercilium; streaked cap; heavily streaked upperparts and breast. Plumage variable.
7a Adult *modesta* Palest race; sandy-rufous above.
7b Adult *nigrita* Darkest race.

8 CRESTED LARK *Galerida cristata*   17 cm   **R**   Page 562
Various open habitats. S Mauritania–Guinea-Bissau to Chad and NE CAR. Not uncommon/locally common.
Stocky; spiky crest; relatively short tail; slightly decurved longish bill. Plumage variable.
8a Adult *senegallensis* Pale grey-brown above; off-white below; streaked breast.
8b Adult *jordansi* Most rufous race; indistinct small spots on breast.

9 RUFOUS-RUMPED LARK *Pinarocorys erythropygia*   18 cm   **M**   Page 559
Open savanna woodland, farmland. Breeds N Ivory Coast to S Chad–CAR. Migrates west and north when not breeding, reaching Sierra Leone, N Niger, N Chad. Vagrant, Senegambia.
Large, slender. Rufous rump and tail edges; heavily streaked breast; well-marked head pattern.

10 THEKLA LARK *Galerida thekla*   17 cm   **V**   Page 563
Dry open country. Palearctic vagrant (rare resident?), NW Mauritania.
Very similar to 8; see text for differences.

PLATE 80: LARKS II

1   BAR-TAILED LARK *Ammomanes cincturus*   15 cm   **R/M**   **Page 560**
Desert. Mauritania to Chad; Cape Verde Is. Local; uncommon/common. Erratic.
1a   Adult Plain, sandy-coloured above; pale bill.
1b   In flight Pale orange-rufous tail with black terminal bar; pale orange-rufous in wings.

2   DESERT LARK *Ammomanes deserti*   15–16 cm   **R**   **Page 560**
Arid country. Mauritania to Mali–Niger; Chad. Locally common.
2a   Adult Greyer than 1; bill stouter and darker. Also compare 3.
2b   In flight Orange-rufous tail with broad, ill-defined blackish terminal bar.

3   DUNN'S LARK *Eremalauda dunni*   14 cm   **R**   **Page 561**
Desert and sub-desert. Mauritania, N Mali, C Niger, N Chad. Uncommon/rare.
3a   Adult Resembles 1, but faintly streaked when fresh; bill stouter, stubbier.
3b   In flight Black outer tail contrasting with pale centre. Compare 1b.

4   LESSER SHORT-TOED LARK *Calandrella rufescens*   13 cm   **P**   **Page 561**
Arid, open country. Coastal Mauritania. Rare.
Resembles 5, but with distinct primary projection; variably streaked breast without black patches;
shorter, stubbier bill.

5   GREATER SHORT-TOED LARK *Calandrella brachydactyla*   13–14 cm   **P**   **Page 560**
Arid and semi-arid country. Mauritania–N Senegal to Chad. Uncommon/locally common.
Streaked above; broad pale supercilium; long tertials almost completely cloaking primaries; variable
dark patch on breast-sides.

6   RED-CAPPED LARK *Calandrella cinerea*   14–15 cm   **R/M**   **Page 561**
Grass- and farmland. C Nigeria (Jos). Uncommon and very local. Vagrant, Gabon and Congo.
Distinctive: rufous crown; white supercilium; rufous patch on breast-sides.

7   CHESTNUT-BACKED SPARROW LARK *Eremopterix leucotis*   12 cm   **R/M**   **Page 563**
Semi-arid country. S Mauritania–Senegambia to Chad. Common. Moves north with rains.
7a   Adult male Black head; white patches on ear-coverts and nape; chestnut upperparts; black
      underparts and underwing.
7b   Adult female Dusky mottled head; pale buff below; breast mottled brownish.

8   BLACK-CROWNED SPARROW LARK *Eremopterix nigriceps*   11–12 cm   **R**   **Page 564**
Arid and semi-arid country. Mauritania–Senegal to C Chad; Cape Verde Is. Uncommon/locally
common.
8a   Adult male White forehead; pale grey-brown upperparts; black underparts.
8b   Adult female Pale sandy-cinnamon above; whitish below; pale cinnamon breast-band.
8c   Adult female in flight Black underwing-coverts diagnostic.

9   TEMMINCK'S HORNED LARK *Eremophila bilopha*   14 cm   **P/R?**   **Page 564**
Desert and sub-desert. NW Mauritania. Rare.
Pale sandy; contrasting face pattern; thin horns on crown-sides.

10   GREATER HOOPOE LARK *Alaemon alaudipes*   18–20 cm   **R/M**   **Page 559**
Sandy desert. Mauritania to Chad; Cape Verde Is. Common/uncommon. Migrant, N Senegal.
10a   Adult Large; sandy; long decurved bill; long legs.
10b   In flight Black-and-white wing pattern diagnostic.

11   THICK-BILLED LARK *Rhamphocoris clotbey*   17 cm   **P**   **Page 559**
Stony sub-desert. NW Mauritania. Rare.
11a   Adult Large, stubby bill; heavily spotted underparts.
11b   In flight Large head; long wings with black-and-white pattern; relatively short tail.

EURASIAN SKYLARK *Alauda arvensis*   18–19 cm   **V**   **Page 563**
Palearctic vagrant, NW Mauritania. **See illustration on plate 147.**
Robust; streaked; short crest. In flight, white trailing edge to wing and white outer tail feathers.

PLATE 81: SWALLOWS I

Distinctive, highly specialised aerial insectivores with slender body, short neck and pointed wings. Bill very short, but gape broad; legs short. Regularly perch on wires and bare branches. Most gregarious when not breeding (some also when breeding), often forming mixed groups with other hirundines. Many are migratory.

1   ROCK MARTIN *Hirundo fuligula* 12–13 cm **R/P**   **Page 572**
Rocky outcrops, cliffs, gorges, villages, towns. Widespread but patchily distributed. Locally not uncommon/common.
- **1a**   **Adult tropical W African races** Dark brown above, paler below. Almost throughout, except where 1c.
- **1b**   **In flight** White patches in tail; no contrasting wing-coverts.
- **1c**   **Adult *obsoleta* group** (Pale Crag Martin) Paler and greyer than 1a. N Mauritania, Mali, N Niger (Aïr), NE Chad (Ennedi).
- **1d**   **In flight** Resembles 2, but much less contrasting underwing-coverts.

2   CRAG MARTIN *Hirundo rupestris* 14–15 cm **P**   **Page 572**
Open habitats. Rare, Mauritania, N Senegal, Mali. Vagrant, Gambia.
- **2a**   **Adult** Grey-brown above; pale buffish below, brownish on lower abdomen.
- **2b**   **In flight** Dark underwing-coverts strongly contrasting with paler flight feathers.

3   COMMON SAND MARTIN *Riparia riparia* 12–13 cm **P**   **Page 567**
Open habitats. Almost throughout, esp. Sahel zone; uncommon/rare to locally common. Vagrant, NE Gabon and NW Congo.
- **3a**   **Adult** Brown above; pure white below; well-defined brown breast-band.
- **3b**   **In flight** White underparts with contrasting breast-band; tail slightly forked.

4   CONGO SAND MARTIN *Riparia congica* 11–12 cm **R**   **Page 567**
Rivers. Congo (Congo R., lower Ubangi R., Sangha R.). Locally common.
- **4a**   **Adult** Resembles 3, but brown breast-band much more diffuse.
- **4b**   **In flight** White underparts with diffuse breast-band; tail almost square.

5   BANDED MARTIN *Riparia cincta* 17 cm **M/R/V**   **Page 568**
Grassy savanna. Patchily distributed; mainly S Congo to N Ivory Coast. Uncommon/rare.
- **5a**   **Adult** Large; brown above, white below; brown breast-band; short white supercilium.
- **5b**   **In flight** Large size conspicuous; slow and erratic, tern-like flight.

6   PLAIN MARTIN *Riparia paludicola* 11–12 cm **R/M**   **Page 567**
Sahel and savanna zones. Widespread but patchily distributed. Locally not uncommon/rare. Vagrant, Ivory Coast, N Congo, Cape Verde Is.
- **6a**   **Adult** Brown above; brownish throat and breast.
- **6b**   **In flight** Greyish-brown throat and breast; rest of underparts white.

7   BRAZZA'S MARTIN *Phedina brazzae* 12 cm **R/M?** DD   **Page 566**
Forested rivers. Congo (from Congo R. to Djambala); probably also SE Gabon. Rare.
- **7a**   **Adult** Dark brown above; white below, heavily streaked dark brown.
- **7b**   **In flight** Streaked underparts; dark brown underwing; square tail.

8   AFRICAN RIVER MARTIN *Pseudochelidon eurystomina* 14 cm **R/M** DD   **Page 565**
Rivers, sandy grassland. Breeds coastal Gabon and Congo, middle Congo R. and lower Ubangi R. Migrates over forested C & NE Gabon.
- **8a**   **Adult** All black; reddish bill and eye; large head.
- **8b**   **In flight** Compact, triangular silhouette produced by broad-based wings and shortish tail.

9   WHITE-THROATED BLUE SWALLOW *Hirundo nigrita* 12 cm **R**   **Page 573**
Forested rivers. Sierra Leone–SE Guinea to SW CAR–Congo. Locally common.
- **9a**   **Adult** Dark glossy purple-blue; white throat.
- **9b**   **In flight** White patches in tail. Flies low and fast over water.

PLATE 82: SWALLOWS II

1 **BARN SWALLOW** *Hirundo rustica*   15–19 cm   **P**   **Page 575**
Various habitats. Throughout; also Cape Verde and Gulf of Guinea Is. Common.
  1a **Adult breeding** Dark rufous forehead and throat; broad dark blue breast-band. Compare 4.
  1b **Adult non-breeding** Duller; rufous on throat may become very pale; no tail streamers.
  1c **Adult breeding in flight** Long tail streamers; whitish underparts.

2 **WHITE-THROATED SWALLOW** *Hirundo albigularis*   14–17 cm   **V**   **Page 574**
Open habitats. Vagrant, N Congo.
  2a **Adult** Rufous forehead; white throat; blue breast-band (sometimes broken).
  2b **In flight** Dark breast-band contrasting with white throat diagnostic. Compare 6 and 7.

3 **GREY-RUMPED SWALLOW** *Pseudhirundo griseopyga*   14 cm   **R/M**   **Page 568**
Grassy and wooded savanna. Widespread but patchily distributed. Rare/uncommon.
  3a **Adult** Small; dark above; white below; long tail streamers.
  3b **In flight from above** Grey rump contrasting with dark blue upperparts.
  3c **In flight from below** No white in tail.

4 **RED-CHESTED SWALLOW** *Hirundo lucida*   15 cm   **R/M**   **Page 574**
Open habitats. S Mauritania–Liberia to SW Niger–N Togo; N Gabon. Uncommon/locally common.
  4a **Adult** As 1a, but rufous reaches to upper breast; breast-band narrower.
  4b **In flight** As 1c, but tail streamers shorter; more white in tail; underparts whiter.

5 **ANGOLA SWALLOW** *Hirundo angolensis*   14–15 cm   **R/M?**   **Page 574**
Open habitats. W & S Gabon, S Congo. Uncommon.
  5a **Adult** As 4a, but with ashy underparts.
  5b **In flight** Dull ashy below, including underwing-coverts. Flight rather slow.

6 **ETHIOPIAN SWALLOW** *Hirundo aethiopica*   13 cm   **R/M**   **Page 573**
Open habitats. Widespread but patchily distributed. Uncommon/locally common.
  6a **Adult** Rufous forehead; buff-white throat; dark blue patch on breast-sides. Compare 7.
  6b **In flight** White below; no breast-band; long tail streamers.

7 **WIRE-TAILED SWALLOW** *Hirundo smithii*   14 cm   **R/M**   **Page 573**
Open habitats, usually near water. Savanna belt; also NE Gabon–Congo. Uncommon.
  7a **Adult** Rufous cap; white underparts; no breast-band.
  7b **In flight** White below; dark blue patches on lower flanks; wire-like tail streamers.

8 **LESSER STRIPED SWALLOW** *Hirundo abyssinica*   15–19 cm   **R/M**   **Page 569**
Open wooded habitats. Almost throughout. Not uncommon/scarce to locally common.
  8a **Adult** *puella* White underparts finely streaked dark. Senegambia to N Cameroon.
    In *bannermani* streaking much finer; rufous areas paler. NE CAR.
  8b **Adult** *maxima* Streaks broader and blacker. SE Nigeria, S Cameroon to SW CAR.
    In *unitatis* streaking intermediate between 8a and 8b. Gabon to Congo.
  8c **In flight from above** Rufous head and rump.
  8d **In flight from below** Streaked underparts; long tail streamers.

  GREATER STRIPED SWALLOW *Hirundo cucullata*   18–20 cm   **V**   **Page 569**
  S African vagrant, S Congo. **See illustration on plate 147.**

9 **RED-RUMPED SWALLOW** *Hirundo daurica*   16–17 cm   **R/M/P**   **Page 570**
  9a **Adult** *domicella* Dark blue cap; rufous collar; creamy-white below. Savanna. Senegambia to Chad.
  9b **Adult** *domicella* **in flight from above** Rufous rump; long tail streamers, often curving inwards.
  9c **Adult** *domicella* **in flight from below** Clear-cut black undertail-coverts; no white in tail.
  9d **Adult** *kumboensis* Pale rufous underparts. Highlands. Sierra Leone and Cameroon.
  9e **Adult** *rufula* Creamy-buff underparts with faint streaks. Palearctic migrant in north.
  9f **Adult** *rufula* **in flight from above** Pale forehead; two-toned rump.

10 **MOSQUE SWALLOW** *Hirundo senegalensis*   21–23 cm   **R/M**   **Page 569**
Open habitats. Almost throughout, except arid north. Rare/locally common.
  10a **Adult** Dark blue cap; rufous collar; pale throat; orange-rufous underparts.
  10b **In flight from above** Large; orange-rufous rump; tail streamers shorter than in 11.
  10c **In flight from below** Rufous undertail-coverts; no white in tail.

11 **RUFOUS-CHESTED SWALLOW** *Hirundo semirufa*   18–21 cm   **R/M**   **Page 568**
Open habitats. Almost throughout, except arid north. Uncommon/locally common.
  11a **Adult** Dark blue hood; no complete collar; orange-rufous underparts.
  11b **In flight from above** Orange-rufous rump; very long tail streamers.
  11c **In flight from below** Orange-rufous undertail-coverts; white patches in tail.

PLATE 83: SWALLOWS III

1 SOUTH AFRICAN CLIFF SWALLOW *Hirundo spilodera*   14 cm   **M**   **Page 572**
Open habitats. C & SE Gabon, Congo. Rare.
1a **Adult** Very dark above; pale rufous/buffish-white below; speckled throat and breast.
1b **In flight from above** Rufous rump; squarish tail.
1c **In flight from below** Throat and breast often appearing darkish; no white in tail.

2 RED-THROATED CLIFF SWALLOW *Hirundo rufigula*   12 cm   **R/M**   **Page 571**
Various habitats, often near rocks and rivers. Gabon, Congo. Locally not uncommon.
2a **Adult** Rufous throat; streaky ear-coverts.
2b **Juvenile** Duller; browner above; rufous areas paler.
2c **Adult in flight from above** Rufous rump.
2d **Adult in flight from below** Squarish tail with white patches.
2e **Juvenile in flight from above** Duller than adult.

3 FOREST SWALLOW *Hirundo fuliginosa*   11 cm   **R/M?**   **Page 571**
Forest. Locally not uncommon, E Nigeria, Cameroon, Eq. Guinea, Gabon. Vagrant, Congo. Endemic.
3a **Adult** Small; dull blackish-brown; rusty tinge on throat.
3b **In flight** All dark; slightly notched tail. Flight fast with much gliding. Compare 5.

4 PREUSS'S CLIFF SWALLOW *Hirundo preussi*   12 cm   **R/M**   **Page 571**
Savanna. Widespread but irregularly distributed. Rare/locally common.
4a **Adult** Blue-black above; pale buff below; small rufous patch behind eye.
4b **Juvenile** Duller; dark brown above; throat and breast washed brownish.
4c **Adult in flight from above** Pale buff rump. Resembles dull or dirty version of 7.
4d **Adult in flight from below** Pale buff; small white patches in tail.
4e **Juvenile in flight from above** Duller than adult.

5 SQUARE-TAILED SAW-WING *Psalidoprocne nitens*   11 cm   **R**   **Page 565**
Forest. Guinea–Sierra Leone to Togo; SE Nigeria to Congo. Common to uncommon/rare.
5a **Adult** Wholly black.
5b **In flight** Notched tail. Flight slow and fluttering. Compare 3.

6 PIED-WINGED SWALLOW *Hirundo leucosoma*   12 cm   **R**   **Page 573**
Wooded savanna, farmbush. Irregularly distributed. Senegambia–Sierra Leone to SW Niger and Nigeria;
scarce/locally not uncommon. Vagrant, W Cameroon.
6a **Adult** Glossy steel-blue above; white below; elongated white wing patches.
6b **In flight from above** Long white patches on inner wing (diagnostic).
6c **In flight from below** White; forked tail with white patches.

7 COMMON HOUSE MARTIN *Delichon urbica*   13–15 cm   **P**   **Page 575**
Open habitats. Throughout. Uncommon/locally common.
7a **Adult** Glossy blue-black above; white below.
7b **In flight from above** Broad white rump (often mottled pale buff-brown in non-breeding).
7c **In flight from below** White; forked tail.

8 MOUNTAIN SAW-WING *Psalidoprocne fuliginosa*   12 cm   **R**   **Page 566**
Forest clearings, montane grassland. Mt Cameroon, Bioko. Locally common. Endemic.
8a **Adult** Dull dark brown; tail fork *c.* 1.5–2.5 cm.
8b **In flight** Grey-brown underwing-coverts.

9 BLACK SAW-WING *Psalidoprocne pristoptera*   13 cm   **R/M**   **Page 566**
Savanna, montane grassland, farmbush. Common/uncommon.
9a **Adult** *petiti* Wholly black with bronze-brown gloss; long tail streamers; tail fork *c.* 2.5–3.5 cm.
E Nigeria to Congo.
9b **Adult** *petiti* **in flight** Whitish underwing-coverts.
9c **Adult** *chalybea* Greenish gloss. N CAR; possibly also N & C Cameroon.
9d **Adult** *chalybea* **in flight** Grey underwing-coverts.

10 FANTI SAW-WING *Psalidoprocne obscura*   17 cm   **R/M**   **Page 565**
Forest zone. Common, Senegambia to W Cameroon. Vagrant, SE Burkina Faso, SW Niger.
10a **Adult** Wholly black; very long tail streamers; tail fork male 5.0–7.5 cm, female 3.0–6.5 cm.
10b **In flight** Black underwing.

PLATE 84: PIPITS

Terrestrial insectivores with rather cryptic, variably streaked brown plumage. Slightly resemble larks, but slimmer. Large species conspicuously long legged. Non-breeding plumage usually similar to breeding; pale tips and fringes to wing feathers broader when fresh. For identification, note extent of streaking on upper- and underparts, colour of outer tail feathers, and calls.

1 TREE PIPIT *Anthus trivialis*   15 cm   **P**   Page 581
Various open wooded habitats. Throughout, south to Gabon and Congo. Common/uncommon.
Vagrant, Cape Verde Is and São Tomé.
Boldly streaked on breast, thinly on flanks; plain rump. Call a single *tzeep* in flight.

2 MEADOW PIPIT *Anthus pratensis*   14.5 cm   **P**   Page 581
Various open habitats. Mauritania. Uncommon/rare.
Bold streaking on breast extending onto flanks; plain rump. Call a sharp *eest-eest-eest*.

3 RED-THROATED PIPIT *Anthus cervinus*   15 cm   **P**   Page 581
Mainly Sahel and savanna belts. Common/scarce. Vagrant, Gabon, Congo, Cape Verde Is.
  3a **Adult breeding** Distinctive brick-red or orange-buff face and throat.
  3b **Adult non-breeding** Boldly streaked above and below; three lines of streaks extending onto
      flanks; streaked rump; moustachial stripe ending in blotch. Call *speeeh*, less rasping than 1.

4 SHORT-TAILED PIPIT *Anthus brachyurus*   12–13 cm   **R**   Page 580
Open grassland. SE Gabon, Congo. Uncommon and local.
Smallest pipit. Heavily streaked; short narrow tail. Skulking; difficult to flush.

5 TAWNY PIPIT *Anthus campestris*   16–17 cm   **P**   Page 579
Sahel and northern savanna belts. Rare/uncommon. Vagrant, Cape Verde Is. Breeding suspected, SW
Mauritania.
Palest, sandy-coloured pipit; contrasting dark centres to median wing-coverts.

6 PLAIN-BACKED PIPIT *Anthus leucophrys*   17 cm   **R**   Page 580
Various open habitats. Widespread. Uncommon/locally common.
  6a **Adult *zenkeri*** Plain dark brown above; warm cinnamon-buff below; buff outer tail. S Mali to
      Ghana east.
      A. l. bohndorffi similar, but paler below. SE Gabon, lower Congo R.
  6b **Adult *gouldii*** Darker above; much paler below; distinct streaks on breast. West of *zenkeri*.
      A. l. ansorgei similar, but greyer above. S Mauritania–Guinea-Bissau.

7 LONG-BILLED PIPIT *Anthus similis*   17–19 cm   **R**   Page 579
Rocky slopes. Widespread, with disjunct range. Uncommon/locally common.
  7a **Adult *bannermani*** Heavily streaked above and below; streaks extending onto flanks; buffish-white
      outer tail. Highlands, Sierra Leone, Mt Nimba, SW Mali, SE Ghana, Nigeria, W Cameroon.
  7b **Adult *asbenaicus*** Paler, more sandy-buff; almost no streaks on underparts. Niger (Aïr), C & E Mali.

8 WOODLAND PIPIT *Anthus nyassae*   17–18 cm   **R**   Page 579
Wooded grassland; grassy hills with exposed rocks. SE Gabon, S Congo. Locally common.
Adult *schoutedeni* Streaking above more diffuse than 7a; below, only streaked on breast.

9 GRASSLAND PIPIT *Anthus cinnamomeus*   16–18 cm   **R/P**   Page 578
Various open habitats, incl. grassland and cultivation. Widespread; patchily distributed.
Adult *lynesi* Large; streaked upperparts and breast; bold face pattern; pale lores; buff-white outer tail.
Highlands, SE Nigeria, W Cameroon. Locally common. Vagrant, N Nigeria, W Chad.
A. r. camaroonensis (Mt Cameroon, Mt Manenguba) similar, but paler below.
Race of those in Liberia, Ghana, NE Nigeria and W Chad unknown.

  RICHARD'S PIPIT *Anthus richardi*   17–20 cm   **P?**   Page 578
Rare Palearctic passage migrant and winter visitor, Mauritania and Mali? **See illustration on plate 147**.
Larger than 9; wing edgings more sandy-buff; underparts whiter; outer tail feathers pure white.

10 LONG-LEGGED PIPIT *Anthus pallidiventris*   *c.* 18 cm   **R**   Page 580
Various grassy habitats. SW Cameroon to Congo. Locally common.
Large; long legs; relatively short tail; prominent supercilium; plain greyish-brown above; buff
outer tail; indistinct streaks on breast.

PLATE 85: WAGTAILS AND LONGCLAWS

Wagtails. Small, slender, terrestrial insectivores with long, frequently pumped tails.

1   WHITE WAGTAIL *Motacilla alba*   18 cm   **P**   **Page 577**
Various open habitats. Mauritania–Senegambia to C Chad; also Sierra Leone, Cape Verde Is. Common/
uncommon. Vagrant elsewhere.
  1a   Adult male breeding Black throat; grey upperparts.
  1b   Adult male non-breeding White face and throat; black breast-band.

2   AFRICAN PIED WAGTAIL *Motacilla aguimp*   19 cm   **R**   **Page 578**
Near water and habitation. SE Senegal–Sierra Leone and S Mali to S Chad, CAR, Congo. Common/
uncommon. Vagrant, Gambia, Guinea-Bissau, São Tomé.
  2a   Adult Black and white; large white wing patch.
  2b   Juvenile Much duller; black replaced by dark grey-brown.

3   GREY WAGTAIL *Motacilla cinerea*   18–19 cm   **P/V**   **Page 577**
Various habitats, normally near water. Mauritania, Senegambia, Mali, W Niger. Generally rare. Vagrant,
Cameroon, Eq. Guinea, N CAR.
  3a   Adult male breeding Very long tail; black throat; wholly yellow below.
  3b   Adult male non-breeding Grey above; yellow on rump and vent.

4   MOUNTAIN WAGTAIL *Motacilla clara*   18–19 cm   **R**   **Page 577**
Along swift, rocky forest streams. Patchily distributed; also Bioko. Scarce/uncommon.
Grey above; white below; narrow blackish breast-band.

5   YELLOW WAGTAIL *Motacilla flava*   16–17 cm   **P**   **Page 576**
Various, generally open, habitats. Throughout. Common/very common.
Illustrated forms are all in breeding plumage. Non-breeding similar to female *flava* but duller.
  5a   Adult male *flava* Blue-grey head; white supercilium; wholly yellow underparts.
  5b   Adult female *flava* Duller; less distinct head pattern; paler below.
  5c   Adult male *flavissima* Yellow-green head.
  5d   Adult male *beema* Pale blue-grey head (paler than 5a); throat sometimes white.
  5e   Adult male *iberiae* Dark grey head; narrow supercilium; white throat.
  5f   Adult male *cinereocapilla* Similar to 5e, but supercilium absent or vestigial.
  5g   Adult male *thunbergi* Top of head slate-grey; head-sides blackish; supercilium indistinct, often
       absent.
  5h   Adult male *melanogrisea* Black head bordered below by white line.
  5i   Adult male *feldegg* As 5h, but without white line.

CITRINE WAGTAIL *Motacilla citreola*   17–18 cm   **V**   **Page 576**
Palearctic vagrant, Senegal. **See illustration on plate 147**.
Resembles 5 but upperparts grey; two conspicuous white or whitish wingbars.

Longclaws. Robust ground-dwellers, with relatively short tail, strong legs and long claw on hind toe.

6   YELLOW-THROATED LONGCLAW *Macronyx croceus*   20–22 cm   **R**   **Page 582**
Various grassy habitats. Throughout, except arid north. Locally common/scarce.
  6a   Adult Yellow below; black necklace.
  6b   Juvenile Duller; necklace buff with dark markings.
  6c   In flight White tail corners.

PLATE 86: CUCKOO-SHRIKES

Unobtrusive arboreal birds. Mainly singly or in pairs, quietly gleaning foliage in search of insects.

1 **BLUE CUCKOO-SHRIKE** *Coracina azurea* 21 cm **R**   **Page 584**
Rainforest. Sierra Leone–SE Guinea to Togo; Nigeria to SW CAR–Congo. Uncommon/rare to common.
1a **Adult male** Deep glossy blue; black face; reddish eye.
1b **Adult female** Duller; less black on face.

2 **GREY CUCKOO-SHRIKE** *Coracina caesia* 23 cm **R**   **Page 584**
Montane forest. SE Nigeria–W Cameroon, Eq. Guinea; not uncommon. Also Bioko; rare.
2a **Adult male** Slate-grey; black lores.
2b **Adult female** No black on lores.
2c **Juvenile** Barred dusky and whitish above and below.

3 **WHITE-BREASTED CUCKOO-SHRIKE** *Coracina pectoralis* 25 cm **R**   **Page 584**
Savanna woodland. Senegambia–Sierra Leone to S Chad–CAR. Uncommon/not uncommon.
3a **Adult male** Pale grey above; lower breast to undertail-coverts white; black lores.
3b **Adult female** Lores paler; throat whitish bordered paler grey.

4 **RED-SHOULDERED CUCKOO-SHRIKE** *Campephaga phoenicea* 20 cm **R/M**   **Page 582**
Forest patches, gallery forest, woodland. Broad savanna belt, S Mauritania–Sierra Leone to S Chad–N Congo. Uncommon/locally not uncommon.
4a **Adult male** Glossy blue-black; scarlet shoulder patch.
4b **Adult male** Orange-shouldered variant. Golden-yellow shoulders also recorded.
4c **Adult female** Greyish olive-brown above; white barred black below.

5 **PURPLE-THROATED CUCKOO-SHRIKE** *Campephaga quiscalina* 20 cm **R**   **Page 583**
Evergreen and gallery forest, forest patches. Patchily distributed, Sierra Leone–SE Guinea to Benin; Cameroon to Congo. Also S Mali, C Nigeria, SW CAR. Uncommon/rare.
5a **Adult male** Black glossed blue-green above; head-sides to breast glossed purple.
5b. **Adult female** Grey head; plain olive upperparts; bright yellow underparts.
5c **Juvenile** Variably barred above and below.

6 **PETIT'S CUCKOO-SHRIKE** *Campephaga petiti* 20 cm **R**   **Page 583**
Montane forest, gallery forest, forest patches. SE Nigeria–Cameroon and S Gabon–Congo. Uncommon.
6a **Adult male** Entirely glossy blue-black; yellow or orange gape wattles usually conspicuous.
6b **Adult female** Mainly yellow; heavily barred blackish above; usually some bars on breast.
6c **Juvenile** More olive above than adult female; spotted below.

7 **EASTERN WATTLED CUCKOO-SHRIKE** *Lobotos oriolinus* 19 cm **R** DD   **Page 584**
Forest. SE Nigeria, S Cameroon, SW CAR, Gabon, N Congo. Very rare.
7a **Adult male** Glossy blue-black head; large gape wattles; yellow tinged orange below.
7b **Adult female** Duller; no orange tinge below.

8 **WESTERN WATTLED CUCKOO-SHRIKE** *Lobotos lobatus* 19 cm **R** VU   **Page 583**
Forest. E Sierra Leone–SE Guinea to Ghana. Rare. Endemic.
8a **Adult male** As 7a but orange-chestnut suffused with yellow below.
8b **Adult female** Yellow below.

PLATE 87: BULBULS I – *Andropadus*

Andropadus bulbuls are small to medium-sized forest inhabitants with mainly featureless olive-green plumage; underparts paler. Essentially frugivorous, hence rarely with mixed-species flocks. Vocalisations an important aid to identification.

1 LITTLE GREENBUL *Andropadus virens* 16.5 cm **R**   Page 586
Forest zone and forest–savanna mosaic. Senegambia to SW Chad, CAR and Congo; Bioko. Very common.
Wholly dirty olive-green without distinguishing markings.

2 LITTLE GREY GREENBUL *Andropadus gracilis* 16 cm **R**   Page 586
Forest zone. W Guinea to SW CAR–Congo. Not uncommon/common.
Small. Head and throat olive-grey; underparts yellowish-olive; narrow white eye-ring.

3 CAMEROON SOMBRE GREENBUL *Andropadus curvirostris* 17 cm **R**   Page 587
Forest zone. Sierra Leone–SE Guinea to SW CAR–Congo; Bioko. Not uncommon/common.
Very similar to 1; voice best distinction. Usually with narrow, broken, white eye-ring.

4 ANSORGE'S GREENBUL *Andropadus ansorgei* 16 cm **R**   Page 587
Forest zone. W Guinea to SW CAR–Congo. Uncommon/locally common.
As 2 but underparts lack any yellow; flanks and vent ginger.

5 CAMEROON MONTANE GREENBUL *Andropadus montanus* 18 cm **R NT**   Page 585
Montane forest. SE Nigeria–W Cameroon. Scarce/locally common. Endemic.
Wholly olive-green.

6 WESTERN MOUNTAIN GREENBUL *Andropadus tephrolaemus* 18 cm **R**   Page 586
Montane forest. SE Nigeria–W Cameroon and Bioko. Common. Endemic.
Grey head and throat; bright olive-green upperparts; yellowish-olive underparts.

7 SLENDER-BILLED GREENBUL *Andropadus gracilirostris* 18 cm **R**   Page 587
Forest zone and adjacent woodland. SW Senegal–Liberia to S CAR–Congo; Bioko. Common.
Olive-brown above; brownish-grey below.

8 YELLOW-WHISKERED GREENBUL *Andropadus latirostris* 17 cm **R**   Page 588
Forest zone and forest–savanna mosaic. SW Senegal, W Guinea to S CAR–Congo; Bioko. Common/very common.
8a Adult Wholly olive-green; bright yellow 'whiskers'; pinkish legs.
8b Immature No or only small 'whiskers'; leg colour as adult. Compare 1.

PLATE 88: BULBULS II – *Pycnonotus, Chlorocichla, Pyrrhurus, Thescelocichla* and *Baeopogon*

1 COMMON BULBUL *Pycnonotus barbatus* 18–20 cm **R**   **Page 596**
Almost all habitats except closed forest and treeless desert; very common.
Mainly brown, darker on head, paler below. Vocal and conspicuous.
  1a *P. b. inornatus* Dark brown breast merges gradually with paler belly. Almost entire region.
  1b *P. b. arsinoe* Breast well demarcated from pale belly. E Chad.
  1c *P. b. tricolor* Undertail-coverts bright yellow. E Cameroon and Congo.
    In *P. b. gabonensis* undertail-coverts are white tinged with variable amount of yellow. C Nigeria and
    C Cameroon to Gabon and S Congo.

2 YELLOW-THROATED LEAFLOVE *Chlorocichla flavicollis* 22.5 cm **R**   **Page 590**
Woodland, thicket, swamp forest. Not uncommon/scarce.
Large. Mainly dark brown.
  2a *C. f. flavicollis* Yellow throat. Senegambia to Cameroon.
  2b *C. f. soror* White throat. Cameroon to CAR–Congo.

3 YELLOW-NECKED GREENBUL *Chlorocichla falkensteini* 18 cm **R**   **Page 589**
Farmbush, forest patches and edges. S Cameroon–Gabon, SW CAR and S Congo. Locally common/rare.
Bright yellow throat contrasting with olive-green upperparts.

4 SIMPLE LEAFLOVE *Chlorocichla simplex* 21 cm **R**   **Page 589**
Forest zone and forest–savanna mosaic; Guinea-Bissau to SW CAR–Congo. Locally common.
White throat; broken white eye-ring.

5 LEAFLOVE *Pyrrhurus scandens* 22 cm **R**   **Page 590**
Forest zone and forest–savanna mosaic. Senegambia to CAR–Congo. Scarce/locally common.
Pale grey head; pale chestnut tail. Vocal and gregarious.

6 SWAMP PALM BULBUL *Thescelocichla leucopleura* 23 cm **R**   **Page 590**
Forest zone and gallery extensions. Senegambia and Guinea-Bissau to SW CAR–Congo. Not
uncommon/locally common.
Large. Creamy belly; outer tail feathers boldly tipped white. Vocal. Partial to palm trees.

7 HONEYGUIDE GREENBUL *Baeopogon indicator* 19 cm **R**   **Page 588**
Forest zone and forest–savanna mosaic. Guinea to S CAR–Congo. Common/not uncommon.
Dark and stocky. Outer tail feathers white tipped blackish (in adult) or wholly white (in immature),
recalling honeyguide (Plate 75).
  7a **Adult male** Whitish eye.
  7b **Adult female** Dark eye.

8 SJÖSTEDT'S HONEYGUIDE GREENBUL *Baeopogon clamans* 19 cm **R**   **Page 589**
Lowland forest. SE Nigeria to SW CAR–Congo. Locally common.
As 7 but underparts paler and buffier; white outer tail feathers lack dark tips.

PLATE 89: BULBULS III – *Phyllastrephus, Calyptocichla, Ixonotus, Neolestes* and *Nicator*

Phyllastrephus bulbuls are generally small to medium-sized, with long, slender bills and russet tails. In family parties in lower and mid-strata of forest. Insectivorous; frequent members of mixed-species flocks. No arresting vocalisations.

1 ICTERINE GREENBUL *Phyllastrephus icterinus* 15–16 cm **R**   **Page 592**
Forest. Sierra Leone–SE Guinea to SW CAR–Congo; Bioko. Not uncommon/very common.
Small. Olive-green above; drab yellowish below; dull rufous tail. Gregarious.

2 XAVIER'S GREENBUL *Phyllastrephus xavieri* 16–18 cm **R**   **Page 592**
Forest. SE Nigeria to Gabon, SW CAR, Congo. Uncommon to locally common.
Extremely similar to 1 but voice diagnostic (see text). Males slightly larger.

3 GOLDEN GREENBUL *Calyptocichla serina* 18 cm **R**   **Page 588**
Forest. Sierra Leone–SE Guinea to SW CAR–Congo; Bioko. Rare/locally not uncommon.
Bright olive-green above; mainly yellow below; pale pinkish bill. Forest canopy.

4 WHITE-THROATED GREENBUL *Phyllastrephus albigularis* 17 cm **R**   **Page 593**
Forest. Sierra Leone–SE Guinea to SW CAR–Congo, also SW Senegal, SE CAR. Scarce/uncommon to locally common.
White throat contrasts with grey head and olive-grey breast; pale yellow belly.

5 BAUMANN'S GREENBUL *Phyllastrephus baumanni* 18 cm **R** **DD**   **Page 591**
Semi-deciduous mid-altitude forest, gallery forest, thicket. Sierra Leone to Togo, Nigeria. Uncommon/rare. Endemic.
Olive-brown above; pale olive-grey below; rusty tail.

6 CAMEROON OLIVE GREENBUL *Phyllastrephus poensis* 18 cm **R**   **Page 591**
Montane forest. SE Nigeria–W Cameroon; Bioko. Locally common endemic.
Rather nondescript. Brown above; rather long rufous tail; pale below with whitish throat.

7 PALE OLIVE GREENBUL *Phyllastrephus fulviventris* 19 cm **R**   **Page 591**
Gallery forest and dense bush. Coastal S Gabon and (possibly) S Congo.
Olive-brown above; dull rufous tail; creamy-white throat does not contrast with rest of underparts.

8 LIBERIAN GREENBUL *Phyllastrephus leucolepis* *c.* 16 cm **R** **CR**   **Page 592**
Rainforest. Liberia. Very rare and local. Endemic.
As 1 but with whitish spots on upperparts.

9 SPOTTED GREENBUL *Ixonotus guttatus* 17 cm **R**   **Page 589**
Forest and forest edge. Sierra Leone–SE Guinea to SW CAR–Congo. Locally common/scarce.
Small. Upperparts spotted white; underparts and outer tail feathers white. Gregarious.

10 GREY-HEADED GREENBUL *Phyllastrephus poliocephalus* 20–23 cm **R** **NT**   **Page 593**
Sub-montane forest. SE Nigeria–W Cameroon. Common endemic.
Large. Grey head; olive-green upperparts; white throat; bright yellow underparts.

11 BLACK-COLLARED BULBUL *Neolestes torquatus* 16 cm **R**   **Page 597**
Lightly wooded savanna. SE Gabon and Congo. Rather uncommon.
Grey top of head; olive-green upperparts and tail; whitish underparts with black breast-band.

12 YELLOW-THROATED NICATOR *Nicator vireo* 17–19 cm **R**   **Page 597**
Forest. Cameroon to SW CAR–Congo. Not uncommon.
As 13 but smaller and with yellow throat; yellow supraloral streak; grey head-sides.

13 WESTERN NICATOR *Nicator chloris* 21–24 cm **R**   **Page 597**
Forest and gallery extensions. Guinea to CAR–Congo; common/scarce. Also S Senegambia (rare).
Large. Olive-green above with bold golden spots; pale grey below; ashy-white throat. Heavy bill.

PLATE 90: BULBULS IV – *Criniger* and *Bleda*

Criniger **Bearded bulbuls.** Medium-sized to large species, distinguished by long white or yellow throat feathers, which are frequently puffed out, forming a 'beard'. Inhabit lower and mid-strata of closed forest, usually occurring in pairs or small groups and frequently joining mixed-species flocks. Vocal.

1   RED-TAILED GREENBUL *Criniger calurus* 20 cm **R**   **Page 595**
Forest. Widespread; not uncommon/common.
   1a   *C. c. verreauxi* Grey head; white 'beard'; central breast and belly yellow; olive-green tail.
         SW Senegal to SW Nigeria; also SW Mali.
   1b   *C. c. calurus* Dull rufous tail. Compare 3 and 5. S Nigeria to SW CAR–Congo; Bioko.

2   WESTERN BEARDED GREENBUL *Criniger barbatus* 22 cm **R**   **Page 595**
Forest. Not uncommon/common.
   2a   *C. b. barbatus* Olive-brown head and upperparts; yellow 'beard'; mottled grey and olive breast; dirty
         olive-yellow belly. Sierra Leone–SE Guinea to W Togo.
   2b   *C. b. ansorgeanus* White chin; very pale yellow throat. S Nigeria.

3   WHITE-BEARDED GREENBUL *Criniger ndussumensis* 18 cm **R**   **Page 596**
Forest. SE Nigeria to SW CAR–Congo. Uncommon to locally common.
Indistinguishable in the field from 1b, but voice as 4.

4   YELLOW-BEARDED GREENBUL *Criniger olivaceus* 18 cm **R** VU   **Page 596**
Forest. E Sierra Leone to S Ghana; also SW Mali. Rare/locally not uncommon. Endemic.
Olive-green head and upperparts; bright yellow throat; olive-green breast and flanks; yellowish belly.
Compare 2a and Icterine Greenbul (Plate 89: 1).

5   EASTERN BEARDED GREENBUL *Criniger chloronotus* 22 cm **R**   **Page 595**
Forest. SE Nigeria to SW CAR–Congo. Common.
Grey head and breast; white 'beard'; bright rufous tail; yellow on underparts pale, confined to belly.
Compare 1b.

Bleda **Bristlebills.** Large, fairly shy skulkers of forest undergrowth, with olive-green upperparts, yellow underparts, and stout bills and legs. Join mixed-species flocks and regularly attend ant columns. Vocal.

6   LESSER BRISTLEBILL *Bleda notata* 19.5–21.0 cm **R**   **Page 594**
Forest. SE Nigeria to SW CAR–Congo; also SE CAR and Bioko. Common.
Entirely olive-green above; yellow spot in front of eye; four outer tail feathers tipped yellow (as 8); bare
blue skin above eye (as 9).

7   GREEN-TAILED BRISTLEBILL *Bleda eximia* 21.5–23.0 cm **R** VU   **Page 594**
Forest. Sierra Leone–SE Guinea to Ghana. Rare. Endemic.
As 6 but without yellow loral spot; yellow tips to outer tail feathers narrower.

8   GREY-HEADED BRISTLEBILL *Bleda canicapilla* 20.5–22.0 cm **R**   **Page 594**
Forest and gallery extensions. SW Senegambia to E Nigeria; also NC Nigeria. Common.
Slate-grey head; olive-green tail feathers tipped yellow on four outer pairs. Compare 6 and 7.

9   RED-TAILED BRISTLEBILL *Bleda syndactyla* 21.5–23.0 cm **R**   **Page 593**
Forest. Sierra Leone–SE Guinea to S CAR–Congo. Rare/common.
Olive-green above; bright rufous tail; sulphur-yellow below; bare blue skin above eye.

PLATE 91: THRUSHES

Large and diverse family of small to medium-sized songbirds (Plates 91–96) with usually strong legs and feet and mostly square or rounded tails. Sexes usually similar, with variable dimorphism, the female being duller. Broad range of vocalisations; some are notable songsters.

1 COMMON ROCK THRUSH *Monticola saxatilis* 18.5 cm **P**   **Page 615**
Various habitats. Mauritania–Sierra Leone to Chad–CAR. Uncommon/rare.
Short, orange-rufous tail; dark wings.
1a **Adult male breeding** Blue head; white patch on back; orange-rufous underparts.
1b **Adult male non-breeding** Dark brown, mottled upperparts; crescent-marked underparts.
1c **Adult female** As 1b but paler. Compare 2b.

2 BLUE ROCK THRUSH *Monticola solitarius* 20 cm **P/R**   **Page 615**
Mainly rocky habitats. Palearctic visitor, Mauritania–Liberia to Chad–CAR; scarce/rare. Year-round in
S Senegal.
2a **Adult male** Wholly dark blue.
2b **Adult female** Dark brown upperparts and tail; barred underparts. Compare 1c.

3 RING OUZEL *Turdus torquatus* 23–24 cm **V**   **Page 618**
Palearctic vagrant, Mauritania.
3a **Adult male** Black; white crescent-shaped breast-band; yellow bill.
3b **Adult female** Browner than male; breast-band less distinct.

4 SONG THRUSH *Turdus philomelos* 23 cm **P**   **Page 618**
Rare Palearctic visitor, Mauritania. Vagrant Senegal, Mali, Chad and Cape Verde Is.
Brown above; boldly spotted below.

5 GREY GROUND THRUSH *Zoothera princei* 21 cm **R**   **Page 616**
Lowland rainforest. Sierra Leone to Ghana; SW Nigeria to Gabon. Rare.
Two black patches on head-sides; two white wingbars.

6 CROSSLEY'S GROUND THRUSH *Zoothera crossleyi* 21.5 cm **R** **NT**   **Page 616**
Montane and submontane forest. SE Nigeria–W Cameroon and SW Congo. Rare/locally not
uncommon.
Resembles 7, but with blackish mask.

7 BLACK-EARED GROUND THRUSH *Zoothera camaronensis* *c.* 18 cm **R**   **Page 616**
Lowland rainforest. Cameroon and Gabon. Rare.
Pattern as 5, but bright orange-rufous on face and underparts.

8 AFRICAN THRUSH *Turdus pelios* 23 cm **R**   **Page 617**
Various wooded habitats; not in rainforest. Widespread. Common/not uncommon.
Typical thrush with yellow bill. Races differ more or less in depth of plumage coloration.
8a **Adult** *chiguancoides* Palest race; upperparts dull grey-brown; underparts pale brown and white;
flanks washed pale orange-buff. Senegal to W Ghana.
T. p. saturatus (W Ghana to Congo), nominate *pelios* (E Cameroon, Chad–N CAR) and *centralis*
(S CAR–Congo) have underparts somewhat deeper coloured, with more distinct streaks on throat.
8b **Adult** *nigrilorum* Darkest race; browner; no orange on flanks. Mt Cameroon.
T. p. poensis (Bioko) similar, but slightly paler.
8c **Juvenile** *chiguancoides* Rufous wash over face and breast; pale tips to coverts; crescentic markings
on breast and flanks.

PLATE 92: FOREST ROBIN, AKALATS, ALETHES, ANT THRUSHES AND FLYCATCHER THRUSHES

1 FOREST ROBIN *Stiphrornis erythrothorax* 13 cm **R** **Page 598**
Lowland forest. Sierra Leone–SE Guinea to CAR–Congo; Bioko. Not uncommon.
- **1a** **Adult** *erythrothorax* Dark olive-brown above; white spot in front of eye; bright orange breast. Sierra Leone to Nigeria.
- **1b** **Juvenile** *erythrothorax* Spotted rufous above; pale buffish throat; blackish breast mottled rufous.
- **1c** **Immature** *erythrothorax* Gradually as adult but duller; throat whitish.
- **1d** **Adult** *xanthogaster* Sooty-grey above; breast paler than 1a. Cameroon to CAR–N Congo.
  - *S. e. gabonensis* (SW Cameroon–SW Congo; Bioko) above as 1d, but below as 1a.
  - *S. e. sanghensis* (SW CAR) has breast yellow-orange; lower underparts pale yellow.

2 BOCAGE'S AKALAT *Sheppardia bocagei* 13 cm **R** **Page 598**
Montane and submontane forests. SE Nigeria–W Cameroon; Bioko. Not uncommon/rare.
- **2a** **Adult** Olive-brown above; orange-rufous below, including head-sides. Secretive.
- **2b** **Juvenile** Dark brown spotted rufous above and below; central belly dirty white.

3 LOWLAND AKALAT *Sheppardia cyornithopsis* 13 cm **R** **Page 599**
Forest. W Guinea to Ivory Coast; S Nigeria to SW CAR–Congo. Rare/locally not uncommon.
- **3a** **Adult** Olive-brown above; orange-rufous below.
- **3b** **Juvenile** Dark brown spotted rufous above and below; belly whitish with some dusky scalloping.

4 FIRE-CRESTED ALETHE *Alethe diademata* 18 cm **R** **Page 603**
Forest, gallery forest, forest patches in savanna. Widespread. Not uncommon.
- **4a** **Adult** *diademata* Dark chestnut above; rusty-orange crown-stripe; white tail corners. S Senegal to Togo.
- **4b** **Juvenile** *diademata* As 4d but with white tail corners; lower underparts whiten with age.
- **4c** **Adult** *castanea* Brighter above; no white in tail. SW Nigeria to S CAR–Congo; Bioko.
- **4d** **Juvenile** *castanea* Spotted orange-rufous above; underparts orange-rufous scalloped blackish on throat and breast.

5 BROWN-CHESTED ALETHE *Alethe poliocephala* 16 cm **R** **Page 603**
Forest, gallery forest, forest patches in savanna. Not uncommon.
- **5a** **Adult** *poliocephala* Grey-black crown; white supercilium; brown head-sides; chestnut upperparts. Secretive. W Guinea to Ghana.
- **5b** **Juvenile** *poliocephala* Upperparts with large orange-rufous spots; dirty white underparts with orange-rufous breast scalloped blackish.
- **5c** **Adult** *compsonota* Head-sides dark grey. SW Nigeria to SW CAR–Congo; Bioko.
  - *A. p. carruthersi* (SE CAR) has crown browner; head-sides brownish.

6 WHITE-TAILED ANT THRUSH *Neocossyphus poensis* 20 cm **R** **Page 604**
Lowland forest. W Guinea to Togo; Nigeria to Congo–CAR; Bioko. Uncommon/locally common.
Dark. White tail corners. Thrush-like jizz. Pumps tail like chat. Terrestrial; secretive.

7 FINSCH'S FLYCATCHER THRUSH *Stizorhina finschi* 18 cm **R** **Page 605**
Lowland forest. Sierra Leone to Ghana; Togo; Nigeria. Not uncommon/rare. Endemic.
White tail corners. Resembles large flycatcher in jizz and foraging behaviour. Flicks outer tail feathers sideways.

8 RED-TAILED ANT THRUSH *Neocossyphus rufus* 22 cm **R** **Page 604**
Lowland forest. S Cameroon to Congo–SW CAR. Not uncommon/rare.
Large, rufous. Rufous outer tail feathers. Thrush-like jizz. Terrestrial; secretive.

9 RUFOUS FLYCATCHER THRUSH *Stizorhina fraseri* 18 cm **R** **Page 605**
Lowland forest. Nigeria to Congo–CAR; Bioko. Not uncommon/common.
Resembles 7 in jizz and actions, but outer tail feathers rufous.

PLATE 93: ROBIN CHATS

Colourful thrushes with orange-rufous underparts and orange-rufous tails with, typically, black central feathers. Most have loud, very melodious and varied songs with much mimicry.

1 WHITE-BELLIED ROBIN CHAT *Cossyphicula roberti* 13 cm **R** Page 600
Montane and submontane forests. SE Nigeria–W Cameroon; Bioko. Uncommon/locally common.
 1a **Adult** Resembles akalats (Plate 92: 2–3) but tail black and red. Quiet and unobtrusive.
 1b **Juvenile** Dark brown spotted rufous above; rufous-buff scalloped black below.

2 MOUNTAIN ROBIN CHAT *Cossypha isabellae* 15 cm **R** Page 601
Montane and submontane forests. E Nigeria–W Cameroon. Not uncommon.
 2a **Adult** *isabellae* Narrow white supercilium; darker above than 2b; contrasting orange-rufous rump. Mt Cameroon only.
 2b **Adult** *batesi* Olive-brown above becoming rufous-brown on rump; lower belly off-white.
 2c **Juvenile** Mainly dark brown spotted rufous.

3 GREY-WINGED ROBIN CHAT *Cossypha polioptera* 15 cm **R** Page 601
Mid-elevation and gallery forests. Guinea–E Sierra Leone to E Ivory Coast; C & E Nigeria–W CAR. Rare/locally common.
 3a **Adult** Long white supercilium underlined by black eye-stripe; uniformly rufous tail.
 3b **Juvenile** Duller; no supercilium; rufous spots on head and wing-coverts.

4 BLUE-SHOULDERED ROBIN CHAT *Cossypha cyanocampter* 16.5 cm **R** Page 601
Forest and gallery forest. SW Mali–Sierra Leone to CAR–N Congo. Uncommon.
 4a **Adult** Long white supercilium; blue shoulder patch.
 4b **Juvenile** Duller; no supercilium; rufous spots on head and wing-coverts.

5 RED-CAPPED ROBIN CHAT *Cossypha natalensis* 17 cm **R/M** Page 602
Wooded habitats. C & E Nigeria, Cameroon, coastal Gabon, Congo, SE CAR. Rare/locally uncommon.
 5a **Adult** Orange-rufous head; beady eye.
 5b **Juvenile** Dark brown spotted rufous above; rufous-brown scalloped dark brown below.

6 WHITE-BROWED ROBIN CHAT *Cossypha heuglini* 20 cm **R** Page 602
Riverine forest, thickets in savanna. Cameroon, SW Chad, CAR, Gabon, Congo. Uncommon/rare to locally not uncommon.
 6a **Adult** Distinctive, long white supercilium.
 6b **Juvenile** Dark brown spotted rufous above; rufous-buff scalloped dark brown below.

7 SNOWY-CROWNED ROBIN CHAT *Cossypha niveicapilla* 20.5–22 cm **R** Page 602
Forest and savanna zones. Throughout. Uncommon/locally common.
 7a **Adult** Large; white crown; rufous hind-collar. Vocal. Compare 8.
 7b **Juvenile** Blackish spotted rufous above; rusty-buff scalloped blackish below.

8 WHITE-CROWNED ROBIN CHAT *Cossypha albicapilla* 26 cm **R/M** Page 602
Narrow savanna belt. Uncommon/locally common.
 8a **Adult** *albicapilla* Very large; white crown; no rufous hind-collar; long tail. Senegambia to N Ivory Coast.
 8b **Adult** *giffardi* Crown mainly black with white crescents. S Burkina Faso to NW CAR.

PLATE 94: BLUETHROAT, REDSTARTS, ROBIN, PALM THRUSH, NIGHTINGALES AND SCRUB ROBINS

1 BLUETHROAT *Luscinia svecica* 14 cm **P/V** Page 600
Sahel belt. S Mauritania–Senegambia to N Nigeria–Chad; uncommon. Vagrant, Ivory Coast, Ghana.
In all plumages: diagnostic rufous sides to basal half of outer tail feathers (conspicuous in flight).
1a Adult male *svecica* Blue throat and breast; rufous spot in centre.
1b Adult female *svecica* Buff-white throat bordered by dappled black breast-band.
1c Adult male *cyanecula* As 1a, but with white spot in centre of blue.

2 COMMON REDSTART *Phoenicurus phoenicurus* 14 cm **P** Page 608
Wooded habitats in broad savanna belt. Uncommon/locally not uncommon. Vagrant, Cape Verde Is.
In all plumages: almost constantly shivering, bright rufous tail.
2a Adult male breeding Blue-grey above; black face; orange-rufous underparts.
2b Adult male non-breeding Pale feather tips partially obscure breeding plumage.
2c Adult female Pale grey-brown above; mainly buffish below. Compare 4b.

3 EUROPEAN ROBIN *Erithacus rubecula* 14 cm **V** Page 599
Palearctic vagrant, coastal Mauritania.
Orange-red face and breast. Compact and rounded.

4 BLACK REDSTART *Phoenicurus ochruros* 14 cm **P/V** Page 607
Semi-arid belt. Coastal Mauritania and N Senegal; rare. Vagrant, S Mali, Niger, Chad, Cape Verde Is.
4a Adult male Dark grey above; white wing panel; bright rufous tail; black face and breast.
4b Adult female Entirely slate-grey (greyer than 2c); tail as male.

5 RUFOUS-TAILED PALM THRUSH *Cichladusa ruficauda* 18 cm **R** Page 605
Palm savanna, thickets, gardens. SW Gabon, Congo, SW CAR. Uncommon/locally common.
Rufous-brown above; rufous tail; pale greyish supercilium and head-sides.

6 THRUSH NIGHTINGALE *Luscinia luscinia* 16.5 cm **V** Page 599
Palearctic vagrant, N Nigeria.
As 7 but less brightly coloured; upperparts and tail darker; indistinctly mottled darkish below.

7 COMMON NIGHTINGALE *Luscinia megarhynchos* 16.5 cm **P** Page 599
Savanna and forest zones. Throughout; not south of Cameroon. Uncommon/locally common. Vagrant,
Cape Verde Is.
Open face; rufous tail.

Scrub robins. Characteristic, fairly long, broad, graduated and white-tipped tails, typically held cocked above
back. Songs melodious.

8 FOREST SCRUB ROBIN *Cercotrichas leucosticta* 15.0–16.5 cm **R** Page 606
Lowland forest. Sierra Leone to Ghana; SE CAR. Scarce.
Note head pattern, white spots on bend of wing.

9 WHITE-BROWED SCRUB ROBIN *Cercotrichas leucophrys* 15 cm **R** Page 606
Wooded savanna, thicket edges. Gabon–Congo. Locally common.
Warm brown above becoming rufous on rump; two white wingbars; streaked breast.

10 BROWN-BACKED SCRUB ROBIN *Cercotrichas hartlaubi* 15 cm **R** Page 606
Wooded savanna. E Nigeria to SW CAR. Scarce/not uncommon.
Cold dark grey-brown above; bright rufous rump; clear-cut blackish terminal band to rufous tail; breast
indistinctly greyish.

11 RUFOUS SCRUB ROBIN *Cercotrichas galactotes* 15 cm **R/P** Page 607
Semi-arid belt. S Mauritania–Senegal to Chad–NE CAR. Locally common.
Plain sandy rufous-brown above; black-and-white tipped tail.

12 BLACK SCRUB ROBIN *Cercotrichas podobe* 18 cm **R** Page 607
Semi-arid belt. S Mauritania–Senegal to N Cameroon–Chad. Not uncommon.
Long, fan-shaped tail boldly tipped white. Rufous in wing visible in flight.

PLATE 95: WHEATEARS

Small, ground-dwelling insectivores. Black-and-white tail pattern an important identification mark. Adult males usually distinctive, but females and immatures often difficult to identify.

1 NORTHERN WHEATEAR *Oenanthe oenanthe* 15–16 cm **P**   Page 609
Broad Sahel belt. Common in north, uncommon to rare further south. Scarce, Cape Verde Is.
1a **Adult male non-breeding** Dull grey-brown above; variable dark mask; whitish supercilium.
1b **Adult female non-breeding** Browner than 1a; no mask.
1c **Adult male** *seebohmi* **breeding** Black face and throat. SW Mauritania, N Senegal, Mali.
1d **Adult tail** T-patterned; terminal band of even width on distal third.

2 HEUGLIN'S WHEATEAR *Oenanthe (bottae) heuglini* 14–15 cm **M/R?**   Page 612
Degraded savanna, dry farmland, rocky hillsides, inselbergs. S Mauritania–C Guinea to Chad–CAR.
Uncommon/rare to locally not uncommon.
2a **Adult** Dark brown above; rufous-buff below.
2b **Adult tail** T-patterned; broad terminal band covering *c.* half of tail.

3 ISABELLINE WHEATEAR *Oenanthe isabellina* 16–17 cm **P**   Page 612
Dry open country. S Mauritania–N Senegal to Chad; common/uncommon. Rare further south.
3a **Adult** Plain, sandy coloured; robust; long legged.
3b **Adult tail** T-patterned; broad terminal band of even width covering *c.* half of tail.

4 BLACK-EARED WHEATEAR *Oenanthe hispanica* 14.5 cm **P**   Page 610
Semi-desert, dry savanna. Mauritania–Senegambia to Chad–NE CAR. Uncommon/rare.
4a **Adult male** *hispanica* **non-breeding** Rich sandy-buff; black mask not connected with black wings.
From Mali west.
4b **Adult male** *hispanica* **non-breeding** Black-throated form. Similar to 4a.
4c **Adult female** *hispanica* **non-breeding** Sandy-brown; indistinct mask; dark brown wings broadly
fringed buffish.
4d **Adult male** *melanoleuca* **non-breeding** As 4e, but has white throat. From Mali east.
4e **Adult male** *melanoleuca* **non-breeding** Black-throated form. upperparts washed grey.
4f **Adult tail** T-patterned; terminal band uneven, sometimes broken.

5 DESERT WHEATEAR *Oenanthe deserti* 14–15 cm **P**   Page 611
Desert and semi-desert. Mauritania–N Senegal to Chad. Common/scarce.
5a **Adult male non-breeding** Sandy; black face and throat connected to black wings.
5b **Adult female non-breeding** Deep sandy-buff; no black on face.
5c **Adult tail** Black, with very little white at base.

6 MOURNING WHEATEAR *Oenanthe lugens* 14.5 cm **V**   Page 611
Desert. N African vagrant, Mauritania, Niger (Aïr).
6a **Adult male non-breeding** Mainly black and white; orange-buff undertail-coverts.
6b **Adult female non-breeding** Dull grey version of 6a.
6c **Adult tail** T-patterned; terminal band of even width on distal third.

7 CYPRUS WHEATEAR *Oenanthe (pleschanka) cypriaca* 13.5 cm **P**   Page 610
Dry, open country. EC Chad. Uncommon.
7a **Adult male non-breeding** Dull brownish crown (paler with wear); black face, throat and
upperparts.
7b **Adult female non-breeding** Dull brownish crown (not paler with wear).
7c **Adult tail** T-patterned; terminal band uneven, black increasing towards outer feathers.

8 BLACK WHEATEAR *Oenanthe leucura* 18 cm **R/P?**   Page 609
Desert. NW Mauritania. Rare.
8a **Adult male** All black except tail base and tail.
8b **Adult female** As male, but browner. Juvenile similar.
8c **Adult tail** T-patterned.

9 WHITE-CROWNED BLACK WHEATEAR *Oenanthe leucopyga* 17 cm **R**   Page 609
Desert. Mauritania, N Mali, Niger, N Chad. Local migrant; common. Vagrant, NE Nigeria.
9a **Adult** Black, with white crown, white tail base and vent.
9b **Juvenile** Dark brown; no white on crown.
9c **Adult tail** White with black on distal half of central feather pair.
9d **Juvenile/adult variant tail** As 9c, but with dark smudges near tip.

PLATE 96: CHATS

1 WHINCHAT *Saxicola rubetra* 12.5 cm **P**   **Page 608**
Open habitats. Almost throughout; uncommon/common. Vagrant, Cape Verde Is and Príncipe.
  1a **Adult male breeding** Broad white supercilium; head-sides blackish; dark tail with white patch at
      sides of base (conspicuous in flight). Adult female similar but paler.
  1b **Adult non-breeding** Very buffish overall; head-sides brownish.

2 COMMON STONECHAT *Saxicola torquata* 12.5 cm **R/P/V**   **Page 608**
Open habitats. Patchily distributed resident; uncommon/locally common. Irregular Palearctic migrant/
vagrant.
  2a **Adult male** *salax* Black head and upperparts; white neck-sides and underparts; chestnut breast
      patch. In flight, white wing patch and rump conspicuous. SE Nigeria east and south; Bioko.
      S. t. moptana (Senegal and Niger Delta) intermediate between 2a and 2b.
  2b **Adult male** *nebularum* Chestnut breast patch more extensive than 1a. Mt Nimba.
  2c **Adult male** *rubicola* **non-breeding** Browner above; rump rusty-brown mottled white. Palearctic
      migrant, coastal Mauritania. Vagrant, Mali, Niger, Chad.
  2d **Adult female** Streaked dark brown above; no supercilium. Compare 1.

3 BLACKSTART *Cercomela melanura* 14 cm **R**   **Page 613**
Rocky areas in arid country. NE Mali, Niger, N Chad; uncommon/locally common. Vagrant, Gambia.
Plain sandy-brown above; black tail.

4 BROWN-TAILED ROCK CHAT *Cercomela scotocerca* 13 cm **R**   **Page 613**
Dry rocky country. Chad. Locally not uncommon.
Small, nondescript; dark brown tail.

5 WHITE-FRONTED BLACK CHAT *Myrmecocichla albifrons* 15 cm **R**   **Page 614**
Savanna belt. Senegambia–Sierra Leone to Chad–CAR. Uncommon/locally common. Vagrant,
S Mauritania.
  5a **Adult male** *frontalis* Entirely black; white patch on forehead.
  5b **Adult male** *limbata* Some white on shoulder. From E Cameroon east.
  5c **Adult female** As male but without any white.

6 FAMILIAR CHAT *Cercomela familiaris* 14 cm **R**   **Page 612**
Rocky areas in savanna. SE Senegal to CAR; uncommon/locally common. Rare, S Mauritania.
Plain brown; conspicuous rufous rump and tail.

7 CLIFF CHAT *Myrmecocichla cinnamomeiventris* 20 cm **R**   **Page 615**
Rocky areas in savanna. Patchily distributed, S Mauritania–Guinea to Chad–CAR. Rare/locally not
uncommon.
  7a **Adult male** *bambarae* Black and rufous with small white shoulder patch.
  7b **Adult female** *bambarae* Pattern as male but duller; no white.
  7c **Adult male** *cavernicola* Distinct shoulder patch; narrow pale line below black breast.
  7d **Adult male** *coronata* White crown; pale area below breast. Mainly in east of range.
  7e **Adult female** *coronata* Head rufous-grey, paler than 7b; entirely rufous below.

8 NORTHERN ANTEATER CHAT *Myrmecocichla aethiops* 19 cm **R**   **Page 614**
Semi-arid belt. S Mauritania–Senegambia to Chad–NE CAR. Locally common/not uncommon.
  8a **Adult** Entirely sooty-brown.
  8b **Adult in flight** Large white patch on primaries.

9 SOOTY CHAT *Myrmecocichla nigra* 16 cm **R**   **Page 614**
Grassland. N & E Nigeria to CAR–Congo. Locally uncommon/common.
  9a **Adult male** Entirely glossy black; white shoulder patches.
  9b **Adult female** Entirely blackish-brown.
  9c **Adult male in flight** White shoulder patches conspicuous.

10 CONGO MOOR CHAT *Myrmecocichla tholloni* 18–19 cm **R**   **Page 613**
Grassland. SE Gabon–Congo. Locally not uncommon.
  10a **Adult** Robust; head-sides and throat mainly dirty white; underparts brown-grey.
  10b **Adult in flight** White wing patch; white rump.

PLATE 97: WARBLERS I – *Melocichla, Schoenicola, Bradypterus, Locustella* and *Bathmocercus*

1 AFRICAN MOUSTACHED WARBLER *Melocichla mentalis* 19–20 cm **R** **Page 620**
Rank herbage in savanna. Throughout. Locally common.
Large and bulky; black malar stripe; broad tail. Distinctive song.

2 BROAD-TAILED WARBLER *Schoenicola platyura* 17 cm **R** **Page 620**
Grassy habitats. Guinea–Sierra Leone, Mt Nimba, SE Nigeria–W Cameroon, Gabon–Congo. Uncommon and local.
Long, broad, blackish tail.

Genus *Bradypterus*. Drab-coloured skulkers of reedbeds, rank herbage, scrub and forest undergrowth. Songs distinctive.

3 DJA RIVER WARBLER *Bradypterus grandis* 18.5 cm **R VU** **Page 618**
Forest edge, tall grassland, marshland. S Cameroon, SW CAR and Gabon. Rare. Endemic.
Resembles 6, but much larger; throat streaked.

4 EVERGREEN FOREST WARBLER *Bradypterus lopezi* 13.0–14.5 cm **R** **Page 619**
Undergrowth of montane forest. Mt Cameroon, Bioko. Not uncommon but highly local.
Darkish. Tail heavily worn.

5 BANGWA FOREST WARBLER *Bradypterus (lopezi) bangwaensis* 14–15 cm **R NT** **Page 619**
Dense vegetation in montane areas. SE Nigeria–W Cameroon. Locally not uncommon. Endemic.
Mainly rufous; throat white.

6 LITTLE RUSH WARBLER *Bradypterus baboecala* 15 cm **R** **Page 618**
Reedbeds, dense aquatic vegetation. Nigeria, Cameroon, W Chad, S Congo. Rare and local.
Dark brown upperparts. Distinctive song.

Genus *Locustella*. Dull-coloured with round, almost graduated tails and long undertail-coverts. Skulk low in dense vegetation.

7 GRASSHOPPER WARBLER *Locustella naevia* 12.5 cm **P** **Page 621**
Dense vegetation. Mauritania–Senegambia, Sierra Leone, Guinea, Mt Nimba, Mali, Ghana. Generally scarce/rare but probably overlooked. Vagrant, Cape Verde Is.
Streaked upperparts; graduated tail; faint supercilium. Secretive.

8 SAVI'S WARBLER *Locustella luscinioides* 14 cm **P** **Page 621**
Dense scrub, herbage near or in water. Patchily distributed, S Mauritania–Senegal to N Nigeria–Chad, rare/scarce. Vagrant, Cape Verde Is.
Plain brown upperparts; broad, rounded tail.

Genus *Bathmocercus*. Secretive species of dense forest undergrowth. Songs distinctive.

9 BLACK-HEADED RUFOUS WARBLER *Bathmocercus cerviniventris* 13 cm **R NT** **Page 620**
Lowland and gallery forest. Sierra Leone–Guinea to Ghana. Rare/uncommon. Endemic.
9a Adult male Black head, throat and centre of breast; upperparts rufous-brown.
9b Immature (female?) Paler than adult; white chin and 'whiskers'.

10 BLACK-FACED RUFOUS WARBLER *Bathmocercus rufus* 13 cm **R** **Page 619**
Lowland and montane forest. Cameroon, Gabon, N Congo, SW CAR. Locally not uncommon.
10a Adult male Black mask extending to centre of breast; upperparts rufous-chestnut.
10b Adult female Upperparts dark grey.

PLATE 98: WARBLERS II – *Acrocephalus, Chloropeta* and *Hippolais*

Genus *Acrocephalus*. Upperparts mainly brown; underparts paler; tail rounded; bill prominent.

1 SEDGE WARBLER *Acrocephalus schoenobaenus* 13 cm **P** Page 622
Reedbeds, moist rank vegetation. Throughout. Locally common.
Bold creamy supercilium; streaked upperparts; unstreaked, rusty rump.

2 AQUATIC WARBLER *Acrocephalus paludicola* 13 cm **P VU** Page 622
Reedbeds, moist herbage. Mauritania, Senegal, Mali; rare. Vagrant, N Ghana.
Creamy-buff crown-stripe; heavily streaked, straw-coloured upperparts; spiky tail.

3 GREAT REED WARBLER *Acrocephalus arundinaceus* 19–20 cm **P** Page 623
Scrub, dense herbage, swamps, etc. Throughout; also São Tomé. Scarce/not uncommon.
Large. Prominent supercilium; stout bill.

4 LESSER SWAMP WARBLER *Acrocephalus gracilirostris* 14–15 cm **R** Page 624
Reedbeds and similar moist vegetation. N Nigeria and Chad. Rare.
Dark brown above; distinct supercilium.

5 GREATER SWAMP WARBLER *Acrocephalus rufescens* 18 cm **R** Page 624
Reedbeds, marshes. Patchily distributed, S Mauritania–N Senegal to Togo, Nigeria to Congo– CAR;
Bioko. Scarce/not uncommon; local.
Dark brown above; greyish-white below; no supercilium.

6 AFRICAN REED WARBLER *Acrocephalus baeticatus* 12.0–12.5 cm **R/M** Page 623
Reedbeds; also drier habitats. Patchily distributed, Mauritania–Senegambia to Chad, Cameroon, Gabon.
Locally not uncommon; status inadequately known through similarity with 7.
As 7, but wings shorter, primaries not projecting beyond rump.

7 EUROPEAN REED WARBLER *Acrocephalus scirpaceus* 12.5–13.0 cm **P** Page 622
Scrub, dense herbage, etc. Mauritania–Liberia to CAR and Gabon. Scarce/common.
Plain brown upperparts; rusty-tinged rump; indistinct supercilium. Note primary projection.

MARSH WARBLER *Acrocephalus palustris* 12.5–13.0 cm **V** Page 623
Palearctic vagrant, N Senegal and NE Nigeria. **See illustration on plate 147.**
Very similar to 7; see text for differences.

Genus *Chloropeta*. Upperparts olive; underparts yellow. Bill broad, flat, flycatcher-like.

8 AFRICAN YELLOW WARBLER *Chloropeta natalensis* 14 cm **R** Page 624
Rank herbage, reeds, dense forest edge. Patchily distributed, SE Nigeria to CAR–Congo. Uncommon/
not uncommon and local.
Bright yellow underparts.

Genus *Hippolais*. Tail square; bill prominent. Arboreal.

9 OLIVACEOUS WARBLER *Hippolais pallida* 12.0–13.5 cm **P/R/M** Page 625
Scrub, various wooded habitats. Palearctic migrant, Mauritania–Guinea to Chad. Vagrant, Cape Verde Is.
Not uncommon resident and partial migrant, N Nigeria, Niger, Chad and N Cameroon.
Pale. Elongated jizz.

10 MELODIOUS WARBLER *Hippolais polyglotta* 13 cm **P** Page 625
Various wooded and scrubby habitats. Mainly in west (east to Cameroon); common/uncommon.
Vagrant, CAR.
Olive-green upperparts; yellow underparts; orangey bill. Compare 12.

11 OLIVE-TREE WARBLER *Hippolais olivetorum* 15.5 cm **V** Page 625
Dry bush country. Palearctic vagrant, N Niger, N Nigeria.
Large; greyish.

12 ICTERINE WARBLER *Hippolais icterina* 13.5 cm **P** Page 626
Open woodland, scrub. Mainly in east; not uncommon. Rarely west of Nigeria.
Resembles 10, but has longer primary projection and pale wing panel.

PLATE 99: WARBLERS III – *Sylvia* 1

Genus *Sylvia*. Palearctic scrub and woodland warblers. Females/immatures of some difficult to separate (see Plate 100). Characteristic vocalisations a sharp *tak* and a harsh churr.

1 GARDEN WARBLER *Sylvia borin* 14 cm **P**   **Page 651**
Various wooded and bushy habitats. Throughout. Generally common.
Plain and featureless. Short, stubby bill.

2 BARRED WARBLER *Sylvia nisoria* 15.5 cm **V**   **Page 650**
Dry bush and woodland. N Senegal, Gambia, NE Nigeria.
2a   **Adult** Robust; crescentic barring on breast-sides and flanks.
2b   **First-winter** Pale fringes to wing feathers and uppertail-coverts; dark crescents on undertail-coverts.

3 LESSER WHITETHROAT *Sylvia curruca* 12.5–13.5 cm **P**   **Page 652**
Dry scrub, Acacia woodland. Mauritania–Senegambia to Cameroon–Chad. Rare/locally common.
Greyish upperparts; whitish underparts; dark ear-coverts.

4 COMMON WHITETHROAT *Sylvia communis* 14 cm **P**   **Page 652**
Dry scrub, wooded savanna, gardens. Mauritania–Senegambia to Chad and CAR–Congo. Common/rare.
4a   **Adult male** Rusty wing panel; greyish head.
4b   **Adult female** Brownish head.

5 BLACKCAP *Sylvia atricapilla* 13 cm **P/R**   **Page 651**
Various wooded habitats. Mauritania–Senegambia to Cameroon; scarce/locally common. Vagrant, Chad.
Common resident, Cape Verde Is.
5a   **Adult male** Plain grey; black cap.
5b   **Adult female** Plain grey; chestnut cap.

6 ORPHEAN WARBLER *Sylvia hortensis* 15 cm **P**   **Page 651**
Dry scrub, wooded savanna. Mauritania–N Senegal, Niger, Chad; uncommon. Rarely further south.
Large, stout. Black or dusky face; white throat.
6a   **Adult male** Blackish hood; pale eyes.
6b   **Adult female** Grey-brown crown.

PLATE 100: WARBLERS IV – *Sylvia* 2

1 SARDINIAN WARBLER *Sylvia melanocephala* 13.5 cm **P**   **Page 653**
Bush, wooded savanna, gardens. Mauritania, Senegal, Niger. Uncommon.
  1a **Adult male** Black hood; white throat; conspicuous, red orbital ring.
  1b **Adult female** Dusky-grey hood; brown upperparts.

2 RÜPPELL'S WARBLER *Sylvia rueppelli* 14 cm **P/V**   **Page 652**
Dry scrub. Chad; not uncommon/common. Vagrant, Mali and Niger.
  2a **Adult male** Black throat; white submoustachial stripe.
  2b **Adult female** Greyish head; pale fringes to wing feathers. Variable.

3 MÉNÉTRIES'S WARBLER *Sylvia mystacea* 13.5 cm **V**   **Page 653**
Dry scrub. Vagrant, N Nigeria.
  3a **Adult male** Dull black crown to ear-coverts; pinkish-tinged underparts.
  3b **Adult female** Sandy grey-brown upperparts; pale fringes to wing feathers.

4 SPECTACLED WARBLER *Sylvia conspicillata* 12.5 cm **P/R**   **Page 654**
Dry scrub. Mauritania–Senegambia; locally common/rare. Vagrant, N Niger. Common resident, Cape
Verde Is.
Resembles miniature Common Whitethroat (Plate 99: 4). Orange-rufous wing panel.
  4a **Adult male** Grey head; black area from bill to eye; white eye-ring; white throat.
  4b **Adult female** Browner; blackish tail contrasts with rest of upperparts.

5 DESERT WARBLER *Sylvia nana* 11.5 cm **R**   **Page 652**
Desert scrub. Mauritania, N Mali, N Niger; rare/uncommon and local. Vagrant, Cape Verde Is.
Small and pale. Rufous tail; yellowish eyes, bill and legs.

6 TRISTRAM'S WARBLER *Sylvia deserticola* 12 cm **P?**   **Page 654**
Desert scrub. Possibly rare migrant to Mauritania.
  6a **Adult male** Slate-grey above; brick-red below; rusty wing panel.
  6b **Immature** Browner above; paler below.

7 SUBALPINE WARBLER *Sylvia cantillans* 12 cm **P**   **Page 653**
Dry scrub, also gardens, mangroves. Mauritania–Senegambia to N Cameroon–Chad; common/scarce.
Vagrant, Sierra Leone, Cape Verde Is.
  7a **Adult male** Blue-grey upperparts; rusty-orange throat and breast; white moustache; red orbital ring.
  7b **Adult female** Pale grey-brown upperparts; pinkish-buff to buffish-white underparts; paler orbital ring.
  7c **Immature** Buffish-brown above; buff below; pale fringes to wing feathers; pale orbital ring.

PLATE 101: WARBLERS V – *Phylloscopus, Hylia, Phyllolais* and *Eremomela*

1 WOOD WARBLER *Phylloscopus sibilatrix* 12 cm **P**   **Page 649**
Forest zone and adjacent savanna. Widespread; common/not uncommon. Vagrant, Bioko.
Throat and upper breast yellow; rest of underparts white. Long primary projection.

2 WILLOW WARBLER *Phylloscopus trochilus* 10.5–11.5 cm **P**   **Page 648**
Various wooded habitats. Throughout; common. Also Cape Verde Is (rare), Bioko (vagrant).
Olive-green upperparts; yellowish underparts and supercilium; pale brownish legs.

3 WESTERN BONELLI'S WARBLER *Phylloscopus bonelli* 11.5 cm **P**   **Page 649**
Dry scrub and bushy savanna. Mauritania–Senegambia to N Cameroon–Chad; common/not
uncommon. Vagrant, S Cameroon, Togo, NE Gabon and Cape Verde Is.
Pale; bland face. Yellowish-green wings and rump contrast with greyish upperparts.

4 CHIFFCHAFF *Phylloscopus collybita* 10–11 cm **P**   **Page 648**
Open, dry and wooded habitats. Mauritania–Senegambia to Chad; common. Vagrant/rare further
south.
More olive-brown and with shorter primary projection than 2; dark legs.

5 GREEN HYLIA *Hylia prasina* 11.5 cm **R**   **Page 656**
Forest. Throughout; also Bioko. Common.
Dark olive-green above; conspicuous yellowish-green supercilium.

6 BLACK-CAPPED WOODLAND WARBLER *Phylloscopus herberti* *c.* 9 cm **R**   **Page 649**
Montane and submontane forest. SE Nigeria–W Cameroon; Eq. Guinea; Bioko. Common/not
uncommon. Endemic.
Black top of head; black eye-stripe; golden-green upperparts.

7 UGANDA WOODLAND WARBLER *Phylloscopus budongoensis* 10 cm **R**   **Page 650**
Forest. S Cameroon, Eq. Guinea, NE Gabon, NW Congo. Locally common.
Well-defined whitish supercilium; plain olive-green above; greyish-white below.

8 RUFOUS-CROWNED EREMOMELA *Eremomela badiceps* 11 cm **R**   **Page 646**
Forest. Sierra Leone–SE Guinea to SW CAR–Congo; Bioko. Common.
8a Adult Bright rufous-chestnut cap; grey upperparts; black gorget.
8b Juvenile Olive above; no cap nor gorget (or just a hint of these); yellowish underparts.

9 GREEN-CAPPED EREMOMELA *Eremomela scotops* 11 cm **R**   **Page 646**
Wooded grassland. SE Gabon and Congo. Uncommon/locally common.
Grey-green above; lemon-yellow below; dark lores; pale eye.

10 BUFF-BELLIED WARBLER *Phyllolais pulchella* 11.5 cm **R**   **Page 638**
Dry *Acacia* woodland. N Nigeria to W Chad. Uncommon/locally common.
Nondescript. Longish, graduated tail.

11 SENEGAL EREMOMELA *Eremomela pusilla* 10 cm **R**   **Page 645**
Savanna woodland. Senegambia–Sierra Leone to W Chad–NW CAR. Common.
Greyish head; bright lemon-yellow lower breast and belly.

12 GREEN-BACKED EREMOMELA *Eremomela canescens* 11 cm **R**   **Page 645**
Wooded grassland. S Chad and CAR; intergrading with 11 in Cameroon. Common.
Clear white supercilium; dark eye-stripe; head more contrasting with upperparts.

13 YELLOW-BELLIED EREMOMELA *Eremomela icteropygialis* 10 cm **R**   **Page 645**
Arid *Acacia* scrub. S Mauritania–N Senegal to Niger–N Nigeria and Chad; also N Ghana. Common/
uncommon.
Pale greyish above; lemon-yellow belly.

14 SALVADORI'S EREMOMELA *Eremomela salvadorii* 11 cm **R**   **Page 645**
Wooded grassland. SE Gabon–Congo. Locally common.
Greenish mantle; brighter yellow belly.

PLATE 102: WARBLERS VI – *Sylvietta*, *Hypergerus* and *Hyliota*

Genus *Sylvietta* (crombecs). Very small size and extremely short tail create distinctive jizz. Often in mixed-species flocks.

1 GREEN CROMBEC *Sylvietta virens* 9 cm **R**   **Page 647**
Forest zone and adjacent savanna woodland. Guinea–Liberia to SW CAR–Congo; also Senegambia, S Mali. Common.
- **1a Adult *virens*** Head brown; upperparts greenish; throat and breast pale brown. East of lower Niger R.
- **1b Adult *flaviventris*** Lower breast and belly yellow. West of lower Niger R.
- **1c Juvenile *flaviventris*** Entire underparts yellow.

2 NORTHERN CROMBEC *Sylvietta brachyura* 9 cm **R**   **Page 646**
Dry and wooded savanna. S Mauritania–Sierra Leone to Chad–CAR. Common/uncommon. Grey above; tawny below; buffish supercilium; dusky eye-stripe.

3 RED-CAPPED CROMBEC *Sylvietta ruficapilla* 10–12 cm **R**   **Page 647**
Wooded grassland. SE Gabon–Congo. Uncommon.
Adult Pale greyish above; rufous on head and breast.

4 LEMON-BELLIED CROMBEC *Sylvietta denti* 8 cm **R**   **Page 647**
Forest. Not uncommon.
- **4a Adult *denti*** Olive-green above; yellow below; breast tinged olive. Nigeria to Congo.
- **4b Adult *hardyi*** Brighter yellow below. Guinea to Ghana.

Genus *Hypergerus*. Large, unmistakable warbler with long, graduated tail.

5 ORIOLE WARBLER *Hypergerus atriceps* 20 cm **R**   **Page 650**
Gallery forest, thicket, mangroves. Senegambia–Sierra Leone to S Chad–CAR. Locally common/rare. Endemic.
Black, scaly head; green upperparts; yellow underparts. Loud, melodious song distinctive.

Genus *Hyliota*. Quite distinct, aberrant warblers with relatively short tails. In canopy of forest or savanna woodland; often in mixed-species flocks. Formerly regarded as flycatchers.

6 VIOLET-BACKED HYLIOTA *Hyliota violacea* 12.5 cm **R**   **Page 655**
Forest. Guinea to Togo; Nigeria to CAR–Congo. Scarce/rare and local.
- **6a Adult male *violacea*** Deep violet-blue above; mainly white below; white patch of variable size on inner greater coverts (lacking in western *nehrkorni*).
- **6b Adult female *violacea*** Throat and breast orange.

7 SOUTHERN HYLIOTA *Hyliota australis* 12.5 cm **R**   **Page 655**
Extremely rare: puzzling record from Rumpi Hills, Cameroon.
- **7a Adult male** As 8a, but upperparts sooty-black; no white edges to tertials and secondaries.
- **7b Adult female** As 8b, but upperparts dark brown.

8 YELLOW-BELLIED HYLIOTA *Hyliota flavigaster* 12.5 cm **R**   **Page 655**
Savanna woodland. Senegambia–Sierra Leone to CAR; also SE Gabon–Congo. Locally not uncommon/scarce.
- **8a Adult male** Glossy blue-black above; white wing patch; pale yellowish-buff below.
- **8b Adult female** Duller and paler than male; upperparts dark grey-brown.

PLATE 103: WARBLERS VII – *Poliolais, Camaroptera, Calamonastes* and *Macrosphenus*

1   WHITE-TAILED WARBLER *Poliolais lopezi*   10 cm   **R**   **NT**   **Page 642**
Montane forest undergrowth. SE Nigeria–W Cameroon and Bioko. Not uncommon; local. Endemic.
Tail short, dark brown above, white below.
1a   **Adult male** *alexanderi* Dark grey with olive wash. Mt Cameroon.
1b   **Adult female** *alexanderi* Forehead and head-sides rufous.
1c   **Adult male** *manengubae/lopezi* Wholly sooty-grey.

Genus *Camaroptera*. Small, with rather long and straight bills; colour of thighs contrasting with belly. Skulk in dense vegetation, mostly low down. Have distinctive mewing calls.

2   YELLOW-BROWED CAMAROPTERA *Camaroptera superciliaris*   11 cm   **R**   **Page 643**
Forest edge. Sierra Leone–SE Guinea to CAR–Congo; Bioko. Uncommon/common.
2a   **Adult** Bright green above; supercilium, ear-coverts, thighs and undertail-coverts bright yellow.
2b   **Juvenile** Throat and breast yellow.

3   OLIVE-GREEN CAMAROPTERA *Camaroptera chloronota*   11 cm   **R**   **Page 643**
Lowland forest and savanna outliers. Throughout; also Bioko. Locally not uncommon. Rare, Senegambia, S Mali.
Skulking, but song loud and distinctive.
3a   **Adult** *chloronota* Dark olive-green above; greyish-white below; ear-coverts greyish. Togo to CAR–Congo.
3b   **Adult** *kelsalli* Ear-coverts tinged rufous. Guinea to Ghana.

4   GREY-BACKED CAMAROPTERA *Camaroptera brachyura*   11.5 cm   **R**   **Page 642**
Dense shrubbery in various habitats. Throughout. One of the most widespread and common warblers in the region.
4a   **Adult** *tincta* Grey, with yellowish-green wings. Forest zone.
4b   **Adult** *brevicaudata* **non-breeding** Ashy-brown above; paler below. Savanna zone.
4c   **Juvenile** Olive-green above; lemon below.

5   MIOMBO WREN WARBLER *Calamonastes undosus*   13 cm   **Page 643**
Savanna woodland. Not (yet?) recorded in our region. Cabinda and lower Congo R.
Wholly dark grey; rather long tail constantly cocked.

Genus *Macrosphenus* (longbills). Aberrant warblers with long straight bill, sharply hooked at tip, short wings and tail, and long, loose feathers on rump and flanks. Occur in dense undergrowth, tangles and vines of forest and forest edge. Song distinctive.

6   KEMP'S LONGBILL *Macrosphenus kempi*   13 cm   **R**   **Page 644**
Forest. Sierra Leone–SE Guinea to Ghana; W and SE Nigeria–SW Cameroon. Uncommon. Endemic.
6a   **Adult** Dark brownish above; orange-rufous flanks and undertail-coverts; yellow eye.
6b   **Juvenile** Olive-green above; yellower below; lemon-yellow throat.

7   GREY LONGBILL *Macrosphenus concolor*   11.5 cm   **R**   **Page 644**
Forest. W Guinea to CAR–Congo; Bioko. Common.
Wholly olive-green; long, straight bill.

8   YELLOW LONGBILL *Macrosphenus flavicans*   13 cm   **R**   **Page 644**
Forest. SE Nigeria to CAR–Congo; Bioko. Uncommon/locally not uncommon.
Mainly olive-green above; whitish throat; rest of underparts bright olive-yellow.

PLATE 104: WARBLERS VIII – *Prinia, Schistolais, Heliolais, Urolais, Drymocichla, Scotocerca, Urorhipis* and *Spiloptila*

Rather slender, active species with long, graduated tails.

1   RIVER PRINIA *Prinia fluviatilis*   *c.* 12 cm   **R**   **Page 636**
Waterside vegetation. NW Senegal, Niger, NE Nigeria, Chad, N Cameroon. Local; distribution
inadequately known. Endemic.
As 2, but greyer above and whiter below. Voice different.

2   TAWNY-FLANKED PRINIA *Prinia subflava*   11–12 cm   **R**   **Page 635**
Various grassy and bushy habitats. Throughout. Common.
2a   Adult breeding Brownish above; buff supercilium; pale tawny flanks.
2b   Adult non-breeding (savanna only) Paler, rusty-brown above; tail longer; bill horn coloured.

3   WHITE-CHINNED PRINIA *Schistolais leucopogon*   14 cm   **R**   **Page 637**
Forest edge, farmbush. SE Nigeria to CAR–Congo. Uncommon/not uncommon.
Entirely grey except black mask and contrasting creamy-white throat.

4   SIERRA LEONE PRINIA *Schistolais leontica*   13 cm   **R**   **VU**   **Page 638**
Forest edge in hilly areas. NE Sierra Leone–Guinea to W Ivory Coast. Uncommon/rare; local. Endemic.
Dark ash-grey; whitish eye; buff flanks and undertail-coverts.

5   BANDED PRINIA *Prinia bairdii*   11.5 cm   **R**   **Page 636**
Forest edges. SE Nigeria to SW CAR–Congo. Uncommon/locally common.
5a   Adult Blackish-brown above with white spots on wings; boldly barred below.
5b   Juvenile Duller; underparts mainly greyish-brown, without barring.

6   RED-WINGED WARBLER *Heliolais erythroptera*   12 cm   **R**   **Page 636**
Wooded grassland. Patchily distributed, Senegambia–Sierra Leone to Chad–CAR. Not uncommon/
uncommon.
6a   Adult breeding Grey head and upperparts; bright rufous wings.
6b   Adult non-breeding Pale vinous-rufous upperparts; horn-coloured bill.

7   GREEN LONGTAIL *Urolais epichlora*   15 cm   **R**   **Page 637**
Montane and submontane forest. SE Nigeria–W Cameroon; Bioko. Locally common. Endemic.
Bright green above; tail long and usually heavily worn.

8   STREAKED SCRUB WARBLER *Scotocerca inquieta*   10 cm   **R**   **Page 635**
Desert scrub. Mauritania. Scarce and local.
Cocked, white-tipped tail; finely streaked crown; broad pale supercilium.

9   RED-WINGED GREY WARBLER *Drymocichla incana*   14 cm   **R**   **Page 638**
Wooded grassland. E Nigeria, Cameroon, CAR. Locally uncommon/rare.
Pale grey; conspicuous rufous wing panel.

10   RED-FRONTED WARBLER *Urorhipis rufifrons*   11 cm   **R**   **Page 642**
Desert scrub and bush. Chad. Uncommon.
Mouse-brown above; rufous forehead.

11   CRICKET WARBLER *Spiloptila clamans*   *c.* 11.5 cm   **R**   **Page 637**
Desert scrub and bush. Mauritania–N Senegal to Chad. Uncommon/locally not uncommon.
Pale; long tail; black-and-white pattern on wing-coverts.
11a   Adult male Hindcrown and nape pale grey.
11b   Adult female Hindcrown and nape cinnamon.

PLATE 105: WARBLERS IX – *Apalis*

Genus *Apalis*. Slender; tail rather long and graduated. Mostly in forest. Active; often in mixed-species flocks. Song typically a rhythmic repetition of a single note.

1 YELLOW-BREASTED APALIS *Apalis flavida* 11.5 cm **R** **Page 639**
Gallery forest, forest patches, bush, mangroves. Irregularly distributed from Gambia–Sierra Leone to S Chad–CAR, and coastal Gabon and Congo. Uncommon/rare.
Adult Grey head; green upperparts; yellow upper breast. Juvenile has green head, pale bill.

2 MASKED APALIS *Apalis binotata* 10 cm **R** **Page 639**
Old clearings and second growth in lowland forest. Cameroon, Eq. Guinea, NE Gabon. Locally not uncommon.
2a **Adult** Black head and throat; white malar stripe.
2b **Juvenile** Greenish head and upperparts; greyish throat; underparts tinged yellow.

3 BLACK-THROATED APALIS *Apalis jacksoni* 11.5 cm **R** **Page 640**
Forest above *c.* 500 m. SE Nigeria–W Cameroon, S Cameroon, Eq. Guinea, SW CAR, NE Gabon, NW Congo. Locally common/rare.
3a **Adult male** Black head and throat; white moustachial stripe; bright yellow underparts.
3b **Adult female** Grey head.

4 BLACK-CAPPED APALIS *Apalis nigriceps* 11.5 cm **R** **Page 640**
Lowland and gallery forest. Sierra Leone–SE Guinea to SW CAR–N Congo; Bioko. Locally common/ rare.
4a **Adult male** Black head and breast-band; golden-yellow mantle; whitish underparts.
4b **Adult female** Grey head and breast-band.

5 BLACK-COLLARED APALIS *Apalis pulchra* 13 cm **R** **Page 639**
Montane forest. SE Nigeria–W Cameroon. Locally common.
Slate-grey head and upperparts; black breast-band; rufous-chestnut flanks. Forages low; cocks tail.

6 GREY APALIS *Apalis cinerea* 13 cm **R** **Page 641**
Montane and submontane forest. SE Nigeria–W Cameroon, N Gabon, Bioko. Locally common.
6a **Adult** Greyish-brown head; grey upperparts; creamy-white underparts.
6b **Juvenile** Tinged olive above and yellow below.

7 BUFF-THROATED APALIS *Apalis rufogularis* 11.5 cm **R** **Page 640**
Lowland forest. S Benin to SW CAR–Congo; Bioko. Not uncommon.
7a **Adult male** Slate-grey head, throat and upperparts; white outer tail feathers.
7b **Adult female** Cinnamon-rufous throat.
7c **Juvenile** Olive-green above; pale yellow below.

8 GOSLING'S APALIS *Apalis goslingi* 11.5 cm **R** **Page 641**
Along rivers in lowland forest. Cameroon, Gabon, Congo and SW CAR. Locally common.
Adult Slate-grey above; pale grey below; throat creamy-white.

9 BAMENDA APALIS *Apalis bamendae* 11.5 cm **R** **Page 641**
Forest patches in highlands. W and C Cameroon. Not uncommon/locally common. Endemic.
Adult Dark brownish-grey above; grey below; forehead to throat rufous.

10 SHARPE'S APALIS *Apalis sharpii* 11.5 cm **R** **Page 640**
Lowland forest. W Guinea to Ghana; not uncommon/locally common. Togo; rare. Endemic.
10a **Adult male** All sooty-grey.
10b **Adult female** Pale grey below; buff throat.
10c **Juvenile** Greyish-olive above; pale lemon below; greyish breast-sides and flanks.

PLATE 106: SMALL CISTICOLAS

Streaked upperparts (warmer coloured in non-breeding season); aerial displays (hence 'cloud-scrapers'); open grassy areas.

1 PECTORAL-PATCH CISTICOLA *Cisticola brunnescens* 10–11 cm **R**   Page 634
Cameroon, Gabon, Congo. Locally not uncommon.
  1a Adult male breeding Top of head russet-brown; black lores; tail blackish above and below.
  1b Adult male non-breeding /adult female Top of head tawny broadly streaked black; faint loral mark.

2 AYRES'S CISTICOLA *Cisticola ayresii* 9–10 cm   Page 635
Status unclear. Records from coastal Gabon and C Congo possibly refer to 1.
Adult male breeding Top of head and rump rufous-brown; very faint loral mark; very short tail plain blackish above and below.

3 PALE-CROWNED CISTICOLA *Cisticola cinnamomeus* 10–11 cm **R**   Page 634
Congo. Locally common.
Adult male breeding Top of head plain rusty buff-brown; blackish around lores and forecrown; back heavily streaked.

4 ZITTING CISTICOLA *Cisticola juncidis* 10 cm **R**   Page 632
Patchily distributed throughout. Locally common/uncommon.
  4a Adult male breeding Top of head brownish-buff heavily streaked black; rump dull rufous-brown; tail black above with black subterminal spots below.
  4b Adult male non-breeding More buff; streaking less heavy.

5 BLACK-BACKED CISTICOLA *Cisticola eximius* *c.* 10 cm **R**   Page 633
S Senegal–Sierra Leone to Burkina Faso, Nigeria; N CAR; N Congo. Locally not uncommon/rare.
  5a Adult male breeding Top of head pale rufous-brown; rump plain rufous, contrasting with rest of upperparts; tail blackish above.
  5b Adult male non-breeding Top of head streaked black.

6 DAMBO CISTICOLA *Cisticola dambo* 10–12 cm   Page 634
SE Gabon. Locally not uncommon.
  6a Adult male breeding Resembles 4, but black better defined; unusually long tail plain black above and below.
  6b Adult male non-breeding Deeper buff; tail blackish edged rusty-buff.

7 DESERT CISTICOLA *Cisticola aridulus* 10–12 cm **R**   Page 633
S Mauritania–N Senegal to N Nigeria, Niger and Chad. Uncommon/not uncommon.
Paler, more sandy coloured than 4.

Plain upperparts not contrasting with top of head; typically sing from perch; savanna.

8 FOXY CISTICOLA *Cisticola troglodytes* 10 cm **R**   Page 632
S Chad, CAR. Local. Status and distribution inadequately known. Upperparts bright russet.

9 RUFOUS CISTICOLA *Cisticola rufus* 10 cm **R**   Page 632
Senegambia–Guinea to S Chad–NW CAR. Uncommon/locally not uncommon. Endemic.
Upperparts dull rust-brown.

10 SHORT-WINGED CISTICOLA *Cisticola brachypterus* *c.* 10 cm **R**   Page 631
Throughout, except arid north. Not uncommon/locally common.
  10a Adult breeding Upperparts dull brown.
  10b Adult non-breeding Upperparts warmer brown with faint dusky streaks.

Plain or streaked upperparts contrasting with rufous top of head; sing from perch; bush.

11 RED-PATE CISTICOLA *Cisticola ruficeps* 13 cm **R**   Page 630
Mali–N Ivory Coast to Chad–CAR. Uncommon/locally common.
  11a Adult breeding Top of head rusty-red; upperparts plain dark brown; undertail-coverts white.
  11b Adult non-breeding Upperparts heavily streaked black.

12 DORST'S CISTICOLA *Cisticola dorsti* 13 cm **R** DD   Page 631
Senegambia, NW Nigeria, N Cameroon, S Chad. Local. Distribution inadequately known. Endemic.
Adult breeding Similar to 11a but undertail-coverts buff. Song different.

13 PIPING CISTICOLA *Cisticola fulvicapillus* 10–11 cm **R**   Page 632
SE Gabon, C Congo. Locally common/uncommon.
Top of head dull rufous; upperparts plain earth-brown.

PLATE 107: MEDIUM-SIZED AND LARGE CISTICOLAS

Plain or faintly dappled upperparts.

1 **ROCK-LOVING CISTICOLA** *Cisticola aberrans* 13–15 cm **R**   **Page 628**
Rocky outcrops, savanna with boulders. Patchily distributed, S Mauritania–Sierra Leone to Chad–N CAR. Locally not uncommon.
Adult breeding Dull rufous top of head; rusty-buff supercilium; longish tail, black spots only visible from below.

2 **WHISTLING CISTICOLA** *Cisticola lateralis* 12.5–14 cm **R**   **Page 627**
Woodland and derived savanna. Senegambia–Liberia to CAR; also Gabon and Congo. Common/not uncommon.
2a **Sooty form** Greyish-brown above; flight feathers edged rufous. Vigorous and melodious song.
2b **Foxy form** Duller rufous-brown.

3 **RED-FACED CISTICOLA** *Cisticola erythrops* 12–14 cm **R**   **Page 626**
Various grassy and bushy habitats. Senegambia–Liberia to Cameroon and CAR; also Gabon and S Congo; locally common. Rare, S Mauritania.
Rufous face. Loud, varied song.

4 **CHUBB'S CISTICOLA** *Cisticola chubbi* 14 cm **R**   **Page 628**
Highlands. SE Nigeria–W Cameroon. Locally common/not uncommon.
Rusty top of head; black lores.

5 **RATTLING CISTICOLA** *Cisticola chiniana* 12–16 cm **R**   **Page 629**
Dry savanna woodland. S Gabon–C Congo. Scarce and local.
Adult breeding Dark rusty-brown top of head; dark brown upperparts; rufous-brown edges to flight feathers.

6 **SINGING CISTICOLA** *Cisticola cantans* 11.5–14 cm **R**   **Page 627**
Bush. Senegambia–Sierra Leone to S Niger, S Chad and CAR. Common/not uncommon.
Chestnut top of head and wing panel.

7 **BUBBLING CISTICOLA** *Cisticola bulliens* 12–15.5 cm   **Page 628**
Grassy habitats. Status unclear; no certain records. Known range: Cabinda to Angola.
Adult breeding As 8 but top of head browner and head-sides paler.

8 **CHATTERING CISTICOLA** *Cisticola anonymus* 12–15 cm **R**   **Page 627**
Lowland forest zone. S Nigeria to SW CAR–Congo. Locally common.
Adult As 5 but occurring in different habitat; wings uniform with rest of upperparts.

Streaked upperparts.

9 **TINKLING CISTICOLA** *Cisticola rufilatus* 13–14.5 cm **R**   **Page 629**
Open grassland with scattered trees. SE Gabon and C Congo. Scarce/locally common.
Adult breeding Rufous top of head and tail; buff supercilium.

10 **STOUT CISTICOLA** *Cisticola robustus* 14–16.5 cm **R**   **Page 630**
Rank grass with shrubs. E Nigeria–W Cameroon (highlands); C Congo. Locally common.
Rufous cap (streaked) and nape (unstreaked).

11 **WINDING CISTICOLA** *Cisticola galactotes* 12–15 cm **R**   **Page 629**
Moist habitats. Throughout, except arid north. Locally common/not uncommon.
11a **Adult breeding** Top of head dull rust-brown; russet-edged flight feathers. Distinctive rasping call.
11b **Adult non-breeding** Warmer coloured.

12 **CROAKING CISTICOLA** *Cisticola natalensis* 12.5–16 cm **R**   **Page 630**
Grassy habitats. Throughout, except arid north. Common/uncommon to rare.
12a **Adult breeding** Large and bulky; upperparts dark earth-brown streaked dusky; almost plain when worn.
12b **Adult non-breeding** Upperparts buffish-brown boldly streaked black.

PLATE 108: FLYCATCHERS I

Insectivorous and arboreal. Most have broad, flattened bills with broad gape and rictal bristles. Catch insects in short sallying flights from perch, by pounding on prey on ground or by picking it from foliage in warbler-like manner. Some join mixed-species flocks.

1 NIMBA FLYCATCHER *Melaenornis annamarulae* *c.* 19 cm **R VU** **Page 657**
Forest. SE Sierra Leone, SE Guinea, Liberia, Ivory Coast. Rare/scarce and local. Endemic.
Very dark plumbeous; robust.

2 NORTHERN BLACK FLYCATCHER *Melaenornis edolioides* *c.* 20 cm **R** **Page 658**
Various types of woodland. S Mauritania–Sierra Leone to S Chad–CAR. Uncommon/not uncommon.
All black; long tail.

SOUTHERN BLACK FLYCATCHER *Melaenornis pammelaina* 19–22 cm **R** **Page 658**
Wooded grassland. Congo. Rare and local. **See illustration on plate 147**.
Similar to 2, but plumage glossed steel-blue.

3 SPOTTED FLYCATCHER *Muscicapa striata* 13.5–14.5 cm **P** **Page 658**
Various wooded habitats. Throughout. Common/not uncommon. Vagrant, Cape Verde and Gulf of
Guinea Is.
Grey-brown; streaked crown and breast.

4 GAMBAGA FLYCATCHER *Muscicapa gambagae* 12–13 cm **R** **Page 659**
Wooded grassland. Distribution patchy; mainly N Ivory Coast–N Togo, NE Nigeria–Chad; also
Mali, Guinea, Burkina Faso. Rare/scarce.
Resembles 3, but smaller; streaks very faint; wings shorter.

5 AFRICAN DUSKY FLYCATCHER *Muscicapa adusta* 10–11 cm **R** **Page 660**
Various wooded habitats in highlands. SE Nigeria, Cameroon, NW CAR, Bioko. Locally not
uncommon/rare.
5a Adult Smaller than 4; underparts darker; throat white.
5b Juvenile Spotted above, mottled below.

6 OLIVACEOUS FLYCATCHER *Muscicapa olivascens* 14 cm **R** **Page 660**
Forest. Sierra Leone–SE Guinea to Ghana; S Nigeria to SW CAR–Congo. Rare/locally not uncommon.
Rather nondescript; mainly olivaceous-brown. Unobtrusive.

7 PALE FLYCATCHER *Melaenornis pallidus* 15–17 cm **R** **Page 658**
Various types of woodland. Throughout, except arid north. Not uncommon.
Medium-sized. Rather nondescript, brownish.

8 RED-BREASTED FLYCATCHER *Ficedula parva* 11.5 cm **V** **Page 665**
Palearctic vagrant, N Senegal.
Bold white sides to base of blackish tail (diagnostic).
8a Adult male breeding Orange-red throat; grey-brown head.
8b Adult female Creamy-buff throat; brown head.

The following three Palearctic migrants are very similar in male non-breeding and female plumages. Note in particular wing and tail patterns, and rump colour (see text for details).

9 COLLARED FLYCATCHER *Ficedula albicollis* 13 cm **P** **Page 664**
Niger, NE Nigeria, Chad, CAR, N Congo. Rare/scarce. Distribution inadequately known.
9a Adult male breeding White collar; large white patch on forehead; whitish rump; all-black tail.
9b Adult male non-breeding/female As 11b but primary patch larger.

10 SEMI-COLLARED FLYCATCHER *Ficedula semitorquata* 13 cm **P** **Page 664**
No certain records. Almost intermediate between 9 and 11.
10a Adult male breeding White half-collar (variable); second upper wingbar; more white in tail than 9
and 11.
10b Adult male non-breeding/female As 9b but usually has white-tipped median coverts.

11 PIED FLYCATCHER *Ficedula hypoleuca* 13 cm **P** **Page 663**
Various wooded habitats. Throughout. Not uncommon/scarce.
11a Adult male breeding White wing patch; small white patch on forehead; very small primary patch;
white edges to tail.
11b Adult male non-breeding/female Brownish above; white wingbar narrower.

PLATE 109: FLYCATCHERS II

1 SWAMP FLYCATCHER *Muscicapa aquatica* 13 cm **R**   **Page 660**
Near water in savanna. Senegambia, Mali, N Ivory Coast to S Chad–CAR. Uncommon/locally common.
Brown above; white below, washed brownish on breast and flanks.

2 CASSIN'S FLYCATCHER *Muscicapa cassini* 13 cm **R**   **Page 660**
By watercourses in forest and gallery forest. Guinea to S CAR–Congo. Common/not uncommon.
Ashy-grey.

3 LITTLE GREY FLYCATCHER *Muscicapa epulata* 9.5 cm **R**   **Page 661**
Forest. Sierra Leone–SE Guinea to Togo; Nigeria; Cameroon to SW CAR–Congo. Scarce.
Small. Greyish; indistinctly streaked below.

4 YELLOW-FOOTED FLYCATCHER *Muscicapa sethsmithi* 10.5 cm **R**   **Page 661**
Forest. SE Nigeria to SW CAR–Congo; Bioko. Locally not uncommon/rare.
Similar to 3 but darker; legs and feet yellow.

5 ASHY FLYCATCHER *Muscicapa caerulescens* 13 cm **R**   **Page 659**
Woodland, forest edge. Sierra Leone–SE Guinea to SW CAR–Congo. Uncommon.
White eye-ring and white supraloral stripe.

6 TESSMANN'S FLYCATCHER *Muscicapa tessmanni* *c.* 13 cm **R**   **DD**   **Page 662**
Forest. Distribution patchy; SE Sierra Leone to SW Ghana; S Cameroon–Eq. Guinea. Rare.
Resembles 7, but throat and belly pure white.

7 DUSKY-BLUE FLYCATCHER *Muscicapa comitata* 12 cm **R**   **Page 661**
Forest zone. Sierra Leone–SE Guinea to CAR–Congo. Not uncommon/rare.
7a   *M. c. comitata* Dark bluish-slate; throat and belly whitish; white eye-ring and supraloral stripe. From
      Cameroon south and east.
7b   *M. c. aximensis* Darker; less marked face; less white below. From Nigeria west.

8 LEAD-COLOURED FLYCATCHER *Myioparus plumbeus* 14 cm **R**   **Page 663**
Wooded savanna and forest edges. Senegambia–Liberia to S Chad–CAR and Congo. Uncommon/rare.
Warbler-like. Black tail with white outer feathers.

9 USSHER'S FLYCATCHER *Muscicapa ussheri* 13 cm **R**   **Page 662**
Forest edges. Guinea to Togo; also Nigeria. Locally not uncommon/rare. Endemic.
9a   Wholly dark. Perches high on dead branches.
9b   **In flight** Martin-like appearance.

10 SOOTY FLYCATCHER *Muscicapa infuscata* 13 cm **R**   **Page 662**
Forest. Nigeria to SW CAR–Congo. Uncommon/locally not uncommon.
Similar to 9; warmer coloured underparts only visible in good light.

11 GREY-THROATED FLYCATCHER *Myioparus griseigularis* 13 cm **R**   **Page 663**
Forest. E Sierra Leone to Ghana; SE Nigeria to CAR–Congo. Rare/locally uncommon.
Wholly grey; no distinctive features.

12 WHITE-BROWED FOREST FLYCATCHER *Fraseria cinerascens* *c.* 17 cm **R**   **Page 657**
Riverine forest. S Senegal to Ghana; Nigeria to CAR–Congo. Uncommon/locally not uncommon.
White supraloral streak. Unobtrusive.

13 FRASER'S FOREST FLYCATCHER *Fraseria ocreata* 18 cm **R**   **Page 657**
Forest. W Guinea to Ghana; Nigeria to CAR–Congo; Bioko. Uncommon/locally not uncommon.
Stout. Dark slate-grey above; white below; dark grey crescentic bars on breast. Vocal.

PLATE 110: MONARCHS

A diverse family of insectivorous, arboreal species with crests or incipient crests, and relatively long, graduated tails. Very active, gleaning most prey from branches in middle and lower levels, often brushing foliage with fanned tail and partially open wings to dislodge insects. Often in mixed-species flocks.

Genus *Terpsiphone* (**paradise flycatchers**). Distinctive genus. Some males have extremely elongated median tail feathers when breeding. Some extremely variable and readily hybridise. Vocal and conspicuous. Call a rasping *zwhee-zwhèh*.

1 AFRICAN PARADISE FLYCATCHER *Terpsiphone viridis* 18 cm **R/M**   **Page 667**
Savanna and forest zones. Throughout, except arid north. Common/rare.
Glossy blue-black crested head. Adult males with very long, ribbon-like tail streamers (+10–18 cm).
Upper- and underparts very variable; two extremes illustrated. Adult females have upperparts and tail always rufous; tail shorter.
  1a   **Adult male rufous 'morph'**
  1b   **Adult male white 'morph'**
  1c   **Adult male** *plumbeiceps* Duller than 1a; no white on wing.

2 RED-BELLIED PARADISE FLYCATCHER *Terpsiphone rufiventer* 18 cm **R**   **Page 668**
Forest and savanna zones. Senegambia to SW CAR–Congo; Bioko. Common.
Rufous underparts. Tail of male + max. 7 cm.
  2a   **Adult male** *rufiventer* Distinct crest, white wing panel, long tail streamers. Senegambia to W Guinea.
  2b   **Adult male** *nigriceps* No crest, no wing panel, much shorter tail. Guinea to SW Benin.
      T. r. fagani (Benin–SW Nigeria) and *ignea* (SE CAR) similar. In *schubotzi* (SW CAR–N Congo) and *mayombe* (S Congo) tail bluish-slate.
  2c   **Adult male** *neumanni* Bluish-slate upperparts and tail. SE Nigeria to S Congo.
      T. r. tricolor (Bioko) similar.

3 RUFOUS-VENTED PARADISE FLYCATCHER *Terpsiphone rufocinerea* 18 cm **R**   **Page 668**
Forest and woodland. SW Cameroon (SE Nigeria?) to S Congo. Common.
Dark greyish-blue underparts; rufous undertail-coverts. Tail of male + 1–11 cm.

4 BATES'S PARADISE FLYCATCHER *Terpsiphone batesi* 18 cm **R**   **Page 668**
Forest. S Cameroon to SW CAR–Congo. Rare/common.
As 3 but less glossy and lacking crest. Tail of male + 1.0–10.5 cm.

5 AFRICAN BLUE FLYCATCHER *Elminia longicauda* 18 cm **R**   **Page 665**
Woodland. Senegambia–Sierra Leone to S Chad–CAR and Congo. Locally not uncommon.
Bright pale blue; short crest; long, graduated tail.

6 CHESTNUT-CAPPED FLYCATCHER *Erythrocercus mccallii* 10 cm **R**   **Page 665**
Lowland forest. W Guinea to SW CAR–Congo; also SW Mali. Not uncommon.
Very small; rufous crown and tail.

7 WHITE-BELLIED CRESTED FLYCATCHER *Elminia albiventris* 11 cm **R**   **Page 666**
Montane forest. SE Nigeria–W Cameroon, Bioko. Not uncommon.
Dark slate; dull black head; white belly.

8 DUSKY CRESTED FLYCATCHER *Elminia nigromitrata* 11 cm **R**   **Page 666**
Lowland forest. Sierra Leone–SE Guinea to Ghana; S Nigeria to SW CAR–Congo; SE CAR. Uncommon/locally not uncommon.
Dark blue-slate; dull black crown.

9 BLUE-HEADED CRESTED FLYCATCHER *Trochocercus nitens* 15 cm **R**   **Page 666**
Lowland and mid-elevation forest. Guinea to Togo; Nigeria to SW CAR–Congo. Rare/not uncommon.
  9a   **Adult male** Glossy blackish-blue with clearly contrasting grey underparts; short glossy crest.
  9b   **Adult female** Much greyer than male; only crest glossy blackish-blue; underparts wholly grey.

PLATE 111: BLACK-AND-WHITE FLYCATCHER, SHRIKE FLYCATCHER AND BATISES

1 BLACK-AND-WHITE FLYCATCHER *Bias musicus* 16 cm **R**   **Page 670**
Forest edge. Guinea to SW CAR–Congo. Generally uncommon. Vagrant, Gambia.
1a **Adult male** Glossy black and white; prominent crest; yellow eye. Vocal.
1b **Adult female** Mainly rufous-chestnut above.
1c **Adult male in flight** Broad rounded wings with white patch; short tail.

2 SHRIKE FLYCATCHER *Megabyas flammulatus* 15–16 cm **R**   **Page 669**
Forest. Sierra Leone–SE Guinea to SW CAR–Congo, Bioko; uncommon. Vagrant, Gambia, SW Mali.
2a **Adult male** Glossy black above; white rump; white below. Moves tail sideways.
2b **Adult female** Earth-brown above; rufous rump; coarsely streaked brown below.

Genus *Batis*. Small, active, arboreal species with relatively large heads and short tails. Flight swift and bouncing, often on whirring wings. Snap bill loudly when catching insects.

3 SENEGAL BATIS *Batis senegalensis* 10.5 cm **R**   **Page 673**
Open woodland. Sahel and savanna zones, east to Niger–N Cameroon. Locally common. Vagrant, Liberia. Endemic.
3a **Adult male** Long broad supercilium.
3b **Adult female** Supercilium and wingbar rusty-buff; pale chestnut breast-band.

4 BLACK-HEADED BATIS *Batis minor* *c.* 11.5 cm **R**   **Page 674**
Open woodland and scrub. Cameroon, S Chad, CAR and S Gabon, Congo. Uncommon/locally common.
4a **Adult male** As 5a, but crown darker grey; breast-band slightly broader.
4b **Adult female** As 5b, but breast-band slightly broader and darker chestnut.
4c **Immature** As adult female but with rusty-buff tinge above, esp. on supercilium and wingbar. Other immature batises have similar plumage.

5 GREY-HEADED BATIS *Batis orientalis* 10 cm **R**   **Page 674**
Open woodland, thorn scrub. NE Nigeria, NE Cameroon, Chad, CAR. Not uncommon to scarce/rare.
5a **Adult male** As 4a, but crown paler; breast-band slightly narrower.
5b **Adult female** As 4b, but breast-band slightly narrower and paler chestnut.

6 CHINSPOT BATIS *Batis molitor* 12 cm **R**   **Page 673**
Wooded savanna. SE Gabon–Congo. Scarce/rare and local.
6a **Adult male** Crown grey; narrow supercilium. Compare 4a (and 5a).
6b **Adult female** Large chestnut throat patch.

7 BIOKO BATIS *Batis poensis* *c.* 12 cm **R**   **Page 675**
Forest. Sierra Leone to Gabon and SW CAR–N Congo. Rare/locally not uncommon. Endemic.
7a **Adult male** *occulta* Crown blackish; distinct supraloral spot; indistinct narrow supercilium.
7b **Adult female** *occulta* Well-defined chestnut breast-band.

8 ANGOLA BATIS *Batis minulla* 10 cm **R**   **Page 674**
Vestigial and fringing forest. SE Gabon–Congo. Uncommon/locally common.
8a **Adult male** No supercilium, broad breast-band.
8b **Adult female** Bright chestnut breast-band broadening on flanks.

9 VERREAUX'S BATIS *Batis minima* 10 cm **R NT**   **Page 675**
Forest. S Cameroon, Eq. Guinea, Gabon. Scarce and local. Endemic.
9a **Adult male** As 7a but supraloral spot indistinct, no supercilium, smaller size.
9b **Adult female** Slate-grey breast-band.

1c

1b

1a

2b

2a

3b

3a

4c

4b

4a

5b

5a

6b

6a

7b

7a

8b

8a

9b

9a

PLATE 112: WATTLE-EYES

Genus *Platysteira*. Distinguished by conspicuous scarlet eye-wattles. Behaviour similar to batises, but jizz different, due to longer tail and larger size.

1 COMMON WATTLE-EYE *Platysteira cyanea* 13 cm **R**   **Page 672**
Various wooded habitats. Throughout, except arid north. Common.
1a Adult male Scarlet eye-wattle, black breast-band, white wingbar.
1b Adult female Throat and upper breast dark chestnut.
1c Immature Olive-grey above; wingbar rusty-buff; buffish wash on breast and flanks.

2 BANDED WATTLE-EYE *Platysteira laticincta* 13 cm **R** EN   **Page 672**
Montane forest. W Cameroon (Bamenda Highlands). Locally not uncommon. Endemic.
2a Adult male As 1a but without wingbar.
2b Adult female Throat and upper breast blue-black.

Genus *Dyaphorophyia*. Small, colourful, very short-tailed species with blue, purplish or green eye-wattles. Active; often joining mixed-species flocks.

3 WHITE-SPOTTED WATTLE-EYE *Dyaphorophyia tonsa* 9.5 cm **R**   **Page 671**
Forest. SE Nigeria to SW CAR–Congo. Rare/locally not uncommon.
3a Adult male As 4c but with short white superciliary streak (partly hidden by eye-wattle).
3b Adult female As 4b but with glossy black crown, partly hidden superciliary streak, longer malar stripe.

4 CHESTNUT WATTLE-EYE *Dyaphorophyia castanea* 10 cm **R**   **Page 670**
Forest. Common/uncommon.
4a Adult male *castanea* Glossy blue-black above; white rump; broad breast-band; purplish eye-wattle. Nigeria to S CAR–Congo; Bioko.
4b Adult female *castanea* Slate-grey head; chestnut upperparts, throat and breast.
4c Adult male *hormophora* White neck collar. Guinea–Sierra Leone to Benin.
4d Juvenile Paler than 4b with broad, diffuse mottled band of grey and chestnut on upper breast.

5 BLACK-NECKED WATTLE-EYE *Dyaphorophyia chalybea* 9 cm **R**   **Page 671**
Undergrowth in lowland and montane forest. Cameroon, Eq. Guinea, Gabon, Bioko. Locally not uncommon.
Adult male Glossy greenish-black above; white underparts with pale yellow cast; emerald-green eye-wattle. Adult female similar but duller.

6 RED-CHEEKED WATTLE-EYE *Dyaphorophyia blissetti* 9 cm **R**   **Page 671**
Undergrowth in lowland forest. Guinea to SW Cameroon. Uncommon/locally not uncommon. Endemic.
6a Adult male Glossy greenish-black above; chestnut patch on cheeks and neck-sides; greenish-blue eye-wattle.
6b Adult female Duller version of 6a.
6c Juvenile Throat and upper breast tawny bordered chestnut.

7 YELLOW-BELLIED WATTLE-EYE *Dyaphorophyia concreta* 10 cm **R**   **Page 672**
Undergrowth in lowland and montane forest. Rare/locally common.
7a Adult male *concreta* Rich chestnut underparts; bright emerald-green eye-wattle. Sierra Leone–SE Guinea to Ghana.
7b Adult female *concreta* Bright yellow underparts; throat and upper breast chestnut.
7c Adult male *graueri* Deep orange-yellow underparts. SE Nigeria to SW CAR–Congo.
7d Adult female *graueri* Throat and upper breast have chestnut wash.
7e Juvenile Greyish-olive above; dirty white washed olive and dull yellowish below.

PLATE 113: ILLADOPSISES AND HILL BABBLER

Unobtrusive forest dwellers, mostly foraging low. Plumages often similar, but voice key to identity. Songs, uttered in duet or in groups, consist of pure, far-carrying whistles that are hard to locate.

1   AFRICAN HILL BABBLER *Pseudoalcippe abyssinica* 14–15 cm   **R**   **Page 679**
Montane forest. E Nigeria–W Cameroon; Bioko. Locally common.
   1a   Adult *atriceps* Black head; dark chestnut above; slate-grey below.
   1b   Adult *monachus* Slate-grey head. Restricted to Mt Cameroon.
      P. a. claudei (Bioko) similar.

2   BROWN ILLADOPSIS *Illadopsis fulvescens* c. 16 cm   **R**   **Page 677**
Forest and forest-scrub mosaic. S Senegal–Liberia to CAR–Congo. Common.
   2a   Adult *fulvescens* Dark fulvous-brown above; paler below; whitish throat. Compare 5. Cameroon to SW CAR–Congo.
      I. f. gularis (from Ghana west) and *ugandae* (SE CAR) similar.
   2b   Adult *moloneyana* Wholly fulvous-brown below, darker than 2a. E Ghana–Togo.
      I. f. iboensis (S Nigeria–W Cameroon) similar.

3   BLACKCAP ILLADOPSIS *Illadopsis cleaveri* 15 cm   **R**   **Page 678**
Forest. Sierra Leone to Ghana; Nigeria to CAR–Congo; Bioko. Not uncommon. Endemic.
   3a   Adult *cleaveri* Dull black cap; whitish or pale greyish lores and supercilium.
   3b   Adult *marchanti* Top of head olive-grey. S Nigeria.
      I. c. poensis (Bioko) similar.

4   SCALY-BREASTED ILLADOPSIS *Illadopsis albipectus* 15 cm   **R**   **Page 677**
Forest. SE CAR. Common.
Breast feathers tipped olivaceous-brown giving scaly appearance.

5   PALE-BREASTED ILLADOPSIS *Illadopsis rufipennis* 15 cm   **Page 676**
Forest. W Guinea to Ghana; SW Nigeria to CAR–Congo; Bioko. Uncommon/locally common.
As 2a but upperparts slightly darker russet-brown; throat pure white; centre of belly whitish.

6   RUFOUS-WINGED ILLADOPSIS *Illadopsis rufescens* 17 cm   **R   NT**   **Page 678**
Forest. S Senegal–Liberia to Togo. Uncommon/rare. Endemic.
Thrush-like. Dark russet-brown above; whitish below; flanks and undertail-coverts olivaceous-grey; strong, pale-coloured legs. Song diagnostic. Compare 7.

7   PUVEL'S ILLADOPSIS *Illadopsis puveli* 18 cm   **R**   **Page 678**
Northern part of forest zone, gallery forest. S Senegambia to Cameroon. Uncommon.
As 6 but paler, more rufous above. Voice different. Compare also Spotted Thrush Babbler (Plate 114: 7).

8   GREY-CHESTED ILLADOPSIS *Kakamega poliothorax* 17 cm   **R**   **Page 679**
Montane forest. SE Nigeria–W Cameroon; Bioko. Uncommon/rare to locally common.
Rich chestnut-brown above; grey below; throat and central belly white.

PLATE 114: BABBLERS AND PICATHARTES

Babblers. Sturdy and thrush-like. Most species occur in small, noisy groups on or close to ground.

1 WHITE-THROATED MOUNTAIN BABBLER *Kupeornis gilberti* 21 cm **R EN** **Page 681**
Montane forest. SE Nigeria–W Cameroon. Locally common/not uncommon. Endemic.
Dark brown; face to breast white.

2 CAPUCHIN BABBLER *Phyllanthus atripennis* 24 cm **R** **Page 681**
Forest edge, gallery forest, thickets. Senegambia–Liberia to Cameroon–CAR. Uncommon/not
uncommon.
2a Adult *atripennis* Very dark chestnut; head to upper breast grey; greenish-yellow bill. West of Ghana.
2b Adult *haynesi* Crown blackish; grey area reduced to head-sides and upper throat. Ghana to
Cameroon.
P. a. bohndorffi (CAR) has chestnut throat.

3 BLACKCAP BABBLER *Turdoides reinwardtii* 25 cm **R** **Page 680**
Savanna zone. Senegambia–Guinea to Chad–CAR. Uncommon/locally common.
3a Adult *reinwardtii* Brown, with black head; contrasting creamy eye. W of Ivory Coast.
3b Adult *stictilaemus* Black of head not sharply defined; underparts darker than 3a. From SE Mali–
Ivory Coast east.

4 BROWN BABBLER *Turdoides plebejus* 24 cm **R** **Page 680**
Bushy habitats. Mauritania–Sierra Leone to S Chad–CAR. Common/not uncommon.
Greyish-brown above; head-sides, rump and underparts paler; faint whitish squamations on breast.

5 ARROW-MARKED BABBLER *Turdoides jardineii* 24 cm **R** **Page 680**
Bushy habitats. S Gabon–Congo. Uncommon and local.
Slightly darker and better marked than 4.

6 DUSKY BABBLER *Turdoides tenebrosus* 24 cm **Page 681**
Wooded savanna; usually near water. No certain records in W Africa. Occurs in Sudan.
Dark brown; scaly forehead, throat and breast.

7 SPOTTED THRUSH BABBLER *Ptyrticus turdinus* 20 cm **R** **Page 679**
Gallery forest, thickets. Cameroon, CAR. Uncommon/locally not uncommon.
Rufous above; white below spotted brown on breast; strong, pale legs. Compare Puvel's Illadopsis (Plate
113: 7).

8 FULVOUS BABBLER *Turdoides fulvus* 25 cm **R** **Page 681**
Arid scrub country. Mauritania–N Senegal to Niger–Chad. Locally not uncommon/rare.
Wholly sandy-buff; long tail.

Picathartes. Strange-looking, slender forest birds with bare head and long, strong legs. Dependent on occurrence
of caves and overhanging rocks (hence alternative name 'rockfowl').

9 YELLOW-HEADED PICATHARTES *Picathartes gymnocephalus* *c.* 38 cm **R VU** **Page 676**
Forest with caves and rocky outcrops. Guinea to Ghana. Scarce and local. Endemic.
Head yellow with large black patch; neck and underparts white.

10 RED-HEADED PICATHARTES *Picathartes oreas* *c.* 38 cm **R VU** **Page 676**
Forest with caves and rocky outcrops. SE Nigeria to Gabon; Bioko. Locally not uncommon. Endemic.
Head blue, black and crimson; neck greyish; underparts yellowish-buff.

PLATE 115: TITS, PENDULINE TITS, SPOTTED CREEPER, WHITE-EYE AND SPEIROPS

Tits. Rather small, arboreal birds with short, relatively stout bills and almost square tails. Active and agile, often hanging upside-down when searching for insects, nuts and seeds. Join mixed-species flocks. Attract attention by rasping calls.

1 WHITE-SHOULDERED BLACK TIT *Parus (leucomelas) guineensis* 14 cm **R**   **Page 683**
Savanna. Senegambia–N Sierra Leone to S Chad–CAR. Not uncommon/uncommon.
Black; white wing patch; conspicuous yellow eye.

2 WHITE-WINGED BLACK TIT *Parus leucomelas* 15 cm **R**   **Page 683**
Savanna. S Gabon–Congo. Not uncommon/uncommon.
As 1 but eye brown; outer tail feathers narrowly edged and tipped white.

3 WHITE-BELLIED TIT *Parus albiventris* 14–15 cm **R**   **Page 682**
Wooded savanna and forest edges in highlands. SE Nigeria–Cameroon. Not uncommon/rare.
Resembles 1 but lower breast and abdomen white; eye dark.

4 RUFOUS-BELLIED TIT *Parus rufiventris* 14–15 cm **R**   **Page 683**
Savanna. Congo. Rare.
Black head; cinnamon-rufous underparts; yellow eye.

5 DUSKY TIT *Parus funereus* 13–14 cm **R**   **Page 682**
Forest. Sierra Leone–SE Guinea to Ghana; Cameroon to SW CAR–N Congo. Rare/scarce.
Entirely black; red eye.

Penduline tits. Tiny, mainly insectivorous and arboreal birds with short tails and short, sharply pointed bills. Active but unobtrusive; often in small parties; some occasionally with other small insectivores.

6 GREY PENDULINE TIT *Anthoscopus caroli* 8.0–8.5 cm **R**   **Page 685**
Savanna. SE Gabon–Congo. Scarce.
Yellowish-olive above; off-white below; yellow forehead.

7 TIT-HYLIA *Pholidornis rushiae* 7.5 cm **R**   **Page 685**
Forest. Sierra Leone–SE Guinea to Ghana; Nigeria to CAR–Congo; Bioko. Not uncommon.
Grey head and breast finely streaked dusky; yellow rump.

8 FOREST PENDULINE TIT *Anthoscopus flavifrons* *c.* 9 cm **R**   **Page 684**
Forest. Liberia to Ghana; Nigeria to Congo. Rare/scarce (probably overlooked).
Olive-green above; olivaceous-grey washed yellow below; golden-yellow forehead (hard to see).

9 SENNAR PENDULINE TIT *Anthoscopus punctifrons* 7.5–8.5 cm **R**   **Page 684**
Dry Acacia savanna. Mauritania–N Senegal to N Cameroon–Chad. Uncommon/locally common.
Yellowish-olive above; buffish-white below; yellowish forehead with blackish dots.

10 YELLOW PENDULINE TIT *Anthoscopus parvulus* 7.5–8.0 cm **R**   **Page 684**
Savanna and Sahel. S Mauritania–Guinea to Chad–CAR. Uncommon/scarce; local.
Olive-yellow above; bright yellow below; bright yellow forehead with blackish dots; variably distinct white wingbar.

Spotted Creeper. Unmistakable, arboreal bird with long, decurved bill and short, strong legs.

11 SPOTTED CREEPER *Salpornis spilonotus* 15 cm **R**   **Page 686**
Savanna. S Senegambia–NW Sierra Leone to Burkina Faso–Togo; Nigeria to S Chad–CAR. Uncommon/rare; local.
Cryptic brown plumage spotted and barred white. Short tail. Clings to tree trunks.

White-eyes. Small, arboreal, warbler-like birds, deriving their name from the distinct white eye-ring shown by most species. Active.

12 YELLOW WHITE-EYE *Zosterops senegalensis* 10–11 cm **R**   **Page 699**
Various wooded habitats. Almost throughout; also Bioko. Common/not uncommon.
Small. Yellowish-olive above; bright yellow below; white eye-ring.

13 MOUNT CAMEROON SPEIROPS *Speirops melanocephalus* 13 cm **R** **VU**   **Page 700**
Forest. Mt Cameroon. Common at 1820–3000 m. Endemic.
Olive-brown and grey; black cap; white throat; narrow white band on forehead and eye-ring.

PLATE 116: SUNBIRDS I

Distinctive passerines with long, slender, sharply pointed, decurved bills (shortest and least decurved in *Anthreptes*, *Deleornis* and *Hedydipna*; this plate). Feed on nectar, insects and spiders. Usually in pairs but may gather at a favourable food source. Active, restless and pugnacious, with rapid and dashing flight.

1 WESTERN VIOLET-BACKED SUNBIRD *Anthreptes longuemarei* 13–14 cm **R**   **Page 686**
Wooded savanna, gallery forest. Senegambia–Sierra Leone to Chad–CAR; also SE Gabon and S Congo. Uncommon/rare.
1a   Adult male Glossy violet above and on throat; white below.
1b   Adult female Brown above; white supercilium; yellow belly; tail glossed violet.

2 BROWN SUNBIRD *Anthreptes gabonicus* 10 cm **R**   **Page 687**
Mangroves, forested rivers. Senegambia–Liberia to Ghana; Nigeria to Congo. Uncommon/locally common.
Adult Grey-brown above; whitish below; narrow white lines above and below eye.

3 VIOLET-TAILED SUNBIRD *Anthreptes aurantium* 13–14 cm **R**   **Page 687**
Forested rivers; also mangroves. Cameroon to S CAR–Congo. Uncommon and local.
3a   Adult male Glossy violet-blue/green above and on throat; pale brownish-buff below.
3b   Adult female Glossy green-blue above; white supercilium; yellow lower underparts.

4 PYGMY SUNBIRD *Hedydipna platura* 9–10 cm **R/M**   **Page 693**
Wooded savanna, Acacia scrub, gardens. Mauritania–Guinea to Chad–CAR. Common.
4a   Adult male breeding Glossy coppery-green upperparts and throat; golden-yellow underparts; very long tail streamers (projecting up to 7 cm).
4b   Adult male non-breeding As 4c; some black on throat; some glossy green on wing-coverts.
4c   Adult female Grey-brown above; yellow below; faint yellowish supercilium.

5 FRASER'S SUNBIRD *Deleornis fraseri* 11–13 cm **R**   **Page 688**
Forest. W Guinea to Togo; Nigeria to SW CAR–Congo; Bioko. Common.
Plain green plumage; pale, straight bill; pale eye-ring. Adult male has pectoral tufts.

6 GREEN SUNBIRD *Anthreptes rectirostris* 10 cm **R**   **Page 687**
Forest zone. Uncommon/locally common.
6a   Adult male *rectirostris* Yellow throat; glossy green breast-band; pale grey lower breast. Compare 8a. Sierra Leone–SE Guinea to Ghana.
6b   Adult male *tephrolaemus* Grey throat; abdomen greyer than 6a. Benin to CAR–Congo; Bioko.
6c   Adult female Warbler-like; olive-green above; mainly olive-yellow below.

7 WESTERN OLIVE SUNBIRD *Cyanomitra obscura* 13–15 cm **R**   **Page 691**
Forest, gallery forest. Throughout; also Bioko and Príncipe. Very common.
Plain olive-green; non-glossy; long decurved bill.

8 COLLARED SUNBIRD *Hedydipna collaris* 10 cm **R**   **Page 693**
Forest and savanna zones. Senegambia–Liberia to CAR–Congo; Bioko. Common.
8a   Adult male Bright glossy green upperparts and throat; yellow underparts. Compare 6a and Variable Sunbird (Plate 117: 9a).
8b   Adult female Entire underparts yellow.
8c   Juvenile Olive above; greyish and lemon-yellow below; yellowish stripes above and below eye.

9 BATES'S SUNBIRD *Cinnyris batesi* 9.5 cm **R**   **Page 698**
Forest. Liberia to Ghana; W Nigeria to SW CAR–Congo; Bioko. Rare/locally common.
Very small, non-glossy. Bill decurved. Compare 7 and 10.

10 LITTLE GREEN SUNBIRD *Anthreptes seimundi* 9.5 cm **R**   **Page 688**
Forest. W Guinea to Togo; Nigeria to S CAR–Congo; Bioko. Uncommon.
Very small, non-glossy. Bill rather straight. Compare 9.

11 URSULA'S SUNBIRD *Cinnyris ursulae* 10 cm **R NT**   **Page 698**
Montane and submontane forest. W Cameroon; locally common. Bioko; rare. Endemic.
Very small. Olive above; greyish below.

12 BUFF-THROATED SUNBIRD *Chalcomitra adelberti* 11.5–12.0 cm **R**   **Page 691**
Forest zone. Sierra Leone–SE Guinea to SE Nigeria–W Cameroon. Not uncommon but local. Endemic.
12a   Adult male Straw-coloured throat; black breast-band; rest of underparts chestnut.
12b   Adult female Olive-brown above; creamy streaked olive-brown below.
12c   Immature male Dark greyish-olive above; dark bib.

PLATE 117: SUNBIRDS II

1 REICHENBACH'S SUNBIRD *Anabathmis reichenbachii* 13–14 cm **R** **Page 688**
Coastal scrub, gardens. Liberia to Ghana; Nigeria to Congo; along Congo R. to SW CAR. Uncommon/ locally common.
- **1a** **Adult male** Glossy blue head and throat; bright yellow lower abdomen; graduated, pale-tipped tail. Adult female similar.
- **1b** **Juvenile** Yellowish-green above; yellow lower abdomen.

2 GREEN-HEADED SUNBIRD *Cyanomitra verticalis* 13–14 cm **R** **Page 690**
Wooded savanna, forest, gardens. Senegambia–Liberia to CAR–Congo. Locally common.
- **2a** **Adult male** Glossy blue-green head and throat; dusky-grey underparts.
- **2b** **Adult female** Wholly pale grey underparts.
- **2c** **Immature male** Forehead and throat blackish-grey.

3 CAMEROON SUNBIRD *Cyanomitra oritis* 12–13 cm **R** **Page 690**
Montane and submontane forest. SE Nigeria–W Cameroon; Bioko. Generally common. Endemic.
Adult male Glossy bluish-purple head and throat. Adult female similar but without lemon pectoral tufts.

4 BLUE-THROATED BROWN SUNBIRD *Cyanomitra cyanolaema* 14–15 cm **R** **Page 690**
Forest. Sierra Leone–SE Guinea to S CAR–Congo; Bioko. Common.
- **4a** **Adult male** Dark; glossy blue forecrown and throat; relatively long tail.
- **4b** **Adult female** White line above and below eye; whitish throat; mottled breast.

5 CARMELITE SUNBIRD *Chalcomitra fuliginosa* 13–14 cm **R** **Page 692**
Mangroves, coastal and riverine scrub. Sierra Leone to Congo. Locally common/rare.
- **5a** **Adult male (fresh)** Dark brown; dark glossy blue forehead; glossy violet throat. Plumage bleaches with wear; then particularly conspicuous.
- **5b** **Adult female** Very pale; throat mottled brownish; breast indistinctly streaked.

6 AMETHYST SUNBIRD *Chalcomitra amethystina* 13.5–14.0 cm **R** **Page 692**
Wooded savanna, forest edge, thickets, gardens. SE Gabon–Congo. Uncommon/locally common.
- **6a** **Adult male** Usually appears black; glossy green crown; glossy violet throat.
- **6b** **Adult female** Olivaceous above; paler below; indistinctly streaked breast.
- **6c** **Immature male** Resembles 6b but throat glossy violet.

7 GREEN-THROATED SUNBIRD *Chalcomitra rubescens* 12–13 cm **R** **Page 691**
Forest, gardens. SE Nigeria to S CAR–Congo; Bioko. Common/not uncommon.
- **7a** **Adult male** *rubescens* Usually appears black; glossy green forecrown and throat.
- **7b** **Adult male** *crossensis* Lacks green throat. SE Nigeria–W Cameroon.
- **7c** **Adult female** Brown above; pale supercilium; dirty yellowish below with streaks.

8 SCARLET-CHESTED SUNBIRD *Chalcomitra senegalensis* 14–15 cm **R/M** **Page 692**
Savanna zone. S Mauritania–Sierra Leone to Chad–CAR. Common/not uncommon.
- **8a** **Adult male** Glossy red lower throat and breast.
- **8b** **Adult female** Grey-brown above; broadly streaked below; dusky throat.

9 VARIABLE SUNBIRD *Cinnyris venustus* 10 cm **R** **Page 696**
Savanna zone. Senegambia to CAR; also S Gabon–S Congo. Uncommon/locally common.
- **9a** **Adult male** Violet forecrown, upper throat and breast; yellow belly.
- **9b** **Adult female** Grey-brown washed olive above; unstreaked, yellowish below.
- **9c** **Immature male** Blackish throat.

10 TINY SUNBIRD *Cinnyris minullus* 9–10 cm **R** **Page 694**
Forest. Sierra Leone to Ghana; Nigeria to Congo; Bioko. Rare/locally not uncommon.
- **10a** **Adult male** As 11a but with shorter, less decurved bill; rump bluish-green.
- **10b** **Adult female** As 11b but less yellow below; bill as 10a.

11 OLIVE-BELLIED SUNBIRD *Cinnyris chloropygius* 10.5 cm **R** **Page 694**
Various habitats in forest zone and moist savanna. S Senegal to S Chad–CAR and Congo; Bioko. Common.
- **11a** **Adult male** Glossy green above and on throat; bright red breast. Compare 10a.
- **11b** **Adult female** Dark olive above; yellowish supercilium; plain, dirty yellow below.

PLATE 118: SUNBIRDS III

1 NORTHERN DOUBLE-COLLARED SUNBIRD *Cinnyris reichenowi* 11.5 cm **R** **Page 694**
Open montane forest, highland thickets, etc. SE Nigeria to NW CAR; Bioko. Common.
- **1a** **Adult male** Resembles Olive-bellied Sunbird (Plate 117: 11); glossy violet uppertail-coverts; darker belly.
- **1b** **Adult female** Dark grey-green above; paler below; no supercilium.

2 CONGO SUNBIRD *Cinnyris congensis* 12–13 cm **R** **Page 695**
Forested banks of large rivers. Congo. Rare and local.
- **2a** **Adult male** Very long tail streamers (projecting up to 7 cm); red breast-band; black belly.
- **2b** **Adult female** Mainly grey-brown, paler on belly.

3 ORANGE-TUFTED SUNBIRD *Cinnyris bouvieri* 11.5–12.0 cm **R** **Page 696**
Edges of montane and lowland forest. SE Nigeria to CAR; also Eq. Guinea, Gabon, Congo. Locally common/uncommon.
- **3a** **Adult male** Glossy purple forehead; glossy purple and chestnut-red breast-bands.
- **3b** **Adult female** Resembles 4b but browner; throat dusky.

4 PURPLE-BANDED SUNBIRD *Cinnyris bifasciatus* 11–12 cm **R** **Page 695**
Forest edges, thickets in savanna, coastal scrub. SW CAR, Gabon, S Congo. Uncommon.
- **4a** **Adult male** Glossy blue and purple breast-bands; black belly.
- **4b** **Adult female** Olive-grey above; pale yellowish streaked dusky below.
- **4c** **Immature male** As 4b with black throat and blotched breast.

5 PALESTINE SUNBIRD *Cinnyris oseus* 10.0–11.5 cm **R/M** **Page 696**
Savanna. CAR; also E Cameroon. Locally common/rare.
- **5a** **Adult male** Iridescent green above; glossy violet-blue forecrown and throat. Appears all black in some lights.
- **5b** **Adult female** Greyish-brown above; dusky-white tinged yellowish below.

6 COPPER SUNBIRD *Cinnyris cupreus* 12–13 cm **R** **Page 698**
Wooded savanna, coastal scrub, gardens. Throughout savanna and forest zones. Common.
- **6a** **Adult male** Glossy coppery and purple. Appears all black in some lights.
- **6b** **Adult female** Olivaceous-green above; olive-yellow below; yellowish supercilium.

7 SPLENDID SUNBIRD *Cinnyris coccinigaster* 14 cm **R** **Page 698**
Wooded savanna, coastal thickets, gardens. Senegambia–Liberia to CAR. Common.
- **7a** **Adult male** Glossy purple head and throat; bright red breast; black belly.
- **7b** **Adult female** Brownish-olive above; pale yellow below; dusky streaks on breast.
- **7c** **Immature male** Black throat and upper breast.

8 BEAUTIFUL SUNBIRD *Cinnyris pulchellus* 9–11 cm **R/M** **Page 695**
Wooded savanna, gardens. S Mauritania–Sierra Leone to Chad–CAR. Common/seasonally common.
- **8a** **Adult male breeding** Long tail streamers (projecting up to 6 cm); glossy green plumage; red breast bordered yellow. Adult male non-breeding as female but retaining tail streamers and glossy green shoulders.
- **8b** **Adult female** Pale ashy-olive above; pale yellow below.

9 SUPERB SUNBIRD *Cinnyris superbus* 16 cm **R** **Page 697**
Forest edge; also wooded savanna. Sierra Leone–SE Guinea to CAR–Congo. Not uncommon/uncommon.
- **9a** **Adult male** Large; conspicuously long bill; dark glossy blue throat and upper breast; dull dark red belly.
- **9b** **Adult female** Plain olive-yellow below; orange-red wash to undertail-coverts.

10 JOHANNA'S SUNBIRD *Cinnyris johannae* 13–14 cm **R** **Page 697**
Forest. Sierra Leone–SE Guinea to Ghana; Nigeria to SW CAR–Congo. Rare/locally not uncommon.
- **10a** **Adult male** Resembles 9a but head and throat glossy green; red on underparts brighter; more compact shape.
- **10b** **Adult female** Heavily streaked below (diagnostic).
- **10c** **Immature**

PLATE 119: TRUE SHRIKES I

Small to medium-sized birds with strong, hooked bills and moderately long to very long tails. Bill and legs usually black(ish). Juveniles typically finely barred above and below; bill paler. Capture prey (insects, reptiles, young birds, small mammals) by pouncing from exposed perch. Some species impale prey on thorns. Flick, swing and fan tail when excited.

1   SOUSA'S SHRIKE *Lanius souzae* 17–18 cm   **R**   **Page 705**
Savanna woodland. SE Gabon and Congo. Scarce.
- **1a   Adult male** Pale grey crown and mantle; white scapulars; narrow tail.
- **1b   Adult female** As male but flanks washed rufous.
- **1c   Juvenile** Paler; narrow wavy bars above and (indistinctly) on breast.

2   RED-BACKED SHRIKE *Lanius collurio* 17–18 cm   **P**   **Page 704**
Open habitats. Scattered records, mostly in east.
- **2a   Adult male** Blue-grey crown and rump; chestnut back; pinkish-white underparts.
- **2b   Adult female** Rufous-brown above; underparts with dusky scaling.
- **2c   Immature** Upperparts scaled; coverts and tertials tipped buffish with subterminal dark markings.

3   EMIN'S SHRIKE *Lanius gubernator* 15–16 cm   **R**   **Page 705**
Savanna woodland. Mali–N Ivory Coast to Cameroon–CAR. Rare/locally uncommon.
- **3a   Adult male** As 2a but rump chestnut; underparts rusty; white wing patch.
- **3b   Adult female** Duller; back and rump grey; underparts paler.
- **3c   Juvenile** Barred dusky above and below.

4   ISABELLINE SHRIKE *Lanius isabellinus* 17–18 cm   **P**   **Page 704**
Open habitats. Mostly Nigeria, Cameroon, Chad. Few records elsewhere. Uncommon/scarce.
- **4a   Adult male** Rather plain, greyish-brown; contrasting rufous tail.
- **4b   Immature** Duller; underparts with indistinct scaling; coverts and tertials tipped buffish with subterminal dark markings.

5   WOODCHAT SHRIKE *Lanius senator* 18–19 cm   **P**   **Page 705**
Open habitats. Widespread. Generally not uncommon in west; rare in Cameroon and CAR. Also Gabon.
- **5a   Adult male *senator*** Chestnut crown and nape; white shoulder patch and rump. Pale fringes to wing feathers indicate fresh plumage.
- **5b   Adult female *senator*** As male but duller and with more white on forehead and lores.
- **5c   Immature *senator*** Variable amount of buff wash and indistinct crescents on scapulars, breast and flanks.
- **5d   Adult male *badius*** No white on primaries; less white on shoulders; narrower black forehead. Worn plumage: no pale fringes to wing feathers.

6   MASKED SHRIKE *Lanius nubicus* 17–18 cm   **P**   **Page 702**
Thorn scrub. Mauritania to Chad–N CAR. Uncommon/rare.
- **6a   Adult male** White forehead; black rump; orange flanks.
- **6b   Adult female** As male but duller.
- **6c   Juvenile** Brown-grey and densely barred above; pale forehead and supercilium.

PLATE 120: TRUE SHRIKES II AND HELMET-SHRIKES

1   COMMON FISCAL *Lanius collaris* 21–23 cm **R**   Page 701
Open habitats. W Guinea to CAR; also S Gabon and Congo; generally common. Rare further north.
1a   Adult Black and white; long, graduated tail.
1b   Immature Rufous-brown barred blackish above; variably barred dusky below.

2   SOUTHERN GREY SHRIKE *Lanius meridionalis* 24 cm **R**   Page 703
Thorn scrub at desert edge and Sahel. Mauritania to Cameroon–Chad. Not uncommon.
Winter visitor/vagrant further south.
2a   Adult *leucopygos* Large; pale grey above; wings black and white.
2b   Immature Duller and tinged buffish.

3   LESSER GREY SHRIKE *Lanius minor* 20–21 cm **P**   Page 703
Open habitats. Extreme E Chad; not uncommon. Mauritania–Senegal to W Chad, Cameroon, Gabon,
Congo, Príncipe; rare/vagrant.
3a   Adult male Black forehead; no white supercilium; long wings.
3b   Immature Duller; no black on forehead.

4   MACKINNON'S SHRIKE *Lanius mackinnoni* 20 cm **R**   Page 702
Edges of lowland and montane forest. SE Nigeria to S Congo. Locally not uncommon.
4a   Adult male All-black wings; white shoulder patch.
4b   Juvenile Narrow dusky bars above and (indistinctly) below.

5   GREY-BACKED FISCAL *Lanius excubitoroides* 25 cm **R**   Page 703
Thorn scrub. NE Nigeria, N Cameroon, Chad, N CAR; rare. Also Mauritania and Mali.
5a   Adult male Large and robust; black forehead; black scapulars. Gregarious and noisy.
5b   Juvenile Brownish, narrowly barred dusky above; smaller mask.

6   YELLOW-BILLED SHRIKE *Corvinella corvina* 30–33 cm **R**   Page 706
Open savanna woodland. S Mauritania–Sierra Leone to Chad–CAR. Locally common.
Adult Long brown tail; yellow bill. Gregarious.

Helmet-shrikes. Gregarious species with rather stout, hooked bills, boldly patterned plumages and brightly
coloured eye-wattles. Brush-like feathers on forehead give 'helmeted' appearance. Conspicuous and vocal,
constantly chattering and often snapping bill.

7   WHITE HELMET-SHRIKE *Prionops plumatus* 19–23 cm **R**   Page 716
Savanna woodland. S Mauritania–Sierra Leone to Chad–CAR. Locally not uncommon.
7a   Adult Long white crest; yellow eye-wattle; black-and-white wings and tail.
7b   Juvenile No crest; no eye-wattle; eye dark.
7c   In flight Broad, rounded wings with white bars.

8   RED-BILLED HELMET-SHRIKE *Prionops caniceps* 20 cm **R**   Page 716
Forest. W Guinea to W Cameroon. Not uncommon.
8a   Adult *caniceps* Greyish-white crown; throat black; breast white merging with deep buff on lower
      underparts; red bill and orbital ring. From Togo west.
      P. c. harterti (Benin to W Cameroon) has head pattern more like *rufiventris*.
8b   Juvenile Blackish bill; no orbital ring; whitish throat.

9   RUFOUS-BELLIED HELMET-SHRIKE *Prionops rufiventris* 20 cm **R**   Page 716
Forest. S Cameroon to SW CAR–Congo and in SE CAR. Not uncommon.
9a   Adult *rufiventris* Greyish-white extending onto head-sides and upper throat; underparts orange-
      chestnut from lower breast.
      P. r. mentalis (SE CAR) deeper coloured on crown and underparts.
9b   In flight Broad, rounded wings with white bar.

Arboreal birds with stout, hooked bills. Many species often encountered in pairs. Most are highly vocal, with loud, ringing calls.

Bush-shrikes (*Malaconotus* and *Telophorus*). Brightly coloured birds of forest and savanna woodland. Forest species generally hard to observe but betray presence by melodious calls and readily respond to playback or whistled imitations.

1 MANY-COLOURED BUSH-SHRIKE *Malaconotus multicolor* 20 cm **R**   **Page 708**
Forest canopy. Sierra Leone–SE Guinea to CAR–Congo. Uncommon/scarce.
1a **Adult male scarlet-breasted form** Black mask; grey crown; green upperparts; tail tip yellow and black.
1b **Adult male orange-breasted form**
1c **Adult male black-breasted form** Belly yellow to scarlet.
1d **Adult female** Forehead and lores white; flanks washed green; tail green tipped yellow.

2 GORGEOUS BUSH-SHRIKE *Telophorus viridis* 19 cm **R**   **Page 710**
Thickets and shrubbery in wooded savanna. SE Gabon–Congo. Locally not uncommon.
2a **Adult male** Dark green; crimson throat; broad black breast-band; black tail.
2b **Immature** Olive-green above; yellow throat; rest of underparts greenish-yellow; some dark crescentic bars on breast; tail washed green.

3 SULPHUR-BREASTED BUSH-SHRIKE *Malaconotus sulfureopectus* 17–19 cm **R**   **Page 709**
Savanna woodland. Senegambia–Sierra Leone to Chad–CAR. Not uncommon.
3a **Adult male** Black mask; yellow forehead and supercilium; orange-washed breast.
3b **Immature** All-grey head.

4 MOUNT KUPE BUSH-SHRIKE *Malaconotus kupeensis* 18–20 cm **R**   **CR**   **Page 708**
Montane forest. W Cameroon. Rare and very local. Endemic.
4a **Adult** Black mask; white throat; grey breast and belly; yellow vent.
4b **Adult** An individual with maroon patch in centre of throat.

5 GREY-HEADED BUSH-SHRIKE *Malaconotus blanchoti* 25 cm **R**   **Page 707**
Savanna woodland. Senegambia–Sierra Leone to Chad–CAR. Locally not uncommon/scarce.
Adult Large; grey head; upperparts and tail green; underparts bright yellow; breast washed orange; massive bill.

6 GREEN-BREASTED BUSH-SHRIKE *Malaconotus gladiator* 25–28 cm **R** **VU**   **Page 708**
Montane forest. SE Nigeria–W Cameroon. Scarce/locally not uncommon. Endemic.
Large. Olive-green with grey head; massive bill.

7 MONTEIRO'S BUSH-SHRIKE *Malaconotus monteiri* c. 25 cm **R** **DD**   **Page 707**
Montane forest. W Cameroon (Mt Cameroon, Mt Kupe). Very rare.
Adult As 5 but white of lores extending above and below eye; underparts uniformly yellow.

8 LAGDEN'S BUSH-SHRIKE *Malaconotus lagdeni* 23–25 cm **R** **NT**   **Page 707**
Forest. Sierra Leone to SW Ghana. Scarce.
Adult Large; deep yellow underparts; massive bill.

9 FIERY-BREASTED BUSH-SHRIKE *Malaconotus cruentus* 25 cm **R**   **Page 707**
Forest. Sierra Leone–SE Guinea to SW CAR–Congo. Uncommon/scarce to locally not uncommon.
9a **Adult** Large; grey head; underparts yellow and scarlet (variable in extent).
9b **Adult** An example of a yellow-breasted individual (rare).

Puffbacks (*Dryoscopus*). Males predominantly black and white with soft elongated feathers on lower back and rump, which are puffed out to produce spectacular white ball in display. Vocal.

1 NORTHERN PUFFBACK *Dryoscopus gambensis* 18–19 cm **R** **Page 712**
Savanna woodland; also forest clearings, mangroves. Throughout. Not uncommon/common.
1a Adult male The only puffback with whitish edges to wing feathers; orange-red eye.
1b Adult female *gambensis* Grey head; earth-brown upperparts; tawny-buff underparts.
1c Adult female *malzacii* Darker than 1b; head brown, as upperparts. CAR.

2 BLACK-SHOULDERED PUFFBACK *Dryoscopus senegalensis* 16–17 cm **R** **Page 712**
Forest edges and second growth. SE Nigeria (rare) to CAR–Congo (not uncommon/common).
2a Adult male Glossy black and pure white. Compare 1a and 4a.
2b Adult female White supraloral streak.
2c Adult male displaying.

3 PINK-FOOTED PUFFBACK *Dryoscopus angolensis* 15–17 cm **R** **Page 711**
Montane and submontane forest. SE Nigeria–Cameroon, Eq. Guinea, S Congo. Uncommon/rare.
3a Adult male Black head; grey upperparts; pinkish legs and feet.
3b Adult female Resembles 1b but more colourful.

4 SABINE'S PUFFBACK *Dryoscopus sabini* 18–19 cm **R** **Page 711**
Forest. Sierra Leone–SE Guinea to SW CAR–Congo. Uncommon.
4a Adult male Larger than 2a, bill longer and heavier.
4b Adult female Tawnier overall than 3b.

Boubous (*Laniarius*). Either all black or with black upperparts and white or red/yellow underparts. Skulk in heavy shrubbery and forest edges. Vocal.

5 MOUNTAIN SOOTY BOUBOU *Laniarius poensis* 18 cm **R** **Page 713**
Montane forest. SE Nigeria–W Cameroon; Bioko. Common.
Entirely black.

6 TROPICAL BOUBOU *Laniarius aethiopicus* 23 cm **R** **Page 713**
Various wooded habitats. NE Sierra Leone–Guinea to Chad–CAR. Not uncommon.
Adult Black above; white below; long white wingbar.

7 SOOTY BOUBOU *Laniarius leucorhynchus* 21.5 cm **R** **Page 712**
Forest zone. Sierra Leone–SE Guinea to S CAR–Congo. Not uncommon/scarce.
7a Adult male All black.
7b Juvenile Bill whitish.

8 SWAMP BOUBOU *Laniarius bicolor* 23 cm **R** **Page 714**
Savanna thickets, coastal scrub, mangroves. Coastal Cameroon–Congo; also along lower Congo R. Locally not uncommon.
Adult Very similar to 6; whiter below; wingbar usually shorter.

9 TURATI'S BOUBOU *Laniarius turatii* 23 cm **R** **Page 714**
Various wooded habitats. Guinea-Bissau, W Guinea, W Sierra Leone. Common. Endemic.
Adult As 6 but without wingbar. Pale pinkish tinge to underparts (often hard to see).

1 BLACK-HEADED GONOLEK *Laniarius erythrogaster* 23 cm **R** **Page 715**
Grassy river banks. NE Nigeria, Cameroon, Chad, CAR. Locally common.
Adult Black above; crimson below.

2 YELLOW-CROWNED GONOLEK *Laniarius barbarus* 23 cm **R** **Page 714**
Thickets, thorn scrub, mangroves. S Mauritania–Sierra Leone to Chad. Locally not uncommon/
common. Also coastal Liberia (rare). Endemic.
2a   **Adult** Black above; crimson below; golden-yellow cap.
2b   **Juvenile** Feathers of upperparts tipped buff; below, yellowish-buff barred black.

3 YELLOW-BREASTED BOUBOU *Laniarius atroflavus* 18–19 cm **R** **Page 715**
Montane forest. SE Nigeria–W Cameroon. Locally not uncommon. Endemic.
Adult Black above; deep yellow below.

4 BOCAGE'S BUSH-SHRIKE *Malaconotus bocagei* 16.5 cm **R** **Page 709**
Wooded savanna, forest edge. Cameroon to SW CAR–Congo. Uncommon.
4a   **Adult male** Black and dark grey above; whitish below; long, white supercilium.
4b   **Juvenile** Upperparts washed greenish and speckled buff; some indistinct dusky barring below.

5 LÜHDER'S BUSH-SHRIKE *Laniarius luehderi* 18–19 cm **R** **Page 713**
Forest zone. SE Nigeria to Congo. Uncommon/not uncommon.
5a   **Adult** Black above; chestnut cap; deep cinnamon throat and breast.
5b   **Immature** Olivaceous-brown above; dirty yellowish below.

6 BRUBRU *Nilaus afer* 13–15 cm **R** **Page 715**
Savanna woodland. S Mauritania–Sierra Leone to Chad–CAR. Not uncommon/locally common.
6a   **Adult male** Small; mottled black and white above; chestnut on flanks.
6b   **Adult female** More brownish-grey above; flanks paler.
6c   **Juvenile** Speckled and mottled above; barred below.

Tchagras (*Tchagra* and *Antichromus*). Mainly brown above with rufous wings and patterned head and tail.
Melodious calls and characteristic display flights attract attention.

7 MARSH TCHAGRA *Antichromus minutus* 16–18 cm **R** **Page 710**
Rank herbage. Sierra Leone–SE Guinea to CAR–Congo. Uncommon and local.
7a   **Adult male** Jet-black cap; rufous-brown upperparts; black tail.
7b   **Adult female** Whitish supercilium.

8 BLACK-CROWNED TCHAGRA *Tchagra senegala* 20–22 cm **R** **Page 711**
Various wooded habitats (not forest). Throughout. Common.
8a   **Adult** Black crown.
8b   **In flight** Rufous wings; black, white-tipped tail.

9 BROWN-CROWNED TCHAGRA *Tchagra australis* 18–19 cm **R** **Page 710**
Forest edge, thicket, scrub. W Guinea to CAR–Congo; also S Mali. Not uncommon/uncommon.
Adult Brown crown. Compare 8.

PLATE 124: ORIOLES

Robust, arboreal birds. In most species occurring in Africa males are bright yellow and black, with black on head, wings and tail, and a strong, reddish bill. Feed on insects and fruit. Mostly in canopy, where often difficult to observe. Voice loud, fluty and melodious. Flight strong and undulating.

1   WESTERN BLACK-HEADED ORIOLE *Oriolus brachyrhynchus*   *c.* 21 cm   **R**   **Page 717**
Lowland and mid-elevation forest. Sierra Leone to CAR–Congo. Not uncommon/common.
  1a   **Adult *brachyrhynchus*** Black head; small white patch on edge of wing; green central tail feathers.
         Sierra Leone–SE Guinea to Benin.
  1b   **Adult *laetior*** Broader yellow collar; yellower mantle. Nigeria to CAR–Congo.
  1c   **Immature** Head and upperparts olive; throat olive streaked yellow; breast streaked black; bill
         dusky.

2   EASTERN BLACK-HEADED ORIOLE *Oriolus larvatus*   *c.* 21 cm   **Page 718**
Open woodland. No certain records. Claims from SW CAR doubtful.
  2a   **Adult** As 1a but flight feathers edged white; tertials edged yellow.
  2b   **Immature** As 3b but with white wing patch.

3   BLACK-WINGED ORIOLE *Oriolus nigripennis*   *c.* 20 cm   **R**   **Page 717**
Lowland and montane forest. Sierra Leone–SE Guinea to CAR–Congo; not uncommon/common.
Bioko; rare.
  3a   **Adult** Similar to 1a, but no white patch on edge of wing; central tail feathers black.
  3b   **Immature** Throat black streaked yellow; breast streaked black.

4   EURASIAN GOLDEN ORIOLE *Oriolus oriolus*   *c.* 24 cm   **P**   **Page 718**
Wooded savanna. Throughout; uncommon/rare. Vagrant, Bioko and Príncipe.
  4a   **Adult male** Golden-yellow; black wings; black, yellow-tipped tail.
  4b   **Adult female** More yellowish-olive above than 4a; streaked dusky below.
  4c   **Immature** Greener above than 4b; whiter below with more distinct streaks.

5   AFRICAN GOLDEN ORIOLE *Oriolus auratus*   *c.* 24 cm   **M**   **Page 718**
Wooded savanna. Senegambia–Sierra Leone to Chad–CAR; locally common. Rare/vagrant further
south.
  5a   **Adult male** Golden-yellow; black mask; black wings and tail broadly edged yellow.
  5b   **Adult female** More yellowish-olive above than 5a; lightly streaked dusky below.
  5c   **Immature** Yellowish-olive above; no mask; heavily streaked below. Compare 4c.

PLATE 125: DRONGOS AND CROWS

Drongos. Arboreal species with black, usually glossy, plumage and stout, slightly hooked bills. Eye red or orange in adults. Hunt from perch, capturing insects on the wing like flycatchers. Conspicuous, bold, pugnacious and vocal. Voice consists of harsh, scolding notes interspersed by varied musical whistles.

1 FORK-TAILED DRONGO *Dicrurus adsimilis* 22.5–25.0 cm **R** **Page 720**
Wooded savanna. S Mauritania–Sierra Leone to Chad–CAR; common. SE Gabon–Congo; locally not uncommon/rare.
Adult Wholly glossy blue-black; forked tail; red eye.

2 VELVET-MANTLED DRONGO *Dicrurus modestus* 24–27 cm **R** **Page 720**
Forest and edges. W Guinea to Congo; also Bioko and Príncipe. Common.
Adult *coracinus* Unglossed velvety black mantle; deeply forked 'fish-tail'.

3 SHINING DRONGO *Dicrurus atripennis* *c.* 21 cm **R** **Page 719**
Forest interior. W Guinea to CAR–Congo. Uncommon/common.
Strongly glossed blue-green; slightly forked tail; red eye. Often in mixed-species flocks.

4 SQUARE-TAILED DRONGO *Dicrurus ludwigii* 19 cm **R** **Page 719**
Forest edge, second growth, thickets. S Senegambia–N Liberia to CAR–Congo. Not uncommon/rare.
Black with slight purplish-blue gloss; slightly notched tail; orange-red eye.

Crows. Medium-sized to large with stout bills and strong legs and feet. Plumage either wholly black, including bill and legs, or black with some white, brown or grey. Omnivorous; foraging mainly on ground. Calls mostly loud and harsh.

5 PIAPIAC *Ptilostomus afer* 35 cm **R** **Page 721**
Sahel and savanna zones. S Mauritania–Sierra Leone to S Chad–CAR. Locally common/uncommon.
5a Adult Long (up to 28 cm), stiff, steeply graduated tail. Usually in small flocks.
5b Immature Pinkish bill with black tip.
5c In flight Drab brown primaries (from above); ashy flight feathers (from below).

6 PIED CROW *Corvus albus* 46–50 cm **R** **Page 721**
Most open habitats; often near habitation. Throughout except most arid areas and rainforest. Common.
6a Black and white below; white collar on hindneck.
6b In flight White on underparts visible from great distance.

7 FAN-TAILED RAVEN *Corvus rhipidurus* 47 cm **R** **Page 721**
Desert. N Mali, N Niger (Aïr) and Chad. Scarce and local.
7a All black; wingtips project well beyond tail.
7b In flight Very short tail; broad wings.

8 BROWN-NECKED RAVEN *Corvus ruficollis* 52–56 cm **R** **Page 721**
Desert and semi-desert. Mauritania–Senegambia to Chad; also Cape Verde Is. Common/rare.
8a Black; brown tinge to head and breast (often difficult to see).
8b In flight Wedge-shaped tail.

WESTERN JACKDAW *Corvus monedula* 33 cm **V** **Page 721**
Palearctic vagrant, N Mauritania. **See illustration on plate 147**.
Small. Black; grey neck; whitish eyes.

PLATE 126: STARLINGS I

Mainly arboreal species with strong and pointed bills and sturdy legs. Omnivorous, most feeding on fruit and insects. Most forage in flocks, roost communally and nest in holes. Calls mostly harsh and grating but also including pleasing sounds.

1 NARROW-TAILED STARLING *Poeoptera lugubris* 20–23 cm (tail 12 cm) **R**   **Page 722**
Forest edges and clearings. Sierra Leone to SW CAR–Congo. Locally not uncommon/scarce.
  1a Adult male Slender; very long, narrow, graduated tail; purple-black plumage; yellow eye.
  1b Adult female Much greyer than 1a.
  1c Adult female in flight Chestnut wing patch.

2 FOREST CHESTNUT-WINGED STARLING *Onychognathus fulgidus* 28–33 cm **R**   **Page 723**
Forest and its outliers. Sierra Leone–SE Guinea to CAR and Congo; Bioko; São Tomé. Not uncommon. Resembles 3 but occurs in different habitat.
  2a Adult male *hartlaubii* Glossy purple-black; head with metallic green reflections.
  2b Adult female *hartlaubii* Head and throat streaked ash-grey.
  2c In flight Chestnut wing patch smaller and darker than 3c.

3 NEUMANN'S STARLING *Onychognathus neumanni* 28–33 cm **R**   **Page 723**
Crags, rocky outcrops. Patchily distributed, Mauritania–Senegal to Chad–CAR. Locally not uncommon/scarce.
  3a Adult male Large; long, graduated tail; glossy purplish-black plumage.
  3b Adult female Head and throat ash-grey streaked blue-black.
  3c In flight Conspicuous chestnut wing patch.

4 WALLER'S STARLING *Onychognathus walleri* 23 cm **R**   **Page 722**
Montane forest. SE Nigeria–W Cameroon; Bioko. Common.
  4a Adult male Tail relatively short; chestnut wing patch; glossy purple-black plumage; head with metallic green reflections.
  4b Adult female Head with some grey streaking.

5 COPPER-TAILED GLOSSY STARLING *Lamprotornis cupreocauda* 20 cm **R** NT   **Page 723**
Forest. Sierra Leone–SE Guinea to Togo. Not uncommon/locally common to rare. Endemic.
  Adult Glossy purple head to upper breast; rest of plumage glossy blue-black; eye yellow.

6 PURPLE-HEADED GLOSSY STARLING *Lamprotornis purpureiceps* 20 cm **R**   **Page 724**
Forest. S Benin to SW CAR–Congo. Uncommon/locally common.
Resembles 5 but mantle and belly glossy blue-green, eye brown.

7 SPLENDID GLOSSY STARLING *Lamprotornis splendidus* 27–30 cm **R/M**   **Page 726**
Forest and its outliers. Senegambia to CAR–Congo; also Bioko and Príncipe. Common/not uncommon. Large. Broad blackish band on wings and tail; eye whitish. Vocal. Swishing wings in flight.

8 WHITE-COLLARED STARLING *Grafisia torquata* 21.5–23.0 cm **R/V**   **Page 727**
Forest/savanna mosaic, grasslands. Cameroon–CAR. Uncommon/locally common. Vagrant, Gabon, N Congo.
  8a Adult male Purple-black; broad white breast-band.
  8b Adult female Greyish-brown slightly glossed blue-purple above; dull greyish below.

9 PURPLE GLOSSY STARLING *Lamprotornis purpureus* 24 cm **R**   **Page 724**
Savanna. Senegambia–Guinea to Chad–CAR. Common/locally uncommon. Rare/vagrant, S Mauritania and Liberia.
Glossy purple head, underparts and tail; relatively long bill and short tail; large yellow eye.

10 CAPE GLOSSY STARLING *Lamprotornis nitens* 23 cm **R**   **Page 724**
Savanna. S Congo (Gabon?). Rare.
All glossy blue-green.

PLATE 127: STARLINGS II AND OXPECKER

1   YELLOW-BILLED OXPECKER *Buphagus africanus* 21–23 cm **R**   **Page 728**
Mainly wooded savanna and bush. S Mauritania–Senegambia to Chad–CAR; also Gabon and Congo.
Uncommon/not uncommon but local.
  1a   **Adult** Yellow, red-tipped bill. Associated with large herbivores.
  1b   **Juvenile** Dusky bill.
  1c   **In flight** Long, pointed wings and tail; contrasting pale rump.

2   WATTLED STARLING *Creatophora cinerea* 21.5 cm **V**   **Page 728**
Savanna. African vagrant, Gambia, Cameroon, Gabon and Congo.
  2a   **Adult male breeding** Head with wattles and bare yellow-and-black skin.
  2b   **Adult male non-breeding** Pale, drab grey; black wings and tail.
  2c   **In flight** Pointed wings; short tail; contrasting greyish-white rump.

3   VIOLET-BACKED STARLING *Cinnyricinclus leucogaster* 16–18 cm **M**   **Page 727**
Various wooded habitats. Throughout. Seasonally common; irregular in some places.
  3a   **Adult male** Brilliant violet; white breast and belly.
  3b   **Adult female** Dark brown above; white streaked brown below.

4   CHESTNUT-BELLIED STARLING *Lamprotornis pulcher* 19 cm **R**   **Page 727**
Semi-arid belt. S Mauritania–Guinea to Niger, N Cameroon, Chad. Locally common/uncommon.
  4a   **Adult** Greyish-brown head; rufous-chestnut belly.
  4b   **Juvenile** Throat and breast ashy-brown.
  4c   **Adult in flight** Creamy wing patch.

5   EMERALD STARLING *Lamprotornis iris* *c.* 20 cm **R**   DD   **Page 726**
Wooded savanna. Guinea, Sierra Leone, Ivory Coast; also SW Mali. Scarce and local. Endemic.
Brilliant emerald-green; head-sides and abdomen glossy purple.

6   BRONZE-TAILED GLOSSY STARLING *Lamprotornis chalcurus* 21.5 cm **R**   **Page 724**
Wooded savanna. Senegambia–Guinea to Chad–CAR. Locally common/uncommon.
Resembles 7 but tail distinctly shorter and purple; eye orange-yellow to reddish-orange. Also compare 9.

7   GREATER BLUE-EARED STARLING *Lamprotornis chalybaeus* 21–24 cm **R**   **Page 725**
Semi-arid belt. S Mauritania–Guinea to N Cameroon–Chad. Common/not uncommon.
  7a   **Adult** Glossy metallic green; belly purple; ear-coverts and rump bluish; eye yellow. Compare 9a.
  7b   **Juvenile** As adult but duller. Darker, more blackish than 9b.

8   LONG-TAILED GLOSSY STARLING *Lamprotornis caudatus* *c.* 51 cm (tail 33 cm) **R**   **Page 726**
Semi-arid belt. S Mauritania–Guinea to Chad–CAR. Not uncommon/common.
Very long, graduated, supple tail.

9   LESSER BLUE-EARED STARLING *Lamprotornis chloropterus* 19–20 cm **R**   **Page 725**
Wooded savanna. Senegambia–Guinea to Chad–CAR. Common/not uncommon.
  9a   **Adult** Very similar to 7a but ear-coverts more contrasting blue-black, tail relatively shorter, rump
       concolorous with upperparts, size smaller, eye yellow to orange-yellow, voice different.
  9b   **Juvenile** Earth-brown head and underparts. Paler below than 7b.

COMMON STARLING *Sturnus vulgaris* 19–22 cm **V**   **Page 728**
Palearctic vagrant, N Mauritania. **See illustration on plate 147.**
Glossy black spotted buff above and white below; long pointed bill; short tail; triangular wings.

PLATE 128: SPARROWS AND WEAVERS

Sparrows. Mostly brown and grey with short, conical bills. Bill of male typically turns from horn coloured to black in breeding season. Forage on or near the ground, feeding largely on seeds. Many species gregarious. Vocal, uttering a variety of rather harsh chirping calls.

1 HOUSE SPARROW *Passer domesticus* 14–15 cm **R**   **Page 729**
Towns. Introduced Mauritania–Senegambia and Cape Verde Is; common. Also Liberia (rare). Claimed from NE Niger and C Chad (vagrant?).
1a   **Adult male breeding** Grey crown and rump; black bib; whitish head-sides.
1b   **Adult female** Dull brown; buff supercilium; horn-coloured bill.

EURASIAN TREE SPARROW *Passer montanus* 14 cm   **Page 729**
Palearctic species claimed from Guinea (Conakry harbour). Ship-assisted or escape? **See illustration on plate 147.**
Resembles 1a, but has chestnut crown and black ear-spot.

2 RUFOUS SPARROW *Passer motitensis* 14 cm **R**   **Page 729**
Arid scrub. EC Chad. Locally common. (Vagrant Mali?)
2a   **Adult male breeding** Grey crown; crown-sides and upperparts rich chestnut; face and underparts whitish.
2b   **Adult female** Paler than 2a; bib dusky-grey.

3 SPANISH SPARROW *Passer hispaniolensis* 15 cm **R/V?**   **Page 729**
Towns, cultivated areas. Cape Verde Is; common. Also Mauritania (introduced?) and NE Chad (Palearctic vagrant?).
3a   **Adult male breeding** Chestnut crown; black bib extending onto breast; heavily streaked flanks.
3b   **Adult female** As 1b but more streaky.

4 DESERT SPARROW *Passer simplex* 13.5 cm **R**   **Page 730**
Desert. Mauritania to N Chad. Uncommon and local.
4a   **Adult male** Pale greyish above; black mask. bib and bill.
4b   **Adult female** Pale sandy-buff; no face markings; bill pale horn.

5 NORTHERN GREY-HEADED SPARROW *Passer griseus* 14 cm **R**   **Page 730**
Various habitats, but mainly towns and villages. Throughout, also Bioko. Common.
Grey head and underparts; rich chestnut upperparts.

6 SUDAN GOLDEN SPARROW *Passer luteus* 14 cm **R/M**   **Page 731**
Thorn scrub. Mauritania–Senegambia to Chad. Common.
6a   **Adult male breeding** Yellow head and underparts; chestnut upperparts; two white wingbars.
6b   **Adult female** Buffish-brown head and upperparts; face and underparts washed yellow.

7 BUSH PETRONIA *Petronia dentata* 13 cm **R**   **Page 732**
Wooded grassland. S Mauritania–Sierra Leone to Chad–CAR. Uncommon/locally not uncommon.
7a   **Adult male** Chestnut supercilium; grey crown.
7b   **Adult female** Buff supercilium; brown crown.

8 YELLOW-THROATED PETRONIA *Petronia superciliaris* 15 cm **R**   **Page 731**
Wooded grassland, cultivation. SE Gabon–Congo. Uncommon/locally common.
8a   **Adult** Broad whitish supercilium.
8b   **Juvenile** Warmer brown above; supercilium brownish-buff.

9 CHESTNUT-CROWNED SPARROW WEAVER *Plocepasser superciliosus* 17 cm **R**   **Page 734**
Thorn scrub, wooded grassland. Senegambia to Chad–N CAR. Uncommon/scarce.
Rufous-chestnut crown; white supercilium; black malar stripe.

10 SPECKLE-FRONTED WEAVER *Sporopipes frontalis* 12 cm **R**   **Page 732**
Dry savanna. S Mauritania–Senegambia to Chad–N CAR. Uncommon/locally common.
Small. Pale rufous nape; forehead and moustachial streak black speckled white.

11 YELLOW-SPOTTED PETRONIA *Petronia pyrgita* 15 cm **R**   **Page 731**
Thorn scrub. S Mauritania–N Senegal to Chad. Scarce/uncommon and local.
Buffish-brown above; creamy-white below; small lemon patch on throat (difficult to see).

PLATE 129: TYPICAL WEAVERS I

Small to medium-sized birds with generally strong, conical bills. Many species gregarious, feeding and roosting in flocks and nesting in colonies. Vocalisations typically consist of characteristic, drawn-out, wheezy, buzzy, chirps and chattering notes.

Genus *Ploceus* (typical weavers). Males in breeding dress mostly bright yellow; females dull and streaky (sparrow-like). In non-breeding dress (if any) males similar to females.

1 LITTLE WEAVER *Ploceus luteolus* 11.5 cm **R**   **Page 736**
Dry wooded savanna. Senegambia to Chad–CAR. Common/not uncommon.
  1a   **Adult male breeding** Small; large black mask; mantle variably streaked dusky.
  1b   **Adult male non-breeding** No mask; crown yellow-green; head-sides to breast buff; mantle buff streaked dusky.
  1c   **Adult female breeding** No mask; head-sides to breast pale yellow; mantle yellow-green streaked dusky.

2 SLENDER-BILLED WEAVER *Ploceus pelzelni* 11 cm **R**   **Page 735**
Mainly coastal zone. Sierra Leone to Congo (along Congo R.). Locally not uncommon.
  2a   **Adult male** As 1a but underparts deeper yellow; upperparts darker green, unstreaked.
  2b   **Adult female** As 1b but brighter yellow; distinct yellow supercilium; unstreaked upperparts.

3 LOANGO WEAVER *Ploceus subpersonatus* 15 cm **R** **VU**   **Page 736**
Narrow coastal strip, Gabon and Congo. Uncommon.
  3a   **Adult male** As 2a but plumage duller and washed brownish, size larger.
  3b   **Adult female** As 2b but darker and larger.

4 COMPACT WEAVER *Ploceus superciliosus* 15 cm **R**   **Page 742**
Grassland. Patchily distributed, S Senegal–Sierra Leone to CAR–Congo. Uncommon/locally not uncommon.
  4a   **Adult male breeding** Black mask; short, stout bill; chestnut forehead; dark upperparts; dark eye.
  4b   **Adult female breeding** As male, but forehead and centre of crown black.
  4c   **Adult non-breeding** Brownish; broad pale supercilium.

5 AFRICAN MASKED WEAVER *Ploceus velatus* 14 cm **R**   **Page 739**
Dry savanna. S Mauritania–Senegambia to Chad–N CAR; São Tomé. Common/not uncommon.
  5a   **Adult male** *vitellinus* **breeding** Black mask (with narrow band on forehead); chestnut wash to crown and lower throat. Compare 6a and 8a.
      P. v. peixotoi (São Tomé): see Plate 144: 4.
  5b   **Adult female** *vitellinus* **breeding** No mask; brownish-olive above; legs pinkish. Compare 6b.

6 HEUGLIN'S MASKED WEAVER *Ploceus heuglini* 14 cm **R**   **Page 739**
Wooded savanna. Senegambia–Ivory Coast to Cameroon–Chad. Uncommon/not uncommon.
  6a   **Adult male breeding** Black mask extending to upper breast; no chestnut wash on head and throat.
  6b   **Adult female non-breeding** As 5b, but less distinct streaking on mantle.

7 ORANGE WEAVER *Ploceus aurantius* 15 cm **R**   **Page 738**
Near water, coastal zone; inland along major rivers. Sierra Leone to CAR–Congo. Locally not uncommon.
  7a   **Adult male** Bright yellow-orange head and underparts. Gregarious.
  7b   **Adult female** Olive above; white below; pale yellow head-sides to upper breast.

8 LESSER MASKED WEAVER *Ploceus intermedius* 14 cm **R**   **Page 738**
Near water. SW Congo. Rare/scarce and local.
  8a   **Adult male breeding** As 5a but mask larger; mantle indistinctly streaked.
  8b   **Adult male non-breeding** No mask; similar to female.
  8c   **Adult female breeding** Yellowish-green above; head-sides yellow; legs bluish-grey.

9 HOLUB'S GOLDEN WEAVER *Ploceus xanthops* 17–18 cm **R**   **Page 737**
Open savanna. S Congo; also S Gabon. Rare/scarce and local.
  9a   **Adult male** Large; mainly golden-yellow; heavy, black bill; yellow eye.
      Adult female is slightly duller. Not gregarious.
  9b   **Juvenile** Olive head; breast and flanks washed buff; horn-coloured bill.

1   BAGLAFECHT WEAVER *Ploceus baglafecht* 15 cm   **R**   **Page 734**
Forest edge, montane (in west) and lowland (in east) areas. E Nigeria, Cameroon, W CAR. Scarce/
uncommon.
  1a   Adult male breeding Bright yellow forehead; black mask with contrasting yellow eye; yellow-green
        mantle.
  1b   Adult female breeding As male, but top of head yellow-green; dusky lores.
  1c   Adult non-breeding Dusky area around eye; underparts whitish washed buff.

2   VILLAGE WEAVER *Ploceus cucullatus* 15.0–17.5 cm   **R**   **Page 740**
The ubiquitous, well-known gregarious weaver, common in towns and villages. Also Bioko and São
Tomé.
  2a   Adult male *cucullatus* **breeding** Black head; chestnut collar; black V bordering mantle. Throughout
        most of range.
        P. c. bohndorffi (Gabon–Congo) has less black on crown, more chestnut on breast.
  2b   Adult male *cucullatus* **non-breeding** Lemon-yellow throat; underparts variable (yellow, or white with
        buffish breast and flanks.
  2c   Adult female breeding Yellow-olive head contrasts with brownish-olive upperparts; lemon-yellow
        below; belly white.
  2d   Adult male *collaris* **breeding** No chestnut collar; no black V on upperparts; dark chestnut breast.
        Coastal Gabon–S Congo.

3   BLACK-HEADED WEAVER *Ploceus melanocephalus* 14–15 cm   **R**   **Page 741**
Near water in semi-arid savanna. S Mauritania–Guinea to Chad–CAR; also Congo. Locally common/not
uncommon.
  3a   Adult male *melanocephalus* **breeding** Black head; golden nuchal collar; yellowish-olive upperparts.
        North of range.
  3b   Adult male *capitalis* **breeding** Chestnut wash on breast. South of range.
  3c   Adult male non-breeding Brownish-olive head; rest of plumage mainly brownish, buff and white.
        Adult female similar but smaller.

4   BLACK-CHINNED WEAVER *Ploceus nigrimentum* *c.* 17 cm   **R**   **Page 734**
Wooded grassland. SE Gabon–C Congo. Locally not uncommon.
  4a   Adult male Black mask; black upperparts; wing feathers edged yellow.
  4b   Adult female As male but head and nape black.

5   SPECTACLED WEAVER *Ploceus ocularis* 16–17 cm   **R**   **Page 737**
Various wooded habitats. SE Nigeria to CAR; Gabon to S Congo. Not uncommon/scarce.
  5a   Adult male As 6a but head yellower; bill more slender.
  5b   Adult female As male but without black on throat.

6   BLACK-NECKED WEAVER *Ploceus nigricollis* 16–17 cm   **R**   **Page 736**
Forest edges, wooded savanna. Senegambia–Liberia to Niger, CAR, Congo; Bioko. Common.
  6a   Adult *brachypterus* **male** Golden-chestnut head; black eye-stripe and throat; yellowish-olive
        upperparts. From W Cameroon west.
  6b   Adult *brachypterus* **female** No black on throat; top of head yellowish-olive.
  6c   Adult *nigricollis* **male** As 6a but upperparts blackish. From W Cameroon east.
  6d   Adult *nigricollis* **female** Top of head black; distinct yellow supercilium.

PLATE 131: TYPICAL WEAVERS III

1 MAXWELL'S BLACK WEAVER *Ploceus albinucha* 15 cm **R**   **Page 741**
Lowland forest. Sierra Leone–SE Guinea to Ghana; Nigeria to SW CAR–N Congo; Bioko. Uncommon/ not uncommon.
1a   **Adult** *albinucha* All black except greyish nuchal patch; greyish-white eye. West of range.
1b   **Adult** *holomelas* No nuchal patch. East of range. Nominate (Bioko) similar.
1c   **Juvenile** Dark sooty-grey, paler below.

2 VIEILLOT'S BLACK WEAVER *Ploceus nigerrimus* 17 cm **R**   **Page 739**
Forest and southern savanna zones. Common. Rare, Gambia.
2a   **Adult male** *nigerrimus* All black. Cameroon to CAR–Congo.
2b   **Adult male** *castaneofuscus* Black and chestnut; yellow eye. W Guinea–SE Nigeria.
2c   **Adult female** *castaneofuscus* Brownish above; buffish-yellow below with some rufous wash.

3 BROWN-CAPPED WEAVER *Ploceus insignis* 14 cm **R**   **Page 742**
Montane and submontane forest. SE Nigeria–W Cameroon; Bioko. Also NE Gabon and SW Congo. Not uncommon.
3a   **Adult male** Golden-yellow; chestnut cap; black mask, wings and tail.
3b   **Adult female** All-black head.
3c   **Juvenile** Head olive mottled black; no black on throat; horn-coloured bill.

4 YELLOW-MANTLED WEAVER *Ploceus tricolor* 17 cm **R**   **Page 741**
Lowland forest. Sierra Leone–SE Guinea to SW Congo; not uncommon. SW CAR; rare.
4a   **Adult** Broad yellow crescent on upper mantle; deep chestnut breast and belly.
4b   **Juvenile** Dull chestnut; wings and tail dull black.

5 PREUSS'S GOLDEN-BACKED WEAVER *Ploceus preussi* 14 cm **R**   **Page 743**
Lowland forest. Sierra Leone–SE Guinea to Ghana; Cameroon to SW CAR–N Congo. Scarce/uncommon.
5a   **Adult male** Golden-chestnut crown; yellow stripe from nape to uppertail-coverts.
5b   **Adult female** Black forehead; no chestnut wash on breast.
5c   **Juvenile** Pattern as adult female, but black on head replaced by yellow-olive; bill horn.

6 DARK-BACKED WEAVER *Ploceus bicolor* 15 cm **R**   **Page 742**
Montane and lowland forest. SE Nigeria to Congo; Bioko. Locally not uncommon.
Adult Black head; blackish-slate upperparts; bright yellow underparts; blue-grey bill.

7 YELLOW-CAPPED WEAVER *Ploceus dorsomaculatus* 14 cm **R**   **Page 743**
Lowland forest. S Cameroon, Gabon, N Congo. Scarce.
7a   **Adult male** Similar to 5a but rump and uppertail-coverts black.
7b   **Adult female** Black head.

8 BATES'S WEAVER *Ploceus batesi* 14 cm **R**   EN   **Page 735**
Lowland forest. Cameroon. Rare. Endemic.
8a   **Adult male** Chestnut head; black upper throat; yellowish-green upperparts and tail.
8b   **Adult female** As adult male but head black, throat yellow.
8c   **Juvenile** As adult female but head olive mottled black; bill horn coloured.

9 BLACK-BILLED WEAVER *Ploceus melanogaster* 14 cm **R**   **Page 737**
Montane forest. SE Nigeria–W Cameroon; Bioko. Locally not uncommon.
9a   **Adult male** Black with yellow head; black throat bordered by yellow band.
9b   **Adult female** As adult male but throat yellow.
9c   **Juvenile** Pattern as adult female; head washed olive; underparts yellowish-olive.

10 BANNERMAN'S WEAVER *Ploceus bannermani* 14 cm **R**   VU   **Page 735**
Montane forest and scrub. SE Nigeria–W Cameroon. Locally not uncommon. Endemic.
Adult Black mask; yellowish-green upperparts and tail.

PLATE 132: MALIMBES

Genus *Malimbus*. Distinctive black-and-red forest weavers. One species also has yellow in its plumage and another is black and yellow, lacking any red. Juveniles duller with horn-coloured bills variably tinged dusky. Calls harsh and rasping.

1 BLUE-BILLED MALIMBE *Malimbus nitens* 17 cm **R**   Page 744
Forest undergrowth, near water. S Senegal–Liberia to SW CAR–Congo. Common.
  1a **Adult** Red patch from throat to upper breast; blue bill.
  1b **Juvenile** Dull red area extending onto throat, head-sides and forehead.

2 CRESTED MALIMBE *Malimbus malimbicus* 17 cm **R**   Page 744
Forest. Guinea to Togo; Nigeria to SW CAR–Congo. Not uncommon/uncommon.
  2a **Adult male** Red head and breast; small nuchal crest; black nape.
  2b **Adult female** As male but without crest; red less extensive below.
  2c **Juvenile** Red area extending onto nape; throat blackish.

3 CASSIN'S MALIMBE *Malimbus cassini* 17 cm **R**   Page 744
Forest. S Cameroon to SW CAR–Congo. Common.
  3a **Adult male** Red head, neck and breast; black mask.
  3b **Adult female** Wholly black (as 10b).
  3c **Juvenile** Pale reddish on head, throat -and breast.

4 IBADAN MALIMBE *Malimbus ibadanensis* 17 cm **R** EN   Page 744
Forest edge, second growth. SW Nigeria (Ghana?). Uncommon and very local. Endemic.
  4a **Adult male** As 3a, but red usually more extensive.
  4b **Adult female** As male, but red on underparts reduced.
  4c **Juvenile** Head to breast dull reddish; blackish mask.

5 RED-VENTED MALIMBE *Malimbus scutatus* 17 cm **R**   Page 745
Forest edge. Sierra Leone to Togo; Benin to SW Cameroon. Not uncommon/common. Endemic.
  5a **Adult male** *scutatus* As 3a but undertail-coverts red. In small noisy groups. West of range.
  5b **Adult female** *scutatus* Head black; underparts as male.
  5c **Adult female** *scutopartitus* Red breast patch vertically separated. East of range.
  5d **Juvenile** As 5b, but red areas pale pinkish and extending onto throat and forehead.

6 RACHEL'S MALIMBE *Malimbus racheliae* 17 cm **R**   Page 745
Forest. SE Nigeria to Gabon. Locally not uncommon. Endemic.
  6a **Adult male** As 5a, but lower part of breast patch and undertail-coverts bright yellow.
  6b **Adult female** Head black; underparts as male.
  6c **Juvenile** As female, but breast patch extending onto throat.

7 RED-HEADED MALIMBE *Malimbus rubricollis* 18 cm **R**   Page 746
Forest and outliers. W Guinea to SW CAR–Congo; Bioko. Not uncommon.
  7a **Adult male** Red top of head, nape and neck-sides.
  7b **Adult female** As male, but forehead and forecrown black.
  7c **Juvenile** Red area paler; some reddish feathers on face and breast.

8 GOLA MALIMBE *Malimbus ballmanni* 17 cm **R** EN   Page 745
Forest. E Sierra Leone, Liberia, W Ivory Coast. Rare/locally common. Endemic.
  8a **Adult male** Yellow nape, breast and undertail-coverts.
  8b **Adult female** As adult male, but without nape patch.
  8c **Juvenile male** As adult male, but top of head pale yellow to orange-brown; nape patch smaller; breast patch extending onto throat.

9 RED-BELLIED MALIMBE *Malimbus erythrogaster* 17 cm **R**   Page 746
Forest. SE Nigeria to CAR /Congo. Not uncommon/rare.
  9a **Adult male** Red head and underparts red; black mask.
  9b **Adult female** As male, but throat also red.
  9c **Juvenile** Head all red; indistinct mask; lower underparts greyish-brown.

10 RED-CROWNED MALIMBE *Malimbus coronatus* 17 cm **R**   Page 746
Forest. S Cameroon to SW CAR–Congo. Not uncommon/uncommon.
  10a **Adult male** Red crown patch.
  10b **Adult female** Wholly black (as 3b).
  10c **Juvenile** Rufous-brown crown patch.

PLATE 133: WEAVERS AND QUELEAS

1 WHITE-BILLED BUFFALO WEAVER *Bubalornis albirostris* 22–24 cm **R**   Page 732
Semi-arid belt. S Mauritania–Guinea to N Cameroon–Chad. Common.
  1a **Adult male breeding** Very large; wholly black; massive whitish bill.
  1b **Adult male non-breeding** Bill black. Adult female similar.
  1c **Juvenile** Blackish-brown; underparts mottled white.

2 RED-HEADED WEAVER *Anaplectes rubriceps* 14–15 cm **R**   Page 747
Wooded savanna. S Senegambia–Guinea to CAR. Uncommon/rare and local.
  2a **Adult male** Red head and bill; black mask. Not gregarious.
  2b **Adult female** Greyish-brown head; red bill.

3 GROSBEAK WEAVER *Amblyospiza albifrons* 18 cm **R**   Page 753
Tall herbage. Sierra Leone to SW CAR; S Gabon to SW Congo. Locally not uncommon.
  3a **Adult male** Heavy bill; white forehead and primary patch.
  3b **Adult female** Heavily streaked below.

4 PARASITIC WEAVER *Anomalospiza imberbis* 13 cm **R**   Page 752
Wooded grassland. Patchily distributed, Sierra Leone to Cameroon. Uncommon/rare and local.
  4a **Adult male** Small; bright yellow head and underparts; short tail; stubby, black bill.
  4b **Adult female** Buff-brown boldly streaked black above; buff below; bill horn.

5 RED-BILLED QUELEA *Quelea quelea* 11–13 cm **M**   Page 747
Open grassy habitats. S Mauritania–Senegambia to Chad–CAR and S Congo. Locally common/
uncommon.
  5a **Adult male breeding** Black mask (incl. frontal band); red bill. Highly gregarious.
  5b **Adult male breeding** Black-faced variant with wholly dark crown.
  5c **Adult male breeding** Black-faced variant without frontal band.
  5d **Adult male breeding** White-faced variant.
  5e **Adult female breeding/male non-breeding** Small; streaky above; long off-white supercilium;
    greyish cheeks. Resembles 7b but with red bill.

6 BOB-TAILED WEAVER *Brachycope anomala* 11 cm **R**   Page 748
Along large rivers. N Congo and S CAR; locally not uncommon. Vagrant, SE Cameroon.
  6a **Adult male** Small; very short tail; black mask.
  6b **Adult female** Plain buffish-brown head and underparts.

7 RED-HEADED QUELEA *Quelea erythrops* 11-12 cm **M**   Page 747
Moist savanna, clearings in forest zone. Senegambia–Liberia to Chad, CAR and Congo.
Resident, São Tomé. Locally common to uncommon.
  7a **Adult male breeding** Red head; black bill. Gregarious.
  7b **Adult female breeding** Face washed with yellow; bill horn.

Genus *Euplectes*. Occur in open grassland and edges of marshy areas. Breeding males conspicuous in plumage and behaviour, and easily identified. In non-breeding season males resemble the nondescript, sparrow-like females, but are often noticeably larger and, in some species, may retain coloured shoulder patches. Females and many males in non-breeding plumage hard to identify; some not safely separable. Most breed during rains, forming mixed-species flocks in off-season.

1   RED-COLLARED WIDOWBIRD *Euplectes ardens* 12–14 cm **R**   **Page 752**
Grassy plains and hill sides. W Guinea to Togo; Nigeria to CAR–Congo. Locally common.
1a   **Adult male breeding** All black; very long, graduated tail (up to 20 cm). In south of range
      (S Gabon, Congo) some have yellow to red crescent-shaped collar on lower throat.
1b   **Adult male non-breeding** Black wing feathers.
1c   **Adult female breeding** Unstreaked below.

2   FAN-TAILED WIDOWBIRD *Euplectes axillaris* male *c.* 18 cm, female *c.* 14 cm **R**   **Page 750**
Rank herbage. Along Niger R. (Mali, Niger); L. Chad area (NE Nigeria–W Chad–N Cameroon),
W Cameroon, NE CAR and Congo (along Congo R.). Locally not uncommon/scarce.
2a   **Adult male breeding** Orange-yellow shoulders above chestnut wing patch.
2b   **Adult male non-breeding** Retains orangey shoulders and black flight feathers.
2c   **Adult female** Lesser coverts black edged orangey.

3   BLACK BISHOP *Euplectes gierowii* 15–16 cm **R**   **Page 749**
Rank herbage. Cameroon and SW CAR. Scarce and local.
3a   **Adult male breeding** Large; orange hindcrown, neck and breast-band; orange-yellow mantle.
3b   **Adult female breeding** As 4b–c but more broadly streaked above; more washed yellowish-buff
      below; breast-sides more distinctly streaked.

4   BLACK-WINGED RED BISHOP *Euplectes hordeaceus* 13–14 cm **R**   **Page 749**
Rank herbage. Senegambia–Liberia to Chad, CAR and Congo; São Tomé. Locally common.
4a   **Adult male breeding** Scarlet crown; black wings; buff undertail-coverts.
4b   **Adult male non-breeding** Black flight feathers (conspicuous in flight).
4c   **Adult female** Streaky above; yellowish supercilium; breast-sides faintly streaked.

5   NORTHERN RED BISHOP *Euplectes franciscanus* 11–12 cm **R**   **Page 749**
Rank herbage. S Mauritania–Sierra Leone to Chad–N CAR. Common/not uncommon.
5a   **Adult male breeding** Black crown; brown wings; short tail concealed by long upper- and
      undertail-coverts.
5b   **Adult male moulting**
5c   **Adult female** Smaller than 4c.

1 YELLOW-MANTLED WIDOWBIRD *Euplectes macrourus* m. 19–21 cm, f. 13–14 cm **R Page 751**
Rank herbage. S Senegambia–Liberia to Chad, CAR and Congo. Common.
> **1a** **Adult male breeding** Yellow mantle and shoulders; long black tail.
> **1b** **Adult male non-breeding** As 1c, but larger, with yellow shoulders.
> **1c** **Adult female breeding** Lesser coverts tipped yellowish.

2 MARSH WIDOWBIRD *Euplectes hartlaubi* male 22 cm, female 16 cm **R**   **Page 752**
Moist grasslands. E Nigeria to Congo. Uncommon/scarce and local.
> **2a** **Adult male breeding** Orange-yellow shoulders; long, broad, black tail.
> **2b** **Adult female breeding** Bulky; heavy bill.
> **2c** **Adult male non-breeding in flight** Orange-yellow shoulders; black flight feathers.

3 WHITE-WINGED WIDOWBIRD *Euplectes albonotatus* m. 14–15 cm, f. 12–13 cm **R**   **Page 751**
Grasslands. S Gabon–Congo, also W CAR; uncommon and local. São Tomé; locally common.
> **3a** **Adult male breeding** Yellow shoulders; white primary patch (forming conspicuous wingbar in
>   flight); long narrow tail (projecting up to 12 cm).
> **3b** **Adult male non-breeding** Sparrow-like but with yellow shoulders and white wing patch.
> **3c** **Adult female breeding** Similar to 1c but with white underwing-coverts.

4 YELLOW BISHOP *Euplectes capensis* male 14 cm, female 11.0–12.5 cm **R**   **Page 750**
Grasslands in montane and hilly areas. SE Nigeria–Cameroon; Bioko. Locally common.
> **4a** **Adult male breeding** Black with yellow shoulders and rump.
> **4b** **Adult male non-breeding** Sparrow-like with yellow shoulders and rump.
> **4c** **Adult female breeding** The only sparrow-like *Euplectes* with dull yellowish rump.

5 YELLOW-CROWNED BISHOP *Euplectes afer* 11 cm **R**   **Page 748**
Moist, rank vegetation; floodplains. S Mauritania–Sierra Leone to Chad, CAR and Congo. Locally
common.
> **5a** **Adult male breeding** Bright yellow and black; small.
> **5b** **Adult female breeding** Broad yellowish supercilium; dusky eye-stripe.

PLATE 136: TWINSPOTS, ANTPECKERS, PYTILIAS AND CRIMSONWING

1 BROWN TWINSPOT *Clytospiza monteiri* 13 cm **R** **Page 760**
Wooded grassland, forest edge. SE Nigeria to Chad–CAR; S Gabon–Congo. Uncommon/locally not uncommon.
1a **Adult male** Rich brown underparts densely spotted white; red throat stripe.
1b **Adult female** As male but has white throat stripe.
1c **Juvenile** Much duller than adult; unspotted orange-brown underparts.

2 RED-FRONTED ANTPECKER *Parmoptila rubrifrons* 10–11 cm **R** **Page 754**
Forest. Sierra Leone to Ghana. Scarce/locally not uncommon.
2a **Adult male** Bright red forehead and forecrown; dark above; rich brown below.
2b **Adult female** Dark earth-brown above; whitish below, densely spotted.
2c **Juvenile** Wholly rufous-brown.

3 DYBOWSKI'S TWINSPOT *Euschistospiza dybowskii* 12 cm **R** **Page 760**
Wooded grassland. SE Senegal–Liberia to S Chad–CAR. Uncommon/scarce and local.
3a **Adult male** Crimson from mantle to uppertail-coverts; slate-grey from head to breast; white spots on black flanks.
3b **Adult female** As male but duller.
3c **Juvenile** Dull slate-grey underparts without spots; dull rusty upperparts.

4 RED-HEADED ANTPECKER *Parmoptila woodhousei* 11 cm **R** **Page 753**
Forest. S Nigeria to SW CAR–Congo. Uncommon/scarce.
4a **Adult male** Orange-rufous face and throat; densely speckled below; some red on forehead.
4b **Adult female** No red on forehead; less densely marked below.
4c **Juvenile** As 2c, but slightly paler.

5 GREEN-WINGED PYTILIA *Pytilia melba* 12–13 cm **R** **Page 756**
Thorn scrub, wooded grassland. Widespread. Uncommon/locally not uncommon.
5a **Adult male** *citerior* Red face; olive-green upperparts; yellow upper breast; red bill. Tail longer than other pytilias. S Mauritania–Guinea to Chad–CAR.
5b **Adult female** *citerior* Grey head; barred grey and white below.
5c **Juvenile** *citerior* Dull olive upperparts; unbarred buffish-brown underparts.
5d **Adult male** *melba* Grey wedge from bill above eye. SE Congo.

6 RED-FACED CRIMSONWING *Cryptospiza reichenovii* 11–12 cm **R** **Page 757**
Montane forest. SE Nigeria–W Cameroon; Bioko. Not uncommon/uncommon.
6a **Adult male** Red eye patch; deep red upperparts.
6b **Adult female** Yellowish-buff eye patch.
6c **Juvenile** Duller than adult; no or indistinct eye patch.

7 YELLOW-WINGED PYTILIA *Pytilia hypogrammica* 12.5–13.0 cm **R** **Page 756**
Wooded grassland. W Guinea to S Chad–CAR. Uncommon/locally not uncommon.
7a **Adult male** Red face; mainly grey body; yellow wings; black bill.
7b **Adult female** No red face; paler, browner grey.

8 ORANGE-WINGED PYTILIA *Pytilia afra* 11 cm **R** **Page 756**
Wooded grassland. S Congo. Locally not uncommon.
8a **Adult male** Red face; olive-green upperparts; orange wings; red bill.
8b **Adult female** Pale grey head; underparts whitish barred olive.

9 GREEN TWINSPOT *Mandingoa nitidula* 10–11 cm **R** **Page 759**
Forest edge. W Guinea to CAR–Congo; Bioko. Uncommon/rare.
9a **Adult male** *schlegeli* Bright green above; red face; lower breast and belly black spotted white; bill black with red cutting edges.
9b **Adult female** *schlegeli* Paler and duller; yellowish-buff face.
9c **Juvenile** Olive and greyish-olive; pale buff face.

10 RED-WINGED PYTILIA *Pytilia phoenicoptera* 12.5–13.0 cm **R** **Page 757**
Wooded grassland. Senegambia–Sierra Leone to Chad–CAR. Uncommon/locally not uncommon.
10a **Adult male** Grey with red wings and tail; black bill.
10b **Adult female** As male but browner grey.

PLATE 137: NEGROFINCHES, OLIVEBACKS, BLUEBILLS AND SEEDCRACKERS

1 GREY-CROWNED NEGROFINCH *Nigrita canicapilla* 15 cm **R**   **Page 754**
Forest zone. Sierra Leone–SE Guinea to CAR–Congo; Bioko. Common/uncommon.
  1a   **Adult** *emiliae* Mainly grey above; black face and underparts. Upper Guinea forest.
  1b   **Adult** *canicapilla* White line separating grey from black on head; white spots on wing-coverts; whitish-grey rump. Lower Guinea forest.
  1c   **Juvenile** Wholly sooty-black.

2 CHESTNUT-BREASTED NEGROFINCH *Nigrita bicolor* 11–12 cm **R**   **Page 755**
Forest zone. Senegambia–Liberia to S CAR–Congo; Príncipe. Locally common/uncommon.
  2a   **Adult** *bicolor* Slate-grey above; dark chestnut below.
  2b   **Juvenile** Pale grey-brown above; rich buff-brown below.

3 WHITE-BREASTED NEGROFINCH *Nigrita fusconota* 10 cm **R**   **Page 755**
Forest zone. Sierra Leone–SE Guinea to Ghana; Nigeria to SW CAR–Congo, Bioko. Uncommon/locally not uncommon.
  Adult Black head and tail; brown upperparts; white underparts.

4 PALE-FRONTED NEGROFINCH *Nigrita luteifrons* 11.5 cm **R**   **Page 754**
Forest zone. Sierra Leone to SW CAR–Congo, Bioko. Locally not uncommon/rare.
  4a   **Adult male** *luteifrons* Grey above; paler forehead tinged buffish.
  4b   **Adult female** Wholly grey with small black mask.
  4c   **Juvenile** Wholly dull grey.

5 WHITE-CHEEKED OLIVEBACK *Nesocharis capistrata* 12–13 cm **R**   **Page 755**
Moist grassland, riparian woodland. Gambia to CAR. Uncommon/rare and local.
  5a   **Adult** Grey head; white face; black throat; yellow flanks.
  5b   **Juvenile** Lacks white cheeks; bill whitish.

6 LITTLE OLIVEBACK *Nesocharis shelleyi* 8–8.5 cm **R**   **Page 755**
Montane forest. SE Nigeria–W Cameroon; Bioko. Locally common/uncommon. Endemic.
  6a   **Adult male** Black head; grey nape; olive upperparts; white stripe on throat-sides; yellowish-olive breast.
  6b   **Adult female** White stripe on throat-sides reduced or lacking; breast grey.

7 GRANT'S BLUEBILL *Spermophaga poliogenys* 14 cm **R**   **Page 758**
Forest. N Congo. Rare.
  7a   **Adult male** As 8a but red restricted to face and forecrown.
  7b   **Adult female** As 8b but head blackish; red restricted to throat and breast.

8 RED-HEADED BLUEBILL *Spermophaga ruficapilla* 15 cm **R**   **Page 759**
Forest, thickets. SE CAR. Common.
  8a   **Adult male** Entirely red head. Compare 7a.
  8b   **Adult female** Entirely red head; red cutting edges to bill.

9 WESTERN BLUEBILL *Spermophaga haematina* 15 cm **R**   **Page 758**
Forest, thickets. Senegambia–Liberia to SW CAR–Congo. Common.
  9a   **Adult male** *haematina* Black with red throat, breast and flanks; heavy blue, red-tipped bill.
  9b   **Adult female** *haematina* Face washed red; red uppertail-coverts; lower underparts densely spotted white.
  9c   **Juvenile** Dark brownish-slate; dull reddish uppertail-coverts.
  9d   **Adult male** *pustulata* Red extends onto lower head-sides; red uppertail-coverts.

10 CRIMSON SEEDCRACKER *Pyrenestes sanguineus* 13–14 cm **R**   **Page 758**
Forest edge, ricefields, farmbush. Senegambia to Ivory Coast. Uncommon/locally common.
  10a   **Adult male** Heavy, triangular, steel-blue bill; red head, breast and flanks; red rump and tail.
  10b   **Juvenile** Wholly brown except for dull red rump and tail.

11 BLACK-BELLIED SEEDCRACKER *Pyrenestes ostrinus* 15 cm **R**   **Page 757**
Forest edge, ricefields, farmbush. E Ivory Coast to CAR–Congo. Uncommon/locally common.
  11a   **Adult male** As 10a but earth-brown areas replaced by blue-black.
  11b   **Adult female** Red areas more restricted; rest of plumage warm brown.

PLATE 138: WAXBILLS, CORDON-BLEUS AND QUAILFINCHES

1 ORANGE-CHEEKED WAXBILL *Estrilda melpoda* 10 cm **R** Page 764
Various grassy habitats. Throughout, except arid north. Common.
1a Adult Bright orange face; red rump; red bill. **1b Juvenile** Duller and buffier; black bill.

2 FAWN-BREASTED WAXBILL *Estrilda paludicola* 11.5 cm **R** Page 764
Moist grassland, grassy clearings. S Gabon, Congo, E CAR. Locally common.
2a Adult Greyish head; red rump; red bill. **2b Juvenile** As adult but duller; black bill.

3 ANAMBRA WAXBILL *Estrilda poliopareia* 11–12 cm **R VU** Page 764
Rank grass along rivers, lagoon sand banks. S Nigeria. Local. Endemic.
Adult Pale, plain grey-brown head; red rump; pale yellowish-buff underparts; red bill.

4 BLACK-CROWNED WAXBILL *Estrilda nonnula* 11 cm **R** Page 765
Grassland, forest regrowth, gardens. SE Nigeria to CAR–Gabon; Bioko. Locally common.
4a Adult male Black cap; white cheeks; red rump; white underparts washed pale grey.
4b Juvenile Duller; greyish-brown above; pale buffish-brown below; black bill.

5 LAVENDER WAXBILL *Estrilda caerulescens* 10 cm **R** Page 763
Wooded grassland. Senegambia–Guinea to Chad–N CAR. Locally common/scarce.
Adult Mainly grey; red rump, tail and undertail-coverts.

6 BLACK-HEADED WAXBILL *Estrilda atricapilla* 10 cm **R** Page 766
Grassy clearings, forest regrowth. S Cameroon to Congo. Locally common.
6a Adult As 4a but darker; underparts smoke-grey; undertail-coverts black.
6b Juvenile Duller and tinged dark brown; black bill.

7 GREY WAXBILL *Estrilda perreini* 11 cm **R** Page 764
Wooded grassland, thickets. S Gabon and Congo. Uncommon/scarce.
Mainly grey; red rump; black tail.

8 BLACK-RUMPED WAXBILL *Estrilda troglodytes* 10 cm **R** Page 765
Savanna belt. S Mauritania–Guinea to Chad–CAR. Locally common.
8a Adult male Broad red eye-stripe; black rump; white undertail.
8b Juvenile Duller; no red on face; darkish eye-stripe; black bill.

9 COMMON WAXBILL *Estrilda astrild* 11 cm **R** Page 765
Various grassy habitats. W Guinea to CAR–Congo; Bioko. Locally common.
9a Adult *occidentalis* Resembles 8a but has brown rump and black undertail.
9b Adult *rubriventris* Heavily tinged pink or pinkish-red. Coastal Gabon–Congo.
9c Juvenile Duller, more buff-brown; eye-stripe paler and smaller; bill blackish.

10 SOUTHERN CORDON-BLEU *Uraeginthus angolensis* 12–13 cm **R** Page 766
Gardens, towns, edges of cultivation. S Congo, São Tomé. Uncommon/locally common.
As 11a but adult male lacks red on cheek.

11 RED-CHEEKED CORDON-BLEU *Uraeginthus bengalus* 13 cm **R** Page 766
Savanna, gardens, villages. S Mauritania–Guinea to Chad–CAR. Common.
11a Adult male Pale blue underparts, rump and tail; bright red cheek patch.
11b Adult female Slightly duller; no red on cheek.

12 AFRICAN QUAILFINCH *Ortygospiza atricollis* 9.5–10.0 cm **R** Page 767
Open grassy habitats. S Mauritania–Liberia to Cameroon–Chad. Locally common.
12a Adult male Dark and dumpy. In flight: short tail with white corners.
12b Adult female Duller; head greyer without black mask.

13 ZEBRA WAXBILL *Amandava subflava* 9–10 cm **R** Page 767
Grassy habitats. Senegambia–Liberia to S Chad, CAR, Congo. Locally common/ scarce.
13a Adult male Bright orange below; red rump. **13b Adult female** Duller; no red supercilium.
13c Juvenile Dull brownish above; rump washed orange; dull buffish below; blackish bill.

14 BLACK-CHINNED QUAILFINCH *Ortygospiza gabonensis* 9.5–10.0 cm **R** Page 768
Open grassy habitats. Eq. Guinea, Gabon, Congo. Uncommon/common and local.
14a Adult male Much as 12a but more streaky above, paler below; bright red bill.
14b Adult female Duller; no black mask.

15 LOCUST FINCH *Ortygospiza locustella* 9–10 cm **R** Page 767
Moist grassland. SE Cameroon, SE Gabon, Congo. Uncommon and local.
15a Adult male Red face to breast; reddish wings (conspicuous in flight). **15b Adult female** Mainly greyish-brown with orange wings; barred flanks. **15c Juvenile** Browner than female; bill dark.

1 BAR-BREASTED FIREFINCH *Lagonosticta rufopicta* 11 cm **R**   **Page 760**
Various open habitats. Senegambia–Sierra Leone to Chad–CAR. Not uncommon.
  1a **Adult** Small white crescents on breast-sides and flanks; pinkish-red bill.
  1b **Juvenile** Dull grey-brown; face and breast tinged pinkish-red; dusky bill.

2 RED-BILLED FIREFINCH *Lagonosticta senegala* 10 cm **R**   **Page 761**
Savanna belt. S Mauritania–Sierra Leone to Chad–CAR. Common.
  2a **Adult male** Plumage largely red; yellowish orbital ring; red bill.
  2b **Adult female** Mainly buff-brown above, paler below; red loral spot and rump; red bill.
  2c **Juvenile** As 2b but lacks red on lores; no white dots on breast; black bill.

3 ROCK FIREFINCH *Lagonosticta sanguinodorsalis* *c.* 10 cm **R**   **Page 762**
Wooded grassland with rocky outcrops. NC & NE Nigeria. Locally not uncommon. Endemic.
  3a **Adult male** Grey crown; bright red mantle; two-toned bill (black and blue-grey).
  3b **Adult female** Bright reddish-brown mantle and back.

4 BLUE-BILLED FIREFINCH *Lagonosticta rubricata* 10.0–11.5 cm **R**   **Page 761**
Forest–savanna mosaic. SW Senegal–Liberia to Chad, CAR, Congo. Not uncommon/rare.
  4a **Adult male** Top of head and upperparts grey-brown; black undertail-coverts; bluish bill.
  4b **Adult female** Paler than male.
  4c **Juvenile** Dull brownish, paler below; dark bill.

5 KULIKORO FIREFINCH *Lagonosticta virata* 10–11 cm **R**   **Page 763**
Rocky areas with grass and bushes. Mali and SE Senegal. Uncommon/rare and local. Endemic.
  Adult male As 4a but slightly duller; fewer spots; bill with black tip, often all-black above.

6 BLACK-BELLIED FIREFINCH *Lagonosticta rara* 10 cm **R**   **Page 761**
Wooded grassland. SE Senegal; W Guinea to Chad–CAR. Not uncommon/rare.
  6a **Adult male** Mainly deep wine-red; black from undertail to lower breast; two-toned bill.
  6b **Adult female** Greyish head; black on underparts as male.
  6c **Juvenile** Dull buff-brown; blackish bill with pinkish base below.

7 REICHENOW'S FIREFINCH *Lagonosticta* (*rhodopareia*) *umbrinodorsalis* 10–11 cm **R Page 762**
Rocky hillsides with grassy areas. N Cameroon and SW Chad. Locally not uncommon.
  Adult male Resembles 4a but top of head greyish.

8 BLACK-FACED FIREFINCH *Lagonosticta larvata* 11.5 cm **R**   **Page 763**
Wooded grassland. Senegambia–Guinea to Chad–CAR. Locally not uncommon.
  8a **Adult male** *vinacea* Black face; pinkish-mauve above; bright pink below. West of range.
  8b **Adult female** *vinacea* Pale grey-brown head; buffish throat; pale pink below.
  8c **Juvenile** *vinacea* Grey-brown above; buff-brown below; no white specks on sides.
  8d **Adult male** *togoensis* Grey above; paler grey below with variable pinkish wash. East of range.
  8e **Adult female** *togoensis* Grey-brown above, paler below; black bill.

9 CUT-THROAT *Amadina fasciata* 11–12 cm **R**   **Page 769**
Sahel zone. S Mauritania–Senegambia to N Cameroon–Chad. Locally common.
  9a **Adult male** Sandy coloured and scaly; red band on throat; chestnut on belly.
  9b **Adult female** No red on throat; face barred; no chestnut on belly.

10 AFRICAN SILVERBILL *Lonchura cantans* 10 cm **R**   **Page 768**
Dry savanna. S Mauritania–Senegambia to Chad–N CAR. Scarce/locally common.
  Adult Pale sandy-brown; black rump; black, pointed tail; blue-grey bill.

11 BRONZE MANNIKIN *Lonchura cucullata* 9 cm **R**   **Page 768**
Grassy habitats, bush, gardens. Throughout; also Bioko, São Tomé and Príncipe. Common.
  11a **Adult** Dark brown above; white below; barred rump and flanks. Small and dumpy.
  11b **Juvenile** Entirely buff-brown.

12 MAGPIE MANNIKIN *Lonchura fringilloides* 11.5–12.0 cm **R**   **Page 769**
Grassy habitats in forest zone. Senegambia to CAR–Gabon. Uncommon/rare and local.
  12a **Adult** Resembles 11a but larger; bill heavier; black patch on breast-sides.
  12b **Juvenile** Dusky-brown above; pale buff-brown below; bill dark.

13 BLACK-AND-WHITE MANNIKIN *Lonchura bicolor* 9.5–10.5 cm **R**   **Page 769**
Grassy habitats in forest zone. Guinea-Bissau to CAR–Congo; Bioko. Common.
  13a **Adult** Glossy black and pure white; heavy, pale grey-blue bill.
  13b **Juvenile** Dull earth-brown above; pale buff-brown below; bill dark.

PLATE 140: INDIGOBIRDS AND WHYDAHS

Small, seed-eating finches, occurring mainly in savanna. Forage on the ground. Polygamous, mostly species-specific brood parasites, laying eggs in nests of estrildids.

Whydahs. Males in breeding plumage have four greatly elongated central tail feathers. All paradise whydahs are very similar but differ mainly in length, width and shape of tail. Status and distribution inadequately known. Parasitise waxbills *Estrilda* and pytilias *Pytilia* species.

1   SAHEL PARADISE WHYDAH *Vidua orientalis*   12.5 cm (+tail of up to *c.* 24 cm)   **R**   **Page 772**
Savanna. S Mauritania–Guinea to N Cameroon–Chad. Uncommon.
  1a   Adult male breeding *aucupum* Very long broad tail streamers; chestnut collar.
  1b   Adult female Boldly striped head; dark bill and legs.
  1c   Juvenile Plain earth-brown above; pale buff-brown to white below.

2   EXCLAMATORY PARADISE WHYDAH *Vidua interjecta*   12.5 cm (+tail of >27 cm)   **R   Page 773**
Savanna. Mali–Guinea to Chad–CAR. Common.
  2a   Adult male breeding Tail streamers longer than 1a and broader than 3.
  2b   Adult female As 1b but with pale orange bill and pinkish legs.

3   TOGO PARADISE WHYDAH *Vidua togoensis*   12.5 cm (+tail of >27 cm)   **R**   **Page 773**
Savanna. Mali–Sierra Leone to N Cameroon–Chad. Status?
  Adult male breeding Tail streamers narrower than 2a; underparts more uniformly yellow; nuchal collar paler.

4   PIN-TAILED WHYDAH *Vidua macroura*   12.5 cm   **R**   **Page 772**
Various open habitats, incl. gardens. Throughout, incl. Bioko, São Tome and Príncipe. Common.
  4a   Adult male breeding Black and white; long ribbon-like tail streamers of up to 20 cm.
  4b   Adult female breeding Boldly striped head; mainly dark brown bill; dark legs.
  4c   Adult male/female non-breeding As 4b but bill pinkish.
  4d   Juvenile Plain pale brown above; pale buffish below; dusky bill.

Indigobirds. Often only distinguishable in the field if song of foster species is known. Songs include non-mimetic chattering and clear, whistled notes mimicking those of hosts. Status and distribution imperfectly known; presence of host an indication of species possibly involved.

5   CAMEROON INDIGOBIRD *Vidua camerunensis*   11.5 cm   **R**   **Page 771**
Adult male breeding Blue gloss; pale purplish legs. Mimics Blue-billed and Black-bellied Firefinches, Brown and Dybowski's Twinspots.
Adult female/adult male non-breeding are as 6b but with pale purplish legs.

The following species are very similar to 5:

JAMBANDU INDIGOBIRD *Vidua raricola*   11.5 cm   **R**
Adult male breeding Blue to green gloss; brown wings. Mimics Zebra Waxbill.

BAKA INDIGOBIRD *Vidua larvaticola*   11.5 cm   **R**
Adult male breeding Greenish-blue gloss. Mimics Black-faced Firefinch.

VARIABLE INDIGOBIRD *Vidua funerea*   11.5 cm   **R**
Adult male breeding Purplish-blue gloss. Mimics Blue-billed Firefinch.

JOS PLATEAU INDIGOBIRD *Vidua maryae*   11.5 cm   **R**
Adult male breeding Green gloss. Mimics Rock Firefinch. C & N Nigeria.

QUAILFINCH INDIGOBIRD *Vidua nigeriae*   11.5 cm   **R**
Adult male breeding Dull green. Mimics African Quailfinch.

WILSON'S INDIGOBIRD *Vidua wilsoni*   11.5 cm   **R**
Adult male breeding Purplish gloss; dark brown wings; purplish legs. Mimics Bar-breasted Firefinch.

6   VILLAGE INDIGOBIRD *Vidua chalybeata*   11.5 cm   **R**   **Page 770**
  6a   Adult male breeding *chalybeata* The only indigobird with orange to red legs; green gloss; black wings. Mimics Red-billed Firefinch.
  6b   Adult female/adult male non-breeding Pale median crown-stripe; dark brown lateral crown-stripes; streaked upperparts; white bill.
  6c   Juvenile Plain dusky-brown crown.

PLATE 141: TRUE FINCHES

Small seed-eaters with stout, conical bills. Arboreal or terrestrial. Several have remarkable, melodious songs.

1 BLACK-FACED CANARY *Serinus capistratus* 11.5 cm **R** Page 774
Woodland, edges of cultivation, gardens. Gabon–Congo. Uncommon/locally common.
1a Adult male Black mask; yellow underparts.
1b Adult female No black mask; some streaks on throat, breast and flanks.

2 WHITE-RUMPED SEEDEATER *Serinus leucopygius* 10.0–11.5 cm **R** Page 774
Sahel and northern savanna. Mauritania–Senegambia to Chad–CAR. Locally common/scarce.
Adult Small; pale greyish-brown; white rump.

3 ORIOLE FINCH *Linurgus olivaceus* 13 cm **R** Page 776
Montane forest. SE Nigeria–W Cameroon; Bioko. Not uncommon.
3a Adult male Black head; golden-yellow underparts; orange-yellow bill.
3b Adult female Olive-green; paler bill.
3c Juvenile As adult female but paler below with faint streaking; dusky bill.

4 BLACK-THROATED SEEDEATER *Serinus atrogularis* 11–12 cm **R** Page 774
Open woodland, cultivation. S Gabon–Congo. Locally not uncommon.
Adult Small; grey-brown; bright yellow rump.

5 YELLOW-FRONTED CANARY *Serinus mozambicus* 11–13 cm **R** Page 775
Open woodland, cultivation, gardens. Throughout; also São Tomé, Annobón. Locally common.
5a Adult male *caniceps* Distinctive head pattern; yellow rump.
5b Adult female *caniceps* Duller; necklace of dark spots.

6 STREAKY-HEADED SEEDEATER *Serinus gularis* 15 cm **R** Page 775
Open woodland, cultivation, scrub. Guinea–Sierra Leone; S Mali–N Ivory Coast to Niger, Cameroon and
CAR. Locally not uncommon/scarce.
6a Adult Mainly brown; broad white supercilium; dark face.
6b Juvenile Duller; streaked underparts.

7 THICK-BILLED SEEDEATER *Serinus burtoni* 18 cm **R** Page 775
Montane forest. SE Nigeria–W Cameroon. Scarce/uncommon to locally common.
Adult Robust; white forehead; heavy bill; double wingbar.

8 COMMON CHAFFINCH *Fringilla coelebs* 15 cm **V** Page 773
Palearctic vagrant. N Mauritania.
8a Adult male *africana* Bluish-grey head; olive-green above; pink below; broad white wingbars.
8b Adult female Mainly grey-brown; white wingbars.
8c In flight White wingbars; white outer tail feathers.

9 BRAMBLING *Fringilla montifringilla* 15.5 cm **V** Page 773
Palearctic vagrant. N Mauritania; also offshore S Senegal.
9a Adult male non-breeding Orangey breast and shoulders.
9b Adult female non-breeding As male but duller; no black on face.
9c In flight White rump.

10 EUROPEAN GREENFINCH *Carduelis chloris* 15 cm **V** Page 776
Palearctic vagrant. Coastal Mauritania.
10a Adult male Robust; olive-green; yellow in wing and tail. Adult female similar but more grey-brown.
10b In flight Yellow patches on primaries and sides of tail.

11 COMMON LINNET *Carduelis cannabina* 14 cm **V** Page 777
Palearctic vagrant. Mauritania and N Senegal.
11a Adult male Greyish head; chestnut upperparts.
11b Adult female More heavily streaked.
11c In flight White patches in wings and tail-sides.

12 TRUMPETER FINCH *Bucanetes githagineus* 12.5 cm **R** Page 777
Stony desert. Mauritania to N Chad. Locally not uncommon.
12a Adult male breeding Grey head; distinct pink tinge esp. on underparts; orange-red bill.
12b Adult female Duller, sandy-grey; horn-coloured bill.

PLATE 142: BUNTINGS

Small, mainly seed-eating species with stout, conical bills. Predominantly terrestrial. Flight typically undulating. Usually singly or in pairs, but some gregarious outside breeding season. Vocalisations usually rather simple.

1 AFRICAN GOLDEN-BREASTED BUNTING *Emberiza flaviventris* 15.5 cm **R** **Page 779**
Dry grassy savanna, open woodland. S Mauritania–Senegal to Chad–CAR; S Congo. Not uncommon/rare.
- **1a** **Adult male** *flavigaster* Bold black-and-white head pattern; white wingbars; grey rump.
- **1b** **Adult female** *flavigaster* As male but head markings browner.
- **1c** **Juvenile** *flavigaster* Duller and browner than adult; head pattern brown and buff.

2 BROWN-RUMPED BUNTING *Emberiza affinis* 14 cm **R** **Page 779**
Open savanna, farmland. Senegambia–Guinea to S Chad–W CAR. Uncommon/locally not uncommon.
- **2a** **Adult male breeding** Head whiter than 1a; no white on wing; brown rump.
- **2b** **Juvenile** Duller; head washed rusty-buff, stripes brown.

3 CABANIS'S BUNTING *Emberiza cabanisi* 16.5 cm **R** **Page 780**
Wooded savanna. Uncommon/locally not uncommon.
- **3a** **Adult male** *cabanisi* Blackish head; long white supercilium; black-and-grey upperparts. S Mali–W Guinea to Chad–CAR.
- **3b** **Juvenile** *cabanisi* Much browner than adult; pale rufous supercilium.
- **3c** **Adult male** *cognominata* Greyish median crown-stripe; buff-grey mantle with dark streaks. SE Gabon and Congo.

4 CINNAMON-BREASTED ROCK BUNTING *Emberiza tahapisi* 14 cm **R/M** **Page 778**
Open savanna with rocky outcrops. Throughout. Locally common.
- **4a** **Adult male** *goslingi* Black-and-white striped head; grey throat; rufous-brown body; rufous wing panel. North of range.
- **4b** **Adult female** *goslingi* Duller; head striped buffish-white and blackish-brown.
- **4c** **Adult male** *tahapisi* Black throat; less rufous in wings. Gabon and Congo.

5 CRETZSCHMAR'S BUNTING *Emberiza caesia* 15.5 cm **V** **Page 778**
Palearctic vagrant. E Chad.
- **5a** **Adult male** Blue-grey head; rusty moustachial stripe and throat; creamy eye-ring.
- **5b** **Adult female** Duller and paler; finely streaked crown and breast.

6 HOUSE BUNTING *Emberiza striolata* 14 cm **R/V** **Page 778**
Rocky wadis, cultivation, villages. Mauritania to Chad; uncommon/locally common. Vagrant, Senegambia.
- **6a** **Adult male** Grey head streaked black; almost uniformly rufous body.
- **6b** **Adult female** Paler and duller; head more grey-brown; no streaks on breast.

EURASIAN ROCK BUNTING *Emberiza cia* 16 cm **V?** **Page 778**
Palearctic vagrant? A juvenile claimed from Chad. **See illustration on plate 147.**

7 ORTOLAN BUNTING *Emberiza hortulana* 16.5 cm **P** **Page 777**
Open upland habitats. Mauritania–Sierra Leone to N Nigeria, Chad. Rare/uncommon, local.
- **7a** **Adult male** Olive-grey head; yellow moustache, throat and eye-ring; pink bill.
- **7b** **Adult female** Slightly duller than male; crown and breast streaked.
- **7c** **Immature** Duller than adult female; flanks streaked.

8 CORN BUNTING *Miliaria calandra* 18 cm **V** **Page 780**
Palearctic vagrant. Coastal Mauritania and N Senegal.
- **Adult** Bulky and streaked; no white in tail; heavy bill.

PLATE 143: ISLAND FORMS I – Cape Verde and Gulf of Guinea Islands

Cape Verde Islands

1 **CAPE VERDE SWIFT** *Apus alexandri* 13 cm; WS 35 cm **R**   Page 509
Most islands. Over all habitats. Common endemic.
Small; relatively short wings; shallowly forked tail.

2 **RASO LARK** *Alauda razae* 12–13 cm **R** CR   Page 563
Raso islet, where the only lark. Endemic. Streaked; rather long bill.

3 **IAGO SPARROW** *Passer iagoensis* 13 cm **R**   Page 730
All islands. Various arid habitats. Common endemic.
  3a   **Adult male** Blackish crown; rich chestnut from eye onto ear-coverts.
  3b   **Adult female** Conspicuous pale, creamy-buff supercilium.

4 **CAPE VERDE WARBLER** *Acrocephalus brevipennis* 13.5 cm **R** EN   Page 624
Santiago and São Nicolau. Well-vegetated valleys. Locally not uncommon endemic.
Greyish head; grey-brown upperparts. Song distinctive.

Gulf of Guinea Islands

5 **ANNOBÓN PARADISE FLYCATCHER** *Terpsiphone* (*r.*) *smithii* 18 cm **R** VU   Page 669
Annobón. Forest and cultivated areas. Common endemic.
Blue-black head; rich orange-rufous body; bluish-slate tail. Tail male + 0.5–1.0 cm.

6 **SÃO TOMÉ OLIVE PIGEON** *Columba thomensis* 37–40 cm **R** VU   Page 473
São Tomé. Forest. Locally common endemic.
Large and dark; yellow bill and feet.
  6a   **Adult male** Mainly maroon; small white spots on wing-coverts.
  6b   **Adult female** Duller; only slightly washed maroon.

7 **SÃO TOMÉ SCOPS OWL** *Otus hartlaubi* 18 cm **R** VU   Page 493
São Tomé, where the only small owl. Most habitats with tall trees. Uncommon endemic.
Broad head with very short ear-tufts. Plumage variable.
  7a   **Pale morph** Mainly grey-brown and ochre. **7b** **Rufous morph** Much darker and rufous.

8 **LEMON DOVE** *Aplopelia larvata* 24–25 cm **R**   Page 475
Bioko (*inornata*; not uncommon), São Tomé (*simplex*; common), Príncipe (*principalis*; common),
Annobón (*inornata*; uncommon). Forest.
Adult male *príncipalis* Pale greyish face; vinous paling to buffish below.
Adult female *príncipalis* and adult male/female *simplex* as *A. l. inornata* (see Plate 51: 8).

9 **SÃO TOMÉ BRONZE-NAPED PIGEON** *Columba malherbii* 28 cm **R**   Page 472
São Tomé, Príncipe, Annobón. Forest. Common/uncommon endemic.
Small; grey bill; red feet.
  9a   **Adult male** Head and underparts grey; glossy nape; blackish upperparts; rufous undertail-coverts.
  9b   **Adult female** Similar, but some ochre mottling below.

10 **PRÍNCIPE GLOSSY STARLING** *Lamprotornis ornatus* 29 cm **R**   Page 726
Príncipe. Forest, cultivation, forest regrowth. Common endemic.
Large; mainly bronzy plumage. Compare Splendid Glossy Starling (Plate 126: 7).

11 **SÃO TOMÉ ORIOLE** *Oriolus crassirostris* 23–24 cm **R** VU   Page 718
São Tomé, where the only oriole. Forest. Uncommon/locally common endemic.
  11a   **Adult male** Black head; whitish underparts; creamy collar; red bill.
  11b   **Juvenile** Greyish head and upperparts; breast streaked blackish; dusky bill.

12 **SÃO TOMÉ GREEN PIGEON** *Treron sanctithomae* 28 cm **R**   Page 470
São Tomé, where the only green pigeon. Forest. Common endemic.
Mainly olive-grey and olive-green.

13 **GULF OF GUINEA THRUSH** *Turdus olivaceofuscus* 24 cm **R** NT   Page 617
São Tomé (common) and Príncipe (extremely rare), where the only thrush. Forested habitats. Endemic.
Dark olive-brown above; pale brown and white below with crescentic bars.

14 **DWARF OLIVE IBIS** *Bostrychia* (*olivacea*) *bocagei* *c.* 50 cm   **R** CR   Page 352
São Tomé (where the only ibis). Primary forest. Rare endemic.
As Olive Ibis (Plate 10: 3) but much smaller.

PLATE 144: ISLAND FORMS II – Gulf of Guinea Islands

1 SÃO TOMÉ SPINETAIL *Zoonavena thomensis* 10 cm **R**   Page 507
São Tomé and Príncipe. Almost all habitats and altitudes. Common endemic.
  1a Adult in flight from above Broad white rump with dark streaks. Small.
  1b Adult in flight from below White belly with dark streaks; square tail.

2 SÃO TOMÉ KINGFISHER *Alcedo* (*cristata*) *thomensis* 13.5 cm **R**   Page 519
São Tomé, where the only kingfisher. Streams, beaches. Common endemic.
  2a Adult As Malachite Kingfisher (Plate 65: 4) but darker; malar area barred dusky.
  2b Juvenile Blackish head-sides, mantle and breast.

3 PRÍNCIPE KINGFISHER *Alcedo* (*leucogaster*) *nais* 13 cm **R**   Page 518
Príncipe, where the only small kingfisher. Streams, beaches; also forest. Common endemic.
  Adult As White-bellied Kingfisher (Plate 65: 3) but no rufous supercilium and less white below.

4 AFRICAN MASKED WEAVER *Ploceus velatus* 14 cm **R**   Page 739
São Tomé. Savanna, open habitats, gardens. Common.
  4a Adult male *peixotoi* breeding Black mask extending in point onto lower throat; chestnut wash on
    crown and lower throat.
  4b Adult female *peixotoi* breeding No mask; brownish-olive above.

5 SÃO TOMÉ PARADISE FLYCATCHER *Terpsiphone atrochalybeia* *c.* 18 cm **R**   Page 254
São Tomé, where the only paradise flycatcher. Forest, edges, plantations. Common endemic.
  5a Adult male Glossy blue-black with very long tail (extending 7.0–11.5 cm).
  5b Adult female Rufous upperparts and tail.

6 SÃO TOMÉ FISCAL *Lanius newtoni* 23 cm **R CR**   Page 702
São Tomé, where the only shrike. Primary forest. Rare endemic.
  Adult As Common Fiscal (Plate 120: 1) but less white on scapulars; rump black; yellow wash to
  underparts.

7 SÃO TOME WEAVER *Ploceus sanctithomae* 14 cm **R**   Page 743
São Tomé. All forested habitats; also towns. Common endemic.
  7a Adult male Rusty-buff head-sides and underparts; blackish crown; double whitish wingbar; slender
    bill.
  7b Adult female Duller; crown and upperparts paler.

8 PRÍNCIPE GOLDEN WEAVER *Ploceus princeps* 18 cm **R**   Page 738
Príncipe, where the only weaver. All habitats with trees; also villages. Common endemic.
  8a Adult male Orange-chestnut head; yellowish-green above; golden-yellow below.
  8b Adult female Yellowish-green top of head; white belly.

9 GOLDEN-BACKED BISHOP *Euplectes aureus* 12 cm **R**   Page 750
São Tomé. Grassland near cultivation. Common.
  9a Adult male breeding Black with orange-yellow mantle, back and rump.
  9b Adult female/adult male non-breeding Rufous-buff upperparts heavily streaked black; lemon wash
    to face.

10 GIANT WEAVER *Ploceus grandis* 22 cm **R**   Page 740
São Tomé. Forest edge, plantations, thickets in savanna. Common endemic.
  10a Adult male Very large; black head; chestnut wash to flanks; large bill.
  10b Adult female Above, greyish-olive streaked dusky; below, brownish-buff and white.

11 PRÍNCIPE SEEDEATER *Serinus rufobrunneus* 11.0–12.5 cm **R**   Page 776
São Tomé and Príncipe. Forest, plantations, dry woodland, towns.
  11a Adult *thomensis* Grey-brown; horn-coloured bill. São Tomé. Common endemic.
  11b Adult *rufobrunneus* Rufous-brown. Príncipe. Uncommon/scarce endemic.

12 SÃO TOMÉ GROSBEAK *Neospiza concolor* 19–20 cm **R CR**   Page 776
São Tomé. Primary forest. Extremely rare endemic.
  Adult Large; dark chestnut; massive bill.

PLATE 145: ISLAND FORMS III – Gulf of Guinea Islands

1 FERNANDO PO SPEIROPS *Speirops brunneus* 13 cm **R** VU **Page 701**
Bioko. Montane forest, Pico Basilé (above 1900 m). Common, but very local endemic.
Rusty-brown above with blackish cap; pale brown below with grey throat.

2 ANNOBÓN WHITE-EYE *Zosterops griseovirescens* 12 cm **R** VU **Page 700**
Annobón. Everywhere with bush and tree cover. Common endemic.
Small. Greyish-olive above; whitish tinged olive-yellow below; white eye-ring.

3 PRÍNCIPE WHITE-EYE *Zosterops ficedulinus* 10.5 cm **R** VU **Page 699**
São Tomé (*feae*, uncommon and local) and Príncipe (nominate; rare). Forest regrowth, plantations.
Endemic.
Adult *feae* Small; greyish-olive to olive-green above; greyish-white tinged yellow below; white eye-ring.
Z. f. ficedulinus slightly paler below, with less grey wash.

4 SÃO TOMÉ SPEIROPS *Speirops lugubris* *c.* 14 cm **R** **Page 700**
São Tomé. Wooded habitats. Common endemic.
Mainly greyish-olive; black cap; white eye-ring; pale bill and legs.

5 PRÍNCIPE SPEIROPS *Speirops leucophaeus* 13.5 cm **R** NT **Page 701**
Príncipe. Forest and regrowth, plantations. Common endemic.
Head and underparts pale grey and white; upperparts brownish-grey.

6 SÃO TOMÉ PRINIA *Prinia molleri* 13 cm **R** **Page 636**
São Tomé. All habitats and altitudes, mainly disturbed and edge situations. Common endemic.
Long, graduated tail; chestnut-brown head; grey upperparts.

7 SÃO TOMÉ SHORT-TAIL *Amaurocichla bocagei* *c.* 11 cm **R** VU **Page 656**
São Tomé. Forest. Rare endemic.
Dark brown above; white upper throat; long straight bill; long legs.

8 DOHRN'S THRUSH BABBLER *Horizorhinus dohrni* *c.* 14 cm **R** **Page 682**
Príncipe. Forest regrowth, plantations. Common endemic.
Olivaceous-grey above; whitish below; olivaceous-grey breast-band and flanks; belly washed pale yellow.

9 PRÍNCIPE SUNBIRD *Anabathmis hartlaubii* 13–14 cm **R** **Page 689**
Príncipe. Forest, cultivation, gardens. Common endemic.
9a Adult male Dark glossy violet-blue throat and upper breast; dark olive above.
9b Adult female Dark olive-green throat and upper breast.

10 NEWTON'S SUNBIRD *Anabathmis newtonii* 10–11 cm **R** **Page 689**
São Tomé. Forest, cultivation, gardens, dry savanna woodland. Common endemic.
10a Adult male Dark glossy violet-blue throat and upper breast; rest of underparts yellow.
10b Adult female Dark olive-green throat and upper breast.

11 BIOKO BATIS *Batis poensis* 12 cm **R** **Page 675**
Bioko. Forest (to 1100 m). Not uncommon.
11a Adult male *poensis* As *B. p. occulta* (Plate 111: 7) but supraloral spot generally smaller, supercilium
indistinct or absent, breast-band broader and white edges to tertials and outer tail feathers
narrower.
11b Adult female *poensis* Chestnut breast-band.

12 GIANT SUNBIRD *Dreptes thomensis* 18–23 cm **R** VU **Page 689**
São Tomé. Forest. Rare to locally common endemic.
Adult male Very large; blackish. Adult female similar but smaller.

PLATE 146: ADDITIONAL VAGRANT AND LOCALISED SPECIES I

1 SOUTHERN GIANT PETREL *Macronectes giganteus* 86–99 cm; WS 185–210 cm **V** VU **Page 328**
Vagrant from southern ocean, offshore Congo.
Very large (like small albatross). Massive, pale bill.
1a Adult dark morph Greyish-brown with whitish head.
1b Adult white morph All white with some black specks.

2 BALEARIC SHEARWATER *Puffinus mauretanicus* 30–40 cm; WS 76–93 cm **V** NT **Page 330**
Vagrant from W Mediterranean, recorded off Mauritania.
Resembles Manx Shearwater (Plate 2: 9) but browish above; dirty greyish-buff below; dark axillaries and
vent.

3 BLACK-BELLIED STORM-PETREL *Fregetta tropica* 20 cm; WS 46 cm **V** **Page 331**
Vagrant from Southern Ocean, recorded off Liberia (Sierra Leone?) and Gulf of Guinea.
White rump; white belly with black stripe on centre; white underwing-coverts.

4 WHITE-BELLIED STORM-PETREL *Fregetta grallaria* 20 cm; WS 46 cm **V?** **Page 332**
Potential vagrant from Southern Ocean (single record north of Cape Verde Is).
Similar to Black-bellied Storm-petrel; belly all white; underwing with more white, dark leading edge
narrower.

5 CAPE CORMORANT *Phalacrocorax capensis* 61–64 cm; WS 109 cm **V** NT **Page 337**
Vagrant from southern African coasts, north to Gabon.
5a Adult breeding Medium-sized; black; yellow-orange gular area.
5b Adult non-breeding/immature Dull brown, underparts paler; yellowish-brown gular area.

6 JACKASS PENGUIN *Spheniscus demersus* c. 60 cm **V** VU **Page 333**
S African vagrant, Gabon and Congo.
6a Adult Black face and throat surrounded by white; black band across breast and down flanks.
6b Juvenile Sooty-grey head and upperparts.

7 WOOD PIGEON *Columba palumbus* 40–45 cm **V** **Page 474**
Palearctic vagrant, Mauritania.
Adult Very large; large white patch on neck-sides; broad white band on wings in flight; broad black
terminal tail band.

8 AMERICAN PURPLE GALLINULE *Porphyrio martinica* 33 cm **V** **Page 415**
American vagrant, recorded off Liberian coast.
Adult Recalls Allen's Gallinule (Plate 33: 3) but larger; red, yellow-tipped bill; yellow legs.

9 HOUBARA BUSTARD *Chlamydotis undulata* 55–65 cm; WS 135–170 cm **R** NT **Page 419**
Semi-desert. Mauritania. Uncommon.
9a Adult Fairly large; black frills on neck-sides.
9b Adult in flight Black flight feathers; white patch on outer primaries.

10 WATTLED CRANE *Bugeranus carunculatus* c. 175 cm; WS 230–260 cm **V?** VU **Page 417**
Marshes and grasslands. Single old record, Guinea-Bissau (vagrant?).
Adult White head and neck; dark grey cap; white feathered wattles below chin; red facial skin.

11 BARBARY PARTRIDGE *Alectoris barbara* 32–35 cm **R** **Page 401**
Rocky slopes. N Mauritania. Rare and local.
Adult Rufous-brown collar with white spots; boldly barred flanks; bright red bill and legs.

PLATE 147: ADDITIONAL VAGRANT AND LOCALISED SPECIES II

1 PRIGOGINE'S NIGHTJAR *Caprimulgus prigoginei* 19 cm **R? CR**   **Page 500**
Forest. SE Cameroon–N Congo?
  1a Adult female Small; darkish; blotched and speckled blackish, rufous, tawny and buff; relatively short tail.
  1b Adult female in flight No white patch in wings; narrow white tail corners.

2 WHITE-BROWED COUCAL *Centropus superciliosus* 40 cm **R?**   **Page 490**
Tall grass and shrubbery. SW Congo. Rare.
Adult The only coucal with a pale supercilium.

3 NUBIAN WOODPECKER *Campethera nubica* 20 cm **R**   **Page 548**
Open woodland. NE CAR. Uncommon.
  3a Adult male Olive-brown upperparts boldly barred and spotted whitish; breast and flanks boldly spotted; red crown and moustachial stripe.
  3b Adult female Forecrown and moustachial stripe black speckled white.

4 EURASIAN SKYLARK *Alauda arvensis* 18–19 cm **V**   **Page 563**
Palearctic vagrant, NW Mauritania.
Robust; streaked; short crest. In flight, white trailing edge to wing and white outer tail feathers.

5 GREATER STRIPED SWALLOW *Hirundo cucullata* 18–20 cm **V**   **Page 569**
S African vagrant, S Congo.
  5a Adult As Lesser Striped Swallow (Plate 82: 8) but streaking much finer; cap darker rufous; ear-coverts white.
  5b In flight Rump paler than in Lesser Striped Swallow; streaking only visible at close range.

6 CITRINE WAGTAIL *Motacilla citreola* 17–18 cm **V**   **Page 576**
Palearctic vagrant, Senegal.
  6a Adult male breeding Bright lemon-yellow head; black nuchal band; grey upperparts; two white wingbars.
  6b First-winter Dull ash-grey above; two white wingbars; broad whitish ear-covert surround; narrow, weak, dusky malar stripe.

7 RICHARD'S PIPIT *Anthus richardi* 17–20 cm **P?**   **Page 578**
Rare Palearctic passage migrant and winter visitor, Mauritania and Mali?
Larger than Grassland Pipit (Plate 84: 9); wing edgings more sandy buff; underparts whiter; outer tail feathers pure white.

8 MARSH WARBLER *Acrocephalus palustris* 12.5–13.0 cm **V**   **Page 623**
Palearctic vagrant, N Senegal and NE Nigeria.
Very similar to European Reed Warbler (Plate 98: 7); typically more uniform olive grey-brown above; less rusty rump; slightly longer wings; 7-8 pale-tipped dark primaries visible; legs usually paler.

9 SOUTHERN BLACK FLYCATCHER *Melaenornis pammelaina* 19–22 cm **R**   **Page 658**
Wooded grassland. Congo. Rare and local.
Similar to Northern Black Flycatcher (Plate 108: 2), but plumage glossed steel-blue.

10 WESTERN JACKDAW *Corvus monedula* 33 cm **V**   **Page 721**
Palearctic vagrant, N Mauritania.
Small crow; black; grey neck; whitish eyes.

11 COMMON STARLING *Sturnus vulgaris* 19–22 cm **V**   **Page 728**
Palearctic vagrant, N Mauritania.
Adult non-breeding Glossy black spotted buff above and white below; long pointed bill; short tail; triangular wings.

12 EURASIAN TREE SPARROW *Passer montanus* 14 cm   **Page 729**
Palearctic species claimed from Guinea (Conakry harbour). Ship-assisted or escape?
Resembles House Sparrow (Plate 128: 1), but has chestnut crown and black ear-spot.

13 EURASIAN ROCK BUNTING *Emberiza cia* 16 cm **V?**   **Page 778**
Palearctic vagrant? A juvenile claimed from Chad.
  13a Adult male Grey head and breast with distinct stripe pattern; white wingbar; grey bill.
  13b Juvenile Dull brown and streaky.

OSTRICH
Family STRUTHIONIDAE (World: 1 species)

A huge, flightless bird, with long, almost bare neck and powerful legs. The largest and heaviest living bird, now endemic to Africa (since extinction in Arabia). Unique in having only two, forward-directed, toes. Sexually dimorphic. Principally herbivorous.

OSTRICH *Struthio camelus*
Autruche d'Afrique

Plate 1 (4)

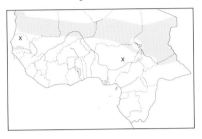

Height male 210–275 cm (83–108"), female 175-190 cm (69–75"). *S. c. camelus*. Unmistakable. At distance more likely to be confused with a large mammal than a bird. **Adult male** Black body, white primaries and tail. In breeding season bill and bare skin (esp. on neck) become redder. **Adult female** Dull greyish-brown; smaller. **Juvenile** Similar to adult female but somewhat darker, more uniformly grey-brown. Adult height attained at *c.* 12 months; full adult plumage of males acquired in *c.* 2 years. Chick has spiky down, buff tipped black, and lines of black spots on neck-sides. Juvenile plumage emerges at *c.* 3 months. **VOICE** Largely silent. Displaying males utter a far-carrying, deep, booming sound, reminiscent of distant roar of lion. [CD5:1a] **HABITS** Usually in pairs or small groups, in open plains, thorn scrub and semi-desert. Very swift and often wary. **SIMILAR SPECIES** None. Small young could be confused with a bustard, but thick legs and unfeathered neck distinctive; also usually accompanied by adults. **STATUS AND DISTRIBUTION** Rare to uncommon resident throughout Sahel belt, from Mauritania to Chad and NE CAR. Population rapidly declining through much of range in W Africa, principally due to excessive hunting and collecting of eggs for food.

ALBATROSSES
Family DIOMEDEIDAE (World: 14 species; Africa: 9; WA: 1)

Huge pelagic seabirds with stout body, extremely long and narrow wings, and short tails. Sexes similar. Adult plumage only attained after several years. Flight continuous and powerful, gliding and banking on stiff, slightly bowed, motionless wings, alternately showing upper- and underside. Spend most of their lives at sea. Normally silent away from breeding colonies. Breed on small oceanic islands, principally in southern hemisphere. Identification best based on underwing pattern and bill coloration.

Almost nothing is known of the occurrence of albatrosses in offshore waters within our region. For more information, reference should be made to a specialist work on the subject (e.g. Harrison 1991).

BLACK-BROWED ALBATROSS *Diomedea melanophris*
Albatros à sourcils noirs

Not illustrated

L 80–95 cm (32–37"), WS 213–246 cm (95"). **Adult** Combination of dark 'saddle' and upperwings, white rump and underparts, and grey tail, typical of smaller albatrosses (or 'mollymawks'). Underwing white *with broad black band on leading edge*. Small dark 'brow'. Bill orange-yellow. **Juvenile** Underwing largely dark, with only ill-defined whitish stripe along centre. Bill greyish-brown with blackish tip. Adult plumage attained in *c.* 5 years. **SIMILAR SPECIES** Yellow-nosed Albatross *D. chlororhynchos* (Albatros à nez jaune) only slightly smaller, but with narrow black border to underwing; bill black with orange-yellow culmen and pinkish-orange tip in adult, all black in juvenile. Gannets and boobies have relatively shorter wings, longer, wedge-shaped tails and conical bills. **STATUS AND DISTRIBUTION** Vagrant from Southern Ocean. An albatross seen off Congo (Aug 1988) was presumably this species or Yellow-nosed. *TBW* category (both spp.): Near Threatened. **NOTE** Sometimes placed in genus *Thalassarche*.

PETRELS AND SHEARWATERS
Family PROCELLARIIDAE (World: *c.* 73 species; Africa: 31; WA: 12)

Medium-sized to very large pelagic seabirds with long, narrow wings and horny-plated, hook-tipped bills with external tubular nostrils ('tubenose'). Sexes usually similar. Fly low over waves on stiff wings, long glides alternating with a few fast flaps. Spend most of their lives at sea and are normally only seen on land at breeding colonies or as storm-driven waifs. Most species breed in burrows or crevices in sea cliffs on oceanic islands. Normally silent away from breeding colonies.

Little is known about the occurrence of pelagic seabirds off the coast of W Africa and new species for the region are likely to be discovered in offshore waters, particularly along upwellings above the continental shelf. In view of the possibility of encountering a species not described in this book, it is recommended that any observer seeking to identify pelagic seabirds refers to a specialist work on the subject (e.g. Harrison 1991).

SOUTHERN GIANT PETREL *Macronectes giganteus*   Plate 146 (1)
Pétrel géant

L 86–99 cm (34–39"); WS 185–210 cm (73–82"). Very large seabird, equal in size to small albatross, but with stouter body and proportionately shorter, narrower wings. Short, wedge-shaped tail; feet project beyond tail in flight. Massive bill pale horn tipped pale greenish, appearing uniformly coloured at sea. Polymorphic. **Adult dark morph** Greyish-brown with slightly paler underparts and contrasting pale, whitish head and neck. **Adult white morph** (up to 15% of individuals) Distinctive; wholly white with a few irregular black specks. **Juvenile/immature dark morph** Wholly sooty-black, progressively fading with age. Attains maturity in *c.* 7 years. **Juvenile/immature white morph** As adult. **HABITS** An aggressive scavenger, occurring singly or in groups. Flight laboured and less graceful than albatross, stiff wingbeats interspersed with short glides. Often appears hump-backed when flapping wings. Occasionally follows ships. Alights on water to retrieve food. **SIMILAR SPECIES** None recorded in our region. Confusion species include Northern Giant Petrel *Macronectes halli* (variable, but white on head typically restricted to face; dark-tipped bill), Sooty Albatross *Phoebetria fusca* (more slender build; black bill) and juvenile Wandering Albatross *Diomedea exulans* (wholly chocolate-brown with white face). Northern and Cape Gannets (q.v.) are slightly smaller and have longer neck and tail, slimmer head and bill, and different flight action, with faster wingbeats. **STATUS AND DISTRIBUTION** Vagrant from Southern Ocean, recorded offshore Congo. *TBW* category: THREATENED (Vulnerable).

CAPE PETREL *Daption capense*   Plate 2 (3)
Damier du Cap

Other names E: Pintado Petrel; F: Pétrel damier
L 38–40 cm (15–16"); WS 81–91 cm. (32–36"). *D. c. capense.* Medium-sized pelagic petrel with *striking black-and-white chequered upperparts and diagnostic white patches on upperwings*. Head black; underparts white; tail principally white with broad black terminal band. **HABITS** Flight typically consists of 5–8 quick wingbeats on stiff wings followed by periods of gliding. In strong winds may tower high above ocean. Usually feeds from surface, sitting gull-like on water and picking at food objects; may also dive. Often found in association with fishing boats and habitually follows ships. **SIMILAR SPECIES** None. No other petrel in the region has similar mottled plumage, though partial albinism in other shearwaters or petrels could produce superficially similar patterns. **STATUS AND DISTRIBUTION** Vagrant from Southern Ocean, recorded off Togo. Also claimed from Gabon. Only likely to be encountered off continental shelf or as storm-driven waif.

ZINO'S PETREL *Pterodroma madeira*   Not illustrated
Pétrel de Madère

Other names E: Madeira Petrel, Freira
L 32–33 cm (12.5–13.0"); WS 78–83 cm (30.5–32.5"). Very similar to, and probably often indistinguishable at sea from, Fea's Petrel (q.v.). Slightly smaller, more slender and paler, appearing less masked due to less contrasting, paler forehead and crown; flanks more heavily mottled grey; wings shorter, broader and blunt tipped; bill slimmer. **HABITS** As Fea's Petrel. **STATUS AND DISTRIBUTION** Potential non-breeding visitor. Breeds only in mountains of Madeira (*c.* 30–40 pairs in 2000). Distribution at sea unknown. *TBW* category: THREATENED (Critical). **NOTE** Formerly treated as race of Soft-plumaged Petrel *P. mollis.*

FEA'S PETREL *Pterodroma feae*   Plate 2 (1)
Pétrel gongon

Other names E: Cape Verde Petrel, Gon-gon
L 33–36 cm (13–14"); WS 84–91 cm (33–36"). Combination of principal plumage marks, shape and flight action distinctive. Dark grey above with pale forehead and lores, dark eye-mask and *pale grey lower rump and tail*. In favourable light, dark outer wing, carpal bar and upper rump (forming variable M mark) contrast with greyer mantle and back. Underparts white with contrasting *blackish-grey underwings* and dusky pectoral patches of variable size (sometimes absent). **HABITS** Singly or in small groups. Flight characteristic, swift and often 'impetuous', with rapid wingbeats, glides and erratic zigzag action (hence name gadfly-petrel). Generally flies relatively low,

but in strong winds may tower high above ocean. Sometimes follows ships; will settle on water to retrieve food. Breeding colonies at inaccessible sites in high mountain ranges. [CD1:1a] **SIMILAR SPECIES** Zino's Petrel (q.v.). Balearic and Manx Shearwaters similar in size, but head smaller, tail shorter, and wings straighter in flight. **STATUS AND DISTRIBUTION** Breeds on Cape Verde Is (Santiago, Fogo, Santo Antão and São Nicolau; total population estimated at 500–1000 pairs in 1995). Movements poorly known. Probably rare non-breeding visitor in offshore waters south to 9°N. A few records off coast of Mauritania, Liberia and Ivory Coast may refer to this species. *TBW* category: Near Threatened. **NOTE** Formerly treated as race of Soft-plumaged Petrel *P. mollis*.

BULWER'S PETREL *Bulweria bulwerii* Plate 2 (2)
Pétrel de Bulwer

L 26–28 cm (10–11"); WS 68–73 cm (27–29"). Medium-sized, long-winged, *all-dark* petrel with pale diagonal band on upperwing. *Long, narrow, graduated tail* (usually held closed and appearing pointed) diagnostic. **HABITS** Flight buoyant and erratic, twisting and weaving close to surface with wings held forward and slightly bowed; usually 2–5 rapid wingbeats followed by a glide. Generally circles low over waves fanning tail to change direction. Rarely follows ships. [CD1:2a] **SIMILAR SPECIES** Sooty Shearwater is much larger, with different flight action and square-ended tail. Great care is needed when identifying all-dark petrels as other species may occur in the region. **STATUS AND DISTRIBUTION** Breeds on Cape Verde Is (Ilhéu de Cima and Raso; probably *c.* 100 pairs) and other Atlantic islands. Disperses south in non-breeding season (Nov onwards); most winter off NE Brazil. Scarce offshore, from Mauritania to Gulf of Guinea.

CORY'S SHEARWATER *Calonectris diomedea* Plate 2 (4)
Puffin cendré

L 45–56 cm (18–22"); WS 110–125 cm (43–49"). Rather featureless large shearwater (about size of Lesser Black-backed Gull) with *ashy-brown upperparts* and *unmarked white underparts*. At close range shows diagnostic *pale yellowish bill* and variable amount of white in uppertail-coverts. Ashy-brown extends onto head- and neck-sides. **Nominate** slightly smaller than *borealis*. **HABITS** Often in flocks. Flight appears lazy with several deep wingbeats followed by long, low glide. Wings typically held bowed downwards. Follows predatory fish, schools of cetaceans and fishing boats to retrieve food scraps. [CD1:2b, 3a] **SIMILAR SPECIES** Cape Verde Shearwater is slimmer with smaller, more angular head and relatively longer tail; head and upperparts darker and greyer brown; bill noticeably slimmer and grey with dark subterminal band (appearing black-tipped at distance). Great Shearwater is similar in size but has contrasting dark cap and white collar, more white in uppertail-coverts, and less clean underwing; flight more rigid, on flatter, slimmer looking wings with more tapering tip. Manx Shearwater is much smaller, has dark, slender bill and more rapid flight. **STATUS AND DISTRIBUTION** Rare to not uncommon non-breeding visitor. Breeds on Atlantic (*borealis*) and Mediterranean islands (nominate), dispersing south in non-breeding season. Status inadequately known as previously not distinguished from Cape Verde Shearwater. Most likely to be encountered Oct–May in offshore waters from Mauritania to Gabon. In Gulf of Guinea, esp. between São Tomé and Príncipe, northward passage of Cory's and/or Cape Verde Shearwaters observed in Mar. Rare in innermost part of Bight of Biafra (1 record, Bioko, Apr).

CAPE VERDE SHEARWATER *Calonectris edwardsii* Plate 2 (5)
Puffin du Cap Vert

L 40–41 cm (16"); WS 90–110 cm (35.5–43.0"). Very similar to Cory's Shearwater but usually smaller (wings 10–15% shorter) and *slimmer* with smaller, more angular head and relatively longer tail; head and upperparts slightly darker and greyer brown; *bill noticeably slimmer and grey with dark subterminal band* (looking black-tipped at distance). **HABITS** As Cory's Shearwater. [CD1:3b] **SIMILAR SPECIES** See under Cory's Shearwater. **STATUS AND DISTRIBUTION** Locally common breeding visitor to Cape Verde Is (estimated at *c.* 10,000 pairs, 1988–93), presumably dispersing southwards in non-breeding season. Distribution at sea inadequately known. Apparently absent from Cape Verde area from late Nov until late Feb. Recorded off Mauritania, Senegal (common in Oct) and Guinea-Bissau. **NOTE** Formerly considered a race of Cory's Shearwater, but treated as separate species by *BWPC*.

GREAT SHEARWATER *Puffinus gravis* Plate 2 (6)
Puffin majeur

L 43–51 cm (17–20"); WS 105–122 cm (41–48"). Large shearwater with *distinctive dark cap*, white collar, brown-grey upperparts and white uppertail-coverts. Underwing largely white with *dark diagonal band on underwing-coverts* (appearing smudgy at distance). White underparts with diffuse dark patch on belly (often difficult to see). Bill dark. At long range, *gleaming white blaze on neck-sides contrasts strongly with dark cap and upperparts*. **HABITS** Flight strong and powerful, with stiff wingbeats interspersed by glides. Wings normally held straight. In strong winds, flight rapid and bounding. Often attends fishing boats to feed on offal. [CD1:4a] **SIMILAR SPECIES** Cory's

Shearwater has more uniform upperparts with ashy-brown neck-sides (no clear-cut dark cap and white collar), and purer white underwings; flight more leisurely, on bowed, broader looking wings held more angled at carpal joint. **STATUS AND DISTRIBUTION** Rare non-breeding visitor from Southern Ocean to offshore waters, from Mauritania to Gabon. Probably not uncommon in Cape Verde waters, Aug–Dec. Southbound passage in Gulf of Guinea mainly late Sep/Oct.

SOOTY SHEARWATER *Puffinus griseus*   Plate 2 (7)
Puffin fuligineux

L 40–51 cm (16–20"); WS 94–109 cm (37–43"). Large, *all-dark* shearwater *with diagnostic silvery underwing-coverts* (often difficult to see at distance). *Long, narrow wings, often angled backwards from carpal joint,* and heavy body give characteristic jizz. **HABITS** Flight strong and direct with usually 2–8 rapid stiff-winged flaps followed by long descending glide. Tends to arc rather higher than other shearwaters, particularly in strong winds. Often around fishing boats in search of food but seldom follows ships. Feeds by plunge-diving or diving from surface. [CD1:6a] **SIMILAR SPECIES** Only likely to be confused with dark Balearic Shearwater (q.v.). At distance could be confused with dark-morph Arctic Skua, but latter has longer tail, pale patch at base of primaries, and quite different flight action. **STATUS AND DISTRIBUTION** Uncommon to rare non-breeding visitor from Southern Ocean to offshore waters of entire region. Possible at any season, though most likely Oct–Jan. Northward passage in Gulf of Guinea, Aug–Sep.

MANX SHEARWATER *Puffinus puffinus*   Plate 2 (9)
Puffin des Anglais

L 30–35 cm (12–14"); WS 76–82 cm (30–32"). Medium-sized shearwater, *black above and white below.* Black of cap and upperparts extends to below level of eye and onto breast-sides. Wings rather straight and not angled backwards from carpal joint. **HABITS** Typical flight a series of rapid stiff-winged strokes followed by shearing over waves, banking from side to side, alternately showing black upperside and white underside. Singly or in small flocks. Occasionally follows ships. May associate with flocks of Sooty Shearwater. [CD1:4b] **SIMILAR SPECIES** Balearic Shearwater (formerly considered conspecific with Manx) very similar, but upperparts browner and underparts washed pale brownish. Little Shearwater similar in plumage pattern, though black of cap rarely extends to level of eye; also more compact and shorter winged. Best distinguished by rapid fluttering flight. See also larger Cory's and Great Shearwaters. Fea's Petrel similar in size, but head larger, tail longer, wings more angled in flight. **STATUS AND DISTRIBUTION** Inadequately known. Non-breeding visitor from N and E Atlantic; scarce but probably regular in offshore waters, particularly during northern autumn and winter. Most likely to be encountered offshore from Mauritania to Liberia. British-ringed birds have been recovered in Ivory Coast and Ghana.

BALEARIC SHEARWATER *Puffinus mauretanicus*   Plate 146 (2)
Puffin des Baléares

L 30–40 cm (12–16"); WS 76–93 cm (30–37"). Slightly larger than Manx Shearwater and lacks striking contrast between upper- and underside, having brownish (not black) upperparts merging into dirty greyish-buff (not pure white) underparts, and grey-brown vent and axillaries. Plumage variable, darkest birds may resemble Sooty Shearwater. **HABITS** As Manx Shearwater, but often closer to shore and thus easier to observe. [CD1:5a] **SIMILAR SPECIES** Manx and Sooty Shearwaters (q.v.). **STATUS AND DISTRIBUTION** Poorly known. Breeds W Mediterranean, dispersing into Atlantic in post-breeding season. Recorded off Mauritania, but also likely to occur further south. *TBW* category: Near Threatened. **NOTE** Originally considered a race of Manx Shearwater *P. puffinus*; forms *yelkouan* and *mauretanicus* subsequently considered a separate species, Mediterranean Shearwater *P. yelkouan*; both forms now generally accorded specific rank, Yelkouan (or Levantine) Shearwater *P. yelkouan* and Balearic Shearwater (cf. *BWPC*).

LITTLE SHEARWATER *Puffinus assimilis*   Plate 2 (8)
Petit Puffin

L 25–30 cm (10–12"); WS 58–67 cm (23–26"). Recalls compact, miniature Manx Shearwater with smaller bill. Underwing margins and tips of wings have less black. Legs and feet blue. In *baroli* black of cap does not usually extend to level of eye, giving bare-faced appearance; undertail-coverts white. *P. a. boydi* has more dark on head-sides and dark undertail-coverts. **HABITS** Distinctive stiff-winged flight usually low, close to waves, consisting of several rapid wingbeats followed by a brief glide with wings held parallel to surface of sea (reminiscent of Common Sandpiper). Rarely follows ships. Often feeds in small rafts pattering across surface of ocean with wings raised like storm-petrel. [CD1:5b] **SIMILAR SPECIES** Manx Shearwater (q.v.). **STATUS AND DISTRIBUTION** Breeds on Cape Verde Is (*boydi*; probably several 1000s) and Atlantic islands (*baroli*). Non-breeding distribution inadequately known; probably scarce but regular offshore to at least 9°N. Recorded off Mauritania and Senegal; also in Gulf of Guinea.

STORM-PETRELS
Family HYDROBATIDAE (World: 20 species; Africa: 10; WA: 6)

Small, swallow-like pelagic seabirds. Largely dark with white rump. Sexes similar. Most have rounded wings and an erratic, bouncing flight, often with legs dangling and webbed feet pattering the surface of the sea. Others, with more pointed wings, have a more swooping flight. Solitary or gregarious at sea. Breed colonially in burrows or crevices on oceanic islands. Field identification difficult and, with experience, best based on flight action. Shape of rump patch and tail, and feet only visible at close range. The family is also known as Oceanitidae.

As species not yet recorded in our region may be encountered, it is recommended that the interested observer consult a specialist work on the subject (e.g. Harrison 1991).

WILSON'S STORM-PETREL *Oceanites oceanicus*   Plate 3 (2)
Océanite de Wilson

Other name E: Wilson's Petrel
L 15–19 cm (6.0–7.5"); WS 38–42 cm (15.0–16.5"). Small, dark brown storm-petrel (appearing black at distance) with *pale diagonal upperwing bar and U-shaped white rump which wraps round to sides of undertail-coverts*. Short, rounded wings lack obvious bend at carpal joint. Legs project beyond *square tail* in flight (sometimes creating illusion of forked tail). Feet have yellow webs, but these are extremely hard to discern at sea. Underwing mostly dark, sometimes with faint greyish band on underwing-coverts. *O. o. exasperatus* larger than **nominate**. **HABITS** Flight direct and purposeful with rapid wingbeats and short glides, often recalling a swallow. 'Walks' on water when feeding, pattering over surface with high-raised wings. Habitually follows ships. [CD1:6b] **SIMILAR SPECIES** European Storm-petrel is smaller, has white bar on underwing-coverts, does not show projecting feet and has more fluttering flight. Madeiran and Leach's Storm-petrels are both larger, have less white on rump, are longer winged and have forked or slightly forked tails. **STATUS AND DISTRIBUTION** Inadequately known. Antarctic breeding race *exasperatus* and subantarctic nominate migrate to N Atlantic in non-breeding season. Probably not uncommon to scarce, though regular, off entire seaboard, Apr–Nov. Regular off Senegambia, Apr–Jun and Aug–Sep. In Gulf of Guinea mainly Jul–Sep. Year-round off Congo.

WHITE-FACED STORM-PETREL *Pelagodroma marina*   Plate 3 (5)
Océanite frégate

L 20–21 cm (8"); WS 41–43 cm (16–17"). Distinctive. Rather large, greyish storm-petrel with *white underparts* and white face with dark patch on ear-coverts. Rump grey. Very long legs project beyond slightly forked tail in flight. Wings broad. *P. m. eadesi* somewhat paler than *hypoleuca*, with more white on neck-sides. **HABITS** Flight erratic, with legs dangling and pendulum-like swinging from side to side, giving impression of bouncing off wave tops. Often gregarious at sea. Rarely follows ships. [CD1:7a] **SIMILAR SPECIES** Unlikely to be confused with any other petrel in the region, though at distance confusion possible with Red Phalarope if seen on surface of sea. **STATUS AND DISTRIBUTION** Inadequately known. Breeds on Cape Verde Is (*eadesi*, c. 5000–10,000 pairs, 1988–93) and extralimital Salvage Is (*hypoleuca*), dispersing south outside breeding season. Few records from the region but probably regular offshore from Mauritania to at least Liberia.

BLACK-BELLIED STORM-PETREL *Fregetta tropica*   Plate 146 (3)
Océanite à ventre noir

L 20 cm (8"); WS 46 cm (18"). *F. t. tropica*. Dark storm-petrel with white central underwing-coverts having *broad dark margins* and *white belly showing diagnostic black stripe down centre that broadens onto vent*. Blackish above with faint pale upperwing bar and white rump. Throat *and upper breast* blackish-brown; undertail-coverts black. Feet project slightly beyond tail in flight. **HABITS** Often accompanies ships, flying in front and alongside in distinctive bounding flight, legs dangling and body rocking from side to side; keeps close to waves, breast bouncing off water. [CD5:1b] **SIMILAR SPECIES** As yet the only storm-petrel recorded from W African waters possessing combination of white rump and white belly and underwing. White-bellied Storm-petrel, a likely vagrant, is very similar but has plain white belly, more white on underwing, and blackish-brown of throat barely extending onto upper breast and more sharply demarcated. Wilson's Storm-petrel lacks white on underparts. **STATUS AND DISTRIBUTION** Vagrant from Southern Ocean. Recorded off Liberia and possibly Sierra Leone. Also Gulf of Guinea (Jul and Sep).

WHITE-BELLIED STORM-PETREL *Fregetta grallaria*   Plate 146 (4)
Océanite à ventre blanc

L 20 cm (8"); WS 46 cm (18"). *F. t. leucogaster.* Resembles Black-bellied Storm-petrel, from which often difficult to separate at sea, but has paler upperparts (black feathers fringed grey) and *entirely white belly. Underwing shows more white, with narrower and better defined dark leading edge.* Dark throat sharply demarcated from white underparts. Tail blackish, darker than upperparts. **HABITS & SIMILAR SPECIES** See Black-bellied Storm-petrel. **STATUS AND DISTRIBUTION** Breeds on S Atlantic and Indian Ocean islands. Not yet recorded from our region, but one observed at 23°48'N 22°01'W, north of Cape Verde Is (Hazevoet 1995), suggests potential vagrancy.

EUROPEAN STORM-PETREL *Hydrobates pelagicus*   Plate 3 (1)
Océanite tempête

Other names E: British Storm-petrel, Storm Petrel
L 15–16 cm (6"); WS 37–41 cm (15–16"). The region's *smallest and darkest* storm-petrel. All black except contrasting square white rump and diagnostic, but variable, *white band on underwing-coverts.* Feet do not project beyond square tail in flight. Narrow, pale diagonal bar sometimes visible on upperwing, esp. in fresh plumage. **HABITS** Flight weak and fluttering, bat-like, with almost continuous wing strokes interspersed by short glides. Often patters surface with wings raised. Usually gregarious. Frequently follows ships. Prone to 'wrecking' in bad weather, when may be seen close inshore. [CD1:7b] **SIMILAR SPECIES** Confusion possible with Wilson's, Madeiran and Leach's Storm-petrels, but distinguished by combination of small size, square tail, feet not projecting in flight, darker upperwing, white band on underwing and fluttering flight. **STATUS AND DISTRIBUTION** Palearctic migrant; breeds in N Atlantic and Mediterranean. Regular offshore from Mauritania to Ivory Coast, sometimes in great numbers. Rare (?) passage migrant in Gulf of Guinea (recorded off São Tomé, Príncipe and, probably, Bioko). Presumed non-breeders present during northern summer.

MADEIRAN STORM-PETREL *Oceanodroma castro*   Plate 3 (4)
Océanite de Castro

Other name E: Band-rumped Storm Petrel
L 19–21 cm (7.5–8.25"); WS 43–46 cm (17–18"). Relatively large storm-petrel with white rump (extending to lateral undertail-coverts) and slightly forked tail (often appearing square). Wings rather angular with obvious bend at carpal joint and pale diagonal bar on upperwing-coverts; underwing all dark. Feet do not project beyond tail in flight. **HABITS** Flight typically buoyant and zigzagging, with deep wingbeats and low shearwater-like glides with wings held horizontally or bowed downwards. Often patters surface with wings held horizontally; frequently sits on water. Does not follow ships. [CD1:8a] **SIMILAR SPECIES** Difficult to separate from Wilson's and Leach's Storm-petrels. Wilson's has long, projecting legs and more noticeable wingbar; Leach's less extensive white rump and more forked tail. **STATUS AND DISTRIBUTION** Inadequately known. Breeds on Cape Verde Is (locally not uncommon; possibly not more than *c.* 1000 pairs) and other Atlantic islands. Probably frequent in offshore waters. Not uncommon over Senegal upwelling where commonest storm-petrel Feb–Mar. Post-breeding dispersal along coast to Gulf of Guinea, Feb–Jul. Recorded off Sierra Leone, Liberia (Nov–Dec, Jul), Ghana (Oct), Congo. Also São Tomé and Príncipe, where there may be undiscovered colonies.

LEACH'S STORM-PETREL *Oceanodroma leucorhoa*   Plate 3 (3)
Océanite cul-blanc

Other name E: Leach's Petrel
L 19–22 cm (7.5–8.5"); WS 45–48 cm (18–19"). *O. l. leucorhoa.* Relatively large, brownish storm-petrel with long angular wings showing obvious bend at carpal joint and *rather long, forked tail* (fork visible from certain angles only). *White rump V-shaped,* less conspicuous and with much less lateral extension than other storm-petrels of the region; *usually has diagnostic dark central division* which, though hard to see, gives rump a smudgy appearance. Upperwing has *prominent dirty grey diagonal band* across coverts. Underwing all dark. Feet do not project beyond tail in flight. **HABITS** Flight buoyant and bounding with deep tern-like wingbeats followed by short shearwater-like glides and characteristically sudden changes of speed and direction. Unlike other storm-petrels does not habitually foot-patter. Rarely follows ships. Prone to being 'wrecked' in bad weather. [CD1:8b] **SIMILAR SPECIES** Most likely to be confused with Madeiran Storm-petrel, which has more prominent white rump without division and much less deeply forked tail. Dark-rumped individuals of Leach's may occur and are extremely difficult to separate from other dark-rumped storm-petrels. **STATUS AND DISTRIBUTION** Inadequately known. Visitor from N Atlantic outside breeding season. Rather few records, from Mauritania (Nov–Dec and Apr–May) to Gulf of Guinea (Nov–Mar), but probably regular along entire W African coast, esp. Senegal upwelling, and in Cape Verde waters.

PENGUINS
Family SPHENISCIDAE (World: 17 species; Africa: 4; WA: 1)

Stocky, flightless seabirds, with flipper-like wings, and short legs and tails. Dense plumage largely blackish above and white below. Sexes similar. Float with head held high and most of body submerged; forage underwater for marine organisms. On land, walk upright with waddling gait. Colonial. One vagrant in W Africa.

JACKASS PENGUIN *Spheniscus demersus*   Plate 146 (6)
Manchot du Cap

Other name E: African Penguin
L *c*. 60 cm (24"). **Adult** Rather small penguin with black face and throat separated from black upperparts by broad white supercilium joining white underparts over neck-sides. Black band across upper breast down flanks to legs. Bill black. Sometimes narrower partial or complete black band between black face and breast-band. **Juvenile** Head and upperparts sooty-grey; underparts white. **VOICE** Usually silent at sea. Occasionally utters its characteristic loud, donkey-like bray away from colony, esp. in panic, during fog or at night. **STATUS AND DISTRIBUTION** S African vagrant, recorded from Gabon (Dec 1956) and Congo (1 captured, Pointe-Noire, Mar 1954). *TBW* category: THREATENED (Vulnerable).

GREBES
Family PODICIPEDIDAE (World: 21–22 species; Africa: 5; WA: 3)

Small to medium-sized waterbirds, adapted for diving and swimming under water. Appear tailless. Sexes similar. Possess breeding and non-breeding plumages. Chicks have stripes on head-sides, neck and upperparts. Flight fast with rapid wingbeats, outstretched neck and head held low (giving hump-backed appearance). Take-off from water with much pattering over surface; land on belly (not feet first). Feed on fish and invertebrates. Build floating nest.

LITTLE GREBE *Tachybaptus ruficollis*   Plate 1 (3)
Grèbe castagneux

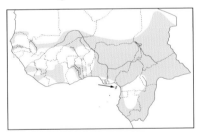

Other name E: Dabchick
L 25–29 cm (9.75–11.5"). Small, compact, mainly dark grebe, with rounded head, short neck, stubby bill and fluffy rear end. **Adult** *capensis* **breeding** Unmistakable. Blackish-brown above with *bright chestnut cheeks and foreneck* and *conspicuous yellowish-white gape patch*. Eye dark brown. **Nominate** somewhat smaller, with slightly less white in secondaries; indistinguishable in field. **Adult non-breeding** Much duller and paler, with chestnut areas becoming pale brownish, gape patch reduced and more white than yellow, crown to upperparts dull brown. **Juvenile** Similar to adult non-breeding but head-sides with irregular white stripes. **VOICE** A loud, descending, whinnying trill *bi-i-i-i-i-i...* uttered frequently and year-round, often in duet. Various clucking notes and *wee-wee-wee* calls audible at close range. Young make high-pitched piping calls. [CD1:10a] **HABITS** Mainly singly or in pairs, occasionally in groups, on almost any type of water, though rarely on fast-flowing streams and open sea. Prefers well-vegetated aquatic habitats, usually avoiding dense areas of tall vegetation unless interspersed with open pools. Swims buoyantly and dives frequently. **SIMILAR SPECIES** Only likely to be confused with young ducklings, which would normally be in company of parent, or with Black-necked Grebe, which is larger and has delicate upturned bill and predominately black-and-white rather than brown-and-buff plumage. **STATUS AND DISTRIBUTION** *T. r. capensis* locally common throughout; also Bioko. Largely resident; some dispersal may occur after rains to utilize seasonal aquatic habitats. Palearctic migrants (of nominate race) reported from NW Mauritania (Banc d'Arguin); presence suspected, N Senegal.

GREAT CRESTED GREBE *Podiceps cristatus*   Plate 1 (1)
Grèbe huppé

L 46–51 cm (18–20"). *P. c. cristatus*. Large, slim grebe with long, slender neck and long, pink, sharp-pointed bill. Hindneck and upperparts dark brownish; face, foreneck and underparts white. In flight has much white on forewing, scapulars and secondaries. **Adult breeding** Unmistakable, with black crown, double-horned crest and distinctive chestnut and black tippets. **Adult non-breeding** (plumage most likely to be encountered in the region)

Largely dark above and white below; triangular head without tippets and with smaller crest, blackish-brown on crown not reaching to level of eye. **HABITS** Could occur in any aquatic habitat, incl. sea. Feeds almost entirely upon fish that are caught by diving. [CD1:10b] **SIMILAR SPECIES** None. Tailless appearance eliminates confusion with cormorants or African Finfoot. **STATUS AND DISTRIBUTION** Palearctic vagrant, Senegal, Mali and Niger. **NOTE** Although only birds of Palearctic origin have been recorded in our region, stragglers of extralimital Afrotropical race *infuscatus* may perhaps occur. This race differs from nominate in lacking white supercilium, blackish-brown of crown being more extensive and descending to level of eye; also appears to lack distinctive non-breeding plumage.

BLACK-NECKED GREBE *Podiceps nigricollis*   Plate 1 (2)
Grèbe à cou noir

Other name E: Eared Grebe
L 28–34 cm (11.0–13.5"). *P. n. nigricollis*. Medium-sized grebe with short, slender, slightly upturned and sharply pointed bill, conspicuous red eyes and high, steep forehead with forecrown peak. In flight has white secondaries. **Adult breeding** Distinctive, with black head, neck, upperparts and upper breast, chestnut flanks and golden ear-tufts. Rest of underparts silky white. **Adult non-breeding** (plumage most likely to be encountered in the region) Mainly black above and white below, with black cap diffusely extending to below eye. Ear-coverts and neck often dusky. **Juvenile** Resembles adult non-breeding but browner overall. **HABITS** Could occur in any still aquatic habitat. Prefers alkaline lakes, estuaries, inshore coastal waters and sewage farms. [CD1:11] **SIMILAR SPECIES** Little Grebe is smaller, brown-and-buff rather than black-and-white in non-breeding plumage, and has stubby bill. **STATUS AND DISTRIBUTION** Palearctic vagrant, which has become more regular winter visitor to N Senegal (Oct–May) since 1990. Also recorded from Mauritania (Mar, May), S Niger (Jan–Feb), N Nigeria (Nov, Apr) and SW Cameroon (Mar).

TROPICBIRDS
Family PHAETHONTIDAE (World: 3 species; Africa: 3; WA: 2)

Medium-sized, graceful, highly aerial seabirds. Adults largely white; tail wedge-shaped with two greatly elongated central feathers. Sexes similar. Young heavily barred black on upperparts, lacking long tail streamers. Adult plumage acquired in 2nd calendar year. Buoyant, rather tern-like, purposeful flight, fluttering wingbeats alternated with long glides. Feed by hovering, then plunging vertically. Mainly pelagic and solitary away from breeding areas. Nest colonially on oceanic islands in caves and crevices.

RED-BILLED TROPICBIRD *Phaethon aethereus*   Plate 3 (7)
Phaéton à bec rouge

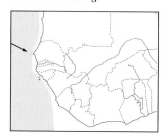

Other name F: Grand Phaéton
L 46–50 cm (18–20"); tail streamers 46–56 cm (18–22"); WS 103 cm (41"). **Adult** Mainly white with conspicuous *red bill*, black streak through eye and long white tail streamers. Upperparts narrowly barred black (sometimes difficult to see in strong sunlight); outer primaries and their coverts black, forming diagnostic band on upperwing. *P. a. mesonauta* differs from **nominate** principally in having rosy cast in fresh plumage. **Juvenile/immature** Bill yellowish with black tip; eye-stripe extending across hindneck, forming dark nuchal collar; upperparts closely barred (appearing grey at distance); tail feathers tipped black. **VOICE** A shrill *keek* or *karreek*. Also loud, piercing calls, only likely to be heard at breeding colonies. [CD1:12] **SIMILAR SPECIES** Adult White-tailed Tropicbird is whiter, with broad black diagonal bar on inner wing and less black on outer wing; it is also smaller and more graceful. For differences between juveniles/immatures, see following species. **STATUS AND DISTRIBUTION** Scarce and local. Breeds on islets off Senegal (Îles de la Madeleine), and on Cape Verde Is (*mesonauta*) (and, extralimitally, on Ascension and St Helena; nominate). Recorded as vagrant offshore Mauritania, Gambia, Guinea-Bissau, Guinea, Gabon and São Tomé.

WHITE-TAILED TROPICBIRD *Phaethon lepturus*   Plate 3 (6)
Phaéton à bec jaune

Other name F: Petit Phaéton
L 38–40 cm (15–16"); tail streamers 33–40 cm (13.0–15.5"); WS 92 cm (36"). *P. l. ascensionis*. **Adult** Pure white with short black streak through eye, black in outer primaries forming patch on upperwing, black diagonal wingbar

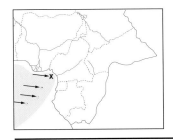

(forming characteristic V), long white tail streamers and yellowish-orange bill. **Juvenile/immature** Similar to juvenile/immature Red-billed Tropicbird, but lack distinct black nuchal collar and are less closely barred above (barring thus more visible at distance); bill pale yellow, tipped black. **VOICE** Similar to Red-billed Tropicbird, but rather higher pitched and faster: *kirrik-kirrik-kirrik*. [CD1:13] **SIMILAR SPECIES** Adult Red-billed Tropicbird (q.v.) has diagnostic red bill and less white appearance; it is also larger with broader based wings. **STATUS AND DISTRIBUTION** Breeds on São Tomé (and presumably Príncipe) and Annobón, where locally common. Elsewhere, mainly recorded as offshore vagrant (Ghana, Liberia, Cape Verde Is).

GANNETS AND BOOBIES
Family SULIDAE (World: 9 species; Africa: 5; WA: 5)

Large, conspicuous seabirds whose stout, tapering bills, cigar-shaped bodies, long, narrow wings and wedge-shaped tails impart distinctive jizz. Sexes similar. Adult plumage acquired over 2–4 years. Flight steady and purposeful with shallow flaps followed by a glide. Flocks often fly in single file. Feed principally by plunge-diving from relatively great height. Breed colonially. In W Africa only likely to be encountered offshore, unless sick or storm-driven.

NORTHERN GANNET *Sula bassana*   Plate 4 (1)
Fou de Bassan

L 87–100 cm (34–40"); WS 165–180 cm (65–71"). The largest seabird likely to be encountered in the region (only vagrant frigatebirds, giant petrels and albatrosses are larger). **Adult** Predominately white with creamy-yellow head and neck, and black wingtips. **Juvenile** Entirely dark brown with small white spots on upperparts, and a whitish, V-shaped rump patch. **Immature** Becomes progressively whiter, producing confusing patterns. Adult plumage acquired over 4 years. **SIMILAR SPECIES** Greatest confusion likely with Cape Gannet (q.v.). In some 1st- or 2nd-year plumages confusion possible with Brown Booby, which is smaller, lacks speckled appearance and has solid brown upperparts lacking white on head and rump. Masked Booby has white, not yellowish head, and all-black secondaries and tail. **STATUS AND DISTRIBUTION** Uncommon to scarce, but regular non-breeding visitor from N Atlantic to at least Guinea-Bissau. Vagrant, Cape Verde Is. **NOTE** Placed in genus *Morus* by some authors.

CAPE GANNET *Sula capensis*   Plate 4 (2)
Fou du Cap

L 84–94 cm (33–37"); WS 165–180 cm (65–71"). **Adult** As Northern Gannet but with all-black secondaries and tail. Usually slightly smaller and has longer black gular stripe, but this rarely noticeable in the field. **Juvenile/immature** Plumage sequence similar to Northern Gannet. Probably inseparable from immature Northern until 3rd year; white secondaries and/or outer tail feathers strongly suggest Northern. **SIMILAR SPECIES** Northern Gannet (q.v. and above). Masked Booby has all-white, not yellowish head, yellowish bill and more extensive black on secondaries. **STATUS AND DISTRIBUTION** Scarce non-breeding visitor from S African waters to Gulf of Guinea (northwest to Ghana), mostly immatures, late May/Jun–Oct. Confusion with Northern Gannet masks true status and range; more information required. *TBW* category: THREATENED (Vulnerable). **NOTES (1)** Some birds may show some white on outer tail feathers, thus recalling Australasian Gannet *S. serrator* (very rare vagrant to S African waters), which has all-black secondaries, variable amount of white in outer tail feathers (typically only 4 central tail feathers black) and much shorter black gular stripe. **(2)** Placed in genus *Morus* by some authors.

MASKED BOOBY *Sula dactylatra*   Plate 4 (3)
Fou masqué

L 81–92 cm (32–36"); WS 152 cm (60"). *S. d. dactylatra*. **Adult** Superficially similar to Northern and Cape Gannets but smaller, with shorter, yellowish bill, small black mask, pure white head, more extensive black on secondaries, black tips to scapulars and more pointed, all-black tail. **Juvenile** Brown head and upperparts, separated by narrow white collar; underparts white. **Immature** Becomes progressively whiter (with back whitening first; white collar broadening), attaining adult plumage over 2 years. **SIMILAR SPECIES** Northern and Cape Gannet (q.v.). Juvenile could be confused with Brown Booby, but white cervical collar diagnostic, underwing pattern different and brown of head extends only to throat, not onto breast. **STATUS AND DISTRIBUTION** Rare vagrant, recorded in Gulf of Guinea (offshore between São Tomé and Príncipe, and from Tinhosas and Santaren I.) and offshore Liberia and Ivory Coast. Nearest breeding colonies on Ascension.

RED-FOOTED BOOBY *Sula sula*   Plate 4 (4)
Fou à pieds rouges

L 66–77 cm (26–30"); WS 91–101 cm (36–40"). *S. s. rubripes.* Small booby with great variety of plumages. Adults have red feet and pink base to bill. **Adult white morph** (typical) Entirely white with black remiges and black carpal patch on underwing. **Adult brown morph** Mainly dull white with brown upperwings. **Juvenile** Similar to adult brown morph, but plumage streaked, bill dark, facial skin purplish and feet yellowish-grey. **Immature** Distinguished from adult intermediate morph by duller bare-part coloration. Adult plumage acquired over 2–3 years. **SIMILAR SPECIES** Brown morph only likely to be confused with juvenile of much larger Northern and Cape Gannets, both of which have crescentic white rumps. Immature Brown Booby has paler underwings and paler, less uniform underparts. **STATUS AND DISTRIBUTION** Rare vagrant, recorded from Cape Verde Is (adult in colony of Brown Boobies, Rombo, Aug 1986) and around Tinhosas and Príncipe (immatures, Jan and Apr 1995). Nearest breeding colony on Ascension.

BROWN BOOBY *Sula leucogaster*   Plate 4 (5)
Fou brun

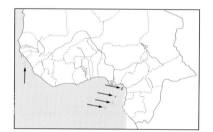

L 64–74 cm (25–29"); WS 132–150 cm (52–59"). *S. l. leucogaster.* **Adult** Distinctive pattern of uniformly dark brown upperparts, head, throat and upper breast contrasting strongly with white underparts, incl. underwing-coverts. Bill yellowish. Feet yellowish-green. **Juvenile/immature** As adult but white of underparts sullied dusky-brown, particularly in juvenile Adult plumage acquired over 2–3 years. **HABITS** Gregarious, mostly in small flocks, but large groups may occur where feeding is productive. Usually encountered inshore. Often perches on buoys, rocks or piers. Rarely seen sitting on surface of sea. Feeding flocks fly in single file and make shallow plunge-dives to catch fish. [CD5:2b] **SIMILAR SPECIES** Confusion possible with juvenile Northern or Cape Gannets, but both these species have white crescentic mark on rump; Brown Booby also smaller, more agile, with faster flight and relatively longer tail. Juvenile Masked Booby has diagnostic white collar on hindneck. Juvenile/immature could be confused with brown morph or immature Red-footed Booby, but Brown Booby usually has indication of dividing line between dark throat and upper breast, and paler underparts. **STATUS AND DISTRIBUTION** Colonies on Cape Verde Is, Alcatraz (Guinea), and Gulf of Guinea Is, where locally common (but only 1 record, in 1895, from Bioko). Uncommon wanderer along region's entire coastline.

CORMORANTS AND SHAGS
Family PHALACROCORACIDAE (World: 39 species; Africa: 8; WA: 3)

Medium-sized to large waterbirds with generally dark plumage, strong bills hooked at tip, long necks and bodies, rather short wings, and long, wedge-shaped tails. Sexes similar. Have breeding and non-breeding plumages. Juveniles duller or paler. Adult plumage acquired in 3rd or 4th year. Swim low in water, with neck erect and bill held up at an angle, often with only head and neck above surface. Catch fish underwater principally by surface-diving. Mostly gregarious and breeding colonially. Fly in line or V-formation, with outstretched neck. Often stand upright with wings held open to dry. Normally silent, except at breeding sites, where they utter various short guttural calls.

GREAT CORMORANT *Phalacrocorax carbo*   Plate 1 (7)
Grand Cormoran

Other name F: Cormoran à poitrine blanche
L 80–100 cm (31–40"); WS 130–160 cm (51–63"). *Large* cormorant with *variable amount of white on throat and breast.* **Adult** *lucidus* has white extending from throat to breast and sometimes belly. *P. c. maroccanus* smaller, with white

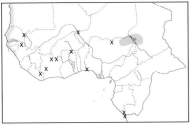

confined to throat and upper neck. In breeding plumage both races have white thigh patch. **Juvenile/immature** Both races virtually identical, brownish above and mainly whitish below. **HABITS** In all aquatic habitats, incl. freshwater lakes, marshes, rivers, mangroves, estuaries and inshore waters. Perches on shorelines, on rocks and in trees. [CD1:15] **SIMILAR SPECIES** Long-tailed Cormorant is much smaller and shorter necked, with much smaller bill and relatively longer tail; adult differs further by lacking any conspicuous white in plumage. See also Cape Cormorant. **STATUS AND DISTRIBUTION** *P. c. lucidus* is locally common resident from Mauritania to Guinea; uncommon, NE Nigeria, N Cameroon and L. Chad. Rare/vagrant elsewhere, with records from Liberia, Ivory Coast, N Ghana, Burkina Faso, W Niger, coastal Nigeria, Gabon and Congo; also Cape Verde Is (former resident?). *P. c. maroccanus* breeds on Moroccan coast and may move south along Mauritanian coast outside breeding season. **NOTE** *P. c. lucidus* sometimes treated as separate species, White-breasted Cormorant.

CAPE CORMORANT *Phalacrocorax capensis*   Plate 146 (5)
Cormoran du Cap

L 61–64 cm (24–25"); WS 109 cm (43"). **Adult breeding** Medium-sized all-black cormorant with yellow-orange gular area and relatively short tail. **Adult non-breeding** Dull brown with slightly paler underparts; gular area dull yellowish-brown. **Juvenile/immature** As adult non-breeding but underparts whiter. **HABITS** A marine species, occurring in coastal waters. [CD5:3a] **SIMILAR SPECIES** Long-tailed Cormorant is smaller and has longer tail. Great Cormorant best distinguished at all ages by larger size and yellow facial skin. **STATUS AND DISTRIBUTION** Vagrant from S African coasts, recorded from Gabon and (probably) Congo. *TBW* category: Near Threatened.

LONG-TAILED CORMORANT *Phalacrocorax africanus*   Plate 1 (6)
Cormoran africain

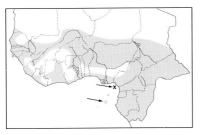

Other name E: Reed Cormorant
L 51–56 cm (20–22"); WS 85 cm (33"). *P. a. africanus.* Small cormorant with short bill, short neck and long tail. **Adult breeding** Almost entirely blackish with green gloss; short crest on forehead; wing-coverts and scapulars silvery grey with black tips producing blotchy effect. Eye red. **Adult non-breeding** Duller; no crest. Some apparently similar to immature but with red eye (Zimmerman *et al.* 1996). **Juvenile/immature** Mainly dull brownish above, pale brownish to buffish-white below with variable amount of buff or brownish streaking; eye brown. **HABITS** In all aquatic habitats except open sea. Often favours smaller rivers and pools than Great Cormorant; also less gregarious, mostly seen singly or in small groups. Rarely on shoreline; prefers to perch on vantage point in or near water. [CD5:3b] **SIMILAR SPECIES** Great Cormorant is much larger, with larger bill and relatively shorter tail; adult differs further by white on underparts, juvenile/immature by being mainly whitish below. See also Cape Cormorant. **STATUS AND DISTRIBUTION** Widespread and usually common resident south of 18°N; also São Tomé. Local movements recorded. Not uncommon dry-season visitor, Liberia. Vagrant, Bioko.

DARTERS
Family ANHINGIDAE (World: 2–4 species; Africa: 1)

Large aquatic birds superficially resembling cormorants but differing in being slimmer and having long, slender bills without hook. Sexes differ; non-breeding plumage duller than breeding. Found mainly on fresh water. Feed singly, spearing fish with bill. Breed colonially in trees.

AFRICAN DARTER *Anhinga rufa*   Plate 1 (5)
Anhinga d'Afrique

L *c.* 80 cm (31"); WS 120 cm (47"). Slim, cormorant-like bird with *long pointed bill* and *slender neck with distinct kink* (obvious in flight and when perched). In flight, long graduated tail often held in fan shape. **Adult male** Mainly black with chestnut foreneck and white stripe below eye extending onto neck-sides; buffish white streaks on wing-coverts and scapulars. **Adult female** Crown and hindneck brown (not black), rest of neck buffish-brown, white

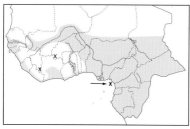

neck-stripe less distinct. **Juvenile** Resembles adult female, but lacks white neck-stripe, has various amounts of buffish-white on head and neck, and buffish underparts suffused with black. Adult plumage acquired in 3rd year. **HABITS** In various aquatic habitats; avoids sea, but frequents coastal lagoons. Swims with only head and neck above water (hence alternative name 'snakebird'). Perches in trees and on rocks, sometimes on shore, often with wings spread like cormorant. May soar on thermals, often at great height. Hunts singly; usually gregarious when roosting and nesting. [CD5:4a] **SIMILAR SPECIES** Great Cormorant is bulkier, with thicker neck, hooked bill and no rufous in plumage. Long-tailed Cormorant is smaller, with shorter bill and neck. African Finfoot has orange-red bill and legs, and quite different habits. **STATUS AND DISTRIBUTION** Widespread, scarce to uncommon resident south of *c.* 18°N. May still be relatively common locally (e.g. Gambia), but has much declined in recent years. Vagrant, Bioko. **NOTE** Variously treated as conspecific with Oriental Darter *A. melanogaster* or as allospecies (cf. Dowsett & Dowsett-Lemaire 1993).

PELICANS
Family PELECANIDAE (World: 7–8 species; Africa 3; WA: 2)

Huge waterbirds with characteristic massive bill and very large, distensible gular pouch. Sexes similar; males average larger. Ungainly on land, but superb flyers. Flight strong, with heavy wingbeats alternated by glides; head drawn back. Often soar. Take-off requires clumsy run across surface. Float high on surface when swimming. Gregarious, breeding colonially, often flying in formation and foraging in groups. Feed almost exclusively on fish.

GREAT WHITE PELICAN *Pelecanus onocrotalus*   Plate 1 (8)
Pélican blanc

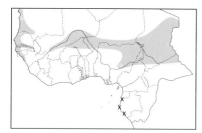

Other name E: White Pelican
L 140–175 cm (55–69"); WS 270–360 cm (106–142"). **Adult non-breeding** Mainly white with diagnostic *black flight feathers*. Bill greyish with pink cutting edges; pouch yellow. Bare facial skin and legs pink. **Adult breeding** Acquires variable pinkish tinge to plumage and small ragged crest on rear of crown; bare-part colours intensify (facial skin pinkish-yellow in male, bright orange in female). **Juvenile** Dull greyish-brown above, whitish below; flight feathers dark brown. Bill, pouch and legs greyish. **Immature** Paler than juvenile, whitening progressively with age, flight feathers darkening, bare parts becoming more yellow. Full adult plumage probably attained in 3rd or 4th year. **HABITS** In large aquatic habitats, incl. estuaries, lagoons, coast and freshwater lakes. Usually prefers alkaline or saline habitats to fresh water. Roosts and nests on ground. Often fishes in coordinated groups. May attend local fisheries to scavenge scraps. Dependent on thermals to move long distances, spiralling to gain height before flying off in V-shaped or curved skeins, often at great altitude. [CD1:18] **SIMILAR SPECIES** Pink-backed Pelican is smaller, generally greyer (not pure white), with grey flight feathers contrasting much less with rest of wing in flight, and has pale dull yellow bill and pouch. At distance, circling flocks could be confused with White Stork, which has similar plumage pattern, but projecting neck and long, trailing legs. **STATUS AND DISTRIBUTION** Scarce or uncommon to locally common resident or intra-African migrant. Breeding colonies in Mauritania, Senegal and Nigeria (Wase Rock); also known to have bred (still breeds?) in Mali, Cameroon and Chad. Could occur in suitable habitat throughout. Movements in W Africa unclear; known to make long-distance migrations in E Africa.

PINK-BACKED PELICAN *Pelecanus rufescens*   Plate 1 (9)
Pélican gris

L 125–132 cm (49–52"); WS 265–290 cm (104–114"). **Adult non-breeding** Whitish with pale grey cast. Blackish primaries and grey secondaries do not exhibit obvious contrast with rest of wing in flight. Pinkish back difficult to see in field. Bill and pouch pale pinkish-yellow; bare facial skin greyish-pink. Legs yellowish. **Adult breeding** Acquires short grey crest on nape, small blackish patch in front of eye, and pinkish tinge to upperparts; bare-part colours intensify. **Juvenile** Brownish above with grey bill, greenish-yellow pouch and greyish-pink legs. **Immature** Becomes progressively paler. Full adult plumage probably attained in 3rd or 4th year. **HABITS** Frequents almost any large aquatic habitat. Prefers freshwater lakes, rivers and swamps; sometimes along coast. Usually roosts and

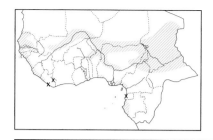

nests in trees, occasionally on ground. Normally fishes singly, not in groups like Great White Pelican. Will use thermals to gain height, but more sedentary than Great White Pelican and less likely to make long-distance movements. [CD5:4b] **SIMILAR SPECIES** Great White Pelican is larger, whiter, with contrasting black-and-white wings in flight, and has differently coloured bare parts. **STATUS AND DISTRIBUTION** Rare or uncommon to locally common. Widespread, but irregularly distributed; sometimes absent from apparently suitable habitat. Principally resident; local movements recorded (e.g. rare dry-season migrant, Liberia).

FRIGATEBIRDS
Family FREGATIDAE (World: 5 species; Africa: 4; WA: 2)

Large, unmistakable seabirds with extremely long, angular wings and very long, deeply forked tails. Bill long and hooked. Highly aerial and pelagic away from breeding colonies. Extraordinary flyers. Flight very buoyant; soar effortlessly for long periods; never settle on sea. Forage in flight by swooping down on prey or by harassing other seabirds (esp. boobies and tropicbirds), in the manner of skuas, forcing them to disgorge their catch. Readily take offal. Usually solitary at sea, but sometimes congregating in large numbers. Sexually dimorphic. Male has bare, scarlet gular pouch, which is spectacularly inflated during display. Period needed to acquire adult plumage unknown, probably 4–6 years. Great care is needed when identifying frigatebirds: plumages extremely variable, differing according to age, sex and colour morph. Note in particular amount of white on underparts and whether there is any extension onto underwing in the form of 'spurs'. Reference to specialist work on seabirds (e.g. Harrison 1991) recommended.

MAGNIFICENT FRIGATEBIRD *Fregata magnificens*   Plate 4 (6)
Frégate superbe

L 89–114 cm (35–45"); WS 217–244 cm (85.5–96.0"). Very similar to Ascension Frigatebird, but sexes possess only one morph. **Adult male** Wholly black with red gular pouch. **Adult female** Breast white; white tips to axillaries form wavy lines. **Immature** Head, breast and belly white; brown breast-band broken in centre; white tips to axillaries. Seven transitional immature plumages described (see Harrison 1991). **HABITS** See family account. In Cape Verdes, breeds on ground within colonies of Brown Boobies on offshore rocks. [CD1:20] **SIMILAR SPECIES** Ascension Frigatebird (q.v.). **STATUS AND DISTRIBUTION** Rare resident, Cape Verde Is (breeds on islets off Boavista). Recorded in offshore waters of Mauritania and Senegambia.

ASCENSION FRIGATEBIRD *Fregata aquila*   Plate 4 (7)
Frégate aigle-de-mer

L 89–96 cm (35–38"); WS 196–201 cm (77–79"). **Adult male** Wholly black with red gular pouch. **Adult female** Typically resembles male but has brown band forming collar and breast-band. Light morph has white breast and belly, and square white 'spurs' on axillaries. **Immature** Similar to light-morph female, but has white head, complete brown breast-band and always has diagnostic square 'spurs' on axillaries. Nine transitional immature plumages described (see Harrison 1991). **HABITS** See family account. **SIMILAR SPECIES** Adult male probably indistinguishable at sea from male Magnificent Frigatebird, but ranges not known to overlap. Adult female Magnificent resembles light-morph female Ascension but lacks 'spurs' and usually has indistinct white wavy lines on axillaries. Immature lacks white 'spurs' and has broken breast-band. **STATUS AND DISTRIBUTION** Vagrant in Gulf of Guinea, where recorded in waters of São Tomé and Príncipe. Only likely to be encountered offshore or as storm-driven waif. Breeds only on islet off Ascension. *TBW* category: THREATENED (Vulnerable).

HERONS, EGRETS AND BITTERNS
Family ARDEIDAE (World: 60–65 species; Africa: 22; WA: 19)

Slender, medium-sized to large wading birds with long necks and legs, and long, straight pointed bills. Flight strong with regular wingbeats and neck retracted, forming an S (unlike storks, spoonbills, ibises and cranes). Wings long and broad; tail short. Sexes similar. Adult plumage attained during 2nd, 3rd or 4th calendar year. Most species usually found in or near water.

GREAT BITTERN *Botaurus stellaris*

Butor étoilé

Plate 7 (7)

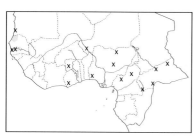

Other name E: Eurasian Bittern

L 70–80 cm (28.5–31.5"); WS 125–135 cm (49–53"). *B. s. stellaris.* Medium-sized, stocky, thick-necked heron with brown, cryptically patterned plumage, *black crown and black malar stripe.* Heavy, dagger-shaped yellowish bill. Flight owl-like with deep wingbeats on broad, rounded wings strongly bowed downwards. **VOICE** Silent in W Africa. [CD1:21] **HABITS** Extremely secretive, solitary and shy. Hides in dense reedbeds, marshes and swamps. Active by day and night. Will 'freeze' in upright posture, bill 'sky-pointing' when threatened, relying on cryptic plumage to avoid detection. **SIMILAR SPECIES** White-crested Tiger Heron lacks black crown and malar stripe, has more barred plumage and occurs in quite different habitat. Juvenile Black-crowned Night Heron much smaller, with white spots on wing-coverts and more uniform streaking below. **STATUS AND DISTRIBUTION** Rare winter visitor from Europe and perhaps N Africa, recorded from S Mauritania, Senegambia, Ghana, SW Niger, Nigeria, Cameroon, CAR and Gabon.

LITTLE BITTERN *Ixobrychus minutus*

Blongios nain

Plate 7 (2)

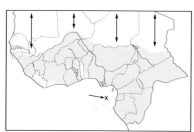

L 27–38 cm (10.5–15.5"); WS 40–58 cm (16–23"). Very small, thick-necked heron with *contrasting pale wing panel striking in flight.* Flies with rapid, shallow wingbeats and glides; legs trail on rising from cover. **Adult male *payesii*** Crown, mantle, back, tail and flight feathers black; face and neck-sides rufous-chestnut; upperwing-coverts pale buff. Lores yellow, becoming orange when breeding. **Nominate** is much paler on face and neck-sides, the rich rufous-chestnut being replaced by pale buff. **Adult female** Streaked brownish, with duller black crown, nape and tail, and browner, less contrasting upperwing-coverts. **Juvenile** Similar to adult female but much more heavily streaked. Following post-juvenile moult resembles adult but has darker, mottled and spotted wing panel. **VOICE** Usually silent. Calls include a hard *ker-ek* and a sharp, rapid *kekekekek.* Breeding male utters deep croaking *hogh* or *woof* at *c.* 2-second intervals. [CD1:22] **HABITS** In reedbeds, marshes, vegetated stream banks and areas of rank vegetation near water, occasionally mangroves. Grounded nocturnal migrants may occur in any habitat incl. desert oases, isolated pools and even arid country devoid of water. Secretive and shy. Will 'freeze' in erect posture with bill 'sky-pointing' when threatened. Most often seen when flushed or when making short flights low over vegetation. **SIMILAR SPECIES** Much smaller than any other heron-like bird except Dwarf Bittern, which is even smaller, uniformly dark above (lacking pale wing panel) and boldly streaked below. Juvenile Green-backed Heron is larger and has all-dark upperparts with white-spotted wing-coverts. Juvenile Black-crowned and White-backed Night Herons are much larger and have white spots on wing-coverts. **STATUS AND DISTRIBUTION** Resident ssp. *payesii* is uncommon in suitable aquatic habitat almost throughout; also Príncipe. Local movements may occur in response to fluctuating water conditions. Vagrant, Cape Verde Is (one juvenile). W Palearctic nominate crosses Sahara on broad front to winter in suitable habitat within the region and further south; probably also breeds locally in Senegal (L. de Guier) and Guinea-Bissau.

DWARF BITTERN *Ixobrychus sturmii*

Blongios de Sturm

Plate 7 (3)

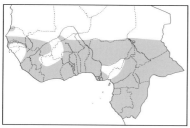

L 25–30 cm (10–12"). **Adult** *Tiny heron* with uniformly *slate-grey upperparts* (incl. face) and *buffish underparts heavily streaked black.* Prominent black streak from base of bill to centre of belly. Lores and orbital skin bluish to yellowish-green. Legs and feet greenish-yellow, becoming orange in breeding season. **Juvenile** Upperparts duller, with rufous-fringed feathers; underparts rufous-brown; legs paler yellow. **VOICE** Normally silent. A loud croak when disturbed. [CD5:5a] **HABITS** Secretive and partially nocturnal. Usually solitary or in pairs at margins of aquatic habitats, incl. vegetated streams and pools, marshes and reedbeds, seasonally inundated areas and mangroves. Will 'freeze' in typical bittern stance when alarmed. Usually seen perched motionless in riverside vegetation or when flushed. Flight slow and heavy with legs dangling. **SIMILAR SPECIES** Confusion only likely with Green-backed Heron, esp. in flight, but latter has contrasting dark crown and unstreaked, pale grey neck, and is larger. Little Bittern always has contrasting pale wing panel. For differences between juvenile/immature see resp. species accounts. **STATUS AND DISTRIBUTION** Uncommon to scarce intra-African migrant, occurring during wet season in suitable habitat throughout most of region, except arid north. Movements poorly understood due to secretive nature; perhaps resident in some areas.

WHITE-CRESTED TIGER HERON *Tigriornis leucolophus*   Plate 7 (6)
Onoré à huppe blanche

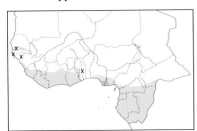

L 66–80 cm (26.0–31.5"). **Adult male** Superficially similar to Great Bittern, but more slender, with barred black and buffish-brown plumage, and diagnostic white crest. When not raised, crest is half-concealed by nape feathers and often difficult to see. Crown may appear black. Flight feathers blackish with white tips to primaries. Barring on tail whiter and more contrasting. **Adult female** Buff barring narrower, resulting in darker appearance. **Juvenile** Much more strongly barred than adult. **VOICE** A far-carrying, single or double, low moaning note, regularly repeated in slow tempo, usually for brief periods. Principally uttered before dawn and after sunset. [CD5:5b] **HABITS** Frequents shaded forest streams and swamps in forested areas; also mangroves. Secretive, shy and largely nocturnal, hence seldom seen. Will 'freeze' in bittern-like posture when threatened. **SIMILAR SPECIES** As usually seen in poor light, confusion theoretically possible with Great Bittern, but latter occurs in quite different habitat, is streaked (not barred), lacks white crest, and is heavier looking. Juvenile Black-crowned and White-backed Night Herons are smaller, streaked (never barred), and have white spots on wing-coverts. **STATUS AND DISTRIBUTION** Uncommon to rare resident in forest zone, from Senegambia to SW CAR and Congo. *TBW* category: Data Deficient.

WHITE-BACKED NIGHT HERON *Gorsachius leuconotus*   Plate 6 (2)
Bihoreau à dos blanc

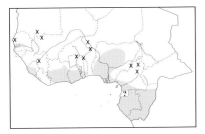

L 50–55 cm (20–22"). **Adult** Medium-sized, stocky heron with black head and *huge eye surrounded by white patch*. Upperparts blackish-brown with white plume-like scapulars forming diagnostic but rather inconspicuous *triangular patch on back*. Short nuchal crest. Throat white; neck and breast rufous-brown; belly white with black streaking. Feet project slightly beyond tail in flight. **Juvenile** Difficult to identify. Dull brown, with speckled upperparts, buffish spots on tips of wing-coverts and heavily streaked underparts. Forehead and crown very dark and unstreaked; head-sides, neck and upper breast streaked buffish-brown. White on scapulars develops quite early. **VOICE** Usually silent. Utters a croaking note when alarmed. Also a short, fast series of hoarse, hushing notes, followed by low trill *taash-taash-taash-taash rr-rrurrr.* [CD5:6a] **HABITS** Singly or in pairs along undisturbed tree-lined rivers and streams; also in mangroves and, less commonly, marshes and reedbeds. Largely nocturnal; secretive and very shy by day, roosting in dense foliage or swampy vegetation. Easily overlooked; seldom seen unless flushed, or when active at dusk or at night. **SIMILAR SPECIES** Adult Black-crowned Night Heron is slightly larger and has quite different plumage. Juveniles very similar; Black-crowned best distinguished by wholly streaked crown and head-sides. See also White-crested Tiger Heron. **STATUS AND DISTRIBUTION** Rare or scarce to locally uncommon resident in forest zone, from Senegambia to Congo. Also recorded from S Mauritania, W Mali, S Burkina Faso, N Ghana, N Togo, SW Niger and N Cameroon. Seasonal movements correlated with rains may occur, but poorly understood.

BLACK-CROWNED NIGHT HERON *Nycticorax nycticorax*   Plate 6 (1)
Bihoreau gris

L 56–65 cm (22.0–25.5"); WS 90–100 cm (35–39"). *N. n. nycticorax.* Medium-sized, stocky heron with relatively short neck and legs. **Adult breeding** Unmistakable. *Crown, mantle and scapulars black* with 2 long white plumes on hindneck. *Wings, rump and tail grey.* Head-sides and underparts white. Eye crimson. Legs and feet pale yellow, becoming red during courtship. Feet extend clearly beyond tip of tail in flight. **Adult non-breeding** Lacks white plumes on hindneck. **Juvenile** Brown above with white spots; buffish below with conspicuous dark brown streaking. Eye yellow. Adult plumage takes 3 years to acquire, the white spotting on upperparts and dark streaking on underparts becoming progressively less marked with advancing age. Black crown acquired during 2nd year. **VOICE** Mostly silent, but utters a distinctive, low, harsh *kwok* in flight. [CD1:23] **HABITS** Gregarious and largely nocturnal,

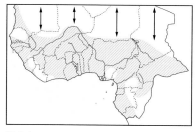

occurring in all well-vegetated aquatic habitats, incl. marshes, lakes, rivers and mangroves. Mostly roosts by day in trees or dense marshy vegetation, leaving roost at dusk in flocks. Occasionally seen feeding in daylight. Breeds in colonies in trees or bushes, usually in company of other herons, cormorants or darters. **SIMILAR SPECIES** Adult White-backed Night Heron markedly different; juvenile more difficult to separate, but distinguished by solid dark (not heavily streaked) forehead and crown, huge eye and, in flight, legs projecting less beyond tail. **STATUS AND DISTRIBUTION** Uncommon to locally common resident in suitable habitats throughout. Migrants from W Palearctic occur alongside residents during northern winter (east to SW Chad). Vagrant, Cape Verde Is.

SQUACCO HERON *Ardeola ralloides* Plate 6 (3)
Crabier chevelu

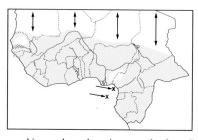

Other name F: Héron crabier

L 42–47 cm (16.5–18.5"); WS 80–92 cm (31.5–36.0"). Small, stocky, short-legged heron, appearing largely *cinnamon or buffish-brown at rest* but with *contrasting white wings, rump and tail in flight*. **Adult breeding** Largely cinnamon with long cream-coloured and black-edged feathers on crown and nape forming a crest. Bill blue tipped black. Legs bright red. **Adult non-breeding** Much duller. Earth-brown above, streaked below, with shorter crest. Bill, legs and feet greenish-yellow. **Juvenile** As adult non-breeding but darker brown; breast more streaked; wings mottled brown. **VOICE** Usually silent. Utters croaking and nasal grating sounds when disturbed and at breeding colony. [CD1:24] **HABITS** Often solitary, occasionally in loose groups of up to 20, in vegetated freshwater habitats incl. marshes, swamps, inundated areas, ricefields and edges of slow-flowing streams. Rarely along coast and in estuaries. Rather secretive, often remaining motionless, when cryptic coloration renders it inconspicuous. **SIMILAR SPECIES** Adult may recall breeding Cattle Egret, but latter appears white at rest and has reddish or yellow bill and legs. Non-breeding adult and juvenile readily separable from all other small herons with brown plumages by conspicuous white wings in flight. **STATUS AND DISTRIBUTION** Not uncommon resident in suitable habitat throughout, except arid north. Migrants from W Palearctic occur alongside residents during northern winter and may outnumber them in most areas. Vagrant, Cape Verde Is, Bioko and Príncipe.

RUFOUS-BELLIED HERON *Ardeola rufiventris* Plate 7 (5)
Crabier à ventre roux

Other name F: Héron à ventre roux

L 38 cm (15"). **Adult male** Small, very dark heron, appearing mostly black, with *dark rufous to maroon wings, belly, rump and tail*. Bill yellow with black tip. Lores and bare orbital skin bright yellow, conspicuous even in flight; eye yellow or yellowish-orange. Legs and feet yellow. **Adult female** Duller, with buffish stripe on throat and duskier bill. **Juvenile** Streaked buffish-brown on head-sides, neck and upper breast. **VOICE** A rasping *kraak* given in flight. **HABITS** In inundated grasslands, marshes and floodplains; occasionally at edges of lakes and rivers. Readily perches in trees. Assumes bittern-like posture at times but bill is held horizontally, not vertically. **SIMILAR SPECIES** Black Heron has all-dark bill, is larger and lacks any rufous in plumage. Juvenile could be confused with juvenile Green-backed Heron but latter is smaller and more slender, with buff tips to upperwing-coverts and a prominent black malar stripe. **STATUS AND DISTRIBUTION** S and E African species, claimed from SW Nigeria (3–4 sight records, 1969 and 1972).

CATTLE EGRET *Bubulcus ibis* Plate 6 (6)
Héron garde-bœufs

L 48–56 cm (19–22"); WS 90–96 cm (35.5–38.0"). *B. i. ibis.* Small, short-necked and largely white egret. Feathering extends below lower mandible to form distinct jowl. Typical stance is hunched with neck sunk into shoulders. In flight, appears stocky with hunched neck and rather short, heavy bill. **Adult breeding** Crown, breast and mantle rich buff. Bill and lores bright orange to coral-red. Legs coral- to dusky-red. **Adult non-breeding** All white with yellow bill and lores, and greenish-yellow to greyish legs. **Juvenile** Bill, legs and feet black. **VOICE** Usually silent. A short, gruff *kok* or *kwok*. [CD1:25] **HABITS** Tame and gregarious, occurring in open fields, grassland, inundated areas, margins of freshwater lakes and rivers, rubbish dumps and around human habitation. Will feed in drier habitats than other egrets. Often with cattle or herds of large game mammals, taking insects disturbed by them.

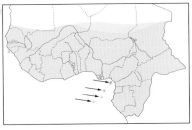

Readily perches and roosts in trees. Flies in loose flocks with distinctive rapid, shallow wingbeats. Once known, distinctive flight action permits identification of flocks at great distance. **SIMILAR SPECIES** Distinguished from other white egrets by small size, stocky build, conspicuous jowl and, in breeding state, combination of buff plumes and colour of bill and legs. Squacco Heron is smaller, appears darkish when standing and shows contrasting dark back and white wings in flight. **STATUS AND DISTRIBUTION** The most common heron throughout, incl. Gulf of Guinea Is. Resident and intra-African migrant; in many areas absent during rains. Common visitor, Cape Verde Is (mainly Dec–Apr); but few breeding records.

GREEN-BACKED HERON *Butorides striatus*   Plate 7 (4)
Héron strié

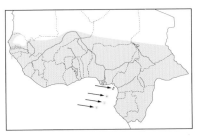

Other names E: Striated Heron; F: Héron vert
L 40 cm (16"); WS 54 cm (21"). *B. s. atricapillus.* **Adult breeding** Small, stocky, short-necked and mainly dark heron with erectile nuchal crest. At distance appears blackish on crown and back. Head-sides and underparts grey; throat white with buff line reaching belly. Greenish tinge to upperparts only visible at close range or in good light, when delicate cream edges to wings and tail feathers also noticeable. Bill all black; lores bright yellow. Legs yellow to reddish-orange. In flight appears largely dark with conspicuous yellow or orange legs (dangling on take-off). **Adult non-breeding** Bill black with yellow-green lower mandible; lores blue or greenish; legs grey-brown with rear of tarsus yellow. **Juvenile** Mainly brown above with white or buff spots on wing-coverts; head and underparts prominently streaked; distinct malar stripe. **VOICE** Mostly silent. Utters a loud, explosive *kyah* when flushed. [CD5:7a] **HABITS** Mainly solitary, in heavily vegetated margins of rivers, streams, lakes and pools, and in mangroves; occasionally floodplains. Shows no preference for salt or fresh water. In tidal habitats often far from cover at low tide. Usually adopts hunched stance when feeding, with body almost horizontal and legs bent. Hunts by stealth, slowly stalking prey or remaining motionless for long periods, often perching on low branches or mangrove roots. Flicks tail vertically. Occasionally 'freezes' in bittern-like posture. **SIMILAR SPECIES** Dwarf Bittern is smaller and has heavily streaked underparts. Juvenile Rufous-bellied Heron is slightly larger with longer neck and distinctive heavier two-tone bill. **STATUS AND DISTRIBUTION** Common resident throughout, except arid north. Also on Gulf of Guinea Is.

Black Herons *Egretta ardesiaca*

BLACK HERON *Egretta ardesiaca*   Plate 6 (5)
Aigrette ardoisée

Other name E: Black Egret
L 48–50 cm (19–20"). Medium-sized, slender and *all-black heron* with conspicuous *deep-orange or yellow feet* and *dark eyes*. 'Canopy-feeding' diagnostic. **Adult breeding** has long, loose plumes on crown, nape and mantle. **Juvenile** As

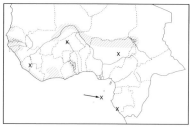

adult but duller. **HABITS** Often in tight groups, but also solitary. Prefers shallow waters of floodplains, marshes, margins of rivers and lakes, mangroves and mudflats. Distinctive feeding posture with wings spread and held forwards over head to form shaded 'umbrella' canopy. Flocks may canopy-feed in unison. Flight swift with rapid wingbeats. [CD5:7b] **SIMILAR SPECIES** Distinguished from dark-morph Western Reef Egret or rare dark-morph Little Egret by darker overall plumage colour, dark, not white throat, black bill and legs, and smaller size. Rufous-bellied Heron is smaller and shorter necked, with yellow legs and feet, and yellow base to bill. **STATUS AND DISTRIBUTION** Uncommon to locally common resident, from S Mauritania and Senegambia to S Nigeria, and from C Mali to N Nigeria, N Cameroon and Chad. Local movements recorded. Vagrant, Burkina Faso, Gabon, Cape Verde Is and São Tomé. **NOTE** Sometimes placed in genus *Hydranassa*.

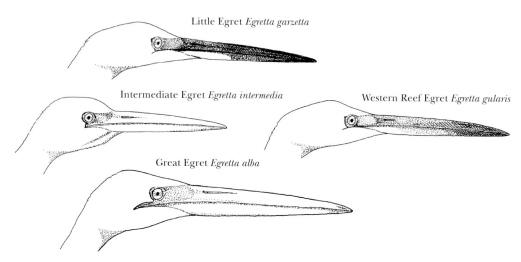

Little Egret *Egretta garzetta*

Intermediate Egret *Egretta intermedia*

Western Reef Egret *Egretta gularis*

Great Egret *Egretta alba*

WESTERN REEF EGRET *Egretta gularis*   Plate 6 (4)
Aigrette à gorge blanche

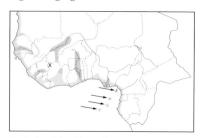

Other names E: Reef Heron; F: Aigrette dimorphe
L 55–65 cm (21.5–25.5"); WS 86–104 (34–41"). *E. g. gularis*. Medium-sized egret with, in breeding plumage, long plumes on nape, scapulars and breast. Dimorphic. **Adult dark morph** Blackish- to slate-grey with *white throat* and *variable amount of white on primary coverts*. Bill rather heavy, slightly drooped at tip, horn-brown or grey with yellowish base to lower mandible; sometimes all black when breeding. Legs greenish-black merging into greenish-yellow feet, yellow often extending up rear edge of legs to tarsal joint or even tibia. Eye yellowish; lores olive or yellowish-olive. **Adult white morph** Entirely white. **Juvenile** Dark morph similar to adult but paler grey with whitish, often mottled belly and white patches on neck and breast. White morph, white with variable amount of grey flecking. **HABITS** Largely confined to coastal areas (mangroves, reefs, tidal pools, mudflats), occasionally inland at freshwater sites. Mostly solitary away from colonies, but occasionally in small, loose parties. Less graceful than Little Egret, feeding by slow stalking, often crouching low in water or on rocks. Occasionally makes sudden rushes with half-opened wings. [CD1:26] **SIMILAR SPECIES** White morph probably often misidentified as Little Egret, which is very similar but distinguishable by combination of bill shape and colour, and colour of legs and feet. Also note habitat and behaviour. Black Heron differs from dark morph by smaller size, darker and more uniform plumage (no white throat), darker bill and legs, and dark eye. Very rare dark morph of Little Egret has more slender, straight and darker bill, and less extensive, deeper yellow on feet. **STATUS AND DISTRIBUTION** Not uncommon to locally common from Mauritania to Gabon; also on Gulf of Guinea Is. Mainly resident; local movements to and from breeding colonies; occasional wanderers noted at inland sites. Rare visitor, Cape Verde Is. White morph locally not uncommon in Guinea, Ivory Coast, São Tomé and Cameroon; probably overlooked elsewhere. **NOTE** Considered conspecific with Little Egret by some authors.

LITTLE EGRET *Egretta garzetta*

Aigrette garzette

Plate 6 (7)

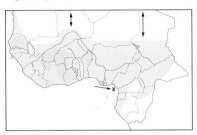

L 55–65 cm (21.5–25.5"); WS 88–95 cm (34.5–37.5"). *E. g. garzetta.* **Adult non-breeding** Medium-sized, all-white egret, distinguished from other white egrets by combination of *slender, straight black bill, black legs* and *contrasting yellow feet*. Lores grey-green. **Adult breeding** has long filamentous plumes on nape, mantle, scapulars and breast; lores and feet become orange-yellow. **Dark morph** (very rare) extremely similar to Western Reef Egret, though usually lacks white chin and throat. **Juvenile** Bill brownish-horn; lores lead-grey; feet grey-green. **HABITS** In all aquatic habitats, both salt and fresh water. Feeds principally by wading through shallow water snapping at prey with quick movements. Often runs quickly on land or in shallow water, opening and closing wings as it does so. Frequently associates with other species of egrets and herons. [CD1:27] **SIMILAR SPECIES** Both morphs of Western Reef Egret (q.v.) distinguished by different bill structure and leg/foot colour. **STATUS AND DISTRIBUTION** Common throughout. Not uncommon resident, Cape Verde Is. Rare migrant, Bioko, São Tomé and Príncipe. Residents augmented by migrants from W Palearctic during northern winter.

INTERMEDIATE EGRET *Egretta intermedia*

Aigrette intermédiaire

Plate 6 (8)

Other name E: Yellow-billed Egret
L 65–72 cm (25.5–28.0"); WS 105–115 cm (41–45"). *E. i. brachyrhyncha.* Medium-sized, all-white egret with yellow bill and mainly black legs and feet. Legs above tarsal joint yellow. Eye and lores yellow. **Adult breeding** has extensive filamentous plumes on foreneck, breast and scapulars (may reach beyond tip of tail), and slight crest on rear crown, but lacks long nape plumes of Little and Western Reef Egrets. Bill becomes red with yellow tip, lores and orbital skin bright green, eye red and tibia crimson. **Juvenile** As adult. **HABITS** In variety of freshwater and coastal habitats, incl. floodplains, inundated areas, margins of rivers and lakes, and mudflats. Often associated with other species of herons and egrets. Feeds by walking slowly in shallow water or on dry grassland, lunging to catch prey. [CD5:8a] **SIMILAR SPECIES** Non-breeding Great Egret may be difficult to separate without direct size comparison, but line of gape reaching to *c.* 1 cm beyond posterior edge of eye diagnostic; neck longer, more kinked than curved; legs and feet project well beyond tail in flight. Adult breeding has black, not yellow, bill. Other white egrets distinguished by combination of size and colour of bill and legs. **STATUS AND DISTRIBUTION** Uncommon to common throughout, except arid north. Mainly resident, though makes local movements; some evidence of migration. Vagrant, Cape Verde Is. **NOTE** Sometimes placed in genus *Mesophoyx*.

GREAT EGRET *Egretta alba*

Grande Aigrette

Plate 6 (9)

Other name E: Great White Egret
L 85–95 cm (33.5–37.5"); WS 140–170 cm (55–67"). Large, slender, long-necked white egret. Largest egret, in size similar to Black-headed Heron. In flight, and often at rest, long neck curved into distinctive S-shaped bulge. Flight slower than other egrets with legs projecting further. **Adult *melanorhynchos* non-breeding** Bill long and yellow; eye yellow. Lores and orbital skin olive-green; *line of gape extending c. 1 cm beyond posterior edge of eye* (diagnostic). Legs and feet black. **Adult breeding** Bill becomes black, eye bright red, lores and orbital skin emerald-green. Long plumes develop on scapulars (extending well beyond tip of tail), foreneck and breast. **Nominate *alba*** has tibia and rear edge of tarsus yellowish. **Juvenile** Bill yellow tipped blackish. **VOICE** A loud, deep, croaking *krraak*. [CD1:28] **HABITS** In most fresh and saltwater habitats. Mostly solitary away from breeding colonies; occasionally in small flocks. Often feeds in deep water when only head and neck visible above surrounding vegetation. Feeds by wading slowly or standing motionless; less active than Western Reef and Little Egrets. **SIMILAR SPECIES** Separated from other white egrets by combination of size, long snaky neck, colour of bill and legs, and lack of long nape plumes in breeding

plumage. Most likely to be confused with smaller Intermediate Egret, which has shorter, less kinked neck, shorter bill, and line of gape ending level with posterior edge of eye. **STATUS AND DISTRIBUTION** Common to not uncommon resident throughout, except arid north (*melanorhynchos*). Vagrant, Cape Verde Is and São Tomé. W Palearctic race *alba* recorded from CAR. **NOTE** Sometimes placed in monotypic genus *Casmerodius* or in *Ardea*.

PURPLE HERON *Ardea purpurea* Plate 5 (3)
Héron pourpré

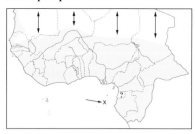

L 78–90 cm (30.5–35.5"); WS 120–150 cm (47–59"). Slender, lightly built heron with long thin bill and snake-like neck. **Adult** *purpurea* Crown black; head-sides, neck and underparts rufous striped with black. Upperparts dark slate-grey tinged purple-chestnut. In flight, long neck kinked into very distinct bulge; upperwing dark with little contrast; underwing-coverts rufous; toes long and often held splayed. Isolated form ***bournei*** is paler, with less black on neck, breast and belly; centre of underparts white and pale chestnut. **Juvenile** Paler, more brownish above, with largely rufous-brown crown and less strongly striped neck and underparts. From 2nd year more like adult, but duller. **VOICE** A loud, harsh *kaark*, similar to, but higher pitched than Grey Heron. [CD1:29] **HABITS** Usually solitary away from breeding colonies, frequenting reedbeds and marshy areas; rarely in mangroves or on mudflats. Skulks in dense aquatic vegetation, where cryptic plumage makes it difficult to observe. Feeds by standing motionless with body almost horizontal, catching fish with rapid strike of head. Will occasionally freeze in bittern-like posture. **SIMILAR SPECIES** Goliath Heron is much larger, with rufous (not black) crown and heavier bill. Purple Heron distinguished from Grey and Black-headed Herons by rufous underparts and very dark upperparts, longer and more snake-like neck, and thin, dagger-shaped bill. In flight more uniform coloration than other herons, with distinct neck bulge and large feet. **STATUS AND DISTRIBUTION** Nominate is uncommon to common resident and migrant from W Palearctic throughout. Vagrant, Cape Verde Is and São Tomé. *A. p. bournei* is rare resident, Cape Verde Is (Santiago, max. *c.* 15 pairs, 1999). **NOTE** *A. p. bournei* sometimes proposed as distinct species, Cape Verde Purple Heron or Bourne's Heron (Héron de Bourne).

GREY HERON *Ardea cinerea* Plate 5 (1)
Héron cendré

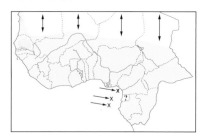

L 90–100 cm (35.5–39.5"); WS 175–195 cm (69–77"). Large, mainly grey-and-white heron. **Adult** *cinerea* Head largely white with diagnostic black stripe from behind eye merging with black plumes on nape. In flight has prominent white carpal patches contrasting with black flight feathers; underwing uniformly dark grey. Bill and lores yellow, becoming orange to vermilion at start of breeding season. Local race *monicae* very pale. **Juvenile** Duller and plainer grey, lacking any black, with grey crown and hindneck, and greyish (not white) throat and upper neck. **VOICE** A loud, croaking *fraank* or *kraak*. [CD1:30,31] **HABITS** In most aquatic habitats, incl. marshes, margins of lakes and rivers, estuaries and shallow sea coasts, occasionally in grassland. Mostly solitary away from breeding colonies, but sometimes in small groups. Feeds by waiting motionless in shallow water and retracting head before making quick stab at prey. Also wades and walks through shallows. **SIMILAR SPECIES** Black-headed Heron has crown and head-sides black (dark grey in juvenile) and underwing-coverts white, contrasting strongly with blackish flight feathers. Purple Heron has rufous plumage, pronounced neck bulge in flight and larger feet. Goliath Heron is much larger and has rufous on head and neck. **STATUS AND DISTRIBUTION** Nominate is common to not uncommon resident and Palearctic migrant throughout, except arid north. Non-breeding visitor to Cape Verde Is (not uncommon, pair bred on Santo Antão, 2000), Bioko (rare), São Tomé and Príncipe (rare). *A. c. monicae* restricted to Banc d'Arguin islands, Mauritania (1000–2000 pairs); non-breeding visitor, Senegal. **NOTE** *A. c. monicae* sometimes proposed as distinct species, Mauritanian Heron (Héron pâle).

BLACK-HEADED HERON *Ardea melanocephala* Plate 5 (2)
Héron mélanocéphale

L 92–96 cm (36–38"). **Adult** Large, mainly black-and-white heron with black head and hindneck, contrasting

white throat and foreneck, and grey belly. Upperparts darker grey than Grey Heron. Bill relatively short and deep at base, black above with greenish-yellow lower mandible. In flight, has distinctive two-tone underwing, white underwing-coverts contrasting with blackish flight feathers. Very rare **dark morph** has underparts entirely black. **Juvenile** Dull grey above with dark grey crown and head-sides, less black on hindneck and variable amount of rufous-buff on underparts. **VOICE** A raucous nasal *kuark*. [CD5:8b] **HABITS** Solitary and largely terrestrial, frequenting grassland and cultivated areas rather than aquatic habitats, though does occur in marshes and margins of rivers and lakes. Often in association with man and found far from water. Feeds by walking slowly over open areas making quick stab at prey once sighted. **SIMILAR SPECIES** Grey Heron has white crown in adult (greyish in juvenile), no black hindneck, uniformly greyish underwings, and is slightly larger. **STATUS AND DISTRIBUTION** Uncommon to common resident throughout, avoiding most arid areas. Moves north with rains, south during dry season. Breeds mainly during rains.

GOLIATH HERON *Ardea goliath*   Plate 5 (4)
Héron goliath

L 135–150 cm (53–59"); WS 210–230 cm (83–91"). The world's largest heron. **Adult** Head, hindneck and underparts rufous-brown; throat white; foreneck and upper breast white streaked with black; upperparts dark grey. Blackish bill powerful and deep. In flight legs held below horizontal. **Juvenile** Basically as adult, but brown parts paler, grey areas suffused with rufous, and whiter head-sides and neck. **VOICE** A far-carrying, raucous *kwaaark*. [CD5:9a] **HABITS** Mostly solitary and rarely encountered away from water. Mainly in freshwater habitats, incl. lakes, marshes and rivers; sometimes in lagoons and estuaries. Feeds by standing motionless for long periods, occasionally by walking and wading. **SIMILAR SPECIES** Large size usually sufficient to prevent confusion with other species. Adult Purple Heron similar in plumage but smaller (almost half the size), much slimmer, with black crown and black lines on head-sides and neck. Juvenile Purple Heron has brownish, not grey upperparts. **STATUS AND DISTRIBUTION** Scarce to uncommon resident in suitable habitat throughout. Some locally dispersive movements recorded. Vagrant, Liberia.

HAMERKOP
Family SCOPIDAE (World: 1 species)

A unique, medium-sized waterbird, endemic to the Afrotropical region. Sexes similar. Builds huge, domed nest of sticks, grass and mud in a tree.

HAMERKOP *Scopus umbretta*   Plate 7 (1)
Ombrette africaine

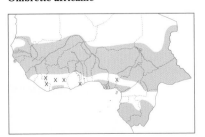

Other name F: Ombrette du Sénégal
L 50–56 cm (20–22"). Unmistakable. **Adult *umbretta*** Uniformly dark brown, with thick backward-pointing crest and stout hook-tipped bill combining to give head distinctive shape. Flight rather buoyant and owl-like, on broad, rounded wings; often rather jerky wingbeats. Glides with neck extended. In soaring flight can appear raptor-like. *S. u. minor* smaller and darker. **Juvenile** Similar to adult. **VOICE** A distinctive nasal trumpeting *yip-purr, yip-yip-yip-purr-purr-yip-yip* uttered when in groups or during display. Flight call a high-pitched nasal *yip* or *wek*. [CD5:9b] **HABITS** Singly or in pairs, occasionally groups, in any aquatic habitat within open woodland, incl. lakesides, river banks, seasonal pools and marshes, but also in drier areas. **STATUS AND DISTRIBUTION** Uncommon to common resident throughout. *S. u. minor* in coastal belt from Sierra Leone to Cameroon. Some dispersal during rains observed in drier areas.

STORKS
Family CICONIIDAE (World: 19 species; Africa: 8; WA: 8)

Large to very large wading and terrestrial birds with long bills, necks and legs. Bill straight in most species, but in some slightly decurved at tip, or with gap between mandibles. Tail short. Sexes usually alike, males average larger; genus *Ephippiorhynchus* unique in exhibiting sexual dimorphism in coloration. Non-breeding plumage similar to breeding, but in some species slightly duller. Full adult plumage attained during 2nd, 3rd or 4th calendar year. Fly with neck outstretched (unlike herons), except Marabou, and legs trailing. Typical flight action consists of soaring and gliding, often at great heights, alternated with some slow wingbeats. Gregarious or solitary. Most species normally silent.

YELLOW-BILLED STORK *Mycteria ibis*   Plate 8 (1)
Tantale ibis

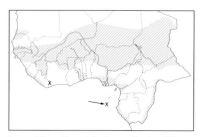

Other name E: Wood Ibis
L 95–105 cm (37.5–41.5"); WS 155–165 cm (61–65"). **Adult** *Large, mainly white stork with black flight feathers and tail, slightly decurved yellow bill, and red legs.* Has bright orange-red bare facial skin and variable amount of pink suffusion on upperwing-coverts, esp. in breeding season. **Juvenile** Largely greyish-brown on head, neck and upperparts; breast and belly white. Bill dull greyish-yellow; facial skin dull orange; legs brownish. By end of 1st year more like adult but duller. **HABITS** In all aquatic habitats, incl. freshwater marshes, lakes and margins of rivers, alkaline lakes, estuaries and mudflats. Gregarious but seldom found in large flocks. Often associated with other waterbirds such as herons, storks and pelicans. Often sits squatting on folded legs. [CD5:10a] **SIMILAR SPECIES** White Stork has red bill and white tail. Greyish juvenile superficially resembles 1st-year Saddle-billed Stork but is much smaller and has slightly down-curved and less massive bill. **STATUS AND DISTRIBUTION** Uncommon to rare in suitable habitats throughout, though mainly in savanna belt. Principally migratory, moving south in dry season and returning north during rains. Also nomadic and subject to irregular movements. Resident in some areas. Breeds Aug–May. Vagrant, São Tomé.

AFRICAN OPENBILL STORK *Anastomus lamelligerus*   Plate 8 (2)
Bec-ouvert africain

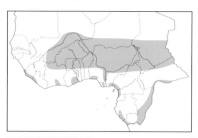

Other name E: Openbill
L 80–94 cm (31.5–37.0"). *A. l. lamelligerus.* Unmistakable. **Adult** *Large, blackish stork with distinctive huge ivory-coloured bill with wide gap between mandibles.* Feathers of neck and upperparts glossed green, bronze and purple. **Juvenile** Dull brown with buff edges to mantle and upperwing-coverts, and shorter, almost straight bill lacking gap. **HABITS** Gregarious, occurring mainly in aquatic freshwater habitats, incl. swamps, marshes, lakes, large rivers, ricefields and floodplains. May disperse when feeding. On migration may occur far from water. Feeds entirely on snails and freshwater mussels. **STATUS AND DISTRIBUTION** Uncommon or rare to locally not uncommon intra-African migrant throughout. Principally dry-season visitor, but appears regular throughout the year at many localities. Breeding records: Nigeria (1) and Sierra Leone (1). Recent decline noted in west of range (Senegal).

BLACK STORK *Ciconia nigra*   Plate 8 (4)
Cigogne noire

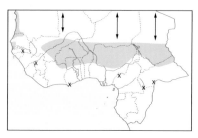

L 95–100 cm (37.5–39.5"); WS 145–155 cm (57–61"). **Adult** *Large, mainly black stork with white breast and belly, and red bill, orbital skin and legs.* **Juvenile** Sooty-brown with bill, orbital skin and legs pale grey-green. Bill and legs become red during 2nd calendar year, when *c.* 10 months old. **HABITS** Solitary or in pairs; occasionally small parties. Usually near water, such as lakes, rivers and marshes, but may occur in dry habitat, incl. grassland, on migration. Will associate with other storks, esp. White Stork and, on wintering grounds, Woolly-necked Stork. [CD1:32] **SIMILAR SPECIES** Abdim's Stork is smaller, with white rump and lower back, yellowish-green bill and legs, and shorter legs projecting less in flight. Immatures more alike but white rump and dusky-greenish legs with pink joints diagnostic of Abdim's. **STATUS AND DISTRIBUTION** Rare to scarce Palearctic migrant, crossing at Gibraltar and following coast to Mauritania, Senegambia and Guinea-Bissau, or crossing Mediterranean and Sahara on broad front to winter in narrow belt of wooded savanna from Mali to C Guinea east to Chad and CAR (most records between 9°30'N and 11°30'N, Oct–Mar).

ABDIM'S STORK *Ciconia abdimii*
Cigogne d'Abdim

Plate 8 (5)

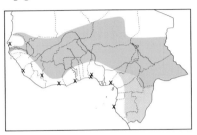

L 75–81 cm (29.5–32.0"). **Adult** Rather *small, largely black stork with white breast, belly, rump and lower back*. Bill greenish-horn with red tip. Bare facial skin grey-blue with red patch in front of eye and under bill. Legs dull green with reddish 'knees' (tarsal joint). During breeding season bare parts become brighter. **Juvenile** Duller, more brownish above and lacks gloss of adult Bill and legs duller and rather darker. **HABITS** Mainly terrestrial, feeding in open grasslands, pastures and cultivated areas. Congregates at swarms of locusts, outbreaks of army worm caterpillars and grass fires. Highly gregarious (flocks on migration often numbering 100s), but spreads out to feed singly. In middle of day flocks often soar to great heights, often in association with other large birds such as Marabou Stork or Pink-backed Pelican. [CD5:11a] **SIMILAR SPECIES** Black Stork lacks white on rump and lower back, and is larger; adult has bright red bill and legs; also less gregarious. **STATUS AND DISTRIBUTION** Widespread intra-African migrant. Generally not uncommon to locally common; rare at limits of range. Occurrence strongly seasonal, breeding in wet season (May–Aug) in broad belt of Sahel and dry savanna, from S Mauritania, Senegal and N Guinea through Mali and N Ivory Coast to Chad and CAR, and moving south at onset of dry season, crossing equator to arrive in southern tropics early in rains. Small wandering groups Dec–May, Gabon.

WOOLLY-NECKED STORK *Ciconia episcopus*
Cigogne épiscopale

Plate 8 (6)

Other name E: White-necked Stork
L 90 cm (*c.* 35"). *C. e. microscelis.* Unmistakable. **Adult** *Large, mainly black stork with white neck, belly and undertail-coverts.* Black on forecrown diffusely extending onto midcrown and head-sides but leaving narrow white forehead. Tail black and slightly forked (often obscured by long undertail-coverts). Black areas strongly glossed bluish and purplish. Bill black with reddish tip and culmen. Legs black. **Juvenile** Duller, more brownish; lacks white forehead. **HABITS** In variety of habitats, incl. wet grassland, usually near water. Mainly solitary, breeding as isolated pairs, but may gather in small flocks (rarely up to a few dozen) prior to migration. [CD5:11b] **SIMILAR SPECIES** No other stork has combination of black body and white neck. **STATUS AND DISTRIBUTION** Uncommon to rare resident and intra-African migrant throughout, north to Senegambia and S Mali to C Chad. Vagrant, S Mauritania. Irregular seasonal movements recorded, north during and after rains.

WHITE STORK *Ciconia ciconia*
Cigogne blanche

Plate 8 (3)

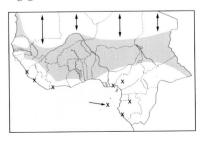

L 100–120 cm (39.5–47.0"); WS 155–165 cm (61–65"). *C. c. ciconia.* **Adult** *Large white stork with black flight feathers and red bill and legs.* **Juvenile** Duller; plumage off-white. Bill blackish with red base and shorter; legs dull red. **HABITS** Mainly in grassland and open savanna where particularly attracted to swarms of locusts and bush fires. Also in wetland areas. Highly gregarious, migrating in large flocks, circling on thermals to gain height before moving on. Roosts in flocks, on ground or in trees, often near water. Spreads out to forage singly in grasslands. [CD1:33] **SIMILAR SPECIES** Yellow-billed Stork has yellow bill, pink in plumage, and black tail. **STATUS AND**

DISTRIBUTION Uncommon to locally not uncommon Palearctic migrant in Sahel belt; rare or scarce further south. Crosses Sahara on broad front. A few, mostly immatures, remain in Africa during northern summer. Vagrant, N Congo and São Tomé.

SADDLE-BILLED STORK *Ephippiorhynchus senegalensis*   Plate 8 (7)
Jabiru d'Afrique

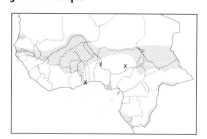

Other names E: Saddlebill; F: Jabiru du Sénégal
L 145–150 cm (57–59"); WS 240–270 cm (95–106"). **Adult** Unmistakable. *Very large black-and-white stork with impressive black and red bill with yellow 'saddle'.* Head, neck, upper and underwing-coverts, scapulars and tail black; rest of plumage white. Legs dark with pinkish-red tarsal joints and toes. In flight, white flight feathers conspicuous. **Male** has brown eyes and 2 small yellow (sometimes red) wattles hanging from base of bill. **Female** has yellow eyes and usually lacks wattles. **Juvenile** Mainly greyish, with dusky bill lacking saddle. **Immature** Duller version of adult; bill duller red and black, with yellow saddle starting to appear. **HABITS** Singly or in pairs in any freshwater or alkaline habitat containing fish, incl. floodplains, large rivers and swamps. Strongly territorial where food supply permanent. Normally shy and wary. [CD5:12a] **SIMILAR SPECIES** Juvenile resembles juvenile Yellow-billed Stork, but is much larger and has heavier bill with slightly upturned lower mandible. **STATUS AND DISTRIBUTION** Rare to uncommon resident. Mainly in Sahel and northern savanna zones, from Senegambia and Guinea-Bissau to Chad, N Cameroon and CAR; also Gabon and Congo. Partially nomadic, moving in response to local conditions.

MARABOU STORK *Leptoptilos crumeniferus*   Plate 8 (8)
Marabout d'Afrique

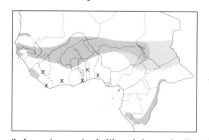

L *c.* 150 cm (59"); WS 230–285 cm (91–112"). Unmistakable. **Adult** *Huge, heavy-bodied* stork with dark slate-grey upperparts, white underparts, *bare pinkish head and neck*, and *massive pale bill*. Has 2 inflatable air-sacs, one long and pendulous hanging below throat, the other smaller and generally hidden between shoulders. Legs black, often heavily stained with white excrement. When breeding, colours of bare parts intensify; upperparts paler bluish-grey; greater wing-coverts broadly edged white. In flight, *uniform dark grey underwings with white axillaries diagnostic.* **Juvenile/immature** First-year is dark brown on upperparts with buff-edged wing-coverts. Second and 3rd year increasingly like adult, gradually gaining paler grey upperparts and more developed markings on bare facial skin. **HABITS** Frequents almost any open habitat, both terrestrial and aquatic. Generally avoids true forest and desert. Often associated with man at fishing villages, slaughterhouses and rubbish dumps. Eats virtually any animal matter. Scavenges at carrion but also catches live food, incl. fish and birds as large as flamingos. Generally gregarious, roosting communally at regular sites. Often sits hunched with legs folded beneath body, sometimes with wings outstretched on ground. Requires thermals to move any distance and frequently soars high to spot carrion. [CD5:12b] **SIMILAR SPECIES** In flight, distinctive underwing pattern and massive size prevent confusion with other storks. Could be confused with distant soaring vultures, but note projecting legs and long bill. Distant soaring Shoebill lacks white belly and axillaries, and has different bill shape. **STATUS AND DISTRIBUTION** Widely but irregularly distributed, mainly in Sahel and northern savanna zones, where rare to locally not uncommon; also S Gabon and Congo. Principally resident; some seasonal movement south during dry season.

SHOEBILL
Family BALAENICIPITIDAE (World: 1 species)

Very large, long-legged wading bird with unique, shoe-shaped bill. Sexes similar; males probably average larger. Sexual maturity probably reached after 3 years or more. Flies with neck retracted. Normally silent. The only member of its family; endemic to Africa.

SHOEBILL *Balaeniceps rex*
Bec-en-sabot du Nil

Plate 9 (3)

L 110–140 cm (43–55"). *Extraordinary, huge, all-grey, stork-like bird with large head and enormous, bulbous bill strongly hooked at tip.* **Adult** Slate-grey with shaggy tuft on rear crown. Bill horn-coloured with irregular dark blotches and streaks. Eye pale yellow to bluish-white. Legs blackish. **Juvenile** Similar, but darker and browner grey; bill smaller and pinkish. **HABITS** Usually solitary. Inhabits extensive freshwater swamps, esp. with papyrus. Shy. Generally found in dense aquatic vegetation, or along channels in swamps, often remaining in one area for long periods. Will rise on thermals to gain height before moving to new areas of swamp. Frequently soars. Flight on take-off laborious, with deep, heavy wingbeats. [CD5:13a] **SIMILAR SPECIES** Unmistakable if seen well. At great distance could be confused with Goliath Heron, but huge swollen bill and lack of any rufous in plumage distinctive. Soaring Marabou Stork has white on belly and axillaries, and differently shaped bill. **STATUS AND DISTRIBUTION** Insufficiently known. Only recorded with certainty as rare non-breeding visitor, N CAR (Manovo-Gounda-St Floris NP). Main range, S Sudan and N Uganda. *TBW* category: Near Threatened.

IBISES AND SPOONBILLS
Family THRESKIORNITHIDAE (World: 32–33 species; Africa: 10–11; WA: 8–9)

Medium-sized to large wading and terrestrial birds with moderately long to long necks and legs; long bill slender and decurved (ibises), or straight with flattened and spoon-shaped tip (spoonbills). Sexes alike or nearly so; males typically larger. Fly with strong wingbeats and neck outstretched; sometimes soar. Gregarious or solitary. Ibises feed in dry habitats, on forest floor or in shallow wet areas by probing in soft mud; spoonbills sweep their bill from side to side in water.

GLOSSY IBIS *Plegadis falcinellus*
Ibis falcinelle

Plate 10 (1)

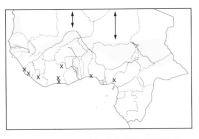

L 55–65 cm (22.0–25.5"); WS 80–95 cm (31.5–37.0"). Small, dark, rather slender ibis. Distinctive flight with rapid wingbeats followed by short glide, legs projecting well beyond tip of tail. **Adult breeding** *Rich purplish-chestnut glossed with green on wings.* Lores pale bluish bordered above and below by narrow white line. **Adult non-breeding** Duller, with variable amount of white streaking on head and neck. **Juvenile** Duller, more sooty-brown, with variable amount of white mottling on head and throat. **Immature** Following post-juvenile moult head and neck streaked with white, but less distinctly than adult. **VOICE** Mostly silent. Occasionally utters a harsh, low *graa-graa-graa* in flight. [CD1:34] **HABITS** Gregarious, occurring in small groups in freshwater habitats incl. marshes, floodplains, inundated areas, and margins of rivers and lakes. Occasionally in coastal lagoons and estuaries. Often associated with herons, egrets and storks. Usually flies in line, sometimes in V formation. **SIMILAR SPECIES** Other dark ibises have less slender appearance with legs not projecting beyond tail in flight. In poor light, could perhaps be mistaken for Eurasian Curlew or Whimbrel, but has longer legs and neck, and upright heron-like walk. **STATUS AND DISTRIBUTION** Not uncommon to rare, locally sometimes common, W Palearctic migrant and resident. Widespread; mainly in Sahel and northern savanna zones, from Mauritania and Senegambia to Chad and N CAR. Breeding recorded in Mauritania and Mali. Vagrant, Cape Verde Is.

HADADA IBIS *Bostrychia hagedash*
Ibis hagedash

Plate 10 (2)

L 76–89 cm (30–35"). *B. h. brevirostris.* Large dark and noisy ibis with rather short legs and dark bill with red on upper mandible. **Adult** Largely dull brown with *white malar stripe* and metallic green gloss on wing-coverts. Flight rather irregular and jerky with one deep wingbeat followed by several shallow flaps; wings broad and rounded; bill held downwards; legs do not project beyond tail. **Juvenile** Similar but duller, lacking green gloss. **VOICE** Vocal. Call a distinctive, far-carrying, nasal *haa! haa-de-dah!* frequently uttered in flight. Also a loud *HAAA!* when

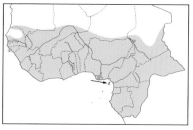

alarmed. [CD5:13b] **HABITS** Along streams and rivers in open woodland, as well as in marshes, moist savanna, mangroves, and at forest edge; occasionally near human habitation. Recorded breeding in rainforest in Cameroon and Gabon; presumably also breeds in Taï NP, Ivory Coast. Usually in pairs, but may flock in non-breeding season. Roosts in trees. **SIMILAR SPECIES** Olive and Spot-breasted Ibises have nuchal crests (though difficult to see in flight) and different, more raucous calls; Spot-breasted is also much smaller, with spotted underparts. Glossy Ibis is more slender, with longer legs projecting beyond tail in flight. **STATUS AND DISTRIBUTION** Not uncommon to uncommon resident throughout, except arid north. Also Bioko.

OLIVE IBIS *Bostrychia olivacea*   Plate 10 (3)
Ibis olive

Other name E: Green Ibis

L 65–75 cm (26.0–29.5"). All-dark forest ibis with *relatively short, coral-red bill*. **Adult** *olivacea* Head dark brown with loose nuchal crest and bare blackish-blue face. Upperparts blackish-brown glossed bronze-green; *wing-coverts metallic coppery and purple-blue*. Underparts dark brown. Legs greenish to dark reddish, not projecting beyond tail in flight. Crest often difficult to see in field, esp. in flight. Races not distinguishable in field; *cupreipennis* has neck and body greener; *rothschildi* has crest glossier purple and more graduated down neck. **Juvenile/immature** As adult but duller, with shorter crest. **VOICE** Call, usually uttered in flight and repeated, a loud, resonant *HAH-hah* or *k-HA-haw* with stress on first syllable. From perch also a single *haaw!* [CD5:14a] **HABITS** Singly, in pairs or small groups along streams and rivers in lowland forest; occasionally also away from water and in mangroves. Shy and rarely seen. Flies above forest at dawn and dusk, uttering call. Feeds in quiet forest glades. Usually silent when flushed from forest floor. **SIMILAR SPECIES** Spot-breasted Ibis best distinguished by call, and, if seen well, by turquoise spots on face, and spotted neck and underparts. Hadada lacks crest, has different call and usually occurs in more open habitats. **STATUS AND DISTRIBUTION** Rare and little-known forest resident, from Sierra Leone to Ghana (nominate), and from Cameroon to Congo (*cupreipennis*). Locally uncommon, Liberia and Gabon. *B. o. rothschildi* endemic to Príncipe, where the only ibis; considered extinct until rediscovered in 1991 (Sargeant 1994).

DWARF OLIVE IBIS *Bostrychia (olivacea) bocagei*   Plate 143 (14)
Ibis de Bocage

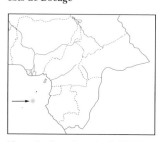

Other name F: Ibis de São Tomé

L *c.* 50 cm (*c.* 21") **Adult** As Olive Ibis (q.v.) but *much smaller*. Dull olive-brown with greenish gloss to wing-coverts. Bill reddish-orange, relatively short and slightly decurved. Legs pinkish. **Juvenile** No information. **VOICE** Call, uttered in flight, described as raucous *kah-gah kah-gah kah-gah* or harsh *karh karh karh...* **HABITS** Singly or in pairs in lowland primary forest. Forages on forest floor. Also along streams, rivers and small oxbows. In areas with steep valley flanks appears to prefer feeding on flatter parts. When flushed, flies (usually silently) to nearby tree and perches on large, horizontal branch at mid-level (3–12 m height), remaining inactive and hidden for a long period if not further disturbed. **SIMILAR SPECIES** None. When perched, small size and dark colour may recall São Tomé Olive Pigeon. **STATUS AND DISTRIBUTION** Rare and little-known forest resident, endemic to São Tomé, where the only ibis. Found in southwest and centre, along R. São Miguel, Xufexufe, Io Grande, Ana Chaves and probably Quija; also in Estação Sousa area. *TBW* category: THREATENED (Critical). **NOTE** Usually considered a race of Olive Ibis, but treated as separate species in *TBW*.

SPOT-BREASTED IBIS *Bostrychia rara*   Plate 10 (4)
Ibis vermiculé

L 47–55 cm (18.5–22.0"). Small, dark forest ibis with *neck and underparts spotted rufous-brown*. **Adult** Head has blackish nuchal crest; *bright turquoise patches in front of and behind eye, and in a line below base of bill*. Upperparts

blackish-brown; wing-coverts metallic green. Crest lax and difficult to see in field, esp. in flight. Bill red. Legs do not project beyond tail in flight. **Juvenile** Duller, with shorter crest and bill, and no turquoise facial markings. **VOICE** Vocal. A fairly loud, raucous and nasal *k-HAH! k-HAH!* or *ah-HAW ah-HAW* and *ah-HAH-hah* with stress on 2nd syllable; uttered in flight and usually frequently repeated. Also a nasal *haw*. [CD5:14b] **HABITS** Singly or in pairs along forest streams and in swamp forest. Usually calls at dusk and during night. **SIMILAR SPECIES** Sympatric but rarer Olive Ibis is noticeably larger, has uniform brown underparts and lacks turquoise facial patches; best distinguished by lower pitched call with stress on 1st syllable. Hadada is much larger, has different call and usually occurs in more open woodland. **STATUS AND DISTRIBUTION** Lowland forest resident, from Liberia to Ghana and from SE Nigeria to SW CAR and Congo. Generally rare or scarce, but locally not uncommon to common in Gabon and Congo.

SACRED IBIS *Threskiornis aethiopicus*   Plate 10 (6)
Ibis sacré

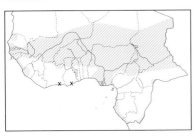

L 65–82 cm (25.5–32.0"); WS *c.* 120 cm (47"). *T. a. aethiopicus.* **Adult** Distinctive, *mainly white* ibis with *bare black head and neck, long, heavy black bill, and long blue-black scapular plumes falling over folded wings and tail.* Black tips to flight feathers visible in flight. Wing edge and flanks often tinged buffish-yellow. Dull grey bare patch at breast-sides extending in line on underwing becomes blood-red when breeding. **Juvenile** Head and neck feathered, blackish with white mottling; throat and foreneck white. Scapulars less developed; primaries more extensively tipped black. **Immature** In subsequent plumages ornamental scapulars develop, black on primaries and greater primary-coverts gradually decreases. Full adult plumage with completely bare head and neck attained in 3–4 years. **HABITS** In variety of habitats incl. margins of rivers and lakes, marshes, floodplains, grasslands, cultivation, rubbish dumps and occasionally estuaries, lagoons and sea coast. Gregarious. Often very tolerant of man. Flocks tend to fly in V formation. [CD5:15a] **SIMILAR SPECIES** Unmistakable. Long curved bill and black tips to flight feathers prevent confusion with white egrets in flight. **STATUS AND DISTRIBUTION** Uncommon to scarce resident. Widespread, mainly in Sahel and savanna zones.

NORTHERN BALD IBIS *Geronticus eremita*   Plate 10 (5)
Ibis chauve

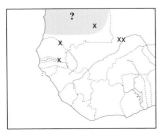

Other name E: Waldrapp
L 70–80 cm (27.5–31.5"); WS 125–135 cm (49–53"). **Adult** Stocky, blackish ibis with shaggy crest, *'bald' reddish head,* and reddish bill and legs. Upperwing-coverts have bronze and violet gloss. Flight with shallow wingbeats interspersed with short glides, legs not projecting beyond tail. **Juvenile** Head feathered and faintly streaked greyish; plumage and bare parts much duller; no elongated feathers on nape. **Immature** In 2nd year nuchal crest and violet gloss on coverts begin to appear, but head still largely feathered. Full adult bare-part coloration probably attained during 3rd year. **HABITS** Prefers dry rocky areas, often along sides of river valleys, also in open fields and coastal lagoons. Flocks fly in V formation. [CD1:35] **SIMILAR SPECIES** Glossy Ibis has more slender appearance, longer neck, feathered head, and longer legs projecting in flight. Hadada lacks bare head and has quite different flight action. **STATUS AND DISTRIBUTION** Very rare Palearctic winter visitor/vagrant, recorded from Mauritania, Senegal, N Mali and Cape Verde Is. Breeds in Morocco in declining numbers (fewer than 100 pairs). *TBW* category: THREATENED (Critical).

EURASIAN SPOONBILL *Platalea leucorodia*   Plate 9 (1)
Spatule blanche

Other name E: White Spoonbill
L 80–90 cm (31.5–35.0"); WS 115–130 cm (45–51"). Large white, long-legged waterbird with *spoon-shaped bill.* Flies with outstretched, slightly sagging neck, and legs projecting well beyond tail. **Adult non-breeding** All white with bare yellow throat patch and black legs. Bill all black (***balsaci***) or tipped yellow (**nominate**). **Adult breeding**

Long, loose nuchal crest and rich yellow throat (often with reddish border). Nominate also has yellow-buff band of variable extent on upper breast. **Juvenile** As adult non-breeding but with black tips to primaries, dull pinkish bill and throat, and dull yellowish-brown legs. **Immature** Bill and legs gradually darken, black on flight feathers decreases, outer primaries retaining some black on tips. Full adult plumage probably attained in 3rd calendar year. **HABITS** Mainly coastal, in shallow water of e.g. lagoons and estuaries. Occasional inland at lakes and marshes. Distinctive feeding action, sweeping bill from side to side while wading. Gregarious. Flocks tend to fly in line or V formation. [CD1:36] **SIMILAR SPECIES** Adult African Spoonbill easily distinguished by bare face and red legs. Juvenile African similar but bill tends to be more yellowish and legs are black. Distinguished in flight from white egrets by extended (not retracted) neck. **STATUS AND DISTRIBUTION** Resident *balsaci* breeds exclusively on Banc d'Arguin, NW Mauritania (1610 pairs in 1984–85) and is locally common on coast; vagrant inland. Nominate is uncommon to rare Palearctic winter visitor from Mauritania to Guinea-Bissau, and patchily in Mali, Burkina Faso, Niger, NE Nigeria and Chad; also Cape Verde Is.

AFRICAN SPOONBILL *Platalea alba* Plate 9 (2)
Spatule d'Afrique

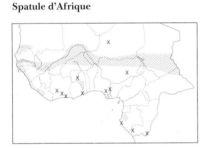

L *c.* 90 cm (35"). **Adult non-breeding** Similar to Eurasian Spoonbill but with *bare red face and red legs*. Bill grey with red edges. **Adult breeding** Loose, fluffy nuchal crest; bare-part coloration intensifies. **Juvenile** Face feathered; primaries and primary-coverts tipped black; bill dull yellowish-horn; legs black. **Immature** Adult features gradually attained. **HABITS** As Eurasian Spoonbill. [CD5:15b] **SIMILAR SPECIES** Adult Eurasian Spoonbill has black legs and bill and lacks bare red facial skin. Juvenile Eurasian similar but bill tends to be more pinkish and legs dull yellowish-brown, not black. **STATUS AND DISTRIBUTION** Widespread but patchily distributed resident. Uncommon to scarce in most areas; mainly vagrant in forest zone.

FLAMINGOS
Family PHOENICOPTERIDAE (World: 5 species; Africa: 2; WA: 2)

Large wading birds with extremely long necks and legs, webbed toes, pink plumage and unique down-curved bills adapted for filter-feeding. Sexes similar; males typically larger. Flight swift and direct with neck outstretched; regular wingbeats sometimes interspersed with short glides. Fly in lines or V formations. Highly gregarious and quite noisy. Occur on shallow brackish, alkaline or saline lakes and lagoons. Feed mainly on microscopic algae and small aquatic invertebrates. Breed irregularly in large colonies on extensive mudflats.

GREATER FLAMINGO *Phoenicopterus ruber* Plate 9 (4)
Flamant rose

Other name F: Grand Flamant
L 127–140 cm (50–55"); WS 140–165 cm (55–65"). *P. r. roseus.* **Adult** Very tall, pinkish-white flamingo with *pale pink bill contrastingly tipped black.* Upper- and underwing-coverts scarlet; flight feathers black. Eye yellow. Legs pink. *Appears largely white at distance,* scarlet wing-coverts only evident in flight. **Juvenile** Greyish-brown with dark grey bill tipped black. Legs grey. **Immature** In subsequent plumages, first becomes whiter, then progressively more pink. Adult plumage attained in 3–4 years. **VOICE** A goose-like *hank-hank* and a nasal *gnaaaa.* [CD1:37] **HABITS** In W Africa confined to coastal habitats such as lagoons and estuaries. Highly gregarious. Wades with head submerged when feeding, or swims and up-ends in

deeper water. Flies in large skeins, usually at night. **SIMILAR SPECIES** Lesser Flamingo is smaller, has more intensely pink plumage, more deeply coloured eye, and mainly dark bill; voice also different. Juvenile Lesser best separated by size, different bill profile and generally darker and more blotchy plumage. **STATUS AND DISTRIBUTION** Locally common in breeding areas (colonies known from Mauritania and N Senegal), uncommon to rare elsewhere (Mauritania to Sierra Leone). Mainly resident, though some dispersal from breeding colonies. Migrants from W Palearctic occur during northern winter. Vagrant, S Niger (Zinder area), N Cameroon and S Congo.

LESSER FLAMINGO *Phoeniconaias minor*   Plate 9 (5)
Flamant nain

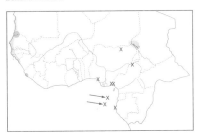

Other name F: Petit Flamant
L 80–90 cm (31.5–35.5"); WS 95–100 cm (37.5–39.0"). **Adult** *Rather small, rose-pink flamingo* with dark red, black-tipped bill (*appearing entirely dark at distance*). Upper- and underwing-coverts blotched deep crimson; flight feathers black. Eye orange-yellow to red. Legs bright red. **Juvenile** Brownish-grey with darker streaking on wing-coverts, breast and mantle. Bill and legs grey. **Immature** Plumage first becomes whiter, then gradually more pink. Adult plumage attained in 3–4 years. **VOICE** A high-pitched *chissik* or *kwirrik*. Feeding flocks utter constant murmuring *murr-err*. [CD5:16a] **HABITS** Normally highly gregarious, but in W Africa also observed singly and in small groups. Principally coastal, frequenting estuaries and lagoons. Feeds mainly at night, either wading in shallow water or swimming in deeper water moving head and neck in rhythmic scything curves. Occasionally associates with Greater Flamingo. **SIMILAR SPECIES** Greater Flamingo is larger, has whiter plumage with contrasting scarlet wing-coverts, conspicuously two-tone pink-and-black bill, and paler eyes. Juvenile stands taller and is paler grey than juvenile Lesser; bill larger with different profile. **STATUS AND DISTRIBUTION** Rare and local. Nomadic. Only known breeding colony in our region in SW Mauritania (bred 1965; breeding suspected, 1998–1999). Mainly coastal, from Senegambia to Sierra Leone; elsewhere reported from S Ghana (1 old record), coastal and N Nigeria, coastal and N Cameroon, SE Niger and W Chad (L. Chad area) and coastal Gabon; also São Tomé and Príncipe. *TBW* category: Near Threatened. **NOTE** Sometimes included in genus *Phoenicopterus*.

DUCKS AND GEESE
Family ANATIDAE (World: 147–157 species; Africa: 53; WA: 33)

Medium-sized to large waterbirds with plump bodies, short, robust legs, webbed feet and flat bills rounded at tip. Wings relatively short and pointed, tail short. Flight fast and direct with neck outstretched. Often gregarious. Generally breed in solitary pairs, near water.

Whistling ducks (*Dendrocygna*) are longer necked and longer legged than other ducks in the region. They are comparatively large and have an upright stance, steep forehead and wedge-shaped bill. In flight their long neck, broad rounded wings and feet projecting beyond the tail, combined with their frequently uttered whistling calls, permit relatively easy identification as members of this subfamily. Sexes similar. No seasonal variation. Juvenile plumage retained for only a few weeks.

Dabbling ducks (*Anas*) feed in water by up-ending or skimming the surface with bill. Rise vertically from the water without any foot-pattering. Upperwing pattern and speculum colour often important identification marks. Sexually dimorphic.

Diving ducks (*Aythya*) feed by diving. Swim under water and run along the surface to take off. Sexually dimorphic.

FULVOUS WHISTLING DUCK *Dendrocygna bicolor*   Plate 11 (2)
Dendrocygne fauve

Other name E: Fulvous Tree Duck
L 45–53 cm (18–21"). Upright-standing brown duck with *bold white streaks on flanks*. In flight, *white U-shaped uppertail-coverts* contrast with blackish wings and tail and fulvous-brown body. **Adult** Head and underparts *fulvous-brown* with indistinct patch on neck-sides finely streaked white, white streaks on flanks and creamy-white undertail-

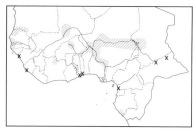

coverts. Upperparts blackish-brown with rusty-buff fringes to feathers forming scaly pattern. Bill and legs grey. **Juvenile** Markedly duller and greyer, with less conspicuous flank streaking and uppertail-covert mark. **VOICE** A clear, slightly nasal whistle *k-wheew* or *ksweeoo*, uttered in flight and usually repeated. [CD5:16b] **HABITS** Gregarious. Frequents well-vegetated freshwater lakes, marshes, ricefields and rivers. Feeds mainly at night. Flight rather slow and laboured with wings making soft muffled sound. Flocks often mix with White-faced Whistling Ducks. **SIMILAR SPECIES** White-faced Whistling Duck shares distinctive shape and stance but has black rump and undertail-coverts and, in adult, white face contrasting with darker plumage. **STATUS AND DISTRIBUTION** Locally common resident and intra-African migrant, from Mauritania, Senegambia and Guinea to Niger, N Cameroon and Chad, and in N CAR and SW Gabon. Important numbers recorded in Senegal delta, central Niger delta, N Nigeria (Hadejia–Nguru wetlands) and N Cameroon (L. Chad and Logone floodplains). Subject to irregular movements.

WHITE-FACED WHISTLING DUCK *Dendrocygna viduata*   Plate 11 (1)

Dendrocygne veuf

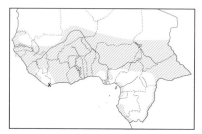

Other name E: White-faced Tree Duck
L 43–48 cm (17–19"). Upright-standing, mainly dark duck. In flight, appears uniformly dark with chestnut forewing. **Adult** Diagnostic *white face and throat* contrast with black rear of head and upper neck. Lower neck, breast and mantle dark chestnut; breast-sides and flanks closely barred black and white; belly and undertail-coverts black. White face may become stained brownish. Bill and legs grey. **Juvenile** Duller, with buffish face and upper neck, blackish crown and hindneck, and chestnut breast-band. **VOICE** Vocal, esp. in flight. A characteristic, clear, sibilant 3-note whistle *swee-swee-sweeoo* or *swee-whee-wheew*, usually repeated. [CD5:17a] **HABITS** Highly gregarious. Occurs in most open freshwater habitats, esp. with some emergent vegetation. Often forages in ricefields. Active at night but also feeds diurnally. Flocks often mix with Fulvous Whistling Ducks. **SIMILAR SPECIES** Fulvous Whistling Duck lacks white face and barred flank patch. Distinguished from juvenile White-faced by lack of black on crown and hindneck, and white (not black) undertail-coverts. In flight, white uppertail distinctive. **STATUS AND DISTRIBUTION** Locally common resident and intra-African migrant throughout. Subject to local movements. Large concentrations recorded in Senegal delta, central Niger delta, SW Niger (along Niger R.), N Nigeria (Hadejia–Nguru wetlands) and N Cameroon.

BLACK-BELLIED WHISTLING DUCK *Dendrocygna autumnalis*   Plate 14 (1)
Dendrocygne à ventre noir

L 48–53 cm (19–21"). Large whistling duck with, in adult, conspicuous *pinkish-red bill*. In flight, mainly dark with long, broad white band on centre of upperwing (diagnostic). **Adult** *autumnalis* Dark brown cap contrasts with pale grey-brown head and neck; conspicuous whitish eye-ring. Upperparts, lower neck and upper breast chestnut; rest of underparts black, mottled white on undertail-coverts. Outer upperwing-coverts and bases of flight feathers white (visible as irregular whitish line when swimming or standing); rest of flight feathers black. Rump, tail and underwing blackish. Legs pinkish-red. *D. a. discolor* has varying amounts of grey on lower breast and upper back. **Juvenile** Duller and browner; bill and legs greyish. **VOICE** Vocal, mostly in flight. A whistled *wee-cheew* or *weechew-weeweewheew...* **HABITS** Normally in freshwater habitats. Mixes with other whistling ducks. **SIMILAR SPECIES** None, but compare other whistling ducks. **STATUS AND DISTRIBUTION** Neotropical vagrant, Gambia (Kotu sewage ponds, Dec 1998; ssp. not specified). First and only record for Africa; possibility of escape cannot be eliminated.

WHITE-BACKED DUCK *Thalassornis leuconotus*   Plate 13 (8)
Dendrocygne à dos blanc

Other names F: Érismature/Canard à dos blanc
L 38–40 cm (15–16"). *T. l. leuconotus.* Small, buff-and-brown barred duck with contrasting white loral patch. Sloping forehead, large-headed and hump-backed appearance, and short tail held flat on water produce distinctive grebe-like shape. In flight, entirely brown and buff except *white lower back* (inconspicuous at distance); feet protrude beyond tip of very short tail. **Adult** Crown and forehead dark brown; head-sides buffish speckled black, with small but conspicuous *white patch at base of bill* extending to chin. Tawny neck collar. Rest of plumage buffish-brown

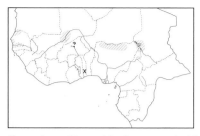

barred dark brown. Bill blackish with yellow-olive sides and prominent hooked nail. Legs and feet greenish-grey. **Juvenile** Similar but duller; facial patch smaller. **VOICE** A clear, sharp, double or 3-note whistle *whit-wee* or *si-wee-wheet*, not unlike call of whistling duck, uttered when swimming and in flight. [CD5:17b] **HABITS** Mainly in pairs or small family groups on vegetation-rich lakes and pools. Most active at dawn and dusk; feeds principally by diving. Spends most of day resting among emergent vegetation when it is easily overlooked. Patters coot-like across surface on take-off, but rarely flies, preferring to swim into cover if disturbed. **SIMILAR SPECIES** Shape, barred brownish plumage and white spot at base of bill distinguish it from all other ducks and grebes. **STATUS AND DISTRIBUTION** Uncommon and local resident, Senegal, Mali, N Nigeria and Chad. Vagrant, Mauritania, Burkina Faso, Togo and N Cameroon. Mainly sedentary. W African population estimated at max. 1000 (Perennou 1992).

BEAN GOOSE *Anser fabalis*   Plate 14 (2)
Oie des moissons

L 66–88 cm (26.0–34.5"); WS 140–175 cm (55–69"). Large, mainly brownish-grey goose with contrasting darker brown head and neck. Vent and uppertail-coverts white. Legs and feet bright orange. *A. f. fabalis* has slender neck and slim, usually long bill with variable amount of orange (almost wholly orange in some). Some have narrow white halter at base of bill. *A. f. rossicus* is slightly smaller and has shorter neck, darker, more contrasting head and neck, and shorter bill with, typically, small orange band near tip (bill always more black than orange). **VOICE** A varied, nasal, quite lively cackle *kayakak* or *kayak*. [CD1:41] **HABITS** Could occur in any open aquatic habitat. **SIMILAR SPECIES** Immature Greater White-fronted Goose (the only similar species recorded from the region) distinguished by pink bill and paler head. **STATUS AND DISTRIBUTION** Palearctic vagrant, Mali (central Niger delta, Jan, and L. Fati, Nov; ssp. unknown). Normal wintering range in W and C Europe. **NOTE** Both forms sometimes considered specifically distinct and named Taiga Bean Goose *A. fabalis* and Tundra Bean Goose *A. serrirostris*.

GREATER WHITE-FRONTED GOOSE *Anser albifrons*   Plate 14 (3)
Oie rieuse

Other name E: White-fronted Goose
L 65–78 cm (26–34"); WS 130–160 (51–63"). *A. a. albifrons*. **Adult** Large, mainly brownish-grey goose with *pink bill, diagnostic white blaze on forehead* and *blackish bars on belly*. Vent and uppertail-coverts white. Legs and feet orange . **Immature** Similar, but without barring on belly; forehead may lack white. **VOICE** A disyllabic *kow-yoo* or *kyo-kyok* suggestive of yelping dogs. **HABITS** Could occur in any open aquatic habitat. **SIMILAR SPECIES** Bean Goose (the only similar species recorded in the region) distinguished from immature White-fronted by dark bill with orange markings and dark wedge-shaped head contrasting with paler body. **STATUS AND DISTRIBUTION** Palearctic vagrant, NW Mauritania (Banc d'Arguin, Nov), Niger (Jan) and N Nigeria (Zaria, dry season; Hadejia–Nguru wetlands, Jan). Normal wintering range in W, C and SE Europe.

BRENT GOOSE *Branta bernicla*   Plate 14 (4)
Bernache cravant

L 55–62 cm (21.5–24.5"); WS 105–120 cm (41–47"). *B. b. bernicla*. Distinctive, small dark goose with *black head, neck and upper breast*, dark grey belly and contrasting white vent and rump. Bill and legs blackish. **Adult** *White flash on neck-sides*. **Juvenile** Similar, but lacks white neck patch and has pale edges to wing-coverts forming bars at rest. **VOICE** A low, guttural *hrot*. [CD1:44] **HABITS** Occurs mainly on mudflats and estuarine waters. Often feeds with other grazing wildfowl, particularly Eurasian Wigeon. **STATUS AND DISTRIBUTION** Palearctic vagrant, NW Mauritania (a single claim, Baie de l'Étoile, Dec) and S Senegal (two photographed, Cap Skirring, Feb). Normally winters on coasts of NW Europe, south to France and occasionally Iberia.

EGYPTIAN GOOSE *Alopochen aegyptiacus*   Plate 11 (8)
Ouette d'Égypte

Other name F: Oie d'Égypte
L 63–73 cm (25–29"); WS 134–154 (53–61"). Unmistakable. Large and bulky with *pink bill* and *long pink legs*. In flight, *white wing-coverts contrast with blackish flight feathers and form diagnostic panels above and below*. **Adult** Head and

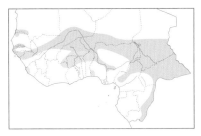

neck pale buffish with distinctive *chocolate-brown eye patch* and narrow dark brown collar on lower neck. Upperparts largely chestnut to grey-brown; rump and tail black. Underparts buffish-grey with chocolate-brown patch on centre of lower breast. **Juvenile** Duller, lacking face and belly markings and collar; head mainly dusky-brown with pale area at base of bill. Bill and legs greyish. **VOICE** A hoarse *taash taash taash...* given in flight. Male utters a harsh, wheezy hiss when displaying. Female gives a harsh, nasal, slightly braying and trumpeting cackle *honk-haah-haah-haah.* [CD1:46] **HABITS** Usually in pairs or small groups in almost any wetland habitat outside forests. Strongly territorial in breeding season. Readily perches in trees. Swims buoyantly with rear end held high, often up-ends but does not usually dive. Moulting flocks may form on larger waters after local breeding has been completed. Flocks disperse at dusk to grazing areas, returning in the dark. **STATUS AND DISTRIBUTION** Locally not uncommon in broad Sahel belt, from S Mauritania to Senegambia and Guinea east to Niger, N Cameroon and Chad, south to CAR, Gabon and Congo. Vagrant, Ghana, Ivory Coast and Benin. Partially migratory over much of range; movements poorly understood.

RUDDY SHELDUCK *Tadorna ferruginea*　　　　　　Plate 14 (5)
Tadorne casarca

L 61–67 cm (24–26"). Large duck with distinctive *plain rusty-orange plumage.* Head paler, orange-buff. Rump and tail black. Bill and legs blackish. In flight, *white wing-coverts contrast with blackish flight feathers* and form conspicuous panels above and below. **Adult male breeding** Narrow blackish neck collar. **Adult male eclipse** Collar indistinct or absent. **Adult female** Similar but lacks collar. **VOICE** Main call a loud honking *aangh.* [CD1:47] **HABITS** Principally on sandy shores, bays and estuaries; also on fresh water near coast and in cultivated fields. **SIMILAR SPECIES** Egyptian Goose has similar white wing panels in flight but is mainly chestnut above and greyish below. **STATUS AND DISTRIBUTION** Palearctic vagrant, claimed from NW Mauritania (Baie de l'Étoile, Mar 1978). Nearest breeding populations in NW Africa (mainly Morocco); these are mostly sedentary, but occasionally undertake short-distance movements.

COMMON SHELDUCK *Tadorna tadorna*　　　　　　Plate 11 (3)
Tadorne de Belon

Other name E: Shelduck
L 58–71 cm (23–28"). Unmistakable. Large, boldly marked, conspicuous duck. **Adult male** Mainly black and white with dark greenish head and neck (appearing black at distance), and chestnut breast-band. Bill bright reddish, with swollen knob on forehead. Legs pink. **Adult female** Slightly smaller; usually has some white markings around eye and base of bill. No knob on bill. **Juvenile** Much duller with brownish-grey upperparts, white throat and no breast-band. Bill and legs greyish. **VOICE** Male utters low whistling calls, female a rapid *ga-ga-ga-ga-ga-ga-gak.* [CD1:48] **HABITS** Mainly on flat sandy shores, bays and estuaries, but also on fresh water near coast and in cultivated fields and grassland. Feeds by swimming in shallow water or dabbling on wet mud. **SIMILAR SPECIES** Unlikely to be confused with any other duck in the region. At long distance could be overlooked as much smaller Pied Avocet, which has similar upperpart pattern. **STATUS AND DISTRIBUTION** Rare Palearctic migrant, Mauritania (Nov–Dec). Vagrant, NW Senegal (Dec–Jan), S Ghana (Jan) and N Niger (Feb). Normal wintering range south to N Africa.

SPUR-WINGED GOOSE *Plectropterus gambensis*　　　　Plate 11 (7)
Oie-armée de Gambie

Other names F: Canard armé, Plectroptère de Gambie
L 75–100 cm (30–39"). *P. g. gambensis.* Huge black-and-white perching duck with long neck and legs. Flight slow and laboured with 'finger-ed' wings producing loud swishing noise. **Adult male** Mainly blackish, glossed iridescent green and purple, with face, shoulder, and lower breast to undertail-coverts white. Bare red skin from base of bill to eye; red frontal knob. Bill and legs pinkish-red. Spur of up to 3 cm projects forward from carpal joint but is normally concealed by body feathers. In flight, has conspicuous *white forewing* above and below. **Adult female** Substantially smaller; white on wing relatively reduced; area of bare facial skin smaller; frontal knob reduced or absent. **Juvenile** Duller and browner; white areas tinged buffish-brown; face entirely feathered. **VOICE** Mainly silent. Male occasionally utters a soft, wheezy *cheweh* or

cherwit. [CD5:18a] **HABITS** Usually gregarious. Frequents flooded areas, marshes, lakes, large rivers and ricefields. Perches in trees. Rather shy and wary. Large flocks gather to moult in dry season. Feeds mainly in early mornings and at night; flocks form staggered lines or V formation when flying to and from roost. **SIMILAR SPECIES** Unlikely to be confused with any other species. White-faced Whistling Duck almost half the size with grey bill. In flight, Egyptian Goose has more extensive white in narrower, unfingered wings and a much shorter neck. **STATUS AND DISTRIBUTION** Uncommon to locally common resident throughout. Marked seasonal movements are poorly understood, but usually related to availability of water. Large concentrations recorded in Senegal delta, central Niger delta, N Nigeria (Hadejia–Nguru wetlands) and N Cameroon.

HARTLAUB'S DUCK *Pteronetta hartlaubii*   Plate 11 (4)
Canard de Hartlaub

Other name F: Ptéronette de Hartlaub
L 56–58 cm (22–23"). Fairly large, mainly dark chestnut-brown duck of forested rivers and streams. Head and neck black. In flight, appears large and bulky with *pale blue wing-coverts forming contrasting panel*. **Adult male** Variable amount of white at base of bill and on forehead. Bill blackish tipped dull pink. **Adult female** Similar, but duller and usually without any white on head. **Juvenile** As adult female but duller still. **VOICE** Mainly silent. Calls include a low, fast *kakakakarrr* uttered in flight, and *whit-whit-whit*. [CD5:18b] **HABITS** In pairs or small groups in well-forested streams and rivers, mangroves and sheltered pools. Occasionally also in more open situations (e.g. reservoirs). Quiet and unobtrusive; most active in evenings and early mornings. Frequently perches in trees; often away from waterside. **SIMILAR SPECIES** Unlikely to be confused with other waterfowl of the region. African Black Duck occurs in similar habitat but is spotted and barred buff above, and has blue-green speculum in flight; it is also noticeably smaller and more local. **STATUS AND DISTRIBUTION** Uncommon to locally not uncommon forest resident, from Guinea to Ghana and from SW Nigeria to SW CAR and Congo. Also SW Mali. *TBW* category: Near Threatened.

KNOB-BILLED DUCK *Sarkidiornis melanotos*   Plate 11 (6)
Canard à bosse

Other names E: Comb Duck; F: Canard casqué
L 56–76 cm (22–30"). *S. m. melanotos.* Unmistakable. Large, bulky duck with mainly *white head, neck and underparts* and iridescent black upperparts. In flight, *white underparts* contrast with *black wings and upperparts*. **Adult male breeding** Bill with fleshy black protuberance ('knob' or 'comb') on upper mandible. Head and neck white speckled black, most densely on crown and hindneck; variable yellowish wash on neck-sides. Upperparts and tail black glossed green and blue, greyish on rump. Underparts white with greyish flanks, black band on vent-sides and yellowish-buff wash to undertail-coverts. Legs greyish-brown. **Adult male non-breeding** Knob reduced; no yellow wash to head. **Adult female** Similar but much smaller, less glossy and without knob; no yellowish wash to head and undertail-coverts. **Juvenile** Dark brown above and buffish-brown below; dark line through eye. Following first moult similar to adult female but duller. **VOICE** Mainly silent. Calls varied; usually uttered in display. [CD5:19a] **HABITS** Frequents a variety of open freshwater habitats, incl. marshes, lakes and rivers. Large flocks occur in dry season. Often perches in dead trees. On land walks with slow swaggering gait. Holds tail clear of water when swimming. Flight powerful, flocks forming lines or V formation. **STATUS AND DISTRIBUTION** Uncommon to locally common resident and intra-African migrant throughout. Marked seasonal movements are poorly understood, but usually related to availability of water. Large concentrations recorded in Senegal delta, central Niger delta, N Nigeria (Hadejia–Nguru wetlands) and N Cameroon. Vagrant, Liberia, S Ivory Coast (Oct–Nov) and São Tomé (Dec).

AFRICAN PYGMY GOOSE *Nettapus auritus*   Plate 11 (5)
Anserelle naine

Other name F: Sarcelle à oreillons
L 30–33 cm (12–13"). *Small,* brightly coloured duck of well-vegetated waters. Mainly dark glossy green above and rusty-orange below. In flight, largely dark with white face and belly; upper- and underwing dark with *conspicuous white patch on inner secondaries*. **Adult male** Head white with black crown and bright metallic green neck-sides bordered by black. Bill bright yellow with black nail. Legs dark grey. **Adult female** Duller. Crown and hindneck

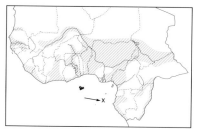

blackish; head-sides and foreneck whitish mottled greyish and has narrow dusky eye-stripe. Bill dull greyish-yellow with dark brown culmen. **Juvenile** Resembles adult female but duller; eye-stripe more prominent. **VOICE** Male utters a soft, rather melodious, whistled *kewheep* or *kewhee-weep* and *khep-khep-kheew*, usually repeated. Female a soft quack. [CD5:19b] **HABITS** Usually in pairs or small groups, frequenting clear freshwater lakes and ponds rich in emergent vegetation, particularly water-lilies. Occasionally on quieter rivers and estuaries. Rarely perches on land, preferring submerged logs or overhanging branches. Spends much time sitting motionless, when easily overlooked among aquatic vegetation. Flies only short distances, usually very low over surface. Feeds by diving and at surface. **SIMILAR SPECIES** Small size and distinctive coloration precludes confusion with other waterfowl. Similar-sized Hottentot Teal lacks white on face, has blue bill and only narrow white trailing edge to inner wing. **STATUS AND DISTRIBUTION** Not uncommon to scarce resident throughout. Vagrant, São Tomé (Jul 1999). Mainly sedentary; nomadic movements in response to changing water levels and habitat availability. Some seasonal movements reported.

EURASIAN WIGEON *Anas penelope* Plate 12 (7)
Canard siffleur

Other name E: Wigeon
L 45–51 cm (18–20"). Medium-sized dabbling duck with rather steep forehead and round head. **Adult male breeding** Head chestnut with creamy-yellow forehead and crown; breast pinkish, upperparts and flanks grey; rear end white and black. Bill rather short, lead-grey with black nail. Legs grey. In flight, has *striking white panel on upper forewing* and *white belly*. **Adult male eclipse** Mainly rich chestnut with contrasting white belly. Retains white wing panel. **Adult female/juvenile** Rather variable. Rusty-brown to brownish-grey with darker smudge around eye. In flight, appears long winged with short pointed tail and relatively small bulbous head; from below has *conspicuous white belly* and grey axillaries. **Immature male** has pale grey secondary coverts and blackish speculum glossed dark green. Some females also have green gloss to dark speculum. **VOICE** Male utters a distinctive melodious, whistling *swheeoooh*. Female has a growling *krrr*, sometimes repeated. [CD1:49] **HABITS** Gregarious, often in very large flocks. Occurs on open shallow freshwater habitats incl. marshes, flooded areas, lakes and ponds; also coastal marshes, lagoons and mudflats. Often feeds by grazing on land. **SIMILAR SPECIES** Vagrant American Wigeon (q.v.). **STATUS AND DISTRIBUTION** Scarce to uncommon Palearctic winter visitor (Nov–Mar), from Mauritania and N Senegal to Chad. Vagrant, N Ghana, SW Nigeria and N Cameroon.

AMERICAN WIGEON *Anas americana* Plate 14 (6)
Canard d'Amérique

Other name F: Canard à front blanc
L 45–56 cm (18–22"). Resembles Eurasian Wigeon in size and structure. In flight, *white axillaries contrast with greyish underwing* (all plumages). **Adult male breeding** Forehead and centre of crown creamy-white; *dark metallic green band from eye to nape* (diagnostic); rest of head speckled greyish-white; body pinkish; rear end white and black. In flight, has prominent white wing panel and white belly. **Adult female/juvenile/adult male eclipse** Similar to Eurasian Wigeon (and perhaps not always safely separable in field), but head, neck and upperparts greyer, more contrasting with rusty-orange breast and flanks. **VOICE** Similar to Eurasian Wigeon but male's whistle more throaty and disyllabic *wheeoh-woh*. [CD5:20a] **HABITS** As Eurasian Wigeon. **SIMILAR SPECIES** Eurasian Wigeon (q.v. and above); in flight, gleaming white (not grey) axillaries best feature. **STATUS AND DISTRIBUTION** N American vagrant, N Senegal (Feb 1975).

GADWALL *Anas strepera* Plate 12 (3)
Canard chipeau

L 46–55 cm (18–22"). *A. s. strepera*. Medium-sized dabbling duck with *diagnostic small white speculum* (conspicuous in flight; also often visible at rest), white belly and white underwing. **Adult male breeding** Mainly greyish, finely vermiculated on breast and flanks with characteristic black rear end and blackish bill. Legs orange-yellow. **Adult male eclipse** As adult female but retains breeding male wing pattern. **Adult female** Mottled brownish with greyer head; bill blackish on culmen with well-defined dull orange sides. White speculum smaller than in male. **Juvenile**

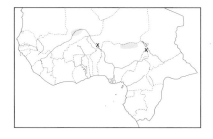

As adult female; white speculum sometimes absent. **VOICE** Mainly silent. Female gives a loud quack, similar to that of female Mallard but slightly higher pitched and harder. [CD1:50] **HABITS** Singly or in small groups, frequenting open freshwater habitats incl. lakes, ponds, marshes and flooded areas. **SIMILAR SPECIES** Female Mallard superficially similar but plumage warmer brown, orange on bill less clear-cut, belly brownish (not white), tail with more white. **STATUS AND DISTRIBUTION** Rare Palearctic winter visitor, N Senegal (Jan–Mar) and N Nigeria. Vagrant, Mali (Nov), SW Niger and N Cameroon.

COMMON TEAL *Anas crecca* Plate 13 (6)
Sarcelle d'hiver

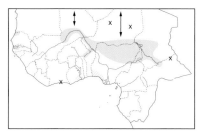

Other names E: Teal, Eurasian Teal
L 34–38 cm (13–15"). *A. c. crecca.* Small dabbling duck, slightly smaller than Garganey. In flight, appears mainly dark and small with narrow pointed wings. Upperwing has short white wingbar bordering dark green speculum, and narrow white trailing edge. **Adult male** Distinctive. *Head chestnut with dark green band from eye to nape narrowly bordered with yellow*, body greyish with white stripe on scapulars; sides of undertail-coverts creamy-yellow bordered with black. Bill grey. Legs dark grey. **Adult female/juvenile/adult male eclipse** Head rather plain with darker crown and indistinct eye-stripe. Body greyish-brown speckled and mottled dark; white stripe at sides of undertail-coverts. Bill brownish-grey, often with dull orange base. **VOICE** Male has a clear, liquid *kreek*, frequently uttered. Female gives a sharp nasal quack and a low growling *trrr.* [CD1:51] **HABITS** In small flocks (mostly of 30–40) in most aquatic habitats incl. sea. Active at night. Often forms tight flocks that twist and turn in flight like waders. **SIMILAR SPECIES** Adult female/ juvenile of similar-sized Garganey best distinguished by more contrasting face pattern with dark eye-stripe and pale loral spot, and longer, stouter, all-grey bill; also by lack of Common Teal's pale mark at sides of tail-base. **STATUS AND DISTRIBUTION** Uncommon or scarce to locally common Palearctic winter visitor (Oct–Mar), from Mauritania and Senegambia to Chad and CAR. Rare, S Ghana and SW Nigeria. Vagrant, S Ivory Coast and Cape Verde Is.

CAPE TEAL *Anas capensis* Plate 13 (1)
Canard du Cap

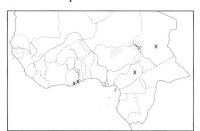

Other name F: Sarcelle du Cap
L 44–48 cm (17–19"). Distinctive, *pale* brownish-grey surface-feeding duck with characteristic *pink bill.* In strong light can appear extremely pale, almost white. In flight, has green speculum with broad white borders at front and rear forming double white wingbars. **Adult** Head pale buffish-grey finely speckled darker. Upperparts brown with buffish feather edges; underparts spotted brown and pale buff. Bill rose-pink with black base and bluish tip. Legs dull yellowish-brown. Eye orange-red to pale brown, sometimes yellow. **Juvenile** Markings on under- and upperparts less distinct; eye dark; black base to bill sometimes lacking. **VOICE** Usually silent. Male has a high-pitched whistled *whee-hee-hew*, female a nasal *rrrhep.* [CD5:20b] **HABITS** In pairs and small groups, frequenting various open aquatic habitats. **SIMILAR SPECIES** Marbled Duck, the only species likely to be confused, has dark bill, dark mask around eye and more prominently buff-spotted upperparts. **STATUS AND DISTRIBUTION** Scarce to uncommon resident and non-breeding visitor, Chad, SE Niger and NE Nigeria (L. Chad area; also westwards, in Kano State). Vagrant, Ghana (Dec, Mar). Nomadic; long-distance movements, presumably in response to changing water levels, recorded.

MALLARD *Anas platyrhynchos* Plate 12 (4)
Canard colvert

L 50–65 cm (20–25"). *A. p. platyrhynchos. Large*, heavily built dabbling duck. In flight, has dark blue speculum prominently bordered with white at front and rear; underwing largely white. **Adult male breeding** Distinctive. *Head dark glossy green*, separated from purplish-brown breast by *narrow white collar*. Rest of body mainly pale grey; rear end black and white with short, upcurled central tail feathers. Bill bright yellow. Legs bright orange. **Adult**

female/juvenile Pale brown, streaked and mottled darker; dark brown crown and line through eye; tail-sides whitish. Bill yellow-orange with variable and irregular dark area on culmen. **Adult male eclipse** Similar to adult female but retains yellow bill; head pattern more contrasting; breast more rufous-brown and only indistinctly marked. **VOICE** Male has a low, nasal *vrreb*. Female gives a loud quack. [CD1:52] **HABITS** Occurs on all aquatic habitats, both fresh and salt. Prefers vegetated pools, lakes and marshes. **SIMILAR SPECIES** Female/juvenile distinguished from other dabbling ducks by combination of large size, upperwing pattern, yellow-orange base and sides of bill, and orange legs. **STATUS AND DISTRIBUTION** Rare Palearctic winter visitor (Sep–Mar), Mauritania. Vagrant, N Senegal, Mali, S Niger (Zinder area) and N Nigeria (Hadejia–Nguru wetlands).

YELLOW-BILLED DUCK *Anas undulata* Plate 12 (2)
Canard à bec jaune

L 51–58 cm (20–23"). Large, mainly greyish-brown dabbling duck with conspicuous yellow bill. In flight, has dark blue speculum narrowly bordered with white at front and rear, white underwing-coverts and dark belly. **Adult** *undulata* Head and neck greyish-black with fine buffish streaking; rest of plumage dark brown with pale fringes giving mottled appearance, esp. on flanks. Bill bright yellow with black stripe on culmen. Legs dull yellowish to reddish-brown. *A. u. rueppelli* somewhat darker and browner; bill deeper yellow. **Juvenile** As adult but pale fringes to feathers broader and more buffy. **VOICE** Male utters low whistles; female a series of quacks. [CD5:21a] **HABITS** In small flocks, on lakes, pools and marshes, occasionally on slow-running rivers. **SIMILAR SPECIES** African Black Duck quite similar in flight, but distinguished by darker body, buff spots on upperparts and longer, barred tail. **STATUS AND DISTRIBUTION** Scarce and local, Cameroon (Adamawa Plateau, Bamenda Highlands); race and status unclear (local breeding population?). Also E Nigeria (Mambilla Plateau).

AFRICAN BLACK DUCK *Anas sparsa* Plate 12 (1)
Canard noirâtre

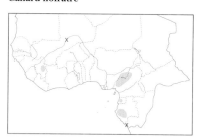

Other name F: Canard noir
L 48–57 cm (19–22"). *Large, blackish* dabbling duck. In flight, has dark blue speculum prominently bordered with white at front and rear, and barred tail; underwing-coverts and axillaries white contrasting with dark flight feathers and underparts. **Adult** *leucostigma* Mainly blackish-brown with pale buffish spots on upperparts and buff bands across rather long tail. Bill dull pink with darker saddle. Legs orange. *A. s. maclatchyi* has pink of bill much reduced. **Juvenile** Duller; buff spots on upperparts much reduced. Bill greyish. **VOICE** Mainly silent. Male utters a very soft, wheezy *wheep*; female a loud quack, sometimes repeated. [CD5:21b] **HABITS** Usually in pairs or family parties on rocky, usually well-wooded streams and rivers in both montane and lowland areas; occasionally on estuaries, lakes and pools. Relatively unobtrusive and easily overlooked. **SIMILAR SPECIES** Unlikely to be confused with any other duck in the region. Yellow-billed Duck is much paler, particularly on underparts, and has bright yellow bill; in flight, lacks barring on tail. **STATUS AND DISTRIBUTION** Rare resident, SE Guinea (race unknown), E Nigeria (Mambilla Plateau) and Cameroon (highlands and Adamawa Plateau; *leucostigma*) and Eq. Guinea and lowland Gabon (*maclatchyi*). Vagrant, Mali (central Niger delta, Nov) and SW Congo. **NOTE** Validity of ssp. *maclatchyi* questionable; this form is not generally recognised.

NORTHERN PINTAIL *Anas acuta* Plate 12 (6)
Canard pilet

Other name E: Pintail
L 51–56 cm (20–26"). *A. a. acuta.* Large, long-necked, *slim* dabbling duck with pointed tail. In flight, has characteristic silhouette with *long neck* and *pointed tail*, broad white trailing edge to speculum and greyish underwing. **Adult male breeding** Distinctive. *Head and hindneck dark brown; stripe on neck-sides*, breast and centre of underparts white. Body vermiculated grey with pale yellowish-buff at rear of flanks and *contrasting black undertail-coverts*. Central tail feathers narrow and extending up to *c.* 10 cm. Speculum dark green glossed with bronze and bordered with pale rufous at front and white at rear. **Adult female/juvenile** Rounded head rather plain, pale brownish (often tinged ginger) with contrasting dark eye and slim, all-grey bill. Body brown with blackish markings. Speculum brownish bordered white. **Adult male eclipse** Similar to adult female, but usually greyer and more finely vermiculated. Dark culmen contrasts with grey bill-sides. Retains green speculum. **VOICE** Male has a clear,

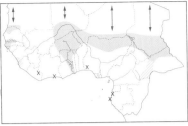

melodious, mellow *krrup*, reminiscent of Common Teal; female has a low quack and Eurasian Wigeon-like growl. [CD1:53] **HABITS** In flocks, frequenting sheltered coastal waters, estuaries and open inland freshwater sites incl. marshes, flooded areas, ricefields, small pools and lakes. Major concentrations of 10,000s or 100,000s reported. **SIMILAR SPECIES** None. Female distinguished from other brownish ducks by combination of size and shape, bland face and all-grey bill; in flight by long neck, pointed tail, white belly and white trailing edge to speculum. **STATUS AND DISTRIBUTION** Palearctic winter visitor (mainly Sep–Feb), from Mauritania to Sierra Leone east to Chad and CAR; common to locally very common south to Senegal, Mali, N Nigeria and Chad, uncommon to scarce further south. Vagrant, Liberia, Ivory Coast, Gabon and Cape Verde Is.

HOTTENTOT TEAL *Anas hottentota* Plate 13 (4)

Sarcelle hottentote

L 30–35 cm (12–14"). The region's smallest dabbling duck. Distinctive combination of *dark crown contrasting with buff head-sides, dusky patch on ear-coverts, and greyish-blue bill*. In flight, has mainly dark upperwing with green speculum bordered by white trailing edge. **Adult male** Upperparts dark brown fringed grey-buff; underparts buffish spotted dark brown. Legs blue-grey. **Adult female** Slightly duller. **Juvenile** As adult female but duller still. **VOICE** A fast, harsh, nasal *kekekekeh*. Male also utters metallic clicking notes in display. [CD5:22b] **HABITS** Mainly in pairs or small groups in shallow freshwater habitats, incl. marshes, lakes and ponds, mainly with floating vegetation or muddy margins. **SIMILAR SPECIES** Common Teal has similar wing pattern but has double white borders to speculum. **STATUS AND DISTRIBUTION** Uncommon resident, S Niger, N Nigeria, W Chad and N Cameroon.

GARGANEY *Anas querquedula* Plate 13 (5)

Sarcelle d'été

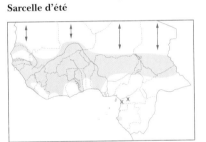

L 37–41 cm (14.5–16"). Small dabbling duck. **Adult male breeding** Distinctive. *Head dark purplish-brown with bold white stripe from above eye curving onto nape*; breast dark brown; flanks pale grey; scapulars pointed, blackish striped white. Bill blackish-grey. Legs grey. In flight, has *pale grey-blue panel* on forewing and green speculum bordered with white at front and rear; dark breast contrasts with whitish belly. **Adult female/juvenile** Mottled dull brown with *pale supercilium bordered by dark crown and eye-stripe*, and *whitish loral spot* and throat usually separated by dusky stripe on ear-coverts (sometimes lacking). Bill grey, rather long. In flight, has dull greyish forewing and white trailing edge (rear border to speculum broader than front bar; reverse of Common Teal). Bill dark grey. **Adult male eclipse** Similar to adult female, but retaining breeding-male wing pattern. **VOICE** Male has a distinctive dry rattling call like scratching fingernail along comb. esp. given in display. Female utters a harsh, nasal *gak* or *kheh* and a soft, high-pitched quack. [CD1:54] **HABITS** In flocks, frequenting mainly open, shallow freshwater habitats incl. marshes, flooded areas and ricefields; on migration also on coastal waters. Large gatherings of 10,000s and 100,000s occur. **SIMILAR SPECIES** Adult female/juvenile of slightly smaller Common Teal best distinguished by less marked, rather plain head, yellow-orange base to bill, and upperwing pattern (darker forewing; front pale wingbar broader than rear one). See also Blue-winged Teal. **STATUS AND DISTRIBUTION** Common to uncommon Palearctic winter visitor almost throughout (mainly Oct–Apr). Major concentrations in Senegal delta, central Niger delta and at L. Chad.

BLUE-WINGED TEAL *Anas discors* Plate 13 (3)

Sarcelle à ailes bleues

Other name F: Sarcelle soucrourou

L 37–41 cm (15–16"). Small vagrant dabbling duck. **Adult male breeding** Distinctive. Dull violet-blue head with *large white crescent in front of eye* and *white patch at sides of black vent*. Underparts warm brown speckled dark. Bill grey, rather heavy and held downwards (like Northern Shoveler). Legs yellowish. In flight, has *bright blue panel on forewing* separated by white wingbar from green speculum, and *no white trailing edge* to secondaries. **Adult female/**

363

juvenile/adult male eclipse Similar to corresponding plumages of Garganey. Head pattern slightly less clearly marked with duller supercilium and no dusky stripe on cheeks, but with *distinct pale loral spot* (usually more prominent than on Garganey) and narrow, broken white eye-ring. In flight, wing pattern duller than in adult male breeding, with white wingbar faint or absent and speculum blacker. **VOICE** Mostly silent. Male has a clear, single or repeated *pying* or *peehp*, like young chicken. **HABITS** Could occur in any freshwater habitat, particularly marshes. **SIMILAR SPECIES** Garganey (q.v. and above); also compare Common Teal. Northern Shoveler (q.v.) has similar wing pattern but differs in other features. **STATUS AND DISTRIBUTION** N American vagrant, N Senegal (Mar).

NORTHERN SHOVELER *Anas clypeata*   Plate 12 (5)
Canard souchet

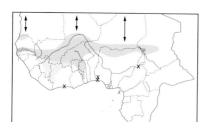

Other name E: Shoveler
L 44–52 cm (17–20"). Medium-sized dabbling duck distinguished in all plumages by *long, broad bill*, producing characteristic front-heavy appearance at rest and in flight. **Adult male breeding** Unmistakable. Head dark green; breast white; *flanks chestnut*; rear end black and white. Bill black. Eye yellow. Legs bright orange. In flight, has conspicuous *blue-grey panel on forewing*, separated by white wingbar from dark green speculum. **Adult female/juvenile** Best distinguished by massive bill. Bill dull brown with orange base and sides. Eye brown, becoming yellow in some females. In flight, wing pattern similar but duller than adult male; underwing-coverts all white, contrasting with dark belly. **Adult male eclipse** Similar to adult female, but head darker; flanks and belly more rufous; brighter breeding-male wing pattern retained. **Adult male sub-eclipse** ('supplementary') Intermediate between eclipse and breeding plumages, with whitish facial crescent. **VOICE** Mostly silent. Male utters a gruff, nasal *took-took*; female has variety of quacks. [CD1:55] **HABITS** In flocks, usually of 20–30, on food-rich shallow wetlands, incl. lakes, marshes, flooded grasslands and ricefields, occasionally on brackish lagoons and tidal mudflats. **SIMILAR SPECIES** No other duck has similar, large, spatulate bill. Vagrant Blue-winged Teal (q.v.) has similar wing pattern. **STATUS AND DISTRIBUTION** Palearctic winter visitor (mainly Oct–Apr). Concentrations of several thousand in Senegal delta, central Niger delta and at L. Chad. Not uncommon to scarce or rare further south.

MARBLED DUCK *Marmaronetta angustirostris*   Plate 13 (2)
Marmaronette marbrée

Other names E: Marbled Teal; F: Sarcelle marbrée
L 39–42 cm (15–16"). Small, slender, *pale* grey-brown duck with *dark eye patch*, rather large head and slim grey bill. Upperparts dark brown with broad buff spots; underparts barred or blotched dark brown; flanks spotted whitish-buff. Legs greyish. In flight, relatively long neck, wings and tail recall female Northern Pintail, but *upperwing lacks speculum or wingbars, and secondaries pale*; underwing whitish. **Adult male** Short hanging crest on back of head. Bill blackish. **Adult female** Crest less distinct; bill duller, usually with indistinct greenish area at base. **Juvenile** Similar to adult female, but spotting duller and more diffuse. **VOICE** Mostly silent. [CD1:56] **HABITS** Gregarious. Frequents shallow and well-vegetated lakes, marshes, ricefields and flooded areas; also on saltpans, lagoons and estuaries. Relatively shy and often remains concealed in emergent vegetation. **SIMILAR SPECIES** No other duck in the region has pale upperwings lacking speculum. Superficially similar Cape Teal has red bill and lacks dark eye patch. White-backed Duck is very different in shape and habits, and has white spot at base of bill. **STATUS AND DISTRIBUTION** Regular Palearctic winter visitor in very small numbers (mainly Dec–Feb), Senegal (mainly Djoudj; also Niokolo-Koba). Vagrant, Mali, NE Nigeria, N Cameroon, Chad and Cape Verde Is. Bred N Senegal (1 pair, 1979). *TBW* category: THREATENED (Vulnerable).

COMMON POCHARD *Aythya ferina*   Plate 13 (7)
Fuligule milouin

Other name E: Pochard
L 42–49 cm (17–19"). Medium-sized diving duck with peaked crown, *sloping forehead and rather long bill* creating distinctive head profile. In flight, has *grey band across flight feathers on upperwing*. **Adult male breeding** Unmistakable combination of *bright chestnut head, black breast and rear end, and pale grey body*. Bill dark grey with pale grey saddle and black tip. Legs grey. Eye orange-yellow to red. **Adult male eclipse** Duller, more female-like, but head plain,

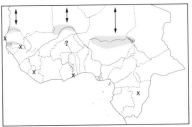

breast darker and eye reddish. **Adult female** Head brownish with diffuse pale loral patch, eye-ring and line behind eye; breast brownish; body greyish-brown. Bill has narrower pale subterminal band. Eye brown or yellowish-brown. In flight, has greyish underwing and belly. **Juvenile** Similar to adult female but browner and usually lacking pale markings on head. **VOICE** Mainly silent. Female may utter a growling *krrrk* when flushed. [CD1:58] **HABITS** Gregarious. On freshwater habitats, incl. lakes, pools, marshes and reservoirs. On coast usually avoids open sea preferring brackish lagoons and estuaries. Principally feeds by diving, sometimes with initial jump. **SIMILAR SPECIES** No other diving duck has similar head profile; in flight, other *Aythya* ducks have white (not grey) wing band. **STATUS AND DISTRIBUTION** Uncommon to rare Palearctic winter visitor (Nov–Mar), Mauritania, Senegambia, Mali and N Nigeria. Vagrant, Liberia, Burkina Faso, Ghana, Niger, Chad, Congo and Cape Verde Is.

FERRUGINOUS DUCK *Aythya nyroca*
Fuligule nyroca

Plate 13 (9)

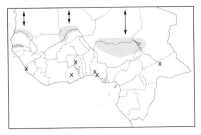

L 38–42 cm (15–16"). Medium-sized, *mainly dark chestnut* diving duck with conspicuous *pure white undertail-coverts*. Bill longish; forehead high; crown peaked. In flight, has *broad white band across flight feathers strongly contrasting with rest of upperwing*; underwing mainly white; well-defined white belly and undertail contrast with dark body. **Adult male** Upperparts blackish. *Eye white* (diagnostic). Bill grey tipped black. Legs grey. **Adult female** Duller, with dark eye. **Juvenile** Duller version of adult female; white of undertail-coverts greyer. **VOICE** Mainly silent. Male utters a soft *wheeoo*. [CD1:59] **HABITS** Singly or in small flocks. Frequents well-vegetated freshwater lakes, pools and marshes; also coastal lagoons. Tends to hide in emergent vegetation more than other diving ducks. Often surface-feeds with head submerged but also dives and up-ends. **SIMILAR SPECIES** Adult female Tufted Duck has different head shape, hint of tuft, yellow eye and white on undertail-coverts often less conspicuous; in flight, white band on upperwing appears less extensive. Juveniles and occasional adult females with dark eyes best distinguished by head and bill shape. **STATUS AND DISTRIBUTION** Uncommon to rare Palearctic winter visitor (Oct–Apr), from Mauritania and Senegambia to Chad. Main wintering areas in Mali and NE Nigeria. Vagrant, Sierra Leone, Ghana, CAR and Cape Verde Is. *TBW* category: Near Threatened.

TUFTED DUCK *Aythya fuligula*
Fuligule morillon

Plate 13 (10)

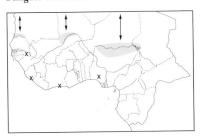

L 40–47 cm (16–18"). Medium-sized diving duck with *drooping crest or short tuft on hindcrown*. Head with steep rounded forehead and flattish crown. Bill rather short and broad, blue-grey tipped black. In flight, has conspicuous white band across upperside of secondaries and inner primaries, and extensive white belly. **Adult male breeding** *Black with striking white flanks and belly and long, drooping crest. Eye yellow.* Legs grey. **Adult male eclipse** Crest short; black parts of plumage blackish-brown; flanks dull brown. **Adult female** Dark brown with slight indication of crest at rear of crown; flanks paler brown. Often has variable amount of white on lores or around base of bill, and whitish undertail-coverts. **Juvenile** Similar to adult female but with brownish-yellow eye. **VOICE** Mainly silent. Female has a growling *krrr*. [CD1:60] **HABITS** Usually in small groups, on any open aquatic habitat incl. sheltered coasts, though favours shallow lakes with vegetated shores. Feeds mainly by diving, occasionally also by up-ending and surface-feeding. **SIMILAR SPECIES** Distinctive head shape and yellow eye distinguish it from other diving ducks known from the region. Vagrant Lesser Scaup also has yellow eye but its crown is peaked. In flight, Ferruginous Duck distinguished by more extensive band on upperwing (reaching outer primaries), and better defined white belly patch. Common Pochard has grey (not white) wingbar. **STATUS AND DISTRIBUTION** Scarce to locally uncommon Palearctic winter visitor (Nov–Mar), from Mauritania and Senegal to N Nigeria and Chad. Rare or vagrant, Gambia, Sierra Leone, S Ivory Coast, S Nigeria, Cameroon and Cape Verde Is.

LESSER SCAUP *Aythya affinis*
Petit Fuligule

Plate 14 (7)

L 38–45 cm (15–18"). Medium-sized diving duck with peaked hindcrown. In flight, has conspicuous white band on upperside of secondaries; inner primaries pale grey; belly white; underwing whitish. **Adult male breeding** Head, breast and rear end black, contrasting with grey upperparts and white underparts. *Upperparts, and often also rear flanks, finely vermiculated dark grey*. Head glossed purple. Bill blue-grey with black nail. Legs grey. Eye yellow. **Adult male eclipse** Dark brown with upperparts vermiculated grey; flanks vermiculated grey and pale brownish. Sometimes a little white around base of bill. **Adult female** Dull brown with white ring at base of bill. Head-sides may possess indistinct pale patch with wear. Belly whitish. **Juvenile** Similar to adult female but with less white at base of bill and brownish-yellow eye. **VOICE** Mainly silent. **HABITS** May occur in any open aquatic habitat; prefers fresh water, but also frequents coastal lagoons and estuaries. **SIMILAR SPECIES** Distinctive head shape with peaked crown distinguish it from other diving ducks. Female Tufted Duck normally has less white at base of bill, and broader black bill tip. **STATUS AND DISTRIBUTION** N American vagrant, Cape Verde Is (São Vicente, Jan–Feb 1999).

COMMON SCOTER *Melanitta nigra*
Macreuse noire

Plate 14 (8)

L 44–54 cm (17–21"). *M. n. nigra*. Medium-sized, compact blackish sea duck. In flight, appears uniformly dark with paler flight feathers. **Adult male** *Entirely black with yellow patch on knobbed black bill*. **Adult female/juvenile** *Dark sooty-brown with characteristic pale cheeks and foreneck*, visible at long range. Bill greyish. **HABITS** On open sea and estuaries. Feeds by diving, usually with preliminary jump forward. [CD1:62] **SIMILAR SPECIES** None; the only dark marine duck recorded from the region. **STATUS AND DISTRIBUTION** Rare or scarce Palearctic winter visitor, Mauritania (mainly Nov–Dec).

OSPREY
Family PANDIONIDAE (World: 1 species)

A comparatively large raptor, exclusively preying on fish caught by plunge-diving. Feet specially adapted to catch slippery prey: soles spiny and hind toe reversible. Sexes similar. Taxonomic position a matter of debate.

OSPREY *Pandion haliaetus*
Balbuzard pêcheur

Plate 15 (2)

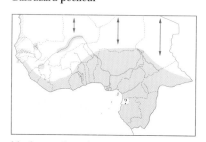

L 52–61 cm (20.5–24.0"); WS 145–173 cm (57–68"). *P. h. haliaetus*. Large, long-winged raptor, with mainly dark upperparts and white underparts. Wings long and narrow, with long 'hand' and only 4 'fingers'. Tail relatively short and square. Flies with shallow, powerful wingbeats interspersed with long glides. Soars and glides on arched wings held angled at carpal joint. **Adult male** Head white with short nuchal crest and broad dark stripe through eye. Some brown streaking on forehead. Upperparts dark brown. Underparts white with pale brown streaks forming indistinct band on upper breast. Bill blue. Eye yellow. Cere and legs blue-grey. In flight, from below, black carpal patches contrast with white underwing-coverts; greater coverts blackish, forming dark line across centre of wing; flight and tail feathers indistinctly barred blackish. **Adult female** Similar, but usually with more prominent breast-band. **Juvenile** Similar, but feathers of upperparts narrowly fringed buffish. Greater coverts white barred dark; flight feathers coarsely barred dark. Eye orange. Adult plumage acquired after *c.* 18 months. **VOICE** Usually silent in Africa. [CD1:69] **HABITS** Usually singly, near water. Perches conspicuously on prominent vantage point at large rivers, lagoons, lakes and reservoirs. Hovers and dives for fish, which is caught in talons; rises shaking water from plumage. **SIMILAR SPECIES** Immature African Fish Eagle occurs in similar habitat and may superficially resemble Osprey, but lacks black band through eye and has dark mottling on lower underparts, short tail and very different flight silhouette. **STATUS AND DISTRIBUTION** Uncommon to locally common Palearctic winter visitor throughout. Immatures frequently remain during rains. Not uncommon resident, Cape Verde Is (all islands).

VULTURES, EAGLES, HAWKS AND ALLIES
Family ACCIPITRIDAE (World: *c.* 235 species; Africa: 76; WA: 62)

A very large and diverse-looking family consisting of most species of diurnal birds of prey. Its members characteristically have hooked bills with yellow cere at base, and powerful talons. Exhibit a wide variety of wing shapes, tail shapes and flight actions, according to habitat and hunting techniques. Sexes more or less alike; females almost always larger. Juveniles often different from adults. Adult plumage attained during 2nd calendar year by small and some medium-sized species (e.g. *Circus* and *Accipiter*); larger species take several years and a series of (often confusing) immature plumages. Normal flight consists of regular wingbeats alternated with gliding. Almost all species carnivorous. Construct own nest, mostly in trees (some on crags or on ground); occasionally adopt nest of others. Conspicuous aerial displays used by many.

Identification often problematic; correct assessment of jizz (combination of structure, proportions, wing attitudes, and flight actions) essential. Main groups occurring in the region include the following.

Vultures (8 species). Medium-sized to huge with long, broad, strongly 'fingered' wings, and largely unfeathered head and neck. The larger species occurring in the region have a long neck with a ruff of feathers at its base, and a relatively short tail. Sexes similar in size, or male larger. Carrion-eaters. Given to soaring, often at great height. In flight, head is tucked between shoulders and appears relatively small.

Harriers (*Circus*, 5 species). Medium-sized and slender with long wings and tails. Sexually dimorphic; females mainly brownish, some presenting identification problems. Flight buoyant with wings characteristically held in a shallow V. Hunt low, quartering the ground and dropping onto prey (frogs and small rodents, occasionally birds). Frequent open country, cultivated areas and marshes. Roost on ground.

Kites (4 species). Medium-sized raptors with graceful flight and, in most species, a forked tail. Prey consists of various small animals; Black Kite also takes carrion.

Snake eagles (*Circaetus*, 5 species). Rather large and eagle-like, with relatively large heads and large yellow eyes (producing rather owl-like appearance), long as well as broad wings, bare tarsi and short toes. Given to soaring and perching for long periods. Drop onto prey, which largely consists of snakes and other reptiles, and kill it on the ground with their powerful feet. Occur in open country and woodland.

Sparrowhawks and goshawks (*Accipiter*, 9 species). Small to medium-sized, with rather short, rounded wings and long tails. Tarsi bare and long. Great size difference between sexes; females much larger. Swift and agile. Hunt by dashing on their prey (mainly birds and small mammals) after a stealthy approach behind cover. Inhabit forest and woodland.

Buzzards (*Buteo*, 3 species). Medium-sized, with broad-winged, eagle-like shape, but generally smaller and with only moderately 'fingered' wings. Tarsi bare. Plumages very variable, sometimes creating identification problems. Mostly seen soaring, or perching on vantage points such as trees and posts. When soaring, wings often held in a shallow V. Prey mainly upon mammals and reptiles caught on ground. Occur in woodland and open country.

Eagles (17 species). Medium-sized to very large with broad 'fingered' wings. Tarsi feathered. Given to soaring. Prey on large and small mammals, birds and reptiles; some take carrion and insects. In general fiercer and more active than buzzards. Occur in open country, woodland and forest.

AFRICAN CUCKOO HAWK *Aviceda cuculoides*   Plate 19 (3)
Baza coucou

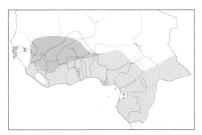

Other names E: African Cuckoo Falcon; F: Faucon-coucou
L 40 cm (16"); WS 91 cm (36"). *A. c. cuculoides.* Plumage pattern and flight silhouette with long wings and longish square tail superficially remini scent of large cuckoo. When perched, wings almost reach tip of tail. **Adult male** Head, upperparts and breast dark grey; rest of underparts white broadly barred chestnut; tail grey with 3 broad black bars. Small chestnut patch below short occipital crest. In flight from below, has conspicuous *chestnut underwing-coverts*; flight feathers whitish with long, narrow, black parallel bars. Eye, cere and legs yellow. **Adult female** Larger and browner, with barring on underparts broader and paler. **Juvenile** Head and upperparts dark brown with narrow buff edges to feathers; short whitish supercilium; underparts white variably marked with brown drop-like spots (occasionally only some coarse barring on flanks); tail brown with 4 broad black bars. Underwing-coverts pale buff speckled dark brown; remiges barred like adult Eye grey; cere and legs yellow. **VOICE** Usually silent. A loud, plaintive *peee-uuw* of variable length; also a sharp *wheet-wheet*. [CD5:25] **HABITS** Frequents woodland and forest clearings. Hunts from perch, catching prey (insects, lizards) on ground or in fast, agile flight. Unobtrusive; often remaining in dense cover. Pairs perform undulating and tumbling display flights, with much calling. **SIMILAR**

SPECIES Adult African Goshawk also has chestnut on underwing-coverts, but has short, rounded wings. This also serves to distinguish juveniles of the two species. **STATUS AND DISTRIBUTION** Uncommon to not uncommon resident in forest and southern savanna belts throughout. Some local movements with rains in savanna zone. **NOTE** Populations in our region here considered to belong to nominate race; ssp. *batesi*, often used for those from the forest zone, is considered a synonym, following e.g. Louette (1981).

EUROPEAN HONEY BUZZARD *Pernis apivorus*   Plate 22 (1)
Bondrée apivore

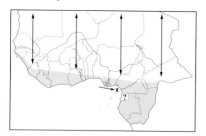

Other name E: Honey Buzzard

L 51–60 cm (20–24"); WS 113–150 cm (44–59"). Medium-sized raptor with highly variable plumage (from blackish through various shades of brown to whitish below), but with characteristic *small-looking head and slender neck, protruding well in flight.* Tail rather long, with rounded corners. Bill small. Soars on flat wings, slightly bowed when gliding. **Adult male** Typically has blue-grey head, brown upperparts, and white underparts with bold, irregular dark bars. In all plumages, *diagnostic tail pattern with dark terminal band and 2 narrower bars near base.* Underwing has dark carpal patches, *prominent broad black trailing edge* and, typically, parallel bands across base of pale grey flight feathers. Primaries with sharply demarcated, small black tips. Cere grey; eye yellow. **Adult female** Usually browner on head and upperparts than male. Tail bands narrower; 2 bars near base wider spaced. Flight feathers usually with more, and more evenly spread, bars. Primaries more diffusely and extensively tipped dusky. **Juvenile** Less distinctive and variable than adult; most appearing dark brown. Flight and tail feathers with 4–5 more evenly spread bars. Secondaries and tail usually darkish; tail often slightly notched. Primaries extensively tipped dusky. Uppertail-coverts often showing whitish U. Cere yellow; eye dark. **VOICE** Usually silent in Africa. [CD1:70] **HABITS** Singly, frequenting forest and wooded savanna. Flies with deep, flexible wingbeats. Principally insectivorous. **SIMILAR SPECIES** Common Buzzard appears more compact, with broad head and neck, shorter tail with sharp corners; soars on raised wings, flies with shallower, stiffer wingbeats. Most dark Common Buzzards differ from dark Honey Buzzard by diffuse pale breast-band (always lacking in latter). Similar-sized eagles never possess contrasting dark carpal patch; adult Ayres's Hawk Eagle moreover has heavily spotted underparts, more evenly barred tail and shorter wings. **STATUS AND DISTRIBUTION** Not uncommon to scarce Palearctic passage migrant and winter visitor throughout. Vagrant, Bioko (Jan, Jun). Some immatures remain during rainy season.

BAT HAWK *Macheiramphus alcinus*   Plate 19 (4)
Milan des chauves-souris

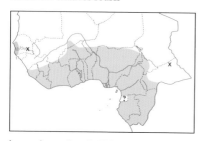

Other name F: Faucon des chauves-souris

L 45 cm (18"); WS 110 cm (43"). *M. a. anderssoni.* Characteristic shape with long, broad and pointed wings and rather long, square tail reminiscent of large dark falcon. **Adult** Mainly blackish-brown with dark stripe on centre of white throat and short occipital crest. Lower underparts occasionally have some white; throat sometimes dark. Small white streaks above and below eye ('eyelids'); white feather-bases may produce 2 small white spots on nape. Eye large, yellow. Cere and legs greyish. **Juvenile** Similar, but underparts with variable amount of white (often mainly white with irregular dark brown breast-band). **VOICE** Calls include a rather hoarse, high-pitched *kwheet-kwheet-wheet-wheet-...* and *kwik-kwikik-kwik-kwikik-...* [CD5:26] **HABITS** Crepuscular; roosts in trees during day and emerges to hunt over open areas at dusk. Catches prey (mainly small bats, also birds) in feet and swallows it whole in flight. Frequents forest edge and wooded savanna; often near rivers. Flight with slow wingbeats, but rapid and graceful when hunting. **SIMILAR SPECIES** Falcons are more slender and noticeably smaller, with narrower wings. **STATUS AND DISTRIBUTION** Uncommon and very local resident in forest and savanna zones throughout.

BLACK-SHOULDERED KITE *Elanus caeruleus*   Plate 19 (1)
Élanion blanc

Other name E: Black-winged Kite

L 31–35 cm (12–14"); WS 75–85 cm (30–33"). *E. c. caeruleus.* Smallish, *very pale* raptor with relatively broad head and long wings. **Adult** Head and upperparts pale blue-grey with small dark eye patch and black 'shoulders' (upperwing-coverts); underparts white. When perched, wings project beyond shortish, square tail. *In flight all-*

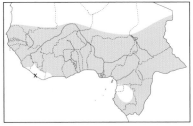

white below except contrasting black primaries; wings long but broad-based. Eye bright red. Cere and legs yellow. **Juvenile/immature** Similar, but washed brown above; feathers edged white; breast washed rufous. Eye initially brownish, then yellowish. **VOICE** Mostly silent. Weak, high-pitched whistling calls and short notes, mostly uttered when breeding. [CD1:71] **HABITS** In various open habitats, except desert. Glides on angled wings raised high in V; hovers frequently and persistently, often with trailing feet. Frequently raises tail when perched. Catches prey (mainly small rodents) by dropping from prominent perch or from hover. **SIMILAR SPECIES** Flight action of distant birds may recall adult male Pallid or Montagu's Harriers, but these are larger with longer wings and tail, and do not hover. **STATUS AND DISTRIBUTION** Common to uncommon resident throughout, except extreme north.

AFRICAN SWALLOW-TAILED KITE *Chelictinia riocourii*   Plate 19 (2)
Élanion naucler

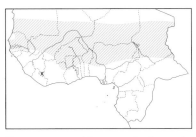

L 35–38 cm (14–15"), incl. tail of up to *c.* 22 cm (9"); WS *c.* 90 cm (35.5"). **Adult** Unmistakable. Small, very slender and graceful raptor with *deeply forked tail*. Top of head and upperparts pale grey; underparts pure white with narrow black patch on coverts, along edge of underwing; tail pale grey and white, projecting well beyond wingtips when perched. Eye bright red. Cere grey. Legs yellow. **Juvenile** Feathers of upperparts edged rufous; pale rufous wash on breast; tail short. **VOICE** Rapid rasping and whistling calls. Usually silent except when breeding. [CD5:27] **HABITS** Gregarious, often in small flocks, in dry open country; occasionally singly or in pairs. Hovers, soars and hangs motionless against wind with spread tail. Catches prey (mainly insects; also lizards and, rarely, small rodents) by dropping from hover; insects also taken in flight. Attracted to bush fires. **STATUS AND DISTRIBUTION** Uncommon to locally common in semi-arid belt, from Mauritania and Senegal through Mali and Burkina Faso to Niger, N Nigeria and Chad; scarce or rare and irregular further south; nomadic and migratory, moving south at start of dry season, returning to breed in Sahel zone during rains.

RED KITE *Milvus milvus*   Plate 17 (2)
Milan royal

L 55–72 cm (22–28"); WS 140–180 (55–71"). Distinctive. Medium-sized, elegant raptor with long, narrow wings and long, deeply forked tail. Flight action as Black Kite, but even more graceful. **Adult** *milvus* Mainly *rufous* narrowly streaked black, with pale greyish head. Wings have conspicuous pale diagonal band across coverts above and strongly contrasting, large whitish 'window' on primaries below. *Tail rusty-red* above, paler below. Bill yellow tipped black. Eye, cere and legs yellow. Island race *fasciicauda* has tail less deeply forked and more barred; inner webs of primaries with darker grey marbling (not white). **Juvenile** Similar to adult, but underparts paler; upperwing-coverts narrowly tipped white; tail browner. **VOICE** Generally silent. Calls shrill and mewing, incl. a drawn-out note followed by a fast rising and falling series *wheeeeuuuh-wheeoweeoweeoweeoweeh* and *peeoweeoh*, often also extended into a series. [CD1:73] **HABITS** Little known in Africa. Form *fasciicauda* breeds on cliffs and rock ledges (Hazevoet 1995). **SIMILAR SPECIES** Black Kite is much darker, duller and less elegant, with only shallowly forked tail. **STATUS AND DISTRIBUTION** Nominate is Palearctic vagrant recorded from Mauritania (Sep, Dec) and Gambia (Dec, Feb, May); also claimed from SW Niger (Jun). Race *fasciicauda* very rare resident, Cape Verde Is (Santo Antão; probably extinct on other islands. Total population estimated at 4–6 (8?) birds in 1996–97; only 2 found in 1999 (Hille & Thiollay 2000)). **NOTE** Poorly known form *fasciicauda* exhibits much individual variation and may be hybrid of Red x Black Kite, or merit specific status as Cape Verde Kite.

BLACK KITE *Milvus migrans*   Plate 17 (1)
Milan noir

L 50–60 cm (19.5–23.5"); WS 130–155 cm (51–61"). Medium-sized, slender, brown raptor with long, *slightly forked tail*. The commonest raptor in our region. Flight buoyant with long, narrow wings angled back at carpal joint. Soars and glides with wings slightly arched and tail constantly twisting. Tail may appear square when spread. **Adult** *parasitus* (Yellow-billed Kite) Entirely chocolate-brown with yellow bill. Some narrow black streaking on head and upperparts. Flight feathers and tail indistinctly barred below. Cere and legs yellow. Eye dark. Palearctic *migrans* has *pale greyish head* narrowly streaked black and *black bill*; wing more contrasting, with broad pale diagonal

369

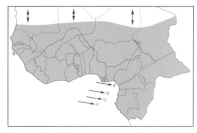

bar across coverts above and pale area at base of primaries below. Eye colour variable, from brown to pale yellowish. **Juvenile** *parasitus* Similar to adult but rather paler below; bill black. Feathers of upperparts have pale tips and underparts some pale streaking. Distinguished from adult nominate by lack of pale greyish on head. Juvenile *migrans* very similar but more streaked on head and underparts. **VOICE** Distinctive. A plaintive, tremulous *keeeey-aarrrr.* [CD1:72; 5:28] **HABITS** Ubiquitous and gregarious. Omnivorous; takes offal, carrion, termites and fish. Scavenger; often searches for road kills. Attracted to bush fires. Swoops onto prey. **SIMILAR SPECIES** Female/immature Eurasian Marsh Harrier have square tails and typically soar and glide with wings in shallow V and less tail twisting. **STATUS AND DISTRIBUTION** Common intra-African migrant throughout, incl. Gulf of Guinea Is (*parasitus*) and uncommon Palearctic migrant (nominate). Rare resident, Cape Verde Is (nominate). **NOTE** *M. m. parasitus* sometimes treated as separate species.

AFRICAN FISH EAGLE *Haliaeetus vocifer*   Plate 15 (1)
Pygargue vocifer

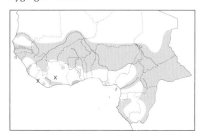

Other name F: Aigle pêcheur
L 63–73 cm (25–29"); WS 190–240 cm (75.0–94.5"). Large eagle associated with water. Soars on broad flat wings. **Adult** *White head, mantle and breast contrast with black of upperparts and chestnut of 'shoulders' and lower underparts.* Short white tail often hidden by wings when perched. Bill greyish or black with yellow base; cere and facial skin yellow. Eye brown. Legs dull yellow. In flight from below, white head and upper breast and white tail contrast with chestnut underwing-coverts and lower underparts and black flight feathers. **Juvenile** Drab mottled brown above; breast whitish variably streaked brown; rest of underparts heavily mottled brown and white, with darker belly. Tail whitish broadly tipped blackish. In flight from below, underwing-coverts contrast with blackish flight feathers; primaries with pale window; *tail whitish with black terminal band* (diagnostic). **Immature** Following first moult, paler than juvenile, with head and underparts mainly whitish streaked brown; often has dark superciliary streak. Terminal dark band on tail narrower. In 3rd–4th year, gradually acquires more adult features: head, mantle and breast white, initially with some black streaking; tail gradually whitens and loses terminal band; chestnut appears on shoulders, belly and underwing. Cere pinkish. Full adult plumage attained in min. 5 years. **VOICE** Vocal. A highly distinctive, far-carrying, ringing *WEEE-AH kleeuw kleeuw kleew,* uttered by both sexes at rest and in flight, with head thrown back; male higher pitched than female. Often duets. [CD5:29] **HABITS** Singly or in pairs near water, incl. large rivers, lagoons, mangrove creeks, lakes and reservoirs. Feeds mainly on fish, which is caught with talons. Also eats carrion and occasionally waterbirds. **SIMILAR SPECIES** Adult Palm-nut Vulture has wing-coverts and lower underparts white; also smaller. Osprey occurs in similar habitat and might superficially resemble immature Fish Eagle, but has black band through eye and longer tail; flight silhouette very different. Immature Palm-nut Vulture has bare yellowish face and is less streaky; also smaller. **STATUS AND DISTRIBUTION** Uncommon or rare to locally common resident in suitable habitat throughout, but mainly in savanna zone. Vagrant, Bioko.

PALM-NUT VULTURE *Gypohierax angolensis*   Plate 15 (3)
Palmiste africain

Other names E: Vulturine Fish Eagle; F: Vautour palmiste
L *c.* 60 cm (24"); WS 140–150 cm (55–59"). Chunky, with broad, rounded wings and very short, rounded tail. **Adult** Readily identified by *black-and-white plumage and bare reddish-pink face.* Bill heavy, yellowish-horn; cere pale bluish. Eye yellow. Legs pinkish. In flight, mainly white with black secondaries and black tail broadly tipped white. Primaries tipped black. **Juvenile** Entirely drab brown with darkish bill, dull yellow face and cere. Eye brown. Legs whitish. In flight, dark brown with blackish-brown flight and tail feathers. **Immature** Brown gradually replaced by white, producing mottled appearance. Adult plumage acquired in 3–4 years. **VOICE** Usually silent. Calls include various low growling and barking notes. [CD5:30] **HABITS** Singly, often near water. Relishes nuts of oil and raphia palms, but also feeds on various small animals, incl. fish, amphibians and crabs, occasionally mammals, birds and large insects; also scavenges. Roosts and nests in trees. **SIMILAR SPECIES** Adult African Fish Eagle has chestnut wing-coverts and

lower underparts; also larger. Adult Egyptian Vulture has slender bill, yellow face and long feathers at back of head; in flight, distinguished by different silhouette with wedge-shaped, entirely white tail and all-black (not white) primaries. Hooded Vulture and juvenile Egyptian Vulture superficially resemble juvenile Palm-nut but have slender bills; Hooded has bare face and throat; Egyptian long feathers on back of head and, in flight, characteristic wedge- or diamond-shaped tail. Juvenile Harrier Hawk is also brown but lacks bare face and has more slender build with longer legs and tail, and different shape in flight. Immature African Fish Eagle streakier and lacks bare yellowish face; also larger. **STATUS AND DISTRIBUTION** Not uncommon to locally common resident in forest and wooded savanna belt throughout. Also Bioko. **NOTE** Exact affinities of this species still debated; may be more related to fish eagles or snake eagles than to vultures (as reflected by alternative name Vulturine Fish Eagle).

EGYPTIAN VULTURE *Neophron percnopterus*   Plate 16 (2)
Vautour percnoptère

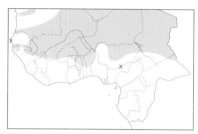

L 55–75 cm (22–29"); WS 155–175 cm (61–69"). *N. p. percnopterus.* Medium-sized. In all plumages, distinctive flight silhouette with *wedge-shaped tail* and small, narrow head. Soars with wings held flat; when gliding, wings slightly arched. **Adult** *Mainly white with bare yellow face, long, narrow bill and long, lanceolate feathers at back of head.* Head, neck, mantle and breast variably washed yellowish or greyish. Face may flush reddish in excitement. In flight from below, black flight feathers contrast with white underwing-coverts. Bill yellow tipped black. Eye deep red. Legs yellow. **Juvenile** Dark brown with broad buffish tips to feathers. Pale tips lost with wear. Bare face and legs greyish; eye brown. **Immature** Gradually acquires white feathers. Adult plumage attained in *c.* 5 years. **VOICE** Usually silent. [CD1:76] **HABITS** Usually singly, in desert, sub-desert and arid savanna. Feeds principally on carrion and offal. Nests on rocks and cliffs. **SIMILAR SPECIES** Adult Palm-nut Vulture superficially resembles adult Egyptian but is more chunky, with heavy bill, no long feathers at back of head, white (not black) primaries and short, rounded black tail tipped white. Juvenile Hooded Vulture has bare neck, broader wings and rounded tail. Juvenile Palm-nut Vulture has heavy bill, feathered head and short, rounded tail. **STATUS AND DISTRIBUTION** Uncommon resident and Palearctic winter visitor in Sahel belt, from Mauritania and Senegambia to Chad and N Cameroon. Not uncommon resident, Cape Verde Is (all islands).

HOODED VULTURE *Necrosyrtes monachus*   Plate 16 (1)
Vautour charognard

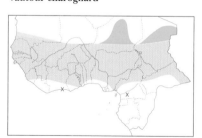

Other name F: Percnoptère brun
L 65–75 cm (26–30"); WS 170–182 (67–72"). *N. m. monachus.* Smallish, dark brown vulture with bare face and throat. In flight, uniformly dark with broad wings and short, rounded tail. **Adult** Bare face and neck pink, flushing red in excitement. Back of head and neck covered in creamy down. Small ruff at base of neck. Bill long, slender and black with pink base. Eye dark. Legs pale bluish. **Juvenile** Similar, but feathers of upperparts tipped buff; bare face and throat whitish (can flush red as in adult); back of head and neck covered in brownish down. Legs brown. **VOICE** Usually silent. [CD5:31] **HABITS** Gregarious and omnivorous scavenger, occurring in various habitats. Often near habitation, scavenging for offal and refuse; also in open country, with other vultures. Nests in trees. **SIMILAR SPECIES** Juvenile Egyptian Vulture has feathered top of head and neck, narrower wings and wedge-shaped tail. Juvenile Palm-nut Vulture has heavy bill, feathered head and short, rounded tail. Juvenile Lappet-faced Vulture has similar silhouette and can be difficult to distinguish at distance, if size cannot be gauged, but has larger head with relatively short bill. **STATUS AND DISTRIBUTION** Locally common resident, from Mauritania to Guinea east to Chad and CAR. Mainly in Sahel and savanna belts; patchily distributed in forest zone; absent from Eq. Guinea, Gabon and Congo.

AFRICAN WHITE-BACKED VULTURE *Gyps africanus*   Plate 16 (4)
Vautour africain

Other name F: Gyps africain
L 80–98 cm (31.5–39.0"); WS 212–218 cm (83–86"). Very large, mainly brownish vulture with, in all plumages, blackish bill. Soars with wings held in shallow V. **Adult** Mainly brown with *white back and rump* (visible when wings are opened and in flight), and bare blackish face. Plumage pales with age; very old birds buffish-white. Skin of

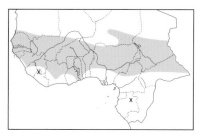

neck blackish, sparsely covered with whitish down. White ruff at base of neck. Bill heavy and blackish; cere grey. Eye dark. Legs blackish. In flight from below, black flight feathers contrast with whitish underwing-coverts. Tail short, rounded and blackish. **Juvenile** Darker brown with narrow whitish streaking; *no white on back*. Head and neck more densely covered with whitish down. Ruff thinner, with brownish, lanceolate feathers. In flight from below, mainly dark brown with long, narrow white bar near leading edge of wing; short white streaking on underwing-coverts visible at close range. **Immature** Gradually becomes plainer and paler; white on back appears in 4th year. Adult plumage acquired in 6–7 years. **VOICE** Usually silent. [CD5:32] **HABITS** Gregarious scavenger favouring lightly wooded savanna. Also near villages. Roosts and nests in trees. **SIMILAR SPECIES** Juvenile Rüppell's Griffon Vulture very similar to juvenile White-backed but is slightly larger, with longer neck and slightly paler bill; in flight, very young birds with single narrow wingbar difficult or even impossible to distinguish. **STATUS AND DISTRIBUTION** Uncommon resident in savanna zones, from S Mauritania to Guinea east to Chad and CAR. Vagrant, Sierra Leone, Liberia and NE Gabon.

RÜPPELL'S GRIFFON VULTURE *Gyps rueppellii*   Plate 16 (5)
Vautour de Rüppell

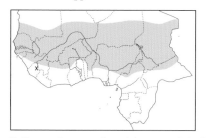

Other name E: Rüppell's Griffon
L 85–107 cm (33.5–42.0"); WS 220–250 cm (87–98"). *G. r. rueppellii*. Very large with long, powerful neck. **Adult** Dark grey-brown body feathers and coverts broadly tipped buffish or whitish, producing scaly appearance. Bare skin on head and neck greyish, sparsely covered with whitish down. White ruff at base of neck. Flight and tail feathers blackish. Bill heavy and yellowish-horn; cere grey. Eye pale yellow. Legs dark grey. In flight from below, appears very dark with long, narrow white bar on leading edge of wing and 2 long and more indistinct white lines formed by white-tipped coverts. **Juvenile** Dull tawny-brown; rather plain, not scaly-looking, above and broadly streaked paler below. Head and neck covered in brownish down; skin initially pinkish, soon darkening. Ruff long and feathery, brownish. Bill blackish. Eye dark. In flight from below, very young birds with single long, narrow white bar near leading edge of wing not safely separable from juvenile White-backed Vulture. **Immature** In flight, white tips to underwing-coverts form additional white lines, permitting separation from young White-backed. Adult plumage acquired in 6–7 years. **VOICE** Usually silent. May utter hisses, grunts and shrieks at carcasses. [CD5:33] **HABITS** Gregarious scavenger generally occurring in dry, open savannas and open habitats in mountainous areas. Dominant over White-backed Vulture at carcasses. Nests and roosts on cliffs. **SIMILAR SPECIES** Juvenile African White-backed Vulture (see above). From distance, young Lappet-faced Vulture similar in flight, but tail more wedge-shaped; head distinctive. **STATUS AND DISTRIBUTION** Not uncommon resident in Sahel belt; locally in Cameroon highlands. Southward movement in dry season reported. Vagrant, Sierra Leone, N Ghana and Togo.

EURASIAN GRIFFON VULTURE *Gyps fulvus*   Plate 16 (6)
Vautour fauve

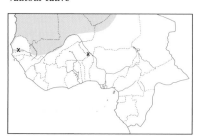

Other name E: European Griffon
L 95–110 cm (37.5–43.0"); WS 230–280 cm (90.5–110.0"). *G. f. fulvus*. Huge. Soars with wings held in shallow V. **Adult** Largely sandy-brown to pale rufous-brown. Skin of head and neck bluish-black densely covered by whitish down. Whitish ruff at base of neck. Greater upperwing-coverts blackish broadly fringed sandy. Bill heavy and yellowish; cere grey. Eye yellowish. Legs blackish. In flight from below, blackish flight feathers contrast with pale brownish underwing-coverts, which have 1–2 paler lines. Tail short, rounded and blackish. **Juvenile** Slightly darker with broad pale streaking on underparts. Ruff pale brownish. Greater upperwing-coverts lack distinct pale fringes. Bill darkish; eye dark. In flight from below, underwing-coverts paler and plainer than adult **Immature** Adult plumage acquired in 6–7 years. **VOICE** Usually silent. [CD1:77] **HABITS** Singly or in small groups. **SIMILAR SPECIES** African White-backed Vulture is darker brown; also smaller; adult has white patch on back. **STATUS AND DISTRIBUTION** Uncommon to rare Palearctic migrant to Mauritania and Senegambia (principally along coast, late Oct–early May) and Mali. Vagrant, SW Niger (Nov).

LAPPET-FACED VULTURE *Torgos tracheliotus*   Plate 16 (7)

Vautour oricou

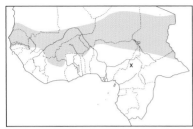

L 98–115 cm (38.5–45.0"); WS 255–290 cm (100–114"). *T. t. tracheliotus*. Huge, mainly dark vulture with bare head and neck and massive bill. Soars with wings held flat or slightly arched. **Adult** Mainly blackish with white streaks on underparts and white 'trousers'. Head and neck deep pink (flushing redder in excitement) with folded skin (lappets). Dark ruff with some white at base of neck. Bill yellowish-horn; cere blue-grey. Eye dark. Legs blue-grey. In flight from below, distinct white bar along leading edge of wing and white thighs contrast with mainly black plumage. **Juvenile** More uniformly darkish. Head and neck dull pinkish-grey. Bill brownish-horn. In flight, appears wholly dark. **Immature** Bare skin on head becomes more reddish; white gradually appears on underparts; white bar on underwing starts appearing in 2nd year; 'trousers' half white in 4th year. Adult plumage acquired in 6–7 years. **VOICE** Usually silent. [CD1:79] **HABITS** Usually singly or in pairs, in dry country. Dominant over other vultures but often stays at periphery of feeding group. Nests in trees. **SIMILAR SPECIES** Hooded Vulture has same flight silhouette and may be difficult to distinguish from juvenile Lappet-faced at distance, but has slim bill; also much smaller. Juvenile/immature White-backed Vulture and Rüppell's Griffon Vulture have more square-tipped tails (slightly more wedge-shaped in Lappet-faced). **STATUS AND DISTRIBUTION** Uncommon resident in Sahel belt, from Mauritania and Senegambia to Chad and CAR. Singles recorded from N Ivory Coast. *TBW* category: THREATENED (Vulnerable). **NOTE** Included in genus *Aegypius* by *BoA*, but majority of recent works use traditional, monotypic *Torgos* (cf. *BWP*, *HBW*, Dowsett & Forbes-Watson 1993).

WHITE-HEADED VULTURE *Trigonoceps occipitalis*   Plate 16 (3)

Vautour à tête blanche

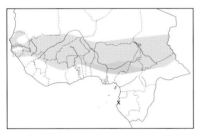

L 78–85 cm (31.0–33.5"); WS 202–230 cm (79.5–90.5"). Distinctive vulture with characteristic peaked hindcrown. **Adult male** Upperparts black; some wing-coverts narrowly fringed white or buff. Face and neck white to pink (flushing redder in excitement); top of head and hindneck densely covered in white down, forming peak or short crest at rear of head. Black ruff at base of hindneck. Breast black; rest of underparts white. *Bill red* tipped black; cere blue. Eye darkish. Legs pink. In flight from below, *white-tipped greater coverts form diagnostic narrow band separating coverts from flight feathers*. **Adult female** Similar but with *inner secondaries white*, conspicuous both perched and in flight. **Juvenile** Mainly dark brown; bare parts duller than adult In flight appears entirely dark with contrasting white band on underwing as adult **Immature** Gradually acquires adult plumage; sexual dimorphism apparent after first moult (2nd year). Full adult plumage attained in *c.* 6 years. **VOICE** Usually silent. **HABITS** Singly or in pairs in dry to lightly wooded savanna. Normally outcompeted by larger vultures. Often singly at small carcass. Nests and roosts in trees. **SIMILAR SPECIES** Hooded Vulture distinguished from juvenile White-headed by slimmer bill and uniformly dark underwing. **STATUS AND DISTRIBUTION** Uncommon resident in Sahel zone, from S Mauritania to Guinea east to Chad and N CAR. Also recorded from coastal Gabon. **NOTE** Included in genus *Aegypius* by *BoA*, but majority of recent works use traditional, monotypic *Trigonoceps* (cf. *HBW*, Dowsett & Forbes-Watson 1993).

SHORT-TOED EAGLE *Circaetus gallicus*   Plate 18 (1)

Circaète Jean-le-Blanc

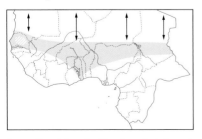

Other name E: Short-toed Snake Eagle
L 59–69 cm (23–27"); WS 162–195 cm (64–77"). Medium-sized, broad-headed raptor with pale underparts and square, rather narrow tail with sharp corners. **Adult male** Head and upperparts typically grey-brown. Below largely white, with upper breast mottled brown or brown streaked whitish, and rest of underparts irregularly blotched and barred brown. Upperwings have variable contrast between greyish-brown coverts and darker brown flight feathers. Underwings white with irregular, narrow dark lines on coverts and 3 narrow dark bands and dusky trailing edge to flight feathers. Tail has 3–4 brown bars with subterminal bar broadest. Some have brown parts darker, creating dark hood, and more distinct markings

below; a few are very pale, without dark hood and almost lacking markings below. Eye orange-yellow. Bill greyish tipped darker; cere grey. Legs pale blue-grey. **Adult female** Similar, but usually with throat and upper breast all dark without pale streaking. **Juvenile** Similar to adult but head and breast more rufous-brown, underwing pattern more indistinct without obvious dusky subterminal band, all bars on tail of equal width (subterminal band not broader), and eye paler. Adult plumage attained in 3rd year. **VOICE** Usually silent in Africa. [CD1:80] **HABITS** Singly in open savanna woodland. Frequently hovers or hangs motionless in wind. Wings held mostly flat when soaring, slightly bowed when gliding. Flies with slow and heavy wingbeats. Feeds mainly on snakes. **SIMILAR SPECIES** Adult Beaudouin's Snake Eagle has all-white underwing-coverts. Subadult Black-breasted Snake Eagle similar to adult Short-toed but differs in all-white underwing-coverts; ranges should not overlap. Pale buzzards distinguished by shorter wings with dark carpal patches (lacking in snake eagles), more rounded tail and smaller size. Osprey has narrower, more angled wings with dark carpal patches, clean white belly, relatively shorter tail with 5–6 bars, and broad dark streak through eye. **STATUS AND DISTRIBUTION** Uncommon Palearctic migrant and dry-season (winter) visitor to Sahel and northern savanna zones, from S Mauritania to Guinea east to Chad. Due to inaccurate descriptions in literature, often not separated with certainty from Beaudouin's Snake Eagle in field; hence respective status is inadequately known in most of their W African range. **NOTE** Sometimes treated as a polytypic species with 2–3 subspecies (*gallicus, beaudouini* and *pectoralis*), but see Clark (1999).

BEAUDOUIN'S SNAKE EAGLE *Circaetus beaudouini* Plate 18 (2)
Circaète de Beaudouin

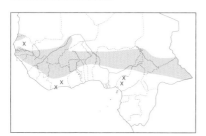

L 62–69 cm (24–27"); WS *c.* 170 cm (67"). **Adult** Similar to Short-toed Eagle but head, upperparts and upper breast typically *darker grey-brown* and lower breast and belly white with variable amount of *narrow* dark bars (sometimes restricted to breast-sides and flanks). Throat may have variable amount of white (considered to relate to sexual differences as in Short-toed Eagle). Tail square and rather narrow with 3 dark bars (subterminal broadest). Underwing white with dark trailing edge and 2 narrow parallel blackish bars across flight feathers. Eye yellow; cere grey. Legs yellowish-grey to blue-grey. Head appears large and owl-like on perched bird, but not so in flight. **Juvenile** Head and upperparts dark brown; underparts more rufous-brown; wings and tail barred darker brown. Underwings have brownish coverts and indistinct narrow dusky bars on flight feathers. Tail has indistinct dusky bands. **VOICE** A clear *kee-u*, uttered in series. Usually silent. [CD5:34] **HABITS** Usually singly, in open savanna woodland. Hunts from prominent perch or in soaring flight interspersed with hovering. Soars with wings held flat. Feeds mainly on snakes; occasionally other small vertebrates. **SIMILAR SPECIES** Even the palest Short-toed Eagle has underwing-coverts with dark markings. Adult Short-toed also has lesser and median upperwing-coverts paler than, and contrasting with, rest of upperwing; Beaudouin's Snake Eagle has more uniformly dark upperwing. Adult Black-breasted Snake Eagle similar to adult Beaudouin's but lacks barring on underparts and is blacker above and on breast. Subadult Black-breasted similar to adult Beaudouin's in having all-white underwing-coverts, but differs in retained juvenile dusky secondaries. Juvenile Black-breasted differs from juvenile Beaudouin's in being more rufous-brown. Juvenile Beaudouin's may be confused with Brown Snake Eagle but differs mainly in narrow dark bars on underside of flight feathers, extensive dark tips to primaries and lack of white bands in tail. Pale buzzards and Osprey have dark carpal patches in flight. **STATUS AND DISTRIBUTION** Uncommon resident in Sahel and northern savanna zones, from S Mauritania to Guinea east to Chad and CAR; some movements recorded, south in dry season, north in rains. Due to inaccurate descriptions in literature, often not separated with certainty from Short-toed Eagle in field; hence respective status is inadequately known in most of their W African range, but Beaudouin's appears to range further south. Wet-season records usually considered Beaudouin's. **NOTE** Considered conspecific with Short-toed Eagle *C. gallicus* by some authors, but see Clark (1999).

BLACK-BREASTED SNAKE EAGLE *Circaetus pectoralis* Plate 18 (3)
Circaète à poitrine noire

Other name E: Black-chested Snake Eagle
L 63–68 cm (25–27"); WS 178 cm (70"). **Adult** Distinctive *pied* raptor with yellow eye. Head, upperparts and breast blackish-brown; rest of underparts pure white. Tail square with 3 dark bars. Underwing white with dark trailing edge and 2 narrow parallel blackish bars on flight feathers. Bill black; cere blue-grey. Legs pale greyish. **Juvenile** Entirely rufous-brown, darker on wings. May fade to buffish within 6 months of fledging; underparts sometimes mottled white. Underwing whitish with rufous-brown coverts, dusky tips to primaries and 3 faint parallel bars on flight feathers (creating dusky appearance, esp. on secondaries, from distance); tail indistinctly barred, appearing grey at distance. **Immature** Blackish-brown and white feathers gradually appear, latter with dark centres, marking underparts with brown blotches. Full adult plumage probably attained in 3rd year. **VOICE & HABITS** As

Beaudouin's Snake Eagle and Short-toed Eagle. [CD5:35] **SIMILAR SPECIES** Adult Short-toed Eagle similar to subadult Black-breasted but mainly distinguished by dark markings on underwing-coverts. Adult Martial Eagle distinguished in flight by dark underwings and tail, and larger size; perched it shows spotted underparts and feathered legs. **STATUS AND DISTRIBUTION** Recorded from SE Gabon and S Congo (Mayombe and Brazzaville). Nomadic and/or regular intra-African migrant; movements inadequately known. **NOTE** Considered conspecific with Short-toed Eagle *C. gallicus* and Beaudouin's Snake Eagle *C. beaudouini* by some authors, but see Clark (1999).

BROWN SNAKE EAGLE *Circaetus cinereus* Plate 18 (4)

Circaète brun

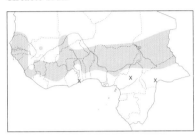

L 66–75 cm (26.0–29.5"); WS *c.* 200 cm (79"). Largest snake eagle. **Adult** *Entirely dark brown.* Broad head with loose feathers at back and large yellow eyes may create owlish appearance when perched. In flight from below, whitish flight feathers contrast with dark brown body and underwing-coverts. Tail dark brown with 3 narrow pale bars. Cere grey. Legs whitish. **Juvenile** Similar to adult, or with white streaks on head or white feathers on lower underparts. **Immature** Acquires variable amount of white mottling on underparts. **VOICE** Usually silent. A guttural *khok-khok-khok-khaw* in soaring display. [CD5:36] **HABITS** In savanna woodland. Perches very upright and for long periods in prominent vantage point (usually top of tree) from which it drops on prey (mainly snakes). Rarely hovers or hunts in flight. Soars on horizontal wings. **SIMILAR SPECIES** Juvenile Beaudouin's may be confused with Brown Snake Eagle but differs mainly in narrow dark bars on underside of flight feathers. Juvenile Black-breasted Snake Eagle appears similar when perched but is more rufous-brown; differs in flight mainly by different underwing pattern. Other all-brown eagles have smaller heads, darker eyes and feathered tarsi; in flight, none has similar contrasting underwing pattern. Perched juvenile Bateleur also has large head but has dark eyes and very short tail. **STATUS AND DISTRIBUTION** Uncommon resident in savanna and Sahel zones, from S Mauritania to Liberia east to Chad and CAR. Some movements recorded, north with rains, south in dry season.

WESTERN BANDED SNAKE EAGLE *Circaetus cinerascens* Plate 18 (5)

Circaète cendré

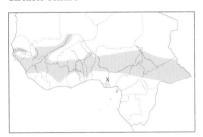

Other name E: Smaller Banded Snake Eagle
L 55–60 cm (22–24"); WS 114 cm (45"). Stocky, shorter tailed than other snake eagles. **Adult male** Head and upperparts grey-brown, darker on wings; below, brown with indistinct whitish barring on lower underparts. *Tail dark brown narrowly tipped white, with single broad white bar across centre* (usually concealed when perched, but conspicuous in flight). Underwing white with 4 dark parallel bars on flight feathers. Eye pale yellow. *Cere and base of bill orange-yellow.* Legs yellow. **Adult female** Similar but usually darker. Some entirely dark brown without noticeable barring on lower underparts. **Juvenile** Much paler than adult Top of head whitish streaked dark; upperparts pale brown, feathers edged buff. Underparts buffish-white, slightly darker on breast and with some brown on belly and thighs. Tail band dirty whitish. **Immature** Variably mottled and streaked brown on underparts, later apparently acquiring brown plumage with varying amounts of barring on lower underparts. **VOICE** Vocal. A loud *kho-kho-kho-kho-...* descending in pitch, uttered in aerial display or from perch. Also a plaintive, nasal *ko-ah ko-ah ...* and *ko-ah ko-koaaah* from perch. [CD5:37] **HABITS** In savanna woodland. Perches unobtrusively for long periods on large tree, often in riverine woodland. Flight direct with rapid shallow wingbeats. Feeds mainly on snakes; also lizards and frogs. **STATUS AND DISTRIBUTION** Uncommon resident in savanna belt, from Senegambia to Sierra Leone east to Chad and CAR. Southward movements recorded locally during rains.

BATELEUR *Terathopius ecaudatus* Plate 15 (5)

Bateleur des savanes

Other name F: Aigle bateleur
L 55–70 cm (21.5–28.0"); WS 170–187 cm (67.0–73.5"). Unmistakable. Stocky, large-headed, almost tail-less eagle. In flight, *unique silhouette with long, broad-based wings curving to a point and very short tail.* Soars with wings held in

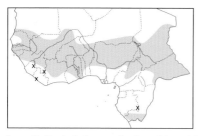

marked V, frequently canting. Rarely flaps once airborne. **Adult male** *Black with chestnut mantle to uppertail-coverts*, grey lesser and median upperwing-coverts, and *chestnut tail*. Loose feathers of head give cowled appearance. Bare facial skin, cere and feet bright red (sometimes fading to yellowish). Bill blue-grey with yellow base. In flight, white underwing-coverts and base of primaries contrast with black inner primaries and secondaries. Feet project beyond tip of tail. Some have creamy or pale brown upperparts. **Adult female** Similar to male but with grey panel on secondaries and inner primaries. In flight, underwing largely white with black trailing edge. **Juvenile** Entirely brown; feathers fringed buff, esp. on head. Bare facial skin and cere greenish-blue. Bill blue-grey. Feet whitish. In flight, appears uniformly brown; silhouette as adult but tail slightly longer, with feet not projecting beyond tip. **Immature** Plumage becomes sooty-brown; tail shortens; face turns yellowish, feet pinkish. Recognisable adult plumage gradually acquired at 5–7 years; full adult plumage attained in 7–8 years. **VOICE** Usually silent. Calls include a far-carrying, barking *kow-aw* and a soft *ko-ko-ko-koaaagh*. [CD5:38] **HABITS** Usually singly, occasionally in small groups, in savanna. Perches in trees. Feeds on small animals, incl. termites, and carrion. **SIMILAR SPECIES** Brown Snake Eagle may appear superficially similar to juvenile/immature Bateleur but has longer (not stumpy) tail and yellow eyes. **STATUS AND DISTRIBUTION** Uncommon to locally common resident in broad savanna belt, from S Mauritania to Guinea east to Chad and CAR; also S Gabon and S Congo.

CONGO SERPENT EAGLE *Dryotriorchis spectabilis* Plate 19 (5)
Serpentaire du Congo

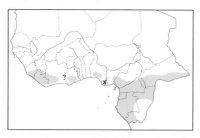

Other name F: Aigle serpentaire du Congo
L 51 cm (20"); WS 85–96 cm (33.5–38.0"). Short-winged and long-tailed forest eagle with relatively large head and long bare legs. **Adult** *spectabilis* Head and upperparts dark chocolate-brown. Underparts white variably tinged pale buff and marked with dark stripe on centre of throat, bold blackish-brown spots on breast and more bar-like spots on belly. Tail long and rounded, brown with 6 broad blackish bars. Wings short and rounded with barred flight feathers. Eye large, dark. Cere and legs yellow. Lower Guinea race *batesi* is paler brown above and less marked below, with spots and bars restricted to flanks. **Juvenile** Head mottled whitish and pale brown. Upperparts brown; dark subterminal spot and narrow white tip to feathers produce scaled appearance. Underparts white with bold blackish spots and variable rufous wash. **VOICE** Vocal. Series of nasal *kow* or *klow* (=*klah*) notes, regularly given by territorial birds, principally in early morning. Uttered from perch or, sometimes, in low, fluttering flight. Also a drawn-out, plaintive *klooooow*, usually given in mid-afternoon. [CD5:39] **HABITS** Inadequately known. Hunts from perch inside forest or along tracks and clearings. Drops onto prey on ground. Feeds on snakes, lizards and amphibians, possibly also small mammals. **SIMILAR SPECIES** Cassin's Hawk Eagle has relatively shorter tail, larger size and feathered legs. **STATUS AND DISTRIBUTION** Scarce to locally not uncommon in forest zone, from Sierra Leone and SE Guinea to Ghana and from W Nigeria to CAR and Congo (common, Odzala NP). Range limits of races inadequately known; nominate thought to range east to W Cameroon.

AFRICAN HARRIER HAWK *Polyboroides typus* Plate 15 (4)
Gymnogène d'Afrique

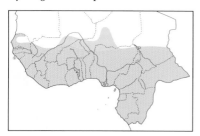

Other names E: Gymnogene; F: Petit Serpentaire
L 60–68 cm (23.5–27.0"); WS *c.* 160 cm (63"). Medium-sized raptor with small head, broad wings and relatively long, slightly rounded tail. Flight with slow, buoyant wingbeats alternated with glides; frequently soars. **Adult** *pectoralis* Largely grey with bare facial skin yellow (flushing orange-red in excitement) and lower underparts finely barred black and white. *Tail black with white band across centre* and narrow white tip. Nape feathers long and loose. Bill black; cere and legs yellow. Eye dark. *In flight, has characteristic pattern of grey wings broadly tipped black, and black tail with distinctive white bar;* on underwing, narrow white line separates grey from broad black trailing edge and tips. **Nominate** is larger, paler and less heavily barred below. **Juvenile** Variably coloured dark brown to tawny; feathers of upperparts narrowly fringed buffish. Tail indistinctly barred blackish. Cere greenish-yellow; bare facial skin greyish. Legs yellow. In flight from below, rufous-brown mottled darker; dark tips to underwing-coverts form more or less distinct line.

Immature Variably mottled whitish on head, streaked buff on underparts, and barred brown on flanks and undertail-coverts. Full adult plumage acquired in *c.* 2–3 years. Young best identified by small head and jizz. **VOICE** A plaintive whistling *sueeeee* or *sueeeeuw* and a high-pitched *hueeeeup-hueeeeup-hueeeeup-*... [CD5:40] **HABITS** Singly or in pairs in forest and wooded savanna. Long legs very flexible at 'knee', enabling insertion into tree holes and crevices. Clings to tree trunks with flapping wings while probing cavities. Feeds on variety of small animals, incl. nestlings, mammals, amphibians, reptiles, fish and insects; also oil-palm husks and eggs. Regularly raids nests, esp. of colonial species, such as weavers. Male performs undulating display flight. **SIMILAR SPECIES** Adult Dark Chanting Goshawk has red cere and legs, no bare face and no white tail bar. Beware of confusing juvenile Harrier Hawk with other brown raptors (e.g. eagles). **STATUS AND DISTRIBUTION** Common resident in forest zone and adjacent savanna throughout (*pectoralis*). Nominate is a dry-season visitor to eastern Sahel (west to N Nigeria).

HEN HARRIER *Circus cyaneus*   Plate 17 (5)
Busard Saint-Martin

Other name E: Northern Harrier
L 43–55 cm (17–21"); WS 97–121 cm (38–48"). *C. c. cyaneus*. Recalls corresponding plumages of Pallid and Montagu's Harriers but has broader, less pointed wings with 5 'fingers' (not 4). **Adult male** Head to breast, upperparts and tail blue-grey; rest of underparts white, sharply demarcated from breast. In flight, large black wingtips and white uppertail-coverts; underwing white with dusky trailing edge. **Adult female** Brown above with white uppertail-coverts. Tail has 4–5 dark bars, subterminal bar broadest. Underparts buffish heavily streaked brown. Underwing has 3 well-defined dark bands (including trailing edge) on secondaries extending onto primaries. **Juvenile** Very similar to adult female, but underparts usually darker and washed rufous; secondaries below darker than primaries with less distinct barring. **VOICE & HABITS** As other harriers. [CD1:81] **SIMILAR SPECIES** Adult female Montagu's and Pallid Harriers very similar to adult and juvenile Hen, but have narrower, more pointed wings. Montagu's is also much paler and more slender, with much more graceful flight. Some adult female Pallid may appear as heavy as Hen but always have narrower wingtips; ear-coverts usually darker with more distinct collar. Juvenile Montagu's and Pallid Harriers easily distinguished from juvenile Hen by much more rufous plumage and unstreaked underparts. **STATUS AND DISTRIBUTION** Palearctic migrant, claimed from Mauritania (a few sightings, Sep–Apr).

PALLID HARRIER *Circus macrourus*   Plate 17 (6)
Busard pâle

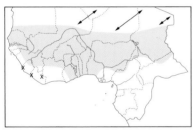

L 40–50 cm (16–20"); WS 95–120 cm (37–47"). **Adult male** Elegant, very pale harrier with distinctive black wedge at wingtip. Pale grey above, mainly white below. **Adult female** Dark brown above with conspicuous white uppertail-coverts ('ring-tail'); tail has 4–5 dark bars. Below, greyish-buff streaked dark brown and chestnut. Distinguished from adult female Montagu's Harrier by darker upperwing, never with dark bar on base of secondaries (often present in Montagu's) and less neatly marked underwing, with paler primaries contrasting with darker secondaries and darkish greater secondary coverts. Also has narrow pale collar (usually absent in Montagu's) and less white above eye. **Juvenile** Upperparts dark brown with paler fringes to coverts; underparts rufous with unstreaked breast. Characteristic head pattern with bold dark cheek patch reaching base of bill, bordered by dark eye-stripe and narrow pale collar (usually absent in juvenile Montagu's); white around eye less extensive than in Montagu's. In flight, pale patch on upperwing usually more distinct than in Montagu's; underwing lacks dark trailing edge to primaries. **VOICE** Usually silent in Africa. [CD1:82] **HABITS** Prefers open habitats, as other harriers. **SIMILAR SPECIES** Montagu's Harrier (q.v. and above). Also Hen Harrier (q.v.). **STATUS AND DISTRIBUTION** Uncommon or scarce to locally not uncommon Palearctic migrant, from Mauritania to Sierra Leone east to Chad and CAR. Rare, Liberia. *TBW* category: Near Threatened.

MONTAGU'S HARRIER *Circus pygargus*   Plate 17 (7)
Busard cendré

L 38–50 cm (15–20"); WS 96–120 cm (38–47"). Most elegant harrier with narrow pointed wings, long tail and very agile, buoyant flight. **Adult male** Head to breast and upperparts blue-grey; rest of underparts white variably streaked chestnut on belly. *In flight, appears tricoloured* with darker grey upperparts and inner wings contrasting with whitish-grey outer wings and extensively black wingtips. Black bars at base of secondaries (1 above, 2 below), white uppertail-coverts and white underwing-coverts marked chestnut. **Adult female** Resembles adult female Pallid Harrier but distinguished by paler upperwing, often with distinct dark bar on base of secondaries (never shown

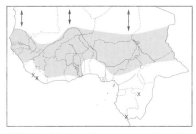

by Pallid) and more uniformly coloured underwing, lacking contrast between primaries and secondaries. Underwing has 3 well-defined dark bands on secondaries with broad pale area between terminal band (trailing edge) and central band; terminal band of equal width and extending onto inner primaries (no dark trailing edge to primaries in Pallid). Facial pattern less distinct; white, claw-shaped area around eye larger. **Juvenile male** Upperparts dark brown with paler fringes to coverts; underparts deep rufous (usually darker, more rufous than juvenile Pallid). Head pattern typically differs from juvenile Pallid by smaller cheek patch (only just reaching base of bill), less black on lores, more white above eye and absence of pale collar (but beware of individual variation). Underwing has dark trailing edge to inner primaries (lacking in Pallid). Eye pale. **Juvenile female** Similar to juvenile male but with dark eye. **VOICE** Usually silent in Africa. [CD1:83] **HABITS** Prefers open habitats, esp. grassland, as other harriers. **SIMILAR SPECIES** Pallid Harrier (q.v. and above). See also Hen Harrier. **STATUS AND DISTRIBUTION** Uncommon or scarce to locally not uncommon Palearctic migrant, from Mauritania to Liberia east to Chad and CAR. Vagrant, Congo and Cape Verde Is. **NOTE** A rare dark morph, occurring in all plumages, is mainly blackish-brown and could be mistaken for dark juvenile/female Eurasian Marsh Harrier, but differs in its lighter, slimmer build and buoyant flight action.

AFRICAN MARSH HARRIER *Circus ranivorus*   Plate 17 (3)

Busard grenouillard

L 44–50 cm (17–20"); WS 110 cm (43"). Brown with, in all plumages, diagnostic combination of *barred tail* and lack of white uppertail-coverts. **Adult male** Brown above with whitish leading edge to wing. Below, paler brown variably streaked dark brown; lower underparts more rufous. Tail brown above, greyish below, both sides barred black. Face outlined by ring of whitish feathers. In flight from below, brownish underwing-coverts; whitish flight feathers distinctly barred black. From above, plain rufous rump and barred tail diagnostic. Eye, cere and legs yellow. **Adult female** Similar, but often darker and more rufous. **Juvenile** Darker brown and plainer than adult with variable amount of whitish on nape and throat and *irregular whitish breast-band*. Eye dark brown. **VOICE** Usually silent. [CD5:41] **HABITS** As other harriers. **SIMILAR SPECIES** Other harriers with barred tails have conspicuous white uppertail-coverts. Adult female/immature Eurasian Marsh Harrier have no barring on wings and tail, and no rufous in plumage; also slightly larger and more heavily built. **STATUS AND DISTRIBUTION** Uncommon resident, Congo (Loudima and Odzala NP). Also claimed from N CAR (Manovo-Gounda-St Floris NP). One uncertain record from Nigeria (Zaria). Main range in S and E Africa.

EURASIAN MARSH HARRIER *Circus aeruginosus*   Plate 17 (4)

Busard des roseaux

Other names E: Western/European Marsh Harrier
L 42–56 cm (17–22"); WS 110–140 cm (43–55"). *C. a. aeruginosus.* Largest and heaviest harrier. Characteristic harrier shape and flight action, with long wings held in shallow V and long tail. **Adult male** Readily identified in flight by *striking tricoloured upperwing*: blue-grey contrasting with large black wingtips and brown mantle and coverts; tail blue-grey. Head to breast buff-brown streaked darker; upperparts and rest of underparts brown. Some very pale, with whitish uppertail-coverts. **Adult female** Mainly dark brown with creamy-white crown, nape, throat and leading edge of wing (absent in some). Often has yellowish patch on breast. **Juvenile** Similar to adult female but darker brown and lacks pale breast patch. Creamy areas usually slightly deeper in colour and restricted to crown and throat, but sometimes also on leading edge of wing; rarely entirely lacking. Adult male plumage acquired in *c.* 3 years. **VOICE** Usually silent in Africa. [CD1:84] **HABITS** Except when migrating, usually flies low over open habitats. Esp. attracted to wet areas, e.g. marshes, reedbeds, ricefields, etc. **SIMILAR SPECIES** Black Kite has shallowly forked tail and flies with wings slightly bowed and with much twisting of tail. Pale adult male Eurasian Marsh Harriers distinguished from other harriers by brown body and upperwing-coverts. **STATUS AND DISTRIBUTION** Not uncommon to locally common Palearctic migrant in suitable habitat throughout. Some remain during rains. Rare winter visitor, Cape Verde Is.

GABAR GOSHAWK *Micronisus gabar*   Plate 20 (2)

Autour gabar

L 28–36 cm (11–14"); WS 60 cm (23.5"). Rather small *Accipiter*-like raptor with prominent white uppertail-coverts.

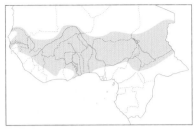

Adult Head, upperparts, throat and upper breast plain grey; rest of underparts narrowly barred grey. Tail grey with 4 broad black bars. Underwing white densely and narrowly barred dark on coverts and remiges. Eye dark red-brown. Cere and legs pinkish-red. **Melanistic morph** is all black except for white barring on upperside of primaries and faint white bars on tail. No white uppertail-coverts. **Juvenile/ immature** Head pale streaked dark brown (or rufous-brown streaked paler) with slight whitish supercilium. Upperparts brown with rufous-buff edges to coverts. Underparts whitish streaked rufous-brown on breast and barred rufous-brown on belly and flanks. Tail brown with 4 black bars. Underwing white with dense, narrow red-brown bars on coverts and blackish bars on remiges. Eye pale yellow. Cere and legs yellowish to orangey. **VOICE** A rapid, high-pitched *kik-kik-kik-...* or *kwk-kwik-kwik-...*; also a slightly slower and more melodious version *kwee-kwee-kwee-...*, reminiscent of Senegal Thick-knee. [CD5:42] **HABITS** In wooded savanna and thornbush. Hunts from perch or on the wing, feeding mainly on small birds, which are usually caught in flight. Often robs nests. **SIMILAR SPECIES** Lizard Buzzard also has white uppertail-coverts but is more stocky, has white throat with black vertical stripe and black tail with single white bar. Shikra is plain grey above, lacks grey on throat and has bright red eye and yellow cere and legs. Dark Chanting Goshawk has similar plumage pattern but is much larger with very long legs. Juvenile Shikra differs from juvenile Gabar in darker brown head and upperparts, unstreaked head, and lack of white on uppertail-coverts. Melanistic Ovambo Sparrowhawk distinguished from melanistic Gabar by yellowish cere and legs; in flight, lacks white barring on primaries and tail. Melanistic Black Sparrowhawk has white throat, unbarred tail, yellow cere and legs, and is much larger. **STATUS AND DISTRIBUTION** Not uncommon to scarce resident in northern savanna and Sahel zones, from S Mauritania and Senegambia through Mali and Ivory Coast to Chad and CAR. **NOTE** Sometimes placed in genus *Melierax*.

DARK CHANTING GOSHAWK *Melierax metabates*   Plate 20 (7)
Autour sombre

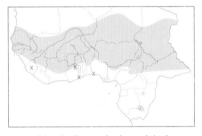

Other name F: Autour-chanteur
L 38–48 cm (15–19"); WS 95–110 cm (37–43"). *M. m. metabates.* Medium-sized, grey raptor with long red legs, long broad wings and longish, rounded tail. **Adult** Head, upperparts, throat and breast plain grey; rest of underparts narrowly barred grey. *Uppertail-coverts white finely barred grey* (appearing greyish from distance). Tail rounded, central tail feathers black, others white with 3 black bars. In flight, dark wingtips above and below. Eye dark. Cere and legs pinkish-red. **Juvenile/immature** Head and upperparts brown with buffish edges to coverts. Underparts variably marked brown and white: blotched, streaked or plain brown on breast and barred brown on belly, flanks and underwing-coverts. Flight feathers barred blackish above and below. Uppertail-coverts white narrowly barred grey-brown. Tail brown broadly barred darker. Eye pale yellow. Cere and legs dull yellowish to orangey. **VOICE** Usually silent except when breeding. Melodious ('chanting') call, a loud, accelerating, piping *wheeeow-whew-whew-whew* or *kleeu-kleeu-klu-klu-klu* uttered by both sexes, from perch or in flight. [CD1:85] **HABITS** Singly, in open woodland and thornbush. Perches prominently and upright on poles or atop trees and bushes. Feeds on variety of small animals, incl. lizards, snakes, mammals and insects; also birds. Catches prey mainly on ground, occasionally in flight. Not shy. **SIMILAR SPECIES** Gabar Goshawk has conspicuous white uppertail-coverts and is noticeably smaller. **STATUS AND DISTRIBUTION** Not uncommon resident in northern savanna and Sahel zones; also SE Gabon and SC and N Congo. Some irregular southward movements during dry season. **NOTE** Race *neumanni*, described as having more vermiculated wings, doubtful and not widely recognised. Would occupy range from L. Chad east to N Sudan.

AFRICAN GOSHAWK *Accipiter tachiro*   Plate 21 (4)
Autour tachiro

L 36–48 cm (15–18"); WS 70 cm (27.5"). Medium-sized forest *Accipiter* with blackish tail possessing 3 white patches. **Adult male** *macroscelides* Head and upperparts (incl. rump) slate-grey. Throat white finely barred grey; rest of underparts and underwing-coverts narrowly barred orange-chestnut and white; flanks and thighs orange-chestnut. Tail blackish with narrow white tip and 3 white patches in centre. Underside of flight feathers barred black and white. Eye, narrow orbital ring, cere and legs yellow. Lower Guinea race *toussenelii* is paler and plainer below. Underparts unbarred greyish-white washed rufous, esp. on flanks and thighs, or slightly barred. Island race *lopezi* slightly richer in colour than *macroscelides*; sometimes with rufous on throat. **Adult female** As resp. males but

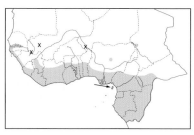

distinctly larger. **Juvenile** *macroscelides/lopezi* Dark brown above with whitish supercilium and rufous edges to feathers. Whitish below, with dark brown stripe on centre of throat, drop-like spots on breast, and short broad bars on flanks and thighs, latter tinged orange-chestnut. Tail barred dark and lighter brown. Eye brown; legs dirty yellow. **Juvenile** *toussenelii* Underparts almost pure white; some dark spots on breast-sides; flanks and thighs variably marked with short broad bars. **VOICE** Distinctive. Series of sharp, abrupt *kwit!* notes, uttered in (usually high) display flight over territory or from perch, esp. at dawn. *A. t. toussenelii* has a slightly longer *kewit!* Resembles call of Velvet-mantled Drongo; often mimicked by robin chats *Cossypha* spp. Juvenile has a plaintive, nasal *we-aaaaauw*. [CD5:43,44] **HABITS** In forest and gallery forest. Unobtrusive, usually keeping to dense cover, except when soaring and calling over territory. Feeds mainly on small birds and mammals; also insects. **SIMILAR SPECIES** Adult Chestnut-flanked Sparrowhawk is only slightly smaller and may be difficult to distinguish from *macroscelides* in small area of overlap, but is darker above and has darker, bolder bars from breast to belly and extensive bright rufous on sides, flanks and thighs; toes dusky. Juvenile very similar to juvenile *macroscelides* and possibly inseparable in field; marks on breast more oval than tear-shaped; bars on breast and flanks broader; toes dusky. **STATUS AND DISTRIBUTION** Common in forest zone and forest outliers in savanna, from Gambia and S Senegal to W Cameroon (*macroscelides*), and from SE Cameroon to CAR and Congo (*toussenelii*); both forms intergrade east of Mt Cameroon. Also Bioko (*lopezi*). **NOTE** All 3 taxa sometimes separated from extralimital eastern and southern forms under the name Red-chested Goshawk *A. toussenelii*, but similar morphology, aerial display and vocalisations suggest they are conspecific.

CHESTNUT-FLANKED SPARROWHAWK *Accipiter castanilius* Plate 21 (3)
Autour à flancs roux

L 28–36 cm (11–14"); WS 60 cm (24"). **Adult male** Head slate-grey darkening to *blackish-grey on upperparts* (incl. rump). Below white with faint, narrow grey bars on throat and *bold, dark brown bars on rest of underparts; breast-sides, flanks and thighs bright rufous. Tail blackish with 3 white spots in centre.* Underside of flight feathers barred black and white. Eye yellow to reddish-brown. Orbital ring, cere and legs yellow; toes dusky. **Adult female** Upperparts browner; flanks darker rufous. Larger (about size of male African Goshawk). **Juvenile** Brown above; white below, with dark brown spots on breast and dark brown bars on flanks and thighs, latter with rufous. Tail barred dark and paler brown. Eye brown; legs dirty yellow; toes dusky. **VOICE** No information. **HABITS** Inadequately known. Hunts from low perch in dense undergrowth of lowland forest, dashing with extreme suddenness and speed onto prey (mainly small birds). Follows ant columns to catch invertebrates and small vertebrates disturbed by them, as well as attending birds. Nest undescribed. **SIMILAR SPECIES** In area of overlap, African Goshawk of race *macroscelides* difficult to separate but has paler grey head, paler upperparts and paler, brighter orange-chestnut barring on underparts; race *toussenelii* paler and much less marked below. Juvenile *macroscelides* very similar to juvenile Chestnut-flanked Sparrowhawk and probably indistinguishable in field. **STATUS AND DISTRIBUTION** Uncommon to scarce forest resident, from W Nigeria to S CAR and Congo.

SHIKRA *Accipiter badius* Plate 20 (1)
Épervier shikra

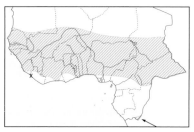

L 26–30 cm (10–12"); WS 58–60 cm (23-24"). *A. b. sphenurus.* Small, pale *Accipiter* of open, wooded country. **Adult** Head and upperparts blue-grey. Underparts white finely barred rufous. Tail grey, outer feathers with 5–6 black bars (visible when tail spread and from below). *In flight uniformly grey above, very pale below,* underwing white with dusky tips to primaries, fine rufous bars on coverts and blackish bars on remiges. Eye orange to red. Cere and legs yellow. **Juvenile/ immature** Head and upperparts brown with buffish edges to coverts. Underparts whitish with dark line on centre of throat, brown blotches on breast and brown bars on belly and flanks. Tail brown with 5 dark bars. Eye, cere and legs yellow. Adult plumage acquired in 2 years. **VOICE** Vocal. A fast, sharp *ki-ki-ki-ki-...* Also *kiwik-kiwik-kiwik-kiwik-...* (male), and plaintive *keeu-keeu-keeu-...* (female). [CD5:46] **HABITS** In savanna, woodland and other open habitats. Usually hunts from perch, dropping onto prey (mostly lizards) on ground; also takes insects and birds (occasionally in flight). Less unobtrusive and more vocal than congeners; favours more open areas and often soars. Not shy. **SIMILAR SPECIES** Adult Ovambo Sparrowhawk has underparts barred grey, uppertail with bold barring and white shaft streaks, and cere and legs red to yellow. Adult Gabar Goshawk has plain grey breast, red cere and legs, and white rump. Juvenile Gabar best distinguished from juvenile Shikra by white rump. Compare also vagrant European and Levant Sparrowhawks. **STATUS AND DISTRIBUTION** Common resident in savanna and Sahel belts throughout; also SE Congo. Moves south to breed in dry season, returning north in rains.

LEVANT SPARROWHAWK *Accipiter brevipes* Plate 20 (6)
Épervier à pieds courts

L 29–39 cm (11.5–15.0"); WS 63–80 cm (25–31"). Medium-sized with rather slender and pointed, *black-tipped wings*. **Adult male** Head (incl. ear-coverts) and upperparts blue-grey. Underparts white narrowly barred rufous. Tail with *unbarred central feathers* and, below, 6 or more narrow dark bars. In flight, appears very pale below, with whitish-looking underwing and contrasting black primary tips. Eye dark. Cere and legs yellow. **Adult female** Only slightly larger. Browner and more barred, with dark stripe on throat. **Juvenile** Dark brown above with rufous edges to feathers. Below white with dark line on throat; rest of underparts with bold drop-shaped spots. Underwing has dark barring, densest on coverts; dark tips on primaries much smaller than in adult and less contrasting. **VOICE** A high-pitched *keewik*. Usually silent on migration. **HABITS** Inadequately known in Africa. Prey, taken in flight or from perch, includes birds, lizards, large insects and small mammals. **SIMILAR SPECIES** Adult Shikra has blunter wings with paler tips, paler upperparts, and smaller size. Juvenile Shikra distinguished from juvenile Levant Sparrowhawk by much less heavily marked underparts. Adult Eurasian Sparrowhawk has more rounded wings lacking black tips and fewer bars on tail. Juvenile Eurasian is barred on breast and belly. **STATUS AND DISTRIBUTION** Palearctic vagrant, claimed from N Cameroon (Waza, Oct 1992).

RED-THIGHED SPARROWHAWK *Accipiter erythropus* Plate 21 (2)
Épervier de Hartlaub

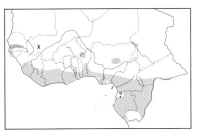

Other name E: Western Little Sparrowhawk
L 23–28 cm (9–11"); WS *c.* 40 cm (16"). Very small (size of wood dove), with distinctive tail pattern. **Adult male** *erythropus* Head and upperparts *grey-black* with *white crescent on uppertail-coverts* (conspicuous in flight) *and 3 small, broken white bars on blackish tail* (middle pair of rectrices lack white). Throat pure white, contrasting with blackish head-sides; rest of underparts greyish-white with pinkish wash (mainly on breast-sides and flanks). Undertail-coverts white; underside of tail barred black and white. Eye orange to red with bright red orbital ring. Cere orange-red. Legs bright orange-yellow. Lower Guinea race **zenkeri** has *deep rufous underparts* and 2 unbroken and more distinct white spots on tail. **Adult female** Similar but noticeably larger with browner upperparts. Eye more brownish-orange. **Juvenile** Head and upperparts brown, many feathers tipped rusty. Uppertail-coverts brown tipped white (forming only indistinct bar). Tail dark brown with 2 irregular buffy bars. Underparts variable, but usually buffy-white barred dark brown; occasionally with some spotting on breast. Tail from below barred greyish-buff and brown. Bare parts duller. **VOICE** Usually silent. A high-pitched *kik-kik-kik-kik-...* or *ki-ki-ki-ki-ki-kiw-kiw*. [CD5:47] **HABITS** Inhabits forest, frequenting second growth at edges and along clearings. Unobtrusive; principally active in early morning or late afternoon. Feeds mainly on small birds, lizards and large insects. Easily overlooked. **SIMILAR SPECIES** Chestnut-flanked Sparrowhawk is barred on underparts, has dark uppertail-coverts and is larger. Little Sparrowhawk is paler above and barred below, and frequents woodland. **STATUS AND DISTRIBUTION** Uncommon to scarce forest resident, from Gambia to Nigeria (nominate) and from S Cameroon to S CAR and Congo (*zenkeri*).

LITTLE SPARROWHAWK *Accipiter minullus* Plate 21 (1)
Épervier minule

L 23–27 cm (9–11"); WS 39 cm (15"). Smallest African *Accipiter*. **Adult** Head and upperparts dark slate-grey. Underparts white narrowly barred chestnut except on throat; breast-sides and flanks washed rufous. *Uppertail-coverts white* (conspicuous in flight). *Tail blackish above with 2 conspicuous white spots in centre* and narrow white tip, white below with 3 dark bars. Eye yellow to orange-yellow. Cere and legs yellow. **Juvenile** Head and upperparts brown with buffish edges to wing-coverts. Underparts whitish boldly marked with brown drop-shaped spots on breast and belly and short broad bars on breast-sides, flanks and thighs. Tail dark brown with pattern as adult; tips of uppertail-coverts white forming narrow line. Eye initially brown. Cere and legs yellow. **VOICE** Vocal when breeding. A rapid *kew-kew-kew-kew-kew* (male) and a high-pitched *kik-kik-kik-kik-kik* (female). [CD5:48] **HABITS** In variety of wooded habitats. Usually in cover, hunting with rapid, twisting flight or perching unobtrusively in shade; occasionally in the open. Feeds mainly on small birds. Easily overlooked but not shy. **SIMILAR SPECIES** Adult Red-thighed Sparrowhawk is blacker above, lacks distinct barring below, and mostly occurs in forest. Juvenile Red-thighed less marked below than juvenile Little. Adult Shikra lacks spots on tail, has red eyes and is larger. Plumage of juvenile African Goshawk similar to juvenile Little Sparrowhawk, but former is much larger and has dark line on throat. **STATUS AND DISTRIBUTION** Resident in E Africa and south of equator; may occur in adjacent areas of our region but not yet recorded with certainty. Claims from Mali (Lamarche 1980) doubtful.

OVAMBO SPARROWHAWK *Accipiter ovampensis*   Plate 20 (3)
Épervier de l'Ovampo

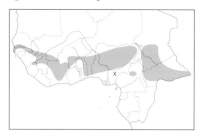

L 33–40 cm (12–16"); WS *c.* 67 cm (26"). Medium-sized *Accipiter* with distinctive white spots on tail shafts. **Adult** Head and upperparts grey; *entire underparts white finely barred grey*. Tail grey above with 4 broad blackish bars, narrow white tip and *rather faint white markings (white shafts of central feathers) in grey parts*. Variable narrow white bar at base of uppertail-coverts (sometimes absent). Underwing white finely barred grey on coverts and more broadly and boldly on remiges. Eye dark red. Cere and legs orange-red to yellow. Rare **melanistic morph** is all black except for 3–4 paler bars on tail and, on underwing, grey remiges barred black. **Juvenile/immature** Two forms. Head and underparts either rufous or whitish variably (usually lightly) streaked dark brown; some barring on flanks. Both forms have brown upperparts with buffish edges to feathers, *whitish supercilium and dark patch on ear-coverts*. Underwing barred blackish on remiges. Eye brownish. Cere and legs yellow. **VOICE** Usually silent except when breeding. Calls include a high-pitched *keep-keep-keep-...* or *kwep-kwep-kwep-...* [CD5:49] **HABITS** Inadequately known. Frequents woodland. Unobtrusive, usually remaining within cover. Dashes on prey (mainly birds) from perch or from soaring position. Flight fast and agile. **SIMILAR SPECIES** Adult Shikra has underparts barred with russet (not grey), no bold barring or white shaft streaks on upperside of tail, and yellow cere and legs. Adult Gabar Goshawk has plain grey breast, red cere and legs, and broad white band on uppertail-coverts. Compare also juvenile Black Sparrowhawk. Melanistic Gabar Goshawk shows white barring on primaries and tail in flight. Melanistic Black Goshawk has white throat, no barring on tail, yellow cere and legs, and much larger size. **STATUS AND DISTRIBUTION** Uncommon to rare intra-African migrant to savanna zone. Breeds south of equator. Recorded from Senegambia, Mali, Guinea, Ivory Coast, Ghana, Togo, Benin, Nigeria, Cameroon, Chad and CAR, mostly in wet season. Status inadequately known; perhaps resident in some areas.

EURASIAN SPARROWHAWK *Accipiter nisus*   Plate 20 (5)
Épervier d'Europe

Other name E: European Sparrowhawk
L 28–41 cm (11–16"); WS 58–80 cm (23–31"). *A. n. nisus*. Small to medium-sized. **Adult male** Head and upperparts dark slate-grey. Underparts white narrowly barred rufous on breast and belly, washed rufous from ear-coverts to flanks. Tail long, narrow-based and square with sharp corners, narrow white tip, broad subterminal band and 4–5 narrower dark bars. Underwing white with narrow rufous bars on coverts and broader blackish bars on remiges. Eye bright yellow to orange. Cere and legs yellow. **Adult female** Much larger and browner with white supercilium and brownish barring on underparts. **Juvenile** Dark brown above with rufous edges to feathers; underparts whitish variably barred brown (markings tending to streaks or spots on throat and upper breast). Eye yellow; cere greenish. Attains adult plumage after *c.* 1 year. **VOICE** Usually silent on migration. [CD1:86] **HABITS** Hunts in open country or woodland with rapid, often low flight using cover and dashing on surprised prey (mainly birds). Normal flight consists of some rapid wingbeats followed by glide on flat wings. **SIMILAR SPECIES** Adult Levant Sparrowhawk has much paler underwing (almost white in male) with contrasting black tips and without heavy barring; from below tail has more bars. Juvenile Levant has entire underparts with bold drop-shaped spots (no barring). Shikra is smaller with paler grey upperparts and unbarred uppertail. Gabar Goshawk easily distinguished by white rump. Poor views may suggest small falcon but shorter, more rounded wings should preclude confusion. **STATUS AND DISTRIBUTION** Scarce Palearctic migrant to Mauritania (Oct–Apr). Vagrant, Gambia, Mali and Chad.

BLACK SPARROWHAWK *Accipiter melanoleucus*   Plate 21 (5)
Autour noir

Other names E: Great Sparrowhawk, Black Goshawk; F: Épervier pie
L 46–55 cm (18–22"); WS *c.* 102 cm (40"). Largest African *Accipiter*. **Adult** *temminckii* Distinctive. *Head and upperparts black contrasting with pure white underparts; mottled black patches on breast-sides, flanks and thighs*. Tail long, dark brown

above, whitish below, barred blackish on both surfaces (terminal band broadest). Underwing white with narrow parallel blackish bars on flight feathers. Eye reddish. Cere and legs yellow. *A. m. melanoleucus* is slightly larger. Rare **melanistic morph** all black with white throat. **Juvenile/immature** Two forms. Upperparts dark brown with rusty edges to feathers. Underparts and underwing either rufous or white, streaked brown (most heavily on breast). Underwing barred blackish on remiges. Eye initially dark. Cere and legs yellow. Attains adult plumage after *c.* 30 months. **VOICE** Usually silent; vocal when breeding. Call *kyip* or *klee-ep* (male) and deeper *chep* (female). Duet consists of alternating series of calls. [CD5:51] **HABITS** In lowland, montane and gallery forest. Usually flies through forest or just above canopy; rarely soars. Often perches inconspicuously in large tree. Feeds mainly on birds, esp. doves and pigeons, caught inside or outside forest. Also attacks poultry. Unobtrusive but not shy. **SIMILAR SPECIES** Adult Cassin's Hawk Eagle also has black markings on breast-sides but has feathered legs and different flight silhouette, with relatively longer wings and shorter tail; also has much black on underwing-coverts. Juvenile Ayres's Hawk Eagle could be confused with rufous juvenile Black Sparrowhawk but lacks bold streaking on underparts, has feathered legs, and different flight silhouette and behaviour. Juvenile African Goshawk could be confused with white juvenile but has drop-shaped spots on underparts and dark stripe on throat. Melanistic forms of Gabar Goshawk and Ovambo Sparrowhawk lack white throat, have barring on tail and are smaller. **STATUS AND DISTRIBUTION** Not uncommon to rare forest resident, from Gambia and S Senegal to Liberia east to S CAR and Congo (*temminckii*); S and E African nominate race reaches Gabon in west but limits poorly known. Single records, S Mali and SW Niger.

LONG-TAILED HAWK *Urotriorchis macrourus*   Plate 19 (6)
Autour à longue queue

L *c.* 57–73 cm (23–29"); WS 90 cm (35"). Unmistakable forest hawk with *very long, graduated tail having irregular white spots and conspicuous white bar at base.* **Adult** Head and upperparts dark slate-grey with contrasting white uppertail-coverts. Below, throat pale grey; *rest of underparts rich chestnut.* Underwing rich chestnut on coverts, white barred black on flight feathers. Undertail-coverts white; tail black tipped and barred white. Eye reddish-yellow. Cere and legs yellow. In rare **variant** chestnut replaced by slate-grey. **Juvenile/immature** Head and upperparts brown with blackish barring on remiges and buff-brown edges to wing-coverts. Underparts white variably marked dark brown (sometimes almost unmarked) with stripe on throat, blotches on upper breast and broad bars on flanks. Tail brown broadly barred blackish-brown. Cere initially dark. **VOICE** Distinctive. Two drawn-out, plaintive whistles *teeu-ieeew teeu-ieeeew*, the second slightly higher pitched, sounding more 'impatient', with variations. Also a soft *klee-klee-klee-klee-klee-klee.* [CD5:52] **HABITS** Mainly in middle strata inside forest, also lower; occasionally near tracks and clearings. Prey probably mainly squirrels and other arboreal mammals; also birds. **SIMILAR SPECIES** White-crested Hornbill occurs in same habitat and has rather similar shape with long graduated tail, but is mainly black and has different behaviour. **STATUS AND DISTRIBUTION** Uncommon or scarce to locally not uncommon forest resident, from E Sierra Leone and SE Guinea to W Togo and from W Nigeria and Cameroon to SW CAR and Congo.

GRASSHOPPER BUZZARD *Butastur rufipennis*   Plate 22 (2)
Busautour des sauterelles

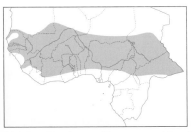

Other name F: Busard des sauterelles
L 41–44 cm (16–17"); WS 102 cm (40"). Medium-sized with long wings and tail, readily identified in flight by *conspicuous rufous, black-tipped 'hand'.* **Adult** Head and upperparts dark grey-brown. Throat white or buff with dark stripe on centre; rest of underparts rufous narrowly streaked darker. Tail brown barred darker, terminal band broadest and most distinct. Distinctive upperwing pattern with rufous primaries and greater primary-coverts; primaries broadly tipped blackish. Underwing has well-defined blackish border to pale rufous flight feathers. Eye, cere, base of bill and legs yellow. **Juvenile** Mainly rufous-brown. Upperparts with rufous edges to feathers. Throat buffish bordered by dark moustachial stripe. Tail has single dark terminal band. Eye brown; bill with less yellow at base. **VOICE** A loud *ki-ki-ki-ki-kee*

when breeding. Otherwise silent. [CD5:53] **HABITS** Singly or in small groups in open country, from thornbush to wooded savanna and cultivation. Perches conspicuously and for long periods atop small trees and bushes. Often near bush fires, catching disturbed insects (mainly grasshoppers) on the wing. Most prey normally captured on ground. Not shy. Flight low and buoyant, rather harrier-like, series of wingbeats interspersed by glides; occasionally soars. **SIMILAR SPECIES** Not easily confused. Perched Fox Kestrel may be superficially reminiscent of juvenile Grasshopper Buzzard but, apart from different jizz, distinguished by more rufous upperparts and graduated (not square) tail. Juvenile Dark Chanting Goshawk has similar upright posture but lacks rufous in wing. **STATUS AND DISTRIBUTION** Common intra-African migrant, breeding locally in semi-arid belt during rains, moving south in dry season; found as far south as northern edge of forest.

LIZARD BUZZARD *Kaupifalco monogrammicus*   Plate 20 (4)
Autour unibande

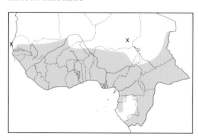

Other name F: Buse unibande
L 35–37 cm (14–15"); WS 79 cm (31"). *K. m. monogrammicus.* Medium-sized, stocky grey raptor with *diagnostic black stripe on centre of white throat and single white band on black tail.* **Adult** Head, upperparts and breast plain grey. *Uppertail-coverts white.* Lower underparts white narrowly barred grey. Underwing white narrowly barred grey; appears all white at distance. Rarely, tail has 2 white bands. Eye dark; narrow red orbital ring. *Cere and legs pinkish-red.* **Juvenile** Similar to adult but mantle and wing-coverts edged buff. Underparts washed buffish. Bare parts paler. Adult plumage attained within 1 year.

VOICE Distinctive. A far-carrying, melodious *KLEEUUu-kluklukluklu.* Also a drawn-out, high-pitched *peeeeeooooo.* [CD5:54] **HABITS** In various types of woodland and cultivation. Perches for long periods in trees or on wires and poles, from which it catches prey (lizards, small snakes, insects, also small mammals and birds) with quick dash. Low flight a series of quick wingbeats followed by a glide; soaring display flight. **SIMILAR SPECIES** Rather similar-sized but more slender Gabar Goshawk has plain grey throat and several tail bars. **STATUS AND DISTRIBUTION** Not uncommon resident in savanna zone and large clearings in forest zone throughout.

COMMON BUZZARD *Buteo buteo*   Plate 22 (4)
Buse variable

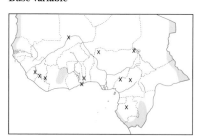

Other name E: Steppe Buzzard
L 45–58 cm (18–23"); WS 110–132 cm (43–52"). Medium-sized, rather compact raptor with broad, rounded wings and tail. Plumage variable. **Adult** *buteo* Very dark to pale brown (occasionally mainly white). Below, usually with diffuse pale breast-band; pale feathers finely cross-barred (not blotched or streaked). In flight from below, flight feathers whitish indistinctly barred grey and tipped blackish (forming dark trailing edge), dark carpal patch; tail densely barred grey with blackish subterminal band. Eastern race *vulpinus* (Steppe Buzzard) usually more rufous-brown; tail paler and often rufous. Underwing pattern more contrasting, with whiter primaries and blacker trailing edge. Barring on flight and tail feathers narrower and more distinct. Pale breast-band often absent. Rare **dark morph** *vulpinus* mainly blackish. Island race *bannermani* similar to nominate; less individual variation. **Juvenile** *buteo* Similar to adult but underparts more streaky; wings and tail lack distinct terminal band. Eye paler than adult Juvenile *vulpinus* usually more regularly streaked below. **VOICE** A loud, mewing *peee-ah.* Usually silent in Africa. [CD1:88] **HABITS** Perches on conspicuous vantage point in wooded habitats. Soars with wings raised in shallow V. Flies with rather shallow, stiff wingbeats. **SIMILAR SPECIES** European Honey Buzzard appears more slender, with slimmer, more protruding head, longer tail with rounded corners, wings held flat when soaring and slightly bowed when gliding. Long-legged Buzzard may resemble very rufous-tailed Steppe Buzzard, but distinguished by paler head and upper breast contrasting with rufous-brown belly, absence of barring on underparts, more uniform rufous underwing-coverts with larger black carpal patch, and unbarred tail. Dark morph very similar to dark morph *vulpinus* and best separated by structural and flight differences. Adult Red-necked Buzzard has tail more rufous and more uniform than Steppe Buzzard and is easily distinguished by white on underwing-coverts and underparts. **STATUS AND DISTRIBUTION** Rare to scarce Palearctic migrant irregularly distributed throughout. Distribution of *vulpinus* and nominate *buteo* unclear; both reported from Liberia; those from Senegal and from Ivory Coast eastwards attributed to *vulpinus.* Rare resident, Cape Verde Is (*bannermani*).

LONG-LEGGED BUZZARD *Buteo rufinus*
Buse féroce

Plate 22 (5)

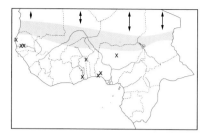

L 50–65 cm (20.0–25.5"); WS nominate 126–163 cm (50–61"), *cirtensis* 115–125 cm (45–49"). **Adult** *rufinus* Resembles Common Buzzard but slightly larger, looking heavier when perched, with larger bill and longer legs. In flight, appears more eagle-like, with longer wings and slightly longer tail. Plumage variable, but typically mainly buffish and orange-rufous. Distinguished from Common Buzzard (esp. race *vulpinus*, Steppe Buzzard) by dark rufous-brown belly and 'trousers' contrasting with pale buff, almost unmarked, head and upper breast. Tail plain (unbarred) orange-rufous, often paling to whitish at base. In flight from below, black carpal patch

and trailing edge to wing contrast with uniformly buff to pale rufous-brown underwing-coverts and whitish, faintly barred flight feathers. N African race *cirtensis* indistinguishable in field, but slightly smaller. **Dark morph** (nominate only) mainly blackish; below, greyish flight feathers narrowly barred darker and with dark trailing edge; tail barred with broad subterminal band. **Juvenile** Similar to adult but wing lacks distinct dark trailing edge and tail more buff-grey with faint barring (rarely with narrow dark terminal band). **VOICE** Usually silent. [CD1:90] **HABITS** Singly. Perches for long periods on exposed vantage point. Soars with wings raised in shallow V (often deeper than Common Buzzard); glides with wings slightly raised. Occasionally hovers. **SIMILAR SPECIES** Adult Steppe Buzzard usually distinguished by darker, more rufous head and breast, often with diffuse pale breast-band, barring on lower underparts, less uniform underwing-coverts with smaller carpal patches (usually restricted to greater coverts), tail typically more rufous, finely barred and with broad subterminal band. Very rufous *vulpinus* with unmarked, entirely rufous underparts and rufous-tinged upperwing-coverts extremely difficult to separate from some rufous Long-legged, but latter usually show contrast on upperwing between rufous lesser coverts and browner median and greater coverts. Juveniles of both species can be very difficult to separate. Rare dark morph *vulpinus* very similar to dark morph Long-legged but has, apart from structural differences, usually less distinct barring on underside of flight feathers and uppertail. **STATUS AND DISTRIBUTION** Scarce to uncommon Palearctic migrant and winter visitor to Sahel belt, from Mauritania and Senegal to Chad. Vagrant, Gambia, Ghana and N Cameroon. Distribution of races unclear; *cirtensis* known from Mauritania, nominate from Niger, both reported from Senegal, Togo and SW Nigeria. **NOTE** Specimen collected in Cape Verde Is in 1897 was provisionally identified as *B. rufinus* but this requires confirmation (Hazevoet 1995).

RED-NECKED BUZZARD *Buteo auguralis*
Buse d'Afrique

Plate 22 (3)

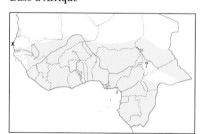

Other names E: Red-tailed Buzzard; F: Buse à queue rousse
L 40–50 cm (16–20"); WS 95 cm (37.5"). Medium-sized raptor with broad wings and relatively short rounded tail. **Adult** Head and upperparts dark brown with chestnut head-sides and nape. Tail bright rufous with narrow black subterminal band. Below, throat and breast dark brown; rest of underparts white variably blotched dark brown. Eye brown. Cere and feet yellow. In flight from above, readily identified by rufous tail; pale panel at base of primaries. From below, white underwing with black tips to flight feathers and black carpal markings continuing as irregular line along coverts.
Juvenile Brown above; tail brown barred black; *underparts entirely white* blotched black on breast and flanks. **VOICE** A loud, drawn-out, mewing *peeee-ah*. [CD5:56] **HABITS** Usually singly, in woodland, forest clearings and edges, and cultivated areas. Given to soaring on raised wings. **SIMILAR SPECIES** Common Buzzard has different underpart pattern. Dark head and breast of Short-toed Eagle and Beaudouin's and Black-breasted Snake Eagles may recall Red-necked Buzzard at distance, but former are much larger and have different flight silhouette.
STATUS AND DISTRIBUTION Not uncommon to scarce resident and intra-African migrant in forest and savanna zones. Moves north at onset of rainy season, returning south at end of rains. Breeds in south of range.

AUGUR BUZZARD *Buteo augur*
Buse augure

Plate 22 (6)

L 55–60 cm (22–24"); WS 132 cm (52"). *B. a. augur.* Large, distinctive buzzard with *very broad wings and conspicuously short, rounded tail.* **Adult male** Top of head and upperparts blackish with secondaries barred whitish; *tail bright rufous* (often with narrow black subterminal band); underparts white. In flight from below, underwings white with blackish trailing edge, primary tips and carpal crescents. **Dark morph** has entire underparts and underwing-coverts black. **Adult female** Similar but with throat black. **Juvenile** Brown above, wing-coverts tipped buff. Tail brown narrowly barred darker. Underparts buffish-white boldly streaked dark on throat, breast-sides and flanks. In flight from below, similar to adult but flight feathers indistinctly barred, dark wing edge less black and tail slightly longer and appearing dusky. **Dark juvenile** has underparts and underwing-coverts plain blackish-brown; flight and tail feathers heavily barred. **VOICE** A loud, rather nasal bark *ewah-ewah-ewah-...* [CD5:57] **HABITS** Perches on conspicuous vantage points. Soars on raised wings. Frequently hovers or hangs motionless in strong wind. In normal range occurs in open highlands. **STATUS AND DISTRIBUTION** Single claim, from N Cameroon, Nov 1993 (Sørensen *et al.* 1996) best regarded as doubtful as nearest breeding areas of this mainly E African species are in E Sudan and NE Congo-Kinshasa.

Genus *Aquila*

Medium-sized to large, mainly brown, eagles of open habitats. Plumage extremely variable; this variation enhanced by effects of wear and moult. Good views essential but some will remain difficult (or almost impossible) to identify. For discussion of plumage variations beyond the scope of this book, see Clark (1999), Forsman (1999), Harris *et al.* (1996) and Porter *et al.* (1981).

LESSER SPOTTED EAGLE *Aquila pomarina*   Plate 23 (4)
Aigle pomarin

L 57–64 cm (22.5–25.0"); WS 145–170 cm (57–67"). *A. p. pomarina.* Medium-sized, well-proportioned, dark brown eagle with relatively long, broad wings and a rather short, wide and rounded tail. Relatively small bill with round nostrils; yellow gape extending to below middle of eye. Rather narrow 'trousers' may produce long-legged appearance. In flight, *wing-coverts typically slightly paler than flight feathers*; below, *2 whitish carpal crescents diagnostic* (if present); above, white patch at base of inner primaries and pale U on uppertail-coverts. Often small white spot on centre of back. Flight quite buzzard-like, with rather fast wingbeats and head not protruding. Soars with wings held flat or 'hand' slightly depressed and tail fanned; when gliding, 'hand' slightly more arched. **Adult** Top of head and lesser and median upperwing-coverts usually paler brown, contrasting with blackish-brown flight feathers. Small and well-defined primary patch. Mantle often darker than coverts, producing 'saddle' effect. Rarely evenly darker and thus very similar to Greater Spotted Eagle. Eye pale yellowish-brown to yellow. **Juvenile** Variable, but typically darker than adult, with rufous-yellow to whitish patch on nape, irregular pale streaking on underparts, pale vent, and whitish tips to greater coverts (and sometimes median coverts). Narrow and regularly spaced barring on flight feathers only visible at close range. Primary patch and white U on uppertail-coverts more distinct. Narrow pale tips to greater coverts, trailing edge of wing and tail quickly abraded. Some dark birds have little or no contrast between coverts and flight feathers. Very rare **pale morph** has pale rufous or buffish underparts. Eye dark brown. **Immature** Adult plumage acquired in *c.* 4–5 years, following two moults. **VOICE** Usually silent. A barking *kow-kow-kow.* [CD1:92] **HABITS** Hunts from perch or on the wing, but in Africa principally feeds on termites, grasshoppers and other insects taken on ground. Also takes small mammals, lizards and nestlings. **SIMILAR SPECIES** Greater Spotted Eagle often difficult to distinguish, but typically darker, more uniformly coloured and heavier, with broader 'hand' and relatively shorter tail, thus appearing more compact; wing-coverts typically slightly darker than flight feathers (the reverse in Lesser Spotted); above, pale primary patch more diffuse; below, single white carpal crescent; 'trousers' more bushy; eye dark. Juvenile and immature Greater Spotted typically darker and more uniformly coloured than same age Lesser Spotted; juvenile also more spotted. Steppe Eagle larger and heavier with slightly longer wings (with more splayed 'fingers') and tail; gape extending to rear of eye, oval (not round) nostrils and bushy 'trousers'; in flight head projects more; bases of primaries barred (unbarred in Lesser and Greater Spotted). Juvenile Steppe has underwing with white band and white trailing edge. Immature Steppe may look very similar to immature Lesser Spotted, with narrow and fairly distinct pale band on greater underwing-coverts; best distinguished by combination of structural differences, absence of whitish carpal crescents and barring on underside of primaries; also lacks dark chin. Tawny Eagle typically paler, with heavier bill and no white on primary bases and uppertail-coverts. **STATUS AND DISTRIBUTION** Rare Palearctic migrant, recorded from NE Nigeria, Chad, N and SE Cameroon, SW CAR and N Congo. Possible record Mali. Mainly in E and S Africa, where passage migrant and winter visitor, Oct–Apr.

GREATER SPOTTED EAGLE *Aquila clanga*   Plate 23 (5)
Aigle criard

Other name E: Spotted Eagle

L 60–70 cm (23.5–27.5"); WS 155–180 cm (61–71"). Large, compact, blackish-brown eagle with broad wings and a relatively short, wide and rounded tail. Strong bill with round nostrils; yellow gape extending to below middle of eye. In flight, *wing-coverts typically as dark or slightly darker than flight feathers* (but sometimes paler); below, *a single whitish carpal crescent*; above, *diffuse white patch on primaries* (mainly formed by pale feather shafts). Soars and glides with wings slightly more arched than Lesser Spotted Eagle. **Adult** Entirely blackish-brown with a *whitish carpal crescent on outermost primaries* (very rarely a second diffuse crescent at base of greater primary-coverts), usually some whitish on uppertail-coverts; sometimes a small white spot on centre of back. Eye dark. **Juvenile** Typically even darker than adult with diagnostic heavily spotted upperwing-coverts and scapulars, and white trailing edge (some paler and less spotted); underparts usually streaked; vent pale. *Wing-coverts typically slightly darker than flight feathers.* Primary patch often more extensive. **Immature** Adult plumage acquired in *c.* 5–6 years. Rare and confusing **pale morph** ('*fulvescens*'), occurring at all ages, is uniformly pale buffish or yellowish-brown to rufous, diffusely streaked darker on underparts. Flight feathers as normal form, greater coverts dark, carpal crescent present. **VOICE** Usually silent. **HABITS** Poorly known in Africa. Perches in trees. Feeds on termites, grasshoppers and other insects; also small mammals, lizards, amphibians and birds. **SIMILAR SPECIES** Adult Lesser Spotted Eagle often difficult to distinguish, but typically paler brown and less heavy with less ample 'hand' and relatively longer tail, thus appearing more slender; upperwing more contrasting and pale primary patch more distinct; underwing has double (not single) whitish carpal crescent; wing-coverts typically slightly paler than flight feathers (the reverse in Greater Spotted); eye pale (not dark). Typical juvenile and immature differ from same age Greater Spotted by paler and more contrasting plumage, with paler lesser and median upperwing-coverts contrasting with blackish greater coverts and secondaries; juvenile also less spotted and (when fresh) has band of pale spots on greater coverts. Adult Steppe Eagle larger and more robust with slightly longer wings (with more splayed 'fingers') and tail; long gape extending to rear of eye, oval (not round) nostrils and more bushy 'trousers'; usually rusty nape patch; more protruding head in flight; bases of primaries barred (unbarred in Lesser and Greater Spotted), no carpal crescents. Pale '*fulvescens*' may be confused with pale Tawny Eagle (q.v.). **STATUS AND DISTRIBUTION** Palearctic vagrant, N Cameroon (Feb/Apr 1973) and Chad (Jan 1971, winter 1996). *TBW* category: THREATENED (Vulnerable).

TAWNY EAGLE *Aquila rapax*
Aigle ravisseur

Plate 23 (2)

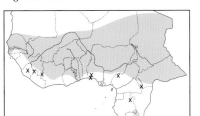

L 62–75 cm (24.0–29.5"); WS 165–185 cm (65–73"). *A. r. belisarius.* Medium-sized, well-proportioned but often rather ragged-looking eagle. Bill powerful with oval nostrils and *yellow gape extending to below centre of eye.* Baggy 'trousers'. Wings fairly long with splayed 'fingers'. Tail broad and rounded. Flight and tail feathers dark greyish with faint dark bars; 'fingers' contrastingly darker. Soars with wings held flat or 'hand' slightly depressed and tail fanned; when gliding, 'hand' slightly more arched. **Adult** Variable. Dark brown to tawny and even pale buffish, with dark-centred body feathers and upperwing-coverts giving streaked appearance. Flight and tail feathers dark with only very indistinct or no dark bars, and no terminal band. Above, creamy-white patch from lower back to uppertail-coverts; pale patch on inner primaries very faint, if present. In pale birds 3 inner primaries usually contrastingly pale, producing 'window' in dark area of underwing. Eye pale brown to yellowish. **Juvenile** Typically paler than adult, rufous-tawny bleaching to creamy-white (a few darker). Wings have narrow white trailing edge; tail tipped white. Pale tips to greater coverts form narrow line on centre of wing, both above and, more irregularly and less distinct, below. Pale 'window' on underside of 3 inner primaries always present. Eye dark. **Immature** Variable. White trailing edge lost. Eye becomes yellow. Adult plumage acquired in *c.* 4–5 years. **VOICE** Usually silent. A barking *kyow* and a guttural *kwork* uttered in display. [CD1:91] **HABITS** In wooded habitats. Hunts from perch or on the wing. Feeds on any animal matter, from small mammals and birds to insects (esp. termites). Takes carrion and robs other large birds of their prey. **SIMILAR SPECIES** Pale adult may appear similar to juvenile Imperial Eagle, but plumage of latter generally warmer coloured and primary patch more distinct. Imperial also often soars with wings slightly raised and glides with wings held level (not bowed). Darker adults similar to Steppe Eagle, but latter has dark trailing edge to wing and dark terminal tail band (lacking in Tawny), and usually has noticeable barring on flight and tail feathers. Greater and Lesser Spotted Eagles appear more compact, with broader, less protruding head; 'trousers' less baggy; pale carpal crescents; nostrils round. Lesser Spotted also has pale primary patch and white U on uppertail-coverts. **STATUS AND DISTRIBUTION** Not uncommon to scarce resident in savanna zones, incl. SE Gabon and Congo. Records suggest southward movement in dry season.

STEPPE EAGLE *Aquila nipalensis*
Aigle des steppes

Plate 23 (3)

L 62–81 cm (24–32"); WS 163–200 cm (64–79"). *A. n. orientalis.* Medium-sized to large, brown eagle with long, broad wings ending in deeply splayed 'fingers'. Flight and tail feathers greyish with rather indistinct but, in good light, noticeable blackish bars (incl. on base of primaries, which are unbarred in spotted eagles); 'fingers' contrastingly black. Head more protruding than in spotted eagles and 'hand' more ample than in Lesser Spotted. Powerful bill with oval nostrils and *long yellow gape extending to below rear of eye*. Chin and throat pale. Bushy 'trousers'. Soars and glides with wings slightly arched, as spotted eagles, but wingbeats slower. **Adult** Dark brown with usually paler patch on nape. Flight and tail feathers greyish (paler than in spotted eagles), and indistinctly barred and tipped dark, producing dark trailing edge to wing and dark terminal band to tail. Carpal area often dark and normally lacks white crescents. Usually whitish patch on base of inner primaries, small white patch on centre of back and some pale on uppertail-coverts (as spotted eagles). Eye dark brown. **Juvenile** Typically paler brown with *diagnostic whitish band on centre of underwing*, broad white trailing edge to wings and broad white terminal band to tail. Above, broad white tips to greater coverts form line on centre of wing, usually prominent whitish patch on primaries and broad white U on uppertail-coverts. **Immature** Variable, but has complete or partial white band on centre of wing and traces of white trailing edge to wings and tail. Older birds may lack pale central wing band, but can be identified by greyish flight and tail feathers with dark barring and dark tips. Adult plumage acquired in *c.* 5–6 years. **VOICE** Generally silent in Africa. **HABITS** Feeds on insects (esp. termites) on ground; also other small animals and carrion. **SIMILAR SPECIES** Both spotted eagles have slightly shorter wings and tail, smaller bill with shorter gape and round (not oval) nostrils, less bushy 'trousers'; less protruding head in flight and differently patterned flight feathers. Adult Greater Spotted Eagle separated from dark Steppe by uniformly dark flight feathers and whitish carpal crescent. Immature Lesser Spotted may look very similar to immature Steppe, with narrow and variably distinct pale band on centre of underwing; apart from structural differences, may be distinguished by 2 whitish carpal crescents, absence of barring on base of primaries and dark chin. Tawny Eagle is usually paler with slightly shorter tail and gape, pale patch on rump and uppertail, wings lacking distinct broad terminal bar and, below, dark carpal patch, flight and tail feathers having only faint or no barring, inner primaries often forming pale wedge below. Juvenile Steppe with vestigial white band on centre of underwing may be confused with juvenile Imperial Eagle (q.v.). **STATUS AND DISTRIBUTION** Palearctic vagrant, recorded from Mali, Niger, NE Nigeria, Chad, Cameroon and N Congo (Odzala NP). Occurs mainly in E and S Africa, where widespread and locally common migrant in open habitats, Sep–Apr. **NOTE** Formerly considered conspecific with Tawny Eagle *A. rapax* by some authors.

EASTERN IMPERIAL EAGLE *Aquila heliaca*   Plate 24 (1)
Aigle impérial

Other name E: Imperial Eagle
L 72–85 cm (28–33"); WS 180–220 cm (71–87"). Large. **Adult** Blackish-brown with pale tawny crown and nape and broad black terminal band to greyish tail. White 'braces' bordering mantle diagnostic but difficult to see. In soaring flight, long and parallel wings held flat and tail usually held closed. **Juvenile** Pale yellowish-brown narrowly streaked dark with paler rump and contrasting dark fight feathers and tail. From distance, streaks on breast and upper belly may appear as broad band. Innermost primaries have pale wedge (visible both from above and below); on upperwing, white tips to greater and median coverts form 2 narrow parallel lines. Wings more S-curved than in adult; tail held slightly spread when soaring. **Immature** Gradually acquires adult plumage over 6–7 years; becomes mottled blackish-brown and yellowish with adult head and tail pattern appearing early. **VOICE** Silent on migration. **HABITS** Feeds on variety of animal prey; also eats carrion. **SIMILAR SPECIES** Steppe Eagle is stockier with shorter, more wedge-shaped and uniformly coloured tail. Juvenile Steppe Eagle usually has white band on underwing and browner flight feathers and tail. **STATUS AND DISTRIBUTION** Vagrant, N Cameroon (imm, Waza NP, Nov 1993). Scarce Palearctic migrant to E Africa (Sudan, Ethiopia, Kenya). *TBW* category: THREATENED (Vulnerable). **NOTE** Formerly considered conspecific with Spanish Imperial Eagle *A. adalberti*.

WAHLBERG'S EAGLE *Aquila wahlbergi*   Plate 23 (1)
Aigle de Wahlberg

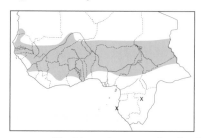

L 55–61 cm (22–24"); WS *c.* 141 cm (55.5"). Smallish, rather slender and variably coloured eagle with slight occipital crest (not always visible), baggy 'trousers' and dark eye. Characteristic silhouette when soaring, with head protruding, *long, parallel-edged wings held flat or slightly downcurved and long, square tail held closed*. **Adult** Commonest morphs uniformly dark to warm brown. Cere and feet yellow. In flight, appears entirely dark, with slightly paler flight feathers contrasting faintly with underwing-coverts. Uncommon **pale morph** has head and underparts white to buffy; upperparts and wing-coverts grey-brown fringed white. In flight, dark flight and tail feathers contrast with white underparts and underwing-coverts. Intermediate plumage variations, from pale buffish-brown to dark brown with white head, may occur. **Juvenile** As adult. **VOICE** Usually silent. A distinctive, clear and loud *kleeeee-ay*. Also a fast *kyip-kyip-kyip-kyip-...* [CD5:58] **HABITS** In wooded savanna and various other wooded habitats. Perches inconspicuously in tree or soars for long periods. Takes mainly birds, rodents and lizards. **SIMILAR SPECIES** Dark morph Booted Eagle has broader wings and tail, pale band across upperwing-coverts, white 'landing lights', and narrow white U on uppertail-coverts. Brown Snake Eagle has large head, yellow eyes, silvery underside to flight feathers and bare tarsi; also larger. Black Kite is slightly smaller but easily distinguished by forked tail. Other brown eagles noticeably larger, with different flight silhouette. **STATUS AND DISTRIBUTION** Not uncommon intra-African migrant in broad savanna belt; more numerous and widespread in dry season. Vagrant, Gabon and N Congo (Odzala NP).

GOLDEN EAGLE *Aquila chrysaetos*　　　　　　Plate 24 (2)
Aigle royal

L 75–90 cm (30–35"); WS 190–225 cm (75–89"). *A. c. homeyeri*. Very large with *rather long tail and long, supple wings held in shallow V when soaring and usually also when gliding*. Secondaries longest in centre, producing slight S-curve at rear edge of wings. **Adult** All-dark brown with rufous-yellow ('golden') crown and nape (usually noticeably contrasting with rest of plumage). In flight, diffuse pale panel on upperwing-coverts; body and underwing-coverts darker than flight feathers. Eye brown. Cere and feet yellow. **Juvenile** Dark chocolate-brown with *broad black terminal band to white tail and variably distinct white wing patches* (both features distinctive and visible from above and below; wing patches occasionally lacking). **Immature** Gradually acquires adult plumage over 6–8 years; distinctive tail pattern retained for 4–5 years. **VOICE** Usually silent. [CD1:94] **HABITS** Hunts on the wing or from prominent perch. Prey mainly medium-sized mammals, also large birds and reptiles; feeds on carrion. **SIMILAR SPECIES** All other *Aquila* eagles soar and glide with wings almost flat or slightly bowed downwards. Adult Imperial Eagle (unlikely to be encountered in same range) is blacker, incl. outer wing above, with more parallel-edged wings and narrower, squarer tail. **STATUS AND DISTRIBUTION** Scarce Palearctic migrant and winter visitor, Mauritania. Also reported from extreme NE Mali (Adrar des Ifôghas), where recently found breeding.

VERREAUX'S EAGLE *Aquila verreauxii*　　　　　Plate 24 (3)
Aigle de Verreaux

Other name E: Black Eagle
L 80–90 cm (31.5–35.5"); WS 190–210 cm (79"). Very large, powerful eagle with, in all plumages, *distinctive flight silhouette produced by long broad wings narrowing towards body and fairly long, often partially spread tail*. Soars with wings held in shallow V. **Adult** Entirely black except for narrow white V on mantle and white lower back and rump. In flight, has *large whitish 'windows' on primaries* (visible from above and below), and white Y on back. Eye dark. Cere, orbital ring and feet yellow. **Juvenile** *Crown and mantle tawny-rufous*; rest of upperparts blackish-brown with feathers edged buff, giving scaled appearance; head-sides to breast black becoming buffish-brown mixed with black on belly and thighs. In flight, primaries have pale buffish 'windows', flight feathers and tail indistinctly barred; from above, rump feathers white edged black, forming narrow white crescent at base of tail. **Immature** Gradually acquires adult plumage in *c.* 3–4 years. **VOICE** Usually silent. [CD5:59] **HABITS** In pairs or singly in rocky hill country and gorges. Glides along rock faces or soars, sometimes to great heights. Soars with wings raised in shallow V and primary tips upturned. Perches for long periods on crags. Feeds mainly on hyrax; also other mammals, birds and reptiles; occasionally carrion. **STATUS AND DISTRIBUTION** Rare resident, Niger (N Aïr) and NE Chad (Ennedi). Vagrant, E Mali (Hombori).

Genus *Hieraaetus*
Small to medium-large eagles of woodland and forest. Tails rather long. Juveniles markedly different from adults. Prey consists largely of birds and small mammals.

BONELLI'S EAGLE *Hieraaetus fasciatus*　　　　Plate 25 (6)
Aigle de Bonelli

L 55–72 cm (22–28"); WS 145–180 cm (57–71"). *H. f. fasciatus*. Medium-sized eagle with relatively small, protruding head and longish tail. **Adult** Above, dark brown with variable amount of white on mantle. Below, dark underwing-coverts form black band of variable width contrasting with white, lightly streaked body. Thighs irregularly streaked

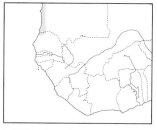

or barred. Tail greyish-white with black terminal band. **Juvenile** Underparts and underwing-coverts usually pale rusty-brown/orange-rufous, palest on belly. From below, flight feathers grey narrowly barred dark; primaries tipped dark; inner primaries often translucent, producing pale panel against light; coverts usually bordered by narrow black line (distinctive when present, but sometimes absent). Tail grey narrowly barred dark, lacking terminal band. Upper wing cinnamon-brown with pale patch on primaries. Rarely juvenile completely dark brown. **Immature** Gradually acquires adult plumage over 3 years, becoming darker brown; underparts streaked dark; broad dark bands on greater coverts and tail starting to show. **VOICE** Usually silent. [CD1:95] **HABITS** Soars less than most other raptors, with wings held flat or slightly arched and tail often closed. Glides on slightly arched wings with carpal joints pressed forward and almost straight rear edge. **SIMILAR SPECIES** From distance, pale European Honey Buzzard may appear similar in shape as it also glides on flat, angled wings, but shorter, rounder edged and sometimes slightly notched tail, less 'fingered' primaries and smaller size should distinguish it. In unfavourable circumstances adult Booted Eagle (pale morph) and Osprey may also be reminiscent of adult Bonelli's, but Booted Eagle is stockier and smaller with white underwing-coverts, and Osprey has black carpal patches. Adult African Hawk Eagle has conspicuous white 'windows' on primaries above and whiter secondaries below. At first glance juvenile/immature Bonelli's may present identification problems, but distinctive silhouette in flight should distinguish them from most other raptors. Juvenile African Hawk Eagle differs in white on primaries above and whiter secondaries below. **STATUS AND DISTRIBUTION** Scarce to rare Palearctic passage migrant and winter visitor, Mauritania and N Senegal (Djoudj NP).

AFRICAN HAWK EAGLE *Hieraaetus spilogaster*   Plate 25 (5)
Aigle fascié

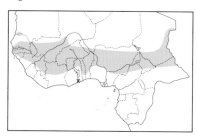

L 60–74 cm (24–29"); WS *c.* 142 cm (56"). Medium-sized eagle with long wings narrowing towards body and longish tail. **Adult male** *Appears black and white* from distance. Top of head and upperparts blackish-brown. Underparts white streaked black; thighs unstreaked. Tail dark grey barred black above with broad black terminal band. In flight from above has conspicuous *white 'windows' on primaries*. From below, mainly white with *black trailing edge* to wing and *broad black mottled bar on greater coverts; tail pale grey with broad black terminal band* and several indistinct narrow dark bars. Eye dark yellow. Cere and feet greenish-yellow. **Adult female** Larger and more heavily streaked below. **Juvenile** Top of head and upperparts dark brown. Underparts rufous with some narrow black streaks on breast-sides. Tail lacks distinct terminal band; dark brown above barred blackish; pale greyish below indistinctly and narrowly barred darker. Underwing-coverts rufous bordered by dark line; flight feathers whitish below with indistinct narrow dark bars and dusky trailing edge. Eye brown. Cere and legs greenish-yellow. **Immature** Gradually acquires adult plumage in 3–4 years. **VOICE** A melodious *klu-klu-klu-kluee* or *kluee-kluee* rather reminiscent of call of Wahlberg's Eagle but shorter. Also various other calls at nest. [CD5:60] **HABITS** Singly or in pairs in wooded savanna. Soars frequently, with wings held flat; perches conspicuously in trees. Preys mainly on large birds (esp. gamebirds) and mammals up to size of hare. **SIMILAR SPECIES** Ayres's Hawk Eagle is noticeably smaller, with more heavily marked underparts, heavily barred underwings and no 'windows' on primaries. Pale adult European Honey Buzzard may have dark underwing-coverts, but these do not form broad band; dark-tipped tail has 2 other distinct dark bars. Bonelli's Eagle has mainly dark underwings with dusky remiges contrasting with white body; above, dark brown with variable white patch on mantle. Underparts of juvenile Black Sparrowhawk of rufous-breasted form similar coloured to juvenile African Hawk Eagle, but former is smaller, has bare legs and different behaviour. Juvenile Ayres's differs from juvenile African Hawk Eagle in being smaller, with paler rufous underparts and much bolder barring on flight feathers below. Juvenile pale morph Booted Eagle has unbarred tail and different underwing pattern. Juvenile Bonelli's Eagle has upperwing lacking obvious white on primaries and underwing with dusky secondaries. **STATUS AND DISTRIBUTION** Uncommon to scarce resident in savanna belt, from S Mauritania to Sierra Leone east to Chad and CAR.

BOOTED EAGLE *Hieraaetus pennatus*   Plate 25 (2)
Aigle botté

L 45–55 cm (18–22"); WS 110–132 cm (43–52"). Small, buzzard-sized eagle with, in all plumages, *small white spot at base of neck* on leading edge of wing ('landing lights'), broad buff diagonal band on upperwing and narrow whitish crescent at base of square tail. Legs heavily feathered (hence name). Two morphs. **Adult pale morph** Brown above, *creamy-white below with contrasting blackish flight feathers*. Tail greyish and unbarred. Eye pale brown.

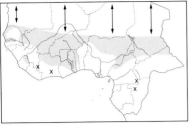

Cere and feet yellow. **Dark morph** entirely dark brown; wing pattern as pale morph. Intermediate, rufous birds occur. **Juvenile** Pale morph has head rufous-brown and underparts and underwing-coverts washed rufous. Dark morph as adult Both morphs have adult pattern in flight but can be aged by even, narrow white trailing edge to wings and tail (uneven in adult); trailing edge to wing noticeably S-curved. Gradually acquires adult plumage over 3 years. **VOICE** Usually silent on migration. [CD1:96] **HABITS** Soars on flat wings with tail often slightly spread; does not hover. Preys mainly on birds up to size of pigeon; also lizards, rodents and insects. **SIMILAR SPECIES** From above, Black Kite may appear similar to dark morph, but is more slender with shallowly forked tail and has flapping and gliding flight. Wahlberg's Eagle is longer winged, less stocky and soars with closed tail. Steppe Buzzard is stockier with dark carpal patches and shorter 'fingers' to more rounded wings; legs bare. Juvenile African and Ayres's Hawk Eagles may be confused with juvenile Booted but have different underwing patterns and barred tails. **STATUS AND DISTRIBUTION** Uncommon Palearctic migrant to Sahel zone, rarer in savanna belt. Also SE Cameroon and N Congo (Odzala NP), principally on passage.

AYRES'S HAWK EAGLE *Hieraaetus ayresii*　　　　　Plate 25 (3)
Aigle d'Ayres

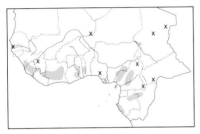

L 45–60 cm (18–23"); WS *c.* 124 cm (49"). Small, stocky eagle with broad, rounded wings and fairly long tail. **Adult male** Top of head and upperparts typically blackish-brown. Underparts white boldly streaked and spotted/blotched with black, incl. thighs. Tail dark grey barred black with broad black terminal band. Short, erectile occipital crest. Plumage variable; variations include: white forehead, short white supercilium, sparsely marked underparts, and feathers of upperparts broadly fringed white. In flight from above uniformly dark; from below, wing-coverts heavily mottled black and white, flight feathers white heavily barred black; tail pale grey with broad black terminal band and several indistinct narrow dark bars. Typically has *small white spot at base of neck* on leading edge of wing ('landing lights'). Eye yellow to orange. Cere and feet greenish-yellow. **Adult female** Larger and usually more heavily marked below. **Juvenile** Upperparts grey-brown with pale fringes to feathers; head and underparts pale rufous to whitish with some narrow black streaks on breast; tail dark brown barred blackish. In flight from below, wing-coverts pale rufous, flight feathers and tail white barred blackish. Eye pale grey-brown. Cere and feet greenish-yellow. **Immature** No information on time required to attain adult plumage. **VOICE** Usually silent. A melodious *whueeep-whip-whip-whip-whip-whueeep* in aerial display; also *whip-whip-whip-...* [CD5:61] **HABITS** Usually singly, in forest and dense woodland. Perches inconspicuously for long periods in cover of leafy tree. Catches medium-sized birds in flight following dive from soaring position or fast dash from perch. **SIMILAR SPECIES** African Hawk Eagle has mainly white, much less heavily marked underparts, pale 'windows' on primaries above, and is larger. Pale adult European Honey Buzzard has longer wings, smaller head, pale tail with dark tip and 2 other dark bars. Juvenile African Hawk Eagle differs from juvenile Ayres's in being larger, with richer rufous underparts, less bold barring on underwing and plain (not scaled) upperparts. Juvenile pale morph Booted Eagle has unbarred remiges and rectrices. Cassin's Hawk Eagle has slightly shorter wings and longer tail; both adult and juvenile paler below. **STATUS AND DISTRIBUTION** Uncommon to scarce resident in forest zone and forest outliers in savanna, from W Guinea to Liberia east to S Chad and Congo. Rare, Gambia. **NOTE** The name *H. dubius*, sometimes used for this species, may actually refer to *H. pennatus* (Dowsett & Dowsett-Lemaire 1993).

LONG-CRESTED EAGLE *Lophaetus occipitalis*　　　　　Plate 25 (1)
Aigle huppard

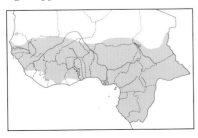

L 53–58 cm (21–23"); WS *c.* 115 cm (45"). Unmistakable. Medium-sized *blackish eagle with long, loose crest* and, *in flight, large white patches ('windows') on primaries both above and below.* **Adult** Blackish-brown with white edge to carpal joint and white thighs. Tail barred black and white. Thighs brown in some. Large yellow gape extends to below eye. Underwing black on coverts and white barred black on flight feathers. Eye, cere and legs yellow. **Juvenile** Similar to adult but with shorter crest and grey eye. Crest attains adult length after 3 months; eye yellow after 9 months. **VOICE** Vocal. A high-pitched, nasal *keeee-aah* and a series of sharp *kikikikikikeeah* with variations, uttered from perch or in soaring display flight. Also a short nasal *kwoh*. [CD5:62] **HABITS** Frequents wooded savanna, cultivation and forest edges. Perches conspicuously and for long periods atop trees or poles. Captures prey (mainly rodents) on ground. Flies with rapid shallow wingbeats interspersed by glides; soars with wings held flat. **STATUS AND DISTRIBUTION** Uncommon to not uncommon resident throughout savanna and forest zones. Apparently sometimes nomadic.

CASSIN'S HAWK EAGLE *Spizaetus africanus*   Plate 25 (4)
Aigle-autour de Cassin

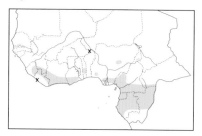

L 50–61 cm (20–24"); WS *c.* 120 cm (47"). Rather small forest eagle with *short rounded wings and longish tail.* **Adult** Mainly black and white. Head and upperparts blackish-brown. *Underparts white marked with black on breast-sides, axillaries* and thighs. *Underwing mainly black on coverts,* barred on flight feathers. Tail dark brown above, whitish below with 3 black bars and a broader subterminal band. Eye yellowish to brown. Cere and feet yellow. **Juvenile** Head pale rufous streaked blackish; upperparts grey-brown, feathers edged buff. Underparts mainly white streaked and spotted blackish; breast-sides and flanks pale rufous. Tail grey-brown barred darker. **Immature** No information on period required to attain adult plumage. **VOICE** Vocal. A loud *ku-ku-wee* or *ku-wee ku-wee* uttered in soaring display flight; also a drawn-out, high-pitched *weeeee-eh* from perch. [CD5:63] **HABITS** Inadequately known. An unobtrusive, infrequently observed forest dweller. Remains perched for long periods within canopy. Displays above forest or over adjacent open areas. Preys mainly on birds and squirrels. **SIMILAR SPECIES** Ayres's Hawk Eagle has slightly longer wings and shorter tail; both adult and juvenile darker below; adult also heavily barred. Black Sparrowhawk also has black markings on breast-sides but has white underwing-coverts, unfeathered legs and different flight silhouette, with relatively shorter wings and longer tail. Congo Serpent Eagle has unfeathered legs, relatively shorter wings and longer tail, and is smaller. **STATUS AND DISTRIBUTION** Scarce to locally not uncommon forest resident, from Sierra Leone and SE Guinea to Togo and from Nigeria to SW CAR and Congo. Also claimed from SW Niger.

CROWNED EAGLE *Stephanoaetus coronatus*   Plate 24 (4)
Aigle couronné

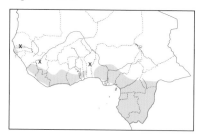

L 80–99 cm (31.5–39.0"); WS 163–209 cm (64–82"). Very large, powerful eagle with broad rounded wings and long tail. **Adult male** Head and throat dark brown; upperparts blackish-brown. Underparts boldly barred and blotched black, rufous and buff. Tail grey above, white below with 3 broad blackish bars. Short, erectile occipital crest. In flight from below, underwings have diagnostic *rufous coverts* bordered by bold black-and-white greater coverts, white flight feathers with broad black trailing edge and 3 narrower parallel black bars. Bill black with orange-yellow gape flanges; cere grey. Eye pale yellow. Feet yellow. **Adult female** Noticeably larger and usually more heavily marked below; flight feathers have only 2 narrow black bars. **Juvenile** Head and underparts white with rufous-buff wash on breast and small black spots on thighs. Upperparts greyish with white edges to feathers. Underwing pattern as adult but wing-coverts pale rufous-buff and flight feathers have 3–4 bars; tail has 4 blackish bars. Eye grey. **Immature** Gradually acquires adult plumage in *c.* 4 years. **VOICE** Vocal. A far-carrying, melodious *kewee-kewee-kewee-...* (male) and a deeper *kowi-kowi-kowi-...* (female), uttered in display flight, esp. in hot midday hours. [CD5:64] **HABITS** In forest and gallery forest. Undulating display flights over territory, often at great height, performed singly or by pair. Perches unobtrusively in canopy. Preys on monkeys and other mammals to size of duiker. **SIMILAR SPECIES** White underparts of juvenile Martial Eagle may be reminiscent of juvenile Crowned in flight, but former has longer wings lacking distinct barring and shorter tail. **STATUS AND DISTRIBUTION** Not uncommon to rare resident in forest zone and forest outliers in savanna, from Gambia to Liberia east to SW CAR and Congo.

MARTIAL EAGLE *Polemaetus bellicosus*   Plate 24 (5)
Aigle martial

L 78–86 cm (31–34"); WS 195–260 cm (77–102"). Very large, powerful eagle with broad, flat-crowned head, long broad wings and *short tail.* **Adult male** Head, upperparts, throat and upper breast dark brown. Underparts white with dark brown speckles. Tail dark brown above, greyish below narrowly barred blackish. Short, erectile occipital

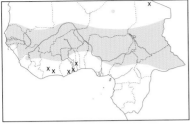

crest. In flight, dark wings and breast contrast with white lower underparts. Flight feathers narrowly and indistinctly barred. Eye yellow. Cere and feet yellowish or blue-grey. **Adult female** Usually more heavily spotted below; larger. **Juvenile** Top of head and upperparts grey with pale edges to feathers; underparts plain white. In flight appears mainly white below with narrowly and indistinctly barred flight feathers and tail. Eye dark brown. **Immature** Acquires adult plumage in *c*. 6–7 years. **VOICE** Usually silent. A rapid *klee-klee-klee-klooee-klooee...*, uttered in display flight and from perch. [CD5:65] **HABITS** Singly or in pairs in wooded savanna. Soars at great height with wings held flat and primary tips slightly upturned. Occasionally hovers. Hunts mostly on the wing, also from prominent perch, preying on mammals, large birds and reptiles; also eats carrion. **SIMILAR SPECIES** White underparts of juvenile Crowned Eagle may be reminiscent of juvenile Martial in flight, but former has shorter and distinctly barred wings and obviously longer tail. In flight, adult Black-breasted Snake Eagle appears mainly white below with contrasting black head and upper breast, and distinct bars on flight feathers and tail; also noticeably smaller; perched it shows unspotted underparts and bare legs; ranges do not overlap in our region. **STATUS AND DISTRIBUTION** Uncommon resident in savanna belt, from S Mauritania to Sierra Leone east to Chad and CAR. Also SE Gabon (Lékoni, status?).

SECRETARY BIRD
Family SAGITTARIIDAE (World: 1 species)

A very large, unique, strange-looking and mainly terrestrial bird of prey with very long, strong legs and short toes adapted for walking and killing prey by impact. Sexes similar. Prey includes reptiles, small rodents and insects, caught on the ground. True relationships unclear. Endemic to Africa.

SECRETARY BIRD *Sagittarius serpentarius*
Plate 15 (6)

Messager serpentaire

Other name F: Serpentaire

L 125–150 cm (47–59"); WS 212 cm (83.5"). Unmistakable. **Adult** *Mainly grey with bare reddish face and long plume-like erectile feathers on nape tipped black.* Rump and belly black. Tail narrowly tipped white with black subterminal band. *Long legs yellow with black-feathered tibia* ('trousers'). Cere bluish grey; legs greyish-pink. In flight, black flight feathers contrast with pale grey wing-coverts and body; very long central tail feathers project beyond long legs, producing distinctive silhouette. **Juvenile** Similar, but facial skin yellow and central tail feathers shorter. Underwing-coverts tinged brownish. **VOICE** Usually silent. [CD5:66] **HABITS** Usually singly, in open savanna. Strides slowly across grassland, searching for prey. Also soars high, like vulture. **STATUS AND DISTRIBUTION** Rare to scarce or uncommon resident in Sahel and dry-season visitor to northern savannas.

FALCONS
Family FALCONIDAE (World: *c*. 60 species; Africa: 20; WA: 16)

Small to medium-sized diurnal birds of prey with long, pointed wings, narrow, longish tail, hooked bill and curved talons. One genus in the region: *Falco* (typical falcons). Sexual dimorphism marked or at least present to some degree; females usually larger. Juveniles most like adult females but often darker and more streaked; adult plumage usually acquired during 2nd calendar year. Aerial hunters; hunt by stooping or hovering. Prey include birds, small rodents, reptiles and insects. Normal flight fast and strong, consisting of fast wingbeats alternating with glides. Occur in open or relatively open habitats.

LESSER KESTREL *Falco naumanni*
Plate 26 (2)

Faucon crécerellette

L 25–33 cm (10–13"); WS 58–74 cm (23–29"). Small falcon resembling Common Kestrel but slightly smaller and

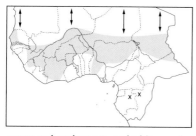

central tail feathers usually slightly protruding. When perched, *wingtips reach black subterminal tail band.* **Adult male** Head pale blue-grey. Upperparts plain rufous-brown with blue-grey greater coverts. Tail grey with broad blackish subterminal band. Underparts pale rufous-buff sparsely marked with small blackish spots. In flight, *tricoloured upperwings exhibit diagnostic blue-grey panel between dark outer wing and rufous coverts.* Underwing mainly white with dark tips and often diffuse trailing edge. **Adult female** Very similar to adult female Common Kestrel of nominate race but usually with more 'open' face pattern and less distinct moustachial stripe, finer streaking on crown and underparts, and whiter underwings with more distinct dusky tip. Claws white (black in Common). **Juvenile** As adult female. **Immature male** Lacks blue-grey wing panel, but usually possesses new plain central tail feathers. Upperparts gradually become less marked. **VOICE** Usually silent away from breeding grounds. [CD2:1a] **HABITS** Often in small or even large groups (of 10s–100s of birds) over various open habitats. Feeds principally on insects, which are caught in the air or on ground. **SIMILAR SPECIES** Adult male Common Kestrel is slightly larger, has dark moustachial stripe, spotted upperparts, and boldly spotted underparts. Adult female/juvenile Common Kestrel very similar to adult female/juvenile Lesser (see above). Also beware confusion of immature male Lesser with adult male Common Kestrel (q.v. and above). **STATUS AND DISTRIBUTION** Uncommon to not uncommon Palearctic migrant (Sep–Apr) from Mauritania and Senegambia to Chad and N CAR, reaching coast in Ivory Coast and Nigeria. Also NE Gabon (on passage; uncommon) and N Congo *TBW* category: THREATENED (Vulnerable).

COMMON KESTREL *Falco tinnunculus*   Plate 26 (1)
Faucon crécerelle

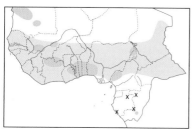

Other name E: Rock Kestrel
L 30–38 cm (12–15"); WS 65–80 cm (26.0–31.5"). Medium-small falcon with rather long wings and tail. When perched, *wingtips do not reach black subterminal tail band.* **Adult male** *tinnunculus* Head grey with *dusky moustachial stripe and dirty white cheeks.* Upperparts rufous-brown spotted black. Tail grey with broad black subterminal band and narrow white tip. Underparts buff densely spotted black. Dark flight feathers contrast with chestnut upperwing-coverts and upperparts in flight; underwing pale, densely barred and spotted dark. Afrotropical *rufescens* overall darker and more heavily marked; head darker slate; underparts more rufous; tail has some barring. Island race *neglectus* has upperparts and tail heavily barred; head and tail with little or no grey. *F. t. alexandri* similar but crown greyish streaked dark; tail grey. **Adult female** *tinnunculus* Head brownish, finely streaked darker. Upperparts and tail brown barred black; tail has broad subterminal band. Underparts and underwing more heavily marked than male. *F. t. rufescens* darker and more heavily marked overall. **Juvenile** As adult female but more boldly streaked below. **VOICE** A sharp, piercing *kee-kee-kee-kee-...*; also a shrill, vibrant *krree-e-e-e-krree-e-e-e-e...* (begging call). [CD2:1b] **HABITS** Singly or in pairs in all open habitats, esp. cultivation and savannas; also rocky outcrops and towns. Flies with fast shallow wingbeats interspersed with glides; frequently hovers with tail fanned. Also frequently soars on slightly blunt-looking wings (recalling small *Accipiter*). Feeds principally on insects; also lizards and small mammals. Hunts from perch or in hovering flight. Residents breed on inselbergs; also on high-rise buildings in towns and cities. **SIMILAR SPECIES** Adult male Lesser Kestrel is similar to adult Common Kestrel of nominate race, but has bluer head without moustachial stripe, unmarked upperparts; blue-grey wing panel, less marked underparts and whiter underwing. Adult female Lesser Kestrel (q.v.) extremely similar to adult Common Kestrel of nominate race. **STATUS AND DISTRIBUTION** Almost throughout. Nominate is not uncommon to common Palearctic migrant; vagrant, Cape Verde Is. Widespread Afrotropical resident *rufescens* uncommon to locally common. *F. t. alexandri* and *neglectus* are common residents, Cape Verde Is (*F. t. alexandri*, Sal to Brava; *neglectus*, Santo Antão to São Nicolau).

FOX KESTREL *Falco alopex*   Plate 26 (3)
Crécerelle renard

Other name F: Faucon renard
L 35–42 cm (14–17"); WS *c.* 90 cm (35"). *Very rufous, long-winged and long-tailed falcon.* **Adult** Entirely rufous (foxy-red) narrowly streaked black. Tail narrowly barred black. In flight from above, mainly blackish flight feathers contrast with rufous coverts. Underwing tipped black; flight feathers whitish indistinctly barred black; underwing-coverts pale rufous-buff. **Juvenile** As adult but dark streaking on upperwing-coverts more prominent and broader black barring on tail. **VOICE** A sharp *kee-kee-kee-kee-...* uttered near breeding sites; otherwise mainly silent. [CD5:67]

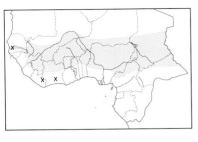

HABITS Singly or in pairs at inselbergs and steep cliffs in savanna. Flight buoyant; rarely hovers. Feeds mainly on insects; also lizards and small mammals. SIMILAR SPECIES Adult female and juvenile Common Kestrel are much more heavily marked above and below, incl. on underwings, and have relatively shorter tail with broad subterminal band. STATUS AND DISTRIBUTION Uncommon to not uncommon and local resident in savanna zone, from S Mauritania to Sierra Leone east to Chad and CAR. Subject to seasonal movements: north with rains, south in dry season. Vagrant, Senegambia. Rare dry-season visitor, Liberia.

GREY KESTREL *Falco ardosiaceus*

Faucon ardoisé

Plate 27 (1)

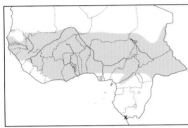

L 30–39 cm (12–15"); WS *c.* 70 cm (27"). Unmistakable. **Adult** *Wholly slate-grey* with bright yellow cere, orbital ring and legs. Feathers have dark shaft streaks, esp. on head. Primaries blackish. **Juvenile** As adult; paler grey on abdomen. VOICE A shrill, rasping, vibrant twitter. Usually silent. [CD5:68] HABITS Singly in savanna, farmland, large clearings and recently burnt areas. Perches on conspicuous vantage point. Never hovers. Preys largely on insects and reptiles, also small mammals and birds. SIMILAR SPECIES Usually none, but see vagrant Sooty Falcon. STATUS AND DISTRIBUTION Not uncommon to scarce resident; almost throughout, except arid north. Rare in southeast, but known from SE Gabon and Congo. Presumed to be mainly sedentary, but seasonal movements reported, north with rains, south in dry season.

RED-NECKED FALCON *Falco chicquera*

Faucon chicquera

Plate 27 (2)

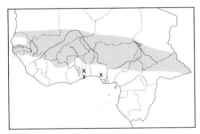

Other name F: Faucon à cou roux
L 29–36 cm (11–14"); WS 65–80 cm (26.0–31.5"). *F. c. ruficollis.* Medium-small falcon. When perched, wingtips do not reach tip of tail. **Adult** *Top of head and nape bright chestnut contrasting with white cheeks*; small dark moustachial stripe. Upperparts blue-grey densely barred black. Tail blue-grey barred black with broader subterminal band and narrow white tip. Underparts white narrowly and densely barred black except on throat; upper breast has pale rufous-buff band. In flight, dark above with blackish primaries; below, dense barring appears greyish from distance; black subterminal tail band. **Juvenile** Duller. Top of head and nape dull brown. Upperparts grey-brown. Underparts pale rufous-buff with black bars on breast-sides and flanks, and some small black spots on breast and belly. VOICE A shrill, harsh, scolding *kheep-kheep-kheep-*... and a soft *k-krrree-up.* [CD5:70] HABITS Singly or in pairs. Typically associated with palm trees, esp. *Borassus* palms. Feeds mainly on birds caught in the air following low, dashing flight; occasionally bats, lizards and small rodents. Esp. active at dusk. SIMILAR SPECIES Adult Lanner Falcon is much larger and has more restricted area of rufous on head, and almost unmarked underparts. STATUS AND DISTRIBUTION Uncommon resident in savanna belt, from S Mauritania to Guinea east to Chad and CAR.

RED-FOOTED FALCON *Falco vespertinus*

Faucon kobez

Plate 27 (5)

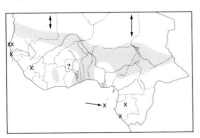

Other name E: Western Red-footed Falcon
L 28–31 cm (11–12"); WS 60–78 cm (24–31"). Rather small, with flight action and silhouette intermediate between Common Kestrel and Eurasian Hobby. **Adult male** *Dark blue-grey with rufous vent and thighs.* Cere, orbital ring and feet reddish. In flight, *silvery grey flight feathers contrast with rest of plumage.* **Adult female** *Crown, nape and underparts orangey*; small black mask around eye; throat and cheeks white. Upperparts slate-grey barred dark (feathers patterned with dark subterminal bands and pale tips). In flight, barred flight and tail feathers with dark trailing edge contrast with orangey underwing-

coverts and body. **Juvenile** Top of head brown with pale forehead and broad pale neck collar; small black mask around eye and short moustachial stripe contrast with white cheeks and throat. Upperparts greyish-brown; feathers fringed rusty-buff. Uppertail barred. Underparts buffish variably streaked dark brown. Cere, orbital ring and feet yellowish. In flight, dark flight feathers contrast with rest of upperparts; underwing with black-barred flight feathers, dark trailing edge, and variably dark-barred underwing-coverts. **Immature male** Has variable amount of blue-grey in plumage and rufous on vent, but underwing barred as in juvenile. **VOICE** Usually silent away from breeding grounds. [CD2:2a] **HABITS** Gregarious, over various open habitats. Feeds principally on insects, caught and eaten on the wing. Regularly hovers; may briefly walk on ground. **SIMILAR SPECIES** Juvenile Eurasian Hobby differs from juvenile Red-footed by darker head pattern, darker and more uniform upperparts and uppertail, more boldly streaked underparts, longer, more pointed wings and shorter tail. **STATUS AND DISTRIBUTION** Rare to uncommon Palearctic passage migrant almost throughout. Vagrant, São Tomé (Nov 1954). Winters in S Africa.

AMUR FALCON *Falco amurensis*   Plate 27 (6)
Faucon de l'Amour

Other name E: Eastern Red-footed Falcon

L 26–32 cm (10–13"); WS 58–75 cm (23.0–29.5"). **Adult male** As Red-footed Falcon but with *white underwing-coverts*. **Adult female** *Crown slate-grey with white forehead and short black moustache contrasting with white cheeks and throat.* Upperparts and tail slate-grey barred blackish. Underparts white barred and blotched black, and washed pale rufous on lower belly, vent and thighs. Cere, orbital ring and feet orange-red. In flight, white below with boldly barred flight feathers, black-mottled underwing-coverts, and broadly barred tail (subterminal band broadest). **Juvenile** Similar to adult female but upperparts browner and fringed buff; underparts more streaked. Soft parts paler. **VOICE** Usually silent. **HABITS** As Red-footed Falcon. **SIMILAR SPECIES** See Red-footed Falcon. **STATUS AND DISTRIBUTION** Palearctic vagrant, Gabon (Lopé, Oct 1993). Passage migrant to E Africa, wintering in S Africa.

ELEONORA'S FALCON *Falco eleonorae*   Plate 27 (8)
Faucon d'Eléonore

L 36–42 cm (14.0–16.5"); WS 84–105 cm (33–41"). Medium-large, very slender falcon with long wings and tail. All plumages possess diagnostic underwing pattern with *darker underwing-coverts contrasting with paler flight feathers*. **Adult** Two morphs, with intermediates. **Pale morph** has blackish top of head and well-defined moustachial stripe, contrasting strongly with white cheeks and throat. Upperparts blackish. *Underparts orange-buff to rufous-orange narrowly streaked blackish.* Tail narrowly barred below. Underwing-coverts blackish contrasting with paler greyish bases to flight feathers. **Dark morph** Entirely brownish-black. **Juvenile** Browner than pale morph. Underparts entirely buffish; streaked dark from breast downwards. In flight, underwing shows dusky tips and trailing edge; pale bases to barred flight feathers contrast with dark brown looking underwing-coverts; tail feathers strongly barred. **VOICE** Usually silent away from breeding grounds. [CD2:2b] **HABITS** Feeds on insects and small birds, usually caught in effortless, fluid flight. **SIMILAR SPECIES** Eurasian Hobby similar to pale morph, but has white (not orangey) underparts, rufous vent and thighs, noticeably smaller size and different silhouette with shorter tail. Peregrine Falcon is compact with different underwing pattern. Sooty Falcon is slightly smaller than dark morph Eleonora's, with shorter tail; lacks contrasting underwing pattern. Adult male Red-footed Falcon has rufous-red vent and thighs; silvery-grey primaries evident in flight. Juvenile Eurasian Hobby differs from juvenile Eleonora's in paler and more boldly streaked underparts. **STATUS AND DISTRIBUTION** Palearctic vagrant, N Mauritania (Nov). Breeds in Mediterranean and along Atlantic coast of Morocco and Canary Is; migrates east through Mediterranean to winter in E Africa and Madagascar (Nov–Mar).

SOOTY FALCON *Falco concolor*   Plate 27 (7)
Faucon concolore

L 32–38 cm (12.5–15.0"); WS 73–90 cm (29–35"). Slender, long-winged and long-tailed falcon. Central tail feathers often slightly protruding, creating *wedge-shaped tail*. When perched, wingtips reach tip of tail. **Adult** *Entirely blue-grey with blackish tips to wings and tail.* Cere, orbital ring and feet yellow. In flight, uniform underwings and tail with dark tips. Darkness of plumage varies; older birds darker (sometimes blackish). **Juvenile** Head pattern as Eurasian Hobby, but *cheeks and throat buffy* (not white). Upperparts grey; feathers narrowly fringed buff. Underparts brownish-buff with diffuse dark streaking. In flight from above, grey with darker flight feathers and tail tip; from below, barred underwing with dusky tips and trailing edge, tail barred with dark, unbarred central tail feathers projecting slightly and broad, dark subterminal tail band. **VOICE** Usually silent away from breeding grounds. **HABITS** Singly or in small flocks. Feeds mainly on insects and small birds, caught on the wing. **SIMILAR SPECIES** Dark morph Eleonora's Falcon is slightly larger with longer tail; overall generally much darker; blackish underwing-coverts contrast with paler flight feathers. Adult male Red-footed Falcon has blunter wings, silvery-grey primaries and rufous vent and thighs. Grey Kestrel is much less slender with noticeably shorter wings and square tail. Juvenile Eurasian Hobby differs from juvenile Sooty by whiter cheeks and throat and whiter, more boldly streaked underparts; in flight lacks dark trailing edge to wings and dark subterminal tail band. Juvenile Eleonora's has paler flight feathers contrasting with darker underwing-coverts, no dark subterminal tail band, and different silhouette. Juvenile Red-footed Falcon has tail barred above and below; also different silhouette. **STATUS AND DISTRIBUTION** Palearctic vagrant, E Chad (May–Jun). Claimed from E Mali (Sep 1979). Breeds from E Libya and Egypt to Jordan, coasts of Red Sea and Arabian Gulf; winters in SE Africa and Madagascar (Oct–May).

MERLIN *Falco columbarius* — Plate 26 (4)
Faucon émerillon

L 24–33 cm (9.5–13.0"); WS 50–69 cm (20–27"). *F. c. aesalon. Smallest falcon of the region.* Compact, with relatively short, pointed wings and moderately long tail. When perched, wingtips do not reach tip of tail. **Adult male** Head pattern rather indistinct, with darkish crown, faint moustachial stripe and diffuse pale nape patch. *Upperparts blue-grey.* Tail blue-grey with broad black subterminal band. Underparts off-white to pale rusty, variably streaked dark. **Adult female** Faint moustachial stripe. Upperparts brown; feathers variably fringed and spotted rufous or buff. Tail barred. Underparts buffish heavily streaked dark brownish. Underwing coarsely barred. **Juvenile** As adult female. Typically has whitish patch on nape. **VOICE** Usually silent away from breeding grounds. [CD2:3a] **HABITS** Frequents open habitats. Preys mainly on small birds, caught after low flight. **STATUS AND DISTRIBUTION** Palearctic vagrant, N Senegal (Nov 1982). Normal southernmost wintering range NW Africa, where scarce.

EURASIAN HOBBY *Falco subbuteo* — Plate 27 (4)
Faucon hobereau

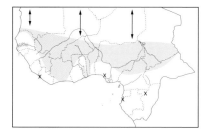

Other names E: Hobby, Northern Hobby
L 29–36 cm (11–14"); WS 70–92 cm (28–36"). *F. s. subbuteo.* Elegant falcon with long, pointed wings and medium-length tail. When perched, wingtips reach or extend beyond tip of tail. Flight action and silhouette may recall swift *Apus*. **Adult** Dark top of head with *well-defined, narrow black moustachial stripe contrasting with white cheeks and throat.* Upperparts dark slate-grey. Underparts whitish boldly streaked blackish; *vent and thighs bright rufous.* In flight, uniformly dark above; *underwing densely barred greyish* (appearing dark from distance). Tail plain above, barred below. **Juvenile** Browner than adult, with paler forehead; feathers of upperparts narrowly fringed buff. Underparts buff-white boldly streaked blackish; no rufous on vent and thighs. **VOICE** Usually silent away from breeding grounds. [CD2:3b] **HABITS** Usually singly; on migration occasionally in flocks, over various open habitats. Feeds mainly on insects and small birds, caught in the air. Insects eaten on the wing. **SIMILAR SPECIES** Adult pale morph Eleonora's Falcon has noticeably longer wings and tail, overall darker plumage with less prominent streaking on orangey underparts, and distinctive, two-toned underwing pattern. Peregrine Falcon is heavier and more stocky, with broader wings and paler rump. Juvenile Red-footed Falcon differs from juvenile Eurasian Hobby in dark flight feathers contrasting with paler upperwing-coverts and relatively longer tail. Juvenile Eleonora's has distinct dusky tip and trailing edge to wings. Juvenile Sooty Falcon has grey upperparts darkening on flight feathers and tip of wedge-shaped tail, and more diffusely streaked, brownish-buff underparts. **STATUS AND DISTRIBUTION** Scarce to uncommon Palearctic passage migrant almost throughout. Winters in S Africa.

AFRICAN HOBBY *Falco cuvierii* — Plate 27 (3)
Faucon de Cuvier

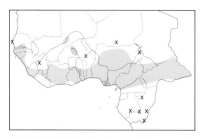

Other name F: Hobereau africain
L 28–31 cm (11–12"); WS *c.* 70 cm (27.5"). Distinctive, swift and slender falcon with long, pointed wings and rufous underparts. **Adult** Head, upperparts and tail blackish-slate. Short moustachial stripe. *Underparts orange-rufous, palest on throat, with some indistinct, fine black streaks.* Cere, orbital ring and feet yellow. In flight from below, mainly orange-rufous; flight feathers narrowly barred dusky. Tail barred below. **Juvenile** Duller. Feathers of crown, nape and upperparts narrowly fringed rufous (often forming variable rusty nuchal patch). Underparts heavily streaked black. Tail more heavily barred below. **VOICE** A shrill *kee-kee-kee-...* and a short, sharp *kik.* [CD5:71] **HABITS** Usually singly or in pairs in wooded savanna, forest–savanna mosaic and derived savanna. Feeds mainly on insects caught and eaten on the wing; also takes

small birds. Esp. active at dusk. **SIMILAR SPECIES** Eurasian Hobby has similar jizz in flight, but rufous underparts of African Hobby should preclude misidentification. **STATUS AND DISTRIBUTION** Uncommon or scarce to locally not uncommon resident in savanna zone and extensive clearings in forest zone almost throughout.

LANNER FALCON *Falco biarmicus*
Faucon lanier

Plate 26 (7)

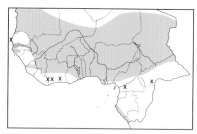

L 38–52 cm (15.0–20.5"); WS 90–115 cm (35–45"). Large falcon with long, relatively blunt-tipped wings. **Adult** *abyssinicus Crown and nape rufous* bordered by black stripe from eye to nape; narrow blackish moustachial stripe contrasts with whitish cheeks. Upperparts blue-grey or grey-brown, paler on rump. Tail narrowly barred dark above and below. Underparts whitish washed pinkish or creamy-buff from breast downwards and variably, finely and sparsely spotted brown, mostly on flanks and thighs. Underwing pale; flight feathers narrowly barred dusky. Cere, orbital ring and feet yellow. N African *erlangeri* and *tanypterus* are paler. **Juvenile** Crown and nape paler, buffish-white to pale rufous; eye-stripe and moustache as adult Upperparts dark brown, often including crown; throat whitish; rest of underparts heavily streaked blackish-brown. Underwing-coverts dark with some pale mottling, contrasting with pale flight feathers. Cere and orbital ring greyish; feet pale yellow. **VOICE** Usually silent except near breeding site. A raucous *kreh-kreh-kreh-...* and a shrill *kirrree-kirrree-*. [CD2:4a] **HABITS** Singly or in pairs in variety of open habitats, principally open savanna. Preys mainly on medium-sized birds, such as pigeons, francolins, waders and domestic chicken, caught on ground and in flight. Often glides and soars on upcurved wingtips. **SIMILAR SPECIES** Peregrine Falcon has shorter, more pointed wings and shorter tail; also darker, with blackish crown and broader moustachial streak; adult Peregrine barred below. See also Saker Falcon. **STATUS AND DISTRIBUTION** *F. b. abyssinicus* is uncommon to not uncommon resident in savanna and Sahel zones; local seasonal movements recorded, south in dry season, north with rains. *F. b. erlangeri* resident N Mauritania; migrant N Senegal. *F. b. tanypterus* E Chad (probably resident).

SAKER FALCON *Falco cherrug*
Faucon sacre

Plate 26 (8)

L 43–60 cm (17–24"); WS 102–135 cm (40–53"). *F. c. cherrug.* Large, powerful falcon, resembling Lanner Falcon but usually heavier. **Adult** *Head conspicuously pale*, whitish-buff with ill-defined eye-stripe and narrow, variably distinct moustachial stripe and crown streaking. Upperparts warm brown to greyish-brown fringed rufous, with contrasting dark brown flight feathers. Tail feathers barred; barring on central pair indistinct or absent. Underparts buff-white variably streaked dark brown, heavier on flanks; 'trousers' usually dark. *Underwing-coverts typically dark, contrasting with paler flight feathers.* Cere, orbital ring and feet pale yellow. **Juvenile** Similar to adult; head pattern usually more pronounced, with more densely streaked crown and more prominent eye-stripe and moustache, underparts more boldly streaked, underwing-coverts darker and even more contrasting. Cere, orbital ring and feet blue-grey. **VOICE** Usually silent away from breeding grounds. [CD2:4b] **HABITS** Frequents open habitats. Preys mainly on medium-sized birds and mammals, caught on ground after low flight. **SIMILAR SPECIES** Adult Lanner Falcon has rufous crown and nape, darker, greyer upperparts, and less marked underparts, Juvenile Lanner difficult to separate, but typically darker above; underwing pattern similar but less contrasting (underwing-coverts darker in juvenile Saker, flight feathers paler); head more strongly patterned, 'trousers' never dark. Some juvenile Peregrine Falcons of ssp. *calidus* (q.v.) similar to some juvenile Saker, but distinguished by lack of rufous tinge to upperparts, less heavily streaked underparts, lack of contrasting dark underwing-coverts and structural differences. **STATUS AND DISTRIBUTION** Rare Palearctic migrant, Mauritania (Jul–May). Vagrant, N Senegal, Mali, Chad and Cameroon.

PEREGRINE FALCON *Falco peregrinus*
Faucon pélerin

Plate 26 (6)

L 33–50 cm (13–20"); WS 80–115 cm (31.5–45.0"). Stocky, powerful falcon with broad-based, pointed wings and relatively short tail. When perched, wingtips reach tip of tail. **Adult** *minor Blackish top of head and broad moustachial stripe contrast strongly with white cheeks and throat.* Upperparts dark slate-grey, paler from back to uppertail-coverts. Tail barred blackish above and below. Underparts buffish-white spotted and densely barred black from lower breast down. Underwing narrowly and evenly barred blackish (appearing grey from distance). Cere, orbital ring and feet yellow. *F. p. brookei* slightly paler above with some rufous on nape; underparts tinged rufous. *F. p. peregrinus* paler and larger; underparts less heavily marked. *F. p. calidus* as nominate but even paler, with narrower moustachial stripe. Island race *madens* tinged brown from crown to mantle; underparts washed pink-buff. **Juvenile** Dark brown

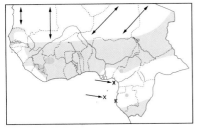

above; feathers of upperparts fringed rufous. Head pattern as ad, but with paler forehead, (often) pale supercilium, usually narrower moustachial streak, and whitish to buff cheeks. Underparts buffish to pale rufous heavily streaked blackish. Bare parts duller. In juvenile *calidus* streaking on underparts narrower (recalling juvenile Barbary Falcon). **VOICE** Usually silent except near breeding site. A loud, raucous *khyeh-khyeh-khyeh-...* becoming shriller when alarmed. [CD2: 5a] **HABITS** Singly or in pairs. Residents frequent inselbergs and cliffs, where they nest; migrants in variety of open habitats. Feeds mainly on birds, caught on the wing; occasionally mammals, reptiles and insects. **SIMILAR SPECIES** Adult Lanner Falcon is less stocky, with longer, blunter wings and longer tail; underparts sparsely marked; crown rufous; moustachial stripe narrower and less prominent. In flight, pale underwing-coverts create paler impression. Juvenile Lanner is paler above than juvenile Peregrine; underparts more boldly streaked; underwing-coverts darker, contrasting with paler flight feathers; moustachial stripe narrower. Eurasian Hobby is smaller and slimmer, with longer, narrower wings; adult is streaked below with rufous thighs and undertail-coverts; rump concolorous with upperparts (paler in Peregrine). Eleonora's Falcon is slimmer with longer wings and much longer tail; underwing-coverts darker than flight feathers; underparts never barred. See also Barbary Falcon. **STATUS AND DISTRIBUTION** Patchily distributed, uncommon to locally not uncommon resident (*minor*), and uncommon to rare Palearctic migrant (*calidus*, *peregrinus* and *brookei*) throughout. *F. p. madens* is rare resident, Cape Verde Is. Distribution of Palearctic races inadequately known; most records refer to *calidus* but may include *peregrinus*. Mediterranean and Middle Eastern ssp. *brookei* reported from Mauritania, but may occur further south and east. Palearctic vagrants reported from Bioko and São Tomé.

BARBARY FALCON *Falco pelegrinoides*   Plate 26 (5)
Faucon de Barbarie

L 32–45 cm (12.5–18.0"); WS 76–100 cm (30–39"). *F. p. pelegrinoides*. Resembles pale Peregrine Falcon in flight action and silhouette. **Adult** Head pattern as Peregrine, but variable amount of *rufous or cinnamon on forehead, sides of crown and nape*; long dark stripe through eye to nape, usually narrower dark moustachial streak contrasting with larger white cheek patch. Upperparts pale blue-grey; primaries darker. Tail barred and darkening towards tip, with broader subterminal band. Underparts, incl. underwing-coverts, pale pinkish-buff or sandy, narrowly and indistinctly barred brown, mostly on flanks and thighs. **Juvenile** Similar to juvenile Peregrine but head pattern with variable amount of buff or rufous on forehead, supercilium and nape (reflecting adult plumage); underparts sandy to buffish-white with narrower and paler brown streaking. **VOICE** Usually silent away from breeding grounds. Calls include a sharp *kek-kek-kek-...* [CD2:5b] **HABITS** As Peregrine Falcon. **SIMILAR SPECIES** Peregrine Falcon (q.v. and above). Adult Lanner Falcon differs by entirely rufous crown and nape and lack of subterminal tail band; for other differences see under Peregrine Falcon. **STATUS AND DISTRIBUTION** Inadequately known. Rare Palearctic migrant or vagrant, recorded from Mauritania, Senegambia, N Burkina Faso, Niger, Chad and N Cameroon. Probably breeds N Mauritania and perhaps elsewhere in southern Sahara. **NOTE** Often considered conspecific with Peregrine Falcon.

QUAILS, PARTRIDGES AND FRANCOLINS
Family PHASIANIDAE (World: 155 species; Africa: 44; WA: 18)

Small to medium-large terrestrial birds with thickset body, small head, short and robust bill, short wings and tail, and strong feet. Plumage often cryptic and intricately patterned. Males typically larger than females. Juvenile often resembles adult female. Occur from forest to subdesert. Usually in pairs or small groups (termed 'coveys'). Only reluctantly take flight; prefer to run. Flight low and rapid on whirring wings, usually not sustained. Vocalisations an important identification aid.

The following genera occur in our region:

Coturnix, **quails** (3 species). Very small and dumpy. Sexually dimorphic. Usually seen when flushed. Reluctant to fly, but migratory and capable of sustained flight.

Francolinus, **francolins** (13 species). Small to medium-large. Identification features include coloration of bare parts and flight feathers, vocalisations, habitat and range. Calls often loud and raucous, drawing attention. Sedentary.

Ptilopachus, **Stone Partridge** (monotypic). Rather small widespread species with bantam-like cocked tail and highly characteristic, melodious voice. Sedentary.

Alectoris, **partridges** (1 species). In our region a very local, semi-desert species, readily identified by head pattern and barred flanks. Sedentary.

COMMON QUAIL *Coturnix coturnix* Plate 29 (6)
Caille des blés

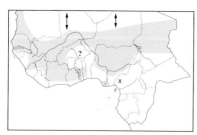

L 16–18 cm (6.25–8.0"); WS 32–35 cm (12.5–13.75"). Small and compact, with variable, but principally brown, streaked plumage. In flight, has plain wings. **Adult male** *coturnix* Top of head blackish or dark brown with narrow creamy median crown-stripe, bordered by long creamy supercilium; head-sides and neck creamy-buff with brownish area through eye, narrow black or brown moustachial stripe and 2 down-curved stripes at neck-sides; black or brown *anchor-shaped mark on throat*. Upperparts brown streaked buff; underparts pale buffish with variable amount of orange-rufous on breast; *breast-sides and flanks boldly streaked buff, black and chestnut*. Rufous morph has variable amount of rufous on head-sides and throat. Island race *inopinata* slightly smaller. **Adult female** As male but duller, lacking anchor on throat; variable amount of small black markings on breast. **VOICE** A far-carrying rhythmic *KWIk-ik-wik* often introduced by a hoarse, nasal *whew-whew whew-whew* (audible only at close range). When flushed, sometimes utters *whreeee*. [CD2:6b] **HABITS** On migration in all types of habitats, but usually favours open grassland. Very skulking and difficult to flush; prefers to run swiftly through grass, or to crouch and remain still. When pressed, flies low on bowed, whirring wings, usually swiftly returning to cover. **SIMILAR SPECIES** Female Harlequin Quail difficult to distinguish in flight but has darker upper- and underparts, and appears less heavily streaked above. Female Blue Quail is darker, lacks striped head pattern and is barred below; also smaller. Little Buttonquail distinguished in flight by contrasting pale upperwing-coverts and slower, not whirring wingbeats; size also smaller. **STATUS AND DISTRIBUTION** Uncommon to scarce or rare Palearctic passage migrant and winter visitor ([Aug] Oct/Nov–Mar), from Mauritania to Sierra Leone and east to Chad and Cameroon; mainly in Sahel zone (nominate). Common resident, Cape Verde Is (*inopinata*).

BLUE QUAIL *Coturnix chinensis* Plate 29 (5)
Caille bleue

L 13–14 cm (4.75–5.5"). *C. c. adansonii*. Very small dark quail. **Adult male** Unmistakable. *Dark slaty-blue* (often appearing black) with *chestnut on wings, flanks and uppertail-coverts*. Boldly contrasting pattern of white malar area and upper breast, and black, laterally extending stripe on throat. **Adult female** Dark brown above, streaked, mottled and vermiculated buff, rufous and black. *Face and underparts pale rufous-brown heavily barred black on breast and flanks* (diagnostic). **Juvenile** Similar to adult female but with pale shaft streaks on wing and rump. **VOICE** Male utters series of 2- and 3-note piping whistles *kee-keew kee-keew kee-kee-kuh*, first note loudest. Also a harsh, growling *HEhéhéhé* (male only?). A squeaky whistle *tir-tir-tir* or shrill *swi* when flushed (*BoA*). [CD5:77] **HABITS** Usually singly or in pairs, in variety of grasslands, at edges of cultivation and in extensive clearings. Wary and normally hard to observe. Flies reluctantly; difficult to flush a second time. Flight short, fast and direct, with whirring wingbeats. Not vocal. **SIMILAR SPECIES** Male Harlequin Quail has white supercilium. Female Harlequin and Common Quails have striped head pattern and lack barring below; also larger. Little Buttonquail is paler with contrasting pale upperwing-coverts distinctive in flight; wingbeats slower, not whirring. **STATUS AND DISTRIBUTION** Uncommon to rare and local resident and intra-African migrant, from Sierra Leone and Guinea to CAR; also coastal Gabon. Vagrant, S Mali, Togo, S Chad and N Congo (Odzala NP). **NOTE** Sometimes considered specifically distinct from Asian forms as African Blue Quail *C. adansonii* (formerly *Excalfactoria*).

HARLEQUIN QUAIL *Coturnix delegorguei* Plate 29 (4)
Caille arlequin

L 15–16 cm (6.0–6.25"). **Adult male** *delegorguei* Small, *dark quail with distinctive black and white head pattern and black-and-chestnut underparts*. Upperparts blackish-brown streaked whitish. Top of head blackish-brown bordered

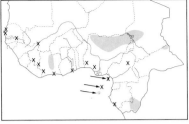

by long whitish supercilium extending onto neck-sides; throat white with black stripe on centre extending laterally and forming 'anchor'. Underparts black in centre and deep chestnut streaked black at sides. Island race *histrionica* is generally darker in both sexes. **Adult female** Resembles female Common Quail but darker, with pale tawny underparts and chestnut-buff undertail-coverts. Variable number of dark spots on breast, esp. on sides (more or less forming collar). **VOICE** An emphatic, rhythmic *kwit kwit-kwit-kwit kwit-kwit-kwit kwit-kwit*, reminiscent of Common Quail, but slightly slower, more deliberate, marginally higher pitched and more monotonous. A squeaky *skreeee* when flushed. [CD5:78] **HABITS** Similar to Common Quail. In grasslands, rank herbage and cultivation. **SIMILAR SPECIES** Male Blue Quail lacks white supercilium and has chestnut on wings and uppertail-coverts. Common Quail rather similar to adult female Harlequin and difficult to distinguish in flight, but browner and more streaked above. Female Blue Quail darker, with barring on underparts. See also Little Buttonquail. **STATUS AND DISTRIBUTION** Nomadic intra-African migrant. Only sparsely reported from our region, with records from N Senegal, Mali, Liberia, Ivory Coast, Ghana, Nigeria, Cameroon, Chad, CAR, Gabon, Congo and Bioko (nominate). Movements inadequately known, correlated with rains in certain areas. Common resident, São Tomé (*histrionica*); vagrant, Príncipe (Aug 1997).

BARBARY PARTRIDGE *Alectoris barbara*   Plate 146 (11)
Perdrix gambra

L 32–35 cm (13.5–14.0"). *A. b. spatzi.* **Adult** Readily identified by *distinctive head and neck pattern and boldly barred flanks.* Crown rufous-brown; face and throat grey-white bordered by white-spotted, rufous-brown collar. Upperparts greyish; *tail corners rufous-brown* (conspicuous in flight). Flanks barred black, rufous-brown and white. Breast pale grey; rest of underparts sandy-buff. Bill and legs bright red. **Juvenile** Very plain, lacking distinct head pattern and flank bars. **VOICE** A rapid series of hoarse, impure *kruk* or *ktchuk* notes interspersed with an occasional *kuk-kow* (or *chukor*), which breaks rhythm, e.g. *kruk kruk kruk kruk kuk-kow kruk kruk...* or *k-tchuk k-tchuk k-tchuk k-tchuk kiak-tchuk k-tchuk...* Also a drawn-out, hoarse squealing *rrrrrhaaaiih.* When flushed, a sharp, repeated *kEEEah* or *ketchew ketchew...* [CD2:7a] **HABITS** Singly or in small flocks on rocky slopes. Wary. Runs fast when disturbed; flies reluctantly. Flight low with rapid wingbeats interspersed with glides on stiff, horizontally held wings. **SIMILAR SPECIES** None; the only partridge within its limited W African range. **STATUS AND DISTRIBUTION** N African species just reaching our region in extreme N Mauritania (Zemmour) and N Chad (Tibesti), where rare and local resident.

STONE PARTRIDGE *Ptilopachus petrosus*   Plate 29 (10)
Poule de roche

L *c.* 25 cm (10"). *P. p. petrosus.* Small dark brown *chicken-like bird with relatively long, broad tail usually held cocked.* Fine feather markings only visible at close range. Legs dark red. Bare skin around eye dull red. **Adult male** Centre of lower breast and belly plain buff. **Adult female** Similar, but patch on underparts paler, more creamy-white. **Juvenile** Similar, but more distinctly barred above and below. **VOICE** Highly distinctive. An abruptly started, rising series of clear, fluty notes with a cheerful, pleasant quality, *weet-weet-weet-weet-...*, often uttered in duet or chorus and principally given at dawn and dusk. [CD5:79] **HABITS** In small groups in variety of habitats, incl. dry wooded grassland, rank herbage, edges of cultivation, rocky outcrops, dry watercourses, erosion gullies, laterite hills, and sandy plains with scrub. Forages on ground. Runs when disturbed; flies reluctantly. **SIMILAR SPECIES** None, but inadequate views may lead to confusion with various francolin species. **STATUS AND DISTRIBUTION** Locally common to uncommon resident, from S Mauritania to Sierra Leone east to Chad and CAR. Reaches coast in Dahomey Gap (west to Accra Plains, Ghana).

LATHAM'S FOREST FRANCOLIN *Francolinus lathami*   Plate 29 (7)
Francolin de Latham

Other names E: Forest Francolin, Latham's Francolin
L 20–25 cm (8–10"). *F. l. lathami.* Small, distinctively patterned francolin. **Adult male** Forehead and head-sides pale grey; top of head dark brown; black eye-stripe bordered by long white supercilium. Upper mantle black spotted white; rest of upperparts largely chestnut with some narrow white shaft streaks. *Underparts black spotted*

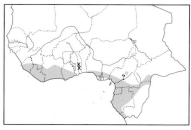

white from lower throat down; lower belly whitish-buff. Legs yellow. **Adult female** Similar but much browner; belly white. **Juvenile** Head-sides brownish; upperparts rufous-brown mottled black; throat white; underparts grey-brown spotted and streaked white; belly whitish. **VOICE** Distinctive nocturnal calls, very different from other francolins, include series of full, rather melodious notes, accelerating at end *krookrookrookrookrookroo*, given for long periods from tree perch or from ground; faster *whotootootootoo*; loud *krok! krokrorrrr krok! krokrorrrr...* Also loud, high-pitched *kirr-kikikiki kirr-kikikiki...* or *kroah-kokoko kroah-kokoko...* uttered in duet, mainly at dusk or in early night [CD5:80] **HABITS** In pairs or singly in mature lowland forest. Keeps to forest interior but occasionally emerges onto tracks. Secretive and usually difficult to observe. Hard to flush; flees danger by running. Rises suddenly and steeply with loud, fast wingbeats when almost stepped upon, flying short distance, then crashing into vegetation. **STATUS AND DISTRIBUTION** Uncommon to locally not uncommon forest resident, from Sierra Leone and SE Guinea to Ghana, and from SW Nigeria to SW CAR and Congo. Rare, Togo.

COQUI FRANCOLIN *Francolinus coqui*   Plate 29 (9)
Francolin coqui

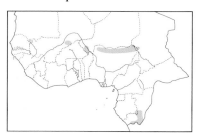

L 20–25 cm (8–10"). **Adult male** *spinetorum Head and neck rufous-orange* paling to buffish on throat; crown and nape rusty-brown; whitish collar with small black bars on lower neck. Underparts pale buff *heavily barred black on breast and flanks*; undertail-coverts deep rusty-buff. Upperparts grey-brown with pale shaft streaks and mix of buff, rufous and black streaks, bars and vermiculations. Bill black with yellow base. Legs yellow; one spur. **Nominate** has barring on underparts extending onto belly. **Adult female** *spinetorum* Face and throat pale buffish; pale supercilium bordered below by narrow black line curving downward on neck; throat with narrow black necklace. Upper breast unbarred vinaceous-grey; lower breast and flanks heavily barred black. Legs spurless. **Nominate** has barring on underparts extending onto belly. **Juvenile** Similar to adult female, but browner above, mottled with rufous; more buffy below. **VOICE** An oft-repeated, rhythmic, shrill *kEE-kwi, kEE-kwi...* or *KO-ki, KO-ki, ...* (hence name coqui), and an equally shrill and rasping *kREEK-krik kREEk-kREEk krikikikikew*, accelerating at end and fading away, or simpler *kREEk, kREEk kREEk-k-krew;* given by male from low perch (large stone, termite mound, or tree stump). [CD5:81] **HABITS** In pairs or small coveys in various types of grassland. Forages and roosts on ground. Flies reluctantly, preferring to crouch when disturbed. Flushes when almost trod upon, flying fast and far. **SIMILAR SPECIES** Similar-sized White-throated Francolin exhibits chestnut wings in flight. Male White-throated lacks barring on underparts; female differs from female Coqui in facial pattern, lacks fine black necklace and has much finer barring on underparts. **STATUS AND DISTRIBUTION** Resident with disjunct distribution in our region: *spinetorum* rare to uncommon, S Mauritania, Mali, W Niger and N Nigeria; nominate locally common, SW Gabon to Congo.

WHITE-THROATED FRANCOLIN *Francolinus albogularis*   Plate 28 (5)
Francolin à gorge blanche

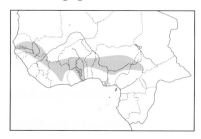

L *c.* 23 cm (9"). Small francolin with *white throat and chestnut wings* (conspicuous in flight). **Adult male** *albogularis* Crown greyish and chestnut bordered by creamy-white supercilium; dusky stripe through eye to ear-coverts; neck sandy- to rusty-buff. Upperparts largely brown-grey with pale shaft streaks and variably marked with black and rusty; mantle and wing-coverts mixed with chestnut. Underparts buff with some chestnut streaks on flanks (sometimes also on breast-sides); undertail-coverts rusty-buff barred blackish. Bill black with yellow base. Legs orange-yellow; one spur. Eastern *buckleyi* has supercilium and head-sides buff to rusty-buff. **Adult female** *albogularis* Similar but with variable fine dark barring on breast and flanks. In *buckleyi* wavy barring extends onto belly. **Juvenile** Similar to adult female but fine barring on underparts often more extensive. **VOICE** A very fast, high-pitched *kulikulikulikulikuli...* or *kuweekuweekuweekuweekuweek...* and a shrill, rasping *kREEK-krik kREEk-krikikikew*, accelerating at end and fading away; similar to Coqui Francolin. Calls mainly at dawn and dusk. [CD5:82] **HABITS** Usually in pairs or small coveys in various types of wooded grassland. Shy and skulking. Hard to flush; prefers to escape danger by slipping away into vegetation. Flight low, fast and direct, on whirring wings. **SIMILAR SPECIES** Similar-sized Coqui and Schlegel's Francolins lack chestnut in wings in flight. Males easily distinguished by ochre-coloured heads and barred underparts. Female Coqui separable from female White-throated by bolder barring on underparts, different head pattern and narrow black necklace. **STATUS AND DISTRIBUTION** Generally uncommon to rare and local resident, from Senegambia to N Cameroon (locally not uncommon, Bénoué plain). Nominate in west of range, *buckleyi* from E Ivory Coast eastwards.

SCHLEGEL'S FRANCOLIN *Francolinus schlegelii*   Plate 29 (8)
Francolin de Schlegel

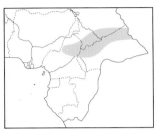

L 21–24 cm (8.25–9.5"). Small. **Adult male** Head mainly rusty-orange (or rufous-yellow) with dusky-brown crown and streak through eye. Upperparts vinous-chestnut with pale shaft streaks and blackish mottling. Underparts whitish narrowly barred black; flanks with some chestnut streaks. Bill black with yellow base. Legs yellow; one spur. **Adult female** Similar but duller, with more irregular barring on underparts. Legs spurless. **Juvenile** Similar to adult female but barred rufous-buff on mantle and scapulars. **VOICE** Loud initial note, followed by accelerating series of more grating ones, fading at end, *KWEEK! kre-kre-krekrekrekew*, lower pitched and less shrill than those of Coqui Francolin. [CD5:83] **HABITS** In pairs or small coveys in wooded grassland. Roosts on ground. Flight silent and short. **SIMILAR SPECIES** Coqui Francolin is allopatric. White-throated Francolin has chestnut on wings, largely brown-grey upperparts and white throat; underparts unbarred in male, finely barred but duller and buffier in female. **STATUS AND DISTRIBUTION** Uncommon to rare and local resident from WC Cameroon through S Chad and N CAR.

RING-NECKED FRANCOLIN *Francolinus streptophorus*   Plate 28 (3)
Francolin à collier

L 30–33 cm (12–13"). Dark brown francolin with *black-and-white barring on neck and upper breast.* **Adult male** Crown dark grey-brown bordered by long white supercilium; head-sides, nape and neck rufous-chestnut. Upperparts dark grey-brown with narrow pale shaft streaks. Throat white. Lower breast to belly dull brown-buff; long flank feathers broadly edged blackish. Bill black; base of lower mandible dull yellowish. Legs yellow. **Adult female** Similar, but upperparts barred rusty-buff; pale shaft streaks broader. **Juvenile** No information. **VOICE** A distinctive, short series of melodious, fluty notes, unlike calls of any other francolin *thuuu, tee whiut tew-tewew.* [CD5:84] **HABITS** Singly, in pairs or small coveys on rocky slopes with short grass. Shy and skulking. Hard to flush; prefers to escape danger by running into cover. Flight reportedly very fast. Roosts on ground. **SIMILAR SPECIES** Double-spurred Francolin lacks barring on neck and has head-sides white streaked black; voice loud and grating. **STATUS AND DISTRIBUTION** Scarce and local resident, Cameroon highlands (Foumban area, 1050–1200 m).

FINSCH'S FRANCOLIN *Francolinus finschi*   Plate 28 (2)
Francolin de Finsch

L *c.* 35 cm (13.75"). **Adult** Medium-sized francolin with *mainly rufous head and wings.* Dusky stripe back from eye, curving down neck. Upperparts dark grey-brown with buff and tawny vermiculations, bars and shaft streaks. Throat white; upper breast brownish-grey; lower breast to belly tawny marked with chestnut and grey. Rufous primaries and outer secondaries conspicuous in flight. Bill black; base of lower mandible yellowish. Legs yellow. **Juvenile** No information. **VOICE** A loud, repeated, rather high-pitched *kwit-e-kwee*, usually given at dusk. [CD5:86] **HABITS** In pairs or small coveys in open grassland. Not particularly shy; only flushes when pressed. Roosts on ground. **SIMILAR SPECIES** Coqui Francolin has more orangey head in male, lacks rufous in wings, and is smaller and more compact (hence faster wingbeats). **STATUS AND DISTRIBUTION** Locally not uncommon to uncommon resident, from SE Gabon (Lékoni area) to Congo (Bateke Plateau).

SCALY FRANCOLIN *Francolinus squamatus*   Plate 28 (7)
Francolin écailleux

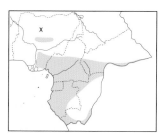

L 30–33 cm (12–13"). Dark francolin with conspicuous *scarlet legs*. **Adult** Dark grey-brown above; slightly paler below. Black and buff vermiculations and bars, whitish throat, and scaly aspect produced by pale fringes to underparts only noticeable under favourable conditions. Bill reddish. **Juvenile** Similar but more rufous-brown; upperparts with black arrowhead markings; underparts barred black and white. **VOICE** A series of loud, grating syllables increasing in volume *kerrrAAK kerrrAAK kerrrAAK...* with variations; also ascending *khiup-khiup-khiup-khiup khiupkiurrr khiupkiurrr...* and fast *kehkehkehkehkeh...* and *khe-kip khe-kip khe-kip...* mostly given at dawn, in early morning and at dusk, occasionally at night; often several together. [CD5:89] **HABITS** Usually in pairs at forest edge and in second growth, gallery forest and thickets. Shy and difficult to observe; occasionally ventures onto tracks. **SIMILAR SPECIES** Double-spurred Francolin has some rather similar calls and should be separated with care in areas of overlap. **STATUS AND DISTRIBUTION** Common to not uncommon resident in forest zone, from Nigeria to S CAR and Congo.

AHANTA FRANCOLIN *Francolinus ahantensis*   Plate 28 (6)
Francolin d'Ahanta

L *c.* 33 cm (13"). Dark francolin with *orange-red bill and legs*. **Adult** Blackish-brown to dark rich brown above, streaked white on hindneck and mantle. Throat white; rest of underparts grey-brown narrowly streaked white. Legs bright scarlet to yellow-orange. Western populations (formerly recognised as ssp. *hopkinsoni*) are generally paler, but differences clinal. **Juvenile** Similar, but upperparts have arrow-shaped black markings on mantle and scapulars. **VOICE** A repeated, loud, grating, rather high-pitched *kee-kee-keRRREE kee-kee-keRRREE...* or *RREEA-kekerrr RREEA-kekerrr...* usually given at dawn, at dawn and in early morning, sometimes at night. [CD5:90] **HABITS** Singly, in pairs or small coveys at forest edge and in second growth, overgrown clearings and cultivation, scrub at edges of gallery forest and thickets. Usually shy and keeping to dense cover, but occasionally emerging onto tracks. Difficult to flush. **SIMILAR SPECIES** Double-spurred Francolin is paler overall with conspicuous pale supercilium, and greenish bill and legs. Latham's Forest Francolin is distinctly smaller with pale head-sides, black throat, dark bill and differently patterned black-and-white underparts; usually keeps to forest interior. Stone Partridge has different jizz with broad, usually cocked tail, darker legs and smaller size. **STATUS AND DISTRIBUTION** Not uncommon or uncommon to locally scarce resident in forest zone, from Senegambia to Liberia east to SW Nigeria; also S Mali (1 record). 'Hopkinsoni' in Senegambia and Guinea-Bissau.

DOUBLE-SPURRED FRANCOLIN *Francolinus bicalcaratus*   Plate 28 (10)
Francolin à double éperon

L 30–35 cm (12.0–13.75"). *F. b. bicalcaratus*. Large, plump francolin, appearing mainly brownish at distance, but with *conspicuous white supercilium*. Plumage coloration varies, darkest in wetter part of range. **Adult male** Forehead black; crown chestnut narrowly bordered with black; long white supercilium bordered below by narrow black eye-stripe; head-sides white finely streaked dark. Neck chestnut marked with black and white. Upperparts grey-brown, edged and vermiculated buff. Underparts buff, densely covered with bold, drop-shaped chestnut-and-black markings on breast and belly. Bill greenish with dusky culmen. Legs greenish with 2 spurs. **Adult female** Smaller; spurs reduced. **Juvenile** Generally duller; bare parts paler; legs with a single or no spur. **VOICE** A series of loud, grating syllables, *rrrraak kerRRAK kerRRAK kerRRAK...*, mostly given at dawn from low perch (termite mound, rock, etc). Contact call *kek kek kek...*; flight call *krrror krrror krrror...* [CD2:7b] **HABITS** In pairs and small coveys (rarely flocks of 12–40) in various types of grasslands, farmbush, cultivation and scrub at forest edge. Principally active in early morning and late afternoon. Often seen on tracks. Flies reluctantly, preferring to run when disturbed. Flight vigorous and short. **SIMILAR SPECIES** Clapperton's Francolin has red eye patch, reddish bill and legs. Heuglin's Francolin has yellow eye patch, bill and legs. Ahanta Francolin is darker with orange-red bill and legs. Scaly Francolin has some rather similar calls (q.v.). **STATUS AND DISTRIBUTION** The most widespread francolin in our region. Common resident from S Mauritania to Liberia east to SW Chad and Cameroon.

HEUGLIN'S FRANCOLIN *Francolinus icterorhynchus*　　　Plate 28 (4)
Francolin à bec jaune

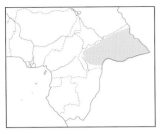

L *c.* 32 cm (12.5"). The only francolin in our region with *combination of yellow bill, eye patch and legs*. **Adult male** Forehead black, becoming chestnut on crown; long, narrow supercilium white freckled brown; head-sides and neck white finely streaked dark brown. Upperparts grey-brown, vermiculated and barred buff; mantle feathers fringed buff. Throat white. Rest of underparts buff, streaked and spotted dark brown on breast and belly, barred on undertail-coverts. Black-tipped bill dusky-yellow above, more orange-yellow below. Bare patch around eye dusky-yellow. Legs orange-yellow. **Adult female** Similar but smaller. **Juvenile** Similar to adult female but barring on upperparts more distinct. **VOICE** A harsh, slow *kerak kerak kek* or faster *kerak-kerak-kerak-kerak-kerrr*, mainly given from low perch in early morning and late afternoon. [CD5:93] **HABITS** Singly, in pairs or small coveys in open and lightly wooded grassland, and cultivation. Flight heavy and usually over short distance. **SIMILAR SPECIES** Double-spurred Francolin lacks yellow eye patch and has greenish bill and legs. Clapperton's Francolin has red eye patch and dusky-red bill and legs. **STATUS AND DISTRIBUTION** Common resident, CAR.

CLAPPERTON'S FRANCOLIN *Francolinus clappertoni*　　　Plate 28 (8)
Francolin de Clapperton

L 30–33 cm (12–13"). Large francolin with *distinctive combination of red eye patch, long white supercilium and black bill*. **Adult male** Forehead dark brown becoming chestnut on crown; head-sides and neck white finely streaked dark brown. Upperparts grey-brown with buff-fringed feathers; uppertail-coverts and tertials barred buff. Throat white; rest of underparts creamy-white streaked and spotted black. Bill black with red base. Legs dusky-red. **Adult female** Similar but noticeably smaller. **Juvenile** Similar to adult female but less distinctly patterned, both above and below. **VOICE** Short series of loud, grating *kerrrAK* or slower, more drawn-out *kerroAH* calls with variations, somewhat resembling those of Double-spurred and Heuglin's Francolins, usually given from low perch (termite mound, small rock, low branch, etc). Also short series of *khek-kheh-kheh-kheh*, sometimes preceded by *kerrak*. [CD5:94] **HABITS** Usually in pairs or small coveys in arid grasslands with scattered trees and bushes; also cultivation and rocky hillsides. Roosts in trees. **SIMILAR SPECIES** Double-spurred and Heuglin's Francolins lack red eye patch, and have differently coloured bills and legs. **STATUS AND DISTRIBUTION** Locally common resident in Sahel zone, from extreme E Mali (Azzawakh) through C Niger, extreme NE Nigeria and N Cameroon to S Chad and N CAR. Also reported from S Mauritania (uncommon and local, Jul–Oct) and C Mali.

MOUNT CAMEROON FRANCOLIN *Francolinus camerunensis*　　　Plate 28 (1)
Francolin du Mont Cameroun

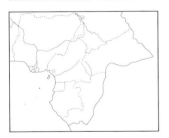

Other name E: Cameroon Mountain Francolin
L *c.* 33 cm (13"). Very dark, *highly local* forest francolin with *bright red bill, eye patch and legs*. **Adult male** Top of head dark brown; head-sides and neck grey with dark centres to feathers; mantle dark brown with grey fringes; upperparts blackish-brown. Throat grey; rest of underparts mainly dark grey with dark-centred feathers. Bare red patch around eye. **Adult female** Crown brown barred black; nape and mantle black streaked pale buff. Upperparts brown heavily blotched and barred with black and rufous-buff. Throat dirty white streaked dusky. Underparts black with mostly U-shaped whitish markings. **Juvenile** Upperparts similar to adult female; underparts barred black and whitish; flank feathers tipped black and white; no bare patch around eye. Bill and legs dusky-red. **VOICE** Short series of loud, trumpeting (not grating) whistles, *KILU KILU KILU*, slower *KEE-ku KEE-ku KEE-ku* or combinations of both, *KILU KILU KEE-ku KEE ku KEE ku KILU KILU KEE kuu KEE kuu...* Usually given at dusk, from concealed perch; also in duet. Alarm an abrupt *khik*. [CD5:95] **HABITS** Usually in pairs or small coveys in montane forest. Prefers dense undergrowth and clearings; absent from grasslands above treeline. Sometimes ventures onto paths. Very shy, seeking cover at slightest disturbance. Occasionally flushes with rapid whirring wingbeats. Most easily observed at dusk. **SIMILAR SPECIES** Scaly Francolin lacks red eye patch and has browner upperparts. **STATUS AND DISTRIBUTION** Endemic to Mt Cameroon (850–2100 m), where uncommon. *TBW* category: THREATENED (Endangered).

RED-NECKED FRANCOLIN *Francolinus afer*
Francolin à gorge rouge

Plate 28 (9)

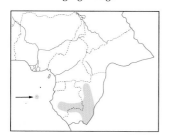

L 35–41 cm (13.75–16.0"). Large francolin with *bare red throat* (diagnostic). **Adult male** *cranchii* Crown and upperparts brown. Head-sides, neck and mantle greyish finely streaked and vermiculated black. Rest of underparts grey with dark vermiculations and broad chestnut streaking on belly. Bill, bare patch around eye, throat and legs red. **Nominate** has white supercilium and moustachial stripe; upper breast pale brown streaked black; rest of underparts blackish-brown, feathers fringed white. **Adult female** Similar but smaller. **Juvenile** Similar but generally duller and browner, with paler bare parts. **VOICE** A loud, grating, rather low-pitched *kAARkukukuw kAARkukukuw...* and a higher pitched series of squealing syllables increasing in volume and ending in a drawn-out grating *koAARRK koAARRK koAARRK-kek koAARRK-kek KeRRRrrr*, with variations and combinations of both; also a fast *koark-koark-koark-koark-...* Mainly given in early morning and late afternoon. [CD5:97] **HABITS** Usually in pairs, occasionally in small coveys, in various types of grassland. Roosts in trees or bushes. **STATUS AND DISTRIBUTION** Locally not uncommon to common resident, S Gabon and Congo (*cranchii*). Nominate introduced São Tomé, where common.

GUINEAFOWL
Family NUMIDIDAE (World: 6 species; Africa: 6; WA: 5)

Afrotropical family of medium-large terrestrial birds with small, bare head, relatively long neck and thickset body. Bare skin of head and neck often brightly coloured. Wings and tail short; feet strong. Plumage principally dark. Sexes similar; males usually larger. Occur from forest to subdesert. Sedentary. Reluctantly take flight; prefer to run. Gregarious when not breeding. Often vocal. Roost in trees. Sometimes treated as a subfamily of Phasianidae.

WHITE-BREASTED GUINEAFOWL *Agelastes meleagrides*
Pintade à poitrine blanche

Plate 30 (2)

L 40–45 cm (16–18"). **Adult** Small black guineafowl with bare, bright reddish head and upper neck, and *broad white collar extending to upper breast*. Tail well developed and held horizontally, reminiscent of domestic chicken rather than guineafowl. Dark greyish legs with 1–2 spurs. **Juvenile** Dull brownish-black with white belly. Head dark reddish-brown with 2 tawny stripes from bill to nape (forming V); some reddish-brown feathers on wings and body (Gatter *et al.* 1988). No spurs. **VOICE** Various quiet, twittering contact and feeding calls, constantly uttered, incl. *pit pit pit ...* and short trill *prrirrr*. [CD5:72] **HABITS** In small groups of 4–20 in primary rainforest and nearby old secondary forest. Forages on ground; roosts in small trees within understorey. Shy. Nest undescribed. **SIMILAR SPECIES** Other guineafowl are larger; forest francolins are much smaller. **STATUS AND DISTRIBUTION** Rare to locally common forest resident, from E Sierra Leone (Gola Forest), through Liberia to W Ivory Coast (Taï NP). Almost extinct, W Ghana (recent sightings from Bia, Mini-Suhien NP, Boin-Tano FR, Tano-Anwia FR, Enchi). *TBW* category: THREATENED (Vulnerable).

BLACK GUINEAFOWL *Agelastes niger*
Pintade noire

Plate 30 (1)

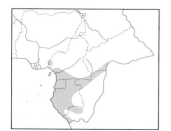

L 40–43 cm (16–17"). **Adult** *Small black guineafowl with bare, bright reddish head and upper neck.* Crown-stripe of short black down. Tail well developed and held horizontally, reminiscent of domestic chicken rather than guineafowl. Dark greyish legs with 1–2 spurs. **Juvenile** Dull black with white belly; no spurs. **VOICE** Various quiet, twittering contact and feeding calls, constantly uttered, incl. hard *pit pit pit...* Song a fast, rising series of rather melodious, whistling notes, *huw hee-huwhee-huwhee-huwhee-huwhee-huwhee-wheet-wheet-wheet...* [CD5:73] **HABITS** In pairs or small groups (usually 2–6) in primary rainforest; principally within forest interior, occasionally in dense undergrowth at edges. Forages on ground, scratching in leaf litter, removing it to rest on bare circular patch of 30 cm diameter during day (good indicator of species' presence). Elusive, but not shy if approached quietly; whistled imitation of song may attract birds close to observer. Nest undescribed. **SIMILAR SPECIES** Plumed Guineafowl is much larger and has distinct crest, finely spotted plumage and dark bare facial skin. **STATUS AND DISTRIBUTION** Scarce and local forest resident, from extreme SE Nigeria to Congo and SW CAR.

PLUMED GUINEAFOWL *Guttera plumifera*   Plate 30 (3)
Pintade plumifère

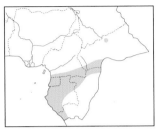

Other name F: Pintade à crête
L 45–51 cm (18–20"). *G. p. plumifera.* **Adult** Medium-large guineafowl with bare *greyish-blue head and straight, stiff, irregular crest.* Plumage black finely speckled bluish-white. Primaries chestnut; outermost secondaries edged white. Legs dark bluish-grey; spurless. **Juvenile** Duller, with shorter crest. Head and neck have some black downy feathers. Plumage largely dusky-grey barred blackish above, spotted and barred whitish below. **VOICE** Vocal. A loud, trumpeting *ku-ku-khep ku-ku-khep ku-ku-khep...* Flock produces discordant chorus of harsh *khep-khep-...* or *kha-kha-kha-...* calls when alarmed. Contact calls low and harsher than those of Black Guineafowl. [CD5:74]

HABITS In groups of 4–20 (occasionally more) in primary rainforest, old second growth and (rarely) large gallery forests. Forages on ground, scratching leaf litter. Perches in trees when disturbed. Shy. **SIMILAR SPECIES** Black Guineafowl is distinctly smaller and has conspicuous reddish head without crest. Crested Guineafowl has curly crest and red throat; only known to overlap in CAR. **STATUS AND DISTRIBUTION** Locally not uncommon to scarce forest resident, from S Cameroon to Congo and CAR (mainly S, but also N in Bamingui-Bangoran NP).

CRESTED GUINEAFOWL *Guttera pucherani*   Plate 30 (4)
Pintade huppée

Other name F: Pintade de Pucheran
L 46–56 cm (18–22"). **Adult** *verreauxi* Large guineafowl with bare blue head, *bushy black crest and bare red throat.* Plumage black finely speckled bluish; unspotted black collar on lower neck. Flight feathers chestnut; whitish outer webs to outermost secondaries form *conspicuous wingbar in flight.* Legs blackish; spurless. *G. p. sclateri* differs in shape of crest: very short in front, longer at rear. **Juvenile** Head tawny-buff with blackish crown and neck. Plumage a mix of black, buff, rufous and blue barring; secondaries dark grey-brown, marked with black, buff and blue. **VOICE** Contact call a soft clucking *chuk.* A sharp, nasal *kak!* interspersed with a hard rattle when alarmed, *kak kak-kak kak-uk kak-uk kurr-r-r-r-k,* resembling that of Helmeted Guineafowl but lower pitched; also sharper *kek-kek* and faster *krrrrrrrrk.* [CD5:75] **HABITS** Usually in small groups of 5–20 in lowland and gallery forest, esp. at edges and in second growth. Forages on ground, occasionally in fruiting trees. Shy and mostly hidden, but emerges on dirt roads or in clearings at dawn or after heavy rain. Flies into trees when disturbed, often cackling loudly. **SIMILAR SPECIES** Plumed Guineafowl has straight crest and lacks red on throat. **STATUS AND DISTRIBUTION** *G. p. verreauxi* is a locally not uncommon resident in forest zone, from Guinea-Bissau to SW Nigeria, and in SC Congo (Bateke Plateau) and CAR. Rare and local, S Mali (1 record, Sagabari). *G. p. sclateri* restricted to forest and forest–savanna mosaic in SE Nigeria and SW Cameroon.

HELMETED GUINEAFOWL *Numida meleagris*   Plate 30 (5)
Pintade commune

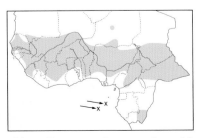

Other name F: Pintade mitrée
L 53–63 cm (20–25"). **Adult** *galeata* Large guineafowl with bare pale bluish head and upper neck, *bony casque* ('helmet') on crown and red gape wattle. Casque reddish at base, paling to horn at top. *Plumage dark grey densely speckled white* except on lower neck and breast. Flight feathers dark grey finely barred white. **Nominate** has larger casque and blue gape wattles. **Juvenile** Head tawny-buff with 2 blackish lateral crown-stripes and no casque; breast buffish scaled dusky. Rest of plumage dull grey-brown speckled buffish-white; upperparts with irregular rusty-buff barring. **Immature** As adult, but with smaller casque, shorter wattles and feathered neck. **VOICE** Commonest call a series of hard, raucous notes

interspersed with a grating rattle *chek-chek-chek krrrrrr chek-chek-chek-...* Contact note a soft, metallic *chink*. Female also utters a rather high-pitched, piping whistle *huu-i huu-i huui- ...*; male responds with single *chek* or *cheenk*. [CD2:6a] **HABITS** In flocks of 10–40 (occasionally more) in wide variety of open habitats. Forages on ground in open areas. Regularly visits waterholes. Runs fast, with half-raised wings (secondaries arched over back). **SIMILAR SPECIES** Crested Guineafowl has feathered crest and occurs in more forested habitats. **STATUS AND DISTRIBUTION** Locally common to scarce resident in broad savanna belt. *N. m. galeata* from S Mauritania to Sierra Leone east to Cameroon; also from SE Gabon to Congo. Nominate *meleagris* (or intermediates formerly recognised as ssp. *strasseni*) from E Cameroon to CAR. Introduced, Cape Verde Is (common), São Tomé (rare) and Annobón.

BUTTONQUAILS
Family TURNICIDAE (World: 16 species; Africa: 3; WA: 3)

Very small, quail-like, terrestrial birds with short, rounded body and wings. Lack hind toe. Females larger and brighter coloured than males. Occur in savanna, grassland and scrub, occasionally forest edge. Secretive and difficult to flush; usually do not fly far. Sex roles reversed.

QUAIL-PLOVER *Ortyxelos meiffrenii*
Turnix à ailes blanches

Plate 29 (3)

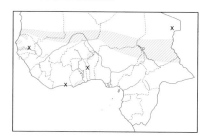

Other names E: Lark Quail, Lark Buttonquail; F: Turnix de Meiffren L 11–13 cm (4.5–5.0"). Very small, short-tailed grassland bird, superficially recalling miniature courser on ground. Mainly sandy-rufous above and whitish below, with *distinctive wing pattern and flight action.* Greater coverts white, contrasting with blackish, white-tipped remiges and forming conspicuous diagonal band on upperwing. *Flight fluttering and erratic, reminiscent of bush lark or butterfly.* **Adult male** Head, upperparts and tail a mix of pale rufous-brown, buff and cream with small black or dusky markings, and broad creamy supercilium. Throat white tinged golden-buff; breast buffish, darker on sides; lower breast and belly cream, becoming white on flanks and undertail-coverts. Bill yellowish-horn or pale greyish with dark culmen; legs whitish-pink to cream. **Adult female** Similar but breast slightly darker rufous-brown. **Juvenile** Paler overall. **VOICE** Described as a very soft low whistle; silent when flushed (Lynes 1925). **HABITS** Singly or in pairs in arid and semi-arid grassland and thorn scrub. Secretive, creeping rapidly through grass, but also running courser-like in the open. Sits tight until almost trodden upon; rises silently, flying off in jerky, undulating flight. **SIMILAR SPECIES** Buttonquails and quails lack contrasting wing pattern and have low whirring flight and different jizz. Quails also have darker upperparts and no, or much less, white on underparts. **STATUS AND DISTRIBUTION** Uncommon, local resident and intra-African migrant in narrow Sahel belt, from S Mauritania and N Senegal through Mali, Burkina Faso, N Benin, Niger, N Nigeria, N Cameroon to Chad; also coastal strip in Ghana. Vagrant, Gambia, coastal Ivory Coast (1 old record, Grand Bassam) and Togo. Dry-season breeder, moving north with rains.

LITTLE BUTTONQUAIL *Turnix sylvatica*
Turnix d'Andalousie

Plate 29 (1)

Other names E: Common/Small/Kurrichane Button-quail; F: Turnix d'Afrique L 14–16 cm (5.5–6.5"). *T. s. lepurana.* Very small, short-tailed, quail-like bird. **Adult male** Head and upperparts cryptically and variably coloured, being marked with shades of brown, rufous and buff; head-sides pale buff finely spotted blackish; median and greater wing-coverts largely pale buff with bold blackish spots. *Breast orangey, spotted black on sides;* rest of underparts whitish. Eye pale yellow. In flight, brown flight feathers contrast with pale coverts. **Adult female** As male, but more brightly coloured and slightly larger. **Juvenile/immature** Paler, less marked and smaller than adult, with mainly pale buffish underparts and small blackish spots extending across breast; eye dark. **VOICE** A strange, low, far-carrying *hooooooo...* resembling distant lowing cow or foghorn. Difficult to locate. Uttered by female mostly at dawn and dusk, also at night and during day. [CD2:8a] **HABITS** Singly or in pairs in various types of grassland and cultivation. Prefers escaping danger by running rather than flying. When flushed, flies low and for only short distance before dropping into cover. Rarely flushed a second time. Secretive, but sometimes permits close approach by quiet observer. Most active in early morning and late afternoon, when often venturing into the open along tracks. **SIMILAR SPECIES** Black-rumped Buttonquail has barred breast and flanks, darker upperparts and black rump. Quails are bulkier, appearing neckless, with shorter bill and differently coloured and patterned head and underparts; flight more rapid on longer, narrower wings. **STATUS AND DISTRIBUTION** Scarce to locally common resident and intra-African migrant almost throughout, except in forest and desert. Wet-season breeding visitor to drier parts of range.

BLACK-RUMPED BUTTONQUAIL *Turnix hottentotta*   Plate 29 (2)
Turnix nain

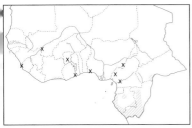

Other name F: Turnix hottentot
L 14–16 cm (5.5–6.5"). *T. h. nana.* **Adult male** Resembles Little Buttonquail, but upperparts darker, *head-sides orange-rufous and sides of orange-rufous breast and flanks barred black and white* (not spotted). Eye pale blue-grey to whitish. In flight, *blackish rump and uppertail-coverts* diagnostic. **Adult female** As male, but more brightly coloured and slightly larger. **Juvenile/immature** Less rufous and barring extending onto entire breast and flanks. **VOICE** Low, resonant hoots similar to those of Little Buttonquail but shorter and delivered in rather fast series *hooo, hooo, hooo, hooo,...* [CD6:1a] **HABITS** Similar to Little Buttonquail, but usually in moister grasslands with shorter herbage (though species may overlap locally); also more secretive and harder to flush. **SIMILAR SPECIES** Little Buttonquail (q.v. and above). **STATUS AND DISTRIBUTION** Inadequately known. Reported from Sierra Leone, S Ivory Coast, Ghana, Nigeria (Lagos and Mambilla Plateau), Cameroon, SW CAR, Gabon and Congo. Apparently vagrant or rare and local resident and intra-African migrant. **NOTE** *T. h. nana* sometimes considered a separate species, principally on basis of differences in plumage pattern, eye colour and migratory habits.

RAILS, FLUFFTAILS, CRAKES, GALLINULES, MOORHENS AND COOTS
Family RALLIDAE (World: *c.* 133 species; Africa: 26; WA: 21)

Diverse family of small to medium-sized terrestrial and aquatic birds, occurring in a variety of habitats. Sexes usually similar or nearly so, except in flufftails. Short tail flicked while walking. Legs and toes moderately or conspicuously long. Wings rounded. Flight over short distances fluttering and seemingly weak, but many are long-distance migrants. Many species skulking and difficult to observe, but voice may attract attention.

    Rails, flufftails and crakes form the most skulking group, comprising many seldom seen species, some of which are definitely rare in the region. The distinctive genus *Sarothrura* (flufftails), confined to Africa and Madagascar, is composed of small, secretive rails of forest or marsh and grassland, with highly characteristic vocalisations. Diagnostic features in males include the amount of chestnut on head, neck and breast, markings on body (spots or streaks) and colour of tail. Females hard to identify in the field (except White-spotted Flufftail); juveniles even more so.

    Gallinules, moorhens and coots include the least secretive species of the family. Frequent open water or clamber within dense waterside vegetation.

NKULENGU RAIL *Himantornis haematopus*   Plate 31 (7)
Râle à pieds rouges

L *c.* 43 cm (*c.* 17"). Largest African rail. **Adult** Large and stout with rather heavy, black bill and conspicuous *long red legs*. Plumage variable; mainly dark brownish with paler fringes to feathers and whitish throat. Lores and orbital ring black. Eye red. **Juvenile/immature** Similar but belly whitish to pale brownish; eye and legs duller. **VOICE** Distinctive antiphonal song a loud, far-carrying, sonorous duet *koKAWkoKAW-hoHO*, repeated for several minutes and uttered mostly after sunset and before sunrise; also at night. Given from ground or from perch up to 20 m high. Solitary birds give 2 loud, honking notes *HO-HO*, repeated at regular intervals. [CD6:1b]

HABITS Skulking, unobtrusive inhabitant of lowland forest. Would mostly remain unrecorded but for characteristic, loud song, from which it derives onomatopoeic name. Forages along small watercourses and among leaf litter. When flushed, readily identified by long red dangling legs. **SIMILAR SPECIES** None. Scaly Francolin has short (not long) red legs, dark, mottled plumage, and different behaviour. **STATUS AND DISTRIBUTION** Uncommon to locally common forest resident, from Guinea to Togo and from S Nigeria to CAR and Congo.

GREY-THROATED RAIL *Canirallus oculeus* Plate 31 (6)
Râle à gorge grise

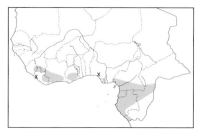

L *c.* 30 cm (12"). Distinctively coloured, slender rail. **Adult** Forehead, forecrown, head-sides and throat grey; crown and nape dark brown merging into olive-brown on rest of upperparts; some large white spots on edge of wing; tail bright rufous-chestnut. Neck, breast and upper belly rich chestnut; lower underparts to undertail-coverts barred brownish and buff. Bill dark with greenish sides. Eye red; yellow-green orbital ring. Legs darkish brown. **Juvenile/immature** Similar but duller, with head and underparts dark brown. **VOICE** Ventriloquial song, uttered at all hours of day and night, consists of series of hollow notes sounding like double drumbeats *pl, pl, pl, pl...*

, interspersed with soft *doo* or *dooah* notes after *c.* 20 s: *doo, pl, doo-doo, pl,...*; the number of *doo* notes increasing (often to 3–4) while the drumbeats fade. Song given from perch 2–6 m high; bird extremely difficult to locate. Calls include a loud, explosive booming of 3–6 notes on a descending scale of half tones *ooe-ooe-ooe* and short bursts of *dook-dook-dook* or muffled *thook-thook-....* (Dowsett-Lemaire & Dowsett 1991). Also hoarse, screeching notes. [CD6:2a] **HABITS** Unobtrusive and rarely seen, occurring singly or in pairs along small watercourses in lowland rainforest. Forages by removing leaf litter with jerky movements of bill. Unlikely to be flushed. **SIMILAR SPECIES** None. Nkulengu Rail, the only rail occurring in same habitat (although less dependent on water), is much larger and has red legs and duller plumage without any white spots. **STATUS AND DISTRIBUTION** Scarce to uncommon forest resident, from E Sierra Leone and SE Guinea to SW Ghana and from S Nigeria to Congo (where locally common in flooded forest). Status inadequately known, due to secretive behaviour.

WHITE-SPOTTED FLUFFTAIL *Sarothrura pulchra* Plate 31 (1)
Râle perlé

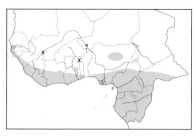

L 15 cm (6"). ¶ Commonest flufftail of the region. Both sexes distinctive. **Adult male** Head, mantle and breast reddish-chestnut; rest of upper- and underparts *black densely spotted white; tail reddish-chestnut*. Bill and legs blackish. Eye brown. **Adult female** Black areas of male more brownish and *entirely barred reddish-buff; tail reddish-chestnut barred blackish*. **Juvenile** Unique in flufftails by being similar to resp. adults, but duller. **VOICE** Vocal. Song a rapid series of short, resonant notes *too-too-too-too-too-...* or *hu-hu-hu-hu-hu* rather reminiscent of Yellow-rumped or Yellow-throated Tinkerbirds. Duets; second bird joining with similar song or with low, growling *krw krw*

krw... When excited (e.g. after playback), a very fast, hard and shrill *kwipipipipipip...* Also a quieter, slower and gradually slowing *kukipipip-pip-pip-pip pip...*, with second bird joining with growling *krw* notes. [CD6:2b] **HABITS** In swampy forest, rank vegetation and shrubby growth along forest trails, edges and abandoned plantations, not always near water; also gallery forest. Skulking and shy, creeping swiftly through dense vegetation without disturbing it and rarely venturing into the open. Much more often heard than seen, but whistled imitation or playback of song often attracts bird very close to source of sound. **SIMILAR SPECIES** Buff-spotted Flufftail is the only other flufftail occurring in forest. Male differs in having buff (not white) spots and barred (not plain) tail; female (q.v.) very different from female White-spotted. **STATUS AND DISTRIBUTION** Common to scarce resident throughout forest zone (up to 1600 m), from Gambia and S Senegal to Nigeria (except extreme SE) and N and C Cameroon (nominate), extreme SE Nigeria to Gabon (*zenkeri*), interior S Cameroon (*batesi*), and S CAR and Congo (*centralis*). Also claimed from SW Niger. **NOTE** Races distinguished by richness of plumage colours, width of barring in female, and size. In *centralis* barring in female narrower and slightly paler, more buffish, than in **nominate**; in *zenkeri*, the most richly coloured race, upperparts blacker; in *batesi* even blacker, chestnut paler.

BUFF-SPOTTED FLUFFTAIL *Sarothrura elegans* Plate 31 (2)
Râle ponctué

L 15 cm (6"). *S. e. reichenovi.* **Adult male** Basic pattern similar to White-spotted Flufftail, but chestnut paler, *upperparts spotted buff* (not white) and *rufous tail barred black*. Underpart spotting white, becoming buff on lower flanks and

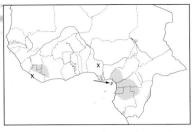

undertail-coverts. **Adult female** Upperparts rich brown finely spotted buff and black. Throat whitish; sides of head and throat finely barred blackish and buff, breast darker, densely marked blackish and buff, grading into coarse tawny and dark brown barring on belly. Tail rufous-brown narrowly barred buff and black. **Juvenile** Mainly sooty-brown with paler belly. **VOICE** Song remarkable and difficult to locate: a drawn-out, low and hollow, foghorn-like *whooooooooooooooo...* lasting 3–4 seconds; mainly uttered at night from concealed perch 1–2 m up, occasionally by day, esp. in wet weather, and from ground. A second bird may join with a very fast, low *tugutugutugutugu...* (rather reminiscent of a running engine). Calls include various moaning, growling, whining and hissing sounds. [CD6:3a] **HABITS** Principally in forest and thick bush, generally preferring edges of tangled growth, but also in forest interior and abandoned and overgrown cultivation, scrub and plantations; not necessarily near water. Skulking and shy like congeners. **SIMILAR SPECIES** White-spotted Flufftail (q.v. and above) is the only other flufftail encountered in same habitat. **STATUS AND DISTRIBUTION** Generally uncommon to rare resident in forest zone, recorded at scattered localities from E Sierra Leone and SE Guinea to Ivory Coast, and from S Nigeria to Congo; also Bioko. Locally common, Gabon. Some records indicate vagrancy or migration.

RED-CHESTED FLUFFTAIL *Sarothrura rufa*   Plate 31 (4)
Râle à camail

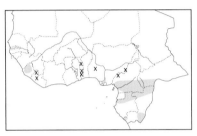

L 15 cm (6"). **Adult male *bonapartii*** Head, mantle and breast bright reddish-chestnut; rest of plumage black densely covered with narrow white streaks, becoming spots on tail. **S. r. *elizabethae*** slightly larger. **Adult female** Upperparts blackish-brown finely and densely barred buffish-brown; head-sides paler brown, becoming buffish on throat; rest of underparts buffish-brown barred blackish; belly paler. **Juvenile** Sooty-black above with some fine white streaking, greyish-black below with paler throat and belly. **VOICE** Song, given mainly at night, less during day, a more or less regular series of single notes *whoah whoah whoah* ... Second bird may join with higher pitched notes *wheea wheea wheea* ... Variations include series of alternating *whoah* and *took* notes or *hoo-du hoo-du hoo-du* ... also in rapid series *whoduduwhoduduwhodudu...* Calls include loud and fast *kuipuipuipuipuip...* and *pulipulipulipulip...* [CD6:3b] **HABITS** Frequents marshy areas, rank herbage and dense vegetation bordering ponds and rivers. Secretive and shy. Difficult to flush; flies reluctantly and only for a few metres. **SIMILAR SPECIES** Chestnut-headed and Streaky-breasted Flufftails (q.v.) occur in similar habitat, but males have less extensive chestnut parts; females have mainly white (not buffish) markings on upperparts. **STATUS AND DISTRIBUTION** Rare resident, locally common in SE Cameroon and Gabon; scattered records from Sierra Leone, Liberia, Burkina Faso, Togo, Nigeria, Cameroon and Congo (*bonapartii*), and SW CAR (*elizabethae*). Probably overlooked.

CHESTNUT-HEADED FLUFFTAIL *Sarothrura lugens*   Plate 31 (3)
Râle à tête rousse

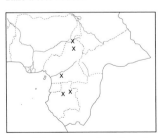

Other name E: Long-toed Flufftail
L 15 cm (6"). *S. l. lugens*. **Adult male** The only flufftail in the region with *chestnut restricted to head*. Throat white; rest of plumage black finely streaked white (densest on breast, broadest on rest of underparts). **Adult female** Head blackish, finely streaked pale brown on crown and speckled on lores and ear-coverts. Upperparts as male. Throat white; underparts spotted and streaked blackish and white. **Juvenile** Dull blackish with some fine white streaking on upperparts; throat and centre of belly whitish. **VOICE** Song a series of moaning *hoo* notes, either delivered at same speed and intensity, or gradually increasing in speed and intensity (as if becoming more excited), then dying away and ending abruptly. Heard all year, but esp. during rains. [CD6:4a] **HABITS** Frequents tall, dense herbage in grassy marshes and clearings. May occur alongside Red-chested and Streaky-breasted Flufftails in open areas, and even with White-spotted and Buff-spotted Flufftails in forested habitat. **SIMILAR SPECIES** Males of other flufftails occurring in similar habitat have more extensive chestnut. In male Streaky-breasted this extends onto upper breast; also has underparts whitish heavily streaked black. In male Red-chested chestnut extends even further, onto mantle and breast. Females hard to distinguish, esp. if seen briefly. Juveniles even more difficult and in most cases probably not identifiable in field. **STATUS AND DISTRIBUTION** Rare to uncommon and local resident, only recorded from Cameroon (Obala, Ngaounyanga) and NE Gabon.

STREAKY-BREASTED FLUFFTAIL *Sarothrura boehmi*   Plate 31 (5)
Râle de Böhm

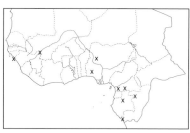

Other name E: Böhm's Flufftail

L 15 cm (6"). **Adult male** Differs from male Chestnut-headed Flufftail in having rufous on head paler and extending onto mantle and upper breast; underparts white heavily streaked blackish, except on centre of belly. **Adult female** Mainly brownish-black with whitish feather fringes giving scalloped and barred appearance; throat and centre of belly white. **Juvenile** Dull sooty-black with whitish throat and centre of belly. Juvenile male has some faint white streaking on upperwing-coverts. **VOICE** Song, uttered at night and during day, a single hollow note on same pitch, repeated in long regular series *whoo whoo whoo whoo whoo ...* (up to 25x). When annoyed (?) a shorter version of the same note, uttered faster and in series of variable speed. Also a rapid, high-pitched, crescendo *kyeh-kyeh-kyeh-kyeh-...* [CD6:4b] **HABITS** Frequents shallow-flooded grassland and open areas of rather short grass dotted with clumps of high herbage, not always near water. May occur alongside Red-chested Flufftail. Flight reportedly stronger and more direct than other flufftails. **SIMILAR SPECIES** Adult males of Chestnut-headed (q.v. and above) and Red-chested Flufftails (q.v.). Female Chestnut-headed distinguished from female Streaky-breasted by streaked upperparts, spotted and streaked underparts, and chestnut wash to head. Female Red-chested paler overall. **STATUS AND DISTRIBUTION** Inadequately known. Resident and intra-African migrant, NE Gabon (locally common) and Congo; a few records from Mali, Nigeria and S Cameroon. One taken at sea *c.* 150 km off Guinea coast. Probably overlooked.

AFRICAN CRAKE *Crex egregia*   Plate 32 (6)
Râle des prés

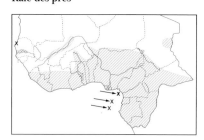

L 20–23 cm (*c.* 9"). **Adult** Upperparts dark brown with dark-centred feathers; head-sides to breast grey; narrow white supercilium; flanks and belly to undertail-coverts boldly barred black and white. Eye bright red; orbital ring pink. Bill pinkish-grey. Legs brownish, sometimes tinged pink. **Juvenile** Browner below, with less distinct barring. Eye dark; bill and legs dusky. **VOICE** A single hard *kluk*, *kruk*, *krw* or *kip*, occasionally uttered in fast series. [CD6:5a] **HABITS** Usually singly or in pairs, on migration sometimes in small groups. Frequents grassy areas in savanna and forest clearings; not in desert and forest. Active all day, but mostly at dawn and dusk. Skulking, but not shy. When flushed, flies reluctantly for short distance with dangling legs. **SIMILAR SPECIES** The rarely recorded Corn Crake is much paler and has bright chestnut wings in flight. All other crakes are much smaller. African Water Rail (rare and local in W Africa) has long, red bill, red legs and plain brown upperparts. **STATUS AND DISTRIBUTION** Common to rare resident and intra-African migrant throughout, except arid north and forest. Also São Tomé and Príncipe (uncommon). Vagrant, Bioko. Wet-season breeder, moving to north of range during rains. **NOTE** Sometimes placed in genus *Crecopsis*.

CORN CRAKE *Crex crex*   Plate 32 (7)
Râle de genêts

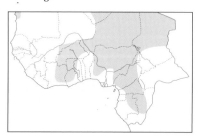

L 27–30 cm (*c.* 10.75–11.75"). Brownish-buff, terrestrial crake with diagnostic *bright chestnut wings*. Upperparts greyish-brown with black feather centres; head-sides and broad supercilium brownish-grey, surrounding brown lores and ear-coverts. Throat whitish; rest of underparts brownish-buff, tinged greyish on breast; flanks and undertail-coverts barred tawny and white. Bill stout, pale pinkish-horn. Legs pinkish-grey. **VOICE** Silent in winter quarters. [CD2:8b] **HABITS** Occurs in grassy areas, usually avoiding wet places. Skulking and rarely seen. When flushed, flies reluctantly over short distance with dangling legs; rusty wings conspicuous. Most active at dawn and dusk. **SIMILAR SPECIES** African Crake is much darker, lacks chestnut on wings, and has lower underparts boldly barred black and white. Francolins have powerful flight on whirring wings and no dangling legs. **STATUS AND DISTRIBUTION** Poorly known due to secretive behaviour. Rare Palearctic winter visitor, recorded from Mauritania, Mali, Ivory Coast, Ghana, Niger, Nigeria, Cameroon, Chad, Gabon and Congo. Main wintering range SE Africa. *TBW* category: THREATENED (Vulnerable).

AFRICAN WATER RAIL *Rallus caerulescens*
Râle bleuâtre

Plate 31 (8)

Other names E: Kaffir Rail, African Rail
L 27–28 cm (*c.* 11"). **Adult** Medium-sized rail with distinctive *long red bill* and red legs. Upperparts plain dark brown; head-sides and underparts grey; flanks to undertail-coverts blackish narrowly barred white. **Juvenile** Basic pattern similar, but browner overall with whitish throat. Bill and legs dark brown, becoming redder with age. **VOICE** Vocal. A fast series of shrill notes, starting with a rapid trill, speeding up, then gradually slowing and descending in intensity *trrrrri-kew-kew-kew-kew-kew-...* Others may join with low clucking and grunting notes. [CD6:5b] **HABITS** Skulking and usually keeping to cover of reedbeds and dense vegetation in swamps or beside ponds and rivers. Walks with jerky movements and cocked tail. Flight low with dangling legs. Most active at dusk. **SIMILAR SPECIES** None. The only rail in the region with long bill. **STATUS AND DISTRIBUTION** Rare and very local resident. Recorded from Gambia, Sierra Leone (Ribi R. area), Cameroon (Ngaoundaba, Ndop, Yaoundé), NE Gabon, NE CAR and S Congo. Vagrant, São Tomé.

LITTLE CRAKE *Porzana parva*
Marouette poussin

Plate 32 (3)

L 18–20 cm (*c.* 7–8"). Slightly larger and more slender than Baillon's Crake, with olive-brown upperparts having some white spots and streaks, inner webs of tertials fringed buff, forming creamy line, and *long primary projection* (almost as long as exposed tertials, with at least 5 primary tips visible). *Bill lime-green with red base.* Eye and narrow orbital ring red. Legs green. **Adult male** Head-sides, neck and underparts blue-grey with rear flanks and undertail-coverts barred dark grey and white. **Adult female** Grey areas of male replaced by brownish-buff, sometimes retaining some grey on supercilium and submoustachial area; throat whitish. **VOICE** Probably silent in winter quarters. Song of male *ik ikik ik*, a rhythmic series *kik-kik-kik-kik-kik-...*, or an accelerating series of notes descending in pitch *kwek, kwek, kwek kwek kwek-kwek-kwek...*, mainly uttered at night. [CD2:9b] **HABITS** In swamps, well-vegetated lake sides, ponds, river banks and ditches. Usually secretive but not shy; sometimes in the open and allowing close approach. Walks on floating vegetation, swims and dives. Only reluctantly flies. **SIMILAR SPECIES** Baillon's Crake is darker and better marked, with red-brown upperparts more densely and conspicuously speckled white, short primary projection, obvious black-and-white barring on flanks extending forward of legs. **STATUS AND DISTRIBUTION** Rare Palearctic winter visitor. Most records from Senegal R. delta (Mauritania/Senegal, Oct–Jan); also Gambia, Liberia, Ivory Coast, Burkina Faso, S Niger (Sep–Jan), N Nigeria (Kano and Hadejia–Nguru wetlands, Dec–Mar). Probably overlooked.

BAILLON'S CRAKE *Porzana pusilla*
Marouette de Baillon

Plate 32 (4)

L 17–19 cm (*c.* 6.75–7.5"). *P. p. intermedia*. Smallest crake. **Adult male** Upperparts rich chestnut-brown densely and irregularly speckled and streaked white. Face and underparts blue-grey; rear flanks, belly and undertail-coverts distinctly barred black and white. Wings short and rounded, *primaries barely projecting beyond tertials*, with rarely more than 3 primary tips visible. *Bill wholly greyish-green.* Eye red. Legs greyish-pink to brownish-olive. **Adult female** As male but often has paler throat and underparts, and brownish ear-coverts. **VOICE** Probably silent in winter quarters. Song of male a dry rattle *tk tk tk tk rrrrrkkkkkkk*, mainly uttered at night. [CD2:10a] **HABITS** In marshes, reedbeds, stream sides and small swampy patches. Very skulking and rarely in the open, but not shy and often tame. Creeps through dense vegetation flicking tail, only reluctantly flying short distances when flushed. Occasionally swims and dives. Most active at dawn and dusk. **SIMILAR SPECIES** Male Little Crake is paler and plainer, with dull muddy brown upperparts, scarcer and less contrasting white speckling, longer primary projection, pale tertial streak, much fainter barring on flanks, and red base to bill. **STATUS AND DISTRIBUTION** Scarce Palearctic winter visitor, Senegal R. delta (Mauritania/Senegal, Oct–Feb). Claimed from Gabon. Probably overlooked.

SPOTTED CRAKE *Porzana porzana*
Marouette ponctuée

Plate 32 (2)

L 22–24 cm (*c.* 9"). Rather small, compact, freckled-looking crake with diagnostic *plain buff undertail-coverts*. **Adult**

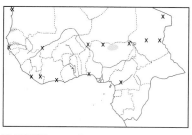

male Upperparts olive-brown streaked black; underparts paler, tinged grey and covered with fine white flecks, which coalesce to form white bars on flanks. Head has broad blue-grey supercilium, brownish ear-coverts and dark lores. Bill yellow with orange-red base. Eye brown. Legs olive-green. **Adult female** Similar but usually browner and more speckled on face and underparts. **Juvenile** As adult but has denser white speckling, no grey on head and whitish throat. **VOICE** A rhythmic series of far-carrying, explosive *hwit!* notes ('whiplash'), also heard in winter quarters, mainly at dusk and at night. Contact calls between wintering birds a hard *whee-up*. [CD2:10b] **HABITS** In marshes, swampy grasslands, edges of reedbeds and densely vegetated ponds. Usually skulking but not particularly shy; also forages in the open. Walks with jerky movements and cocked tail, like congeners (thereby showing diagnostic undertail). Flies reluctantly; also swims and dives. Mainly active at dusk. **SIMILAR SPECIES** Little and Baillon's Crakes are smaller with unspeckled underparts and barred undertail-coverts. Similar-sized Striped Crake is more slender with striped upperparts and russet undertail-coverts. African Crake is larger, more elongated, with bold black-and-white barring on lower underparts. **STATUS AND DISTRIBUTION** Generally rare Palearctic winter visitor (Nov–Apr), patchily recorded from Mauritania, Senegal (common in Senegal R. delta), Mali, Liberia, S Ivory Coast, Nigeria, SW Niger, Cameroon and Chad. Probably overlooked. Main wintering area S Africa.

STRIPED CRAKE *Aenigmatolimnas marginalis*   Plate 32 (5)
Marouette rayée

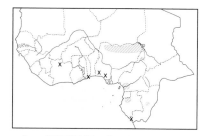

L 18–21 cm (*c.* 7.0–8.25"). Medium-small crake with diagnostic *russet rear flanks and undertail-coverts*. Upperparts dark brown with white-edged feathers forming lines and giving streaked appearance. Bill greenish and relatively heavy. Eye golden-brown or dark brown (not red), orbital ring pale green or yellow. Legs and very long toes greenish. **Adult male** Head-sides and breast to flanks buffish-brown with very faint pale barring on lower flanks; throat and belly off-white. **Adult female** Similar but head grey; breast to flanks pale grey, faintly streaked white. **Juvenile** Like adult male but duller and browner overall, lacking white streaks; head-sides and breast tinged rufous. **VOICE** Usually silent. In breeding season utters a long dry rattle (rather reminiscent of an engine) *rrtktktktktktktktktk...*, mainly at night. Call a fast *krw-krw-krw-...* [CD6:6] **HABITS** In seasonally flooded grassland, swampy areas and edges of ricefields and drainage ditches with waist-high herbage, avoiding permanent marshes with higher vegetation. Highly secretive and very rarely seen. Very hard to flush, prefers creeping unobtrusively through vegetation. Bobs tail in typical crake fashion. **SIMILAR SPECIES** *Porzana* crakes (q.v.) never have russet undertail-coverts and always possess barred flanks. African Crake is larger and lacks white streaks on upperparts. Juvenile Allen's Gallinule is much larger with longer, brownish legs and scaly (not streaked) upperparts. **STATUS AND DISTRIBUTION** Poorly known. Scarce to rare resident and intra-African migrant, patchily recorded from N Ivory Coast, Ghana, Togo, Nigeria, Cameroon, Gabon and Congo. Breeds in wet season.

BLACK CRAKE *Amaurornis flavirostris*   Plate 32 (1)
Râle à bec jaune

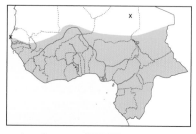

Other name F: Marouette noire
L 19–23 cm (*c.* 7.5–9.0"). **Adult** Distinctive, smallish, *all-black* crake with conspicuous *yellow bill* and *red legs*. Eye and orbital ring red. **Juvenile** Duller, with whitish throat and dusky bill and legs. **VOICE** Noisy. A variety of clucking, purring and growling sounds, uttered in stereotyped duet. [CD6:7] **HABITS** Singly or in pairs, occasionally in small groups, frequenting a variety of swampy places and edges of lakes, ponds, rivers and reedbeds. Less shy than other crakes and often seen in the open, walking on floating vegetation. **SIMILAR SPECIES** Common and Lesser Moorhens are larger and have white undertail-coverts. **STATUS AND DISTRIBUTION** Common resident throughout, except arid north. Local migrant in Sahel zone, appearing during rains. The commonest species of its family. **NOTE** Previously placed in genus *Limnocorax*.

ALLEN'S GALLINULE *Porphyrio alleni*

Talève d'Allen

Plate 33 (3)

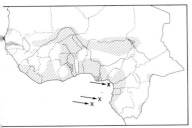

L 25–26 cm (*c.* 10"). **Adult** Rather small and elegant, glossy purplish-blue and dark green gallinule with white undertail, *bluish frontal shield*, red bill and *long red legs and feet*. Eye red to red-brown or yellow-brown. Bare parts brighter when breeding. At start of breeding, frontal shield turquoise-blue in male, apple-green in female; subsequently, shield becomes blue in both adults; following breeding, shield dark grey or dark blue. **Juvenile** Mainly warm buffish-brown with dark-centred upperpart feathers giving scaly appearance. Throat and belly white; undertail-coverts buff. Bill and legs dusky; frontal shield olive-brown; eye brown. **VOICE** Various hard, dry calls, uttered singly or in series, e.g. *kuk kuk kuk kuk kk* and *kip-kip-kip-kip-kirrrr*. A sneering, quacking *hah*. In flight, a high-pitched *kli-kli-kli-...* [CD6:8] **HABITS** Frequents weedy pools, ricefields, marshes, reedbeds and rank vegetation by lakes and rivers. Usually skulking; mostly seen when it dashes from dense cover or clambers in tangled vegetation. Occasionally feeds in the open. Jerks tail when moving; walks on floating plants. Most active at dawn and dusk. **SIMILAR SPECIES** Purple Swamphen is much larger and has red frontal shield. See also American Purple Gallinule. If seen in bad light, also compare Common Moorhen and Lesser Gallinule, esp. to distinguish juveniles. **STATUS AND DISTRIBUTION** Uncommon to locally common resident and intra-African migrant throughout, except arid north. Vagrant, Bioko, São Tomé, and Annobón. Movements complex and inadequately known; northward migrations with rains reported; also local movements in response to seasonal habitat changes. **NOTE** Sometimes placed in *Porphyrula* as it possesses characteristics of both *Porphyrio* and *Gallinula*.

AMERICAN PURPLE GALLINULE *Porphyrio martinica*

Talève violacée

Plate 146 (9)

Other name F: Talève pourprée

L 33 cm (*c.* 13"). **Adult** Superficially resembles Allen's Gallinule (q.v.), but larger, with red, *yellow-tipped bill*, blue-white frontal shield and *yellow legs*. Undertail wholly white. **Juvenile** Upperparts unpatterned (not scaly) and more olive than juvenile Allen's; throat pale (not white); lower flanks, belly and undertail white. Bill dull grey and green; frontal shield grey; legs dull yellow. **HABITS** As Allen's Gallinule. **STATUS AND DISTRIBUTION** Vagrant from the Americas. One record 90 km off coast Liberia (Jun). The species is known for its many records of long-distance vagrancy. **NOTE** Sometimes placed in genus *Porphyrula*.

PURPLE SWAMPHEN *Porphyrio porphyrio*

Talève sultane

Plate 33 (4)

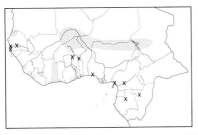

Other names E: Purple Gallinule; F: Talève poule-sultane

L 45–50 cm (*c.* 18–20"). *P. p. madagascariensis*. Unmistakable. **Adult** *Very large*, glossy purple and blue-green rail with white undertail, *massive bright red bill, red frontal shield* and long red legs. Eye red. **Juvenile** Paler and greyer, with duskier bare parts. **VOICE** Various nasal, groaning, trumpeting, grunting, clucking, hooting and wailing calls. [CD2:11a] **HABITS** Solitary, in pairs or small groups in marshes, reedbeds, and shallow ponds and lakes with rank herbage. Rather shy, usually keeping to dense vegetation, but occasionally foraging in the open. **SIMILAR SPECIES** Allen's Gallinule is much smaller and has a bluish frontal shield. **STATUS AND DISTRIBUTION** Locally common to rare resident with patchy distribution. Recorded from S Mauritania and Senegambia, from C Mali (central Niger delta) through Burkina Faso, S Niger and N Nigeria to Chad, from Cameroon, CAR and N Congo, and from coastal Sierra Leone to Nigeria. Local movements in response to seasonal habitat changes.

COMMON MOORHEN *Gallinula chloropus*

Gallinule poule-d'eau

Plate 33 (1)

L 30–36 cm (*c.* 13–14"). **Adult** *Mainly blackish* with ragged white line along flanks, conspicuous white-sided undertail, *red, rounded frontal shield* and *red, yellow-tipped bill*. Head sooty-black becoming slaty on neck, mantle and underparts; upperparts dark brown. Legs and long toes yellow-green with red 'garters'. Eye red. Resident *meridionalis* has upperwing-coverts slaty blue-grey without olive tinge; in **nominate** wing-coverts are concolorous with upperparts and tinged olive. **Juvenile** Mainly greyish-brown, with whitish throat and belly, white flank line and white on

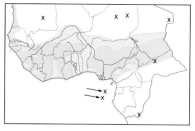

undertail as adult, greenish-brown bill and dull green legs. Eye grey brown. **VOICE** Most characteristic calls include a sudden, single *yerrrrp* and a short hard *kik*, sometimes in rapid series *kikikikikik-kik kik-kik...* [CD2:11b] **HABITS** Singly, in pairs or small groups, frequenting a variety of freshwater habitats with fringing vegetation Rather shy, dashing into cover if alarmed. Tail cocked when walking Swims readily with jerky movements of head and tail. Short flight low and laboured with legs dangling; patters along surface in taking off. **SIMILAR SPECIES** Lesser Moorhen is smaller and has yellow bill with red culmen. Allen's Gallinule has purplish plumage lacking white flank line, all-red bill and blue-green frontal shield. **STATUS AND DISTRIBUTION** *G. c. meridionalis* locally not uncommon to rare resident, patchily distributed throughout. Also São Tomé and Príncipe. Formerly bred Annobón (extinct?). Nominate is Palearctic winter visitor recorded from Mauritania, Senegambia, Guinea, C Mali, Burkina Faso, Niger, N Nigeria and Chad. Former resident, now vagrant, Cape Verde Is.

LESSER MOORHEN *Gallinula angulata* Plate 33 (2)
Gallinule africaine

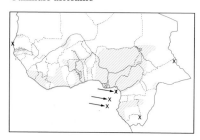

L 23 cm (*c.* 10"). **Adult male** Resembles Common Moorhen, but has *bright yellow bill with red culmen* and is noticeably smaller. Frontal shield red and pointed. Legs and feet yellowish-green. Eye red. **Adult female** Somewhat paler, with smaller frontal shield. **Juvenile** Upperparts mainly dark brown; head-sides, neck and breast pale greyish-brown or brownish-buff; throat and centre of belly whitish, becoming pale greyish on flanks. Bill dusky-yellow with darkish culmen; frontal shield yellowish to orangey; legs dull greenish. **VOICE** Calls include short, sharp *khup* and *prp*, also uttered in series. [CD6:9] **HABITS** Singly or in pairs in variety of wet habitats, incl. temporarily flooded places, ricefields, swamps and rank vegetation. Flight and behaviour as Common Moorhen, but much more skulking. **SIMILAR SPECIES** Common Moorhen distinguished by red bill with yellow tip and larger size. Juvenile Common has largely greyish-brown plumage with much darker underparts lacking any brownish-buff. **STATUS AND DISTRIBUTION** Locally common to uncommon resident and intra-African migrant throughout, except arid north. Also São Tomé and Príncipe (rare). Vagrant, Bioko. Some move north during rains.

EURASIAN COOT *Fulica atra* Plate 33 (5)
Foulque macroule

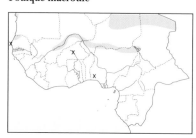

Other name E: Common Coot
L 36–38 cm (*c.* 15"). *F. a. atra.* **Adult** Plump, thickset, *all-black* waterbird with *white bill and frontal shield.* Legs and lobed toes grey-green. **Immature** Duller, tinged with brown, some retaining whitish on throat like juvenile. **VOICE** A loud, sharp *kut* or *khuk* (hence name). [CD2:12a] **HABITS** Gregarious, often in large flocks. Prefers large, still water bodies. Dives frequently. Takes off in long pattering run across water with fluttering wings. **SIMILAR SPECIES** Virtually unmistakable, but compare Common Moorhen. **STATUS AND DISTRIBUTION** Rare to locally common Palearctic winter visitor to desert oases and Sahel. Recorded from Mauritania, N Senegal (delta), Gambia, Mali (central Niger delta), Burkina Faso, Niger, Nigeria and Chad. Breeding recorded, N Senegal (2 pairs, Jan 2001).

CRANES
Family GRUIDAE (World: 15 species; Africa: 6; WA: 4)

Large to very large terrestrial and wading birds with long neck and legs, relatively short, straight bills and mainly grey plumage. Sexes similar; male usually larger. Juveniles tinged brown; adult plumage attained after 1.5–3.0 years. Flight slow and laboured with narrow neck outstretched and long legs projecting. Groups fly in V or line formation. Utter far-carrying honking and trumpeting calls.

COMMON CRANE *Grus grus*
Plate 34 (2)

Grue cendrée

L 100–120 cm (40–47"); WS 180–220 cm (71–87"). *G. g. grus.* **Adult** *Large;* mainly grey with greatly elongated, bushy tertials hanging over tail. *White stripe from eye* reaching head- and neck-sides and contrasting with black head and upper two thirds of neck. *Small red patch on crown visible at close quarters.* Bill horn-coloured. Legs black. In flight, all grey with contrasting black remiges. **Juvenile** Browner overall; head and neck brownish, lacking adult pattern. **First-winter** As juvenile but whitish patch (often tinged buffish) behind eye in some. **VOICE** A far-carrying, nasal trumpeting call, usually uttered in flight. [CD2:13a] **HABITS** Usually in flocks. Frequents open habitats. Wary. **SIMILAR SPECIES** In flight, Demoiselle Crane (q.v.) difficult, at great distance even impossible, to distinguish. Grey and Black-headed Herons are smaller with longer bill; fly with head retracted on deeply bowed, rounded wings (without 'fingers'). **STATUS AND DISTRIBUTION** Palearctic vagrant, N Mauritania (wintering flock, Dec–Feb) and N Nigeria (Dec–Mar). Claimed from Niger (Dec).

WATTLED CRANE *Bugeranus carunculatus*
Plate 146 (11)

Grue caronculée

L c. 175 cm (69"); WS 230–260 cm (91–102"). Very large. **Adult** *White head and neck with dark grey cap, white feathered wattles below chin and red bare facial skin extending onto base of bill and on wattles.* Upperparts grey; underparts black. Greatly elongated, pointed tertials cover tail, almost reaching ground. Legs black. **Juvenile** Similar but with smaller wattles, white crown, shorter tertials and no red facial skin. **VOICE** Usually silent. [CD6:10] **HABITS** Typically in pairs or small groups in marshes and moist grasslands. Wary. **SIMILAR SPECIES** Normally unmistakable, but compare other cranes and Grey Heron. **STATUS AND DISTRIBUTION** A single old record from Guinea-Bissau (Cufada lagoon, March 1948, 3 together); either vagrants or last representatives of relict and now extinct population (Hazevoet 1997). Present range Ethiopia and C and S Africa. *TBW* category: THREATENED (Vulnerable). **NOTE** Sometimes placed in genus *Grus.*

DEMOISELLE CRANE *Anthropoides virgo*
Plate 34 (1)

Grue demoiselle

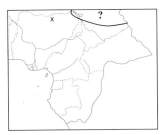

L 90–100 cm (35–39"); WS 155–180 cm (61–71"). **Adult** Smaller and more elegant than Common Crane, with *long white plumes from eye contrasting with black head and foreneck,* and plainer, paler grey upperparts. Greatly elongated, narrow, pointed tertials fall neatly over tail. Bill greyish tinged reddish on tip. Legs greyish to black. In flight as Common Crane, but *bill shorter, forehead steeper, black on neck reaching breast, and inner primaries tinged greyish;* neck, tail and legs also relatively shorter, but this often difficult to assess. **Juvenile** Wholly pale ash-grey (incl. head). **First-winter** Dull, washed-out version of adult. Paler than 1st-winter Common. **VOICE** A clear trumpeting call, higher pitched than that of Common Crane. [CD2:13b] **HABITS** Singly, in pairs or flocks in *Acacia* grassland; also cultivated areas and margins of lakes and rivers. **SIMILAR SPECIES** Common Crane (q.v. and above). **STATUS AND DISTRIBUTION** Scarce Palearctic winter visitor to C and S Chad. Formerly locally common in NE Nigeria (Chad basin); only 1 record since 1972 (1 adult, Hadejia–Nguru wetlands, Dec 1988–Feb 1989).

BLACK CROWNED CRANE *Balearica pavonina*
Plate 34 (3)

Grue couronnée

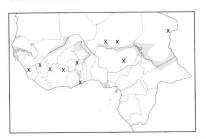

Other name E: Northern Crowned Crane

L c. 100 cm (40"); WS 180–200 cm (71–79"). Unmistakable. **Adult male** *pavonina* breeding *Very dark, slate-grey and black, with large white wing panel and straw-coloured crest.* Large bare area on head-sides, upper part white, lower pink; small pink wattle on throat. Innermost wing-coverts straw-coloured; elongated, dark chestnut bushy feathers cover tail. Bill and legs black. In flight, conspicuous white upper- and underwing-coverts contrast with dark remiges and body; head and legs held slightly below horizontal, giving hump-backed appearance. *B. p. ceciliae* has bare cheek patch more extensively red (not pink), with only upper quarter white (approximately half in nominate). **Adult male non-breeding** Long lanceolate, dark slate-grey feathers of neck and upper mantle less developed. **Adult female** Crest slightly smaller. **Immature** Similar to adult but washed rusty-brown; crest shorter, no bare pinkish area on head-sides. Adult plumage attained after *c.* 1 year, full development of facial pattern and throat wattle may require another year. **VOICE** A far-carrying, mellow trumpeting *honk* or *ka-wonk*, usually uttered in flight. [CD6:11] **HABITS** In pairs, small family groups or, outside breeding season, larger flocks. Frequents dry and moist open habitats, incl. wet plains, ricefields, and margins of lakes and rivers. Roosts in trees. **STATUS AND DISTRIBUTION** Rare to locally not uncommon resident with local seasonal movements. Patchily distributed with records from most countries; main breeding areas Senegambia and Chad basin (nominate), with some limited populations in area between. Eastern *ceciliae* reaches E and C Chad. Decreasing and threatened in parts of its range. *TBW* category: Near Threatened.

FINFOOTS
Family HELIORNITHIDAE (World: 3 species; Africa: 1; WA: 1)

Medium-sized, inconspicuous, grebe-like waterbirds with long neck and small head. Only one representative of the family in Africa, endemic to the Afrotropical region.

AFRICAN FINFOOT *Podica senegalensis*   Plate 33 (8)
Grébifoulque d'Afrique

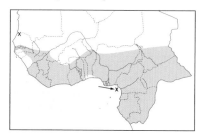

Other name F: Grébifoulque du Sénégal
L *c.* 50 cm (20"). Distinctive waterbird with *darkish upperparts, long stiff tail, red bill and bright orange-red legs and feet.* **Adult male *senegalensis* breeding** Forehead to hindneck black with green gloss, separated from slate-grey throat and foreneck by whitish line from eye down neck-side. Upperparts dark brown glossed green, finely spotted white; lower breast and belly white; flanks barred dark brown and white. *P. s. camerunensis* is much darker above and below, with blue gloss and very few or no spots on upperparts, and no white line on neck-sides. **Adult male *senegalensis* non-breeding** Throat and foreneck white. **Adult female** Smaller, forehead to hindcrown brown; throat and foreneck white. **Juvenile** As adult female but paler, much browner with few white spots; lower neck, breast and flanks tawny buff. Bill dark. **VOICE** Normally silent. Calls include dry cackles and guttural notes. [CD6:13] **HABITS** In pairs along quiet forested rivers, streams, lakes and lagoons with dense vegetation overhanging water. Swims low in water, usually near water's edge. Pumps head and neck back and forth with each leg-stroke. Occasionally creeps with remarkable agility through undergrowth of muddy banks. Unobtrusive and easily overlooked. **SIMILAR SPECIES** Swimming African Darter has longer, more slender neck and bill; also larger and lacking red on bill. **STATUS AND DISTRIBUTION** Uncommon to locally common resident throughout forest zone; also in wooded savanna. *P. s. camerunensis* confined to S Cameroon, Gabon and Bioko. Both races, sometimes considered colour morphs, occur sympatrically, with intermediates, in S Cameroon. Nowhere numerous, but perhaps more common in suitable habitat than records suggest.

BUSTARDS
Family OTIDIDAE (World: 25–26 species; Africa: 20; WA: 7)

Medium-sized to very large terrestrial birds with long neck and legs, short bill and cryptically coloured upperparts. Large species sexually dimorphic; males much larger than females. Juveniles usually much like female; adult plumage probably attained at 2–6 years. Occur in a variety of open-country habitats. Omnivorous, though probably mainly feed on insects. Walk slowly and deliberately and only reluctantly take flight. Flight powerful with neck outstretched, broad wings showing variable amount of white. Males perform distinctive courtship displays in which neck feathers are fluffed out. Largely silent outside breeding season. In W Africa, many species subject to seasonal movements, migrating north in response to rains. Breed mainly in rainy season. All species declining because of hunting and trapping.

DENHAM'S BUSTARD *Neotis denhami*   Plate 34 (7)
Outarde de Denham

Other name E: Stanley's Bustard
L male *c.* 100 cm (40"), female *c.* 80 cm (32"); WS 170–250 cm (67–98"). *N. d. denhami.* **Adult male *denhami*** Large,

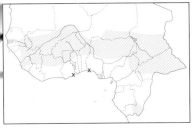

with diagnostic *rufous hindneck and bold white markings on blackish wings*. Top of head black with pale median crown-stripe. Head-sides greyish; foreneck and upper breast blue-grey becoming whitish on belly. Upperparts dark brown; tail broadly barred black and dirty white. Legs yellowish. In flight, mainly dark above with contrasting white on wing. *N. d. jacksoni* has darker rufous hindneck. **Adult female** As adult male, but smaller, with head-sides, throat and foreneck tinged and vermiculated buffish and blackish-brown. **Juvenile** As adult but duller. **VOICE** Mostly silent. Occasionally utters a guttural barking and booming sound. **HABITS** Usually singly, sometimes small parties, in wooded savanna and farmland; attracted to freshly burnt grassland. Usually wary. Male performs 'balloon' display, puffing out neck and breast feathers; then particularly conspicuous. **SIMILAR SPECIES** Similar-sized Arabian Bustard has entire neck pale grey and lacks conspicuous white patches on wing. Nubian Bustard is smaller and much paler, without rufous on hindneck. **STATUS AND DISTRIBUTION** Scarce to locally not uncommon in broad zone, from S Mauritania to Sierra Leone east to Chad and CAR (nominate); also SE Gabon and Congo (*jacksoni*). Subject to seasonal movements, moving north with rains (May–Jun). *TBW* category: Near Threatened.

NUBIAN BUSTARD *Neotis nuba* Plate 34 (9)
Outarde nubienne

L male *c.* 70 cm (28"), female *c.* 50 cm (20"); WS 140–180 cm (55–71"). *N. n. agaze.* **Adult male** Fairly large, pale bustard with diagnostic *rufous, black-edged crown* (ending in short crest at nape) and *black throat* contrasting with whitish head-sides. Neck pale blue-grey; upperparts and tail tawny-buff with black vermiculations. Breast rich tawny, rest of underparts white. Bill pale yellow with black tip to upper mandible, or pale green with dusky base. Legs creamy-white. In flight, black remiges with large white area on primaries and greater primary-coverts; secondaries tipped white and white at base; tail with white patches at sides of base; underwing largely white. **Adult female** As adult male, but smaller; head markings duller, throat patch smaller. **Juvenile** As adult female, but head markings duller still and black on throat reduced to stripe. **VOICE** Described as a shrill *maqur* (Mackworth-Praed & Grant 1970). **HABITS** Occurs in dry thorn scrub and savanna at desert edge. Behaviour and ecology little known. **SIMILAR SPECIES** White-bellied Bustard is considerably smaller, with black-and-grey crown, white supercilium, less black on throat, and no white on upperwing. Denham's Bustard is larger, much darker and has barred tail without white base. **STATUS AND DISTRIBUTION** Uncommon to rare resident in arid and semi-arid belt, in Mauritania, Mali, Niger and Chad. Vagrant, N Nigeria (May 1959) and N Cameroon. Northward movement with rains. *TBW* category: Near Threatened. **NOTE** Sometimes considered monotypic (cf. *HBW*).

HOUBARA BUSTARD *Chlamydotis undulata* Plate 146 (9)
Outarde houbara

L 55–65 cm (22–26"); WS 135–170 cm (53–67"). *C. u. undulata.* **Adult male** Fairly large, mainly sandy coloured with *conspicuous black frills on neck-sides*. Crown has tuft of white erectile feathers; neck greyish-white becoming white on belly; upperparts vermiculated brown; tail has 4 blue-grey bars. Flight feathers black with large white patch on outer primaries. **Adult female** As male, but smaller and with shorter crown and neck plumes. **Juvenile** As adult female, but crown feathers and neck frills even less developed. **VOICE** Generally silent. [CD2:14b] **HABITS** In semi-desert. Wary. Male performs spectacular 'trotting' display with raised neck plumes and head drawn back onto mantle. **SIMILAR SPECIES** Nubian Bustard, the only other bustard occurring in its range, is more tawny and lacks black frills on neck. **STATUS AND DISTRIBUTION** Uncommon resident, Mauritania (north of 20°N). *TBW* category: Near Threatened.

ARABIAN BUSTARD *Ardeotis arabs* Plate 34 (8)
Outarde arabe

L male *c.* 100 cm (38"), female *c.* 80 cm (30"); WS 205–250 cm (81–98"). *A. a. stieberi.* **Adult male** *Large*, with greyish, thick-looking neck and breast finely and densely barred blackish (appearing all grey at distance). Crown buffish broadly bordered black, with short black nuchal crest; supercilium creamy-white. Upperparts mainly dull earth-brown and tawny; underparts whitish. Wing-coverts tipped white. Eye, bill and legs yellowish; culmen darkish. In flight, upperwing has irregular, narrow white barring; outer primaries and primary-coverts mostly lack white,

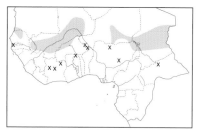

forming black patch. Old birds have much pale grey bloom on remiges. **Adult female** As adult male, but smaller, with greyer upperparts. **Juvenile** As adult but duller. **VOICE** Generally silent. [CD2:15b] **HABITS** Singly, in pairs or family parties in semi-desert and arid grassy plains; also *Acacia* woodland. Reported to migrate in flocks, sometimes with Denham's Bustard. Wary. **SIMILAR SPECIES** Similar-sized Denham's Bustard has rufous hindneck and darker upperparts with much more white on wings. **STATUS AND DISTRIBUTION** Uncommon to rare resident and intra-tropical migrant in Sahel belt, from S Mauritania and N Senegal to N Cameroon and Chad; also CAR (Bamingui-Bangoran NP). Vagrant, N Ivory Coast and N Ghana. Moves north with rains.

SAVILE'S BUSTARD *Eupodotis savilei*   Plate 34 (4)
Outarde de Savile

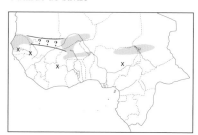

Other names E: Pygmy Bustard, Crested Bustard; F: Outarde houpette
L 41 cm (16"). *Smallest bustard of the region*, with relatively short neck and legs. **Adult male** Crown grey; nape with tuft of rufous feathers (erected in display); head-sides creamy-white; neck deep buff; upperparts and tail tawny with black vermiculations. Throat black; foreneck grey; rest of underparts black with white patch on breast-sides. Bill and legs yellowish, culmen brown. In flight, tawny upperwing-coverts separated from blackish flight feathers by white band. **Adult female** As adult male, but crown dark brown and buff, throat white, neck wholly cinnamon-buff and breast with broad white band. **Juvenile** As adult female, but with buff tips to primaries. **VOICE** Distinctive. Contact call a clear, whistled *tuit-thit*. Male advertisement consists of a short whistled note followed by a rapid series of short whistles *tuit! tutututututututu*, or an accelerating series of clear whistles *thut thut-thut-thut-thututututut*; also a series of frog-like notes in same rhythm. [CD6:14] **HABITS** Singly or in pairs in arid and semi-arid habitats, incl. grassy plains and lightly wooded savanna. Secretive; flies reluctantly and for short distance only, preferring to avoid detection by remaining immobile in cover. Behaviour poorly known; display unrecorded. **SIMILAR SPECIES** Adult male Black-bellied Bustard, the only other bustard with black on underparts, has black extending onto foreneck and is obviously larger and more slender. **STATUS AND DISTRIBUTION** Locally not uncommon to rare resident with patchy distribution in Sahel belt. Recorded from S Mauritania, Senegal, Gambia (rare), Mali, Burkina Faso, N Ivory Coast (rare), Niger (scarce), NE Nigeria and Chad. Generally considered sedentary, but seasonal movements suspected. **NOTES (1)** Often treated as race of extralimital Buff-crested Bustard *E. ruficrista* but vocalisations very different and not known to perform similar rocket-flight display. **(2)** Sometimes placed in genus *Lophotis* (cf. *HBW*).

WHITE-BELLIED BUSTARD *Eupodotis senegalensis*   Plate 34 (6)
Outarde du Sénégal

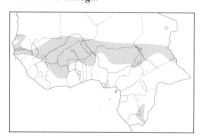

L *c.* 50 cm (20"). *Fairly small* and slender, *white-bellied* bustard. **Adult male** *senegalensis* Distinctive head pattern and greyish-blue neck. Forehead and forecrown black becoming *greyish-blue* on hindcrown, bordered black on nape; head-sides and throat whitish except black patch on throat which extends onto sides of upper neck. Upperparts and tail tawny-buff with dark vermiculations; tail has 2 narrow dark bars. Breast tawny; rest of underparts white. Bill and legs yellowish. In flight, black flight feathers with some whitish on inner webs of primaries contrast with tawny-buff upperwing-coverts; underwing whitish with black trailing edge. *E. s. mackenziei* has tawny-buff hindneck. **Adult female** As adult male, but lacks distinctive markings: crown dusky-brown, throat white, neck mainly pale tawny-buff. **Juvenile** Similar to adult female. **VOICE** Vocal. A loud, sonorous, rather nasal, *kuk-kwarrak* often repeated and occasionally interspersed with variations, e.g. *kuk-wrik-e-haah!* [CD6:15] **HABITS** Singly, in pairs or small family groups in open savanna, grassland, bush bordering cultivation and thorn scrub. Most active at dawn and dusk. Wary. **SIMILAR SPECIES** Nubian Bustard is considerably larger and has rufous crown. Female Black-bellied Bustard differs from female White-bellied in larger size, more slender build, uniform dull buffish-brown neck with fine vermiculations and mainly black underwing. **STATUS AND DISTRIBUTION** Locally not uncommon to rare resident in Sahel belt, from S Mauritania, Senegal, Mali and N Ivory Coast to S Niger, Nigeria,

N Cameroon, Chad and N CAR (nominate), and savanna in SE Gabon and Congo (*mackenziei*). Northward movement with rains reported from Chad. **NOTE** *E. s. mackenziei* and extralimital S African *E. s. barrowii* sometimes treated as separate species, Barrow's Bustard *E. barrowii*.

BLACK-BELLIED BUSTARD *Eupodotis melanogaster*

Plate 34 (5)

Outarde à ventre noir

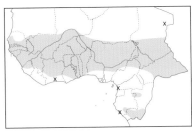

L *c.* 60 cm (24"). *E. m. melanogaster.* Medium-sized bustard, with long, thin neck and long legs. **Adult male** Head and hindneck buff with fine dark markings; black stripe from eye ending in short nuchal crest. Upperparts and tail buffish-brown with dark blotches and vermiculations; tail has dark bars. *Throat mottled blackish merging into black line bordered with white on foreneck; rest of underparts black.* Bill and legs yellowish, darker on culmen. In flight, upperwing has large area of white (larger than in any other bustard): greater and lesser coverts white, primaries white tipped black, except blackish outer one, secondaries mostly blackish with some white; underwing mainly black with white patch on primaries. **Adult female** Mainly brownish-buff with dark blotches and vermiculations. Throat and belly whitish. In flight, much less white on upperwing, greater and lesser coverts buff, flight feathers blackish above and below with whitish spots on inner primaries and buff tips to secondaries; underwing-coverts blackish spotted white and buff. **Juvenile** As adult female, but wing feathers edged buff. **VOICE** Generally silent. Male advertising call, given from regularly used prominence, a hoarse, nasal sound followed after a short pause by a low, drawn-out growl leading into a cheerful, popping or hiccuping note *vwok, rrorr-WIK!*; head stretched out and retracted during call. [CD6:16] **HABITS** Singly or in pairs in open savanna; also farmland and derived savanna. Male display includes flight with deliberate wingbeats, alighting after glide with raised wings. **SIMILAR SPECIES** Male Savile's Bustard distinguished from male Black-bellied by much smaller size, richer coloured upperparts and greyish foreneck. Female White-bellied Bustard differs from female Black-bellied in smaller size, richer coloration, esp. on breast, and mainly white underwing. **STATUS AND DISTRIBUTION** Not uncommon to rare resident in savanna and Sahel, from S Mauritania to Sierra Leone east to Chad and CAR, and in S Gabon and Congo. Also recorded near coast in Ivory Coast and SW Cameroon. Mainly sedentary, but northward movement with rains reported from Mali and Nigeria.

JACANAS
Family JACANIDAE (World: 8 species; Africa: 2; WA: 2)

Distinctive family of small to medium-sized waterbirds with long legs and extremely long toes and claws, enabling them to walk on floating vegetation. Sex roles reversed in many species (e.g. African Jacana), the larger female being polyandrous and leaving parental care mostly or entirely to the male. Flight usually low, on rounded wings and with large feet trailing.

AFRICAN JACANA *Actophilornis africana*

Plate 33 (7)

Jacana à poitrine dorée

Other names E: Lily-trotter; F: Jacana
L 30 cm (12"). Combination of *bright chestnut upperparts* and *long greyish legs with extraordinary long toes* diagnostic. In flight, huge trailing feet, rounded chestnut wings with black primaries and short tail. **Adult** Crown and hindneck black, bill and large frontal shield bright blue. Head-sides, throat and neck white bordered by golden-yellow upper breast; rest of underparts chestnut. **Juvenile** White supercilium contrasting with dark brown crown, hindneck and eye-st ripe; frontal shield small and grey (invisible in field). Underparts white with breast-sides washed yellow; flanks chestnut. **VOICE** Distinctive.

Various nasal, strident and grating sounds, incl. a husky drawn-out whining *kyowrrr*, a shorter, repeated *kreep-kreep-kreep...* and a high-pitched *weep-weep-weep...* (reminiscent of some flufftail calls). [CD6:17] **HABITS** Occurs on lakes, pools, marshes, lagoons and backwaters with floating and emergent vegetation, also temporarily flooded areas (even small) and ricefields. Usually in small, loose groups, occasionally in pairs or large gatherings. Not shy. Walks on floating vegetation (esp. water lilies, hence alternative name). Swims and dives. Vocal. **SIMILAR SPECIES**

Unlikely to be confused, but compare juvenile with much smaller Lesser Jacana. **STATUS AND DISTRIBUTION** Common resident throughout, except arid north.

LESSER JACANA *Microparra capensis*
Jacana nain

Plate 33 (6)

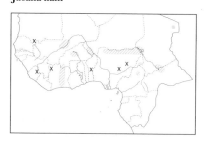

L 16 cm (6.25"). *Small*, inconspicuous, crake-like jacana. **Adult** Much smaller than African Jacana, with *rufous crown and nape*, white supercilium, dark eye-stripe and no frontal shield. Upperparts mainly brown; underparts white; neck-sides tinged pale golden-yellow. Tail short, rufous-chestnut. Bill pale brownish-olive. Legs pale olive. *In flight, white trailing edge contrasts with black flight feathers.* Pale brown panel in centre of upperwing; underwing largely black. Short, rounded wings and trailing feet may recall crake. **Juvenile** As adult. **VOICE** Varied. Calls include a soft, rapid *whoop-whoop-whoop-whoop-whoop*, chattering *kikikikikikikiki*, softer *ku-ku-ku*, plaintive *shree shree shree shree* or *see sree srrr*, and soft *chrr-chrr-chrr*. [CD6:18] **HABITS** Frequents various wetlands with floating vegetation (water lilies) or emergent grass; also temporally flooded areas. Usually singly, but may gather in numbers. Shy and easily overlooked. Flies readily, briefly raising wings upon landing. **SIMILAR SPECIES** Juvenile African Jacana is almost twice as large with longer bill, golden wash on underparts and black crown and hindneck; in flight, no white trailing edge to wing. When only briefly seen, confusion with crakes possible, but no crake in the region has white underparts. **STATUS AND DISTRIBUTION** Widespread but local resident. Locally common (e.g. Mali, Nigeria) to rare. Mainly sedentary, but some seasonal wandering may occur. Recorded from C Mali, Ivory Coast, Burkina Faso, N Nigeria/S Niger to N Cameroon, Chad and CAR. Vagrant, Sierra Leone and S Mauritania (Kankossa, Nov).

PAINTED-SNIPES
Family ROSTRATULIDAE (World: 2 species; Africa: 1; WA: 1)

Medium-sized, unobtrusive marsh-dwelling birds resembling true snipes, but brighter coloured, shorter billed, with broad, rounded wings and entirely different flight action. Sex roles reversed, the more brightly coloured and larger female being polyandrous and leaving parental care to the male.

GREATER PAINTED-SNIPE *Rostratula benghalensis*
Rhynchée peinte

Plate 36 (8)

L 23–26 cm (9–10"). *R. b. benghalensis*. Distinctive, rather plump, short-tailed, *rail-like* wader. In flight, *rounded wings, slow wingbeats and dangling legs characteristic*. **Adult male** Head, neck and upper breast olive-brown with striking golden-buff eye-ring extending as stripe behind eye, and narrow golden-buff median crown-stripe. Upperparts olive-grey narrowly barred black and spotted golden-buff, esp. on wing-coverts; golden-buff lines on mantle form conspicuous V. Lower breast to undertail-coverts white; white line extending onto neck-sides and joining golden-buff scapular line. Tail grey narrowly barred black and finely spotted golden-buff. Bill long and brownish, slightly decurved, with pinkish-brown or dark brown tip. Legs olivaceous. **Adult female** Similarly patterned but more colourful than male. Head, neck and upper breast chestnut becoming black on breast. Eye-ring white. Upperparts dark glossy bottle-green narrowly barred black. **Juvenile** Similar to adult male but plumage less neatly patterned; grey of upper breast not sharply delimited from rest of underparts. **VOICE** Usually silent, except when breeding. Song, uttered by female, a low, soft, sonorous *hooOOoo*, likened to sound made by blowing across top of empty bottle; usually given at night, occasionally in late afternoon or at dusk. When flushed, may give a loud *kek*. [CD6:19] **HABITS** Singly or in pairs in wetlands, incl. overgrown ricefields, edges of lakes, pools and streams, and swamps. Unobtrusive; mostly active at dawn and dusk. **SIMILAR SPECIES** Unlikely to be confused. Other waders are more slender and have more rapid flight. **STATUS AND DISTRIBUTION** Uncommon to locally common resident and intra-African migrant, from S Mauritania to Liberia east to Chad and CAR. Also recorded from Gabon. Seasonal, short-distance movements, responding to habitat requirements.

OYSTERCATCHERS
Family HAEMATOPODIDAE (World: 11 species; Africa: 2; WA: 1–2)

Large, conspicuous and distinctively shaped, rather plump waders with pied or all-black plumage, long orange-red bill and pinkish legs. Sexes similar, female slightly larger. Mainly coastal. Vocal and gregarious.

EURASIAN OYSTERCATCHER *Haematopus ostralegus*
Huîtrier pie

Plate 35 (5)

L 40–45 cm (16.0–17.5"). *H. o. ostralegus.* Unmistakable. Large, bulky, black-and-white wader with *long, stout orange-red bill* and *sturdy, pinkish legs.* In flight, has broad white wingbar and white wedge from lower back to uppertail-coverts. **Adult breeding** Head, upperparts and breast black; underparts and underwing white. Eye and orbital ring red. **Adult non-breeding** Similar, but with white half-collar on throat and neck-sides ('chinstrap'). **Immature** Duller, with white half-collar and dark-tipped bill. Upperparts brownish-black, eye brownish-red, orbital ring dull orange, legs dull pinkish. **VOICE** Distinctive. A loud and vigorous, high-pitched *KLEEP!* or *K-PEEP!* Also a sharp *kip kip kip...* [CD2:16b] **HABITS** Locally in large flocks, where rare in pairs or even singly, frequenting coastal creeks, sandy beaches and estuaries. **STATUS AND DISTRIBUTION** Palearctic passage migrant and visitor to W African coast. Locally common, Mauritania and Senegambia (mainly Aug/Sep–Apr/May; recorded in all months); uncommon to rare south to Nigeria; vagrant, Eq. Guinea and Gabon. An inland record from N Mali (Lac Faguibine). Vagrant, Cape Verde Is.

AFRICAN BLACK OYSTERCATCHER *Haematopus moquini*
Huîtrier de Moquin

Plate 35 (6)

L 42–45 cm (16.5–17.5"). **Adult** As Eurasian Oystercatcher, but plumage entirely black. **Immature** Duller, with dark-tipped bill; plumage browner, eye brownish, orbital ring dull orange, legs greyish-pink. **VOICE & HABITS** Similar to Eurasian Oystercatcher, but usually prefers rocky coasts. **SIMILAR SPECIES** Extralimital Canarian Black Oystercatcher *H. meadewaldoi* very similar but, in flight, has pale bases to primaries on underwing; last definite record in 1913. **STATUS AND DISTRIBUTION** Sedentary S African species; seasonal short-distance movements recorded. Unconfirmed record from Gabon. Entirely black oystercatchers also reported from Senegal (near Dakar, Feb 1970, one; Basse Casamance, Dec 1975, two); their identity remains uncertain, as African Black Oystercatcher not known north of 12°S in Angola, melanism unknown in Eurasian Oystercatcher and Canarian Black Oystercatcher presumed extinct.

AVOCETS AND STILTS
Family RECURVIROSTRIDAE (World: 7–10 species; Africa: 2; WA: 2)

Fairly large, pied and graceful waders with slender, upturned bills and long legs (avocets) or thin, straight bills and extremely long legs (stilts). Sexes similar. Juveniles relatively similar to adults.

BLACK-WINGED STILT *Himantopus himantopus*
Échasse blanche

Plate 35 (7)

L 35–40 cm (14.0–15.5"). *H. h. himantopus.* Unmistakable. Graceful black-and-white wader with *long neck, needle-like bill* and *extraordinary long legs.* In flight, legs extend 14–17 cm beyond tail; wings pointed, all black above and below; white wedge from back to tail. **Adult male** Head and body mainly white, with wings and lower mantle glossy black. Crown and hindneck have variable amount of blackish (sometimes wholly white). Bill black. Eye red. Legs reddish-pink. **Adult female** Black areas slightly duller. Dark areas on head usually less extensive (but much overlap and not useful character for sexing). **Juvenile** Duller. Top of head and neck washed brownish;

upperparts brownish with feathers narrowly fringed buffish; secondaries and inner primaries narrowly tipped white. Legs dull pink. **VOICE** Vocal in breeding season, otherwise rather silent. Calls varied, incl. sharp, quickly repeated *kyik kyik kyik* ... or *ket ket*..., grating *kreet kreet kreet*..., more drawn-out *krrrrrt* and high-pitched *kip-kip-kip-...* [CD2:17a] **HABITS** Singly or in small groups in various aquatic habitats, incl. coastal lagoons, shallow lakes, pools, inland reservoirs, ricefields and swamps. Feeds on land and in shallow water, picking food from mud, surface of water or floating vegetation; also probes and occasionally immerses head in water. **STATUS AND DISTRIBUTION** Locally common to uncommon resident and Palearctic visitor, from S Mauritania to Liberia east to Chad and CAR; also coastal Gabon. Vagrant, Congo. Seasonal local movements reported but poorly understood. Locally not uncommon resident, Cape Verde Is.

PIED AVOCET *Recurvirostra avosetta* — Plate 35 (8)
Avocette élégante

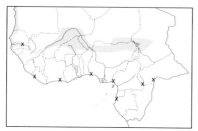

L 42–45 cm (16.5–17.5"). Unmistakable. Elegant, white-and-black shorebird with *slender, upturned black bill* (diagnostic) and grey-blue legs. **Adult male** Plumage mainly white, with black from top of head extending to just below eye and on hindneck, and black areas on side of mantle, across wing-coverts and on wingtip. Bill black. **Adult female** Bill slightly shorter and more upturned; black area on head occasionally less neatly demarcated and tinged brownish. **Immature** Black areas replaced by brownish. **VOICE** A loud, clear *kluit kluit* or *klup klup*... [CD2:17b] **HABITS** In flocks, frequenting mudflats, estuaries and lake margins. Typically feeds by sweeping bill from side to side through water or soft mud; also pecks items from mud. Frequently swims and up-ends like duck. **STATUS AND DISTRIBUTION** Widespread Palearctic visitor (mainly Aug–Mar; some present all year). Major concentrations in Mauritania, Senegambia, Guinea and L. Chad area; not uncommon Mali (central Niger delta) and N Nigeria; scarce to rare elsewhere. Vagrant, Cape Verde Is.

THICK-KNEES AND STONE-CURLEWS
Family BURHINIDAE (World: 9 species; Africa: 4; WA: 4)

Medium-sized to large, rather strange-looking waders with cryptically patterned plumage, large head and eyes, relatively short, stout bill and longish, yellow or greenish legs. Lack hind toe. Prominent tarsal or 'knee' joint is origin of name 'thick-knee'. Sexes similar. Juveniles similar to adults. Crepuscular and nocturnal. Vocal at night; calls melodious and far carrying. In our region often encountered on roads at night. Wing markings are important identification features.

STONE-CURLEW *Burhinus oedicnemus* — Plate 35 (3)
Oedicnème criard

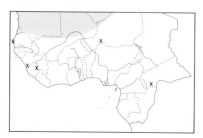

L 40–45 cm (16.0–17.5"). **Adult male** *oedicnemus* Sandy-brown streaked dark brown above; underparts white washed buff and streaked dark brown on breast; undertail-coverts cinnamon-buff. *Closed wing has horizontal white bar bordered above and below by black, with broad greyish panel below.* Bill yellow tipped black. Eye large, yellow. Legs yellow. In flight, has black flight feathers with contrasting white patches on outer and innermost primaries, and broad pale panel across wing-coverts bordered above by narrow black-bordered white bar. N African *saharae* slightly paler, more rufescent, but differences clinal, and much individual variation. **Adult female** Very similar, but white covert bar less strongly bordered with black (noticeable when pair seen together). **Juvenile** Whitish supercilium and black covert bars less prominent. **VOICE** Varied, incl. loud, melodious *kur-LEE*, reminiscent of Eurasian Curlew, and plaintive *tlueeEE*. Mostly silent in W Africa. [CD2:18a] **HABITS** In arid, stony areas. On migration, singly or in small groups. Rests in shade of shrub during day; mainly active at dusk. Calls esp. at dusk and at night. **SIMILAR SPECIES** Senegal Thick-knee has relatively longer bill with more extensive black tip, and single black horizontal bar bordering more distinct, pale grey wing panel. **STATUS AND DISTRIBUTION** Uncommon to rare Palearctic visitor (Sep–Mar; both races), recorded from Mauritania, Senegal and Mali. Vagrant, Guinea, Sierra Leone, Niger and CAR. Movements poorly known. Rare resident, N Mauritania (*saharae*).

SENEGAL THICK-KNEE *Burhinus senegalensis*

Oedicnème du Sénégal

Plate 35 (1)

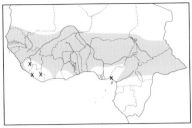

L 32–38 cm (12.5–15.0"). **Adult** Upperparts sandy or tawny-brown streaked dark brown; underparts white streaked dark brown on lower throat and breast; undertail-coverts cinnamon-buff. Head sandy-brown streaked dark brown; white eye-ring extending as short streak behind eye; white streak from lores, below eye, to rear ear-coverts. *Closed wing has broad, pale greyish panel bordered above by single horizontal black bar.* Bill largely black with yellow base; relatively longer and heavier than that of other thick-knees. Eye large, yellow. Legs yellow. In flight, has black flight feathers with prominent white patches on outer and innermost primaries, and broad pale panel across wing-coverts. Feet do not project beyond tail. **Juvenile** Median coverts and tertials tipped buff. **VOICE** A series of clear, piping notes, accelerating and increasing in volume, with last few notes fading away *pi pi pi-pi-pi-pi-pi-PII-PII-PII-pii-pii-pii-pii-piu.* [CD6:20] **HABITS** Singly, in pairs or small groups on river banks, sandbanks and lake shores, and in mangroves. Mainly active at night, when also venturing onto tracks and roads. **SIMILAR SPECIES** Water Thick-knee has conspicuous horizontal white bar across wing. Compare also Stone-curlew. **STATUS AND DISTRIBUTION** Common to not uncommon resident and intra-African migrant, from S Mauritania to Sierra Leone east to Chad and CAR; rare, Liberia. Local movements in response to changes in water levels reported.

WATER THICK-KNEE *Burhinus vermiculatus*

Oedicnème vermiculé

Plate 35 (2)

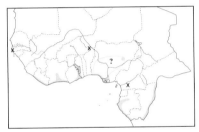

Other name E: Water Dikkop

L 38–41 cm (15–16"). **Adult** *buettikoferi* Upperparts brown streaked dark brown; fine vermiculations only visible at close range. Underparts white streaked dark brown on lower throat and breast; undertail-coverts rufous-buff. Head pattern as Senegal Thick-knee. *Closed wing has broad, pale greyish panel bordered above by narrow horizontal white bar,* latter conspicuous and highlighted by blackish bar above it. Bill mainly black with greenish-yellow base. Eye large, yellow. Legs yellowish; also greyish or greenish. In flight, has black flight feathers with contrasting white patches on outer and innermost primaries, and broad pale panel across wing-coverts; feet project slightly beyond tail. **Nominate** similar but paler and greyer above. **Juvenile** Grey wing panel spotted buff; upperparts and tail more vermiculated. **VOICE** A series of plaintive, piping, whistled notes, first accelerating and rising in pitch and volume, then dying away with slower, drawn-out, plaintive notes, *pi-pi-pi-pi-pee-pee-PEE-PEE-PEE-PEE-PEE-PEE-PEE-peeu-peeeu-peeeu-peeeu-peeeu.* Also very rapid *pipipipipipipipipipipi.* [CD6:21] **HABITS** Singly, in pairs or small groups, frequenting riverbanks, lake shores, estuaries, mangroves, lagoons and beaches. Rests by day in light woodland or scrub near water. Mainly active at night, when also foraging far from water. **SIMILAR SPECIES** Senegal Thick-knee lacks white horizontal bar across lesser wing-coverts. **STATUS AND DISTRIBUTION** Locally common to rare resident. *B. v. buettikoferi* recorded from S Senegal (rare), Liberia (common on coast), Ivory Coast (common along coast and major rivers), Ghana (uncommon), Burkina Faso, SW Niger, Nigeria (uncommon), Cameroon and Gabon; nominate in Congo. Local movements in response to changes in water levels reported.

SPOTTED THICK-KNEE *Burhinus capensis*

Oedicnème tachard

Plate 35 (4)

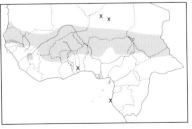

Other name E: Spotted Dikkop

L c. 43 cm (17"). *B. c. maculosus.* Largest and tallest African thick-knee, readily separated from others by lack of any bars or panel on wing. **Adult** Upperparts, incl. wing-coverts, warm tawny-brown densely spotted blackish; underparts warm buff becoming whitish on belly; head, neck and underparts streaked dark brown; undertail-coverts rufous-buff. Head pattern less distinct than other thick-knees, with white areas smaller. Bill largely black with yellow base. Eye large, yellow. Legs yellow. In flight, has black flight feathers with contrasting white patches on outer and innermost primaries. **Juvenile** Rather duller, with streaked upperparts. **VOICE** A series of loud piping notes, rising in pitch and volume, then dying away, *pi-pi-pi-pi-plee-plee-PLEEW-PLEEW-WHEEW-WHEEW-wheew-wheew-...* and an accelerating, slightly rising series of similar notes *piu-piu-piu-piupiupiupiu...*; lower pitched than calls of Senegal and Water Thick-knees. Also harsh notes and rapid *pi-pi-pi-...* in alarm. [CD6:22] **HABITS** Mainly singly or in pairs in arid areas and dry, open woodland. Rests by day on stony ground in the open or under shrub. Mostly active at night. **SIMILAR SPECIES** Other thick-knees easily distinguished by upperpart and wing pattern. **STATUS AND DISTRIBUTION** Uncommon resident, from S Mauritania to Senegambia east to Chad and NE CAR. Vagrant, Guinea and Gabon. Some local movements reported.

COURSERS AND PRATINCOLES
Family GLAREOLIDAE (World: 17 species; Africa: 12; WA: 8)

Small to medium-sized plover-like birds in two subfamilies.

Coursers are terrestrial, ground-feeding species with long legs and cryptically coloured plumage, occurring in dry habitats.

Pratincoles are principally aerial-feeding, gregarious and more water-dependent species with short legs, long, pointed wings and graceful flight, reminiscent of terns. The aberrant, strikingly patterned Egyptian Plover is usually considered closest to the coursers. Sexes similar in all species.

EGYPTIAN PLOVER *Pluvialis aegyptius*
Pluvian fluviatile

Plate 36 (7)

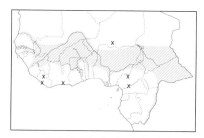

Other names E: Crocodile Bird; F: Pluvian d'Égypte
L 19–21 cm (7.5–8.25"). *Uniquely patterned, chunky, riverine wader.* **Adult** Crown and head-sides black with long white supercilium; mantle black with long feathers extending in point over back; *rest of upperparts blue-grey.* Throat white; rest of underparts creamy to deep peach-buff with black breast-band. Bill black. Legs blue-grey. In flight, has striking wing pattern produced by *white area on flight feathers and coverts with contrasting diagonal black band and black tips to primaries*; underwing pure white with black diagonal band. **Juvenile** Rather duller, with some rusty on crown, mantle, and lesser and median coverts. **VOICE** Harsh *chreek-chreek-chreek* or *cherk-cherk-cherk.* [CD6:23] **HABITS** Singly, in pairs or small groups on sandbars in large rivers, where breeding. Also in larger flocks and in various other aquatic habitats when not breeding. **STATUS AND DISTRIBUTION** Locally not uncommon to rare resident and intra-African migrant, along major rivers in Sahel, northern savannas and Congo. Irregular local movements in response to changes in water levels; longer distance movements reported from Nigeria and Chad (north with rains). Vagrant, Liberia and S Ivory Coast.

CREAM-COLOURED COURSER *Cursorius cursor*
Courvite isabelle

Plate 36 (15)

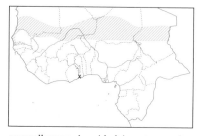

L 21–24 cm (8.25–9.5"). Pale desert courser. **Adult** *cursor* Entirely *pale sandy-cream*, paling to whitish on lower underparts. White supercilium and black stripe from behind eye form V on nape and delimit *grey hindcrown*. Bill relatively long and decurved, mainly blackish. Legs whitish. In flight, has contrasting black outer wing and black underwing; white tips to secondaries form narrow trailing edge. Island race *exsul* described as slightly darker. **Juvenile** Duller, with plainer head (grey on hindcrown absent, eye-stripe indistinct), and faintly scaled upperparts. **VOICE** Mostly silent. A short, sharp *kwit* or *krit* usually given in flight. [CD2:18b] **HABITS** Singly, in pairs or small groups in arid plains, stony and sandy desert and semi-desert; occasionally on bare farmland and saltpans. Forages by running fast, then stopping briefly to pick insects from ground or look around. Bobs head. Runs when approached, flying away when pressed; may crouch. Flight jerky. **SIMILAR SPECIES** Temminck's Courser is smaller and darker, with rufous hindcrown and black patch on belly; in flight, has black secondaries. **STATUS AND DISTRIBUTION** Uncommon Palearctic visitor (Aug–May) and resident, from Mauritania and N Senegal east through N Mali, N Burkina Faso and Niger to C Chad (nominate). Rare, Gambia (Dec–Feb). Vagrant, coastal Ghana (two, May 1996). Not uncommon resident, Cape Verde Is (*exsul*).

TEMMINCK'S COURSER *Cursorius temminckii*

Courvite de Temminck

Plate 36 (14)

L 19–21 cm (7.5–8.25"). Small savanna courser with *blackish patch on centre of chestnut belly*. **Adult** Crown entirely *rufous* bordered by white supercilium and broader black stripe back from eye, forming V on nape. Upperparts grey-brown. Throat whitish; breast soft brown washed chestnut, becoming deep chestnut on lower breast and belly; lower underparts white. Bill slightly decurved, mainly dark with paler base to lower mandible. Legs whitish. In flight, has contrasting black outer wing and secondaries, and black underwing. **Juvenile** Duller, with plainer head (crown buff, eye-stripe brownish), faintly scaled upperparts and smaller, more diffuse belly patch. **VOICE** Mostly silent. A nasal, metallic *het-het-het-herr het-het-het-herr...* rather like toy trumpet or squeaking hinge, uttered in high, circular display flight. [CD6:24] **HABITS** Singly, in pairs or small groups in open and burnt grassland, dry farmland, semi-arid savanna and similar open habitats. Prefers to run when disturbed; may also crouch or fly short distance. **SIMILAR SPECIES** Cream-coloured Courser is larger and paler, with grey hindcrown and pale belly; in flight, has no black on secondaries. **STATUS AND DISTRIBUTION** Uncommon to locally not uncommon resident, from S Mauritania to Guinea east to Chad and CAR, and in Gabon and Congo. Vagrant, Sierra Leone. Nomadic, with seasonal movements in response to rainfall and burning of grasslands; some possibly longer distance migrants.

BRONZE-WINGED COURSER *Rhinoptilus chalcopterus*

Courvite à ailes bronzées

Plate 36 (13)

Other name E: Violet-tipped Courser

L 25–29 cm (10.0–11.5"). Large, *long-legged* courser with *relatively big, distinctively patterned head, black breast-band and pinkish-red legs*. **Adult** Upperparts plain grey-brown. Forehead and supercilium creamy-white; lores and ear-coverts dusky; whitish stripe behind eye bordered above by chestnut line and below by ear-coverts; neck-sides and throat white with dark malar stripe. Breast plain grey-brown paling below to buffish and bordered by narrow black band; rest of underparts buffish-white. Bill black with reddish base below. Reddish orbital ring around large dark eye. In flight, has black flight feathers contrasting with brown coverts, white uppertail-coverts and black tail-band; underwing-coverts buffish-white. Glossy violet tips to primaries usually not visible in field. **Juvenile** Upperparts faintly scaled and breast-band mottled. **VOICE** A short series of 3–4 remarkable syllables (difficult to transcribe) *hu thu-WHEH-hep*, often rhythmically repeated. Also a harsh sound in flight. [CD6:25] **HABITS** Usually in pairs, in wooded savanna. Active at dusk and at night. Rests in shade by day. Runs when approached, flying only short distance when pressed, then running again. **SIMILAR SPECIES** When briefly seen in flight, may recall lapwing, esp. African Wattled and Brown-chested, but these have broad white diagonal wingbars (not just short, barely visible white line); head and underpart patterns very different. **STATUS AND DISTRIBUTION** Uncommon to scarce resident and probably intra-African migrant, from S Mauritania to Guinea east to Chad and CAR; also Gabon and Congo. Movements poorly understood. **NOTE** Placed in genus *Cursorius* by BoA; it has, however, been argued that including all coursers in a single genus disguises natural relationships and we therefore prefer to retain the long-established genus *Rhinoptilus* (cf. Dowsett & Dowsett-Lemaire 1993 and *HBW*).

COLLARED PRATINCOLE *Glareola pratincola*

Glaréole à collier

Plate 36 (9)

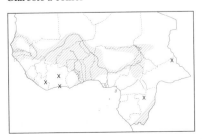

Other name E: Red-winged Pratincole

L 24–28 cm (9.5–11.0"). Distinctively shaped, short-legged, mainly brown wader. *Graceful flight with long, pointed wings and strongly forked tail recalls terns*. **Adult breeding** Head and upperparts plain grey-brown; lower rump and uppertail-coverts white; tail mainly black. Narrow white crescent below eye. Throat creamy-yellow bordered by black line; breast grey-brown becoming white on belly. Bill stubby with decurved tip; black with red base below. Legs blackish. In flight, has *chestnut underwing-coverts and axillaries*, conspicuous white rump, blackish flight feathers and white-tipped secondaries forming narrow trailing edge to inner wing. **Adult non-breeding** Duller, with indistinct throat outline; breast mottled grey-brown.

Juvenile Feathers of upperparts have narrow black subterminal bar and buff fringe producing scaly aspect. Paler than adult below, with indistinct throat pattern and blotched breast. Tail shorter. **VOICE** A variety of high-pitched, tern-like calls incl. short, sharp *kik* and *kirrik* often repeated and given in combination; also a piercing, rattling *krrrrrret*. [CD2:19a] **HABITS** In small or large flocks. Breeds in loose colonies on variety of flat, open areas with short or sparse vegetation, incl. farmland, generally near water. Catches insects on the wing. Most active at dawn and dusk. **SIMILAR SPECIES** Black-winged Pratincole is darker brown, less contrasting above and has all-dark underwing lacking white trailing edge. **STATUS AND DISTRIBUTION** Widespread, uncommon and local resident, intra-African migrant and Palearctic visitor. Breeding areas inadequately known; nesting reported on coast from Senegal to Ghana, along Niger R. in Mali and Niger, and along major rivers in Nigeria and N Cameroon. Vagrant, Cape Verde Is. Nomadic; movements unclear.

BLACK-WINGED PRATINCOLE *Glareola nordmanni* Plate 36 (10)
Glaréole à ailes noires

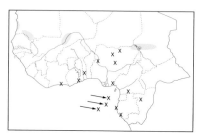

L 24–28 cm (9.5–11.0"). Very similar to Collared Pratincole and difficult to separate at rest. **Adult breeding/non-breeding** In flight, darker and more uniform above, with less contrasting flight feathers; *underwing all dark; no white trailing edge to inner wing.* At rest, slightly longer legs and usually shorter tail. Bill with less red at base. **Juvenile** Similar to juvenile Collared but differs from it as adult. **VOICE** Similar to Collared Pratincole but slightly lower pitched and drier. [CD2:19b] **HABITS** In small or large flocks, in W Africa also singly, frequenting same habitats as Collared Pratincole and often associating with it. **SIMILAR SPECIES** Collared Pratincole (see above); beware: chestnut underwing-coverts can be surprisingly hard to see, and narrow white trailing edge to inner wing may be lost with wear. **STATUS AND DISTRIBUTION** Rare to locally uncommon Palearctic migrant (Sep–Apr) to major lakes and rivers in Sahel zone. Scattered records elsewhere. Recorded from Mauritania, Mali, Nigeria, S Cameroon, Chad (Ndjamena and NE) and N Congo. Vagrant, Ivory Coast, Ghana, Togo, Gabon, São Tomé, Príncipe and Annobón. Winters mainly in S Africa. *TBW* category: Data Deficient.

ROCK PRATINCOLE *Glareola nuchalis* Plate 36 (12)
Glaréole auréolée

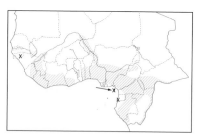

L 18–20 cm (7–8"). Small, dark pratincole, associated with emergent rocks in rivers. **Adult** *liberiae* Mainly dark grey-brown with white stripe behind eye continuing as *chestnut collar across nape.* Belly white. Narrow white crescent below eye. Bill red tipped black. Legs bright red. In flight, has contrasting white uppertail-coverts, shallowly forked, blackish tail with white outer feathers, and blackish flight feathers. **Nominate** has white nuchal collar. **Juvenile** Duller than adult, without nuchal collar and eye-stripe, and with buff-tipped feathers of upperparts producing scaly aspect. Bill duller. **VOICE** Calls include a short, sharp *kip* and *kwip*, a frequently repeated, clear *kweee*, a shrill, rasping *krrreep krrrree*, and a soft, high-pitched, whistled *wheedidi whee wheedidi*. [CD6:26] **HABITS** In pairs or small groups on exposed rocks in rivers or, occasionally, lakes. Mainly active at dawn and dusk, spending most of day resting on rocks or, during high water, on overhanging branches. **SIMILAR SPECIES** Collared and Black-winged Pratincoles are larger, have black legs and no nuchal collar, and frequent different habitats. **STATUS AND DISTRIBUTION** Common to uncommon resident and intra-African migrant, from Sierra Leone and SE Guinea to CAR and Congo; also Bioko. Western race *liberiae* meets nominate in W Cameroon (where intermediates occur). Nominate also reported from Guinea and Togo. Irregular movements in response to changes in water levels; migrations poorly understood.

GREY PRATINCOLE *Glareola cinerea* Plate 36 (11)
Glaréole grise

L 18–20 cm (7–8"). Unmistakable. *Beautiful, small and pale pratincole associated with large sand banks.* Striking pattern in flight. **Adult** Pale grey above with broad pale chestnut collar across nape. Long black eye-stripe curves around white ear-coverts and is bordered above by white supercilium. Underparts white washed pale cinnamon on breast. Bill orange-red tipped black. Legs bright orange-red. In flight, has pure white areas on flight feathers contrasting with black outer wing and trailing edge, and shallowly forked, white tail with black bar. Underwing largely white

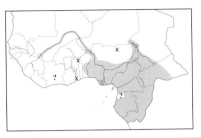

with black primary-coverts and primary tips. **Juvenile** Duller; head plainer, plumage mixed with sandy-buff. **VOICE** Calls include hoarse *zri* or *kree* frequently uttered in series. [CD6:27] **HABITS** In small or large groups on sand banks in large rivers. Principally active at dusk, spending most of the day resting. **STATUS AND DISTRIBUTION** Locally common to uncommon intra-African migrant on large rivers in Mali and Niger, and from Benin to S Chad, CAR and Congo. Vagrant, Guinea, S Ghana and N Togo; also claimed from Ivory Coast. Moves in response to changes in water levels; migrations inadequately known.

PLOVERS AND LAPWINGS
Family CHARADRIIDAE (World: 67 species; Africa: 31; WA: 25)

Small to medium-sized waders with large, rounded head, relatively short bill and long legs. Many have distinct breeding and non-breeding plumages. Feeding action typically consists of short, swift runs interrupted by abrupt stops followed by a pause or a dip.

The **small plovers** (*Charadrius*) have pointed wings and largely white underparts; head and breast markings are important field marks.

Golden and Grey Plovers (*Pluvialis*) are highly migratory, medium-sized and rather plump birds with pointed wings and spotted or spangled plumages. The 3 species of golden plover are very similar in non-breeding plumage and require extremely careful observation to be identified.

Lapwings (*Vanellus*) are medium-sized, round-winged, mostly well-marked and easily identified species; several in the region sport wattles or spurs. Distinctive black-and-white wing pattern in flight. Flight rather slow and heavy.

LITTLE RINGED PLOVER *Charadrius dubius*   Plate 38 (7)
Pluvier petit-gravelot

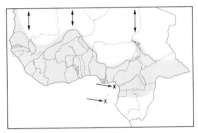

L 15–17 cm (6.0–6.75"). *C. d. curonicus*. Resembles corresponding plumages of Common Ringed Plover, but slightly smaller and slimmer, with narrower bill and more tapering rear end. Call different. In flight, *no wingbar*. **Adult male breeding** Distinguishing features include white line separating black frontal band from brown crown, black bill, *yellow orbital ring* and dull pinkish or yellowish-brown legs. **Adult female breeding** Black head markings and breast-band mixed with brown; breast-band reduced. **Adult non-breeding** Following moult Jul–Nov, duller, with head markings and breast-band brown; forehead and supercilium buff. **Juvenile** Resembles adult non-breeding, but upperparts scaled buff; breast-band often incomplete. **VOICE** A piping, rather plaintive *PEE-uu*. [CD2:20b] **HABITS** Usually in small groups, frequenting various open habitats, incl. edges of lakes, ponds and rivers, farmland, airfields, and golf courses; less frequently estuaries, coastal lagoons and marine shores. **SIMILAR SPECIES** Common Ringed Plover is more compact, has stubby bill and prominent white wingbar in flight. In breeding plumage, lacks white line above black frontal band, and has black-tipped orange bill, orange legs and no yellow orbital ring. Non-breeding has white forehead and supercilium. **STATUS AND DISTRIBUTION** Not uncommon to locally common Palearctic passage migrant and visitor throughout (mainly Oct–Apr), except extreme south. Rare, Cape Verde Is. Vagrant, Bioko and São Tomé.

COMMON RINGED PLOVER *Charadrius hiaticula*   Plate 38 (5)
Pluvier grand-gravelot

Other names E: Ringed Plover; F: Grand Gravelot
L 18–20 cm (7–8"). Resembles Little Ringed Plover, but more compact, with *stubby bill* and *orange legs*; call different. In flight, *prominent white wingbar*. **Adult male breeding** Brown above, white below, with white forehead, black mask, narrow white collar and black breast-band. Distinguishing features include orange bill with black tip and orange legs. Arctic *tundrae* slightly smaller and slimmer than European **nominate**, and moults flight feathers in winter quarters, not near breeding grounds as nominate. **Adult female breeding** Black head markings and breast-band mixed with brown; breast-band reduced. **Adult non-breeding** Duller, with head markings and breast-band brownish; forehead and supercilium remain white. Bill dark, usually with some dull orange at base. Legs often duller orange. **Juvenile** Resembles adult non-breeding, but upperparts scaled buff; breast-band reduced and

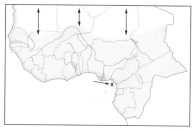

often incomplete. Legs dull orange or yellowish. **VOICE** A distinctive, mellow *too-EE* rising at end, and a shorter *tuwee*. [CD2:21] **HABITS** Usually in small groups on tidal mudflats and inland shores; also frequents various other open habitats, incl. farmland, areas with short grass and estuaries. **SIMILAR SPECIES** Little Ringed Plover is more slender and lacks wingbar; in breeding plumage, has white line above black frontal band, black bill, yellow orbital ring and dull-coloured legs. Non-breeding has plainer head with buff forehead and supercilium. **STATUS AND DISTRIBUTION** Common to not uncommon Palearctic passage migrant and visitor throughout (mainly Sep–May); also Cape Verde Is and Bioko.

SEMIPALMATED PLOVER *Charadrius semipalmatus*   Plate 42 (1)
Pluvier semipalmé

L 17–19 cm (*c.* 7"). Very similar to corresponding plumages of Common Ringed Plover but slightly slimmer, with rather stubbier bill and *small webs between all 3 front toes* (not only between middle and outer). **Adult breeding** Black breast-band usually narrower and often of more even width; supercilium behind eye indistinct or lacking. **Adult non-breeding** Duller pattern as Common Ringed Plover; white post-ocular streak sometimes joins white forehead. **Juvenile** Small white wedge from throat-side to above corner of gape (juvenile Common Ringed normally dark from lores to gape); breast-band not broken in centre. **VOICE** Different from Common Ringed. A short, rising *chewit*, almost monosyllabic and with stress at end, recalling subdued Spotted Redshank; also a more drawn-out *chewee*. **HABITS** As Common Ringed Plover. **SIMILAR SPECIES** Common Ringed Plover (q.v. and above). **STATUS AND DISTRIBUTION** N American vagrant, Cape Verde Is (Sal, Mar 1999).

KITTLITZ'S PLOVER *Charadrius pecuarius*   Plate 38 (4)
Pluvier pâtre

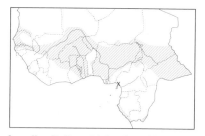

Other name F: Pluvier de Kittlitz
L 14–16 cm (5.5–6.0"). *C. p. pecuarius*. Small, rather long-legged plover with *warm buff breast*. In flight, dark leading edge to wing, white patch on dark outer wing extending as narrow white line on inner wing, white sides to tail, and toes projecting beyond tail. **Adult breeding** *Distinctive head pattern* with white forehead, broad white supercilium meeting on nape, and black band across forecrown extending through eye to neck-sides. Upperparts dark brown; feathers with paler fringes. Underparts white variably washed creamy-buff on breast and flanks. Bill black. Legs blackish. **Adult non-breeding** Duller, with head markings washed out and brownish; supercilium orange-buff; breast paler buff with diffuse brownish patch on sides. Legs greyish. **Juvenile** Resembles adult non-breeding, but upperparts scaled buff. **VOICE** Mostly silent. Calls include *tuweet* or *pipeep*, a hard *trip* and a dry rattling *trrrr*. [CD6:28] **HABITS** In pairs or small groups in various open habitats, incl. areas with short grass, edges of reservoirs and lakes, mudflats, beaches and saltpans; not always near water. **SIMILAR SPECIES** Slightly larger White-fronted Plover has paler upperparts and different head pattern, with white forehead extending as supercilia that do not meet on nape; toes do not project beyond tail in flight. Kentish Plover has paler, more uniform upperparts, lacks creamy-buff tinge to head and breast, and has supercilia that do not meet on nape. **STATUS AND DISTRIBUTION** Widespread, uncommon to locally not uncommon resident and intra-African migrant throughout, north to S Mauritania. Rare or vagrant, Liberia, Ivory Coast and Togo.

THREE-BANDED PLOVER *Charadrius tricollaris*   Plate 38 (2)
Pluvier à triple collier

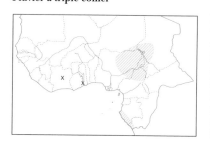

L 17–18 cm (7"). *C. t. tricollaris*. Similar to Forbes's Plover but smaller, with shorter legs and *white forehead*. In flight, narrow white line across wings. **Adult** Crown dark brown bordered by white stripe from forehead to nape. Head-sides paler brown-grey, becoming pale grey on throat and whitish-grey towards well-defined black upper breast-band. Upperparts dark grey-brown. Bill pinkish-red with black tip. Legs pinkish-red. **Juvenile** Duller, with brownish forehead; upperparts narrowly scaled buff. **VOICE** Calls include a piercing *peeweet* or *peep*, a more rasping *kreep* and a shrill *wik-wik*. [CD6:29] **HABITS** Singly or in pairs, occasionally in small groups, along rivers, muddy pools, lakes and reservoirs; more rarely in coastal habitats. **SIMILAR SPECIES** Forbes's Plover lacks white forehead; upper breast-band merges with brown of throat; lower breast-band noticeably broader; band across nape also broader; legs longer. **STATUS AND DISTRIBUTION** Uncommon and local resident, C and N Nigeria, Cameroon, W Chad and Gabon. Rare, Congo. Vagrant, Ivory Coast and Ghana. Local movements reported.

FORBES'S PLOVER *Charadrius forbesi* Plate 38 (3)

Pluvier de Forbes

L *c.* 20 cm (8"). Rather long-tailed plover with *2 blackish breast-bands* and *red orbital ring*. In flight, appears dark, without white wingbar. **Adult** Head to throat brown with white band from eye across nape. Upperparts dark brown. Brown of throat merges into black upper breast-band. Bill pinkish-red with black tip or mainly dark with variable amount of pink at base. Legs pinkish. **Juvenile** Duller, with buffish band across nape; upperparts scaled buff. **VOICE** A plaintive *pee-oo*, sometimes repeated, and a sharp *pee-pee-pee-...* [CD6:30] **HABITS** Singly or in pairs, occasionally small groups, frequenting grassland, farmland, recently burnt ground, and edges of muddy pools, lakes and rivers. In breeding season also on rocky outcrops. **SIMILAR SPECIES** Three-banded Plover is smaller, with shorter legs, white forehead and greyer brown upperparts; throat becomes pale grey towards upper breast-band; band across nape and lower breast-band narrower. **STATUS AND DISTRIBUTION** Locally not uncommon to rare throughout, north to Senegambia and S Niger (Zinder area; vagrant?). Seasonally migratory. Breeds during rains.

KENTISH PLOVER *Charadrius alexandrinus* Plate 38 (6)

Pluvier à collier interrompu

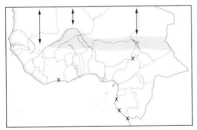

L 15–17 cm (6.0–6.75"). *C. a. alexandrinus.* Small, pale plover with *narrow dark patch at breast-sides* and blackish legs. Rather rounded with relatively large head and short rear end. In flight, white wingbar and broad white sides to rump and tail. **Adult male breeding** White forehead and supercilium, black bar across forecrown, *variable amount of rufous on crown* and black mask. Upperparts sandy-brown, separated from head by narrow white collar. Underparts white with narrow black patches on breast-sides. Bill black. **Adult female breeding** Duller, with head markings and breast patches grey-brown; crown has little or no rufous. **Adult non-breeding** After moult Jul–Sep, similar to adult female breeding. Male usually has slightly darker mask and breast patch. **Juvenile** Resembles adult non-breeding but paler, with indistinct buff supercilium; upperparts narrowly scaled buff; breast patches paler and reduced. **VOICE** A short *kip* or *kitip*. Alarm a hard *pirrrrr.* [CD2:22] **HABITS** Singly or in small groups on sandy beaches, tidal mudflats, lagoons, saltpans and inland shores. Runs very fast. **SIMILAR SPECIES** White-fronted Plover separated by absence of dark breast patches, rusty tinge to upperparts, less white underparts, greenish-grey legs and relatively longer tail (projecting beyond wingtips). **STATUS AND DISTRIBUTION** Scarce resident, coastal Mauritania and Senegal; common, Cape Verde Is. Common Palearctic visitor (mainly Sep–Apr), Mauritania; rare to uncommon elsewhere. Vagrant, Gabon and Congo. Distribution inadequately known.

WHITE-FRONTED PLOVER *Charadrius marginatus* Plate 38 (10)

Pluvier à front blanc

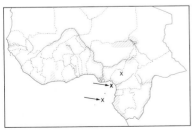

L *c.* 18 cm (7"). *C. m. mechowi.* Small plover with *large white frontal patch.* In flight, white wingbar and broad white sides to rump and tail. **Adult male breeding** Large white forehead and supercilium, black bar across forecrown, rusty grey-brown crown and hindneck and black eye-stripe. Upperparts grey-brown with rusty tinge; separated from head by narrow white collar. Underparts white with *variable creamy-buff to orange wash on breast.* Primary tips do not reach tip of tail. Bill black. Legs dark grey-green. **Adult female breeding** Less black on forecrown. **Adult non-breeding** Similar to adult female breeding but duller; head pattern brownish; breast coloration duller and reduced. **Juvenile** Resembles adult non-breeding but upperparts scaled sandy-buff; underparts white, sometimes with creamy wash. **VOICE** A short, low *wit* or *twirit.* Alarm a dry *trrrr.* [CD6:31] **HABITS** Singly, in pairs or small groups on sandy beaches, large sand banks, and sandy shores of rivers and lakes. Runs very fast. **SIMILAR SPECIES** Kentish Plover has distinct lateral breast patches, lacks rusty tinge above and below, and has blackish legs; also more rounded, with shorter rear end. Adult Kittlitz's Plover has darker upperparts and different head pattern with black stripe extending on neck-sides; toes project beyond tail in flight. Juvenile Kittlitz's Plover has buff supercilium extending to nape. **STATUS AND DISTRIBUTION** Locally not uncommon to rare resident throughout. Mainly sedentary, but seasonal movements recorded. Vagrant, Bioko and São Tomé.

GREATER SAND PLOVER *Charadrius leschenaultii* — Plate 38 (8)
Pluvier de Leschenault

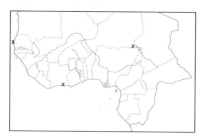

Other name F: Pluvier du désert
L 20–23 cm (8–9"). Long-legged, medium-sized plover with relatively long and heavy bill. In flight, white wingbar and white sides to tail; feet project beyond tail. **Adult *crassirostris* non-breeding** Grey-brown above with white forehead and supercilium; no white nuchal collar. Underparts white with large grey-brown patch on breast-sides. Bill black. Legs greyish. *C. l. columbinus* has shorter, straighter and more slender bill. **Adult breeding** Not observed in W Africa. Male has black mask, white forehead bordered by black frontal bar and bright orange-rufous nape and breast; female much duller. **Juvenile** Similar to adult non-breeding, but upperparts scaled buff. **VOICE** A trilling *trrr* and a longer *tirirrilip*, often frequently repeated. [CD2:23] **HABITS** In W Africa usually singly, on intertidal mudflats; also grassland. Associates with other waders. **SIMILAR SPECIES** Caspian Plover is smaller and slimmer, with bolder supercilium, broad, unbroken breast-band, faint wingbar and no white sides to tail. **STATUS AND DISTRIBUTION** Palearctic vagrant, Senegal (Nov-Dec), Liberia (Jan), Ivory Coast (Feb), NE Nigeria (Aug) and Gabon (Feb). Probably also Gambia (Jan). Not subspecifically identified (immature specimen, 2 Aug, L. Chad, could be either). *C. l. crassirostris* winters mainly on E and S African coasts (Aug-early May); westernmost *columbinus* winters Red Sea, Gulf of Aden and SE Mediterranean.

CASPIAN PLOVER *Charadrius asiaticus* — Plate 38 (9)
Pluvier asiatique

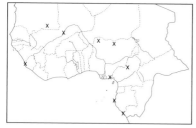

L 19–21 cm (7.5–8.0"). *Long-legged, elegant plover.* At rest, wings project beyond tail. In flight, only faint white wingbar and no white on sides of tail; feet project beyond tail. **Adult non-breeding** Grey-brown above with white forehead, lores and supercilium; grey-brown area behind eye. Underparts white with *broad grey-brown breast-band.* Bill black, slim. Legs greyish to yellowish. **Adult male breeding** Plumage more contrasting; cheeks white; broad breast-band bright chestnut narrowly bordered below by black. **Adult female breeding** Much duller, similar to adult non-breeding with some chestnut mixed in breast-band. **Juvenile** Similar to adult non-breeding, but upperparts scaled buff. **VOICE** A short, sharp *chip* or *kwit*; also a soft, repeated *tik.* [CD6:32] **HABITS** Normally in groups, but in W Africa singly, in grassland. Often far from water. **SIMILAR SPECIES** Mainly coastal Greater Sand Plover is larger and heavier, with longer, heavier bill, less distinct supercilium, patches on breast-sides, bold wingbar and white sides to tail. **STATUS AND DISTRIBUTION** Palearctic vagrant from C Asia (Nov–Apr), recorded from Mauritania, Mali, Sierra Leone, Nigeria, Cameroon, Gabon and Congo. Winters mainly in E and S Africa.

EURASIAN DOTTEREL *Charadrius morinellus* — Plate 39 (5)
Pluvier guignard

Other name E: Dotterel
L 20–22 cm (8.0–8.75"). Rather plump and compact, recalling golden plovers, but has *broad pale supercilia meeting in V on nape, narrow pale breast-band* and *yellowish legs.* In flight, plain upperwing; white shaft to outer primary forms narrow pale leading edge. **Adult non-breeding** Brown-buff with whitish to creamy-buff supercilium; upperparts heavily scaled creamy-buff. Underparts buff mottled darker on breast and with paler narrow breast-band. Bill black. **Adult male breeding** Supercilium white, contrasting with dark grey crown. Upper breast grey bordered below by black line and narrow white band; lower breast and flanks orange-rufous becoming blackish on belly; vent and undertail white. **Adult female breeding** Similar but brighter. **Juvenile** Similar to adult non-breeding, but dark feather centres of upperparts contrast strongly with pale buffish fringes. **VOICE** Mostly silent. A rattling call in flight, rather reminiscent of Ruddy Turnstone. [CD2:24] **HABITS** Could occur anywhere but prefers open inland habitats. **SIMILAR SPECIES** Juvenile may recall golden plover, but latter lack pale supercilium and breast-band, and have dark legs. **STATUS AND DISTRIBUTION** Palearctic vagrant; singles in Mauritania (Oct) and Gambia (Mar).

AMERICAN GOLDEN PLOVER *Pluvialis dominica*   Plate 39 (3)
Pluvier bronzé

Other name F: Pluvier doré américain
L 24–28 cm (9.5–11.0"). Noticeably smaller and slimmer than Grey Plover, with finer bill. In flight, indistinct wingbar and *grey underwing*. At rest, 4–5 primary tips project beyond relatively short tertials (exposed primaries approximately equal in length to tertials; tips of latter fall well short of tail tip); wings project clearly beyond tail. **Adult non-breeding** Plumage overall *cold grey*. Conspicuous white supercilium bordering dark grey-brown crown with pale streaks; head-sides, throat and neck paler brown-grey; dusky crescent in front of eye and dusky smudge on ear-coverts. Upperparts dark brown-grey spotted and spangled off-white. Underparts greyish-white, breast and flanks mottled. Bill black. Legs blackish. **Adult male breeding** Head-sides and entire underparts (incl. undertail-coverts) black; well-defined white band ('shawl') extending from forehead and supercilium to breast-sides, where bulging. Crown and upperparts black spangled golden. **Adult female breeding** Similar but black area mixed with some white. **Immature** Similar to adult non-breeding but plumage neater, esp. on upperparts. **VOICE** A rather mellow *tlu-ee*, with stress on first syllable, and *tlu-ee-uh* or *tlu-uu-ee*. [CD2:25] **HABITS** Singly, frequenting coastal mudflats, ricefields, grassland and ploughed farmland. **SIMILAR SPECIES** Pacific and European Golden Plovers (q.v.). Grey Plover is larger and has white rump, black axillaries and bold wingbar; juvenile also differs from juvenile American Golden in stouter bill, less contrasting crown and less distinct supercilium. **STATUS AND DISTRIBUTION** N American vagrant (Sep–May), N Senegal, Gambia, Sierra Leone, Ivory Coast, Togo, Nigeria and Cape Verde Is. Not specifically identified 'golden plovers' reported from Liberia, Ghana, Nigeria, Gabon and São Tomé. **NOTE** Previously considered conspecific with Pacific Golden Plover and known as Lesser Golden Plover.

PACIFIC GOLDEN PLOVER *Pluvialis fulva*   Plate 39 (2)
Pluvier fauve

L 23–26 cm (9–10"). Very similar to American Golden Plover, having indistinct wingbar and *grey underwing* in flight. Primaries shorter with, at rest, only 3 (rarely 4) tips projecting beyond tertials (exposed primaries approximately a quarter to half length of tertials); tertials longer with tips approximately level with tail tip. Wings project slightly beyond tail. Legs slightly longer than American; toes project beyond tail in flight. **Adult non-breeding** Plumage has warm buff-brown tones. Conspicuous buff-white supercilium bordering buff crown with dark streaks; head-sides and neck pale buff; dusky crescent in front of eye and dusky smudge on ear-coverts. Upperparts dark brownish spotted and spangled golden-buff. Throat pale buff becoming brownish-white on breast. Underparts may retain black feathers of breeding plumage. Bill black. Legs blackish. **Adult male breeding** Head-sides and underparts black; white band extending from forehead and supercilium to flanks; undertail-coverts variably mixed black and white. Crown and upperparts black spangled golden. **Adult female breeding** Similar but black area mixed with some white. **Immature** Similar to adult non-breeding but plumage neater, esp. on upperparts, breast and flanks golden-buff faintly mottled darker. **VOICE** A clear, rapid *chu-it* or *kuweet*, reminiscent of Spotted Redshank, and more drawn-out and plaintive *klee-wee* and *klu-wee-up*. **HABITS** Usually singly, frequenting beaches, mudflats, ricefields and freshwater ponds; also farmland. A small group of 5–7 was found in Gabon. Associates with other waders. **SIMILAR SPECIES** American Golden Plover (q.v. and above). European Golden Plover is slightly larger and more compact, with relatively shorter legs; supercilium indistinct; in flight, has white underwing. **STATUS AND DISTRIBUTION** Rare winter visitor from N Siberia, coastal Gabon (Oct–May). Vagrant, coastal Ivory Coast (Oct–Dec). **NOTE** Previously considered conspecific with American Golden Plover and known as Lesser Golden Plover *P. dominica*.

EUROPEAN GOLDEN PLOVER *Pluvialis apricaria*   Plate 39 (1)
Pluvier doré

Other names E: Eurasian/Greater Golden Plover
L 26–29 cm (10.25–11.5"). Slightly smaller than Grey Plover, with finer bill and golden-yellow tones to plumage. Similar to American and Pacific Golden Plovers but *white underwing* visible in flight. Also slightly larger and more chunky, with shorter legs. At rest, wings project only just beyond tail; 3–4 primary tips project beyond tertials; tertial tips almost level with tail tip. **Adult non-breeding** Supercilium usually rather ill defined; dusky smudge on rear of ear-coverts. Upperparts dark brown spotted and spangled golden-yellow. Underparts yellow-buff mottled brownish on breast and flanks, becoming whitish on belly. Bill black. Legs blackish to greenish-grey. **Adult breeding** Head-sides and underparts black, sometimes mottled; white band extending from forehead and supercilium to flanks; undertail-coverts white. **Immature** Similar to adult non-breeding but plumage neater, breast slightly more streaked, belly greyer. **VOICE** A soft, melodious *tluuee*. [CD2:26] **HABITS** In W Africa mainly singly, on tidal mudflats. **SIMILAR SPECIES** American and Pacific Golden Plovers (q.v.) are slightly smaller and slimmer, with relatively longer legs; supercilium more prominent and whitish; in flight, grey underwing. American Golden

Plover is also overall much greyer. **STATUS AND DISTRIBUTION** Rare Palearctic winter visitor, Mauritania. Vagrant, Senegambia. Winters mainly in N Africa; occurrence south of Senegambia unproven.

GREY PLOVER *Pluvialis squatarola*   Plate 39 (4)
Pluvier argenté

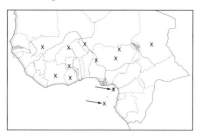

L 27–30 cm (10.5–11.75"). *Rather large, chunky, coastal wader* with relatively large head, large dark eye, and *stout blunt bill*. In flight, bold white wingbar, square white rump and uppertail-coverts, and whitish underwing with *contrasting black axillaries* ('armpits'). **Adult male breeding** Head-sides to belly black, bordered by white from forehead to breast-sides; lower underparts white. Crown and upperparts spangled black and white. Bill black. Legs blackish. **Adult female breeding** Similar but black area admixed with some white. **Adult non-breeding** Mainly dull pale greyish; upperparts mottled darker; lower underparts white. Supercilium subdued; diffuse dusky patch through eye. **Immature** Similar to adult non-breeding but plumage tinged pale yellowish-buff; upperparts neatly chequered; underparts finely streaked grey. **VOICE** Distinctive. A far-carrying, clear, rather plaintive, drawn-out *tlueeee* or *tleee-oo-ee*. [CD2:27] **HABITS** Singly or in loose groups; principally on tidal mudflats. **SIMILAR SPECIES** The 3 species of golden plovers are smaller and lack white rump and black axillaries; wingbar much fainter. Juvenile American Golden Plover also differs from juvenile Grey in its smaller bill, more contrasting crown and more distinct supercilium. **STATUS AND DISTRIBUTION** Not uncommon to common Palearctic visitor along entire coast (mainly Oct–Apr); also Cape Verde Is and São Tomé. Rare inland. Vagrant, Mali, Burkina Faso, Niger, Chad and Bioko.

AFRICAN WATTLED LAPWING *Vanellus senegallus*   Plate 37 (9)
Vanneau du Sénégal

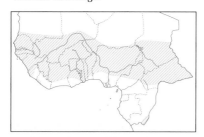

Other names E: Wattled Plover; F: Vanneau caronculé
L *c.* 34 cm (13.5"). Large, mainly pale brown with *conspicuous yellow wattles* and *long yellow legs*. In flight, broad, diagonal white band on inner wing contrasts with black flight feathers. **Adult** *senegallus* Forehead and forecrown white bordered blackish; neck streaked; throat black. Lower belly to undertail-coverts whitish. Bill yellow. Large wattles yellow with red base. Yellow orbital ring. Southern *lateralis* has belly darkening to blackish between legs and bill with black tip. **Juvenile** Head pattern duller, with white forecrown reduced, throat streaked and wattles very small. **VOICE** A shrill *kwip-kwip-kwip-...* and more nasal *ke-weep ke-weep...* [CD6:33] **HABITS** Singly, in pairs or small groups on damp grassland, grassy edges of lakes and rivers, marshes, open and lightly wooded grassland, usually (but not always) near water; also mangrove fringes. **SIMILAR SPECIES** White-headed Lapwing, the only other plover with yellow wattles, is easily distinguished by more contrasting plumage with much white, esp. in flight. Spotted Thick-knee lacks wattles and has variegated plumage. **STATUS AND DISTRIBUTION** Common to uncommon resident and intra-African migrant, from S Mauritania to Liberia east to Chad and CAR (nominate). Vagrant, S Congo (*lateralis*). Movements poorly understood.

WHITE-HEADED LAPWING *Vanellus albiceps*   Plate 37 (8)
Vanneau à tête blanche

Other names E: White-crowned Lapwing/Plover
L 28–32 cm (11.0–12.5"). Medium-sized waterside lapwing with *white band on wing, conspicuously large yellow wattles* hanging from base of bill and long greenish-yellow legs. In flight, mainly white wings with black outer 3 primaries and squarish black area on coverts. **Adult** Unmistakable. White stripe from forehead and nape; rest of head and neck grey separated from brown upperparts by white line. Black wing-coverts broadly bordered by white band. Underparts white. Bill yellow tipped black. Yellow orbital ring around yellowish eye. Long carpal spur. **Juvenile** White on crown reduced, upperparts mottled buff, outer coverts brownish and wattles very small. **VOICE** A hard, sharp, high-pitched *kip, kip, kip,...* and *kwip, kwip, kwip,...* [CD6:34] **HABITS** Singly, in pairs or small groups on sand banks, mud or rocks in rivers; also forages in grassy areas near water. **SIMILAR SPECIES** African Wattled Lapwing, the only other plover with

yellow wattles, is easily distinguished by more uniform, brown plumage. **STATUS AND DISTRIBUTION** Common to not uncommon resident and intra-African migrant throughout, except arid north. Seasonal movements related to water levels.

BLACK-HEADED LAPWING *Vanellus tectus*   Plate 37 (4)
Vanneau à tête noire

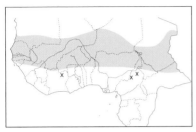

Other name E: Black-headed Plover
L *c.* 25 cm (10"). *V. t. tectus.* Rather small *dry-country lapwing* with distinctive black-and-white head pattern, wispy black crest, black-tipped red bill, and red legs. In flight, large white area on forewing extends as diagonal band on inner wing. **Adult** Top of head and neck black; white area on throat and nape connected under eye. White forehead mostly obscured by small reddish wattles at base of bill. Upperparts brown. Underparts white with black stripe on centre of breast; breast-sides washed brown. Eye yellow. **Juvenile** Duller; crest shorter; feathers of head and upperparts fringed buff; wattles very small. **VOICE** A short, piercing *kir* and a shrill *kwairr.* [CD6:35] **HABITS** Usually singly, in pairs or small groups in dry grassland with scattered bushes and patches of bare ground. Mostly active in early morning, in evening and at night, spending most of day resting in shade. Not shy. **SIMILAR SPECIES** In flight, only Long-toed and White-headed Lapwings have more white in wings. **STATUS AND DISTRIBUTION** Common to uncommon resident, from Senegambia to Guinea east through Mali, N Ghana, N Nigeria, Niger and N Cameroon to Chad (north to Ennedi). Rare, N Ivory Coast, N Togo and N CAR. Seasonal movements, north with rains, recorded.

SPUR-WINGED LAPWING *Vanellus spinosus*   Plate 37 (6)
Vanneau à éperons

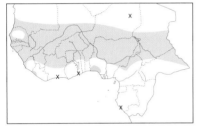

Other names E: Spur-winged Plover; F: Vanneau armé
L 25–28 cm (10–11"). Distinctively patterned, waterside lapwing. In flight, diagonal white band from carpal joint to inner secondaries contrasts with black of flight feathers and brown coverts. **Adult** Top of head black; head-sides and neck pure white contrasting with broad black stripe on throat; breast to upper belly black; rest of underparts white. Upperparts pale grey-brown. Bill and legs black. Long carpal spur. **Juvenile** Duller; black areas tinged brownish and flecked white; upperparts scaled buff. **VOICE** Alarm a sharp, metallic *kit*, often rapidly repeated. Territorial call a loud, harsh *ti-ti-terr-it.* [CD6:37] **HABITS** Usually in pairs or small groups at variety of wetlands, incl. marshes, rivers, lakes, reservoirs, estuaries, coastal saltpans and irrigated farmland. Vocal. **SIMILAR SPECIES** None, but wing pattern may recall African Wattled Lapwing (larger, browner; legs yellow) or Brown-chested Lapwing (different head and underpart pattern). Black-headed Lapwing has much white on outer wing. **STATUS AND DISTRIBUTION** Common to locally uncommon resident, from S Mauritania to Sierra Leone east to Chad and N CAR. Vagrant, Liberia and Gabon.

BROWN-CHESTED LAPWING *Vanellus superciliosus*   Plate 37 (2)
Vanneau à poitrine châtaine

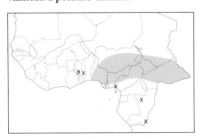

Other name E: Brown-chested Plover
L *c.* 23 cm (9"). Small and beautifully patterned lapwing. In flight, diagonal white band across wing separates black flight feathers from brown coverts. **Adult** Chestnut forehead and forecrown mostly hidden by *bright yellow wattle extending from base of bill to eye*; rest of crown black bordered by narrow black line; head-sides, neck and upper breast pale grey merging into *broad, dark chestnut breast-band* (diagnostic); rest of underparts white. Upperparts brown; uppertail-coverts white; tail white with black terminal band. Bill black. Legs dark grey. **Juvenile** Duller; head and breast dull brownish-grey with pale rusty forehead and supercilium, and no black cap; upperparts scaled rusty; wattles very small. **VOICE** Calls harsh and shrill, easily identifiable as being from a lapwing. [CD6:38] **HABITS** In pairs in short grassland, bare and recently burned ground near rivers and lakes, and other open areas in or near lightly wooded grassland. Also in small groups when not breeding. Active at night. **SIMILAR SPECIES** Lesser Black-winged Lapwing is much plainer, with blackish breast-band, white forehead, greyish crown and no wattles; in flight, has broad white trailing

edge to wing. **STATUS AND DISTRIBUTION** Uncommon and local intra-African migrant. Breeds Nigeria, Cameroon and (probably) CAR. Vagrant, E Ghana, Togo and Congo.

LESSER BLACK-WINGED LAPWING *Vanellus lugubris*   Plate 37 (1)
Vanneau terne

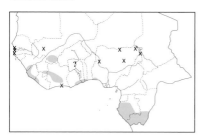

Other name E: Senegal Plover
L 22–26 cm (9–10"). Small, slim and rather plain-looking with *white forehead*. In flight, *white secondaries form broad, contrasting trailing edge to wing* (diagnostic). **Adult** Head and neck brownish-grey; upperparts grey-brown. Breast brownish-grey becoming blackish on lower border; rest of underparts white. Uppertail-coverts white; tail white with black terminal band and white outer feathers. Bill black. Eye orange-yellow. Legs dark brownish. **Juvenile** Head pattern duller; wing feathers and scapulars fringed buff. **VOICE** A melodious *thi-HUwit*, shorter *thu-WIT* or *thi-whoo*, and longer *tihi-hooee*. [CD6:39] **HABITS** Singly, in pairs or small groups in coastal savannas and lightly wooded grassland. **SIMILAR SPECIES** Brown-chested Lapwing is more colourful, with chestnut breast-band, black crown and yellow wattles; in flight, has diagonal band across wing. **STATUS AND DISTRIBUTION** Patchily distributed intra-African migrant. Uncommon to locally not uncommon from Guinea to Nigeria and in Gabon and Congo. Rare/vagrant, Senegambia.

WHITE-TAILED LAPWING *Vanellus leucurus*   Plate 37 (3)
Vanneau à queue blanche

Other name E: White-tailed Plover
L 26–29 cm (10.5–11.5"). **Adult** Long legged and slender with rather plain, greyish-brown plumage, palish face, and *all-white tail*. Lower breast to undertail-coverts white. Bill black. Eye dark brownish. Legs yellow. In flight, broad diagonal white band across wing, from primary-coverts to secondaries. **VOICE** Usually silent. Calls include a high-pitched *pet-ee-wit* and shorter *pee-wik*, recalling Northern Lapwing but quieter, less strident. [CD2:28] **HABITS** Singly or in small groups in marshes and damp grassland, and near rivers and lakes. Frequently forages in shallow or deep water. **STATUS AND DISTRIBUTION** Palearctic vagrant, S Niger (Jan), NE Nigeria (Jan–Feb) and Chad (L. Chad area, Oct). Winters mainly in Sudan and Eritrea.

LONG-TOED LAPWING *Vanellus crassirostris*   Plate 37 (7)
Vanneau à ailes blanches

Other names E: Long-toed Plover; F: Vanneau à face blanche
L *c.* 31 cm (12"). *V. c. crassirostris*. Large lapwing frequenting aquatic habitats and distinctly patterned with white face and foreneck contrasting with black hindcrown, hindneck, neck-sides and breast. In flight, *mainly white wing-coverts* contrasting with black flight feathers diagnostic. **Adult** Upperparts grey-brown. Belly to undertail-coverts white. Rump white; tail mainly black. Bill red tipped black. Eye reddish with red orbital ring. Legs bright pinkish-red. **Juvenile** Duller; feathers of dark areas tipped buff; white wing-coverts mottled buff. **VOICE** An often repeated, metallic *kik-k-k-k* and a plaintive *wheet*. [CD6:41] **HABITS** In pairs or small groups at lakes, ponds and river edges, and in swampy areas and short grass near water. Relatively long toes enable foraging on floating vegetation. **STATUS AND DISTRIBUTION** Not uncommon but local resident, L. Chad (Nigeria and Chad; claimed from Cameroon); also recorded from CE Nigeria (Pandam).

NORTHERN LAPWING *Vanellus vanellus*   Plate 37 (5)
Vanneau huppé

L 28–31 cm (11–12"). Rather thickset, relatively short-legged lapwing with wispy black crest. In flight, *strikingly broad, rounded wings*; wings all dark above, tipped white on outer 3–4 primaries. Characteristic, flappy flight action. **Adult non-breeding** Top of head black; head-sides and throat mainly white with dusky mark below eye. Upperparts dark glossy green (appearing black at distance), feathers tipped buff. Tail white with broad black subterminal band; outer feathers white. Underparts white with broad black breast-band; undertail-coverts cinnamon. Bill black. Eye dark brown. Legs dull pinkish-red. **VOICE** A hoarse, shrill *cheew-ep* or *pee-witt*. [CD2:29] **HABITS** Singly or in groups, frequenting various open habitats, incl. beaches, mudflats and ricefields. Flocks of 100 and 1000 reported from Mauritania. **SIMILAR SPECIES** The only other lapwing with crest, Black-headed Lapwing, has very different plumage pattern and coloration. **STATUS AND DISTRIBUTION** Scarce Palearctic visitor, Mauritania (mainly Nov–Feb). Vagrant, N Senegal, Gambia and Cape Verde Is.

SANDPIPERS AND ALLIES
Family SCOLOPACIDAE (World: 86–87 species; Africa: 50; WA: 40)

Varied family of small to large waders. Typically more slender than plovers, with longer neck and bill. Wings relatively long and pointed. Many species have distinct breeding and non-breeding plumages. Sexes mostly similar. Juveniles separable. Mostly gregarious outside breeding season; often encountered in mixed species gatherings. Flocks of smaller species often fly in tight formation, performing swift manoeuvres in unison. Occur at sea shores, lagoons, lakes, marshes and other wet habitats. Identification of larger species in our region usually easy, but small ones, especially calidrids, may be problematic and require careful observation. Important identification features include wing, rump and tail patterns, bill shape and length, and call. As yet unrecorded vagrants may appear, reference to a specialist work (e.g. Hayman *et al.* 1986) is recommended when faced with an unusual wader.

Understanding moult greatly facilitates identification. (As all species visiting W Africa are long-distance migrants from the Palearctic or Nearctic, we refer for convenience to northern hemisphere seasons.)

First moult is post-juvenile (mainly Sep–Nov): a partial moult, replacing head and body feathers and resulting in first-winter plumage. Subsequently, there are two moults per year: a partial one from winter to summer plumage (mainly Jan–Apr), replacing head and body feathers and a variable number of inner wing-coverts, and a complete one following the breeding season (mainly Aug–Nov), resulting in adult non-breeding or second-winter plumage. Moults mainly take place in or near wintering areas and may result in distinctly mixed plumages.

Juveniles (Aug–Oct) are neatly patterned with upperpart feathers having crisp pale fringes. Difference to plainer, worn plumage of adults is often striking, especially on tertials and greater coverts.

First-winter plumage (Oct–Mar) is very similar to adult non-breeding, but as post-juvenile moult is partial and does not include wings, first-winters can, at least initially, be distinguished by their wing pattern (esp. tertials and greater coverts). The distinguishing paler fringes are gradually lost through wear, however, and by midwinter first-winters are usually difficult to age.

First-summer plumage (May–Jul) is often only partially similar to adult summer (=breeding) plumage, or resembles winter plumage. Many first-summers do not migrate north with adults but remain in their winter quarters, usually moulting early into second-winter plumage, which is identical to adult non-breeding plumage.

Adult breeding plumage (Mar/Apr–Sep/Oct), is, in many species, distinctly more colourful than plainer non-breeding plumage (Sep–Mar); just prior to 'spring' migration, the feathers often have crisp white fringes, but by the time the birds return to their wintering grounds, the plumage is highly worn and generally darker above.

Species in our region can be grouped as follows.

Stints and small sandpipers (*Calidris*, 13 species). Mostly small and very active. Feed by pecking busily.

Snipes (*Gallinago* and *Lymnocryptes*, 3 species). Small to medium-sized with long, straight bill and cryptic plumage. Feed by probing rapidly and almost vertically into soft ground.

Godwits (*Limosa*, 2 species). Large with long legs and necks and long, straight or slightly upturned bills.

Curlews (*Numenius*, 2 species). Large to very large with long legs and necks and long, decurved bills. Plumage mainly brown and without seasonal variation.

Shanks and allies (*Tringa*, *Actitis* and *Xenus*, 11 species). Medium-sized, mostly elegant waders with relatively long bills; legs long in *Tringa*, shorter in *Actitis* and *Xenus*.

Phalaropes (*Phalaropus*, 3 species). Fairly small to medium-sized, graceful waders. Swim buoyantly and forage on water with spinning action, these features distinguishing them from all other waders. Females in breeding plumage brighter than males; sex roles reversed. The most pelagic of all waders, seen flying far out above or swimming on open sea.

Five species do not fall into the above categories: Ruff, Buff-breasted Sandpiper, Broad-billed Sandpiper, Upland Sandpiper and Ruddy Turnstone. In addition, vagrant dowitchers have reached our region, but were not specifically identified.

RED KNOT *Calidris canutus*   Plate 40 (10)
Bécasseau maubèche

Other name E: Knot
L 23–25 cm (9–10"). *C. c. canutus.* Largest *Calidris.* Stocky, rounded, coastal wader with straight, relatively short and thick bill. Legs relatively short, greyish or greenish. In flight, narrow white wingbar, whitish-grey rump and

grey tail. **Adult non-breeding** Pale grey above with indistinct whitish supercilium and dusky eye-stripe; feathers narrowly fringed white. Underparts white with grey-streaked breast and flanks. **Adult breeding** Upperparts rufous and black; wing-coverts grey fringed white. Face and underparts deep rufous; lower underparts white, variably marked black. **Juvenile** Similar to adult non-breeding but feathers of upperparts have narrow whitish fringes and narrow dark subterminal crescents, producing scaly appearance. Breast and flanks have V-shaped markings and spots; initial ochre-buff tinge soon wears. **VOICE** A low, hoarse *nut, nut, ...*; also various rather nasal sounds. [CD2:30] **HABITS** Dense flocks in main wintering area, elsewhere singly or in small groups, favouring tidal mudflats and estuaries. **SIMILAR SPECIES** Curlew Sandpiper is distinctly smaller and slimmer, with longer, decurved bill; white rump in flight. Dunlin is even smaller and has longer bill and dark-centred rump and tail. **STATUS AND DISTRIBUTION** Palearctic passage migrant and visitor (mainly Sep–Apr) to entire coast. Winters mainly NW Mauritania (Banc d'Arguin), where very common; common, Guinea-Bissau and Guinea; uncommon to scarce elsewhere. Rare, Mali. Vagrant, Cape Verde Is.

SANDERLING *Calidris alba* Plate 40 (7)
Bécasseau sanderling

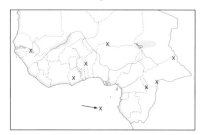

L 20–21 cm (8"). Active, rounded wader with relatively short, straight black bill and black legs. *Lacks hind toe* (diagnostic). In flight, bold white wingbar on blackish upperwing. **Adult non-breeding** *Palest wader*. Very pale grey above with white face and *distinctive blackish patch at bend of closed wing* (sometimes obscured by breast-side feathers). Underparts white. **Adult breeding** Head and breast rufous marked black; upperparts black fringed rufous and grey; wing-coverts grey fringed whitish. **Juvenile** Contrastingly marked black and white above. Crown streaked blackish; upperparts blackish spangled and spotted white and creamy-buff, producing chequered appearance. Breast-sides washed buff. **VOICE** A short, sharp *twik* or *krit*; repeated notes sometimes accelerating into trill. [CD2:31] **HABITS** Usually in small groups, on sandy beaches; also sandy river banks, lagoons, saltpans. Runs very fast at edge of surf, nervously picking at small items. Occasionally inland, where single birds with less restless behaviour can be confusing. **SIMILAR SPECIES** Red Phalarope (q.v.) is the only other small wader with similarly contrasting wing pattern. Little Stint much smaller, with shorter, finer bill. **STATUS AND DISTRIBUTION** Common to not uncommon Palearctic and Nearctic passage migrant and visitor to entire coast (mainly Sep–Apr; records in all months). Inland records from Mali, Ghana, Togo, Niger, Nigeria, Chad and CAR. Also Cape Verde Is (common) and Annobón.

SEMIPALMATED SANDPIPER *Calidris pusilla* Plate 42 (4)
Bécasseau semipalmé

L 13–15 cm (5–6"). Similar to Little Stint but has half-webbed toes (visible under good conditions). Also rather dumpier, with slightly heavier, thick-tipped bill. **Adult non-breeding** Slightly paler and more uniform than Little; best distinguished by palmations and call. **Adult breeding** Grey-brown above, lacking pale V and rusty-orange of Little Stint; scapulars with dark centres; wing-coverts and tertials with relatively paler centres (always paler than Little). Underparts white with streaks on breast forming band; some extending onto flanks. **Juvenile** Dull greyish above; feathers fringed grey-buff to off-white. Differs from Little in more uniform plumage usually lacking rufous tones; no distinct white Vs on mantle (sometimes faint pale V); wing-coverts and tertials with paler centres; lower 2 rows of scapulars pale grey with dark anchor-shaped mark at tips. **VOICE** A short, low *chrup* or *krrit*; also a slightly higher pitched, more Little Stint-like *tip* or *kit*. **HABITS** Frequents mudflats and sandy beaches; also muddy edges of freshwater pools. **SIMILAR SPECIES** Little Stint (q.v. and above). **STATUS AND DISTRIBUTION** N American vagrant, NW Mauritania (Banc d'Arguin, Jan 1994) and Cape Verde Is (Boavista, Mar 1999).

LITTLE STINT *Calidris minuta* Plate 40 (3)
Bécasseau minute

L 12–14 cm (5–6"). Very small wader with *relatively short, straight black bill* and *black legs*. In flight, narrow white wingbar and grey outer tail feathers. **Adult non-breeding** Grey above; feathers often with dark shaft streaks. Underparts white with grey breast-sides (sometimes forming breast-band). **Adult breeding** Head, breast and upperparts variably tinged rusty-orange with darkish-marked crown and creamy-buff V on mantle. **Juvenile** Crisp

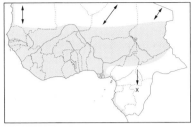

plumage, usually tinged rusty-orange above. Crown streaked blackish contrasting with whitish forehead and supercilium, and pale greyish collar. Upperparts blackish-brown fringed deep buff and rufous; usually has prominent white Vs on mantle and scapulars. Underparts white with breast-sides streaked dark and washed rusty-orange. **VOICE** A sharp, high-pitched *kip* or *chit*. [CD2:32] **HABITS** Singly or in small loose flocks, frequenting coastal and inland mudflats and wetlands, incl. lagoons, estuaries, edges of pools and rivers, rice-fields, etc. Forages actively, briskly picking at items on mud surface. **SIMILAR SPECIES** Similar-sized Temminck's Stint distinguished by plain brownish head, brownish upperparts, usually entirely brownish breast, and shorter, greenish legs; also forages more unobtrusively in crouching posture and towers high when flushed (Little Stint flies off low). Sanderling distinguished at distance by paler overall plumage and broad white wingbar in flight. Dunlin is larger and has longer bill. **STATUS AND DISTRIBUTION** Common Palearctic passage migrant and visitor throughout (mainly Sep–Apr; records in all months). Also Cape Verde Is (not uncommon).

TEMMINCK'S STINT *Calidris temminckii*   Plate 40 (6)
Bécasseau de Temminck

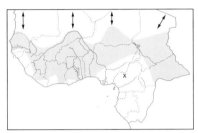

L 13–15 cm (5–6"). Similar in size to Little Stint, but slightly more elongated and much plainer, with (typically) complete grey-brown breast-band and shorter, brownish to olive-yellow legs. Crouching posture and plain plumage reminiscent of miniature Common Sandpiper. In flight, narrow white wingbar and white outer tail feathers. **Adult non-breeding** Dull grey-brown above. Underparts white with grey-brown breast-band (sometimes only at sides). **Adult breeding** Head, breast and upperparts grey-brown; variable number of scapulars and mantle feathers have dark centres and dull cinnamon and greyish fringes. **Juvenile** Grey-brown above; narrow buffish fringes bordered by dark subterminal line produce scaly appearance at close range. **VOICE** A distinctive, high-pitched, dry trill *tirrr*, often drawn-out into *tirrr-r-r-r-r*. [CD2:33] **HABITS** Singly, occasionally in small groups, at muddy edges of pools, lakes, rivers, marshes and ricefields. Unobtrusive. When flushed, flies high and erratically (like snipe). **SIMILAR SPECIES** Similar-sized Little Stint in non-breeding plumage distinguished by greyer head and upperparts with whitish face and supercilium, grey on breast usually restricted to sides, and black legs; also forages actively in more upright posture. **STATUS AND DISTRIBUTION** Rare to locally not uncommon Palearctic passage migrant and visitor (mainly Sep–Apr) from Mauritania to Liberia east to Chad and CAR. Vagrant, Cape Verde Is.

LONG-TOED STINT *Calidris subminuta*   Plate 42 (3)
Bécasseau à longs doigts

L 14–15 cm (6"). Small stint with *yellowish to greenish legs*. Very similar to Least Sandpiper, but slightly larger, with *longer neck* and *longer legs* (esp. tibia) *and toes* (esp. middle); fine-tipped bill usually has pale base to lower mandible. Forehead dusky, usually extending to base of bill (on Least pale supercilia normally join on forehead); dark streaks on breast narrower and extending onto flanks. Often in upright posture. In flight, narrower white wingbar than Little Stint and darker underwing than Least Sandpiper; feet project slightly beyond tail (unlike other stints). **Adult non-breeding** Dull brown-grey above as Least but has more distinct dark centres to scapulars and more contrasting, broad pale fringes (giving scalloped appearance when fresh). **Adult breeding** Usually more rusty above than Least (recalling breeding Little Stint); dusky loral stripe narrower, less prominent and often broken; malar area less heavily streaked. **Juvenile** Upperparts have blackish-centred, rufous-fringed feathers and whitish V on mantle as in juvenile Least, but distinguished by rather contrasting, brown-grey, whitish-fringed coverts (juvenile Least has rufous-fringed coverts). **VOICE** A soft *chrrrp*, reminiscent of Curlew Sandpiper. **HABITS** Frequents estuaries, freshwater and brackish pools, lakes and marshes. May adopt crouching posture when feeding, as Least Sandpiper, but stands very erect when alarmed, looking conspicuously long legged and long necked. **SIMILAR SPECIES** Least Sandpiper (q.v. and above). Temminck's Stint has rather uniform grey-brown breast in all plumages, tail projecting beyond wingtips, and different call. Little Stint has black legs; non-breeding plumage plainer above. **STATUS AND DISTRIBUTION** E Palearctic vagrant, coastal Congo (Nov 1990).

LEAST SANDPIPER *Calidris minutilla*   Plate 42 (2)
Bécasseau minuscule

L 13–14 cm (5–6"). Small stint with *greenish to yellowish legs* and *streaks on breast usually forming complete band*. Tip of tail falls approximately level with wingtips. Supercilium reaches base of all-black bill and is bordered below by dusky loral stripe. In flight, narrow white wingbar and white outer tail feathers. **Adult non-breeding** Dull brown-grey above with diffuse dark feather centres. Underparts white diffusely streaked brown-grey on breast. **Adult breeding** Crown heavily streaked; upperparts blackish-centred with brown-grey to rufous fringes. Breast coarsely streaked dark brown, clearly demarcated from white lower underparts. **Juvenile** Blackish-centred, rufous-fringed upperparts recall juvenile Little Stint, but whitish V on mantle narrower and less distinct, primary projection shorter and legs yellowish (not black). **VOICE** A high-pitched, rising *krreeep*. **HABITS** Normally favours muddy edges of inland waters. Often forages in crouched posture. When flushed, flies high like Temminck's Stint. **SIMILAR SPECIES** Long-toed Stint (q.v.). Temminck's Stint has rather uniform grey-brown breast in all plumages, tail projecting beyond wingtips, and different call. **STATUS AND DISTRIBUTION** N American vagrant, Cape Verde Is (São Vicente, Mar 1996).

WHITE-RUMPED SANDPIPER *Calidris fuscicollis*   Plate 40 (1)
Bécasseau à croupion blanc

Other name F: Bécasseau de Bonaparte
L 15–18 cm (6–7"). Long bodied, in size between Little Stint and Curlew Sandpiper, with relatively short legs, long primary projection and *wingtips extending clearly beyond tail*. Prominent white supercilium. Bill slightly decurved with dirty orange-pink base to lower mandible. In flight, faint white wingbar and *white uppertail-coverts contrasting strongly with dark tail*. **Adult non-breeding** Greyish above. Underparts white with greyish, finely streaked breast. **Adult breeding** Dull greyish above with some rusty on crown, mantle and scapulars; feathers with dark centres. Breast heavily streaked black; streaking extending onto flanks. **Juvenile** Brighter. Crown tinged rusty; mantle and upper rows of scapulars fringed rusty; usually indistinct Vs on mantle. **VOICE** A distinctive, high-pitched, squeaky, insect- or bat-like *tzreet* or *tzeeeht*, often rapidly repeated. [CD2:34] **HABITS** Favours brackish and freshwater pools, lagoons, estuaries and ricefields. **SIMILAR SPECIES** Curlew Sandpiper is the only other *Calidris* that also has white uppertail-coverts, but white band broader; also larger, with longer, obviously decurved bill. Baird's Sandpiper has similar shape but tends to have slightly straighter bill lacking pale patch at base of lower mandible; plumage more buffish-grey lacking white on uppertail-coverts; breast streaking does not extend onto flanks in non-breeding plumage; supercilium usually less prominent. Juvenile Baird's lacks rufous tones of juvenile White-rumped, and has more uniform upperparts. Dunlin is slightly larger, with longer bill, longer legs and shorter wings. **STATUS AND DISTRIBUTION** N American vagrant, Ivory Coast (Oct–Nov 1988) and Ghana (Dec 1985).

BAIRD'S SANDPIPER *Calidris bairdii*   Plate 40 (2)
Bécasseau de Baird

L 14–17 cm (5.5–6.5"). Long bodied, resembling White-rumped Sandpiper in shape and size, but has less prominent supercilium, usually shorter, straighter, all-black bill and *lacks white band on uppertail-coverts*. Primary projection long; wingtips extending clearly beyond tail; legs relatively short. In flight, faint white bar on long wings and dark-centred rump and uppertail-coverts. **Adult non-breeding** More buffish above than White-rumped; breast also more buffish and darker, with streaking less distinct; no streaks on flanks. **Adult breeding** Grey-buff above with some dark feather centres. Breast buffish finely streaked brown. **Juvenile** Head buffish with indistinct pale supercilium and pale supraloral spot. Upperparts grey-buff with white-fringed feathers giving uniform scaly appearance (without white Vs). Breast buff finely streaked brown. **VOICE** A soft, rolling *prrreet* or *kirrrp*, lower and softer than Curlew Sandpiper. **HABITS** Frequents tidal mudflats, edges of brackish and freshwater pools, saltpans and sandy areas with short grass. **SIMILAR SPECIES** White-rumped Sandpiper (q.v. and above). **STATUS AND DISTRIBUTION** Possible N American vagrant, claimed from Mauritania (Nov 1987), N Senegal (Dec 1965) and Gambia (Nov 1976).

PECTORAL SANDPIPER *Calidris melanotos*   Plate 40 (5)
Bécasseau à poitrine cendrée

Other name F: Bécasseau tacheté
L 19–23 cm (7.5–9.0"). Slightly larger than Curlew Sandpiper and somewhat smaller than female Ruff, with relatively long neck and small head; *densely streaked neck and upper breast sharply demarcated from white lower breast and belly*; bill slightly decurved and pale based; *legs greenish-yellow to yellowish-brown*. In flight, indistinct narrow pale wingbar and white sides to dark-centred rump and uppertail-coverts. **Adult non-breeding** Greyish-brown above with diffuse dark feather centres; pale, finely streaked supercilium. **Adult breeding** Dark-centred feathers of upperparts fringed rufous and buff; indistinct pale Vs on mantle and scapulars. **Juvenile** Similar to adult breeding, but brighter and more neatly scaled above; white Vs on mantle and scapulars usually prominent; fringes to upperparts brighter rufous. **VOICE** A rolling, hoarse *krrrt* or *chrrp*, somewhat like call of Curlew Sandpiper but lower pitched and rather harsher. [CD2:35] **HABITS** Frequents muddy edges of pools, marshy places, wet

grasslands, lagoons and mudflats. **SIMILAR SPECIES** In flight, much fainter wingbar than other *Calidris* and long wings may suggest small Ruff, but flies with faster wingbeats and feet do not project beyond tail. **STATUS AND DISTRIBUTION** Vagrant from N America and E Siberia (Oct–Apr), recorded in Mauritania, N Senegal, Sierra Leone, Liberia, Ivory Coast, Ghana, Gabon and Príncipe.

CURLEW SANDPIPER *Calidris ferruginea* Plate 40 (11)
Bécasseau cocorli

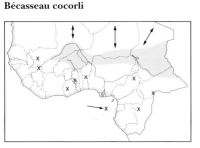

L 18–23 cm (7–9"). *Rather elegant* with longish, evenly decurved black bill and longish black legs. In flight, clear white wingbar and *broad white band on uppertail-coverts*. **Adult non-breeding** Grey above with prominent white supercilium. Underparts white with breast-sides washed grey. **Adult breeding** Distinctive, but usually only partial in Africa. Head and underparts deep rusty-red; upperparts dark brown fringed rusty-red; wing-coverts grey. **Juvenile** Resembles adult non-breeding but upperparts uniformly scaly and washed buff; breast lightly streaked and washed cinnamon-buff. **VOICE** A rather pleasing, soft, rippling *chirrup* or *chirririp*, less grating than Dunlin. [CD2:36] **HABITS** In small or large groups on tidal mudflats, estuaries, muddy shores, lagoons and saltpans. Inland, mainly in small groups on muddy shores of freshwater pools, lakes and reservoirs. **SIMILAR SPECIES** Dunlin is slightly smaller and less elegant, with less evenly decurved bill, shorter neck and shorter legs. **STATUS AND DISTRIBUTION** Common or not uncommon to locally very common Palearctic passage migrant and visitor to entire coast (mainly Aug–Apr; records in all months). Smaller numbers locally inland. Also Cape Verde Is (not uncommon), Bioko and Príncipe (rare).

PURPLE SANDPIPER *Calidris maritima* Plate 42 (6)
Bécasseau violet

L 20–22 cm (8.0–8.75"). *Stocky, darkish wader with short, yellowish legs.* Tail projects beyond wingtips. In flight, mainly dark above with narrow white wingbar and white sides to rump and uppertail. **Adult non-breeding** Head, upperparts and breast slaty brown-grey; no supercilium. Rest of underparts white. Bill slightly decurved with variable yellowish base. **Adult breeding** Mantle and scapulars fringed whitish and rufous; breast and flanks mottled and streaked blackish. Legs greenish-brown. **Juvenile** Feathers of upperparts with neat whitish fringes produce scaly appearance; crown and mantle with rufous; breast streaked. **VOICE** A short *kwit*. [CD2:37] **HABITS** Normally favours rocky areas on coasts, where readily associates with Ruddy Turnstones. **STATUS AND DISTRIBUTION** Palearctic vagrant, NW Mauritania (Banc d'Arguin, Apr 1985).

DUNLIN *Calidris alpina* Plate 40 (9)
Bécasseau variable

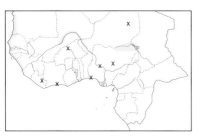

L 16–22 cm (6.5–8.5"). *Rather dumpy* with variably longish, black bill, slightly decurved at tip, and black legs. In flight, narrow white wingbar and dark-centred rump and uppertail-coverts. **Adult non-breeding** Plain drab grey above with *indistinct* white supercilium. Underparts white with breast finely streaked grey. **Adult breeding** Mantle and scapulars with variable amount of rufous; wing-coverts grey. Diagnostic *large black patch on belly*. Races differ in brightness of rufous on upperparts, extent of belly patch and mean bill length: *arctica* least rufous, belly patch relatively small, bill shortest; *schinzii* intermediate, upperparts with dull chestnut, belly patch mixed with white; **nominate** richest rufous, belly patch largest and blackest, bill longest. **Juvenile** Head, nape and upper breast washed rusty-buff; no obvious supercilium. Upperparts with dark feather centres and buffish-white and rufous fringes, usually with narrow white Vs on mantle and scapulars. Breast *densely streaked; belly-sides variably spotted*, contrasting with unmarked flanks. **VOICE** A distinctive, shrill, rasping *krrreet* or *treerrp*. [CD2:38] **HABITS** Usually in small or large groups on tidal mudflats; singly or in small groups on muddy shores of inland waters. Forages actively, picking at small items on mud surface or shallowly probing; also wades in shallows. **SIMILAR SPECIES** Curlew Sandpiper has longer wings (wingtips extending beyond tail), longer legs, more evenly decurved bill, more prominent supercilium and white rump. Juvenile Curlew Sandpiper further differs from juvenile Dunlin in more neatly patterned upperparts and absence of coarse markings on belly and flanks. Red Knot is distinctly larger and more robust, with shorter, straight bill and paler grey non-breeding plumage. **STATUS AND DISTRIBUTION** Palearctic passage migrant and visitor (mainly Sep–Apr; records in all months). Mainly Mauritania

(Banc d'Arguin); not uncommon from Senegambia to Guinea; rare along rest of coast. Small numbers or singles occasionally inland, mainly in central Niger delta, Mali. Also Cape Verde Is (scarce). Mainly *schinzii*, with smaller numbers of *arctica* and some nominate *alpina*.

BROAD-BILLED SANDPIPER *Limicola falcinellus* Plate 40 (8)
Bécasseau falcinelle

L 16–18 cm (6–7"). Slightly smaller than Dunlin, with slightly shorter, greyish legs and longish, black bill with downward kink at tip. In flight, dark upperwing with narrow white wingbar, and dark-centred rump and uppertail-coverts. **Adult non-breeding** Grey above, resembling Dunlin, but traces of breeding plumage crown pattern (with 'split supercilium') usually visible; upperparts have dark shaft streaks and dark carpal area ('shoulder'; often hidden). Underparts white with rather indistinct streaks and spots on breast-sides extending onto flanks. In flight, broad dark leading edge to wing. **Adult breeding** Dark crown, dark eye-stripe, pale supercilium and narrow pale lateral crown stripe. Upperparts blackish-brown with whitish and buff feather fringes and narrow white Vs on mantle and scapulars; soon darken with wear, Vs becoming indistinct. Underparts white with heavily streaked breast and arrow-like markings on flanks. **Juvenile** Resembles adult breeding but upperparts paler brown and more neatly patterned; wing-coverts fringed buff; breast streaked lighter and flanks unmarked. **VOICE** A dry *trrreet* or *krr-r-r-r-p*. Also a short, Little Stint-like *trett* or *dritt*. [CD2:39] **HABITS** Favours mudflats and muddy edges of pools and lakes. **SIMILAR SPECIES** Compare Dunlin (q.v.). **STATUS AND DISTRIBUTION** Palearctic vagrant (Aug–Apr), Mauritania, N Mali, NE Nigeria, C Chad, Cameroon and Gabon.

BUFF-BREASTED SANDPIPER *Tryngites subruficollis* Plate 40 (4)
Bécasseau roussâtre

L 18–20 cm (7–8"). Slightly smaller than female Ruff, with distinctly small, round head, short, fine black bill and yellowish legs. In flight, no white above but white underwing has conspicuous *dusky half crescent on greater primary-coverts*; feet do not project beyond tail. Appears plover-like, but has distinctly finer bill. **Adult** Head and underparts plain sandy-buff with finely streaked crown, dark beady eye in plain face, and finely spotted breast-sides. Upperparts scaly, dark fringed sandy-buff. **Juvenile** Similar to adult but more neatly scalloped above, with fringes to scapulars and coverts whiter. **VOICE** Usually silent. **HABITS** Favours grassy areas, mudflats, and edges of pools, lagoons and marshes. Rather deliberate, plover-like in its movements, often foraging in short vegetation. Often tame. **SIMILAR SPECIES** Juvenile female Ruff is larger, with longer bill, usually indistinct dusky eye-stripe, no or less distinct spots on breast-sides, and duller legs; flight pattern different, with clear white wingbar and white patches at tail-sides. **STATUS AND DISTRIBUTION** N American vagrant (Nov–Jan and Apr), Senegal, Gambia, Sierra Leone, Ghana and Gabon. *TBW* category: Near Threatened.

RUFF *Philomachus pugnax* Plate 41 (4)
Combattant varié

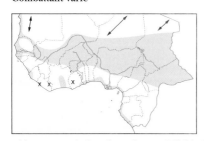

Other name F: Chevalier combattant
L male 26–32 cm (10.5–12.5"), female 20–25 cm (8–10"). Distinctly shaped wader with relatively small head, longish neck and slightly curved, medium-long bill. Plumage pattern, bare-part coloration and size variable; males considerably larger. *Feathers of mantle appear loose and are often fluffed.* In flight, narrow but clear white wingbar and *white oval area on sides of dark-centred uppertail*; feet project beyond tail. Flying flocks often identifiable by marked size difference between sexes. **Adult non-breeding** Grey-brown above with scaly upperparts; often whitish feathers at base of bill. Underparts white with grey-mottled wash on breast. Bill black with orange base. Legs orange-red to yellowish. Some males have head and neck white. **Adult male breeding** (Apr–Jun) Unmistakable with spectacular head tufts and neck feathers ('ruff'), which vary from black through chestnut and cinnamon to white, and are barred, vermiculated or unmarked. When moulting, often has white on head and black blotching on breast. **Adult female breeding** Mainly brown variably marked with black and buff. **Juvenile** Head, neck and breast warm buff; upperparts with dark brown feather centres and buff fringes give strongly scaled appearance. Legs dull greenish to yellowish-brown. **VOICE** Usually silent. [CD2:40] **HABITS** Singly or in small groups in marshes, muddy edges of pools, lakes and lagoons, ricefields, flooded grassland and estuaries. Wingbeats slightly but distinctly slower than other similar-sized waders, giving more 'relaxed' impression. **SIMILAR SPECIES** Common Redshank (slightly smaller than male Ruff, larger than female) has longer bill without white at base, plain (not scaled) upperparts, different flight pattern, and distinctive call. **STATUS AND DISTRIBUTION** Palearctic passage migrant and visitor throughout (mainly Sep–Apr; some year-round). Common to locally very common from Mauritania to Guinea east to Chad and N CAR; rare to uncommon elsewhere. Also Cape Verde Is (scarce).

JACK SNIPE *Lymnocryptes minimus*

Bécassine sourde

Plate 36 (5)

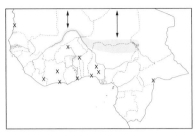

L 17–19 cm (6.75–7.5"). Small and skulking with relatively short bill and legs, and 2 bold creamy-yellow stripes on upperparts. Head marked with buff, split supercilium and dusky crescent around ear-coverts. Lacks median crown-stripe. Breast and flanks *streaked* (not barred) dark brown; belly white. Flight low, revealing wedge-shaped tail. **Juvenile** Similar. **VOICE** Usually silent, but occasionally utters quiet, hoarse *gatch* when flushed. [CD2:41] **HABITS** Singly in marshy areas and at muddy edges of pools, ricefields and small streams. Unobtrusive and hard to detect. Flushes silently when almost trod upon; does not usually fly far before dropping into cover. **SIMILAR SPECIES** Other snipes are distinctly larger with longer bills and flush at longer distances. **STATUS AND DISTRIBUTION** Rare to locally uncommon Palearctic winter visitor (mainly Nov–Mar), Mauritania, Senegambia, Mali (central Niger delta), N Ghana and N Nigeria. Scattered records elsewhere. Vagrant, Cape Verde Is. Possibly under-recorded due to secretive habits.

COMMON SNIPE *Gallinago gallinago*

Bécassine des marais

Plate 36 (6)

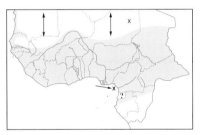

L 25–27 cm (10.0–10.5"). *G. g. gallinago*. Cryptically patterned, medium-sized, stocky wader with boldly striped head and very long, straight bill. Flight fast and zigzagging, revealing narrow white trailing edge to inner wing and white edges to tail. **Adult** Mainly brown, with buff median crown-stripe and buff stripes on mantle-sides and scapulars. Breast brown-buff densely marked with lines of dark brown spots; flanks barred dark brown; belly white. Underwing has pale, unmarked central area. Legs greenish. **Juvenile** Similar. **VOICE** A hoarse *rretch*, uttered when flushed and in flight. [CD2:42] **HABITS** Singly or in small groups, favouring muddy edges of pools, lakes and watercourses, and marshes. Often wades. Forages actively by thrusting bill vigorously in and out of mud ('sewing machine' action). Secretive and wary. Squats when threatened; flushes explosively at *c.* 10 m distance, climbing rapidly in frantic, tilting flight and usually flying far before dropping again. **SIMILAR SPECIES** Great Snipe (q.v.). **STATUS AND DISTRIBUTION** Uncommon to common passage migrant and visitor throughout (mainly Sep–Mar). Also Cape Verde Is (rare). Vagrant, Bioko. **NOTE** Although those recorded in Cape Verde Is are likely to be migrants from the Palearctic, the possibility of Nearctic *G. g. delicata* (Wilson's Snipe: darker, with narrower white trailing edge to inner wing and uniformly dark-barred underwing) cannot be eliminated.

GREAT SNIPE *Gallinago media*

Bécassine double

Plate 36 (4)

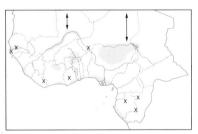

L 27–29 cm (10.5–11.5"). Slightly larger and bulkier than Common Snipe with slightly shorter bill, white wingbars formed by bold tips to coverts, and entirely barred flanks, belly and underwings. Flight low and straight, distinctly slower and heavier than Common Snipe, with 2 narrow white lines delimiting dark midwing panel on greater coverts; broad white corners to tail conspicuous on take-off or landing. **Juvenile** Similar, but with less white on tail. **VOICE** A hoarse, gruff *shrt* or *krrrt*, occasionally given in flight. [CD2:43] **HABITS** Singly in marshy areas and at muddy edges of pools and small streams. Less wary than Common Snipe. Flushes silently but with audible sound of wings at *c.* 5 m distance; usually dropping quickly into cover. **SIMILAR SPECIES** Common Snipe (q.v. and above); in flight, note in particular narrow white trailing edge to inner wing, narrower white edges to tail, white belly, flight action and call. **STATUS AND DISTRIBUTION** Widespread, uncommon to rare Palearctic passage migrant and visitor (mainly Oct–Apr); locally common in C Mali and N Nigeria. *TBW* category: Near Threatened.

SHORT-BILLED DOWITCHER *Limnodromus griseus*

Bécassin roux

Plate 42 (7)

L 25–29 cm (10.0–11.5"). Medium-sized wader with very long, straight, blunt-tipped bill, prominent supercilium and moderately long, greenish legs. In flight, whitish trailing edge to wing, white oval ('cigar') on back and barred tail; feet barely project beyond tail. **Adult non-breeding** Grey above with whitish supercilium and dusky eye-stripe; white below with grey upper breast and grey barring on flanks. **Adult breeding** Dark brown above with pale rusty supercilium; upperparts fringed pale rufous. Face, neck and underparts orange-rufous; belly whitish to orange-rufous; breast-sides, flanks and undertail-coverts with variable number of spots and bars. **Juvenile** Paler and less heavily patterned below; upperparts neatly fringed rusty-buff to pale buff; centres of tertials, greater coverts and scapulars irregularly striped and barred ('marbled') buff. **VOICE** Diagnostic flight call a soft, fast, rattling *tiu-tuk-tuk* or *chu-tu*, reminiscent of Ruddy Turnstone. **HABITS** Could occur in any wet habitat. Probes in soft mud with 'sewing machine' action, like Common Snipe. **SIMILAR SPECIES** Long-billed Dowitcher *L. scolopaceus*, a N American species that could occur as vagrant to our region, is extremely similar and best distinguished by call: a short, sharp, rather high-pitched *keek* or *kyip*, often quickly doubled or trebled when alarmed. Juvenile/1st-winter differs from juvenile/1st-winter Short-billed by plain (not 'marbled') dark centres to tertials, greater coverts and scapulars. **STATUS AND DISTRIBUTION** N American vagrant, claimed from Ghana (Oct 1976; not accepted by *BoA*). An unidentified dowitcher was reported from Gambia (Dec 1978).

BLACK-TAILED GODWIT *Limosa limosa*   Plate 39 (6)
Barge à queue noire

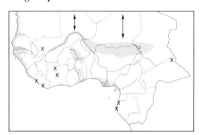

L 37–44 cm (14.5–17.0"). *L. l. limosa*. Large, slender wader with *very long, straight bill*, long neck, and long dark legs. In flight, distinctive pattern of *broad white wingbar* and *square white uppertail contrasting with broad black terminal band*. Underwing white with dusky trailing edge. Legs project well beyond tail. **Adult non-breeding** Head, upperparts, breast and flanks plain grey; rest of underparts white. Short white supercilium. Bill pinkish with black tip. **Adult male breeding** Head, neck and upper breast rufous-orange; breast and flanks variably barred. Upperparts dark brown mixed with rufous and grey. Bill pinkish-orange tipped dark. **Adult female breeding** Similar, but usually duller; bill slightly longer. **Juvenile** Neck and breast variably washed pale rusty-orange; upperparts and wing-coverts neatly fringed pale rusty-orange. **VOICE** Calls include a short *kip* or *kip-kip-kip* and a sharp *weeka-weeka*. [CD2:45] **HABITS** Singly, in small or large groups on coastal lagoons, tidal mudflats, saltpans, and freshwater edges. Forages by probing deep in mud; often wades up to belly, even submerging head. Flight rather fast with rapid wingbeats. **SIMILAR SPECIES** Bar-tailed Godwit is stockier, with slightly upturned, somewhat shorter bill and shorter tibia; easily distinguished in flight, when plain brown upperwing, white wedge on back and barred tail become evident. **STATUS AND DISTRIBUTION** Palearctic passage migrant and visitor throughout (mainly Sep–Apr; some year-round). Common from Mauritania to Guinea east through Mali to NE Nigeria and Chad; uncommon to rare elsewhere. Vagrant, Cape Verde Is (Aug–Jan).

BAR-TAILED GODWIT *Limosa lapponica*   Plate 39 (7)
Barge rousse

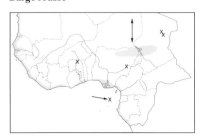

L 37–41 cm (14.5–16.0"). *L. l. lapponica*. Large, long billed and long legged. Resembles Black-tailed Godwit but more stocky, with somewhat shorter, *slightly upturned bill* and shorter legs. In flight, *plain brown upperwing*, white wedge onto back and *barred tail*. Underwing white. Feet project only slightly beyond tail. **Adult non-breeding** Head, upperparts and breast pale brown-grey with dark shaft streaks; rest of underparts whitish. Long white supercilium. Bill pinkish with black tip. **Adult male breeding** Head, neck and underparts dark rufous; lower underparts sometimes whitish. Upperparts blackish-brown fringed dark rufous; wing-coverts pale grey. Bill black. **Adult female breeding** Much paler, with noticeably longer bill; head and breast washed pale orange-buff; rest of underparts off-white. **Juvenile** Head and breast washed buffish; prominent dusky loral area; upperparts, wing-coverts and tertials dark brown, fringed off-white and notched buff. **VOICE** Generally silent. Flight call a low, nasal *kirruk* or *kvip*, and a sharper, also nasal, *keweep*. [CD2:46] **HABITS** In small groups or singly on tidal mudflats and sandy shores; occasionally at inland freshwater habitats. In large flocks on main wintering grounds. Probes deep in mud. **SIMILAR SPECIES** Black-tailed Godwit (q.v. and above); non-breeding Black-tailed appears plainer (Bar-tailed more streaky). Distant Whimbrel difficult to separate from non-breeding Bar-tailed Godwit in flight if bill not seen, but has more uniform upperwing, wings slightly broader and wingbeats slower; flies with head held up. **STATUS AND DISTRIBUTION** Common to scarce Palearctic passage migrant and visitor (mainly Sep–Apr; some year-round) to entire coast. Locally very common from Mauritania to Guinea; a few inland records, Ghana, Nigeria and Chad. Also Cape Verde Is (uncommon). Vagrant, Príncipe (Jan 1995).

WHIMBREL *Numenius phaeopus*

Plate 39 (8)

Courlis corlieu

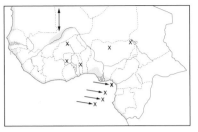

L 40–46 cm (16–18"). Large, streaky brown wader with long, decurved bill. In flight, **nominate** has white wedge on back and barred tail; upperwing rather dark and uniform. *N. p. hudsonicus* has entirely brownish upperparts, lacking white wedge, and dark (not white) underwings. **Adult male** Head pattern distinctive. Crown dark brown with pale stripe on centre; pale supercilium bordered below by dusky eye-stripe. Bill black with pinkish base. **Adult female** Bill slightly longer. **Juvenile** Similar, but more neatly patterned; upperparts, wing-coverts and tertials dark brown fringed with buff notches. **VOICE** Distinctive. A clear, loud, rapid *bi-bi-bi-bi-bi-bi-bi*. Occasionally also a Eurasian Curlew-like *kurrlee* and a hard, rasping *krrreep*. [CD2:47] **HABITS** Singly or in small, loose groups on tidal mudflats, sandy and rocky shores, estuaries and lagoons; occasionally on dry ground inland. Associates with Eurasian Curlews and other waders. **SIMILAR SPECIES** Eurasian Curlew is larger and paler, with longer and evenly decurved bill, longer legs and no bold stripes on head. Distant Bar-tailed Godwit difficult to separate in flight if bill not seen, but dark outer wing contrasts with paler inner wing, wings slimmer and wingbeats quicker, neck longer and almost drooping. **STATUS AND DISTRIBUTION** Common Palearctic passage migrant and visitor (mainly Aug–Apr; some year-round) to entire coast. Rarely inland except in Mali, where locally uncommon. Also Cape Verde Is (common), Bioko (rare), São Tomé and Príncipe (common) and Annobón. N American *hudsonicus* is vagrant to Cape Verde Is (São Nicolau, Feb 1991) and Sierra Leone.

SLENDER-BILLED CURLEW *Numenius tenuirostris*

Plate 42 (9)

Courlis à bec grêle

L 36–41 cm (14–16"). Small curlew, differing from Eurasian Curlew and Whimbrel by shorter, straighter, thinner, all-black bill and whiter underparts with clearly defined streaks and spots. Breast and flanks have *rounded spots* (diagnostic). Head has darker crown and indistinct median crown-stripe, pale supercilium and dusky loral stripe. **Juvenile** Underparts streaked; spots acquired during first winter. **VOICE** A short *kur-lee*, higher pitched than Eurasian Curlew. Alarm a sharp *kwee*. [CD2:48] **HABITS** In winter, frequents grassy areas near lagoons and estuaries, and inland marshes. **SIMILAR SPECIES** Compare similar-sized and darker Whimbrel and distinctly larger Eurasian Curlew. **STATUS AND DISTRIBUTION** A single, uncertain sighting from L. Chad (not accepted by *BoA*). There is a (remote) possibility that this extremely rare Palearctic species visits the region. *TBW* category: THREATENED (Critical).

EURASIAN CURLEW *Numenius arquata*

Plate 39 (9)

Courlis cendré

Other names E: Curlew, European Curlew
L 50–60 cm (20–23"). Largest, streaky brown wader, with *very long, evenly decurved bill*. In flight, white wedge on back and barred tail; dark outer wing contrasts with paler inner wing. Eastern *orientalis* has even longer bill than western **nominate**; underwing-coverts and axillaries largely plain white (white with variable brown barring in nominate), contrasting more with typically more strongly streaked flanks. **Adult male** Entirely streaked and barred dark grey-brown. Bill black with pinkish base. **Adult female** Bill slightly longer. **Juvenile** Similar; bill often shorter. **VOICE** Distinctive. A clear, liquid, bubbling *kur-lee* and *kwurrrr-lee*; also a shorter *kwee-kwee*. [CD2:49] **HABITS** Singly or in small groups on tidal mudflats and estuaries; inland on lakeshores and river banks. Forages by probing in mud or picking items from surface. Flies with rather slow wingbeats, recalling gull. **SIMILAR SPECIES** Whimbrel is smaller and darker, with shorter bill (more abruptly kinked at tip), shorter legs and boldly striped head pattern. **STATUS AND DISTRIBUTION** Common to scarce Palearctic passage migrant and visitor (mainly Aug–Apr; some year-round) to entire coast. Rare to locally uncommon inland. Rare, Cape Verde Is (nominate), São Tomé and Príncipe. Available data suggest nominate winters south to NW Mauritania (Banc d'Arguin), with *orientalis* occupying (most of) wintering grounds elsewhere.

UPLAND SANDPIPER *Bartramia longicauda*

Plate 42 (5)

Maubèche des champs

Other name F: Bartramie
L 28–32 cm (11.0–12.5"). Distinctively shaped wader with *small head, long thin neck, shortish straight bill* with yellowish base to lower mandible and *long tail* extending well beyond wingtips. Dark-streaked crown with narrow pale median stripe; plain face with contrasting dark beady eye. Throat and upper breast streaked brown; flanks with

arrow-shaped markings. Legs yellowish. In flight, uniformly dark above; feet do not project beyond long tail. **Adult** Scapulars, wing-coverts and tertials barred blackish and fringed buff. **Juvenile** Upperparts more neatly patterned, with pale fringes and dark subterminal crescents. **VOICE** A liquid *kwee-lip* and a clear, bubbly *kwip-ip-ip-ip*. [CD2:50] **HABITS** Favours grassy areas, foraging in plover-like fashion. **STATUS AND DISTRIBUTION** N American vagrant, NW Mauritania (May 1986) and Gabon (Oct–Nov 1988).

SPOTTED REDSHANK *Tringa erythropus*   Plate 41 (7)
Chevalier arlequin

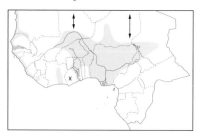

L 29–32 cm (11.5–12.5"). Elegant wader with *long red legs* and *long fine bill* with reddish base to lower mandible and slight but distinct downward kink at tip. In flight, white wedge on back and plain wings. **Adult non-breeding** Pale grey above; white below with pale grey wash on breast. White supercilium highlighted by dusky eye-stripe. Legs red. **Adult breeding** Unmistakable. Entirely sooty-black with white spots and notches on upperparts. Underwing and back white. Legs dusky red. In moult has black-blotched plumage. **Juvenile** Head greyish-brown with prominent, short whitish supercilium and dark eye-stripe; upperparts greyish-brown finely speckled white. Underparts whitish, densely and finely streaked, mottled and barred grey-brown. Legs orange-red. **VOICE** Distinctive. A loud, abrupt *chu-wit!* Alarm a short *chip*. [CD2:51] **HABITS** Singly or in small flocks, favouring estuaries, lagoons, mangroves, saltpans and various freshwater habitats. Often wades deeply; frequently swims and up-ends. **SIMILAR SPECIES** Common Redshank is less elegant, with shorter bill and legs; bill red with black tip; in flight, white panel on secondaries. Does not wade as deeply in water. Common Greenshank has similar pattern in flight, but wings darker, contrasting with pale head; rump and tail whiter. **STATUS AND DISTRIBUTION** Not uncommon to scarce Palearctic passage migrant and visitor (mainly Oct–Apr), from Mauritania to Liberia east to Chad. Rare, Cape Verde Is.

COMMON REDSHANK *Tringa totanus*   Plate 41 (6)
Chevalier gambette

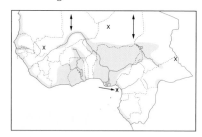

Other name E: Redshank
L 27–29 cm (10.5–11.5"). *T. t. totanus.* Medium-sized wader with *orange-red legs* and *red-based bill*. In flight, distinctive *broad white trailing edge to wing* (visible above and below) and white wedge on back. **Adult non-breeding** Rather plain grey-brown above; white eye-ring; underparts white washed pale grey-brown and finely streaked on breast; flanks sparsely barred. **Adult breeding** Acquires some dark barring on upperparts; underparts densely streaked brown. **Juvenile** Upperparts grey-brown fringed with buff spots and notches; underparts streaked brown, heavily on foreneck and breast, sparsely on flanks. Bill base and legs duller than adult. **VOICE** Vocal. A mournful, drawn-out *tiuuu* and a clear, ringing *tiu-lu tiu-lu-lu*, recalling Common Greenshank but distinctly higher pitched. [CD2:52] **HABITS** Singly or in small groups, frequenting tidal mudflats, estuaries, lagoons, mangroves, saltpans and various freshwater habitats. Forages mainly by walking and picking on mud. **SIMILAR SPECIES** Spotted Redshank, the only similar-sized wader with red legs, is more elegant, with longer legs and longer, finer bill; wings uniform. Non-breeding Spotted Redshank paler, greyer and with more prominent supercilium. Ruff sometimes has orange-red legs but has scaly upperparts and often small whitish area at base of slightly decurved bill; flight pattern different. **STATUS AND DISTRIBUTION** Palearctic passage migrant and visitor throughout (mainly Aug–Apr; some year-round). Common to very common from Mauritania to Guinea; uncommon to scarce elsewhere. Also Cape Verde Is (uncommon). Vagrant, Bioko.

MARSH SANDPIPER *Tringa stagnatilis*   Plate 41 (10)
Chevalier stagnatile

L 22–25 cm (9–10"). Recalls Common Greenshank, but distinctly smaller and more delicate, with *straight, very fine, needle-like dark bill* and relatively longer legs. In flight, plain dark upperwing and long white wedge on back; feet project well beyond tail. **Adult non-breeding** Plain pale grey above with *white forehead and supercilium* (creating distinct white-faced appearance); underparts white. Often has dark carpal area (shoulder). Legs dull greenish. **Adult breeding** Crown and neck finely streaked brown-grey; upperparts brown-grey blotched black; breast spotted; flanks with some V-shaped markings. Legs often tinged yellowish. **Juvenile** Crown and hindneck finely streaked

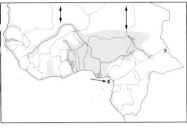

brown-grey; upperparts brown-grey narrowly fringed white. **VOICE** A clear *keeuw* or *kyu*, singly or repeated up to 3–4 times, reminiscent of Common Greenshank, but less ringing. Alarm a sharp, rapidly repeated *chip* or *kiup*. [CD2:53] **HABITS** Singly or in small groups, frequenting tidal mudflats, estuaries, lagoons, and muddy edges of various freshwater habitats. Forages preferably on soft mud, picking small prey from surface. **SIMILAR SPECIES** Common Greenshank is larger, less elegant, with stouter, slightly upturned bill. **STATUS AND DISTRIBUTION** Uncommon to locally not uncommon Palearctic passage migrant and visitor throughout (mainly Aug–Apr;

some year-round). Rare, Bioko. Vagrant, Cape Verde Is.

COMMON GREENSHANK *Tringa nebularia*   Plate 41 (8)
Chevalier aboyeur

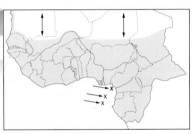

Other name E: Greenshank
L 30–34 cm (12.0–13.5"). Largest *Tringa*. Medium-sized wader with *dull greenish legs* and long, stout, *slightly upturned, greyish bill*. In flight, plain dark upperwing and broad white wedge on back; feet project slightly beyond tail. **Adult non-breeding** Head, hindneck and breast-sides white finely streaked grey; upperparts pale grey; underparts white. **Adult breeding** Head, neck and breast heavily streaked and spotted dark grey; upperparts blotched black. **Juvenile** Darker than adult. Head, neck and breast streaked dark grey; upperparts grey-brown fringed white. **VOICE** Distinctive. A loud, ringing *tiu-tiu-tiu*.

Occasionally also a rasping *kruip-kruip-kruip* when flushed, and a rapid *chip-chip-chip*. [CD2:54] **HABITS** Singly or in small groups frequenting tidal mudflats, estuaries, lagoons, mangroves, saltpans and various freshwater habitats. Forages at water's edge or in shallow water, picking from surface, probing, or scything in Pied Avocet-like manner. **SIMILAR SPECIES** Marsh Sandpiper is smaller, more elegant, with finer, straight bill and relatively longer legs. Spotted Redshank has similar pattern in flight, but more barred lower rump and uppertail-coverts. **STATUS AND DISTRIBUTION** Common Palearctic passage migrant and visitor throughout (mainly Aug–Apr; some year-round). Also Cape Verde Is (not uncommon), Bioko (rare), São Tomé and Príncipe (not uncommon).

LESSER YELLOWLEGS *Tringa flavipes*   Plate 41 (9)
Petit Chevalier

Other name F: Petit Chevalier à pattes jaunes
L 23–25 cm (9–10"). Resembles Common Redshank but distinctly slimmer, with finer, black bill and longer, *yellow legs*. In flight, dark upperwing and square white rump; feet project well beyond tail. **Adult non-breeding** Brown-grey above with white supraloral streak and narrow white eye-ring; wing-coverts speckled off-white and dark. Underparts white with foreneck and upper breast washed and finely streaked brownish-grey. **Adult breeding** Head, neck and upper breast heavily streaked; upperparts with blackish blotches. **Juvenile** Upperparts speckled pale buff. **VOICE** A clear *tew* or *tew-tew* (occasionally up to 4 notes), reminiscent of Common Greenshank but more subdued. **HABITS** Singly, frequenting tidal mudflats, lagoons and various freshwater habitats. **SIMILAR SPECIES** Marsh Sandpiper and Common Greenshank easily distinguished in flight by large white wedge on back. Wood Sandpiper has thicker, slightly shorter bill, longer supercilium delimiting dark cap, shorter primary projection and shorter legs. **STATUS AND DISTRIBUTION** N American vagrant (Nov–Mar), Gambia, Ghana, Nigeria and Cape Verde Is.

SOLITARY SANDPIPER *Tringa solitaria*   Plate 42 (10)
Chevalier solitaire

L 18–21 cm (7–8"). Similar to Green Sandpiper but slightly smaller and slimmer. Primary projection usually longer; wingtip protrudes clearly beyond tail, creating more attenuated rear. Best distinguished in flight, when *dark band on centre of rump and tail* is visible; tail-sides barred. **VOICE** A high-pitched *tewit-weet* or *pleet-weet-weet*, resembling call of Green Sandpiper. [CD6:43] **HABITS** As Green Sandpiper. **STATUS AND DISTRIBUTION** N American vagrant, Gambia (Jan 1999) and Cape Verde Is (Boavista, Mar 1997).

GREEN SANDPIPER *Tringa ochropus*   Plate 41 (1)
Chevalier cul-blanc

L 21–24 cm (8.5–9.5"). Medium-sized wader with contrasting, black-and-white-looking plumage. Resembles Wood

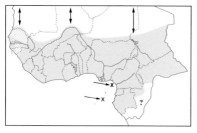

Sandpiper but rounder, with darker, rather plain-looking upperparts, darker breast sharply contrasting with white belly, shorter white supercilium (only extending to eye) and more distinct white eye-ring; bill slightly longer, legs shorter, greyish-green. In flight, plain *blackish upper- and underwings* contrast strongly with pure white belly and rump; tail broadly barred blackish; feet barely project beyond tail. **Adult non-breeding** Rather plain, dark grey-brown above. **Adult breeding** Darker above, with more distinct streaking on head and breast and more contrasting speckling. **Juvenile** Similar to adult non-breeding; upperparts finely speckled buffish. **VOICE** Distinctive. A loud, clear, ringing *tlooweet weet-weet!* uttered when flushed and in flight. [CD2:55] **HABITS** Usually singly in various freshwater habitats, foraging inconspicuously at muddy edges of ditches, pools, lakes and rivers. Bobs rear as Common Sandpiper. Flight fast with jerky wing action, reminiscent of Common Snipe. **SIMILAR SPECIES** Wood Sandpiper is slimmer, with longer legs, distinct supercilium and more distinct speckling on upperparts; also more active, less wary and has different call. Common Sandpiper is smaller and browner, with different flight action and call. **STATUS AND DISTRIBUTION** Common to not uncommon Palearctic passage migrant and visitor throughout (mainly Sep–Apr; some year-round). Rare, Cape Verde Is. Vagrant, Bioko and São Tomé.

WOOD SANDPIPER *Tringa glareola*   Plate 41 (2)
Chevalier sylvain

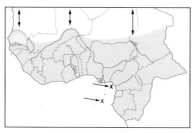

L 19–21 cm (7.5–8.0"). Smallish, slender wader with *long, prominent supercilium* and *densely speckled upperparts*. Bill medium-long, straight, dark. Legs greenish to dull yellowish. In flight, plain dark upperwings, square white rump and pale grey underwing; feet clearly project beyond tail. **Adult non-breeding** Dark-streaked crown bordered by pale supercilium; narrow whitish eye-ring; upperparts grey-brown extensively speckled whitish. Underparts white washed and streaked grey-brown on foreneck and breast. **Adult breeding** Darker and more distinctly speckled above. **Juvenile** Similar to adult non-breeding but more neatly marked above. **VOICE** Distinctive. A loud, ringing *chiff-iff-iff* or *chipipip*. Alarm a sharp *chip-chip-chip...*, often repeated. [CD2:56] **HABITS** In small groups or singly, frequenting various freshwater habitats, incl. ricefields, muddy edges of ditches, pools and lakes, marshes, flooded grassland and sewage outlets. Bobs rear as Common Sandpiper. **SIMILAR SPECIES** Green Sandpiper is plumper, more rounded; upperparts more contrasting with underparts and plainer; also more wary and has different call. **STATUS AND DISTRIBUTION** Common Palearctic passage migrant and visitor throughout (mainly Sep–Apr; some year-round). Also Cape Verde Is (uncommon), Bioko (rare) and São Tomé (rare).

TEREK SANDPIPER *Xenus cinereus*   Plate 41 (5)
Chevalier bargette

Other name F: Chevalier de Térek
L 22–25 cm (9–10"). Rather stocky wader, readily identified by *long, gently upturned bill* and *shortish, reddish-pink to orange-yellow legs*. In flight, distinctive wing pattern with dark leading edge and outer wing and whitish trailing edge; rump and tail grey. **Adult non-breeding** Rather plain pale grey above with dark carpal area; white below with pale grey breast-sides. Bill black with dull yellow or orange base. **Adult breeding** Prominent dark stripe on scapulars ('braces'). Bill black. **Juvenile** Similar to adult but has narrow buffish fringes to upperparts. **VOICE** A loud, melodious, ringing *eeb-eeb-eeb* or *du-du-du*. [CD2:57] **HABITS** Usually singly, occasionally several, frequenting tidal mudflats, estuaries and muddy edges of lakes and rivers. Often forages fast and actively, running in low horizontal posture over mudflats, but also walks slowly, picking at prey on surface. Associates with other waders. Occasionally bobs rear like Common Sandpiper. **STATUS AND DISTRIBUTION** Palearctic vagrant (Aug–May), NW Mauritania, Gambia, Mali, Ivory Coast, Ghana, Togo, Nigeria, Chad, N Cameroon and Gabon. Winters mainly on E and S African coasts.

COMMON SANDPIPER *Actitis hypoleucos*   Plate 41 (3)
Chevalier guignette

L 19–21 cm (7.5–8.0"). Smallish, rather compact wader readily identified by horizontal posture and *constant nervous bobbing of rear body*. Neck short; rear elongated with tail extending well beyond wingtips. Bill straight and dark; legs rather short, grey-green to dull yellowish-brown. Flight action distinctive: *low over water with rapid, fluttering wingbeats interspersed by short glides on stiff, bowed wings, revealing white wingbar*; rump and central tail dark,

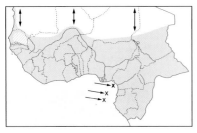

sides of tail white. Feet do not project beyond tail. **Adult non-breeding** Rather plain brown above with narrow white eye-ring and indistinct pale supercilium; white below with sharply demarcated brown breast; white extends onto breast-sides and forms wedge in front of closed wing. Narrow buff tips and blackish subterminal bars to wing-coverts visible at close range. **Adult breeding** Upperparts have blackish shaft streaks and irregular narrow bars. **Juvenile** Similar to adult non-breeding but narrow barring above more distinct. **VOICE** Distinctive. A rapid, clear, very high-pitched, piercing *tsee-wee-wee* or *hee-dee-dee* given in flight. Alarm a drawn-out *sweeeee-eet*. [CD2:58] **HABITS** Usually singly, frequenting muddy edges of pools, lakes, rivers, streams, ricefields, marshes, lagoons, mangroves and beaches. **SIMILAR SPECIES** Green Sandpiper is larger and darker, with shorter tail and no pectoral wedge; in flight, has different pattern and action. See also vagrant Spotted Sandpiper. **STATUS AND DISTRIBUTION** Common Palearctic passage migrant and visitor throughout (mainly Aug–Apr; some year-round). Also Cape Verde Is (not uncommon), Bioko (common), São Tomé and Príncipe (common).

SPOTTED SANDPIPER *Actitis macularia*   Plate 42 (8)
Chevalier grivelé

L 19–21 cm (7.5–8.0"). Very similar to Common Sandpiper but with *shorter tail*, which projects noticeably less beyond wingtips, and paler, yellowish legs. In flight, *shorter wingbar*, not extending onto inner secondaries (in Common, wingbar extends across all secondaries), and usually less white on tail-sides. **Adult non-breeding/juvenile** Patches on sides usually less prominent and less well-defined than in Common; foreneck and breast lack fine streaking. **Adult breeding** Unmistakable, with underparts boldly spotted black, pinkish bill tipped black and pinkish legs. **VOICE** A piping *peet!* or *tweet-weet-weet*, reminiscent of Green Sandpiper and lower pitched than Common Sandpiper. **HABITS** As Common Sandpiper. **STATUS AND DISTRIBUTION** N American vagrant, coastal Cameroon (Limbe area, Apr 2000) and Cape Verde Is (São Vicente, Feb–Mar 1999).

RUDDY TURNSTONE *Arenaria interpres*   Plate 38 (1)
Tournepierre à collier

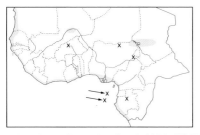

Other name E: Turnstone
L 21–25 cm (8–10"). *A. i. interpres*. Stocky, coastal wader with stubby, wedge-shaped, black bill and *short orange legs*. In flight, striking pied pattern with white wingbar, white stripe at base of wing and another on back, and white, black-banded tail. **Adult non-breeding** Mottled blackish-brown above and on breast; underparts white. **Adult breeding** Head and underparts white streaked black on crown with complex black pattern on face and breast. Upperparts rufous-chestnut and black. **Juvenile** Similar to adult non-breeding but duller; upperparts have pale fringes producing scaly appearance. **VOICE** A low, nasal *tuk* and a rattling *tukatukatuk*. [CD2:59] **HABITS** Usually in small groups frequenting rocky coasts, shorelines, mangroves, saltpans and muddy edges of lagoons; rarely inland waters. Forages actively, taking variety of prey, poking at rocks and tossing aside seaweed, pebbles and shells with bill. Tame. **STATUS AND DISTRIBUTION** Common to uncommon Palearctic passage migrant and visitor to entire coast (mainly Sep–Apr). Vagrant inland, in Mali, Burkina Faso, NE Nigeria and Chad. Also Cape Verde Is (common), São Tomé and Príncipe.

WILSON'S PHALAROPE *Phalaropus tricolor*   Plate 36 (2)
Phalarope de Wilson

L 22–24 cm (9"). Medium-sized, with relatively small head and long, needle-thin, black bill. In flight, plain upperparts and wings and white rump and uppertail-coverts. **Adult non-breeding** Plain, very pale grey above with white supercilium, small grey smudge in front of eye and grey line behind eye curving onto neck; white below. Legs medium-long and yellowish. **Juvenile** Upperparts dark brown with buff fringes. **Adult breeding** Unlikely to be encountered in our region. Distinctive pattern of grey, chestnut and black on head and upperparts. Legs black. **VOICE** Usually silent. [CD2:60] **HABITS** Frequents fresh and brackish water. Forages actively by running in crouched posture on muddy margins or sand bars, stabbing at small prey; swims less than other phalaropes. **SIMILAR SPECIES** Other phalaropes are smaller, shorter necked and shorter billed. Marsh Sandpiper may superficially appear similar, but has much longer, less yellow legs, white wedge on back and different, slimmer shape. **STATUS AND DISTRIBUTION** N American vagrant, Ivory Coast (Mar–Apr 1989). **NOTE** Sometimes placed in monotypic genus *Steganopus*.

RED-NECKED PHALAROPE *Phalaropus lobatus*   Plate 36 (1)
Phalarope à bec étroit

L 18–19 cm (7.0–7.5"). Resembles Red Phalarope in shape and behaviour, but more dainty, with *needle-thin, all-black bill*. In flight, more contrasting white wingbar. **Adult non-breeding** As Red Phalarope but upperparts slightly darker with more distinct white fringes. **Adult breeding** Head, neck and breast-sides grey with pale rusty-brown band from behind eye down neck-sides; throat and small spot above eye and on shoulder white; upperparts grey with 2 rusty-yellow Vs. Male duller than female. **Juvenile** Resembles juvenile Red Phalarope, but more boldly patterned, with distinct rusty-yellow Vs on dark brown upperparts. Post-juvenile moult starts in winter quarters. **VOICE** Usually silent. A short, sharp *kip* or *tsik*. Also a nasal chatter. [CD2:61] **HABITS** As Red Phalarope. **SIMILAR SPECIES** Red Phalarope has slightly thicker, blunter bill; for other differences see above. Wilson's Phalarope (q.v.) is noticeably larger; lacks dark eye patch in non-breeding plumage. **STATUS AND DISTRIBUTION** Palearctic and Nearctic vagrant (Oct–Apr), NW Mauritania, N Senegal, Gambia, Sierra Leone and Cape Verde Is. W Eurasian population winters in Arabian Sea and is uncommon to scarce in E Africa; no regular wintering area known in Atlantic.

RED PHALAROPE *Phalaropus fulicaria*   Plate 36 (3)
Phalarope à bec large

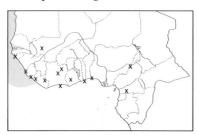

Other name E: Grey Phalarope
L 20–22 cm (8.0–8.5"). Small and active with shortish, straight bill. In flight, long white wingbar (broadest on inner wing). **Adult non-breeding** Head white with black eye patch and black hindcrown; upperparts plain, pale blue-grey; underparts white. Bill black, often with small yellowish patch at base. Appears very pale in flight. **Adult breeding** Rarely seen away from breeding areas. Distinctive combination of black crown, white head-sides and rufous-chestnut neck and underparts. Male duller than female. **Juvenile** Mainly dark brown above with whitish forehead and head-sides, darkish eye patch, buff fringes to upperparts and white fringes to coverts. Breast washed pale rusty, soon fading to white. Post-juvenile moult starts early; plain, pale grey feathers soon appear on upperparts. **VOICE** A sharp, high-pitched *kip* or *tsik*. [CD2:62] **HABITS** Mainly pelagic outside breeding season. On land in various aquatic habitats, incl. pools, lakes and sewage ponds. Forages by spinning buoyantly on water, picking rapidly at small items on surface. Tame. **SIMILAR SPECIES** Red-necked Phalarope has thinner, all-black bill; adult non-breeding slightly darker and less plain above. At sea, flying Sanderling has broader white wingbar and usually occurs in more compact flocks. **STATUS AND DISTRIBUTION** Rare Palearctic and Nearctic visitor (Jul–Apr), from Mauritania to Cameroon. Also Cape Verde Is (uncommon, Oct–May). Winters mainly at sea off W and SW Africa.

SKUAS
Family STERCORARIIDAE (World: 7 species; Africa: 5; WA: 5)

Strong, medium-sized to large, gull-like seabirds with mainly dark brown or brown-and-white plumage, and stout, dark bill, hooked at tip. Wings long and pointed in smaller *Stercorarius*, broader and blunter in larger *Catharacta*. Sexes similar; female slightly larger. Flight swift, purposeful and agile. Predatory and piratical. Feed mainly by harassing other seabirds, especially gulls and terns, to force them to disgorge or drop their catch. Generally silent away from breeding grounds. In W Africa, pelagic passage migrants and visitors. Their variable and confusing plumages render identification often problematic; with experience jizz distinctive.

Most likely to be confused with immature gulls, but skuas have a more powerful, steadier and faster flight, and usually appear darker (gulls have pale uppertail-coverts and lack contrasting pale wing patches). Smaller skuas are also more elegant and have more pointed wings. Features useful to separate the *Stercorarius* species include structure and flight action; also presence or absence of tail projection and of prominent pale areas in plumage.

POMARINE SKUA *Stercorarius pomarinus*   Plate 43 (3)
Labbe pomarin

Other name E: Pomarine Jaeger
L 46–51 cm (18–20"); WS 125–135 cm (49–53"). Largest and most powerful of the smaller skuas (slightly smaller than Lesser Black-backed Gull). In flight, *noticeably heavier than Arctic Skua, with broader based wings and more rounded breast*. Bill slightly heavier and *distinctly two-toned*. Pale primary patch as Arctic but much smaller than in Great. **Adult breeding** Readily identified by *elongated, spoon-shaped central tail feathers* (projecting up to 11 cm beyond rest

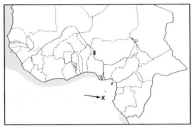

of tail). Bill blue-grey tipped black. **Pale morph** has blackish cap extending below gape; head-sides yellowish; upperparts blackish-brown. Underparts white with *dark-mottled breast-band* (often incomplete or absent in males), dark flanks and dark undertail-coverts. Rare **dark morph** entirely blackish-brown except pale primary patches. **Adult non-breeding** Variable. Most have head-sides darkish and upper- and undertail-coverts barred; elongated tail feathers shorter and often lacking. Underwing-coverts all dark. **Juvenile** Blackish-brown to mid-brown (appearing very dark at distance) with paler barring on underwings, uppertail-coverts and vent, *which are typically paler than, and contrast with, rest of plumage.* Pale crescent at base of primary-coverts visible at fairly close range, creating *distinctive double pale patch on underwing.* Head usually unstreaked but sometimes lightly barred and rarely with paler nuchal band (typical of Arctic). Upperparts scaly. Blunt central tail feathers do not, or only just, project beyond rest of rectrices but create triangular tail shape. Bill bluish or pale brownish tipped black. **Immature** Adult plumage gradually acquired in 3–5 years. **HABITS** Chases seabirds up to size of large gull; attacks shorter than those of Arctic, with slower, more laboured flight. Unlike Arctic, often attacks victim itself. Also frequently feeds on self-caught fish and carrion. [CD2:63] **SIMILAR SPECIES** Arctic Skua (q.v. and above). Great Skua distinguished from dark and immature Pomarine by more compact and heavier body, broader wings with blunter tip and generally distinctly larger and whiter patches, and relatively shorter tail; also lacks barring on uppertail-coverts, vent or underwing (these areas thus never contrast); top of head often slightly darker, forming cap. **STATUS AND DISTRIBUTION** Holarctic passage migrant and visitor (mainly Oct–Apr; records in all months). Winters mainly off W African coast between *c.* 20° and 8°N. Not uncommon from Mauritania to Ghana; rare or vagrant elsewhere. Also Cape Verde and Gulf of Guinea seas (uncommon/scarce).

ARCTIC SKUA *Stercorarius parasiticus* Plate 43 (2)
Labbe parasite

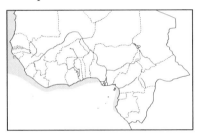

Other name E: Parasitic Jaeger
L 41–46 cm (16–18"); WS 110–120 cm (43–47"). Agile and rather elegant with *fast, falcon-like flight.* Noticeably slimmer than Pomarine Skua, with *more slender wings (narrower at base),* smaller head and slimmer bill. Pale primary patch as Pomarine, but typically lacks pale crescent at base of primary-coverts. **Adult breeding** Distinguished by *elongated, narrow, pointed central tail feathers* (projecting up to 10 cm beyond rest of tail). Plumage variation as Pomarine. **Pale morph** has blackish cap *not* extending below gape; head-sides paler yellowish; upperparts dark greyish-brown. Underparts white with *pale grey breast-band* (variable; often incomplete or absent), greyish flanks and blackish undertail-coverts. Bill blackish, sometimes with paler base, but rarely as two-toned as Pomarine. **Dark morph** entirely blackish-brown to grey-brown except pale primary patches. Intermediates have dark cap contrasting with paler head-sides. **Adult non-breeding** Variable. Mostly with head-sides darkish and upper- and undertail-coverts barred; elongated tail feathers often absent. Underwing-coverts all dark. **Juvenile** Very variable. Blackish-brown to warm rufous-brown. Barring on underwings, uppertail-coverts and vent more irregular and wavy than Pomarine. Head narrowly streaked dark and *often has warm rufous nuchal band* (diagnostic); upperparts scaly, frequently tinged rusty; primary tips usually edged buffish (visible on closed wing). *Central tail feathers pointed* and only slightly projecting beyond rest of tail. Bill blue-grey or pinkish-grey tipped black. **Immature** Juvenile plumage gradually lost in 2nd and 3rd calendar years. **HABITS** Mostly reliant on kleptoparasitism. Chases terns and small gulls for up to several minutes. Unlike Pomarine, rarely attacks victim. [CD2:64] **SIMILAR SPECIES** Pomarine Skua (q.v. and above). Long-tailed Skua is smaller and slimmer, more elongated, with narrower wings possessing only 2 pale shafts on leading edge of wing, and more tern-like flight. **STATUS AND DISTRIBUTION** Holarctic passage migrant and visitor (mainly Sep–Apr; records in all months). Winters mainly off Namibia and W South Africa. Not uncommon from Mauritania to Ghana; rare or vagrant elsewhere. Also Cape Verde and Gulf of Guinea seas (rare/scarce).

LONG-TAILED SKUA *Stercorarius longicaudus* Plate 43 (1)
Labbe à longue queue

Other name E: Long-tailed Jaeger
L 48–53 cm (19–21"); WS 105–115 cm (41–45"). Small and slim (about size of Black-headed Gull but often appearing larger) with *buoyant, almost tern-like flight.* Head smaller than Arctic Skua, bill shorter, body and tail more elongated. Upperwing has 2 pale shafts on dark primaries, forming pale line on leading edge. **Adult breeding**

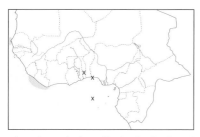

Central tail feathers greatly elongated, very narrow and pointed (projecting up to 24 cm beyond rest of tail). Well-defined, neat black cap; head-sides and breast pale creamy becoming grey on belly; *upperparts greyish with contrasting black flight feathers.* Bill blackish. No dark morph. **Adult non-breeding** Variable. Most have darkish head-sides and dark barring or spotting on breast (creating band), flanks, and upper- and undertail-coverts; elongated tail feathers much shorter and often absent. Underwing-coverts all dark. **Juvenile** Very variable. Palest have whitish head and belly, darkest mainly sooty-black. Typically cold brownish-grey, *always lacking warm rufous-brown tones* often shown by juvenile Arctic. Dark and pale barring on underwings, uppertail-coverts and vent. Upperparts scaly; fringes whitish (more contrasting and straighter than in Arctic). *Central tail feathers narrow but rather blunt tipped* and *clearly projecting* beyond rest of tail (projection up to 4 cm). Bill greyish tipped black. **Immature** Juvenile plumage gradually lost in 2nd and 3rd calendar years. **HABITS** Mostly reliant on kleptoparasitism during migration, briefly chasing terns and small gulls, but frequently feeds on self-caught fish. Also catches insects and picks items from surface. [CD2:65] **SIMILAR SPECIES** Arctic Skua (q.v. and above). **STATUS AND DISTRIBUTION** Holarctic passage migrant (mainly Aug–Sep and Mar–Apr). Winters mainly off Namibia and W South Africa. Rare to scarce or vagrant, off Mauritania, Senegal, Liberia, Togo and Nigeria; also Cape Verde and Gulf of Guinea seas. Highly pelagic, thus rarely observed from coast; migration routes poorly known.

GREAT SKUA *Catharacta skua*   Plate 43 (4)
Grand Labbe

L 51–58 cm (20–23"); WS 135–145 cm (53–57"). Largest skua. *Heavy, compact* and entirely dark with relatively short tail and *broad wings* with *conspicuous white patches at base of primaries above and below.* All plumages similar. **Adult** Coarsely streaked yellowish-buff. Sometimes has dark cap slightly contrasting with rest of head. Bill heavy, blackish. Legs black. **Juvenile** Separable at close range. Darker and more uniform, less streaky than adult, generally more chestnut-brown below; white wing crescent often slightly narrower and less conspicuous above. **HABITS** Chases variety of seabirds, up to size of gannets; also takes carrion and fish offal. [CD2:66] **SIMILAR SPECIES** Dark immature gulls (e.g. similar-sized Lesser Black-backed Gull) are less heavy and stocky, with more relaxed flight action, slimmer wings (broader at base in Great Skua), without contrasting wing patches. For differences with dark and immature Pomarine, see that species. **STATUS AND DISTRIBUTION** Palearctic passage migrant and visitor (mainly Sep–Mar). Winters mainly in Atlantic south to W Africa. Uncommon, Mauritania. Rare or vagrant, Senegal, Liberia, Ivory Coast, Ghana and Nigeria; also Cape Verde seas and Annobón. Some records may refer to South Polar Skua.

SOUTH POLAR SKUA *Catharacta maccormicki*   Plate 43 (5)
Labbe de McCormick

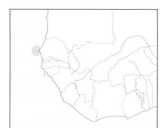

L 51–54 cm (20–21"); WS 125–135 cm (49–53"). Very similar to Great Skua but slightly smaller and more slender, with relatively smaller head and slightly slimmer bill. Plumage variable but *unstreaked*, and typically with paler nape (often hard to discern at sea) and lacking dark cap of Great. Occasionally has indistinct, small pale area at base of bill. **Adult dark morph** Wholly cold greyish-brown with paler nape. Darkest birds virtually impossible to separate from some juvenile Great Skuas. **Adult intermediate morph** Head and underparts pale brown; upperparts blackish-brown. **Adult pale morph** Distinctive. Head, mantle and underparts plain, pale buff-brown. **Juvenile** Similar to adult of respective morph. **HABITS** Feeds mostly by fishing. Less reliant on kleptoparasitism than other skuas, and short attacks, on e.g. shearwaters, apparently less successful. **SIMILAR SPECIES** Great Skua (q.v. and above). **STATUS AND DISTRIBUTION** Poorly known. Breeds in Antarctica. Immatures migrate to northern hemisphere in southern winter, performing clockwise migration. Southwestward migrants observed in numbers off Senegal. Presumably regular passage migrant off W African coast; status unclear due to confusion with Great Skua.

GULLS
Family LARIDAE (World: *c.* 51 species; Africa: 23; WA: 17)

Robust, medium-sized to large birds of coast, oceans and inland waters with predominantly white plumage and grey or black wings and mantle. Wings long and narrow, tail usually slightly rounded, bill powerful. Distinct breeding, non-breeding and immature plumages. Adult plumage reached in 2nd–4th year; period of immaturity generally related to size. Tail white in adults; dark tail band indicative of immature. Sexes similar, male slightly larger. Flight strong and buoyant with slow wingbeats and much gliding and soaring. Often swim, rarely dive. Usually vocal and gregarious, breeding colonially. Most are omnivorous aquatic and littoral scavengers, fishers and predators. Several are long-distance migrants.

In W Africa only one species resident, rest migrants or vagrants, mostly from Palearctic or Nearctic, one from southern hemisphere. At least two Palearctic species have bred in the region.

To identify gulls, note wing and head patterns, and bare-part coloration. Understanding moult greatly facilitates identification. First moult is post-juvenile: a partial moult, replacing head and body feathers and resulting in first-winter plumage. Subsequently, there are two moults per year: a partial one from winter to summer plumage, mainly replacing head and body feathers (flight feathers and all or most tail feathers thus still juvenile and much abraded in first-summer plumage), and a complete one after the breeding season, resulting in adult or immature winter plumages. Sabine's and Franklin's Gulls (q.v.) are the only exceptions to this rule. For detailed plumage descriptions of all species recorded in W Africa, except Kelp Gull, see Grant (1986).

MEDITERRANEAN GULL *Larus melanocephalus*   Plates 44 (12) & 46 (9)
Mouette mélanocéphale

L 36–38 cm (14–15"); WS 92–100 cm (36.0–39.5"). Slightly larger and heavier than Black-headed Gull, with stouter bill and longer legs. Adult plumage acquired in 3rd winter. **Adult breeding** *Very white-looking* with black hood extending on nape, white crescents above and below eye, blood-red bill (with black band near tip), and red legs. In flight striking: the only gull in the region with *all-white primaries and underwing*. **Adult non-breeding** Hood replaced by dark wedge-shaped patch on ear-coverts, extending diffusely over crown. **First-winter** Head as adult non-breeding, upperparts very pale grey, outer primaries and coverts blackish forming broad dark leading edge, brown carpal bar, black secondary bar, broad blackish tail band. Bill blackish, usually with paler base. Legs variable, blackish to orange. **First-summer** Hood variably developed, wing and tail pattern faded. **Second-winter** As adult non-breeding, but variable amount of black in tips of outer primaries. Bill colour variable (pink to reddish, dark tipped or with black subterminal band); legs usually dark reddish. **Second-summer** Hood fully developed, sometimes with some white flecks. [CD2:67] **SIMILAR SPECIES** First-winter Common Gull distinguished from 1st-winter Mediterranean by dirty brown head markings (no distinct patch on ear-coverts), darker grey upperparts, less contrasting upperwing pattern, less clean white underwing, broader tail band, more slender bill with clear-cut black tip, and pink legs. Black-headed differs, at all ages, in structure and wing pattern. **STATUS AND DISTRIBUTION** Scarce to rare Palearctic winter visitor, coastal Mauritania (mainly Nov–Mar) and Senegambia (Nov–Feb; also May); mainly immatures

LAUGHING GULL *Larus atricilla*   Plates 44 (5) & 46 (6)
Mouette atricille

L 38–41 cm (15–16"); WS 100–125 cm (39.5–49.0"). Rather small and slim (only slightly larger than Black-headed Gull) with relatively long wings, legs and bill, the latter often appearing drooping. Adult plumage acquired in 3rd winter. **Adult breeding** Black hood; white crescents above and below eye; ashy-grey upperparts (darker than Common Gull); black wingtips (*lacking mirrors*). Bill and legs dull red. **Adult non-breeding** Head mainly white with variable dusky markings on ear-coverts and over crown and nape. Bill and legs blackish. **First-winter** Hindneck, breast and flanks extensively tinged dirty grey; grey upperparts; brown inner wing-coverts; blackish flight feathers and tail band; underwing with darkish markings. **First-summer** Variably developed hood; wing markings faded. **Second-winter/-summer** As adult breeding/non-breeding, but black on wing more extensive, and often have traces of secondary bar and/or tail band. In summer plumage hood may still be only partially developed. [CD6:44] **SIMILAR SPECIES** Franklin's Gull (q.v.) is smaller with more rounded head, more conspicuous eye-crescents, shorter bill, and shorter wings appearing more rounded at tip. Immature differs by dark half-hood and white hindneck and underparts; 1st-winter has paler upper and underwing, narrower tail band and white outer tail feathers. Flight more buoyant. **STATUS AND DISTRIBUTION** American vagrant, coastal Senegambia (Dec–Apr). One may have paired with Grey-headed Gull (Saloum delta, Mar 1983). Also claimed from Mauritania (Apr).

FRANKLIN'S GULL *Larus pipixcan*
Mouette de Franklin

L 32–36 cm (12.5–14.0"); WS 85–95 cm (33.5–37.5"). Small and rather dumpy, with shortish wings, tail, bill and legs, and conspicuous white eye-crescents in all plumages. In winter, darker headed than any other similar-sized gull (but beware transitional stages of moult). The only gull to have two complete moults per year (only post-juvenile moult is partial); adult plumage normally acquired in 2nd summer. **Adult breeding** Black hood; upperparts slate-grey; primaries prominently tipped white; black subterminal tip separated by white band from rest of upperwing. Bill and legs red. **Adult non-breeding** *Blackish rear of head* (forming 'half-hood') and *broad white eye-crescents* diagnostic. Bill and legs darkish. **First-winter** Half-hood; brown inner wing-coverts; blackish flight feathers and tail band (the latter not extending to outer tail feathers). **First-summer** As adult non-breeding, but black on primaries extending along leading edge and not isolated by white band, white tips smaller. **Second-winter** Even more like adult non-breeding (many may be inseparable), but typically has more extensive black wingtips. [CD6:45] **SIMILAR SPECIES** Laughing Gull (q.v.) is larger, with longer head and bill, and longer, more pointed-looking wings. First-winters have hindneck, breast and flanks extensively washed brownish-grey, wholly blackish primaries, dark-marked underwings and broader tail band. **STATUS AND DISTRIBUTION** N American vagrant, coastal Senegambia (Jan–Apr). One paired with Grey-headed Gull (Saloum delta, May 1983).

LITTLE GULL *Larus minutus*
Mouette pygmée

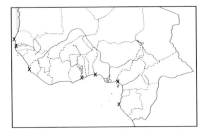

L 25–27 cm (10.0–10.5"); WS 75–80 cm (30.5–31.5"). Distinctive, tiny gull (the smallest of its family) with rounded-looking wings and buoyant, surface-picking feeding flight, recalling marsh tern. Adult plumage acquired in 3rd year. **Adult breeding** The only black-hooded gull lacking white eye-crescents; upperparts pale grey; *underwings blackish with white trailing edge*. Bill and legs red. **Adult non-breeding** Head white with dark crown/nape and blackish ear-spot. Bill blackish. Legs dark red. **First-winter** Head white with blackish 'cap' and ear-spot. In flight, wings more pointed than adult, with blackish W above (formed by black outer primaries and carpal bar); underwings white; tail with narrow black band. **First-summer** Variably developed hood. **Second-winter/-summer** As adult breeding/non-breeding, but variable blackish markings on tips of outer primaries; summer plumage with hood often flecked white. [CD2:68] **SIMILAR SPECIES** First-winter Black-legged Kittiwake differs from 1st-winter Little Gull in having black on primaries more clear-cut, all-white secondaries and no dark crown; flight steadier. **STATUS AND DISTRIBUTION** Scarce Palearctic migrant, Mauritania. Vagrant, coastal Senegambia, Sierra Leone, Ghana, Nigeria, Cameroon and Gabon.

SABINE'S GULL *Larus sabini*
Mouette de Sabine

L 27–32 cm (11.0–12.5"); WS 90–100 cm (35.5–39.5"). Rather small with diagnostic *tricoloured, triangular-patterned upperwing*, slightly forked tail (often hard to see), and elegant, tern-like flight. Unique among gulls in having a *partial* moult in Nov–Dec (in wintering areas) and a *complete* one in Feb–Apr (before northward migration). Adult plumage acquired in 2nd winter. **Adult breeding** Dark grey hood bordered by black line; yellow-tipped black bill; darkish legs. In flight, striking grey, white and black wing pattern. **Adult non-breeding** Head largely white with blackish patch of variable extent on nape. **Juvenile** Head, upperparts, wing-coverts and breast-sides greyish-brown; clear-cut black tail band; all-black bill. **First-winter** Acquires grey upperparts and adult non-breeding head pattern. **First-summer** As adult breeding, but hood only partially developed, yellow bill tip smaller or lacking, and sometimes having traces of tail band. **HABITS** Highly pelagic and only occasionally seen from land. Gregarious; usually in small flocks. Often with Red Phalaropes. **SIMILAR SPECIES** First-year Black-legged Kittiwake differs in dark nape collar, diagonal bar across wing, and less dainty appearance. **STATUS AND DISTRIBUTION** Arctic passage migrant to W African seas (mainly Aug–Nov and Mar–Jun). Recorded along entire coast; also Cape Verde seas. Numbers difficult to assess due to its pelagic habits. Winters mainly off Namibia and South Africa. **NOTE** Sometimes placed in monotypic genus *Xema*.

GREY-HEADED GULL *Larus cirrocephalus*

Mouette à tête grise

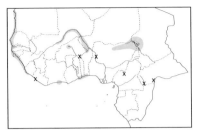

L 39–42 cm (15.5–16.5"); WS 100–115 cm (39.5–45.5"). *L. c. poiocephalus.* Medium-sized gull with diagnostic upperwing pattern in adult and *all-dusky underwing* at all ages. Adult plumage usually acquired in 3rd winter. **Adult breeding** Pale grey hood; ashy-grey upperparts. Eye pale. Bill and legs red. In flight, extensive black wingtips with white mirrors, bordered by broad white diagonal bar. **Adult non-breeding** Head largely white, often having some grey, with faint dusky ear-spot. **Juvenile** Head, upperparts and breast-sides have greyish-brown markings; brown carpal bar, black outer primaries, secondaries and subterminal tail band; white patch on middle of outer wing. Eye dark. Bill dull pinkish tipped dusky. Legs pink. **First-winter** Head mainly white with pale dusky 'headphones'; ashy-grey mantle and back. **First-summer** Variably developed hood; dark areas on wings and tail much faded. **Second-winter** As adult non-breeding but black on primary tips more extensive, white mirrors small or lacking, inner wing with dusky trailing edge. **Second-summer** As 2nd-winter, but hood usually fully developed. **VOICE** Vocal. Calls include harsh *kaarr* and short *kok-kok*. [CD2:69] **HABITS** Frequents coastal areas, wetlands, large rivers and lakes. Gregarious; rarely singly. **SIMILAR SPECIES** Black-headed Gull is smaller, with narrower, pointed wings having white leading edge; upperparts paler grey. Slender-billed Gull has similar size and structure, but is always much whiter, with a complete white leading edge to upperwing. **STATUS AND DISTRIBUTION** Common to rare resident and wanderer along coast, from Mauritania to Benin, and inland along major rivers from Mali to L. Chad; also recorded from SW Cameroon and SW CAR.

BLACK-HEADED GULL *Larus ridibundus*

Mouette rieuse

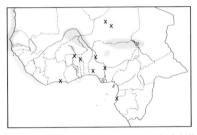

L 34–37 cm (13.5–14.5"); WS 100–110 cm (39.5–43.5"). Rather small and slender, with rounded head and thin bill. In flight slim, pointed wings with conspicuous white leading edge to outer wing (most extensive on adults). From below, primaries appear mainly dusky. Adult plumage acquired in 2nd winter. **Adult breeding** Diagnostic dark brown hood (looking blackish at distance) reaching only to nape; white crescents above and below eye. Bill and legs dark red. **Adult non-breeding** Head white with dusky ear-spot and 2 faint dusky bands over crown ('headphones'). **First-winter** Differs from adult non-breeding in brown carpal bar, blackish secondary bar and subterminal tail band, and dull pinkish bill with blackish tip. **First-summer** Hood variably developed (often white flecked); wing and tail markings faded. **VOICE** Vocal. Calls include harsh *krreeah* and *kwarr*, and short *kak, kok-kok-kok*. [CD2:70] **SIMILAR SPECIES** Grey-headed Gull is larger with broader wings; upperwing lacks complete white leading edge and has extensive black wingtips; underwing wholly dusky. Slender-billed Gull has characteristic, elongated profile with flat forehead and longer bill, neck and tail; in winter also whiter head. **STATUS AND DISTRIBUTION** Uncommon to rare Palearctic visitor (mainly Oct–Apr; records in all months) to coasts (mainly from Mauritania to Ghana) and Cape Verde Is. Also inland along Niger R. in Mali and Niger, and in Nigeria and Chad. Some may oversummer.

BONAPARTE'S GULL *Larus philadelphia*

Plate 47 (3)

Mouette de Bonaparte

L 28–30 cm (11–12"); WS 90–100 cm (35.5–39.5"). Small, elegant version of Black-headed Gull, with white underwing and mainly black bill. Plumage sequence as Black-headed (q.v.); adult plumage acquired in 2nd winter. **Adult breeding** Hood black (not dark brown); white crescents above and below eye more prominent; upperparts slightly darker grey. Bill black (not dark red), sometimes with reddish base. Legs more orange-red. In flight, wings also have white leading edge, but outer primaries white and translucent (not dusky) from below, with narrow black border. **Adult non-breeding** Ear-spot more distinct; hindneck and breast-sides tinged grey (not white). **First-winter** Upperwing markings more blackish and better defined, with neat black trailing edge to flight feathers. Bill appears all dark (not pale pinkish with dark tip). Legs pink. **First-summer** Variably developed hood; wing and tail markings faded. **HABITS** Food-picking from surface in dipping, buoyant flight recalls Little Gull or marsh tern. **SIMILAR SPECIES** Black-headed Gull (q.v. and above). Little Gull is smaller and has neat dark cap in winter plumages. **STATUS AND DISTRIBUTION** N American vagrant, N Senegal (Jan 1986). Also claimed from Mauritania (Apr 1987).

SLENDER-BILLED GULL *Larus genei*
Goéland railleur

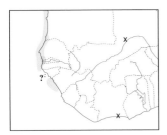

L 42–44 cm (16.5–17.5"); WS 100–110 cm (39.5–43.5"). Medium-sized, white-headed and long-necked gull with distinctive *elongated head*, produced by gently sloping forehead and longish bill. Adult plumage acquired in 2nd winter. **Adult breeding** Head all white; upperparts pale grey; underparts washed pink. Pale eye. Dark reddish bill (looking black at distance) and legs. In flight, white leading edge to outer wing. **Adult non-breeding** Faint dusky ear-spot (often lacking). **First-winter** Very faint dusky ear-spot, pale brown carpal bar, blackish secondary bar and tail band. Dark eye. Pale bill (often with darkish tip) and legs. **First-summer** As 1st-winter, but wing and tail markings faded. [CD2:71] **SIMILAR SPECIES** Black-headed Gull also has white leading edge in flight, but is slightly smaller and has rounder head, and shorter bill and neck. First-winter Black-headed differs in having darker wing markings, ear-spot, bill and legs. Grey-headed Gull has darker grey upperparts and different wing pattern. **STATUS AND DISTRIBUTION** Locally not uncommon resident and intra-African migrant, breeding in Mauritania and Senegal. Also Palearctic winter visitor, to coasts of Mauritania to Guinea. Vagrant, Mali, Ivory Coast, Niger, Nigeria and Cape Verde Is.

AUDOUIN'S GULL *Larus audouinii*
Goéland d'Audouin

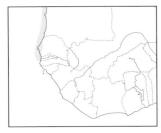

L 48–52 cm (19.0–20.5"); WS 115–140 cm (45–55"). Smallest, *palest and most elegant of the large gulls* in W Africa. Adult plumage acquired in 4th year. **Adult** Very pale with sharply contrasting black wingtips. In flight, *obviously paler than Yellow-legged Gull* (lacking prominent white leading and trailing edges to wing and white mirrors on wingtips), with *dark eye* and *dark-looking bill* (red, with black subterminal band and yellow tip) imparting less fierce expression. Legs olive-grey. **First-year** Whitish head contrasts with brownish upperparts; breast greyish-brown becoming white on belly. Diagnostic dark grey legs. In flight from above, flight feathers blackish, greater coverts dark, forming a second bar, tail black, contrasting with white U on lower rump and uppertail-coverts. From below, contrasting dark and pale bars on underwing-coverts, dark tail band. **Second-year** Upperparts obviously paler than corresponding Yellow-legged, with narrower tail band, dark legs and usually reddish-based bill. **Third-year** Much like adult, but black of wingtips extends onto greater primary-coverts and mirror on outer primary smaller or lacking. [CD2:72] **SIMILAR SPECIES** Yellow-legged Gull (q.v. and above). Also compare 1st-year with 1st-year Lesser Black-backed. **STATUS AND DISTRIBUTION** Rare migrant from Mediterranean to coastal Mauritania, Senegal and Gambia (Sep–May). *TBW* category: Near Threatened.

RING-BILLED GULL *Larus delawarensis*
Goéland à bec cerclé

Plate 47 (4)

L 43–47 cm (17.0–18.5"); WS 120–155 cm (47–61"). Intermediate in size and structure between Common and Yellow-legged Gulls, with rather flat head and stout, 'parallel', blunt-looking bill. Adult plumage acquired in 3rd winter. Plumage sequence and patterns similar to Common; differences as follows. **Adult breeding** Upperparts paler grey; white mirrors smaller. Eye pale. Bill yellow with clear black subterminal band. **Adult non-breeding** As breeding, but with dusky head markings. **First-winter** Usually heavily spotted on head, neck and breast-sides; narrow (not rather wide) pale edges to tertials. Heavy, pinkish bill with dark tip. In flight, grey on upperparts paler, tail band not clear-cut. **First-summer** Paler 'saddle', not contrasting with greater covert panel. **Second-winter** Single small white mirror (or none); tail often showing traces of subterminal band. Eye usually pale. **Second-summer** Bill colour and pattern nearing adult. **SIMILAR SPECIES** Common Gull (see above). Compare also Yellow-legged Gull. **STATUS AND DISTRIBUTION** N American vagrant, Senegal (Saloum delta, Oct 1985). Also claimed from Mauritania (Apr 1988).

COMMON GULL *Larus canus*
Goéland cendré

Other name E: Mew Gull
L 40–42 cm (16.0–16.5"); WS 110–130 cm (43.5–51.0"). *L. c. canus.* Medium-sized, fairly elegant gull with rounded head, slim bill and dark eye. Adult plumage acquired in 3rd winter. **Adult breeding** Upperparts rather dark blue-grey. Bill and legs yellow-green. In flight, black wingtips with relatively large white mirrors. **Adult non-breeding**

Head has dusky markings; bill pale with dark subterminal band. **First-winter** Head, breast and flanks variably mottled and streaked grey-brown; rest of underparts white. Blackish leading edge to outer wing, secondary bar and tail band. Bill pinkish with dark tip. Legs greyish-pink. **First-summer** Faded version of 1st-winter (blue-grey mantle and scapulars forming contrasting 'saddle'). **Second-winter** Uniformly grey wings, black leading edge to outer wing, small white mirrors. Bill greyish-green with dark tip. Legs greyish-pink. **Second-summer** Similar but head whiter, black areas faded. [CD2:73] **SIMILAR SPECIES** Adult Yellow-legged Gull is larger, with flatter crown, stout bill, yellow eye and smaller white mirrors on black wingtips. Adult Ring-billed Gull has paler grey upperparts and smaller white mirrors. First-year has paler grey upperparts and ill-defined tail band. Second-years have only one small mirror and often show traces of tail band. First-winter Mediterranean Gull differs from 1st-winter Common by dark eye patch, paler upperparts, more contrasting upperwing pattern and whiter underwing; bare-part coloration also different. **STATUS AND DISTRIBUTION** Palearctic vagrant to coastal Mauritania (Dec–May) and Senegambia (Sep, Dec–Jan, Jul). Probably under-recorded.

KELP GULL *Larus dominicanus*   Plates 44 (10) & 45 (4)
Goéland dominicain

L 58 cm (23"); WS 128–142 (50.5–56.0"). *L. d. vetula*. Large, heavy looking gull, with massive head, stout bill and dark eye. Adult plumage acquired in 4th year. **Adult** Upperparts black; white mirror on outermost primary. Bill yellow with orange-red spot on gonys. Legs olive-grey to olive-yellow. **First-year** Mainly dark brown. Outer wing and secondaries dull black; rump and uppertail-coverts barred and mottled brown and white; tail blackish. Bill blackish. Legs pinkish-brown. **Second-year** Head mainly white; 'saddle' dark brown; underparts white with brownish streaks; tail white at base. Bill and legs gradually becoming more yellowish. **Third-year** As adult, but retaining some traces of immaturity, esp. on wings. [CD6:46] **SIMILAR SPECIES** Lesser Black-backed Gull is more slender, with longer primary projection and less heavy bill. Adult has blackish-grey upperparts and yellow eye and legs; immature very similar and best separated by structure. Great Black-backed (q.v.) is larger and appears even bulkier and more powerful. **STATUS AND DISTRIBUTION** Scarce migrant from southern hemisphere, recorded from Gabon, Senegambia and Mauritania (Banc d'Arguin). Probably overlooked. Bred Senegal, 1983.

LESSER BLACK-BACKED GULL *Larus fuscus*   Plates 44 (8) & 45 (3)
Goéland brun

L 52–67 cm (20.5–26.5"); WS 135–155 cm (53–61"). Large but rather elegant, with slightly narrower, longer looking wings than other large gulls in W Africa. Adult plumage acquired in 4th year. Three races occur. **Adult breeding** Diagnostic combination of slate-grey (*graellsii*), blackish-grey (*intermedius*) or blackish (**nominate**) upperparts and yellow legs. Black wingtips with 1–2 small white mirrors. In flight from below, dusky-grey band across flight feathers (darker than Yellow-legged Gull). **Nominate** somewhat smaller and slimmer than other races. Eye yellow with narrow red orbital ring. Bill yellow with orange-red spot on gonys. **Adult non-breeding** Head and breast-sides have extensive grey-brown streaking (*graellsii*), or little or no streaking (**nominate**); *intermedius* moderately streaked. **First-winter** Mainly dark grey-brown. In flight from above, outer wing uniformly dark, rear of inner wing with 2 dark bars formed by blackish-brown secondaries and greater coverts (diagnostic); broad blackish tail band contrasting with whitish rump. Bill black. Legs dull pink. **First-summer** Head and underparts whiter; rest of plumage more faded. **Second-year** Adult-like mantle colour appears; head and body gradually become white, bare parts turn paler with yellowish tinge, bill has black subterminal band. **Third-year** As adult, but retaining some traces of immaturity on wings and tail; bare-part colours paler, bill with darkish spot on gonys. [CD2:74] **HABITS** Frequents coastal areas, major rivers, and lakes. Migrates singly or in small groups. **SIMILAR SPECIES** Slightly bulkier Yellow-legged Gull has similar size and plumage sequences, but is, in most plumages, distinguished by much paler upperparts. First-years may be more difficult to separate, esp. at distance, but upperwing has more contrasting pattern and tail band is less clear-cut. Compare also Great Black-backed and Kelp Gulls. **STATUS AND DISTRIBUTION** Common to uncommon Palearctic migrant to entire coast and inland (e.g. along Niger R.; trans-Saharan migration probable); also Cape Verde Is. Some oversummer throughout.

YELLOW-LEGGED GULL *Larus cachinnans*   Plates 44 (7) & 45 (5)
Goéland leucophée

L 55–67 cm (21.5–26.5"); WS 138–155 cm (54.5–61.0"). Large, powerful gull. Adult plumage acquired in 4th year. **Adult** *michahellis* **breeding** Fierce looking, with ashy-grey upperparts. Contrasting white leading and trailing

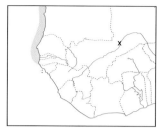

edges to wings; black wingtips with small white mirrors. Eye yellow with narrow red orbital ring. Bill yellow with orange-red spot on gonys. Legs yellow. **L. c. atlantis** is slightly darker grey above; black on wingtips more extensive. **Adult michahellis non-breeding** Similar, but head very faintly streaked (streaking lost from Dec/Jan onwards); legs sometimes yellowish-pink. **L. c. atlantis** more heavily streaked on head and neck. **First-winter** Mainly brownish, with pale head and underparts and contrasting black bill; legs brownish-pink. In flight, blackish-brown flight feathers contrast with rest of upperparts; dark tail band. **L. c. michahellis** has indistinct paler area on inner primaries, but in **atlantis** all primaries dark (thus resembling Lesser Black-backed). **First-summer** Head and underparts become paler; rest of plumage more faded. **Second-year** Upperparts start showing adult-like ashy-grey colour ('saddle'); head and body become whiter; tail band more clear-cut. Eye pale; bill black with pinkish base; legs yellow tinged. **Third-year** As adult, but with traces of immaturity on wings and tail; bare parts paler, bill with dark subterminal band. [CD2:76-78] **SIMILAR SPECIES** Adult non-breeding Lesser Black-backed Gull of ssp. *graellsii* is much more streaked on head and neck than adult non-breeding Yellow-legged *michahellis* but approaches heavier streaked *atlantis*. First-year Lesser Black-backed may be difficult to separate from corresponding Yellow-legged, but note proportionately narrower and more pointed wings, imparting slimmer appearance; upperparts generally darker, inner wing with 2 dark bars, whitish rump more contrasting. Compare also Common and Ring-billed Gulls. **STATUS AND DISTRIBUTION** Uncommon Palearctic visitor (mainly Oct–May; records in all months) to coastal Mauritania and Senegambia. Rare or vagrant further south and Cape Verde Is. Possibly under-recorded. **NOTE** Previously considered conspecific with Herring Gull *L. argentatus* (cf. *BoA*), but now generally treated as separate species (cf. *BWPC, HBW*). Form *atlantis* sometimes included in Lesser Black-backed Gull *L. fuscus*. Taxonomy of the *L. argentatus/L. cachinnans/L. fuscus* complex much debated and requires further analysis.

GREAT BLACK-BACKED GULL *Larus marinus*   Plate 47 (5)
Goéland marin

L 64–78 cm (25–31"); WS 150–165 cm (59–65"). Very large, powerful gull, recalling Lesser Black-backed but markedly larger and bulkier, with massive bill and pink legs. In flight appears heavier and more lumbering. Plumage sequence as Lesser Black-backed; adult plumage acquired in 4th year. **Adult** Diagnostic combination of blackish upperparts and pink (not yellow) legs. Large white mirrors on outer 2 primaries (Lesser Black-backed has only one subterminal mirror on outer primary). Eye yellow with red orbital ring. Bill pale yellow with orange-red spot on gonys. Head remains white in non-breeding plumage. **First-year** Differs from 1st-winter Lesser Black-backed by whiter head and upper breast contrasting with more coarsely marked grey-brown plumage. In flight from above, dark flight feathers contrast with coverts. Bill black. Legs dull pink. **Second-year** Head and body gradually become white; adult-like mantle colour appears. Eye gradually turns pale. Bill with pale base and black tip. **Third-year** As adult, but some traces of immaturity on wings and tail; bill with darkish spot on gonys. [CD2:80] **SIMILAR SPECIES** Compare Lesser Black-backed, immature Yellow-legged and Kelp Gulls. **STATUS AND DISTRIBUTION** Palearctic vagrant, Mauritania (south to Nouakchott; Oct–May).

BLACK-LEGGED KITTIWAKE *Rissa tridactyla*   Plates 44 (4) & 46 (4)
Mouette tridactyle

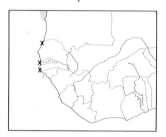

Other name E: Kittiwake
L 38–40 cm (15–16"); WS 95–120 cm (37.5–47.0"). Medium-sized, slender gull with rather pointed wings and a square or slightly notched tail. Adult plumage acquired in 2nd winter. **Adult breeding** Darkish grey and white, with diagnostic clear-cut black wingtips. Bill greenish-yellow. Legs black and rather short. **Adult non-breeding** Hindcrown and neck greyish; blackish ear-spot. **First-winter** Neatly defined, broad black W on wings (formed by black outer primaries and carpal bar); clear-cut black terminal tail band; blackish ear-spot; blackish collar (or traces of it) on hindneck. Bill black. **First-summer** Variable; a more or less faded version of 1st-winter. **HABITS** Highly pelagic. Often follows ships. [CD2:81] **SIMILAR SPECIES** First-winter Little Gull also has black W on upperwings, but this is less well defined; it is also much smaller, with more rounded wings, a darkish 'cap', and a more buoyant flight. At long range, Sabine's Gull (q.v.) may also be confused with 1st-winter Kittiwake, esp. when latter's carpal bar is difficult to see, but Sabine's always has a clear-cut contrast between the black, grey and white triangles. **STATUS AND DISTRIBUTION** Palearctic migrant just reaching coastal Mauritania, where rarely observed but probably regular offshore. Vagrant, Senegambia (Nov–Apr). Scarce, Cape Verde Is.

TERNS

Family STERNIDAE (World: 44 species; Africa: 23; WA: 17)

Slender, small to medium-sized marine and freshwater birds with long, pointed wings, usually long, forked tail and slender, pointed bill. Distinct breeding and non-breeding plumages. Sexes similar. Flight graceful, more buoyant and agile than gulls, on faster and deeper wingbeats and without any gliding or soaring. Feed by plunge-diving or plucking food from water; some (esp. marsh terns and Gull-billed Tern) also catch insects in the air. Seldom swim. Gregarious; esp. vocal at nesting colonies. Several are long-distance migrants. Most immatures of migratory Palearctic species spend first, sometimes also second, summer in winter quarters.

Main identification features include size, wing pattern, bill shape and colour, and call. Post-breeding moult of head, body and wing-coverts complete; pre-breeding partial. Moult of flight feathers peculiar.

In W Africa, 5 species are resident, 7 are Holarctic/Palearctic migrants, 4 have both resident and Palearctic populations, one is an intra-African migrant and one a possible vagrant. They belong to one of the following four genera.

Sterna, typical terns (11 species). Plumage usually pale grey above (dark in Sooty and Bridled Terns) and white or pale grey below. Most have contrasting black cap. In non-breeding plumage black on head reduced or flecked white, lesser coverts dark (forming dusky carpal bar), fork in tail shallower. Adult plumage usually acquired during 2nd or 3rd winter.

Gelochelidon, Gull-billed Tern. Similar to typical terns, but with comparatively short and stout bill.

Chlidonias, marsh terns (3 species). Small and compact, with short, shallowly forked tail. Breeding plumage dark grey or black above and below. In this plumage easy to identify, but whiter non-breeding and immature plumages may present problems. Adult plumage usually acquired during 2nd summer. Do not plunge for food.

Anous, noddies (2 species). Distinctive pelagic terns with wholly dark brown plumage, contrasting white or pale grey cap, and spatulate, slightly forked tail. No seasonal variation. Rarely soar high above ocean or plunge-dive; normally fly *c.* 3 m above sea. Separation at sea requires very careful observation.

GULL-BILLED TERN *Gelochelidon nilotica*   Plate 49 (1)
Sterne hansel

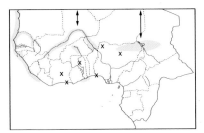

L 35–43 cm (14–17"); WS 86–103 cm (34–40.5"). *G. n. nilotica.* Fairly large, comparatively thickset and white-looking tern with *stout, all-black bill*, no crest, short shallowly forked tail and longish black legs. **Adult breeding** Black cap (from Mar); upperparts and tail very pale grey; upper- and underwing with dark trailing edge to primaries. **Adult non-breeding** Head mainly white with variable blackish streak behind eye; crown feathers moulted at once, resulting in mottled cap (never a white forehead) in transitional stages (late Jul to mid-Oct); outer primaries gradually darken through wear. **Juvenile** As adult non-breeding but upperparts have variable V-shaped markings (usually less distinct than other terns), and faint brownish tinge to crown and upperparts. **Immature** As adult non-breeding but initially with some traces of immaturity on wings (brownish tertials) and tail feathers (dark subterminal patch); these lost by Feb. Primaries worn and darker. **First-summer** has varying amount of black on head (never wholly black cap). **VOICE** A harsh, raucous *ger-vik* and *khaak.* Juvenile utters a high-pitched *pe-eep* or a fast *pe-pe-eep.* [CD2:82] **HABITS** Frequents larger lakes and rivers; also coasts. Not on open sea. Attracted to bush fires. Does not habitually plunge-dive (unlike Sandwich Tern). Flight with steady wingbeats. **SIMILAR SPECIES** Sandwich Tern has similar size but appears slimmer with more protruding head and neck, and long slender bill with, in adult, diagnostic yellow tip. Trailing edge to primaries absent on upperwing and greyer, less clear-cut on underwing; rump and tail white (not grey; but difficult to see in strong light). Call higher pitched (but beware, juveniles have very similar calls). Similar-sized Lesser Crested Tern has slightly darker upperparts and orange bill. **STATUS AND DISTRIBUTION** Locally common to uncommon resident breeder, Mauritania (Banc d'Arguin and near Nouakchott) and N Senegal (Senegal delta). Locally common to rare Palearctic winter visitor (mainly Sep–Apr), both inland (e.g. central Niger delta to Chad) and on coast, from Mauritania to Nigeria. Vagrant, Cape Verde Is. Crosses Sahara. Immatures mostly remain in winter quarters during first summer. **NOTE** Sometimes included in genus *Sterna.*

CASPIAN TERN *Sterna caspia*   Plate 49 (6)
Sterne caspienne

L 47–54 cm (18.5–21.0"); WS 130–145 cm (51–57"). Largest tern (about size of large gull) with *massive red bill*, thick neck, conspicuous *dark wedge on tip of underwing* (formed by blackish outer primaries) and short, shallowly

459

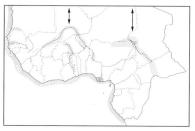

forked tail. Flight gull-like with slow, heavy wingbeats. **Adult breeding** Large black cap ending in short rough crest; blood-red, yellow-tipped bill with black subterminal band; black legs. **Adult non-breeding** *Forehead* and crown closely streaked white (appearing greyish). **Juvenile** Crown more buff than adult non-breeding; upperparts with variable scaly pattern; dusky lesser coverts; tail tipped dusky. Bill orange-red with extensive dark tip. Legs darkish. **Immature** Differs from adult non-breeding in having dusky leading edge and secondary bar on upper wing; outer primaries and primary-coverts darken with wear. **VOICE** A deep, hoarse *kraah-ap* or *rrha-ak* somewhat resembling call of Grey Heron. [CD2:83] **HABITS** Mainly on coasts, lagoons, large rivers, lakes and inland deltas. Often with other terns. Flies high; occasionally soars. Flight powerful with steady wingbeats. Catches prey by plunge-diving, submerging completely. **SIMILAR SPECIES** Royal Tern is smaller, less heavily built, with more slender, unmarked, orange bill; in flight, primaries white with blackish trailing edge. **STATUS AND DISTRIBUTION** Resident and Palearctic winter visitor. Breeds from Mauritania to Guinea; also Eq. Guinea and Gabon. Common to not uncommon migrant along entire coast south to Gulf of Guinea and locally inland (e.g. central Niger delta), uncommon further south to Congo. Vagrant, Cape Verde Is. Crosses Sahara. **NOTE** Sometimes placed in monotypic genus *Hydroprogne*.

ROYAL TERN *Sterna maxima*   Plate 49 (5)
Sterne royale

L 45–50 cm (18–20"); WS 125–135 cm (49–53"). *S. m. albididorsalis*. Second largest tern. **Adult breeding** Black cap ending in short, shaggy crest; *deep orange bill*; black legs. In flight from below, *dark trailing edge to outer primaries*. Black cap retained for only short period, in early breeding season. **Adult non-breeding** Black on head reduced to 'shawl' from eyes across hindcrown; bill pale orange. **Juvenile** Head pattern similar to adult non-breeding; upperparts indistinctly scaled. In flight, boldly patterned upperwing with dark outer wing and pale grey inner wing marked with 3 contrasting dark bars (on leading edge, greater coverts and secondaries); tail tipped dark. Bill yellowish-orange. Legs dull yellowish. **Immature** Initially as juvenile but with uniformly grey upperparts; gradually more like adult non-breeding. **VOICE** A high-pitched, shrill *kee-err* and a loud, sharp *krryuk*, resembling call of Sandwich Tern but deeper. [CD2:84] **HABITS** Marine; mainly on shores, lagoons and in harbours. Singly or in small groups. **SIMILAR SPECIES** Caspian Tern has heavier build and much heavier, dusky-tipped red bill; in flight from above, black on outer primaries, from below, black wedge on wing point. Non-breeding Caspian never has white forehead. Lesser Crested Tern is much smaller, with thinner bill. **STATUS AND DISTRIBUTION** Resident and intra-African migrant; common along entire coast. Also recorded in Gulf of Guinea waters. Breeds Mauritania (Banc d'Arguin) and Senegambia. Vagrant, Cape Verde Is.

GREATER CRESTED TERN *Sterna bergii*   Plate 49 (4)
Sterne huppée

Other name E: Swift Tern

L 46–49 cm (18–19"); WS 125–130 cm (49–51"). Large, crested tern with long, *yellow, slightly decurved bill* and *dark grey upperparts*. Legs black. **Adult breeding** Narrow white forehead (unlike other large terns); black cap ending in shaggy crest. Bill lemon-yellow. Upperparts darker than other large terns in the region. **Adult non-breeding** Black of cap receding (but more extensive than in Royal and Lesser Crested Terns) and speckled white, crest much reduced. **Immature** Head pattern as adult non-breeding; dark leading and trailing edges to upper wing contrast with paler mid-wing panel. Bill dusky-yellow. **VOICE** A loud strident *errik*, a staccato *rrh-rrh-rrh* and a low *rak*. **HABITS** Marine; in coastal habitats and offshore, often with other terns. Flight powerful with steady wingbeats. **SIMILAR SPECIES** Compare Royal, Caspian and Lesser Crested Terns. **STATUS AND DISTRIBUTION** Not recorded with certainty in our region. Claimed from Nigeria (single sight record, Lagos, Feb 1969). Occurs in E and S Africa.

LESSER CRESTED TERN *Sterna bengalensis*   Plate 49 (3)
Sterne voyageuse

L 35–37 cm (14.0–14.5"); WS 92–105 cm (36.0–41.5"). Fairly large, elegant crested tern with *orange bill*. Legs black. **Adult breeding** Black cap ending in shaggy crest; silvery-white primaries contrasting with rest of grey

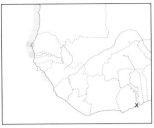

upperparts; underwing white with narrow dark trailing edge to outer primaries; rump and tail grey. Bill orange-red. **Adult non-breeding** Forehead and most of crown white; bill paler orange. **Immature** Similar to adult non-breeding but with dark outer primaries and dusky secondary bar. Bill dusky-orange. **VOICE** A rasping *errik* or *kirrik* and *krrr-eep*, resembling call of Sandwich Tern but higher pitched. [CD2:85] **HABITS** Marine; mainly occurring inshore. Gregarious; often foraging with other terns. Catches prey by plunge-diving; usually submerging completely. **SIMILAR SPECIES** Royal Tern is larger, with stouter bill. Similar-sized Sandwich Tern has yellow-tipped black bill. Both have slightly paler upperparts and white rump and tail. **STATUS AND DISTRIBUTION** Generally scarce Mediterranean visitor (mainly Sep–Apr) to coasts, from Mauritania to Guinea; not uncommon, Gambia. Vagrant, Ghana.

SANDWICH TERN *Sterna sandvicensis*   Plate 49 (2)
Sterne caugek

L 36–41 cm (14–16"); WS 95–105 cm (37.5–41.5"). *S. s. sandvicensis.* Fairly large, white-looking crested tern with long, slender black bill (yellow-tipped in adult), long wings and short, forked white tail. Legs black. **Adult breeding** Black cap ending in shaggy crest; upperparts very pale grey (usually appearing white), with outer primaries forming darkish wedge; underwing with rather diffuse trailing edge to outer primaries; rump and tail white. **Adult non-breeding** Black cap gradually receding to form 'shawl' from eyes over nape (appears early, while breeding). **Immature** Initially differs from adult non-breeding in darkish carpal bar, secondaries and outer tail feathers, but these lost by late winter. **VOICE** Distinctive. A loud, rasping *kerrik* or *keerr-wit.* [CD2:86] **HABITS** Marine; mainly occurring inshore. Gregarious, often foraging with other terns. Typically plunge-dives from great heights (5–10 m), but also from lower. **SIMILAR SPECIES** Gull-billed Tern is more thickset, with shorter, deeper all-black bill, shorter neck and shorter, broader wings, imparting different, more gull-like jizz. Adult breeding lacks crest; adult moulting to non-breeding has cap mottled white (never only a white forehead); call lower (juveniles, however, have very similar calls). Lesser Crested Tern has orange bill and slightly darker upperparts. Medium-sized terns have dark bills in some plumages but differ in more slender build, longer wings, dark carpal bar and (in adults) red legs. **STATUS AND DISTRIBUTION** Common Palearctic winter visitor to entire coast. Rare, Cape Verde Is, São Tomé and Príncipe. Most immatures remain in winter quarters during first summer, some also during second.

ROSEATE TERN *Sterna dougallii*   Plate 46 (3)
Sterne de Dougall

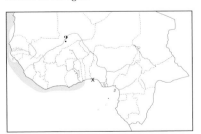

L 33–38 cm (13–15"); WS 72–80 cm (28.5–31.5"). *S. d. dougallii.* Medium-sized, slender sea tern; more delicate and shorter winged than Common Tern. Underwing lacks dark trailing edge to primaries in all plumages. **Adult breeding** Very pale grey, *appearing almost white at distance* (recalling Sandwich Tern), with black cap, extremely long tail streamers (moulted Aug–Sep) and black bill with red base. Underparts white tinged pink. On upperwing, 2–4 outer primaries gradually darken to form narrow dark wedge; *underwing all white.* Legs bright red. **Adult non-breeding** Narrow white forehead; upperparts pale grey (paler than in Common and Arctic Terns); underparts lack pink wash; no tail streamers. Bill black; legs dusky-red. **Juvenile** All-dark cap; brown scaling on mantle and scapulars forms dark 'saddle' (reminiscent of juvenile Sandwich); dusky carpal bar; no dusky trailing edge to underwing. Bill and *legs black.* Post-juvenile moult started before southward migration. **Immature** As adult non-breeding; legs gradually become orange-red. **VOICE** Distinctive. A shrill *cher-vrik* (reminiscent of Sandwich Tern). Also a guttural, rasping *rrraakh.* [CD2:87] **HABITS** Marine; more pelagic than Common Tern. Gregarious; often with other terns. Plunge-dives; also dips to surface. Direct flight less graceful than Common and Arctic, with faster wingbeats (almost as Little Tern). **SIMILAR SPECIES** Common and Arctic Terns (q.v. and above). **STATUS AND DISTRIBUTION** Uncommon to rare Palearctic winter visitor on coasts, from Mauritania to Nigeria. Immatures usually remain in winter quarters. Bred Senegal (one pair, 1980). Vagrant, Cape Verde Is.

COMMON TERN *Sterna hirundo*
Sterne pierregarin

Plate 46 (1)

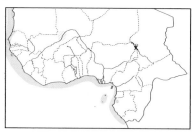

L 31–35 cm (12–14"); WS 77–98 cm (30.5–38.5"). *S. h. hirundo.* Medium-sized, elegant sea tern with, in all plumages, diagnostic underwing pattern of *dusky-tipped outer primaries forming diffuse trailing edge.* **Adult breeding** Black cap; slender, orange-red bill *with black tip;* red legs. Tail has long streamers, not extending beyond wingtips at rest. In flight, upperwing never uniformly grey, but darker area starting as small dusky wedge on middle primaries and gradually, with wear, extending to all outer primaries. Rump white. **Adult non-breeding** White forehead; inner wing with *dusky carpal and fainter secondary bar,* outer wing more uniformly pale once worn outer primaries moulted early winter. Rump and tail tinged grey; no tail streamers. Bill dusky. **Juvenile** White forehead and scaly upperparts with gingery tinge; dark carpal bar; dark grey secondary bar. Bill pinkish to orange-yellow with dark tip, gradually becoming more blackish. Legs dull reddish or yellowish-orange. Post-juvenile moult of Palearctic birds usually started in winter quarters. **Immature** As adult non-breeding; in 1st-winter/-summer juvenile flight feathers darken with wear before being gradually replaced Jan–Jul (outer primaries blackish Feb–Jun, contrasting with fresh inner feathers). **VOICE** A shrill, drawn-out *KEEEarrh* and a sharp *kip.* [CD2:88] **HABITS** Mainly coastal. Gregarious. Forages by hovering and plunge-diving from low height. **SIMILAR SPECIES** Arctic Tern is even more elegant looking, with narrower wings and longer tail; upperwings have uniformly grey flight feathers; underwings white with neat black edge to wingtip. First-winter never has dusky secondary bar. Roseate Tern has paler, white-looking upperparts and longer tail; underwings lack dusky trailing edge. Sandwich Tern is larger, bulkier and whiter, with longer bill and relatively shorter tail. **STATUS AND DISTRIBUTION** Common resident, intra-African migrant and Palearctic winter visitor along entire coast. Rare, Cape Verde Is, Bioko, São Tomé and Príncipe. Few inland records, Mali and Cameroon. Breeds annually Mauritania (Banc d'Arguin); occasionally Senegal (Saloum delta), Guinea-Bissau (Bijagos Is), Nigeria (Dodo R. mouth) and Gabon (Ogooué R.). Most immatures from Palearctic remain in winter quarters during first summer.

ARCTIC TERN *Sterna paradisaea*
Sterne arctique

Plate 46 (2)

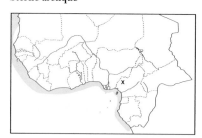

L 33–35 cm (13–14"); WS 75–85 cm (29.5–33.5"). Medium-sized, very elegant sea tern with, at all ages, diagnostic *well-defined narrow black trailing edge to primaries* on underwing. **Adult breeding** Black cap; wholly deep red bill and red legs (both shorter than Common Tern). Upperparts uniformly grey (never having dark wedges on wings as Common); underparts pale grey (usually slightly darker than Common, but this difficult to see); *underwing pure white (looking wholly translucent against light).* Rump and tail white; tail with long streamers, extending beyond wingtips at rest. **Adult non-breeding** White forehead; black bill; faint dusky carpal bar; no tail streamers. **Juvenile** Much whiter looking than juvenile Common, lacking gingery tinge to plumage; upperwing has only faint carpal bar, *white secondaries* and paler primaries; rump white. **Immature** Initially similar to juvenile; 1st-summer very bleached. Second-summer similar to adult non-breeding. **VOICE** A strident *keeAARRGH* resembling call of Common Tern, but harsher and with stress on second syllable; a rapid *ki-ki-ki-ki.* [CD2:89] **HABITS** Both inshore and offshore; more pelagic than Common Tern. Hovers and plunge-dives from low height; feeding action more hesitant than Common. Also catches crustaceans by surface-picking. **SIMILAR SPECIES** Common and Roseate Terns (q.v. and above). **STATUS AND DISTRIBUTION** Uncommon to rare Holarctic winter visitor along entire coast and in Cape Verde seas. Vagrant, Bioko. Most immatures remain in or near winter quarters during first summer.

BRIDLED TERN *Sterna anaethetus*
Sterne bridée

Plate 48 (8)

L 30–32 cm (12.0–12.5"); WS 77–81 cm (30.5–32.0"). Medium-sized sea tern with *dark upperparts* and very buoyant, graceful flight. Long, slender black bill. Legs black. **Adult breeding** Black cap and *white forehead extending beyond eye as narrow supercilium;* dark grey-brown upperparts and tail; long, deeply forked tail with white outer tail feathers; underparts white washed greyish on breast and belly. Underwing-coverts white contrasting with dark flight feathers. **Adult non-breeding** White streaking on crown; mantle with pale grey fringes. **Juvenile** Paler and browner above, with pale forehead and faint narrow supercilium; pale-fringed upperparts, coverts and tertials creating scalloped

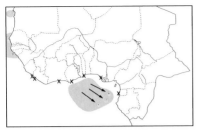

appearance; underparts white with grey wash on breast-sides and flanks. **Immature** As adult but with traces of immaturity on wings and tails. **VOICE** A variety of harsh, grating calls. At colony a yapping *wep-wep* or *yuk-yuk* and *kerrr.* [CD2:90] **HABITS** Mainly pelagic. Forages singly, in small or large flocks; also with other terns. Flies low over sea, hovering frequently and surface-picking; also plunge-dives. **SIMILAR SPECIES** Sooty Tern is more pelagic and has blacker upperparts, broader white forehead not extending beyond eye, and more contrasting underwing; juvenile differs in dark head and underparts. **STATUS AND DISTRIBUTION** Uncommon, local resident and intra-African migrant along coast. Breeds Mauritania (Banc d'Arguin), São Tomé (Sete Pedras) and Annobón; also Senegal (a few pairs, Îles de la Madeleine and Langue de Barbarie). Disperses offshore after breeding; recorded from Guinea-Bissau to Eq. Guinea and (probably) Gabon.

SOOTY TERN *Sterna fuscata*   Plate 48 (7)
Sterne fuligineuse

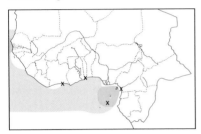

L 33–36 cm (13–14"); WS 82–94 cm (32.5–37.0"). *S. f. fuscata.* Medium-sized slender sea tern with *blackish upperparts* and long forked tail. Legs and feet black. **Adult breeding** Highly contrasting pattern of blackish upperpart and tail, and pure white underparts. *Forehead white, not extending as supercilium.* Underwing white with contrasting blackish flight feathers. **Adult non-breeding** As adult breeding but faded; upperpart feathers variably fringed whitish. **Juvenile** Distinctive. Wholly blackish-brown with whitish lower belly and undertail-coverts; upperparts have white fringes. **Immature** Gradually becomes paler on head and underparts. **VOICE** A nasal *wekawek* ('wide-awake') and *kwèè-eh*; also a low *krrrr.* [CD2:91] **HABITS** Mainly pelagic. Usually gregarious; flocks often joined by other terns. Does not plunge-dive but flies low over water and snatches prey from or just below surface. **SIMILAR SPECIES** Bridled Tern (q.v.) is smaller, more slender, with less contrasting plumage; juvenile much paler with white underparts. Brown Noddy could be confused with juvenile at long range, but has wedge-shaped tail and no pale areas on lower underparts. **STATUS AND DISTRIBUTION** Local resident and intra-African migrant. Breeds in large colonies of up to 200,000 pairs, Tinhosas Is (Príncipe) and, irregularly, in isolated pairs, Senegal. Disperses widely after breeding; recorded from Mauritania to Cameroon and Bioko.

DAMARA TERN *Sterna balaenarum*   Plate 48 (2)
Sterne des baleiniers

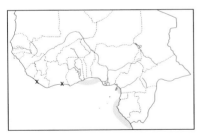

L 23 cm (9"); WS 51 cm (20"). *Small* and compact with slightly decurved black bill, pale grey upperparts, rump and uppertail, and darkish legs. **Adult breeding** Combination of small size and all-black cap diagnostic. In flight, outer 2–3 primaries form dusky leading edge contrasting with rest of wing. At distance appears almost white. **Adult non-breeding** Similar, but forehead white. **Juvenile** White forehead; upperpart feathers have brownish fringes. In flight, dusky carpal bar and leading edge to outer wing. **Immature** As adult non-breeding but with dusky carpal bar. **VOICE** A high-pitched *rrrik* and *rrreekikik.* [CD6:47] **HABITS** Coastal. Singly or small parties on sandy shores, estuaries, lagoons and in sheltered bays. Forages by hovering and plunge-diving. **SIMILAR SPECIES** Little Tern appears less dumpy, with shorter, straight bill, slightly darker upperparts, and, in ssp. *guineae*, white (not grey) rump and tail. Adult breeding further differs in well-defined white forehead and yellow bill and legs; other plumages by dark leading edge to upperwing. **STATUS AND DISTRIBUTION** Rare to scarce migrant from SW African coasts (Apr–Jan), recorded from coastal Congo to Liberia. *TBW* category: Near Threatened.

LITTLE TERN *Sterna albifrons*   Plate 48 (1)
Sterne naine

L 22–24 cm (8.5–9.5"); WS 48–55 cm (19–22"). Small with narrow pointed wings, short tail and *white forehead in all plumages.* Wingbeats faster than other terns. **Adult *guineae* breeding** Black cap; well-defined broad white forehead tapering to point above eye; thin, straight bill *yellow* with little or no black on tip. Legs yellow. In flight, dark outer primaries, forming narrow wedge on leading edge of wing, contrast with rest of pale grey upperparts. Rump

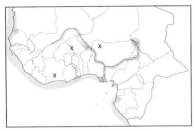

white. **Nominate** differs in having bill with clear black tip and grey rump. Vagrant *sinensis* has black-tipped bill and whitish shafts to primaries. **Adult non-breeding** Black on head recedes to form 'shawl' from eyes across nape; crown streaked grey; lores white. Bill black. Legs brownish. In flight, dark carpal bar and leading edge to outer wing. **Juvenile** Forehead and upperparts initially tinged sandy-buff, but whitening with wear; feathers of mantle and scapulars fringed brownish. In flight, upperwing has dusky leading edge to entire wing (formed by dark carpal bar, primary-coverts and outer primaries) contrasting with pale grey mid-wing panel and almost white secondaries and inner primaries. Bill blackish with dull yellowish base. Legs dull yellow. **Immature** Much as adult non-breeding. **VOICE** A short, shrill *kitik* and *kree-ik*. [CD2:92] **HABITS** Singly or with other terns on coast, lagoons and creeks, and inland on major rivers and lakes with exposed beaches. Forages by hovering and plunge-diving from moderate height, also by surface-picking. **SIMILAR SPECIES** Damara Tern is dumpier, with longer, slightly down-curved bill (black at all ages) and paler, more uniform upperparts. Marsh terns *Chlidonias* have shorter bill and smaller head. **STATUS AND DISTRIBUTION** *S. a. guineae* is local resident and common to uncommon partial intra-African migrant; breeds on coast in Mauritania, Senegambia, Ghana, Nigeria, Cameroon and, inland, along Niger R. and major tributaries, L. Chad and Ogooué R. (Gabon). *S. a. albifrons* is common to uncommon Palearctic winter visitor to entire coast; rare, Cape Verde Is. *S. a. sinensis* presumably vagrant; recorded from Ghana.

WHISKERED TERN *Chlidonias hybridus*　　　　　Plate 48 (3)
Guifette moustac

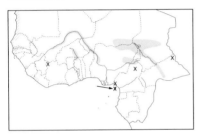

L 23–25 cm (9–10"); WS 74–78 cm (29–31"). *C. h. hybridus.* Largest marsh tern, with distinctive *Sterna* jizz. **Adult breeding** (mainly early Apr–Jun/Jul) Unmistakable. Black cap; head-sides white, contrasting with grey throat and neck; upperparts and tail grey. Underparts dark grey becoming blackish-grey on belly; undertail-coverts white. Bill blood-red. **Adult non-breeding** (Oct–Nov onwards) Very pale. Black on head reduced to U-shaped band from eyes across nape and some streaking on crown. Upperparts and tail pale grey. Underparts white. Bill black. **Juvenile** Head pattern as adult non-breeding; dark brown 'saddle' (soon replaced); dark bill and legs. No dark carpal bar.

Immature Gradually more like adult non-breeding (much individual variation; post-juvenile moult usually completed Oct–Jan). **VOICE** Low grating and rattling sounds, incl. a harsh *kirrk* and a nasal *airkh*. [CD2:93] **HABITS** Frequents wetlands. Often gregarious. Forages low over water or aquatic vegetation, hovering and surface-picking; also catches insects in flight. **SIMILAR SPECIES** Compare non-breeding and immature White-winged and Black Terns, and larger, structurally different, Common and Arctic Terns. **STATUS AND DISTRIBUTION** Widespread, common to scarce Palearctic passage migrant (mainly Aug–Oct and Apr) and winter visitor; also Bioko. Immatures probably remain in winter quarters during first summer; extent of oversummering not known.

BLACK TERN *Chlidonias niger*　　　　　Plate 48 (4)
Guifette noire

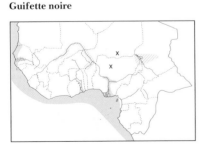

L 22–24 cm (8.5–9.5"); WS 64–68 cm (25–27"). *C. n. niger.* Small. Unmistakable in breeding plumage; in other plumages *small dark breast patches* diagnostic. **Adult breeding** (Mar/Apr–Jun) Head and underparts black; upperparts, upper- and underwings slate-grey; rump and tail grey; vent and undertail-coverts white. Bill black. Legs blackish-red. **Adult non-breeding** Plain grey above, white below. Black cap extending to ear-coverts; small dark patches on breast-sides. Moulting birds blotched below. **Juvenile** As adult non-breeding but with dark grey-brown 'saddle' (mantle and scapulars), and broad, dark carpal bar. **Immature** Post-juvenile moult generally started in winter quarters. Gradually more like adult non-breeding (much individual variation). **VOICE** Mainly silent. A nasal *kyay* or *kleeah* and *kyek*, a grating *kerr* and a short *kik*. [CD2:94] **HABITS** Mostly marine, frequenting lagoons, estuaries, harbours and other sheltered habitats. Gregarious; offshore regularly in large flocks. Forages low and erratically over water, hovering and surface-picking; also catches insects on the wing. **SIMILAR SPECIES** Compare non-breeding and immature White-winged and Whiskered Terns. White-winged has whiter head, shorter, stubbier bill, white rump and no breast patches; non-breeding has paler grey upperparts, juvenile more contrasting 'saddle' (diagnostic). Whiskered has more *Sterna*-like jizz, differently patterned head markings and no breast patches;

non-breeding has paler upperparts. **STATUS AND DISTRIBUTION** Common Palearctic passage migrant (mainly Sep–Oct and Mar–May) and winter visitor to entire coast and Gulf of Guinea waters. Records from Mali (uncommon) and Niger and Chad (rare) suggest some cross Sahara. Immatures remain in winter quarters during first summer.

WHITE-WINGED TERN *Chlidonias leucopterus*   Plate 48 (5)
Guifette leucoptère

Other name E: White-winged Black Tern
L 20–23 cm (8–9")' WS 63–67 cm (25.0–26.5"). Small. Unmistakable in breeding plumage; no breast patches in other plumages. **Adult breeding** (late Apr–Jul) Jet-black head, body and underwing-coverts contrast with pale grey wings and white rump, vent and tail. Upperwing has blackish outer primaries and inner secondaries. Bill and legs bright red. **Adult non-breeding** Black plumage confined to head, where less extensive than on Black Tern and reduced to small patch behind eye and streaking over rear crown (forming 'head-phones'). Upperparts pale grey (paler than Black) with contrasting white rump. Often retains traces of breeding plumage late into moult (esp. on wing-coverts), and in early winter upperwings still have dark outer primaries and inner secondaries. **Juvenile** Dark brown 'saddle' contrasts strongly with pale grey wings (paler than juvenile Black) and white rump. Carpal bar narrower and less conspicuous than Black. Head as adult non-breeding **Immature** Post-juvenile moult generally started in winter quarters. Gradually more like adult non-breeding (much individual variation). **VOICE** Various squealing and grating sounds, incl. *khee-ep*, a harsh *krrek* and a softer *kverr-kek*; also a fast *kikikik*. [CD2:95] **HABITS** Frequents wetlands; rare on coast (unlike Black). Forages low over water, hovering and surface-picking/-dipping; also forages far from water. **SIMILAR SPECIES** Compare non-breeding and immature Black and Whiskered Terns. Black has slightly longer bill, more black on head, greyer upperparts, grey (not white) rump and diagnostic breast patches. Whiskered has longer bill, *Sterna* jizz, differently shaped head markings and grey rump. **STATUS AND DISTRIBUTION** Common to uncommon Palearctic visitor (late Jul–Apr) throughout. Particularly numerous in Senegambia, Mali and Nigeria. Immatures remain in winter quarters during first summer.

BLACK NODDY *Anous minutus*   Plate 48 (10)
Noddi noir

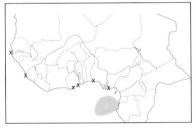

L 35–39 cm (14.0–15.5"); WS 66–72 cm (26.0–28.5"). *A. m. atlanticus.* **Adult** Medium-sized, wholly blackish-brown pelagic tern with contrasting white cap. White crescent below eye. Bill long and slender, black. Legs black. **Juvenile** White well defined and restricted to forehead; upperparts and coverts fringed buffish. **VOICE** Mainly silent away from colonies and communal roosts. [CD6:48] **HABITS** As Brown Noddy, with which it often associates at feeding areas and roosts. **SIMILAR SPECIES** Brown Noddy is more heavily built with stouter bill and two-toned wing; wingbeats slower. Typically has less contrasting, greyer and smaller cap, and browner upperparts, but differences in overall coloration, extent, colour and demarcation of cap, and size generally of little use in separating the two species at sea. **STATUS AND DISTRIBUTION** Locally common resident at breeding grounds in Gulf of Guinea. Breeds Tinhosas Is (Príncipe; 5000 pairs) and Annobón (70,000 pairs). Vagrant elsewhere; recorded from Gambia to Gabon. Record from Mauritania either Black or Brown Noddy.

BROWN NODDY *Anous stolidus*   Plate 48 (9)
Noddi brun

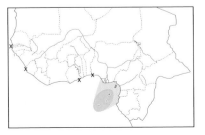

Other name E: Common Noddy
L 38–40 cm (15–16"); WS 77–85 cm (30.5–33.5"). *A. s. stolidus.* Medium-sized dark brown pelagic tern with characteristic spatulate and slightly forked tail. Bill and legs black. **Adult** Contrasting greyish-white forehead and forecrown becoming ash-grey on rear crown and nape. White crescent below eye. *Underwing greyish-brown bordered by contrasting dark brown band*, narrow on leading edge, broad on trailing edge. Upperwing-coverts contrast slightly with darker flight feathers (visible only in good light). **Juvenile** Greyish on head confined to forehead; narrow whitish line above base of bill to over

eye; upperparts and coverts with pale fringes. When heavily worn, no trace of grey on cap, pale line indistinct. **VOICE** A low grating *krrrah.* Mainly silent away from colonies and communal roosts. [CD6:49] **HABITS** Highly pelagic and gregarious, occurring mainly offshore. **SIMILAR SPECIES** Black Noddy is more graceful with proportionately smaller head, slimmer neck, longer, more slender bill, narrower wings, more uniformly blackish plumage and usually more contrasting and extensive white cap; wingbeats more rapid. Juvenile Sooty Tern has forked tail, whitish lower underparts and slightly shorter bill; it is also slightly larger and has dashing flight. At distance may be confused with dark morph *Stercorarius* skuas (q.v.) but these have quite different, more vigorous flight and pale wing panels. **STATUS AND DISTRIBUTION** Locally common resident on breeding grounds in Gulf of Guinea; breeds Tinhosas Is (Príncipe; 10,000s), São Tomé (100s) and Annobón (1000s). Vagrant elsewhere; recorded from Gambia, Sierra Leone, Ghana, Nigeria, Gabon and Bioko.

SKIMMERS
Family RYNCHOPIDAE (World: 3 species; Africa: 1; WA: 1)

Robust, medium-sized, tern-like birds with unique bill shape and feeding technique. Fish by flying close to surface with tip of much longer lower mandible slicing through water. Wings long and narrow, tail short and forked. Sexes similar, male slightly larger. Flight fast and agile. Breed colonially.

AFRICAN SKIMMER *Rynchops flavirostris*   Plate 48 (6)
Bec-en-ciseaux d'Afrique

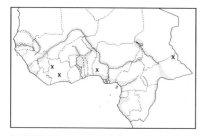

L 36–42 cm (15"); WS 125–135 cm (49–53"). Conspicuous black-and-white species with long wings and distinctive, *scissor-like bill.* **Adult breeding** Crown, hindneck, upperparts and tail black; forehead, underparts and outer tail feathers white. *Bill and feet orange-red.* In flight, has white trailing edge to inner primaries and secondaries on both surfaces; underwing greyish-white with blackish primaries. **Adult non-breeding** Upperparts browner with whitish collar on hindneck; bill and legs paler. **Juvenile** Duller. Browner above with buffish fringes to feathers; bill dusky with yellowish base (becoming more orange with age), shorter than in adult; legs dull yellowish.

VOICE A loud, shrill *kik* and *kree,* often repeated. [CD6:50] **HABITS** Gregarious, but may occur singly, esp. outside breeding season. Curious feeding action requires large expanses of quiet water; breeding dependent on large sandbars in broad rivers and lakes. Disperses when not breeding, then also on estuaries, coastal lagoons, dams and saltpans. Spends much of day resting on sandbars, often in small flocks, foraging mainly at dawn and dusk. **STATUS AND DISTRIBUTION** Locally not uncommon to rare resident and partial intra-African migrant. Breeding known from mid Senegal R., mid Niger R., lower Benue, Ogooué, Logone and Chari R., and Congo R. Widespread outside breeding season; recorded from various coastal sites and large rivers. Movements dependent on height and turbulence of water. *TBW* category: Near Threatened.

AUKS
Family ALCIDAE (World: 22–23 species; Africa: 4; WA: 2)

Small to medium-sized seabirds with elongated bodies and strong, highly variable, bills. Plumages typically dark above and pale below. Wings short and narrow, used for both flight and swimming underwater; tail short. Sexes similar. Feed by diving. Flight direct and low with rapid, whirring wingbeats. Both species recorded in our region are Palearctic vagrants.

COMMON GUILLEMOT *Uria aalge*   Plate 47 (1)
Guillemot marmette

Other names E: Guillemot, Common Murre; F: Guillemot de Troïl
L 38–43 cm (15–17"). *U. a. albionis.* Medium-sized auk with *slender, pointed, black bill.* In flight, has narrow white rump-sides; feet project beyond short, rounded tail. **Adult breeding** Head, neck and upperparts dark grey-brown; short white wingbar. Underparts white with variable dark streaking on flanks. **Adult non-breeding** Head-sides and

foreneck white with dark line from eye across ear-coverts. **Immature** As adult non-breeding but with smaller bill and little or no streaking on flanks. **HABITS** Spends non-breeding season at sea, when rarely seen from shore. [CD2:96] **SIMILAR SPECIES** Could be confused with Razorbill at long range, when distinctive bill shape hard to discern. Razorbill has black (not grey-brown) upperparts, pure white (not streaked) flanks and white (not dusky) 'armpits'; in flight, has broad (not narrow) white sides to black central rump band; feet do not project beyond longer, pointed tail. **STATUS AND DISTRIBUTION** Palearctic vagrant, NW Mauritania (2 records, Jan–Feb).

RAZORBILL *Alca torda* — Plate 47 (2)
Petit Pingouin

L 37–39 cm (15"). *A. t. islandica.* Medium-sized auk with *heavy, blunt bill* and pointed, often raised tail. **Adult breeding** Head, neck and upperparts black; short white wingbar; white line from bill to eye. Underparts pure white. Bill has white vertical line. **Adult non-breeding** Head-sides and foreneck white (lacking dark line on ear-coverts, unlike Common Guillemot). **Immature** As adult non-breeding but with smaller bill and dusky head-sides. **HABITS** Spends non-breeding season at sea; not usually seen from shore. [CD2:97] **SIMILAR SPECIES** Common Guillemot (q.v.) **STATUS AND DISTRIBUTION** Palearctic vagrant, NW Mauritania (3 records, Nov–Dec).

SANDGROUSE
Family PTEROCLIDAE (World: 16 species; Africa: 12; WA: 5)

Medium-sized pigeon-like terrestrial birds of arid habitats. Plumage cryptic, making them difficult to see on the ground, wings and tail pointed, bill and legs short. Sexually dimorphic, females less distinctly marked. Flight swift and direct, reminiscent of pigeons. Vocal (calls being good identification clues) and often gregarious. May cover considerable distances daily to drink at favourite waterholes, where sometimes gathering in large flocks, usually at dawn or dusk.

CHESTNUT-BELLIED SANDGROUSE *Pterocles exustus* — Plate 50 (1)
Ganga à ventre brun

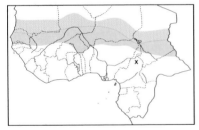

Other name F: Ganga sénégalais
L 29–33 cm (11.5–13.0"). *P. e. exustus. Long-tailed,* pale yellow-buff with *entirely dark underwing* and *dark belly.* Above, blackish flight feathers contrast with pale upperwing-coverts. **Adult male** Head to mantle and upper breast plain; upperparts washed grey and lightly scaled brown. Underparts with narrow black breast-band and *dark chestnut belly* (becoming black in centre). Tail with finely pointed elongated central feathers. Bill and feet bluish-grey. Orbital ring pale greenish. **Adult female** Upperparts densely barred dark brown. Head-sides and throat plain sandy-orange. Crown, neck and upper breast finely and densely streaked and spotted brown, separated from plain lower breast by single or double brown line. Belly dark brown narrowly barred yellow-buff on flanks and tarsus. Tail with dark brown bars and finely pointed elongated central feathers. **Juvenile** Yellow-buff with rufous wash and fine dark vermiculations on upperparts; no narrow breast-bands; tail short. **VOICE** In flight, a repeated guttural *kwit-gurut, kwit kwit-gurut...* or *kvitt-kerr-kerr...* [CD6:51] **HABITS** Usually in pairs or small flocks in bare semi-desert and sandy arid scrub. Visits waterholes in early morning and before sunset in small or large flocks (occasionally numbering 1000s). **SIMILAR SPECIES** Spotted Sandgrouse also has long tail but has pale underwing-coverts and pale belly with narrow black stripe down middle. **STATUS AND DISTRIBUTION** Common to not uncommon resident in Sahel belt, from Mauritania and Senegal through Mali, Burkina Faso and Niger to Chad. Southward movements in dry season, with records from Gambia, N Togo, N Nigeria, N Cameroon and N CAR.

SPOTTED SANDGROUSE *Pterocles senegallus* — Plate 50 (2)
Ganga tacheté

L 29–35 cm (11.5–13.75"). *Long-tailed,* pale yellow-buff with *pale underwing-coverts* and *pale belly with dark streak on centre* (only visible at close range). In flight, from below, dark flight feathers (primaries greyish, secondaries black) contrast with *pale underwing-coverts; upperwing largely pale* with dusky trailing edge. **Adult male** Cheeks, neck-sides and throat deep yellow-orange bordered by pale blue-grey supercilium, neck and breast. Crown pale rufous-buff. Upperparts mainly pale yellow-buff; scapulars and wing-coverts greyish with large sandy tips. Bill and

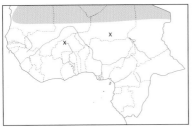

legs bluish-grey. **Adult female** Crown, upperparts and upper breast spotted black; cheeks, neck-sides and throat orange-yellow. **Juvenile** Resembles adult female, but upperparts barred and streaked dark; throat whitish; breast with small U-shaped marks; tail short. **Immature** As respective adult, but duller and paler. **VOICE** In flight, a frequently repeated, distinctive, staccato *wikow wik wikow...*, in chorus *wikowikowikowikowikow...* [CD3:2a] **HABITS** In pairs and small flocks in open, flat, patchily vegetated stony desert. Visits waterholes in early morning, where 100s may congregate. **SIMILAR SPECIES** Chestnut-bellied Sandgrouse also has long tail, but is easily separated in flight by entirely dark underwing and belly. Crowned Sandgrouse has short tail and contrasting pattern on both upper- and underwing; also lacks black stripe on belly. Juvenile Crowned distinguished from juvenile Spotted by barred underparts without stripe on belly. **STATUS AND DISTRIBUTION** Uncommon resident, Mauritania, N Mali, Niger and Chad. Vagrant, Burkina Faso (Ouagadougou area).

CROWNED SANDGROUSE *Pterocles coronatus*   Plate 50 (3)
Ganga couronné

L 25–29 cm (10.0–11.5"). *P. c. coronatus. Short-tailed,* pale yellow-buff with, in flight, blackish flight feathers contrasting with pale coverts on both surfaces. **Adult male** Black vertical mark from forecrown to upper throat; crown pale rufous bordered pale blue-grey from above eye to nape. Cheeks, neck-sides and lower throat orange-yellow. Upperparts pale yellow-buff washed grey with *well-defined sandy spots* to feather tips, from mantle to coverts. Bill and legs bluish-grey. **Adult female** Densely and finely spotted and barred with black both above and below. Cheeks, neck-sides and throat pale orange-yellow. **Juvenile** Resembles adult female but throat whitish and barring coarser. **Immature** Resembles respective adult, but duller. **VOICE** In flight, a fast guttural chatter *klak-klagarrarra klak-klak-klak-klagarra...* (or *chaga-chagarra...*). [CD3:2b] **HABITS** In pairs or small flocks in desert, esp. in stony areas. Gathers at waterholes in early morning. **SIMILAR SPECIES** Spotted Sandgrouse appears more slender, with long tail, and has paler upperwing pattern; also has black central belly stripe. Juvenile Spotted distinguished from juvenile Crowned by plain belly with blackish central streak. Female/juvenile Lichtenstein's may recall female/juvenile Crowned but have face and throat finely vermiculated black. **STATUS AND DISTRIBUTION** Uncommon resident, N Mauritania, N Niger and N Chad. Vagrant, Mali.

LICHTENSTEIN'S SANDGROUSE *Pterocles lichtensteinii*   Plate 50 (4)
Ganga de Lichtenstein

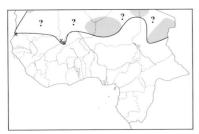

L 22–26 cm (8.75–10.25"). *P. l. targius.* Stocky, smallish, finely and densely barred sandgrouse. In flight, blackish flight feathers contrast with paler upperwing-coverts; underwing dark grey-brown, with little or no contrast. **Adult male** Buff to rufous-buff, speckled dark on head and throat, barred on rest of upper- and underparts. Forehead white traversed by black band; white supercilium with black spot above eye. Wings barred golden-buff, white and black. Breast unbarred buff-yellow traversed by 2 narrow black bands (one on centre and one separating breast and belly). Bill orange. Bare pale yellow eye patch. Feet pale yellowish. **Adult female** Head and throat finely streaked and speckled black; rest of plumage entirely covered by fine, closely spaced, wavy bars. Bill slate-grey or pinkish-grey. Bare pale yellowish to greyish eye patch. **Juvenile** Resembles adult female but duller. **Immature** Resembles respective adult, but young male has partial (or no) black-and-white facial marks and reduced, duller wingbars and breast-bands. **VOICE** A clear, liquid, sharp *k-kwio k-kwio....* or *kliuw kliuw...* (also transcribed as *KWEtal* or *kuitl*). Alarm a dry croaking *krre-krre-krre-krre-...* [CD3:3a] **HABITS** In pairs and small flocks in rocky and scrubby desert with scattered bushes. Gathers at freshwater source at dusk. **SIMILAR SPECIES** Adult female/juvenile Crowned may recall female/juvenile Lichtenstein's but have face and throat plain yellow-buff; underwing pattern different, more contrasting. **STATUS AND DISTRIBUTION** Uncommon and patchily distributed resident/migrant, Mauritania, Mali, Niger and Chad. Vagrant, N Senegal (Jul and Sep).

FOUR-BANDED SANDGROUSE *Pterocles quadricinctus*   Plate 50 (5)
Ganga quadribande

Other name F: Ganga de Gambie

L 25–28 cm (10–11"). Rather small, short-tailed sandgrouse. In flight, blackish flight feathers contrast with yellowish-buff upperwing-coverts and grey underwing-coverts. **Adult male** Head, upperparts and tail mainly yellowish-buff

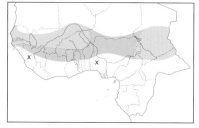

with white forehead and forecrown traversed by black band and rufous-buff crown streaked black. Head-sides, neck and breast washed olive; upperparts barred black and chestnut; rump and tail barred black. Median coverts have narrow black, white-edged subterminal bars. Breast traversed by 3 bands: a white one bordered chestnut above and black below. Belly finely barred black and white; undertail-coverts rufous-buff barred black. Bill dull yellow. Bare yellow eye patch. Feet yellow. **Adult female** Lacks male head markings and breast-bands. Hindneck rufous-buff barred black; head-sides, throat and upper breast deep buff; lower breast to undertail-coverts barred buff and black. Lesser and median coverts have narrow black bars. Bill brownish. **Juvenile** Resembles adult but more rufous, black barring duller and narrower, feathers of upperparts fringed buff, primaries broadly tipped rufous or buff. **VOICE** In flight, a far-carrying whistled twittering *kik-krrr-reee* (also transcribed *wur-wulli* or *pirrou-ee*). [CD6:52] **HABITS** Usually in pairs or small groups in dry wooded grassland, cultivation and open sandy scrub. Flocks gather at dusk and fly to waterholes, where sometimes congregate in numbers. Apparently largely nocturnal. **SIMILAR SPECIES** Chestnut-bellied Sandgrouse has entirely dark underwing and belly, and long tail. Lichtenstein's Sandgrouse is a desert dweller with densely barred plumage, incl. neck and upper breast; male has different breast-band pattern. **STATUS AND DISTRIBUTION** Locally common to not uncommon in broad savanna belt, from S Mauritania to Guinea-Bissau east through Mali and N Ivory Coast to Chad, N Cameroon and CAR. Resident in centre of range; non-breeding migrant to Sahel during rains, breeding migrant in south in dry season. Commonest sandgrouse in our region, with more southerly range than other species. **NOTE** Name 'Four-banded' is a misnomer, as male has only three breast-bands.

PIGEONS AND DOVES
Family COLUMBIDAE (World: *c.* 308 species; Africa: 37; WA: 27)

Small to medium-sized arboreal and terrestrial birds with compact body and small head. Flight strong and fast. Sexes similar in most. Occur from rainforest to desert. Feed on fruit and/or seeds. Calls distinctive and an excellent identification clue. Often produce loud, characteristic clapping sound on taking wing or in display (caused by rigid flight feathers). Most species in the region sedentary.

AFRICAN GREEN PIGEON *Treron calva*
Colombar à front nu

Plate 51 (3)

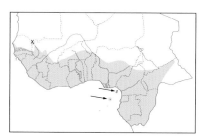

Other names E: Green Fruit Pigeon; F: Pigeon vert à front nu
L 25–28 cm (10–11"). Plump, heavily built green pigeon with bicoloured bill. **Adult** Head and underparts yellowish-green darkening to olive-green on upperparts; purplish patch on shoulder; greater coverts tipped and edged yellow, forming narrow but conspicuous wingbar; secondaries black edged yellow near tip. Feathers of lower belly and vent fringed bright yellow; undertail-coverts chestnut edged buff. Tail feathers blackish with broad grey terminal band, except all-grey central pair. Bill whitish-horn with *large red base.* Feet yellow. Races differ in shade of green and extent of red cere. *T. c. sharpei* slightly greyer, less yellow-olive than *nudirostris*, with cere larger, extending well onto forehead; **nominate** darker and greener than *sharpei*; *uellensis* slightly larger than *sharpei*, with larger cere. Island race *virescens* deeper and duller green than nominate, with more slender bill. **Juvenile** Duller and lacking purplish shoulder patch. **VOICE** Distinctive. Curious and not pigeon-like, starting with a rather soft, fluty trill, suddenly rising, then descending, *hru-hru-wrih-hu-hrruu hrruuu* and ending in throaty chuckling, creaking, barking and growling notes. [CD6:53] **HABITS** Usually in parties, in forest and wooded savanna. Feeds on figs and other fruit, clambering among branches like a parrot, occasionally hanging upside-down. Rarely on ground, but may locally visit saltpans. Easily overlooked when perching quietly in thick foliage. **SIMILAR SPECIES** Bruce's Green Pigeon separated by lemon-yellow lower breast and belly. *Poicephalus* parrots have relatively large head with heavy bill and screeching voice. **STATUS AND DISTRIBUTION** Common resident throughout forest and savanna zones. *T. c. nudirostris* from Senegambia to Guinea; *sharpei* from Sierra Leone to Nigeria and N Cameroon; *calva* from SE Nigeria and Bioko to Congo; *uellensis* SE CAR; *virescens* Príncipe.

SÃO TOMÉ GREEN PIGEON *Treron sanctithomae*

Plate 143 (12)

Colombar de Sao Tomé

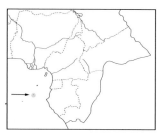

Other name F: Pigeon vert de São Tomé
L 28 cm (11"). Resembles African Green Pigeon, but duller, with somewhat parrot-like bill and small cere. **Adult** Head to mantle and underparts olive-grey; rest of upperparts olive-green; purplish patch on shoulder; greater coverts tipped yellow, forming conspicuous wingbar. Feathers of lower belly and vent fringed bright yellow; undertail-coverts chestnut edged buff. Tail blackish with broad slate-grey terminal band. Bill heavy with hooked tip, whitish-horn with red cere. Feet orange. **Juvenile** Duller and greyer; purplish shoulder patch much reduced. **VOICE** Similar to African Green Pigeon but even more curious and complex, starting with a drawn-out, accelerating, grating rattle *kek kek kek kek kek kek* or *krrrrrrrrrrek* followed by a fluty *hru-hru-wrih-hu-hruu hruuu* and ending with rasping, barking and growling notes. [CD6:54] **HABITS** Usually in canopy of primary and secondary forest, to at least 1600 m. **STATUS AND DISTRIBUTION** Common resident, endemic to São Tomé, where the only green pigeon. **NOTE** Sometimes treated as conspecific with African Green Pigeon (e.g. by Dowsett & Dowsett-Lemaire 1993); if Madagascar Green Pigeon is also included within this species, scientific name becomes *T. australis.*

BRUCE'S GREEN PIGEON *Treron waalia*

Plate 51 (2)

Colombar waalia

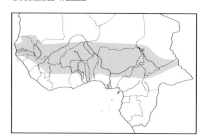

Other names E: Yellow-bellied Fruit Pigeon; F: Pigeon à épaulettes violettes
L 28–30 cm (11–12"). Plump green pigeon with *bright yellow belly.* **Adult** *Head, throat and upper breast pale olive-grey*; upperparts olive-green; lower breast and belly bright yellow, sharply demarcated from upper breast; purplish patch on shoulder; greater coverts tipped and edged yellow, forming narrow wingbar; secondaries edged yellow near tip. Vent mainly *whitish* streaked olive-grey; undertail-coverts chestnut edged buff. Tail feathers slate-grey with broad grey terminal band, except all-grey central pair. Bill whitish-horn with *pinkish-blue cere.* Feet orange. **Juvenile** Duller with no or smaller purplish shoulder patch. **VOICE** Curious and not pigeon-like, starting with a rapid series of hard, cracking notes (resembling slow opening of a creaking door), followed by fluty whistles, suddenly rising, then descending and ending in some abrupt yapping grunts *k k k-k-k-kkkkkkrrrrrrr whuuuuu-wiwhuwheeew errrr whrreh errrr whrreh whrreh whrreh.* [CD6:55] **HABITS** In small flocks in dry wooded savanna and thorn scrub. Partial to figs. Easily overlooked when perched in leafy canopy. **SIMILAR SPECIES** African Green Pigeon distinguished by green breast and belly, yellow (not whitish) vent, yellow legs and red cere at base of bill. **STATUS AND DISTRIBUTION** Not uncommon resident in dry savanna belt, from S Mauritania and Senegambia to Guinea east to Chad and CAR. Local movements, related to rains or ripening of figs, reported.

BLUE-HEADED WOOD DOVE *Turtur brehmeri*

Plate 51 (9)

Tourtelette demoiselle

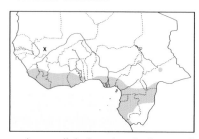

Other name F: Tourterelle à tête bleue
L 25 cm (10"). Dark, rufous-chestnut forest dove with blue-grey head. **Adult** *infelix* Wing has 2–6 brilliant metallic green spots (on tertials and inner greater coverts). Bill purplish tipped greenish. Feet red. In **nominate** wing spots are iridescent golden-copper. **Juvenile** Duller, with indistinct dark barring on upperparts and no metallic spots on wing. **VOICE** A series of plaintive *hoo*s, starting rather hesitantly, accelerating, then fading; very similar to song of Tambourine Dove, but notes slightly modulated, more 'bouncing', esp. at end. [CD6:56] **HABITS** Singly or in pairs, on or near rainforest floor. Shy and unobtrusive, usually in forest interior, but occasionally venturing onto paths and tracks. **STATUS AND DISTRIBUTION** Not uncommon forest resident, from Sierra Leone and SE Guinea to SW CAR and Congo. Also reported from SW Mali and N CAR. Western *infelix* east to coastal Cameroon; nominate from S Cameroon further east and south.

TAMBOURINE DOVE *Turtur tympanistria*

Plate 51 (7)

Tourtelette tambourette

Other name F: Tourterelle tambourette
L 20–22 cm (8–9"). Small with *contrastingly white underparts.* In flight, rufous flight feathers and 2 blackish bands

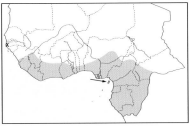

on lower back. **Adult male** Striking *white face* (forehead and area around eye) contrasts with dark brown hindcrown and nape; blackish patch on ear-coverts. Upperparts dark brown. Wing has *c*. 6 dark metallic blue spots (on tertials and innermost coverts). Underparts pure white. Bill and feet dark purplish. **Adult female** Duller face pattern and throat to breast and flanks grey. **Juvenile** Duller and browner, with upperpart feathers fringed and barred rufous and dark brown, giving scaly impression; throat and upper breast dusky-greyish with small dark brown bars; no wing spots. **VOICE** A series of *hoo*s, starting rather slowly, then accelerating *ho, ho, ho ho hohoho hohohohohohohohoho*; similar to song of Blue-headed Wood Dove, but 'flatter' (not modulated), less 'bouncing'. [CD6:57] **HABITS** Singly or in pairs, in various types of forest, dense woodland and plantations. Forages on ground; occasionally venturing onto tracks and in clearings. Shy; mostly seen in fast, direct flight. **SIMILAR SPECIES** Blue-spotted and Black-billed Wood Doves lack contrasting white face and underparts. **STATUS AND DISTRIBUTION** Common resident throughout forest zone and gallery extensions, from Guinea to Congo; also Bioko. Rare, S Senegal.

BLUE-SPOTTED WOOD DOVE *Turtur afer*   Plate 52 (11)
Tourtelette améthystine

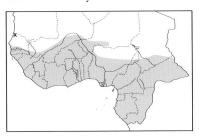

Other names E: Red-billed Wood Dove; F: Tourterelle à bec rouge L 20 cm (8"). *T. a. afer*. Small with *red, yellow-tipped bill*. In flight, *conspicuous rufous primaries*, 2 blackish bands on lower back and broad dark subterminal tail band. **Adult** Top of head blue-grey. Upperparts grey-brown; underparts vinous-pink, paler on abdomen. Wing has 5–6 dark metallic blue spots (on tertials and inner greater coverts); underwing rufous. Feet dark red. **Juvenile** Duller and browner, with upperpart feathers tipped and barred tawny-buff; wing spots absent or smaller and duller; bill dusky. **VOICE** Two muffled, plaintive introductory syllables followed by a series of *hoo*s, initially rather hesitant and irregular, then accelerating, finally fading *oh-wuh oh-wuh oh-ho-ho-ho-hohohohoho*. [CD6:58] **HABITS** Singly or in pairs at forest edges and in clearings, gardens, cultivation, gallery forest and dense woodland. Often seen foraging on ground in open situations and near habitation. **SIMILAR SPECIES** Black-billed Wood Dove has dark bill and occurs in drier habitats (with some overlap). Emerald-spotted Wood Dove has green wing spots and only overlaps in coastal Gabon and Congo. **STATUS AND DISTRIBUTION** Common resident throughout, except arid north. Partial migrant in some areas, with southward movements in dry season, north with rains.

BLACK-BILLED WOOD DOVE *Turtur abyssinicus*   Plate 52 (12)
Tourtelette d'Abyssinie

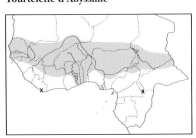

Other name F: Tourterelle à bec noir L 20 cm (8"). **Adult** Very similar to Blue-spotted Wood Dove but with *black bill*. Top of head blue-grey becoming whitish on forehead. Upperparts somewhat greyer brown; underparts slightly paler and pinker, white on belly. **Juvenile** Very similar to juvenile Blue-spotted and probably inseparable in field, but overall paler, with greyer head. **VOICE** A series of soft, plaintive *hoo*s, similar to Blue-spotted Wood Dove, but initial notes more hesitant and drawn out. [CD6:60] **HABITS** Singly or in pairs in dry savanna woodland, open scrub, cultivation, thickets and second growth. Congregates in flocks at waterholes in dry areas. Behaviour as Blue-spotted Wood Dove. **SIMILAR SPECIES** Blue-spotted Wood Dove (q.v. and above); beware confusion with juvenile Blue-spotted, which also has darkish bill. **STATUS AND DISTRIBUTION** Common resident in Sahel and northern savannas, from S Mauritania, Senegambia and Guinea through Mali and N Ivory Coast to Ghana, reaching coast in Dahomey Gap (E Ghana to W Nigeria), and east through N Cameroon to Chad and N CAR. Possibly non-breeding migrant in some areas. Vagrant, Liberia.

EMERALD-SPOTTED WOOD DOVE *Turtur chalcospilos*   Plate 52 (10)
Tourtelette émeraudine

Other name F: Tourterelle émeraudine L 20 cm (8"). **Adult** Similar to Blue-spotted Wood Dove but *bill dusky* and *wing spots brilliant green* and slightly

larger. **Juvenile** Similar to juvenile Blue-spotted but with whitish tips to greater and median coverts. **VOICE** A series of soft, plaintive *hoo*s, very similar to Blue-spotted Wood Dove but slightly longer and higher pitched. [CD6:59] **HABITS** Singly or in pairs in open scrub and second growth. Behaviour as Blue-spotted Wood Dove. **SIMILAR SPECIES** Blue-spotted Wood Dove has red, yellow-tipped bill and dark blue wing spots. **STATUS AND DISTRIBUTION** Uncommon resident, coastal Gabon and Congo. Mainly in E and S Africa.

NAMAQUA DOVE *Oena capensis* Plate 52 (9)
Tourtelette masquée

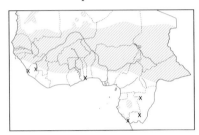

Other names E: Long-tailed Dove; F: Tourterelle à masque de fer L 22–26 cm, incl. tail (8.5–10.0"); WS 28–33 cm (11–13"). Small and slim with *long, steeply graduated, pointed tail*. In flight, *conspicuous rufous primaries* and 2 black bands, separated by pale buff, on lower back. **Adult male** Head mainly grey with *black mask* extending from forehead to breast. Upperparts brown-grey. Upperwing has 3–5 metallic blue spots; underwing mainly rufous. Belly white. Bill dark red broadly tipped yellow. Feet purple-red. **Adult female** Face pale greyish without black; bill grey with reddish base. **Juvenile** Resembles adult female, but upper breast barred dark grey and buff; upperwing-coverts coarsely spotted and barred white, buff and dark grey. Bill grey. **VOICE** A short, soft initial note, followed by a plaintive, drawn-out, emphatic and slightly rising syllable: *oh-whoooooah oh-whoooooah* ... , frequently repeated. [CD3:3b] **HABITS** Singly or in pairs in *Acacia* savanna, thornbush, dry grassland and cultivation, and near habitation. Forages on bare ground; often on roads and tracks. Rises with clattering wings; flight fast, low and direct. **STATUS AND DISTRIBUTION** Common resident and intra-African migrant in broad Sahel belt; common to uncommon dry-season breeding visitor to Sudan and Guinea savannas. Scarce resident, Gabon; rare (vagrant?), Congo. Vagrant, Sierra Leone and Cape Verde Is.

WESTERN BRONZE-NAPED PIGEON *Columba iriditorques* Plate 52 (1)
Pigeon à nuque bronzée

L 28 cm (11"). Medium-sized, mainly dark forest pigeon with bright crimson feet. *Broad pale terminal tail band* conspicuous on landing. Gloss on hindcrown to mantle usually hard to see. **Adult male** Head and neck dark blue-grey; hindcrown and nape glossed green; hindneck and upper mantle glossed copper and vinous-pink; upperparts blackish-slate. Tail blackish-slate; all feathers except central pair broadly tipped buff. Breast and belly dark vinous; undertail-coverts chestnut. Red orbital ring. Bill blue-grey tipped whitish; cere dark red. **Adult female** Crown cinnamon-rufous; rest of head and underparts chestnut washed grey-brown. Hindneck and mantle glossed violet. **Juvenile** Duller and browner. **VOICE** One or two soft, barely audible, hesitating introductory notes abruptly followed, after a short pause, by 5–6 clear coos, last 2 descending in scale *ehuu ehuu KOOOW KOOOW KOOW koo-koo*. [CD6:61] **HABITS** Singly or in pairs in lowland and riparian forest. Usually high in trees or at mid-level; occasionally lower. Unobtrusive; mainly noticed by voice. **SIMILAR SPECIES** Lemon Dove has whitish face and brown upperparts, and frequents lower strata. Blue-headed Wood Dove mainly chestnut and occurs low down. **STATUS AND DISTRIBUTION** Uncommon to scarce forest resident, from Sierra Leone and SE Guinea to Congo. **NOTE** Sometimes treated as conspecific with extralimital Eastern Bronze-naped Pigeon *C. iriditorques* (cf. Dowsett & Dowsett-Lemaire 1993).

SÃO TOMÉ BRONZE-NAPED PIGEON *Columba malherbii* Plate 143 (9)
Pigeon de Malherbe

L 28 cm (11"). Small, dark pigeon with glossy nape, rufous undertail-coverts and bright red feet. **Adult male** Head and underparts grey; nape and upper mantle glossy green or purple; upperparts and tail blackish-slate; vent and undertail-coverts ochreous. Bill grey with pale tip. **Adult female** Similar but with ochre-mottled underparts. **Juvenile** As adult female but duller, with browner underparts. **VOICE** Two or three raucous initial notes followed by a fast,

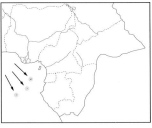

descending and accelerating series of bouncing coos, ceasing abruptly *krohuu, krohuu-krooh koo-koo-koo-koo-kookookookookookoo*. Also a deep, raucous sustained cooing *krurrr-rrurrrrrrrrrrh*. Calls include a low *krrreu* and a raucous *krrrek*. [CD6:62] **HABITS** Singly or in small parties in forest, forest regrowth and plantations with tall trees. Usually high in trees or in mid-stratum, but also within dense undergrowth and on ground. **SIMILAR SPECIES** Lemon Dove has whitish face and frequents lower strata. São Tomé Olive Pigeon much larger, maroon, with yellow bill. **STATUS AND DISTRIBUTION** Common to uncommon resident, endemic to São Tomé, Príncipe and Annobón. **NOTE** Sometimes treated as conspecific with Western and Eastern Bronze-naped Pigeons under name *C. iriditorques* (cf. Dowsett & Dowsett-Lemaire 1993).

CAMEROON OLIVE PIGEON *Columba sjostedti*   Plate 51 (4)
Pigeon du Cameroon

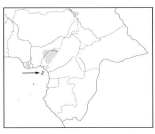

L 36–40 cm (14–16"). Large, *very dark* pigeon with *yellow eye and bill*. **Adult** Head dark blue-grey. Upper- and underparts mainly vinous-maroon; mantle, breast and belly profusely speckled white; some white speckles on wing-coverts. Tail blue-black. Bill yellow with red base. Feet dark purple. **Juvenile** Much duller. Not speckled but feathers of lower breast and belly fringed white; upperparts and upper breast feathers fringed pale grey and rufous. Bill dusky. **VOICE** A deep, drawn-out growl, followed by a quavering series of low, muffled coos; also a bleating sound in display flight (both vocalisations very similar to those of extralimital and closely related Olive Pigeon *C. arquatrix*). [CD6:64] **HABITS** Singly, in pairs or small groups in montane forest and forested gullies, thickets and gallery forest in highland areas. Mainly in canopy. **SIMILAR SPECIES** White-naped Pigeon has large white or whitish nape patch, plain wing-coverts, pale terminal tail band and brownish-red bill. **STATUS AND DISTRIBUTION** Common resident in highlands of SE Nigeria (Obudu and Mambilla Plateaux) and W Cameroon (Mt Cameroon to Mt Oku); also Bioko. **NOTE** Sometimes treated as conspecific with Olive Pigeon *C. arquatrix* (cf. Dowsett & Dowsett-Lemaire 1993).

SÃO TOMÉ OLIVE PIGEON *Columba thomensis*   Plate 143 (6)
Pigeon de Sao Tomé

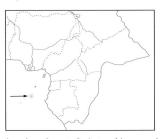

Other name E: Maroon Pigeon
L 37–40 cm (14.5–16.0"). Large, dark, mainly maroon pigeon with yellow bill and feet. **Adult male** Head dark slate-grey; upper- and underparts maroon darkening to blackish-slate on rump and tail; small white spots on wing-coverts; undertail-coverts grey. **Adult female** Duller; breast and wing-coverts dark brownish-grey only slightly washed maroon. **Juvenile** Mainly dark brown, lacking any white spotting. **VOICE** A quavering series of very low, muffled coos (reminiscent of Cameroon Olive Pigeon and extralimital Olive Pigeon *C. arquatrix*). [CD6:66] **HABITS** Mainly in canopy of mid- to high-altitude primary forest, but also in lowland forest, cultivated areas, forest regrowth and bamboo forest. Quiet and inconspicuous, thus easily overlooked when perched. Tame, permitting close approach. **SIMILAR SPECIES** São Tomé Bronze-naped Pigeon is much smaller, grey and slate-black, with grey bill. Lemon Dove even smaller, with whitish face, and occurs mainly low down. **STATUS AND DISTRIBUTION** Locally common but little-known resident, endemic to São Tomé. *TBW* category: THREATENED (Vulnerable). **NOTE** Sometimes treated as conspecific with extralimital Olive Pigeon *C. arquatrix* (cf. Dowsett & Dowsett-Lemaire 1993).

WHITE-NAPED PIGEON *Columba albinucha*   Plate 51 (5)
Pigeon à nuque blanche

L 36 cm (14"). *Large, very dark pigeon resembling Cameroon Olive Pigeon* but with *large white or whitish nuchal patch. Tail with broad pale grey terminal band.* Eye yellow with orange-red outer ring. Bill brownish-red tipped dirty yellowish or bright red. Feet bright red. **Adult male** Nuchal patch (from eye to hindcrown and nape) white. Throat pale grey. White speckles confined to lower breast and belly. **Adult female** Nuchal patch pale greyish. **Juvenile** Much duller. Head grey; nuchal patch absent or indistinct. Not speckled but feathers of lower breast fringed white, those of mantle and upper breast narrowly fringed rusty. Bill dusky; eye brownish. **VOICE** Resembles Cameroon Olive Pigeon. Described as a deep, rather quavering, deliberate note followed by 3–4 coos in decreasing volume (Eisentraut 1968). In flight, a bleating *meeeh* (Prigogine 1971). **HABITS** Usually in pairs or small groups, in

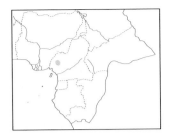

canopy of transitional and montane forest, at 1000–2100 m. Visits farmbush with coffee plantations near primary transitional forest. Occasionally at mid-level or lower. Spends long periods perched motionless. Hard to see when sitting quietly in foliage. **SIMILAR SPECIES** Cameroon Olive Pigeon lacks white nape patch and has speckled wing-coverts, wholly dark tail and yellow bill. Western Bronze-naped Pigeon also has broad pale terminal tail band, but is noticeably smaller and differs in other plumage characteristics. **STATUS AND DISTRIBUTION** Rare and very local forest resident in highlands of W Cameroon (Rumpi Hills, Mt Kupe, Bakossi Mts, Mt Manenguba). Perhaps at other montane localities. *TBW* category: Near Threatened.

AFEP PIGEON *Columba unicincta*　　　　　　　　Plate 51 (6)
Pigeon gris

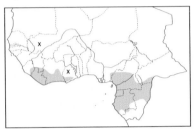

Other name E: Grey Wood Pigeon
L 36–40 cm (14–16"). Large, *pale forest pigeon* with conspicuous red eye. In flight, blackish wings and pale grey tail band contrasting with broad black terminal band. **Adult** Head pale grey; upperparts darker grey, feathers narrowly fringed pale grey. Throat white; breast pale pinkish becoming pale grey on flanks; rest of underparts white. Bill grey-blue tipped paler. Eye and orbital ring crimson. Feet blue-grey. **Juvenile** Upperparts darker; breast brownish-vinous. **VOICE** A deep, drawn-out *hoooo*, uttered in series of up to 20, preceded by a soft and barely audible vibrant guttural *oooorrr*. [CD6:67] **HABITS** Singly or in small groups of 3–5 in canopy of moist evergreen and riparian forest. Often perches conspicuously on bare branch atop large tree. May locally visit saltpans in large numbers. **SIMILAR SPECIES** None. All other forest pigeons are much darker. **STATUS AND DISTRIBUTION** Not uncommon to scarce forest resident, from Sierra Leone and SE Guinea to Ghana and from SE Nigeria to SW CAR and Congo.

WOOD PIGEON *Columba palumbus*　　　　　　　　Plate 146 (8)
Pigeon ramier

L 40–45 cm (16–18"); WS 68–77 cm (27–30"). Very large and heavy looking. In flight, broad white transverse band on wing and broad black terminal tail band. **Adult** Mainly grey above with *large white patch on neck-sides*. Breast vinous-pink becoming pale grey on belly. Neck-sides glossed green and purple. Eye pale yellow. Bill yellow with red base. Feet pinkish-red. **Juvenile** Duller; lacks white neck patch. Eye and bill dark. **VOICE** A short series of 5 muffled coos with stress on second note (sometimes rendered *who COOKS for-you, oh...*). [CD3:4a] **HABITS** Feeds on ground and in trees. Flight fast and direct. Wings make clattering noise on take-off. **SIMILAR SPECIES** None, but may sometimes recall large falcon in flight. **STATUS AND DISTRIBUTION** Palearctic vagrant, Mauritania (Nouakchott, Apr 1981, one). Nearest breeding population is in NW Africa (*excelsa*); nominate (differing only by smaller mean size) is winter visitor from Europe to NW Africa.

SPECKLED PIGEON *Columba guinea*　　　　　　　　Plate 51 (10)
Pigeon roussard

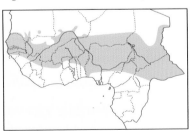

Other name F: Pigeon de Guinée
L 35–40 cm (14–16"). *P. g. guinea*. Large grey-and-chestnut pigeon with *triangular white spots on wing-coverts*. In flight, pale grey rump and uppertail-coverts contrast with dark upperparts; tail grey with broad black terminal band. **Adult** Head and neck grey with conspicuous bare red patch around yellow eye, and vinous-chestnut streaking on neck. Upperparts vinous-chestnut; underparts grey. Bill blackish; cere whitish. Feet dark pinkish. **Juvenile** Duller and browner; bare eye patch brownish. **VOICE** A fast, rising series of *oo* notes, increasing in pace and volume *oo-oo-oo-oo-oo-oo-oo-oo-OOH-OOH-OOH-....* Also *rroh rroh rroh*. [CD6:68] **HABITS** Usually in flocks in open country, incl. savanna, cultivation, gardens; also villages and towns, where often perches on buildings. Feeds on ground. **SIMILAR SPECIES** None, but inadequate views of pale-rumped feral pigeons may be confusing. **STATUS AND DISTRIBUTION** Common resident in Sahel and savanna belts, from S Mauritania to Guinea east to Chad and CAR. Reaches coast in Dahomey Gap. Local seasonal movements probable, as numbers increase in dry season in some areas.

ROCK DOVE *Columba livia*
Pigeon biset

Plate 51 (11)

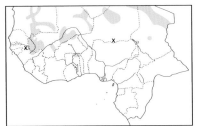

L 31–34 cm (12–13"); WS 62–68 cm (24.5–27.0"). **Adult *gymnocyclus*** *Dark slate-grey with pure white rump*, 2 broad black wingbars and white underwings. Neck and upper mantle glossed metallic green and purple. Eye yellowish with narrow red orbital ring. Bill dark grey with white cere. Feet dark red. Saharan *targia* much paler; mainly soft grey (incl. rump) with contrasting black wingbars and black terminal tail band. **Juvenile** Duller; little or no iridescence on neck; wingbars less distinct. **VOICE** A low, drawn-out, emphatic *rrooh* or *o-o-orr* repeated several times. [CD3:6] **HABITS** In flocks on coastal cliffs, rock ledges, escarpments, gorges, and rocky hills and mountains; also buildings. Feeds on ground. Flight fast and powerful. **SIMILAR SPECIES** Rock Dove is ancestor of domestic pigeon; feral population of latter highly variable, some very similar to genuine Rock Dove. **STATUS AND DISTRIBUTION** Locally common to not uncommon resident, Mauritania, coastal Senegal, Guinea, Mali, Burkina Faso, N Ghana and N Togo (*gymnocyclus*) and central Sahara (Hoggar, Aïr, Tibesti, Ennedi: *targia*). Common resident, Cape Verde Is. **NOTE** Birds from Cape Verde Is are very dark and have been separated as ssp. *atlantis*; position unclear, perhaps derived from feral pigeons (*BWP*, Hazevoet 1995).

LEMON DOVE *Aplopelia larvata*
Pigeon à masque blanc

Plate 51 (8) & 143 (8)

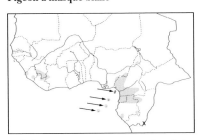

Other names E: Cinnamon Dove; F: Tourterelle à masque blanc
L 24–25 cm (9.5–10.0"). Rather small, dark forest pigeon with *whitish face*. **Adult male *inornata*** Forehead, forecrown and throat whitish; rest of crown to upper mantle greyish glossed green and violet; upperparts blackish-brown. Tail feathers blackish-brown with broad grey terminal band except on central pair. Underparts grey, paler on lower belly. Dark red to brown eye with red orbital ring. Bill black. Legs dark purple-red. Island race *simplex* similar but slightly paler; *principalis* has face pale greyish and underparts vinous becoming buffish on belly. **Adult female *inornata*** Similar to adult male but upperparts browner and *underparts cinnamon* becoming buff or whitish on lower belly and vent; *principalis* similar, slightly paler on breast. Female *simplex* paler and browner than male; head and breast greyish-buff; nape and neck-sides more rufous glossed golden and green. **Juvenile** Duller and browner with black-and-rufous barring. **VOICE** A monotonous series of 10–50 similar, low notes, often slightly rising in volume *hoot hoot hoot hoot hoot hoot* ... Also a slower series of lower notes *ooorr, ooorr, ooorr, ooàrr, ooorr,* ... and a faster, slightly higher pitched *whoopwhoopwhoopwhoopwhoop...* [CD6:63] **HABITS** Singly or in pairs in various lowland and montane forests. Largely terrestrial. Usually elusive and easily overlooked. **SIMILAR SPECIES** Bronze-naped pigeons are canopy dwellers lacking white face. Blue-headed Wood Dove is bright chestnut with contrasting blue head. **STATUS AND DISTRIBUTION** Rare to locally uncommon forest resident, from Sierra Leone and N Liberia to SE Guinea and W Ivory Coast, from SE Nigeria (Obudu and Mambilla Plateaux) to Congo, and on Bioko and Annobón (*inornata*). Also São Tomé (*simplex*) and Príncipe (*principalis*), where common. **NOTES (1)** Placed in *Columba* by BoA, but we prefer to follow the more widely accepted view and retain this ground-feeding, essentially seed-eating dove within its traditional genus *Aplopelia*. See also Dowsett & Dowsett-Lemaire 1993 (*pro Aplopelia*) and *HBW* (*pro Columba*). **(2)** São Tomé form *simplex* accorded specific rank by *HBW* on basis of alleged vocal differences.

RED-EYED DOVE *Streptopelia semitorquata*
Tourterelle à collier

Plate 52 (5)

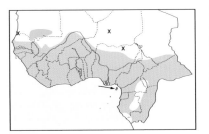

L 30–34 cm (12–13"). Large, dark 'collared' dove with grey crown, dark grey-brown upperparts and vinous-pink underparts. In flight, dark band across central tail. **Adult** Black collar on hindneck. Underparts become grey on lower belly. Eye dark red with narrow dark red orbital ring (visible at close range). Bill black. Feet purple-red. **Juvenile** Upperparts fringed rufous-buff; breast tinged rufous; black collar indistinct. **VOICE** A rapid series of 6 short, rather nasal coos, with stress on 5th, *ho-hu ho-hu HOO-ho* (or *hooHOOhuhu HOOhu*) with variations. Also *rroorr-huhu rroorr-huhuhu* ... and *rroorr-rrhuh rroorr-rrhuh...* Occasionally a nasal *ehèèh* on landing. [CD6:69]

HABITS Singly or in pairs, occasionally small groups, in variety of habitats, incl. woodland, gardens, clearings, forest edge and mangrove. Not in forest. Feeds mainly on ground, sometimes in trees. Wings make clattering noise on take-off. **SIMILAR SPECIES** African Mourning Dove is smaller and overall paler, has white in tail but no black band, and a yellow eye with red orbital ring. **STATUS AND DISTRIBUTION** Common and widespread resident throughout forest and savanna zones; also Bioko. North to extreme S Mauritania (uncommon), SW Niger, S Chad; absent from extreme N Nigeria, and N and SE Cameroon.

AFRICAN MOURNING DOVE *Streptopelia decipiens* Plate 52 (4)
Tourterelle pleureuse

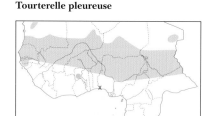

L 28–30 cm (11–12"). Medium-sized, dry-country 'collared' dove. In flight, tail has white corners. **Adult** *shelleyi* Face and crown grey with *yellow to pale orange eyes* and *narrow but conspicuous red orbital ring*. Hindneck and neck-sides vinous-pink; black collar on hindneck bordered white. Upperparts grey-brown. Tail feathers grey-brown tipped white above, except central pair. Breast soft vinous-pink becoming grey on belly. Bill black. Feet purple-red. *S. d. logonensis* has slightly greyer breast. **Juvenile** Browner, esp. on head; eye pale brownish. **VOICE** An emphatic note followed by a quavering, prolonged coo, *whoh! kho-o-o-o-o-o-o* and a rapidly repeated, 3-syllable *whoh whoh-hoo*. On landing, a resonant, quavering gargle *arh-r-r-r-r-rw* or *krrrowrrru*. [CD6:70] **HABITS** Singly or in small groups in riparian *Acacia* woodland and arid wooded grasslands near water. Feeds mainly on ground. Moves daily between roosting, foraging and drinking areas, often in large flocks. **SIMILAR SPECIES** Red-eyed Dove is larger and overall darker, tail lacks white but has black band, and has red eyes. **STATUS AND DISTRIBUTION** Common resident in broad Sahel belt, from Mauritania and Senegambia to Chad (north to Ennedi) and N CAR. *S. d. shelleyi* east to S Niger and Nigeria; *logonensis* from N Cameroon east. Local seasonal movements recorded.

VINACEOUS DOVE *Streptopelia vinacea* Plate 52 (7)
Tourterelle vineuse

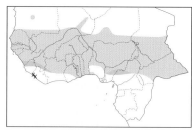

L 25 cm (10"). The region's smallest and commonest 'collared' dove. Head and underparts pale vinous-pink; eye dark. In flight, tail has white corners; underwing pale grey. **Adult** Black collar on hindneck (bordered grey on upper edge). Upperparts grey-brown. Tail feathers grey-brown tipped white above, except central pairs. Underparts become white on belly. Bill black. Eye dark brown. Feet purple-red. **Juvenile** Similar but with pale fringes to wing-coverts. **VOICE** A rather high-pitched, far-carrying, fast, 3-syllable *wheh heh-heh wheh heh-heh wheh heh-heh...* and faster *wheh-heho wheh-heho wheh-heho...* (or *ho-huhu ho-huhu ho-huhu...*), uttered throughout the day and even at night. Also fast series of rolling *horrrh-horrrh-horrrh-horrrh-...*. Nasal call on landing. [CD6:71] **HABITS** Singly, in pairs or flocks in a variety of savanna habitats, incl. cultivation. Feeds on ground. **SIMILAR SPECIES** Red-eyed Dove is much larger and darker, with black tail band and red eye. African Mourning Dove has grey head with yellow eye and red orbital ring. African Collared Dove is paler overall. **STATUS AND DISTRIBUTION** Common to very common resident in broad savanna and Sahel belt, from Mauritania to Sierra Leone east to Chad and CAR. Also coastal savannas. Seasonal movements recorded, north with rains, south in dry season. Rare dry-season visitor, Liberia.

RING-NECKED DOVE *Streptopelia capicola* Plate 52 (6)
Tourterelle du Cap

Other name E: Cape Turtle Dove
L 25 cm (10"). *S. c. tropica*. Pale 'collared' dove with dark eye. In flight, tail has white corners; underwing grey. **Adult** Head bluish-grey washed pink; black collar on hindneck. Upperparts grey-brown. Tail feathers grey-brown broadly tipped white, except central pairs. Underparts pale pinkish-grey becoming white on belly. Bill black. Eye blackish-brown. Feet dark purplish. **Juvenile** Feathers of upperparts fringed buff. **VOICE** A rather high-pitched, rhythmic, monotonous and far-carrying *ku-KOORRRR-ku ku-KOORRRR-ku ku-KOORRRR-ku...* uttered throughout the day and even at night. Also a faster *kuKOORRRR kuKOORRRR kuKOORRRR...* Harsh, nasal call on landing.

[CD6:72] **HABITS** In pairs and flocks in open wooded grassland. Feeds mainly on ground, mostly in early morning and late afternoon. Often on tracks and roads. **SIMILAR SPECIES** Red-eyed Dove, the only sympatric 'collared' dove, is much larger and darker, with black tail band and dark red eye. **STATUS AND DISTRIBUTION** An E and S African species, reaching our region in SE Gabon and S Congo, where locally scarce to not uncommon resident.

AFRICAN COLLARED DOVE *Streptopelia roseogrisea*   Plate 52 (3)
Tourterelle rieuse

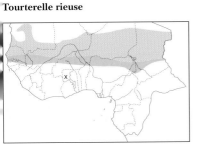

Other name E: Rose-grey Dove
L 27–29 cm (10.5–11.5"); WS 45–50 cm (17.75–19.5") *S. r. roseogrisea.* Pale, fairly uniform, medium-sized 'collared' dove. In flight, pale wing-coverts contrast with dark primaries; tail has white corners; underwings white. **Adult** Head and underparts pale grey-pink; black collar on hindneck (bordered white at upper margin). Upperparts pale grey-brown. Tail pale grey-brown; all feathers, except central pairs, tipped white. Underparts become white on belly. Bill black. Eye dark red with narrow white orbital ring. Feet purple-red. **Juvenile** Duller, with indistinct collar and pale fringes to feathers of upper-

parts. **VOICE** A single, emphatic note followed, after a brief pause, by a sustained, rolling purr *whooh rrrwhrrrooh* or *ooh krrruuuuu.* Excitement call a fast, nasal jeering *heh-heh-heh-heh-heh-heh-...* [CD6:73] **HABITS** In pairs or flocks in thornbush, arid farmland and at desert edge. Feeds on ground, often with other doves. Congregates in large numbers at water and roosts. **SIMILAR SPECIES** Similar-sized African Mourning Dove easily separated by much darker plumage and yellow eye with red orbital ring. Vinaceous Dove slightly smaller and darker, with grey underwings and without white orbital ring. **STATUS AND DISTRIBUTION** Common resident and partial migrant in Sahel belt, from Mauritania and Senegambia through Mali, N Burkina Faso, Niger and N Nigeria to N Cameroon and Chad. Vagrant, N Ghana. Local seasonal movements recorded at northern and southern edges of range, south in dry season, north with rains.

EUROPEAN TURTLE DOVE *Streptopelia turtur*   Plate 52 (1)
Tourterelle des bois

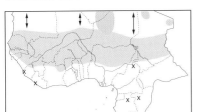

L 26–28 cm (10–11"); WS 47–55 cm (19.0–21.5"). Slender, medium-sized dove with black-and-white streaked patch on neck-sides and patterned upperparts. In flight, blackish flight feathers separated from rufous shoulders by blue-grey band; tail has contrasting white terminal band. **Adult** *turtur* Top of head and hindneck blue-grey. Scapulars and most wing-coverts broadly fringed rufous with contrasting black centres; outer wing-coverts grey. Rump and uppertail-coverts mainly blue-grey. Tail blue-grey above, black below; all feathers, except central pair, broadly tipped white. Underparts pinkish becoming white on belly. Bill blackish. Eye yellow. Feet dark pink. *S. t. arenicola* paler and slightly smaller; *hoggara* sandier, with crown buffier and wing-coverts more broadly fringed orange-buff. **Juvenile** Paler and duller; no neck patch. **VOICE** A deep purring *rrrurrr rrrurrr rrrurrr...* Not recorded on migration. [CD3:8] **HABITS** Locally highly gregarious on migration. Frequents dry savannas and farmland, feeding on ground. Flight fast and direct with clipped wingbeats and tilting from side to side. **SIMILAR SPECIES** No other dove has white-striped black neck patch. Slightly larger and bulkier Adamawa Turtle Dove is darker overall with solid black neck patch; in flight, appears more overall grey-brown above and dark below. **STATUS AND DISTRIBUTION** Common Palearctic winter visitor to Sahel belt, from S Mauritania and Senegambia east to Chad (European nominate and N African *arenicola*). Vagrant, Sierra Leone, Liberia, N Ivory Coast (Comoé NP), NE Gabon and N Congo (Odzala NP). Rare passage migrant, Cape Verde Is. Saharan race *hoggara* resident in mountains of Niger and Chad (Hoggar, Aïr, Tibesti, Ennedi).

ADAMAWA TURTLE DOVE *Streptopelia hypopyrrha*   Plate 52 (2)
Tourterelle de l'Adamaoua

L 30–31 cm (*c.* 12"). Fairly large and *dark* with *broad black patch on neck-sides.* **Adult** Head, neck and breast blue-grey, paler on forehead and throat. Upperparts dark brown fringed buff. Scapulars and most wing-coverts broadly fringed rufous with contrasting black centres; outer wing-coverts fringed blue-grey. Lower underparts dark cinnamon-pink becoming grey on vent. Tail blackish; all feathers, except central pair, broadly tipped pale grey. Bill blackish. Eye orange with reddish orbital ring. Feet purple-red. **Juvenile** Paler and duller. **VOICE** A low, hard purring *rrrurrr rr-rrurr,* resembling that of European Turtle Dove, but sharper and deeper; sometimes with a 4th

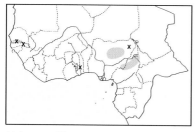

syllable. [CD6:74] **HABITS** Singly (when breeding) or in flocks in various habitats, incl. open woodland on rocky, hilly terrain, wooded gorges, riparian woodland, cultivation and gardens. Feeds mostly on ground. Associates with other doves, e.g. Vinaceous Dove and Speckled Pigeon; in Senegambia with European Turtle Dove and African Mourning Dove. **SIMILAR SPECIES** Slightly smaller and slimmer European Turtle Dove is overall paler, with black-and-white striped neck patch and pinkish (not grey) breast. **STATUS AND DISTRIBUTION** Locally common to uncommon resident in C and E Nigeria (Jos-Bauchi and Mambilla Plateaux), N Cameroon (Adamawa Plateau and Bénoué NP) and SW Chad. Local seasonal movements reported. Also recorded from Togo, where perhaps vagrant. Recently discovered in Senegal (Djoudj area and Niokolo-Koba NP) and Gambia; status unclear.

LAUGHING DOVE *Streptopelia senegalensis* Plate 52 (8)
Tourterelle maillée

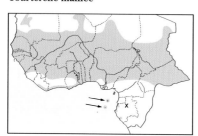

L 23–25 cm (9–10”); WS 40–45 cm (16–18”). *S. s. senegalensis.* Slender and small with mainly rufous-brown upperparts, *blue-grey outer wing-coverts*, and *black-speckled rufous necklace*. In flight, blackish flight feathers separated from rufous shoulders by blue-grey panel; tail has white corners; undertail white. **Adult male** Head vinous-pink. Mantle rufous-brown; back to uppertail-coverts blue-grey. Tail feathers dark grey, tipped white, except central pairs. Underparts vinous-pink becoming white on belly. Bill dusky-brown. Eye dark brown. Feet purple-red. **Adult female** Paler head and breast, and less prominent necklace. **Juvenile** Duller and browner, without necklace. **VOICE** A single, emphatic note followed by a rather soft series of 5–7 hurried notes, first rising, then falling *ho ho-hu-hu-ho* or *hoo koHUHUhu-hoo*, likened to a gentle laugh or chuckle. [CD3:9] **HABITS** Singly or in pairs, occasionally small groups, in villages and towns, farmland, bush and open woodland. Feeds on ground. Not shy. May congregate in numbers at water. Towering display flight with clattering wingbeats, gliding down with open wings and fanned tail. **SIMILAR SPECIES** European Turtle Dove has chequered upperparts, no necklace but a black-and-white neck patch, and less grey on wing. **STATUS AND DISTRIBUTION** Widespread and common resident. Commonest in Sahel; also in oases and Saharan mountains in W and C Mauritania, NE Mali (Adrar), NW Niger (Aïr) and N and E Chad (Tibesti, Ennedi); not in forest. Also São Tomé and Príncipe (introduced?).

PARROTS AND LOVEBIRDS
Family PSITTACIDAE (World: *c.* 330 species; Africa: 19; WA: 9)

Very small to large, mostly arboreal birds with compact body and stout, strongly hooked and powerful bills, also used in climbing. Legs short, feet strong and zygodactyl, used like hands in feeding. Sexes similar or nearly so. Flight strong, fast and direct. Noisy, especially in flight. Calls include high-pitched screeches (parrots) and shrill twitterings or chirrupings (lovebirds). Rather gregarious. Feed mostly on fruits and seeds. Nest in tree holes. Several W African species sold as pets.

GREY PARROT *Psittacus erithacus* Plate 53 (9)
Perroquet jaco

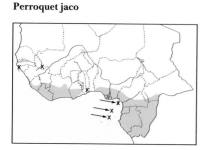

Other name F: Perroquet gris
L 28–39 cm (11.0–15.25”). Large, stocky parrot with unique grey plumage. **Adult** *erithacus* All grey with *contrasting paler grey rump* and *bright scarlet tail* (both esp. conspicuous in flight). Bare facial area white. Pale fringes to feathers give scaly appearance. Primaries blackish-grey. Upper- and undertail-coverts scarlet. Bill blackish. Eye yellow. Distinctive western *timneh* smaller and darker with dark maroon tail edged brownish. Uppertail-coverts dark grey tinged reddish. Upper mandible pinkish-ivory tipped black. **Juvenile** Eye grey. *P. e. erithacus* also has tail dark red at tip. **VOICE** Vocal. The

only parrot in the region to utter loud clear whistles, besides variety of high-pitched screeches, harsh and grating calls, and imitations. [CD6:76] **HABITS** In pairs or small flocks. Mainly in lowland forest; also gallery forest, wooded savanna and mangroves. Congregates in numbers at regular roosts. Partial to oil-palm fruits; may travel long distances in search of fruiting trees or minerals in saltpans. Generally wary. **SIMILAR SPECIES** Red-fronted Parrot, seen from distance or against light, distinguished by faster wingbeats, and less raucous and slightly softer calls lacking whistles. **STATUS AND DISTRIBUTION** Locally common to scarce resident. *P. e. timneh* from Guinea to Ivory Coast (west of Comoé R.); also Guinea-Bissau and S Mali. Nominate from E Ivory Coast (east of Comoé R.) to S CAR and Congo; also Bioko, São Tomé and Príncipe. Suffers from heavy trapping for pet trade. **NOTE** The two forms may be separate species (very little interbreeding; voice reportedly 'fairly distinctive'. *HBW*).

BROWN-NECKED PARROT *Poicephalus robustus*   Plate 53 (8)
Perroquet robuste

L 28–33 cm (11–13"). *P. r. fuscicollis*. **Adult male** Large, green parrot with brownish-grey head and conspicuously large, horn-coloured bill. Leading edge of wing and thighs red. Flight and tail feathers blackish. **Adult female** Red forehead and forecrown. **Juvenile** Duller; head brownish-olive; no red on wings and thighs. **VOICE** A strident, harsh *zzkeek*. [CD6:77] **HABITS** Singly, in pairs or small groups in savanna woodland; also mangroves (Gambia). Undertakes regular daily flights from roosts to feeding areas. Generally wary. **SIMILAR SPECIES** Red-fronted Parrot has green head, darker upperparts and wing-coverts, and contrasting yellowish-green rump and uppertail-coverts. **STATUS AND DISTRIBUTION** Patchily distributed, scarce to locally uncommon resident, Senegambia, Guinea-Bissau, SW Mali, N Guinea, C Ivory Coast (esp. Lamto) and Ghana. Rare, Sierra Leone, Liberia, Togo and EC Nigeria. Local movements reported.

RED-FRONTED PARROT *Poicephalus gulielmi*   Plate 53 (7)
Perroquet à calotte rouge

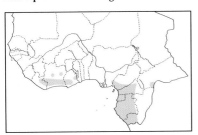

Other name F: Perroquet à front rouge
L 26–30 cm (10.25–12.0"). **Adult** *fantiensis* Large, green parrot with variable amount of orange on forehead, leading edge to wing and thighs. Upperparts and wings black fringed dusky-green; rump and uppertail-coverts yellowish-green, flight feathers and tail black. Bill mostly horn-coloured, with variable amount of black. Cere and orbital ring pale buffish-yellow. In **nominate** orange replaced by orange-red. **Juvenile** Duller, without orange or orange-red; head (esp. crown) dusky-brown; feathers of upperparts and wings have narrower dusky-green fringes; underparts more bluish-green.
VOICE High-pitched screeches. [CD6:78] **HABITS** In pairs or small groups in lowland forest. **SIMILAR SPECIES** Brown-necked Parrot is duller with greyish head and heavier bill. Ranges overlap in C Ivory Coast, but habitat normally different. Grey Parrot, seen from distance or against light, distinguished by slower wingbeats, more raucous and louder calls, including whistles. **STATUS AND DISTRIBUTION** Rare to uncommon resident, from Liberia to Ghana (*fantiensis*), from SE Nigeria to Congo and in S CAR (nominate).

MEYER'S PARROT *Poicephalus meyeri*   Plate 53 (5)
Perroquet de Meyer

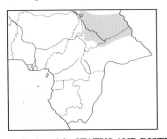

Other name E: Brown Parrot
L 21–23 cm (8.25–9.0"). *P. m. meyeri*. **Adult** Medium-sized parrot with ash-brown head, upperparts and tail, conspicuous yellow patch on crown and leading edge of wing, and green to blue-green rump and underparts. Underwing-coverts and thighs yellow. Bill black. **Juvenile** Duller; yellow lacking on crown and thighs, reduced or lacking on upper- and underwing; feathers of upperparts and wings fringed green. **VOICE** Harsh, high-pitched shrieks. [CD6:79] **HABITS** In pairs or small parties in savanna woodland. Generally wary. **SIMILAR SPECIES** Niam-niam Parrot is slightly larger and appears greener above and below. Unaccompanied juvenile may be difficult to distinguish. **STATUS AND DISTRIBUTION** Common resident, S Chad to N CAR.

SENEGAL PARROT *Poicephalus senegalus*
Perroquet youyou

Plate 53 (4)

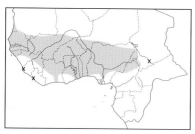

L 23 cm (9"). **Adult** Rather small green parrot with grey head and conspicuous yellow to orange belly. Breast and thighs green; underwing-coverts yellow. Eye yellow. **Nominate** has breast-sides, flanks, underwing- and undertail-coverts yellow darkening to orange on central belly. In *versteri* central underparts deep-orange to scarlet. Poorly defined *mesotypus* intermediate. **Juvenile** Duller; underparts mostly greenish. Eye dark brown. **VOICE** Various harsh shrieks. [CD6:80] **HABITS** In pairs or small groups in savanna woodland and farmland with scattered trees. **SIMILAR SPECIES** No other parrot in the region has yellow or orange on underparts. **STATUS AND DISTRIBUTION** Common to uncommon resident in broad savanna belt. Nominate from Senegambia and Guinea to NW Nigeria; *versteri* further south, from W Ivory Coast to SW Nigeria; *mesotypus* in NE Nigeria and adjacent parts of Niger, N Cameroon and SW Chad, and NE CAR. Seasonal movements reported, north with rains (e.g. in S Mauritania). **NOTE** Ssp. *mesotypus* sometimes considered indistinguishable from, or only one end of cline within nominate (*HBW*).

NIAM-NIAM PARROT *Poicephalus crassus*
Perroquet des niam-niam

Plate 53 (6)

L *c.* 25 cm (10"). **Adult** Little-known, largely grass-green parrot with greyish-brown head and upper breast. Flight feathers blackish-brown with dark green outer webs; underwing-coverts green. Tail blackish-brown edged dark green. Upper mandible dusky-horn, lower mandible yellowish-horn. Eye yellow. **Juvenile** Grey-brown areas suffused with yellowish; belly paler; innermost secondaries edged yellow. **VOICE** A short, sharp screech. [CD6:81] **HABITS** In pairs or small flocks in wooded savanna and forest–savanna mosaic; often near water. **SIMILAR SPECIES** Compare slightly smaller Meyer's and Senegal Parrots. Unaccompanied juvenile may be difficult to distinguish. **STATUS AND DISTRIBUTION** Local resident, SW Chad and CAR. Also claimed from N Cameroon (1 sight record, Waza). Status inadequately known.

RED-HEADED LOVEBIRD *Agapornis pullarius*
Inséparable à tête rouge

Plate 53 (2)

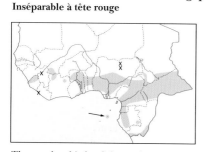

L 13–15 cm (5–6"). *A. p. pullarius.* Very small, bright green parrot with orange-red to yellow face and bill. Short tail with green central feathers; others mainly red with subterminal black bar and green tip. In flight, striking cobalt-blue rump, triangular wings and short tail. **Adult male** Face orange-red; underwing-coverts black. **Adult female** Face orange-yellow; underwing-coverts green. **Juvenile** Face yellow. Young male distinguishable by black underwing-coverts. **VOICE** A clear, loud *kl-eee*, uttered in flight and at rest. [CD6:82] **HABITS** In pairs or small groups in grassy savanna woodland. Feeds mainly on grass seeds. Flight fast and direct. **SIMILAR SPECIES** The two lovebirds of the region are separated by habitat. Black-collared occurs exclusively in mature lowland forest. **STATUS AND DISTRIBUTION** Locally common to uncommon resident in Guinea, Sierra Leone, N Ivory Coast and from Ghana to S Chad, CAR and Congo. Feral birds, Liberia (Monrovia, rare). Also São Tomé (probably introduced, common). Local movements reported.

BLACK-COLLARED LOVEBIRD *Agapornis swindernianus*
Inséparable à collier noir

Plate 53 (1)

L 13 cm (5"). Tiny, bright green forest parrot. **Adult** Head separated from upperparts by narrow black collar on nape followed by broad band of dirty orange-yellow (**nominate**) or deep orange to orange-red (*zenkeri*) extending onto upper breast-sides. Rump and uppertail-coverts purplish-blue. Tail very short; lateral feathers mainly red with subterminal black bar and green tip. Bill black. **Juvenile** Lacks black collar and yellowish or orange band,

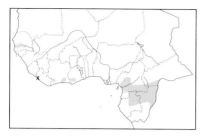

thus appearing greener; bill horn-coloured. **VOICE** A distinctive shrill chirruping *srleeee* or *tchirrrlu*, uttered in flight. [CD6:83] **HABITS** In small flocks in canopy of lowland forest. Blends remarkably well with foliage, where difficult to spot. Feeds on figs and various fruits; may visit favoured fruiting trees daily at dawn. **SIMILAR SPECIES** Red-headed Lovebird occurs in different habitat and has orange-red face. **STATUS AND DISTRIBUTION** Scarce to rare forest resident, from Liberia to Ghana (nominate); locally more common from S Cameroon to SW CAR, E Gabon and N Congo (*zenkeri*).

ROSE-RINGED PARAKEET *Psittacula krameri*   Plate 53 (3)
Perruche à collier

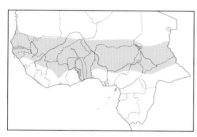

Other name E: Ring-necked Parakeet

L 38–42 cm (15.0–16.5") incl. tail of up to 25 cm (10"). *P. k. krameri.* Distinctive, slender, all-green parrot with *long pointed tail.* **Adult male** Chin and malar area black extending as narrow black line on necksides, where bordered below by narrow rosy-pink line that continues as inconspicuous collar over nape. Bluish wash on nape. Upper mandible red; lower blackish. Eye yellow. **Adult female** Head all green; tail usually shorter. **Juvenile** As adult female, but with slightly more yellowish tinge to plumage, shorter tail and paler, pinkish bill. Most males acquire head markings at *c.* 3 years. **VOICE** Vocal. Loud, screeching calls, incl. a shrill *kee-ak.* [CD6:84] **HABITS** Usually in small flocks in savanna-like habitats, incl. wooded grassland, riparian forest, farmland, gardens and mangroves. **STATUS AND DISTRIBUTION** Not uncommon to common resident in most of range, from S Mauritania to Guinea east through Mali and N Ivory Coast to Chad and N CAR, reaching coast in Dahomey Gap; also coastal savannas in Liberia (uncommon) and Ivory Coast (Abidjan–Grand Bassam). Local seasonal movements reported. Introduced, Cape Verde Is (Santiago, <10 birds, 1995).

TURACOS
Family MUSOPHAGIDAE (World: 23 species; Africa: 23; WA: 10)

Medium-sized to large arboreal birds with crested heads and long tails. Sexes similar. Flight weak, over only short distances and consisting of some flaps followed by a glide. Agile in trees, characteristically running and bounding along branches (except *Crinifer*). Vocal. Feed mainly on fruit. Endemic to Africa.

Four genera occur in our region. *Tauraco*, the largest genus, consists of mainly green species. Their loud, raucous calls are among the most characteristic sounds of the African forest; although overall very similar, they are, with practice, separable in most cases. The two *Musophaga* are mainly glossy violet, have a frontal shield and vividly coloured bare skin around the eye, and utter pleasant, rolling sounds. The flashing crimson flight feathers are the most conspicuous feature of all *Tauraco* and *Musophaga* species in flight. Colour pattern and shape of crest, facial ornamentation and bill colour are important field marks. *Crinifer* species (plantain-eaters) are conspicuous, though they are mainly grey and white, lacking any green or red. They have various cackling and yapping calls. *Corythaeola* consists of a single, huge and quite unmistakable forest species, which has a series of impressive calls.

GREAT BLUE TURACO *Corythaeola cristata*   Plate 54 (11)
Touraco géant

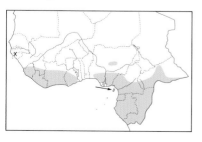

Other name E: Blue Plantain-eater

L 70–75 cm (28–30"). Unmistakable. **Adult** Very large and spectacular, mainly greyish turquoise-blue with large, erect, black crest and bright yellow, reddish-tipped bill. Tail long and strikingly patterned with broad subterminal black band and yellowish-green outer feathers. Lower belly and undertail-coverts chestnut. No red in wings. **Juvenile** Duller, with smaller crest. **VOICE** Impressive and unmistakable. When one starts calling, other group members join in, producing a roaring sound, including a fast series of explosive *kok* notes and a deep guttural *krraou.* [CD6:85] **HABITS** In parties

481

of six or more in canopy of forest and gallery forest. Poor flyer but agile climber, hopping and running along branches. **STATUS AND DISTRIBUTION** Common resident where not persecuted, in forest zone and forest–savanna mosaic in Guinea-Bissau, from Guinea to Togo and from Nigeria to CAR and Congo; also Bioko.

GREEN TURACO *Tauraco persa*
Touraco vert

Plate 54 (6)

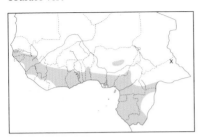

Other name E: Guinea Turaco
L 40–43 cm (16–17"). The region's typical and most widespread turaco. **Adult** Head, mantle and breast bright green; back, wing-coverts and tail glossy violet-black. *Crest entirely green.* Small white patch in front of eye. Bill orange-red; red orbital ring. Flashing crimson flight feathers striking in flight. Western *buffoni* has black line running from lores to below eye. **Nominate** has thinner black line bordered below by larger, white line. **Juvenile** As adult but crimson on wing more limited. **VOICE** Two or three rising notes *woop-woop* followed by a regular series of loud, raucous *khaw* notes.
A calling bird often incites all others within earshot to join a rapidly swelling chorus. [CD6:86] **HABITS** In pairs or family parties in lowland and gallery forest. In areas of overlap with Yellow-billed Turaco always in more secondary habitat. Reaches lower montane forest in Bamenda Highlands, Cameroon, where comes into contact with Bannerman's Turaco. Noisy and active. Runs and hops along branches in manner typical of family. **SIMILAR SPECIES** Yellow-billed Turaco has crest tipped black and white or red (but beware juvenile, which has all-green crest!), lacks white patch in front of eye and has yellow bill; call also different. Black-billed is very similar but has black bill and white-tipped crest. The other three *Tauraco* species have white or red crests. **STATUS AND DISTRIBUTION** Common resident in forest zone and forest outliers in savanna, from S Senegambia to S CAR and Congo. The two races meet in Ivory Coast (*buffoni* in west, nominate in east).

BLACK-BILLED TURACO *Tauraco schuetti*
Touraco à bec noir

Plate 54 (4)

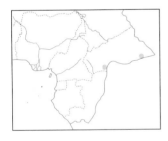

L *c.* 40 cm (16"). *T. s. schuetti.* Very similar to Green Turaco, but has black (not orange-red) bill and shorter, white-tipped crest. **VOICE** As Green Turaco, but slightly slower. [CD6:88] **HABITS** In lowland forest and savanna woodland. Behaviour as other green turacos. **SIMILAR SPECIES** May come into contact with Green and perhaps Yellow-billed Turacos (q.v.). **STATUS AND DISTRIBUTION** Common resident, S CAR (Bangui, Zémio).

YELLOW-BILLED TURACO *Tauraco macrorhynchus*
Touraco à gros bec

Plate 54 (5)

Other name E: Verreaux's Turaco, Black-tipped Crested Turaco
L 40–43 cm (16–17"). **Adult** *macrorhynchus* Similar to Green Turaco, but *crest tipped black with white subterminal line*, no white patch in front of eye, and upperparts more glossy dark blue. Bill mainly yellow, appearing larger than orange-red bill of Green Turaco. Eastern *verreauxii* has crest tipped red. **Juvenile** Duller, with all-green crest. **VOICE** Different from Green Turaco: starts abruptly with single loud, harsh, barking note, followed by a series of raucous *khaw* notes, starting rather fast, then progressively slowing down. [CD6:89] **HABITS** As Green Turaco, but more exclusively forest based.
SIMILAR SPECIES Beware confusion (esp. juvenile) with Green Turaco. **STATUS AND DISTRIBUTION** Uncommon to locally common resident in forest zone, from Sierra Leone and SE Guinea to Ghana (nominate), and from Nigeria to Congo and Bioko (*verreauxii*).

WHITE-CRESTED TURACO *Tauraco leucolophus*

Plate 54 (1)

Touraco à huppe blanche

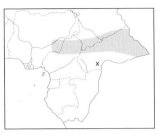

L *c.* 40 cm (16"). **Adult** Differs from all other turacos in conspicuous *pure-white head and crest*. Lores, forehead, forecrown and area behind eye blue-black; mantle and breast bright green; back, wings and tail dark violet-blue. Flight feathers largely crimson. Bill pale yellow. **Juvenile** No information. **VOICE** A single, drawn-out *whooap* followed, after short pause, by a regular series of raucous *khaw* notes. [CD6:90] **HABITS** As other *Tauraco* species, but occurs in more open and drier habitats. Rather shy. **STATUS AND DISTRIBUTION** Locally common resident in wooded savanna and gallery forest in E Nigeria, Cameroon, S Chad and N CAR.

BANNERMAN'S TURACO *Tauraco bannermani*

Plate 54 (2)

Touraco de Bannerman

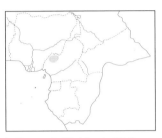

L *c.* 43 cm (17"). **Adult** Highly localised and endangered turaco with *brilliant crimson crest*. Head-sides grey; upperparts and breast green; tail dark violet-blue. Belly and undertail-coverts dark grey washed greenish. Flight feathers mainly bright crimson. Bill yellow with red-brown culmen. Red orbital ring. **Juvenile** No information. **VOICE** A single introductory *whoop* followed, after short pause, by a regular series of raucous *khaw* notes. Softer and higher pitched than Green Turaco. [CD6:91] **HABITS** As other green turacos, but occurs exclusively in montane forest. **SIMILAR SPECIES** Green Turaco has green crest, violet-black upperparts and different call. **STATUS AND DISTRIBUTION** Common resident in remaining montane forest patches in Bamenda Highlands, W Cameroon. *TBW* category: THREATENED (Endangered).

RED-CRESTED TURACO *Tauraco erythrolophus*

Plate 54 (3)

Touraco pauline

L *c.* 43 cm (17"). **Adult** Similar to Bannerman's Turaco but differs in having white or greyish-white face, all-yellow bill and some white-tipped feathers in red crest. **Juvenile** No information. **VOICE** A single *whoop* followed, after short pause, by a regular series of raucous *khaw* notes. [CD6:92] **HABITS** Little known. In evergreen and gallery forest. **STATUS AND DISTRIBUTION** No certain records from our region. Claimed from S Cameroon (sight record Kribi, 1977). Occurs in Angola.

VIOLET TURACO *Musophaga violacea*

Plate 54 (9)

Touraco violet

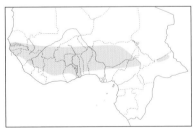

Other name E: Violet Plantain-eater
L 45–50 cm (18–20"). **Adult** All-purple and blue-black turaco with heavy, deep yellow and reddish-tipped bill. Crown crimson; bare orbital skin scarlet. White line over ear-coverts, bordering orbital skin and crown. Bright crimson in wings conspicuous in flight. **Juvenile** Much duller, with all-dark head and blackish bill. **VOICE** A melodious rolling series of far-carrying *koorroo* notes; often joined by second bird in asynchronous duet and eliciting response from all others within earshot. Also a harsh note. [CD6:94] **HABITS** In pairs or small parties in gallery forest and forest edge, moving like other green turacos. Not shy. **SIMILAR SPECIES** Ross's Turaco has a crest, yellow (not scarlet) orbital skin, and much more limited distribution. **STATUS AND DISTRIBUTION** Widespread and locally not uncommon resident, from S Senegambia and Guinea to N Cameroon and N CAR.

ROSS'S TURACO *Musophaga rossae*

Plate 54 (7)

Touraco de Lady Ross

Other name E: Lady Ross's Turaco
L 51–54 cm (20–21"). **Adult** Similar to Violet Turaco, but distinguished by *crimson crest*, and conspicuous *all-yellow*

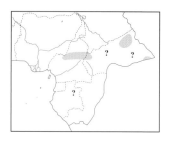

bill and orbital skin. **Juvenile** Duller. Bill and orbital skin blackish, culmen yellow. Crown black with red patch in centre. **VOICE & HABITS** Similar to Violet Turaco. [CD6:95] **STATUS AND DISTRIBUTION** Locally not uncommon resident, Cameroon (Adamawa Plateau, south to Mbam Mts) and CAR. Also reported from NE Gabon (M'Passa).

WESTERN GREY PLANTAIN-EATER *Crinifer piscator*   Plate 54 (8)
Touraco gris

L *c.* 50 cm (20"). **Adult** Mainly *grey*, long-tailed turaco with conspicuous *yellow bill* and *spiky nuchal crest*. Lower breast to undertail-coverts white, streaked blackish. In flight has striking white wing patch. **Juvenile** As adult but crest shorter. **VOICE** A series of loud, high-pitched and rolling *kow-kow-kow-...* ending in chatter; a cackling *kak-kak-kak-kalak-kalak...* [CD6:98] **HABITS** In pairs or small parties, frequenting treetops in gallery forest along watercourses, savanna woodland and cultivated areas. Less agile than other turacos but a far better flyer. Conspicuous and not shy. **SIMILAR SPECIES** Could be confused with Eastern Grey Plantain-eater (q.v.) in E Chad and E CAR. **STATUS AND DISTRIBUTION** Common resident in wooded savanna belt, from S Mauritania to Sierra Leone east to Chad and CAR; also coastal Liberia and Congo along Congo R.

EASTERN GREY PLANTAIN-EATER *Crinifer zonurus*   Plate 54 (10)
Touraco à queue barrée

L *c.* 50 cm (20"). **Adult** Very similar to Western Grey Plantain-eater, but distinguished by white on outer tail feathers, smaller white wing patch, scalloped (rather than spotted) mantle and wing-coverts, and paler yellow bill. **Juvenile** As adult but crest shorter. **VOICE** A variety of cackling and yapping notes, similar to Western Grey Plantain-eater but slower and lower pitched. [CD6:99] **HABITS** As Western Grey Plantain-eater. **STATUS AND DISTRIBUTION** Common resident in open wooded savanna, SE Chad and SE CAR.

CUCKOOS AND COUCALS
Family CUCULIDAE (World: *c.* 136 species; Africa: 25; WA: 21)

Mainly medium-sized. Feet zygodactyl. Sexes usually similar. Three subfamilies occur in W Africa.

Typical cuckoos (Cuculinae) are arboreal with long, pointed wings and long tails, reminiscent of a small raptor in flight. Feed on insects and hairy caterpillars and are brood parasites. Many are rather inconspicuous and some quite difficult to observe, but all have distinctive calls. Fourteen species, of which 11 are endemic to the Afrotropics.

The **Yellowbill**, endemic to Africa and the only Afrotropical representative of the malkohas (Phaenicophaeinae), is arboreal, principally insectivorous, and non-parasitic.

Coucals (Centropodinae) are semi-terrestrial, stoutly built with rounded wings and long, broad tails. Flight clumsy and not sustained. Largely insectivorous, but also take vertebrates, such as lizards and nestlings. Non-parasitic. Utter rapid series of deep, hollow hoots and liquid bubbling notes (like water poured from a bottle). Six species, 5 endemic to the Afrotropics. Accorded family status (Centropodidae) by some authors (e.g. Sibley & Monroe 1990).

JACOBIN CUCKOO *Oxylophus jacobinus*
Coucou jacobin

Plate 55 (6)

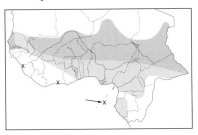

Other name E: Black-and-white Cuckoo
L 33 cm (13"). **Adult** *pica* Pied, crested cuckoo, black above and white below. Long graduated tail black, broadly tipped white. White wing patch conspicuous in flight. Southern *serratus* has some fine streaking on throat and breast; also melanistic birds with black or grey underparts and all-black tail. **Juvenile** Duller, with shorter crest and buffish underparts. **VOICE** A series of clear *keew* or *pweep* notes (as Levaillant's Cuckoo but slightly higher pitched) ending in staccato chatter. [CD7:1] **HABITS** Usually unobtrusive. Singly or in pairs in dry savanna woodland; small groups may form on migration. Not in forest. Parasitises mainly Common Bulbul. **SIMILAR SPECIES** Levaillant's Cuckoo is larger and has boldly striped throat and upper breast. **STATUS AND DISTRIBUTION** Widespread but patchy; locally common. *O. j. pica* breeds in Sahel zone from S Mauritania and Senegambia to Chad and CAR. Intra-African migrant, some moving south, crossing equator, in dry season. Vagrant, São Tomé. S African *serratus* winters in E Africa, northwest to Sudan, and has been reported from Gabon and Chad. **NOTE** Sometimes placed in genus *Clamator*.

LEVAILLANT'S CUCKOO *Oxylophus levaillantii*
Coucou de Levaillant

Plate 55 (7)

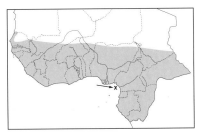

Other name E: African Striped Cuckoo
L 38–40 cm (*c.* 15"). **Adult** Large, long-tailed and crested cuckoo with black upperparts and white underparts, boldly streaked black on throat and upper breast. Long graduated tail black broadly tipped white. White wing patch conspicuous in flight. **Juvenile** Duller, with shorter crest and buffish underparts. **VOICE** A series of clear, loud *KEEow* notes, ending in staccato chatter (like machine gun burst). [CD7:2] **HABITS** In dense savanna woodland, forest edge, second growth and adjacent cultivated areas in forest zone. Solitary and unobtrusive, foraging within cover, often low. Parasitises babblers. **SIMILAR SPECIES** Jacobin Cuckoo is smaller and more slender with proportionately shorter tail, and lacks streaking on underparts. **STATUS AND DISTRIBUTION** Widespread and common to uncommon throughout; absent from arid north. Intra-African migrant breeding mainly during rains and moving south in dry season. Vagrant, Bioko. **NOTE** Sometimes placed in genus *Clamator*.

GREAT SPOTTED CUCKOO *Clamator glandarius*
Coucou geai

Plate 55 (5)

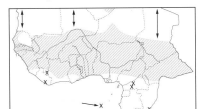

L 35–40 cm (14–16"). *Large, elongated cuckoo, boldly spotted white above.* **Adult** *Pale grey crest*; upperparts grey-brown boldly spotted white. Long graduated tail blackish, all feathers except central pair broadly tipped white. Throat and breast creamy-buff; rest of underparts white. **Juvenile** *Black cap* with small crest; upperparts darker brown; *primaries rufous* (conspicuous in flight); underparts richer coloured. **VOICE** Noisy in breeding season. A loud *kweeow kweeow kweeow*, an excited *kiu-ku-ku-ker* and a variety of harsh and chattering calls. [CD3:10] **HABITS** Inconspicuous and usually solitary. Mainly in semi-arid open woodland. Easily overlooked when not calling, despite size and open savanna-type habitat. Parasitises mostly Pied Crow. **STATUS AND DISTRIBUTION** Widespread, scarce to locally common resident and intra-African migrant in broad Sahel and savanna belts, from S Mauritania to Liberia east to Chad and CAR. Vagrant, SE Cameroon, Gabon and Congo, also Annobón. Breeds during rains (Apr–Aug), moving south in dry season, when Palearctic migrants winter south of Sahara (south to 10°N).

THICK-BILLED CUCKOO *Pachycoccyx audeberti*
Coucou d'Audebert

Plate 56 (6)

L *c.* 36 cm (14"). *P. a. brazzae*. **Adult** Large and rather bulky cuckoo with dark, slate-grey upperparts (becoming slate-brown with age) and all-white underparts. Bill slightly heavier than other cuckoos, hook-tipped, dark above, yellowish below. **Juvenile** Upperparts mainly dark brown with large buffish-white spots on feather tips; white

underparts slightly washed buff. **VOICE** A series of clear, far-carrying *whuee-di* or *hwee-wik*. Also a not particularly loud *weedidi weedidi weedidi* uttered in flight. [CD7:3] **HABITS** Unobtrusive and mainly solitary. In wooded savanna, gallery forest and forest edge. Restless, often flying from treetop to treetop. Undertakes noisy, buoyant display flight over long distances. Parasitises helmet-shrikes. **SIMILAR SPECIES** Jacobin, Levaillant's and Great Spotted Cuckoos also have white underparts, but all are crested and have longer, more slender tails. **STATUS AND DISTRIBUTION** Rare and patchily distributed resident known from Sierra Leone, Guinea, Liberia, Ivory Coast, Ghana, Togo, Burkina Faso, Nigeria, Cameroon, CAR, Gabon and Congo. Some seasonal movements recorded.

RED-CHESTED CUCKOO *Cuculus solitarius* Plate 56 (5)
Coucou solitaire

L 28–31 cm (*c.* 12"). **Adult male** *solitarius* Mainly slate-grey with chestnut breast (occasionally barred), and buffish-white belly and undertail-coverts boldly barred blackish. Tail tipped white. Poorly differentiated island race *magnirostris* has slightly longer and heavier bill. **Adult female** Similar, but often has chestnut of breast paler, less extensive and barred. **Juvenile** Upperparts blackish-brown, feathers narrowly fringed white; white spot on nape. Tail dark brown spotted and tipped white. Throat and breast as upperparts, rest of underparts buffish-white, boldly barred blackish-brown. **VOICE** A frequently uttered, loud and descending 3-note *WHIT-whit-teew* (stress on first syllable). [CD7:4] **HABITS** Solitary and unobtrusive in forest and wooded savanna. Highly vocal when breeding, usually calling from hidden perch. Also calls at night. Parasitises mainly Turdidae. **SIMILAR SPECIES** Chestnut-breasted race of Black Cuckoo has similar basic pattern, but is much darker, with chestnut extending to throat and neck-sides; call different. African and Common Cuckoos with buffish wash on breast may resemble subadult with barred throat and little chestnut, but have paler grey upperparts. **STATUS AND DISTRIBUTION** Not uncommon to uncommon resident and intra-African migrant, from Senegambia to Liberia east to CAR and Congo (nominate); also Bioko (*magnirostris*). Visitor to northern savannas, Mar–Dec.

BLACK CUCKOO *Cuculus clamosus* Plate 56 (7)
Coucou criard

L 28–31 cm (*c.*12"). Two races, with intermediates. **Adult** *gabonensis* Head, upperparts and tail glossy blue-black. Throat and breast dark chestnut with or without dark barring; rest of underparts buffish-white heavily barred blackish-brown. Tail graduated, narrowly tipped white. Typical adult *clamosus* all black with narrowly white-tipped tail. Underparts may be partially or wholly barred with white to pale rufous and have some chestnut on throat and breast. **Juvenile** (both races) entirely dull sooty-black. **VOICE** A series of 3 hesitant notes, each with more emphasis and higher pitched than the preceding, final note sometimes repeated after a short pause: *who whuu whee*, *wheee*. Also a wild, bubbling trill, rising and falling and gradually dying away *lululululululuWHIRlulululuWHIRlulululuWHIRlulu WHIRlu...* [CD7:5] **HABITS** Usually solitary and unobtrusive in forest, forest patches in savanna and woodland, remaining in dense cover, often high. Calls also at night. In S Africa recorded to parasitise mainly shrikes. **SIMILAR SPECIES** Red-chested Cuckoo has similar pattern to *C. c. gabonensis* but is much paler, has grey throat and different call. **STATUS AND DISTRIBUTION** Uncommon to locally common resident and intra-African migrant. *C. c. gabonensis* is strictly forest based, occurring from E Sierra Leone and SE Guinea to Togo and from Nigeria to CAR and Congo; probably non-migratory *C. c. clamosus* is a wet-season visitor recorded from Gambia and S Senegal to Chad, CAR and Congo. Status unclear.

COMMON CUCKOO *Cuculus canorus* Plate 56 (2)
Coucou gris

Other names E: European/Eurasian Cuckoo
L 32–36 cm (12.5–14.0"). Slender, grey-backed cuckoo with long, graduated tail **Adult male** Head, breast and

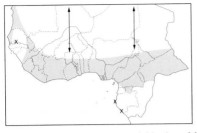

upperparts usually grey. Lower breast to undertail-coverts white barred grey. Tail blackish spotted and tipped white, outer feathers with incomplete white bars (visible from below). Eye and orbital ring yellow. Bill blackish with yellow base. *C. c. canorus* has slightly longer wings than *bangsi*. **Adult female** Upperparts tinged brown; barring on underparts browner; breast tinged rusty; sometimes faint barring on upper breast. Very rarely entirely rufous barred black above ('hepatic' form). **Immature** Variably brownish or greyish, but always appearing very barred. Feathers of upperparts tipped whitish; crown partially barred; white patch on nape; rump and uppertail-coverts plain. Underparts variably tinged buffish; throat and breast barred. **VOICE** Silent in Africa. [CD3:11] **HABITS** Usually solitary and unobtrusive, like congeners. Occurs in wooded and derived savanna and cultivated areas in forest zone. Forages mostly high in trees and bushes. Flight often low and hawk-like. Flies with rapid shallow wingbeats (only occasionally and briefly interspersed by glides), wings mostly held below horizontal (unlike small raptors) while head, body and tail are very straight, with bill pointing forward. **SIMILAR SPECIES** Adult African Cuckoo is extremely similar but has more yellow on bill. Immatures also similar, but African has crown entirely (not partially) barred, and rump and uppertail-coverts barred (not plain). Subadult Red-chested Cuckoo with barred throat and very pale and limited rufous on breast may appear very similar but has darker upperparts. **STATUS AND DISTRIBUTION** Widespread but scarce Palearctic passage migrant or winter visitor (both races). Vagrant, Cape Verde Is. Possible old record from São Tomé.

AFRICAN CUCKOO *Cuculus gularis*   Plate 56 (1)
Coucou africain

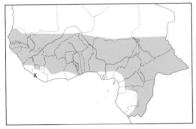

L 31–33 cm (12–13"). **Adult male** Very similar to Common Cuckoo but differs in *mainly yellow, black-tipped bill*, and *call*; also has complete white bars on outer tail feathers (visible from below). **Adult female** Similar, but breast may be washed buff; grey of upper breast sometimes indistinctly barred. **Juvenile/immature** Like young Common Cuckoo but crown entirely (not incompletely) barred; rump and uppertail-coverts barred (not plain). **VOICE** A frequently repeated, far-carrying *oo-OO*, with stress on second, slightly higher pitched syllable (as flat, inverted Common Cuckoo song). Also a hard, harsh *krèk*. [CD7:6] **HABITS** Unobtrusive, like congeners. Occurs in wooded savanna. Often calls from bare branch, perching upright; occasionally at night. May recall small hawk. Parasitises Fork-tailed Drongo. **SIMILAR SPECIES** Common Cuckoo (q.v. and above). **STATUS AND DISTRIBUTION** Uncommon to common intra-African migrant throughout, except lowland forest and arid north. Moves north with rains. Probably breeds throughout range.

DUSKY LONG-TAILED CUCKOO *Cercococcyx mechowi*   Plate 56 (3)
Coucou de Mechow

L *c.* 33 cm (13"). Slender, small-bodied cuckoo with very long, broad tail. **Adult** Head, upperparts and tail blackish-grey; wings finely barred rufous. Throat to belly buffish-white boldly barred blackish; undertail-coverts rich buff. Tail tipped white; outer rectrices barred. **Juvenile** Head, upperparts and throat dark slate-grey barred rufous. Tail blackish-brown barred rufous. Breast to undertail-coverts whitish barred blackish. **VOICE** Two song types, each with distinct western and eastern versions. In Upper Guinea (east to W Cameroon), first type consists of 3 rising notes, *hu hee wheeu*; second, a less frequently uttered whinnying series of rather plaintive notes, first accelerating, then slowing and descending *tiutiutiutiutiutiui-tiu-tiu-tiu-...*, reminiscent of *Halcyon* kingfisher. In Lower Guinea, first type is faster, with 3 similar, less melodious notes *wheet-wheet-wheet*; second, a fast, descending *wheewheewheewheewheewhee...*, almost twice as fast as equivalent in Upper Guinea. [CD7:9–10] **HABITS** Shy and seldom seen, frequenting lower strata of lowland forest. Calls also at night. Hosts unknown; possibly include Forest Robin and Brown Illadopsis. **SIMILAR SPECIES** Adult Olive Long-tailed Cuckoo has dark greyish-brown upperparts, paler undertail-coverts and different song. Juvenile has streaked underparts and never has blackish throat. **STATUS AND DISTRIBUTION** Uncommon or rare to locally common forest resident, from Sierra Leone and SE Guinea to Ghana and Togo, and from S Nigeria to S CAR and Congo.

OLIVE LONG-TAILED CUCKOO *Cercococcyx olivinus* Plate 56 (4)
Coucou olivâtre

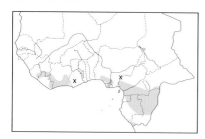

L *c.* 33 cm (13"). Slender and very long tailed. **Adult** Upperparts and tail dark greyish-brown, underparts buffish-white boldly barred blackish-brown (finer on throat). Undertail-coverts plain pale buff. **Juvenile** Upperparts dark brown with rufous fringes. Underparts whitish *streaked* dark brown; streaking progressively replaced by barring (last on throat). **VOICE** Usual song a series of 3 notes *whit tiuw-tiuw* (first note inaudible at distance). Also a long, rising series *teeru-teeru-teeru-teeru-...,* appearing ever more impatient, ceasing abruptly. [CD7:11] **HABITS** Shy and seldom seen, although vocal, occurring in or below canopy of lowland forest. Often calls at night. Hosts unknown; possibly include Pale-breasted Illadopsis and Rufous Ant Thrush. **SIMILAR SPECIES** Dusky Long-tailed Cuckoo has darker upperparts and richer coloured undertail-coverts, ventures lower and has different song. Juvenile has barred (not streaked) underparts and a blackish throat. Three-note songs of Red-chested and Olive Long-tailed Cuckoos similar, but in Red-chested first note always loud and clearly audible; in Olive Long-tailed first note quiet and two following at same pitch. **STATUS AND DISTRIBUTION** Rare to locally common forest resident, from Sierra Leone and SE Guinea to Ghana and from Nigeria and S Cameroon to SW CAR and Congo.

AFRICAN EMERALD CUCKOO *Chrysococcyx cupreus* Plate 55 (3)
Coucou foliotocol

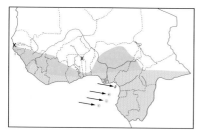

L 20 cm (8"). **Adult male** *cupreus* Unmistakable. Strikingly beautiful, with bright glossy green upperparts, throat and breast and deep golden-yellow belly. Undertail-coverts white barred green; tail green with outer rectrices spotted white. Bill bluish; blue-green orbital ring. Poorly differentiated island race *insularum* stated to have slightly shorter tail and smaller white area on outer rectrices. **Adult female** Upperparts glossy brown densely barred russet and green. Underparts white heavily barred bronzy-green. **Juvenile** Very similar to and easily confused with female, but crown and nape feathers green (not bronzy-brown) fringed white. Juvenile males tend to have green parts more brilliant. **VOICE** A far-carrying, clear and melodious *ptiu, tiu-ut,* quite unmistakable and one of the characteristic sounds of African rainforest. Also an explosive, stuttering *tiuw tu-tu-tu...* Female utters a clear *tiuw* and *huu tu-tu.* [CD7:13] **HABITS** Solitary, in rainforest, gallery forest and densely wooded savanna. Often heard, but much less seen, as usually well hidden in canopy. Occasionally forages lower. Known hosts include bulbuls, illadopsises, flycatchers, sunbirds and weavers. **SIMILAR SPECIES** Female and juvenile very similar to juvenile Klaas's Cuckoo, but latter has whitish (although sometimes indistinct) post-ocular patch and finer barring on underparts, giving whiter bellied appearance. **STATUS AND DISTRIBUTION** Widespread and fairly common intra-African migrant in forest zone and adjacent dense woodland, from S Senegambia to CAR, Congo and Bioko (nominate). Also São Tomé, Príncipe and Annobón (*insularum*). Probably resident in parts of forest zone.

YELLOW-THROATED CUCKOO *Chrysococcyx flavigularis* Plate 55 (4)
Coucou à gorge jaune

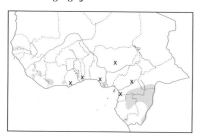

L 18 cm (7") Seldom seen, darkish forest cuckoo. White outer tail conspicuous in flight. **Adult male** Head, upperparts and breast-sides dark metallic green (sometimes appearing brown). Conspicuous *golden-yellow stripe on throat;* lower breast and belly buffy narrowly barred and vermiculated dark greenish-brown. Tail dark, outer rectrices white with dark subterminal bar. Eye yellowish. Feet yellow-green. **Adult female** Head-sides and underparts buffish narrowly barred dark greenish-brown. Female of eastern *parkesi* has underparts, esp. undertail-coverts, browner, less creamy-white than **nominate** (Dickerman 1994). **Juvenile** As adult female but upperpart feathers fringed russet; head and underparts more tawny. **VOICE** A rapid series of *c.* 10 pure whistled notes on same pitch *tiu tee-tee-tee-tee-tee-...* Also a distinctive, loud, clear, double whistle *hee-huu.* [CD7:14] **HABITS** Extremely unobtrusive and little known, frequenting canopy of lowland forest. Calls year-round. Hosts unknown. **SIMILAR SPECIES** Females and juveniles of other *Chrysococcyx* cuckoos have whiter underparts. **STATUS AND DISTRIBUTION** Rare to uncommon forest resident. Rare in west of range, where recorded from Sierra Leone to Togo, and in Nigeria (nominate); uncommon to scarce from S Cameroon to Congo (*parkesi*).

KLAAS'S CUCKOO *Chrysococcyx klaas*
Coucou de Klaas

Plate 55 (1)

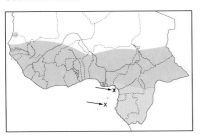

L 18 cm (7"). **Adult male** Head and upperparts bright metallic green with *small white streak behind eye*. Tail green with *outer rectrices mainly white*. Underparts clean white with *green patch on breast-sides* forming incomplete collar. Eye dark. **Adult female** Variable, but typically has upperparts more bronzy and flanks finely barred olive-brown. Post-ocular patch buffy, less conspicuous than in male and sometimes indistinct. Some females similar to males. **Juvenile** Upperparts mainly bronzy-brown finely barred green and russet. Underparts finely barred olive-brown; slight buffish post-ocular patch (sometimes very indistinct). **VOICE** Unmistakable. A plaintive *huee-ti huee-ti* or *whee-ee chew whee-ee chew*. [CD7:15] **HABITS** Solitary and unobtrusive; usually going unnoticed but for call. Occurs in variety of habitats, incl. savanna woodland, forest (mainly edges and clearings) and cultivated areas; not in very arid country. Parasitises mainly warblers, sunbirds, flycatchers and Yellow White-eye. **SIMILAR SPECIES** Didric Cuckoo lacks white outer tail feathers and green breast patches, has broad white eye-stripe, barred flanks, red eye and entirely different call. Juvenile Klaas's is very similar to female/juvenile Emerald Cuckoo and only safely separable if reasonable views are obtained; latter always lack post-ocular patch and have bolder barring on underparts. **STATUS AND DISTRIBUTION** Widespread and generally common throughout; also Bioko and São Tomé. Resident in rainforest and southern savannas, some migrating north to 16.5°N during rainy season.

DIDRIC CUCKOO *Chrysococcyx caprius*
Coucou didric

Plate 55 (2)

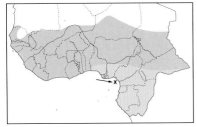

Other name E: Diederik Cuckoo

L 19 cm (7.5"). **Adult male** Head and upperparts metallic green with bronze reflections. *White streak in front and behind eye*; short green malar stripe. Wings with some white spots and bars. Tail bronzy-green, outer rectrices spotted white. Underparts white barred bronzy-green on flanks and thighs. *Orbital ring and eye red*. **Adult female** Similar but variable, with upperparts more bronzy-brown and white areas sometimes tinged buff. **Juvenile** Highly variable. Upperparts dull green (some with rusty barring) or russet. Underparts white with streaked throat and breast and lightly barred flanks, or spotted from breast to belly. *Bill coral-red*. **VOICE** Unmistakable. A plaintive *deea deea deedrik* from perch or in flight. Also a rapid *di-di-di-di-di*. Female a *deea-deea-deea*. [CD7:16] **HABITS** Solitary or in pairs in variety of open habitats, incl. savanna, cultivated areas, gardens, scrub, etc. Not in forest. Conspicuous (unusual for a cuckoo). Flight direct with rapid wingbeats. Parasitises mainly weavers. **SIMILAR SPECIES** Male Klaas's Cuckoo is more uniform green above, with green patches on breast-sides, all-white outer tail feathers, white on head restricted to small patch behind dark eye, and no barring on flanks. **STATUS AND DISTRIBUTION** By far the commonest and most widespread of the four glossy *Chrysococcyx* cuckoos, occurring throughout sub-Saharan Africa; also Bioko. In forest zone mainly resident; many migrate north in rainy season to breed, locally to 17°N.

YELLOWBILL *Ceuthmochares aereus*
Malcoha à bec jaune

Plate 57 (1)

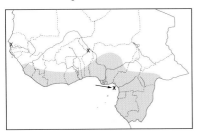

Other names E: Green Coucal; F: Coucal à bec jaune

L 33 cm (13"). Shy and dark with *long, graduated tail* and *conspicuous yellow bill*. **Adult *flavirostris*** Top of head, upperparts and tail dark slate-grey with purplish-blue gloss. Underparts dark grey, paler than upperparts. Bill bright yellow. Eye reddish; orbital ring yellowish. **Nominate** has gloss on upperparts more greenish-blue. **Juvenile** Duller, with upper mandible brownish. **VOICE** A series of loud, explosive *kuk* notes, first uttered slowly, but rapidly gathering speed *kuk, kuk, kuk kuk kukkukkukkukukukkkkkkrrrrrr*. Single *kuk* notes resemble those of Western Nicator. Also a soft, plaintive *mweeeew*. [CD7:17] **HABITS** Singly or in pairs in forest, even where heavily degraded. Frequents all levels, though mostly mid-stratum and lower canopy, unobtrusively creeping through dense leafy vegetation with creepers. Joins mixed-species flocks. **STATUS AND DISTRIBUTION** Widespread resident in forest zone from Gambia (rare), S Senegal (uncommon) and Guinea to CAR and Congo (common). Also reported from SW Niger. *C. a. flavirostris* occurs west of lower Niger R., Nigeria; nominate east.

BLACK-THROATED COUCAL *Centropus leucogaster*

Coucal à ventre blanc

Plate 57 (6)

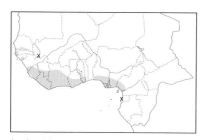

L 48–58 cm (19–23"). **Adult** *leucogaster* Large coucal with *blue-black head, mantle, throat and upper breast*; rest of underparts white. Wings deep chestnut; back to uppertail-coverts black barred buff. Tail black. Thighs and undertail-coverts have variable amount of pale tawny. Eastern *efulensis* has head and mantle glossed green. **Juvenile** Duskier, with barred upperparts and tail. **VOICE** A series of 10–20 deep, resonant *hoots*; similar to Senegal Coucal's in structure, but lower and slower, never accelerating at end. Songs of duetting pair have similar pitch but different rhythm (one almost half as fast as the other). [CD7:18] **HABITS** Singly or in pairs in lower strata of lowland forest and at forest edges. Rather shy and retiring. **SIMILAR SPECIES** No other coucal has combination of black throat and white belly. **STATUS AND DISTRIBUTION** Not uncommon forest resident in S Senegal and Guinea-Bissau, from Guinea to Togo, and in S Nigeria (nominate), and Cameroon and coastal Eq. Guinea and Gabon (*efulensis*). Also recorded from S Mali (south of Mandingues Mts). Claim from SW Niger doubtful.

GABON COUCAL *Centropus anselli*

Coucal du Gabon

Plate 57 (5)

L 48–58 cm (19–23"). **Adult** Very dark, large coucal with *tawny underparts*. Head and mantle glossy blue-black; rest of upperparts blackish-brown; wings dark rufous-brown. Tail black with rufous-brown barring at base. **Juvenile** Much as adult but head and mantle dull black with pale feather shafts, rest of upperparts and tail barred, *throat barred*. **VOICE** A series of deep, resonant *hoots*, similar to those of Black-throated Coucal. Songs of duetting pair have different pitch and rhythm. Vocalisations often higher pitched and more varied than other coucals, including more drawn-out syllables and series of rapid notes with 'yelping' quality. [CD7:19] **HABITS** Shy, occurring in undergrowth and edges of lowland forest. **SIMILAR SPECIES** No other coucal in our area has tawny underparts. Juvenile unique in having barred throat. **STATUS AND DISTRIBUTION** Common forest resident, S Cameroon, SW CAR, Eq. Guinea, Gabon and Congo.

WHITE-BROWED COUCAL *Centropus superciliosus*

Coucal à sourcils blancs

Plate 147 (2)

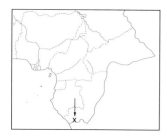

L *c.* 40 cm (15"). *C. s. loandae*. The only coucal with a pale supercilium. **Adult** Top and head-sides blackish-brown with creamy supercilium; nape, neck-sides and mantle streaked black and cream; rest of upperparts rufous-brown; rump and uppertail-coverts blackish narrowly barred buff. Tail black glossed green. Underparts buffish-white finely barred brownish on flanks. **Juvenile** Top and sides of head streaked buff; upperparts barred dusky; tail narrowly barred buff. **VOICE** An accelerating series of hollow *hoots*, similar to, but normally faster than Senegal Coucal. [CD7:25] **HABITS** Singly or in pairs in tall grass and shrubbery. **STATUS AND DISTRIBUTION** Mainly S and E African species, recorded once in extreme SW Congo (Djéno, Jul 1991).

BLACK COUCAL *Centropus grillii*

Coucal de Grill

Plate 57 (2)

L 30–35 cm (12–14"). **Adult breeding** Rather small, all-black coucal with chestnut wings. Broad, black tail sometimes has some narrow buff bars. **Adult non-breeding** Top of head to mantle dusky streaked buffish; rest of upperparts and tail tawny-chestnut barred blackish; rump and uppertail-coverts buff narrowly barred black. Underparts buffish; flanks and undertail-coverts with some dusky barring. Bill dusky above, horn-coloured below. **Juvenile** Similar to adult non-breeding but more heavily barred (e.g. barring of upperparts extending to forehead and face). **VOICE** Advertisement call, given in breeding season, a long series of variable speed *wok-wok, wok-wok,*

wok-wok, ... or *po-op po-op po-op*... and faster *popopopopopopop*..., from exposed perch. Also 'water-bottle' song, similar to that of Senegal Coucal but higher pitched, faster and not rising at end. [CD7:21] **HABITS** Prefers moist, rank grasslands and marshy areas; occasionally shrub and cultivation. Rather shy. **SIMILAR SPECIES** No other coucal has entirely black underparts. 'Epomidis' morph of Senegal Coucal also appears very dark, but has back, breast and belly chestnut (not black). Juvenile separated from other juvenile coucals by pale overall appearance and very streaky head and mantle. **STATUS AND DISTRIBUTION** Uncommon to locally common throughout, except arid north. Irregularly distributed. Resident in permanent moist places in and near forest zone; savanna population migratory, moving north to 13°N during rains.

SENEGAL COUCAL *Centropus senegalensis* Plate 57 (3)
Coucal du Sénégal

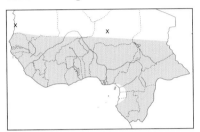

L 36–40 cm (14–16"). *C. s. senegalensis*. Commonest and most easily seen coucal. **Adult** Head, upper mantle and tail glossy black; rest of upperparts rufous-brown. Underparts whitish washed buff (mostly on flanks). Eye red. Dark and very distinctive morph ('*epomidis*') has head and throat entirely black, breast and belly rufous-brown as upperparts, and lower back and rump black. **Juvenile** Dusky and tawny version of adult, with barred upperparts and tail; pale feather shafts give streaky appearance to head and upper mantle. **VOICE** A characteristic, accelerating series of hollow *hoots*, first rising in pitch, then dying away, sounding like water being poured from a bottle. Often in duet, both songs delivered in similar rhythm but different pitch. Speed and pitch variable, slowest and deepest resembling song of Black-throated Coucal. Calls include nasal *gook* (alarm) and fast, dry *k-t-k-t-k-t-k-t-k-...* [CD7:22–23] **HABITS** Singly or in pairs in gardens, roadsides, plantations and variety of open habitats with tall grass and bushes. Clambers in low vegetation, on ground, or flies clumsily for short distance, crash-landing into bush. **SIMILAR SPECIES** Blue-headed Coucal (q.v.) is larger and heavier looking; head and mantle strongly glossed blue, upperparts darker rufous. Juvenile Blue-headed has black tail. **STATUS AND DISTRIBUTION** Common resident throughout, except in heavy forest. Dark morph in areas with highest rainfall (east to Nigeria).

BLUE-HEADED COUCAL *Centropus monachus* Plate 57 (4)
Coucal moine

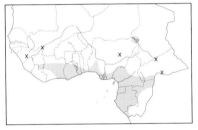

L *c.* 45 cm (18"). *C. m. fischeri.* **Adult** Very similar to Senegal Coucal, esp. in poor light, but head and mantle strongly glossed dark blue, and rest of upperparts darker, more chestnut. Also slightly larger and heavier looking. **Juvenile** Duller, with pale feather shafts giving streaky appearance to dull black head and upper mantle; upperparts rufous-brown barred black; uppertail-coverts faintly barred buff; tail black; underparts whitish washed rufous-buff, esp. on breast. **VOICE** Very similar to Senegal Coucal. Songs of duetting pair have different pitch but same rhythm. [CD7:24] **HABITS** Frequents swampy places, densely vegetated river banks, forest edge and savanna woodland. **SIMILAR SPECIES** Senegal Coucal (q.v.). **STATUS AND DISTRIBUTION** Uncommon to common resident, from Guinea to CAR and Congo. Also reported from Mali. Status and distribution in west (from Nigeria west) obscured due to resemblance to Senegal Coucal.

BARN AND GRASS OWLS
Family TYTONIDAE (World: 16 species; Africa: 3; WA: 2)

Medium-sized nocturnal birds of prey with distinctive heart-shaped, pale facial disc. Sexes similar, females larger. Flight buoyant and silent. Perch upright.

BARN OWL *Tyto alba* Plate 58 (1)
Effraie des clochers

Other name F: Chouette effraie
L 33–36 cm (13–14"). *Pale owl* with conspicuously *heart-shaped face*, black beady eye and no ear-tufts. **Adult** *affinis*

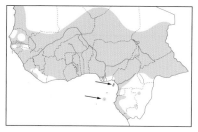

Upperparts pale grey mixed with golden-buff and speckled white and black. Facial disc whitish. Underparts white sparsely speckled black and variably washed pale golden-buff. Flight feathers and tail barred dark grey. **Nominate** slightly paler overall. Island races *detorta* and *thomensis* darker; upperparts dark greyish; underparts golden-brown; facial disk cinnamon (*detorta*) or pale brownish (*thomensis*). **Juvenile** Upperparts darker grey; underparts washed golden-buff. **VOICE** A hoarse, drawn-out screech *shreeeeeeeeee*, uttered in flight. [CD3:12] **HABITS** In various habitats, often associated with man; avoids closed forest. Nocturnal and crepuscular. **SIMILAR SPECIES** African Grass Owl is much darker, occurs in open grassy areas and has a much more restricted distribution. **STATUS AND DISTRIBUTION** Uncommon to not uncommon resident throughout mainland and Bioko (*affinis*), Cape Verde Is (*detorta*) and São Tomé (*thomensis*). Nominate, Niger (Aïr). **NOTE** Birds from Bioko, described as being slightly darker than *affinis*, are sometimes recognised as distinct race *poensis* (cf. *HBW*).

AFRICAN GRASS OWL *Tyto capensis* Plate 58 (2)
Effraie du Cap

Other name F: Chouette effraie du Cap
L 36–40 cm (14–16"). *Two toned owl* with conspicuously *heart-shaped face* and no ear-tufts. **Adult** Facial disc white. Top of head and upperparts dark brown finely speckled white. Underparts whitish to pale rufous-buff, speckled brown; specks densest on breast. Tawny-buff patch at base of outer primaries (conspicuous in flight). Tail dark brown with 2 outer feathers white. Eye dark. **Juvenile** Facial disc dusky-russet to pale tawny-buff; upperparts lack white specks; underparts rufous-buff. **VOICE** Usually silent. A hoarse, drawn-out screech, recalling Barn Owl but softer. [CD7:26] **HABITS** In dambos, dense montane grassland and open grasslands, often near water. Mainly nocturnal. When flushed, flies short distance with dangling legs before dropping into cover of high grass. **SIMILAR SPECIES** Barn Owl is much paler, occurs in broader range of habitats and is widespread. Marsh Owl occurs in similar habitat, but is paler above with much larger tawny-buff primary patches and white trailing edges to wings, and brown below. **STATUS AND DISTRIBUTION** Rare to locally not uncommon resident, W Cameroon and Congo.

TYPICAL OWLS
Family STRIGIDAE (World: *c.* 190 species; Africa: 30; WA: 23)

Small to large, mainly nocturnal birds of prey with characteristic large, rounded heads and large, forward-facing eyes in flat facial disc. Plumage mostly cryptic, many species with 'ear-tufts'. Sexes similar, females often larger. Calls distinctive; usually most vocal just after dark and before dawn.

SANDY SCOPS OWL *Otus icterorhynchus* Plate 59 (7)
Petit-duc à bec jaune

Other name E: Cinnamon Scops Owl
L 18–22 cm (7.0–8.75"). Variably coloured, small forest owl with short ear-tufts. **Adult** *icterorhynchus* Mainly cinnamon-brown, speckled white; speckles bordered black; underparts paler, with heavier markings. Scapulars with bold white spots, forming row on shoulder; greater coverts spotted white; primaries with white spots on outer webs. Plumage variable; some birds more rufous. Eyes pale yellow. Lower Guinea race *holerythrus* more rufous, with fewer markings on breast and no white spots on wing-coverts. **Juvenile** Paler and almost unmarked, with upperparts indistinctly barred and underparts plain. **VOICE** A single, drawn-out, descending whistle *wheeoo*, repeated with pauses of a few seconds. In east of range, could be confused with call of Allen's Squirrel Galago *Galago alleni*, which is slightly more variable. [CD7:27] **HABITS** Inadequately known. Frequents high evergreen forest and forest edge. **SIMILAR SPECIES** African Scops Owl, which may occur at forest edge, has greyer plumage and different voice. **STATUS AND DISTRIBUTION** Uncommon to rare forest resident, from Liberia to Ghana (nominate) and from Cameroon to SW CAR and Congo (*holerythrus*).

EUROPEAN SCOPS OWL *Otus scops*
Petit-duc scops

Plate 59 (5)

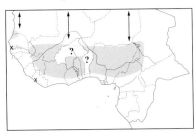

Other names E: Eurasian/Common Scops Owl; F: Hibou petit-duc L 18–20 cm (7–8"). **Adult** *scops* Very similar to slightly smaller African Scops Owl and not safely separable in field. In hand, wing formula different: outermost (10th) primary longer than 5th. ***O. s. mallorcae*** more uniformly grey than nominate, with dark markings slightly more distinct and shorter wings. **VOICE** A pure, whistled *pyuu*. Normally silent in Africa. [CD3:13] **HABITS** In various bushy and wooded habitats; avoids forest. **SIMILAR SPECIES** African Scops has outermost primary shorter than 5th; in field best distinguished by voice. **STATUS AND DISTRIBUTION** Inadequately known due to identification issue. Rare to not uncommon Palearctic migrant (Sep–Apr) in broad savanna belt; recorded from Mauritania, N Senegal, Gambia, Guinea, Mali, Liberia, Ivory Coast, Ghana, Togo, Niger, Nigeria, Chad and Cameroon. Mostly nominate *scops*, but *O. s. mallorcae*, which occurs in Iberia and possibly NW Africa, may be resident at some Saharan oases or visitor south to Niger.

AFRICAN SCOPS OWL *Otus senegalensis*
Petit-duc africain

Plate 59 (6)

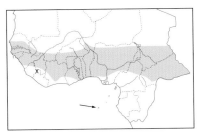

L 16–18 cm (6.25–7.0"). *Cryptically and variably coloured small owl with prominent ear-tufts.* Short tail does not project beyond wingtips at rest. **Adult** *senegalensis* Mainly grey, with fine blackish streaks and vermiculations; underparts paler, with heavier black streaks. Face grey outlined with black. Scapulars with bold white spots, forming row on shoulder; median and greater coverts tipped white; primaries barred white. Variable amount of buff or rufous in plumage; some birds mainly brownish. Eyes pale yellow. Island race *feae* darker and more broadly streaked below. **Juvenile** As adult. **VOICE** A single, short, vibrant *prr-u-u-p*, often repeated, with pauses of 5–10 seconds. [CD7:28] **HABITS** In wooded savanna; occasionally gardens and mangroves. Nocturnal. Roosts by day in dense cover or against tree trunk, where cryptic plumage make it very hard to discover. Generally starts calling at dusk, but may do so up to one hour before. **SIMILAR SPECIES** Slightly larger European Scops Owl (q.v.) not safely distinguishable in field. **STATUS AND DISTRIBUTION** Nominate is uncommon to locally common resident throughout, except in forest zone and arid north. *O. s. feae* restricted to Annobón. **NOTE** Formerly considered conspecific with European Scops Owl.

SÃO TOMÉ SCOPS OWL *Otus hartlaubi*
Petit-duc de Sao Tomé

Plate 143 (7)

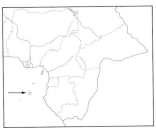

L 18 cm (7"). Variably coloured small owl, with short ear-tufts. **Adult pale morph** Mainly dark grey-brown and ochre, paler below, finely vermiculated and barred brown, ochre, buff and white. Throat white; scapulars with white spots, forming row on shoulder; median and greater coverts tipped white. Bill and eyes yellow. **Adult rufous morph** Overall much darker and rufous. **Juvenile** As adult but paler. **VOICE** A soft *tuh* or *whu* and a higher pitched, single, purring *prr-u-u-p* or *pwu-huhu*, given at intervals of 15–20 seconds, reminiscent of African Scops Owl song but more mellow and higher pitched. Calls include a nasal *tchèèp* and a drawn-out *shrweeee*. [CD7:29] **HABITS** In most habitats with tall trees, incl. high- and low-altitude primary and secondary forest (but not shade forest), occurring in dense foliage at all levels. Sings frequently from 1 hour before dusk to 1 hour after dawn; occasionally during day. Not shy. **SIMILAR SPECIES** No other owls, except larger and very dissimilar Barn Owl, occur on São Tomé (and none on Príncipe). **STATUS AND DISTRIBUTION** Not uncommon resident, endemic to São Tomé. Perhaps also Príncipe. *TBW* category: THREATENED (Vulnerable).

WHITE-FACED OWL *Ptilopsis leucotis*
Petit-duc à face blanche

Plate 59 (8)

Other names E: White-faced Scops Owl
L 23–28 cm (9–11"). Smallish owl with *contrasting white face boldly outlined with black, large orange eyes*, and *distinct ear-tufts*. **Adult** *leucotis* Mainly soft grey, with long, narrow blackish streaks and fine vermiculations; underparts

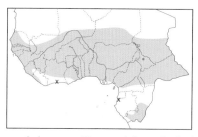

paler, with more obvious blackish streaks. Scapulars with bold white spots, forming row on shoulder; primaries finely barred black. Southern *P. l. granti* generally darker and greyer with more distinct black markings. Individual variation in both races. **Juvenile** Paler and browner; face pale grey-brown; underparts less heavily marked; eyes yellow. **VOICE** Nominate *leucotis* utters 2 rather low, melodious, fluting syllables *whoh whoow*, repeated at intervals of 4–8 seconds. Also a lower and slightly faster *whoh thohoow*. *O. l. granti* has a rapid, muffled bubbling followed by a clear, melodious hoot *kdkdkdkdkdkd-whOOw*. [CD7:30] **HABITS** In various wooded habitats, mainly open wooded savanna. Nocturnal. Roosts by day in trees and bushes, often in rather open situations. **SIMILAR SPECIES** African Scops Owls is noticeably smaller with grey face only narrowly bordered black, smaller ear-tufts and yellow eyes. **STATUS AND DISTRIBUTION** Not uncommon resident, from S Mauritania to Liberia east to S Chad and CAR (nominate), and in Gabon and Congo (*granti*). **NOTE** Genus *Ptilopsis* merged with *Otus* by BoA, but molecular-biological studies suggest it to be very different; eyes also larger. Considered to contain two monotypic species, Northern and Southern White-faced Owl, by some authors, on basis of differences in DNA and vocalisations (cf. *HBW*).

MANED OWL *Jubula lettii*   Plate 60 (2)
Duc à crinière

Other name F: Hibou à bec jaune
L 34–37 cm (13.5–14.5"). Rufous-brown forest owl with bushy ear-tufts and elongated crown and nape feathers. **Adult** Top of head and upperparts dark rufous-brown; crown and nape irregularly speckled white; face rufous boldly bordered black. Underparts rufous becoming paler with long blackish-brown streaks from lower breast down. Scapulars with pale rufous spots; flight and tail feathers rufous with some blackish bars. Bill yellow. Eyes deep yellow to crimson. Feet yellow. **Juvenile** Almost unmarked pale rufous; head and neck almost white; upperpart feathers finely barred rufous and tipped white. **VOICE** Song unknown. Probably consists of a single hoot followed by a series (Dowsett-Lemaire 1996). **HABITS** Inadequately known. In lowland forest, esp. favouring sites with dense lianas. **SIMILAR SPECIES** None. Eagle owls are all noticeably larger and heavier, with differently shaped ear-tufts. **STATUS AND DISTRIBUTION** Rare forest resident, recorded from Liberia to Ghana and from S Cameroon to Congo. May be locally less rare than records suggest.

EURASIAN EAGLE OWL *Bubo bubo*   Plate 60 (8)
Grand-duc d'Europe

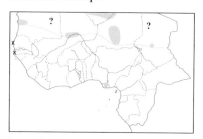

Other name F: Hibou grand-duc
L 46–50 cm (18–20"). *B. b. ascalaphus*. Large owl with distinct ear-tufts and orange or deep yellow eyes. **Adult** Mainly tawny-rufous to sandy-ochre above with dark and pale blotches and streaks. Face plain tawny or pale ochre, indistinctly bordered dusky. Flight feathers and tail rufous barred dusky-brown. Underparts pale tawny to sandy coloured, boldly blotched blackish-brown on upper breast; rest of underparts finely and indistinctly barred dark brown. Bill black. **Juvenile** Ear-tufts smaller. **VOICE** A far-carrying, deep, sonorous hooting *WHOO-oo*, second syllable lower pitched; repeated at intervals of 10–15 seconds. During courtship, a trisyllabic *WHOO-hoo-hoo*. Duets; female higher pitched. [CD3:14] **HABITS** In desert and subdesert, esp. with rocky outcrops and in oases. Mainly nocturnal. Calls at dusk. **STATUS AND DISTRIBUTION** Uncommon resident in dry belt, in Mauritania, Mali, Niger and Chad. Vagrant, Senegal. Seasonal movements recorded. **NOTE** Sometimes considered a separate species, Desert (or Pharaoh) Eagle Owl *B. ascalaphus* (Grand-duc ascalaphe), with birds from our region, which are generally paler than those in N Africa, recognised as race *desertorum* (cf. *HBW*).

SPOTTED EAGLE OWL *Bubo africanus*   Plate 60 (6)
Grand-duc africain

L 43–48 cm (17–19"). Large, greyish owl with prominent, two-toned ear-tufts. **Adult** *cinerascens* Top of head and upperparts dark grey-brown finely barred and vermiculated; face darkish bordered black. Underparts whitish finely and densely barred grey-brown; throat white; breast with dark blotches. Tail and flight feathers dark grey-brown barred buff. *Eyes dark brown.* Bill black tipped grey. Toes mainly feathered; bare tips dark horn. **Nominate**

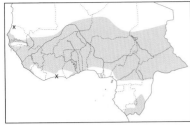

greyer, with *yellow eyes* and blackish-horn bill. **Juvenile** Browner with shorter ear-tufts. **VOICE** Male utters a short, abrupt hoot, followed after a short pause by a second, lower and drawn-out hoot *HO!, hoooo*. Female has trisyllabic *ho, hohooo*, middle note higher. Often duets. [CD7:31] **HABITS** In various habitats, from rocky outcrops in desert to well-wooded savanna and forest edge; often near villages and towns. Calls mainly at dawn and dusk. Often on dirt roads at night. **SIMILAR SPECIES** Akun Eagle Owl is darker with yellow eyes and bare, pale yellow feet. **STATUS AND DISTRIBUTION** Uncommon to not uncommon resident throughout. *B. a. cinerascens* in most of range; nominate in Gabon and Congo. **NOTE** Ssp. *cinerascens* sometimes treated as a species, Greyish (or Vermiculated) Eagle Owl (cf. *HBW*).

FRASER'S EAGLE OWL *Bubo poensis* Plate 60 (3)
Grand-duc à aigrettes

L 38–45 cm (15.0–17.75"). Large, dark rufous forest owl with prominent ear-tufts and narrowly barred plumage. **Adult** Top of head and upperparts rufous-brown barred blackish-brown; face rufous bordered black. Underparts pale rufous becoming white on belly and undertail-coverts; entirely and finely barred blackish-brown, with dusky blotches on breast. Tail and flight feathers dark rufous barred blackish. Eyes dark; eyelids pale blue. Bill pale blue-grey. Feet bluish-grey. Rufous tone and density of barring variable. **Juvenile** Very pale rufous, almost white, with narrow dark bars. **VOICE** A plaintive, drawn-out *whuah* or *whooaah* and an often long, low purring or grunting trill, rising and falling in tempo and volume, ceasing abruptly, recalling sound of small engine *kudakudakudakudakdkdkdkdkdkd rdrdrdrd...* [CD7:33] **HABITS** In evergreen forest, incl. edges and clearings. Calls mainly at and just after dusk and before dawn. Nocturnal. **SIMILAR SPECIES** Akun Eagle Owl has yellow eyes and is darker and more slender. **STATUS AND DISTRIBUTION** Uncommon to not uncommon forest resident, from Sierra Leone to Ghana and from Nigeria to CAR and Congo; also Bioko. **NOTE** Here treated as monotypic, following *HBW* and others.

SHELLEY'S EAGLE OWL *Bubo shelleyi* Plate 60 (5)
Grand-duc de Shelley

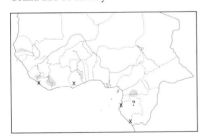

L 54–61 cm (21.25–24.0"). Very large, dark forest owl, with boldly barred underparts and prominent ear-tufts. **Adult** Top of head and upperparts dark brown, usually with some white feathers on crown; face dirty white barred blackish-brown and bordered black. Underparts whitish heavily barred blackish-brown. Tail and flight feathers dark brown barred blackish. Eyes dark. Bill pale creamy-horn. **Juvenile** Mainly white barred brown; flight and tail feathers as adult. **VOICE** A single, loud, high-pitched, plaintive scream *KEEEEOOOOUW!* [CD7:34] **HABITS** Little known. In evergreen forest, incl. edges and clearings. Reported to call mainly at and just after dusk and before dawn. Nocturnal. **SIMILAR SPECIES** Other forest eagle owls are noticeably smaller. **STATUS AND DISTRIBUTION** Rare forest resident, from SE Sierra Leone (Gola Forest) to W Ivory Coast and in Ghana, S Cameroon, Gabon and S Congo.

VERREAUX'S EAGLE OWL *Bubo lacteus* Plate 60 (7)
Grand-duc de Verreaux

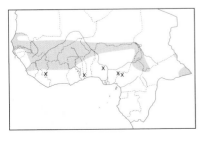

Other names E: Giant/Milky Eagle Owl
L 58–65 cm (23–26"). Very large, greyish owl with pink eyelids. **Adult** Top of head and upperparts grey-brown with fine vermiculations; face whitish bordered black at sides; black bristles at base of bill; broad, bushy ear-tufts rarely raised. Underparts whitish finely vermiculated grey-brown. Tail and flight feathers broadly barred dark brown. Eyes dark. Bill pale creamy-horn. **Juvenile** Browner, with ear-tufts limited to small bumps. **VOICE** A short, irregular series of low, discontented grunts. [CD7:35] **HABITS** In wooded savanna, often in riverine woodland with large trees. Calls mainly at dusk;

also not infrequently during day. **SIMILAR SPECIES** Spotted Eagle Owl is distinctly smaller, with more prominent ear-tufts, dusky blotches on breast and no pink eyelids. **STATUS AND DISTRIBUTION** Uncommon to rare resident in broad savanna belt, from S Mauritania to N Liberia east to Chad and CAR. Distribution patchy.

AKUN EAGLE OWL *Bubo leucostictus*
Grand-duc tacheté

Plate 60 (4)

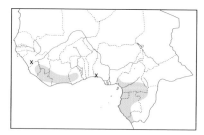

L 40–46 cm (15.75–18.0"). Large, dark forest owl with prominent two-toned ear-tufts and *yellow eyes*. **Adult** Top of head and upperparts dark brown indistinctly barred paler. Face dusky bordered black; ear-tufts dark brown at front, white at back; conspicuous white V from bill to ear-tufts. Underparts whitish with long, rather narrow bars and large dark blotches, esp. on breast and upper belly. Tail and flight feathers dark brown barred paler. Bill pale greenish-yellow. Eyes pale yellow; eyelids dark brown. Feet pale yellow. Brown tone and density of markings variable. **Juvenile** Almost entirely white, with narrow, widely spaced rufous-brown bars. **VOICE** Reminiscent of Fraser's Eagle Owl, with similar low grunting trill, but even more plaintive wail *wheeaah!*; also utters kind of barking and series of low *ro, ro, ro,...* [CD7:32] **HABITS** In tall evergreen forest, incl. edges and clearings. Calls mainly just after dusk and before dawn. Nocturnal. **SIMILAR SPECIES** Fraser's Eagle Owl has dark eyes and is overall more barred. Spotted Eagle Owl paler and greyer with, in northern ssp. *cinerascens*, dark eyes. **STATUS AND DISTRIBUTION** Rare to locally not uncommon forest resident, from Sierra Leone and SE Guinea to Ghana and from Nigeria to SW CAR and Congo.

PEL'S FISHING OWL *Scotopelia peli*
Chouette-pêcheuse de Pel

Plate 58 (6)

L 55–63 cm (22–25"). *Very large, strikingly orange-rufous owl* with large round head and no ear-tufts. **Adult male** Upperparts and tail narrowly barred blackish; underparts paler, variably marked with dark brown drop-shaped spots, streaks and bars. No distinct facial disk; face unmarked. Eyes large, dark brown. Bill black; cere grey. Bare legs and feet yellowish-white. **Adult female** Larger; usually paler. **Juvenile** Initially covered in whitish to pale buffish down, with pale rufous wing and tail feathers narrowly barred blackish. Retains pale yellowish head until attaining adult plumage at *c.* 15 months. **VOICE** A far-carrying, deep sonorous hoot, sometimes preceded and often followed by a low grunt *hooommm-hut*. Pair may duet, male starting with low grunting *uh-uh-uhu...* reaching higher *hoommm*, answered by deeper hoot of female (Steyn 1982). Young utter single, initially rising, then falling scream *weeeeaaow*. [CD7:36] **HABITS** Frequents large rivers bordered by forest, and swamps. Largely nocturnal. Roosts by day in shady spot in large tree near water. Flushes with noisy wingbeats. Catches fish from low perch. **SIMILAR SPECIES** Rufous Fishing Owl is smaller and darker rufous, without obvious barring on upperparts and with dark streaking on underparts. Vermiculated Fishing Owl also smaller, with brown upperparts and brown-streaked underparts. **STATUS AND DISTRIBUTION** Rare to locally not uncommon resident on suitable waterways throughout forest and savanna zones.

RUFOUS FISHING OWL *Scotopelia ussheri*
Chouette-pêcheuse rousse

Plate 58 (5)

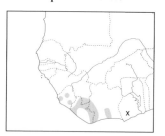

Other name E: Ussher's Fishing Owl

L 43–51 cm (17–20"). Large, rufous owl with streaked underparts. Round head flattened at sides, without ear-tufts nor distinct facial disk. **Adult** Top of head rufous with fine darker streaking; upperparts plain rufous. Face unmarked pale cinnamon-rufous. Underparts pale cinnamon-rufous with narrow dusky rufous streaks. Flight feathers and tail rufous barred dusky. Eyes large, dark brown, strongly contrasting with pale face. Bill black, cere creamy. Bare legs and feet yellowish-white. **Juvenile** Paler. Entire head, mantle, wing-coverts and underparts very pale buff; flight feathers similar to adult, strongly contrasting with rest of plumage. Bill yellow-grey, darker at tip, cere dirty yellow-green. **VOICE** A single, deep, drawn-out wailing hoot, repeated at 1-minute intervals. [CD7:37] **HABITS** Inadequately known. In riverine forest and mangroves. **SIMILAR SPECIES** Pel's Fishing Owl is noticeably larger with barred (not plain) upperparts. **STATUS AND DISTRIBUTION** Rare to scarce resident, recorded from Sierra Leone, Liberia, SE Guinea (1 record, 1951), Ivory Coast and Ghana. *TBW* category: THREATENED (Endangered).

VERMICULATED FISHING OWL *Scotopelia bouvieri* Plate 58 (7)
Chouette-pêcheuse de Bouvier

Other name E: Bouvier's Fishing Owl

L 43–51 cm (17–20"). **Adult male** Above, dusky-brown mixed with rufous, marked with dark streaks on top of head and fine vermiculations on upperparts. Face pale dusky-brown; underparts buffish-white to pale rufous with long, bold, dark brown streaks. Flight feathers and tail broadly barred dusky. Eyes dark brown. Bill yellow tipped dusky. Bare legs and feet yellowish. **Adult female** Less rufous above. **Juvenile** Mainly white washed pale rufous; underparts white with indistinct narrow brown streaks; flight feathers more rufous than adult. **VOICE** A deep, drawn-out, croaking wail *krooOOoah.* Also a series of short, abrupt notes *kroh! woh!-woh! woh!-woh! woh!-woh!* Scream of young similar to that of young Pel's Fishing Owl, but lower pitched. [CD7:38] **HABITS** Along forest streams and marshy pools, and in swamp forest. Secretive. **SIMILAR SPECIES** Pel's Fishing Owl is larger with orange-rufous plumage barred black above. **STATUS AND DISTRIBUTION** Rare to locally not uncommon or common forest resident, S Nigeria, S Cameroon, CAR, Gabon, and Congo (locally common).

PEARL-SPOTTED OWLET *Glaucidium perlatum* Plate 59 (4)
Chevêchette perlée

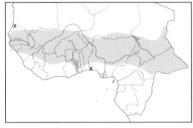

L 19–21 cm (7.5–8.25"). *G. p. perlatum.* Small owl with rounded head, no ear-tufts and rather long tail. Two dark 'eye-spots' form false 'face' on nape. **Adult** Head, upperparts and tail brown finely spotted white. Face whitish edged brown. Underparts whitish heavily streaked rufous-brown. Scapulars with bold white spots, forming row on shoulder; median and greater coverts tipped white; flight and tail feathers with white spots forming bars. Rufous in plumage variable. Eyes yellow. Bill yellowish-horn. **Juvenile** Similar to adult but lacks white pearly spots on crown and mantle. **VOICE** A distinctive, rhythmically rising series of short notes followed by a loud whistling climax *hu-hu-hu-hu-HU-HU-HU TEEEUW TEEEUW TEEEUW...* [CD7:39] **HABITS** In savanna woodland. Partially diurnal; regularly calls by day. **SIMILAR SPECIES** African and European Scops Owls generally greyer, more slender, and have ear-tufts. **STATUS AND DISTRIBUTION** Uncommon to locally common resident in broad savanna belt, from S Mauritania to Guinea east to Chad and CAR.

RED-CHESTED OWLET *Glaucidium tephronotum* Plate 59 (3)
Chevêchette à pieds jaunes

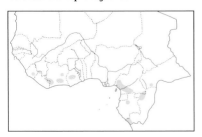

Other name E: Yellow-legged Owlet

L 20–24 cm (8.0–9.5"). Small forest owl with rounded head, no ear-tufts and rather long tail. **Adult** *tephronotum* Head dark grey with greyish face, yellow cere and yellow eyes. Upperparts dark slate-grey; tail dusky-brown *with 3 bold white spots on centre*. Underparts white with large rufous-brown spots; breast-sides and flanks rufous-chestnut. Bill greenish-yellow. *G. t. pycrafti* darker above with less rufous on breast-sides and flanks; spots on underparts blackish. **Juvenile** Underparts lack rufous-chestnut and have smaller or no spots. **VOICE** A rhythmic, rather slow series of whistles *huut huut huut huut ...* (repeated at intervals of *c.* 1 second); length and speed of series variable. [CD7:40] **HABITS** In mature evergreen forest and forest edge. Crepuscular and nocturnal. **SIMILAR SPECIES** Chestnut-backed Owlet is larger with narrow dark barring on underparts and white barring on wings and tail. **STATUS AND DISTRIBUTION** Rare to uncommon forest resident, from Sierra Leone to Ghana (nominate) and from S Cameroon to SW CAR and Congo (*pycrafti*).

CHESTNUT-BACKED OWLET *Glaucidium sjostedti*   Plate 59 (1)
Chevêchette à queue barrée

Other name E: Sjöstedt's Barred Owlet
L 24–28 cm (9.5–11.0"). Forest owl with large rounded head and finely barred head and underparts. **Adult** Head and mantle dusky-brown finely and densely barred white; white eyebrows. Rest of upperparts *deep chestnut*. Flight feathers and rather long tail dusky-brown narrowly barred white; greater upperwing-coverts irregularly tipped white. Throat whitish; *rest of underparts cinnamon-rufous*, paler on belly and undertail-coverts, *finely barred dark brown*. Bill pale yellow. Eyes yellow. **Juvenile** Underparts paler, with indistinct barring restricted to breast. **VOICE** A short, rapid series of similar hoots *ho-ho-ho-ho* and a descending and accelerating series of vibrant notes *krroow krroow krroow-rroo-rroo-rroo*. [CD7:43] **HABITS** In tall evergreen forest interior. Nocturnal. Calls mainly at dawn and dusk, but may commence up to one hour before dusk (esp. on overcast days) and continue after dawn. **SIMILAR SPECIES** Red-chested Owlet is much smaller and has white underparts spotted blackish and rufous breast-sides and flanks. African Barred Owlet has whitish underparts broadly barred dusky on upper breast only. **STATUS AND DISTRIBUTION** Rare to common forest resident, from SE Nigeria to S Congo and in SW CAR.

AFRICAN BARRED OWLET *Glaucidium capense*   Plate 59 (2)
Chevêchette du Cap

Other name F: Chevêchette à poitrine barrée
L 20–21 cm (8.0–8.5"). Small, dumpy owl with rounded head. **Adult** *etchecopari* Head dark brown barred white (becoming more speckled on forehead); face grey. Upperparts dark brown sparsely barred tawny (mantle and back almost unmarked), with white spots on scapulars forming row on shoulder. Underparts white; upper breast broadly barred dusky-brown on buff-brown; rest of underparts marked with largish dusky-brown spots. Outer median and greater coverts tipped white; flight feathers barred whitish; tail fairly long, very narrowly barred tawny. Eyes yellow. Bill yellowish. *G. c. castaneum* has mantle and back plain chestnut. **Juvenile** As adult. **VOICE** A fairly rapid, rhythmic series of 6–18 plain whistled hoots *hoot-hoot-hoot-hoot-hoot-...*, repeated at intervals of *c.* 2 seconds or slightly faster (Dowsett & Dowsett-Lemaire 1993). Also similar, but slightly accelerating series with more vibrant notes. [CD7:41] **HABITS** In lowland evergreen and gallery forests. **SIMILAR SPECIES** Chestnut-backed Owlet is larger and has longer tail and more colourful plumage; underparts cinnamon-rufous with dense, narrow dark barring. **STATUS AND DISTRIBUTION** Generally rare to locally not uncommon forest resident, recorded in Liberia and Ivory Coast (*etchecopari*) and from S Cameroon to SW CAR and Congo (where locally common) (probably *castaneum*). **NOTE** *G. c. etchecopari* and *castaneum* sometimes considered specifically distinct from other forms of *G. capense* as Chestnut Owlet (Chevêchette châtaine) *G. castaneum* (cf. *HBW*). For discussion, see Dowsett & Dowsett-Lemaire (1993).

LITTLE OWL *Athene noctua*   Plate 59 (9)
Chevêche d'Athéna

Other name F: Chouette chevêche
L 21–25 cm (8.25–10.0"). *A. n. saharae*. The only small owl without ear-tufts in its range. Flat headed, with 'frowning' face. **Adult** Mainly sandy-brown above, streaked white on head, spotted and dappled white on upperparts. Underparts white broadly streaked brownish. Flight and tail feathers barred whitish. Plumage coloration variable. Eyes yellow. Bill pale horn. **Juvenile** Less boldly marked. **VOICE** Commonest call a sharp *keeow* and *woow*. Advertising call of male a rather low, mellow, whistled *ko-ooep*. Call of female higher pitched and more nasal. Alarm, a loud, sharp *kyip kyip*. Various other calls incl. *kyow* and *WHE-w*. [CD3:15] **HABITS** In stony desert, esp. in montane areas. Partially diurnal. Flight low and undulating. Bobs head when excited. **SIMILAR SPECIES** Migrant European Scops Owl is much greyer and slimmer, with obvious ear-tufts. **STATUS AND DISTRIBUTION** Uncommon and local in Sahara (north of 17°N), in Mauritania, N Mali, N Niger (Todora Mts, Sep 1922) and N Chad. **NOTE** Subspecific boundaries unclear due to individual variation and existence of intermediate populations. Proposed race *solitudinis* (cf. *BoA*) here treated as synonym of *saharae* (following *HBW*), which itself is sometimes considered a colour morph of extralimital *glaux*.

AFRICAN WOOD OWL *Strix woodfordii*

Plate 60 (1)

Chouette africaine

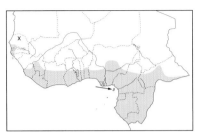

Other name F: Hulotte africaine

L 33–35 cm (13–14"). *S. w. nuchalis.* Medium-sized forest owl with round head and barred underparts. **Adult** Dark brown above, speckled white; scapulars and wing-coverts spotted white. Face pale brownish, with large dark eyes emphasised by dark brown surround and broad whitish eyebrows. Underparts barred rufous-brown and white. Flight and tail feathers barred. Bill yellowish. **Juvenile** White facial markings less distinct. **VOICE** A characteristic, single, drawn-out *whoOOow* or *oowhEEoo* and a rhythmic *hu-hoo, hu-hoo hoo-hu-hoo.* Duets; female higher pitched. Male may answer female *whoOOow* by low *hooo.* [CD7:44] **HABITS** In evergreen and gallery forests, dense woodland and plantations. Often in pairs. Nocturnal. **SIMILAR SPECIES** None. Eagle owls are larger and have ear-tufts. Similar-sized Maned Owl has yellow eyes, bushy feathers on head, and differently coloured plumage. **STATUS AND DISTRIBUTION** Generally common forest resident, from Senegambia (rare) to Liberia east to CAR and Congo; also Bioko.

SHORT-EARED OWL *Asio flammeus*

Plate 58 (3)

Hibou des marais

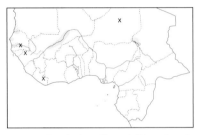

L 35–40 cm (14–16"); WS 95–110 cm (37–43"). *A. f. flammeus.* **Adult** Above, sandy-buff blotched and streaked dark brown and tawny. Face whitish narrowly edged dark, with lemon-yellow eyes surrounded by black. Small, short ear-tufts in centre of forehead usually hard to see. Underparts buffish streaked dark brown, mainly on breast. Flight and tail feathers barred dark brown. In flight, long wings with dark carpal patch and contrasting sandy-buff patch at base of primaries; underwing pale with black tips and dark carpal patch. **VOICE** Usually silent in Africa. A high-pitched yelp and a hooting note, often repeated. [CD3:18] **HABITS** In various open habitats. Crepuscular, nocturnal and often diurnal. Hunts low; buoyant flight with rigid wingbeats alternating with gliding and banking, and wings held in shallow V, recalling harrier. Roosts on ground. **SIMILAR SPECIES** Marsh Owl is darker and plainer above and has dark brown eyes. **STATUS AND DISTRIBUTION** Rare to uncommon Palearctic migrant, Mauritania, Senegambia, N Guinea and Mali. Vagrant, Niger, Chad, Liberia and Cape Verde Is.

MARSH OWL *Asio capensis*

Plate 58 (4)

Hibou du Cap

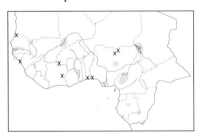

Other name F: Hibou des marais africain

L 30–35 cm (*c.* 12–14"); WS 80–95 cm (31.5–37.0"). *A. c. capensis.* Rather plain, dark brown owl with round head and dark eyes. **Adult** Above, dark brown. Face pale brown to buffish, with large dark brown eyes emphasised by blackish surround. Small, short ear-tufts in centre of forehead rarely visible. Dark brown breast, becoming buffish on lower underparts. In flight, contrasting large pale patch at base of primaries; underwing pale with dark carpal patch and black tips to primaries. Bill blackish. **Juvenile** Face darker and bordered black. **VOICE** Various harsh croaks and rasping calls, uttered in flight and on ground. [CD3:19] **HABITS** Singly, in pairs or small groups in open grassland. Mainly crepuscular and nocturnal, occasionally diurnal. Hunts from low perch or in low, buoyant flight with wings held in shallow V and sudden banking, recalling harrier. Roosts on ground. Inquisitive, often circling around intruder when flushed. **SIMILAR SPECIES** Short-eared Owl occurs in similar habitats but has paler and more variegated plumage, and yellow eyes; also slightly larger. Grass Owl has paler, heart-shaped face. **STATUS AND DISTRIBUTION** Local resident and partially intra-African migrant. Uncommon to rare, S Mauritania, coastal Gambia, Guinea, Burkina Faso, Ivory Coast, Benin, Nigeria, L. Chad area (W Chad, N Cameroon), SE Gabon and S Congo. Locally not uncommon, Mali (central Niger delta), Nigeria (Jos and Mambilla Plateaux) and W Cameroon (highlands). Also claimed from coastal S Senegal.

NIGHTJARS

Family **CAPRIMULGIDAE (World: 89 species; Africa: 24; WA: 16)**

Rather small to medium-sized, crepuscular and nocturnal birds with very cryptic plumage, long wings, moderately long tail, tiny bill and huge gape for catching insects on the wing. No nest, normally laying eggs (usually two) on bare ground, leaf litter or rocks. Rarely seen during day, except when flushed from underfoot. Identification often problematic; important features include overall coloration, markings on scapulars and wing-coverts, presence or absence of nuchal collar, and amount of white (male) or buff (female) in wings and tail, but best identified by voice. Typically sing at or just after dusk and just before dawn; often also throughout moonlit nights. Songs churring or reeling, whistling, yelping and chuckling. Often encountered on dirt roads at night; orange reflection of eyes in headlights of car visible at considerable distance. Many species poorly known.

BATES'S NIGHTJAR *Caprimulgus batesi*   Plate 61 (2)
Engoulevent de Bates

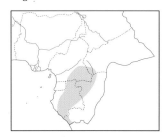

L 29–31 cm (11.5"). Very large, dark nightjar with relatively long tail. **Adult male** Top of head grey-brown with black blotches and streaks; head-sides and variably distinct collar on hindneck ochreous narrowly barred dark brown. Upperparts dark brown blotched, streaked and spotted tawny and buff. Lesser coverts black; flight feathers dark brown narrowly barred tawny or buff, except outer primaries. Tail dark brown with *c.* 5 regularly spaced blackish bars; outer tail feathers tipped white. Throat with white patch; breast dark brown variably marked buff or rufous; belly and undertail-coverts buff barred brown. In flight, *small* white patch towards wingtip (on 4 outer primaries) and white tips to outer 2 tail feathers. **Adult female** Generally paler and more rufous; no white wing patch; outer tail feathers tipped buff. **VOICE** Song, usually given from tree perch, an initial clear, yelping note, followed after a short pause by a rapid series of 2–12 (usually 5–6) similar notes, *whow! whow-whow-whow-whow whow! whow-whow...* resembling Freckled Nightjar but more monotonous. Occasionally sings from ground. Flight call a low, guttural *ugh-ugh-ugh.* [CD7:45] **HABITS** Frequents lowland forest, gallery forest and forest edge. Roosts by day on ground or on lianas. When flushed, flies a short distance before dropping to ground or perching on low branch. **SIMILAR SPECIES** Brown Nightjar is much smaller and lacks white in wings and tail. **STATUS AND DISTRIBUTION** Uncommon to locally not uncommon forest resident, S Cameroon, SW CAR, Gabon and N and S Congo. **NOTE** An unknown nightjar resembling a small Bates's Nightjar was found in Bafut-Ngemba FR (Bamenda, W Cameroon), on 9 Apr 1990. Despite measurements and photographs taken in the hand, it has not been identified (Robertson 1992).

BROWN NIGHTJAR *Caprimulgus binotatus*   Plate 61 (1)
Engoulevent à deux taches

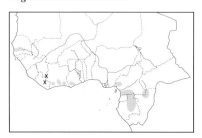

L 21–23 cm (8.25–9.0"). Smallish, very dark nightjar without white spots on wings or tail. **Adult** Head, upperparts and tail very dark brown with row of contrasting buff spots on scapulars and small white patch on throat-sides. Small 'ear-tufts' at rear of crown. Underparts dark brown finely mottled and barred tawny or rufous. Tail rather stiff and 'tent-shaped'. **VOICE** Song, given from perch or in flight, occasionally on ground, a long series of *kyup* or *kliou* notes. [CD7:46] **HABITS** Frequents lowland rainforest and forest edge. Roosts by day on lianas or high branches. Hawks insects from perch, in or above canopy. Typically sings at dusk and just before dawn. Nest unknown, but one perched on low palm leaf was apparently brooding young (Carroll & Fry 1987). **SIMILAR SPECIES** Bates's Nightjar is much larger with more variegated plumage; male has white in wings and tail. **STATUS AND DISTRIBUTION** Inadequately known. Rare forest resident, recorded from Liberia, Ivory Coast, S Ghana, W and S Cameroon, SW CAR, Gabon and N Congo. **NOTE** Sometimes placed in monotypic genus *Veles* because of lack of rictal bristles, slightly curved primaries and unique ∧-shaped tail.

PRIGOGINE'S NIGHTJAR *Caprimulgus prigoginei*   Plate 147 (1)
Engoulevent de Prigogine

Other name E: Itombe Nightjar

L 19 cm (7.5"). Small, rather dark nightjar with relatively short tail. Known only from single female specimen.

Adult female Top of head brown, rufous and buff with black blotches and streaks. No collar on hindneck but tiny white spot on nape-sides. Upperparts brown blotched blackish and mixed with rufous, tawny and buff. Lesser coverts rufous-brown barred tawny; flight feathers dark brown marked rufous and buff. Tail dark brown irregularly barred buff and rufous, and narrowly tipped buffish, with outer tail feathers tipped white. Small buffish-white patch on throat-sides; breast rufous-brown barred blackish; belly and undertail-coverts buff barred blackish. **VOICE** A song, consisting of a long, regular series of rapid notes *chuk-chuk-chuk-chuk-chuk-chuk-*, similar to song of Swamp Nightjar, and short *rek, rek* call notes, are possibly from this species. **HABITS** Unknown. Presumed to occur in forest. **SIMILAR SPECIES** Brown Nightjar is darker and lacks white in tail. Bates's Nightjar is much larger, much darker and has long tail. **STATUS AND DISTRIBUTION** Specimen taken in Itombwe forest, E Congo-Kinshasa, Aug 1955. Vocalisations described above were tape-recorded in forest in SE Cameroon and N Congo-Brazzaville, and proved similar to recordings made in Itombwe. It has been suggested that they may be attributable to Prigogine's Nightjar (Dowsett & Dowsett-Lemaire 2000) *TBW* category: THREATENED (Endangered).

SWAMP NIGHTJAR *Caprimulgus natalensis*   Plate 62 (6)
Engoulevent du Natal

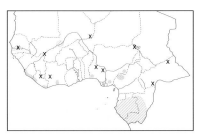

Other names E: Natal/White-tailed Nightjar; F: Engoulevent à queue blanche
L 20–24 cm (8.0–9.5"). Medium-sized, mainly grey-brown, heavily marked and spotted nightjar of wet grasslands. Tail rather short. **Adult male** *natalensis* Top of head blackish speckled grey and buff; *head-sides dark brown bordered by long, pale supercilium and moustachial stripe*. Indistinct collar on hindneck buffish or tawny. Upperparts brown to grey-brown with *large, irregular blackish markings* on scapulars and wing-coverts. Throat has white, triangular patch; *breast brown with rounded buff spots*, becoming buff barred dark brown on belly. In flight, *white patch towards wingtip* and *white-edged tail*. Coastal *accrae* more uniformly grey-brown; blackish markings on crown, scapulars and wing-coverts larger. **Adult female** Similar, but has buffish patch on wings and buff-edged tail. **VOICE** Song, mainly given from ground, a long, regular series of rapid notes *chuk-chuk-chuk-chuk-chuk-chuk-...* or *tuk-tuk-tuk-tuk-tuk-...*, resembling knocking on dry wood, and a faster *chukukukukukukuk*. Also a sharp note, often repeated a few times, *kip kip-kip-kip*; a highly distinctive, variable series of melodious, tremulous notes *whoa-whulululu whoa-whuwhu whoa-whulululu whoa!* in flight, and a fast *rrrrukukukukukukuk-whoalulululu.* [CD7:47] **HABITS** Frequents damp or locally dry grasslands and edges of swamps and coastal lagoons. Roosts on ground in rank vegetation by day. Sits on tracks and low perches. Typically sings at dusk and just before dawn. **SIMILAR SPECIES** Square-tailed and Long-tailed Nightjars also have white-edged tails but differ in pale trailing edge to wings; songs churring. **STATUS AND DISTRIBUTION** Locally rare to not uncommon and patchily distributed resident. Nominate throughout, from E Gambia, Mali and NE Nigeria to CAR, Gabon and Congo, except where replaced by *accrae*, from coastal Liberia (and presumably Sierra Leone) to W Cameroon. Seasonal local movements reported.

LONG-TAILED NIGHTJAR *Caprimulgus climacurus*   Plate 61 (8)
Engoulevent à longue queue

L 28–43 cm (11–17"), incl. central tail feathers of 20.0–26.5 cm. Slim, small nightjar with *long, graduated tail* and variable, relatively contrasting plumage. **Adult male** *climacurus* Above, mainly grey-brown to pale buff-brown with conspicuous *clear white bar on wing-coverts* (formed by broad white tips to lesser coverts). Head-sides dark rufous bordered by narrow white moustachial stripe and, usually, pale supercilium. Broad tawny or buffish collar on hindneck. Scapulars broadly fringed buff. White patch on 5 outer primaries and white tips to secondaries often visible at rest. Tips of outer 2 tail feathers white; outer tail feather entirely edged white. White triangular throat patch. Breast mainly greyish or pale brown; rest of underparts buffish, narrowly barred brown on belly and flanks. In flight, white patch towards wingtip and white trailing edge to inner wing. *C. c. sclateri* darker and more rufous. **Adult female** Wingbar, patch on outer primaries and trailing edge buff or buffish-white. Central tail feathers much shorter; edge of outer tail feather barred dark brown and buff. **VOICE** Song, given from ground or tree perch, a monotonously sustained, hard reeling with a single, fast rhythm *rrerrrrrrrrrrrrrrrrrrr...* Female has similar, slightly lower pitched *rrorrrrrrr...* In flight, a repeated *chiong-chiong ...* [CD7:48] **HABITS** In wide variety of habitats, from semi-desert to derived savanna in forest zone. Roosts by day on ground. When flushed, flies short distance before dropping onto leaf-littered ground; occasionally perches lengthwise on low branch. Typically sings at dusk and just before dawn, but periodically throughout clear night. **SIMILAR SPECIES**

Square-tailed Nightjar also has white wingbar but is easily distinguished by short tail. **STATUS AND DISTRIBUTION** Widespread and common resident and intra-African migrant throughout. Nominate largely migratory, breeding from Sahel to northern Guinean savannas (north to S Mauritania, C Mali, Niger and Chad), some migrating to southern Guinean savannas and forest zone in off season, possibly south to Congo. *C. c. sclateri* probably mainly sedentary and partially migratory, breeding in southern Guinean savannas and derived savannas in forest zone, from Guinea to Cameroon and in Congo; outside breeding season also from S Cameroon south. Movements poorly understood.

SQUARE-TAILED NIGHTJAR *Caprimulgus fossii* — Plate 62 (5)
Engoulevent du Mozambique

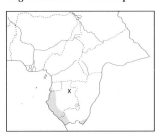

Other names E: Mozambique/Gabon Nightjar
L 23–24 cm (9.0–9.5"). *C. f. fossii.* **Adult male** Top of head greyish streaked black in centre of crown and bordered by narrow pale supercilium. Head-sides grey mixed with rufous, bordered below by white malar stripe. Rufous or tawny collar on hindneck. Upperparts mainly brownish-grey or grey-brown with row of large tawny spots on blackish-brown scapulars and *white bar on wing-coverts* (formed by white-tipped lesser coverts); median and greater coverts boldly tipped buff; secondaries tipped white (often visible at rest). Inner primaries and secondaries rufous barred blackish. Small white patch on throat-sides; underparts greyish mixed with rusty-buff, becoming buffish narrowly barred brown on belly and flanks. In flight, white patch towards wingtip, white trailing edge to wing and white-edged tail. **Adult female** Wingbar, wing patch, trailing edge and tail edge buffish. **VOICE** Song, given from ground or, sometimes, tree perch, a far-carrying, sustained reel alternating with a more sputtering rattle *rrerrrrrrr-rreheheheh-rrerrrrrrr-rreheheheh-...* Flight call a yelping *whaoop!* or *wowaw!* [CD7:49] **HABITS** Frequents grassland, open woodland, forest edge and farmland. Roosts by day on ground. **SIMILAR SPECIES** Long-tailed Nightjar also has white wingbar, which is even more conspicuous, but is easily distinguished by long tail. **STATUS AND DISTRIBUTION** Common resident, Gabon and S Congo. Also recorded from Eq. Guinea. Claims from Guinea, Ghana and Ivory Coast require substantiation.

FIERY-NECKED NIGHTJAR *Caprimulgus pectoralis* — Plate 61 (4)
Engoulevent musicien

L 23–25 cm (9.0–9.75"). *C. p. fervidus.* **Adult male** Top of head greyish streaked black in centre of crown. Head-sides dark rufous, bordered below by white malar stripe. Rufous collar on hindneck. Upperparts mainly grey-brown with double row of black patches on scapulars, upper row fringed buff, lower row larger, fringed rufous. Upperwing-coverts tipped buff; some lesser coverts tipped whitish, forming narrow bar. Inner primaries and secondaries rufous barred blackish. Outer 2 tail feathers broadly tipped white. White patch on throat-sides bordered below by black; underparts dark rufous mixed with grey and blackish, becoming buffish narrowly barred brown on belly and flanks. In flight, white patch towards wingtip and white corners to tail. **Adult female** Wing patch and tail tips buffish. **VOICE** Song, given from tree perch, a clear, melodious drawn-out note followed by a tremulous whistling *kyoo-yiup kyuiurrrr,* very similar to Black-shouldered Nightjar but first note more modulated, often preceded by fast *whoa-whoa-whoa-whoa-...* (slightly slower than Black-shouldered Nightjar). [CD7:51] **HABITS** Frequents woodland, gallery forest and forest edge. Perches in trees. **SIMILAR SPECIES** Black-shouldered Nightjar is generally more rufous with dark brown lesser coverts ('shoulders'). **STATUS AND DISTRIBUTION** A mainly S African species, reaching our region in Congo and SE Gabon, where locally common. Seasonal movements reported.

BLACK-SHOULDERED NIGHTJAR *Caprimulgus nigriscapularis* — Plate 61 (3)
Engoulevent à épaulettes noires

L 23–25 cm (9.0–9.75"). Resembles Fiery-necked Nightjar, but more rufous-brown with *contrasting blackish shoulder.* **Adult male** Top of head brown streaked black in centre of crown. Head-sides and hindneck collar dark rufous. Upperparts mainly dark brown tinged rufous. Double rows of black patches on scapulars edged rusty-buff. *Lesser coverts blackish-brown* speckled rufous; median and greater coverts have rusty-buff spot near tips. Inner primaries and secondaries blackish barred rufous. Outer 2 tail feathers broadly tipped white. Throat with white patch on either side or single triangular patch, bordered below by black; breast dark rufous-brown mixed with grey and blackish; belly and flanks rusty-buff narrowly barred brown. In flight, small white patch towards wingtip and white

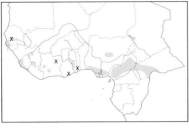

corners to tail. **Adult female** White wing patch and tail corners slightly smaller. **VOICE** Song, given from tree perch, a clear, melodious drawn-out note followed by a tremulous whistling *kiiiuu-iup kiiurrrr* or *tiyuu-iup tyurrrr*, sometimes preceded by a fast *whoap-whoap-whoap-whoap-...* or *whoawhoawhoawhoa...* Flight call a low short *chuk* often repeated rapidly 2–3 times. [CD7:52] **HABITS** Frequents woodland, gallery forest and forest edge. Roosts by day in tree. **SIMILAR SPECIES** Fiery-necked Nightjar is overall paler, greyer brown with more uniform wing-coverts; wing patches slightly larger. **STATUS AND DISTRIBUTION** Inadequately known. Not uncommon to rare, from S Senegal to CAR. Claimed from Gambia. Probably resident. **NOTE** Often considered conspecific with Fiery-necked Nightjar.

PLAIN NIGHTJAR *Caprimulgus inornatus*   Plate 62 (4)
Engoulevent terne

L 22–23 cm (8.75–9.0"). Rather slender nightjar with *uniform plumage* lacking distinctive markings. Plumage coloration varies from cinnamon-rufous or sandy-buff to grey-brown, grey or dark brown (matching soil colour). **Adult male** Centre of crown has small dark streaks. *Row of small black spots on scapulars.* Wing-coverts tipped paler than rest of plumage, sometimes forming indistinct bars. White patch on (4) outer primaries sometimes visible at rest. Inner primaries and secondaries blackish barred cinnamon or buff. Outer 2 tail feathers broadly tipped white. Underparts mainly cinnamon-buff or brownish; belly and flanks narrowly barred brown. In flight, small white patch towards wingtip and white corners to tail. **Adult female** Patch on outer primaries and tips to outer 2 tail feathers buff. **VOICE** Song, given from ground or low bush, a monotonous, sustained, hard reeling *rrorrrrrrrrrrrrrrrr... rrerrrrrrrrrrrrrrrrrrrr...* similar to but noticeably lower pitched and slower than that of Long-tailed Nightjar. In flight, a chuckling *kwakow*. On ground, a low *chuk*. [CD7:55] **HABITS** In variety of open and wooded habitats, incl. arid scrub, wooded grassland, derived savanna and large forest clearings. Roosts by day on ground. When flushed, usually flies short distance before landing again on ground or, sometimes, in tree. **SIMILAR SPECIES** Egyptian Nightjar may resemble greyish forms of Plain Nightjar, but has white on throat and no white in wings and tail. Freckled Nightjar is larger and much darker, and only occurs in rocky habitats. **STATUS AND DISTRIBUTION** Generally common to not uncommon resident and intra-African migrant, from Sahel to forest zone. Breeding visitor to Sahel, from S Mauritania and N Senegal through Mali, N Burkina Faso and Niger (north to Aïr) to C Chad (north to Ennedi). Dry-season visitor to south, from Liberia to S Chad and CAR; occasional breeding in this area reported. Present all months, Lagos (and S Nigeria generally?). Vagrant, Gabon.

FRECKLED NIGHTJAR *Caprimulgus tristigma*   Plate 62 (3)
Engoulevent pointillé

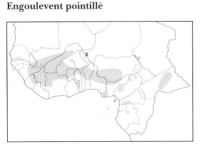

L 26–28 cm (10.25–11.0"). Distinctive, large, dark nightjar of *rocky habitats. Entirely greyish-black, dark brownish-grey or blackish-brown finely speckled white or pale buff*, blending with bare rock on which it generally rests, making detection difficult. **Adult male** *sharpei* Outer 2 tail feathers broadly tipped white. Throat with small white patch on either side or single triangular patch. In flight, small white patch towards wingtip and white corners to tail. Local race *pallidogriseus* paler overall. **Adult female** White wing patch smaller and no white in tail. **VOICE** Song a series of yelping notes *whaow! whaow! whaow! ...* and *aow-whaow! aow-whaow! aow-whaow! ...*, also a more barking *wah!-wah!-wah!-...* Flight call *wok*. [CD7:56] **HABITS** On rocky outcrops and inselbergs with vegetation. Roosts by day on bare rocks and rocky hillsides. **STATUS AND DISTRIBUTION** Uncommon to locally not uncommon resident, from Mali, Guinea and N Sierra Leone to SW Niger and Nigeria, and in Cameroon, C CAR and Gabon (*sharpei*). Perhaps locally migratory. *C. t. pallidogriseus* described from C Nigeria (Jos Plateau); status uncertain.

GOLDEN NIGHTJAR *Caprimulgus eximius*   Plate 62 (1)
Engoulevent doré

L 23–25 cm (9.0–9.75"). Unique, golden-buff nightjar of arid country. **Adult male** *simplicior* Entirely cinnamon-

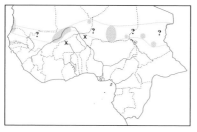

buff covered with irregular white spots, each bordered and speckled black. Outer 2 tail feathers broadly tipped white. Short white malar stripe; white throat patch. Belly plain buff. In flight, conspicuous large white patch towards wingtip and white corners to tail. **Nominate** slightly darker. **Adult female** Wing patch and tail corners smaller and buffish. **VOICE** No information. **HABITS** In arid country with sparsely vegetated, rocky and sandy soils. Roosts by day on ground; hard to flush. **SIMILAR SPECIES** Egyptian Nightjar is pale grey and lacks white in wings and tail. **STATUS AND DISTRIBUTION** Uncommon to rare resident in Sahel zone, from S Mauritania and N Senegal through C and NE Mali, N Burkina Faso and S Niger/NE Nigeria to C Chad (*simplicior*); also NE Chad (Ennedi: nominate).

EGYPTIAN NIGHTJAR *Caprimulgus aegyptius*   Plate 62 (2)
Engoulevent du désert

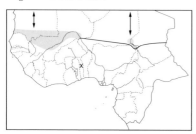

L 24–27 cm (9.5–10.5"); WS 53–58 cm (21–23"). Rather uniform, pale greyish desert nightjar. **Adult *saharae*** Entirely pale sandy-grey above. Indistinct buffish nuchal collar. Scapulars and tertials with some small black-and-buff markings; wing-coverts with indistinct buff spots. Short whitish malar stripe. Throat with white patch on either side or single triangular patch. Lower underparts buff, narrowly barred brown on belly and flanks. In flight, no white in wings and tail; outer wing long and mainly dark brownish. **Nominate** darker, less sandy and greyer. **VOICE** Mostly silent in W Africa. Song, given on ground, a long, very rapid, regular series of *krow* notes, *krowkrowkrowkrowkrowkrowkrowkrow...*, resembling sound of running engine. [CD3:20] **HABITS** On open sandy ground in desert and semi-desert; often near water. Roosts by day on ground, occasionally in groups (up to 50 recorded in Senegal). **SIMILAR SPECIES** Greyish forms of Plain Nightjar may resemble Egyptian Nightjar, but always lack white on throat; males have white patch near wingtip and white tail corners. Golden Nightjar also occurs in Sahel, but is differently coloured and has large white or whitish patches in wings. **STATUS AND DISTRIBUTION** Uncommon to locally not uncommon Palearctic winter visitor to Sahel zone, (Oct) Nov–Feb (Apr). *C. a. saharae* recorded in S Mauritania, N Senegal, Mali and N Nigeria. Nominate in E Mauritania, C Mali, N Togo (probably this ssp.) and Chad. Also S Niger (ssp. not specified).

RED-NECKED NIGHTJAR *Caprimulgus ruficollis*   Plate 61 (7)
Engoulevent à collier roux

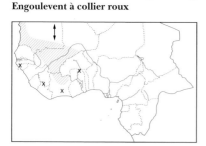

L 30–34 cm (12.0–13.5"); WS 60–65 cm (23.5–25.5"). Large, grey-brown nightjar with broad rufous collar on hindneck. **Adult male *ruficollis*** Top of head greyish with rufous-edged black streaks in centre of crown. Head-sides dusky-rufous, bordered below by white malar stripe. Upperparts mainly greyish, narrowly streaked black. Scapulars with row of black, buff-edged patches. Upperwing-coverts tipped buff, forming pale bars. Inner primaries and secondaries barred rufous and blackish. Outer 2 tail feathers broadly tipped white. Throat rufous with white patch on either side or single triangular patch. Breast vermiculated grey; rest of underparts buff narrowly barred brown. In flight, white patch towards wingtip and white corners to tail. *C. r. desertorum* overall paler, with rufous breast and paler nuchal collar. **Adult female** Wing patch and tail tips buffish. **VOICE** Song, occasionally given on 'spring' passage, a series of rapid, mechanical-sounding *kotok-kotok-kotok-kotok-kotok-kotok-...* accelerating in long *kotokotokotokotokotokoto...* then slowing down again; also slower *ko-tuk ko-tuk ko-tuk ko-tuk ...* [CD3:21] **HABITS** In various open and wooded habitats. Roosts by day on ground and in trees. **SIMILAR SPECIES** European Nightjar is darker, less variegated and smaller, with shorter wings. **STATUS AND DISTRIBUTION** Rare to uncommon Palearctic winter visitor (Oct–Mar), recorded in Mauritania, N Senegal, Gambia, Guinea-Bissau, Mali, Liberia, N Ivory Coast, Burkina Faso, N Ghana and NE Nigeria. Winter quarters of both races probably similar.

EUROPEAN NIGHTJAR *Caprimulgus europaeus*   Plate 61 (6)
Engoulevent d'Europe

L 24–28 cm (9.5–11.0"); WS 52–60 cm (20.5–23.5"). Rather large and dark greyish. **Adult male *europaeus*** Top of

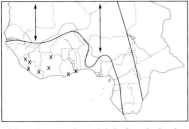

head grey-brown with black streaks in centre of crown. Head-sides dark rufous freckled black, bordered below by white moustachial stripe. Upperparts grey narrowly streaked black. *Scapulars with buff line* formed by row of pointed, black, buff-edged patches. Upperwing-coverts tipped buff, forming *contrasting pale bar on lesser coverts*. Inner primaries and secondaries barred rufous-buff and blackish. Outer 2 tail feathers broadly tipped white. Small white patch on throat-sides. Breast dark grey-brown; rest of underparts buff narrowly barred brown. In flight, white patch towards wingtip and white corners to tail. *C. e. meridionalis* overall paler, with slightly larger white wing patches. **Adult female** Lacks white wing patch and tail corners; throat patches smaller and buffy. **VOICE** Normally silent in W Africa. Song a far-carrying, hard 'two-geared' rattle *rrerrrrrrrrrr-rrorrrrr-rrerrrrrrrrr-rrorrrrr-...* Flight call a short *kweek*. [CD3:22] **HABITS** In various open and lightly wooded habitats. Roosts by day on ground or in trees. Occasionally perches in bushes between foraging bouts. **SIMILAR SPECIES** Red-necked Nightjar is paler, more variegated and larger, with longer wings. **STATUS AND DISTRIBUTION** Uncommon to rare Palearctic winter visitor. Main wintering grounds apparently in N Cameroon (locally not uncommon), Nigeria (uncommon, Oct–Apr) and Gabon (uncommon); generally rare elsewhere, with records from S Mauritania (passage migrant, Oct–Nov and May–Jun), NW Senegal, Gambia (uncommon, late Oct–early Feb), Mali, Sierra Leone, Liberia, Ivory Coast, Ghana, Niger, and Congo. Probably overlooked. Winter quarters of both races overlap. **NOTE** Races intergrade, geographical variation clinal.

RUFOUS-CHEEKED NIGHTJAR *Caprimulgus rufigena*　　　　　Plate 61 (5)
Engoulevent à joues rousses

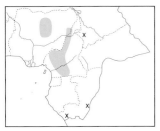

L 23–24 cm (9.0–9.5"). *C. r. rufigena*. Medium-sized, grey-brown to pale grey nightjar with narrow rufous to buffish nuchal collar. **Adult male** Top of head grey-brown with black streaks in centre of crown. Head-sides dark rufous freckled black, bordered below by small white moustachial stripe. Upperparts grey-brown narrowly streaked blackish. *Scapulars with buff line* formed by row of pointed, black, buff-edged patches. Upperwing-coverts tipped buff, forming variably distinct pale bars. Inner primaries and secondaries barred rufous and blackish. Outer 2 tail feathers broadly tipped white. Throat with small white patch on either side or single large patch. Breast dark grey-brown mixed with buff; rest of underparts buff narrowly barred brown on belly and flanks. In flight, white patch towards wingtip and white corners to tail. **Adult female** Lacks white tail corners; wing patch smaller and usually buffy. **VOICE** Probably silent in W Africa. Song, given from ground, a monotonous, sustained, dry reeling *rrerrrrrrrrrrrrrrrrrrrrrr...* sometimes preceded by coughing *k-hoop k-hoop*. Also *chiop chiop* or *kiup kiup* and low, fast *dongdongdongdong...* like a low note on a string instrument. [CD7:57] **HABITS** In various types of wooded grasslands and bush. Roosts by day on ground. When flushed, flies short distance before settling on ground again, usually at base of tree; occasionally lands on low branch. **SIMILAR SPECIES** European Nightjar is rather similar but larger, with conspicuous pale bar on wing-coverts and no nuchal collar. Fiery-necked Nightjar has broader, more distinct rufous nuchal collar. Square-tailed Nightjar has white- or buff-edged tail. **STATUS AND DISTRIBUTION** Intra-African migrant. Non-breeding visitor to Cameroon (main 'wintering' range; Apr–Aug) and Nigeria (rare). Vagrant, Chad. Rare passage migrant, Congo. Breeds in S Africa (Sep–Jan). **NOTE** Ssp. *damarensis* possibly also migrates from S African breeding grounds to our region, but occurrence not proven. Compared to nominate, it is paler grey above, with black markings more contrasting; nuchal collar narrower.

STANDARD-WINGED NIGHTJAR *Macrodipteryx longipennis*　　　　　Plate 62 (7)
Engoulevent à balanciers

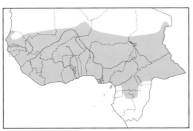

L 21–22 cm (8.5"). Smallish, richly coloured nightjar with broad rufous nuchal collar and *no white in wings and tail*. **Adult male breeding** Unmistakable. In flight, *extremely elongated 2nd innermost primaries with bare shaft and blackish vane at tip* ('standards', length 45.0–53.5 cm), gives impression of bird being closely pursued by two small fluttering bats. Top of head grey-brown with rufous-edged black feathers in central crown. Head-sides rufous finely barred black, bordered above by pale buffish supercilium; sometimes an indistinct line of small white feathers below eye. Upperparts mainly grey-brown. *Scapulars have broad buff edges* to black marks, forming irregular pale line. Upperwing-coverts tipped buff, forming variably distinct pale bars. Flight feathers barred dark rufous and blackish. Tail short. Underparts deep buff finely barred brown; breast slightly darker with rufous-buff spots; vent plain. **Adult male non-breeding/adult female** Similar, but without 'standards'. **VOICE** Song, given from ground, low perch or in flight, a fast, high-pitched, shrill, insect-like *tsikitsikitsikitsikitsiki...* and slower *tseepetseepetseep etseepetseepe...* Also shorter bouts, e.g. *tsikitsikit tsikitsikit, tseepit tseepit, tsikik ...* Flight call a low *kuk*. [CD7:58] **HABITS** In variety of open, wooded habitats, incl. wooded grassland, coastal plains, farmland, dry scrub, and open bush in and around villages and towns. Roosts by day on ground. Starts foraging at dusk. **SIMILAR SPECIES** Female Pennant-winged Nightjar distinguished from female Standard-winged by slightly darker plumage above, with broader and darker rufous nuchal collar, paler buff and barred lower underparts, larger rufous bars in primaries, and larger size. **STATUS AND DISTRIBUTION** Not uncommon to common intra-African migrant. Breeds from Senegambia to Liberia east to CAR. South of that zone, present in breeding season (but not yet proved to breed) from S Ivory Coast to coastal Nigeria and SW Cameroon. Moves to northern savannas and Sahel in rainy season, north to S Mauritania, Mali (to *c.* 18°N), S Niger and Chad (north to Ennedi). Also rare non-breeding visitor, Gabon and Congo.

PENNANT-WINGED NIGHTJAR *Macrodipteryx vexillarius*   Plate 62 (8)
Engoulevent porte-étendard

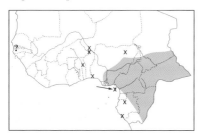

L 24–28 cm (9.5–11.0"). Medium-large, brownish nightjar with broad rufous nuchal collar. **Adult male breeding** Unmistakable. In flight, *uniquely shaped wings* with black remiges, *broad white band on primaries and extremely elongated whitish 2nd innermost primary* ('pennant', length 48–78 cm). Inner primaries and secondaries tipped whitish forming contrasting trailing edge. Top of head grey-brown with blackish feathers in central crown. Head-sides rufous speckled black, bordered by pale buffish supercilium and malar stripe. Upperparts mainly grey-brown. Scapulars blackish with broad buff edges, forming irregular pale line. Upperwing-coverts have some large buffish spots. Tail slightly notched, without white. Throat has white triangular patch; breast rusty-buff indistinctly spotted buffish; rest of underparts pale buff finely barred brown, becoming white on central belly. **Adult male non-breeding** Similar, but pennants broken or short. **Adult female** Very similar to adult female/adult male non-breeding Standard-winged Nightjar but upperparts slightly darker, nuchal collar broader and darker rufous, throat patch white, lower underparts paler buff and more heavily barred, primaries with larger rufous bars; also larger. **VOICE** Probably silent in W Africa. Song, given from low perch or in flight, a fast, high-pitched, shrill, insect-like *tsitsitsitsitsitsitsitsitsi...* recalling Standard-winged Nightjar. Flight calls *wheeeo* and *chup*. [CD7:59] **HABITS** In various types of wooded grasslands and bush. Roosts by day on ground. When flushed, may fly into tree, perching lengthwise on branch. May start foraging well before dusk. **SIMILAR SPECIES** Female Standard-winged Nightjar (q.v.). **STATUS AND DISTRIBUTION** Intra-African migrant. Uncommon to scarce and local non-breeding visitor (Feb/Apr–Aug/Oct) from E Nigeria and Cameroon east to S Chad and CAR. Locally common to scarce passage migrant, Gabon and Congo. Vagrant, SW and N Nigeria, SW Niger, SE Burkina Faso, Togo and Bioko. Claims from S Ghana (Feb 1933) and S Senegal (Dec, Bouet 1961) doubtful; those from Gambia (Apr 1981, Nov 1977) require confirmation. Breeds in S Africa (Aug–Dec/Mar).

SWIFTS
Family APODIDAE (World: 92–98 species; Africa: 23; WA: 20–21)

Highly aerial, small to medium-sized anchor-shaped birds with mostly dark plumage. Flight swift and effortless on long, stiff wings. Sexes similar; juveniles similar to adults or duller. Catch insects on the wing; never perch. Often gregarious and vocal. Some difficult to identify: size, silhouette, rump pattern, tail shape, habitat and locality are useful for identification.

Two distinct groups occur in W Africa.

Spinetails (6 species) have distinctively shaped, notched wings, with curved wingtips, bulging midwings and shorter secondaries pinched-in at body, and square tails (with bare shafts extending beyond feather tips, hence name, but invisible in the field).

Typical swifts (13–14 species) and **swiftlets** (1 species, genus *Schoutedenapus*) have sickle-shaped wings and generally forked tails.

SÃO TOMÉ SPINETAIL *Zoonavena thomensis*
Martinet de Sao Tomé

Plate 144 (1)

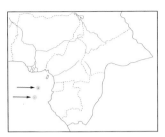

L 10 cm (4"). Small. Distinctive within limited range. **Adult** Crown and upperparts glossy black with broad, well-defined dark-streaked white rump. Head-sides, throat and breast blackish-brown becoming whitish on rest of underparts, latter streaked dark. Tail square. **Juvenile** Breast slightly paler; undertail-coverts pale brown. **VOICE** A frequently uttered, very high-pitched, bat-like squeak. **HABITS** In pairs or small parties (of up to 30) in almost all habitats and at all altitudes. Forages either low down (lower than other swifts) or high above forest canopy. Flight rather weak and fluttering. Occasionally associates with Little Swift. **SIMILAR SPECIES** The only sympatric swifts, African Palm and Little, have different jizz and lack white belly. **STATUS AND DISTRIBUTION** Common resident, endemic to São Tomé and Príncipe.

SABINE'S SPINETAIL *Rhaphidura sabini*
Martinet de Sabine

Plate 63 (2)

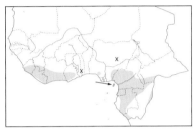

L 10.5–11.5 cm (4.0–4.5"). Easily identified by small size and extensive white on upper- and underparts. **Adult** Black above with white rump and uppertail-coverts. White belly and undertail-coverts sharply demarcated from black breast. *Upper- and undertail-coverts very long and almost completely obscuring tail* (of which only black corners occasionally visible). Secondaries shorter than primaries, producing slightly notched wing. Indistinct black shaft streaks on white parts usually not visible in field. **Juvenile** Uppertail-coverts shorter, falling *c.* 1 cm short of tail tip. **VOICE** Series of weak, high-pitched notes *pit ptrrrit pit-pit-pit-pit prrritt...* somewhat reminiscent of African Palm Swift. [CD7:60] **HABITS** Singly, in pairs or small groups over clearings and edges of rainforest (to 1700 m); often near water. Flight fluttering. Associates with other spinetails, esp. Cassin's. **SIMILAR SPECIES** No other spinetail has similar amount of white on rear upperparts. Cassin's also has white belly, but is noticeably larger with white on upperparts limited to narrow crescent at base of short tail. **STATUS AND DISTRIBUTION** Locally common to uncommon resident, irregularly distributed in rainforest zone, from Sierra Leone and SE Guinea to Togo, and from Nigeria to CAR and Congo; also Bioko (scarce). Local movements reported.

BLACK SPINETAIL *Telacanthura melanopygia*
Martinet de Chapin

Plate 63 (5)

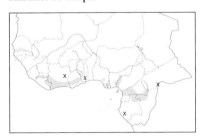

Other name E: Chapin's Spinetail
L 15–17 cm (6–7"). *Large, all-blackish spinetail* with dusky mottled throat and upper breast (only visible at close range and in favourable light). Wings with slight notch, imparting typical spinetail shape. Tail relatively long and square, often spread in circling flight, when appearing rounded. Silhouette resembles large, slender Mottled Spinetail. **VOICE** Usually silent. Various harsh and rasping ('teeth-gnashing') calls e.g. *chrtt* or *kritt*, and *chrewitit chreo.* [CD7:61] **HABITS** Singly, in pairs or small groups above clearings and canopy of lowland rainforest. Flight fast and powerful. Associates loosely with other swifts, esp. spinetails. Nest unknown. **SIMILAR SPECIES** *Apus* swifts have forked (not square) tails and evenly tapering (not 'paddle-shaped') wings. **STATUS AND DISTRIBUTION** Rare to locally uncommon rainforest resident (and Afrotropical migrant/wanderer?), from SE Sierra Leone to Togo, and from S Nigeria to SW CAR, Gabon and N Congo.

MOTTLED SPINETAIL *Telacanthura ussheri*
Martinet d'Ussher

Plate 63 (1)

L 13–14 cm (5.0–5.5"). Blackish with *broad white rump* patch and narrow white band of variable extent on lower belly (usually difficult to see in field). Throat pale mottled brownish. Tail square and relatively long (imparting *Apus*-like appearance). *T. u. sharpei* blacker than **nominate**, with darker throat. **VOICE** Distinctive rasping twittering, *chrwit chrwitt-itit chrrr-r-r-r-itititt* and variations. [CD7:62] **HABITS** Usually in pairs or small groups above wooded

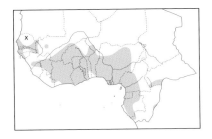

savanna (esp. with *Borassus* palms and baobabs) and open areas in rainforest zone. Associates with other swifts. **SIMILAR SPECIES** Little Swift is superficially similar, but has different jizz, with less protruding head, shorter wings and more fluttery flight; also well-defined white throat, no white on lower belly and different call. **STATUS AND DISTRIBUTION** Widespread but irregularly distributed, scarce to locally common resident, from Senegambia to Liberia east to SW Niger and Nigeria (nominate), and from Cameroon to SW CAR and Congo (*sharpei*). Vagrant, Bioko.

CASSIN'S SPINETAIL *Neafrapus cassini* Plate 63 (3)
Martinet de Cassin

L 12–13 cm (*c* 5"). Characteristic silhouette with long, *distinctly notched wings* (innermost secondaries shorter than rest of flight feathers) and *very short tail*. Upperparts and wings black; tail slightly rounded with *narrow white band* at base. White belly and undertail-coverts sharply demarcated from greyish throat and upper breast. **VOICE** Usually silent. Various harsh and rasping ('teeth-gnashing') calls. [CD7:63] **HABITS** Singly, in pairs or small groups, above clearings and canopy of lowland rainforest. Flight fluttering, loose jointed and bat-like. Associates with other spinetails, esp. Sabine's. **SIMILAR SPECIES** Sabine's Spinetail has similarly patterned underparts but white on upperparts much more extensive, with tail appearing entirely white; noticeably smaller (even without direct comparison), tail longer, flight less bat-like. **STATUS AND DISTRIBUTION** Locally not uncommon resident with irregular distribution in rainforest zone, from Sierra Leone to Ghana and from SW Nigeria to SW CAR and Congo. Also Togo and Bioko (rare).

BÖHM'S SPINETAIL *Neafrapus boehmi* Plate 63 (4)
Martinet de Böhm

L 9–10 cm (3.5–4.0"). *N. b. boehmi* (?). Highly distinctive, *tiny swift with extremely short tail* (appears tail-less) and very fluttering, bat-like flight. Plumage pattern as Cassin's Spinetail, but white rump patch larger; greyish throat and upper breast paler and less sharply demarcated from white belly. **VOICE** Rather silent. A high-pitched rippling twitter *CHIRRrrititititititrew*, also single *chit* or *tit* notes. [CD7:64] **HABITS** Singly or in small groups above clearings and canopy of evergreen forest and wooded savanna. Associates with other swifts and hirundines. **SIMILAR SPECIES** Very small size and short-tailed shape diagnostic. **STATUS AND DISTRIBUTION** Recorded from NE Gabon (Makokou, Mar, Jul/Aug, Nov; status?). Normal range extralimital, in E and S Africa (nearest W Angola), where resident.

SCARCE SWIFT *Schoutedenapus myoptilus* Plate 63 (9)
Martinet de Shoa

L 16.5 cm (6.5"). *S. m. poensis*. Small, dark brown swift with pale throat, sharply tapering wings, *longish tail* and *diagnostic voice*. Upperparts uniformly blackish-brown; black eye patch. Throat greyish merging into dark brown underparts. Tail deeply forked but usually held closed and appearing pointed. In favourable light, contrast between dark head and broad pale throat patch may be noticeable at long range and create slightly capped appearance. **VOICE** Rapid series of high-pitched nasal twitterings *kri-kri-kri trihihihihi* (reminiscent of Cardinal Woodpecker), interspersed by characteristic metallic clicks *tik-tik-tik-...* [CD7:65] **HABITS** Gregarious and noisy. Typically over montane forest, but on Bioko observed below 100 m, near habitation. Extralimitally, partial to crags and cliffs (500–2450 m). Flight fluttering. Nest unknown. **SIMILAR SPECIES** African Palm Swift easily distinguished by much paler, grey plumage, and much longer wings and tail. Fernando Po Swift is larger and blacker, with shorter, less deeply forked tail, much smaller or no throat patch, and different voice. Other species recorded from Bioko (Little Swift, Mottled, Cassin's and Sabine's Spinetails) easily distinguished by different jizz and white rumps. **STATUS AND DISTRIBUTION** Rare to scarce and local resident (or short-distance migrant/wanderer?), W Cameroon (Mt Cameroon, Mar 2001; Mt Menenguba, 1900–2200 m, Feb 1999) and Bioko (8 specimens, 1902–33; 1 possible sighting, Dec 1966–Jan 1967 (Pérez del Val 1996); three, Feb 2001). Probably under-recorded.

AFRICAN PALM SWIFT *Cypsiurus parvus* Plate 60 (15)
Martinet des palmiers

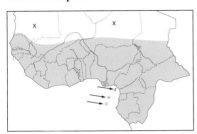

L 16 cm (6.25"), incl. tail of up to 9 cm (3.5"); WS 33–35 cm (13–14"). Distinctive. **Adult *parvus*** Very slender, *all-greyish* swift with *very long and narrow wings and tail.* Deeply forked tail often held closed, appearing thin and pointed. *C. p. brachypterus* darker. **Juvenile** As adult but with shorter tail. **VOICE** Vocal. A frequently uttered, rapid, high-pitched, sibilant twittering *srrit-itititititit.* [CD7:66] **HABITS** In pairs or small groups, typically foraging low around palm trees. Frequently in towns. Flight fast and graceful, with quick wingbeats and sudden twists and turns. Occasionally in mixed flocks with other swifts. **STATUS AND DISTRIBUTION** Common resident throughout, except arid north (north to *c.* 16°50'N). Nominate from S Mauritania to Guinea east to Chad; *brachypterus* from Sierra Leone to CAR and Congo, also Bioko and (presumably this ssp.) São Tomé and Príncipe.

CAPE VERDE SWIFT *Apus alexandri* Plate 143 (1)
Martinet du Cap-Vert

L 13 cm (5"); WS 34–35 cm (13.75"). Small and rather pale, with relatively short wings and only slightly forked tail (shallowest of its genus). Mainly dark grey-brown with slightly paler rump and uppertail-coverts, and ill-defined, greyish-white throat patch. **VOICE** As Common Swift but higher pitched and weaker. **HABITS** In all habitats and at all altitudes within its restricted range, esp. favouring cliffs, rock faces and ravines, as well as towns and villages. Often in large flocks. Flight weak and fluttering. **SIMILAR SPECIES** The only other swifts recorded in Cape Verde Is, Common, Pallid and Alpine Swifts, are all larger and longer winged, with more deeply forked tails. **STATUS AND DISTRIBUTION** Common resident, endemic to Cape Verde Is; reported from most islands, most common on Fogo and Bravo.

AFRICAN BLACK SWIFT *Apus barbatus* Plate 63 (11)
Martinet du Cap

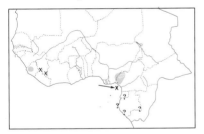

Other name E: African Swift
L 16 cm (6.25"). Medium-sized, dark blackish-brown swift (appearing black in field) with forked tail and small, ill-defined greyish throat patch. Darkest of the dark swifts. Very difficult to separate from Common Swift. In good light, best separated *from above*, where *paler area on inner wing* (secondaries and greater coverts) contrasts with rest of plumage. Minor differences include slightly bulkier body, slightly blunter wings and shorter, fuller tail. *A. b. glanvillei* has some blue gloss on mantle, lacking in *serlei*. **VOICE** Typical *Apus* scream but more toneless (due to higher frequency) than Common Swift's, sounding like harsh, buzzing trill *szrirrrzr.* [CD7:67] **HABITS** Singly or in flocks, occurring mainly over wet highlands with cliffs and gorges, but may also range over any lowland habitat. Gregarious. Flight strong and fast. Associates with other swifts. **SIMILAR SPECIES** Common Swift (q.v. and above). Fernando Po Swift is even darker. Similar-sized Pallid Swift is distinctly paler, with broad white throat patch and pale forehead. Mottled Swift has browner plumage with whitish throat and greyish-looking underparts; also larger. **STATUS AND DISTRIBUTION** Locally uncommon to rare resident, recorded from Sierra Leone (*glanvillei*), N Liberia, W Ivory Coast and Mt Nimba (race uncertain) and W Cameroon (*serlei*); possibly Gabon. (Map includes *A. (b.) sladiniae.*) **NOTE** Paucity of specimens considered to impair validity of races in W Africa (*BoA*). Fernando Po Swift often regarded as conspecific.

FERNANDO PO SWIFT *Apus (barbatus) sladeniae* Plate 63 (12)
Martinet de Fernando Po

L 16 cm (6.25"). Very similar to African Black Swift (q.v.) but even darker with little or no whitish-grey on throat. **VOICE, HABITS & SIMILAR SPECIES** See African Black Swift. **STATUS AND DISTRIBUTION** Rare and local resident, recorded from SE Nigeria (Obudu Plateau), W Cameroon (Bakossi Mts), Bioko and, extralimitally, Angola. Dark swifts recently observed in Eq. Guinea and on São Tomé and Príncipe tentatively identified as *sladeniae*. Status unclear; formerly considered resident on Bioko from 6 specimens collected in 1903–1904, but unrecorded since. *TBW* category: Data Deficient. **NOTE** Usually considered a race of African Black Swift, but treated as separate species by *TBW* and Dowsett & Dowsett-Lemaire (1993).

PALLID SWIFT *Apus pallidus*
Martinet pâle

Plate 63 (6)

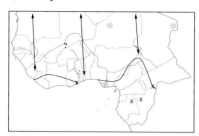

L 16 cm (6.25"); WS 42–46 cm (16.5–18.0"). Medium-sized, strongly resembling Common Swift but *paler*, more greyish-brown; *inner wing contrasting with darker mantle, forewing and outer wing*; whitish throat patch larger and extending over lores to forehead, making dark eye patch stand out. Slightly broader and blunter wings, slightly shallower tail fork and less agile flight with slower wingbeats impart subtly different jizz. **Nominate** slightly paler and greyer than Palearctic *brehmorum*. **VOICE** A high-pitched, shrill scream similar to that of Common Swift but often slightly deeper and more disyllabic. [CD3:23] **HABITS** Gregarious, occurring over variety of habitats, often at great height. Associates with other swifts. Breeds on rocky escarpments and cliff faces. **SIMILAR SPECIES** Common Swift (q.v. and above). For differences from darker African Black Swift and vagrant Plain Swift, see those species. Vagrant Nyanza Swift has even more contrasting wing pattern, but rest of plumage more uniform. **STATUS AND DISTRIBUTION** Nominate is locally common breeder, Niger (S Aïr) and NE Chad (Ennedi), uncommon NW Mauritania (Banc d'Arguin); probably also E Mali (Bandiagara escarpment). *A. p. brehmorum* locally common to rare Palearctic passage migrant and winter visitor (Sep–Apr), from coastal Mauritania and Senegambia to Liberia east to Chad and Cameroon; also SW CAR, NE Gabon and N Congo. Vagrant, Cape Verde Is. **NOTE** Race *illyricus* (E Adriatic coast) possibly also winters in our region, but occurrence not yet proven. It is slightly darker than *brehmorum*.

NYANZA SWIFT *Apus niansae*
Martinet du Nyanza

Plate 63 (7)

L 15 cm (6"). Medium-sized, rather uniformly dark brownish swift with strikingly *contrasting paler area on inner wing* (secondaries and greater coverts). Whitish throat patch small and ill defined; tail forked. **VOICE** Similar to Common Swift. **HABITS** In normal range occurs near rocky outcrops, gorges, crags, mountains and above towns; also lowlands. Gregarious. **SIMILAR SPECIES** African Black Swift has similar, but less distinct, plumage pattern, with lesser coverts appearing slightly paler than mantle and outer wing; also much darker overall. Pallid Swift also has less contrasting wing pattern and paler head and throat. Common Swift has similarly shaped, narrow and pointed wings, but tail is slightly more deeply forked with more pointed outer feathers; overall plumage darker. **STATUS AND DISTRIBUTION** One specimen, from N Congo (Oka, 1 Feb), appears to be this species (presumably nominate). Normal range E Africa.

PLAIN SWIFT *Apus unicolor*
Martinet unicolore

Plate 63 (13)

L 14–15 cm (5.5–6.0"); WS 38–39 cm (15.0–15.25"). Medium-sized, uniformly dark grey-brown swift with paler brown throat and forked tail. Separated from very similar Common and Pallid Swifts by slightly smaller size, slimmer appearance, darker, brownish throat (indistinct and often difficult to see in field; sometimes appearing pale in strong light), faster wingbeats and more erratic flight. From below, semi-translucent flight feathers an important field mark. **VOICE** A high-pitched, shrill scream similar to that of Common Swift. **HABITS** Usually in small groups at coast. Associates with hirundines and other swifts. **SIMILAR SPECIES** Apart from above-mentioned differences, Common Swift has slightly darker plumage and slightly shallower forked tail. Pallid Swift has paler, more contrasting plumage, white throat and forehead, and slightly broader wings with blunter tips. **STATUS AND DISTRIBUTION** Palearctic vagrant (winter visitor?), NW Mauritania (Banc d'Arguin). Common breeder in extralimital Canary Is and Madeira. Winter quarters unknown.

COMMON SWIFT *Apus apus*
Martinet noir

Plate 63 (8)

Other name E: Eurasian Swift
L 16–17 cm (6.25–6.75"); WS 42–48 cm (16.5–19.0"). *A. a. apus.* Medium-sized, mainly uniformly blackish-brown swift with narrow pointed wings, forked tail and small, ill-defined whitish throat patch. **VOICE** A high-pitched, shrill, screaming *srreeee*. [CD3:24] **HABITS** Highly gregarious, occurring over all habitats, often at great height. Flight strong and fast. Very mobile, covering long distances in search of good feeding areas. Attracted to centres of low atmospheric pressure. Presumed to remain on the wing, even for roosting, during entire sub-Saharan stay. **SIMILAR SPECIES** Other dark swifts often

difficult to distinguish. Pallid Swift (q.v.) very similar but separable under favourable conditions. African Black Swift is darker and has limited distribution. Mottled Swift has more powerful flight with slower wingbeats; larger size and more robust appearance impart different jizz which, once known, enables identification of distant or single birds. **STATUS AND DISTRIBUTION** Common throughout. Mainly Palearctic passage migrant; winter visitor to S Cameroon, SW CAR, Gabon and Congo (main wintering range south of equator). Also Cape Verde Is (uncommon) and São Tomé (rare).

BATES'S SWIFT *Apus batesi*
Martinet de Bates

Plate 63 (16)

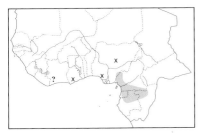

L 14 cm (5.5"). Rather small, slender, all-black glossy swift with fairly deeply forked tail (often held closed to a point). Indistinct grey-brown throat patch. **VOICE** Usually silent. A high-pitched trill *trrrriiiiirrr* near nest. **HABITS** Singly or small flocks above canopy or clearings of rainforest, esp. with crags, and adjacent areas. Foraging flight fluttering with shallow, rapid wingbeats occasionally interspersed by short glides. Associates loosely with other swifts and hirundines. Uses old nests of Forest Swallow and possibly Mottled Spinetail; in Gabon also suspected to use Lesser Striped Swallow nests (Christy & Clarke 1994). **SIMILAR SPECIES** African Black and Common Swifts are distinctly larger and browner. Scarce Swift has similar size but is noticeably paler and greyer. Dark morph Horus Swift is less slender, with shorter tail. **STATUS AND DISTRIBUTION** Rare to locally uncommon in forest zone; probably resident. Recorded from Liberia, Ghana, Nigeria, W and S Cameroon (locally not uncommon), Eq. Guinea, N Gabon and N Congo. Claimed from Sierra Leone and Ivory Coast.

WHITE-RUMPED SWIFT *Apus caffer*
Martinet cafre

Plate 63 (14)

L 14–15 cm (5.5–6.0"); WS 34–36 cm (13.5–14.25"). Rather small, slender black swift with long outer tail feathers forming *deeply forked tail* (usually held closed, tapering to thin point), *relatively narrow U-shaped white rump patch* extending to rear flanks, and white throat. **VOICE** Rather silent. Low, harsh, twittering notes. [CD3:25] **HABITS** Singly, in pairs or small flocks in savanna and farmland. Flight fast and graceful; also fluttering. Associates with other swifts and swallows. Often uses nests of hirundines, mainly Lesser Striped and Rufous-chested Swallows, and Little Swifts, ousting original occupants. **SIMILAR SPECIES** Only Horus Swift has similar combination of white rump and forked tail, but rump patch much broader and tail much shorter; also less elegant, more like Little Swift in jizz. **STATUS AND DISTRIBUTION** Inadequately known. Principally resident, but also partially migratory and probably dispersive; wet-season visitor to Sahel belt on edge of Sahara. Recorded throughout. Uncommon and local, Senegambia, Ghana, Nigeria, Chad, S Gabon and Congo; rare or scarce and patchily distributed elsewhere. Northernmost records NW Mauritania (Banc d'Arguin).

HORUS SWIFT *Apus horus*
Martinet horus

Plate 63 (10)

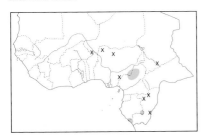

L 15 cm (6"). *A. h. horus.* Rather small, black swift with forked tail, *broad white rump patch* extending onto rear flanks and *white throat.* Tail may appear rather square when banking. **Dark morph** has smaller, greyer throat patch and dark brown rump. **VOICE** Rather silent when not breeding. Call a remarkable trill, lower pitched and sounding more pleasant than calls of Little and White-rumped Swifts, *krrweepeeo krrweepeeo...*, *przweew przweew ...* or *preeeooo preeeooo...* with variations. [CD7:68] **HABITS** In small groups above grassland and wooded savanna. Forages low. Flight fluttering. Associates with other swifts and swallows. Nests in old burrows of bee-eaters, kingfishers and martins. **SIMILAR SPECIES** Only White-rumped Swift has similar combination of white rump and forked tail, but rump patch much smaller and U-shaped, tail much longer and deeply forked; also appears slimmer, more graceful. Little Swift has square tail. Dark morph Horus resembles smaller Bates's Swift (which has longer, slimmer tail) and larger African Black Swift; best identified by accompanying and normally more numerous white-rumped birds. **STATUS AND DISTRIBUTION** Resident and/or intra-African migrant. Rare to scarce and local, possibly overlooked, SW Niger ('W' NP), Nigeria (Sokoto, Zaria, Obudu and Mambilla Plateaux), Cameroon (Plaine Tikar to Adamawa Plateau), S Chad and SE Gabon (Lékoni area). Locally not uncommon, Congo. Dark morph recorded in Gabon and Congo. **NOTE** Dark morph sometimes treated as subspecies *A. h. toulsoni* or (with extralimital dark race *fuscobrunneus*) as separate species, Loanda Swift *A. toulsoni*.

LITTLE SWIFT *Apus affinis*   Plate 63 (9)
Martinet des maisons

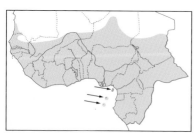

L 12–13 cm (4.75–5.0"); WS 33–35 cm (13.0–13.75"). *A. a. aerobates* Small, rather stocky, blackish swift with *square tail* (sometimes appearing slightly notched) and *broad white rump patch* extending onto rear flanks; throat white. Island race *bannermani* darker, with lightly streaked throat (usually not visible in field). **VOICE** Vocal. A frequently uttered, rapid, high-pitched twittering. [CD3:26] **HABITS** Highly gregarious; mainly in towns, but also forages over adjacent open areas. Associates with other swifts. **SIMILAR SPECIES** Mottled Spinetail has different jizz, with more protruding head, longer wings and different, more powerful flight action; also greyer throat, some white on vent and different call. Horus Swift has forked tail. **STATUS AND DISTRIBUTION** Common resident throughout, except arid north (*aerobates*). Also Bioko, São Tomé and Príncipe (*bannermani*).

MOTTLED SWIFT *Tachymarptis aequatorialis*   Plate 63 (17)
Martinet marbré

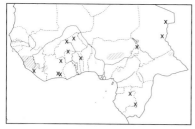

L 23 cm (9"). Very large, robust, dark swift with forked tail. *T. a. lowei* Upperparts dark brownish; throat greyish-white to white; rest of underparts scalloped whitish, appearing greyish in field. **Nominate** darker, with smaller, greyer throat patch. **VOICE** A harsh, shrill call, infrequently uttered. [CD7:69] **HABITS** Gregarious, but also occurring singly. Frequents crags, rocky outcrops and escarpments but may travel some distance to forage and disperses widely after breeding; then recorded over variety of habitats. Associates with other swifts, esp. Common Swift. **SIMILAR SPECIES** Similar-sized Alpine Swift easily distinguished by white underparts. Single or distant Common and African Black Swifts separated by different jizz imparted by smaller size, less bulky body, and faster wingbeats; closer birds show uniform underparts. **STATUS AND DISTRIBUTION** Inadequately known. Rare to uncommon and local, patchily distributed resident or intra-African migrant, recorded from Guinea to Togo and Nigeria (*lowei*), Cameroon (nominate), E Chad, NE CAR and Gabon (ssp. uncertain). **NOTES (1)** Sometimes placed in genus *Apus*. **(2)** Ssp. *furensis* (Sudan; upperparts greyish-brown, throat patch large and white) may occur.

ALPINE SWIFT *Tachymarptis melba*   Plate 63 (18)
Martinet alpin

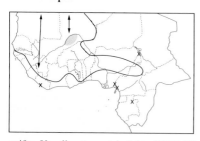

L 20–22 cm (8.0–8.75"); WS 54–60 cm (21.25–23.75"). Distinctive. *Very large*, mainly brown swift with *white throat and belly separated by dark brown breast-band*; tail shallowly forked. Size and powerful flight may recall small falcon. **Nominate** has dark grey-brown upperparts. NW African *tuneti* distinctly paler. **VOICE** A rapid, rolling, twittering *chit rititititititit chetetet* ... rising and falling in pitch. Reminiscent of Little Swift but slower, lower pitched and less harsh. [CD3:27] **HABITS** Gregarious. Migrants and winter visitors occur over variety of habitats; breeds on subdesert crags. Flight fast and more powerful than Common Swift, with slower wingbeats. Associates with other swifts. Usually at great height. **SIMILAR SPECIES** Similar-sized Mottled Swift easily distinguished by dark underparts. **STATUS AND DISTRIBUTION** Not uncommon to rare Palearctic passage migrant and winter visitor (both races), from Mauritania to Liberia east to Cameroon and NE Gabon. Also Cape Verde Is (rare). Winter range inadequately known. Breeds in C Mali (Bandiagara escarpment), where common (ssp. unknown). **NOTE** Sometimes placed in genus *Apus*.

MOUSEBIRDS
Family COLIIDAE (World: 6 species; Africa: 6; WA: 2)

Small, arboreal, buffish-brown or greyish birds with long, stiff, graduated tails and short, erectile crests. Sexes similar. Always in small parties. Flight on short, whirring wings, fast and direct with long glides. Climb well. Feed mainly on fruit and leaves. Sedentary. Endemic to Africa.

BLUE-NAPED MOUSEBIRD *Urocolius macrourus*   Plate 64 (1)
Coliou huppé

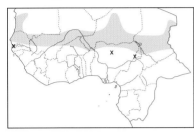

Other name F: Coliou à nuque bleue
L 33–38 cm (12–15"), incl. tail of up to 28 cm (11"). **Adult *macrourus*** Greyish with conspicuous *azure-blue nape*, very long, slender tail, red on bill and bare red patch around eye. Upperparts brownish-grey; rump blue-grey. Lower underparts buff. Bill black with base of upper mandible red. Eye red. Feet dull purplish-red. *U. m. laeneni* paler. **Juvenile** Paler, without blue on nape; crest shorter; bill dusky; bare facial skin pink. **VOICE** A far-carrying, long, clear whistle *pwheeeeeeer*; also shorter notes including *kwee, pwee-u* and *kruw.* [CD7:71] **HABITS** In small, vocal parties in arid open country with scrub, thorn trees and cultivated areas. Sedentary or locally nomadic. **STATUS AND DISTRIBUTION** Nominate is common to uncommon resident in Sahel zone, from Mauritania and Senegal through Mali, Burkina Faso, N Benin, Niger, N Nigeria and N Cameroon to Chad; rare, Gambia. *U. m. laeneni* restricted to Aïr, Niger, where common.

RED-BACKED MOUSEBIRD *Colius castanotus*   Plate 64 (3)
Coliou à dos marron

Other name F: Coliou à dos roux
L *c.* 33 cm (13"). Resembles Speckled Mousebird but distinguished by *chestnut lower back and rump*, absence of bare whitish spot behind eye, and yellow-and-green (not brown) eye. **Juvenile** Similar but wing-coverts have indistinct pale brown scallops; eye brown. **VOICE** Unknown. **HABITS** Like Speckled Mousebird. **STATUS AND DISTRIBUTION** Not recorded with certainty from our region. Locally common resident, Angola. Claimed from coastal strip at mouth of Congo R.; may range north: more information required. Claim from SW Gabon no longer accepted.

SPECKLED MOUSEBIRD *Colius striatus*   Plate 64 (2)
Coliou rayé

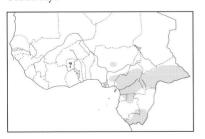

Other name F: Coliou strié
L 30-36 cm (12-14"), incl. tail of up to 22 cm (8.5"). **Adult *nigricollis*** Wholly brown with very long tail, short, bushy crest, black face and bare bluish-white patch behind eye. Bill black above with blue-grey patch on culmen, bluish-white below. Eye brown. Feet bright red. *C. s. leucophthalmus* has yellowish-white crest; hindneck and back lightly barred dark brown; eye white. **Juvenile** Tail and crest shorter; pale brown stripe down back; feet dusky. **VOICE** Sibilant twittering calls *tsiu, whseet-whut, tsi-ui* etc, constantly uttered at rest and in flight. Alarm a buzzing *tzik-tzik.* [CD7:72] **HABITS** In small, vocal parties of up to 20 birds, actively foraging by clambering mouse-like through bushes, often hanging vertically on branches. Inhabits a variety of open habitats, woodland, forest edges, clearings, gardens and scrub. Not in forest. Sedentary, but some local movements reported. **SIMILAR SPECIES** In our region quite unmistakable since no overlap with other mousebirds (Red-backed Mousebird, formerly claimed from SW Gabon, may occur in coastal strip north of Congo R.) **STATUS AND DISTRIBUTION** Common to locally not uncommon resident, from E Nigeria to CAR and Congo. Claimed from N Ghana (Mole NP). *C. s. nigricollis* almost throughout; *leucophthalmus* in SE CAR.

TROGONS

Family **TROGONIDAE (World: 36–39 species; Africa: 3; WA: 3)**

Medium-sized, arboreal forest birds with brightly coloured plumage and long, broad tails. Bill with broad gape. All three African species are quite similar – males mainly green above and red below, females duller – but differ vocally. They feed on insects, which they catch on vegetation during swift, swooping flight, reminiscent of drongos. Nest in tree-holes. Easily overlooked despite their brilliant colours as they perch motionlessly for long periods at mid-level in the forest shade. The 3 species are endemic to Africa.

BAR-TAILED TROGON *Apaloderma vittatum*   Plate 64 (5)
Trogon à queue barrée

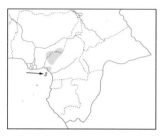

Other name F: Couroucou à queue barrée
L 28 cm (11"). Darker than the other two trogons. **Adult male** Head and throat glossy black; rest of upperparts dark green. Tail purple-blue with 3 outer rectrices graduated and *closely barred black and white.* Upper breast dark green bordered below by narrow bands of purple, blue and dark green; remaining underparts crimson. Bill greenish-yellow; bare skin at gape and below eye orange-yellow. **Adult female** Differs in having head and throat dark brown washed green, margin g through cinnamon on lower breast into pink on belly and undertail-coverts. Bill dark horn above, greenish-yellow below. **Juvenile** As adult female, but inner secondaries and greater and median coverts with white or buff terminal spots. **VOICE** Distinctive. A series of 7–14 high-pitched, yelping *kew* or *kiup* notes, starting softly and increasing in volume. Also a clear, drawn-out whistle, descending in scale *whueeeeuw* (easily imitated). Silent when not breeding. [CD7:73] **HABITS** Singly or in pairs. Confined to moist montane forest. Behaviour as other trogons (see family account). **SIMILAR SPECIES** Narina's and Bare-cheeked Trogons lack barring on tail and have different facial skin coloration; voices different. **STATUS AND DISTRIBUTION** Uncommon to locally not uncommon forest resident in highlands of SE Nigeria (Obudu and Mambilla Plateaux), W Cameroon (Mt Cameroon to Tchabal Mbabo) and Bioko.

NARINA'S TROGON *Apaloderma narina*   Plate 64 (4)
Trogon narina

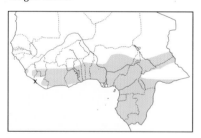

Other name F: Couroucou narina
L 29–34 cm (11.5–13.25"). **Adult male** Upperparts, throat and breast bright green; rest of underparts crimson. Tail white below (outer 3 remiges white and graduated) and mainly blackish-blue above. Wing-coverts finely barred grey and white. Bill yellow. Head-sides with 3 areas of bare skin: at gape, below eye and above eye. In western race *constantia* bare skin is vivid yellow, in *brachyurum* green. **Adult female** Duller, with forehead, throat and breast grey washed green; belly to undertail-coverts pink; bare skin bluer. Female *brachyurum* has fine vermiculations on breast (lacking in *constantia*). **Juvenile** Resembles adult female, but secondaries and wing-coverts tipped with large white and buff spots. **VOICE** Distinctive but easily passing unnoticed. A series of paired, moaning, soft and dove-like hoots, with stress on second syllable and repeated up to 14 times, starting hesitantly, then slightly rising in volume and giving impression of being forced out *who-ot-WHO poe-WHO poe-WHO poe-WHO...* [CD7:74] **HABITS** Singly or in pairs in old secondary forest; also in gallery forest in savanna. Unobtrusive. When singing, throat swells, revealing blue skin, while tail pumps rhythmically down (as if to force notes out). **SIMILAR SPECIES** Bare-cheeked Trogon, whose range overlaps with ssp. *brachyurum* from Cameroon to Congo, is very similar but has patches of bare facial skin yellow (not green) and larger; tail of perched individual seen from front freckled brownish at base along centre (pure white in Narina's); voice and habitat preferences also differ. Bar-tailed Trogon, which overlaps in Cameroon and SW Nigeria, is darker with barred tail; voice different. **STATUS AND DISTRIBUTION** Generally uncommon or scarce, but locally common forest resident, from Sierra Leone and SE Guinea to Benin (*constantia*) and from Nigeria to Chad and Congo (*brachyurum*).

BARE-CHEEKED TROGON *Apaloderma aequatoriale*   Plate 64 (6)
Trogon à joues jaunes

Other name F: Couroucou à joues jaunes
L 28 cm (11"). **Adult male** Very similar to Narina's Trogon, but distinguished by larger, *yellow* and more conspicuous

patches of bare skin at gape, below eye and above eye. **Adult female** Forehead, throat and breast dull to rusty brown merging through grey and pinkish into red on belly and undertail-coverts. **Juvenile** As adult female, but duller, with lower breast narrowly barred green and upperwing-coverts with large white or buff subterminal spots. **VOICE** A series of 6–8 plaintive *hoo* notes, with first note longer and stressed, the following being shorter and gradually dying away; reminiscent of Blue-spotted Wood Dove. [CD7:75] **HABITS** In primary lowland rainforest. Behaviour much as other trogons. **SIMILAR SPECIES** Ssp. *brachyurum* of Narina's Trogon has facial skin apple-green (not yellow) and different voice; also prefers disturbed forest and old second growth with dense lianas. Bar-tailed Trogon, which only overlaps in W Cameroon, is darker with barred tail and different call. **STATUS AND DISTRIBUTION** Not uncommon to rare forest resident, from SE Nigeria to SW CAR and Congo.

KINGFISHERS
Family ALCEDINIDAE (World: *c.* 92 species; Africa: 18–20; WA: 14–16)

Small to large and often brightly coloured birds with characteristic compact silhouette, formed by large head, long, dagger-shaped bill, short body and very short legs. Sexes similar, except in giant and pied kingfishers (Cerylinae); females often larger. Occur in a variety of habitats. Aquatic species feed mainly on fish, which they capture by plunge-diving; terrestrial species subsist on insects and reptiles, on which they swoop from a perch. Flight fast and direct. Nest in tree or earth holes.

CHOCOLATE-BACKED KINGFISHER *Halcyon badia* Plate 66 (1)
Martin-chasseur marron

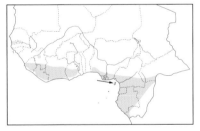

L 21 cm (8.25"). Shy forest kingfisher; often heard, but rarely seen. **Adult** *Head and upperparts dark chocolate-brown* with bright azure rump and wing panel; tail dull azure with dusky tip. *Underparts clean white.* Bill bright red. Western *obscuridorsalis* slightly darker than **nominate** (Dickerman *et al.* 1994). **Juvenile** Breast faintly washed buff with thin black fringes to feathers; flanks have some dusky streaks; bill dusky-black with orange tip. **VOICE** Song distinctive, starting with a single, weak *wheet* (only audible at close range) followed after a short pause by a series of 12–17 fluty *hu* notes, initially slightly increasing in volume and speed, then slowing down, becoming more plaintive and trailing off *wheet, huhuhuhuHuHuHUhuhu-hu-hu-hu...* [CD7:76] **HABITS** Singly in lowland rainforest, perching quietly at mid-height, flying to seize prey on ground or catching insects in mid-air. Attends ant columns. Not associated with water. Habitually uses semi-suspended earthen termitaria for nesting. **SIMILAR SPECIES** Song of Blue-breasted Kingfisher has similar structure and speed but single introductory note is abrupt *chiup!* and those in following series are shorter, more clipped, lacking melancholy quality of Chocolate-backed. **STATUS AND DISTRIBUTION** Uncommon to locally fairly common forest resident, from Sierra Leone and Guinea to Ghana and from SW Nigeria to CAR and Congo; also Bioko. *H. b. obscuridorsalis* west of Dahomey Gap, nominate east.

BROWN-HOODED KINGFISHER *Halcyon albiventris* Plate 66 (2)
Martin-chasseur à tête brune

L 22 cm (8.75"). *H. a. prentissgrayi.* Medium-sized. **Adult male** Dirty brown blackish-streaked head separated, by buffish collar, from blackish-brown mantle, scapulars and wing-coverts; azure-blue wing panel, back, rump and tail. Underparts dirty buffish, lightly streaked grey-brown. Bill heavy, bright red tipped dusky. Feet red. **Adult female** Mantle browner; underparts more tawny and streaked. **Juvenile** Duller, with dusky, whitish-tipped bill. **VOICE** A series of 3–5 weak, descending notes *kweep eep eep eep*. Alarm a loud, harsh chatter *chrrit-chrrit-chrrit-...* Displaying birds utter a loud, excited *kik-kik-kik-kik-...* [CD7:77] **HABITS** Singly or in pairs in open woodland, thickets, scrub and cultivation; not associated with water. Pairs display by facing each other and rapidly flicking wings open and shut, calling excitedly. **SIMILAR SPECIES** Grey-headed Kingfisher is neater, not streaky, with chestnut belly, and has large pale window in wing in flight. Juveniles may appear very similar, but Grey-headed not streaky. **STATUS AND DISTRIBUTION** Not uncommon resident, S Gabon and Congo.

GREY-HEADED KINGFISHER *Halcyon leucocephala* Plate 66 (4)
Martin-chasseur à tête grise

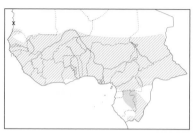

Other name E: Chestnut-bellied Kingfisher
L 22 cm (8.75"). **Adult** *leucocephala* Diagnostic combination of pale grey head and breast, *bright chestnut belly and undertail-coverts* and black mantle, scapulars and wing-coverts. Flight feathers, rump and tail violet-blue; throat white. Bill and feet bright red. In flight, upper-wing has pale blue panel at base of primaries (white on underwing); underwing-coverts chestnut. Southern *pallidiventris* has paler grey head and pale chestnut belly. Island race *acteon* has head even paler, greyish-white. **Juvenile** Duller, with dusky whitish-tipped bill, scaly collar on breast and hindneck extending onto head-sides and buff or pale chestnut belly and undertail-coverts. **VOICE** Call a loud, scolding and explosive *CHeK!* or *KHE!*, frequently uttered in short series. Song a weak descending trill *chichichichi-chi-chiu*, usually given in flight. [CD7:78] **HABITS** Singly or in pairs in woodland, bush and cultivated areas. On migration in a variety of habitats, except dense forest and very arid regions. Hunts from perch, sitting quietly for long periods, occasionally pouncing on insects on ground. Displaying pairs circle above trees, calling, and perch conspicuously on topmost branches, opening wings to reveal underwing pattern. **SIMILAR SPECIES** Normally unmistakable. In Congo, Brown-headed Kingfisher occurs in similar habitat, but lacks chestnut belly and is more streaky. **STATUS AND DISTRIBUTION** Nominate is fairly common resident and intra-African migrant throughout, from derived savanna to Sahel zone. Migrates north in rains, reaching 15.0°–15.5°N, south to open areas in forest zone in dry season. *H. l. pallidiventris* reported as non-breeding migrant in S Congo. *H. l. acteon* is common resident, Cape Verde Is (Santiago, Fogo and Brava).

BLUE-BREASTED KINGFISHER *Halcyon malimbica* Plate 66 (5)
Martin-chasseur à poitrine bleue

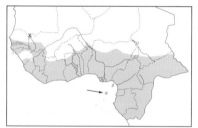

L 25 cm (10"). Large. **Adult** *malimbica* Mainly bright azure-blue with *black back, scapulars and upperwing-coverts* and *pale azure-blue breast* becoming grey or whitish on rest of underparts. Crown grey. Black streak through eye produces stern appearance. Flight feathers have broad black tips. Tail black below. Bill heavy, upper mandible red, lower mandible black. Feet red. *H. m. torquata* has blue areas paler and greener; *forbesi* intermediate. Island race *dryas* like *forbesi* but has brown crown and larger bill. **Juvenile** Duller. Head and underparts with buffish wash. **VOICE** Song distinctive and far carrying, starting with single, abrupt *chiup!*, followed after a short pause by a series of piping *pu* notes, initially accelerating and ascending slightly, then descending and slowing down: *chiup! pu-pupupuPUPUUpuu-puu puu puu*. Song of *H. m. dryas* slower, more melancholic. Alarm a raucous *chup, chup-chup-chup* or *kiah, kiah-kiah*. [CD7:79] **HABITS** Singly or in pairs in forest and dense woodland. Shy; more often heard than seen. Mainly keeps in deep shade below canopy. Feeds primarily on insects, also small vertebrates, hunted from perch. **SIMILAR SPECIES** Commoner and more easily seen Woodland Kingfisher has more azure-blue upperparts (no black on back and scapulars), lacks blue on breast and is smaller; absence of narrow black mask produces gentler appearance. Also has different song and occurs in more open habitats. Song of Brown-backed Kingfisher similar in structure and speed to Blue-breasted but single introductory note is a soft *wheet* and those in following series are fluty and plaintive. **STATUS AND DISTRIBUTION** Uncommon to common in forest and densely wooded savanna. *H. m. torquata* from S Senegambia to W Mali; *forbesi* from Sierra Leone to E Nigeria; nominate from Cameroon to CAR and Congo; *dryas* on Príncipe. Vagrant, S Mauritania. Resident, but range expands slightly during rains.

WOODLAND KINGFISHER *Halcyon senegalensis* Plate 66 (3)
Martin-chasseur du Sénégal

Other name E: Senegal Kingfisher
L 22 cm (8.75"). Widespread, medium-sized kingfisher. **Adult** *senegalensis* Head and mantle grey; rest of upperparts bright azure-blue with contrasting *black upperwing-coverts*. Underparts greyish-white; undertail black. Upper mandible red, lower mandible black. Feet black. In flight striking blue above with black wing-coverts, mainly

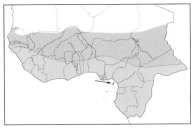

black primaries and dusky-tipped secondaries. Forest-zone race *fuscopilea* has crown darker, brownish-grey, and mantle and breast greyer. **Juvenile** Bill dusky; grey parts of head suffused with azure-blue; supercilium, breast and flanks washed yellowish-buff; head-sides and breast with narrow grey crescents giving dusky appearance; azure upperparts suffused with grey. **VOICE** Vocal. Distinctive and easily learnt song vigorous and explosive, consisting of a single sharp initial note, followed after a short pause by a hard descending trill *PTIK TIRRRRrrrrrrr*. Alarm a fast *kee-kee-kee-kee-...* [CD7:80] **HABITS** In pairs in woodlands, clearings, cultivated areas, gardens and mangroves; not true forest. Perches on high branch or other, often exposed, vantage point, hunting insects and small vertebrates. Conspicuous, attracting attention by loud song. In display flicks wings open while singing, revealing contrasting underwing pattern, and turns around repeatedly. Often interacts aggressively with other birds. **SIMILAR SPECIES** Blue-breasted Kingfisher is larger with black (not blue) mantle and scapulars, and blue breast; occurs in more wooded habitats and has different song. **STATUS AND DISTRIBUTION** Common throughout savanna and forest zones. Sedentary (*fuscopilea*: Sierra Leone to Congo and S CAR; also Bioko) or migratory (nominate: Senegambia to Chad and N CAR), esp. at northern extremity of range; wet-season breeder in Sahel zone north to 17°N. Races intergrade in forest–savanna contact zone.

STRIPED KINGFISHER *Halcyon chelicuti*   Plate 66 (6)
Martin-chasseur strié

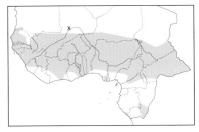

L 17 cm (6.75"). Small, *rather drab savanna kingfisher* with distinctive song. **Adult male** *chelicuti* Forehead and crown buff-grey heavily streaked blackish-brown; blackish streak through eye; head-sides and throat whitish. Mantle and wings greyish-brown, some blue of flight feathers evident; back, rump and uppertail-coverts azure; tail dull blue. Underparts dirty white, finely streaked blackish-brown. Upper mandible dusky, lower mandible reddish. Feet dark reddish. Underwing white with dark carpal patch, broad grey trailing edge and broad black subterminal band. *H. c. eremogiton* has underparts almost unstreaked. **Adult female** Underwing has smaller carpal patch and lacks subterminal black band on remiges. **Juvenile** Bill blackish tipped whitish; underparts more buffy; blue areas of wings and tail paler and more limited, or lacking. **VOICE** Song a series of loud, far-carrying *KEE-RRRUU KEE-RRRUU KEE-RRRUU...* [CD7:81] **HABITS** Singly or in pairs in dry savanna woodland and bush. Perches un-obtrusively on branch at mid-height. Feeds on insects (mainly grasshoppers) and, occasionally, small vertebrates, nearly all taken on ground. Returns to perch to consume prey. Displaying birds face each other perching upright, flicking wings open and shut, and calling excitedly. **STATUS AND DISTRIBUTION** Generally not uncommon resident in Sahel and savanna zones south of 17°N, from S Mauritania and Senegambia to Chad and CAR, and in S Gabon and Congo. Nominate almost throughout; *eremogiton* in arid zone from C Mali (Mopti) east, intergrading with nominate in S Sahel.

AFRICAN DWARF KINGFISHER *Ceyx lecontei*   Plate 65 (1)
Martin-pêcheur à tête rousse

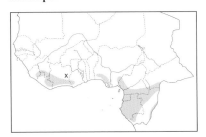

L 10 cm (4"). Smallest kingfisher; tiny and secretive. **Adult** Distinguished from other small kingfishers by *wholly orange-rufous crown and black forehead* (difficult to see from below). Ear-coverts and hind-neck washed violet; white streak on neck-sides; back to tail and wings brilliant violet-blue. Throat white; rest of underparts orange-rufous. Bill and feet coral-red. Bill has horizontally flattened, square tip (sometimes visible in field). **Juvenile** Crown and upperparts blackish, the feathers tipped blue; bill blackish tipped white (not square ended); orange-rufous parts of adult rufous-buff; pale rufous collar on hindneck; dusky moustachial area and white throat. **VOICE** A high-pitched *tseep*. Mainly silent. [CD7:82] **HABITS** Singly in dense rainforest and its edge, perching at all levels, mainly low. Unobtrusive. Insectivorous. Occasionally in mixed-species flocks and attending ant columns. **SIMILAR SPECIES** African Pygmy Kingfisher is very similar but has dark blue crown and occurs in more open habitat. Juveniles of both species even more similar, but Pygmy has violet wash to ear-coverts and less pronounced moustachial streak. **STATUS AND DISTRIBUTION** Rare to uncommon forest resident, from Sierra Leone and SE Guinea to Ghana and from SW Nigeria to CAR and Congo. **NOTE** Formerly placed in genus *Ispidina*.

AFRICAN PYGMY KINGFISHER *Ceyx pictus*   Plate 65 (2)
Martin-pêcheur pygmée

L 12 cm (4.75"). **Adult** *pictus* Very small with brilliant violet-blue upperparts, orange-rufous underparts and coral-red bill and feet. Head mainly orange-rufous with blue-black crown, broad orange-rufous supercilium, ear-coverts strongly washed violet and white patch on neck-sides. Throat white. Forest-zone race *ferruginus* slightly darker. **Juvenile** Duller, with black, white-tipped bill, blackish moustachial stripe and lores, and mottled upperparts. **VOICE** A single sharp, high-pitched *tseet*, uttered in flight; also shorter *tsip*. [CD7:83] **HABITS** Singly in variety of habitats, incl. forest edge, woodland, clearings, cultivation and gardens; not usually near water. In Gabon also in forest. Hunts for insects from low perch. **SIMILAR SPECIES** Malachite Kingfisher is slightly larger, has crest, lacks supercilium and always occurs near water. African Dwarf Kingfisher has all-rufous crown and black forehead, and is a true forest species. Juvenile Dwarf very similar to juvenile Pygmy, but lacks violet wash on ear-coverts and has more pronounced moustachial streak. White-bellied Kingfisher has white underparts with chestnut sides and occurs near small streams in dense forest. **STATUS AND DISTRIBUTION** Generally common throughout, from Senegambia to Chad and Congo; uncommon at periphery of range. Breeding visitor north of 12°–13°N during rainy season (north to 17°N in Mauritania), and resident and partial migrant south of 9°N. Nominate in savanna; *ferruginus* in forest zone from Guinea-Bissau to Congo. **NOTE** Formerly placed in genus *Ispidina*. Forest-zone race originally named *I. p. ferrugina*; in genus *Ceyx* becomes *ferruginus* (not *ferrugineus* as in *HBW*).

WHITE-BELLIED KINGFISHER *Alcedo leucogaster*   Plate 65 (3)
Martin-pêcheur à vent blanc

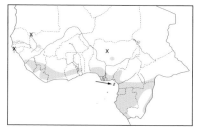

L 13 cm (5"). **Adult** *leucogaster* Small and unobtrusive with violet-blue upperparts and coral-red bill and feet of small congeners, but with *clear white underparts* from throat to belly and *chestnut head-sides, breast patches and flanks*. Western **bowdleri** has rufous area above lores more extensive; southern **leopoldi** has superciliary area blue, not rufous. **Juvenile** Duller, with black, white-tipped bill and dusky-mottled moustachial area, ear-coverts, breast and mantle. **VOICE** A high-pitched, vigorous *(t)seee*, given in flight; reminiscent of African Pygmy Kingfisher but louder, more piercing. **HABITS** In dense primary and secondary forest. Shy and difficult to observe. Usually seen flying low and fast over water or perching low above small stream. **SIMILAR SPECIES** Malachite Kingfisher has same size but lacks white on underparts and does not enter forest. For differences from African Pygmy and African Dwarf Kingfishers, see those species. **STATUS AND DISTRIBUTION** Uncommon to scarce forest resident, from Guinea-Bissau and Guinea to Togo (*bowdleri*), Nigeria to Gabon, CAR and Bioko (nominate) and E Congo (nominate, presumably intergrading with *leopoldi* in Congo R. basin). Also S Mauritania and SW Mali. **NOTES (1)** Placed in genus *Corythornis* in *BoA*, but subsequently in *Alcedo* by same author (Fry & Fry 1992). **(2)** Príncipe Kingfisher *A.* (*l.*) *nais* often regarded as conspecific with White-bellied Kingfisher.

PRÍNCIPE KINGFISHER *Alcedo* (*leucogaster*) *nais*   Plate 144 (3)
Martin-pêcheur de Principe

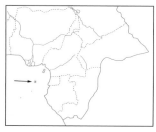

L 13 cm (5"). **Adult** Differs from mainland White-bellied Kingfisher (nominate race) in lacking rufous supercilium and having paler and bluer crown; white on underparts restricted to throat and central belly; rest of underparts pale rufous (not chestnut). **Juvenile** Crown strongly barred; speckled moustachial streak; bill dark. **VOICE** A high-pitched, sharp *(t)seee* or *tsee-eet*, uttered in flight, and a short *tsip*. **HABITS** Frequents watercourses and beaches, where it fishes for shrimps. Also in secondary forest, away from water. Not particularly shy and easily observed, unlike its mainland relative. **STATUS AND DISTRIBUTION** Common resident, endemic to Príncipe, where the only small kingfisher. **NOTE** Usually treated as race of White-bellied Kingfisher (cf. *BoA*), but also as separate species, *A. nais*, or as race of Malachite Kingfisher.

MALACHITE KINGFISHER *Alcedo cristata*   Plate 65 (4)
Martin-pêcheur huppé

L 13 cm (5"). **Adult** *galerita* Small, *aquatic* kingfisher with glossy violet-blue upperparts, orange-rufous underparts

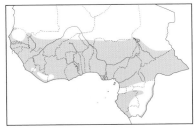

and coral-red bill and feet. Forehead and crown feathers long and violet-blue barred black, occasionally raised. White patch at neck-sides and white throat, like all other small kingfishers. **Nominate** has paler underparts. **Juvenile** Duller and duskier, esp. on ear-coverts and breast, with black bill (turns red from *c.* 3 months). **VOICE** A shrill *pseek*, given in flight. Song in duet *tsrrr-tsrrrr tsip-tsip tsrr tsip t-t-t...* [CD7:84] **HABITS** Singly or in pairs on lakes, pools, marshes, lagoons and slow-running rivers. Dives for fish, shrimps and insects from low perch; sometimes takes prey on land. Flight fast and direct, low over surface of water. Not shy. **SIMILAR SPECIES** African Pygmy Kingfisher has same coloration but is smaller, lacks crest, has broad rufous supercilium restricting blue on head to crown and distinctive violet wash to ear-coverts; also occurs in drier habitats. Other small blue-and-rufous kingfishers are forest dwellers. Shining-blue Kingfisher is larger, has deeper blue upperparts and black bill, occurs mainly along forested rivers and lakes, and is much scarcer. **STATUS AND DISTRIBUTION** Widespread and often common resident in suitable habitat throughout, except arid north; *galerita* in west of range to approximately Ghana, nominate in east. Partially migratory in at least some areas. **NOTES** (1) Placed in genus *Corythornis* in *BoA*, but subsequently in *Alcedo* by same author, who also revised ranges of races (Fry & Fry 1992). (2) São Tomé Kingfisher *A. (c.) thomensis* often regarded as conspecific with Malachite Kingfisher.

SÃO TOMÉ KINGFISHER *Alcedo (cristata) thomensis* Plate 144 (2)
Martin-pêcheur de Sao Tomé

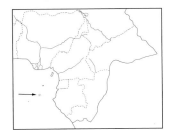

L 13.5 cm (5.25"). **Adult** Distinguished from mainland Malachite Kingfisher by overall darker plumage, smaller rufous loral spot and dusky barring on malar area. **Juvenile** Very dark, with head-sides, mantle and breast blackish; bill black. **VOICE** As Malachite Kingfisher. **HABITS** Frequents streams and rivers with shallow pools, and beaches. **STATUS AND DISTRIBUTION** Common resident, endemic to São Tomé, where the only kingfisher. **NOTE** Usually treated as race of Malachite Kingfisher (cf. *BoA*), but also as separate species, *A. thomensis*.

SHINING-BLUE KINGFISHER *Alcedo quadribrachys* Plate 65 (5)
Martin-pêcheur azuré

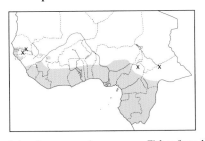

L 16 cm (6.25"). Unobtrusive *aquatic* forest kingfisher with *black bill.* **Adult male** Head, upperparts and patch at breast-sides brilliant, deep ultramarine-blue with white patch on neck-sides and rufous loral spot. Throat white; rest of underparts deep orange-rufous. Feet red. **Nominate** has mantle to uppertail-coverts bright purple-blue; eastern *guentheri* has these paler, bright azure-blue, contrasting with rest of upperparts. **Adult female** Bill has some dark red at base of lower mandible. **Juvenile** Duller, with dusky mottling on breast and bill tipped whitish. **VOICE** A shrill, high-pitched *tseep*, uttered in flight. [CD7:86] **HABITS** Singly or in pairs along forested rivers, lakes, lagoons and mangroves. Fishes from low perch above water. Shy and easily overlooked; mostly seen in low, fast flight over water. **SIMILAR SPECIES** All other adult blue-and-rufous kingfishers are smaller with red bills. Juvenile Malachite and White-bellied Kingfishers have black bills and similar plumage pattern, but head-sides are orange-rufous or chestnut (not blue), and White-bellied has broad white streak from throat to undertail-coverts. **STATUS AND DISTRIBUTION** Uncommon or scarce to common in forest and savanna zones, from Senegambia to WC Nigeria, also SW Mali and Burkina Faso (nominate), and from SW Nigeria to CAR and Congo (*guentheri*). In Cameroon to 850 m.

COMMON KINGFISHER *Alcedo atthis* Plate 65 (7)
Martin-pêcheur d'Europe

Other name E: European Kingfisher
L 15–16 cm (6.0–6.25"). *A. a. atthis.* **Adult male** The only aquatic kingfisher with combination of mainly brilliant blue upperparts, orange-rufous underparts and *broad green-blue malar stripe.* Rufous loral spot and ear-coverts; white patch on neck-sides; white throat. Bill black. Feet coral-red. **Adult female** Bill usually reddish at base.

VOICE A high-pitched, shrill *tzeee*, uttered in flight. [CD3:28] **HABITS** Frequents rivers, small streams and lakes with vegetated banks and reedbeds; in winter also estuaries, harbours, fishponds and rocky coasts. **SIMILAR SPECIES** No other kingfisher has been reported at northern limit of our region. **STATUS AND DISTRIBUTION** Palearctic species; claimed as vagrant from NW Mauritania (Nouadhibou, Aug).

GIANT KINGFISHER *Megaceryle maxima*   Plate 66 (7)
Martin-pêcheur géant

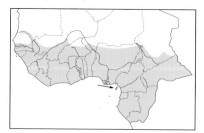

Other names E: African Giant Kingfisher; F: Alcyon géant
L 42–46 cm (16.5–18.0"). Largest African kingfisher, with slate-grey upperparts speckled white, bushy crest and massive black bill. **Adult male** *maxima* Throat white; *breast chestnut; belly white barred black.* Underwing-coverts white. Forest race *gigantea* darker, upperparts less heavily speckled; belly more heavily barred. **Adult female** *Breast white very densely spotted black; belly to undertail-coverts chestnut.* Underwing-coverts chestnut. **Juvenile male** As adult male, but breast-sides speckled black; flanks with some chestnut feathers. **Juvenile female** As adult female, but breast less densely spotted; spots crescentic and mainly at sides. **VOICE** A loud *KEK!* uttered singly or in series. [CD7:87] **HABITS** Aquatic. Solitary or in pairs along wooded rivers, lagoons and lakes, and in mangroves; also on rocky and sandy seashores. Perches quietly and often concealed over water, searching for fish and crabs. Shy. Flies low over water when disturbed. **SIMILAR SPECIES** Unmistakable if seen reasonable well, but beware of confusing flying bird with Green-backed Heron, which has similar size, upperpart coloration and call. **STATUS AND DISTRIBUTION** Not uncommon resident in savanna and forest zones throughout, north to S Mauritania; also Bioko. Ranges of races poorly defined; *gigantea* in forest zone from Liberia east, intergrading with nominate in forest–savanna contact zone.

PIED KINGFISHER *Ceryle rudis*   Plate 65 (6)
Martin-pêcheur pie

Other name F: Alcyon pie
L 25 cm (10"). *C. r. rudis.* Conspicuous and unmistakable aquatic kingfisher. **Adult male** Head and upperparts black and white with bushy crest on hindcrown. Underparts clean white with double black breast-band (upper band broad, lower narrow). Bill long, black. **Adult female** Single, incomplete breast-band (limited to patch on breast-sides). **Juvenile** As adult female but feathers of lores, throat and breast fringed brown; breast patches greyish-black; gape pink. **VOICE** Vocal. Calls include a high-pitched, chattering *kwik-kwik* uttered in flight or on perch, *chikrr-chekrr...* (threat call) and *kittle-te-ker* (when flying from perch). [CD7:88] **HABITS** In pairs or small parties in open areas along rivers, lagoons, mangroves, lakes and creeks. Tame. Feeds almost exclusively on fish, caught by plunge-diving from low perch or hovering flight. Flight fast and direct, with uneven wingbeats occasionally interspersed by short glides. Nests solitarily or in small colonies. **STATUS AND DISTRIBUTION** Common in suitable habitat throughout, south of 17°N, incl. Bioko. Probably mainly sedentary, but seasonal changes in abundance reported.

BEE-EATERS
Family MEROPIDAE (World: 24–26 species; Africa: 18–19; WA: 14)

Slender, small to medium-sized, mostly brightly coloured birds with long, pointed, slightly decurved bills and triangular wings. Sexes similar. Feed mainly on wasps and bees, either caught in short sallies from perch or in lengthy hawking flights. Some migratory and gregarious, others sedentary and occurring in pairs or family parties. Nest in burrows excavated in banks or level ground. Most species in this distinctive and beautiful family are easy to identify. Fourteen are endemic to Africa, 11 of these occur in W Africa.

BLACK-HEADED BEE-EATER *Merops breweri*   Plate 67 (3)
Guêpier à tête noire

L 25–28 cm (10–11"). Unmistakable; large and robust. **Adult male** The only bee-eater with *dull black head*, grass-

green upperparts and *cinnamon-buff underparts*. Tail cinnamon edged green, mostly covered by elongated, green, central tail feathers. Underwing-coverts buff. Bill rather heavy. Eye crimson. **Adult female** Tail streamers shorter. **Juvenile** Duller, with greenish crown and no streamers; throat and head-sides black suffused with green; breast without cinnamon and washed green on sides. **VOICE** Mainly silent. Calls include *churuk churuk* and a melodious *chiuk* or *chiok* in flight; alarm call *wik*. [CD7:89] **HABITS** Mostly solitary or in pairs, foraging in forest canopy (usually gallery and swamp forest) and breeding in adjacent savanna. Flight slow and 'sailing'. Rather inconspicuous. **STATUS AND DISTRIBUTION** Scarce to rare resident. Mainly Gabon and Congo; also recorded in CAR, SE Nigeria (Igalaland plateau) and C Ivory Coast (Marahoué NP); also SE Ghana (1952, Afram R., area now inundated). Sedentary, but local movements reported.

BLUE-HEADED BEE-EATER *Merops muelleri* Plate 67 (2)
Guêpier à tête bleue

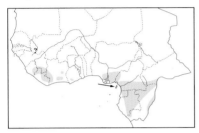

L 19 cm (17.5"). Small, dark, unobtrusive forest bee-eater. **Adult** Mainly deep purplish-blue with rich chestnut mantle, back and wings and small scarlet throat patch surrounded by black (sometimes hard to see in forest shade). Eye red. In flight appears all dark with rufous wings. Western **mentalis** has short, blunt central tail feathers; **nominate** has white forehead, pale blue crown and lacks streamers. **Juvenile** Duskier. **VOICE** Mainly silent. Contact call a frequently uttered, discreet *slip* or *sip*. Infrequently a high-pitched *kee-klip* or *ptii-wit*. Also bill-snapping and soft, muffled, hoarse little sounds. [CD7:90] **HABITS** In pairs or trios along shady tracks or in small openings within primary and old secondary forest. Perches on lianas and thin bare branches, usually in mid-strata, but occasionally very low, from where it forages in short, sweeping sallies. Wags tail in short arc. Confiding and readily approachable. **SIMILAR SPECIES** More widespread and commoner Black Bee-eater occurs in more open areas (thus more conspicuous), has larger scarlet throat patch and striking azure-blue rump and lower underparts. **STATUS AND DISTRIBUTION** Scarce to rare and local forest resident, from Sierra Leone and SE Guinea to W Ghana, and S Nigeria to W Cameroon (*mentalis*), and from S Cameroon and Bioko to SW CAR and Congo (nominate). Claimed from SW Mali. The two races meet near Douala, Cameroon. **NOTE** Birds from Bioko are slightly larger, with slightly longer wings than *mentalis*, and are sometimes recognised as distinct ssp. *marionis*.

BLACK BEE-EATER *Merops gularis* Plate 67 (1)
Guêpier noir

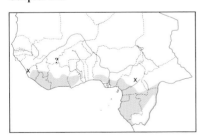

L 20 cm (8"). **Adult** Small and square tailed with *mainly jet-black upperparts, azure rump, belly and undertail-coverts, scarlet throat and black azure-streaked breast*. Eye dark red. In flight brilliant azure rump is conspicuous. Scarlet throat can be surprisingly hard to see in bad light. **Nominate** has blue forehead and supercilium. *M. g. australis* has head entirely black with only a few blue feathers on forehead; blue streaks on underparts sometimes tipped scarlet (virtually impossible to see in field). **Juvenile** Duskier and lacking scarlet throat (reddish feathers may appear 1 month after fledging). **VOICE** A rather loud and distinctive *klip* or *wik*, not infrequently uttered. [CD7:91] **HABITS** Mostly in pairs, occasionally in trios, frequenting clearings and forest edges. Perches conspicuously and usually quite high on dead branches, from where short flycatching sallies are made. Returns to same or nearby perch to consume catch. **SIMILAR SPECIES** Much rarer Blue-headed Bee-eater lacks striking azure rump, has chestnut wings and mantle, and is more restricted to forest. **STATUS AND DISTRIBUTION** Scarce to locally common resident and partial migrant in forest zone, from Sierra Leone and SE Guinea to Nigeria (nominate) and from Cameroon to SW CAR and Congo (*australis*). Some intergradation in SE Nigeria and W Cameroon. Claimed from SE Mali.

LITTLE BEE-EATER *Merops pusillus* Plate 68 (2)
Guêpier nain

L 15–17 cm (6.0–6.5"). Smallest bee-eater. **Adult *pusillus*** Top of head and upperparts grass-green. Throat rich

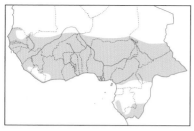

yellow bordered by black mask and gorget; breast chestnut becoming cinnamon to buff on belly and vent. Eye red. In flight, wings mainly cinnamon with broad black trailing edge; square or slightly notched tail cinnamon with black tips and green central feathers. *M. p. meridionalis* has short, narrow blue supercilium. **Juvenile** Paler, with pale greenish breast and no gorget; eye brown. **VOICE** A quiet *slip* or *sip* and a hard, clipped *tsip* or *klip*, uttered singly or in series. Also a high-pitched, sibilant *siddle-iddle-ip, d'jee* in high-intensity greeting display. [CD7:92] **HABITS** In pairs or family parties in grassland, open savanna, plantations and variety of open-country habitats; fond of moist areas. Forages low, hawking in short sallies from perch on long grass stem or low shrub, occasionally higher. **SIMILAR SPECIES** Blue-breasted Bee-eater is very similar, but has white spot on neck-sides, dark blue (not black) gorget, and is slightly larger; see also distribution and voice. **STATUS AND DISTRIBUTION** Generally common resident throughout, except arid north; uncommon to scarce at north edge of range. Nominate from S Mauritania to Liberia east to Chad and CAR; *meridionalis* in S Gabon and Congo. Local movements reported.

BLUE-BREASTED BEE-EATER *Merops variegatus*   Plate 68 (1)
Guêpier à collier bleu

L 17 cm (6.5"). **Adult** *variegatus* Very similar to Little Bee-eater but has *purple-blue* (not black) *gorget* (visible in good light) and *white spot on neck-sides*, at corner of yellow throat. Although latter is only visible at short range, it usually makes yellow throat appear larger and brighter; from behind it stands out clearly against black mask and green neck. Highland race *loringi* has blue supercilium. **Juvenile** Lacks gorget and has pale greenish breast. **VOICE** A quiet *prru* or *tup*, uttered singly or in series, *prru-tup prru-tup-tup ptup prru...* [CD7:93] **HABITS** Prefers more upland and often moister areas than Little Bee-eater, but behaviour similar. The two may occur together. **SIMILAR SPECIES** Easily confused with Little Bee-eater (q.v. and above). **STATUS AND DISTRIBUTION** Local and uncommon resident in highlands of SE Nigeria (Obudu and Mambilla Plateaus) and W Cameroon (*loringi*), and from S Cameroon through wet lowlands of Gabon, Congo and SW CAR (nominate).

SWALLOW-TAILED BEE-EATER *Merops hirundineus*   Plate 68 (5)
Guêpier à queue d'aronde

L 20–22 cm (8.0–8.5"). *M. h. chrysolaimus.* Medium-sized greenish bee-eater easily distinguished by *strongly forked tail.* **Adult male** Top of head green with azure-blue forehead and supercilium; upperparts green; rump and tail pale blue. Throat yellow bordered by black mask and narrow blue gorget; breast green becoming blue on belly and undertail-coverts. In flight appears mainly green with forked, white-tipped tail; wings have some rufous and black trailing edge. **Adult female** Tail less deeply forked; gorget narrower. **Juvenile** Throat greenish-white; no gorget. **VOICE** A subdued *weeerp-weeerp, tip-tip,* or *diddle-diddle-ip,* typically bee-eater-like in quality. [CD7:95] **HABITS** In pairs or small parties in open savanna woodland. Feeds in short sallies from perch. Rather silent and unobtrusive. Often more mobile than other small bee-eaters, moving from one spot to another when foraging. May roost in tightly packed ranks of 6–10 (rarely up to 30) when not breeding. **STATUS AND DISTRIBUTION** Rather scarce and local in savanna belt, from Senegambia to Sierra Leone east to S Chad and N CAR. Vagrant, S Mauritania and Liberia. At least partially migratory; movements poorly understood. Resident in north, but some move south for dry season.

RED-THROATED BEE-EATER *Merops bulocki*   Plate 68 (3)
Guêpier à gorge rouge

L 20–22 cm (8.0–8.5"). *M. b. bulocki.* The only bee-eater in its range with *green upperparts and red throat.* Tail square. **Adult** Top of head, upperparts and tail mainly grass-green. Black mask; throat bright scarlet (rarely yellow); hindneck, breast and belly cinnamon-buff; vent and undertail-coverts deep ultramarine-blue. In flight, wings green above and buff below with black trailing edge; tail warm buff. **Juvenile** Dull version of adult. **VOICE** Vocal. Calls include *wip,* querulous *kirrup, kwirrup* or *krrip,* delivered in trilling series in greeting, and nasal *kweep,* uttered in flight and from perch. [CD7:96] **HABITS** In small to large groups in open savanna woodland and bushy

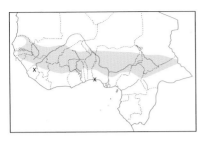

pastures. Mainly forages from perch. Breeds colonially in sand banks. **SIMILAR SPECIES** White-fronted Bee-eater has similar plumage pattern but does not overlap in our region. Yellow-throated morph separated from other bee-eaters with yellow throats by lack of gorget. **STATUS AND DISTRIBUTION** Locally common resident in savanna belt, from S Mauritania to Guinea east to S Chad and N CAR. Vagrant, Sierra Leone. **NOTE** Species name spelt *bullocki* in *BoA*, but see Dowsett & Dowsett-Lemaire (1993) for maintaining traditional and widely used *bulocki*.

WHITE-FRONTED BEE-EATER *Merops bullockoides*   Plate 68 (4)
Guêpier à front blanc

L 22–24 cm (8.5–9.5"). *M. b. bullockoides.* Unmistakable. **Adult** Forehead white; crown and hindneck buff; upperparts and square tail green. Black mask bordered below by white chin and malar stripe; throat scarlet; breast and belly cinnamon-buff; vent, upper- and undertail-coverts deep blue. In flight, has black trailing edge to wing (like many congeners). **Juvenile** Duller. **VOICE** Vocal. Call a distinctive nasal *wèèh* or *wèèh-up*, very different from other bee-eaters. Alarm a sharp *waark*. [CD7:97] **HABITS** In parties in lightly wooded grassland. Breeds colonially. **SIMILAR SPECIES** None in our region: no overlap with Red-throated Bee-eater, its northern counterpart. **STATUS AND DISTRIBUTION** Locally common resident, Gabon and S Congo. Local movements reported.

WHITE-THROATED BEE-EATER *Merops albicollis*   Plate 68 (6)
Guêpier à gorge blanche

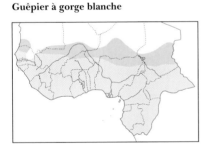

L 19–21 cm (8"), excl. tail streamers projecting up to *c.* 12 cm (4.75"). Graceful, conspicuous bee-eater with largely pale green upperparts, *long tail streamers* and *strikingly patterned, pied head.* **Adult** Unmistakable when perched: *black crown* (unique among bee-eaters), mask and gorget contrast with white forehead, supercilium and throat; hindneck ochreous; back and wings pale green; rump and tail pale blue; breast whitish-green; belly white. Eye red. In flight, green and pale rufous wings (buffish below) have black trailing edge. **Juvenile** Duller, with pale yellow throat and no tail streamers; eye brownish. **VOICE** Vocal. A pleasing, melodious and far-carrying *pruuee, trooee* or *pruik*, reminiscent of European Bee-eater but distinctly higher pitched and delivered in longer series. Sometimes interspersed with shorter notes (e.g. *kwik kwik kwik kearrlup kearrlup...*). Frequently uttered in flight and on perch. [CD7:98] **HABITS** Usually hunts from low perch on breeding grounds, from medium-high to high perch in non-breeding zone. Highly gregarious. **STATUS AND DISTRIBUTION** Seasonally common to not uncommon throughout, except in desert. Breeds during rains in narrow band at south edge of Sahara (May–Oct), from S Mauritania and N Senegal through Mali to Niger, NE Nigeria, N Cameroon and Chad. Moves south in dry season. Migration and non-breeding range include entire moist savanna and rainforest zones, from Guinea to Congo.

LITTLE GREEN BEE-EATER *Merops orientalis*   Plate 68 (8)
Guêpier d'Orient

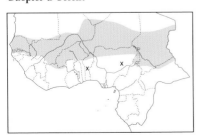

Other name E: Green Bee-eater
L 16–18 cm (6.5–7.0"), excl. tail streamers projecting up to 9.5 cm (3.75"). **Adult male** *viridissimus* Small, *all-green* bee-eater with very long tail streamers, black mask and thin black gorget. Eye red. In flight, has rufous flight feathers above and below, and black trailing edge (like many bee-eaters). *M. o. flavoviridis* similar but throat often yellow. **Adult female** Gorget usually even narrower and tail streamers shorter. **Juvenile** Paler, lacking gorget and with shorter or no streamers. **VOICE** A hard, rapid and slightly buzzing trill *trrri-trrri-trrri* or *kree-kree-kree-...* Alarm a staccato *ti-ti-ti* or *ti-ik*. [CD8:1a] **HABITS** Mostly in arid country, occurring singly, in pairs or small parties in subdesert and woodland. Makes short flycatching

sallies from low perch. **SIMILAR SPECIES** Swallow-tailed Bee-eater has similar coloration but a yellow throat and forked tail without streamers. **STATUS AND DISTRIBUTION** Common to uncommon in Sahel and northern savanna zones, from S Mauritania and Senegambia through Mali and N Ivory Coast to N Benin, Niger, N Nigeria, N Cameroon, Chad and N CAR (*viridissimus*). *M. o. flavoviridis* in NE Chad (Ennedi). Resident and partial migrant, dispersing mostly south in dry season, after breeding. **NOTE** Ssp. *flavoviridis* perhaps invalid (possibly *viridissimus* in worn or very fresh plumage: *HBW*).

BLUE-CHEEKED BEE-EATER *Merops persicus*   Plate 68 (7)
Guêpier de Perse

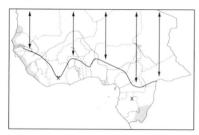

L 24–26 cm (9.5–10.0"), excl. tail streamers projecting up to 10.5 cm (4"). Large, slender, bright green bee-eater with very long tail streamers. **Adult male** *chrysocercus* All green, with black mask bordered blue, and yellow chin merging into chestnut of throat and upper breast. Streamers extend 7.0–10.5 cm beyond other tail feathers. In flight, entirely golden-green above, underwing rufous with narrow dusky trailing edge. In worn plumage appears paler, more bluish. **Nominate** has shorter tail streamers, extending 4.5–6.7 cm; upperparts grass-green. **Adult female** Tail streamers shorter. **Juvenile** Duller and lacking streamers. **VOICE** A far-carrying *prri-ip*, *dirrip* or *pririk*, very similar to call of European Bee-eater but harder, less liquid and shorter. Alarm *pik-pik-pik* or *dik-dik-dik*. [CD3:29] **HABITS** A highly gregarious desert-edge species, wintering in bushy grassland and open farmland. Perches on power lines and low bushes from which it makes fast sallies in pursuit of prey. Also forages high in the air, in continuous flight. Flight graceful and soaring, like European Bee-eater, rising on a series of rapid wingbeats, stalling, then gliding. Migrates by day at great height, sometimes with European Bee-eater. **SIMILAR SPECIES** At distance and in flight, European Bee-eater can be hard to separate, but differs in chestnut-and-golden upperparts, yellow throat and paler underwing with broader trailing edge. **STATUS AND DISTRIBUTION** Widespread intra-African and Palearctic migrant. *M. p. chrysocercus* breeds locally near Mauritanian coast (May–Oct, not uncommon), Senegal, C Mali, Nigeria (Kainji Dam) and W and E shores of L. Chad. These and those breeding in NW Africa winter in savanna from Gambia to Sierra Leone and coastal Liberia (not uncommon), Ivory Coast and Ghana (rare) to Nigeria, Cameroon and CAR (uncommon); vagrant, Cape Verde Is. Nominate is winter visitor (mainly from Asia) to E and S Africa, reaching our region in Congo and Gabon.

EUROPEAN BEE-EATER *Merops apiaster*   Plate 67 (4)
Guêpier d'Europe

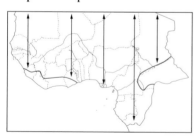

L 23–25 cm (9–10"), excl. tail streamers projecting 0–3 cm (0–1"). Large, vocal bee-eater with unique coloration. **Adult male** Top of head and upperparts chestnut and golden; tail blue-green with short streamers. Yellow throat bordered by black mask and thin black gorget; rest of underparts turquoise-blue. In flight, scapulars and rump form golden V; wings chestnut and green with broad black trailing edge. Underwing pale rufous. **Adult female** Often separable by less chestnut in wing and green wash to scapulars. **VOICE** A melodious, liquid, remarkably far-carrying *pruuip*, *pru-ik* or *kwirip*. (Calls of Blue-cheeked and Rosy Bee-eaters very similar and only separable with practice). Also *pik-pik*. [CD3:30] **HABITS** In flocks in open woodland and cultivated areas. Hawks insects from perch on tree, shrub or wire, or in continuous twisting flight. Often perches on power lines. Flight graceful, wheeling; rapid wingbeats alternating with gliding. **SIMILAR SPECIES** At distance and in flight from below could be confused with Blue-cheeked Bee-eater, but latter is bright green, has longer tail streamers and more rufous underwing with narrower blackish trailing edge. **STATUS AND DISTRIBUTION** Locally common to scarce Palearctic passage migrant and winter visitor throughout, except in forest zone, from Mauritania (Jul and Mar–Apr) and Senegambia to NE Liberia (Mt Nimba) east to Chad, CAR and Congo (Sep–May). Vagrant, Cape Verde Is.

ROSY BEE-EATER *Merops malimbicus*   Plate 67 (6)
Guêpier gris-rose

L 22–25 cm (9–10"). Unmistakable. **Adult** Top of head and upperparts slate-grey contrasting conspicuously with vivid reddish-pink underparts; black mask bordered below by white line. Tail dusky-carmine with short streamers. Underwing dark grey. Eye red. **Juvenile** Duller, without streamers. **VOICE** A rather hoarse *pru* or *krrp*, shorter,

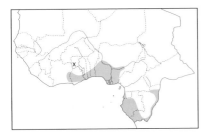

more abrupt, less melodious than call of European Bee-eater. Alarm *wik*. [CD8:2b] **HABITS** Aerial and gregarious, foraging above gallery forest, moist savanna woodland and rainforest, often near water. Nests on sandbars in large rivers, often in huge colonies. **SIMILAR SPECIES** None, but see Carmine Bee-eater. **STATUS AND DISTRIBUTION** Not uncommon intra-tropical migrant. Breeds in Nigeria (R. Niger and tributaries, Apr–Jun), S Gabon (Gamba) and coastal Congo (Conkouati Reserve). Disperses after breeding and then occurs from Nigeria to Burkina Faso, Ghana and (very rarely) W Ivory Coast, and from Eq. Guinea and Gabon to Congo.

NORTHERN CARMINE BEE-EATER *Merops nubicus*   Plate 67 (5)
Guêpier écarlate

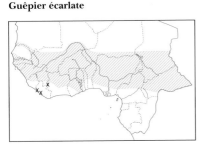

L 24–27 cm (9.5–10.5"), excl. tail streamers projecting up to 12 cm (4.75"). Unmistakable. **Adult** Large, conspicuous, carmine-and-pink bee-eater with blue head and long tail streamers. Rump, upper- and undertail-coverts pale blue, strongly contrasting with rest of plumage. Underwing buffy. Tail grey below. **Juvenile** Duller, with mainly brownish upperparts and short streamers. **VOICE** A rather unmusical *klienk*, *klunk* or *terk*, often repeated and sometimes followed by *ki-ki-ki-ki-...* [CD8:3a] **HABITS** Aerial and gregarious, occurring in dry and open savanna. Forages in 'sailing', twisting flight. Also feeds from perch. Partial to grasshoppers and locusts and therefore attracted to bush-fires. Uses mammals, large terrestrial birds, people and vehicles as beaters and swoops after insects flushed from grass. Nesting colonies usually number 100–1000 nests, occasionally up to 10,000. **SIMILAR SPECIES** None, but see Rosy Bee-eater. **STATUS AND DISTRIBUTION** Not uncommon to locally common throughout savanna belt. Breeds from S Mauritania and Senegambia through Mali and N Ghana to Chad and CAR. Disperses south in dry season, after breeding, almost reaching north edge of rainforest zone.

ROLLERS
Family CORACIIDAE (World: 12 species; Africa: 8; WA: 7)

Medium-sized, robust, conspicuous birds with colourful plumage, large heads and stout bills, slightly hooked at tip. Sexes similar; juveniles duller. Hunt from perch, swooping to ground to catch large insects or small vertebrates (*Coracias*), or pursuing insects in the air (*Eurystomus*). Spectacular display flights (hence English name). Vocal and pugnacious. Voice harsh. Breed in tree holes.

RUFOUS-CROWNED ROLLER *Coracias naevius*   Plate 69 (3)
Rollier varié

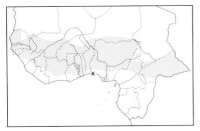

Other name E: Purple Roller
L 35–40 cm (14.0–15.5"). *C. n. naevius*. Large, stocky and compara-tively dull roller with *conspicuous white forehead and supercilium, dull brownish-pink underparts heavily streaked white* and square tail. **Adult** Crown and nape vinaceous-rufous; small white patch on nape. Upperparts mainly olive-brown; rump pale purple; upperwing-coverts pinkish-brown; flight feathers purplish. Tail dark purple-blue with brownish-green central feathers. **Juvenile** Duller, more olive. **VOICE** Various cackling, muffled and nasal notes, uttered singly or in rapid series, less harsh than other rollers and sometimes reminiscent of Green Wood-hoopoe. [CD8:4a] **HABITS** Usually singly, perching on high vantage point. Frequents various open woodlands, from dry *Acacia* savanna to wooded savanna and cultivation. Less vocal than other rollers. **SIMILAR SPECIES** None. All other *Coracias* are brighter, bluer and have azure-blue in wings; *Eurystomus* rollers much smaller, with bright yellow bills and no white supercilium. **STATUS AND DISTRIBUTION** Generally uncommon in Sahel and savanna zones, from S Mauritania to Guinea east to S Chad and CAR. Rare, Sierra Leone and Liberia. Mainly sedentary in centre of range, but wet-season visitor to Sahel zone and mainly dry-season visitor to southern Guinean woodlands.

BLUE-BELLIED ROLLER *Coracias cyanogaster*
Rollier à ventre bleu

Plate 69 (4)

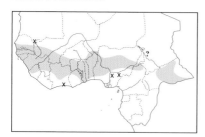

L 28–30 cm (11–12"); streamers projecting up to 6 cm (2.25"). Strikingly coloured and unmistakable, with pale head and breast contrasting with purple, black and blue of rest of plumage. **Adult** Head, neck and breast pale pinkish-fawn or buffish-white. Back black; rump and uppertail-coverts purplish-blue; wings purplish-blue with broad azure-blue bar. Tail azure-blue, forked, with streamers. Belly to undertail-coverts purplish-blue. **Juvenile** Duller, lacking streamers. **VOICE** A fast, sharp *keh-keh-keh-keh-k-r-r-r-r-r* and an accelerating, descending, scolding 'laugh' *HEheheheheh...* uttered in flight and from perch. [CD8:4b] **HABITS** In pairs or small groups in wooded savanna, edges of gallery forest, derived savanna and cultivation in or near rainforest. Noisy. Perches conspicuously atop trees, electricity poles, etc. Flight direct with rapid, shallow wingbeats. **STATUS AND DISTRIBUTION** Locally common, mainly in savanna bordering forest, from Senegambia to Sierra Leone east to CAR, becoming increasingly scarce in east. Rare, S Mauritania and SW Niger. Resident in south, with numbers augmented by those from further north during dry season.

ABYSSINIAN ROLLER *Coracias abyssinicus*
Rollier d'Abyssinie

Plate 69 (7)

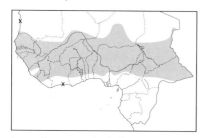

L 28–30 cm (11–12"); streamers projecting up to 12 cm (5"). Slender and brightly coloured with *very long outer tail feathers*. **Adult** Bright turquoise-blue with chestnut mantle, scapulars and tertials; forehead and throat whitish; rump and uppertail-coverts deep violet-blue. Forewing violet-blue, forming shoulder patch (sometimes concealed) when perched; flight feathers deep violet-blue. **Juvenile** Duller; breast and crown washed brownish; no tail streamers. **VOICE** A harsh, explosive, scolding screech *kwrèèèèh* or *kèèèèèhh* uttered from perch; a sharp *kek*, mainly in flight. In display flight a rapid series of similar harsh notes. [CD8:5a] **HABITS** Singly or in pairs, gathering in small groups at bush fires. In various dry wooded habitats, incl. *Acacia* savanna, wooded grassland, farmland and large gardens. Noisy and conspicuous, perching on poles, power lines and bare branches. Not shy. **SIMILAR SPECIES** European Roller lacks tail streamers and is slightly larger and more robust; distinguished from juvenile Abyssinian by blackish flight feathers. **STATUS AND DISTRIBUTION** Common in broad belt of Sahel and savanna north of forest zone, from S Mauritania to Liberia east to Chad and CAR. Intra-African migrant, moving south during dry season.

EUROPEAN ROLLER *Coracias garrulus*
Rollier d'Europe

Plate 69 (5)

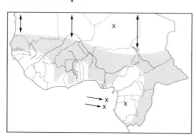

L 32 cm (12.5"). *C. g. garrulus*. **Adult** Mainly turquoise-blue with chestnut mantle, scapulars and tertials, *blackish flight feathers* (deep violet below) and *square tail*. Back to uppertail-coverts and forewing purplish-blue. **Juvenile** Paler and duller until winter moult (Nov–Dec). **VOICE** Mostly silent in winter. A harsh, short *wrek*, often developing into *wrek-kekekek-erek...* Also a drawn-out *wrèèèèèèh*. [CD3:31] **HABITS** Singly or in groups in wooded savanna, thorn-bush and cultivated areas. Perches on bare branches, power lines and poles. Flight strong and direct. **SIMILAR SPECIES** Abyssinian and Lilac-breasted Rollers have long tail streamers and are slightly smaller and more slender. Juvenile and moulting adult Abyssinian differ in having uniformly pale blue wing-coverts and upperside of flight feathers dark blue (not black); they appear bluer, are noisier and have a more agile flight. **STATUS AND DISTRIBUTION** Uncommon to rare Palearctic passage migrant throughout, incl. São Tomé and Príncipe; winter visitor to narrow belt from S Senegambia to SE Nigeria. Vagrant, Cape Verde Is (Raso).

LILAC-BREASTED ROLLER *Coracias caudatus*
Rollier à longs brins

Plate 69 (6)

L 28–30 cm (11–12"); streamers projecting up to 8 cm (3"). *C. c. caudatus*. **Adult** Distinctive combination of *white-*

streaked lilac throat and breast and *long tail streamers*. White forehead and supercilium; top of head to mantle greenish-blue; mantle, scapulars and tertials chestnut; rump purple-blue. Belly to undertail-coverts azure-blue. Forewing violet-blue, forming shoulder patch when perched; flight feathers deep violet-blue. **Juvenile** Duller, more buffish, without tail streamers. **VOICE** Loud, harsh rattling notes *wrek, wrek*... or sharp *keh! keh!*..., in display flight developing into a series. [CD8:5b] **HABITS** As Abyssinian and European Rollers. Frequents wooded grassland. **SIMILAR SPECIES** Abyssinian Roller has all-blue underparts; no overlap in our region. **STATUS AND DISTRIBUTION** An E and S African species reaching SC Congo, where not uncommon and presumably resident, and SE Gabon (Lékoni).

BLUE-THROATED ROLLER *Eurystomus gularis*   Plate 69 (2)
Rolle à gorge bleue

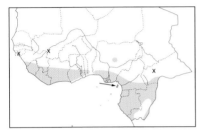

L 25 cm (10"). **Adult** *gularis* Resembles Broad-billed Roller but slightly darker and with *diffuse blue throat patch* (sometimes hard to see). Head and upperparts, incl. greater coverts, dark chestnut; uppertail-coverts blackish. Tail azure-blue and black. *Entire underparts* (incl. vent and undertail-coverts) *dark rufous-brown*. **E. g.** *neglectus* more richly coloured; underparts washed lilac. **Juvenile** Duller; throat and upper breast dark brown; lower breast and belly greyish-blue mottled dusky. Bill yellow with dusky culmen. **VOICE** Various strident notes, e.g. *khlee* or *gleek* and *grwree*, sometimes developing into a series, e.g. *khleep-khleep-khleep-*... or *kikikikik*... [CD8:6b]

HABITS Singly or in pairs in clearings, forest edge and gallery forest, perching conspicuously on bare branch from canopy, occasionally hawking for insects. Gathers in feeding flocks in late afternoon, often with Broad-billed Roller. Fast, dashing flight on pointed wings reminiscent of falcon. **SIMILAR SPECIES** Beware confusion, esp. in bad light, with Broad-billed Roller, which is very similar, but lacks throat patch and has deep lilac (not dark chestnut) underparts with azure-blue vent and undertail-coverts; more difficult to see are dark blue (not dark chestnut) greater coverts and greyish-blue (not blackish) uppertail-coverts. **STATUS AND DISTRIBUTION** Uncommon to not uncommon resident in forest zone and forest outliers in savanna, from Guinea-Bissau to S Nigeria (nominate) and from Cameroon and Bioko to CAR and Congo (*neglectus*). Intergrades in extreme SE Nigeria/SW Cameroon.

BROAD-BILLED ROLLER *Eurystomus glaucurus*   Plate 69 (1)
Rolle violet

L 29–30 cm (11.5"). Relatively small, thickset and dark roller with *bright yellow, triangular bill*. **Adult** *afer* Top of head and upperparts chestnut; head-sides and underparts deep lilac. Lateral uppertail-coverts, vent and undertail-coverts grey-blue or dull azure-blue. Flight feathers and greater coverts dark blue. Tail rather short and shallowly forked; central feathers blackish, rest azure-blue tipped black and with dark blue subterminal band. Southern *suahelicus* brighter, with uppertail-coverts all blue and undertail-coverts pale blue. **Juvenile** Much duller; upperparts, throat and upper breast mainly dirty brown; lower breast and belly dull greenish-blue mottled brownish. Bill yellow with dusky tip and culmen. **VOICE** Vocal. A sharp, nasal *kek*, sometimes developing into harsh 'laughing' *kekekekekek-k-k-k-r-r-r-r*; also variety of other guttural, grating and growling notes, uttered in series. [CD8:7a] **HABITS** Mostly singly or in pairs, in clearings, forest edge, along rivers, in woodland and more open habitats with scattered large trees. Perches on tall treetops, occasionally hawking insects high in the air. Assembles in wheeling feeding parties in late afternoon and at dusk. Pugnacious. Pointed wings combined with powerful and aerobatic flight may recall falcon. **SIMILAR SPECIES** Easily confused with Blue-throated Roller, which differs mainly in small blue throat patch and entirely dark rufous-brown (not lilac) underparts; more difficult to see are chestnut (not dark-blue) greater coverts and blackish (not dull blue) uppertail-coverts. Blue-throated is also more exclusively associated with forest and has different, shrieking voice. **STATUS AND DISTRIBUTION** Common throughout, except arid north. Mainly *afer*, from Senegambia to Liberia east to Chad and CAR, with S African *suahelicus* intergrading between 5°N and equator. Resident and intra-African migrant, expanding north during rains (wet-season breeder in north of range). Vagrant, Cape Verde Is, São Tomé and Príncipe.

WOOD-HOOPOES
Family PHOENICULIDAE (World: 8 species; Africa: 8; WA: 4)

Medium-sized, slender birds with largely dark glossy plumage, slender, decurved bills and long graduated tails. Legs short, wings rounded. Sexes similar; juvenile duller. Arboreal and agile, often hang upside-down probing bark for insects. Larger species (genus *Phoeniculus*) conspicuous, noisy and usually gregarious; smaller taxa (*Rhinopomastus*) quieter, mostly solitary or in pairs. Nest in tree holes. Endemic to Africa.

FOREST WOOD-HOOPOE *Phoeniculus castaneiceps*   Plate 70 (4)
Irrisor à tête brune

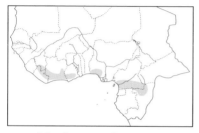

L 28 cm (11"). **Adult male** *castaneiceps* Small and slender wood-hoopoe with chestnut head and all-dark, iridescent green plumage. Bill dark horn. Feet black. Eastern *brunneiceps* has head bottle green, pale brownish or buffish-white with feathers narrowly fringed darker. **Adult female** Similar; in *brunneiceps* head is pale brownish or whitish. **Juvenile** Duller and smaller; *brunneiceps* similar to adult female. **VOICE** A fast twittering chatter (recalling White-headed Wood-hoopoe), often uttered by several birds together, and a series of up to 20 plaintive notes *kweep-wheep-wheep-wheep-....-wheew* (similar to Black Wood-hoopoe). [CD8:7b] **HABITS** Singly, in pairs (mostly) or small family groups, foraging high up in dense rainforest and semi-deciduous forest. Occasionally in mixed-species flocks. **SIMILAR SPECIES** White-headed Wood-hoopoe (the only other forest species) is larger and less slender; adult has bright red bill and feet. Black Wood-hoopoe occurs in savanna and differs from dark-headed male *brunneiceps* by white in wings and tail. **STATUS AND DISTRIBUTION** Scarce to uncommon and local forest resident. Nominate in SE Guinea, Liberia, Ivory Coast, S Ghana and SW Nigeria; *brunneiceps* in S Cameroon, SW CAR and N Congo.

WHITE-HEADED WOOD-HOOPOE *Phoeniculus bollei*   Plate 70 (2)
Irrisor à tête blanche

L 30–35 cm (12–14"). **Adult male** *bollei* Medium-large wood-hoopoe with black plumage glossed violet-green, conspicuous *buffish-white head* and bright red bill and feet. Long, graduated tail without spots. Isolated race *okuensis* has buffish-white on head much reduced. **Adult female** Bill slightly smaller. **Juvenile** Duller, with dusky bill and legs, and variable head colour: buffish-white mottled dusky or all dark gradually whitening with age. **VOICE** A fast, rippling, chattering twitter, often uttered by several birds together. Also a frequent, high-pitched *kuk*, uttered singly or in a rapid series. [CD8:8a] **HABITS** Gregarious, foraging high up in groups of 2–10 (sometimes more) in lowland, montane and large gallery forest; occasionally woodland. Noisy, but less so than Green Wood-hoopoe. **SIMILAR SPECIES** Green Wood-hoopoe occurs in savanna and is larger with, in adult, white wingbars and tail spots, and all-dark head; juvenile has white in tail and primary tips narrowly fringed white. Forest Wood-hoopoe occurs in same habitat but is smaller, more slender, and lacks red on bill and feet. **STATUS AND DISTRIBUTION** Uncommon forest resident, from SE Guinea and Liberia to S Ghana and from S Nigeria to S CAR (nominate); *okuensis* restricted to montane forest around Mt Oku, W Cameroon. Also claimed from SE Mali.

GREEN WOOD-HOOPOE *Phoeniculus purpureus*   Plate 70 (1)
Irrisor moqueur

Other names E: Red-billed Wood-Hoopoe; F: Moqueur
L 35–40 cm (14–16"). *P. p. guineensis/senegalensis*. Largest wood-hoopoe. **Adult male** All black with violet-green iridescence. Tail long and graduated with white subterminal spots on outer 2–3 rectrices. Double white wingbar conspicuous in flight. Bill bright red or black with red tip. Feet red. **Adult female** Bill slightly smaller. **Juvenile** Dull black with shorter all-black bill and black feet; all-black primaries with narrow white tips; white tail spots smaller. **VOICE** A single, loud, cracked *whak* or *kuk* and loud, high-pitched cackling, usually started by one and soon accelerating and developing into a resonant 'laughter' when other group members join in. [CD8:8b] **HABITS** Noisy and gregarious, typically in groups of 3–12 (or more), flying in single file between trees in open wooded habitats. Breeds cooperatively. **SIMILAR SPECIES** White-headed Wood-hoopoe is the only other wood-hoopoe

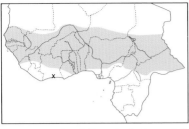

with red bill and feet, but it occurs in forest and has whitish head, uniform wings and tail, and faster, higher pitched and quieter calls. Confusion between juvenile Green and Black Wood-hoopoes (which occurs in same habitat, but is not gregarious) unlikely, due to presence of adults. **STATUS AND DISTRIBUTION** Uncommon to locally common resident in broad savanna and Sahel belts, from S Mauritania to Sierra Leone east to Chad and CAR. Reaches coast between E Ghana and W Nigeria. Absent from forest and arid zones. **NOTE** Two races described from our region, *guineensis* (S Mauritania and N Senegal to N Ghana and further east) and *senegalensis* (Gambia and S Senegal to S Ghana), the latter duller with variably coloured bill, from red with extensive black base to all black. Strong intergradation between the two races.

BLACK WOOD-HOOPOE *Rhinopomastus aterrimus* Plate 70 (3)
Irrisor noir

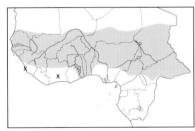

Other names E: Lesser Wood-Hoopoe; F: Petit Moqueur noir
L 23 cm (9"). **Adult male** *Small* violet-black wood-hoopoe with *black bill and feet*. White wingbar obvious in flight. **Nominate** has white primary-coverts and all-dark tail; southern *anchietae* lacks white on primary-coverts and has white subterminal spots on outer 2 tail feathers. **Adult female** Bill slightly smaller. **Juvenile** Duller, with shorter bill and tail. **VOICE** A series of plaintive, fairly loud or subdued notes *kwheep-kwheep-kwheep-...* or *pooee-pooee-pooee...* [CD8:9b] **HABITS** Singly or in pairs in savanna trees and bushes, occasionally low, even on ground. Joins mixed-species flocks. **SIMILAR SPECIES** The only other savanna species is Green Wood-hoopoe, which is much larger (q.v.). **STATUS AND DISTRIBUTION** Uncommon to locally not uncommon in savanna and Sahel zones to desert edge, from S Mauritania to Sierra Leone east to Chad and CAR (nominate). Also S Gabon and SC Congo (*anchietae*). Migrates south to Guinea savanna in dry season. **NOTE** Taxonomy much debated. Often placed in genus *Phoeniculus* (cf. *BoA*) or, previously, monospecific *Scoptelus*, but DNA studies in particular suggest best included in *Rhinopomastus* (*HBW*).

HOOPOES
Family UPUPIDAE (World: 2 species; Africa: 1; WA: 1)

Very distinctive, medium-sized birds with slender, decurved bills, broad rounded wings and erectile, fan-shaped crests. Sexes similar. Feed on ground; nest in holes. Most populations migratory.

HOOPOE *Upupa epops* Plate 70 (5)
Huppe fasciée

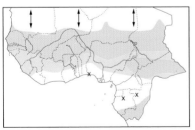

L 25–28 cm (10–11"); WS 42–48 cm. Unmistakable. **Adult male** Rufous-sandy or cinnamon with *black-tipped crest* and *boldly patterned black-and-white wings and tail*. Five races occur in our region, best distinguished by wing pattern and tone of body plumage. **Nominate** is palest (cinnamon-buff or sandy) with broad white band across primaries and 4 relatively neat white bands across coverts and secondaries; *senegalensis* has slightly darker, more cinnamon-rufous head and more white on secondaries, creating large white patches. More deeply coloured *africana* is rich cinnamon and lacks white in primaries; secondaries largely white (resembling *senegalensis*). Poorly defined *waibeli* has colour and wing pattern intermediate between *senegalensis* and *africana* (but still has white across primaries). *U. e. major* as nominate but slightly duller, with less white in secondaries, narrower tail band and more streaked belly; bill longer and thicker at base. **Adult female** Similar but slightly duller and smaller; *africana* rather paler, with less white in secondaries (like nominate). **Juvenile** Similar to adult female but duller, with shorter crest and bill. **VOICE** A low and soft, but far-carrying, trisyllabic *hoop-oop-oop*. [CD3:32] **HABITS** Singly or in pairs in wooded savanna and dry scrub. Forages by walking briskly about, vigorously probing ground with long bill. Highly conspicuous in flight when pattern and irregular flapping of broad wings suggest huge butterfly. Surprisingly easily overlooked when on the ground. Crest not often raised, except on landing and when alarmed.

STATUS AND DISTRIBUTION Uncommon to not uncommon in broad Sahel and savanna belt, and in savanna south of forest zone in Gabon and Congo. Nominate is not uncommon Palearctic passage migrant and winter visitor south of Sahara, from Mauritania to Sierra Leone east to Cameroon and Chad, where largely resident *senegalensis* (which moves to southern limit of its range in dry season) also occurs. Vagrant, Cape Verde Is (nominate); rare, coastal Liberia (*senegalensis*). *U. e. waibeli* occupies zone south of breeding range of *senegalensis* in Cameroon (Adamawa Plateau), S Chad and CAR. *U. e. africana* in Gabon (where nominate has also been recorded; scarce) and Congo (rare). *U. e. major* in NE Chad (Ennedi). NOTE African breeding races (or ssp. *africana* only) occasionally treated as separate species (African Hoopoe), but morphological, behavioural and vocal similarities do not suggest specific distinctness.

HORNBILLS
Family BUCEROTIDAE (World: *c.* 57 species; Africa: 23–24; WA: 13)

Medium-sized to very large birds, with large decurved bills surmounted by a horny casque of variable size. Most are black and white, with a variable amount of bare skin around the eye and on the malar area or throat; tail long. All have long eyelashes. Sexes dissimilar, separable by shape and pattern of casque, bare-part and/or plumage coloration; males usually larger. Juveniles as adults but with smaller bills and no casque. Most species arboreal, frugivorous and insectivorous. Fly with neck outstretched. Wingbeats distinctly audible, especially in large *Bycanistes* and *Ceratogymna*, in part because lack of underwing-coverts permits air to rush between bases of remiges. Nest in tree holes; entrance sealed from within by female (except in ground hornbills), leaving a narrow slit through which she is fed by her mate throughout incubation and until the young are partially grown. Female breaks out as young grow; latter re-seal entrance. Most species vocal, with distinctive calls.

Two subfamilies (sometimes considered families): ground hornbills *Bucorvinae* and true hornbills *Bucerotinae*. In Africa, true hornbills contain four genera: *Tockus* (small to medium-sized with small casque usually limited to low ridge), *Bycanistes* (medium-sized to very large with, in most species, prominent casque), *Ceratogymna* (very large with prominent casque and throat wattles) and monotypic *Tropicranus* (medium-sized with very long tail, distinctive voice and behaviour).

ABYSSINIAN GROUND HORNBILL *Bucorvus abyssinicus*   Plate 72 (7)
Bucorve d'Abyssinie

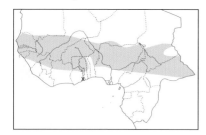

Other names F: Calao terrestre d'Abyssinie, Grand Calao d'Abyssinie
L 90–110 cm (35–43"). Unmistakable. *Huge, black, terrestrial hornbill,* revealing all-white primaries in flight. Adult male Bare skin around eye blue; inflatable bare skin on throat and neck red with blue on upper throat. Bill long and black with short, high, open-ended casque and reddish patch at base of upper mandible. Adult female Smaller, with bare skin entirely dark blue. Juvenile Dark sooty-brown with smaller bill, undeveloped casque and grey-blue bare skin. Immature Gradually as adult. In male, bare skin, incl. around eye, becomes dull pinkish. Adult coloration acquired in *c.* 3 years. VOICE A deep, far-carrying, booming *uu-uh, uh-uh-uh,* uttered from ground or perch, esp. at dawn. [CD8:10a] HABITS Usually in pairs or small family groups, in open grassland. Walks in search of prey; only occasionally flies (e.g. when disturbed). STATUS AND DISTRIBUTION Uncommon or rare to locally not uncommon resident in savanna belt, from S Mauritania to Sierra Leone east through NE Ivory Coast and Burkina Faso to Chad and CAR. NOTE Placed, with extralimital Southern Ground Hornbill *B. leadbeateri* (=*cafer*), in separate family Bucorvidae by some authors (Sibley & Monroe 1990, Kemp 1995).

WHITE-CRESTED HORNBILL *Tropicranus albocristatus*   Plate 72 (6)
Calao à huppe blanche

Other name E: Long-tailed Hornbill
L 70–80 cm (27.5–31.5"), incl. tail of *c.* 45 cm (18"). Slender hornbill with diagnostic *white crest* and *very long, strongly graduated, white-tipped tail.* Adult male *albocristatus* Black with white crown and head-sides; white feathers elongated, with black shafts and tips giving greyish appearance. Bill blackish with long, low casque and creamy patch on sides of upper mandible. Eye creamy with blue orbital ring. Bare pinkish throat patch. *T. a. macrourus* similar but white of head extends to throat and neck. Eastern *cassini* has white restricted to crown and white tips to flight feathers, greater wing-coverts and scapulars. Adult female Smaller, with shorter casque. Juvenile Similar

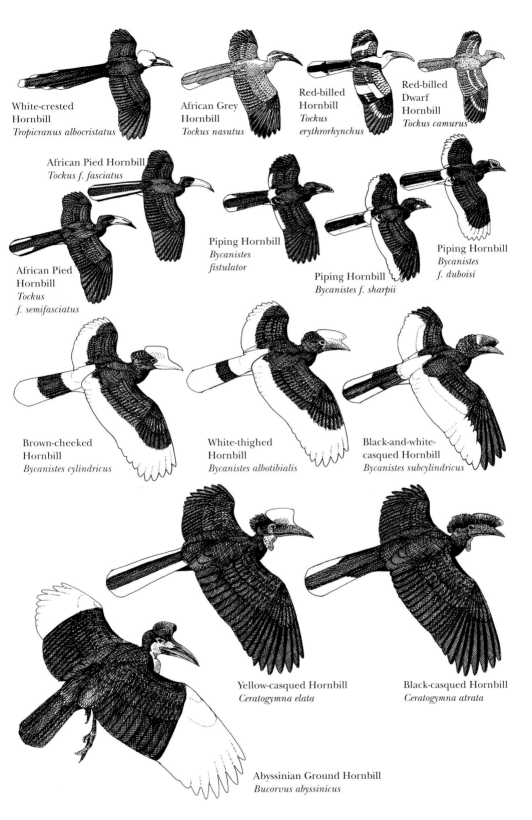

White-crested
Hornbill
Tropicranus albocristatus

African Grey
Hornbill
Tockus nasutus

Red-billed
Hornbill
*Tockus
erythrorhynchus*

Red-billed
Dwarf
Hornbill
Tockus camurus

African Pied Hornbill
Tockus f. fasciatus

African Pied
Hornbill
*Tockus
f. semifasciatus*

Piping Hornbill
*Bycanistes
fistulator*

Piping Hornbill
Bycanistes f. sharpii

Piping Hornbill
*Bycanistes
f. duboisi*

Brown-cheeked
Hornbill
Bycanistes cylindricus

White-thighed
Hornbill
Bycanistes albotibialis

Black-and-white-
casqued Hornbill
Bycanistes subcylindricus

Yellow-casqued Hornbill
Ceratogymna elata

Black-casqued Hornbill
Ceratogymna atrata

Abyssinian Ground Hornbill
Bucorvus abyssinicus

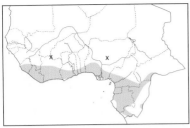

to adult female but smaller, with darkish bill and no casque; eye pale blue. **VOICE** Unlike other hornbills. A series of rather soft, plaintive sounds culminating in a distinctive, drawn-out wail *ooooooooaah!*; also harsh clucks. [CD8:11a] **HABITS** In pairs or small parties in mature and secondary forest, also relict forest patches. Forages in dense foliage. Joins mixed-species flocks and associates with monkeys for the small animals they disturb. **SIMILAR SPECIES** Beware confusion with fleetingly glimpsed Long-tailed Hawk, which also has long, white-tipped tail and occurs in same habitat. **STATUS AND DISTRIBUTION** Not uncommon to locally common forest resident, from Sierra Leone and SE Guinea to W Ivory Coast (nominate); E Ivory Coast to Benin (*macrourus*); and Nigeria to S CAR and Congo (*cassini*). **NOTE** Placed in *Tockus* by BoA, but we prefer to retain this very distinctive hornbill in its traditional, monotypic genus *Tropicranus* (cf. Dowsett & Dowsett-Lemaire 1993, *HBW*).

BLACK DWARF HORNBILL *Tockus hartlaubi* Plate 71 (2)
Calao de Hartlaub

Other name F: Calao pygmée à bec noir
L 35–39 cm (13.75–15.25"). Small, black forest hornbill with *broad greyish-white supercilium* from bill to nape. Tail graduated; all feathers tipped white except central pair. Underparts grey, becoming whitish on belly. White spot in centre of outer primary (not always visible on perched bird). **Adult male** *hartlaubi* Bill black with distinct dark red tip; casque limited to low ridge. Bare blue-pink throat patch. *T. h. granti* has white tips to wing-coverts and inner secondaries; red on bill more extensive. **Adult female** Smaller; bill all blackish or dark dusky-red with almost no casque ridge; throat patch greyish and inconspicuous. **Juvenile** Similar to adult female. **VOICE** Rather silent. A fairly loud *ee-ep ee-ep* (also uttered singly or in series of 3), frequently repeated. Also a series of rather soft notes *kwu-wu-wu-wu-...* ending in rising *kwee-kukwee-kukWEE*. [CD8:11b] **HABITS** Singly, in pairs or small family parties in lowland and gallery forest. Favours liana-rich areas at mid-level and lower canopy. Often perches silently for long periods. Makes short sallies to take insects from leaves or in mid-air. Follows monkeys for the insects they disturb. Unobtrusive. **STATUS AND DISTRIBUTION** Uncommon forest resident. Nominate from S Sierra Leone and SE Guinea to Togo and from S Nigeria and Cameroon to Congo. *T. h. granti* in SW CAR and (probably) Congo.

RED-BILLED DWARF HORNBILL *Tockus camurus* Plate 71 (1)
Calao pygmée

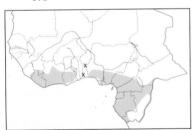

Other name F: Calao pygmée à bec rouge
L 34–39 cm (13.25–15.25"). Small forest hornbill with *red bill*. Head, upperparts, tail, throat and upper breast *rufous-brown*. Wing-coverts tipped white, forming 2 bars; flight feathers edged pale rufous-brown to whitish. Tail feathers tipped white except central pair. Lower breast to undertail-coverts white. **Adult male** Bill red; casque limited to low ridge. Eye whitish or cream, with brownish orbital ring. **Adult female** Smaller, with red bill tipped black and almost no casque ridge. **Juvenile** Similar to adult female but with paler, more orange bill. **VOICE** A distinctive, far-carrying series of slightly mournful, melancholy calls, first rising, then falling, *koo-kio-kio-kio-kio-kio-kio-kio*. [CD8:12a] **HABITS** Usually in small, vocal groups in lowland forest, gallery forest and relict forest patches. At mid-levels and lower canopy. Gleans insects from leaves and branches. Often joins mixed-species flocks. Unobtrusive when not calling. **STATUS AND DISTRIBUTION** Not uncommon forest resident, from S Sierra Leone and SE Guinea to Ghana, and from Benin to S CAR and Congo.

RED-BILLED HORNBILL *Tockus erythrorhynchus* Plate 71 (7)
Calao à bec rouge

Other name F: Petit Calao à bec rouge
L 40–48 cm (15.75–19.0"). Medium-small, black-and-white savanna hornbill with *slender red bill*. **Adult male** Crown and nape dark grey; head-sides and neck white. Upperparts blackish-brown with white streak on central back and *white spots on wing-coverts*. Tail black with white outer feathers; others broadly tipped white, except 2 central pairs. Underparts white. Bill orange-red with basal half of lower mandible black; no casque. Eye brown with pale yellowish to pink (**nominate**) or black (**kempi**) orbital ring. **Adult female** Smaller, with smaller bill often lacking black. **Juvenile** Bill

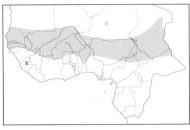

smaller and duller, yellowish-brown. **VOICE** A long series of clucking notes, increasing in tempo and volume (as if getting more excited) *uk uk uk uk-uk-uk-UK-UK-UK-UK-UK-uhWUK uhWUK UK-UK-uhWUK...* Several birds may join in excited clucking chorus. [CD8:12b] **HABITS** In pairs or small parties in open savanna woodland and thorn scrub. May congregate in large feeding flocks outside breeding season. Forages mostly on ground. **STATUS AND DISTRIBUTION** Common to not uncommon resident in savanna belt, from S Mauritania to Guinea east through Mali and N Ivory Coast to Chad and N CAR. Vagrant, Sierra Leone. *T. e. kempi* from W Mali (east to inner Niger delta) west; nominate in rest of range. **NOTE** Ssp. *kempi* recently described by Tréca & Erard (2000).

AFRICAN PIED HORNBILL *Tockus fasciatus*   Plate 71 (3)
Calao longibande

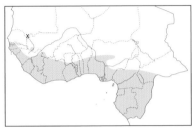

Other names E: Black-and-white-tailed Hornbill, Allied Hornbill
L 48–55 cm (19.0–21.5"). Medium-sized, slender black-and-white hornbill with cream-yellow bill. **Adult male** *semifasciatus* Black, with lower breast to undertail-coverts white. Outer tail feathers 3 and 4 broadly tipped white. Bill cream-yellow with black tip extending as line on cutting edges and end of low casque ridge. Dark brown eye surrounded by dark blue skin. **Nominate** has white in tail much more extensive, forming 2 white lines (feather pairs 3 and 4 almost entirely white). Bill has dark red tip extending on underside of lower mandible. **Adult female** Smaller, with slightly smaller bill. Nominate has bill often darker tipped; throat patch pinkish. **Juvenile** Bill smaller and entirely cream-coloured. In both races white in tail restricted to tips of feather pairs 3–4. **VOICE** Vocal. A series of high-pitched, scolding whistles varying in pitch and tempo *pyi-pyi-pyi-pyi-PYI-PYI-PYI-...*; in display ending with drawn-out *pieeu.* [CD8:13a] **HABITS** In pairs or small family parties in forest and various adjacent wooded habitats. May congregate in large feeding flocks outside breeding season. Arboreal. Flight flapping and buoyant on broad, rounded wings. **SIMILAR SPECIES** Similar-sized Piping Hornbill distinguished by darker, shorter and heavier bill (with obvious casque in eastern races), and white in wings; flight fast and slightly undulating. **STATUS AND DISTRIBUTION** Common forest and woodland resident, from Senegambia to W Nigeria (*semifasciatus*), and E Nigeria to CAR and Congo (nominate). Intermediates recorded, Nigeria (Owerri). Rare, extreme S Mauritania.

CROWNED HORNBILL *Tockus alboterminatus*   Plate 71 (5)
Calao couronné

L 48–55 cm (19.0–21.5"). **Adult male** Brownish-black with lower breast to undertail-coverts white; white streaks from bill to nape form long broad stripe. Tail feathers broadly tipped white, except central and outermost pairs. *Bill red* with yellow band at base and low casque ridge. Eye yellow with black orbital ring; bare throat patch black. **Adult female** Smaller, with slightly smaller bill; throat skin greenish. **Juvenile** Bill smaller, duller and without ridge; wing-coverts edged pale; throat skin dull yellowish; eye grey. **Immature** Adult bare-part coloration acquired after 8–10 weeks. **VOICE** A series of shrill whistles varying in pitch and tempo, recalling Pied Hornbill. [CD8:13b] **HABITS** In pairs or small family parties in coastal forest patches. Arboreal. Flight action as Pied Hornbill, flapping and buoyant on broad, rounded wings. **SIMILAR SPECIES** Compare Pied Hornbill. **STATUS AND DISTRIBUTION** Unrecorded in our region, but extends north from Angola along coastal Congo-Kinshasa and Cabinda, and may occur in coastal Congo.

AFRICAN GREY HORNBILL *Tockus nasutus*   Plate 71 (6)
Calao à bec noir

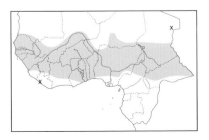

Other names F: Petit Calao à bec noir, Calao nasique
L 45–51 cm (17.75–20.0"). *T. n. nasutus.* Medium-sized, dull greyish savanna hornbill. **Adult male** Head and neck dark grey with prominent white supercilium from bill to nape; upperparts and tail *brown-grey.* Wing feathers edged pale buff-grey. Tail feathers broadly tipped white, except central pair. Throat and upper breast brown-grey; rest of underparts white. *Bill black with triangular creamy patch at base of upper mandible,* low, short casque and narrow grooves at base of lower mandible. Bare skin around eye dark grey. **Adult female** Bill smaller with dark red tip and rest of upper mandible pale yellow.

Juvenile Similar to adult but with smaller, entirely dusky-grey bill. **VOICE** A rather melancholy *pee-o*, often uttered in series, and a series of far-carrying, rhythmic, piping notes *pee-pee-pee-pee-PEE-PEE PEE-pyew PEE-pyew PEE-PEE-pyew-pee...* [CD8:15] **HABITS** In pairs or small family parties in savanna woodland. Arboreal. Flight undulating and flapping with glides on broad, rounded wings. **SIMILAR SPECIES** Red-billed Hornbill has much whiter and more contrasting plumage, and red bill. **STATUS AND DISTRIBUTION** Common resident and intra-African migrant in broad savanna belt, from S Mauritania to Sierra Leone east to Chad and CAR. Reaches coast from Ghana to W Nigeria. Vagrant, Liberia. Moves north with rains, after breeding.

PIPING HORNBILL *Bycanistes fistulator*   Plate 71 (4)
Calao siffleur

Other names E: White-tailed Hornbill, Laughing Hornbill
L 50–60 cm (19.5–23.5"). Medium-sized, compact black-and-white forest hornbill. **Adult male** *fistulator* Head to upper breast, upperparts and tail black; uppertail-coverts white. Tail feathers broadly tipped white, except central pair. Wings black with broad white tips to central secondaries. Underparts white. Bill dusky and heavily grooved, with horn-coloured base and tip. Eye brown with dark blue orbital ring. *B. f. sharpii* has white extending onto inner primaries; outer tail feathers all white. Bill has raised casque ridge. *B. f. duboisi* has white extending even further, onto outer primaries; central tail feathers narrowly tipped white (often abraded); casque well developed. **Adult female** Smaller, with smaller bill. **Juvenile** Bill smaller and all dusky. **VOICE** Vocal. A distinctive, loud, harsh, nasal laughing. Also a shrill piping *peep-peep-peep.* [CD8:17] **HABITS** In pairs or family trios in canopy of lowland and gallery forest. May congregate in feeding flocks outside breeding season. Arboreal. Flight slightly undulating with fast, shallow wingbeats interspersed by glides. Wings make rushing sound in flight. **SIMILAR SPECIES** Similar-sized Pied Hornbill distinguished by more slender, cream-coloured bill without obvious casque, broad, entirely black wings, and flapping flight. Brown-cheeked, White-thighed and Black-and-white-casqued Hornbills much larger, with much larger casques, more white in wings and tail (latter also differently patterned), and less white on underparts. **STATUS AND DISTRIBUTION** Not uncommon to common forest resident, from S Senegal to Liberia east to W Nigeria (nominate); E Nigeria to Congo (*sharpii*); and SC Cameroon to CAR (*duboisi*). Also S Mali (1 record, Kangaba). Intermediates occur in contact zones; birds with characters of *sharpii* reported in Ghana and Togo. **NOTE** Placed in genus *Ceratogymna* in *BoA*, but subsequently in traditional *Bycanistes* by same author (*HBW*).

BLACK-AND-WHITE-CASQUED HORNBILL *Bycanistes subcylindricus*   Plate 72 (5)
Calao à joues grises

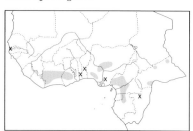

Other name E: Grey-cheeked Hornbill
L 65–78 cm (25.5–30.75"). Large black-and-white forest hornbill with *blackish bill and two-toned casque.* **Adult male** *subcylindricus* Head and upperparts black; head feathers tipped grey; rump and uppertail-coverts white. *Central tail feathers black*, rest white with broad black band on centre. Wings black with black-based white inner primaries and secondaries; greater coverts broadly tipped white. Underparts to upper belly black; lower belly, thighs and undertail-coverts white. Bill and high casque blackish with base of casque pale horn. Bare orbital skin dull pinkish. *B. s. subquadratus* has horn-coloured area extending to about middle of casque; much individual variation. **Adult female** Smaller, with smaller bill and casque. Bare facial skin becomes red in breeding season. **Juvenile** Similar to adult but bill smaller, lacking casque. **VOICE** Calls loud and nasal, including an abrupt *keh!* or *heh!* and a slow series *hah-hah-hah-hah-hah-...* Calls of subspecies may differ; more research required. [CD8:18] **HABITS** In pairs or small family parties in canopy of mature lowland forest. Favours edges and large clearings; also in adjacent woodland. Several may gather at fruiting trees, often with other large hornbills. Flight strong and direct, with loud, swishing wingbeats interspersed by glides. **SIMILAR SPECIES** Brown-cheeked and White-thighed Hornbills have all-pale and differently shaped casque, pale or dusky bill, different tail pattern without all-black central feathers, and less white on flight feathers. Piping Hornbill is much smaller, with smaller, differently shaped casque, less white in wings and tail (latter also differently patterned), and much more white on underparts. **STATUS AND DISTRIBUTION** Locally common to rare forest resident, from Sierra Leone to W Nigeria (nominate), and from E Nigeria (east of Niger R.) to CAR and N Congo (*subquadratus*). **NOTE** Placed in genus *Ceratogymna* in *BoA*, but subsequently in traditional *Bycanistes* by same author (*HBW*).

BROWN-CHEEKED HORNBILL *Bycanistes cylindricus*   Plate 72 (1)
Calao à joues brunes

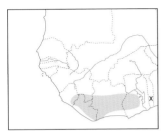

L 65–77 cm (25.5–30.5"). Large, stout black-and-white forest hornbill. **Adult male** Head and upperparts black; lower rump and uppertail-coverts white. *Tail white with broad black band on centre.* Wings black with outer half of flight feathers white. Underparts to upper belly and upper thighs black; lower belly and undertail-coverts white. Head-sides and upper throat dark brownish (hard to see in field). *Bill pale horn-coloured with high, grooved casque.* Orbital ring red. **Adult female** Smaller, with smaller bill and casque. Orbital ring pinkish. **Juvenile** Similar to adult but bill smaller, lacking casque. **VOICE** Calls include a high-pitched *pfeet!* or *khlee!* and harsh notes, uttered singly or in series at variable speed. Distinctly higher pitched and less raucous than calls of White-thighed Hornbill. [CD8:20] **HABITS** In pairs or family trios in canopy of mature lowland forest. May gather in small groups at fruiting trees, often with other hornbills. Flight strong and direct, with loud, swishing wingbeats interspersed by glides. **SIMILAR SPECIES** White-thighed Hornbill (q.v.) similar but no known overlap. Black-and-white-casqued Hornbill has mainly dark bill with differently shaped casque, different tail pattern with black central feathers, and more white on flight feathers, except all-black outer primaries. Piping Hornbill much smaller, with smaller, differently shaped and coloured casque, less white in wings and tail (latter also differently patterned), and more white on underparts. **STATUS AND DISTRIBUTION** Uncommon to locally not uncommon forest resident, from Sierra Leone and SE Guinea to Ghana. Record from Togo probably refers to this species (rather than to White-thighed Hornbill). *TBW* category: Near Threatened. **NOTE** Placed in genus *Ceratogymna* in *BoA*, but subsequently in traditional *Bycanistes* by same author (*HBW*).

WHITE-THIGHED HORNBILL *Bycanistes albotibialis*   Plate 72 (2)
Calao à cuisses blanches

L 65–77 cm (25.5–30.5"). Resembles Brown-cheeked Hornbill. **Adult male** Differs from Brown-cheeked in all-white thighs and *narrower black band on tail.* No brown on head-sides. *Bill largely dusky with differently shaped, longer, horn-coloured casque ending in point.* Orbital ring pale yellow. **Adult female** Smaller, with smaller bill and casque. Orbital ring pinkish. **Juvenile** Similar to adult but bill smaller, lacking casque. **VOICE** A loud, harsh, barking *gak!* or *rrhoak!* uttered singly or in slow, descending series. Also an abrupt *kekh!* [CD8:19] **HABITS** As Brown-cheeked Hornbill. **SIMILAR SPECIES** See Brown-cheeked Hornbill. **STATUS AND DISTRIBUTION** Uncommon to locally common forest resident, from S Nigeria to SW CAR and Congo. Records in Benin probably refer to this species (or Brown-cheeked Hornbill). **NOTE** Often considered conspecific with Brown-cheeked Hornbill (e.g. *BoA*), but morphological and vocal differences suggest they may be better treated as separate species.

BLACK-CASQUED HORNBILL *Ceratogymna atrata*   Plate 72 (4)
Calao à casque noir

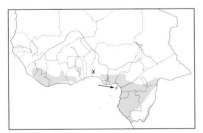

Other names E: Black-casqued Wattled Hornbill; F: Grand Calao à casque noir
L 70–90 cm (28.0–35.5"). *Very large, black* forest hornbill with *black bill.* **Adult male** All black except *broad white tips to all but central pair of tail feathers.* Shaggy crest. Bill and massive cylindrical casque blackish. Bare skin around eye and inflatable bare throat wattles light blue. Eye red. **Adult female** Smaller, with rufous head and neck, and smaller bill and casque. **Juvenile** Similar to adult female but bill smaller, without casque; no wattles. Young male acquires all-black head at end of first year. **VOICE** A far-carrying, resonant, strikingly plaintive trumpeting, often uttered in flight. Female calls are much higher pitched than male's. More raucous and plaintive than calls of Yellow-casqued Hornbill. [CD8:21] **HABITS** In pairs or small family parties in canopy of mature lowland forest. Several may gather at fruiting trees, often with other large hornbills. Flight strong and direct, with loud, swishing wingbeats interspersed by glides. **SIMILAR SPECIES** Male Yellow-casqued Hornbill has differently shaped, cream-coloured casque and different tail pattern with all-white outer feathers. In addition, female Yellow-casqued also distinguished from female Black-casqued by creamy-horn bill, paler rufous head and white bases to neck feathers. Other large hornbills have more white in plumage. **STATUS AND DISTRIBUTION**

Uncommon or rare to locally common forest resident, from Sierra Leone and SE Guinea to Ghana, and from S Nigeria to S CAR and Congo; also Bioko (where the only hornbill). Distribution in Upper Guinea now patchy.

YELLOW-CASQUED HORNBILL *Ceratogymna elata*   Plate 72 (3)
Calao à casque jaune

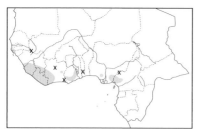

Other names E: Yellow-casqued Wattled Hornbill; F: Grand Calao à casque jaune
L 70–90 cm (28.0–35.5"). *Very large, black* forest hornbill with high *creamy casque*. **Adult male** All black except *white tail with black central feather pair*. Shaggy crest; neck feathers have white bases and brown tips. Bill and lower part of casque blackish; upper part of casque creamy-horn. Bare skin around eye and inflatable bare throat wattles pale blue with line of dark brown feathers on central throat. Eye red. **Adult female** Smaller, with rufous head and neck; neck feathers with white bases; bill and casque smaller and all creamy-horn.

Juvenile Similar to adult female but bill smaller, without casque; no wattles. Young male presumably acquires all-black head at end of first year. **VOICE** A far-carrying, resonant, nasal trumpeting, often uttered in flight. Less raucous and plaintive, more fluting and slower than calls of Black-casqued Hornbill. [CD8:22] **HABITS** In pairs or small family parties in canopy of mature lowland forest; often at edges. Penetrates gallery forest and forest patches in savanna. Several may gather at fruiting trees, often with other large hornbills. Flight strong and direct, with loud, swishing wingbeats interspersed by glides. **SIMILAR SPECIES** Male Black-casqued Hornbill has differently shaped, blackish casque and different tail pattern with white restricted to distal third of all but central feather pair. In addition, female Black-casqued distinguished from female Yellow-casqued by blackish bill, darker rufous head and lack of white bases to neck feathers. Other large hornbills have more white in plumage. **STATUS AND DISTRIBUTION** Rare to uncommon and local forest resident, from SW Senegal to Togo, and from SW Nigeria to W Cameroon. Distribution now patchy. *TBW* category: Near Threatened.

BARBETS
Family CAPITONIDAE (World: *c.* 83 species; Africa: 41–42; WA: 21)

Small to medium-sized stocky birds, with large heads and heavy bills. Zygodactyl. Sexes generally similar. Flight usually direct. Most are arboreal, mainly frugivorous and nest in self-excavated tree holes. Their calls are a characteristic sound in forest and savanna and typically consist of a single note monotonously repeated. Tinkerbirds derive their English name from the supposed likeness of their calls to the sound of a hammer on an anvil. Non-migratory.

GREY-THROATED BARBET *Gymnobucco bonapartei*   Plate 74 (4)
Barbican à gorge grise

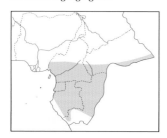

Other name F: Barbican de Bonaparte
L 16.5–17.5 (6.5–7.0"). *G. b. bonapartei.* **Adult** Dark, brown barbet with *dark greyish head and throat* and short but conspicuous brush-like brownish nasal tufts. *Bill black.* Eye brown to red. **Juvenile** More uniform chocolate-brown; nasal tufts absent or very short; bill yellowish with dusky tip; eye dark. **VOICE** A hoarse *whewp*, uttered singly or in loose series, also accelerating into rattling *kripipipipppp...*; also a nasal *nyaaa* and a buzzy *spszsz.* [CD8:23] **HABITS** Mostly in pairs or small parties in various types of lowland and lower montane forest and its edge. Less noisy than other *Gymnobucco.* Often perches quietly in canopy. **SIMILAR SPECIES** Naked-faced and Bristle-nosed Barbets have pale bills and bare faces; Bristle-nosed also has longer nasal tufts. **STATUS AND DISTRIBUTION** Locally common in forest zone, from Cameroon to Congo and S CAR. **NOTE** Eastern *cinereiceps*, with yellowish eyes, does not appear to occur in CAR (cf. Chapin 1939, Schouteden 1963; *contra BoA*).

SLADEN'S BARBET *Gymnobucco sladeni*   Plate 74 (3)
Barbican de Sladen

L 17–18 cm (6.75–7.0"). **Adult** Dark brown with bare blackish face, *blackish bill, conspicuous buff to brownish-buff* *nasal tufts*, and *grey throat and neck-sides*. Some bristles at chin and gape. Underpart feathers indistinctly fringed yellowish. Flight feathers dark brown narrowly edged yellowish on secondaries and tertials. *Eye reddish.* **Juvenile** Face feathered, nasal tufts small and dark, bill yellowish-horn at base and feathers of underparts with more yellowish fringes. **VOICE & HABITS** Similar to other *Gymnobucco* barbets. **SIMILAR SPECIES** Grey-throated Barbet has short dark nasal tufts and greyish head with feathered face. **STATUS AND DISTRIBUTION** Recorded once in lowland forest, S CAR (south of Bangui).

BRISTLE-NOSED BARBET *Gymnobucco peli*   Plate 74 (2)
Barbican à narines emplumées

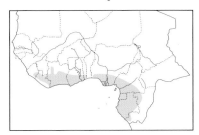

L 17–18 cm (6.75–7.0"). **Adult** Brown with bare blackish face and *conspicuous buff to brownish-buff nasal tufts*. Some bristles at chin and gape. Rump and uppertail-coverts paler; belly feathers fringed yellowish. Flight feathers dark brown narrowly edged yellowish on secondaries and tertials. *Bill yellowish to pale brown.* **Juvenile** Head feathered; nasal tufts small or absent; bill yellowish-horn or dusky. **VOICE** A hard *KYEW!*, shorter and higher pitched than Naked-faced Barbet. Noisy, rattling *kreepipippppp...* in social interaction. Also a rhythmic, continuous series of low-pitched, hollow *hoops*. [CD8:24] **HABITS** Similar to Naked-faced Barbet (q.v.). **SIMILAR SPECIES** Naked-faced Barbet very similar but has shorter, brown nasal tufts. Grey-throated Barbet has short dark nasal tufts, dark bill, greyish head and feathered face; compare distributions. **STATUS AND DISTRIBUTION** Uncommon to locally common in lowland forest, from Sierra Leone to Congo. Less common than sympatric Naked-faced Barbet.

NAKED-FACED BARBET *Gymnobucco calvus*   Plate 74 (1)
Barbican chauve

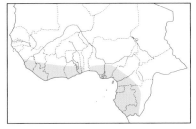

L 17–18 cm (6.75–7.0"). **Adult** *calvus* Brown with bare, blackish face, brownish feather tufts on chin and *sparse short bristles at gape and nostrils*. Bill dull yellowish, reddish-horn or dusky-brown. *G. c. congicus* has slightly paler grey throat. **Juvenile** Head more feathered; tufts shorter; bill with dark blotches. **VOICE** A sharp, explosive *KYEW!*, uttered singly or in loose series. In social interaction a rattling and rasping *kirrrrrr...* and *kreepipipipppp...* Also a series of 7–10 accelerating *hoops*. [CD8:25] **HABITS** Noisy and gregarious, in lowland and lower montane forest, typically in second growth. Nests and roosts colonially in large dead trees, sometimes with other *Gymnobucco* species. **SIMILAR SPECIES** Less common, sympatric Bristle-nosed Barbet very similar but has conspicuous pale nasal tufts, slightly smaller, paler bill, underparts suffused with yellowish-green and secondaries narrowly edged yellowish. Grey-throated Barbet has short dark nasal tufts, dark bill, greyish head and no bare face; compare distributions. **STATUS AND DISTRIBUTION** Widespread and locally common in forest zone, from SW Guinea to Gabon (nominate) and Congo (*congicus*).

SPECKLED TINKERBIRD *Pogoniulus scolopaceus*   Plate 73 (2)
Barbion grivelé

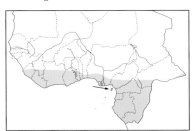

Other name F: Petit Barbu grivelé
L 13 cm (4.5"). Small, plump and rather cryptically coloured barbet with contrasting *pale eye*. **Adult** *scolopaceus* Head and upperparts largely dark brown with yellow fringes producing scaly appearance; rump yellower. Throat whitish; rest of underparts dirty yellowish with dusky mottling. Bill black. Eastern *flavisquamatus* has more yellow speckling on head and stronger marked underparts. In island race *stellatus* upperparts more olive-green and whitish with duller markings; underparts paler. **Juvenile** Base of lower mandible yellowish. **VOICE** Usual call a series of *kwip* notes, first a single note, repeated a few times, then a doubled *kwip-ip*, also repeated (reminiscent of Common Quail), then tripled, gradually 4–5 rapid notes. Also a variety of other calls, always delivered in rhythmic series: *poop-poop-poop-...* resembling other tinkerbirds; an insect- or toad-like *kwibibibbbbbt*; trills *kukkrrrrr...*, and a hoarse, high-pitched *hyep*. [CD8:28] **HABITS** Singly or in pairs in forest (up to 1800 m), clearings and bush. Occasionally joins mixed-species flocks. Usually quite tame and foraging

lower than most other barbets. **STATUS AND DISTRIBUTION** Common resident in forest zone, from Sierra Leone and SE Guinea to SE Nigeria (nominate), east to SE CAR and Congo (*flavisquamatus*); also Bioko (*stellatus*).

WESTERN GREEN TINKERBIRD *Pogoniulus coryphaeus*   Plate 73 (4)
Barbion montagnard

Other name F: Petit Barbu montagnard
L 9 cm (3.5"). *P. c. coryphaeus.* **Adult** Very small barbet with conspicuous *yellow stripe from crown to rump* contrasting with black upperparts. White moustachial stripe; *underparts wholly dusky-grey.* Wing feathers edged yellow, forming bar on median coverts. Bill blackish. **Juvenile** Duller above, paler below, esp. on belly; base of lower mandible yellowish. **VOICE** Rather metallic *kwip* or *pwip* in descending series at varying speed. Also a trill *kirrrrik.* Vocalisations identical to those of E African counterpart, Moustached Green Tinkerbird *P. leucomystax.* [CD8:27] **HABITS** Restricted to forest and forest patches in montane areas. Usually forages singly in middle and higher strata of fruiting trees, hanging tit-like on branches. Inconspicuous and easily overlooked when not calling. Apparently more vocal in fruiting season of e.g. mistletoes. **SIMILAR SPECIES** Combination of yellow stripe on upperparts, white moustachial stripe and dark grey underparts is unique among tinkerbirds. **STATUS AND DISTRIBUTION** Locally common forest resident in highlands of SE Nigeria (Obudu and Mambilla Plateaux) and W Cameroon.

RED-RUMPED TINKERBIRD *Pogoniulus atroflavus*   Plate 73 (7)
Barbion à croupion rouge

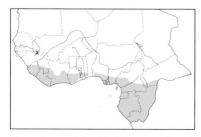

Other name F: Petit Barbu à croupion rouge
L 13 cm (5"). The only tinkerbird with *red rump.* **Adult** Head, upperparts and tail glossy black; *head-sides have 3 narrow yellow stripes* (above eye, below eye and on moustachial area); rump and uppertail-coverts bright red; flight feathers and greater coverts edged yellow; yellow tips to median coverts form wingbar. Underparts yellow, washed olive-grey from breast to undertail-coverts. Bill black. **Juvenile** Duller; crown and back feathers tipped olive; underparts paler; bill pale at base. **VOICE** A rhythmic series of *poop* calls (occasionally distorted into a nasal *klunk*), resembling those of Yellow-rumped Tinkerbird, but slower, lower pitched (*c.* 1 per second) and in much longer series. Also a series of trills *kirr kirr kirr...* or *krukukkk...* [CD8:29] **HABITS** Singly in lowland forest and clearings, occurring at all levels, but usually high. **SIMILAR SPECIES** Yellow-rumped and Yellow-throated Tinkerbirds have yellow rumps, 2 (not 3) well-marked (not narrow) stripes on head-sides, and are slightly smaller. **STATUS AND DISTRIBUTION** Uncommon to not uncommon forest resident, in S Senegal and from SW Guinea to SE CAR and Congo; also reported from SW Mali.

YELLOW-THROATED TINKERBIRD *Pogoniulus subsulphureus*   Plate 73 (5)
Barbion à gorge jaune

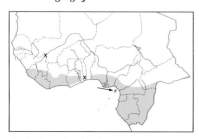

Other name F: Petit Barbu à gorge jaune
L 10 cm (4"). Glossy black above with 2 conspicuous facial stripes (above and below eye), yellow-edged wing feathers and lemon-yellow rump. **Adult** *chrysopyga* Very similar to Yellow-rumped Tinkerbird. *Facial stripes white; throat white;* breast pale greyish or yellow-olive washed grey, becoming pale yellowish on belly. Bill black. Eastern *flavimentum* has *pale yellow facial stripes* and *yellow throat.* **Nominate** as *flavimentum* but underparts greyer. **Juvenile** Duller, with some yellow-tipped back feathers and pale base to bill. **VOICE** Rhythmic series of *poop* notes similar to Yellow-rumped Tinkerbird's but higher pitched, noticeably faster (5 per second versus 3 per second) and usually in longer series. Also an even series of *krrrw* notes and an accelerating *kwip kwip-kwipkwipkwipipip.* [CD8:30] **HABITS** As Yellow-rumped, but usually in heavier forest. **SIMILAR SPECIES** Yellow-rumped Tinkerbird best distinguished from western *chrysopyga* by slower, more monotonous song, given in short series of 3–6 notes. **STATUS AND DISTRIBUTION** Common forest resident, from Sierra Leone and SE Guinea to Ghana (*chrysopyga*); S Nigeria (Togo?) to Congo and SE CAR (*flavimentum*); and Bioko (nominate). Also S Mali. **NOTE** Western race originally named *Barbatula chrysopyga;-* pyga being a Latin noun here used in appostion, it should remain unchanged. Thus the correct name is *P. s. chrysopyga* (not *chrysopygius* as in *BoA*) (R.J. Dowsett *in litt.*).

YELLOW-RUMPED TINKERBIRD *Pogoniulus bilineatus* Plate 73 (3)
Barbion à croupion jaune

Other names E: Golden-rumped/Lemon-rumped Tinkerbird; F: Petit Barbu à croupion jaune
L 10 cm (4"). **Adult** *leucolaima* Small barbet with glossy black upperparts, lemon-yellow rump and 2 conspicuous white facial stripes (above and below eye). Throat white; breast pale greyish becoming pale yellowish on belly. Flight feathers and greater coverts edged yellow; yellow-tipped coverts form 2 wingbars. Bill black. Island race *poensis* has slightly paler rump and underparts. **Juvenile** Duller, with some yellow-tipped crown and back feathers, and pale base to bill. **VOICE** One of the most characteristic sounds of forest and woodland: a monotonous, far-carrying, rhythmic series of 3–6 (sometimes more) *kok* or *poop* notes, uttered with short pause between each series and endlessly repeated, even during the heat of the day. Also a series of *krrw* notes and rattling *kkkkkk*. [CD8:31] **HABITS** Alert and inquisitive, foraging actively in canopy or mid-level of various forest and woodland types; also gardens. Singly or in pairs. Sporadically flycatches. Aggressive to other tinkerbirds. **SIMILAR SPECIES** Race *chrysopyga* of Yellow-throated Tinkerbird very similar and best distinguished by faster calls in longer series. **STATUS AND DISTRIBUTION** Common resident, from Senegambia to Liberia east to CAR and Congo (*leucolaima*); also Bioko (*poensis*).

YELLOW-FRONTED TINKERBIRD *Pogoniulus chrysoconus* Plate 73 (1)
Barbion à front jaune

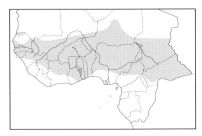

Other name F: Petit Barbu à front jaune
L 11.5 cm (4.5"). *P. c. chrysoconus.* **Adult** Small barbet with *golden-yellow forecrown*, black-and-white streaked upperparts and head-sides, and lemon-yellow underparts. Rump and uppertail-coverts pale yellow. Bill black. **Juvenile** Similar, but forecrown black. **VOICE** A monotonous series of *poops* repeated for long periods without pause. Also series of hoarse *kwèp* notes and *kwrrrr kwrrrr ...* trills in variable series. [CD8:32] **HABITS** Usually singly or in pairs in wooded grassland, savanna and bush, actively darting at insects or gleaning from foliage in warbler fashion. Creeps on tree trunk like woodpecker. Attracted to mistletoe berries. Often in mixed-species flocks. Not in forest. **SIMILAR SPECIES** None. Song of Yellow-rumped Tinkerbird distinguished by short series of 3–6 notes (Yellow-fronted continuous); notes also more 'popping'. **STATUS AND DISTRIBUTION** Common resident in broad savanna belt, from SW Mauritania to Guinea east to Chad and CAR.

YELLOW-SPOTTED BARBET *Buccanodon duchaillui* Plate 73 (6)
Barbican à taches jaunes

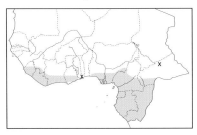

L 15 cm (6"). Distinctive. **Adult** Head black with *bright red forehead to central crown* and conspicuous *yellow stripe from eye to neck-sides.* Upperparts black spotted yellow; wing feathers edged yellow; underparts black barred yellow from lower breast to undertail-coverts and with yellow stripe on centre of lower breast and belly. Bill black. **Juvenile** Duller, *forehead and crown black*, bill yellowish tipped black. **VOICE** Main call (only in east of range?) a characteristic purring *rrurrrrrr...* (lasting 1–2 seconds), unique among barbets, uttered by adult and juvenile Also (only in west of range?) a series of 7–10 accelerating *oop* notes similar to a song of Hairy-breasted Barbet. [CD8:33] **HABITS** Usually singly, in canopy of lowland and lower montane forest. Several may gather in fruiting tree. Occasionally joins mixed-species flocks. Inconspicuous and easily overlooked. **SIMILAR SPECIES** Hairy-breasted Barbet is noticeably larger with heavier bill. **STATUS AND DISTRIBUTION** Common to uncommon forest resident, from Sierra Leone and SE Guinea to Togo (where rare) and from SW Nigeria to SW CAR and Congo; also N CAR (Bamingui area).

HAIRY-BREASTED BARBET *Tricholaema hirsuta* Plate 73 (8)
Barbican hérissé

L 16.5 cm (6.5"). Stout, medium-sized barbet. **Adult** *hirsuta* Black head and throat with bold white supercilium

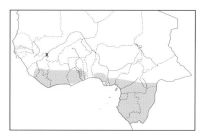

(back from eye) and moustachial stripe. Upperparts black spotted yellow; flight feathers and coverts black with yellow fringes. Underparts greenish-yellow, streaked black on breast, spotted on belly. Bill black. *T. h. ansorgii* has narrower moustachial stripe, whitish throat finely streaked black and yellow-spotted nape; upperparts more brownish-black. *T. h. flavipunctata* lacks facial stripes, has freckled head with yellow-spotted crown and whitish-speckled head-sides; upperparts brownish-black; belly dull yellow with indistinct brown markings. *T. h. angolensis* as *flavipunctata* but overall browner, esp. on belly. **Juvenile** Yellow spots on upperparts broader; black markings on underparts more bar-like, esp. on flanks; bill horn coloured with black base. **VOICE** Deep *hoops*, uttered in continuous, rhythmic series at variable speed (most often *c.* 1–2 per second, but also much faster). Also accelerating *hoop hoop-hoophoopoopoopoopoop*. [CD8:34–35] **HABITS** Singly or in pairs in lowland and lower montane forest (up to 1800 m) and clearings; occasionally gathering in numbers at fruiting trees or joining mixed-species flocks. Inconspicuous, far more often heard than seen. **SIMILAR SPECIES** Sympatric Yellow-spotted Barbet has red forehead, single yellow stripe on head-sides and is smaller. **STATUS AND DISTRIBUTION** Locally common in forest zone, in S Senegal and from Sierra Leone and SE Guinea to Ghana (*hirsuta*), through Togo and W Nigeria (where intergrading with *flavipunctata*), S Nigeria, Cameroon, N and C Gabon (*flavipunctata*), E Cameroon, CAR (*ansorgii*), S Gabon, W and C Congo (*angolensis*). One record from SW Mali.

VIEILLOT'S BARBET *Lybius vieilloti*   Plate 74 (8)
Barbican de Vieillot

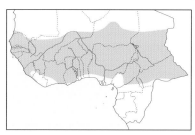

L 15 cm (6"). **Adult** *buchanani* Diagnostic combination of red face, pale yellow underparts, lower back and uppertail-coverts, and largely drab greyish-brown upperparts and tail. Red spots from throat to central belly. Bill black. *L. v. rubescens* usually darker. **Juvenile** Duller, with less red; bill paler at base and tip. **VOICE** Distinctive. Duet consisting of a rhythmic series of far-carrying and melodious *poop* calls, preceded by chattering or purring notes (slightly reminiscent of Green Wood-hoopoe) *kekkekkekkkk-urrrrr poop poop poop-eh-poop epoop poop-eh...* [CD8:38] **HABITS** In pairs or small groups in savanna, wooded grassland and thorn scrub; also cleared forest areas. Breeding behaviour poorly known (breeds in pairs or small groups?). Parasitised by Lesser Honeyguide. **STATUS AND DISTRIBUTION** Common to uncommon in Sahel and savanna zones, from S Mauritania to Sierra Leone and Liberia, east to Chad and CAR. *L. v. buchanani* in north of range, *rubescens* in south, but races poorly defined.

WHITE-HEADED BARBET *Lybius leucocephalus*   Plate 74 (5)
Barbican à tête blanche

L 15.0–16.5 cm (6.0–6.5"). *L. l. adamauae.* **Adult** The only barbet with *all-white head and breast*. Upperparts and tail brownish-black; wing-coverts spotted white; rump and uppertail-coverts white. Underparts white except flanks and belly, which are brownish-black with dull white feather tips. Bill black. **Juvenile** Head and underparts blotched dusky-brown; bill brownish-horn. **VOICE** A high-pitched *pyup* or *kyip*, uttered singly or in a series. Duet may end in *ki-ki-ki-ki-ki-...* [CD8:39] **HABITS** Singly, in pairs or small groups in riverine woodland with fig trees, open country and cultivated areas; also gardens. Often perches quietly atop tree. **STATUS AND DISTRIBUTION** Locally common to scarce resident, from NC Nigeria to Cameroon, S Chad and CAR.

BLACK-BILLED BARBET *Lybius guifsobalito*   Plate 74 (7)
Barbican guifsobalito

Other name F: Barbican à bec noir

L 15 cm (6"). **Adult** *Black with red face, throat and upper breast.* Flight feathers edged yellowish; wing-coverts edged white. Bill black. **Juvenile** Duller, with red areas more orange and less extensive. **VOICE** Duet consists of a series of *hik-kup-oot* sometimes introduced by harsh *ch-ch-ch, cha-cha-cha.* Interactive calls include *kek*, a nasal *kaw* and grating notes. [CD8:40] **HABITS** Usually in pairs or small groups in open savanna woodland. Noisy and conspicuous, often perching atop trees. Occasionally flycatches. **SIMILAR SPECIES** Malimbes are also black and red but occur in forest. **STATUS AND DISTRIBUTION** One record, N Cameroon (Waza NP, Feb 1993). Main range in E Africa.

BLACK-BACKED BARBET *Lybius minor*   Plate 74 (6)
Barbican de Levaillant

L 16.5 cm (6.5"). Easily identified by unique coloration. **Adult *minor*** Head and mantle typically grey-brown; forehead to central crown red; rest of upperparts and relatively long tail blackish. Underparts white with thighs to flanks black and central lower breast to belly pinkish. Bill whitish-horn; orbital skin pale violet. Some intergrade with extralimital (?) ***macclounii*** and have white V on blackish upperparts and white head-sides. **Juvenile** Duller; no red on head; bill darker horn. **VOICE** Calls include *kyek*, uttered singly or in series, *kik-ik-ik-ik, drrrrr* and other grating notes. Song a buzzy trill *krrrrriiiii*, slightly reminiscent of Double-toothed Barbet. [CD8:42] **HABITS** Mostly in pairs or family groups in open savanna woodland and gallery forest. Forages at all levels. Rather inconspicuous. Often perches quietly and for long periods in top of tree. **STATUS AND DISTRIBUTION** Scarce to locally common resident, S Gabon and Congo.

DOUBLE-TOOTHED BARBET *Lybius bidentatus*   Plate 74 (13)
Barbican bidenté

Other name E: Tooth-billed Barbet
L 23 cm (9"). Large black-and-red barbet with *heavy, whitish bill* and bare yellow orbital skin. **Adult male *bidentatus*** Top of head, upperparts and tail blue-black with small white patch on lower back (conspicuous in flight); *dark red band on wing-coverts*. Head-sides and underparts bright red with elongated white patch bordering black of flanks and undertail-coverts. Legs brownish. Eastern ***aequatorialis*** has wingbar narrower and whitish or pinkish. **Adult female** Similar but has some thin black streaks just above white side patch (virtually impossible to see in field). **Juvenile** Duller; throat to breast blackish-grey. **VOICE** Usually silent. A loud *KEK!* or *KZEK!* reminiscent of Green Wood-hoopoe, uttered singly or in series. Song a long purring, frog-like *errrrrrrrr* (lasting up to 5 seconds or more). Also other grating and buzzing notes. [CD8:43] **HABITS** Singly, in pairs or family parties in various types of woodland, forest edge and cleared areas. Attracted to fruiting fig trees; catches insects on the wing and by clinging to tree trunks, inspecting bark. Flicks tail when excited. **SIMILAR SPECIES** Bearded Barbet has black breast-band, no red on wing, heavier, more yellowish bill and yellowish legs; also slightly larger and has a generally more northern distribution (but some overlap). **STATUS AND DISTRIBUTION** Widespread and not uncommon resident, in Mali (north to 15°N) and from Guinea-Bissau to Sierra Leone east to Cameroon, Gabon and Congo (nominate), and in CAR (*aequatorialis*).

BEARDED BARBET *Lybius dubius*   Plate 74 (12)
Barbican à poitrine rouge

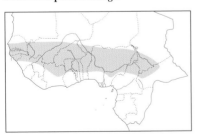

L 25 cm (10"). Large black-and-red barbet with *heavy yellowish bill* and bare yellow orbital skin. **Adult male** Top of head, upperparts and tail blue-black with small white patch on lower back (conspicuous in flight). Head-sides and underparts bright red with *black band on lower breast*, whitish flanks and black undertail-coverts. Legs yellowish. **Adult female** Similar but has small black spots on white flanks. **Juvenile** Duller, with dull black breast and dusky base to bill. **VOICE** Harsh, low, grating notes (such as *khroaa*), uttered singly or in series. [CD8:44] **HABITS** Singly, in pairs or small, conspicuous groups in open woodland, wooded grassland and cleared areas with suitable trees (esp. large fig trees). **SIMILAR SPECIES** Double-toothed Barbet has wingbar, no breast-band, paler bill and dark legs. See also Black-breasted Barbet, which it may meet southeast of L. Chad. **STATUS AND DISTRIBUTION** Locally not uncommon resident, from Senegambia and Guinea-Bissau to N Cameroon, SE Chad and C CAR.

BLACK-BREASTED BARBET *Lybius rolleti*   Plate 74 (11)
Barbican à poitrine noire

L 26.5 cm (10.5"). Largest barbet in the region. **Adult male** *Mainly black with massive ivory bill.* White patch on

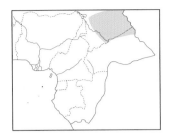

lower back. Centre of lower breast and belly bright red; flanks white. Orbital skin bluish; legs greyish. **Adult female** Similar but has fine black streaks on white flanks. **Juvenile** Duller, with smaller and smoother bill. **VOICE** Low rasping and growling notes, and sharp *kak!* reminiscent of Green Wood-hoopoe, uttered singly or in series. [CD8:45] **HABITS** Social and conspicuous, in wooded grassland and cultivated areas. Whirring, often noisy flight. **SIMILAR SPECIES** No known overlap with Double-toothed or Bearded Barbet (but see those species). **STATUS AND DISTRIBUTION** Uncommon to locally common resident, S Chad and N CAR.

YELLOW-BILLED BARBET *Trachylaemus purpuratus*   Plate 74 (10)
Barbican pourpré

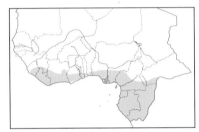

L 24 cm (9.5"). Large, long-tailed barbet with glossy blue-black upperparts, conspicuous *yellow bill* and *yellow bare orbital patch.* **Adult** *goffinii* Forehead to mid-crown reddish (extending above eye and onto ear-coverts); white line on scapulars (often concealed). Black throat and upper breast speckled silvery-pink, bordered by reddish band; lower breast and belly yellow. **Nominate** has red on head duller; throat and upper breast more streaked than speckled white; yellow on underparts blotched black. Intermediate ***togoensis*** similar to *goffinii* in head and rump colour, but throat and underparts as nominate. **Juvenile** Duller; throat and upper breast all dark. **VOICE** A long series of low-pitched *hoops*, resembling those of Hairy-breasted Barbet but louder, more popping and usually delivered slower (often *c.* 1 per second); *goffinii* slower than nominate. Duet may start with a curious mewing call. [CD8:46] **HABITS** Quiet, inconspicuous and easily overlooked, occurring singly or in pairs in forest, second growth and adjacent cultivated areas. Occasionally in mixed-species flocks. Excavates nest in large dead tree, sometimes in association with *Gymnobucco* barbets. **STATUS AND DISTRIBUTION** Not uncommon resident in forest zone, from Sierra Leone and SE Guinea to Ghana (*goffinii*), E Ghana to SW Nigeria (*togoensis*) and SE Nigeria to S CAR and Congo (nominate). **NOTE** Placed in *Trachylaemus*, following e.g. Zimmerman *et al.* (1996), to differentiate from open-country social species in *Trachyphonus*, which have different plumage, behaviour and vocalisations.

YELLOW-BREASTED BARBET *Trachyphonus margaritatus*   Plate 74 (9)
Barbican perlé

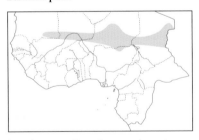

L 20 cm (8"). *T. m. margaritatus*. Large, slender and distinctive terrestrial barbet with long, unnotched, *pinkish to red-brown bill*. **Adult male** Head yellow with black cap forming short crest; upperparts and tail mainly brown spotted white; rump lemon-yellow; uppertail-coverts red. Underparts yellow, paler on flanks and belly; breast has black patch in centre bordered by variably developed band of brown spots; undertail-coverts red. **Adult female** Similar but lacks breast patch. **Juvenile** Duller; no white spots on back. **VOICE** Vocal. Duet starts with clear *pwewp* notes and evolves in buoyant, rolling *pwewp-up kwewp-up tew-kwip-to...* [CD8:48] **HABITS** Conspicuous, occurring in pairs or small parties near dry watercourses in *Acacia* woodland, dry wooded grassland, thorn bush and desert edge. Forages on ground. The only barbet in our region to nest in self-excavated hole in earth bank. **STATUS AND DISTRIBUTION** Locally common resident in Sahel zone, from S Mauritania through Mali, Niger and extreme NE Nigeria to C Chad.

HONEYGUIDES
Family INDICATORIDAE (World: 17 species; Africa: 15; WA: 11)

Small to medium-sized, drab-coloured and generally inconspicuous birds with dark-tipped white outer tail feathers often conspicuous in flight. Sexes similar, except in Greater Honeyguide. Feed on wax, insects and fruit and are brood parasites. Their English name is derived from the habit of two *Indicator* species 'guiding' man and mammals to beehives by alternately calling, flying and waiting (to feed on remnants of broken hive).

CASSIN'S HONEYBIRD *Prodotiscus insignis*

Indicateur pygmée

Plate 75 (11)

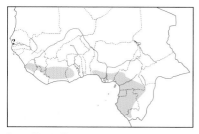

Other name E: Cassin's Honeyguide

L 11–12 cm (4.5"). Small and flycatcher-like with *all-white outer tail* and *fine pointed bill*. **Adult** *flavodorsalis* Head and upperparts olive to yellowish-green with rump feathers tipped bright yellow; narrow white eye-ring. Tail blackish with outer 2 feather pairs entirely white. Underparts grey, paler on belly and undertail-coverts. White feathers from flanks to rump-sides fluffed out in display, but normally concealed. Tail frequently spread in undulating flight, with white outer feathers conspicuous. Bill and legs blackish. **Nominate** more olive, lacking golden-yellow wash on upperparts; underparts darker. **Juvenile** Duller, more dusky above; bill and legs pale horn. **VOICE** A distinctive buzzy *tsrrr-tsrrr-...*, uttered in flight. **HABITS** In canopy or mid-strata of evergreen and gallery forest, frequenting edges, clearings and second growth. Usually singly. Active and restless, gleaning insects from leaves and branches like warbler and hawking insects from perch like flycatcher. Occasionally joins mixed-species flocks. Parasitises flycatchers, warblers, sunbirds and probably white-eyes. **SIMILAR SPECIES** Adult Wahlberg's Honeyguide has white outer tail feathers tipped dark and browner upperparts; juvenile has all-white outer tail but is distinctly browner above. Other small honeyguides have dark-tipped outer tail feathers and stumpy bills. **STATUS AND DISTRIBUTION** Scarce to uncommon resident, from Guinea and Sierra Leone to Togo, and from S Nigeria to CAR and Congo. Claimed from Senegal. Western *flavodorsalis* east to SW Nigeria; nominate mainly east of Niger R.

WAHLBERG'S HONEYBIRD *Prodotiscus regulus*

Indicateur de Wahlberg

Plate 75 (10)

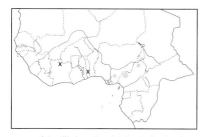

Other names E: Brown-backed/Sharp-billed Honeyguide

L 12–13 cm (4.25–5.0"). *P. r. camerunensis*. Small and flycatcher-like, mainly dull grey-brown with fine pointed bill. **Adult** Head and upperparts grey-brown. Tail blackish with outer 3 feather pairs white broadly tipped blackish. Underparts grey becoming whitish on belly and undertail-coverts. White feathers from flanks to rump-sides fluffed out in display, but normally concealed. Tail frequently spread in undulating flight, revealing inverted T pattern. Bill and legs blackish. **Juvenile** Outer tail feathers entirely white; also paler above and yellower below; gape orangey. **VOICE** A buzzy *tsrrr-tsrrr-...*, uttered in flight; also high-pitched *tsip* notes. Song a long, buzzy *tsrrrrrrrrrrrrrrrrrrrrr...* [CD8:50] **HABITS** As Cassin's Honeyguide, but occurs in woodland and bush (not evergreen forest) and forages lower and in smaller trees. Perches upright on bare branch in treetop. Parasitises hole-nesters and species with globular nests (e.g. Grey-backed Camaroptera, Scarlet-chested Sunbird, Yellow-throated Petronia and cisticolas). **SIMILAR SPECIES** Cassin's Honeyguide differs in olive-green upperparts and smaller size, from adult Wahlberg's also by all-white outer tail feathers. **STATUS AND DISTRIBUTION** Rare resident with patchy distribution, recorded from Liberia (Mt Nimba), N Ivory Coast (once), Togo (once), SE Nigeria (4 records, Enugu), Cameroon and W CAR.

ZENKER'S HONEYGUIDE *Melignomon zenkeri*

Indicateur de Zenker

Plate 75 (9)

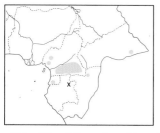

L 14–15 cm (5.0–5.5"). Darkish and bulbul-like with *yellowish legs*. **Adult** Head and upperparts warm olive-brown with olive feather edgings giving slightly streaked appearance. Tail graduated, mainly dark, inner webs of outer feathers greyish-white (often difficult to see); outer tail feathers of different lengths may recall miniature Lyre-tailed Honeyguide. Entire underparts, incl. undertail-coverts, dirty greyish-olive tinged yellow, slightly paler on throat and lower underparts. Bill *bicoloured*, grey-brown to blackish above, *yellow below*. Legs dull pinkish-yellow or greenish-yellow. **Juvenile** Dark greenish above; dusky grey-green below; tail whiter. **VOICE** Song a far-carrying, deliberate, rhythmic series of identical, short, whistled notes *wEEu wEEu wEEu wEEu ...*, first rising, then gradually fading. [CD8:51] **HABITS** Poorly known. Occurs in evergreen and riparian forest, frequenting edges and secondary situations. Hosts unknown, possibly barbets. **SIMILAR SPECIES** Yellow-footed Honeyguide best separated from below by whitish undertail-coverts and largely white tail; above brighter olive; bill completely yellowish. Some *Andropadus* greenbuls may appear superficially similar but have longer and square tails without any white. **STATUS AND DISTRIBUTION** Generally rare forest resident, recorded from W and S Cameroon, Eq. Guinea (Mt Alen, not uncommon), NE Gabon (M'Passa), N Congo and SE CAR (Ouossi R.).

YELLOW-FOOTED HONEYGUIDE *Melignomon eisentrauti* — Plate 75 (8)
Indicateur d'Eisentraut

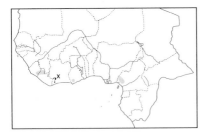

L 14–15 cm (5.0–5.5"). Rather bulbul-like with slender yellowish bill and yellowish legs. **Adult** Top of head olive, head-sides greyish; upperparts yellowish-olive. Tail graduated but square-tipped, dark in centre, outer 4 feather pairs white tipped blackish (*tail thus appearing mainly white from below*). Underparts grey slightly washed olive on breast *becoming off-white on vent and undertail-coverts*. Bill yellowish-brown to olive-yellow. Legs olive-yellow tinged pinkish on sides. **Juvenile** Paler; tail whiter with smaller and duller dark tips. **VOICE** A series of *c.* 13 clear, emphatic notes, slightly descending and slowing down at end, *tuu-i tuu-i tuu-i tuu-i ... tuu tuu tuu*. **HABITS** Poorly known. Frequents mid-strata and canopy of primary and secondary evergreen lowland forest, incl. lower montane slopes. Searches leaves and branches, occasionally hanging upside-down (thus superficially recalling warbler). Hosts unknown. **SIMILAR SPECIES** Zenker's Honeyguide is darker overall with dark bill, more olive underparts (incl. undertail-coverts) and much less white in tail (which may appear rather uniformly dark). Other honeyguides have differently shaped bills. Small *Andropadus* bulbuls may appear surprisingly similar but have longer and square tails without any white. Honeyguide Greenbul and Sjöstedt's Honeyguide Greenbul have white in tail but both are larger and have longer, ungraduated tails. **STATUS AND DISTRIBUTION** Rare forest resident, recorded from SE Sierra Leone, Liberia (Wonegizi Mts and Mt Nimba), Ivory Coast (Marahoué NP), SW Ghana (Kakum FR) and W Cameroon (easternmost localities near Mt Cameroon and Mamfe). Unconfirmed sightings from SW Ivory Coast (Taï NP). Probably more widespread. *TBW* category: Data Deficient.

LYRE-TAILED HONEYGUIDE *Melichneutes robustus* — Plate 75 (5)
Indicateur à queue en lyre

L 16.5–17.5 cm (6.5–7.0"). Rarely seen honeyguide with *diagnostic sound* and graduated *lyre-shaped tail*. **Adult male** Head and upperparts dark olive-brown with darker centres to feathers. Underparts olivaceous becoming creamy-white on belly and undertail-coverts; flanks variably streaked blackish. Tail exceptionally shaped: 2 central feather pairs long, broad, curved outwards and blackish, 4 outer pairs short, narrow and white, central pair of undertail-coverts long and blackish, projecting between rectrices. Bill strong, moderately long. **Adult female** Similar but tail slightly shorter and browner; longest undertail-coverts do not or barely project between central rectrices. **Juvenile** Entirely sooty blackish-brown washed olive. Tail similar to adult but much shorter. **VOICE** Calls include *prrr prrr ...* in flight and a short harsh note likened to slow version of Greater Honeyguide chatter. In display flight a highly distinctive, far-carrying, accelerating series of 10–30 nasal sounds (probably caused by air passing through tail feathers during rapid descent), first rising in pitch with increasing resonance, then fading *heyih heyih heyih-heyih-heiheiheihei...*; extremely difficult to locate. [CD8:52] **HABITS** In lowland and lower montane forest and its outliers in savanna. Usually in canopy, but descends to lower strata, even near ground. Spectacular and unique display high above forest, performed by both sexes, consists of undulating flight followed by fast descent in zigzags or spirals while making remarkable sound. Hosts unknown, *Gymnobucco* barbets suspected. **SIMILAR SPECIES** Not fully grown tail of Velvet-mantled Drongo may superficially resemble Lyre-tailed Honeyguide, but overall shape of bird and behaviour should preclude confusion even if coloration is not discernable. **STATUS AND DISTRIBUTION** Rare to uncommon resident, from Sierra Leone and SE Guinea to Ivory Coast and from E Nigeria to S CAR (east to Ouossi R.) and Congo.

SPOTTED HONEYGUIDE *Indicator maculatus* — Plate 75 (2)
Indicateur tacheté

L 16.5–17.5 cm (6.5–7.0"). Distinctive but unobtrusive honeyguide with *mainly spotted underparts*. **Adult** *maculatus* Head and upperparts olive-green. Underparts olive spotted creamy-green on breast and upper belly, paler and streaked on throat and undertail-coverts. Tail graduated, outer 4 feather pairs white broadly tipped dusky. Eastern *stictothorax* has head-sides finely streaked yellowish-green; spots on underparts paler and more distinct; vent

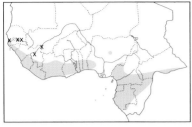

yellower. **Juvenile** Top of head dark grey speckled pale yellowish-green on forehead; throat and breast greyer, making spots more distinct; spotting also more extensive. **VOICE** A purring *brrrrrrr...* identical to that of extralimital Scaly-throated Honeyguide *I. variegatus* and slightly reminiscent of song of Double-toothed Barbet. [CD8:53] **HABITS** In evergreen and gallery forest; locally to 1650 m. At all levels, from canopy to undergrowth, even on ground. Inconspicuous. Hosts unknown, *Gymnobucco* barbets and Buff-spotted Woodpecker suspected. **SIMILAR SPECIES** Honeyguide Greenbul and Sjöstedt's Honeyguide Greenbul lack spots on underparts and have slender bills. **STATUS AND DISTRIBUTION** Uncommon or rare to locally not uncommon forest resident, from Senegambia through Ivory Coast (north to Comoé NP) to SW Nigeria (nominate), and from SE Nigeria and Cameroon to SE CAR and Congo (*stictothorax*). Also C Nigeria (Kagoro; nominate?).

GREATER HONEYGUIDE *Indicator indicator* Plate 75 (1)
Grand Indicateur

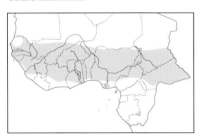

L 18.0–19.5 cm (7–8"). Stout with tail pattern typical of genus. **Adult male** Grey-brown crown, *whitish ear patch, black throat* and *pink bill* diagnostic. Upperparts dark grey-brown with white edges to wing-coverts and white streaks on rump and uppertail-coverts. Breast greyish-white, sometimes washed yellowish, becoming white on rest of underparts; flanks have some blackish streaking. Yellow lesser coverts form patch at base of leading edge of wing, visible in flight. **Adult female** Duller than male and lacking its face pattern; rather bulbul-like. Head dull greyish-brown. Bill dark. **Juvenile** Top of head and upperparts brown with darker mask; some white on rump and uppertail-coverts. *Throat and breast pale lemon to orange-yellow*, becoming white on rest of underparts. Bill dark. **VOICE** Song a distinctive, far-carrying series of *WHIT-birr WHIT-birr...* ending with a single *WHIT*; uttered by male. 'Guiding' call a loud, fast chattering. [CD8:55] **HABITS** In various wooded habitats. May guide humans and Ratels *Mellivora capensis* to beehives by fluttering from tree to tree, uttering chattering call. Flight undulating, conspicuously revealing white outer tail feathers. Occasionally joins mixed-species flocks. Parasitises hole-nesters or species with deep nests, incl. bee-eaters, kingfishers, hoopoes, woodpeckers, swallows, starlings and others. **SIMILAR SPECIES** Spotted Honeyguide is similar sized but distinguished from female Greater by spots on underparts. Bulbuls have more slender bills and different behaviour. **STATUS AND DISTRIBUTION** Uncommon or rare to locally common throughout savanna zone, from S Mauritania to Sierra Leone and NW Liberia east to Chad and CAR, and in SE Gabon and SW Congo.

LESSER HONEYGUIDE *Indicator minor* Plate 75 (3)
Petit Indicateur

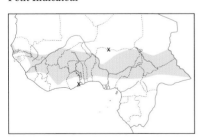

L 14–15 cm (5.5–6.0"). Inconspicuous, with dark-tipped white outer tail feathers and stubby bill. **Adult male** *senegalensis* Head grey, usually with dusky submoustachial stripe and sometimes white on lores. Upperparts olive-green with blackish feather centres and dull greenish-yellow fringes producing rather indistinct streaking. Underparts grey becoming whitish on throat and belly; flanks have some dusky streaks. Bill dark; base of lower mandible pinkish. *I. m. riggenbachi* is slightly greener olive above; streaking more distinct; underparts darker grey. **Adult female** Submoustachial stripe weaker. **Juvenile** Greener above and greyer below; loral and submoustachial marks faint or lacking; throat streaked dusky. **VOICE** Song, uttered from traditional perches, a far-carrying, deliberate, rhythmic series of 10–30 identical notes *prwit prwit prwit prwit ...* or *wrip wrip wrip wrip ...* introduced by faint *pee-yew* or *tyeew* (only audible at close range); rhythm variable. [CD8:56] **HABITS** Usually singly, in various types of woodland. Gleans foliage and branches; catches insects in flight. Frequents beehives. Displays include bounding, circling flight often with fanned tail and 'drumming' sound. Parasitises hole-nesters or species with deep nests, incl. bee-eaters, kingfishers, barbets, woodpeckers, swallows, starlings and others. **SIMILAR SPECIES** Thick-billed Honeyguide extremely similar (usually more distinctly streaked above and darker grey below), with identical song, but confined to forest; not safely distinguishable in field where they meet. Least Honeyguide has distinct submoustachial stripe, white lores, bold streaking on upperparts and flanks, and is smaller, with smaller bill. Willcocks's Honeyguide also smaller with even shorter, stubbier bill, is greener above and darker below, and lacks submoustachial stripe. **STATUS AND DISTRIBUTION** Uncommon resident in savanna zone, from Senegambia to Sierra Leone east to N Cameroon and Chad (*senegalensis*), and from C Cameroon to CAR (*riggenbachi*).

THICK-BILLED HONEYGUIDE *Indicator conirostris* Plate 75 (4)
Indicateur à gros bec

L 14–15 cm (5.5–6.0"). As Lesser Honeyguide but a forest species, with darker grey underparts, brighter greenish-yellow fringes to upperparts and slightly stouter bill. **Adult male *ussheri*** Head dusky-grey variably washed olive, usually with dusky submoustachial stripe and sometimes narrow whitish stripe on lores (both often hard to see in field). Upperparts olive-green with blackish feather centres and greenish-yellow fringes producing more or less streaked appearance. Underparts grey becoming whitish on belly; flanks streaked dusky. Bill dark; base of lower mandible grey to pinkish. **Nominate** slightly greyer above and darker below. **Adult female** Submoustachial stripe even fainter. **Juvenile** Greener above; no loral mark; throat streaked dusky; dark tail tips reduced. **VOICE** Song, uttered from traditional perches, a far-carrying, deliberate, rhythmic series of 10–30 identical notes *wrip wrip wrip wrip* ... introduced by faint *tyeew* (only audible at close range); indistinguishable from that of Lesser Honeyguide. [CD8:57] **HABITS** Usually singly, in lowland, lower montane and dense gallery forest. Foraging behaviour as Lesser Honeyguide. Only known host is Grey-throated Barbet; other *Gymnobucco* barbets suspected. **SIMILAR SPECIES** Lesser Honeyguide extremely similar (usually less distinctly streaked above and paler grey below), with identical song, but occurs in more open woodland; not safely distinguishable in field where they meet. Least Honeyguide has bolder flank streaking and is smaller. Willcocks's Honeyguide is also smaller and has strong green wash on breast, less distinct streaks on upperparts, and no submoustachial stripe. **STATUS AND DISTRIBUTION** Uncommon forest resident, from SE Sierra Leone to S Ghana (*ussheri*) and from S Nigeria to S CAR and Congo (nominate). **NOTE** Often treated as conspecific with Lesser Honeyguide and they probably interbreed in Nigeria and Cameroon (and elsewhere?).

LEAST HONEYGUIDE *Indicator exilis* Plate 75 (7)
Indicateur menu

Other names E: Western Least Honeyguide; F: Indicateur minule
L 11–13 cm (4.25–5.0"). *I. e. exilis*. Small dark forest honeyguide. **Adult** Head greyish to greenish-olive with narrow white line on lores and blackish submoustachial stripe. Upperparts olive-green, with blackish feather centres and yellow-green fringes producing distinct streaked appearance. Underparts grey becoming whitish on belly; flanks have distinct dark streaks. Tail pattern typical of genus. Bill stubby and dark with pale grey to pinkish base to lower mandible. **Juvenile** Darker; no loral mark; submoustachial stripe and flank streaks fainter; dark tail tips reduced. **VOICE** Song a far-carrying, deliberate, rhythmic series of 10–30 identical notes *wrEEu wrEEu wrEEu wrEEu* ... resembling that of Lesser and Thick-billed Honeyguides but less abrupt and less snapping. [CD8:58] **HABITS** In lowland, lower montane and gallery forest; also forest edges. Mostly high in canopy, hopping through tall trees, inspecting foliage and clinging to tree trunks. Flight undulating with white of outer tail feathers conspicuous. Hosts unknown; barbets suspected, with possible preference for tinkerbirds. **SIMILAR SPECIES** Thick-billed and Lesser Honeyguides have fainter flank streaking, less distinct facial markings and are larger. Similar-sized Willcocks's Honeyguide lacks loral and submoustachial stripes. **STATUS AND DISTRIBUTION** Uncommon to rare forest resident, from Guinea-Bissau to Togo (patchy) and from S Nigeria to CAR and Congo; also Bioko.

WILLCOCKS'S HONEYGUIDE *Indicator willcocksi* Plate 75 (6)
Indicateur de Willcocks

L 11–13 cm (4.25–5.0"). Small, unobtrusive and poorly known honeyguide with bland face. **Adult *willcocksi*** Head olive; no submoustachial stripe; pale orbital ring; dark speckles on forehead visible in favourable conditions. Upperparts olive-green with blackish feather centres and yellow-green fringes producing streaked appearance. Underparts olive-grey *with strong olive wash on breast* becoming whitish on belly; flanks have dusky streaks. Tail pattern typical of genus. Bill stubby and dark, with pale grey to pinkish base to lower mandible. Northern *hutsoni* slightly paler and greyer, intergrading with nominate; western *ansorgei* paler and greyer still. **Juvenile** Greener above; dark tail tips reduced; legs pale. **VOICE** Song a long rhythmic series of identical notes, resembling that of

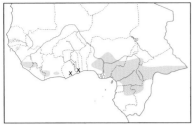

other small *Indicator* species in structure but interspersed with a distinctive snapping note, and lacking soft introductory syllable *p-wEEw-Pk p-wEEw-Pk p-wEEw-Pk...* or *huwEEw-TK huwEEw-TK huwEEw-TK* (*Pk* and *TK* = snapping note). [CD8:59] **HABITS** In gallery forest, forest patches and dense woodland in savanna; also in lowland and lower montane forest. Foraging behaviour as other small *Indicator*. Hosts unknown. **SIMILAR SPECIES** Similar-sized Least Honeyguide has loral and submoustachial stripes, distinctly bolder flank streaks and lacks green wash to breast. Lesser Honeyguide is less green above and paler and greyer below. Thick-billed is darker, less olive below and more streaked above; also larger. **STATUS AND DISTRIBUTION** Rare to uncommon forest resident in Guinea-Bissau (*ansorgei*), from Sierra Leone to Ghana and from S Nigeria and S Cameroon to Congo (nominate), and from C Nigeria and N Cameroon to S Chad and CAR (*hutsoni*).

WRYNECKS, PICULETS AND WOODPECKERS
Family PICIDAE (World: *c.* 215 species; Africa: 30–31; WA: 21)

Very small to large birds with straight, pointed or chisel-tipped bills and zygodactylous feet. Most are arboreal, largely insectivorous and non-migratory. Often in pairs. Three subfamilies occur in W Africa.

Wrynecks (Jynginae) are relatively small with cryptic plumages and soft tails, not used as brace when clinging to trees. Sexes similar. Feed mainly on ants. Nest in natural cavities and old woodpecker or barbet holes. Two species, one resident (endemic to Africa) and one Palearctic migrant.

Piculets (Picumninae) are tiny woodpecker-like species with short bills and very short tails, not used as brace when clinging to bark. Sexes differ in forehead colour. Nest in self-excavated holes in trees and vines. One African endemic resident.

Woodpeckers (Picinae) are small to fairly large species that characteristically cling to trees, using stiff tail as prop, and chip at wood with strong bill. Sexes differ in head pattern; males usually have (more) red. Nest in self-excavated holes in trees or termitaria. Flight strong and bounding, with brief bursts of flapping interspersed by glides on completely folded wings. All 18 species are resident and endemic to Africa.

EURASIAN WRYNECK *Jynx torquilla*   Plate 78 (1)
Torcol fourmilier

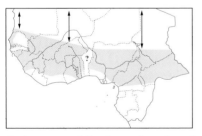

Other name E: Northern Wryneck
L 16–18 cm (6.5"). *J. t. torquilla*. Cryptically patterned, elongated species. **Adult** Head, upperparts and fairly long tail a mix of grey, brown and buff. Dark stripes from eye onto neck-sides, from centre of crown onto back and on edge of scapulars. Densely barred underparts buffish-cream on throat, becoming whitish on lower breast and belly. Bill rather short and pointed. **Immature** Similar. **VOICE** Usually silent in winter quarters, but may occasionally utter series of loud *kièh-kièh-kièh-...*, recalling small falcon. [CD3:33] **HABITS** Singly, in open habitats. Unobtrusive and easily overlooked. Often feeds on ground. Perches on branches; also clings to tree like woodpecker. Shallow bounding flight, with short series of wingbeats interspersed by glides on closed wings. **SIMILAR SPECIES** Could be confused with Red-throated Wryneck where overlap, but latter has rufous breast patch. **STATUS AND DISTRIBUTION** Uncommon to rare Palearctic passage migrant and winter visitor (mainly Oct–Mar), from Mauritania to N Liberia east to Chad and CAR.

RED-THROATED WRYNECK *Jynx ruficollis*   Plate 78 (2)
Torcol à gorge rousse

Other name E: Rufous-breasted Wryneck
L 18–19 cm (7"). **Adult** *ruficollis* Mainly dark brown above with black markings, and broken black stripe from crown to mantle. Below pale buffish-white with distinctive *chestnut breast patch* extending to chin. Head-sides and throat barred blackish and buff-white. Tail barred black. In *pulchricollis* chestnut patch only reaches lower throat. **Juvenile** Chestnut patch duller and slightly smaller; flanks more barred. **VOICE** A series of 2–12 hard *kweeh* notes,

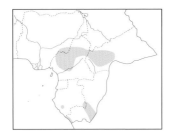

uttered atop bushes or trees; delivery slower than Eurasian Wryneck. [CD8:61] **HABITS** Singly or in pairs in scrub, open woodland, clearings and forest edges. Unobtrusive. Forages mainly on ground. Clings to trees and moves jerkily like woodpecker, but also perches like passerine on branches and in exposed positions. **SIMILAR SPECIES** Slightly smaller Eurasian Wryneck is greyer and lacks chestnut breast patch. **STATUS AND DISTRIBUTION** Uncommon to locally common resident with two disjunct populations: from SE Nigeria and Cameroon to SW CAR (*pulchricollis*), and from S Gabon through Congo (nominate). Often erratic, disappearing after breeding.

AFRICAN PICULET *Sasia africana* — Plate 78 (3)
Picumne de Verreaux

Other name F: Picule
L 8–9 cm (3.0–3.5"). *Tiny*, unobtrusive woodpecker with short, pointed bill, *very short tail*, yellowish-olive upperparts and grey underparts. Short white streak behind eye and duller one below grey ear-coverts. Eye red, with red orbital ring. Feet red. **Adult male** has reddish forehead. **Adult female** Forehead olive, as crown. **Juvenile** Head-sides and underparts mixed with dull rufous. Sexes as adults. **VOICE** A very fast, high-pitched, piercing little trill *tsiririririri*, reminiscent of bat or insect. [CD8:62] **HABITS** Very inconspicuous and difficult to observe as it usually keeps to dense forest undergrowth, foraging singly, in pairs or small groups, in lower and mid-strata; occasionally at forest edge. Often feeds on very thin twigs. Moves constantly and taps often; tapping rapid and relatively loud. Occasionally in mixed-species flocks. Flight swift and darting. **SIMILAR SPECIES** When glimpsed in clearings or at edges may initially recall crombec *Sylvietta*, but plumage characteristics and behaviour preclude confusion. **STATUS AND DISTRIBUTION** Forest resident. Rare and local in SE Liberia, Ivory Coast and SE Ghana; uncommon to locally common from S Cameroon to SW CAR and Congo. Probably overlooked.

FINE-SPOTTED WOODPECKER *Campethera punctuligera* — Plate 76 (2)
Pic à taches noires

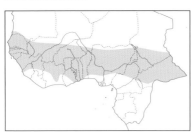

Other name F: Pic ponctué
L 21 cm (8"). *C. p. punctuligera*. Medium-sized woodpecker with green upperparts and *pale yellow underparts finely spotted black*, most densely on breast. Head-sides and throat white. Tail golden-yellow barred brown. **Adult male** has forehead to nape red and *red moustachial stripe*. **Adult female** Red on head confined to hindcrown and nape; forehead to mid-crown black streaked white; moustachial area spotted black. **Juvenile** Similar to adult female but darker green above and buffier below; forehead to mid-crown plain black; moustachial area blacker. **VOICE** A ringing series, often starting with a few *kip* calls: *kip-kip-kip-kieeh-kieeh-kieeh*, often uttered simultaneously by pair. [CD8:63] **HABITS** Singly, in pairs or small family groups in open woodland, foraging on trunks and branches, regularly descending to base of tree or ground. Occasionally in mixed-species flocks. **SIMILAR SPECIES** Golden-tailed Woodpecker is duller green above, and densely streaked below and on head-sides. Nubian Woodpecker has similar voice but is browner, more boldly marked white above and boldly spotted below; only recorded in NW CAR. **STATUS AND DISTRIBUTION** Locally common to uncommon resident in broad woodland belt between desert and forest, from S Mauritania to Sierra Leone east to Chad and CAR.

NUBIAN WOODPECKER *Campethera nubica* — Plate 147 (3)
Pic de Nubie

L 20 cm (8"). *C. n. nubica.* Medium-sized woodpecker with olive-brown upperparts *boldly barred and spotted whitish* and yellowish-white underparts *boldly spotted black on breast and flanks*. Head-sides white *heavily streaked black on ear-coverts*. Tail dull yellow barred brown. **Adult male** has forehead to nape red and *red moustachial stripe*. **Adult female** Red on head confined to hindcrown and nape; forehead to mid-crown and moustachial stripe black speckled white. **Juvenile** Similar to adult female but darker and browner above, more heavily spotted and barred below, with spots or streaks often extending onto throat; forehead to mid-crown plain black or speckled white. **VOICE**

Similar to Fine-spotted Woodpecker. An accelerating series of ringing, strident notes *kieeh, kieeh kieeh kieeh kieeh-kieeh-kieeh kieeh kiee.* [CD8:64] **HABITS** Singly or in pairs in open woodland, foraging on trunks, branches and ground. Pairs maintain vocal contact. **SIMILAR SPECIES** Two similar-sized woodpeckers with red or white-speckled black moustachial stripes occur. Golden-tailed Woodpecker is greener, less marked above and densely streaked (not spotted) below. Fine-spotted Woodpecker is much greener above, only finely spotted below and mainly white on head-sides. **STATUS AND DISTRIBUTION** Uncommon resident, N CAR (Manovo-Gounda-St Floris NP).

GOLDEN-TAILED WOODPECKER *Campethera abingoni*   Plate 76 (1)
Pic à queue dorée

L 20 cm (8"). *C. a. chrysura.* Medium-sized woodpecker with olive-green upperparts *finely spotted and barred yellowish-white*, and pale yellowish underparts *boldly streaked blackish*, most densely on breast. Head-sides and throat white streaked blackish. Tail deep yellow with broad brownish bars and golden-yellow shafts. **Adult male** has dusky-red forehead, crown and moustachial stripe, and red nape. **Adult female** Forehead, crown and moustachial stripe black speckled white; red confined to nape. **Juvenile** As adult female but duller and more heavily streaked below. **VOICE** A single, drawn-out *k-heeeeew.* Drums softly and infrequently. [CD8:66] **HABITS** Singly or in pairs in gallery forest and woodland, foraging on trunks and branches. Rather unobtrusive. Sometimes joins mixed-species flocks. **SIMILAR SPECIES** No similar-sized woodpecker within its range has combination of greenish upperparts, barred (not plain) mantle, and streaked (not spotted or barred) underparts; Cardinal Woodpecker is much smaller and has different head pattern. Compare Fine-spotted and Nubian Woodpeckers, which frequent same habitat and are the only other woodpeckers in which males also have red moustachial stripe. **STATUS AND DISTRIBUTION** Inadequately known. Rare to uncommon resident. Patchy distribution includes S Mauritania, Senegambia, Guinea, Mali, N Ivory Coast, NW Ghana, N Benin, N Cameroon, S Chad, CAR, SE Gabon and Congo. Perhaps more widespread.

LITTLE GREEN WOODPECKER *Campethera maculosa*   Plate 76 (6)
Pic barré

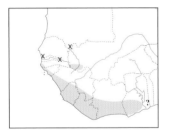

Other name E: Golden-backed Woodpecker
L *c.* 16 cm (6.25"). Rather little-known medium-sized forest woodpecker. Head-sides and throat buffish with fine dark specks. Upperparts plain dark golden-olive or bronzy-green. Tail black. Underparts pale yellowish-green with bold dark bars. **Adult male** has dusky-red forehead and crown, and red nape. **Adult female** Forehead to crown black finely spotted buffish. **Juvenile** As adult female but upperparts greener, and underparts paler and more irregularly barred. **VOICE** A drawn-out, plaintive *huweeeeh,* very similar to that of Green-backed Woodpecker, and a shorter, harsh *whee.* Also a hard *kewik* (in response to playback). [CD8:67] **HABITS** Singly or in pairs in lowland forest, clearings and edges. **SIMILAR SPECIES** No other woodpecker in its range has combination of plain upperparts and barred underparts, except in SC Ghana where overlap with Green-backed Woodpecker. Latter has greener upperparts, green tail, and head-sides and throat more densely spotted on less buff background; female has red nape. **STATUS AND DISTRIBUTION** Uncommon to locally common resident in forest zone, from Senegal to C Ghana; also SW Mali. Rare, S Mauritania. Claimed from W Togo.

GREEN-BACKED WOODPECKER *Campethera cailliautii*   Plate 76 (5)
Pic à dos vert

Other name E: Little Spotted Woodpecker
L *c.* 16 cm (6.25"). *C. c. permista.* Appears rather small and darkish. Upperparts and tail plain green; head-sides and throat buffish-grey or greenish-grey densely speckled and barred brownish-black; underparts yellowish-green densely and boldly barred black. **Adult male** has forehead to nape red. Eye reddish-brown. **Adult female** Forehead and forecrown black spotted buffish, red confined to hindcrown and nape. **Juvenile** Similar to adult female but eye grey. **VOICE** A drawn-out, plaintive *huweeeeh* or *wheeeee,* often given in series of 3–5. [CD8:68] **HABITS** Singly or in pairs in various forest

types, forest edge and adjacent woodland. Forages inconspicuously on trees, often high. **SIMILAR SPECIES** No other woodpecker in its range has combination of plain green upperparts and boldly barred underparts, except in SC Ghana where overlap with Little Green Woodpecker; latter has more olive upperparts, buffier head-sides and throat, blackish tail and, in female, no red on nape. **STATUS AND DISTRIBUTION** Uncommon to common resident, mainly in forest zone, from Ghana to S CAR and Congo.

TULLBERG'S WOODPECKER *Campethera tullbergi*　　　　Plate 76 (3)
Pic de Tullberg

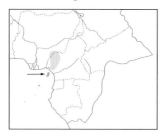

Other name E: Fine-banded Woodpecker

L 18–20 cm (7–8"). *C. t. tullbergi.* Medium-sized. Upperparts and tail *plain bright green* with some red on bend of wing; head-sides and throat pale greyish-yellow finely speckled dusky; rest of underparts pale greenish-yellow with fine spots on breast and bolder spots and bars on flanks and undertail-coverts. **Adult male** has forehead to nape red. **Adult female** Red confined to nape; forehead and crown black speckled white. **Juvenile** Similar to adult female, with crown olive-black speckled white. Young male soon acquires red on crown. **VOICE** A single *kreeer* or *kweeh.* [CD8:69] **HABITS** Confined to moist montane forest, where it forages mostly high in trees. Occasionally joins mixed-species flocks. **SIMILAR SPECIES** No other woodpecker in montane habitat has combination of green upperparts, finely spotted face and underparts, and red on 'shoulder' (unique among woodpeckers). **STATUS AND DISTRIBUTION** Uncommon forest resident in highlands of SE Nigeria (Obudu Plateau) and W Cameroon; also Bioko.

BUFF-SPOTTED WOODPECKER *Campethera nivosa*　　　　Plate 76 (8)
Pic tacheté

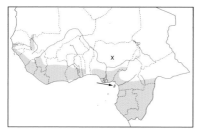

L 14–16 cm (5.5–6.25"). Small, dark woodpecker. **Adult male** *nivosa* Forehead, crown and upperparts plain dark olive; red patch on nape; head-sides and throat densely streaked olive-brown on dirty white; underparts dark olive with dense yellowish-white spots on breast becoming more bar-like on flanks, belly and undertail-coverts. Tail blackish. Doubtful race *maxima* slightly larger; island race *poensis* has more bar-like breast markings and whiter abdomen. Eastern *herberti* has upperparts greener, less olive; underparts yellower with more bar-like markings on breast and broader bars on belly. **Adult female** Lacks red nape patch. **Juvenile** Similar to adult female, with browner, more barred breast and greyer top of head. **VOICE** Mostly silent. Occasionally utters a soft *peeer.* Drums softly. [CD8:70] **HABITS** Singly or in pairs, quietly searching branches and vines at lower levels of primary and secondary forest. Frequently joins mixed-species flocks. Unobtrusive, but not shy. Nest usually excavated in ant nests and termitaria. **SIMILAR SPECIES** Brown-eared Woodpecker occurs in same habitat and also has dark olive plumage with spots on underparts, but is easily distinguished by brown ear patch and larger size. **STATUS AND DISTRIBUTION** Locally common resident in lowland and gallery forest, from SW Senegambia to S CAR and Congo. Rare, SW Mali. Also claimed from S Mauritania. Nominate throughout, except extreme N Ivory Coast (*maxima*), Bioko (*poensis*), CAR and perhaps E Congo (*herberti*).

BROWN-EARED WOODPECKER *Campethera caroli*　　　　Plate 76 (7)
Pic à oreillons bruns

L 18–19 cm (7.0–7.5"). Medium-sized, dark olive woodpecker with *brown ear patch.* **C. c. arizelus** Forehead to nape blackish-brown or olive; head-sides (except ear-coverts) and throat olive densely spotted buff-white; rest of underparts dark olive densely spotted yellowish-green; tail blackish. In **nominate** upperparts are more bronzy-green; underparts more heavily spotted. **Adult male** has feathers of hindcrown tipped red, nape red. **Adult female** Lacks red on crown and nape. **Juvenile** Similar to adult female, with paler, whiter spots becoming more bar-like on belly. **VOICE** A distinctive single *huuwEEEEuu,* drawn-out, plaintive and remarkably prolonged for a woodpecker. [CD8:71] **HABITS** Unobtrusive inhabitant of primary and secondary forest, occurring mainly at mid-levels and frequently joining mixed-species flocks. **SIMILAR SPECIES** Buff-spotted Woodpecker shares same habitat and also has plain dark olive upperparts and spots on underparts, but lacks brown ear patch and is noticeably smaller. **STATUS AND DISTRIBUTION** Forest resident with apparently disjunct distribution. *C. c. arizelus* locally not uncommon from Sierra Leone and SE Guinea to Ivory Coast; rare, Ghana and perhaps Guinea-Bissau. Nominate locally common to uncommon from Cameroon to S CAR and Congo; rare, S Nigeria.

LITTLE GREY WOODPECKER *Dendropicos elachus*   Plate 77 (1)
Pic gris

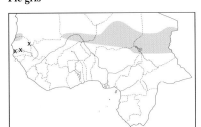

Other name F: Petit Pic gris
L 11–13 cm (4.5–5.0"). Very small, *pale greyish-brown* woodpecker with conspicuous *red rump* and white-barred back and wings. Forehead, crown, ear-coverts and moustachial stripe greyish-brown; white line from eye around ear-coverts to bill; throat white finely streaked brown. Underparts whitish with pale brown spotting or barring on breast even paler on flanks and belly. Head markings fade. **Adult male** has red hindcrown and nape. **Adult female** lacks red. **Juvenile** As adult female, but even duller. **VOICE** A sharp, very rapid rattling *kree-kree-kree-kree-...* similar to that of Cardinal Woodpecker, but harder and faster. [CD8:72] **HABITS** Relatively little known. Occurs in arid areas with small, scattered trees south of desert. **SIMILAR SPECIES** Brown-backed Woodpecker has darker, more contrasting plumage, esp. on head, no red on rump and plain (unbarred) back. **STATUS AND DISTRIBUTION** Uncommon to rare resident in Sahel zone, in SW Mauritania and N Senegal and from C Mali through Niger, NE Nigeria and N Cameroon to Chad. A few old records from Gambia.

SPECKLE-BREASTED WOODPECKER *Dendropicos poecilolaemus*   Plate 77 (5)
Pic à poitrine tachetée

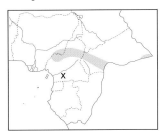

Other name E: Uganda Spotted Woodpecker
L 14–15 cm (5.5–6.0"). *Small* woodpecker with distinctive *fine spotting on breast.* Head-sides and throat whitish with some indistinct dark streaking and speckling, and faint malar stripe. Upperparts yellowish-green with pale barring on back and wings; reddish-tipped feathers give uppertail-coverts and rump orange cast. Underparts pale dirty yellowish. Some have fine bars on breast, broad pale grey bars on sides and indistinct streaks on belly. **Adult male** has forehead and forecrown brown, hindcrown and nape red. In **adult female** these are black, becoming brown on forehead. **Juvenile** Similar to adult but greyer, lacking red on rump and with red crown in both sexes.
VOICE A hard rapid series of *k-ret* notes. [CD8:73] **HABITS** At forest edges, in clearings and in light woodland, foraging at all heights, often low. **SIMILAR SPECIES** Similar-sized Gabon Woodpecker has underparts entirely and much more boldly spotted; also more restricted to forest. Cardinal Woodpecker has streaked underparts. **STATUS AND DISTRIBUTION** Rare to locally common resident, from extreme E Nigeria (Serti) through C Cameroon to SW CAR. Also extreme SE CAR.

GABON WOODPECKER *Dendropicos gabonensis*   Plate 76 (4)
Pic du Gabon

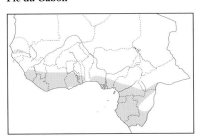

L *c.* 15 cm (6"). Small, plain-backed woodpecker. **Adult male** *lugubris* Forehead, crown, head-sides and malar stripe dark olive-brown, nape red. Upperparts golden olive-green; underparts greenish-yellow heavily streaked olive-brown. Tail black. In **nominate** forehead and forecrown are brownish, hindcrown and nape red; head-sides and throat whitish with dark streaks. Upperparts green; entire underparts boldly spotted black. Tail olive tipped blackish. Intermediate *reichenowi* has streaked underparts as *lugubris* and head markings more like *gabonensis* but malar streak narrow and rather indistinct. **Adult female** Similar to resp. males, but *lugubris* has nape concolorous with crown; in **nominate** forehead to nape blackish. **Juvenile** (all races) Similar to adult but duller and with red crown patch and black nape (both sexes). **VOICE** A diagnostic, rapid, high-pitched trill, *krititititititti* (also rendered *trrreeeeee* or *br-r-r-r-r-ip*), reminiscent of police whistle. Also a series of loud *kree* notes, resembling that of Cardinal Woodpecker, *kwik-ik-ik...* and *kreek-rrek-rrek-rrek...* [CD8:74] **HABITS** Singly or in pairs at forest edges, in clearings and in canopy of secondary forest. Forages actively at all heights. **SIMILAR SPECIES** Similar-sized Cardinal and Speckle-breasted Woodpeckers have barred (not plain) upperparts and much less marked underparts and head-sides. **STATUS AND DISTRIBUTION** Uncommon to locally common resident in lowland forest (below 1400 m), from Sierra Leone and SE Guinea to Togo (*lugubris*), and from S Nigeria and SW Cameroon (*reichenowi*) to Gabon and Congo (nominate). Also SW CAR. Race *reichenowi* intergrades with *lugubris* in SW Nigeria (Lagos area). **NOTE** Race *lugubris* sometimes considered a separate species, Melancholy Woodpecker (Pic à raies noires).

CARDINAL WOODPECKER *Dendropicos fuscescens* Plate 77 (6)
Pic cardinal

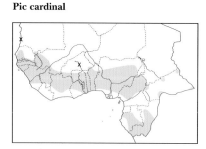

L 14–15 cm (5.5–6.0"). Small woodpecker with green upperparts, faintly barred mantle and streaked underparts. **Adult male** *lafresnayi* Forehead and forecrown brownish; hindcrown and nape red; head-sides and throat whitish with some faint dark streaking and narrow malar stripe; rest of underparts pale yellowish lightly streaked black-ish; some orange on uppertail-coverts. Race *sharpii* has more heavily streaked underparts. **Adult female** Forehead to nape brownish-black. **Juvenile** Dull version of adult, with red crown (brighter in male) and black nape. **VOICE** A rapid series of harsh, rattling notes *kree-kree-kree-kree...* Other rapidly uttered notes include *kwik-ik-ik-ik...* [CD8:75] **HABITS** Usually in pairs, in various types of woodland. Occasionally in mixed-species flocks. **SIMILAR SPECIES** Similar-sized Speckle-breasted Woodpecker has breast finely spotted, not streaked. **STATUS AND DISTRIBUTION** Scarce or uncommon to locally common resident in broad belt, from S Mauritania to Sierra Leone east to Nigeria (*lafresnayi*) and from Cameroon to CAR and Congo (*sharpii*).

BEARDED WOODPECKER *Dendropicos namaquus* Plate 77 (8)
Pic barbu

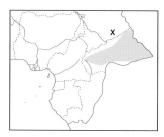

L 23–25 cm (9–10"). *D. n. schoensis*. Large, dark woodpecker with bold head markings. **Adult male** Forehead black finely speckled white; hindcrown red; nape black; head-sides white with broad black eye-stripe reaching malar stripe. Upper- and underparts mainly dark greyish-olive with narrow whitish bars; rump and uppertail-coverts tinged yellowish. Tail grey-brown barred yellowish. **Adult female** Similar, but hindcrown black. **Juvenile** As adult, with both sexes having red on crown mixed with black and spotted white. **VOICE** A loud and rapid *wik-wik-wik-wik...* Drums loudly and rather slowly (single taps clearly noticeable); both sexes drum. [CD8:76] **HABITS** Singly or in pairs in various types of woodland provided some large trees are present. **SIMILAR SPECIES** None. Yellow-crested Woodpecker, which has plain upperparts and, in male, yellow crown patch, may marginally meet Bearded at forest edges. **STATUS AND DISTRIBUTION** Not uncommon resident, CAR. Rare, SE Chad (Am Timan). **NOTE** Sometimes included within genus *Thripias*.

FIRE-BELLIED WOODPECKER *Dendropicos pyrrhogaster* Plate 77 (9)
Pic à ventre de feu

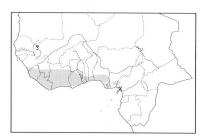

L *c.* 24 cm (9.5"). Readily identified, strikingly marked, large wood-pecker (the largest within its range). Combination of *bright red rump and central underparts* and *pure white head-sides with broad black eye-stripe and black malar stripe* diagnostic. Upperparts plain golden olive-green; breast-sides and flanks spotted and barred black and white. Tail blackish. **Adult male** has red crown and nape. **Adult female** Crown and nape all black. **Juvenile** As adult but duller, with red crown in both sexes (more extensive in male). **VOICE** A hard, short *kweep* or *kwip*, accelerating into rattling *kwipibibibibib*. Drumming loud, in rather fast and short bursts. [CD8:77] **HABITS** Usually in pairs, in various types of forest, second growth and edges. Frequent loud drumming betrays presence. **STATUS AND DISTRIBUTION** Not uncommon resident in forest zone, from Guinea and Sierra Leone to W Cameroon. Also reported from SW Mali (rare). **NOTE** Sometimes placed within genus *Thripias*.

YELLOW-CRESTED WOODPECKER *Dendropicos xantholophus* Plate 77 (7)
Pic à couronne d'or

L 20–23 cm (8–9"). Large and mainly dark olive with boldly marked black-and-white head. Upperparts generally plain, except for some white spotting on upper mantle; underparts densely spotted and barred white. Tail black.

In **adult male** hindcrown is yellow; in **adult female** black. **Juvenile** As adult but duller; both sexes have some yellow on hindcrown (more extensive in male). **VOICE** Calls include *kreeerr, kreeerr, kreeerr* followed by short, descending series of *kweek* notes. Drumming loud and rather slow; given by both sexes. [CD8:78] **HABITS** In pairs or singly, in forest canopy. Taps frequently and loudly. **SIMILAR SPECIES** None, but may meet Bearded Woodpecker in S CAR and Fire-bellied Woodpecker in extreme W Cameroon or SE Nigeria (see those species). **STATUS AND DISTRIBUTION** Rare to locally common resident, from extreme SE Nigeria to SW CAR and Congo. **NOTE** Sometimes placed within genus *Thripias*.

ELLIOT'S WOODPECKER *Dendropicos elliotii*   Plate 77 (3)
Pic d'Elliot

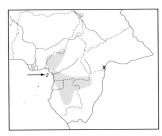

L 19–21 cm (7.5–8.0"). Medium-sized woodpecker with diagnostic combination of *plain green upperparts, black forehead and forecrown*, and *unmarked buffish-green head-sides*. Tail darkish. **Nominate** has pale yellowish underparts heavily streaked blackish. In montane *johnstoni* underparts mainly unmarked except for some faint streaking on breast. **Adult male** has hindcrown and nape red. **Adult female** Forehead to nape black, lacking red. **Juvenile** As adult but duller, with both sexes having some red on hindcrown and nape. **VOICE** Calls include shrill *kree-kree-kree* or softer *kiwik-kiwik-kiwik*. Rather silent. Drumming similar to Yellow-crested Woodpecker's but softer (*BoA*). [CD8:79] **HABITS** Singly or in pairs in forest (up to 2300 m). Forages at all levels; joins mixed-species flocks. **SIMILAR SPECIES** No other woodpecker within its range has similar combination of field marks. **STATUS AND DISTRIBUTION** Uncommon to locally common forest resident. Nominate from lowland Cameroon to Gabon, SW CAR and NW Congo; *johnstoni* confined to montane areas of SE Nigeria, Cameroon and Bioko, interbreeding with *elliotii* at base of Mt Cameroon and Mt Kupe. **NOTE** Sometimes placed within genus *Mesopicos*.

GREY WOODPECKER *Dendropicos goertae*   Plate 77 (4)
Pic goertan

Other name F: Pic gris
L *c.* 20 cm (8"). Fairly large woodpecker with diagnostic *grey head and underparts*. **Adult male** *goertae* Hindcrown and nape red; upperparts golden olive-green with some greyish barring on wings; uppertail-coverts and rump red; tail darkish. Underparts mainly grey with orange-red or yellow-orange patch on central belly to undertail-coverts. In pale race *koenigi* belly patch is yellow and much reduced, or even lacking. **Adult female** has plain grey head, lacking any red. **Juvenile** As adult but duller; both sexes have some red on crown. **VOICE** Calls include shrill, loud *krreet-krreet-krreet*, descending in volume and sometimes preceded by *kik-kik-kik-* ; fast and hard *kwik-wik-wik-wik* resembling that of Bearded Woodpecker but higher pitched; *kee-krrirrt-krrirrt...* and variations. [CD8:80] **HABITS** In various types of woodland, gallery forest and forest edge. Avoids closed forest. Usually in pairs or small family parties. Forages at all levels. **STATUS AND DISTRIBUTION** Not uncommon to common resident, from S Mauritania to Liberia east to Chad and CAR; also S Gabon and S Congo. Some local movements. Nominate throughout except at desert edges, from E Mali and Niger to Chad (*koenigi*). **NOTE** Sometimes placed within genus *Mesopicos*.

BROWN-BACKED WOODPECKER *Picoides obsoletus*   Plate 77 (2)
Pic à dos brun

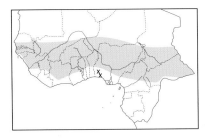

Other name F: Petit Pic à dos brun
L 13–14 cm (5.0–5.5"). *P. o. obsoletus.* Small, mainly greyish-brown woodpecker with distinctive combination of *plain back and spotted wings*. **Adult male** Forehead, forecrown, ear-coverts, and moustachial streak grey-brown; nape bright red. Upperparts and tail grey-brown, heavily spotted and barred white on wings and tail. Underparts whitish faintly streaked brownish. **Adult female** Nape grey-brown. **Juvenile** As adult; both sexes have some red on nape. **VOICE** A rapid *ki-ki-keeew-keeew-keeew* and *krreet-krreet-krreet*. [CD8:82] **HABITS** A rather unobtrusive species occurring in woodland. Usually in pairs.

SIMILAR SPECIES Little Grey Woodpecker is smaller, paler, and has barred back and red rump. **STATUS AND DISTRIBUTION** Uncommon to not uncommon resident in broad savanna belt, from Senegambia to Sierra Leone east to Chad and CAR. Rare, S Mauritania.

BROADBILLS
Family EURYLAIMIDAE (World: 15 species; Africa: 4; WA: 3)

Small to medium-sized, rather thickset, arboreal forest birds with broad gapes and large heads. Species in our region (genus *Smithornis*) are small, unobtrusive, streaky brown insectivores with short tails. Sexes similar but separable. Juveniles similar to adults. Possess unique short circular display flight accompanied by distinctive trilling or rattling sounds.

GREY-HEADED BROADBILL *Smithornis sharpei* Plate 78 (6)
Eurylaime à tête grise

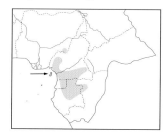

L 15 cm (6"). **Adult male *zenkeri*** Large, with *dark grey head and broad orange-rufous breast-band* (usually divided by white in centre). Lores orange-rufous; upperparts and tail largely dark brown; feathers of mantle and back with white bases (usually concealed at rest but puffed out in display). Throat white bordered by orange-rufous on sides and by deep orange-rufous breast; flanks and belly white streaked black; undertail-coverts rufous. Bill black above, white below. **Nominate** typically has paler, blue-grey head and less heavy streaking on underparts. **Adult female** Duller; orange on breast paler and streaked dusky. **Juvenile** Similar to adult female but head concolorous with upperparts; breast duller and heavily streaked blackish; wing-coverts tipped rufous. **VOICE** A thin, plaintive *theew* or *huiiii*. Loud, fast, strident rattle, produced during display flight, resembles that of congeners, but is rather shorter and noticeably lower pitched than that of Rufous-sided Broadbill. [CD8:83] **HABITS** Singly or in pairs in lower and middle strata of lowland and montane forest. Quiet and inconspicuous, like congeners; performs similar display flight. Feeds and displays lower than other broadbills. **SIMILAR SPECIES** Rufous-sided Broadbill distinguished by smaller size, double white wingbar, orange on breast brighter and restricted to sides. Juvenile Rufous-sided differs from juvenile Grey-headed by orange on breast, narrower streaking and brighter upperparts. African Broadbill has no orange on underparts. **STATUS AND DISTRIBUTION** Inadequately known. Locally not uncommon to rare forest resident recorded from E Nigeria (base of Mambilla Plateau), W and S Cameroon, Eq. Guinea, N and C Gabon, and N Congo (*zenkeri*); also Bioko (nominate). Esp. in mid-altitude forest (e.g. common on Mt Kupe), much rarer in lowland forest.

RUFOUS-SIDED BROADBILL *Smithornis rufolateralis* Plate 78 (5)
Eurylaime à flancs roux

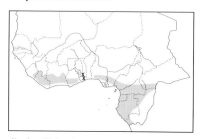

L 11.5 cm (4.5"). *S. r. rufolateralis.* Small, with conspicuous rufous-orange patches on breast-sides. **Adult male** Head black with buff lores. Upperparts and tail mainly dark brown and black; feathers of back and rump with white bases (usually concealed at rest but *puffed out in display*); double wingbar formed by white tips to wing-coverts. Underparts white streaked black on throat-sides, breast and flanks; rufous-orange patch on breast-sides and neck. Bill flat, very broad at gape, black above, whitish below. **Adult female** Duller; head and upperparts brown. **Juvenile** Like adult female but upperparts more rufous. **VOICE** A thin, plaintive *theew* or *huiiii* often followed by display flight. A remarkable, loud, fast, strident rattling *tttt-trrrrrrrrree*, resembling sound made when attempting to pronounce *pffrrrrrru* by vigorously expelling air between vibrating protruded lips; produced during display flight and believed to be caused by vibration of outer primaries. [CD8:84] **HABITS** Singly or in pairs in lower and middle storey of primary and old secondary forest interior. Quiet and unobtrusive when not displaying. Display, performed from horizontal branch or liana, starts with 1–2 abrupt jumps, followed by short, very fast, elliptical flight accompanied by strange rattle (resembling mechanical toy). Both sexes display (male more frequently), mainly in early morning, starting before dawn, and in late afternoon and at dusk. **SIMILAR SPECIES** African Broadbill produces similar but slightly lower pitched sound; lacks orange breast patches and occurs in more open and disturbed forested habitats. Juvenile African distinguished from juvenile Rufous-sided by lack of orange on underparts and streaked upperparts. Grey-headed Broadbill differs in larger size, more extensive and darker orange on underparts, no white spots on wing-coverts. Juvenile Grey-headed very similar to juvenile Rufous-sided but differs by dirty orange-buff (not dull orange) on breast and darker, less rufous upperparts without any white spots on wing-coverts. **STATUS AND DISTRIBUTION** Uncommon to locally not uncommon forest resident, recorded from Sierra Leone to Ghana and Togo (rare) and from S Nigeria to SW CAR and Congo.

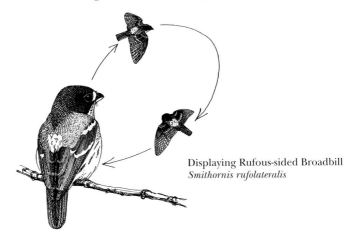

Displaying Rufous-sided Broadbill
Smithornis rufolateralis

AFRICAN BROADBILL *Smithornis capensis*   Plate 78 (4)
Eurylaime du Cap

L 13 cm (5"). **Adult male** *camarunensis* Crown black; head-sides greyish. Upperparts dark rusty-brown streaked black; feathers of lower back and rump with white bases (usually concealed at rest but puffed out in display). Underparts white, heavily streaked black on breast and flanks, and strongly washed rusty-buff on breast. Bill flat, very broad at gape, black above, white below. Western *delacouri* is very similar but slightly paler, with greyer nape. **Adult female** Duller; crown black to greyish-brown streaked black. **Juvenile** Similar to adult female but crown brown indistinctly streaked black. **VOICE** A weak plaintive *hweee*. A remarkable, loud, fast, strident rattle, similar to that of Rufous-sided Broadbill but lower pitched (sometimes likened to sound of an old-fashioned automobile horn: Zimmerman *et al.* 1996); produced during display. [CD8:85] **HABITS** Singly or in pairs in a variety of wooded habitats, incl. undergrowth of secondary forest, gallery forest, dense thickets, forest edges and plantations. Mainly in lower storey, up to 6 m above ground. Unobtrusive and easily overlooked if not displaying. **SIMILAR SPECIES** Rufous-sided Broadbill produces similar but slightly higher pitched sound, has orange breast patches and prefers interior of primary or mature secondary forest. Juvenile Rufous-sided distinguished from juvenile African by orange breast patches and unstreaked upperparts. Grey-headed Broadbill has orange-rufous on breast. **STATUS AND DISTRIBUTION** Patchily distributed, generally scarce to rare resident, recorded from Sierra Leone, Liberia, Ivory Coast (north to Comoé NP), Ghana and Togo (*delacouri*) and from E Nigeria (Mambilla Plateau), S Cameroon, CAR, Gabon, N Congo (*camarunensis*).

PITTAS
Family PITTIDAE (World: 31–32 species; Africa: 2; WA: 2)

Medium-sized, generally brilliantly coloured but secretive terrestrial forest birds with long, strong legs and short tails. Sexes similar in African species. Juveniles darker.

GREEN-BREASTED PITTA *Pitta reichenowi*   Plate 78 (8)
Brève à poitrine verte

L 19–21 cm (7.5–8.25"). **Adult** Similar to African Pitta but differs mainly in *pure white throat* and *green (not cinnamon-*

buff) breast. **Juvenile** Darker with paler supercilium. Upperparts blackish-brown; 1–2 lines of azure-blue spots on wing. Breast brownish-olive; belly pinkish. Bill brown-red with black band across centre. **VOICE** A pure, rather plaintive whistle *huu ... huu ... huu ...* uttered in series of variable speed and resembling call of Fire-crested Alethe. A short *prrrt* in flight is probably mechanical and often the first indication of the species' presence. [CD8:86] **HABITS** Poorly known. Singly in old secondary forest, forest edge and thickets. **SIMILAR SPECIES** African Pitta (q.v. and above). Juvenile African usually lacks olive on breast and has fewer azure spots in wing. **STATUS AND DISTRIBUTION** Inadequately known. Rare forest resident, recorded from S Cameroon and NE Gabon. **NOTE** Occasionally treated as conspecific with African Pitta (cf. Dowsett & Dowsett-Lemaire 1993). For discussion, see Lambert & Woodcock (1996).

AFRICAN PITTA *Pitta angolensis*   Plate 78 (7)
Brève de l'Angola

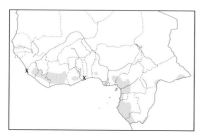

Other names E: Angola Pitta; F: Brève à poitrine fauve
L 17–22 cm (6.75–8.75"). Brightly coloured but seldom seen, thrush-like forest bird. **Adult** *pulih* Head black with long broad golden-buff supercilium. Upperparts dark green; rump and uppertail-coverts azure-blue; tail short, black. Wings blackish with azure-blue tips to coverts and white patch at base of primaries (conspicuous in flight). Throat white washed pink; breast cinnamon-buff; belly and undertail-coverts red. Bill strong, black with red spot on culmen. Legs pale. Westernmost populations (west of Dahomey Gap) often lack azure-blue tips to outer greater coverts. **Nominate** similar but tips to outer greater coverts usually tinged violet; also larger. Eastern *longipennis* has azure-blue of median wing-coverts and rump tinged violet; red on underparts less extensive; also larger and heavier. **Juvenile** Much darker. Upperparts brownish; usually some blue spots in wing. Underparts buff-brown; belly washed pink. Bill brown-red with black band near tip. **VOICE** A low frog-like croak. In display a loud, abrupt, rising and rather melodious whistle *prrruueep!* or *krrooit!* introduced by wing noise. Only the mechanical sounding wingbeats *prrt* are audible at long range. [CD8:87] **HABITS** Singly on the floor of evergreen and semi-deciduous forest. Forages by waiting motionless, then hopping to catch invertebrate prey in leaf litter. When disturbed, flies onto low branch and remains motionless, or disappears in low fast flight. Display, performed at dawn and dusk on breeding grounds, from thick horizontal branch 3–8 m above ground, consists of a quick vertical jump with rapid wingbeats, accompanied by liquid call. **SIMILAR SPECIES** Green-breasted Pitta differs mainly in green (not cinnamon-buff) breast; indistinguishable from behind or when flushed. Juvenile Green-breasted has olive on breast and more azure spots in wing. **STATUS AND DISTRIBUTION** Inadequately known. Rare forest resident with disjunct distribution. *P. a. pulih* from Sierra Leone and SE Guinea to Togo, and from S Nigeria to W Cameroon; records suggest local movements or migration. *P. a. angolensis* known to breed extralimitally in N Angola and suspected to breed in S Congo (display observed, Nov), but northern limits of breeding range unclear; probably migrates north in dry season: recorded from Gabon, S Cameroon and SW CAR. E African migrants (*longipennis*) reach SE CAR. May be less rare than records suggest. **NOTE** For discussion regarding taxonomy and distribution, see Lambert & Woodcock (1996).

LARKS
Family ALAUDIDAE (World: 85–91 species; Africa: 67–70; WA: 23)

Small to fairly small, cryptically coloured terrestrial birds with variable bill shape and long tertials almost entirely covering primaries. Sexes similar, except in sparrow larks. Juveniles similar to adults (or adult females) but feathers of upperparts often have pale fringes, giving a spotted or scaly appearance; this plumage completely (not partially) moulted immediately after fledging. Feed on insects, seeds and plants. Walk or run on the ground. Occur in various open habitats, from desert to savanna. Nest on the ground. Mainly sedentary, but some are migratory or nomadic.

   Their inconspicuous plumage pattern and coloration, which often varies within a species to match the dominant shade of the soil, make them often hard to identify. Atypically in passerines, larks have only one annual, post-breeding, moult. Fresh and worn plumages often strikingly different: with wear, pale feather tips and fringes abrade, leaving only dark centres (which fade from blackish to brownish), resulting in a more uniform plumage.

SINGING BUSH LARK *Mirafra cantillans*
Alouette chanteuse

Plate 79 (1)

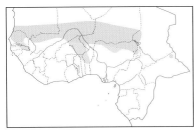

L 13 cm (5"). *M. c. chadensis*. Small, sandy-coloured lark with short, stout bill and white outer tail feathers (conspicuous on landing). **Adult** Crown sandy lightly streaked brown, bordered by creamy-white supercilium. Upperparts sandy grey-brown lightly streaked dusky; flight feathers edged rufous. Tail feathers dark brown with central pair sandy grey-brown and outer pair almost entirely white. Underparts dirty white, breast washed buff and faintly streaked brown. Bill pale horn. Legs pinkish-white. **Juvenile** Similar, but mottled dark brown above; feathers of upperparts and wings have broader pale fringes. **VOICE** Song a variable, rapid medley of short phrases consisting of varied notes; same phrase often repeated several times before starting another. Uttered from ground or low perch, or in fluttering display flight. [CD8:88a] **HABITS** In dry open grassland with scattered bushes and semi-arid thornbush. Usually unobtrusively creeps about on ground and often flushed from underfoot; hovers before dropping into grass again after low, hesitant flight. **SIMILAR SPECIES** Kordofan Lark has sandy-rufous (not sandy-brown) plumage with more contrastingly tricoloured tail. Flappet Lark darker and more streaked above, with rufous on tail and less stubby bill. Sun Lark has distinct supercilium and bolder breast streaking. **STATUS AND DISTRIBUTION** Locally not uncommon resident and intra-African migrant in Sahel zone, from S Mauritania and Senegambia to Niger, N Nigeria, N Cameroon and W Chad. Moves north during rains to breed in subdesert.

WHITE-TAILED BUSH LARK *Mirafra albicauda*
Alouette à queue blanche

Plate 79 (2)

L 13 cm (5"). Small lark *with blackish, scaly-looking upperparts*. In flight, *rufous wing panel and short tail with conspicuous white outer feathers*. **Adult male** Crown buff-brown densely streaked black; pale supercilium and eye-ring. Upperparts blackish, feathers fringed buffish, imparting scaly appearance. Flight feathers blackish-brown edged buff and rufous, basal half with rufous inner webs (visible in flight). Tail blackish with pure white outer feathers. Throat whitish; rest of underparts pale buff, mottled and streaked blackish on breast. Short, stout bill, dark above, pale below. Legs pinkish-brown. **Juvenile** Similar. **VOICE** Song, given in high display flight, consists of a series of harsh notes and some whistles, lacking Singing Bush Lark's trills. [CD8:89] **HABITS** Singly or in pairs, in open grassland. Flushes silently from short distance and drops back into grass after hovering. Unobtrusive, but not particularly shy, often permitting close approach. **SIMILAR SPECIES** Singing Bush Lark has similar structure, but is much paler. **STATUS AND DISTRIBUTION** Rare and local resident, restricted to L. Chad area, Chad.

KORDOFAN LARK *Mirafra cordofanica*
Alouette du Kordofan

Plate 79 (3)

L 14 cm (5.5"). Small, pale sandy-rufous lark with distinctive *tricoloured tail pattern* (rufous, black and white). **Adult** Top of head and upperparts pale cinnamon-rufous with some indistinct dusky streaking; when fresh, feathers fringed buff with narrow blackish subterminal crescents. Tail feathers pale cinnamon-rufous on central 2 pairs, adjacent pairs blackish and outer pair largely white. Underparts whitish, with pale cinnamon-rufous wash and faint dusky cinnamon speckles on breast. Short, stout bill whitish-horn. Legs pinkish. **Juvenile** Feathers of upperparts and wings more broadly fringed buff; breast speckled dull blackish. **VOICE** Song a sustained series of short, varied phrases which, unlike e.g. Singing Bush Lark, are not repeated. Phrases include various short trills, harsh and mellow chirps, melodious whistled notes and imitations of other species. Given from ground, tops of bushes or in high display flight. [CD8:88b] **HABITS** In open arid areas with red sandy soil (matching plumage) and low, scattered scrub. Nest undescribed. **SIMILAR SPECIES** Dunn's Lark is paler above, unmarked below, and has black-and-rufous tail, lacking white. Singing Bush Lark has less cinnamon plumage with shorter, sandy grey-brown, dark brown and white tail, lacking black and rufous. **STATUS AND DISTRIBUTION** Uncommon and widespread to rare and local resident and intra-African migrant, from Mauritania (widespread) to N Senegal (rare) east through N Mali (uncommon) and N Burkina Faso (local) to W Niger (Tahoua; rare); also E Chad (Abéché area; scarce).

RUFOUS-NAPED LARK *Mirafra africana*

Alouette à nuque rousse

Plate 79 (6)

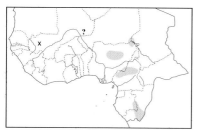

L 16–18 cm (6.25–7.0"). *Large, chunky lark* with short crest. In flight, *conspicuous rufous wing patch and relatively short tail*. Plumage variable; rufous nape distinctive at close range. **Adult** Top of head and upperparts cinnamon-buff to rich rufous-brown, lightly or heavily streaked and mottled black. Flight feathers broadly edged rufous, forming panel on closed wing. Tail blackish with buff or rufous outer feathers. Underparts creamy-buff to rich cinnamon, paler on throat and lightly or heavily streaked on breast. Bill elongated and slightly decurved, blackish above, pinkish below. *M. a. henrici* is dark and boldly marked; upperparts mainly blackish with sandy-rufous fringes to feathers; underparts rich cinnamon, breast boldly streaked; *batesi* similar but slightly paler; *bamendae* similar but breast lightly streaked; *stresemanni* rufous-brown lightly streaked black above; rich cinnamon below, breast lightly streaked; *malbranti* dull brown lightly streaked black above; warm buffish below, throat whitish, breast almost unstreaked. **Juvenile** Similar. **VOICE** Song from low perch a variable, simple phrase of 2–3 clear, high-pitched whistles, endlessly repeated, e.g. *swhee-swhuu* ... ; *tsuwee-tsweeoo* ... ; *swhu-wheeuu* ... ; *tsee-tsluu* ... ; *treelee-treeloo* ... ; *tiu twiliu* ... etc. Song flight more complex, including whistled, chirping and trilling notes, occasionally with imitations of other species. *M. a. malbranti* utters more vibrating or rapidly modulated notes, e.g. *trlip-trlip trlip-trlip...* and *writ-tirru writ-tirru... writ-tiwhiu writ-tiwhiu* ... [CD8:90-91] **HABITS** Singly or in pairs in various open habitats, incl. montane grassland, open *Acacia* scrub and fallow fields. Not easily flushed; prefers to run. When it does take to the wing, usually soon returns to cover. **SIMILAR SPECIES** None. Flappet Lark is much smaller. **STATUS AND DISTRIBUTION** Local and uncommon to common resident with patchy distribution. *M. a. henrici* Sierra Leone, Mt Nimba (SE Guinea, N Liberia, W Ivory Coast) and (this ssp.?) SW Mali; *batesi* SE Niger (L. Chad area), Nigeria (Jos-Bauchi Plateau) and (possibly this ssp.) E Mali; *bamendae* W Cameroon and possibly E Nigeria (Mambilla Plateau; Gashaka-Gumti NP); *stresemanni* C Cameroon (Adamawa Plateau); *malbranti* SE Gabon and Congo.

FLAPPET LARK *Mirafra rufocinnamomea*

Alouette bourdonnante

Plate 79 (5)

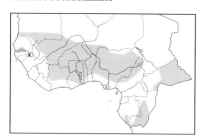

L 13–14 cm (5.0–5.5"). Small lark *with diagnostic wing-clapping display flight.* Plumage variable; all forms have rufous wing patch. **Adult** *buckleyi* Top of head and upperparts cinnamon-brown streaked blackish; pale buff or whitish supercilium. Tail feathers rufous-brown on central pair, rest blackish with cinnamon outer pair. Throat whitish or buff; rest of underparts creamy-buff to orange-buff, spotted and streaked blackish and rusty on breast. Bill dark above, pale below. *M. r. serlei* deeper rufous above and below; *tigrina* similar, but slightly less reddish above than *serlei*; *schoutedeni* rather pale brown, lightly marked above. **Juvenile** Generally duller. **VOICE** Most characteristic sound a dry, rattling *prrrrrrp,* produced by rapid wing-flaps during undulating, cruising display flight and occasionally accompanied by a few thin notes. Rarely heard song consists of short phrases of melodious whistled notes, uttered from ground or perch. [CD8:93] **HABITS** Singly or in pairs, in various grassy habitats. **SIMILAR SPECIES** No other species has similar display flight. Rufous-naped Lark is much larger and bulkier. Singing Bush Lark is paler, with pale stubby bill. Sun Lark has slight crest and no rufous in tail. **STATUS AND DISTRIBUTION** Uncommon to common and often local resident throughout, except arid north and forest zone. *M. r. buckleyi* S Mauritania to Guinea east through Ghana to S Niger, Nigeria, N Cameroon and W Chad (rare); *serlei* SE Nigeria; *tigrina* E Cameroon to E CAR; *schoutedeni* Gabon, S CAR and Congo.

RUSTY BUSH LARK *Mirafra rufa*

Alouette rousse

Plate 79 (4)

L 13–14 cm (5.0–5.5"). Small, rufous lark with comparatively longer tail than other bush larks, *lacking any white.* Plumage variable; all forms have rufous wing patch. **Adult** *rufa* Top of head and upperparts rufous-brown variably streaked black and buff (sometimes almost plain). Tail feathers wholly rufous-brown on central pair, rest blackish with narrow buff edges to outer pair. Supercilium and throat pale buff; rest of underparts creamy-buff, variably washed rufous and streaked blackish on breast. Bill dark above, pale below. *M. r. nigriticola* darker, less streaked above and deeper buff below, with heavier, more reddish streaking on breast. **Juvenile** Generally duller. **VOICE**

Song, uttered in cruising display flight, described as 'pleasing' (Bannerman 1936). **HABITS** Singly or in pairs in dry open bush country with rocks. Perches on bushes. **SIMILAR SPECIES** Kordofan Lark is paler, with tricoloured tail and uniformly pale bill. Dunn's Lark is even paler above, with white, unstreaked underparts. Flappet Lark has richer coloured underparts and shorter tail. **STATUS AND DISTRIBUTION** Locally not uncommon resident, from NE Mali to W Niger (*nigriticola*) and in E Chad (Abéché area; *rufa*). Vagrant, N Togo.

RUFOUS-RUMPED LARK *Pinarocorys erythropygia*   Plate 79 (9)
Alouette à queue rousse

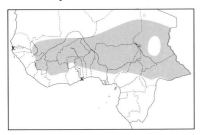

L 18 cm (7"). Distinctive. Rather large, slender lark with well-marked head pattern, rufous rump contrasting with dark upperparts, long rufous-edged tail and heavily marked breast. **Adult** Top of head and upperparts dark brown. Head-sides and supercilium whitish with dark brown eye-stripe and crescent from behind eye across ear-coverts to bill, joining narrow malar stripe. Primaries edged rufous. Rump, uppertail-coverts and base of tail rufous; tail feathers dark brown edged rufous (except central pair). Underparts pale buffish, heavily streaked dark brown on breast; flanks and undertail-coverts rufous. Bill pale horn tipped dusky. Legs long, pinkish-white. **Juvenile** Paler above, feathers tipped cinnamon; breast streaking paler brown. **VOICE** Flight call *wree*. Song a short, often repeated series of clear, far-carrying whistles with some variation, e.g. *seeuu... seeuu.. seeuu...* and *seeuu-seeuu-teeu-teeu*, also *wree-tseeu wree-tseeu..., tee-tseeuu tee-tseeuu...* and *trrru trrru trrru...*; uttered in soaring display flight high above territory. [CD8:94] **HABITS** Singly, in open savanna woodland and farmland. Prefers recently burnt areas. Flight flapping with legs dangling. Frequently perches in trees. **STATUS AND DISTRIBUTION** Not uncommon but local intra-African migrant, breeding from N Ivory Coast and N Ghana to S Chad and CAR, migrating west and north in non-breeding season (Mar/Apr–Oct/Nov, reaching Sierra Leone, N Niger and N Chad. Vagrant, Senegambia.

GREATER HOOPOE LARK *Alaemon alaudipes*   Plate 80 (10)
Sirli du désert

Other name E: Hoopoe-Lark
L 18–20 cm (7–8"). *A. a. alaudipes*. Unmistakable. Large, elongated, pale sandy lark with *long, black, decurved bill and long grey-white legs*. In flight, *conspicuous black-and-white pattern on wing* diagnostic. **Adult male** Top of head and upperparts pale sandy-buff tinged greyish; head-sides marked with long creamy supercilium bordered below by narrow dusky eye-stripe, and dusky moustachial stripe. Tail feathers sandy-buff on central pair, rest blackish narrowly fringed whitish. Underparts white, breast creamy variably spotted blackish. **Adult female** Similar, but smaller. **Juvenile** Upperparts browner, feathers have buff tips and fringes, and dusky subterminal bars. **VOICE** Song a piping, accelerating and ascending series of pure drawn-out whistles, occasionally interspersed with a short buzzy trill. [CD3:37] **HABITS** Singly or in pairs in sandy desert. Not easily flushed; prefers to run, often over long distances. Not shy. Spectacular display flight: male rises from low perch with fluttering wings and spread tail, rolls over and dives back to same perch on closed wings. **STATUS AND DISTRIBUTION** Common to uncommon resident, Mauritania, NW Mali, Niger, Chad and Cape Verde Is. Partial migration recorded, e.g. Senegal (where rare/vagrant).

THICK-BILLED LARK *Rhamphocoris clotbey*   Plate 80 (11)
Alouette de Clot-Bey

L 17 cm (6.75"). Unmistakable. Medium-sized, sandy-coloured lark with *conspicuous large, stubby bill and heavily spotted underparts*. In flight, large head, rather long wings with bold black-and-white pattern above, and relatively short tail. **Adult male** Head distinctly patterned with *white eye-ring and black on sides*, with white spot at base of lower mandible and on cheeks. Top of head and upperparts sandy grey-buff. Tail sandy-buff with blackish subterminal band (except on centre) and white outer feathers. Underparts white boldly spotted black from breast to belly and flanks. Black flight feathers with broad white trailing edge to secondaries and inner primaries;

underwing wholly blackish with white trailing edge. Bill yellow-horn tipped dusky. Legs greyish. **Adult female** Facial pattern and markings on underparts duller. **Juvenile** Upperparts and head-sides cinnamon. Underparts creamy-white, breast washed cinnamon, without black streaking. **VOICE** Call in flight *prit* or *blit-blit*. Song a short jingle of clear, sweet and tinkling notes. [CD3:38] **HABITS** Singly, in pairs or small flocks in stony subdesert. **STATUS AND DISTRIBUTION** Rare Palearctic wanderer to NW Mauritania.

BAR-TAILED LARK *Ammomanes cincturus*   Plate 80 (1)
Ammomane élégante

Other name E: Bar-tailed Desert Lark

L 15 cm (6"). Plain, warm sandy-coloured lark. In flight, pale orange-rufous in wings and well-defined black terminal bar to pale orange-rufous tail conspicuous. **Adult *arenicolor*** Head and upperparts orange-sandy to sandy-buff. Underparts whitish washed buff on breast. Flight feathers pale orange-rufous, tipped black on outer primaries. Bill pale. Legs whitish to pinkish-grey. **Nominate** slightly deeper coloured. **Juvenile** Black in tail and on primary tips reduced and less sharply demarcated. **VOICE** Calls include *trrlup* and *wreelup*, often repeated in series; on take-off *trup trup trup...* Song a monotonously repeated, clear, high-pitched, drawn-out whistle *tsu-wheeh, tsu-wheeh, tsu-wheeh...* slightly reminiscent of creaking iron gate; uttered during undulating display flight or, occasionally, from ground. [CD3:40] **HABITS** Singly or in pairs in sand or stone deserts with very sparse vegetation. Also in small flocks when not breeding. **SIMILAR SPECIES** Desert Lark is greyer, with black of tail ill defined and gradually merging with rufous; also has stouter and darker bill, and longer primary projection. Dunn's Lark best distinguished by different tail pattern: mainly blackish with pale sandy-cinnamon central feathers; also has slightly streaked upperparts, almost no primary projection and swollen bill. Female Black-crowned Sparrow Lark lacks wing and tail pattern. **STATUS AND DISTRIBUTION** Uncommon to common but local resident or local migrant, Mauritania, WC Mali, Niger (mainly Ténéré) and Chad (*arenicolor*). Common resident, Cape Verde Is (nominate). Erratic.

DESERT LARK *Ammomanes deserti*   Plate 80 (2)
Ammomane isabelline

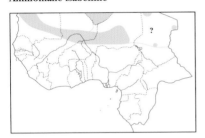

L 15–16 cm (6.0–6.25"). ¶ Plain grey-brown to sandy-coloured lark. Plumage coloration varies geographically with colour of soil. Spread tail has *dull rufous base and broad, ill-defined, blackish terminal band.* **Adult** Head and upperparts pale buff-grey to sandy-cinnamon; darkest birds grey-brown to slate-grey. Tail rather dark (darker than upperparts); outer webs of feathers rufous. Usually has ill-defined short pale supercilium and narrow pale eye-ring. Throat whitish; rest of underparts sandy-buff to deep tawny-buff, paler on belly and vent; upper breast usually diffusely streaked. Flight feathers darker than upperparts with rufous outer webs. Bill yellowish-horn with dusky culmen and tip. Legs yellowish-brown. **Juvenile** Wing feathers narrowly tipped pale buffish. **VOICE** A weak *chup* or *chewup*. Song a repeated, simple series of trilling syllables *treewrrurrip, treewrrurrip, ... , wrulp-whrrreew, wrulp-whrrreew, ..., treewrrurrip...* uttered from perch or in short, undulating display flight. [CD3:41] **HABITS** Singly, in pairs or small, loose flocks, in arid areas with sparsely vegetated, rocky and sloping terrain. Not shy. **SIMILAR SPECIES** Pale forms may resemble Bar-tailed Lark, but latter distinguished by well-defined black tail band and pale orange-rufous flight feathers. Dunn's Lark best separated by different tail pattern: mainly blackish with pale sandy-cinnamon central feathers; also has slightly streaked upperparts, almost no primary projection and pale, swollen bill. **STATUS AND DISTRIBUTION** Locally common resident, from Mauritania to Mali and W and S Niger (*geyri*), in N Burkina Faso (unknown ssp.), N Niger (Aïr: *mya*), and Chad (Tibesti: *algeriensis*; Ennedi: *kollmanspergeri*; from Ndjamena eastwards: *erythrochrous*). **NOTE** Plumage variation complex, with clinal intergradation and variation within populations according to local soil types. *A. d. geyri* pale sandy grey-brown above, sandy-buff below; *algeriensis* tinged cinnamon-pink above, buff below; *mya* as *algeriensis* but larger and with heavier bill; *kollmanspergeri* darker and more reddish-brown above than *geyri*, more rufous below; *erythrochrous* similar but darker rufous below. Unknown ssp. in Burkina Faso dark slate-grey above (Fishpool *et al.* 2000).

GREATER SHORT-TOED LARK *Calandrella brachydactyla*   Plate 80 (5)
Alouette calandrelle

Other name E: Short-toed Lark

L 13–14 cm (5.0–5.5"). ¶ Small, pale lark with *streaked upperparts, broad pale supercilium* and *long tertials almost*

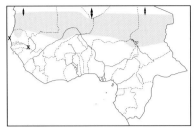

completely covering primaries. Plumage coloration variable. **Adult** Top of head and upperparts pale sandy-brown to sandy-grey streaked darker; crown often tinged rufous. Tail feathers blackish, broadly edged cinnamon on central pair, mainly white on outer pair. *Underparts whitish and unstreaked, with variable narrow, dark crescent-shaped patch on breast-sides;* breast variably washed pale buff and occasionally with some faint streaks, esp. below patches. Wing feathers dark brown tipped and edged buffish; dark centres to median coverts may form wingbar. Bill pale horn. Legs pinkish. Crown feathers may be raised. **Juvenile** Breast lacks patches and has some dusky spots or streaks. **VOICE** A hard, dry *prrt, prrt-trrt.* Song in wintering grounds a weak rattling jumble of harsh and buzzy notes. Full song, given in bouncing flight, consists of series of short, rapid, varied phrases, often including imitations. [CD3:42] **HABITS** In open arid and semi-arid country. Highly gregarious; often in large flocks. **SIMILAR SPECIES** Lesser Short-toed Lark has distinct primary projection, streaked breast without patch on sides, shorter bill and more indistinct head pattern with less pronounced supercilium. **STATUS AND DISTRIBUTION** Uncommon to locally common Palearctic winter visitor in wide Sahel belt, from Mauritania and N Senegal to Niger, extreme NE Nigeria and Chad. **NOTE** Races differ mainly in tone of upperparts and are very difficult to identify; their distribution in our region is very poorly known. **Nominate**, presumably the most widespread race, is warm brown and prominently streaked above; *rubiginosa* is more rufous with finer streaking; *longipennis* greyish-sandy with fine streaks.

RED-CAPPED LARK *Calandrella cinerea*   **Plate 80 (6)**
Alouette cendrille

L 14–15 cm (5.5–6.0"). *C. c. saturatior.* The only lark in the region with *rufous crown bordered by white supercilium* and *rufous patch on breast-sides.* In flight, dark tail contrasts with rufous rump. **Adult** Upperparts dark brown streaked blackish; rump more rufous-brown; uppertail-coverts rufous; tail blackish with white margins. Underparts white with rufous patch on breast-sides and variable amount of rufous on flanks. Wing feathers blackish fringed brown. Crown feathers may be erected to form short crest. Bill and legs blackish. **Juvenile** Darker. **VOICE** Calls include a sparrow-like *chirrup,* a hard *pit* and a sharp *tsreee.* Flocks produce a soft twittering. Song, given in high display flight, a short, varied jumble of high-pitched notes and trills, including imitations of other birds. [CD8:96] **HABITS** Singly or in pairs in grassland, fields and recently burnt areas. Male in display rises almost vertically, then flies out to wind with slowly flapping wings, rising and dipping, singing for up to 10 mins, finally descending gradually and dropping on closed wings, opened just prior to landing. **STATUS AND DISTRIBUTION** Uncommon and very local resident in C Nigeria (Jos Plateau). Vagrant, Gabon and Congo.

LESSER SHORT-TOED LARK *Calandrella rufescens*   **Plate 80 (4)**
Alouette pispolette

L 13 cm (5"). *C. r. minor.* Similar to Greater Short-toed Lark, but with distinct primary projection (3 tips exposed), variably streaked breast without black patches, shorter, more stubby bill, and more indistinct head pattern with less pronounced supercilium. Plumage coloration variable. **Adult** Head and upperparts pale sandy-grey to sandy rufous-buff streaked darker. Tail feathers blackish, with central pair broadly edged cinnamon and outer pair mainly whitish. Underparts whitish washed pale buff and finely streaked dark on breast, streaks extending more faintly on flanks. Wing feathers dark brown to blackish, broadly tipped and edged buffish. Bill pale horn. Legs pinkish. **Juvenile** Feathers of upperparts have pale buff tips and black subterminal bars. **VOICE** A rippling *prrrt* and *trr chrrr.* Song, given from ground and in flight, a continuous jumble of varied notes, including many imitations of other birds. [CD3:43] **HABITS** In dry open country, often in small flocks. **SIMILAR SPECIES** More widespread Greater Short-toed Lark lacks primary projection, has virtually unstreaked underparts with dark breast patches, longer, more pointed bill and prominent pale supercilium sharply delimiting crown (producing capped impression). Juveniles best separated by primary projection. **STATUS AND DISTRIBUTION** Rare Palearctic winter visitor to coastal Mauritania. Also claimed from Nigeria (Jos Plateau).

DUNN'S LARK *Eremalauda dunni*   **Plate 80 (3)**
Alouette de Dunn

L 14 cm (5.5"). *E. d. dunni.* Warm sandy-coloured lark with wholly pale, stubby bill. In flight, black outer tail

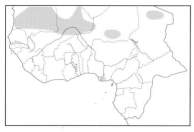

contrasts with pale centre. **Adult** Head and upperparts pale orange-sandy or sandy-cinnamon, with slightly darker centres producing faint, diffuse streaking (hard to see when worn). Underparts whitish washed sandy-cinnamon on breast-sides. Flight feathers slightly darker than upperparts, diffusely tipped dusky on outer primaries; tertials long, almost completely covering primaries. Bill pale creamy-horn. Legs pinkish-white. **Juvenile** Darker; no dusky primary tips. **VOICE** Calls include *chiup* and *chruip* or *prrip*. Song, given from ground or low perch, a fast, regularly repeated *wit-wit-wtrrreedridridridrree*; in display flight variable, with whistling and rattling notes. [CD3:44]

HABITS Singly or in pairs in desert and subdesert with low grasses. In small flocks when not breeding, often with other lark species, esp. sparrow larks. Not shy. Male rises in display, singing, then hovers, finally descending slowly on outstretched wings. **SIMILAR SPECIES** Desert Lark is plain and greyer above, with blackish distal half of tail gradually merging with rufous base; also has darker, more pointed bill and longer primary projection. Bar-tailed Lark most easily separated by different tail pattern: pale sandy-rufous with black terminal band; also has plain upperparts, short primary projection and much smaller bill. Kordofan Lark has tricoloured tail with white outer feathers. Female Black-crowned Sparrow Lark lacks wing and tail pattern. **STATUS AND DISTRIBUTION** Uncommon to rare resident at southern edge of Sahara and in N Sahel, in Mauritania, N Mali, C Niger and N Chad. Distribution irregular; seasonal fluctuations recorded.

SUN LARK *Galerida modesta*   Plate 79 (7)
Cochevis modeste

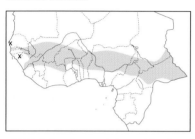

L 14 cm (5.5"). Small, brown lark with prominent supercilium surrounding streaked cap, and heavily streaked upperparts and breast. Crown feathers may be raised to form short crest. Plumage coloration varies geographically and with wear. **Adult** *modesta* Head and upperparts sandy-rufous heavily streaked blackish; neck-sides and hindneck less streaked. Supercilium buff-white to sandy-rufous, joining nape and bordered below by narrow black eye-stripe. Underparts dirty white washed tawny on breast and streaked blackish on throat-sides and breast. Tail blackish edged pale rufous. Flight feathers blackish narrowly edged buff to rufous. Bill blackish above, whitish below. Legs pale brownish. *G. m. bucolica* darker; *strumpelli* even darker, with pale brownish underparts more heavily streaked on breast; *nigrita* darker still. **Juvenile** Feathers of crown and upperparts tipped whitish; wing-coverts broadly fringed whitish (forming 2 wingbars). **VOICE** Song a short, fast, grating warble, frequently repeated and uttered from ground or bushes, or in flight. A second song type, given in hovering display flight, is much more varied and sustained, with whistling, grating and buzzy notes, and imitations of other species. Also a lisping, trisyllabic *thli-thli thli.* [CD8:97] **HABITS** Singly or in pairs in various open grassy habitats, often with rocky areas or lateritic soils. In small flocks outside breeding season. **SIMILAR SPECIES** Flappet Lark has indistinct (not heavy) streaking on breast, and pale feather fringes produce scaly rather than streaked impression. Singing Bush Lark paler with faint speckling on breast, less sharply defined supercilium and wholly pale bill. **STATUS AND DISTRIBUTION** Locally not uncommon to common resident in savanna belt, from S Senegambia, Guinea and Sierra Leone to Cameroon, S Chad and CAR. Nominate almost throughout; *nigrita* Guinea and Sierra Leone; *strumpelli* Cameroon Plateaux; *bucolica* SE CAR. Local movements recorded.

CRESTED LARK *Galerida cristata*   Plate 79 (8)
Cochevis huppé

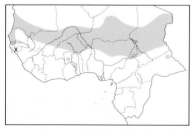

L 17 cm (6.75"). Rather stocky with *distinct spiky crest*, relatively short tail, and slightly decurved, longish bill. Plumage coloration and markings vary geographically and with wear. **Adult** *senegallensis* Upperparts pale grey-brown variably streaked dusky; distinct pale supercilium; underparts off-white with buff wash and dark streaks on breast; tail edged buff. Bill dusky above; paler below. Legs pale pinkish-brown. *G. c. balsaci* and *alexanderi* very similar but *balsaci* whiter below, *alexanderi* more cinnamon-brown above with faint streaks. *G. c. jordansi* more rufous and less marked than other races, plain rufous-brown above, rufous-buff below with indistinct small spots on breast; *isabellina* like *alexanderi* but upperparts more cinnamon, less streaked. **Juvenile** Crest shorter; feathers of crown and upperparts tipped whitish; wing-coverts broadly fringed whitish or buff. **VOICE** A liquid *doo-leeoo* and *diui*, with variations. Song loud, varied and sustained, with fast clear whistles and throaty twitters, occasionally including imitations of other birds. Also a short, fast and clear *tee-titee-titiu*. Uttered from perch or ground and in flight. [CD3:46] **HABITS** In pairs or family parties in various open habitats, incl. farmland, roadsides, savanna, areas with low scrub and bare soil, and sea shores. **SIMILAR SPECIES** See Thekla Lark. No other species has similar crest. **STATUS AND DISTRIBUTION** Not uncommon to locally common resident and partial migrant in broad belt, from S Mauritania, Senegambia and Guinea-Bissau to Niger, also N Ghana (*senegallensis*), N Nigeria, N Cameroon, Chad and NE CAR (*alexanderi*). *G. c. balsaci* coastal Mauritania; *jordansi* N Niger (Aïr); *isabellina* possibly NE Chad.

THEKLA LARK *Galerida thekla*   Plate 79 (10)
Cochevis de Thékla

L 17 cm (6.75"). *G. t. aguirrei* (?). Very similar to Crested Lark and field identification usually hard (sometimes impossible); difficulty increased by variation in plumage coloration and markings in both species. **Adult** Upperparts rufous-brown to pale cinnamon-rufous streaked blackish; underparts off-white to pale buff with prominent dark streaks on breast. Combination of following features generally useful to separate from Crested Lark: slightly shorter, more spread crest rising from forecrown (longer and more spiky, rising more steeply from mid-crown in Crested); shorter bill with almost straight culmen (longer, with more gradually decurved culmen in Crested); often better marked head pattern, with more striking pale supercilium (more diffuse in Crested), and darker edges to ear-coverts clearly delimited by pale half-collar; bolder, neater streaking on whiter breast (usually finer and more diffuse in Crested); cinnamon-tinged rump contrasting with back and tail (no contrast in Crested); almost no primary projection in fresh plumage (slight in Crested, with at least 2 tips visible); slightly more compact appearance (Crested may look elongated in comparison). Often perches atop bushes (Crested usually on ground). **Juvenile** Similar to juvenile Crested. **VOICE** Commonest call a fluting *too-tee-tweeeu* or *too-tee-tu-teeu*. Song similar to that of Crested Lark but softer and lower pitched. [CD3:47] **HABITS** Singly or in pairs in dry open country. **SIMILAR SPECIES** Crested Lark (q.v. and above). **STATUS AND DISTRIBUTION** Palearctic vagrant (rare resident?), NW Mauritania (presumably *aguirrei*). **NOTE** Sometimes considered conspecific with *G. malabarica* from India, with which it forms a superspecies.

EURASIAN SKYLARK *Alauda arvensis*   Plate 147 (4)
Alouette des champs

L 18–19 cm (7.0–7.5"). ¶ Robust, streaked lark with short crest. In flight, white trailing edge to wing and white outer tail feathers. **Adult** Crown and upperparts brown heavily streaked black; diffuse pale supercilium; underparts whitish with rufous-buff wash and dense blackish streaks on breast. Bill dark horn above; paler below. Legs pinkish-brown. **VOICE** Flight call a characteristic rippling *trruwee* or *chirrup*. [CD3:49] **HABITS** In our region only found singly in arid areas. Flight fluttering. **STATUS AND DISTRIBUTION** Palearctic vagrant, NW Mauritania (ssp. unknown). **NOTE** Geographical variation slight and general coloration variable, with intergradation and individual variation within populations. Race of majority of non-breeding visitors to N Africa unknown. Possible vagrants to our region include *harterti*, nominate, *cantarella* and *sierra*.

RASO LARK *Alauda razae*   Plate 143 (2)
Alouette de Razo

Other name F: Alouette des Îles du Cap-Vert
L 12–13 cm (4.75–5.0"). **Adult male** Head and upperparts dull grey streaked blackish; tail blackish with white outer feathers. Throat white; rest of underparts off-white washed cream-buff with narrow black streaks on breast. Wings short and rounded, without pale trailing edges. Bill relatively long, greyish-horn. **Adult female** Bill noticeably shorter. **Juvenile** As adult but buffier. **VOICE** Call on take-off reportedly similar to rippling *trruwee* or *chirrup* flight call of Eurasian Skylark. Song, given in vertical display flight or from ground or perch, loud, clear and sustained, reminiscent of Eurasian Skylark but more repetitive and less varied. [CD3:50] **HABITS** Inadequately known. Mainly on flat ground, sparsely vegetated with herbage and low bushes. Outside breeding season in flocks of up to 25. Not shy. **STATUS AND DISTRIBUTION** Endemic resident, restricted to arid islet of Raso, Cape Verde Is, where the only lark. Population *c.* 50–250, fluctuating in response to climate. *TBW* category: THREATENED (Critical).

CHESTNUT-BACKED SPARROW LARK *Eremopterix leucotis*   Plate 80 (7)
Moinelette à oreillons blancs

Other names E: Chestnut-backed Finch Lark; F; Alouette-moineau à oreillons blancs
L 12 cm (4.75"). *E. l. melanocephala*. Small dark lark with distinctive male plumage. **Adult male** *Head black with large*

white patch on ear-coverts and small white half-collar on nape. Upperparts rich chestnut becoming pale buffish on rump and with whitish patch on shoulder. *Underparts and underwing black* with whitish flanks. Bill pale horn. **Adult female** Much duller and variable, lacking black-and-white pattern of male. Crown brownish mottled darker; head-sides buffish mottled dusky on ear-coverts. Upperparts dull chestnut. Underparts pale buff with throat and breast mottled brownish and centre of belly dark brown. **Juvenile** Similar to adult female, but feathers of head and upperparts broadly tipped white. **VOICE** A short, hard *chip-chip* or *twit-twit*. Song a short, rapid phrase with harsh, grating notes *krr-krr-krr-chreep krr-krr-krr-chreep-chreep...*, frequently repeated and given from ground or low perch and in flight. [CD8:98] **HABITS** In pairs in various open areas with bare soil and short grass or scrubs; partial to recently burnt ground. Often in small flocks when not breeding. Forages actively, shuffling low on ground; may perch on bushes or trees. **SIMILAR SPECIES** No other lark in the region has chestnut upperparts. Black-crowned Sparrow Lark (q.v.) much paler. **STATUS AND DISTRIBUTION** Common resident and intra-African migrant in semi-arid belt, from S Mauritania and Senegambia to Niger, N Nigeria, N Cameroon and Chad. Moves north with rains. Erratic; often in fluctuating numbers.

BLACK-CROWNED SPARROW LARK *Eremopterix nigriceps*   Plate 80 (8)
Moinelette à front blanc

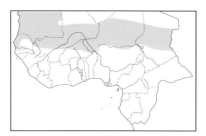

Other names E: Black-crowned Finch Lark; F: Alouette-moineau à front blanc
L 11–12 cm (4.25–4.75"). *E. n. albifrons/ nigriceps*. Small, compact lark with distinctive male plumage and, in both sexes, black underwing-coverts and black outer tail-feathers. **Adult male** Bold pied head pattern with black area from crown through eye joining black throat, and black half-collar on neck-sides, contrasting with white forehead and large white cheek patches. Upperparts pale grey-brown. Underparts and underwing black with white patch on breast-sides. Tail blackish broadly edged buff-white and with pale grey-brown central feathers. Bill pale horn or pale bluish-grey. **Adult female** Very different. Head and upperparts pale sandy-cinnamon faintly streaked on crown; patch around eye and neck-sides whitish. Underparts whitish with pale cinnamon breast-band; underwing-coverts black. Tail as male but central feathers pale sandy-cinnamon. **Nominate** usually slightly darker and more rufous. **Juvenile** Similar to adult female, but feathers of head and upperparts broadly tipped pale buff with faint narrow dark subterminal bar; breast diffusely spotted dusky. **VOICE** An abrupt, buzzing *eezp* or *jeep*; also a quiet *djib*. Song, usually given in flight, a pleasant, simple, persistently repeated, rapid phrase of high, clear whistles *chiWEE-chreep* or *chichiWEE-chireep*, and variations. [CD8:99] **HABITS** In pairs in dry open areas with sparse vegetation. In small flocks when not breeding. **SIMILAR SPECIES** Adult female may superficially resemble other small sandy-coloured larks, but distinguished by black underwing-coverts and smaller size. **STATUS AND DISTRIBUTION** Uncommon to locally common resident in semi-arid and arid belts, in Mauritania, Senegal, Mali, Burkina Faso, Niger, N Nigeria and C Chad (*albifrons*). Locally common, Cape Verde Is (nominate).

TEMMINCK'S HORNED LARK *Eremophila bilopha*   Plate 80 (9)
Alouette bilophe

L 14 cm (5.5"). Pale sandy-coloured lark with distinctive, contrasting adult face pattern. **Adult male** Unmistakable. Crown, nape and upperparts pale orange-sandy or sandy-cinnamon; *narrow black band on forecrown ending in delicate 'horns' on sides*; forehead, head-sides and underparts white with black patch from lores to cheeks and black breast-band; breast-sides and flanks washed orange-sandy. Tail blackish with white edges and pale orange-sandy centre. Underwing white. Bill and legs dark greyish. **Adult female** Black areas often duller and slightly reduced; horns shorter. **Juvenile** Lacks black markings and is faintly spotted and scalloped white on upperparts. **VOICE** Calls include *tsip*, *seeuu* and *sweeeup*. Song, uttered in flight and on ground, varied but rather simple, with clear whistles, dry grating notes, melodious twittering and soft warbling. [CD3:52] **HABITS** Singly or in pairs in desert and subdesert. In small flocks outside breeding season, regularly in company of other larks. **STATUS AND DISTRIBUTION** Rare resident or Palearctic migrant, NW Mauritania.

SWALLOWS AND MARTINS
Family HIRUNDINIDAE (World: *c.* 75 species; Africa: 38; WA: 31)

Distinctive, small, highly specialised aerial insectivores with slender body, short neck and pointed wings. Bill very short, but gape broad; legs short. In some, tail deeply forked with long streamers, often with white 'windows' (white patches on inner webs of feathers). Sexes similar; in long-tailed species, females have shorter outer tail feathers. Juveniles similar but duller. Regularly perch on wires and bare branches. Most species gregarious when not breeding (some also when breeding), often forming mixed groups with other hirundines. Many are migratory. Main genera are:

Psalidoprocne, saw-wings (4 species): entirely or largely black hirundines, deriving their name from serrated outer edge of outer primary (in males only). Tail square or forked with very long streamers. Flight slow and fluttering. Nest in self-dug burrows (3 spp.) or on ledge (1 sp.).

Riparia, sand martins (4 species): small to medium-sized with brown upperparts and brown or white underparts. Tail square or slightly forked. Nest in self-dug burrows in sandy ground.

Hirundo, swallows (19 species): largest genus, including different-looking groups. All construct mud nests plastered under overhanging surface or on ledge. Crag martins (2 spp.; sometimes placed in *Ptyonoprogne*) are rather plain coloured with slightly notched tails; cliff swallows (sometimes placed in *Petrochelidon*) have buff or rufous rumps and squarish or slightly notched tails.

Remaining genera *Pseudochelidon*, *Phedina*, *Pseudhirundo* and *Delichon*, each have only one representative in the region.

AFRICAN RIVER MARTIN *Pseudochelidon eurystomina*   Plate 81 (8)
Pseudolangrayen d'Afrique

Other name F: Hirondelle de rivière
L 14 cm (5.5"). Robust, all-black large-headed swallow with bright reddish bare parts. In flight, broad-based wings and shortish tail produce character-istic *compact, triangular silhouette*. **Adult** Black, glossed metallic blue or green. *Bill bright orange-red tipped yellow*. Eye scarlet with pink orbital ring. Feet pinkish. **Juvenile** Duller, blackish-brown above, greyish-brown below; bare parts duller. **VOICE** Distinctive. Call a hard, grating *dzreh dzreh ...* or *chèrr chèrr...* [CD9:1] **HABITS** Gregarious. Forages over large rivers, forest, savanna and human habitation. Perches on wires, roofs, branches or ground. Breeds in large colonies in burrows excavated in sand banks and in grassy savanna with sandy soil. **STATUS AND DISTRIBUTION** Highly local and seasonal. Breeds coastal Gabon (Port Gentil area and Gamba, Sep–Nov), coastal Congo (Conkouati Reserve, Sep–Nov), and on sand banks in middle Congo R. (between Lukolela and Basoko) and lower Ubangi R. (Jan–May). Migration observed, NE and C Gabon (Makokou and Lopé; Dec–Mar east and Jun–Sep west). *TBW* category: Data Deficient.

SQUARE-TAILED SAW-WING *Psalidoprocne nitens*   Plate 83 (5)
Hirondelle à queue courte

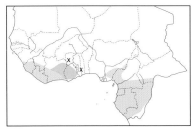

Other name F: Hirondelle hérissée à queue courte
L 11 cm (4.25"). *P. n. nitens.* **Adult** Small, entirely blackish swallow with notched tail. Pale greenish gloss only visible under favourable circumstances. **Juvenile** Duller. **VOICE** Calls include soft, rather hoarse *pzzuit, pseeru, pruruit* and *psit.* [CD9:2] **HABITS** In pairs or small flocks in lowland forest. Forages above canopy, in clearings and along tracks and edges. In west of range often with Fanti Saw-wing. Flight slow and fluttering, typical of genus. Pairs breed solitarily, in burrow excavated in earth bank. **SIMILAR SPECIES** Forest Swallow very similar, but voice different and flight faster, with much gliding and back less hunched. Other saw-wings have forked tails. **STATUS AND DISTRIBUTION** Common to uncommon or rare forest resident from Guinea and Sierra Leone to Togo and from SE Nigeria to CAR and Congo.

FANTI SAW-WING *Psalidoprocne obscura*   Plate 83 (10)
Hirondelle fanti

Other names F: Hirondelle hérissée, Hirondelle de Fanti
L 17 cm (6.75"). **Adult male** Small, all-blackish swallow with very long, strongly forked tail (depth of fork *c.* 5.0–

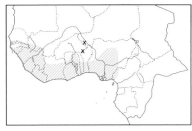

7.5 cm). Underwing dark. Greenish gloss visible in favourable circumstances. **Adult female** Tail streamers typically shorter (depth of fork *c.* 3.0–6.5 cm). **Juvenile** Browner; tail shorter. **VOICE** Mainly silent. Call a soft *sheep.* [CD9:3] **HABITS** In pairs or small flocks in forest clearings, along tracks and edges, in farmbush, woodland and savanna. Flight slow and fluttering, typical of genus. Pairs breed solitarily, in burrow excavated in earth bank or similar site. **SIMILAR SPECIES** Black Saw-wing, which replaces Fanti in east of region (with some overlap in E Nigeria/W Cameroon: ssp. *petiti*), has shorter tail and white underwing-coverts. **STATUS AND DISTRIBUTION** Common resident and intra-African migrant, from Senegambia and SW Mali to Nigeria and W Cameroon. Locally and partially migratory, some moving north in wet season. Vagrant, SE Burkina Faso and SW Niger.

BLACK SAW-WING *Psalidoprocne pristoptera*   Plate 83 (9)
Hirondelle hérissée

Other name F: Hirondelle hérissée bleue
L 13 cm (5"). **Adult male *petiti*** Small, entirely blackish swallow with long, forked tail (depth of fork *c.* 2.5–3.5 cm) and *whitish underwing-coverts.* Plumage has bronze-brown gloss. *P. p. chalybea* has *grey* underwing-coverts and greenish gloss. **Adult female** Tail streamers usually shorter. **Juvenile** Duller, browner; underwing-coverts dusky whitish; tail shorter. **VOICE** Mainly silent. Call a soft, nasal *sheeu.* [CD9:4] **HABITS** In pairs or small flocks in savanna, montane grassland, edges of gallery forest and farmbush. Forages mainly low, with slow, fluttering flight, typical of genus. Typically perches on bare branches of tall trees, but also in smaller savanna trees. Joins mixed flocks of hirundines and swifts. Pairs breed solitarily, in burrow excavated in earth bank or similar site. **SIMILAR SPECIES** Fanti Saw-wing, which replaces Black in west of region (with some overlap in E Nigeria/W Cameroon), has longer tail and black underwing-coverts. Mountain Saw-wing has shorter tail, and is browner and less slender. **STATUS AND DISTRIBUTION** Common to uncommon resident and intra-African migrant. *P. p. petiti* from SE Nigeria (Obudu and Mambilla Plateaux) to Congo; *chalybea* in N CAR (Bamingui-Bangoran NP) and, possibly, N and C Cameroon. Movements inadequately known; presence in some areas related to season. **NOTE** The two forms in our region are sometimes considered specifically distinct from *pristoptera* as Petit's Saw-wing *P. petiti* and Shari Saw-wing *P. chalybea.*

MOUNTAIN SAW-WING *Psalidoprocne fuliginosa*   Plate 83 (8)
Hirondelle brune

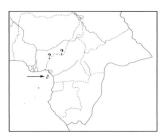

Other names E: Cameroon Mountain Rough-winged Swallow; F: Hirondelle hérissée brune
L 12 cm (4.75"). Drabbest saw-wing. **Adult** Small, entirely dull dark brown swallow with *moderately forked tail* (depth of fork *c.* 1.5–2.5 cm). Underwing-coverts paler, more grey-brown or smoky-brown. **Juvenile** Paler brown. **VOICE** Call a soft *see-su.* [CD9:5] **HABITS** In pairs or small flocks in forest clearings and edges, farmbush and montane grassland, from sea level to 3000 m. Joins mixed flocks of hirundines and swifts. Pairs usually breed solitarily; moss nest placed on ledge of cliff or cave. **SIMILAR SPECIES** Black Saw-wing of ssp. *petiti* is more slender, with longer tail and more contrasting, whitish underwing-coverts. Fanti Saw-wing is blacker, with much longer tail and black underwing. Brownish juveniles may appear very similar but are generally in company of parents; note build, tail length and colour of underwing-coverts. **STATUS AND DISTRIBUTION** Locally common resident, Cameroon (Mt Cameroon) and Bioko; possibly also E Nigeria (sight records, Obudu and Mambilla Plateaux). Local seasonal movements recorded.

BRAZZA'S MARTIN *Phedina brazzae*   Plate 81 (7)
Hirondelle de Brazza

Other name E: Congo Martin
L 12 cm (4.75"). Small, brown-and-white martin with *streaked underparts* and *square tail.* **Adult** Head, upperparts, tail and underwing dark brown. Underparts white heavily streaked dark brown; breast variably washed brownish. **Juvenile** Feathers of upperparts tipped and fringed rufous; streaking on underparts less distinct. **VOICE** No

information. **HABITS** In small groups along forested rivers. Sometimes forages with other hirundines. Nests in burrows excavated in vertical river banks or cliffs, usually in small colonies. **SIMILAR SPECIES** No other martin within its range has streaked underparts. **STATUS AND DISTRIBUTION** Rarely recorded, local resident (migrant?), Congo (from Congo R. to Djambala). Probable sightings, SE Gabon (Lékoni). *TBW* category: Data Deficient. **NOTE** Placed in genus *Phedinopsis* by some authors on basis of morphological traits and nesting habits, which are different from Mascarene Martin *Phedina borbonica* (Dowsett & Dowsett-Lemaire 1993).

PLAIN MARTIN *Riparia paludicola*   Plate 81 (6)
Hirondelle paludicole

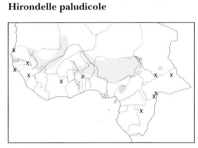

Other names E: Brown-throated/African Sand Martin
L 11–12 cm (4.5"). Small martin with brownish throat and breast. **Adult** *paludibula* Head, upperparts, tail and underwing brown. Throat and breast greyish-brown becoming white on belly and undertail-coverts. Tail slightly forked, almost square. Localised *newtoni* has darker upperparts and white on underparts more extensive (also whiter when fresh, but pale brown wash appears with wear). **Juvenile** Feathers of upperparts have narrow pale fringes. **VOICE** Calls include a rasping *chtrrr* and harsh *steeh*. Song a soft twitter. [CD3:53] **HABITS** Usually in small flocks, over or near rivers and lakes, but outside breeding season also away from water. Occasionally singly or in large flocks. Weak, fluttering flight. Associates with other hirundines. Dry-season breeder, nesting in burrows dug in sand banks, generally in small colonies. **SIMILAR SPECIES** Other martins in range have contrasting white throats and well-marked brown breast-bands. Rock and Crag Martins are larger with white in tail; Crag also has dark underwing-coverts contrasting with flight feathers. **STATUS AND DISTRIBUTION** Locally not uncommon to rare resident in Sahel and savanna zones. Patchily distributed, S Mauritania, Senegambia, Guinea, Mali and N Ghana to Niger, Nigeria, Chad and CAR (*paludibula*). Also E Nigeria (Mambilla Plateau) and W Cameroon highlands (*newtoni*). Vagrant, Ivory Coast, N Congo and Cape Verde Is. Local movements (post-breeding dispersal) recorded.

CONGO SAND MARTIN *Riparia congica*   Plate 81 (4)
Hirondelle du Congo

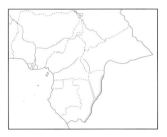

Other name F: Hirondelle de rivage du Congo
L 11–12 cm (4.5"). Small martin with very restricted range; similar to Common Sand Martin but has *more diffuse brown breast-band*. **Adult** Head, upperparts, tail and underwing brown; broad breast-band greyish-brown. Tail slightly forked, almost square. **Juvenile** Feathers of upperparts have narrow pale fringes; breast-band more indistinct. **VOICE** No information. **HABITS** In flocks, foraging low over rivers and riverine vegetation. Sometimes forages with other swallows. Nests colonially in burrows excavated in river banks, usually in small colonies. **SIMILAR SPECIES** Common Sand Martin has well-defined breast-band. **STATUS AND DISTRIBUTION** Locally common resident, Congo (along Congo R. and lower Ubangi R., also recorded along Sangha R.).

COMMON SAND MARTIN *Riparia riparia*   Plate 81 (3)
Hirondelle de rivage

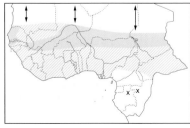

Other names E: European Sand Martin, Bank Swallow
L 12–13 cm (5"). *R. r. riparia*. Small martin with *pure white underparts* and *well-defined brown breast-band*. **Adult** Head, upperparts, tail and underwing brown; breast-band broadest at sides. Tail slightly forked. **Immature** Similar. **VOICE** Call a dry, rasping *chrrp*. [CD3: 54] **HABITS** Over various open habitats. Gregarious; often associating with other hirundines. **SIMILAR SPECIES** Congo Sand Martin is very similar but has more diffuse breast-band. Banded Martin also has distinct brown breast-band, but is easily separated by much larger size. **STATUS AND DISTRIBUTION** Uncommon or rare to locally

common Palearctic passage migrant and winter visitor, from Mauritania to Liberia east to Chad and CAR; mainly in Sahel zone. Vagrant, NE Gabon, NW Congo and Cape Verde Is.

BANDED MARTIN *Riparia cincta* Plate 81 (5)
Hirondelle à collier

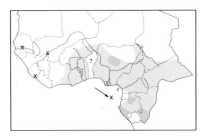

L 17 cm (6.75"). ¶ Conspicuously *large* martin with *well-defined breast-band* and square tail. **Adult *parvula*** Head, upperparts and tail dark brown; short white supercilium. Underparts white with broad brown breast-band (width *c.* 18 mm); underwing-coverts white. **Juvenile** Upperwing-coverts and tertials fringed rufous-buff. **VOICE** Mostly silent. Call *chirp.* [CD9:6] **HABITS** Singly or in small flocks. Forages low over grassland, with slow and erratic, tern-like flight. Associates with other hirundines. Pairs breed solitarily, in burrow excavated in bank or similar site. **SIMILAR SPECIES** All other hirundines with brown upperparts are much smaller and none has superciliary stripe. **STATUS AND DISTRIBUTION** Inadequately known. Resident, S Gabon and Congo. Uncommon to rare intra-African migrant, from Gabon to SW Niger, Ghana and N Ivory Coast; also N CAR and S Chad. Vagrant, Gambia, SW Mali, Sierra Leone, Príncipe. Main breeding grounds in E and S Africa. **NOTE** Wintering grounds of races unknown. ***R. c. parvula*** known to migrate north to (at least) Cameroon; other races possibly reaching our region include **xerica** (as *parvula* but breast-band narrower, *c.* 12 mm), **nominate** (uppertail-coverts slightly paler, breast-band *c.* 15 mm) and **suahelica** (darker above, breast-band as *parvula*).

GREY-RUMPED SWALLOW *Pseudhirundo griseopyga* Plate 82 (3)
Hirondelle à croupion gris

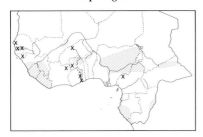

L 14 cm (5.5"). *P. g. melbina.* Small, slim swallow with deeply forked tail and *greyish rump contrasting with dark blue upperparts.* **Adult** Top of head to nape brown-grey; lores and ear-coverts dark brown, sometimes bordered by narrow pale supercilium. Upperparts dark glossy blue; rump and uppertail-coverts brown-grey. Underparts, incl. underwing-coverts, white. **Juvenile** Duller, with pale fringes to feathers of upperparts and shorter tail. **VOICE** Mostly silent. Call a soft, nasal *chwèèp.* [CD9:7] **HABITS** Singly, in pairs or small flocks in grassland, wooded savanna and similar open areas. Flight fluttering without long glides. Usually forages low. Associates with other hirundines. Pairs breed solitarily; sometimes in small loose colonies. Nests in rodent burrow (occasionally in old kingfisher or bee-eater burrow) in sandy ground. **SIMILAR SPECIES** No other hirundine has combination of greyish rump and long tail streamers. Moulting birds with short tail could recall Preuss's Cliff Swallow, but the latter has white in tail. **STATUS AND DISTRIBUTION** Patchily distributed resident and migrant. Generally rare or scarce to uncommon, but sometimes locally not uncommon. Movements inadequately known. Recorded from Gambia, Guinea, Sierra Leone, Liberia, Ivory Coast, Mali, Burkina Faso, N Ghana, Benin, SW Niger, Nigeria, NW Cameroon, N CAR, coastal Eq. Guinea and Gabon, and Congo. **NOTE** Placed in its own, monotypic, genus on basis of nesting habits.

RUFOUS-CHESTED SWALLOW *Hirundo semirufa* Plate 82 (11)
Hirondelle à ventre roux

Other name E: Red-breasted Swallow
L 18–21 cm (7.0–8.25"). *H. s. gordoni.* Large dark blue and rufous swallow with *dark blue hood* and *very long tail streamers.* **Adult male** Top of head to below eye and onto ear-coverts glossy dark blue connecting, on hindneck, with similarly coloured upperparts; rump orange-rufous; uppertail-coverts dark blue; tail dark blue with white patches. Underparts pale orange-rufous, palest on throat (becoming buff or whitish with wear); underwing-coverts pale rufous-buff. **Adult female** Tail streamers shorter. **Juvenile** Duller; brownish above, paler below; tail shorter. **VOICE** A soft, gurgling *trlrrrrr* or *chip-chip-chleeeeurrrr* with variations; also a soft *dee-uuuh*, a short *chip* and a high-pitched *weet-weet.* [CD9:8] **HABITS** In pairs or small flocks in farmbush, degraded savanna, grassland, villages and towns. Flight slow and buoyant on broad wings, with much gliding. Forages at all heights. Associates with other hirundines. Pairs breed solitarily; mud nest is bowl shaped with tunnel entrance plastered to overhanging rock, human habitation, hollow tree or similar place. **SIMILAR SPECIES** Mosque Swallow is even larger and heavier, with pale head-sides and rufous nuchal

collar surrounding dark blue cap, shorter tail streamers, no white in tail and usually whiter underwing-coverts. Red-rumped Swallow has rufous head-sides and collar, black (not rufous) undertail-coverts and no white in tail. **STATUS AND DISTRIBUTION** Uncommon to locally common resident and intra-African migrant almost throughout, except arid north. From Senegambia to Liberia east to SW Chad, SW CAR and Congo. Wet-season breeding visitor in some areas.

MOSQUE SWALLOW *Hirundo senegalensis*   Plate 82 (10)
Hirondelle des mosquées

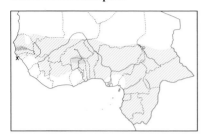

L 21–23 cm (8.25–9.0"). Largest and heaviest African swallow. **Adult male *senegalensis*** Top of head, upperparts, uppertail-coverts and tail glossy dark blue; lores and head-sides whitish to pale rufous, becoming dark rufous on hindneck and neck-sides, forming collar (sometimes broken by blue feather tips); rump orange-rufous. Throat and upper breast whitish to pale rufous, darkening to orange-rufous on rest of underparts; underwing-coverts pale rufous-buff to white, contrasting with blackish flight feathers. ***H. s. saturatior*** has deeper rufous underparts. **Adult female** Tail streamers shorter. **Juvenile** Duller; dark brownish glossed blue above, paler below; tail shorter. **VOICE** Calls include a nasal *nyaa*, a distinctive, piping, reedy note, recalling sound of tiny trumpet, and various guttural notes. Song consists of various rather slow, short series of nasal, whining and creaking sounds, introduced by a few chuckles and repeated several times. [CD9:9] **HABITS** Singly, in pairs or small flocks in various open habitats; mostly in broad savanna belt but also open areas in forest zone; frequently around villages and towns. Flight slow, with much gliding; sometimes recalling small falcon. Typically forages high over treetops. Associates with swifts and other hirundines. Pairs breed solitarily; mud nest bowl shaped with tunnel entrance plastered against overhanging surface of natural or artificial sites. **SIMILAR SPECIES** Rufous-chested Swallow best separated by dark blue cap extending to below eye and onto ear-coverts (forming hood); also has incomplete rufous collar, longer tail streamers and white in tail; difference in colour of underwing-coverts less reliable, but usually more rufous. Red-rumped Swallow has black undertail-coverts. **STATUS AND DISTRIBUTION** Rare to locally common resident and partial intra-African migrant almost throughout, except arid north. Wet-season breeding visitor in some areas (mostly in north of range). Unrecorded from Guinea-Bissau to S Ivory Coast, except as vagrant in Guinea and Liberia. Nominate from S Mauritania and Senegambia to N Ghana, N Nigeria, N Cameroon and S Chad; *saturatior* from S Ghana to S Cameroon, CAR and Congo.

LESSER STRIPED SWALLOW *Hirundo abyssinica*   Plate 82 (8)
Hirondelle striée

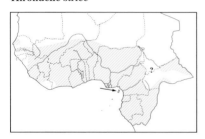

Other name F: Hirondelle à gorge striée
L 15–19 cm (6.0–7.5"). The region's only widespread rufous-rumped swallow with *rufous head* and *streaked underparts*. **Adult male *puella*** Head rufous-chestnut; upperparts glossy dark blue with rufous rump. Tail has long outer feathers and small white patches. Underparts white finely streaked dark. ***H. a. unitatis*** is more heavily streaked below; in *maxima* streaks even heavier, broader and blacker; in *bannermani* much finer and narrower, and rufous areas paler. **Adult female** Tail streamers shorter. **Juvenile** Duller; crown speckled blackish; wing-coverts and tertials tipped buffish; rump paler; tail shorter. **VOICE** Call a wheezy *cheeew*. Song varied and nasal, including liquid, squeaky, buzzy and gurgling notes. [CD9:10] **HABITS** Singly, in pairs or small flocks in various open wooded habitats; also forest edges and clearings; occasionally mangroves. Flight fluttering with much gliding. Forages low or at mid-height, often over water. Associates with other hirundines. Pairs usually breed solitarily; mud nest bowl shaped with tunnel entrance plastered to overhanging surface of natural or artificial sites. **SIMILAR SPECIES** Greater Striped Swallow (q.v.). **STATUS AND DISTRIBUTION** Generally not uncommon to locally common resident and partial intra-African migrant. Wet-season breeding visitor to north of range. *H. a. puella* from Senegambia (where scarce) to Liberia east to NE Nigeria and N Cameroon; *maxima* from SE Nigeria and S Cameroon to SW CAR and Bioko; *bannermani* in NE CAR; *unitatis* from Gabon to Congo.

GREATER STRIPED SWALLOW *Hirundo cucullata*   Plate 147 (5)
Hirondelle d'Afrique

Other name F: Hirondelle à gorge striée
L 18–20 cm (7–8"). Resembles Lesser Striped Swallow, but noticeably larger, with streaking on underparts much

finer (only visible at close range), presenting much paler appearance, darker rufous cap, white (not rufous) ear-coverts, and paler rump. **VOICE** Calls include *chrrp* and plaintive *wheep*. Song a few *chwurp* or *chrrp* notes followed by a pleasant, descending trill *tirrrrrr*. [CD9:11] **HABITS** In open grassy habitats. Forages low or at mid-height. Associates with other hirundines. Flight slower than Lesser Striped Swallow. **SIMILAR SPECIES** Lesser Striped Swallow (q.v. and above). **STATUS AND DISTRIBUTION** Vagrant, S Congo (Nov). Breeds S Africa (Aug–Apr), moving north in non-breeding season.

RED-RUMPED SWALLOW *Hirundo daurica* Plate 82 (9)
Hirondelle rousseline

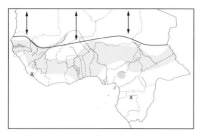

L 16–17 cm (6.25–6.75"). Medium-sized blue-and-rufous swallow with *dark blue cap, rufous collar*, deeply forked tail and *diagnostic clear-cut black undertail-coverts*. **Adult male *domicella*** Crown glossy dark blue, bordered by rufous forehead and narrow supercilium; head-sides buffish, becoming rufous on hindneck and neck-sides, forming distinct collar; upperparts and uppertail-coverts glossy dark blue with contrasting orange-rufous rump. Tail blackish with long, inwardly curving outer feathers. Underparts creamy-white slightly washed rufous on sides and flanks; underwing-coverts buff-cream, contrasting with blackish flight feathers. *H. d. kumboensis* has pale rufous underparts; *rufula* has two-toned rump, with lower rump paler; underparts pale creamy-buff with faint, narrow dark streaks; dark blue of crown does not reach bill. Rufous and buff parts of all forms pale with wear. **Adult female** Tail streamers shorter. **Juvenile** Duller; feathers of upperparts tipped buff; rump paler; tail shorter. **VOICE** Calls include a soft *djuit* or *chreet*. Song a varied, quiet twittering with nasal notes and rasping trills. [CD3:55] **HABITS** Singly, in pairs or small flocks in various types of savanna. Flight slower than Barn Swallow. Associates with other hirundines. Pairs breed solitarily or in small, loose colonies; mud nest bowl shaped with tunnel entrance plastered against variety of natural or artificial sites. **SIMILAR SPECIES** Mosque and Rufous-chested Swallows are larger, with rufous undertail-coverts; Rufous-chested also has dark blue cap extending onto ear-coverts (forming hood) and incomplete rufous collar, longer tail streamers and white in tail. **STATUS AND DISTRIBUTION** Uncommon to locally common resident, partial intra-African migrant and Palearctic migrant in Sahel and savanna zones. *H. d. domicella* widespread, from Senegambia and Guinea to Chad; *kumboensis* restricted to highlands of Sierra Leone (Birwa Plateau) and W Cameroon (Bamenda Highlands); *rufula* Palearctic passage migrant and winter visitor, south to N Senegal through Mali to Chad. Vagrant, NE Gabon and Cape Verde Is. **NOTE** Ssp. *domicella* sometimes treated as separate species (West African Swallow; Hirondelle ouest-africaine).

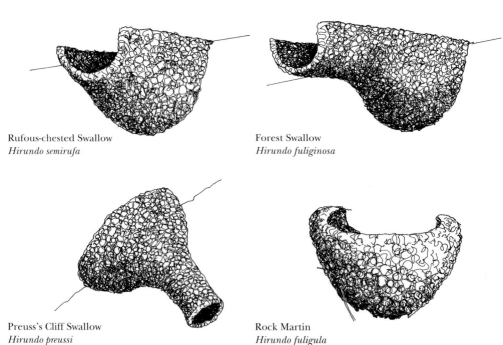

Rufous-chested Swallow
Hirundo semirufa

Forest Swallow
Hirundo fuliginosa

Preuss's Cliff Swallow
Hirundo preussi

Rock Martin
Hirundo fuligula

FOREST SWALLOW *Hirundo fuliginosa*

Plate 83 (3)

Hirondelle de forêt

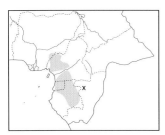

Other name E: Dusky Cliff Swallow
L 11 cm (4.25"). Small, all-dark forest swallow. **Adult** Entirely dull blackish-brown, with rusty tinge on throat. Tail slightly notched (almost square). Looks wholly blackish against sky. **Juvenile** Similar. **VOICE** Mostly silent. Call a clear, sharp *whit, whit* and a double, 'sucking' *pritchi* or *pitchri*. [No published recording] **HABITS** Usually in pairs, occasionally up to 5, along forest tracks and clearings. Flight fluttering and fast, with much gliding, typical of small *Hirundo*. Pairs breed solitarily. Mud nest bowl shaped with long tunnel entrance, plastered to overhanging rocks in forest, often near those of Red-headed Picathartes; occasionally on artificial surfaces. **SIMILAR SPECIES** Square-tailed Saw-wing is very similar but voice different and flight noticeably slower, with back more hunched. Mountain Saw-wing is slightly larger, with grey underwing-coverts. **STATUS AND DISTRIBUTION** Inadequately known. Locally not uncommon resident or intra-African migrant, SE Nigeria (Oban Hills), Cameroon, Eq. Guinea and Gabon. Vagrant, Congo.

PREUSS'S CLIFF SWALLOW *Hirundo preussi*

Plate 83 (4)

Hirondelle de Preuss

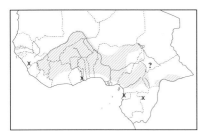

Other name F: Hirondelle de rochers à dos noir
L 12 cm (4.75"). The only swallow in its range with *pale buff rump and underparts* contrasting with blue-black upperparts. **Adult** Top of head, ear-coverts and upperparts glossy blue-black; rump creamy or pale buff. Tail slightly forked, blackish with small white patches on inner webs. Underparts (incl. underwing-coverts) pale buff. In field, small rufous patch behind eye and greyish-white streaks on nape and mantle (forming small, irregular patch) only visible in favourable circumstances. **Juvenile** Duller; dark brown above; throat and breast washed brownish. **VOICE** Call *prrp-prrp*, given in flight. [CD9:13] **HABITS** In small or large flocks in grassy and wooded savanna, often near rocky outcrops and rivers. Flight slow with much gliding. Forages at all heights, often very high. Associates with other hirundines. Breeds in dense colonies of a few to 2500 nests. Mud nest bowl shaped with downward-angled entrance tunnel plastered under variety of natural or artificial surfaces, esp. bridges. **SIMILAR SPECIES** Common House Martin has similar plumage pattern, but appears 'cleaner', with pale areas white (not pale buff); tail more forked. **STATUS AND DISTRIBUTION** Irregularly distributed, rare to locally common resident and intra-African migrant, from Guinea-Bissau, Guinea, Sierra Leone, Mali and N Ivory Coast to Niger, Nigeria and Cameroon; also Eq. Guinea, Congo and S CAR. Vagrant, Chad. Movements inadequately known.

RED-THROATED CLIFF SWALLOW *Hirundo rufigula*

Plate 83 (2)

Hirondelle à gorge fauve

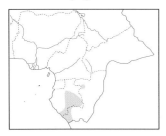

Other names E: Angolan Cliff Swallow; F: Hirondelle de rochers à gorge fauve
L 12 cm (4.75"). Small swallow with *rufous rump and throat*. **Adult** Top of head and upperparts glossy blue-black; ear-coverts streaky; rump and uppertail-coverts rufous. Tail slightly forked, blackish with small white patches on inner webs. Throat and upper breast rufous becoming orange-buff on rest of underparts. In field, greyish-white streaks on mantle only visible in favourable circumstances. **Juvenile** Duller; dark brown above; rufous areas paler. **VOICE** Call a rasping *prrp-prrp*, given in flight. Song a rapid twittering. [CD9:14] **HABITS** Usually in flocks over various open habitats, often near rocky outcrops and rivers; also in forest at bridges over large rivers. Flight slow with much gliding. Associates with other hirundines. Breeds in dense colonies. Mud nest bowl shaped with downward-angled entrance tunnel plastered under variety of natural or artificial surfaces, esp. bridges. **SIMILAR SPECIES** South African Cliff Swallow distinguished by pale throat, speckles on breast forming indistinct pectoral band, square tail lacking spots and less blue upperparts. **STATUS AND DISTRIBUTION** Locally not uncommon resident and intra-African migrant, Gabon and Congo. Colonized Gabon in late 1970s; northward range expansion ongoing. Seasonal movements inadequately known.

SOUTH AFRICAN CLIFF SWALLOW *Hirundo spilodera*

Hirondelle sud-africaine

Plate 83 (1)

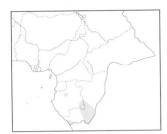

Other name F: Hirondelle de rochers sud-africaine

L 14 cm (5.5"). Medium-sized swallow with *rufous rump* and *square, unmarked tail.* **Adult** Head sooty-brown with narrow rufous forehead and lores; upperparts glossy blue-black; rump pale rufous. Tail blackish, square or slightly notched. Underparts pale rufous to buffish-white (darkest on breast, palest on belly), and variably speckled and mottled black on throat and breast. In field, often appears dark throated or having ill-defined darkish breast-band; greyish-white streaks on mantle only visible in favourable conditions. Rufous areas pale with wear. **Juvenile** Duller; dark brown above; rufous areas paler. **VOICE** Call *prrp-prrp*, given in flight. [CD9:15] **HABITS** Usually in flocks over various open habitats. Flight slow with much gliding. Usually forages low. Associates with other hirundines. **SIMILAR SPECIES** Red-throated Cliff Swallow distinguished by wholly blue-black cap, plain rufous throat, and white spots in tail. **STATUS AND DISTRIBUTION** Rare intra-African migrant, C and SE Gabon (Lopé, Lékoni) and Congo. Breeds in S Africa (Aug–Apr); winters mainly in lower Congo basin (May–Aug).

ROCK MARTIN *Hirundo fuligula*

Hirondelle isabelline

Plate 81 (1)

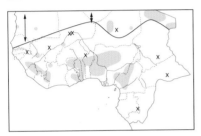

Other name F: Hirondelle de rochers africaine

L 12–13 cm (5"). ¶ Small, compact, *all-brown* hirundine with *white patches in almost square tail.* Coloration and size variable: tropical W African races darkest and smallest, those from N Africa and Sahara (*obsoleta* group) palest and slightly larger. **Adult** (dark races) Head, upperparts and tail dark earth-brown. Underparts paler dusky-brown, variably washed pale rufous, esp. on throat. Tail slightly notched, with two small white patches near tip (conspicuous when tail spread). Birds from *obsoleta* group paler and greyer. **Juvenile** Similar, but feathers of upperparts narrowly tipped buff. **VOICE** Mostly silent. Calls include *wik* and a high-pitched *sree*. Song a soft twitter. [CD3:56] **HABITS** Singly, in pairs or small flocks around inselbergs, rocky outcrops, cliffs and gorges, and in villages and towns. Often far from water. Flight slow, with much gliding. Associates with swifts and other hirundines. Breeds in pairs or small colonies; nest a mud cup attached to rocky surface, bridge or similar place. **SIMILAR SPECIES** Crag Martin has similar tail pattern, but distinguished by darker and more contrasting, blackish underwing-coverts and larger size. Plain Martin lacks white in tail, has white abdomen and fluttering flight. **STATUS AND DISTRIBUTION** Patchily distributed, locally not uncommon or common resident, from desert to forest zone. Dark races throughout, except arid north, where replaced by *obsoleta* group in N Mauritania, Mali, N Niger (Aïr) and NE Chad (Ennedi). **NOTES (1)** Sometimes placed in genus *Ptyonoprogne*. (2) Northern populations sometimes considered a separate species, Pale Crag Martin *H. obsoleta* (Hirondelle du désert), but broad cline in coloration and size suggests all are conspecific. (3) Races occurring in our region, from darkest brown to palest: tropical W Africa: *bansoensis* (Cameroon, Nigeria to Sierra Leone and probably lower Gambia R.), *fusciventris* (S Chad, CAR), *pusilla* (S Mali to S Niger, C Chad); *obsoleta* group: *buchanani* (Niger: Aïr), *arabica* (NE Chad: Ennedi), *spatzi* (Mali?), *presaharica* (N Mauritania).

CRAG MARTIN *Hirundo rupestris*

Hirondelle de rochers

Plate 81 (2)

Other name E: Eurasian Crag Martin

L 14-15 cm (5.5-6.0"). Stocky, brown hirundine with white patches in tail and *blackish-brown underwing-coverts contrasting strongly with grey-brown flight feathers.* **Adult** Head and upperparts grey-brown with darker flight feathers. Underparts pale buffish becoming brown on lower belly and undertail-coverts; upper throat indistinctly speckled brown. Tail slightly notched, with white subterminal patches (noticeable when tail spread). **VOICE** Mostly silent. Calls include a soft *chrp* or *prrt*. [CD3:57] **HABITS** Normally associated with cliffs and gorges, but on migration also over low, open areas. Flight slow, with frequent gliding. **SIMILAR SPECIES** Rock Martin has similar tail pattern, but races of *obsoleta* group separated by paler and greyer plumage, with paler underwing-coverts contrasting much less with flight feathers. **STATUS AND DISTRIBUTION** Rare Palearctic winter visitor, Mauritania, N Senegal, E Mali. Vagrant, Gambia. **NOTE** Sometimes placed in genus *Ptyonoprogne*.

WIRE-TAILED SWALLOW *Hirundo smithii*   Plate 82 (7)
Hirondelle à longs brins

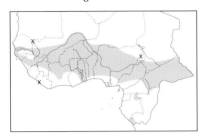

L 14 cm (5.5"). *H. s. smithii.* Slender, blue-and-white swallow with *rufous cap and very thin, wire-like, tail streamers.* **Adult male** Top of head rufous bordered by black lores and blue-black ear-coverts; upperparts and tail glossy violet-blue. Tail has white patches on inner webs and very long, thin outer feathers. Underparts pure white with blue-black patch on breast-sides and lower flanks. **Adult female** Tail streamers shorter. **Juvenile** Duller; crown browner; tail shorter. **VOICE** Call a quiet *chit.* Song a soft twittering. Mostly silent. [CD9:18] **HABITS** In pairs or small flocks in open habitats, usually near water. Flight fast with much gliding. Forages low or at mid-height. Occasionally with other hirundines. Pairs breed solitarily; mud nest an open cup, plastered to overhanging surface of natural or artificial site. **SIMILAR SPECIES** Ethiopian Swallow has rufous restricted to forehead, tail streamers shorter and less wire-like, and no patches on lower flanks. **STATUS AND DISTRIBUTION** Uncommon resident and partial intra-African migrant in broad savanna belt throughout; also NE Gabon and Congo.

WHITE-THROATED BLUE SWALLOW *Hirundo nigrita*   Plate 81 (9)
Hirondelle à bavette

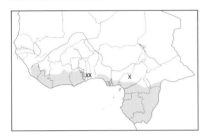

Other name F: Hirondelle noire
L 12 cm (4.75"). Unmistakable. **Adult** Beautiful, dark glossy *purple-blue swallow with contrasting white throat and white patches in tail.* Flight feathers blackish; tail slightly notched (almost square), with white conspicuous when tail spread in flight. May appear black at distance or against light. **Juvenile** Duller, browner, with less white on throat and in tail. **VOICE** Calls include a vigorous *weetch,* a hard *vwhit, vwhit* and a soft *whit,* the last also uttered in flight. Song a soft, dry trill *prl-trrrrrr* mixed with disharmonic notes. **HABITS** Usually in pairs, over rivers, lakes and lagoons with forested banks; also mangroves. Flight low and fast, with much twisting and banking and some gliding; perches on branches in water, snags and rocks. Mud nest an open cup, plastered to surface of natural or (occasionally) artificial site overhanging water (usually hollow logs fallen in water). **STATUS AND DISTRIBUTION** Locally common resident in forest zone, from Sierra Leone and SE Guinea to SW CAR and Congo.

PIED-WINGED SWALLOW *Hirundo leucosoma*   Plate 83 (6)
Hirondelle à ailes tachetées

L 12 cm (4.75"). The only hirundine with *long white patches on upper wing.* **Adult** Head, upperparts and tail glossy steel-blue; inner greater upperwing-coverts, inner secondaries and outer webs of tertials white, forming elongated patch on inner wing. Tail forked, with white patches on inner webs. Entire underparts (incl. underwing-coverts) pure white. **Juvenile** Duller and browner; tail shorter. **VOICE** Mostly silent. Call a low *chut.* [CD9:19] **HABITS** Singly or in pairs, occasionally in small flocks, over wooded savanna and farmbush. Flight fast with much banking and turning. Forages low, often flying between trees and bushes, like saw-wings. Associates with other hirundines. Pairs breed solitarily. Mud nest an open cup, plastered to surface of artificial site. **STATUS AND DISTRIBUTION** Irregularly distributed, scarce to locally not uncommon resident; seasonal movements in some areas. Recorded from Senegambia to Sierra Leone east to SW Niger and Nigeria. Vagrant, W Cameroon.

ETHIOPIAN SWALLOW *Hirundo aethiopica*   Plate 82 (6)
Hirondelle d'Ethiopie

L 13 cm (5"). *H. a. aethiopica.* Small blue-and-white swallow with *rufous forehead, buff-white throat and no breast-band.* **Adult male** Head and upperparts glossy steel blue with rufous forehead and forecrown. Tail blue-black with white patches on inner webs and long outer feathers. Underparts white with variable buff tinge on throat and dark blue patch on breast-sides. **Adult female** Tail streamers shorter. **Juvenile** Duller and browner, lacking rufous on forehead; tail shorter. **VOICE** Call a soft *chit* or *cheep.* Song a melodious twittering. [CD9:21] **HABITS** In pairs or small

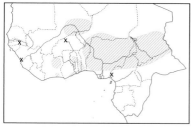

flocks over various open habitats, often near water; also villages and towns. Flight fast with much gliding. Forages low or at mid-height. Occasionally with other hirundines. Pairs breed solitarily; mud nest an open cup, plastered to overhanging surface of natural or artificial site. **SIMILAR SPECIES** Wire-tailed Swallow has entire top of head rufous, very long, thin tail streamers and patches on lower flanks. Red-chested Swallow has rufous throat. **STATUS AND DISTRIBUTION** Uncommon to locally common resident and partial intra-African migrant. Recorded from S Senegal, Guinea, Ivory Coast, Ghana, E Mali, Burkina Faso, Togo, Benin, S Niger, Nigeria to N Cameroon, Chad and N CAR.

WHITE-THROATED SWALLOW *Hirundo albigularis*   Plate 82 (2)
Hirondelle à gorge blanche

L 14–17 cm (5.5–6.75"). Blue-and-white swallow with *rufous forehead, white throat* and *blue breast-band*. **Adult male** Head and upperparts glossy dark blue with rufous forehead and forecrown. Tail blue-black with white patches on inner webs and long outer feathers. Throat pure white bordered by dark blue breast-band (narrow, sometimes broken, in centre); rest of underparts off-white. **Adult female** Tail streamers shorter. **Juvenile** Duller and browner, with little or no rufous on forehead; breast-band brownish; tail shorter. **VOICE** Calls include a sharp *chit* and various nasal and squeaky notes. [CD9:22] **HABITS** In open habitats, often near human habitation. Flight fast and agile. Forages low or at mid-height. Associates with other hirundines. **SIMILAR SPECIES** Wire-tailed Swallow has entire top of head rufous, no breast-band, very long, thin tail streamers, and patches on lower flanks. Ethiopian Swallow has blue on breast restricted to sides, not forming band. Barn, Red-chested and Angolan Swallows have rufous throat. **STATUS AND DISTRIBUTION** Vagrant from S Africa, recorded in N Congo (Odzala NP, Aug). Claim from Cameroon rejected (Dowsett & Dowsett-Lemaire 2000).

ANGOLA SWALLOW *Hirundo angolensis*   Plate 82 (5)
Hirondelle de l'Angola

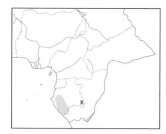

L 14–15 cm (5.5–6.0"). *Similar to Red-chested Swallow, but with ashy underparts.* **Adult male** Head, upperparts and tail glossy dark blue; forehead, throat and upper breast dark rufous bordered by narrow blue breast-band (narrowest or sometimes broken in centre). Tail has large white patches on inner webs and elongated outer feathers. Rest of underparts dull ash-grey; undertail-coverts tipped whitish. **Adult female** Outer tail feathers shorter. **Juvenile** Duller and browner; forehead and throat paler, more rufous-buff; outer tail feathers shorter. **VOICE** Song a soft twittering. [CD9:23] **HABITS** Singly or in pairs; occasionally small flocks. Frequents various open habitats; also villages. Flight rather slow. Forages low or at mid-height. Pairs usually breed solitarily; occasionally in small colonies. Mud nest an open cup, plastered to overhanging surface of natural or artificial site. **SIMILAR SPECIES** Red-chested Swallow has lower breast to undertail-coverts pure white. Barn Swallow has rufous below restricted to throat, broader breast-band, whitish underparts, longer tail streamers and smaller white patches on tail. **STATUS AND DISTRIBUTION** Uncommon resident (and partial local migrant?), W and S Gabon and S Congo.

RED-CHESTED SWALLOW *Hirundo lucida*   Plate 82 (4)
Hirondelle de Guinée

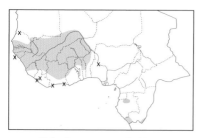

Other name F: Hirondelle à gorge rousse
L 15 cm (6"). **Adult male** *lucida* Very similar to Barn Swallow but has shorter outer tail feathers, more white in tail, dark rufous of throat extending to upper breast, narrower breast-band (narrowest in centre; sometimes broken), and whiter underparts. In flight, more compact silhouette with broad-based, more triangular wings. *H. l. subalaris* has rufous areas darker; rufous on forehead also more extensive. **Adult female** Outer tail feathers shorter. **Juvenile** Duller and browner; forehead and throat paler; outer tail feathers shorter. **VOICE** Similar to Barn Swallow. [CD9:24] **HABITS** Singly, in pairs or small flocks over open habitats and villages and towns. Flight fast and agile. Forages low or at mid-height. Often with other hirundines. Pairs usually breed solitarily or in small loose colonies. Mud nest an open cup,

plastered to overhanging surface of natural or artificial site, often under bridges. **SIMILAR SPECIES** Adult Barn Swallow (see above); juveniles more difficult to separate, but Barn has broader breast-band and less white in tail. Angolan Swallow has ashy underparts. Ethiopian Swallow has more rufous on forehead, white throat and no breast-band (only patches at sides). **STATUS AND DISTRIBUTION** Uncommon to locally common resident and partial intra-African migrant. Two widely disjunct populations: nominate from S Mauritania to Liberia east to SW Niger and N Togo; vagrant W Nigeria; *subalaris* along Congo R. and N Gabon. Movements inadequately known; wet-season visitor to some areas.

BARN SWALLOW *Hirundo rustica*
Hirondelle rustique

Plate 82 (1)

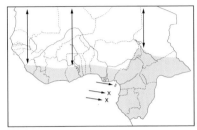

Other names E: European Swallow; F: Hirondelle de cheminée
L 15–19 cm (6.0–7.5"). *H. r. rustica*. The region's most widespread and often commonest swallow. **Adult male** Head, upperparts and tail glossy dark blue (duller when worn); *forehead and throat dark rufous bordered by dark blue breast-band*. Tail with small white patches on inner webs and very long outer feathers. Lower breast to undertail-coverts whitish. **Adult female** Tail streamers shorter. **Immature** Duller and browner; forehead and throat paler, more rufous-buff; tail streamers shorter. **VOICE** Call a vigorous *whit-whit*; also a sharp *siflit*. Song, occasionally uttered in winter quarters, a fast, varied twitter including clear notes, rasping sounds and gurgling rattles. [CD3:58] **HABITS** In small or large flocks in all habitats, from desert to forest. Flight fast and agile with much banking and turning. Forages low or at mid-height. Often with other hirundines. **SIMILAR SPECIES** Red-chested Swallow has shorter tail streamers, more white in tail, rufous of throat extending to upper breast, narrower breast-band and whiter underparts. Juvenile Red-chested separated from juvenile Barn by narrower breast-band and more white in tail. Angolan Swallow has ashy underparts. Ethiopian Swallow has more rufous on forehead, white throat and no breast-band (only patches at sides). **STATUS AND DISTRIBUTION** Common to locally very common Palearctic passage migrant and winter visitor throughout, also Cape Verde Is, Bioko, São Tomé and Príncipe (mainly Aug–May, rarely oversummering).

COMMON HOUSE MARTIN *Delichon urbica*
Hirondelle de fenêtre

Plate 83 (7)

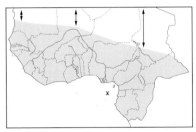

L 13–15 cm (5–6"). *D. u. urbica*. Small, rather compact, black-and-white martin with *conspicuous white rump*. **Adult breeding** (from Apr) Head, upperparts and tail glossy blue-black; rump and entire underparts white; underwing with pale buff-grey underwing-coverts and dark silvery flight feathers. Tail forked. **Adult non-breeding** (from Aug) White of rump, throat and flanks often mottled pale buff-brown. **Immature** Duller and browner. **VOICE** Call a hard *prrt prrpt*. [CD3:59] **HABITS** In small or large flocks over various open habitats. Flight fluttering with much gliding. Forages at all heights, often very high. Often associates with other hirundines, esp. cliff swallows. **SIMILAR SPECIES** Preuss's Cliff Swallow has similar plumage pattern, but duller, with pale areas buffish (not white); tail less forked. **STATUS AND DISTRIBUTION** Uncommon to locally common Palearctic passage migrant and winter visitor throughout, also Cape Verde Is (mainly Aug/Sep–Mar/May, rarely oversummering). Old record, Príncipe.

WAGTAILS, PIPITS AND LONGCLAWS
Family MOTACILLIDAE (World: *c.* 65 species; Africa: 38; WA: 17)

Small, mostly slim and mainly terrestrial insectivores with slender pointed bills and long tertials reaching to tips of primaries. Occur in open habitats, often near water. Walk. Many species migratory. Taxonomic status of several species/subspecies complex and requiring further research.
Three genera in our region:
Motacilla, wagtails (6 species): long tail and unstreaked black, grey, white and yellow plumage. Derive name from habit of wagging tail. Non-breeding plumage of Palearctic species different, less colourful and contrasting than breeding; sexes also different.

Anthus, pipits (10 species): rather cryptic, variably streaked brown plumage; larger species conspicuously long legged. Non-breeding plumage usually similar to breeding; pale tips and fringes to wing feathers broader when fresh. Identification features include extent of streaking on upper- and underparts, colour of outer tail feathers and calls.

Macronyx, longclaws (1 species): robust, with relatively short tail and round wings. Derive name from long claw on hind toe of strong legs. Sedentary.

YELLOW WAGTAIL *Motacilla flava* Plate 85 (5)
Bergeronnette printanière

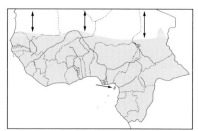

L 16–17 cm (6.5"). Commonest and most widespread wagtail, with comparatively short tail and, in all plumages, generally some yellow on underparts. Eight races occur, differing mainly in head pattern of males in breeding plumage. Non-breeding plumages usually indistinguishable. **Adult male *flava* breeding** Head blue-grey with white supercilium; upperparts olive-green; wing feathers blackish with greenish-yellow fringes to coverts (forming double wingbar). Tail black with white outer feathers. Underparts bright yellow. Legs blackish. Moult to breeding plumage starts Jan–Feb, completed Mar–Apr. *M. f. iberiae* has head darker grey; supercilium narrower, sometimes absent in front of eye; throat white; breast-sides tinged olive. *M. f. cinereocapilla* similar, but supercilium absent or restricted to streak behind eye. *M. f. thunbergi*: top of head slate-grey; supercilium indistinct, often absent; head-sides blackish. *M. f. feldegg*: head black; breast-sides olive. *M. f. melanogrisea* similar to *feldegg*, but with white line bordering black head-sides. *M. f. beema* similar to *flava* but head paler, pale blue-grey; distinct white supercilium and streak below eye; throat sometimes white. *M. f. flavissima*: head yellow-green; supercilium yellow, often extending to forehead. **Adult female *flava* breeding** Similar, but duller. Slightly browner above with less distinct head pattern; paler yellow below, often becoming whitish on throat and breast, upper breast sometimes mottled buff. **Adult non-breeding** Similar to adult female breeding, but duller. **First winter** Variable. Generally similar to adult non-breeding, but often even more washed out. **VOICE** Call a rather loud *pseew*. [CD3:60] **HABITS** Singly and in small to large flocks in variety of habitats, generally open areas, incl. short grassland, lawns, airfields, marshes, dams, ricefields and farmland, but also forest clearings and tracks. Often wags tail. Assembles, often in large numbers, at roost. Flocks may include several races. **SIMILAR SPECIES** Grey Wagtail, the only other wagtail (except vagrant Citrine) with yellow in plumage, has longer tail, greyer upperparts, no pale tips to upperwing-coverts and pale brownish-pink legs. **STATUS AND DISTRIBUTION** Common to very common Palearctic passage migrant and winter visitor throughout. *M. f. flava, thunbergi* and *iberiae* throughout, incl. Bioko; *flavissima* mainly from Mali west, also recorded from Liberia, Ivory Coast (probably), NE Gabon and N Congo; *cinereocapilla* Mali to Nigeria; *feldegg* Nigeria to Chad, also N Congo; *melanogrisea* NE Nigeria (L. Chad; vagrant); *beema* Chad to NE Nigeria (L. Chad). Vagrant, Cape Verde Is. **NOTE** Several forms, esp. *feldegg* ('Black-headed Wagtail'), treated as separate species by some authors.

CITRINE WAGTAIL *Motacilla citreola* Plate 147 (6)
Bergeronnette citrine

L 17–18 cm (7"). ¶ Resembles Yellow Wagtail but has, in all plumages, *grey upperparts and 2 conspicuous white or whitish wingbars* (narrowing with wear). Tail and legs slightly longer. **Adult male breeding** Easily identified by bright lemon-yellow head bordered by black nuchal band. White wingbars and white edges to tertials prominent when fresh. Underparts yellow, usually becoming whitish on undertail-coverts; flanks dusky-grey. Tail black with white outer feathers. Legs black. **Adult female breeding/adult non-breeding** Crown greyish washed yellow; broad yellow supercilium extending around greyish ear-coverts and joining yellow throat; no black on nape. Yellow coloration duller, paling from breast downwards; undertail-coverts whitish. **First-winter** Above, dull ash-grey (lacking olive tones). Characteristic head pattern with broad whitish ear-covert surround, palish lores (often tinged buff) and narrow, weak, dusky malar stripe. Underparts whitish (occasionally tinged buff on breast) washed grey on breast-sides and flanks. Bill all dark. Yellowish tinge to head gradually appears in Dec–Jan. **VOICE** Call a sharp, rasping *tsrreep*. **HABITS** Often near water. **SIMILAR SPECIES** Yellow Wagtail (esp. of race *flavissima*) separated from adult female/adult non-breeding Citrine by different head pattern (lacking broad yellow ear-covert surround and usually with dark loral line), olive-tinged (not dull ashy-grey) upperparts with narrower, less distinct wingbars, and yellow undertail-coverts; call less rasping (that of eastern races, such as *feldegg*, rather similar but usually softer). In addition, 1st-winter Yellow has slightly paler base to lower mandible (all dark in Citrine). Grey Wagtail has different head pattern, no conspicuous wingbars, yellow undertail-coverts and longer tail. **STATUS AND DISTRIBUTION** Palearctic vagrant, Senegal (Jan 1999). **NOTE** Race of vagrants to Africa unknown; *werae* slightly smaller and paler than nominate; doubtfully separable, except in adult male breeding plumage.

GREY WAGTAIL *Motacilla cinerea*   Plate 85 (3)
Bergeronnette des ruisseaux

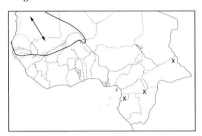

L 18–19 cm (7.0–7.5"). *M. c. cinerea.* Slimmer, longer tailed than Yellow Wagtail with *grey upperparts, contrasting blackish wings and, in all plumages, yellow on rump and vent.* **Adult male breeding** (from Feb) Head and upperparts grey with narrow, white supercilium and eye-ring; rump yellow-olive; wings brownish-black with white fringes to tertials. Tail black with 3 outer feathers white. Throat black bordered by white moustachial stripe; rest of underparts yellow. Legs pale brownish or pinkish. **Adult female breeding** Similar, but throat whitish variably mottled black. **Adult male non-breeding** Similar, but throat white; yellow on underparts paler, brightest on breast and vent. **Adult female non-breeding** Very similar to male non-breeding, but yellow on underparts paler. **Immature** As adult female non-breeding, but upper breast buffier. **VOICE** Call a disyllabic *chzizik,* similar to White Wagtail's but shorter, more metallic. [CD3:61] **HABITS** Singly in various habitats, normally near water. Active. Often wags tail. **SIMILAR SPECIES** Yellow Wagtail has olive-tinged upperparts and blackish legs. **STATUS AND DISTRIBUTION** Generally rare Palearctic passage migrant and winter visitor, recorded from Mauritania (regular on coast), Senegambia, Mali (mainly SW, where uncommon; also along Niger R.), W Niger (Mari), SE Cameroon, Eq. Guinea and N CAR.

MOUNTAIN WAGTAIL *Motacilla clara*   Plate 85 (4)
Bergeronnette à longue queue

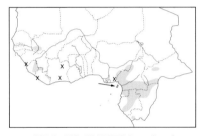

L 18–19 cm (7.0–7.5"). *M. c. chapini.* Slender wagtail of fast-flowing, rocky streams. **Adult** Head and upperparts grey with long white supercilium; wing feathers blackish fringed white. Tail black with white outer feathers. Underparts white with *narrow blackish breast-band.* **Juvenile** Browner above; underparts tinged pale buffy; breast-band reduced or absent. **VOICE** Call a loud, hard, ringing trill *tsrrrrup* and a drawn-out, high-pitched *tseeeet.* Song a sustained medley of vigorous, rather melodious phrases with high-pitched, sibilant notes and ringing trills, e.g. *tsu-tseeee-seeeuu, tsrrrup-tsrrrup-tsrrrup, tseeeee, tsrrrrreeeuw, tsee-tsu-tsee-uw, trrrip-trrrip, tsweu-tswee-tsee-uw...* [CD9: 26] **HABITS** In pairs along swift, boulder-strewn streams in lowland and montane forest. Active. Often wags tail. **SIMILAR SPECIES** Palearctic Grey Wagtail similar in shape and habits, but has yellow on underparts. **STATUS AND DISTRIBUTION** Patchily distributed, scarce to uncommon resident. Recorded from SW Mali, Guinea, Sierra Leone, Liberia and Ivory Coast, and from E Nigeria to CAR, Congo and Bioko. Claimed from Ghana. Seasonal movements recorded in some areas (e.g. Ivory Coast).

WHITE WAGTAIL *Motacilla alba*   Plate 85 (1)
Bergeronnette grise

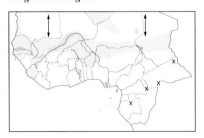

L 18 cm (7"). *M. a. alba.* Grey, black and white wagtail with white face. **Adult male breeding** (from Feb) Forehead, forecrown and head-sides white; hindcrown and nape black, sharply contrasting with grey upperparts; wing feathers black broadly fringed white. Tail black with white outer feathers. Throat to upper breast black; rest of underparts white. **Adult female breeding** Similar, but black on rear of head progressively merging into grey upperparts. **Adult male non-breeding** Throat white (sometimes tinged cream) bordered by black breast-band; black of crown duller, mixed with grey. **Adult female non-breeding** Usually entire top of head (incl. forehead) grey; head-sides variably tinged greyish and cream. **Immature** As adult female non-breeding, but entire face strongly tinged cream-yellow (becoming whitish with wear). **VOICE** Call a sharp *chissik;* alarm *chik.* [CD3:62] **HABITS** Singly or in small or large flocks in variety of habitats, incl. fields, gardens, villages, sea shores, edges of lakes and rivers, dams. Often wags tail. Gathers in large roosts. Defends winter territories. **SIMILAR SPECIES** African Pied Wagtail is black and white, lacking any grey, with black face and white wing patch. **STATUS AND DISTRIBUTION** Common to uncommon Palearctic passage migrant and winter visitor (Sep–Apr), from Mauritania to Sierra Leone east to Niger, N Nigeria, N Cameroon and C Chad. Rare, Cape Verde Is. Vagrant, N Togo, CAR and Gabon.

AFRICAN PIED WAGTAIL *Motacilla aguimp*
Bergeronnette pie

Plate 85 (2)

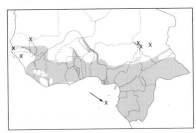

L 19 cm (7.5"). *M. a. vidua*. Black-and-white wagtail with large white wing patch. **Adult male breeding** Head and upperparts black with long, broad white supercilium, white patch on neck-sides, and long white patch on wing (conspicuous in flight); tertials edged white. Tail black with white outer feathers. Underparts white with broad black breast-band. **Adult female breeding** Similar, but black less intense. **Adult non-breeding** Upperparts duller, more slate-grey. **Juvenile** Similar pattern, but much duller, with black replaced by dark grey-brown. **VOICE** Call a clear, whistled *tluwsee*. Song a sustained and varied series of vigorous, quite melodious, short phrases of clear piping and whistling notes. [CD9: 27] **HABITS** Singly, in pairs or family parties near water and habitations. Tame. **SIMILAR SPECIES** White Wagtail has white face and grey upperparts; only overlaps in Mali and SE Niger. Mountain Wagtail has grey upperparts, narrow breast-band and no white in wing. **STATUS AND DISTRIBUTION** Common to uncommon resident, from SE Senegal, Guinea, Sierra Leone, Liberia and S Mali east to S Chad, CAR and Congo. Vagrant, Gambia, Guinea-Bissau and São Tomé.

GRASSLAND PIPIT *Anthus cinnamomeus*
Pipit africain

Plate 84 (9)

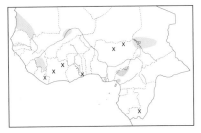

Other name E: African Pipit
L 16–18 cm (6–7"). Large pipit with streaked upperparts and breast, and buff-white outer tail. *Bold face pattern, with prominent buffish supercilium*, pale buff lores, buff ear-coverts streaked dark, and narrow but distinct blackish moustachial and malar stripes. **Adult *lynesi*** Top of head and upperparts dark brown streaked blackish; hindneck paler; rump almost plain; wings feathers blackish-brown, broadly tipped and edged cinnamon-buff when fresh. Tail blackish with buffish-white outer feathers. *Underparts warm cinnamon-buff* (paler on throat and belly) with *short dark streaking confined to upper breast*. Bill dark with yellowish base to lower mandible. Legs pale pinkish-brown; hind claw very long (longer than hind toe). Ssp. *camaroonensis* similar, but underparts paler; wings feathers tipped and edged pale buff. **Juvenile** Streaking heavier; wing-coverts and tertials fringed whitish. **VOICE** Call a hard *chip* when flushed. Song a short, rapid series of 3–5 identical notes *shree-shree-shree-shree-...*, uttered from low perch or in display flight, and a longer series during steep dive at end of display. [CD3:63] **HABITS** Singly, in pairs or loose flocks in various open habitats, incl. grassland and cultivation. Attracted to recently burnt areas. Forages on ground; perches on low vantage points. **SIMILAR SPECIES** Long-billed Pipit is less distinctly streaked and has different voice; also somewhat larger with slightly longer bill and slightly shorter hind claw. Plain-backed Pipit is plain above, with plainer face pattern and buff outer tail feathers. Immature Tawny Pipit has dark loral stripe and more wagtail-like jizz, with slightly shorter legs and less upright carriage; hind claw relatively short. See also Richard's Pipit. **STATUS AND DISTRIBUTION** Locally common resident in highlands of SE Nigeria and W Cameroon (*lynesi*; *camaroonensis* known from Mt Cameroon and Mt Manenguba); *lynesi* also vagrant, N Nigeria and W Chad (single records). Also recorded from Liberia, Ivory Coast, Ghana and Congo (race unknown) and NE Nigeria and W Chad (unnamed local population?). **NOTE** Taxonomic status of *A. cinnamomeus* complex; more research needed. Often considered conspecific with widespread *A. novaeseelandiae* (cf. *BoA*) or Palearctic Richard's Pipit *A. richardi* (cf. Dowsett & Dowsett-Lemaire 1993). Sub-Saharan forms here tentatively treated as a separate species, following Clancey (1990) and others. Our region's two subspecies are sometimes considered a distinct species, Cameroon Pipit (Pipit du Cameroun) *A. camaroonensis* (cf. Sibley & Monroe 1990).

RICHARD'S PIPIT *Anthus richardi*
Pipit de Richard

Plate 147 (7)

L 17–20 cm (7–9"). Very similar to Grassland Pipit but larger, with longer legs and hind claw, more sandy-buff wing-feather edgings, whiter underparts and pure white outer tail feathers. **VOICE** Typical call a rather loud, abrupt, grating *pshreep* or *pshriu* (sometimes reminiscent of sparrow). [CD3:63] **HABITS** Mainly singly in various open habitats. Flight strong, undulating; often hovers briefly before landing. **SIMILAR SPECIES** Grassland Pipit (q.v. and above). **STATUS AND DISTRIBUTION** Potential non-breeding Palearctic visitor. Presence unproven, but probable in view of rare reports from coastal and SE Mauritania (passage Oct–Nov and Mar–Apr) and Mali (Sep–Apr).

TAWNY PIPIT *Anthus campestris* Plate 84 (5)
Pipit rousseline

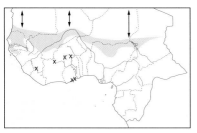

L 16–17 cm (6.5"). *A. c. campestris*. Slim, *pale* pipit with white outer tail feathers. **Adult** Easily identified by *pale sandy-brown upperparts with contrasting dark centres to median wing-coverts*. Head sandy with prominent creamy supercilium, dark lores and narrow dark malar stripe. Underparts plain cream-white, faintly streaked on breast-sides. Upperparts appear plain, but indistinct dusky streaks noticeable in favourable conditions. Wing feathers blackish-brown, broadly tipped and edged sandy-buff when fresh; tail blackish with white outer feathers. Bill dark horn with pale pinkish base to lower mandible. **First-winter** Similar; but sometimes with traces of streaked juvenile plumage on upperparts and breast. **VOICE** Calls a sparrow-like *chlip* or *shlup* and *cheeup* or *treeup*. [CD3:64] **HABITS** Singly in open habitats, incl. short dry grassland and fields. Forages on ground, walking rather horizontally and creeping through short grass like wagtail; also wags tail. Occasionally perches on low vantage point. **SIMILAR SPECIES** All other pipits in the region are darker. See Grassland Pipit for differences with immature Tawny. **STATUS AND DISTRIBUTION** Rare to uncommon Palearctic winter visitor to Sahel and northern savanna belts, from S Mauritania to Guinea-Bissau east through C and S Mali and N Ivory Coast to S Niger, N Nigeria, N Cameroon and C Chad. Vagrant, Cape Verde Is and coastal Ghana. Breeding suspected SW Mauritania (Nouakchott).

LONG-BILLED PIPIT *Anthus similis* Plate 84 (7)
Pipit à long bec

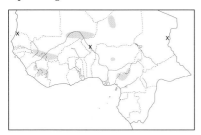

L 17–19 cm (6.75–7.5"). Large pipit with two quite distinct races. **Adult *bannermani*** Top of head and upperparts dark brown streaked blackish; prominent buffish supercilium; narrow dark malar stripe; wing feathers blackish-brown, broadly tipped and edged cinnamon-buff when fresh. Tail blackish with buffish-white outer feathers. Underparts warm buff (paler on throat and belly), boldly streaked dark on breast and flanks. Bill dark horn with pale pinkish base to lower mandible. *A. s. asbenaicus* much paler, more sandy-buff with almost unstreaked breast. **Juvenile** Darker; upperpart feathers pale tipped; breast more spotted than streaked. **VOICE** Call a rather soft *chee* or *djeep*, rapidly repeated once or twice. Song a simple series of 3 (occasionally 2) detached, sparrow-like notes repeated at intervals *chriu shree chewee* or *tsreep shree shruw*, uttered from ground, tree or large rock, or in slow, fluttering display flight. [CD9: 28] **HABITS** Singly or in pairs in montane grassland with rocky slopes. **SIMILAR SPECIES** Grassland Pipit best distinguished by voice; face pattern usually bolder. Plain-backed Pipit has similar vocalisations, but sings much less often than Long-billed and song usually consists of only 1–2 notes repeated at intervals; *A. l. zenkeri* distinguished from *A. s. asbenaicus* by plain upperparts. See also Woodland Pipit. **STATUS AND DISTRIBUTION** Uncommon to locally common resident with disjunct range. *A. s. asbenaicus* C Niger (Aïr), C and E Mali; *bannermani* highlands of Sierra Leone, Guinea, N Liberia, W Ivory Coast (Mt Nimba), SE Ghana, C and SE Nigeria (Jos and Obudu Plateaux, Gashaka-Gumti NP) and W Cameroon, also SW Mali. Vagrant, SW Niger and E Chad. **NOTE** Taxonomic status of Long-billed Pipit in Africa complex; more research needed. Some forms may constitute separate species.

WOODLAND PIPIT *Anthus nyassae* Plate 84 (8)
Pipit forestier

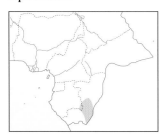

L 17–18 cm (6.75–7.0"). *A. n. schoutedeni*. **Adult** Resembles ssp. *bannermani* of Long-billed Pipit but streaking on upperparts more diffuse; streaks on underparts confined to breast; throat, belly and outer tail feathers whiter. **Juvenile** Darker; upperpart feathers pale-tipped; breast more heavily marked with spots rather than streaks. **VOICE** Song a simple, sustained series of detached chirruping notes, e.g. *tswee twuree shree srwee chruee...* [CD9:29] **HABITS** Singly or in pairs in wooded grassland; also grassy hills with exposed rocks. Forages on ground, walking half upright. Often perches on bushes or low trees, bill pointing upwards. **SIMILAR SPECIES** See Long-billed Pipit. Grassland Pipit best distinguished by voice; face pattern usually bolder. Also compare Plain-backed Pipit. **STATUS AND DISTRIBUTION** Locally common resident, SE Gabon (Lékoni) and S Congo. **NOTE** Taxonomic status unclear; more research required. Often considered conspecific with Long-billed Pipit (cf. *BoA*); here tentatively treated as distinct species, following Clancey (1985).

PLAIN-BACKED PIPIT *Anthus leucophrys*
Pipit à dos uni

Plate 84 (6)

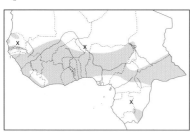

Other name F: Pipit à dos roux

L 17 cm (6.75"). Large pipit with *plain, unstreaked upperparts*. Four races; two cinnamon-buff below, two others pale creamy-buff. **Adult zenkeri** Top of head and upperparts dark brown; prominent buffish supercilium; rather indistinct dark moustachial and malar stripes; wing feathers blackish-brown broadly tipped and edged cinnamon. Tail blackish with buff outer feathers. Underparts warm buff tinged cinnamon, paler on throat and indistinctly streaked dark on breast. Bill dark with yellowish base to lower mandible. *A. l. bohndorffi* similar, but underparts paler. *A. l. gouldii* darker above, with less cinnamon on wings; underparts much paler, with distinct breast streaking. *A. l. ansorgei* like *gouldii* but paler and greyer above. **Juvenile** Browner; breast streaking heavier; upperpart feathers and wing-coverts narrowly fringed whitish. **VOICE** Call a soft *chee-chee* or *tsip-tsip* on take-off. Song simple and repetitive, consisting of a rather slow and monotonous series of single or alternating *swree* and *chirrup* notes, given from ground or low perch. [CD9: 30] **HABITS** Singly or in pairs in short grassland, cultivation, ricefields and burnt ground. Forages on ground, walking rather upright; wags tail. Occasionally perches on low vantage point. **SIMILAR SPECIES** Grassland, Long-billed and Woodland Pipits have streaked upperparts and distinct dark malar stripes. Vocalisations of Long-billed Pipit similar to Plain-backed, but song usually consists of 3 notes. Compare also Tawny Pipit. **STATUS AND DISTRIBUTION** Uncommon to locally common resident; local movements recorded in some areas. Widespread, from S Mauritania, Senegambia and Guinea-Bissau (*ansorgei*) to Sierra Leone, Liberia and Ivory Coast (*gouldii*) and Mali to Ghana, east to CAR (*zenkeri*); also in SE Gabon (Lékoni) and Congo (*bohndorffi?*).

LONG-LEGGED PIPIT *Anthus pallidiventris*
Pipit à longues pattes

Plate 84 (10)

L *c.* 18 cm (7"). *Large, rather plain pipit*, with prominent supercilium, *pale underparts* and *relatively short tail emphasising conspicuously long*, pale yellowish-pink *legs*. **Adult** Top of head and upperparts plain greyish-brown; clear whitish supercilium bordered below by narrow black eye-stripe; short and narrow dark moustachial and malar streaks. Tail blackish-brown with buffish outer feathers. Underparts mainly whitish, distinctly tinged buff on breast, paler buff on flanks; breast with variably distinct, diffuse dark streaking. Wing feathers blackish-brown edged and tipped buff. Bill appears long and largely yellowish, with dark culmen and tip. **Juvenile** Crown streaked; supercilium buff; feathers of upperparts and wings tipped and fringed tawny; breast mottled black. **VOICE** Flight call *psee-ip* or *ch-seep*; contact call a vigorous *psip*. Song simple and consisting of frequently repeated chirrupy notes. [CD9: 32] **HABITS** In pairs or small groups in various grassy habitats, incl. forest clearings. Normally on ground, foraging with upright posture and constantly wagging tail. Perches on poles, wires, etc; only occasionally in trees. Not shy. **SIMILAR SPECIES** No other large pipit in range appears so pale and plain. **STATUS AND DISTRIBUTION** Locally common resident, from SW Cameroon to Congo.

SHORT-TAILED PIPIT *Anthus brachyurus*
Pipit à queue courte

Plate 84 (4)

L 12–13 cm (5"). *A. b. leggei*. Small, dark, heavily streaked pipit with short narrow tail (conspicuous in flight) and distinctive voice. **Adult** Head and upperparts olive-brown streaked or blotched blackish, rump plain. Tail blackish-brown with whitish outer feathers. Underparts whitish, breast and flanks washed buff and boldly streaked black. Wing feathers tipped and edged buff. Bill dark above, pinkish below. **Juvenile** Similar, but feather tips and fringes more tawny. **VOICE** Call a nasal *tseep*. Song a series of nasal *chirrup* notes accompanied by buzzing wing-snaps in cruising display flight. [CD9: 33] **HABITS** Usually singly or in pairs, in short, moist open grassland. Skulking; not easily flushed, prefers to escape by running. Flight low, jerky and not sustained, soon dropping into cover of grass. **STATUS AND DISTRIBUTION** Uncommon and local in SE Gabon and Congo. Local movements recorded. Movements inadequately known (intra-African migrant?).

TREE PIPIT *Anthus trivialis*
Pipit des arbres

Plate 84 (1)

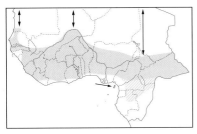

L 15 cm (6"). *A. t. trivialis*. Slim, small pipit with narrowly streaked flanks. **Adult** Top of head and upperparts olive-brown streaked blackish with almost plain rump; supercilium warm buff; wings dark brown with buff edges to flight feathers and white fringes to median coverts. Tail dark brown with largely white outer feathers. Underparts whitish washed creamy-buff on throat, breast and flanks, with blackish malar stripe and coarse blackish streaking on breast extending thinly on flanks. **First-winter** Similar. **VOICE** Call a single *tzeep* or *teez* in flight. [CD3:66] **HABITS** Singly, in twos or small flocks in various wooded habitats, from Sahel and Guinea savanna to forest clearings and farmbush. Forages on ground, often moving stealthily and slowly wagging tail when pausing. Frequently perches in trees or on bushes. **SIMILAR SPECIES** Red-throated Pipit in non-breeding and first-winter plumages always separable by streaked (not plain) rump and broader streaking on flanks. Meadow Pipit has streaking on underparts extending heavily onto flanks; voice different. **STATUS AND DISTRIBUTION** Common to uncommon Palearctic passage migrant and winter visitor throughout, south to Gabon and Congo; also Bioko (rare). Vagrant, Cape Verde Is and São Tomé.

MEADOW PIPIT *Anthus pratensis*
Pipit farlouse

Plate 84 (2)

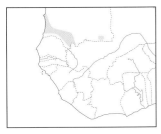

L 14.5 cm (5.75"). *A. p. pratensis*. Small pipit with rather featureless face, extensively streaked flanks, and distinctive call. **Adult** Top of head and upperparts brown heavily streaked blackish (and tinged greenish when fresh), with almost plain rump; wing feathers dark brown edged and fringed buffish. Tail dark brown with largely white outer feathers. Underparts whitish (strongly washed buff when fresh), with coarse blackish streaking on breast and flanks. Head pattern less distinct than Tree and Red-throated Pipits, with more subdued supercilium and eye-stripe (imparting more plain-faced look); blackish malar stripe. **First-winter** Similar. **VOICE** Call a sharp, high-pitched *eest-eet-eest* (usually 1–3 notes). [CD3:68] **HABITS** Singly in various open habitats. Forages on ground mouse-like, creeping gait. Flight hesitant, jerky. **SIMILAR SPECIES** Tree Pipit has streaking on flanks much narrower (sometimes hard to see in field), creamy-buff breast and flanks contrasting with whitish belly, bolder head pattern (with more distinct supercilium), and white-fringed median coverts more contrasting with rest of upperparts. Red-throated Pipit in non-breeding and first-winter plumages much more boldly streaked, with streaked (not plain) rump, creamy stripes on upperparts, and white fringes of median coverts forming contrasting wingbar. **STATUS AND DISTRIBUTION** Uncommon to rare Palearctic winter visitor to coastal and S Mauritania (inland to Tagant and Karakoro).

RED-THROATED PIPIT *Anthus cervinus*
Pipit à gorge rousse

Plate 84 (3)

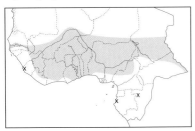

L 15 cm (6"). Small, *boldly streaked* pipit. **Adult male breeding** *Diagnostic brick-red or orange-buff face and throat.* Top of head and entire upperparts (incl. rump) brown heavily streaked blackish, often with 2 pale lines on mantle-sides ('tramlines' or 'braces'; lacking when worn); wings dark brown with whitish fringes to median coverts (forming wingbar) and buff edges to greater coverts and flight feathers. Tail dark brown with white outer feathers. *Underparts have heavy, coarse blackish streaking on breast extending in 2 broad lines along entire flanks.* Head pattern less distinct than Tree and Red-throated Pipits, with more subdued supercilium and eye-stripe (imparting more plain-faced appearance); blackish malar stripe ends in ill-defined patch on neck-sides. **Adult female breeding** Usually with less bright face and throat, and more streaked underparts. **Adult non-breeding** Colour of face and throat variable but always duller, much subdued or replaced by warm buff. **First-winter** Similar to adult non-breeding, but with head-sides and breast dirty whitish to warm buff; upperparts have pale 'braces'. **VOICE** Call a drawn-out, high-pitched *speeeh*, similar to, but longer, less rasping and more piercing than call of Tree Pipit. [CD3:67] **HABITS** Often in small loose flocks, in various types of grassland and cultivation. Forages on ground, wagging tail when pausing. Occasionally perches on low vantage point. **SIMILAR SPECIES** Tree Pipit distinguished by plain rump, narrow streaking on flanks and lack of creamy stripes on upperparts. Meadow Pipit less boldly streaked, with plain rump. **STATUS AND DISTRIBUTION** Common to scarce Palearctic passage migrant and winter visitor, mainly in Sahel and savanna belts, smaller numbers further south. Vagrant, Gabon, Congo and Cape Verde Is.

YELLOW-THROATED LONGCLAW *Macronyx croceus* Plate 86 (6)
Sentinelle à gorge jaune

Other name F: Alouette sentinelle
L 20–22 cm (8.0–8.75"). Distinctive. Robust, with black necklace on yellow underparts. White tail corners conspicuous in flight. **Adult male** Head and upperparts brown streaked blackish; broad yellow supercilium. *Tail* relatively short, brown edged white *with distal half of 2 outer feathers white*. Underparts bright yellow with black moustachial stripe extending and broadening into complete breast-band; variable dark streaking below necklace. **Adult female** Yellow underparts often duller, washed buffish below less well-defined necklace. **Immature** Yellow underparts duller, washed buff; necklace ill defined; streaking on breast and flanks more extensive; white tail corners duller. **Juvenile** Underparts buff with necklace of dark spots. **VOICE** Most characteristic calls include a far-carrying, melodious and repeated whistle *teeuwheee* or *twee-eu*, uttered by both sexes and by male in display flight, sometimes followed by *tiri-tiri-ti* or variation. [CD9: 36] **HABITS** Singly or in pairs in variety of grassy habitats. Forages on ground, but often perches on vantage point (hence French name 'Sentinelle'). Flight distinctive: low and slow, bursts of stiff fluttering alternated with glides, with spread tail. **STATUS AND DISTRIBUTION** Locally common to scarce resident throughout, except arid north and forest.

CUCKOO-SHRIKES
Family CAMPEPHAGIDAE (World: 72–82 species; Africa: 10; WA: 8)

Small to medium-sized, unobtrusive arboreal birds of forest and woodland. Sexually dimorphic. Juveniles resemble adult females. Quietly glean foliage in search of insects, including hairy caterpillars. Usually occur singly or in pairs. Not related to cuckoos or shrikes. All African species endemic. Three genera occur.

Campephaga (3 species): slender with rather long, slightly rounded, ample tails; partially puffed-out feathers of lower back and rump may produce hump-backed appearance. Sexually highly dimorphic; males largely glossy black with skin at gape sometimes swollen into yellow or pinkish wattles.

Lobotos (2 species): distinctive and colourful with large gape wattles; tails rather shorter than in *Campephaga*. Sexes similar.

Coracina (3 species): rather robust, with fairly stout bills and almost square tails. Sexes similar.

RED-SHOULDERED CUCKOO-SHRIKE *Campephaga phoenicea* Plate 86 (4)
Échenilleur à épaulettes rouges

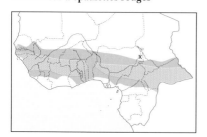

L 20 cm (8"). **Adult male** Entirely glossy blue-black with *large, conspicuous scarlet shoulder patch*. Tail relatively long and graduated. Bill black; swollen gape wattles yellow or pinkish. Shoulder patch sometimes orange or golden-yellow. **Adult female** Top of head, upperparts and tail grey-brown slightly washed olive; wing feathers and outer tail feathers edged yellow; rump and uppertail-coverts barred black. Underparts white barred black and variably washed yellow on breast-sides and flanks; underside of tail feathers broadly tipped yellow on inner webs. **Juvenile** Similar to adult female but upperparts barred black; underparts more heavily barred, bars spade-shaped on belly; underside of tail feathers only narrowly tipped yellow. **VOICE** Calls include a single clear loud whistle *heeew*, a double *huu-tseew* and a short *tsuk*. Song, a jumble of high-pitched and scratchy notes, somewhat reminiscent of sunbird, often uttered in duet. [CD9:39] **HABITS** Singly or in pairs in forest patches, gallery forest, woodland and thickets in savanna. Usually in upper and mid-storeys, occasionally lower, quietly foraging in foliage. Occasionally in mixed-species flocks. Mostly silent and inconspicuous, despite brilliant plumage of male. **SIMILAR SPECIES** Male Purple-throated and Petit's Cuckoo-shrikes lack shoulder patches; females and juveniles have yellow underparts. **STATUS AND DISTRIBUTION** Uncommon to locally not uncommon resident and intra-African migrant in broad savanna belt, from S Mauritania to Sierra Leone east through S Mali and Ivory Coast to SW Niger, Cameroon, S Chad, CAR and N Congo. Wet-season breeder, moving north with rains.

PETIT'S CUCKOO-SHRIKE *Campephaga petiti*

Plate 86 (6)

Échenilleur de Petit

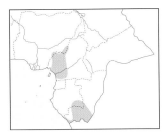

L 20 cm (8"). **Adult male** *Entirely glossy blue-black* with yellow or orange gape wattles usually conspicuous. **Adult female** *Mainly yellow.* Top of head yellowish-green; upperparts olive-yellow heavily barred blackish; wing feathers black edged yellow. Underparts yellow with some barring on breast-sides (barring variable: sometimes almost entirely lacking, rarely more extensive). **Juvenile** Similar to adult female but darker, more olive above; top of head indistinctly barred; underparts with variable number of spade-shaped spots on breast, flanks and undertail-coverts. **VOICE** A short whistled *seep.* Song a high-pitched *psiuu, tsi-tsi*; also a high, scratchy warbling *sueet-sueet, siueet-seet-seet-sireet* (Zimmerman *et al.* 1996). **HABITS** Singly or in pairs in lowland and montane forest, gallery forest and forest patches in savanna. Usually in canopy, occasionally lower, quietly and unobtrusively foraging in foliage. **SIMILAR SPECIES** Male Purple-throated Cuckoo-shrike differs from male Petit's by purple gloss on head-sides and breast and smaller gape wattles; female has grey head and plain olive upperparts; juvenile has olive upperparts and bars (not spots) on underparts. **STATUS AND DISTRIBUTION** Uncommon forest resident, SE Nigeria (Obudu Plateau, Gotel Mts), Cameroon, S Gabon and Congo.

PURPLE-THROATED CUCKOO-SHRIKE *Campephaga quiscalina*

Plate 86 (5)

Échenilleur pourpré

L 20 cm (8"). *C. q. quiscalina.* **Adult male** Entirely black with top of head and upperparts glossed blue-green and *head-sides, throat and breast glossed purple.* Bill black; gape yellow to reddish (*often inconspicuous*). **Adult female** *Head mainly grey; upperparts plain olive; tail green-brown edged and tipped yellow. Throat white; rest of underparts bright yellow.* **Juvenile** Similar to adult female but with variable amount of barring above and below; head browner; wing-coverts tipped black and yellow. **VOICE** A frequently repeated, vigorous, far-carrying *tsee-up*; female may utter similar but shorter and less vigorous *tseeu* note. Song a rather fast and vigorous series of melodious syllables *slueet-slueet-swit-wit slueet-slueet-swit-wit tluw-tluweew* ...[CD9:41] **HABITS** Singly or in pairs in evergreen and gallery forest and forest patches in savanna. Mostly foraging unobtrusively in canopy. Occasionally in mixed-species flocks. **SIMILAR SPECIES** Male Petit's Cuckoo-shrike differs from male Purple-throated by lack of purple gloss on head-sides and breast; female has top of head yellowish-green (not grey) and barred upperparts; juvenile is yellower above and has spots (not bars) on underparts. Male Purple-throated may superficially resemble glossy starling or drongo, but these are vocal and conspicuous, and have less slender outline and differently shaped tails (relatively shorter and narrower in glossy starlings, forked or almost square in drongos). **STATUS AND DISTRIBUTION** Patchily distributed, uncommon to rare forest resident, from Sierra Leone and SE Guinea to Benin and from Cameroon to Gabon and Congo. Isolated records from S Mali, C Nigeria and SW CAR.

WESTERN WATTLED CUCKOO-SHRIKE *Lobotos lobatus*

Plate 86 (8)

Échenilleur à barbillons

L 19 cm (7.5"). **Adult male** *Head glossy blue-black with large yellow-orange to orange gape wattles.* Upperparts olive becoming orange-chestnut on rump; flight feathers black with yellow edges to secondaries and tertials. Tail feathers black tipped and edged yellow except dark green central pair. *Underparts orange-yellow to orange-chestnut suffused with yellow,* paler on belly. **Adult female** Duller; underparts yellow; gape wattles smaller. **Juvenile** Similar to adult female but slightly barred above and below; flight feathers and greater coverts tipped white. **VOICE** A short *tsik*, uttered in flight. **HABITS** Singly or in pairs in canopy of evergreen and semi-deciduous forest. Silent and unobtrusive. Joins mixed-species flocks. **SIMILAR SPECIES** Black-winged and Western Black-headed Orioles have stout red bills and lack gape wattles. **STATUS AND DISTRIBUTION** Rare forest resident, recorded from E Sierra Leone, SE Guinea, Liberia (21 localities), Ivory Coast (Taï NP, Marahoué NP, Mopri) and W Ghana. *TBW* category: THREATENED (Vulnerable).

EASTERN WATTLED CUCKOO-SHRIKE *Lobotos oriolinus*   Plate 86 (7)
Échenilleur loriot

Other name E: Oriole Cuckoo-shrike
L 19 cm (7.5"). **Adult male** *Head glossy blue-black with large yellow-orange to orange gape wattles.* Upper mantle bright yellow becoming olive on rest of upperparts; rump yellower olive; flight feathers black with yellow edges to secondaries and tertials. Tail feathers black tipped and edged yellow except dark green central pair. *Underparts bright yellow slightly washed orange.* **Adult female** Duller; underparts yellow; gape wattles smaller. **Juvenile** Similar to adult female but slightly barred above and below; flight feathers and greater coverts tipped white. **VOICE** Mainly silent. **HABITS** Singly or in pairs in canopy of primary or tall secondary rainforest. Silent and unobtrusive. Joins mixed-species flocks. **SIMILAR SPECIES** Orioles with black heads occurring in same range differ by stout red bills and absence of gape wattles. **STATUS AND DISTRIBUTION** Very rare forest resident, recorded from SE Nigeria (Ikpan), S Cameroon, SW CAR, Gabon and N Congo. *TBW* category: Data Deficient.

GREY CUCKOO-SHRIKE *Coracina caesia*   Plate 86 (2)
Échenilleur gris

L 23 cm (9"). *C. c. pura.* Fairly large grey cuckoo-shrike of highland forests. **Adult male** *Entirely slate-grey* with *black lores* and paler grey eye-ring. Wings and tail darker, blackish-grey. **Adult female** Lores concolorous with rest of head; underparts paler. **Juvenile** Similar to adult but barred dusky and whitish above and below. **VOICE** A thin, sharp, high-pitched and drawn-out *seeeeeu.* Song a spluttering jumble of short, high-pitched notes interspersed with call note. [CD9:42] **HABITS** Mainly singly or in pairs in canopy of mature montane forest, old second growth and stands of tall trees in forest clearings. Quiet and unobtrusive as other members of family, often perching motionless for long periods. Gleans insects from foliage and branches and pursues them in the air. **STATUS AND DISTRIBUTION** Not uncommon forest resident, SE Nigeria (Obudu Plateau), W Cameroon (Mt Cameroon, Rumpi Hills, Mt Kupe, Bamenda Highlands) and Eq. Guinea (Mt Alen); rare, Bioko.

WHITE-BREASTED CUCKOO-SHRIKE *Coracina pectoralis*   Plate 86 (3)
Échenilleur à ventre blanc

L 25 cm (9.75"). Large, grey-and-white cuckoo-shrike with relatively stout bill. **Adult male** Head, upperparts, throat and upper breast pale grey, sharply delimited from pure white rest of underparts. Lores black; white eye-ring. Flight feathers black broadly edged grey on secondaries and tertials; tail feathers blackish except grey central pair. Bill black. **Adult female** Lores paler and upper throat whitish bordered paler grey. **Juvenile** Similar but barred black-and-white above; spotted blackish below. **VOICE** A rather thin, high-pitched whistle *seeeu,* easily passing unnoticed. Various other calls, incl. weak *tsip-tsip* and *tsitsitsi,* louder *wreewree-tiuptiup,* vigorous and abrupt *tsiu,* and rolling *wrrllu-wrrllu-...* [CD9:43] **HABITS** Singly or in pairs in well-wooded savanna. Forages mainly in canopy, gleaning insects from foliage and branches and pursuing them in the air. Regularly joins mixed-species flocks. **STATUS AND DISTRIBUTION** Uncommon to not uncommon resident in broad savanna belt, from Senegambia to Sierra Leone east to S Chad and CAR. Northward movements with rains reported.

BLUE CUCKOO-SHRIKE *Coracina azurea*   Plate 86 (1)
Échenilleur bleu

L 21 cm (8.25"). Uniquely coloured forest bird. **Adult male** Entirely deep glossy blue with black on face and upper throat; flight feathers and tail largely black. Eye reddish. Unmistakable if seen well, but may appear all black in poor light. **Adult female** Slightly duller, with less black on face. **Juvenile** Similar to adult female but has white edges to secondaries and belly feathers; tail tipped white. **VOICE** Calls include a hoarse, nasal *chwee-ep* or *chuee* and a short *chup;* also a series of identical *tuk* notes, first accelerating, then slowing down. Song consists of

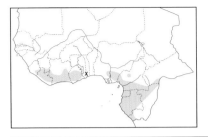

a variable series of loud and clear whistling notes, e.g. *pooeet-pooi-pooeet-peeoo*, a repeated *chup-peeeo* or *peeeoo*, often interspersed with call notes. [CD9:44] **HABITS** Singly or in pairs in primary and mature secondary rainforest; sometimes in small family groups. Usually in canopy, occasionally lower. Regularly joins mixed bird parties. Hops and runs on larger branches; catches insects in the air or snatches them from foliage and branches. Vocal. **STATUS AND DISTRIBUTION** Uncommon to common forest resident, from Sierra Leone and SE Guinea to Ghana and Togo (rare) and from S Nigeria to SW CAR and Congo.

BULBULS
Family PYCNONOTIDAE (World: *c.* 130 species; Africa: 58; WA: 38)

Small to medium-sized (13–26 cm), mostly arboreal birds. Sexes similar, males usually slightly larger; juveniles/immatures similar to, but generally duller than, adults. At all ages, a single complete annual moult shortly after breeding season; post-juvenile moult may start when not yet fledged. Family is well represented in Africa; all species but one are endemic. Most are inconspicuously coloured in various shades and combinations of dull green, grey, brown and yellow, and are associated with forest habitat. Several have a reddish-brown tail. Frugivorous and insectivorous; occasionally flycatch. Mainly sedentary.

The main genera are the following:

Andropadus (8 species): small to medium-sized forest inhabitants with largely featureless olive-green plumage; underparts paler. Essentially frugivorous, hence rarely with mixed-species flocks. Vocalisations important for identification.

Chlorocichla (3 species): medium-sized to large, with contrasting white or yellow throat. Prefer thickets and scrub, mainly outside forest. Mostly in pairs or noisy family groups.

Baeopogon, honeyguide greenbuls (2 species): medium-sized, with white outer tail feathers, giving them superficial resemblance to honeyguides (Indicatoridae). Voice distinctive.

Phyllastrephus (8 species): generally small to medium-sized, with long, slender bills and russet tails. In family parties within lower and middle strata of forest. Insectivorous; frequent members of mixed-species flocks. No arresting vocalisations.

Bleda, bristlebills (4 species): large skulkers of forest undergrowth, with olive-green upperparts, yellow underparts, and stout bills and legs. Rictal bristles well developed. Regularly attend ant columns. Rather shy but vocal.

Criniger, bearded greenbuls (5 species): medium-sized to large, distinguished by their long white or yellow throat feathers which are frequently puffed out, forming a 'beard'. Long, hair-like feathers spring from nape, falling over mantle. Inhabit lower and middle strata of closed forest, usually occurring in pairs or small groups and frequently joining mixed-species flocks. Rather shy but vocal.

Nicator, nicators (2 species): medium-sized to large with bold yellow spots on wings and heavy, hooked bill. Vocalisations distinctive.

CAMEROON MONTANE GREENBUL *Andropadus montanus*   Plate 87 (5)
Bulbul concolore

L 18 cm (7"). **Adult** The only uniformly olive-green bulbul within its restricted range. Rump and uppertail-coverts slightly brighter than rest of upperparts. Underparts slightly paler, rather more yellowish, esp. on belly and undertail-coverts. Underwing-coverts and axillaries sulphur-yellow. **Immature** Inadequately known. A probable immature male had yellowish-grey feet (*BoA*). **VOICE** Song a soft and unobtrusive, rather subdued, nasal, husky babble ending with a faster, rather cheerful, chuckling phrase *churp-churp-churp-chipurchipurcherr*. Calls include a low *kerr* or *purrr*, easily passing unnoticed, and a rapidly repeated, nasal *chup*. [CD9:45] **HABITS** Unobtrusive, occurring singly or in pairs in montane forest, second growth and edge situations. Forages low, but occasionally up to 10 m. **SIMILAR SPECIES** Western Mountain Greenbul has grey head and brighter plumage. Cameroon Olive Greenbul is browner with rusty tail, and frequently flicks wings. **STATUS AND DISTRIBUTION** Scarce to locally common resident in SE Nigeria (Obudu Plateau, Mt Gangirwal) and W Cameroon (Mt Cameroon, Rumpi Hills, Mt Kupe, Bakossi Mts, Mt Manenguba, Bamenda Highlands). *TBW* category: Near Threatened.

WESTERN MOUNTAIN GREENBUL *Andropadus tephrolaemus*   Plate 87 (6)
Bulbul à gorge grise

Other names E: Mountain Greenbul; F: Bulbul à tête grise
L 18 cm (7"). Restricted-range species with distinctive plumage combination.
Adult *tephrolaemus* *Head slate-grey with narrow white eye-ring*; throat paler grey;
upperparts and tail bright olive-green. Underparts yellowish-olive becoming
yellow on belly. *A. t. bamendae* duller, less yellow below, with dark olive wash
on breast. **Juvenile** Duller; head washed olive; underparts darker. **VOICE**
Song a monotonous, steady series of notes, same form repeated continuously,
e.g. *whup-wheep-whip-whupchipup* and *whup-wheep-whup-wheep-whup-wheep-...*;
also faster series of nasal, chirruping notes. Call a nasal, scolding *whee-up* or
dzut-dzuwi. [CD9:47] **HABITS** Singly, in pairs or small groups (usually with
juveniles) in all types of montane forest, incl. small relict patches. Vocal. Forages at all levels, catching insects in
the air or by gleaning foliage or bark, occasionally clinging to tree trunks like woodpecker. Congregates at certain
fruiting trees; attends ant swarms. **SIMILAR SPECIES** Cameroon Montane Greenbul is uniformly dull olive-
green. **STATUS AND DISTRIBUTION** Common resident on Mt Cameroon and Bioko (nominate) and in
SE Nigeria (Obudu and Mambilla Plateaux, Gotel Mts) and rest of Cameroon highlands (*bamendae*); variation
clinal. **NOTE** Here treated as specifically distinct from E African forms, included in *A. nigriceps* (Eastern Mountain
Greenbul), following Dowsett & Dowsett-Lemaire (1993).

LITTLE GREENBUL *Andropadus virens*   Plate 87 (1)
Bulbul verdâtre

L 16.5 cm (6.5"). *Skulking, wholly dirty olive-green bulbul with distinctive,
easily learnt song.* Absence of any contrast in plumage is a distinguishing feature. **Adult** *virens* Head and upperparts dark olive-green; tail
and uppertail-coverts washed brownish. Throat pale olive-green
becoming dark olive on breast and flanks and pale yellowish in centre
of belly. No eye-ring. Bill blackish, often with more or less yellow
gape. Legs usually pale yellowish-brown, sometimes darker. Western
erythropterus darker and greyer above with browner flight feathers.
Island race *poensis* similar to nominate, but greener above and
slightly brighter yellow on belly (Dickerman 1994). **Immature**
Similar, but lower mandible, tip of upper mandible and gape yellow. **VOICE** Song starts with a few subdued
chuckling notes, followed by a rapidly ascending series forming a pleasant bubbling warble, increasing in volume
and abruptly ending on a clear, high, rising note. Call a dry *kuk-kuk-kuk...* [CD9:49] **HABITS** Singly and in pairs in
variety of habitats, incl. forest edge, rank scrub, abandoned cultivation, thicket, even gardens. Sings all day and
nearly year-round, from deep cover, usually 2–3 m above ground; only ceases for 1–2 months when moulting. Shy
and often frustratingly hard to see. **SIMILAR SPECIES** Cameroon Sombre Greenbul is very similar and best
separated by voice and eye-ring. Young Yellow-whiskered Greenbul has distinctive pinkish legs. **STATUS AND
DISTRIBUTION** Very common (except at periphery of range) resident in forest zone and adjacent forest–savanna
mosaic, from Gambia to Nigeria, north to extreme SW Mali (Mandingo Mts) and 10°N in Nigeria (*erythropterus*),
and from Cameroon to extreme SW Chad, CAR and Congo (nominate); also Bioko (*poensis*).

LITTLE GREY GREENBUL *Andropadus gracilis*   Plate 87 (2)
Bulbul gracile

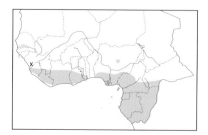

L 16 cm (6.25"). **Adult** *extremus* Small bulbul with *olive-grey head and
throat*, narrow but usually obvious *white eye-ring*, olive-green upperparts and *yellowish-olive underparts*, brightest in centre. Tail and
uppertail-coverts washed chestnut. Bill black. Eastern *gracilis* duller
and greyer below, with little green or yellow. **Juvenile** Brighter above
and below; eye-ring and base of bill yellowish. **VOICE** Song consists
of 4–5 rapid, jaunty notes *wheet wu-wheet-wu-wheet*; also a quieter *tehu-
tehee-tee.* Call a short *tyuk.* [CD9:50] **HABITS** Singly, in pairs or family
groups, frequenting upper mid-level of variety of forest habitats incl.
abandoned cultivation, and preferring edges. Not shy; easily seen.
SIMILAR SPECIES Ansorge's Greenbul is very similar but lacks any yellow tones to underparts and has distinctive
dry rattling trill. **STATUS AND DISTRIBUTION** Not uncommon to common in forest zone, from W Guinea to
Togo and in SW Nigeria, west of Niger R. (*extremus*) and from SE Nigeria to SW CAR and Congo (nominate).

ANSORGE'S GREENBUL *Andropadus ansorgei*
Bulbul d'Ansorge

Plate 87 (4)

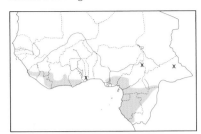

L 16 cm (6.25"). *A. a. ansorgei*. Small and very similar to Little Grey Greenbul, with olive-grey head and throat, narrow, but usually obvious *white eye-ring* and olive-green upperparts, but *underparts lack any yellow*: lower breast and belly mainly olive-greyish, becoming ginger on flanks and around vent. Tail and uppertail-coverts washed chestnut. **VOICE** Three-note song *wheet-whuut-whit* or *tiu-wheet-tweet*, resembling that of Little Grey Greenbul, but lacking its sprightliness; last syllable may be dropped. Also a distinctive, rapid, harsh and flat trill *rititititititit* or *tchitchitchitchi*. [CD9:51] **HABITS** Singly, in pairs or family parties in mature forest, usually frequenting crown of low trees at upper mid-level. May occur alongside Little Grey Greenbul, though usually higher. Not shy and fairly easily observed. **SIMILAR SPECIES** Little Grey Greenbul distinguished by yellow on underparts and different voice. **STATUS AND DISTRIBUTION** Uncommon to locally common forest resident, from W Guinea to Togo and from Nigeria to SW CAR and Congo. Status and distribution obscured by confusion with more common and widespread Little Grey Greenbul.

CAMEROON SOMBRE GREENBUL *Andropadus curvirostris*
Bulbul curvirostre

Plate 87 (3)

Other name E: Plain Greenbul
L 17 cm (6.75"). One of the most difficult forest bulbuls to identify on plumage characters alone. **Adult *leoninus*** Mainly olivaceous-brown, with darker, reddish-brown uppertail-coverts and tail and slightly paler underparts; flanks washed ginger. Usually has narrow whitish eye-ring, often broken or faint. **Nominate** overall somewhat paler, with olivaceous-grey throat contrasting slightly with darker head and olive-brown breast (not apparent in *leoninus*): a field mark in good light. **Immature** Belly yellower; gape yellow. **VOICE** Song of western *leoninus* unarresting but distinctive *tiuwhee-tiu trriiiiii* or *su-hi-oo trriiii*, with stress on final harsh trill, which is often also given separately. Also a hard *wrrrit* and longer *wrrrrrititit*. Nominate has quite different, 3-note *wheet-tiuwhee-tuu* and *wheet-tu-twhee*, also a harsh trill *trriiii*. [CD9:52–53] **HABITS** Unobtrusive, occurring singly, in pairs or small family groups in mature and degraded forest and edges. Frequents lower mid-level. May join other bulbuls at fruiting trees and in mixed-species flocks. Usually skulking and hard to observe. **SIMILAR SPECIES** Little Greenbul is extremely similar but more olivaceous-green; best separated by voice and lack of eye-ring. Slightly more slender bill of Cameroon Sombre is not a useful field character. Young Yellow-whiskered Greenbul distinguished by pinkish legs. Little Grey and Ansorge's Greenbuls noticeably smaller, with whiter eye-rings and different habits. **STATUS AND DISTRIBUTION** Not uncommon to common resident in forest zone, from Sierra Leone and SE Guinea to S Benin and from SW Nigeria to SW CAR and Congo; also Bioko. *A. c. leoninus* in west of range, meeting nominate in Ghana.

SLENDER-BILLED GREENBUL *Andropadus gracilirostris*
Bulbul à bec grêle

Plate 87 (7)

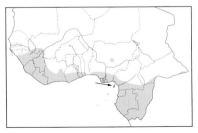

L 18 cm (7"). *A. g. gracilirostris*. A conspicuously two-toned, rather slim, canopy dwelling bulbul. **Adult** Head, upperparts and tail olive-brown. Underparts pale brownish-grey, with throat, centre of belly and undertail-coverts paler, more greyish-white. Bill black. Legs black to dark brown. **Juvenile** Similar. **VOICE** A distinctive, clear, drawn-out whistle *tseeeeu*, frequently uttered. Also a seasonal song (used for only a few weeks) consisting of 4–5 whistled notes *whee-ti-twheew-ti-twhee* or *whuut-hEET whuut-hEET whuut-hEET whuut-hEET*, with slight variations. [CD9:54] **HABITS** Singly, in pairs or small family groups. Largely restricted to canopy of various forest types and cultivation with large trees. Occasionally joins mixed-species flocks. **SIMILAR SPECIES** No other bulbul frequenting canopy has uniform pale grey underparts. **STATUS AND DISTRIBUTION** Common resident in forest zone and adjacent woodland, from SW Senegal and Guinea-Bissau through Liberia and SW Mali to S CAR and Congo; also Bioko.

YELLOW-WHISKERED GREENBUL *Andropadus latirostris*   Plate 87 (8)
Bulbul à moustaches jaunes

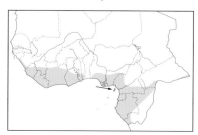

L 17 cm (6.75"). **Adult** *congener* Common, olive-green forest bulbul with *bright yellow malar stripe* and distinctive *orange-brown or yellow-brown legs*. Head and upperparts dark olive-green; tail darker, washed rufous. Underparts paler greyish-olive, becoming yellowish on centre of belly. Bill blackish. Yellow 'whiskers' often puffed out. **Nominate** paler and greener above; tail more rusty. **Juvenile/immature** No or only rudimentary whiskers, but *pinkish legs* key to identity. Bill blackish with dark orange-yellow gape and base. **VOICE** Song a monotonous series of *c.* 12 *chruk* notes, slightly increasing in volume; uttered all day for long periods and one of the characteristic sounds of many forests. Calls include a repeated *chuk* and a rapid rattling *ditditditdit...* (alarm). [CD9:55] **HABITS** Rather shy, occurring singly in various forest types and abandoned cultivation with dense scrub. Found at all heights. Rarely in mixed-species flocks. **SIMILAR SPECIES** Adult with yellow whiskers unmistakable. Little and Cameroon Sombre Greenbul distinguished from immature by darker legs. **STATUS AND DISTRIBUTION** Common to very common resident in forest zone and adjacent forest–savanna mosaic. *A. l. congener* in SW Senegal, from W Guinea to Togo and in SW Nigeria, west of Niger R.; nominate from SE Nigeria to SW CAR and Congo; also SE CAR and Bioko.

GOLDEN GREENBUL *Calyptocichla serina*   Plate 89 (3)
Bulbul doré

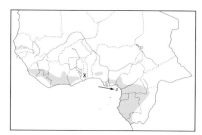

L 18 cm (7"). **Adult** Distinctive, medium-sized bulbul with olive-green upperparts, yellow underparts becoming bright golden-yellow on centre of belly, and *pale pinkish bill*. Yellowish-olive ear-coverts and pale bill give it an open-faced appearance. **Immature** Duller. **VOICE** Song a clear, short *tiup-chiweew*, with stress on last syllable; many variations, e.g. *whit-tu-tiup, tiu-tu-tip, whit-tuti-tiheew, whu-whit whu-hu-wheet*, etc. Calls include a short *tsip* or *tyip* and, in flight, a thin, high-pitched *see*. [CD9:56] **HABITS** Singly, in pairs or small family parties in canopy of primary and secondary lowland forest; prefers edges. Occasionally lower at fruiting bushes. Sometimes joins mixed-species flocks. **SIMILAR SPECIES** No other lowland forest bulbul with yellow underparts inhabits canopy. Pinkish bill conspicuous and unique among bulbuls. **STATUS AND DISTRIBUTION** Rare to locally not uncommon resident in forest zone, from Sierra Leone and SE Guinea to Togo, and from S Nigeria to SW CAR and Congo; also Bioko.

HONEYGUIDE GREENBUL *Baeopogon indicator*   Plate 88 (7)
Bulbul à queue blanche

L 19 cm (7.5"). **Adult male** *leucurus* Dark, stocky bulbul with 3 pairs of outer tail feathers conspicuously white, tipped blackish-brown (thus superficially resembling large honeyguide). Top of head and upperparts brownish-olive; head-sides and throat greyish; rest of underparts dark olive-grey, becoming creamy-white on lower belly and undertail-coverts. Eye whitish. **Nominate** more washed with olive below; lower underparts slightly darker, buffish. **Adult female** Similar, but eye dark. **Immature** Outer tail feathers all white. **VOICE** Wide range of calls unlike those of any other bulbul. Main song a clear, vigorous thrush-like whistle comprising a series of melodious drawn-out notes; often abbreviated as a hurried *vik-vik-view* or *tiu-liuuw*. Also a single mewing note. [CD9:57] **HABITS** Singly, in pairs or small groups in upper mid-level and canopy of forest, preferring edges. Vocal; sings throughout day. Flight direct, not undulating like honeyguide. **SIMILAR SPECIES** Sjöstedt's Honeyguide Greenbul has buffish underparts and white outer tail feathers in all plumages; best distinguished by harsh, less melodious voice. Honeyguides (Plate 75) also have white in outer tail, but shape and pattern different; plumage characteristics also differ; bill stubby. **STATUS AND DISTRIBUTION** Generally common resident in forest zone and adjacent forest–savanna mosaic, from Guinea to Togo (*leucurus*) and from Nigeria to SW and SE CAR and Congo (nominate).

SJÖSTEDT'S HONEYGUIDE GREENBUL *Baeopogon clamans* Plate 88 (8)
Bulbul bruyant

L 19 cm (7.5"). **Adult** Resembles Honeyguide Greenbul but underparts paler and buffier; white outer tail feathers lack darkish tips. Lores and ear-coverts grey; centre of throat greyish. Eye dark. **Immature** Duller. **VOICE** A loud, hard, nasal *whEw!*, frequently uttered and often running into a short, rapid, nasal babble *teeturuteetutwhee*. [CD9:58] **HABITS** Singly, in pairs or small groups, frequenting mid-levels and lower canopy (sometimes lower) in interior of primary and old secondary lowland forest, often near streams. Occasionally joins mixed-species flocks. Calls frequently, spreading tail to display white feathers. **SIMILAR SPECIES** Honeyguide Greenbul is more uniformly dark and has white outer tail feathers tipped blackish (but dark tips lacking in immature!); male has whitish eye. Most easily separated by quite different, melodious voice. **STATUS AND DISTRIBUTION** Locally common resident, from SE Nigeria to SW CAR and Congo.

SPOTTED GREENBUL *Ixonotus guttatus* Plate 89 (9)
Bulbul tacheté

L 17 cm (6.75"). Small and distinctive. **Adult** Top of head and upper-parts dark olive-brown with *white spots on wing-coverts, inner secondaries and rump*. Underparts and outer tail feathers white faintly tinged yellow. Spots on upperparts may be difficult to see as usually high in trees, but white underparts, gregarious behaviour and voice render identification easy. **Immature** Similar. **VOICE** A distinctive dry chirping or 'ticking' call, resembling noise made by electrical spark, constantly uttered. [CD9:59] **HABITS** In noisy, mobile monospecific groups of up to 50 (usually 7–15), restlessly gleaning insects from leaves and branches, in canopy of primary and secondary forest; also at forest edges, in forest–savanna mosaic and man-made habitats. Frequently raises one wing alternately. **SIMILAR SPECIES** Unlikely to be confused. Western and Yellow-throated Nicator also have spots on wings, but these are yellow; they are solitary, skulking species principally frequenting lower and mid-levels and have entirely different voices. **STATUS AND DISTRIBUTION** Locally common to scarce resident in forest zone, from Sierra Leone (Gola Forest) and SE Guinea to Ghana, and from S Nigeria to SW CAR and Congo; also N CAR (Bamingui-Bangoran NP).

YELLOW-NECKED GREENBUL *Chlorocichla falkensteini* Plate 88 (3)
Bulbul de Falkenstein

Other name E: Falkenstein's Greenbul
L 18 cm (7"). **Adult** Head, upperparts and tail bright olive-green. Throat bright yellow, extending onto neck-sides; rest of underparts very pale grey. Eye dark red. **Immature** Similar; gape yellow. **VOICE** Song a string of nasal notes *kik-kuk-ku-KWEE-uk-wik-e-wik-kup*, with slight variations. Song and chattering calls similar in quality to those of Simple Leaflove. [CD9:62] **HABITS** In pairs or small groups in dense vegetation within a variety of shrubbery habitats, incl. small forest patches, farmbush and fallow fields. Not in forest. Occasionally joins mixed-species flocks. Skulking, but occasionally vocal. Sometimes raises wings alternately like Spotted Greenbul, thereby revealing bright yellow underwing-coverts. **SIMILAR SPECIES** Yellow-throated Leaflove has brown upper- and underparts and is stouter. **STATUS AND DISTRIBUTION** Locally common to rare resident; three apparently disjunct populations from S Cameroon to Gabon, in SW CAR (near Bangui), and in S Congo.

SIMPLE LEAFLOVE *Chlorocichla simplex* Plate 88 (4)
Bulbul modeste

Other name E: Simple Greenbul
L 21 cm (8.25"). **Adult** Large bulbul with conspicuous *white throat* and *broken eye-ring* contrasting with rest of plumage. Head, upperparts and tail dark brown. Underparts olive-brown, becoming whitish in centre of belly; undertail-coverts pale cinnamon. **Immature** Similar. **VOICE** Song a frequently uttered, subdued nasal chattering. Calls include a scolding *wherr* and a short, clipped *kwit!* [CD9:63] **HABITS** In pairs or small family groups in

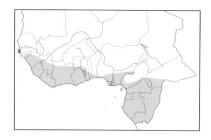

habitats with dense shrubs (farmbush, forest regrowth, thickets). Rather shy and skulking, but often emerges in the open. **SIMILAR SPECIES** Common Bulbul superficially similar in general colour, but lacks white throat and white crescents around eye. White-throated race of Yellow-throated Leaflove lacks eye-crescents and has different voice. **STATUS AND DISTRIBUTION** Locally common resident in forest zone and forest–savanna mosaic, from Guinea-Bissau to SW CAR and Congo.

YELLOW-THROATED LEAFLOVE *Chlorocichla flavicollis*   Plate 88 (2)
Bulbul à gorge claire

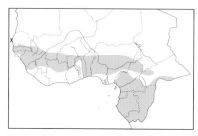

Other name F: Grand Bulbul à gorge jaune
L 22.5 cm (*c.* 9"). **Adult male** Large bulbul with dark olive-brown head, upperparts and tail. **Nominate** has bright yellow throat contrasting with dark head and olive-grey underparts. Eye greyish-buff. In distinctive eastern race *soror* throat is white, faintly tinged yellow; underparts paler. **Adult female** Similar, but eye greyish-white. **Immature** Slightly darker. **VOICE** Contact calls include loud, nasal *chow* and *kyip*. Song a string of *chows* interspersed with shorter notes; in quality reminiscent of Swamp Palm Bulbul. [CD9:64] **HABITS** Singly, in pairs or noisy family parties, mainly in lower and mid-levels of dense woodland and thickets; also low swamp forest. Fairly shy, but vocal. **SIMILAR SPECIES** White-throated race likely to be confused only with Simple Leaflove, which also has white throat and shares same habitat, but has distinctive broken white eye-ring. **STATUS AND DISTRIBUTION** Not uncommon to scarce resident. Nominate from Senegambia to Sierra Leone through S Mali and N Ivory Coast to Nigeria, N Cameroon and (this race?) NW CAR; *soror* from NC Cameroon to CAR and Congo. **NOTE** The two races may be species: they are separated by only 50 km on Adamawa Plateau and no intergrades are known.

SWAMP PALM BULBUL *Thescelocichla leucopleura*   Plate 88 (6)
Bulbul des raphias

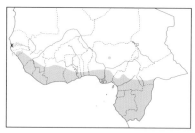

Other names E: White-tailed Greenbul; F: Bulbul à queue tachetée
L 23 cm (9"). Large, vocal and easily identified. **Adult** Top of head, upperparts and tail dark olive-brown; head-sides speckled white; *outer tail feathers broadly tipped white*. Throat off-white; upper breast pale grey-buff; *lower breast and belly creamy-white*. **Immature** Throat and breast pale yellow. **VOICE** A raucous nasal cackling like a tape-recorded conversation being played too fast. One bird starts and is immediately copied by other members of the group. Also a variety of other loud, nasal, scolding calls. [CD9:65] **HABITS** Gregarious, occurring in small, mobile and noisy parties and associated with palm trees, particularly *Raphia*, in lowland and gallery forest. **SIMILAR SPECIES** Unlikely to be confused, but beware Leaflove and Yellow-throated Leaflove when attempting to identify this species on call alone. **STATUS AND DISTRIBUTION** Not uncommon to locally common resident in forest zone and its gallery extensions, from Senegambia and Guinea-Bissau to SW CAR and Congo. Isolated population, C Nigeria (Jos Plateau).

LEAFLOVE *Pyrrhurus scandens*   Plate 88 (5)
Bulbul à queue rousse

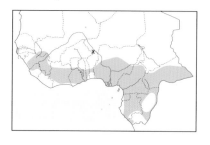

L 22 cm (8.5"). Large, handsome bulbul with distinctive coloration. **Adult** *scandens* Head pale grey; upperparts pale brownish-olive; wings brown edged rufous on flight feathers; uppertail-coverts and tail pale chestnut. Throat white; breast and flanks greyish-buff becoming creamy-buff on lower breast and belly. Eastern *orientalis* is rather greyer above and slightly paler below. **Immature** Top of head has rusty wash. **VOICE** Song starts with some subdued, nasal notes, then bursts into loud, resonant, pleasant conversational cackling, including *kyop-kee-kyop-kyop-ke-kyop*—..., uttered for long periods by family groups. [CD9:66] **HABITS** In small, noisy groups in forest fringing rivers and swampy places within forest; also forest patches in savanna. Occurs at all levels, though mainly in canopy and middle stratum, working its way through dense liana tangles. Joins mixed-species flocks. Easily located and identified by its incessant call. **SIMILAR SPECIES** Caution is needed when attempting to identify this species on voice alone: Swamp Palm Bulbul and Yellow-throated Leaflove sound quite similar. **STATUS AND DISTRIBUTION** Scarce to locally common resident in forest zone and adjacent forest–savanna mosaic, from Senegambia and S Mali to CAR and Congo. The two races meet in Cameroon.

PALE OLIVE GREENBUL *Phyllastrephus fulviventris*   Plate 89 (7)
Bulbul à ventre roux

L 19 cm (7.5"). **Adult** Top of head and upperparts dark olive-brown; rump, uppertail-coverts and tail dull rufous. Throat creamy-white; rest of underparts pale tawny-yellow washed pale olive-brown on breast and flanks. Broken eye-ring white, merging with pale streak from lores to just behind eye; ear-coverts pale olive-brown with narrow whitish streaks. Bill dark above, paler below. Legs pale pinkish to greyish. **Immature** Unknown. **VOICE** A nasal, chattering note. Alarm call described as a loud, sharp *tsik-tschirr-tschirr* (*BoA*: Heinrich 1958). [CD9:69] **HABITS** Skulking and little known, occurring in gallery forest, dense bush and forest patches in savanna. Forages at all levels, but principally in undergrowth, occasionally on ground. **SIMILAR SPECIES** Could occur alongside White-throated Greenbul, which is greener above and has white throat contrasting with greyish head and breast. **STATUS AND DISTRIBUTION** Inadequately known. Reported from coastal S Gabon and (possibly) S Congo. Locally common resident in extralimital Cabinda and along lower Congo R. in Congo-Kinshasa.

BAUMANN'S GREENBUL *Phyllastrephus baumanni*   Plate 89 (5)
Bulbul de Baumann

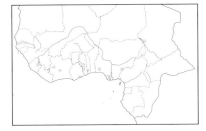

L 18 cm (7"). Relatively little-known, medium-sized bulbul, lacking diagnostic features. **Adult** Top of head and upperparts dark olive-brown; rump, uppertail-coverts and tail rusty; head-sides grey-brown with narrow pale streaks. Underparts pale olive-grey; undertail-coverts pale sandy-rufous. Throat may be slightly paler than rest of underparts but is never contrasting. Eye brown to tawny. Bill pale horn with dark culmen. **Immature** Unknown. **VOICE** Calls include series of loud *week* and *wik* notes, and a scolding *chèrr*. Song consists of 2-4 slightly nasal, rising notes *whu whee wheew* followed by some scolding *wik* or *chewik* notes. **HABITS** Usually in pairs, sometimes small groups of 3–6, frequenting lower levels of semi-deciduous mid-altitude forest (*c.* 500–1100 m), often on hills and mountain slopes, and, at lower levels, gallery forest and thickets. Not in lowland rainforest. Joins mixed-species flocks. Skulking and rather shy, but occasionally noisy. Flicks wings. **SIMILAR SPECIES** In dense undergrowth most likely to be confused with illadopsis species (q.v.); compare also *Andropadus* greenbuls, esp. *A. gracilirostris*. White-throated Greenbul is more colourful, with grey head, olive upperparts, rufous-brown tail, contrasting white throat and yellow underparts washed olive (Baumann's does not have any yellow). **STATUS AND DISTRIBUTION** Inadequately known. Uncommon to rare resident in forest–savanna mosaic. Recorded from Sierra Leone, Liberia, Ivory Coast, Ghana, SW Togo and Nigeria. Status and distribution obscured by identification problems; possibly more widespread than records suggest (cf. Fishpool 2000). *TBW* category: Data Deficient.

CAMEROON OLIVE GREENBUL *Phyllastrephus poensis*   Plate 89 (6)
Bulbul olivâtre

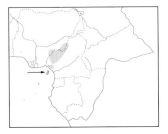

L 18 cm (7"). Rather nondescript bulbul with rather long, fine, darkish bill. **Adult** Olive-brown above with greyish head-sides. Tail relatively long and washed rufous. Underparts pale dirty olive-grey with whitish throat, becoming pale rufous-brown on undertail-coverts. **Immature** Upperparts have chestnut wash; dull brownish band on breast; some rusty feathers on rest of underparts. **VOICE** Calls include a low, grating *chrrr-chrrr-chrrr-....* and a hard, dry *trrr-trrr-trrr-...* Song a series of unmusical notes *chewp chop chop chip chip chewp chip cher...* and a loud, scolding *kweep-kwep-kwerr-kweep-krrr-kweep-kerrr...* with variations. [CD9:72] **HABITS** Singly, in pairs and small groups in montane forest and forested ravines on plateaux at 1000–2200 m. Forages in dense cover, mainly low down, systematically gleaning branches and leaves. Frequently flicks tail. Shy and unobtrusive. Joins mixed-species flocks when attending ant swarms. **SIMILAR SPECIES** Other montane bulbuls have quite different plumages. Western Mountain Greenbul is olive-green with grey head and yellowish belly. Cameroon Montane Greenbul is uniformly olive-green. Grey-headed Greenbul has bright yellow underparts. **STATUS AND**

DISTRIBUTION Locally common forest resident, SE Nigeria (Obudu and Mambilla Plateaux, Danko FR) and W Cameroon (Mt Cameroon, Rumpi Hills, Mt Kupe, Mt Nlonako, Bakossi Mts, Mt Manenguba, Bamenda Highlands); also Bioko.

ICTERINE GREENBUL *Phyllastrephus icterinus*   Plate 89 (1)
Bulbul icterin

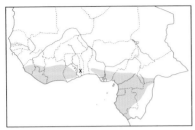

L 15–16 cm (6.0–6.25"). Small, slender bulbul. **Adult** Head and uperparts dull olive-green; head-sides paler with some yellowish. Tail dull rufous. Underparts drab yellow washed olive on breast and flanks, brighter yellow on throat, belly and undertail-coverts. Bill blackish above, grey or yellowish below, gape pale yellow. **Immature** Greener above; throat and breast slightly washed brownish; tail feathers more pointed. **VOICE** Song a rather fast nasal chatter, slowing at end, *chirrrrup-chup-chup-chup-chup-chup*, easily passing unnoticed. Alarm a nasal trill. [CD9:73] **HABITS** Gregarious and usually in mixed-species flocks, gleaning insects from leaves, shoots and tangles at lower and mid-levels of forest interior. **SIMILAR SPECIES** Yellow-bearded Greenbul distinguished by tufted yellow throat feathers clearly demarcated from olive upper breast, bare skin around eye and woodpecker-like behaviour. Xavier's Greenbul (q.v.) extremely similar and in some cases probably inseparable in field, but voice diagnostic and males noticeably larger; in favourable conditions, greenish (not rusty) uppertail-coverts and duller rufous tail may be visible; often flock together, but tending to forage at higher levels. **STATUS AND DISTRIBUTION** Not uncommon to common forest resident, from Sierra Leone and SE Guinea to Ghana and from S Nigeria to SW CAR and Congo; also Bioko. Rare, Togo.

XAVIER'S GREENBUL *Phyllastrephus xavieri*   Plate 89 (2)
Bulbul de Xavier

L 16–18 cm (6.25–7.0"). **Adult** *xavieri* Extremely similar to and very difficult to separate from Icterine Greenbul. Best distinguished by call. Less useful characters include slightly heavier, deeper based bill with upward kink in lower mandible (more dagger-like in Icterine), greenish (not rufous) uppertail-coverts, and less rufous tail. Olive wash on breast perhaps tends to be less extensive (sometimes lacking in centre) in Xavier's, thus producing slightly yellower underparts (with less or no contrasting throat); Icterine has more complete olive breast-band. Larger size occasionally noticeable by direct comparison of large male Xavier's and small female Icterine (female Xavier's and male Icterine overlap in size). *P. x. serlei* slightly duller, less yellow below. **Immature** Lower mandible and gape more yellow. **VOICE** Diagnostic. A short, nasal *kwah, kwah, kwah,... kwahkwah* and a more drawn-out, squeaky *kwèèèh* and *kèèh*, uttered in shorter or longer series of either similar or combined notes (*kwah-kèèh*). [CD9:74] **HABITS** In pairs or family parties in interior and edge of primary and secondary forest, also large trees in clearings. Often in mixed-species flocks of insectivores, foraging at all levels. **SIMILAR SPECIES** Icterine Greenbul (q.v. and above). **STATUS AND DISTRIBUTION** Uncommon to locally common forest resident in extreme SE Nigeria, Cameroon, SW CAR, Gabon and Congo; nominate almost throughout, *serlei*-restricted to area north and west of Mt Cameroon. Status inadequately known due to confusion with Icterine Greenbul.

LIBERIAN GREENBUL *Phyllastrephus leucolepis*   Plate 89 (8)
Bulbul du Libéria

Other names E: White-winged Greenbul; F: Bulbul ictérin tacheté
L *c*. 16 cm (6.25"). **Adult** Similar to Icterine Greenbul but with whitish spots on wings. Head and upperparts olive-green; *2 conspicuous bars on brown wing formed by cream-coloured subterminal spots on flight feathers and row of whitish to creamy-grey spots on wing-coverts.* Tail brighter chestnut than in Icterine. Underparts drab yellowish washed with olive on breast and flanks. **Immature** Unknown. **VOICE** No information. **HABITS** Only observed singly or in pairs, within transition zone between evergreen and semi-deciduous forest. Joins mixed-species flocks and forages mainly at mid-levels (4–8 m). Flicks both wings concurrently, creamy bars then conspicuous. **SIMILAR SPECIES** Unlikely to be confused. Spotted Greenbul has white underparts and different behaviour. Western Nicator is

larger with heavy bill and white underparts except yellow throat and undertail-coverts. **STATUS AND DISTRIBUTION** Very rare resident, known from only two isolated patches of rainforest near Zwedru, Grand Gedeh County, Liberia. *TBW* category: THREATENED (Critical).

WHITE-THROATED GREENBUL *Phyllastrephus albigularis* Plate 89 (4)
Bulbul à gorge blanche

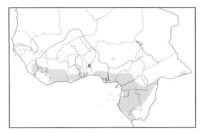

L 17 cm (6.75"). *P. a. albigularis*. **Adult** Rather small; *white throat contrasting with grey head and olive-grey breast*. Upperparts dark olive-green; uppertail-coverts and tail olive-rufous. Grey of breast mixed with some pale yellow, which increases on lower breast, belly and undertail-coverts. Eye whitish to dark brown. Bill black to dark brown with pale cutting edges. **Immature** Gape and tip of bill yellow. **VOICE** A variable, rapid series of clear, scolding notes, usually rising in pitch, then dying away, often preceded by loud, rolled *turrr*. [CD9:75] **HABITS** In pairs or family groups in primary and secondary forest. Prefers edges with dense undergrowth and lianas. Usually forages very low, but also in middle and upper levels. Joins mixed-species flocks, but less frequently than Icterine or Xavier's Greenbuls; often in small monospecific groups. Rather unobtrusive. **SIMILAR SPECIES** Unlikely to be confused if seen well. Icterine and Xavier's Greenbuls have yellow underparts. Red-tailed Greenbul is much larger, has white throat feathers often puffed out, rest of underparts yellow, and different behaviour. Compare also Baumann's Greenbul. **STATUS AND DISTRIBUTION** Scarce to uncommon or locally common forest resident, from Sierra Leone and SE Guinea to SW CAR and S Congo; also SW Senegal, C Nigeria (Jos Plateau) and SE CAR.

GREY-HEADED GREENBUL *Phyllastrephus poliocephalus* Plate 89 (10)
Bulbul à ventre jaune

L 20–23 cm (*c.* 8–9"). **Adult male** Largest and brightest montane bulbul, easily distinguished by combination of grey head, olive-green upperparts and tail, white throat and bright yellow underparts. Narrow white eye-ring. **Adult female** Similar, but smaller. **Immature** Unknown. **VOICE** A loud, abrupt, rather harsh *churp*, often repeated, and a fast *churp-p-p* or *chrititit*. [CD9:77] **HABITS** Usually in small, noisy groups within tall, mature rainforest at mid-elevations. Sometimes singly. A characteristic member of mixed-species flocks where often the predominant species. Forages at all levels, but mainly in middle and upper strata. Frequently flicks wings and tail. **SIMILAR SPECIES** Western Mountain Greenbul is overall duller, with grey, not white throat, and has different voice and behaviour. **STATUS AND DISTRIBUTION** Common forest resident in highlands of SE Nigeria (Obudu Plateau) and W Cameroon (Mt Cameroon, Rumpi Hills, Mt Kupe, Mt Nlonako, Bakossi Mts, S Bamenda Highlands). *TBW* category: Near Threatened.

RED-TAILED BRISTLEBILL *Bleda syndactyla* Plate 90 (9)
Bulbul moustac

Other name F: Bulbul moustac à queue rousse
L 21.5–23.0 cm (8.5–9.0"). *B. s. syndactyla*. Distinctive, handsome and robust bulbul, largest of its genus. **Adult** Sulphur-yellow underparts with olivaceous wash to breast and flanks contrasting strikingly with olive-green to olive-brown head and upperparts. *Tail bright rufous*. Conspicuous half crescent of bare blue skin above eye; pale yellow-olive spot in front of eye. Bill blackish above, blue or grey below. Legs greyish to brownish. **Juvenile/immature** Wholly russet but swiftly assumes adult-like plumage; bill black with tip, cutting edges and gape yellow; legs dull yellowish. **VOICE** Character-istic song ends with a series of pure, vibrant syllables on same pitch *turrruuu turrruuu turrruuu*. A variant consists of a series of tremulous, melancholic notes descending in pitch. Variety of calls includes a nasal *kyow* or *pyeeuw* and a hard *chup*. [CD9:78] **HABITS** Shy and secretive, occurring singly, in pairs or small family groups in dense forest undergrowth. Regularly joins mixed-species flocks. Very partial to ant columns. **SIMILAR SPECIES** Grey-headed, Green-tailed and Lesser Bristlebills have olive-green tails tipped yellow, and different voice. **STATUS AND DISTRIBUTION** Rare or uncommon to common forest resident, from Sierra Leone and SE Guinea to Ghana and from S Benin to S CAR and Congo.

GREEN-TAILED BRISTLEBILL *Bleda eximia*

Plate 90 (7)

Bulbul à queue verte

Other name F: Bulbul moustac à tête olive
L 21.5–23.0 cm (8.5–9.0"). **Adult** Similar to Lesser Bristlebill but spot in front of eye indistinct, yellowish-green; yellow tips to outer 3–4 tail feathers narrower (absent or barely noticeable on T3, broadening to *c.* 10 mm on outer pair). **Juvenile/immature** No information. **VOICE** Song a series of pure, whistled notes *hee-huu-huu-hu-heeu* and *hu-hu-heeuu-heeuu*, often slightly vibrating and rather similar in tone to notes of Red-tailed Bristlebill. Calls include an abrupt, nasal *kyop*, sometimes followed by a fast *kiuwkiuwkiuwkiuwkiuw...* [CD9:79] **HABITS** Skulking and very unobtrusive. Usually singly or in pairs in undergrowth of primary forest and old second growth. Regularly joins mixed-species flocks. Attends ant swarms, often with other bristlebills. **SIMILAR SPECIES** Grey-headed Bristlebill has similar tail pattern but head slate-grey (not olive-green), without bare blue skin above eye; yellow tips to tail feathers more or less of equal length (not broadening outwardly); more often occurs at forest edges and around clearings. **STATUS AND DISTRIBUTION** Rare forest resident, from Sierra Leone and SE Guinea to Ghana. *TBW* category: THREATENED (Vulnerable).

LESSER BRISTLEBILL *Bleda notata*

Plate 90 (6)

Bulbul jaunelore

L 19.5–21.0 cm (7.75–8.25"). **Adult** *notata* Head, upperparts and tail olive-green; *bright yellow spot in front of eye*; outer 4 tail feathers broadly tipped yellow. Underparts bright yellow washed olive on breast-sides and flanks. Crescent of bare blue skin above eye. Bill blackish above, bluish below. Legs greyish. In eastern *ugandae* loral spot is dull greenish-yellow, yellow tips to outer tail feathers 5–10 mm wider (12–15 mm on T3, broadening to 22–32 mm on outer pair). **Juvenile/immature** Mostly dark russet but very soon like adult; bill blackish with yellow cutting edges and gape; legs dull yellowish to pale grey. **VOICE** Song tremulous and descending in pitch, slightly resembling that of Red-tailed Bristlebill but lacking its melancholic quality. Calls include a hard, oft-repeated *chup* and *chiup*, sometimes followed by a rattle *trrrrttttt*, and *wheew*. Combinations may form series e.g. *chup chup whEE chuw chuw...* [CD9:81] **HABITS** Singly, in pairs or small family groups (pair with one young) in undergrowth of primary forest and old second growth. Regularly joins mixed-species flocks. Very partial to ant swarms. Skulking but vocal. **SIMILAR SPECIES** No other bulbul within its range has similar head and tail markings. Range of Grey-headed Bristlebill not known to overlap. **STATUS AND DISTRIBUTION** Common forest resident, from extreme SE Nigeria to SW CAR, Congo and Bioko (nominate); also SE CAR (*ugandae*). **NOTE** Often treated as conspecific with Green-tailed Bristlebill, but considered specifically distinct on basis of acoustic and morphological data; forms superspecies with parapatric Grey-headed Bristlebill (Chappuis & Erard 1992).

GREY-HEADED BRISTLEBILL *Bleda canicapilla*

Plate 90 (8)

Bulbul fourmilier

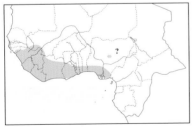

L 20.5–22.0 cm (8.0–8.75"). **Adult** *canicapilla* Head slate-grey with pale grey loral spot; upperparts and tail olive-green; 3 outer tail feathers with broad, wedge-shaped yellow tips (conspicuous in flight). Throat pale yellow; rest of underparts bright sulphur-yellow washed olive on breast-sides and flanks. Bill blackish above, blue or greyish below. Legs bluish to grey. Western *morelorum* overall paler. **Juvenile/ immature** Unknown. **VOICE** Very varied. Song far carrying with cheerful, ringing quality, typically consisting of a loud initial note, a brief pause, then a series on descending scale. Other characteristic notes include a loud *CHEEup* or *CHRIup*, and *kyuw* or *peeuw*, often repeated in long series. [CD9:80] **HABITS** Shy and skulking, but vocal. In pairs or small parties in forest undergrowth (mostly at edges and clearings), gallery forest and woodland thickets. Often in mixed-species flocks. Attends ant columns. **SIMILAR SPECIES** Green-tailed and allopatric Lesser Bristlebills have similar tail pattern (though yellow tips broaden outwardly), but head is olive-green, concolorous with rest of upperparts (not always easy to see in forest undergrowth) with bare blue skin above eye; also less frequent in edges and second growth; Lesser Bristlebill further distinguished by yellow (not grey) loral spot. **STATUS AND DISTRIBUTION** Generally common forest resident from W Gambia and SW Senegal (*morelorum*) east to E Nigeria (nominate). Isolated population, C Nigeria (Jos Plateau).

WESTERN BEARDED GREENBUL *Criniger barbatus*   Plate 90 (2)
Bulbul crinon

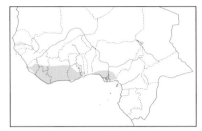

Other names F: Grand Bulbul huppé, Bulbul crinon occidental
L 22 cm (8.5"). Large, mainly olive-green bulbul with conspicuous yellow throat. **Adult *barbatus*** Head and upperparts olive-brown. Tail dull olive-chestnut. Throat bright yellow, contrasting with mottled grey and olive breast and belly; centre of belly olive-yellow. Yellow throat feathers frequently puffed out. Grey feathers of breast have pale shafts, giving rather untidy appearance. Eastern **ansorgeanus** has white chin, very pale yellow throat and browner underparts. **Immature** Gape yellow. **VOICE** Most common song a frequently uttered, a clear, far-carrying whistle *teruu twEEEur*. Calls include a hard *tsyuk* or *tsik* and a loud *KYUW*. [CD9:82] **HABITS** Noisy, usually occurring in pairs or family parties in lower and mid-levels of closed forest. Frequently joins mixed-species flocks, often with other *Criniger* species. Attends ant columns. **SIMILAR SPECIES** Smaller Yellow-bearded Greenbul also has bright yellow throat feathers, but these are less frequently puffed out and appear somewhat shorter; head, breast and belly olive-green (lacking grey tones); narrow blue orbital ring. **STATUS AND DISTRIBUTION** Not uncommon to common forest resident, from Sierra Leone and SE Guinea to W Togo (nominate) and in S Nigeria (*ansorgeanus*)

EASTERN BEARDED GREENBUL *Criniger chloronotus*   Plate 90 (5)
Bulbul à dos vert

Other name F: Bulbul crinon oriental
L 22 cm (8.5"). **Adult** Large bulbul with *grey head, conspicuous white throat and bright rufous tail*. Upperparts olive-green. Breast pale grey washed brownish, becoming yellowish-olive on flanks and pale dull yellow on belly; undertail-coverts yellowish-cinnamon. White throat feathers frequently puffed out. **Immature** Gape yellow. **VOICE** A rather soft, mournful, quavering 2-note song, quite different from Western Bearded Greenbul. Alarm a weak chatter. [CD9:83] **HABITS** In small noisy groups in lower and mid-levels of primary rainforest. Shy. Occasionally joins mixed-species flocks and follows ant columns. **SIMILAR SPECIES** Slightly smaller Red-tailed Greenbul also has white throat, but tail dull rufous, underparts bright yellow with olivaceous breast; voice different. **STATUS AND DISTRIBUTION** Fairly common forest resident, from SE Nigeria (Cross R. area) to SW CAR and Congo.

RED-TAILED GREENBUL *Criniger calurus*   Plate 90 (1)
Bulbul à barbe blanche

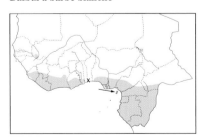

L 20 cm (8"). **Adult *verreauxi*** Fairly large, handsome bulbul with *pure white throat contrasting strongly with grey head and bright yellow breast and belly*. Upperparts and flanks dull olive-green. Tail olive-green with slight brownish wash. Conspicuous blue-grey orbital ring. Throat feathers frequently puffed out. **Nominate** has top of head and upperparts slightly paler olive-green and *reddish-brown tail*. **Immature** Upperwing-coverts fringed dull cinnamon. **VOICE** Song a cheerful rising *chup-chup-chwirulup*, somewhat reminiscent of Common Bulbul. Also a 3-note *tsik-tyu-tyip* and rapid series *tyu-tyutyutyu-tyip-tyip-tyip...* Alarm *tsik*. [CD9:84] **HABITS** Vocal, occurring in pairs or small parties (4–8) in lower and mid-levels of primary and secondary forest. Frequent member of mixed-species flocks, often with other *Criniger* species. **SIMILAR SPECIES** No confusion possible in west of range. In east, the only other white-throated *Criniger* species are Eastern Bearded Greenbul, which has olive-grey underparts, and White-bearded Greenbul, which is extremely similar and only safely distinguishable by voice (q.v.). **STATUS AND DISTRIBUTION** Not uncommon to common forest resident: *verreauxi* from Guinea to SW Nigeria, also SW Senegal and SW Mali; nominate from S Nigeria to SW CAR and Congo, also Bioko.

WHITE-BEARDED GREENBUL *Criniger ndussumensis* Plate 90 (3)
Bulbul de Reichenow

L 18 cm (7"). **Adult** Very similar to sympatric Red-tailed Greenbul and probably not safely distinguishable on plumage characters alone (not even in the hand!). When foraging together, more slender (and slightly shorter) bill of White-bearded sometimes noticeable at close range. Also reported to have more pronounced whitish spot in front of eye, darker flanks (more heavily washed olive) and cinnamon-buff rather than yellow undertail-coverts. **Immature** No information. **VOICE** Song consists of 3 short, harsh syllables, uttered on same, relatively low pitch, with slight emphasis on last syllable *whut-chruw-chruw* or *chuk-ker-chyer*, indistinguishable from allopatric Yellow-bearded Greenbul; similar in structure to Red-tailed Greenbul but lower pitched and lacking its cheerfulness. Alarm *tsik* (similar to Red-tailed). [CD9:86] **HABITS** In pairs or small family parties (pair with one juvenile) in lower and mid-levels of forest. A usual member of mixed-species flocks, almost always with other *Criniger* species, esp. *calurus*. Often clings to tree trunks while foraging. Habits inadequately known due to identification problems. **STATUS AND DISTRIBUTION** Uncommon to locally not uncommon forest resident, from SE Nigeria to SW CAR and Congo. Status inadequately known. **NOTE** Treated as race of Yellow-bearded Greenbul by some authors.

YELLOW-BEARDED GREENBUL *Criniger olivaceus* Plate 90 (4)
Bulbul à barbe jaune

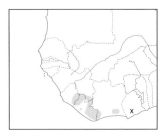

Other name E: Yellow-throated Olive Greenbul
L 18 cm (7"). **Adult** Rather small bearded bulbul with olive-green head and upperparts, bright yellow throat and warm olive-green breast and belly becoming more yellowish in centre. Tail has slight rufous wash. Narrow blue orbital ring. Ear-coverts olive-green streaked yellowish. Throat feathers less often puffed out than in sympatric *Criniger* and appear slightly shorter. **Immature** Similar, but information scant. **VOICE** Song consists of 3 short, harsh syllables, uttered on same, relatively low pitch, with slight emphasis on last syllable *whut-chruw-chruw*, similar in structure to song of Red-tailed Greenbul, but lower pitched and lacking its cheerfulness; indistinguishable from allopatric White-bearded Greenbul's. [CD9:85] **HABITS** Quiet and unobtrusive, occurring singly, in pairs or family parties in lower and mid-levels of forest, often in mixed-species flocks with other two *Criniger* species in its range. Often clings to tree trunks. **SIMILAR SPECIES** Western Bearded Greenbul is noticeably larger and has mottled greyish-olive breast, no bare skin around eye and different call. Icterine Greenbul also has bright yellow throat, but lacks tufted throat feathers clearly demarcated from olive upper breast, bare orbital skin and trunk-clinging behaviour. **STATUS AND DISTRIBUTION** Generally rare forest resident, from E Sierra Leone to SW Ghana. Locally not uncommon, Ivory Coast (Yapo Forest). *TBW* category: THREATENED (Vulnerable).

COMMON BULBUL *Pycnonotus barbatus* Plate 88 (1)
Bulbul des jardins

Other names E: Common Garden Bulbul; F: Bulbul commun
L 18–20 cm (7–8"). Africa's ubiquitous, most conspicuous and familiar bird. **Adult male** Earth-brown, with blackish-brown head and throat, pale brown breast and whitish belly. Peaked crown gives head distinctive shape. Races vary principally in degree of darkness of head and throat, and colour of undertail-coverts. In *inornatus* brown of breast merges gradually with dirty white belly; in *arsinoe* dark breast more demarcated from and contrasting with pale belly. Undertail-coverts vary from white (*inornatus* and *arsinoe*), white tinged with variable amount of yellow (*gabonensis*) to bright yellow (*tricolor*). **Adult female** As male but smaller. **Juvenile** Duller and paler, with upperparts tinged rufous. **VOICE** Song a cheerful phrase of 3–6 notes, e.g. *chuk chuk twirulup* or *chuk twee tu twuri* or *kwik kweek kwuyu*, etc.; most adequately rendered by popular *quick quick, doctor, quick!* Calls include a variety of chattering and ringing notes. [CD3:71] **HABITS** In almost all habitats, incl. gardens and other man-made environments, but absent from closed forest (though penetrates edges) and treeless desert. Pugnacious and noisy. Usually in pairs or small (family?) groups. **SIMILAR SPECIES** Unlikely to be misidentified, but compare Simple Leaflove. **STATUS AND DISTRIBUTION** Very common throughout, ranging north to 16–18°N; recorded even further north in Mauritania

(Adrar region), Niger (Aïr) and Chad (Tibesti). *P. b. inornatus* almost throughout the region; *gabonensis* C Nigeria and C Cameroon to Gabon and S Congo; *tricolor* E Cameroon and Congo (also SE Gabon?); *arsinoe* E Chad.

BLACK-COLLARED BULBUL *Neolestes torquatus*　　　　　　　　Plate 89 (11)
Bulbul à collier noir

L 16 cm (6.25"). Distinctive, handsome, rather shrike-like species. **Adult** Crown and hindneck grey; upperparts and tail olive-green with some yellow on bend of wing. Creamy-white throat bordered by black collar from lores through eye, curving around ear-coverts and broadening on breast; rest of underparts whitish-grey. **Juvenile** Dull version of adult, with crown and hindneck olive-green, concolorous with rest of upperparts, and wing-coverts fringed buff. **VOICE** A rapid subdued babble, somewhat reminiscent of Common Bulbul. Also a quavering *twee-dududu*. [CD9:87] **HABITS** Singly or in pairs in lightly wooded savanna. Rather unobtrusive, usually foraging low in dense herbaceous vegetation. Sings from tops of low trees and bushes. **STATUS AND DISTRIBUTION** Rather uncommon resident in SE Gabon and Congo. **NOTE** Treated variously as a bulbul or a 'shrike' (Laniidae, Malaconotidae or Prionopidae), but recent review, incorporating biology, anatomy and DNA, suggests it is a primitive member of the Pycnonotidae (Dowsett *et al.* 1999).

WESTERN NICATOR *Nicator chloris*　　　　　　　　　　　　　Plate 89 (13)
Bulbul nicator

Other names E: West African Nicator; F: Pie-grièche nicator, Nicator à gorge blanche, Nicator vert
L 21–24 cm (8.25–9.5"). **Adult male** Top of head, upperparts and tail olive-green *with bold golden spots on tips of median and greater wing-coverts and tertials*; flight feathers narrowly edged yellowish; outer tail feathers tipped yellow. Underparts pale grey becoming greyish-white on upper throat and centre of belly; undertail-coverts yellow. White supraloral spot; *head-sides ('cheeks') olive-yellow*, narrow but distinct yellow eye-ring. Bill black, heavy and hooked, recalling bush-shrike. **Adult female** As male but smaller. **Immature** Primaries tipped yellow. **VOICE** Distinctive. A powerful, melodious and very varied song comprising an explosive crescendo of notes, clear whistles, guttural rattles, etc., somewhat reminiscent of Common Nightingale. Occasionally mimics other species (e.g. Yellow-throated Nicator, Bates's Paradise Flycatcher, Yellow-bellied Wattle-eye, Chestnut-backed Owlet). Calls include a loud, abrupt *tok!*, often in long, accelerating series *tok... tok... tok tok tok tok-tok-tok-toktoktoktiuktiuktiuktiuk...*, and various harsh whistled notes. [CD14:1a] **HABITS** A skulking but inquisitive species of tangles, dense foliage and thick cover, in forest and its gallery extensions. Usually singly; sometimes joins mixed-species flocks. Mostly at mid-level (to 25 m), but may descend to lower levels (down to 1–2 m or even on ground). Far more often heard than seen. **SIMILAR SPECIES** Yellow-throated Nicator is smaller, with yellow (not greyish-white) throat, yellow supraloral streak and grey head-sides; voice also different (q.v.). Single loud *tok* note resembles *kuk* of Yellowbill, but latter 'drier' and less powerful. **STATUS AND DISTRIBUTION** Generally common resident in forest and its outliers in savanna, becoming scarce at edges of range, from Guinea to CAR and Congo; also Gambia (rare) and S Senegal. **NOTE** Treated variously as a bulbul or a 'shrike' (Laniidae or Malaconotidae), but recent reassessment of taxonomic position suggests it is a bulbul (Fishpool in prep.).

YELLOW-THROATED NICATOR *Nicator vireo*　　　　　　　　　Plate 89 (12)
Bulbul à gorge jaune

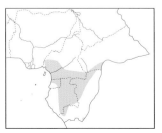

Other name F: Nicator à gorge jaune
L 17–19 cm (6.75–7.5"). **Adult male** *Small version of Western Nicator*, with *yellow throat* (sometimes neither conspicuous or extensive). Forehead, forecrown, lores and ear-coverts *grey*, separated by *small but conspicuous yellow supraloral streak*. Hindcrown, upperparts and tail olive-green with bold golden spots on tips of median and greater wing-coverts and tertials; flight feathers narrowly edged yellowish; outer tail feathers tipped yellow. Breast to belly pale grey; undertail-coverts yellow. **Adult female** As male but smaller. **Juvenile** Forehead and forecrown green. **VOICE** Song a loud, far-carrying series of resonant, explosive notes e.g. *ko-kwee-ko-ko-ko-kwee-kuk-kuk*, with variations. Calls include angry *gwrrrrrrr*. [CD14:1b] **HABITS** Shy, skulking and solitary, inhabiting middle and lower levels of lowland

forest with dense creepers. Betrays presence by distinctive, loud song. **SIMILAR SPECIES** Western Nicator is noticeably larger and has greyish-white (not yellow) throat, white supraloral spot, olive-yellow head-sides and different voice. **STATUS AND DISTRIBUTION** Not uncommon forest resident, from Cameroon to SW CAR and Congo.

THRUSHES, CHATS AND ALLIES
Family TURDIDAE (World: *c.* 315 species; Africa: 125; WA: 57)

A large and diverse family of small to medium-sized songbirds, usually with strong legs and feet and mostly square or rounded tails. Sexes often similar, with variable dimorphism, females being duller (e.g. in wheatears and some chats). Juveniles more or less spotted. Occur in a variety of habitats, from desert to rainforest. Feed on invertebrates and fruit. Broad range of vocalisations; some species are notable songsters. In W Africa, 18 species are of Palearctic origin.

FOREST ROBIN *Stiphrornis erythrothorax*   Plate 92 (1)
Rougegorge de forêt

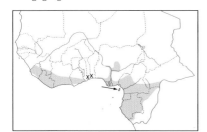

L 13 cm (5"). Small, inconspicuous, short-tailed forest species. **Adult** *erythrothorax* Top of head, upperparts and tail dark olive-brown. Head-sides slate-grey with *distinct white spot in front of eye. Throat and breast bright orange*; rest of underparts white washed grey on breast-sides and flanks. *S. e. gabonensis* similar, but sooty-grey above; *xanthogaster* similar to *gabonensis* but orange area distinctly paler (esp. on throat) and less well demarcated from rest of underparts, which have pale cream tinge. *S. e. sanghensis* has throat and breast yellow-orange; rest of underparts pale yellow (Beresford & Cracraft 1999). **Juvenile** Head, mantle and upperwing-coverts speckled and spotted rufous; throat whitish tinged buff; breast blackish mottled rufous. **Immature** Spots on upperparts gradually lost, those on wing-coverts last. Breast becomes orange but throat remains whitish until full adult plumage assumed. **VOICE** Distinctive. Song fast, sweet and melodious, consisting of variable, high-pitched whistled motifs. Varies between races (and/or populations?): in nominate and *gabonensis* motifs typically short (*gabonensis* similar to Bocage's Akalat), in *xanthogaster* semi-continuous (rather similar to White-bellied Robin Chat). Calls include low, sharp *karrrr* and double whistle *whi-whiuuu*. [CD9:89] **HABITS** Singly or in pairs within lowland rainforest, riverine forest, thickets and relict forest patches in savanna. Unobtrusive, quietly hopping on forest floor or in undergrowth. Occasionally joins mixed-species flocks at ant columns. Shy and easily overlooked, but song attracts attention. **SIMILAR SPECIES** Lowland Akalat has longer, rufous tail and lacks supraloral spot. **STATUS AND DISTRIBUTION** Not uncommon forest resident. Nominate from Sierra Leone and SE Guinea to S Benin and Nigeria; *gabonensis* in SW Cameroon, W Gabon and Bioko; *xanthogaster* from SC Cameroon and interior of Gabon to N Congo and SE CAR; *sanghensis* in SW CAR. **NOTE** Form *sanghensis* originally described as phylogenetic species, Sangha Forest Robin. *TBW* category: Data Deficient.

BOCAGE'S AKALAT *Sheppardia bocagei*   Plate 92 (2)
Rougegorge de Bocage

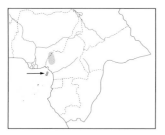

Other names E: Rufous-cheeked Robin-chat, Bocage's Robin, Alexander's Akalat; F: Cossyphe à joues rousses
L 13 cm (5"). Small and very secretive robin-like forest species. **Adult** *granti* Top of head to lores olive-grey; upperparts olive-brown becoming rufous on uppertail-coverts; tail rusty-brown. Head-sides and underparts orange-rufous; centre of belly white. Base of feathers in front of eye white, but this supraloral stripe is usually concealed. Island race *poensis* has top of head darker, blackish washed olive; underparts deeper orange-rufous. **Juvenile** Dark brown spotted rufous above and below; centre of belly dirty white. **VOICE** Song a quiet, rather mournful series of 7–10 sweet whistled notes, *hu-hee-hu-hluwee-hu hlu-whee-hu-hee* ... Calls include quiet twittering and a sibilant ratchet note reminiscent of Dusky Flycatcher; alarm 1–3 high-pitched *seeep* notes (*BoA*). [CD9:91] **HABITS** Usually singly in undergrowth of montane and submontane forest (600–1700 m). Commonest at lower elevations. Occurs at lower altitudes than Mountain Robin Chat. On or near ground. Attends ant columns. Observed in small mixed-species flocks on Bioko (Pérez del Val 1996). Very unobtrusive and shy. **SIMILAR SPECIES** White-bellied Robin Chat has orange area not extending onto

head-sides and black-and-rufous tail. Slightly larger Grey-winged Robin Chat distinguished by white supercilium, black eye-stripe, loud robin chat song and different forest habitat. See also Mountain Robin Chat. **STATUS AND DISTRIBUTION** Generally not uncommon forest resident, SE Nigeria (Obudu Plateau; rare), W Cameroon (Mt Cameroon, Rumpi Hills, Mt Kupe, Mt Nlonako, Bakossi Mts) (*granti*) and Bioko (*poensis*). **NOTE** Sometimes considered specifically distinct as *S. poensis* (='*insulana*', Alexander's Akalat; Rougegorge d'Alexander) from extralimital southern population on basis of vocal and minor morphological differences.

LOWLAND AKALAT *Sheppardia cyornithopsis*   Plate 92 (3)
Rougegorge merle

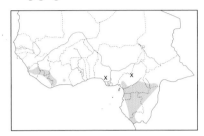

Other names E: Akalat; F: Merle rougegorge
L 13 cm (5"). Small and very secretive robin-like forest species. **Adult** *houghtoni* Head and upperparts dark olive-brown, becoming more rufous on uppertail-coverts and tail. Underparts mainly orange-rufous; belly white. **Nominate** very similar; flanks washed olive; undertail-coverts paler, sometimes white. **Juvenile** Dark brown spotted rufous above and below; belly whitish with some dusky scalloping. **VOICE** Song consists of 2 soft whistles in alternation *whee, whiu, whee, whiu*, or series of *whiu* notes, repeated at 2-second intervals or less. When more excited, whistles may be slightly purred *pur, peer, purr,...* Also a faster series of weak *whiu-whiu-whiu-...*(in response to playback). Strongly reminiscent of, though not identical to, rolled whistles of extralimital Equatorial Akalat *S. aequatorialis* (Dowsett-Lemaire 1997). Call a short whistle *tiee*; alarm a low *krrr*. [CD9:92] **HABITS** Singly or in pairs within undergrowth of primary and secondary forest, forest patches in savanna and seasonal swamp forest. Occasionally joins mixed-species flocks. Very unobtrusive and shy. **SIMILAR SPECIES** Bocage's Akalat occurs at higher elevations. **STATUS AND DISTRIBUTION** Rare to scarce or locally not uncommon forest resident, from W Guinea to Ivory Coast (*houghtoni*), and from S Cameroon to SW CAR and Congo (nominate). One record from S Nigeria (Okumu, Jul 1988).

EUROPEAN ROBIN *Erithacus rubecula*   Plate 94 (3)
Rougegorge familier

Other name F: Rougegorge
L 14 cm (5.5"). *E. r. rubecula/witherbyi*. **Adult** Small, compact and rounded with conspicuous warm orange-red face and breast. Top of head, upperparts and tail olive-brown; lower underparts dirty buffish becoming whitish in centre of belly. Races intergrade and differ principally in measurements (*witherbyi* smaller). **VOICE** Calls include a dry *tik* often repeated in fast series *tik-tikikikikkk* and a thin, high-pitched *seeeeh*. [CD3:77] **HABITS** Singly, in various habitats. Forages on ground, hopping with upright posture. **STATUS AND DISTRIBUTION** Palearctic vagrant, coastal Mauritania (Nov–Dec and Mar).

THRUSH NIGHTINGALE *Luscinia luscinia*   Plate 94 (6)
Rossignol progné

Other name E: Sprosser
L 16.5 cm (6.5"). **Adult** Very similar to Common Nightingale but typically overall less brightly coloured; upperparts duller and slightly darker brown; tail darker, less rufous, usually less contrasting with upperparts; breast and flanks variably marked with indistinct dark brown mottling; white throat often more contrasting with head-sides due to ill-defined, dusky-brown malar stripe; eye-ring usually less distinct. **VOICE** A low *krrrr*, a high-pitched, in-drawn *heet* and a short *tak*. Song similar to Common Nightingale's but somewhat slower, more monotonous, deeper and harder. [CD9:94] **HABITS** Singly in savanna scrub and thickets. General habits similar to Common Nightingale. **SIMILAR SPECIES** Common Nightingale (q.v. and above). **STATUS AND DISTRIBUTION** Palearctic winter visitor to E Africa. Vagrant, N Nigeria (Kano, Sep–Oct).

COMMON NIGHTINGALE *Luscinia megarhynchos*   Plate 94 (7)
Rossignol philomèle

L 16.5 cm (6.5"). *L. m. megarhynchos*. Small, plain thrush with *open face* and *rufous tail*. **Adult** Head and upperparts warm brown; uppertail-coverts and tail rufous. Underparts pale grey-brown becoming whitish on throat and belly. Whitish eye-ring. **VOICE** Song loud, rich and melodious, including a variety of clear whistles, vigorous chuckles and piping notes, which are often repeated; characteristic are a rapid, accelerating series of full *tchiok* notes and in-drawn whistles *heeeet-heeeet-heeeet-...*; often uttered in winter quarters. Calls include a high *hueet*, a hard *tak* and a low *karrr*. [CD3:78] **HABITS** Singly in savanna scrub, farmbush, gardens and overgrown clearings.

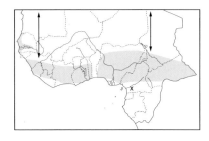

Usually skulking in dense cover, but attracting attention by remarkable song. Feeds on ground, standing erect and hopping with drooped wings, often flicking wings and tail; cocks tail when excited. **SIMILAR SPECIES** Thrush Nightingale (rare vagrant) often difficult to distinguish, but slightly darker, with faintly mottled breast. Female Common Redstart distinctly smaller and slimmer, with dark central tail feathers; also more arboreal. **STATUS AND DISTRIBUTION** Uncommon to locally common Palearctic passage migrant and winter visitor (Sep–Apr) almost throughout; in WC Africa unknown south of Cameroon. Vagrant, Cape Verde Is.

BLUETHROAT *Luscinia svecica*   Plate 94 (1)
Gorgebleue à miroir

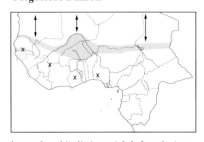

Other name F: Gorgebleue
L 14 cm (5.5"). Unobtrusive with, in all plumages, dark brown upperparts, whitish supercilium and *diagnostic rufous sides to basal half of outer tail feathers* (conspicuous in flight); rest of tail and central feathers dark brown. **Adult male breeding** (from Feb) Unmistakable. *Throat and upper breast bright blue* with white (**cyanecula**) or rufous (**nominate**) central spot, bordered below by black and rufous bands (these separated by white line); rest of underparts buff-white. **Adult male non-breeding** Throat buff-white bordered by dappled black breast-band and malar streak, the latter variably suffused blue; rufous breast-band indistinct. **Adult female** As non-breeding male, but usually lacks blue suffusion in band surrounding throat or rufous on breast. **First-winter** Similar to adult female, but outer greater coverts tipped buff. **VOICE** Calls include a hard *tak* and *hueet* (like *Phylloscopus* warbler). Song loud and remarkably varied, including melodious whistles, liquid trills and harsh, discordant notes. [CD3:79] **HABITS** Usually in scrub near water and in damp places. Skulking. Forages on or near ground, occasionally leaving cover. Hops or runs; stands upright on long legs with cocked tail. **SIMILAR SPECIES** No other chat has similar tail pattern. **STATUS AND DISTRIBUTION** Uncommon Palearctic winter visitor (Sep–Mar) to Sahel belt, from S Mauritania and Senegambia to N Nigeria and Chad. Vagrant, NW Ivory Coast, S Ghana and S Nigeria. Both races occur throughout, though *cyanecula* generally more common in west, nominate in east.

WHITE-BELLIED ROBIN CHAT *Cossyphicula roberti*   Plate 93 (1)
Cossyphe à ventre blanc

L 13 cm (5"). *C. r. roberti*. Smallest robin chat. Akalat-like in size and plumage but with *black-and-red tail*. **Adult** Head and upperparts olive-brown; *narrow white supercilium* bordered by black lores. Tail orange-rufous with black central rectrices. Throat, breast and flanks orange-rufous; flanks washed olive; belly and undertail-coverts white. **Juvenile** Head and upperparts dark brown spotted rufous; underparts rufous-buff (paler on centre of belly) scalloped black. **VOICE** Song a short, fast series of 6 rather high-pitched whistled notes *tsu-ti-tu-ti-tu-tu*, frequently repeated without pause. Alarm call a fast *ti-ti-ti-ti-ti-*... [CD9:95] **HABITS** Usually singly, in primary and secondary montane and submontane forest (650–2000 m). From ground to mid-storey; mainly at 2–4 m. Catches insects on the wing or gleans them from foliage. Quiet and unobtrusive. **SIMILAR SPECIES** Bocage's Akalat lacks supercilium (but shows white supraloral stripe when excited!) and has orange-rufous extending onto head-sides and almost covering entire underparts, and rusty-brown tail. Mountain Robin is noticeably larger and has underparts entirely, or almost entirely, orange-rufous. **STATUS AND DISTRIBUTION** Uncommon to locally common forest resident, SE Nigeria (Obudu Plateau), W Cameroon (Mt Cameroon, Rumpi Hills, Mt Kupe, Mt Nlonako, Bakossi Mts) and Bioko.

Genus *Cossypha*

Small to medium-sized, colourful thrushes. All species in our region have orange-rufous underparts and orange-rufous tails with black central feathers (except Grey-winged Robin Chat). Most have loud, very melodious and varied songs with many imitations.

MOUNTAIN ROBIN CHAT *Cossypha isabellae*   Plate 93 (2)
Cossyphe d'Isabelle

L 15 cm (6"). Smallish *Cossypha* of montane forest. **Adult *batesi*** Top of head and upperparts olive-brown becoming rufous-brown on rump; blackish head-sides bordered by *narrow white supercilium.* Tail orange-rufous with dark brown central rectrices. Underparts orange-rufous becoming off-white on lower belly. **Nominate** darker above with contrasting orange-rufous rump; *entire* underparts orange-rufous. **Juvenile** Dark brown spotted rufous above and below; centre of belly dirty off-white. **VOICE** Mainly silent. Song a fairly loud, rather tuneless 2-note trill *tsri—tsrrrr* repeated in rapid succession, and slight variations, e.g. *tsri-tsrrrr-tsi-tsee.* Call a guttural *grrr.* [CD9:96] **HABITS** Usually singly in undergrowth of montane and submontane forest (800–2700 m). Commonest at higher elevations. Generally replaces marginally sympatric Bocage's Akalat at higher altitudes. On or near ground. Attends ant columns. Quiet and unobtrusive but not shy. **SIMILAR SPECIES** White-bellied Robin Chat is distinctly smaller and has more white on lower underparts. Bocage's Akalat is also smaller, has orange-rufous extending onto head-sides, rusty-brown tail and no supercilium, and generally occurs at lower altitudes. Juvenile smaller than juvenile Mountain Robin Chat; tails similar to adults. **STATUS AND DISTRIBUTION** Not uncommon forest resident, SE Nigeria (Obudu Plateau, Gotel Mts, Chappal Hendu) and W Cameroon (Rumpi Hills, Mt Kupe, Bakossi Mts, Mt Manenguba, Bamenda Highlands) (*batesi*); nominate confined to Mt Cameroon.

GREY-WINGED ROBIN CHAT *Cossypha polioptera*   Plate 93 (3)
Cossyphe à sourcils blancs

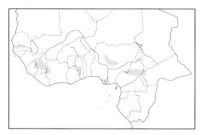

L 15 cm (6"). Small, unobtrusive *Cossypha* with *uniformly rufous tail.* **Adult *nigriceps*** Crown black; *long white supercilium underlined by black eye-stripe*; upperparts olive-brown with slaty wing-coverts; head-sides and underparts bright orange-rufous becoming whitish on central belly. **C. p. tessmanni** reportedly darker, esp. on cheeks. **Juvenile** Similar but duller and spotted rufous on head and upperwing-coverts; no supercilium. **VOICE** Song varied and melodious, including imitations. Higher pitched than song of other robin chats that include imitations. [CD10:1a] **HABITS** Inhabits edges and clearings of mid-elevation forest, dense gallery forest and forest patches in savanna. Mainly on or near ground, but also higher. Flicks wings and tail when perched. Shy. **SIMILAR SPECIES** Bocage's Akalat is slightly smaller and lacks supercilium. Other robin chats have two-toned tail with dark central feathers. **STATUS AND DISTRIBUTION** Patchily distributed, rare to locally common resident, recorded from Guinea, E Sierra Leone, Liberia and E Ivory Coast (Mt Nimba, Mt Tonkoui), C and E Nigeria (incl. Mambilla Plateau, Gashaka-Gumti NP), Cameroon (Bamenda Highlands, Adamawa Plateau) and W CAR. *C. p. tessmanni* known only from E Cameroon ('Upper Kadei R.'). **NOTE** Sometimes placed in genus *Sheppardia.*

BLUE-SHOULDERED ROBIN CHAT *Cossypha cyanocampter*   Plate 93 (4)
Cossyphe à ailes bleues

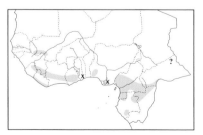

L 16.5 cm (6.5"). *C. c. cyanocampter.* Shy, medium-sized robin chat with *conspicuous supercilium and blue shoulder patch.* **Adult** Head black with long white supercilium; upperparts olive-brown becoming rufous on rump; tail orange-rufous with black central rectrices; wings blackish-brown with lesser and median coverts blue (someti mes largely concealed); underparts orange-rufous, paler on belly. **Juvenile** Duller, with head and wing-coverts spotted rufous; no supercilium. **VOICE** Song very varied, melodious and sustained, including perfect imitations of other birds and human whistles. Lower pitched and slower than songs of Grey-winged and Snowy-crowned Robin Chats. Readily responds to human whistles. Call a loud *trurr.* [CD10:1b] **HABITS** Singly in dense undergrowth of primary and secondary forest, gallery forest, forested ravines and seasonal swamp forest. On or near forest floor. Very skulking and difficult to observe. **SIMILAR SPECIES** White-browed Robin Chat is larger, lacks shoulder patch, frequents forest edges and has very different song without mimicry. **STATUS AND DISTRIBUTION** Uncommon forest resident, from SW Mali, E Guinea and Sierra Leone to Togo and from S Nigeria to SW CAR , Gabon and N Congo; also reported from NE CAR.

WHITE-BROWED ROBIN CHAT *Cossypha heuglini* Plate 93 (6)
Cossyphe de Heuglin

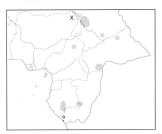

Other name E: Heuglin's Robin

L 20 cm (8"). Retiring but highly vocal robin chat with distinctive melodious song. **Adult** *heuglini* *Head black with long white supercilium*; upperparts dark olive-brown to slate-grey separated from head by narrow orange-rufous collar. Rump, uppertail-coverts and tail orange-rufous with central rectrices brownish. Underparts orange-rufous. *C. h. subrufescens* has central rectrices black. **Juvenile** Head and upperparts blackish-brown spotted rufous; no supercilium; underparts rufous-buff scalloped blackish-brown; legs pale. **VOICE** Song loud, varied and melodious, with simple, whistled phrases characteristically starting quietly, then increasing in volume. Does not usually include imitations. Members of pair duet or sing antiphonally, female uttering high-pitched *tseeeee*. Calls include a harsh, rattling *tsrek-tsrek* (alarm). [CD9:99] **HABITS** Singly or in pairs in riverine forest and thickets in savanna. Forages mainly on ground, often in the open. **SIMILAR SPECIES** Blue-shouldered Robin Chat is smaller, has blue shoulder patch, inhabits dense forest interior and has very different song with much mimicry. In Gabon and Congo, Red-capped Robin Chat may share same habitat but mainly distinguished by orange-rufous head. **STATUS AND DISTRIBUTION** Patchily distributed, uncommon or rare to locally not uncommon resident, recorded from Cameroon, SW Chad and CAR (nominate), Gabon and Congo (*subrufescens*). Also claimed from extreme NE Nigeria.

RED-CAPPED ROBIN CHAT *Cossypha natalensis* Plate 93 (5)
Cossyphe à calotte rousse

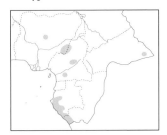

L 17 cm (*c.* 7"). *Only robin chat with orange-rufous head*; beady eye conspicuous. **Adult** *larischi* Head and underparts orange-rufous, crown and mantle darker, more olive-brown. Rump, uppertail-coverts and tail orange-rufous with black central rectrices. Wings blue-grey. *C. n. intensa* slightly paler above. **Juvenile** Head and upperparts dark brown spotted rufous; underparts rufous-buff scalloped dark brown; legs pale. **VOICE** Song rich and melodious, consisting of whistled phrases and often including many imitations. Calls include a rather plaintive *whuh ti-eh* (second syllable nasal) and a slightly trilled *prree prrup*, monotonously repeated; alarm a guttural *grrr*. [CD10:2a] **HABITS** Various wooded habitats. Forages mainly on or near ground; occasionally also in mid-strata and canopy. Attends ant columns. When perched typically flicks wings and tail, then slowly lowers tail, often fanning it briefly. Difficult to detect when not singing. **SIMILAR SPECIES** In Gabon and Congo, White-browed Robin Chat may share same habitat but mainly distinguished by head pattern and darker mantle. Snowy-crowned Robin Chat is much larger with longer tail, much darker above and has conspicuous white crown. **STATUS AND DISTRIBUTION** Patchily distributed, rare to locally uncommon resident and intra-African migrant, recorded from C and E Nigeria (Kagoro, Chappal Waddi), Cameroon, coastal Gabon and Congo (*larischi*) and SE CAR (*intensa*).

SNOWY-CROWNED ROBIN CHAT *Cossypha niveicapilla* Plate 93 (7)
Cossyphe à calotte neigeuse

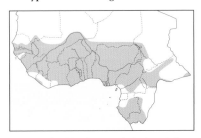

Other names E: Snowy-headed Robin Chat; F: Petit Cossyphe à tête blanche

L 20.5–22.0 cm (8.0–8.75"). Large robin chat with *white crown and rufous hind-collar*. **Adult** *niveicapilla* Forehead and head-sides black; upperparts slate-grey, separated from head by narrow orange-rufous collar. Rump, uppertail-coverts and tail orange-rufous with black central rectrices. Underparts orange-rufous. *C. n. melanota* has upperparts black. **Juvenile** Head and upperparts blackish spotted rufous; underparts rusty-buff scalloped blackish, paler on throat. Spots retained longest on head. **VOICE** Song rich, melodious and sustained, with characteristic fast delivery and including many imitations of other birds and human whistles. Calls include a drawn-out, often repeated *heeee* (contact) and a guttural *krrr* (alarm). [CD10:2b] **HABITS** Singly or in pairs, inhabiting thickets in savanna, forest edges and clearings, overgrown farmland, gallery forest and gardens. Forages mainly on or near ground. Skulking. **SIMILAR SPECIES** White-crowned Robin Chat is much larger and lacks collar on nape; song less varied. **STATUS AND DISTRIBUTION** Uncommon to locally common resident in forest and savanna zones, north to S Mauritania and central Niger delta in Mali to S Niger, S Chad and

N CAR; also Gabon and Congo. Western populations (east to Nigeria) nominate; rest *melanota*, but intergrades occur from S Ghana east.

WHITE-CROWNED ROBIN CHAT *Cossypha albicapilla*   Plate 93 (8)

Cossyphe à calotte blanche

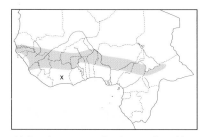

Other name F: Grand Cossyphe à tête blanche
L 26 cm (11"). Largest robin chat. *Large and long-tailed with large white or black-and-white crown.* **Adult** *albicapilla* Forehead to nape white with some indistinct dark scaling (disappearing with wear); head-sides (to chin) and upperparts black. Rump, uppertail-coverts and tail orange-rufous with black central rectrices. Underparts orange-rufous. Eastern *giffardi* has crown mainly black with white crescents. **Juvenile** Unknown. **VOICE** Song varied, sustained and fast, including scratchy notes and trills, but without imitations. Call a high-pitched, penetrating *sweeuee.* [CD10:3a] **HABITS** Singly in thickets in savanna, riverine scrub and large overgrown gardens. Forages mainly on ground. Skulking but not shy. **SIMILAR SPECIES** Snowy-crowned Robin Chat is noticeably smaller and has narrow rufous hind-collar, smaller white crown patch and no black on chin; song richer. **STATUS AND DISTRIBUTION** Uncommon to locally common resident in narrow savanna belt, from Senegambia to Sierra Leone and S Mali to N Ivory Coast (nominate), east through S Burkina Faso and N Ghana to Cameroon (Adamawa Plateau) and NW CAR (*giffardi*). Local seasonal movements reported, north with rains, south in dry season.

FIRE-CRESTED ALETHE *Alethe diademata*   Plate 92 (4)

Alèthe à huppe rousse

L 18 cm (7"). **Adult** *diademata* Top of head and upperparts dark olive-chestnut with *rusty-orange central crown-stripe;* head-sides grey. Tail blackish with *outer three feathers broadly tipped white.* Underparts white washed grey on breast-sides and flanks. Crown feathers can be raised, forming short crest. Eastern *castanea* lacks white in tail and is brighter chestnut above; tail dull dark brown edged chestnut. **Juvenile** Blackish above with orange-rufous spots (small on head, large on upperparts). Underparts pale orange-rufous scalloped blackish on throat and breast. Nominate has white tail corners. **Immature** With age, spots gradually disappear from upperparts; lower underparts become whitish; white feathers appear on throat and breast. **VOICE** Song consists of 3 sweet ascending whistles, *huu hee hueee;* third note sometimes absent. Also a subdued song with mixture of whistles and imitations (e.g. Black Cuckoo, Emerald Cuckoo, Chocolate-backed Kingfisher, Forest Robin, Finsch's Flycatcher Thrush, malimbe sp.). Call a single whistled *huu*, like first note of song, often repeated in long, regular series *huu... huu... huu... huu...* [CD10:3b] **HABITS** Singly or in pairs in primary and old secondary forest, gallery forest and forest patches in savanna. Mainly on or near forest floor. Frequently attends ant columns, where most commonly seen; joins mixed-species flocks. Secretive. **SIMILAR SPECIES** Brown-chested Alethe has long white supercilium, brown-washed underparts and no crown-stripe. Juvenile Brown-chested very similar to juvenile *castanea* and not safely distinguishable in field. **STATUS AND DISTRIBUTION** Not uncommon forest resident, from S Senegal to Togo (nominate) and SW Nigeria to S CAR, Congo and Bioko (*castanea*). **NOTE** Nominate sometimes treated as separate species, White-tailed Alethe.

BROWN-CHESTED ALETHE *Alethe poliocephala*   Plate 92 (5)

Alèthe à poitrine brune

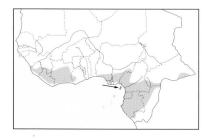

L 16 cm (*c.* 6"). Secretive terrestrial forest species. **Adult** *poliocephala* Grey-black crown separated from brown head-sides by long white supercilium. Upperparts chestnut; tail blackish-brown, rather short. Underparts white washed brownish-buff on breast and flanks; flanks tinged grey. Legs long, pale. *A. p. compsonota* has head-sides dark grey; in *carruthersi* crown browner, head-sides brownish, upperparts duller chestnut **Juvenile** Top of head blackish spotted and streaked orange-rufous; head-sides orange-rufous mottled dusky; upperparts have large orange-rufous spots. Throat dirty white washed orange and indistinctly scalloped dusky; breast orange-rufous scalloped blackish; rest of underparts dirty white. **Immature** With age, spots gradually disappear from upperparts; underparts become whiter. Fully grown immature similar to adult, but often has some orange spots on tips of greater coverts. **VOICE** Song a series of 5–8 clear whistled notes, descending in scale; similar to song of Yellow Longbill, but weaker and given closer to ground. Calls include a soft *karr-karr*. Generally silent. [CD10:4b] **HABITS** Singly, in pairs or small family groups in primary and old secondary forest, gallery forest and forest patches in savanna. Mainly on or near forest floor. Frequently attends ant columns; joins mixed-species flocks. Difficult to detect; much shyer than Fire-crested Alethe. **SIMILAR SPECIES** Fire-crested Alethe lacks supercilium but has crown-stripe and is whiter below; nominate has white tail corners. Juvenile Fire-crested very similar to juvenile Brown-chested and, except for nominate with white in tail, not safely distinguishable in field. Blackcap Illadopsis (Plate 113) rather similar in size and movements but principally distinguished by less white supercilium; ear-coverts dusky grey. Compare songs of Yellow Longbill and Sabine's Puffback. **STATUS AND DISTRIBUTION** Not uncommon to scarce forest resident, from W Guinea to Ivory Coast (north to Comoé NP) and Ghana (nominate) and from SW Nigeria to SW CAR, Congo and Bioko (*compsonota*); also SE CAR (*carruthersi*).

RED-TAILED ANT THRUSH *Neocossyphus rufus* — Plate 92 (8)
Néocossyphe à queue rousse

Other name F: Grive fourmilière à queue rousse
L 22 cm (8.75"). *N. r. gabunensis.* Shy, rufous terrestrial forest species with typical thrush-like jizz. **Adult** Head, throat and upper breast olive-brown washed grey; upperparts olive-brown, becoming more rufous on wings, rump and tail. Tail relatively long, *outer feathers rufous.* Lower breast to undertail-coverts rufous. Legs pale pinkish-horn. **Juvenile** Similar but duller; unspotted. **VOICE** Common call a drawn-out, descending, sibilant whistle *pseeeuuw,* sometimes with additional note *pseeeuuw-pyeew,* also a lower *wruuw.* A sharp, dry *prrt prrt* given around ant swarms and on take-off. Song consists of 2 clear, rather high-pitched whistles followed by a drawn-out, descending and accelerating trill *tseee wheh tsisisisisisisrrru.* [CD10:5] **HABITS** In pairs or singly in lowland forest. Mainly on forest floor or just above it. Regularly attends ant columns; occasionally in mixed-species flocks. Skulking. **SIMILAR SPECIES** Rufous Flycatcher Thrush is arboreal and has darker tail, shorter bill and is smaller. White-tailed Ant Thrush and Finsch's Flycatcher Thrush have white in outer tail feathers and darker upperparts. **STATUS AND DISTRIBUTION** Not uncommon to rare forest resident, S Cameroon, SW CAR, Eq. Guinea, Gabon and Congo. **NOTE** Taxonomic position of *Neocossyphus* is debatable; considered not to belong to thrush family by some authorities, on basis of morphological and other differences (*BoA*).

WHITE-TAILED ANT THRUSH *Neocossyphus poensis* — Plate 92 (6)
Néocossyphe à queue blanche

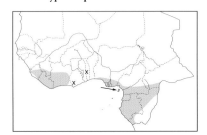

Other name F: Grive fourmilière à queue blanche
L 20 cm (8"). Dark, secretive forest species with *conspicuous white tail corners* and typical thrush-like jizz. **Adult** *poensis* Head and upperparts slate-brown. Tail blackish with 3 outer feathers broadly tipped white. Throat slate-brown becoming rufous on upper breast; lower breast to undertail-coverts dull rufous. Dark chestnut bases to flight feathers form indistinct panel on closed wing and broad bar in flight. Legs pale pinkish-horn. *N. p. praepectoralis* slightly browner above. **Juvenile** Unknown. **VOICE** A clear, drawn-out whistle *huweeeeet* or *tlooeeet* and a descending, sibilant *tseeeuw.* Common call around ant swarms, on take-off and in flight, a sharp, dry *prrt prrt.* Song rich and *Turdus*-like, but subdued and rarely heard. [CD10:6] **HABITS** In pairs or singly in lowland forest. Mainly on forest floor or just above it. Frequently attends ant columns; occasionally in mixed-species flocks. Wags tail up and down like a chat. **SIMILAR SPECIES** More conspicuous Finsch's Flycatcher Thrush is mainly arboreal, has more upright stance, different tail movement (flicks outer tail feathers sideways) and voice, and browner upperparts with more rufous rump. Red-tailed Ant Thrush lacks white in tail and is more rufous above. **STATUS AND DISTRIBUTION** Uncommon to locally common forest resident, from W Guinea to Togo, and from Nigeria and Bioko to Congo (nominate) and CAR (*praepectoralis*). **NOTE** Taxonomic position of *Neocossyphus* is debatable; may not belong to thrush family (*BoA*).

RUFOUS FLYCATCHER THRUSH *Stizorhina fraseri*　　Plate 92 (9)
Stizorhin de Fraser

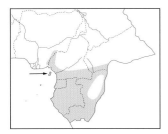

Other names E: Rufous Ant Thrush, Rufous Flycatcher, Fraser's Rusty Thrush; F: Grive fourmilière rousse

L 18 cm (7"). Rufous, flycatcher-like forest species. **Adult *rubicunda*** Head and upperparts olive-brown; rump rufous; underparts rich rufous. Tail dark olive-brown with *rufous outer feathers*. Rufous bases to secondaries produce diagonal bar on underwing. **Nominate** only very slightly darker. **Juvenile** Similar but duller and darker; unspotted. **VOICE** Varied. Calls include vigorous, high-pitched whistling *tsweetweetweetweet*, rapid *trrwit-rrwit-rrwit*, scolding *swit-swit-swit-swit-...*, and various hoarse notes. Song a short, rather slow, slightly rising series of 4 whistled notes, *trwee tu-trwee-twee* or *hee hu-hwee hee-i* with variations. [CD10:7] **HABITS** Singly or in pairs in middle and lower strata of primary lowland forest and old second growth. Flicks outer tail feathers sideways, in scissor-like fashion. Forages like flycatcher, catching insects on the wing and snatching prey from bark or foliage. Joins mixed-species flocks; occasionally near ant columns. **SIMILAR SPECIES** Red-tailed Ant Thrush is largely terrestrial and shyer, with more rufous tail, longer bill and larger size. White-tailed Ant Thrush has white in outer tail feathers and darker upperparts. **STATUS AND DISTRIBUTION** Not uncommon to common forest resident, from S Nigeria and S Cameroon to S CAR and Congo (*rubicunda*); also Bioko (nominate). **NOTE** Taxonomic position of *Stizorhina* (sometimes included in *Neocossyphus*) subject to conjecture; may not belong to thrush family (*BoA*).

FINSCH'S FLYCATCHER THRUSH *Stizorhina finschi*　　Plate 92 (7)
Stizorhin de Finsch

Other names E: Finsch's Ant Thrush; F: Grive fourmilière de Finsch

L 18 cm (7"). Dark, rufous forest species resembling large flycatcher, with *conspicuous white tail corners*. **Adult** Head and upperparts dark olive-brown, more rufous on rump. Tail slate-brown with outer 3 feathers broadly tipped white. Underparts rich rufous strongly washed olive on breast. **Juvenile** Similar but duller; unspotted. **VOICE** Call a harsh croaking *truwee-trueet* and *wreet wreet wreet...* Song a rather slow, slightly rising series of 4 drawn-out, melodious, whistled notes *hooee, hooee hooee-huEE*, slower and lower pitched than song of Rufous Flycatcher Thrush. Often silent. [CD10:8] **HABITS** Singly or in pairs in mid-strata of lowland forest. Flicks outer tail feathers sideways, in scissor-like fashion. Forages like flycatcher, catching insects on the wing or snatching them from foliage; often perches motionless with upright posture. Joins mixed-species flocks; occasionally attends ant columns. **SIMILAR SPECIES** Less conspicuous White-tailed Ant Thrush is terrestrial, has different tail movement (wags tail up and down) and voice, and more sooty upperparts. **STATUS AND DISTRIBUTION** Not uncommon to rare forest resident, from Sierra Leone and Guinea to S Ghana and Togo and in S Nigeria. **NOTES** (1) Treated as conspecific with Rufous Flycatcher Thrush by some authorities. Apparent hybrids between *finschi* x *fraseri*, based on plumage, have been reported from SE Nigeria; these produced songs at intermediate speed (Dowsett & Dowsett-Lemaire 1993). (2) *Stizorhina* sometimes merged with *Neocossyphus* (cf. *BoA*), but structural and behavioural differences suggest they are best treated as separate genera (Dowsett & Dowsett-Lemaire 1993).

RUFOUS-TAILED PALM THRUSH *Cichladusa ruficauda*　　Plate 94 (5)
Cichladuse à queue rousse

L 18 cm (7"). Small and local thrush with conspicuous rufous tail. **Adult** Top of head and upperparts rufous-brown; rump and tail bright rufous; supercilium, head-sides and neck dull pale greyish; dark eye-stripe. Underparts dirty white washed pale creamy-buff with greyish breast, upper belly and flanks; lower belly dirty buff. Bill black. Legs blue-grey. **Juvenile** Crown indistinctly streaked dusky; underparts mottled dusky, esp. on breast; bill horn. **VOICE** Song consists of loud melodious whistled phrases, including imitations; often in duet. Alarm a harsh *chrr*. [CD10:10] **HABITS** In palm savanna, thickets, plantations and gardens. Vocal. **STATUS AND DISTRIBUTION** Uncommon to locally common resident, SW Gabon (coast, Gamba area) and Congo (Congo R. valley) north to SW CAR.

Genus *Cercotrichas*

Scrub robins. Rather small with relatively long, broad, graduated and white-tipped tails, typically held cocked above back. Sexes similar. Juveniles like adult (unspotted). Species occurring in our region placed in genus *Erythropygia* by some authorities.

FOREST SCRUB ROBIN *Cercotrichas leucosticta*   Plate 94 (8)
Agrobate du Ghana

Other names E: Northern Bearded Scrub-Robin; F: Robin-agrobate de forêt; Rouge-queue du Ghana
L 15.0–16.5 cm (6.0–6.5"). The only scrub robin in our region occurring in forest. **Adult** *colstoni* Long white supercilium and moustachial stripe contrast with dark brown head and upperparts; black malar stripe; 3–5 white spots on bend of wing; rufous-brown rump and uppertail-coverts. Tail blackish-brown tipped white on 2–3 outer feathers. Throat whitish contrasting with broad brownish-grey breast-band; rest of underparts whitish with flanks tinged rufous-buff. **Nominate** is paler below, with poorly defined narrow greyish-buff breast-band and rest of underparts strongly washed rufous-buff; *collsi* is more olive above and has clean white underparts with pure grey breast-band and olive-grey flanks. **Juvenile** Similar, but scalloped black above and below, lesser wing-coverts spotted dark rufous, underparts rufous-buff. **VOICE** Song very melodious and sweet, consisting of a variable, high-pitched, whistled phrases. No imitations. Calls include a single *chuk* and a fast, high-pitched *chit-chit-chit*. [CD10:12] **HABITS** Very shy and difficult to observe, inhabiting undergrowth of lowland forest. Frequently flicks tail up and down. Forages on ground, often near ant columns. **SIMILAR SPECIES** Brown-backed and White-browed Scrub Robins inhabit open habitats, are slightly smaller and have different songs; ranges not known to overlap. **STATUS AND DISTRIBUTION** Scarce forest resident, recorded from Sierra Leone, Liberia, Ivory Coast (*colstoni*), Ghana (nominate) and SE CAR (*collsi*). Also claimed from Burkina Faso. **NOTE** Precise ranges of *colstoni* and nominate unknown; differences may be clinal (Tye 1991).

BROWN-BACKED SCRUB ROBIN *Cercotrichas hartlaubi*   Plate 94 (10)
Agrobate à dos brun

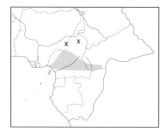

L 15 cm (6"). **Adult** Long white supercilium and 2 white wingbars contrast with cold *dark grey-brown upperparts*; *bright rufous rump and tail*, the latter with *clear-cut broad blackish subterminal band* and white tips to all feathers except central pair. Underparts white with grey wash or smudges and some variable fine streaking on breast forming indistinct band; flanks washed tawny. Bill mainly blackish. **Juvenile** Supercilium and wingbars rusty-buff; some rusty-buff tipped feathers on nape; breast slightly scalloped. **VOICE** Song loud and clear, consisting of short melodious whistled phrases constantly repeated; delivered from prominent perch and in short flight. [CD10:14] **HABITS** In wooded savanna with elephant grass, bushes, banana trees and cultivation. Constantly raises and lowers tail. Song attracts attention. **SIMILAR SPECIES** White-browed Scrub Robin has similar song but differs in brighter, rufous-brown back merging (not contrasting) with rufous rump, breast washed brownish (not grey) and distinctly streaked, wing feathers edged rufous. **STATUS AND DISTRIBUTION** Scarce to not uncommon resident, E Nigeria (Gashaka) to SW CAR.

WHITE-BROWED SCRUB ROBIN *Cercotrichas leucophrys*   Plate 94 (9)
Agrobate à dos roux

L 15 cm (6"). *C. l. munda.* **Adult** Long white supercilium and 2 white wingbars contrast with *warm brown upperparts* becoming *rufous on rump and tail*; central tail feathers mainly rufous with variable dark tips, other rectrices with blackish outer webs and white tips. Underparts whitish with buffish-brown wash and distinct dark brown streaking on breast; flanks washed tawny. Bill blackish with basal half of lower mandible pale horn. **Juvenile** Spotted above (rusty-buff and blackish on head, rufous-brown and black on upperparts); whitish below mottled rusty-buff and blackish on breast; flanks washed tawny. **VOICE** Song loud and clear, consisting of short melodious whistled phrases constantly repeated; delivered from prominent or hidden perch. Alarm call a hard *chrrr.* [CD10:15] **HABITS** In wooded savanna and thicket edges. Frequently flicks tail up and down and droops wings. Forages mainly on ground, hopping, running and searching leaf litter. **SIMILAR SPECIES** Similarly patterned Brown-backed Scrub Robin has more contrasting plumage with darker upperparts, bright rufous rump and tail, clear-cut black band and whiter underparts with pale greyish (not brownish) wash on breast and some narrow (not distinct) dark streaking. **STATUS AND DISTRIBUTION** Locally common resident in coastal and SE Gabon, and in Congo.

RUFOUS SCRUB ROBIN *Cercotrichas galactotes*   Plate 94 (11)
Agrobate roux

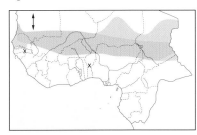

Other names E: Rufous Bush Chat, Rufous Bush Robin, Rufous-tailed Scrub Robin; F: Agrobate rubigineux
L 15 cm (6"). Distinctive. **Adult** *minor* Plain sandy rufous-brown above with rufous rump and fan-shaped rufous tail boldly tipped black and white. Creamy-white supercilium bordered below by narrow blackish eye-stripe. Underparts plain sandy-white. **Nominate** has broader subterminal black band to tail. **Juvenile** Similar. **VOICE** Song quite loud and sustained, consisting of short melodious whistled phrases, regularly repeated; delivered from prominent perch or in butterfly-like display flight. Calls include a hard *tek tek*, a hoarse *tseeeip*, a low *chrrr* and *deeu*. [CD3:80] **HABITS** A dry-country species, occurring in open grassland with shrubs, degraded savanna, farmland, etc. Mainly on ground or in low bushes, constantly fanning and cocking tail. **STATUS AND DISTRIBUTION** Locally common resident in semi-arid belt (*minor*), from S Mauritania and Senegal to N Nigeria, N Cameroon, Chad and NE CAR. In Sep–Apr numbers increased by Palearctic migrants (nominate).

BLACK SCRUB ROBIN *Cercotrichas podobe*   Plate 94 (12)
Agrobate podobé

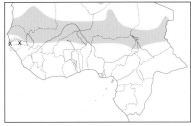

Other names E: Black Bush Robin; F: Merle podobé
L 18 cm (7"). *C. p. podobe.* Unmistakable. **Adult** Slender, long-legged, *black* chat with white-tipped undertail-coverts and *long fan-shaped tail boldly tipped white.* Underwing-coverts and inner webs of flight feathers rufous (visible in flight). **Juvenile** Duller, more sooty-brown; undertail-coverts have indistinct pale tips. **VOICE** Song, delivered from prominent perch, far carrying and melodious, consisting of fairly short, varied phrases, resembling song of Rufous Scrub Robin; also a sustained babble of sweet, fluty and scratchy notes, recalling subsong of Olivaceous Warbler. [CD10:16] **HABITS** In arid savanna, thorn scrub with clumps of palm trees, dry brush and oases. Mainly on ground or in low bushes, fanning and cocking tail. Sings frequently and almost year-round. Not shy. **STATUS AND DISTRIBUTION** Not uncommon resident in semi-arid belt, from Mauritania and Senegal to Niger, N Nigeria, N Cameroon and Chad.

BLACK REDSTART *Phoenicurus ochruros*   Plate 94 (4)
Rougequeue noir

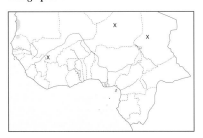

L 14 cm (5.5"). *P. o. gibraltariensis.* **Adult male breeding** Top of head and upperparts dark grey with white wing panel; rump and tail bright rufous. Face and underparts black becoming grey on flanks and belly; undertail-coverts pale rufous. **Adult male non-breeding** Black face and breast partially obscured by grey feather tips. Breeding plumage acquired with wear. **Adult female** Entirely slate-grey; tail as male. **First-winter male** As adult female. **VOICE** Call *tsip*, often followed by *tak-tak-tak*. [CD3:81] **HABITS** In our region rather similar to Common Redstart. Less vocal. **SIMILAR SPECIES** Female Common Redstart paler and warmer coloured than female Black Redstart, with pale brown upperparts and buff (not drab grey) underparts. **STATUS AND DISTRIBUTION** Rare Palearctic winter visitor, coastal Mauritania (late Aug/Oct–Apr) and N Senegal. Vagrant, S Mali (Bamako), N Niger (Aïr), Chad and Cape Verde Is.

COMMON REDSTART *Phoenicurus phoenicurus*
Plate 94 (2)
Rougequeue à front blanc

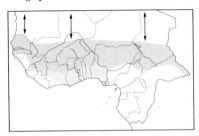

Other names E: Redstart; F: Rougequeue
L 14 cm (5.5"). *P. p. phoenicurus*. Dainty, slim chat with upright stance and, in all plumages, *bright rufous tail, which is almost constantly shivering* (giving impression that it is about to fall off). **Adult male breeding** Unmistakable. Top of head and upperparts blue-grey; rump and tail bright rufous but for brown central feathers. Black face and throat bordered above by white streak from forehead to eye. Rest of underparts orange-rufous, paler on belly. **Adult male non-breeding** Narrow buffish and white feather tips partially obscure full breeding plumage, which is gradually revealed through wear. **Adult female** Pale grey-brown above, buffish below, often tinged pale orange on breast and flanks; throat whitish. Dark eye with pale eye-ring appears large in open face. **First-winter male** As adult male non-breeding, but buffish and white feather tips broader, thus concealing underlying coloration more. **VOICE** Call *huweet* (recalling Willow Warbler but deeper and farther carrying), sometimes followed by *tuk tuk*. Also a soft *tsip*. Rarely sings in winter quarters. [CD3:82] **HABITS** Singly in various wooded habitats, mainly *Acacia* savanna and woodland. Catches insects in the air or on ground, sallying from concealed perch, often in shrubs or canopy of small trees. **SIMILAR SPECIES** Familiar Chat differs from adult female Common Redstart by duller rufous tail with dark tip; does not quiver tail. Adult female Black Redstart (a vagrant) also resembles female Common Redstart, but is distinctly greyer. **STATUS AND DISTRIBUTION** Uncommon to locally not uncommon Palearctic winter visitor (Sep–Apr) in broad savanna belt, from SW Mauritania, Senegambia and Sierra Leone to N Cameroon, Chad and N CAR. Rarely further south. Vagrant, Cape Verde Is.

COMMON STONECHAT *Saxicola torquata*
Plate 96 (2)
Tarier pâtre

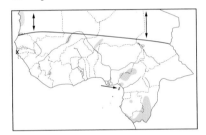

Other name F: Traquet pâtre
L 12.5 cm (5"). Small chat with relatively large, round head and short tail. **Adult male *salax* breeding** Head, throat, upperparts and tail black; neck-sides, rump and uppertail-coverts white; long white wing panel mainly formed by inner coverts; underparts white with centre of breast chestnut. *S. t. moptana* has chestnut on breast paler and more restricted; in *nebularum* chestnut more extensive. Palearctic *rubicola* browner above; feathers of upperparts and wings with more brownish-buff fringes; rump and uppertail-coverts rusty-brown mottled white. **Adult female** Head and upperparts streaked dark brown and buff *without supercilium*; white rump; breast orange-buff becoming buff and whitish on rest of underparts. **Juvenile** Head and upperparts mottled dark brown and buff; throat greyish; rest of underparts warm buff mottled dusky. **VOICE** Call (*wheet*) *trek-trek*. Song consists of short clear, twittering and scratchy phrases. [CD3:84] **HABITS** Usually in pairs, in various open habitats, incl. farmland, grassland, forest edge and marshy areas. Sits upright on prominent, often low perch, frequently flicking wings and tail. **SIMILAR SPECIES** Whinchat distinguished from female Common Stonechat by broad buff supercilium. **STATUS AND DISTRIBUTION** Patchily distributed, uncommon to locally common resident. *S. t. moptana* Senegal delta, inner Niger delta (C Mali) and (this race?) SW Niger; *nebularum* highlands of Sierra Leone, SE Guinea, N Liberia and W Ivory Coast (Mt Nimba); *salax* SE Nigeria (Obudu and Mambilla Plateaux), Cameroon (W and Adamawa Plateau), Bioko, C and S Gabon and Congo. Also irregular Palearctic winter visitor (*rubicola*), coastal Mauritania, and vagrant, N Mali, NE Niger and E Chad.

WHINCHAT *Saxicola rubetra*
Plate 96 (1)
Tarier des prés

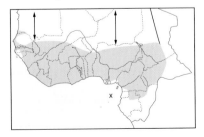

Other name F: Traquet tarier
L 12.5 cm (5"). Small, streaky buffish-brown chat with, in all plumages, *broad buff or white supercilium*. **Adult male breeding** (from Jan/Feb) Top of head and upperparts brown heavily streaked black; head-sides blackish, bordered by broad white supercilium above and white stripe below. Wings blackish-brown edged buff with variable white line on innermost coverts ('shoulders') and small white patch on edge of wing (primary-coverts). Tail blackish-brown with white patch at sides of base (conspicuous in flight). Underparts orange-buff becoming buffish-white on belly. **Adult female breeding** Paler

version of male; supercilium buff; head-sides brownish; wings with almost no white markings; tail base with less prominent white sides. **Adult male/adult female non-breeding** (from Aug) Very buffish overall. Head-sides brown mottled buff, bordered above by prominent, buff supercilium; underparts creamy-buff with some dark speckles on breast. Some males can be sexed by white line on inner coverts. **VOICE** Usually silent. Call (*whu*) *tek-tek*. [CD3:86] **HABITS** Uses prominent, low perches in open areas, incl. farmland, gardens, degraded savanna, wooded grassland and forest clearings. **SIMILAR SPECIES** Adult female Common Stonechat lacks supercilium. **STATUS AND DISTRIBUTION** Uncommon to common Palearctic passage migrant and winter visitor (Sep–May), wintering mainly from S Mauritania to Liberia east to SW Chad, W CAR, N Gabon and N Congo. Vagrant, Cape Verde Is and Príncipe.

Genus *Oenanthe*

Small, ground-dwelling insectivores with mainly sandy-brown or black-and-white plumage. Usually silent. Generally in arid or semi-arid habitats. Black-and-white tail pattern (prominent in flight) an important identification mark. Non-breeding plumage of Palearctic species acquired after complete moult before migration to winter quarters. Breeding plumage acquired after partial moult in winter quarters before return migration; may initially resemble non-breeding dress, but full breeding plumage rapidly acquired by wear of pale feather fringes, revealing underlying coloration. First-winter plumage similar to adult female non-breeding; acquired after post-juvenile moult in natal area, involving head and body feathers and some inner wing-coverts, while juvenile tail feathers and most or all juvenile wing feathers are retained with, occasionally, a few spotted juvenile scapulars. First-summer plumage similar to respective adult breeding; acquired after moult of some or all head and body feathers, just before return migration (Clement 1987).

WHITE-CROWNED BLACK WHEATEAR *Oenanthe leucopyga*   Plate 95 (9)
Traquet à tête blanche

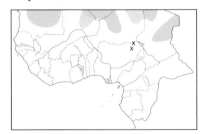

L 17 cm (6.75"). *O. l. aegra*. Large, black wheatear with white tail lacking T-pattern but having broad black central line. **Adult** Black with *white crown* and white lower rump, tail-coverts and vent. Tail feathers white, often with small dark smudge near tip; distal half of central pair black. **First-winter** As adult but crown black; dark smudges near tip of tail often larger. **Juvenile** As 1st-winter but dark brown (not black). **VOICE** Calls include a hard *chak* and whistled notes. Song rich and variable, consisting of short phrases with clear whistles, slurred notes and, occasionally, imitations and a few scratchy sounds. Also a quieter, more continuous song, including more harsh notes. [CD3:87] **HABITS** Singly or in pairs in rocky areas within desert; also oases and near habitation. At all altitudes in central Saharan massifs. Perches on rocks and bushes. Tame and inquisitive. **SIMILAR SPECIES** Black Wheatear distinguished from young by broad black terminal bar on tail. **STATUS AND DISTRIBUTION** Common desert resident and local migrant, Mauritania, N Mali, Niger and N Chad. Vagrant, NE Nigeria.

BLACK WHEATEAR *Oenanthe leucura*   Plate 95 (8)
Traquet rieur

L 18 cm (7"). *O. l. syenitica*. Largest wheatear. Black with white, T-patterned tail. **Adult male** Entirely black except for white at base of tail; tail white broadly tipped black, distal half of central pair black. **Adult female** Similar but browner; underparts rather mottled. **Juvenile** Similar to adult female but underparts more uniform. **First-winter** As juvenile but underparts darker, more like adult male. **VOICE** Call a loud *pee-pee-pee* (alarm); also *chak*. Song a rich and melodious warble. [CD3:88] **HABITS** Singly or in pairs in rocky areas with little or no vegetation. Rather shy. **SIMILAR SPECIES** Young White-crowned Black Wheatear lacks broad terminal bar on tail. **STATUS AND DISTRIBUTION** Rare resident and/or Palearctic migrant, NW Mauritania.

NORTHERN WHEATEAR *Oenanthe oenanthe*   Plate 95 (1)
Traquet motteux

L 15–16 cm (6.0–6.25"). Most widespread wheatear with, in all plumages, *white lower rump and tail with characteristic black inverted T* (formed by broad terminal tail band of even width, and black on distal two-thirds of central pair) **Adult male** *oenanthe* breeding *Top of head to back blue-grey* (diagnostic); black mask bordered by white supercilium;

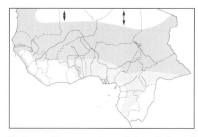

wings black. Underparts sandy-buff, richer on breast, becoming whitish on belly. Underwing-coverts dark grey. *O. o. libanotica* very similar, but typically paler. Greenland race *leucorhoa* larger; brown-grey above; deeper (rather orange-buff), more extensively and uniformly coloured below. Distinctive N African *seebohmi* has face and throat black. **Adult male non-breeding** Above, buff-brown variably washed greyish; variable darkish mask (sometimes very indistinct) bordered by whitish supercilium; wing feathers black broadly fringed orange-buff. In *seebohmi* black throat partially or wholly obscured by pale grey feather fringes. **Adult female breeding** Duller and browner than adult male breeding above, with variably distinct mask (sometimes lacking); supercilium buffish. **Adult female non-breeding/first-winter** Similar to adult male non-breeding but lack mask; wing feather centres dark brown. **First-summer male** As adult male breeding but upperparts browner; dark mask mottled brown; wings and tail browner. **VOICE** Call a hard *chak* (alarm), less often *wheet-chak*. Song, given from ground or low perch, consists of variable short phrases including whistles, scratchy notes and imitations; occasionally uttered in winter quarters. [CD3:89-90] **HABITS** Singly in various open habitats, from subdesert to degraded savanna. Forages mainly by running short distances, then pecking at prey. Bobs head, flicks wings and tail. Flies low. Defends individual winter territory. **SIMILAR SPECIES** Adult female/1st-winter easily confused with some other wheatears. Isabelline Wheatear paler, more sandy-coloured in all plumages, with generally broader and shorter stemmed tail band; wings more concolorous with rest of upperparts; blackish alula contrasting strongly; head broader, bill stronger (imparting subtly different head shape). Black-eared and Desert Wheatears in all plumages distinguishable by different tail pattern. Adult male *seebohmi* differs from other black-throated wheatears by greyer upperparts, but non-breeding sometimes very similar to sandy-coloured species. **STATUS AND DISTRIBUTION** Palearctic winter visitor (Sep–Apr) in broad Sahel belt. *O. o. seebohmi* SW Mauritania (common), Senegal and Mali (uncommon to rare); also reported from N Cameroon (vagrant?). Other races common from Mauritania and N Senegal to Niger, N Nigeria, N Cameroon and Chad; uncommon further south, from Gambia and S Senegal to CAR. Rare, coastal Guinea to Liberia. Vagrant, NE Ivory Coast, N Ghana, Gabon, N Congo. Scarce but regular, Cape Verde Is (Nov–Apr). Senegambia main wintering area for *leucorhoa*.

CYPRUS WHEATEAR *Oenanthe (pleschanka) cypriaca*   Plate 95 (7)
Traquet de Chypre

Other name F: Traquet pie
L 13.5 cm (5"). Small, chat-like wheatear. T-patterned tail with uneven terminal bar, black increasing on outer feathers. **Adult male breeding** Top of head to nape white, contrasting with black face, throat and upperparts; rump white. Breast rusty-buff becoming white on undertail-coverts. **Adult female breeding** Similar to adult male breeding but crown and nape dark brown-grey, bordered by white supercilium. **Adult non-breeding** Top of head to nape dull brownish bordered by narrow whitish supercilium; black face and throat variably mottled by buffish feather tips; feathers of upperparts and wings tipped and fringed buffish. Underparts buff, more rusty on breast (rapidly bleaching in winter quarters). **First-winter** Similar to adult non-breeding but feathers of upperparts, wings and throat have broader buffish tips and fringes. **VOICE** Call a hard *chak*. [CD10:17] **HABITS** Usually singly in open country with scattered bushes. Behaviour as other wheatears. **SIMILAR SPECIES** Adult Black-eared Wheatear of black-throated forms distinguished by pale back and less extensive black on throat. Mourning Wheatear has orange-buff (not white) undertail-coverts. **STATUS AND DISTRIBUTION** Uncommon Palearctic migrant, EC Chad (Nov–Mar). **NOTE** Variously treated as race of Pied Wheatear *O. pleschanka* (e.g. *BoA*), or more frequently in recent years as separate species (e.g. *BWPC*).

BLACK-EARED WHEATEAR *Oenanthe hispanica*   Plate 95 (4)
Traquet oreillard

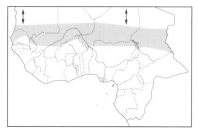

L 14.5 cm (5.75"). Rather slim and long-tailed wheatear with variable plumage. T-patterned tail *with uneven, sometimes broken, terminal bar: generally narrower than Northern Wheatear on 2nd–4th pair of rectrices, broader on outer pair.* Two races, both with white-throated and black-throated forms in both sexes. **Adult male *hispanica* breeding** Top of head to back rich sandy-buff (paler with wear); black mask highlighted by white forehead and narrow white supercilium above, and by white throat and neck-sides below; wings black. Breast warm sandy-buff becoming whitish on rest of underparts. Underwing-coverts black. Black-throated form has black mask extending to throat. Eastern *melanoleuca* more black and white, with black-throated form commoner than white-throated. **Adult male** *hispanica* **non-breeding** Similar, but basic pattern overlaid by buffish feather tips; greater coverts and inner secondaries broadly tipped and fringed rich buff, forming pale wing panel. Black-throated form has black feathers of throat narrowly tipped white. *O. h. melanoleuca* has upperparts washed grey. **Adult female** *hispanica* **breeding** Similar but much less strongly marked than male. Top of head and upperparts rich sandy-brown; mask indistinct, often bordered by pale supercilium; wings dark brown. Black-throated form has black face and throat variably mottled buffish. *O. h. melanoleuca* generally browner, less sandy; supercilium generally absent. **Adult female** *hispanica* **non-breeding/1st-winter** Similar, but mask and wing feathers mostly obscured by broad, rich buff tips and fringes. In black-throated form dark feather bases on throat visible with wear. *O. h. melanoleuca* more grey-brown above. **First-summer male** Similar to adult male breeding, but darker above and duller. Black-throated forms have black face and throat variably mottled buff. **VOICE** Call a hard *chak*. Song, sometimes given in winter quarters, a rapid jumble of scratchy notes and whistles, occasionally including imitations. [CD3:91] **HABITS** Singly in semi-desert, *Acacia* savanna and dry farmland. Perches more on shrubs than Northern and Isabelline Wheatears. **SIMILAR SPECIES** Northern and Desert Wheatears distinguished in all plumages by tail pattern. Adult male Desert further differs from male black-throated forms by sandy (not black) scapulars, and black on throat connected to black of wing. **STATUS AND DISTRIBUTION** Uncommon to rare Palearctic migrant (Sep–Apr) in Sahel belt, from Mauritania and Senegambia to Chad and NE CAR. Nominate in west; *melanoleuca* in east; both races meet in Mali.

MOURNING WHEATEAR *Oenanthe lugens* Plate 95 (6)
Traquet deuil

L 14.5 cm (5.75"). *O. l. halophila*. Vagrant with T-patterned tail having broad black terminal band of even width. In flight, whitish inner webs to primaries and outer secondaries produce pale panel. **Adult male breeding** Top of head to upper mantle whitish; face to throat and upperparts black; back to uppertail-coverts white. Breast to belly white; *undertail-coverts orange-buff*. **Adult male non-breeding** Similar but duller, with some white tips to wing and tail feathers. **Adult female breeding** Dull grey version of male. Top of head and upperparts greyish buff-brown; face and throat variably dark; rest of underparts white, usually washed buffish on breast and undertail-coverts. Throat sometimes lacks black. **Adult female non-breeding** Similar but duller; wing feathers tipped and fringed buffish. **First-winter** As adult female non-breeding but even duller; wing-coverts more broadly tipped buffish, forming 2 pale bars. **VOICE** Call a hard *chak*. Song, sometimes given in winter, resembles that of other wheatears. [CD3:92] **HABITS** Singly in rocky desert. Behaviour typical of genus. **SIMILAR SPECIES** No other wheatear in our region has similar combination of dark throat, rusty-tinged undertail-coverts, tail band of even width and (in flight) pale wing panel. **STATUS AND DISTRIBUTION** N African vagrant, Mauritania (Sep–Dec) and N Niger (Aïr, Nov).

DESERT WHEATEAR *Oenanthe deserti* Plate 65 (5)
Traquet du désert

L 14–15 cm (5.5–6.0"). Sandy-coloured wheatear with *diagnostic black tail in all plumages*. **Adult male** *homochroa* **breeding** Top of head to back pale sandy-cream; *face and throat black* bordered above by whitish supercilium and *connected to black wings at shoulder*; rump and uppertail-coverts white. Breast to undertail-coverts pale sandy-buff. **Nominate** virtually inseparable in field, but upperparts tinged greyer. **Adult male non-breeding** Similar but slightly greyer above; black feathers of face and throat narrowly tipped white; wing feathers tipped and fringed white or pale sandy. **Adult female breeding** Top of head and upperparts deep sandy-buff, sometimes with greyish tinge, duller than male; supercilium pale sandy, indistinct; ear-coverts variable, pale sandy to brownish; wings dark brownish tipped and fringed pale sandy; underparts pale sandy-buff. Throat occasionally marked dusky. **Adult female non-breeding** Similar to adult female breeding, but greyer above; some throat feathers narrowly tipped brownish; wing feathers more broadly tipped and fringed sandy-buff. **First-winter male** Similar to adult male non-breeding, but face and throat feathers more extensively tipped whitish, obscuring black to varying degree; wing feathers browner and more broadly tipped and fringed sandy-buff. **VOICE** Calls include *hweee* and hard *tuk*. Song, often given in winter quarters, a characteristic, plaintive, descending *swee-you* or *deedjrruu*, frequently repeated. [CD3:94] **HABITS** Usually singly in semi-desert and desert with some vegetation. Defends individual winter territory. Frequently wags tail. **SIMILAR SPECIES** Adult male Black-eared Wheatear of black-throated form distinguished from male by T-patterned tail, black scapulars, and gap between black of throat and wings. Isabelline Wheatear differs from female by noticeably larger size and paler wing. Blackstart lacks white rump and has black (not white) uppertail-coverts. **STATUS AND DISTRIBUTION** *O. d. homochroa* common to scarce N African winter visitor (Sep–Apr) to Sahara and N Sahel, from Mauritania and N Senegal to Niger, N Nigeria (rare) and Chad. Unsubstantiated records from Gambia. Recorded breeding, Mauritania (Adrar). Nominate, vagrant to Niger and Chad.

HEUGLIN'S WHEATEAR *Oenanthe (bottae) heuglini* Plate 95 (2)
Traquet de Heuglin

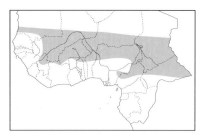

Other names E: Red-breasted Wheatear; F: Traquet à poitrine rousse L 14–15 cm (5.5–6.0"). Distinctive, long-legged wheatear with *rufous-buff underparts.* T-patterned tail with *broad* black terminal band (*c.* half of tail). **Adult breeding** Top of head and upperparts dark brown; narrow white supercilium bordered by dark eye-stripe; rump and uppertail-coverts white. Underparts rich rufous-buff to brick-red on breast and flanks, paler on throat and centre of belly. **Adult non-breeding** Similar, but wing feathers tipped and fringed rufous-buff. **Juvenile** Similar, but spotted dark rufous-buff above and scalloped dusky below; wing feathers broadly fringed rufous. **VOICE** Call a sharp *chak.* Song consists of very varied phrases, including imitations. [CD10:19] **HABITS** Usually in pairs in degraded savanna, dry farmland, recently burnt ground, rocky hillsides and inselbergs, up to 2300 m. Often wags rear body. Has short display flight with spread tail. Rather shy. **SIMILAR SPECIES** No other wheatear or chat has combination of dark brown upperparts, rufous-buff underparts and white rump. **STATUS AND DISTRIBUTION** Uncommon or rare to locally not uncommon intra-African migrant between Sahel and savanna zones, from S Mauritania to Chad. Rare, C Guinea, N Ivory Coast, N Ghana, N Togo, N Benin and Burkina Faso. Possibly partially resident in Mali, Nigeria, Chad and CAR. Breeds in savanna zone, moving north in rainy season. **NOTE** Variously treated as race of Red-breasted Wheatear *O. bottae* (e.g. *BoA*) or as separate species, *O. heuglini,* on basis of differences in size, habitat and behaviour.

ISABELLINE WHEATEAR *Oenanthe isabellina* Plate 95 (3)
Traquet isabelle

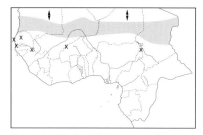

L 16–17 cm (6.25–6.75"). *Robust, long-legged, plain sandy-coloured wheatear.* T-patterned tail with *broad* black terminal band. **Adult breeding** Top of head and upperparts sandy-brown; supercilium pale creamy; wing feathers brown broadly fringed sandy-brown (thus barely contrasting with upperparts when fresh), with *contrasting blackish alula;* rump and uppertail-coverts white. Underparts pale sandy, paler on throat and belly. Underwing pale grey to white. **Adult non-breeding** Very similar but wing paler; primaries and tail feathers tipped pale buff. **First-winter** As adult non-breeding; broader pale buff tips to primaries and tail feathers. **VOICE** Calls a hard *chak* (alarm) and *wheew.* Song, uttered in winter quarters, consists of variable short phrases including whistles, scratchy notes and imitations; similar to Northern Wheatear's. [CD3:95] **HABITS** Usually singly in open country with scattered bushes. Runs rather than hops. Defends individual winter territory. **SIMILAR SPECIES** Non-breeding Northern Wheatear has duller, slightly darker, greyer brown upperparts, darker wings and generally narrower tail band (black covering *c.* one-third of length, but *c.* half in Isabelline). Some adult female/1st-winter Northern very similar to Isabelline, but distinguished by darker centres to wing-coverts. Desert and Black-eared Wheatears are noticeably smaller and have different tail pattern. **STATUS AND DISTRIBUTION** Common to uncommon Palearctic winter visitor (Oct–Apr) to Sahel zone, from S Mauritania and N Senegal to Mali, Niger and Chad. Rare, Gambia, SE Senegal and N Cameroon.

Genus *Cercomela*

Small, plain-coloured, ground-feeding chats with long, slender legs; upperparts variously brown or grey, underparts paler. Tail colour an important identification feature.

FAMILIAR CHAT *Cercomela familiaris* Plate 96 (6)
Traquet familier

Other name F: Traquet de roche à queue rousse
L 14 cm (5.5"). *C. f. falkensteini.* Rather nondescript, plain brown chat with *conspicuous rufous rump and tail.* **Adult** Head and upperparts grey-brown with narrow pale eye-ring and rusty ear-coverts; rump and uppertail-coverts

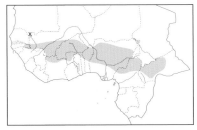

bright rufous. Tail feathers bright rufous tipped dark brown, except entirely dark brown central pair. Underparts pale greyish becoming whitish on central belly. **Juvenile** Similar, but spotted buff above and has dusky barring below, giving scaly appearance. **VOICE** Song a soft series of whistled and chattering notes. Calls include a high *whee*, often followed by a hard *chak-chak*; also a harsh *cher-cher*. [CD10:20] **HABITS** Singly or in pairs around inselbergs, bare rocky ground and erosion gullies. On landing, often raises and lowers tail and flicks wings. **SIMILAR SPECIES** Adult female Common Redstart has brighter rump and tail lacking dark tip; also constantly quivers tail. **STATUS AND DISTRIBUTION** Uncommon to locally common resident in savanna zone, from SE Senegal through S Mali and N Ivory Coast to CAR. Rare, S Mauritania.

BROWN-TAILED ROCK CHAT *Cercomela scotocerca*   Plate 96 (4)
Traquet à queue brune

Other name F: Traquet de roche à queue brune
L 13 cm (5"). *C. s. furensis.* Small, nondescript and very localised chat. **Adult** Head and upperparts plain grey-brown, more rufous on rump; tail dark brown; underparts pale buffish, becoming off-white on lower belly. Narrow pale eye-ring. **Juvenile** Unknown. **VOICE** Song a rapid, chirruping phrase, frequently repeated with slight variations. [CD10:21] **HABITS** Singly or in pairs in dry rocky country with low shrubs. Flicks wings and tail. Tame. **SIMILAR SPECIES** Blackstart and Familiar Chat have differently coloured tail; ranges not known to overlap. **STATUS AND DISTRIBUTION** Locally not uncommon resident, Chad (east of 19°E).

BLACKSTART *Cercomela melanura*   Plate 96 (3)
Traquet à queue noire

Other name F: Traquet de roche à queue noire
L 14 cm (5.5"). **Adult** *ultima* Head and upperparts plain sandy-brown with *black uppertail-coverts and tail.* Underparts paler, becoming whitish on belly. Flight feathers dark brown. ***C. m. airensis*** reportedly paler, with cinnamon tinge, but coloration variable; races doubtfully distinct. **Juvenile** Similar; unspotted; wing-coverts and tertials tipped and fringed paler. **VOICE** Song, delivered from prominent perch, a short, rather pleasant warble, frequently repeated after a short pause, with only slight variations; includes *cherlu*, which is also uttered singly. Calls include a harsh note and a high-pitched *hiih* (alarm). [CD10:22] **HABITS** Singly or in pairs in rocky hills, mountains and ravines within open, arid scrub country. Constantly flicks tail open, often simultaneously drooping and flicking wings half-open. **SIMILAR SPECIES** Brown-tailed Rock Chat has dark brown tail and usually prefers thicker scrub; ranges not known to overlap. **STATUS AND DISTRIBUTION** Uncommon to locally common resident, NE Mali (from Hombori area east), W and NE Niger (common in Aïr) and N Chad (Tibesti; south to Faya). Vagrant, Gambia.

Genus *Myrmecocichla*

Small to medium-sized, mainly ground-dwelling chats with strong legs. Most have mainly black or grey plumage. Occur in open habitats. Endemic to Africa.

CONGO MOOR CHAT *Myrmecocichla tholloni*   Plate 96 (10)
Traquet du Congo

Other name F: Traquet-fourmilier du Congo
L 18–19 cm (7.0–7.5") Medium-sized, *robust, dark chat* with contrasting *white wing patch and white rump.* **Adult** Crown, upperparts and tail dark brown-grey (appearing blackish in poor light); crown feathers tipped paler, giving scalloped appearance; head-sides mainly dirty white; narrow pale collar formed by whitish neck-sides and pale dirty brown nape; rump and uppertail-coverts white; primaries white at base (forming wing patch). Throat and undertail-coverts dirty white; rest of underparts dirty brown-grey scalloped paler. Bill and legs strong, black.

Juvenile Crown plain brown-grey (no scallops); *no white wing patch*; wing-coverts and secondaries tipped dirty white; underparts plain soft brown-grey except dirty white throat. Bill smaller and paler with pale horn cutting edges. **VOICE** Song consists of short, clear, melodious whistles interspersed with rolled *chiurrr*, repeated with slight variations. Alarm a sharp but weak *peep*. [CD10:23] **HABITS** In pairs or small parties in open grassland with few or no bushes and trees. Mainly on ground, but readily uses perches. Characteristic low display flight with vibrating wings. Wary. **STATUS AND DISTRIBUTION** Locally not uncommon resident, SE Gabon (Lékoni) and Congo (Bateke Plateau).

NORTHERN ANTEATER CHAT *Myrmecocichla aethiops*   Plate 96 (8)
Traquet brun

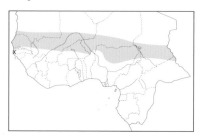

Other name F: Traquet-fourmilier brun du nord
L 19 cm (7.5"). *M. a. aethiops*. Long-legged, blackish-brown chat with large white wing patches visible in flight. **Adult** Wholly sooty-brown (appearing black in poor light); basal two-thirds of inner webs of primaries white above and below (usually concealed at rest). **Juvenile** Similar, but warmer brown; throat and breast feathers edged buff. Gape yellow during first 3 months. **VOICE** Song, delivered from perch, a sustained, varied mixture of clear whistles and short, hard trills. Calls include penetrating *tseeu* and *heeh*. [CD10:24] **HABITS** In pairs or small parties in open grassland and farmland with scattered bushes, trees and, usually, termitaria. Mainly on ground, but readily uses perches. Tame and conspicuous. **SIMILAR SPECIES** Sooty Chat is noticeably smaller, lacks white wing patch and has, in male, conspicuous white shoulder patch; see also distribution. White-fronted Black Chat is even smaller with, in male, white forehead. **STATUS AND DISTRIBUTION** Locally common to uncommon resident in semi-arid belt, from S Mauritania and Senegambia to N Cameroon, Chad and NE CAR.

SOOTY CHAT *Myrmecocichla nigra*   Plate 96 (9)
Traquet commandeur

Other name F: Traquet-fourmilier noir
L 16 cm (6.25"). **Adult male** Entirely glossy black with conspicuous *white shoulder patches*. **Adult female** Entirely blackish-brown. **Juvenile** Similar to adult female. **VOICE** Song consists of clear whistles occasionally interspersed with a short, low, hard trill; delivered from perch or, occasionally, in short flight. Sometimes imitates other species. Both sexes sing. Calls include various whistles. [CD10:25] **HABITS** Singly or in pairs in open grassland with termitaria and few or no bushes; also farmland, by roads and on recently burnt areas. Slowly moves tail up and down. Fairly tame. **SIMILAR SPECIES** Northern Anteater Chat is noticeably larger, has white wing patch and lacks shoulder patch; see also distribution. White-fronted Black Chat is smaller with, in male, white forehead. **STATUS AND DISTRIBUTION** Uncommon to common and local resident, N and E Nigeria (mainly Gashaka-Gumti NP and Mambilla Plateau), Cameroon, CAR, SW and SE Gabon, and Congo. A few records from Senegal, Guinea and Liberia.

WHITE-FRONTED BLACK CHAT *Myrmecocichla albifrons*   Plate 96 (5)
Traquet à front blanc

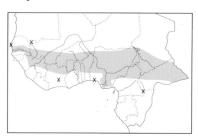

Other name F: Traquet noir à front blanc
L 15 cm (6"). Small, slim chat. **Adult male** *frontalis* Entirely black except conspicuous white patch on forehead (usually extending to forecrown). Flight feathers whitish below. Eastern *limbata* also has some white on shoulder. **Adult female** Similar, but lacks white. **Juvenile** Similar to adult female but browner and slightly mottled. **VOICE** Song consists of short phrases with sharp and rolling notes, including frequently repeated *uwheetirr*; sometimes imitates other species. Calls include a high-pitched, penetrating *heet*. [CD10:26] **HABITS** Singly or in pairs in open savanna woodland; also cultivated

areas and clearings. Partial to recently burnt ground. Mostly perches in small trees and bushes, from where it pounces on prey on ground. Frequently flicks tail. Male has butterfly-like display flight. **SIMILAR SPECIES** Sooty Chat is noticeably larger; male lacks white forehead and has conspicuous white shoulder patch. Northern Anteater Chat is even larger, also without forehead patch but large white patch in wing. **STATUS AND DISTRIBUTION** Uncommon to locally common resident in broad savanna belt, from Senegambia to Sierra Leone east to Chad and CAR. *M. a. limbata* from E Cameroon east. Vagrant, S Mauritania.

CLIFF CHAT *Myrmecocichla cinnamomeiventris*   Plate 96 (7)
Traquet à ventre roux

Other names E: Mocking Chat, White-crowned Cliff-Chat; F: Traquet de roche à ventre roux
L 20 cm (8"). Fairly large chat with characteristic coloration. Three races recognised in our region. **Adult male** *cavernicola* Black above with white shoulder patch and rufous rump. Below, black breast separated from rufous rest of underparts by narrow pale line. *M. c. bambarae* has white on shoulder restricted to lesser coverts, uppertail-coverts black (not rufous) and no paler line separating black breast from rest of underparts. *M. c. coronata* has white crown (sometimes white restricted to a few feathers) and pale area between breast and rest of underparts. **Adult female** *cavernicola/bambarae* Similar to male but duller and lacking white on shoulder; black areas more sooty; rufous areas browner, less bright; no pale line bordering breast. *M. c. coronata* has much paler, rufous-grey head and entire underparts dull chestnut. **Juvenile male/female** Like respective adults but duller; male has less white on shoulder. **VOICE** Song melodious, rich and far carrying, consisting of clear, vigorous and varied whistles or rapid imitations; often in duet. Calls include a high-pitched, penetrating *seeeo* and a more rapid *seeu-seeu*. [CD10:28-29] **HABITS** Usually in pairs on rocky outcrops, inselbergs, escarpments and cliffs in savanna. Often slowly raises and lowers tail, fanning it on upswing. Flight low with strong flapping and gliding. **STATUS AND DISTRIBUTION** Local and patchily distributed, rare to not uncommon resident. Recorded from S Mauritania, Senegal, Guinea, Mali, NW Ivory Coast, SE Burkina Faso, Ghana, N Togo, N Benin, Nigeria, Cameroon, Chad and CAR. *M. c. cavernicola* and *bambarae* mainly in west of range, *coronata* in east. **NOTES (1)** Sometimes included in genus *Thamnolaea*. **(2)** *M. c. coronata* sometimes treated as separate species, White-crowned Cliff Chat. This form occurs alongside black-crowned males at several localities in Burkina Faso, Togo, Benin, Nigeria and Cameroon.

COMMON ROCK THRUSH *Monticola saxatilis*   Plate 91 (1)
Monticole merle-de-roche

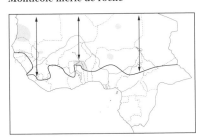

Other names E: Rock Thrush; Mountain Rock Thrush, Rufous-tailed Rock Thrush, European Rock Thrush; F: Merle de roche
L 18.5 cm (7.25"). Chat-like, short-tailed, rather long-billed thrush with distinctive rufous tail in all plumages. **Adult male breeding** (from Apr) Head and upperparts greyish-blue; back has pure white patch; wings blackish-brown. Underparts orange-rufous. **Adult male non-breeding** (after complete moult Jul–Sep) Mottled dark brown above; wing feathers fringed pale buff. Underparts rufous, partially obscured by white feather tips and black crescentic barring; central throat pale. **Adult female** Resembles non-breeding male but paler brown above, no white on back, underparts buffish with dark scaling. **VOICE** Mostly silent in winter quarters. Call a short *chak*. Song consists of clear, fluty, short, warbling phrases, each starting with same notes. [CD3:96] **HABITS** Solitary and shy. Occurs in a variety of habitats, incl. rocky plateaux, savanna, recently burnt bush, cultivated areas, even gardens. **SIMILAR SPECIES** Female Blue Rock Thrush differs from female Common in being darker brown, unspotted above, incl. tail; also larger and longer billed. **STATUS AND DISTRIBUTION** Uncommon to rare and local Palearctic migrant and winter visitor (Sep–Apr), from Mauritania to Sierra Leone east to Chad and CAR. More regularly recorded in east (Chad, Cameroon, Nigeria).

BLUE ROCK THRUSH *Monticola solitarius*   Plate 91 (2)
Monticole merle-bleu

Other name F: Merle bleu
L 20 cm (*c.* 8"). Larger, with longer bill and tail than Common Rock Thrush. **Adult male** Unmistakable. Wholly deep dark blue (appearing blackish at distance). Bill and legs black. In fresh plumage (Sep–Dec) slate-grey of head obscured by brownish feather tips; upperparts have pale brown fringes, wing-coverts small white tips and underpart feathers buffish tips. **Adult female** *solitarius* Upperparts, incl. tail, dark brown (tinged bluish in some);

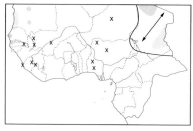

head-sides grey-brown finely mottled and streaked buff. Underparts buffish, throat and breast scaled, rest barred blackish-brown. *M. s. longirostris* has underparts paler, greyer and less barred. **Juvenile** Resembles adult female, but slightly paler brown; upperparts spotted buffish; underparts appear more spotted; barring on lower underparts paler and less sharply defined. **VOICE** Mainly silent. Calls include *tak-tak* (contact), *chuk-chuk* (alarm) and a high-pitched *tsee*. Song (occasionally uttered in winter quarters) consists of far-carrying, melodious, oft-repeated, short phrases, slightly reminiscent of Common Blackbird *Turdus merula*. [CD3:97] **HABITS** Mostly solitary, frequenting mainly rocky habitats on plateaux, coastal hills and mountain slopes; also on buildings (Senegal). Shy. **SIMILAR SPECIES** Female Common Rock Thrush has rufous tail, paler brown, spotted upperparts, and pale-centred throat; also smaller. **STATUS AND DISTRIBUTION** Scarce to rare Palearctic migrant and winter visitor (Oct–early Apr), patchily distributed from Mauritania to Liberia east to Chad and CAR. Year-round at Cap de Naze (S Senegal); breeding suspected. Nominate throughout, except in extreme east, where *longirostris* (east of *c*. 20°E).

BLACK-EARED GROUND THRUSH *Zoothera camaronensis*　　　Plate 91 (7)
Grive du Cameroun

Other name F: Grive terrestre du Cameroun

L *c*. 18 cm (7"). *Z. c. camaronensis*. Small and rather short-tailed, distinctively coloured ground thrush. **Adult** Top of head and upperparts russet-brown with *2 conspicuous white wingbars* (formed by broad white spots on greater and median coverts); *head-sides orange-rufous with 2 black streaks* (one below eye and one on ear-coverts). Underparts orange-rufous becoming white on central belly and undertail-coverts. Tail brown. Bill black and rather heavy. Legs whitish-pink. **Juvenile** Similar to adult, but crown and mantle have pale buff shaft streaks; tertials fringed orange; breast and belly slightly mottled dusky; base of lower mandible pale. **VOICE** Calls include a thin, high-pitched *ssreee* (probably alarm) and *tssrrr*. Song unknown. [CD10:33] **HABITS** Singly or in pairs in lowland rainforest. Forages on ground. Very shy and difficult to observe. **SIMILAR SPECIES** Crossley's Ground Thrush occurs at higher elevations and is distinctly larger. Grey Ground Thrush, the only other lowland ground thrush, has similar pattern but lacks distinctive orange-chestnut underparts and is larger. **STATUS AND DISTRIBUTION** Rare forest resident in W, C and S Cameroon, Eq. Guinea and NE Gabon.

GREY GROUND THRUSH *Zoothera princei*　　　Plate 91 (5)
Grive olivâtre

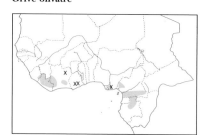

Other name F: Grive terrestre grise

L 21 cm (*c*. 8.25"). **Adult** *batesi* Compact with rather short wings and tail, *2 black vertical patches on whitish head-sides* (one below eye and one on ear-coverts), and *2 conspicuous irregular white bars on wings* (formed by broad white tips to coverts). Upperparts olive-brown becoming more russet on back and rump; tail rufous-brown. Throat dirty white, breast and flanks brownish, central belly and undertail-coverts white. **Nominate** greyer above. **Juvenile** Similar to adult, but head and mantle heavily streaked rufous, tertials tipped rufous, underparts initially rufous (gradually lost with age), breast mottled dusky. **VOICE** Calls include a sharp, high-pitched *ssrii* or *sssirrr* (reminiscent of White-browed Forest Flycatcher and Ashy Flycatcher; Brosset & Erard 1977) and a thin, high-pitched *seeep* (alarm). Song unknown. [CD10:34] **HABITS** Singly or in pairs in lowland rainforest. Forages on ground, hopping on forest floor, searching leaf litter. Joins mixed-species flocks; sometimes attends ant swarms. Extremely shy and rarely observed. **SIMILAR SPECIES** Black-eared Ground Thrush has bolder white wingbars, and orange-rufous face and underparts. **STATUS AND DISTRIBUTION** Rare forest resident, from Sierra Leone to Ivory Coast (north to Comoé NP) and SW Ghana (nominate) and from S Nigeria (1 possible record) and Cameroon to NE Gabon (*batesi*).

CROSSLEY'S GROUND THRUSH *Zoothera crossleyi*　　　Plate 91 (6)
Grive de Crossley

Other name F: Grive terrestre de Crossley

L 21.5 cm (8.5"). *Z. c. crossleyi*. **Adult** Mainly orange- and rufous-chestnut ground thrush with *blackish mask* from

lores to ear-coverts. Upperparts and tail rufous-chestnut, more russet on rump and uppertail-coverts; *bright orange-rufous collar*; wings dark brown with large white terminal spots on coverts forming 2 rather irregular bars. Underparts bright orange-rufous; centre of belly to undertail-coverts white. Bill black. Legs pinkish-white. **Juvenile** Head-sides and breast mottled dusky and rufous; belly wholly orange-rufous. **VOICE** Alarm a thin, high-pitched *seeep*. Song varied and far carrying, consisting of short, melodious phrases, repeated several times at regular intervals, until a new motif or variant is introduced. [CD10:31] **HABITS** Singly or in pairs in wetter montane and submontane forests, often in ravines. Sings from ground or from high, hidden perch. Shy and difficult to observe. **SIMILAR SPECIES** Other ground thrushes inhabit lowland forest and have 2 black streaks on head-sides. **STATUS AND DISTRIBUTION** Rare to locally not uncommon forest resident, SE Nigeria (Obudu and Mambilla Plateaux, Gotel Mts) and W Cameroon (Mt Cameroon, Rumpi Hills, Mt Kupe, Mt Nlonako, Bakossi Mts; 1000–2300 m); also in SW Congo (Mayombe; 500–600 m). Altitudinal migration recorded (to 200 m on Mt Cameroon). *TBW* category: Near Threatened.

GULF OF GUINEA THRUSH *Turdus olivaceofuscus*   Plate 143 (13)
Merle de Sao Tomé

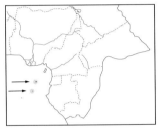

Other names E: São Tomé Thrush; F: Grive de São Tomé
L 24 cm (*c.* 9.5"). Distinctive. Long billed and leggy. **Adult** *olivaceofuscus* Head, upperparts and tail dark olive-brown, throat whitish spotted brown, breast pale brown merging into white with brown crescentic bars on rest of underparts. Bill dark horn. *T. o. xanthorhynchus* more distinctly marked below; bill bright yellow. **Juvenile** Duller; wing-coverts tipped pale buff; bars on underparts less distinct. **VOICE** Song, a few short, full whistles followed by sharper, higher pitched notes, e.g. *tyoo-heeo hwit hwit tyoo-heeo heeo hwit hwit hwit ...* with a few slight variations; monotonously repeated in unhurried, rather rhythmical series. Alarm, soft *whup* notes. [CD10:39] **HABITS** In all forested habitats, incl. plantations and patches of dry woodland in savanna. Rarely far from cover of tall trees. Usually on or near ground but occasionally high in trees. Most commonly observed towards dusk. Wary but sometimes permits close approach. **SIMILAR SPECIES** None. The only thrush within its restricted range. **STATUS AND DISTRIBUTION** Endemic to São Tomé and Príncipe. Nominate widespread and common on São Tomé, but Príncipe race *xanthorhynchus* not seen since 1928, although discovery of thrush anvils in 1996 indicate that it survives in forested south-west. *TBW* category: Near Threatened.

AFRICAN THRUSH *Turdus pelios*   Plate 91 (8)
Merle africain

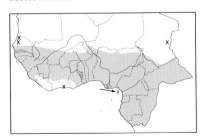

Other name F: Grive grisâtre
L 23 cm (9"). **Adult** Typical thrush, ashy-brown above, paler below, with *conspicuous yellow bill*. Throat variably streaked dusky; breast pale ashy-brown becoming whitish on belly. Eye dark. Legs olive-brown. Six races in W Africa differ in plumage coloration. Western *chiguancoides* palest, with whitish throat and pale orange-buff wash to flanks and underwing-coverts; gradually merges into *saturatus*, which is overall slightly deeper coloured and has more heavily streaked, brownish throat. **Nominate** like *chiguancoides* above, but has orangey flanks and more distinct streaks on throat; *centralis* as nominate, but upperparts and breast slightly darker. Remaining races darker and browner, with very restricted range; *nigrilorum* darkest, without orange on flanks; *poensis* slightly paler. **Juvenile** Similar, but wing-coverts tipped pale chestnut; breast and flanks mottled greyish-brown. **VOICE** Song far carrying, melodious and variable, with series of short repetitions of simple motifs, e.g. *toolee toolee toolee weetyuuwee weetyuuwee teewit teewit teewit swrreep swrreep leepoo leepoo leepoo churp churp churp...* etc. Call a hard *chuk*, often repeated in a dry series *chukukukukuk...* [CD10:40] **HABITS** Singly or in pairs in variety of wooded habitats, incl. cultivation with some trees, gardens, wooded savanna and patches of montane forest. Not in rainforest. Forages on ground, like congeners. **STATUS AND DISTRIBUTION** Widespread and common to not uncommon resident. *T. p. chiguancoides* Senegal to N Ghana; *saturatus* Ghana to W Congo; *pelios* Chad, E Cameroon and N CAR; *centralis* S CAR and E Congo; *nigrilorum* Mt Cameroon; *poensis* Bioko.

SONG THRUSH *Turdus philomelos*
Grive musicienne

Plate 91 (4)

L 23 cm (9"). *T. p. philomelos.* Upperparts warm brown; boldly spotted underparts cream-white washed buff on breast and flanks. Legs pinkish. In flight has orange-buff underwing-coverts. Flight fast and direct. **VOICE** In flight, a short dry *tsik.* [CD3:99] **STATUS AND DISTRIBUTION** Rare Palearctic winter visitor, Mauritania (Nov–Feb). Vagrant, N Senegal, N Mali, E Chad and Cape Verde Is.

RING OUZEL *Turdus torquatus*
Merle à plastron

Plate 91 (3)

L 23–24 cm (9.0–9.5"). *T. t. torquatus.* **Adult male** Easily identified by black plumage with white crescent-shaped breast-band and whitish edges to wing feathers. Bill mainly yellow with blackish tip. Legs brown. **Adult female** Browner with, typically, less distinct, brownish-white breast-band (absent in 1st year). Bill less yellow. In fresh plumage (autumn) both sexes have pale fringes to underparts feathers, producing scaly effect. **VOICE** Call a loud *chak* or *chakchakchak.* [CD4:2a] **STATUS AND DISTRIBUTION** Palearctic vagrant, Mauritania (Nouakchott, Jan–Feb). **NOTE** Race unspecified, but probably nominate, from N and W Europe (*T. t. alpestris* from C, S and E Europe has feathers of upperparts with much broader pale fringes, giving paler overall appearance; winters in N Africa and Sudan).

WARBLERS
Family SYLVIIDAE (World: *c.* 390 species; Africa: *c.* 205; WA: 111–112)

A large, varied family of principally small species, reaching its highest diversity in Africa; some genera of uncertain affinity. Sexes similar, except in *Sylvia* and *Apalis*; juveniles unspotted. Mainly insectivorous. Broad range of vocalisations. Occur in all habitats. In W Africa, 26 species are of Palearctic origin. Some establish feeding territories in their winter quarters and show remarkable site fidelity over years. Correct identification should usually be based on a combination of plumage characters, habitat, behaviour, voice and distribution.

Genus *Bradypterus*

Drab-coloured skulkers occurring in reedbeds, rank herbage, scrub and forest undergrowth. Songs distinctive.

LITTLE RUSH WARBLER *Bradypterus baboecala*
Bouscarle caqueteuse

Plate 97 (6)

Other names E: African Sedge Warbler; F: Bouscarle des marais
L 15 cm (6"). Dark brown swamp dweller with broad, steeply graduated tail, characteristic of its genus. **Adult** *centralis* Top of head and upperparts dark olive-brown washed rufous on rump and uppertail-coverts; narrow buffish supercilium; brownish cheeks. Tail broad, often worn and appearing tatty. Underparts whitish faintly streaked dusky on upper breast; flanks and undertail-coverts dull rufous-brown. Bill blackish-horn with pale base to lower mandible. Legs brownish to greyish-pink. *B. b. chadensis* lacks streaks on throat. **Juvenile** Underparts tinged yellowish. **VOICE** Song a distinctive, loud series of identical, dry, single notes, accelerating at end and ceasing abruptly *kruk, kruk kruk kruk-krukukukukukuk*; likened to sound of stick held in spokes of wheel turning faster and faster, then stopped. Alarm a nasal *meew.* [CD10:43] **HABITS** Singly or in pairs in reedbeds and dense aquatic vegetation. Secretive and hard to see; may, however, perch in the open, esp. in early morning. Display flight low and bouncing on whirring wings, only over short distances before diving into cover. **SIMILAR SPECIES** *Acrocephalus* warblers occur in same habitat, but have much paler brown upperparts. **STATUS AND DISTRIBUTION** Rare and local resident, recorded from Nigeria (Onitsha, Kazaure, Kano, Obubra) and Cameroon (*centralis*), W Chad (*chadensis*) and S Congo (ssp.?). Claimed from N Ivory Coast. Perhaps overlooked due to skulking behaviour.

DJA RIVER WARBLER *Bradypterus grandis*
Bouscarle géante

Plate 97 (3)

Other names E: Ja River Scrub-Warbler; F: Bouscarle du Dja
L 18.5 cm (7.25"). Little known; resembles Little Rush Warbler but much larger, with streaks on underparts more

distinct and not confined to upper breast but extending to entire throat, and breast-sides and flanks washed darker brown. Eye brown. Bill blackish with pale base to lower mandible. Legs greyish. **Juvenile** No information. **VOICE** Song consists of 4 introductory notes, increasing in speed, followed by a drawn-out trill *sweep-sweep-sweepswip-rrrrrurrrr*. [CD10:48] **HABITS** Skulks near or on ground, creeping mouse-like through dense rank herbage. Occurs in contact zones between forest and open areas such as savannas and river borders. Very difficult to see but song reveals presence. **SIMILAR SPECIES** Compare Little Rush Warbler. Bangwa and Evergreen Forest Warblers occur at higher altitudes and in forest. **STATUS AND DISTRIBUTION** Rare resident, recorded from a few localities in S Cameroon, SW CAR and Gabon. Perhaps commoner than records suggest. *TBW* category: THREATENED (Vulnerable).

EVERGREEN FOREST WARBLER *Bradypterus lopezi* Plate 97 (4)
Bouscarle de Lopes

Other names E: Cameroon Scrub-Warbler; F: Bouscarle brune
L 13.0–14.5 cm (5.0–5.75"). Sombre-coloured skulker with pale buff supercilium and long, strongly graduated, usually heavily worn tail. **Adult *camerunensis*** Top of head, upperparts and tail dark chestnut-brown to olive-brown. Underparts paler, with throat whitish (sometimes pale brown) and belly becoming whitish. Eye brown. Bill blackish-brown with paler lower mandible. Legs brownish. ***B. l. manengubae*** more olive below. **Nominate** more cinnamon below, esp. on throat and belly. **Juvenile** Throat and belly olive-yellow. **VOICE** Song a vigorous, rapid, rhythmic series of 3–9 (sometimes more) identical single or double notes, uttered by both sexes. Speed of delivery and notes variable, e.g. *chip-chip-chip...*, *chip-o-chip-o-chip...*, *chipipipipipip*, *chupchup-chupchup-chupchup-...*, *twee-tweet-twee-twee-...*, *tweetotweetotwee...*, *tsuweew-tsuweew- tsuweew-...* etc. In duet female utters a weak, high-pitched, drawn-out whistle *wheee-ooo*. Calls include a hard *chrrk*, a vibrant *pirr* and a loud *klik* (alarm). [CD10:49–50] **HABITS** Usually in pairs, in undergrowth of montane forest. Secretive, like congeners, keeping close to ground and moving restlessly through understorey, tangles and thickets by tracks and clearings. Very difficult to observe, but betrays presence by loud song. **SIMILAR SPECIES** In poor light and seen from above, Bangwa Forest Warbler may appear similar, but bright cinnamon underparts with white throat and belly should prevent misidentification. **STATUS AND DISTRIBUTION** Not uncommon but highly local forest resident in montane areas of Cameroon (Mt Cameroon: *camerunensis*; Mt Manenguba: *manengubae*; Kumbo and Mt Oku: undescribed form (Stuart & Jensen 1986)) and Bioko (nominate).

BANGWA FOREST WARBLER *Bradypterus (lopezi) bangwaensis* Plate 97 (5)
Bouscarle du Bangwa

L 14–15 cm (5.5–6.0"). **Adult** Secretive and very rufous looking. Upperparts and tail russet-brown; pale supercilium. Underparts bright rufous-cinnamon with white throat and belly. Eye brown. Bill blackish-brown, lower mandible paler. Legs brownish. **Juvenile** Duller, with some yellow on underparts. **VOICE** Song a loud, rhythmic series of identical notes, increasing in volume; speed variable (sometimes reminiscent of slow machine-gun fire). Female sometimes forms duet with high-pitched whistled notes. Alarm or excitement call a rattling *krrr*. [CD10:51] **HABITS** Usually in pairs, in montane areas. Occurs in shrubbery along forest tracks and clearings, edges of riparian forest, rank herbage in forest plantations, tall thick grass, scrub, bracken and brambles. Usually keeps low to ground in dense vegetation. **SIMILAR SPECIES** Evergreen Forest Warbler (q.v.) may, in poor light, appear similar, but is much more sombre coloured, esp. on underparts; also more restricted to forest. **STATUS AND DISTRIBUTION** Locally not uncommon to common resident in highlands of SE Nigeria (Obudu and Mambilla Plateaux) and W Cameroon (Mt Manenguba, Bamenda Highlands). *TBW* category: Near Threatened. **NOTE** Considered a race of *B. lopezi* by *BoA* and lumped with extralimital *B. cinnamomeus* (Cinnamon Bracken Warbler) by Sibley & Monroe (1990, 1993). Here treated as distinct species, following Dowsett & Forbes-Watson (1993) and *TBW*. More fieldwork in limited area of overlap with *lopezi* required.

BLACK-FACED RUFOUS WARBLER *Bathmocercus rufus* Plate 97 (10)
Bathmocerque à face noire

Other name F: Rousselette à face noire
L 13 cm (5"). *B. r. rufus*. Distinctive but secretive warbler, exhibiting marked sexual dimorphism. **Adult male**

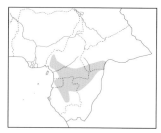

Forehead, head-sides, throat and central breast black; *upperparts, tail and breast-sides rufous-chestnut;* flanks and belly grey. Eye dark brown. Bill black. Legs greyish. **Adult female** Rufous-chestnut replaced by grey (dark olivaceous-grey on upperparts and tail, paler on underparts); breast-sides with some creamy-buff. **Immature male** Head dull brownish-grey; upperparts dull greyish-chestnut (dull rufous on edge of wing); tail dull rufous-chestnut. Underparts wholly olive-grey. Bill dark above, horn-coloured below. Legs paler than adult. **Immature female** Uniformly drab, with head, upperparts and tail dull brownish-grey. **Juvenile** Mainly olive-green. **VOICE** Song consists of distinctive, monotonous series of penetrating, high-pitched whistles which are hard to locate, e.g. *HEEEEET, HEEEEET, HEEEEET, ... ; HEET-HEET-HEET-HEET-... ; HEE-HEE hu-EE HEE hu-EE HEE hu-EE HEE ...* (duet; sounding like unoiled wheelbarrow); in another duet male whistles while female utters dry rattling *chrrk* notes. Male also sings alone. Call *chip*. Members of a party utter muffled notes. [CD11:30] **HABITS** In dense undergrowth of lowland and montane forest (up to 1700 m). Behaviour as Black-headed Rufous Warbler. Very vocal. **STATUS AND DISTRIBUTION** Locally not uncommon forest resident, Cameroon, Gabon, N Congo and SW CAR.

BLACK-HEADED RUFOUS WARBLER *Bathmocercus cerviniventris*   Plate 97 (9)
Bathmocerque à capuchon

Other names F: Rousselette/Fauvette aquatique à capuchon
L 13 cm (5"). Rather thickset with distinctive coloration and strongly graduated, often cocked tail. **Adult male** Head and throat to central breast black; upperparts and breast-sides rufous-brown; flanks and belly cinnamon. Eye chestnut. Bill black. Legs dark greyish. **Adult female** Probably inseparable from male (differences insufficiently known). **Immature** Generally paler than adult with whitish throat. **Juvenile** Unknown. **VOICE** Song simple but variable, consisting of endlessly repeated series of 2–3 clear, high-pitched whistles, e.g. *pee-pee hwu, pee-pee hwu, pee-pee hwu...* and *peet-whuu whut, ...; tiuuu-tiuu whu, ...; pwhee-whu, ... ; piuu-thee, ...* etc. [CD11:29] **HABITS** Usually in pairs in lowland and gallery forest. Frequents dense undergrowth along paths and glades, often near forest streams and creeks. Difficult to observe as it keeps low to ground in densest vegetation, but presence revealed by song. **STATUS AND DISTRIBUTION** Rare to uncommon, extremely local forest resident, Sierra Leone (uncommon), Guinea (1 record), Liberia (uncommon/not uncommon), Ivory Coast (5 localities) and Ghana (1–2 records). *TBW* category: Near Threatened.

AFRICAN MOUSTACHED WARBLER *Melocichla mentalis*   Plate 97 (1)
Mélocichle à moustaches

Other names E: Moustached Grass-Warbler; F: Grande Fauvette à moustaches
L 19–20 cm (7.5–8.0"). *M. m. mentalis.* Large, plump *warbler* with *fairly long, broad tail.* **Adult** Head and upperparts plain earth-brown except rufous-brown forehead and whitish supercilium; *black malar stripe* contrasts with white throat; rest of underparts buff. Tail blackish-brown with paler tips. *Eye yellowish.* Bill heavy, blackish above, pale horn below. Legs greyish. **Juvenile** Lacks rufous on forehead; wing-coverts tipped buff; underparts somewhat mottled; eye dark. **VOICE** Song a cheerful and vigorous *tup-tup-twiddle-diddle-dee.* Call a hoarse, querulous *pièh-pièh-pièh-...* [CD11:38] **HABITS** In rank herbage in savanna and dry coastal areas. Usually hidden in tall grass or scrub but loud, distinctive song attracts attention. Sometimes quite confiding, singing in the open. **SIMILAR SPECIES** Broad-tailed Warbler is noticeably smaller, lacks malar stripe and has entirely different vocalisations. **STATUS AND DISTRIBUTION** Locally common to uncommon resident, from S Senegal to Ivory Coast east to S Chad and CAR; also S Gabon and Congo. Scarce wet-season visitor, S Mauritania.

BROAD-TAILED WARBLER *Schoenicola platyura*   Plate 97 (2)
Graminicole à queue large

Other names E: Fan-tailed Grassbird; F: Fauvette à large queue
L 17 cm (7"). *S. p. alexinae.* Brown with *long, broad, blackish-brown tail.* **Adult** Upperparts brown; indistinct buffish supercilium. Underparts whitish washed buffish on breast and flanks; elongated, brown undertail-coverts and tail

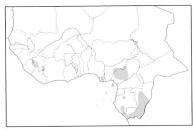

feathers tipped buffish. Bill strong, blackish above, greyish below. Eye brown. Legs pinkish. **Juvenile** Underparts yellowish. **VOICE** A slow, deliberate series of similar, high-pitched whistles *heeet, heeet, heeet, ...* In song flight a dry *tzit...tzit...tzit...* [CD11:57] **HABITS** Singly or in pairs within rank vegetation and tall grass in moist areas. Usually hidden but occasionally clambers to perch in very upright position atop grass stem. Song flight up to 15 m above ground in broad circles, tail fanned and jerking up and down, wings snapping. Flushes with difficulty, 'tail-heavy' bird flying only short distance on jerky wings before diving into cover, rarely flushing a second time. **SIMILAR SPECIES** African Moustached Warbler is noticeably larger, has black malar stripe and different song. **STATUS AND DISTRIBUTION** Locally uncommon to common resident, recorded from Sierra Leone (Tingi Mts) and adjacent Guinea, Liberia/Guinea/Ivory Coast (Mt Nimba), SE Nigeria (Obudu and Mambilla Plateaux), W Cameroon (highlands), Gabon and Congo. **NOTE** Sometimes considered specifically distinct from Indian population, as *S. brevirostris,* but differences appear minor (*BoA*).

Genus *Locustella*

Dull-coloured Palearctic migrants, with rounded, almost graduated tails and long undertail-coverts. Skulk low in dense vegetation. Buzzing or reeling song (hence name).

GRASSHOPPER WARBLER *Locustella naevia* Plate 97 (7)

Locustelle tachetée

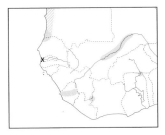

L 12.5 cm (5"). *L. n. naevia*. Secretive, darkish warbler with faint supercilium and graduated, rounded tail. Upperparts olive-brown with diffuse dark streaks. Underparts dirty white to pale yellowish; breast often washed brownish with some faint streaking; long dark streaks on rear flanks and undertail-coverts. Legs usually pinkish. **VOICE** Song (exceptionally heard in winter quarters) a distinctive, monotonous, insect-like reeling (hence name), uttered on constant high pitch; often compared to sounds of line running off angler's reel, or of small muffled alarm clock. Call a short *twit* or *pit.* [CD4:3b] **HABITS** Solitary and unobtrusive, keeping low to ground, walking or running mouse-like through dense vegetation. Flies reluctantly and hides for long periods. Flight flitting, low and only over short distances. Bobs tail in characteristic *Locustella*-fashion, thereby showing streaky undertail-coverts. In winter quarters recorded in rank herbage and regrowth. **STATUS AND DISTRIBUTION** Palearctic migrant (Sep–Apr) with main wintering range in W Africa. Locally common to not uncommon in W Mauritania and Liberia/Guinea (Mt Nimba); scarce in Senegambia, Sierra Leone and Mali. Vagrant, Ghana and Cape Verde Is. Probably overlooked due to skulking behaviour.

SAVI'S WARBLER *Locustella luscinioides* Plate 97 (8)

Locustelle luscinioïde

L 14 cm (5.5"). *L. l. luscinioides*. Resembles African and European Reed Warblers, but larger and with broad, rounded tail and curved primaries (straight in all *Acrocephalus*). Upperparts plain warm brown; narrow pale supercilium. Underparts whitish becoming buffish-brown on flanks; breast with patchy olive-brown wash; *undertail-coverts very long, cinnamon-buff, often with paler tips.* Bill rather long, black above, mainly yellowish below. Legs brownish or pinkish. **VOICE** Song (exceptionally uttered in winter quarters) a far-carrying insect-like reel similar to that of Grasshopper Warbler but faster, lower pitched and more buzzing. Call a sharp *tswik.* [CD4:4] **HABITS** Solitary. In winter quarters in dense scrub and herbage near or in water. Behaviour skulking and furtive, typical of its genus. **SIMILAR SPECIES** Compare *Acrocephalus* warblers, esp. African and European Reed Warblers (smaller; plumage paler and brighter brown, tail less rounded; undertail-coverts shorter; legs darker). **STATUS AND DISTRIBUTION** Rare to scarce Palearctic winter visitor (Oct–Apr) patchily recorded from narrow belt in S Mauritania and Senegal to S Mali, N Ghana, N Nigeria and N Cameroon. Vagrant, Chad and Cape Verde Is.

Genus *Acrocephalus*

Robust, dull-coloured warblers with mainly brown upperparts, paler underparts, rounded tails and prominent bills. Occur in reedbeds, marshes and rank vegetation; Palearctic migrants also in less typical habitats, far from water.

AQUATIC WARBLER *Acrocephalus paludicola*   Plate 98 (2)
Phragmite aquatique

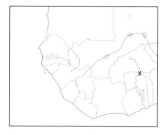

L 13 cm (5"). Boldly marked yellowish-buff *Acrocephalus*. **Adult** Bold creamy-buff supercilium and sharply defined median crown-stripe; black stripe behind eye; lores usually pale. Upperparts (incl. rump) yellowish-buff heavily streaked blackish-brown (becoming greyish-brown with wear), mantle bordered by 2 broad buffish longitudinal stripes ('tramlines'); tail spiky. Underparts buffish-white; breast-sides and flanks washed buff and finely streaked. Bill dark brown-horn, base of lower mandible pale pinkish. Legs pale pinkish. **First-winter** Bright straw coloured with no or very little streaking on underparts. **VOICE** Song (perhaps exceptionally uttered in winter) includes call alternated with short series of rapidly repeated notes e.g. *krrr hweeweeweeweeweewee, krrr diidiidiidiidii, krrr dudududududu kirrr....*etc. Does not mimic. Call a sharp *krrr krrr.* [CD4:5] **HABITS** Solitary, occurring low in beds of reeds and sedges by lakes and pools, wet grasslands and similar habitats. Flight low and jerky. **SIMILAR SPECIES** First-winter Sedge Warbler may also have median crown-stripe but this is always less sharply defined; also has dark lores (producing 'sterner' expression), pale rufous-brown, unstreaked rump and more rounded tail feathers. **STATUS AND DISTRIBUTION** Rare Palearctic winter visitor (Sep–Mar), recorded from Mauritania, Senegal (Senegal delta) and Mali. Vagrant, N Ghana. Winter quarters inadequately known. *TBW* category: THREATENED (Vulnerable).

SEDGE WARBLER *Acrocephalus schoenobaenus*   Plate 98 (1)
Phragmite des joncs

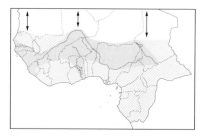

L 13 cm (5"). Easily distinguished from other small *Acrocephalus*, except Aquatic Warbler, by *conspicuous creamy-buff supercilium* bordered by dark eye-stripe, and *streaked upperparts.* **Adult** Dark, heavily streaked crown. Upperparts brown delicately streaked blackish; rump unstreaked rusty-brown contrasting with dark brown tail. Underparts buffish-white washed rusty-buff on breast-sides and flanks. Bill blackish, base of lower mandible yellowish. Legs greyish. **First-winter** Often has ill-defined creamy median crown-stripe and faintly speckled or spotted breast; also paler and more buff than adult; legs paler, more yellowish-grey. **VOICE** Song a fast series of extremely varied rasping and musical notes, most repeated several times, including clear trills and imitations. Call *tuk*. Alarm *trrr*. [CD4:6] **HABITS** Solitary. In reedbeds, rank vegetation by rivers and lakes, inundated grassland and similar habitats. Often difficult to observe in dense vegetation, but not shy and frequently forages in the open. Inquisitive and attracted to imitation of call. **SIMILAR SPECIES** Aquatic Warbler is yellower and more heavily marked, with creamy median crown-stripe, pale lores, buffish, streaked rump, and spiky, more graduated tail. **STATUS AND DISTRIBUTION** Locally common Palearctic winter visitor (Sep–May) throughout.

EUROPEAN REED WARBLER *Acrocephalus scirpaceus*   Plate 98 (7)
Rousserolle effarvatte

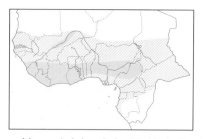

Other name E: Eurasian Reed Warbler
L 12.5–13.0 cm (*c.* 5"). *A. s. scirpaceus.* Plain brown above with *contrasting rusty-tinged rump.* Rather flat forehead and crown in combination with long thin bill produce tapering profile. Pale supercilium faint and short; narrow whitish eye-ring. Underparts whitish with rusty-brown wash on breast-sides and flanks. Bill dark brown above, pinkish to yellowish-horn below. Legs greyish-brown. **VOICE** Song a series of slow, grating and nasal syllables repeated 2–5 times e.g. *kerr-kerr-kerr kirrik-kirrik-kirrik peet-peet-peet trer-trer-trer tjetje-tjetje...*, occasionally including imitations. Less rasping, slower and less varied than Sedge Warbler's. In winter quarters song is usually more subdued. Call *krrrr* or *chrrr* and

chrrrt. Alarm a sharp *tek, tek* and a scolding rattle *krrrt.* [CD4:7] **HABITS** Frequents dense dry herbage, riverside thickets with tall grass, scrub, cassava plantations and gardens. Solitary. Skulking but not shy; attracted to imitations of its call. **SIMILAR SPECIES** African Reed Warbler is virtually indistinguishable in field. Savi's Warbler is larger, with broad, rounded tail and curved wing. See also Marsh Warbler. **STATUS AND DISTRIBUTION** Scarce to common Palearctic passage migrant and winter visitor (Sep–May) almost throughout, from Mauritania to Liberia east to CAR and Gabon.

AFRICAN REED WARBLER *Acrocephalus baeticatus*   Plate 98 (6)
Rousserolle africaine

L 12.0–12.5 cm (*c.* 5"). Very similar to European Reed Warbler and virtually indistinguishable in field but slightly smaller with shorter, rounder wing not extending beyond rump. Plumage slightly brighter coloured (***cinnamomeus***), or a shade darker with slight grey tinge (***guiersi***). **Juvenile** As adult. **VOICE** Song and calls indistinguishable from those of European Reed Warbler (q.v.). [CD10:52] **HABITS** Breeds in reedbeds and similar wet or moist habitats; migrants also frequent drier habitats and may occur alongside European Reed Warbler. **SIMILAR SPECIES** See European Reed Warbler. **STATUS AND DISTRIBUTION** Patchily distributed, locally not uncommon resident and intra-African migrant, recorded from Mauritania, Senegambia, Mali (central Niger delta), Niger, Nigeria, Chad, Cameroon and Gabon. *A. b. cinnamomeus* throughout range; *guiersi* described from N Senegal. Status inadequately known due to similarity to *A. scirpaceus*. **NOTE** Treated as conspecific with *A. scirpaceus* by some authors on basis of vocal and other evidence (cf. Dowsett & Dowsett-Lemaire 1993).

MARSH WARBLER *Acrocephalus palustris*   Plate 147 (8)
Rousserolle verderolle

L 12.5–13.0 cm (*c.* 5"). Very similar to European Reed Warbler and extremely difficult to separate when not singing. Typically somewhat paler with more uniformly coloured, olive grey-brown upperparts (lacking warm brown tones) and distinctly less rusty-tinged rump (usually more concolorous with rest of upperparts). Underparts whiter, with breast-sides and flanks buffish rather than tinged rusty-brown. Wings appear slightly longer, with narrow pale tips to 7–8 primaries and narrow pale edgings to secondaries and tertials when fresh (but these rather distinctive features disappear with wear!). More rounded head and slightly shorter bill create more gentle impression. Narrow eye-ring usually more distinct. Legs usually paler (esp. in 1st-winters), often orangey to pale yellowish (darker, greyish-brown or brownish-pink in European Reed Warbler). **VOICE** Song extremely rich and varied, including many imitations and liquid trills. Call a sharp *tek* or *chek* and a low *tuk.* Alarm a grating *krrr.* [CD10:53] **HABITS** As European Reed Warbler. **SIMILAR SPECIES** European and African Reed Warblers (q.v. and above). **STATUS AND DISTRIBUTION** Palearctic passage migrant and winter visitor in E and S Africa. Vagrant, N Senegal (1 record, Djoudj NP, Jan 1994) and NE Nigeria (two, Sep-Oct 2000). Previous claims from Nigeria do not adequately eliminate European Reed Warbler.

GREAT REED WARBLER *Acrocephalus arundinaceus*   Plate 98 (3)
Rousserolle turdoïde

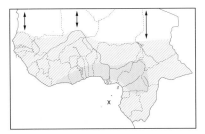

L 19–20 cm (*c.* 7.5"). *A. a. arundinaceus.* Like very large and bulky European Reed Warbler, with much stouter bill, angular head and more prominent supercilium. **Adult** Upperparts plain brown; tail broad and rounded. Underparts whitish washed brownish-buff on breast-sides and flanks. Bill dark brown with pinkish base to lower mandible. Legs brownish-grey. **First-winter** More rusty above and orange-buff below. **VOICE** Song a rather slow series of far-carrying, low, raucous, grating notes interspersed with high squeaks, repeated 2–3 times e.g. *krr-krr karra-karra-karra-keet-keet, kruruk-kruruk kreek-kreek...* In winter quarters utters both subdued and full song. Call a hard *krek.* [CD4:9] **HABITS** Frequents variety of habitats, incl. scrub, wooded grassland, dense herbage, cassava plantations and swamps. Solitary and not shy. **SIMILAR SPECIES** Less heavy Greater Swamp Warbler is darker overall and greyer below, lacks clear supercilium and has more slender bill, shorter primary projection and different song. Lesser Swamp Warbler is much smaller. **STATUS AND DISTRIBUTION** Palearctic winter visitor (Sep–May) throughout; also São Tomé. Scarce in west, more common further east.

GREATER SWAMP WARBLER *Acrocephalus rufescens*
Rousserolle des cannes

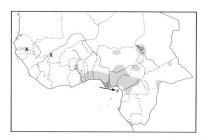

L 18 cm (*c.* 7"). **Adult** *rufescens* Robust, recalling Lesser Swamp Warbler but larger and heavier, with darker, cold brown upperparts, faint and greyish supercilium, and longer bill. Throat and central belly white, rest of underparts pale greyish-brown washed tawny on flanks. Bill dark brown with yellowish base to lower mandible. Legs dark greyish. *A. r. senegalensis* and *chadensis* slightly paler above, whiter below; former greyer on head. **Juvenile** Plumage tones warmer; upperparts more tawny-brown; underparts washed pale tawny. **VOICE** Song consists of loud croaking, churring and gurgling notes, very similar to song of Lesser Swamp Warbler, but slightly lower pitched and more guttural. Calls include a harsh *chrr*, a low *kreeok* and *chok!* [CD10:55] **HABITS** Occurs in reedbeds and marshes; also adjacent sugarcane plantations. Solitary and secretive but inquisitive. **SIMILAR SPECIES** Lesser Swamp Warbler has white supercilium and whiter underparts with rufous on flanks; also smaller. Great Reed Warbler has warmer brown upperparts, obvious and pale supercilium, stouter bill and longer primary projection. **STATUS AND DISTRIBUTION** Patchily distributed resident, recorded from S Mauritania, NW Senegal, Gambia (*senegalensis*), Mali, Ghana, Togo, Nigeria, Cameroon, N CAR, Gabon, Congo, Bioko (nominate) and L. Chad area (*chadensis*). Not uncommon but local in west of range, more common and widespread in east.

Plate 98 (5)

LESSER SWAMP WARBLER *Acrocephalus gracilirostris*
Rousserolle à bec mince

L 14–15 cm (*c.* 5.5–6.0"). *A. g. neglectus*. Noticeably larger than European and African Reed Warblers, with darker brown upperparts, distinct whitish supercilium and whiter underparts; breast-sides and flanks dull rusty. Bill blackish-horn, base of lower mandible yellowish or pink. Legs dark grey or black (darker than in most other *Acrocephalus*). **Juvenile** Upperparts warmer, more tawny. **VOICE** Song consists of a few introductory notes followed by a fast trilling or bubbling series e.g. *chok chok djujujujujuju* or *klok klik djidji gligligligligligli* and *twip prr-titititititti trrr chroorreee* and variations. Calls include a hard *chuk!*, a low *cheruk* and a harsh *chaa*. [CD10:54] **HABITS** Singly or in pairs in reedbeds and similar vegetation along lakes, pools, marshes, etc. Forages low. Inquisitive and highly vocal in breeding season, when easier to observe. **SIMILAR SPECIES** European and African Reed Warblers (q.v.). Greater Swamp Warbler lacks clear supercilium and has darker upperparts, greyish-brown breast-sides and flanks, longer bill and larger size; song slightly lower pitched and more throaty. **STATUS AND DISTRIBUTION** Rare resident, recorded from N Nigeria (Jekara) and Chad (north side of L. Chad). Claim from N CAR doubtful.

Plate 98 (4)

CAPE VERDE WARBLER *Acrocephalus brevipennis*
Rousserolle du Cap-Vert

Plate 143 (4)

Other names E: Cape Verde Cane Warbler; F: Rousserolle des Îles du Cap-Vert
L 13.5 cm (5.25"). **Adult** Top of head brown-grey becoming paler and greyer on head-sides and neck, and merging into greyish-brown of upperparts; rump brighter. Short, faint pale supercilium. Underparts white washed greyish on breast-sides and flanks. Wings notably short and rounded. Bill long, dark above, pale below. Legs slate-grey. **Juvenile** As adult but plumage warmer, less grey; above tinged rufous or cinnamon; below more creamy-white washed buffish on flanks. **VOICE** Song short and distinctive with harsh quality and including loud liquid bubbling and trills, reminiscent of closely related Greater Swamp Warbler and/or Common Bulbul. Call a low, hard *kruk* or *kerr*. [CD4:10] **HABITS** Favours well-vegetated valleys, esp. with patches of reeds, up to 500 m (mostly lower); also in sugarcane and banana plantations, and gardens, usually near water. Secretive when not breeding. Occasionally in small parties, foraging in fruiting fig trees. **SIMILAR SPECIES** None within its restricted range. **STATUS AND DISTRIBUTION** Endemic resident, Cape Verde Is, confined to Santiago (local; max. 500 pairs) and São Nicolau (8 territories in 1998). Apparently extinct on Brava. *TBW* category: THREATENED (Endangered).

AFRICAN YELLOW WARBLER *Chloropeta natalensis*
Chloropète jaune

Plate 98 (8)

Other names E: Yellow Flycatcher-Warbler; F: Fauvette jaune commune
L 14 cm (5.5"). **Adult male** *batesi* Upperparts and tail yellowish-brown, crown slightly browner, rump yellower;

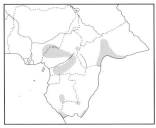

faint, short supercilium. Wing feathers olive-brown edged yellowish. *Underparts bright yellow*, breast-sides, belly and flanks washed olive. Eye brown. Bill dark above, pinkish or pale horn below. Legs blackish. **C. n. major** has crown and upperparts slightly paler and greener. **Adult female** Underparts duller. **Juvenile** As adult female, but more buffy. **VOICE** Song consists of a few dry notes followed by a fast series of throaty and varied notes, e.g. *tsk-tsk-tsk-tsk hihihihihi* or *chip-chip-chip reetreetreetreet* and *trrp-trrp chi-i-i-i-i-i* etc. Alarm a sharp *tsk!* or *chrr!* [CD10:57] **HABITS** Singly or in pairs within reeds, rank herbage near water or dense forest-edge vegetation, up to 2150 m. Secretive, foraging low in vegetation, but may clamber up vertical stem to sing from exposed perch. Dives into vegetation and creeps away if disturbed. Habits much like *Acrocephalus* warblers. Posture often upright. Raises crown feathers when excited. **SIMILAR SPECIES** Icterine Warbler and smaller Willow Warbler have paler upperparts and much less yellow underparts. **STATUS AND DISTRIBUTION** Patchily distributed, locally uncommon to not uncommon resident, from SE Nigeria (Obudu and Mambilla Plateaux) to Cameroon and CAR (*batesi*), and in Gabon and Congo (*major*).

Genus *Hippolais*

Plain coloured and arboreal, with prominent, rather broad, flattened bills and square tails. Palearctic migrants, except one local resident race.

OLIVACEOUS WARBLER *Hippolais pallida*   Plate 98 (9)
Hypolaïs pâle

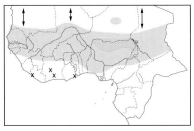

L 12.0–13.5 cm (4.75–5.25"). Strikingly pale and rather elongated with long bill and flat crown. Upperparts dull pale olive-grey to greyish-brown; short pale supercilium faint and typically not extending behind dark eye. Underparts whitish washed pale buff on breast-sides and flanks. Bill long, dark brown above, yellowish or pinkish below. Legs bluish-grey to pinkish-brown. Races distinguished by subtle differences in size, bill length and plumage tone but often hard or impossible to separate in field. Widespread *opaca* mostly greyish-olive above; eastern *elaeica* colder grey-brown above, whiter below; N African *reiseri* paler and sandier. Resident *laeneni* quite distinct: even paler than *reiseri* and greyer, esp. on rump; also smaller. **VOICE** Song a harsh chattering phrase, repeated constantly; somewhat reminiscent of European and African Reed Warblers. Calls include a frequently uttered hard *chek!* or *tak!* and *churr-churr.* [CD4:11] **HABITS** In wooded savanna, scrub, overgrown plantations and gardens. Resident race breeds in *Acacia* woodland. Solitary and restless, frequently pumping tail downwards. Often hides within foliage. **SIMILAR SPECIES** European and African Reed Warblers are browner with rusty-tinged rump, rounded tail and long undertail-coverts. Garden Warbler is more compact, with rounder head and much shorter bill. **STATUS AND DISTRIBUTION** Not uncommon to rare Palearctic passage migrant and winter visitor (late Aug–May; most *H. p. opaca*, also *reiseri* and *elaeica*), from Mauritania to Guinea east to Chad. Some remain year-round. Vagrant, Cape Verde Is (*opaca*, Sep). Race *laeneni* not uncommon resident and partial migrant in N Nigeria, Niger, Chad and N Cameroon; northernmost breeding grounds (Aïr, N Niger) vacated Nov–Jan.

OLIVE-TREE WARBLER *Hippolais olivetorum*   Plate 98 (11)
Hypolaïs des oliviers

L 15.5 cm (6"). Large, greyish warbler with long, heavy bill, rather flat, sloping crown, pale supraloral streak and pale wing panel. Upperparts dirty brownish-grey, wings darker with buff-white fringes to inner secondaries and tertials (forming pale wedge). Tail dark edged whitish. Underparts dirty white, occasionally washed pale yellowish on throat and breast. Bill greyish above, yellowish-pink below. Legs bluish-grey. **VOICE** Call a hard *tsek* or *tuk.* [CD10:59] **HABITS** In winter quarters frequents *Acacia* savanna and dry bush country. Unobtrusive. **SIMILAR SPECIES** Olivaceous Warbler is smaller with a less powerful bill. **STATUS AND DISTRIBUTION** Palearctic vagrant, N Niger (N Aïr, Aug) and N Nigeria (Kano, Oct). Claimed from N Cameroon (Maroua, Jan). Principal wintering grounds in E and S Africa.

MELODIOUS WARBLER *Hippolais polyglotta*   Plate 98 (10)
Hypolaïs polyglotte

L 13 cm (5"). Very similar to Icterine Warbler (q.v.) but typically browner green above and brighter yellow below;

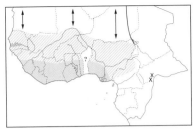

wings shorter (primary projection *c.* half of tertial length); *wing panel lacking or indistinct*; crown rounder. Bill mainly orange-pink. Legs brownish (grey in Icterine). **VOICE** Song a rapid, sustained and varied babble, faster, less harsh and more even than Icterine Warbler's, sometimes including imitations. Call a chattering, sparrow-like *tchèèèrr* or *tret-tret-...* [CD4:12] **HABITS** In scrub, overgrown gardens and plantations, and wooded savanna. Mostly solitary and rather skulking, but often not shy, feeding in the open. **SIMILAR SPECIES** Icterine Warbler (q.v. and above). **STATUS AND DISTRIBUTION** Common to uncommon Palearctic passage migrant and winter visitor (Aug–Apr). Mainly in west of region, east to Cameroon. Vagrant, CAR.

ICTERINE WARBLER *Hippolais icterina*   Plate 98 (12)
Hypolaïs ictérine

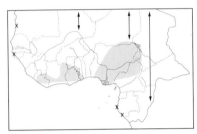

L 13.5 cm (5.25"). Bright olive-green and yellow. *Pale lores* and faint supercilium confer open-faced expression. **Adult** Differs from Melodious Warbler in *peaked* (not rounded) *crown* and *longer wings* (primary projection almost equal tertial length). Tail square. In autumn, prior to moult, may appear very washed out with pale grey-green upperparts and creamy underparts. Following moult in winter quarters has *pale wing panel* (formed by yellow fringes to inner secondaries and tertials). Bill relatively long (emphasised by sloping forehead) and mainly orangey. Legs grey. **First-winter** Duller, with greyish-green upperparts and pale underparts washed yellow. **VOICE** Song a very varied, vigorous and sustained jumble of harsh, melodious and discordant, often repeated notes, including a distinctive nasal and creaky *GEE-a* (recalling sound made by squeezing rubber toy) and imitations. In winter quarters often a quieter and less varied subsong. Calls a short hard *tek* and *deedeweet* or *hipperueet*. [CD10:60] **HABITS** Occurs in open woodland, gardens and scrub. Solitary and appearing rather lively and alert. Raises crown feathers when excited. **SIMILAR SPECIES** Melodious Warbler (q.v. and above). Willow and Wood Warblers are both smaller with strong facial patterns (lacking pale lores), smaller bills and rounded crowns; also have more restless behaviour. Garden Warbler has dull brown upperparts, pale buffish underparts and short stubby bill. European and African Reed Warblers have browner plumage and rounded tail. **STATUS AND DISTRIBUTION** Palearctic passage migrant and winter visitor (late Aug–mid-May). Mainly in east of region, but also (rarely) recorded from Mauritania, Mali, Guinea-Bissau, Ivory Coast, Ghana, Togo. Winters mainly south of equator.

Genus *Cisticola*

Large genus of similar-looking species, occurring in grassy habitats (hence the name 'grass warblers'). Plumage in various shades of brown, plain or patterned with black on upperparts, mainly whitish and variably tinged buff or brown on underparts. Tail strongly graduated, darkish coloured above, paler and greyer below, mostly with white (or pale) tips and black (or dusky) subterminal spots forming band (most conspicuous from below). Males usually larger than females; in larger species this difference is particularly important and noticeable in the field. Juveniles are browner and more rufous, like adult non-breeding, and often have yellow on underparts. Usually, but not invariably, breed during rains, which is best season to observe them, as males are singing and displaying; in dry season all are inconspicuous. In northernmost and southernmost areas distinct breeding and non-breeding plumages occur, but in central areas a perennial dress, with only one annual moult, is usual. In breeding plumage, males have darker or blackish bills; in non-breeding dress, the tail in some is slightly longer. Occur singly or in pairs, after breeding in small family groups. In the dry season some may form small parties, sometimes with other cisticolas.

Among the hardest species to identify on plumage characters alone, but vocalisations and aerial displays, if any, diagnostic. For correct identification, note size (small, medium or large), upperpart pattern (streaked or plain), head pattern (colour of crown and head-sides, presence or absence of streaking and supercilium), tail colour and pattern. Then check against known range.

RED-FACED CISTICOLA *Cisticola erythrops*   Plate 107 (3)
Cisticole à face rousse

L 12–14 cm (4.75–5.50"). *C. e. erythrops*. Medium-sized, plain-backed cisticola with *diagnostic rufous face*. **Adult breeding** Forehead, head-sides, incl. lores and supercilium, rufous; crown, upperparts and tail dark olivaceous-brown. Tail has broad whitish tips and black subterminal band. Underparts rusty-buff becoming whitish on throat and central belly. **Adult non-breeding** Upper- and underparts more russet. (This plumage perennial in Cameroon.)

Juvenile As adult breeding but duller and with face whitish (at most with slight rusty tinge). **VOICE** Song a rhythmic succession of loud, varied notes, e.g. *ch-ch-ch trweet-trweet-trweet WEET-WEET-WEET plik-up plik-up plik-up WEET WEET WEET* ...; uttered from low perch and usually in duet. Alarm call *cheap cheap cheap*. [CD10:90] **HABITS** Singly or in pairs in variety of grassy and bushy habitats. Forages low, skulking in rank herbage. **SIMILAR SPECIES** Singing Cisticola has rufous-brown crown, less rufous head-sides and rufous wing panel. **STATUS AND DISTRIBUTION** Locally common to uncommon resident, from Senegambia to Liberia east to Cameroon and S CAR (also Bamingui-Bangoran NP). Rare visitor, S Mauritania. Local, Gabon and S Congo.

SINGING CISTICOLA *Cisticola cantans*   Plate 107 (6)
Cisticole chanteuse

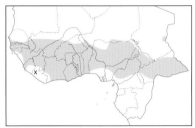

L 11.5–14.0 cm (4.5–5.5"). Medium-sized, plain-backed cisticola with chestnut crown and wing panel. **Adult *swanzii* breeding** Forehead to nape dull rufous contrasting with greyish-brown to earth-brown upperparts; head-sides pale brown; lores whitish; flight feathers edged rufous. Tail has dirty white tips and black subterminal band. Underparts pale buff becoming whitish on throat and central belly. *C. c. concolor* has upperparts paler, underparts rather brighter; in *belli* crown is richer rufous; *adamauae* intermediate between *swanzii* and *belli*. **Adult non-breeding** Similar, but slightly richer coloured. **Juvenile** Upperparts duller and more rusty-brown. **VOICE** Song, uttered from perch, a single, repeated, variable note, such as *krwleep krwleep krwleep...* or *p-lip p-lip p-lip ...* and *kwiplip kwiplip kwiplip ...* etc. Occasionally duets, female uttering low grating notes. Calls include a thin *tsit tsit* and a drawn-out *cheerr cheerr*. [CD10:91] **HABITS** In thick bushes and rank undergrowth; generally in drier areas than Red-faced Cisticola. Singly or in pairs; outside breeding season in small parties, sometimes forming mixed flocks with other cisticolas. **SIMILAR SPECIES** Red-faced Cisticola occurs in similar habitat, but adult has head-sides rufous, no rufous wing panel and different song. Juvenile has top of head and wing edging concolorous with upperparts (not more rufous); pale tips to tail feathers broader. **STATUS AND DISTRIBUTION** Common to not uncommon resident, from Senegambia to Sierra Leone east to Chad and CAR. *C. c. swanzii* from S Nigeria west; *concolor* from N Nigeria to Chad and N CAR; *adamauae* Cameroon and W CAR; *belli* S CAR.

WHISTLING CISTICOLA *Cisticola lateralis*   Plate 107 (2)
Cisticole siffleuse

L 12.5–14.0 cm (4.75-5.5"). Medium-sized, plain-backed cisticola, uniformly coloured above. **Adult male *lateralis* breeding** Head, upperparts and tail sooty greyish-brown to brown (browner in east); flight feathers edged rufous (difficult to see; most visible on grey individuals). Underparts whitish washed greyish on breast-sides, becoming buff or grey-buff on flanks and undertail-coverts; thighs rusty-brown. Tail tipped greyish-white with black subterminal band only visible from below. Eastern *antinorii* browner, less sooty above; southern *modesta* similar to *antinorii* but with greyer crown. **Adult male non-breeding** Above rufous-brown; below rich buff becoming whitish on throat and belly. **Adult female breeding/non-breeding** As male but smaller. **Juvenile** Upperparts dull rufous-brown; head-sides and throat tinged yellow. **VOICE** Song a short, vigorous and melodious whistled phrase, uttered from prominent perch. Those east of Nigeria have a more monotonous *thup thup thuthuthuthuw*. Call *thup*, often repeated. Alarm harsh and rasping. [CD10:87] **HABITS** Singly or in pairs in savanna woodland, overgrown farmland with scattered trees, bushy forest edge and large forest clearings with low scrub and tall trees. **STATUS AND DISTRIBUTION** Common to not uncommon resident in savanna zone and degraded habitats within forest zone. Nominate from Senegambia to Liberia east to Benin and across S Nigeria (also at Kagoro, C Nigeria) to Cameroon (Adamawa Plateau south to edge of forest zone) and CAR (also Bamingui-Bangoran NP, rare) where grading into *antinorii*; *modesta* in Gabon and Congo (rare).

CHATTERING CISTICOLA *Cisticola anonymus*   Plate 107 (8)
Cisticole babillarde

L 12–15 cm (4.75–6.0"). Medium-sized, plain-backed cisticola with dull rufous cap. **Adult** Forehead to nape dull

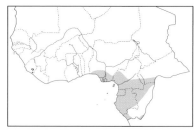

chestnut slightly contrasting with uniformly dark brown upperparts. Head-sides, incl. lores, rusty-buff; no supercilium. Underparts buff becoming creamy-white on throat and central belly. Tail tipped greyish-white, dark brown above with indistinct subterminal dusky spots, greyish-brown below, subterminal spots distinct. **Juvenile** Head-sides to upper breast yellow. **VOICE** Song, uttered from low perch, a loud repeated phrase of 2–3 harsh notes followed by a bubbling trill *ch-ch-twurrrrlp* and (fast, abrupt) *tetete-tchrr*, with variations. Similar to song of Rattling Cisticola. Various harsh call notes. [CD10:89] **HABITS** A noisy and easily observed species, occurring in pairs or small parties in grassy patches with scattered bushes in forest clearings, low second growth and near tracks, farmland and rivers; also in wet grassland bordered with thickets. Absent from savanna belt. **SIMILAR SPECIES** Bubbling Cisticola is like a pale version of Chattering, without rusty tinge on face. Rattling Cisticola has very similar plumage and song but is a savanna species with rufous-brown edges to flight feathers and no rusty tinge to face. **STATUS AND DISTRIBUTION** Locally common resident in lowland forest zone, from S Nigeria to SW CAR and Congo. Birds resembling this species claimed to be locally not uncommon in SE Sierra Leone.

BUBBLING CISTICOLA *Cisticola bulliens* Plate 107 (7)
Cisticole murmure

L 12.0–15.5 cm (4.75–6.0"). *C. b. septentrionalis*. Medium-sized, plain-backed cisticola with relatively long, protruding bill. **Adult male breeding** Forehead to nape earth-brown tinged rufous, not contrasting with rest of earth-brown upperparts. Head-sides, incl. lores, whitish. Underparts pale buff becoming whitish on throat and central belly. Tail earth-brown with broad white tips and black subterminal spots. **Adult male non-breeding** Warmer coloured overall. Top of head paler and more rusty; upperparts paler with indistinct dappling. **Adult female** As male, but noticeably smaller. **Juvenile** As adult non-breeding, but duller and head-sides to upper breast slightly tinged yellow. **VOICE** Song, uttered from elevated perch, described as 3 sharp notes followed by a bubbling flourish *di-di-di drrreee*; last part occasionally in flight. Apparently very similar to song of Chattering Cisticola. **HABITS** In brackish meadows with coarse grass and bushes, open grassy woodland, scrub and palm groves. **SIMILAR SPECIES** Chattering Cisticola has rusty-buff (not whitish) face. Rattling Cisticola very like Chattering but flight feathers edged rufous-brown. **STATUS AND DISTRIBUTION** Unclear. Not recorded with certainty from our region. Claims from S Gabon (Mouila; Lékoni) require confirmation. Known range: Cabinda and Lower Congo R. to Angola.

CHUBB'S CISTICOLA *Cisticola chubbi* Plate 107 (4)
Cisticole de Chubb

L 14 cm (5.5"). Distinctive, *vocal highland cisticola* with rusty cap and plain back. **Adult *adametzi*** Forehead to nape rich rusty-brown; rest of upperparts plain dull brown; lores black. Tail dark brown with white tips and dark subterminal spots. Underparts brownish-grey to greyish-brown on breast, becoming buff on flanks and undertail-coverts and whitish on throat and central belly. *C. c. discolor* has almost plain tail, with dark spots obscure, almost absent. **Juvenile** Duller, with upperparts slightly more rufous (cap less contrasting). **VOICE** Song an easily recognised, loud and explosive duet of rapidly repeated phrases, such as *switch-a-bee switch-a-bee switch-a-bee* ...or *sweet-a-pitoo sweet-a-pitoo...* and *sweet-wit sweet-wit...* [CD10:92-93] **HABITS** In pairs or family parties in dense herbage of forest clearings and edge, grassy areas of abandoned cultivation with scattered bushes and trees, and bracken. **STATUS AND DISTRIBUTION** Locally common to not uncommon resident in highlands of SE Nigeria (Obudu and Mambilla Plateaux) and W Cameroon (Mt Cameroon to Bamenda Highlands). *C. c. discolor* confined to Mt Cameroon. **NOTE** Population in our region sometimes considered a separate species, *C. discolor* (Brown-backed Cisticola; Cisticole à dos brun), but treated as conspecific by *BoA* on vocal evidence.

ROCK-LOVING CISTICOLA *Cisticola aberrans* Plate 107 (1)
Cisticole paresseuse

Other names E: Lazy Cisticola; F: Cisticole des rochers

L 13–15 cm (5–6"). Medium-sized, plain-backed cisticola with rufous crown. **Adult *petrophilus* breeding** Forehead to nape dull rufous becoming pale rusty on head-sides, supercilium rusty-buff; rest of upperparts grey-brown. Tail relatively long, dark grey-brown tipped greyish with black subterminal spots only visible from below. Underparts pale rusty-buff, darkening on breast-sides, flanks and undertail-coverts. *C. a. admiralis* slightly darker above, with top of head less rufous, more uniform with upperparts. **Adult non-breeding** Brighter coloured overall. **Juvenile**

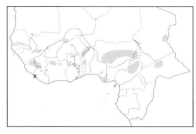

Similar to adult non-breeding. **VOICE** Song, uttered from perch, a slow series of various squeaky, metallic notes resembling squeezing of rubber toy (*kwee-et* or *tu-whee-a* and variations), interspersed with short dry trills and/or short, rapid series of clicking, sucking notes (*tzk-tzk-tzk-tzk* or *tswik-tswik-tswik*); also various harsh notes and short, buzzing or dry, rattling trills. Alarm a sharp *krrrt* or *prrrip*. [CD10:86] **HABITS** Singly or in pairs on rocky outcrops in wooded grassland, and in open woodland with scattered boulders and low bushes. Longish tail, often flicked and held cocked, give prinia-like impression. Low in herbage or on ground, running mouse-like. **STATUS AND DISTRIBUTION** Patchily distributed, locally not uncommon resident, in S Mauritania, Mali, Guinea, Sierra Leone and NW Liberia east to Cameroon, Chad and N CAR. *C. a. petrophilus* from Ghana west; *admiralis* from Togo east.

RATTLING CISTICOLA *Cisticola chiniana*   Plate 107 (5)
Cisticole grinçante

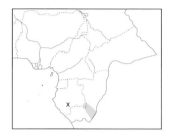

Other name F: Cisticole à crécelles
L 12–16 cm (4.75–6.25"). *C. c. fortis*. Medium-sized, robust cisticola. **Adult male breeding** Forehead to nape plain rusty-brown; lores dusky; upperparts dark brown (appearing plain in most, very faintly dappled dusky in some); flight feathers edged rufous-brown. Tail dusky-brown with white tip and black subterminal band. Underparts buffish-white becoming white on throat and central belly; breast-sides and flanks grey. **Adult male non-breeding** Top of head more rufous. **Adult female** As male, but noticeably smaller. **Juvenile** As adult breeding, but more rusty-brown and head-sides and upper breast slightly tinged yellow. **VOICE** Song, uttered from low perch, a loud repeated phrase of 1–4 harsh or high-pitched notes followed by a trill *chi-chi-chrrrr* or *seep seep titrrrrip*, also *cheecheecheechee trrrr* and other variations. [CD10:78] **HABITS** Singly or in pairs in dry savanna woodland. **SIMILAR SPECIES** Chattering Cisticola difficult to separate but occurs in lowland forest belt and has uniform upperparts lacking rufous-brown edgings to flight feathers. Bubbling Cisticola as Chattering but paler. **STATUS AND DISTRIBUTION** Scarce and local resident, S Gabon (Lékoni, Mouila) and C Congo (Brazzaville, Djambala).

TINKLING CISTICOLA *Cisticola rufilatus*   Plate 107 (9)
Cisticole grise

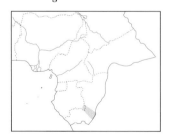

Other name F: Cisticole des taillis
L 13.0–14.5 cm (5.0–5.75"). *C. r. ansorgei*. Medium-sized, streaked cisticola, distinguished in all plumages by *plain rufous crown and ear-coverts*, buff supercilium and *rufous tail*. **Adult male breeding** Forehead to nape and head-sides bright rufous; upperparts buff-brown streaked blackish; tail bright rufous with white tip and black subterminal band. Underparts buff becoming white on throat and central belly. **Adult male non-breeding** Head-sides paler, less well marked; upperparts more rufous. **Adult female breeding/non-breeding** As male, but with shorter tail. **Juvenile** As adult non-breeding, but duller. No yellow on underparts. **VOICE** Song, uttered from perch atop tree or bush, a short series of clear piping whistles *hweee-hweee-hweee-...* often followed by bill-snapping and trill *tzk-tzk-tzk chirrrrrr*; also snaps wings. [CD10:77] **HABITS** In open grassland with scattered trees and bushes. Secretive and shy, diving into cover at least disturbance. **STATUS AND DISTRIBUTION** Local resident, SE Gabon (scarce) and C Congo (Brazzaville and Bateke Plateau; locally common).

WINDING CISTICOLA *Cisticola galactotes*   Plate 107 (11)
Cisticole roussâtre

Other name E: Greater Black-backed Cisticola
L 12–15 cm (4.75–6.0"). Medium-sized, streaked cisticola occurring in moist habitats. **Adult male breeding** *amphilectus* Forehead dull rusty-brown becoming greyish-brown on crown and nape; faint buffish supercilium; head-sides rusty-buff; upperparts brownish-grey broadly streaked blackish; rump plain greyish; flight feathers edged russet. Tail ash-grey with white tips and black subterminal band. Underparts creamy-buff. Eastern *zalingei* has upperparts rather grey sparsely streaked dusky. **Adult male non-breeding** (rarely assumed in *amphilectus*). Forehead to nape indistinctly streaked black; upperparts tawny broadly streaked black; underparts warm buff,

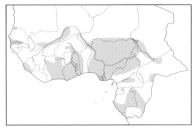

paler on throat and belly and, in *zalingei*, narrowly streaked black. **Adult female** As male, but slightly smaller. **Juvenile** Similar to adult non-breeding, but crown broadly streaked black; throat to upper breast pale yellow. **VOICE** Song a monotonously repeated, single, drawn-out rasping note *zrrrreeeeeee* uttered from low perch in herbage. [CD10:95] **HABITS** The typical cisticola of wet places, such as rank grass and sedge of marshes, temporary water pans and similar moist habitats. Singly or in pairs. **STATUS AND DISTRIBUTION** Locally common to not uncommon resident throughout, except arid north. *C. g. zalingei* from N Nigeria east to N CAR; *amphilectus* in rest of range. Race in Mali and Niger unclear.

STOUT CISTICOLA *Cisticola robustus* Plate 107 (10)
Cisticole robuste

L 14.0–16.5 cm (5.5–6.5"). Medium-sized, well-marked, streaked species with bright rufous cap and nape. **Adult male** *santae* Forehead and crown rufous streaked black *strongly contrasting with unstreaked rufous nape*, buff supercilium and head-sides; upperparts brownish, heavily streaked black. Flight feathers dusky with pale buff edges. Tail blackish tipped whitish with no visible subterminal spots on upper surface. Underparts buff becoming whitish on throat and central belly; undertail-coverts tawny. Those in Congo (presumed *nuchalis*) paler, less richly coloured above, particularly on crown. **Adult female** As male, but noticeably smaller. **Juvenile** Rusty version of adult non-breeding, with yellow of underparts pervading entire head. **VOICE** A short, piping ripple *tsri tsri tsrrrrrrr*, usually uttered from low perch (grass stalk or top of low bush); occasionally in low display flight. Alarm call *tsip-tsip-tsip*. [CD10:76] **HABITS** Singly or in pairs in rank grass with shrubs. In highlands up to 2300 m. **SIMILAR SPECIES** Croaking Cisticola has top of head concolorous with rest of upperparts and shorter, thicker and decurved bill; also bulkier and exhibiting considerable seasonal colour variation. **STATUS AND DISTRIBUTION** Locally common resident in highlands of E Nigeria (Gangirwal, Mambilla Plateau) and W Cameroon (*santae*); local in Congo (Gamboma: matching description of *nuchalis*).

CROAKING CISTICOLA *Cisticola natalensis* Plate 107 (12)
Cisticole striée

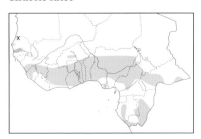

L 12.5–16.0 cm (5.0–6.25"). *C. n. strangei*. Large, bulky cisticola with stout bill. **Adult male breeding** Top of head and upperparts dark earth-brown mottled dusky (almost plain when heavily worn); tail ash-brown with white tips and subterminal spots. Underparts creamy-buff. Bill black. **Adult male non-breeding** Strikingly different from worn breeding plumage. Upperparts buffish-brown boldly streaked black; tail longer and tipped buff; underparts warmer buff. Bill more horn-coloured. **Adult female breeding/non-breeding** As male, but noticeably smaller. **Juvenile** Rusty version of adult non-breeding, with yellow underparts. **VOICE** Song a loud *klink klunk* in display flight at medium height (*c.* 3 m or more); when perched give slower *kluuu klink!* and *kzeee klunk!* Occasionally uttered by individuals in non-breeding plumage. Croaking alarm call. [CD10:75] **HABITS** Usually singly in rank grass with scattered low bushes and at edges of cultivation. **SIMILAR SPECIES** Stout Cisticola is also robust, but has contrasting rufous nape, less stout bill and is generally smaller. In worn breeding plumage sometimes confused with Whistling Cisticola, but upperparts never quite so plain and flight feathers lack rufous edges. **STATUS AND DISTRIBUTION** Common to uncommon almost throughout, except arid north. Rare, Senegambia, SW Mali and Liberia.

RED-PATE CISTICOLA *Cisticola ruficeps* Plate 106 (11)
Cisticole à tête rousse

L 13 cm (5"). Medium-sized with conspicuous *plain reddish cap*. Upperparts plain in breeding, streaked in non-breeding plumage. **Adult *guinea* breeding** Forehead to nape plain rufous contrasting with plain grey-brown upperparts; whitish supraloral spot; head-sides plain buffish-brown; rufous-brown wing panel (produced by pale fringes to greater coverts and inner secondaries); uppertail-coverts rufous-brown; tail dark brown with white tips

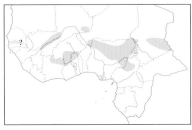

and black subterminal spots. Underparts mainly white faintly washed buffish, esp. on flanks and (darker) on thighs; undertail-coverts white. **Nominate** has paler brown upperparts, enhancing contrast between cap and mantle; wing panel greyish-white and buff. **Adult non-breeding** Top of head brighter rufous; upperparts heavily streaked black; rump plain buff-brown; uppertail-coverts bright rufous; tail longer and broadly edged rufous. Nominate has upperparts buff and more heavily streaked. **Juvenile** Dull version of adult non-breeding with underparts variably tinged yellow. **VOICE** Song consists of a drawn-out, high-pitched note followed by 2 fast phrases of alternatively rising and falling notes, the first progressively falling in pitch, the second ascending (similar to some phrases of Whistling Cisticola). Uttered from top of small tree. [CD10:79] **HABITS** In grassy thorn scrub and open grassy areas with scattered trees or bushes. **SIMILAR SPECIES** Dorst's Cisticola (q.v.). In non-breeding dress Red-pate is the only streaked cisticola with plain rufous crown within its range and habitat. **STATUS AND DISTRIBUTION** Inadequately known. Uncommon to locally common resident in Sahel belt in Mali, N Ivory Coast and N Ghana, and from SW and SE Niger through Nigeria to N Cameroon (*guinea*), and in Chad and CAR (nominate). Intergrade in L. Chad area. **NOTE** Birds previously identified as Red-pate Cisticola may be Dorst's Cisticola in certain parts of range.

DORST'S CISTICOLA *Cisticola dorsti*   Plate 106 (12)
Cisticole de Dorst

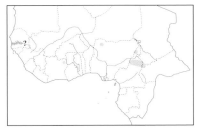

L 13 cm (5"). **Adult breeding** Very similar to Red-pate Cisticola, but has *buff* (not white) *vent and undertail-coverts* (reducing contrast between thighs and rest of underparts), and less sharply defined black-and-white pattern to undertail. Upperparts also appear more uniform and wing panel less conspicuous; head-sides slightly more rusty with less contrasting lores. **Adult non-breeding and juvenile** Unknown. **VOICE** Song a drawn-out trill of variable speed, often preceded by a short vibrant note (analogous to corresponding note of Red-pate Cisticola) and followed by a simple motif, repeated several times, e.g. *chirrrrrrr tsu-wheet tsu-wheet tsu wheet* ... Uttered from atop stem or small bush (much lower than Red-pate). [CD10:80] **HABITS** In dry grassland with clumps of thicket or shrub layer; also cassava plantations and old fields. **SIMILAR SPECIES** Red-pate Cisticola (q.v. and above). **STATUS AND DISTRIBUTION** Inadequately known. A recently described species (Chappuis & Erard 1991) recorded from Senegambia, N Ivory Coast, NW Nigeria (Gusau), N Cameroon and S Chad (Bekao, Baïbokoum). Perhaps more widespread. *TBW* category: Data Deficient.

SHORT-WINGED CISTICOLA *Cisticola brachypterus*   Plate 106 (10)
Cisticole à ailes courtes

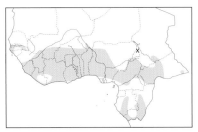

Other name E: Siffling Cisticola
L *c.* 10 cm (4"). *C. b. brachypterus.* Small. **Adult breeding** Head and upperparts plain dull brown. Tail dark brown with pale grey tips and indistinct dusky subterminal spots. Lores, head-sides and underparts whitish washed buff, esp. on flanks. **Adult non-breeding** Upperparts warmer brown diffusely streaked dusky. **Juvenile** Rusty version of adult non-breeding, with yellow underparts. **VOICE** Song, uttered from treetop or other perch (such as pole or wire), a continuously repeated short, rapid series of 2–3 high-pitched notes descending in pitch e.g. *see-se-swu see-su-swu* ... ; also a fast *tsuw-tsuw-tsuw-tsuw...*, *tsurup-tsee tsurup-tsee...* and variations. From Cameroon east additional notes are added to the series. Also a fast, hesitant *tsitsitsutsitutsiti....* in flight or from perch. [CD10:81] **HABITS** In various types of wooded grassland, incl. grassy edges of farmland with scattered trees, scrub at forest edge, second growth and grassy clearings in woodland. Outside breeding season in small parties, sometimes with other cisticolas. **SIMILAR SPECIES** Rufous Cisticola generally has warmer, more rufous plumage (but some may appear very similar), and always has plain upperparts with wings more uniform with mantle. **STATUS AND DISTRIBUTION** Not uncommon to locally common resident throughout, except arid north. **NOTE** A small cisticola resembling Short-winged Cisticola, but with contrasting white throat and different song, occurs in SE Gabon (Lékoni); it appears to constitute an undescribed species.

RUFOUS CISTICOLA *Cisticola rufus*   Plate 106 (9)
Cisticole rousse

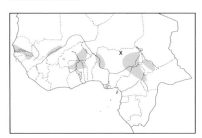

L 10 cm (4"). Small and plain backed. **Adult breeding** Head and upperparts plain dull rust-brown; rump and uppertail-coverts paler and more rufous. Tail dark rust-brown with pale tips and dark subterminal smudges. Underparts white washed pale rusty-brown, most heavily on flanks; thighs dull rusty-brown. **Adult non-breeding** Slightly paler and brighter. **Juvenile** Dull version of adult non-breeding, with underparts variably tinged yellow. **VOICE** Song, delivered from treetop, 2–3 thin descending notes e.g. *see see hu, see see hu, ...* and *see-hu see-hu see-hu.* Also a rapid, hesitant series *tsutititsutititu...* uttered in flight or from perch. [CD10:82] **HABITS** In dry grassy savanna with scattered trees. **SIMILAR SPECIES** Short-winged Cisticola is less rufous (but warmest coloured birds may appear confusingly similar), with rufous fringes to wing feathers slightly contrasting with duller mantle; song very similar. Foxy Cisticola has much brighter rufous upperparts and plain black tail. **STATUS AND DISTRIBUTION** Uncommon to locally not uncommon resident with patchy distribution in dry savanna zone. Recorded from Senegambia, Guinea, Mali, Burkina Faso, N Ghana, N Togo, Nigeria, N Cameroon, S Chad and NW CAR.

FOXY CISTICOLA *Cisticola troglodytes*   Plate 106 (8)
Cisticole russule

Other name F: Cisticole grisâtre

L 10 cm (4"). *C. t. troglodytes.* Small, plain and bright rufous-brown. **Adult** Head and upperparts bright russet; underparts rusty-buff becoming whitish on central belly. Tail russet-brown tipped whitish, with subterminal black spots visible from below. No non-breeding plumage. **Juvenile** Dull-coloured version of adult, with yellow underparts. **VOICE** Call a soft *tsit, tsit-tsit.* Song, uttered from low treetop and in display flight, a rapid series of similar *tsit* or *tsee* notes. [CD10:83] **HABITS** In wooded grassland. **SIMILAR SPECIES** Rufous Cisticola has less rufous upperparts and pale-tipped tail with black subterminal smudges visible on both surfaces. **STATUS AND DISTRIBUTION** Inadequately known. Resident, S Chad and CAR (recorded in W and in Manovo-Gounda-St Floris NP). Claimed from NE Mali (1 reportedly collected east of Azzawakh, Aug).

PIPING CISTICOLA *Cisticola fulvicapillus*   Plate 106 (13)
Cisticole à couronne rousse

Other name E: Neddicky

L 10–11 cm (4.0–4.25"). *C. f. dispar.* Small, plain-backed species with dull rufous crown. **Adult breeding** Rufous cap contrasts with plain earth-brown upperparts and tail. Indistinct supercilium and head-sides buff. Underparts whitish tinged grey on flanks. Tail has rather indistinct dusky subterminal spots. **Adult non-breeding** Upperparts slightly more russet. **Juvenile** As dull adult non-breeding, with face and breast tinged yellow. **VOICE** Song a simple, monotonous repetition of a single, penetrating, but not very loud note *whee whee whee whee...*, or faster *whip-whip-whip-...* uttered from perch on top of tree or tall bush. Alarm call a rapid *tikitikitikitiki-...* or *krrrrrrrr* 'like running fingernail across teeth of comb' (Maclean 1985). [CD10:84] **HABITS** Singly or in pairs in shrubs and rank grass in woodland, second growth at forest edge and along fields and roads. When not breeding sometimes in small parties; occasionally in mixed-species flocks. **SIMILAR SPECIES** None. Other small, plain-backed species in its range have top of head concolorous with upperparts. **STATUS AND DISTRIBUTION** Local resident, SE Gabon (uncommon) and C Congo (locally common).

ZITTING CISTICOLA *Cisticola juncidis*   Plate 106 (4)
Cisticole des joncs

Other name E: Fan-tailed Cisticola

L 10 cm (4"). Small and streaked. **Adult male *uropygialis* breeding** Forehead, crown and upperparts brownish-buff heavily streaked blackish, contrasting with buffish supercilium and head-sides, pale, vaguely streaked nape, and dull, mostly plain rufous-brown rump and uppertail-coverts. Tail dusky black tipped white; black subterminal

spots esp. conspicuous from below. Underparts mainly buffish, becoming whitish on throat and central belly. Southern *terrestris* is slightly darker and browner. **Adult male non-breeding** More buff, less blackish upperparts. **Adult female** As male non-breeding **Juvenile** Rusty version of adult non-breeding; underparts tinged pale yellow. **VOICE** Song a distinctive monotonous repetition of a single harsh *tsip*, *zit* or *tslik*, from perch or, more often, in high, undulating display flight, one note at each dip; without wing-snapping. Alarm call a rapid *tik-tik-tik.-...* or *chik-chik-chik*. [CD4:36] **HABITS** Singly or in pairs in both wet and dry grassy areas without trees, in savanna, edges of farmland and similar habitats. Unobtrusive outside breeding season, foraging mostly low in rank grass or on ground. **SIMILAR SPECIES** Desert Cisticola is paler and subterminal spots on tail only visible from below; also occurs in drier areas. Adult male non-breeding and female Black-backed Cisticola are brighter, overall more rufous, with pale supraloral spot (no supercilium) and orangey rump and flanks. **STATUS AND DISTRIBUTION** Patchily distributed throughout, locally common to uncommon. *C. j. uropygialis* throughout, except Eq. Guinea, Gabon and Congo, where *terrestris*.

DESERT CISTICOLA *Cisticola aridulus* Plate 106 (7)
Cisticole du désert

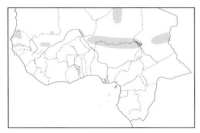

L 10–12 cm (4.0–4.75"). *C. a. aridulus*. Small and streaked. **Adult** In all plumages very similar to Zitting Cisticola but noticeably paler, more sandy coloured, with plain blackish tail tipped white; black subterminal tail spots only visible from below. **Juvenile** Rusty version of adult non-breeding; no yellow on underparts. **VOICE** Song, uttered from low perch or in low, bouncing display flight, a rapid repetition of a clear, high-pitched note (*teehu*), interspersed with dry clicking bill-snaps (*tk;* only audible at short range) and occasional wing-snaps: *teehu (tk)-teehu (tk)-teehu (tk)-tktktktk-teehu (tk)-....* Dry clicks also used in alarm. [CD10:74] **HABITS** Singly or in pairs in short dry grass and fallow farmland in semi-arid country. **SIMILAR SPECIES** Zitting Cisticola is slightly richer coloured, overall more buff, has longer looking tail with subterminal black spots forming band visible from above, and occurs in moister areas. **STATUS AND DISTRIBUTION** Uncommon to not uncommon resident in Sahel belt, from S Mauritania and N Senegal to N Nigeria, Niger and Chad.

BLACK-BACKED CISTICOLA *Cisticola eximius* Plate 106 (5)
Cisticole à dos noir

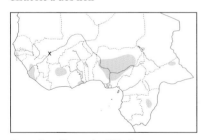

Other name E: Black-necked Cisticola
L *c*. 10 cm (4"). Small and streaked. **Adult male *occidens* breeding** Forehead and crown pale rufous-brown; head-sides pale greyish-buff; pale supraloral spot. Upperparts brown boldly streaked black; conspicuous orange-rufous rump and uppertail-coverts. Flight feathers brownish-black with pale fringes. Tail brownish-black above, whitish below with black subterminal spots. Underparts whitish tinged buff on breast-sides; lower flanks orange-rufous. Local race *winneba* paler; **nominate** darker. Undescribed race from Congo has golden-yellow crown and rump (Dowsett-Lemaire 1997). **Adult male non-breeding** Crown heavily streaked black; upperparts more rufous. **Adult female** Forecrown indistinctly streaked black. **Juvenile** Upperparts more buffish; underparts yellow. **VOICE** Song a rather sharp, vibrant *tsree-tsree-tsree-tsree-...* uttered from perch or in undulating display flight high above territory, sometimes accompanied by a rhythmic noise, sounding like wing-snapping, and occasionally including a distinctive, sharp, dissonant *chereet-chereet* (slightly reminiscent of Icterine Warbler); latter also used in alarm. Song of those in Congo include a disyllabic *dzuah, dzuah,...* uttered in flight, and *tchik-pee, tchik-pee,...* given on landing (Dowsett-Lemaire 1997). [CD10:72] **HABITS** Singly or in pairs in moist open grassland, wet dambos and grassy valleys. Active and conspicuous when breeding, but secretive and easily overlooked in dry season. **SIMILAR SPECIES** The only other small, streaked cisticola in range is Zitting, which is overall less rufous, with brownish-buff (not rufous-brown) and always streaked top of head, buffish supercilium, rather dull rufous-brown (not orange-rufous) rump and no orange-rufous on lower flanks; song also different. **STATUS AND DISTRIBUTION** Locally not uncommon to rare resident, from S Senegal to Sierra Leone east to Burkina Faso and Nigeria (*occidens*), and in N CAR (nominate) and N Congo (Oyo to Odzala NP; undescribed race). *C. e. winneba* in coastal Ghana.

DAMBO CISTICOLA *Cisticola dambo*   Plate 106 (6)
Cisticole dambo

Other names E: Black-tailed Cisticola; F: Cisticole des dambos
L 10–12 cm (4.0–4.75"). Small and streaked, with *relatively long, black tail.*
Adult male breeding Resembles Zitting Cisticola, but has better defined, neater black markings; more buff on head-sides, breast, flanks and undertail-coverts, and distinctly longer, whitish-tipped black tail (lacking subterminal black spots). **Adult male non-breeding** Deeper, rusty-buff above and below; tail slightly longer, blackish edged rusty-buff. **Adult female breeding/non-breeding** As male. **Juvenile** Underparts tinged pale yellow. **VOICE** Song, uttered in display flight at great height, a rasping, rather piercing *hree-ep, hree-ep, hree-ep...* accompanied by wing-snapping. [CD10:73] **HABITS** Singly or in pairs in open grassy plains and damp meadows (dambos). **SIMILAR SPECIES** Other small 'cloud-scraper' cisticolas may occupy same habitat, but none has long tail. **STATUS AND DISTRIBUTION** Locally not uncommon resident, SE Gabon (Lékoni). Race unknown. **NOTE** Formerly only known from E Angola to SE Congo-Kinshasa (nominate) and NW Kasai, Congo-Kinshasa (*kasai*). *C. d. kasai* described as having narrower black feather centres to upperparts, brighter rusty-buff underparts and rustier tail tips.

PECTORAL-PATCH CISTICOLA *Cisticola brunnescens*   Plate 106 (1)
Cisticole brune

Other name F: Cisticole brunâtre
L 10–11 cm (4.0–4.25"). Small and streaked. **Adult male *lynesi*** Forehead and crown plain russet-brown; lores black contrasting with buff supercilium and head-sides. Upperparts black with russet fringes to feathers (producing streaked appearance); rump and uppertail-coverts rufous, vaguely streaked black. Tail black edged russet and tipped white. Throat and central belly white; rest of underparts tinged buff; some blackish streaks on breast-sides. *C. b. mbangensis* slightly paler, less streaky above. **Adult female** Forehead and crown tawny broadly streaked black; lores dusky; rump and uppertail-coverts streaked black. **Juvenile** Underparts yellow. **VOICE** Song, usually given in jerky display flight, a rhythmic repetition of a single, buzzing note, accelerating to a series when diving to ground *tzit-tzit-tzit-tzit-... -tzitzitzitzitzit...*; accompanied by barely audible dry clicks sounding like wing-snaps, but also heard from perched singers. [CD10:69] **HABITS** Singly or in pairs in open plains with short or burnt grass, dry dambos and, in highlands, areas with short grass and scattered low bushes on stony ground. **SIMILAR SPECIES** Ayres's Cisticola is very similar, but male breeding has less contrasting lores, no or only very faint streaks on breast-sides, shorter tail and different song; female extremely difficult to separate from female Pectoral-patch. **STATUS AND DISTRIBUTION** Locally not uncommon to common resident, Cameroon (*lynesi*, Bamenda Highlands; *mbangensis*, Adamawa Plateau), and (race unknown) Gabon and Congo (Odzala NP, Bateke Plateau).

PALE-CROWNED CISTICOLA *Cisticola cinnamomeus*   Plate 106 (3)
Cisticole châtain

L 10–11 cm (4.0–4.25"). *C. c. midcongo.* Small and streaked. **Adult male** Forehead and crown plain rusty buff-brown; area from *base of bill to around eye blackish*, forming diagnostic face pattern. Upperparts black with russet fringes to feathers (producing streaked appearance); rump and uppertail-coverts rufous. Tail black edged russet and tipped white. Throat and central belly white; rest of underparts tinged rufous-buff; some blackish streaks on breast-sides. **Adult female** Forehead and crown tawny broadly streaked black and bordered by buff supercilium; lores dusky; rump and uppertail-coverts streaked black. **Juvenile** As adult female but more russet above and washed yellow below. **VOICE** Song, usually given in jerky display flight, a short series of thin, high-pitched notes *eeeyip eeeyip eeeyip eeeyip...* followed by a series of lower, slightly vibrating notes *rreee rreee rreee rreee...* as bird dives; finally returns to ground silently, without dry clicks. [CD10:70] **HABITS** Singly or in pairs in short, dry or moist grassland. **SIMILAR SPECIES** Compare Pectoral-patch and Ayres's Cisticolas. Adult female not safely distinguishable in field from other small, streaky cisticolas. **STATUS AND DISTRIBUTION** Locally common resident, Congo (Bateke Plateau). **NOTE** Often treated as race of Pectoral-patch Cisticola.

AYRES'S CISTICOLA *Cisticola ayresii*

Cisticole gratte-nuage

Plate 106 (2)

Other name E: Wing-snapping Cisticola
L 9–10 cm (3.5–4.0"). *C. a. gabun.* Very small and streaked, with short tail. **Adult male** Forehead and crown russet-brown. Buffish supercilium; head-sides pale brownish; small faint dusky loral streak. Upperparts blackish with broad russet fringes to feathers (producing streaked appearance); rump and uppertail-coverts rufous streaked blackish. Tail blackish tipped white. Underparts white tinged rusty-buff on breast and flanks. **Adult female** Forehead and crown heavily streaked black with rusty-brown fringes to feathers. **Juvenile** Underparts washed yellow. **VOICE** Song, given in display flight at great height, a series of thin, high-pitched whistles repeated many times, interspersed by dives accompanied by rapid *tiktiktik...* and occasional wing-snaps. [CD10:71] **HABITS** Singly or in pairs in open grassland. **SIMILAR SPECIES** Pectoral-patch Cisticola is very similar, but generally has stronger loral mark (producing more contrasting head-sides), more distinct blackish streaking on breast-sides, slightly longer tail, stronger legs and feet, and different song; females extremely difficult to separate. **STATUS AND DISTRIBUTION** Unclear. Formerly reported to be local resident, coastal Gabon and C Congo (Gamboma), but vocalisations heard in Port-Gentil area, Gabon, type locality of *gabun*, identified as being from *C. brunnescens*, while *C. ayresii* has yet to be found there. Further research should elucidate if *C. 'a.' gabun* belongs to *brunnescens*, or if *ayresii*, *brunnescens* and *cinnamomeus* occur sympatrically.

STREAKED SCRUB WARBLER *Scotocerca inquieta*

Dromoïque vif-argent

Plate 104 (8)

Other name F: Dromoïque du désert
L 10 cm (4"). *S. i. theresae.* Secretive small desert warbler with cocked tail. **Adult** Upperparts dark grey-brown; crown finely streaked dark brown; broad pale supercilium; thin black eye-stripe. Underparts whitish with rufous flanks and vent; very finely streaked breast. Tail graduated, dark brown tipped white. Eye pale yellow to brown. Bill pinkish with darker culmen. Legs yellowish-brown to pale pinkish. **Juvenile** Duller. **VOICE** Song a distinctive dry *dzit dzit* followed by a melodious, liquid *deedle doleedle doleed.* [CD4:25] **HABITS** A desert species occurring singly or in pairs within dry wadis with low scrub. Skulking but not particularly shy. Very active, scurrying mouse-like on ground and hopping into and from bushes. **STATUS AND DISTRIBUTION** Scarce and local resident, Mauritania (Adrar).

Genus *Prinia*

Rather slender, active species with long, graduated tails. Frequent grassy and bushy habitats.

TAWNY-FLANKED PRINIA *Prinia subflava*

Prinia modeste

Plate 104 (2)

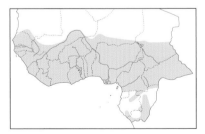

Other names F: Prinia/Fauvette-roitelet commune
L 11–12 cm (4.25–4.75"). ¶ Very common warbler with brownish upperparts, distinct *buffish supercilium*, creamy underparts and rather long, graduated tail. **Adult breeding** Lores blackish. Breast-sides, flanks and undertail-coverts pale tawny. Tail has dark subterminal band and whitish tip. Eye brown to reddish-brown. Bill black. Legs pinkish-brown. **Adult non-breeding** (this plumage only assumed in dry season in savanna zone) Paler brown with rusty tinge, noticeably longer tail and horn-coloured bill. **Juvenile** Underparts tinged pale yellow; bill yellowish. **VOICE** Song, uttered from prominent perch, a monotonous, rhythmic series of a single note; speed of delivery and tone variable, e.g. sharp *tzreep tzreep tzreep tzreep...* and fast *plip-plip-plip-plip-...* Alarm a harsh *zbeee.* [CD11:2] **HABITS** In pairs or small parties in wide variety of grassy and bushy habitats, incl. overgrown plantations, gardens, road edges and scrub at forest edges. Active and vocal, foraging low in rank herbage and shrubbery, with frequent jerky movements of cocked tail. Not shy. **SIMILAR SPECIES** Unstreaked cisticolas mainly distinguished by relatively shorter tails (imparting a less slender appearance) and different vocalisations; also by combination of different head pattern and/or underparts coloration. River Prinia is paler and has slightly more slender appearance, with longer tail. **STATUS AND DISTRIBUTION** Common resident throughout. **NOTE** Plumage variation strongly clinal, with much inter-gradation; palest populations in driest zones. Races described from our region include *pallescens* (Sahel zone; Mali to Chad), *subflava* (northern savannas; Senegal to CAR) and *melanoryncha* (southern savannas and forest zone).

RIVER PRINIA *Prinia fluviatilis*   Plate 106 (1)
Prinia aquatique

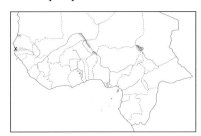

L *c.* 12 cm (4.75"). **Adult** Similar to sympatric Tawny-flanked Prinia, but distinguished by slightly greyer upperparts, whiter, less buff underparts, and slightly longer tail. **Juvenile** No information. **VOICE** Song a rapid, rhythmic series of a single, high-pitched note. Not reminiscent of Tawny-flanked Prinia; notes more drawn-out, rhythm much less variable. [CD11:3] **HABITS** Inhabits waterside vegetation, such as reedbeds, rank herbage and bushes. Very active. Sings very low, near ground (Tawny-flanked usually, but not always, higher). **SIMILAR SPECIES** In addition to different plumage coloration, tail length and voice, Tawny-flanked Prinia also has different habitat preferences, frequenting drier areas where both species meet. **STATUS AND DISTRIBUTION** Insufficiently known. Recorded from NW Senegal, Gambia, SW Niger and NE Mali (along Niger R., from Tillabéri to Goa) and L. Chad area (Chad, SE Niger, NE Nigeria, N Cameroon). Probably resident throughout contact zone of Sahel–northern savanna.

SÃO TOMÉ PRINIA *Prinia molleri*   Plate 145 (6)
Prinia de Sao Tomé

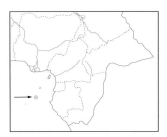

L *c.* 13 cm (5"). Conspicuous, noisy warbler with rather long, graduated tail. **Adult** Face rufous becoming rufous-brown on crown and nape; upperparts grey; underparts creamy-white with rufous thighs. Tail feathers tipped white with dark subterminal spots (visible from below). Bill dark brown above, whitish below. Legs pinkish. **Juvenile** Upperparts, throat and breast washed rufous. **VOICE** Call a nasal *dzik* or *dzi-dzi-dzi-dzi-...* Song a monotonous, rhythmic series of loud high-pitched single or double notes *tsee-tsee-tsee-...* or *tissee-tissee-tissee...* and *tsiwhip-tsiwhip-tsiwhip...*, with many variations. [CD11:4] **HABITS** Singly, in pairs or small groups, occurring in all habitats at all altitudes, but particularly common in disturbed and edge habitats (even in towns), and rare in dense forest regrowth or undisturbed forest. Vocal and conspicuous, often using exposed perches and snapping wings in flight. **SIMILAR SPECIES** None. No other prinias (nor cisticolas) occur on the island. **STATUS AND DISTRIBUTION** Endemic to São Tomé, where common resident.

BANDED PRINIA *Prinia bairdii*   Plate 104 (5)
Prinia rayée

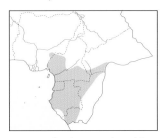

L 11.5 cm (4.5"). *P. b. bairdii.* **Adult** Distinctive dark prinia with boldly and densely barred black-and-white underparts. Head, upperparts and tail blackish-brown (darkest on head); wing-coverts, tertials and strongly graduated tail tipped white. Central belly to undertail-coverts white (not barred). Eye brownish. Bill and legs black. **Juvenile** Duller, more brownish, with greyish-brown underparts (becoming whitish on belly), only slightly barred on flanks. Eye pale brown. **VOICE** Song a loud, shrill, fast series of single notes *plee-plee-plee-plee-...* and more rapid *plipliplipliplipli...*, uttered from cover; also *wheet-wheet-trlrrrr.* [CD11:5] **HABITS** Mostly in pairs, within tangled undergrowth at forest edges, old clearings and wet grassland in forest/marsh ecotone. Restless; moving with much wing-flicking and cocking of tail. Often forages on or just above ground. **STATUS AND DISTRIBUTION** Uncommon to locally common resident, in highlands of SE Nigeria (Obudu Plateau) and lowlands from W Cameroon to SW CAR and Congo.

RED-WINGED WARBLER *Heliolais erythroptera*   Plate 104 (6)
Prinia à ailes rousses

Other name F: Fauvette à ailes rousses

L 12 cm (5"). Prinia-like warbler with rufous wings and long, graduated tail. **Adult *erythroptera* breeding** Diagnostic combination of *grey head and upperparts, bright rufous wings*, and white throat merging into buff on rest of underparts (darkening on undertail-coverts). Tail rufous-brown with white tips and black subterminal spots. Central belly white. Eye yellowish to brown. Bill fairly long, blackish. Legs pinkish. *H. e. jodoptera* more tawny above; bill brownish. **Adult *erythroptera* non-breeding** Upperparts pale vinous-rufous; bill horn-coloured. Some may breed in this

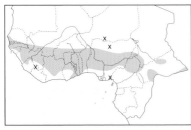

plumage. *H. e. jodoptera* darker, tawnier. **Juvenile** Similar to adult non-breeding, but paler or duller and virtually without blackish tail markings. **VOICE** Song a rapid, monotonous series of high *pseep* or *thu-weet* notes; in duet second bird utters dry chatter. Calls include thin *tseek* notes. [CD11:25] **HABITS** In pairs or small family groups in wooded grasslands, restlessly foraging low in rank herbage. May perch conspicuously on bushes and fly to treetops. **SIMILAR SPECIES** Tawny-flanked Prinia lacks rufous in wings and has pale supercilium. Cisticolas have different jizz and no plain-backed species has such bright rufous wings. **STATUS AND DISTRIBUTION** Not uncommon to uncommon and patchily distributed resident in savanna zone, from Gambia and S Senegal to Sierra Leone and N Liberia east to N Cameroon and Chad (nominate), Cameroon (south of Adamawa Plateau) and CAR (*jodoptera*). **NOTE** Sometimes included in genus *Prinia*.

GREEN LONGTAIL *Urolais epichlora*   Plate 104 (7)
Prinia verte

Other name F: Fauvette verte à longue queue
L 15 cm (6"). **Adult** *epichlora* Easily identified by *bright green upperparts, long, strongly graduated and usually heavily worn tail*, and *restricted range*. Deep yellow supraloral streak; buffish underparts becoming whitish on belly; undertail-coverts and thighs yellow. Eye pale brown. Bill black. Legs brown. *U. e. cinderella* slightly larger. Island race *mariae* has head-sides yellower, mantle greener, tail much longer. **Juvenile** Similar. **VOICE** Commonest song a long rhythmic series of a single sharp note *tsip-tsip-tsip-*... or *djib-djib-djib-*... Also a frequent ascending *peeeep-peep-peep-peep-pip-pip-pip*. Sings frequently. [CD11:24] **HABITS** In pairs, family groups and mixed-species flocks in montane and submontane forest (above 800 m); also second growth and isolated patches in savanna or farmland. Very active, foraging mainly at middle and upper levels, but also lower. **STATUS AND DISTRIBUTION** Locally common resident in highland areas of SE Nigeria (Obudu Plateau, Chappal Wadi) and W Cameroon (*epichlora/ cinderella*); also Bioko (*mariae*). **NOTE** Validity of ssp. *cinderella* questionable; this form not generally recognised.

CRICKET WARBLER *Spiloptila clamans*   Plate 104 (11)
Prinia à front écailleux

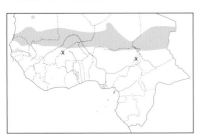

Other names E: Scaly-fronted Warbler; F: Prinia grillon
L *c.* 11.5 cm (4.5"). Distinctive, delicately coloured, pale and long-tailed dry-country warbler. **Adult male** Forehead and forecrown black and white (giving streaked or scaly impression); hindcrown and nape pale grey; upperparts sandy-cinnamon with wing-coverts contrastingly black fringed white; rump pale yellow. Flight feathers grey with pinkish outer webs. Tail long and graduated, all feathers with broad white tips and black subterminal bars. Underparts pale creamy-buff. Eye pale brown. Bill greyish tipped black. Legs pale pinkish-yellow. **Adult female** As male, but hindcrown and nape cinnamon, concolorous with mantle. **Juvenile** Browner; crown streaked dark brown. **VOICE** Song consists of dry, rhythmic, insect-like trills and series of single, high-pitched notes, both delivered in fast or slower tempo. Duets. [CD11:22] **HABITS** Frequents thorn scrub at desert edge and northern savanna. Usually in small, restless parties, constantly jerking tail in all directions. Not shy, often perching in the open. When disturbed, dives for cover and prefers to escape by running rather than flying. **STATUS AND DISTRIBUTION** Uncommon to locally not uncommon in dry Sahel and adjacent Sudan zone, from Mauritania and N Senegal to Niger, NE Nigeria and Chad. Also N Cameroon (rare/vagrant?). Mainly resident; some movements reported in north of range (May–Oct).

WHITE-CHINNED PRINIA *Schistolais leucopogon*   Plate 104 (3)
Prinia à gorge blanche

L 14 cm (5.5"). *S. l. leucopogon*. **Adult** *All grey* with long, graduated tail; black lores and ear-coverts contrast with conspicuous *creamy-white throat*. Forecrown feathers have black centres, giving scaly appearance (only visible at very close range). Flight feathers dark brownish. Eye reddish-brown. Bill black. Legs pinkish. **Juvenile** Similar; bill with horn-coloured tip and yellow gape. **VOICE** A rather harsh *chi-chik*, constantly uttered but easily passing

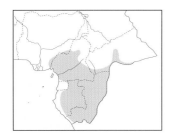

unnoticed. Also *djuweet* or *kluweet*. Song, always in duet or group, a jumble of *chi-chik* calls. [CD11:6] **HABITS** In pairs or small parties within dense shrubbery at forest edge and overgrown cultivation. Restless; not easily seen in dense vegetation. **SIMILAR SPECIES** Gosling's Apalis (q.v.) has similar plumage but is distinctly smaller, occurs in different habitat and has different voice. **STATUS AND DISTRIBUTION** Uncommon to not uncommon resident, from SE Nigeria (Calabar to Mambilla Plateau) to CAR and Congo. **NOTE** Formerly included in genus *Prinia*.

SIERRA LEONE PRINIA *Schistolais leontica* — Plate 104 (4)
Prinia du Sierra Leone

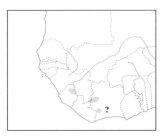

Other name E: White-eyed Prinia
L 13 cm (5"). **Adult** *Dark ash-grey with contrasting whitish eye*, deep buff lower underparts (flanks, vent, undertail-coverts), and rather long, graduated tail. Flight feathers dark brown; tips of rump feathers tinged olive; tail feathers narrowly tipped white; belly whitish. Bill black. Legs brownish-pink. **Juvenile** No information. **VOICE** Call *psit* or *pit*. Song, given in unsynchronized duet, a rapid, high-pitched *sipsipsipsipsip...* with second bird uttering lower, nasal *bur-bur-bur-bur-...* Other calls include *psew-psew-psew-...* , *psip-psip-psip-...* and *bzee, bzee* (*BoA*). [CD11:7] **HABITS** Inhabits dense vegetation at forest edge, along streams and in gorges and gullies within hilly areas, esp. at 700–1600 m. Behaviour much as other prinias: active, occurring singly, in pairs or small parties of up to 9. **SIMILAR SPECIES** White-chinned Prinia has creamy throat and does not overlap. **STATUS AND DISTRIBUTION** Uncommon to rare and extremely local resident, recorded from E Sierra Leone (Loma Mts, Tingi and Kambui Hills), Mt Nimba (Guinea/Liberia/Ivory Coast), other ranges in Nimba County, Liberia (Mt Kitoma, Mt Bele) and Ivory Coast (Taï, Man, Sipilou; Lamto?). *TBW* category: THREATENED (Vulnerable). **NOTE** Formerly included in genus *Prinia*.

RED-WINGED GREY WARBLER *Drymocichla incana* — Plate 104 (9)
Prinia grise

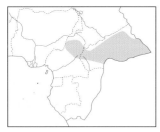

Other name F: Fauvette grise à ailes rousses
L 14 cm (5.5"). Easily identified by pale grey plumage (darker on wings and tail; paler, becoming whitish, on underparts and supraloral area) and rufous outer webs to primaries forming conspicuous wing panel. Thighs buffish to chestnut. Eye greenish-grey. Bill black. Legs yellowish-pink. **Juvenile** Similar, but bill dark brown above, pale horn below. **VOICE** Song includes a series of loud, clear *kweeup kweeup kweeup...* and a repeated *kwup* ending in a chatter. [CD11:27] **HABITS** In pairs or family parties in small trees and bushes in wooded savanna, often along streams, and edges of gallery forest. **STATUS AND DISTRIBUTION** Local and uncommon to rare resident, E Nigeria (1 record, Yim R., near Gumti village), Cameroon (Adamawa Plateau) and CAR. **NOTE** Claims from SE Guinea (birds observed in grass tussocks in flooded marshy area; Halleux 1994) require substantiation (confusion with e.g. Sierra Leone Prinia cannot be eliminated).

BUFF-BELLIED WARBLER *Phyllolais pulchella* — Plate 101 (10)
Phyllolaïs à ventre fauve

Other name F: Apalis à ventre jaune
L 11.5 cm (4.5"). Nondescript small warbler with apalis- or prinia-like tail. **Adult** Upperparts pale olive-brown; lores, head-sides and underparts pale yellowish-buff. Dark, graduated tail with white outer feathers and tip. Eye pale brown. Bill greyish-pink. Legs pinkish-brown. **Juvenile** Upperparts slightly darker, underparts yellower. **VOICE** Song a dry, ascending, insect-like rattling *zrrrrrt zrrrrrt*. Calls *cht-cht-cht-cht* and *tzr tzr chit*. [CD11:26] **HABITS** Inconspicuous. In pairs or small parties in dry *Acacia* woodland. Frequents treetops. Joins mixed-species flocks. **SIMILAR SPECIES** Poorly seen *Phylloscopus* may superficially recall this species, but all differ in absence of white in tail, better marked head pattern, longer wings and, in some, dark legs. **STATUS AND DISTRIBUTION** Uncommon to locally common resident, N Nigeria, N Cameroon and W Chad (L. Chad area).

Genus *Apalis*

Slender, with rather long, graduated tails. Very active, foraging like *Phylloscopus* warblers. Mostly in forest, occupying all levels, from undergrowth to canopy. Often join mixed flocks of small insectivores. Song typically consists of rhythmic repetitions of a single note. Endemic to Africa.

BLACK-COLLARED APALIS *Apalis pulchra*   Plate 105 (5)
Apalis à col noir

Other names F: Apalis/Fauvette forestière à collier noir
L 13 cm (5"). *A. p. pulchra*. Combination of *black breast-band* and *rufous flanks* distinctive. **Adult** Head and upperparts slate-grey. Tail feathers tipped white except central pair, outer feathers mainly white. Throat creamy-white bordered by broad black breast-band; rest of underparts white with conspicuous rufous-chestnut flanks. Eye brownish. Bill black. Legs brownish-pink. **Juvenile** Breast-band dusky grey. **VOICE** Broad repertoire includes rhythmic series of mewing *pew-pew-pew-pew-...* or *pwee-pwee-pwee-...*, rapid *kewkewkewkewkewkew...* and slower *pweet pweet pweet pweet*. [CD11:20] **HABITS** Usually in small parties, in dense undergrowth of montane forest and its edges. Restless; actively forages low in dense tangles, flicking tail up and down, and sideways, and often holding it cocked at an acute angle. Occasionally moves higher. Not shy. **STATUS AND DISTRIBUTION** Locally common forest resident in highlands of SE Nigeria (Mambilla Plateau; Gotel Mts) and W Cameroon (Bamenda Highlands).

YELLOW-BREASTED APALIS *Apalis flavida*   Plate 105 (1)
Apalis à gorge jaune

L 11.5 cm (4.5"). *A. f. caniceps*. **Adult** Head grey; upperparts and tail green; underparts white with yellow breast-band. Eye reddish-brown. Bill black. Legs pinkish grey. **Juvenile** Head green, concolorous with upperparts; throat and breast yellow. Bill pale. **VOICE** Male song a dry, rhythmic *churup-churup-churup-churup-...* very similar to one of the songs of Grey-backed Camaroptera. In duet, female utters a sharp *kep-kep-kep-kep-....* [CD11:8] **HABITS** Occurs in gallery forest and thickets in wooded savanna, forest patches and edges, bush, and mangroves (Nigeria, Gabon). In pairs or small family parties, from low down to treetops. Often in mixed-species flocks. **SIMILAR SPECIES** Grey-backed Camaroptera (q.v.) has some very similar forms of song. **STATUS AND DISTRIBUTION** Patchily distributed, generally uncommon and local resident, recorded from Gambia, Guinea, Sierra Leone, N Ivory Coast (rare), Ghana, Burkina Faso, Togo, Benin, Nigeria, Cameroon, S Chad, CAR, W Gabon and S Congo.

MASKED APALIS *Apalis binotata*   Plate 105 (2)
Apalis masquée

L 10 cm (4"). Greenish with blackish head. **Adult male** Head dark slate-grey; upperparts and tail green. Throat to centre of upper breast black bordered by white malar stripe (often reduced to short streak at base of neck) and yellowish-green breast-sides. Rest of underparts white with yellowish flanks and undertail-coverts. Eye pale brown. Bill black. Legs pinkish. **Adult female** As male, but always has a complete white malar stripe. **Juvenile** Head grey-green; throat greyish; rest of underparts tinged yellow. **VOICE** Song a fast series of sharp, dry notes *tiree-tiree-tiree-...* or *trièk-trièk-trièk...*; with variations. Female often joins duet with fast *tatatatatata...* [CD11:9] **HABITS** In pairs in old clearings, edges of second growth and degraded forest patches near cultivation. Frequents bushes, small trees and dense vine tangles. **SIMILAR SPECIES** Black-throated Apalis has broad moustachial stripe and bright yellow underparts. **STATUS AND DISTRIBUTION** Uncommon to locally not uncommon resident in lowland and submontane areas (up to 1000 m), W and S Cameroon (Rumpi Hills, Mt Kupe, Mt Manenguba, and from Yaoundé south), Eq. Guinea (Mt Alen) and NE Gabon.

BLACK-THROATED APALIS *Apalis jacksoni*   Plate 105 (3)
Apalis à gorge noire

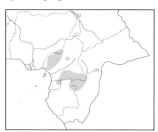

Other name F: Apalis à moustaches blanches

L 11.5 cm (4.5"). Colourful montane apalis. **Adult male *bambuluensis*** Easily identified by mainly *black head and throat with conspicuous white moustachial streak*, and bright yellow underparts. Upperparts olive-green; flight feathers blackish, edged white on secondaries and tertials; tail dark grey tipped and edged white. Eye brown. Bill black. Legs pinkish. ***A. j. minor*** has top of head sooty-grey; upperparts paler, brighter green. **Adult female** Duller, with grey head. **Juvenile** As adult female, but even duller, with head yellowish-green (greyer in young male); moustachial streak tinged yellow; bill paler. **VOICE** Very different from other apalises and difficult to transcribe. A rapid *ku-kree ku-kree ku-kree...* Duets. [CD11:11] **HABITS** Singly or in pairs in canopy of montane forest, incl. edges; locally also in lower altitude forest. Sometimes with other small insectivores. **SIMILAR SPECIES** Masked Apalis is less brightly coloured and has narrow malar stripe. **STATUS AND DISTRIBUTION** Very local forest resident. *A. j. bambuluensis* in highlands of SE Nigeria (Mambilla Plateau) and W Cameroon; *minor* also at lower altitudes (down to *c.* 500 m) in S Cameroon, Eq. Guinea (Mt Alen), SW CAR (Ngotto), NE Gabon (Mékambo area) and NW Congo (Odzala NP). Common to rare.

BLACK-CAPPED APALIS *Apalis nigriceps*   Plate 105 (4)
Apalis à calotte noire

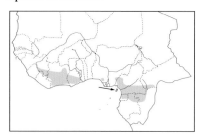

Other name F: Apalis à tête noire

L 11.5 cm (4.5"). *A. n. nigriceps.* **Adult male** *Head black; mantle golden-yellow becoming yellowish-green on rest of upperparts.* Tail slate-grey tipped white. *Throat pure white bordered by black crescent-shaped breast-band;* rest of underparts greyish-white. Eye reddish-brown. Bill black. Legs pinkish. **Adult female** Pattern as male but head and breast-band grey; throat creamy-white. **Juvenile** As adult female but head tinged green; breast-band lacking; underparts tinged yellow. **VOICE** Song a monotonous, rhythmic series of a single note *tzrrrrr tzrrrr tzrrrr* or *turrrirrrp turrrirrrp turrrirrrp....* Sings persistently throughout the day. Duets. [CD11:14] **HABITS** In pairs or small parties in canopy of lowland and gallery forest. Usually in mixed flocks of small insectivores. Very active. **STATUS AND DISTRIBUTION** Locally common to rare resident in lowland forest and its outliers in savanna, from NE Sierra Leone and SE Guinea to Ghana and from S Nigeria to Gabon, SW CAR and N Congo; also Bioko.

SHARPE'S APALIS *Apalis sharpii*   Plate 105 (10)
Apalis de Sharpe

L 11.5 cm (4.5"). **Adult male** Very dark with *wholly sooty-grey plumage*, underparts becoming soft grey from breast down. Tail narrowly tipped white. Eye red-brown. Bill black. Legs reddish-brown. **Adult female** Dark grey above; *throat deep buff;* rest of underparts pale grey becoming white on central belly. **Juvenile** Greyish-olive above; underparts pale lemon; breast-sides and flanks greyish. Bill black above, pink-horn below. **Immature male** Similar to adult female, but throat paler buff and rest of underparts greyish with pale yellowish-green patch on belly. **VOICE** Song a monotonous, fast, rhythmic *cherit-cherit-cherit-...* or *tirrit-tirrit-tirrit-...* given throughout the day; very similar to Buff-throated Apalis. Duets. [CD11:15] **HABITS** A forest species, foraging actively in canopy, sometimes lower near edges. Usually in pairs, with mixed parties of small insectivores. **STATUS AND DISTRIBUTION** Not uncommon to locally common forest resident, from W Guinea to Ghana. Rare, Togo.

BUFF-THROATED APALIS *Apalis rufogularis*   Plate 105 (7)
Apalis à gorge rousse

L 11.5 cm (4.5"). **Adult male *rufogularis*** *Entire head, upperparts, throat and upper breast dark slate-grey;* 4 outer tail feathers white (conspicuous in flight); rest of underparts creamy-white, washed greyish on flanks. Eye red-brown. Bill black. Legs brownish-pink. *A. r. sanderi* has head darker; throat black. **Adult female** Similar, but *throat and*

upper breast cinnamon-rufous. **Juvenile** Upperparts dull olive-green; underparts pale yellow; bill with pale base to lower mandible. **VOICE** Song a fast, monotonous, rhythmic *truit-truit-truit-...* or *cheerk-cheerk-...* with variations. [CD11:16] **HABITS** In pairs or small family parties within canopy of lowland forest. Regularly joins mixed flocks of small insectivores. Active and restless, like congeners. **STATUS AND DISTRIBUTION** Not uncommon to locally common forest resident, from S Benin to SW CAR and Congo; also Bioko. Also reported from C Nigeria, NE Cameroon and NE CAR. *A. r. sanderi* in SW Nigeria (and presumably S Benin); nominate in rest of range.

BAMENDA APALIS *Apalis bamendae* Plate 105 (9)
Apalis du Bamenda

L 11.5 cm (4.5"). **Adult** Greyish with *rufous forehead, head-sides and throat.* Crown and upperparts dark grey tinged brownish. Rest of underparts grey, paler on central belly. Eye yellowish-brown. Bill black. Legs pinkish. **Juvenile** Dull version of adult with crown and upperparts more olive-brown, rufous on face and throat paler, and underparts dirty cream. **VOICE** Song consists of a simple short note, monotonously repeated in a relatively slow rhythm, resembling song of Gosling's Apalis. Another song uses a 3-note motif *tswee-tit-tit tswee-tit-tit ...* or *tsu-twit-twit tsu-twit-twit ...* Also a fast *tsutititititititititit.* [CD11:18] **HABITS** Little known. Found in canopy of (even small) forest patches in savanna within montane and submontane areas (680–2050 m). **SIMILAR SPECIES** None, but to some observers song perhaps reminiscent of Grey-backed Camaroptera. **STATUS AND DISTRIBUTION** Not uncommon to locally common resident in highland areas of W and C Cameroon (Bamenda Highlands, Adamawa Plateau).

GOSLING'S APALIS *Apalis goslingi* Plate 105 (8)
Apalis de Gosling

L 11.5 cm (4.5"). **Adult** Head and upperparts slate-grey. Tail shorter than other apalises; dark brownish, all feathers, except central pair, narrowly tipped white. Throat distinctly pale buff or creamy-white; rest of underparts pale grey becoming whitish on central belly. Eye red-brown. Bill black. Legs pinkish. **Juvenile** Upperparts tinged greenish; throat and centre of underparts pale yellow. **VOICE** Song a short, fast, rhythmic series of a single note *twit-twit-twit-twit-...* and faster *twitititititititit,* slightly reminiscent of Banded Prinia. [CD11:19] **HABITS** In pairs or small family parties in forest bordering rivers. Mostly forages at mid-level. **SIMILAR SPECIES** White-chinned Prinia (q.v.) has similar plumage pattern, but is distinctly larger, occurs in different habitat and has different voice. **STATUS AND DISTRIBUTION** Locally common forest resident, S Cameroon, NE Gabon, SW CAR and N Congo.

GREY APALIS *Apalis cinerea* Plate 105 (6)
Apalis cendrée

Other names F: Apalis grise, Fauvette forestière à tête brune
L 13 cm (5"). **Adult** *funebris* Head dark ashy-grey to dusky grey-brown; upperparts grey; 3 outer tail feathers mainly white; underparts creamy-white, flanks washed greyish. Eye pale brown. Bill black. Legs pinkish. *A. c. sclateri* very similar; top of head slightly more grey-brown; underparts without greyish wash. **VOICE** Song a monotonous series of a single, rather croaking note (difficult to transcribe) *kwek-kwek-kwek-kwek-...* Often preceded by a sweet, descending trill *pirrrrrrrr* (reminiscent of ringing telephone), or *trrrr-tik-tik* with variations. Also duets, male (presumably) uttering *kwek* notes, female series of *pirrrrr* trills. [CD11:12] **HABITS** In canopy and mid-levels of montane and mid-altitude forest, occasionally descending to undergrowth. Singly, in pairs or most commonly in small parties, or with mixed flocks of small insectivores. Frequently fans and flicks tail, like congeners. **STATUS AND DISTRIBUTION** Locally common forest resident in highlands of SE Nigeria (Obudu and Mambilla Plateaux, Gotel Mts), W Cameroon, N Gabon (where also partial migrant or vagrant to lowland forest) and Bioko. *A. c. sclateri* confined to Mt Cameroon and Bioko.

RED-FRONTED WARBLER *Urorhipis rufifrons*
Fauvette à front roux

Plate 104 (10)

Other name F: Apalis à front roux

L 11 cm (4.25"). *U. r. rufifrons*. Small, pale dry-country warbler *with contrasting rufous forehead*. **Adult** Forehead and crown have variable amount of rufous; rest of head and upperparts plain pale mouse-brown. Long, graduated tail black broadly tipped white. Underparts creamy-white with cinnamon thighs. Eye pale brown. Bill black or dull pinkish-brown. Legs dark pinkish. **Juvenile** Similar, but forehead has only faint trace of rufous. **VOICE** Song a simple series of chirping notes *tik tik tik tik ...* and a rhythmic *tsyep-tsyep-tsyep-tsyep-...* Alarm a sharp *seep-seep* or *tzii* and trill *spispihehehe*. [CD11:23] **HABITS** Inhabits arid bush and desert scrub. Usually in pairs actively foraging in low bushes, with tails constantly moving in all directions. Not shy. **STATUS AND DISTRIBUTION** E African species reaching westernmost limits of range in Chad. Uncommon resident. **NOTE** Formerly placed in genus *Spiloptila*; sometimes in *Apalis* (Red-faced Apalis).

WHITE-TAILED WARBLER *Poliolais lopezi*
Poliolaïs à queue blanche

Plate 103 (1)

Other name F: Camaroptère à queue blanche

L 10 cm (4"). Small, short-tailed warbler with marked sexual dimorphism. Three races occur. In all forms tail feathers mainly white except 2 central pairs, which are dark brown. **Adult male *alexanderi*** Head blackish-grey; upperparts blackish-grey with olive wash, paler on rump. Throat slate-grey; rest of underparts pale olive-green becoming pale greyish on central breast and belly. Thighs olive-brown; undertail-coverts buff. Eye pale brown. Bill blackish. Legs brown. In **nominate** and *manengubae* upperparts dark sooty-grey; underparts sooty-grey with ashy-grey tinge on belly (*manengubae* has edges of flight feathers and tertials, and thighs dark olive-green, but this hardly visible in field). **Adult female** (all races very similar) Forehead and head-sides rufous-chestnut merging into dark greyish-brown crown and upperparts. Underparts dusky-grey washed rufous on throat and upper breast; flanks, thighs and undertail-coverts rusty-brown. **Juvenile** Head and upperparts olive-brown; underparts olive-green becoming yellower in central breast and belly. **VOICE** Song a regular series of clear, high-pitched, single notes, each repeated several times, e.g. *hweet hweet hweet hweet hweet tsuit tsuit tsuit tsee tsee tsee tsee tsee tsuui tsuui tsuui ...* often lasting for several minutes. Calls include a clear *hee-huuw hee-huuw...* Alarm a plaintive, high-pitched *peep*, uttered in series and reminiscent of Grey-backed Camaroptera. [CD11:28] **HABITS** Frequents undergrowth thickets in montane forest, incl. edges. Behaviour similar to camaroptera, restlessly moving within dense vegetation, where difficult to observe. Usually in pairs or small family parties. Not shy. **STATUS AND DISTRIBUTION** Not uncommon local forest resident in highlands of SE Nigeria (Obudu Plateau: *manengubae*), W Cameroon (Mt Cameroon: *alexanderi*; Rumpi Hills, Mt Kupe, Mt Nlonako, Mt Manenguba: *manengubae*) and Bioko (*lopezi*). *TBW* category: Near Threatened. **NOTE** Scientific name sometimes spelt *lopesi* (e.g. *BoA*), but see Dowsett & Dowsett-Lemaire (1993).

Genus *Camaroptera*

Small, with rather long straight bill; colour of thighs contrasting with belly. Skulk in dense vegetation, mostly low. Distinctive mewing calls. Pouch-like nest sewn into leaves of a shrub. Endemic to Africa.

GREY-BACKED CAMAROPTERA *Camaroptera brachyura*
Camaroptère à tête grise

Plate 103 (4)

Other names E: Bleating Warbler; F: Camaroptère à dos gris

L 11.5 cm (5"). **Adult *tincta*** Small *grey* warbler with contrasting *yellowish-green wings*, formed by greenish scapulars, wing-coverts and edges to flight feathers. Central belly whitish. Thighs orange-buff. Eye pale brown. Bill black. Legs pinkish. Paler savanna race ***brevicaudata*** assumes non-breeding plumage in dry season (Oct–May) when upperparts become ashy-brown and underparts paler; bill black above, pale horn below. **Juvenile** Head, upperparts and tail dull olive-green; underparts pale lemon-yellow. **VOICE** Calls include a distinctive mewing note. Song variable, but always consists of rhythmic repetitions (variable in speed) of a single note, e.g. *churrup-churrup-churrup-...*, *churp-churp-churp-...*, *chrup chrup chrup ...*, *kechup-kechup-kechup-...*, *chupchupchupchup...*etc. [CD11:32] **HABITS** Singly or in pairs within dense shrubbery in a variety of habitats, incl. gardens, from desert edge to coast. Very active; skulking. Tail often held cocked at sharp angle over back. **SIMILAR SPECIES** Yellow-breasted Apalis (q.v.) has some very similar song types. **STATUS AND DISTRIBUTION** One of the commonest and most widespread resident African species, occurring throughout. *C. b. tincta* in forest zone, from Liberia to Congo; *brevicaudata* in savanna zone, from Senegambia to Sierra Leone east to Chad and CAR.

YELLOW-BROWED CAMAROPTERA *Camaroptera superciliaris*   Plate 103 (2)
Camaroptère à sourcils jaunes

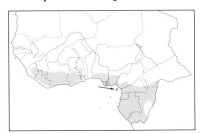

Other name F: Camaroptèr à sourcils
L 11 cm (4.5"). Short tailed and rather long billed with distinctive colour pattern. **Adult** Top of head and upperparts bright green; bright yellow supercilium and ear-coverts; black lores; underparts greyish-white becoming whiter on belly; thighs and undertail-coverts yellow. When singing, throat of male swells and reveals bare blue patch at sides. Eye pale brown. Bill black. Legs pinkish or greyish-brown. **Juvenile** Throat and upper breast bright yellow. **VOICE** Song a distinctive double nasal note, sounding like *koa-koa* or *kweh-kweh*. Female may respond with nasal *maaaah*. [CD11:34] **HABITS** Singly or in pairs in dense shrubbery and tangles at forest edges and in overgrown farmland. Wing-snapping often heard. **STATUS AND DISTRIBUTION** Uncommon to common resident in forest zone, from Sierra Leone and SE Guinea to CAR and Congo; also Bioko.

OLIVE-GREEN CAMAROPTERA *Camaroptera chloronota*   Plate 103 (3)
Camaroptère à dos vert

L 11 cm (4.5"). Secretive, short tailed and rather long billed. **Adult** *chloronota* Top of head and upperparts dark olive-green; lores and ear-coverts grey. Underparts slate-grey becoming greyish-white on undertail-coverts and washed dusky-greenish on breast. Thighs cinnamon. Eye pale brown. Bill black above, paler below. Legs yellowish-brown. Western *kelsalli* has lores and ear-coverts washed rufous and underparts uniform grey. Island race **granti** deeper grey below. *C. c. toroensis* has forehead, head-sides and throat pale brownish or rufous-buff. **Juvenile** Upperparts more yellowish-green; underparts pale lemon-yellow washed grey on flanks. **VOICE** Song a remarkable, loud and sustained series of a single note uttered at same pitch and without interruption for up to several minutes before ceasing abruptly. Unmistakable. [CD11:35] **HABITS** Singly or in pairs within dense undergrowth at forest edges. Skulking and usually difficult to see, but distinctive song attracts attention. **SIMILAR SPECIES** Adult Grey-backed Camaroptera has conspicuously bright green wings and longer tail; juveniles very similar but jizz different. **STATUS AND DISTRIBUTION** Locally not uncommon resident in lowland forest and adjacent savanna zone, from Guinea and Sierra Leone to Ghana (*kelsalli*) east to Cameroon, CAR and Congo (nominate); also Bioko (*granti*) and SE CAR (*toroensis*). Rarely recorded in Senegambia and S Mali.

MIOMBO WREN WARBLER *Calamonastes undosus*   Plate 103 (5)
Camaroptère du miombo

L 13 cm (5"). **Adult** *cinereus* Distinctive, *wholly dark grey* with rather *long tail*. Wings browner grey; throat and belly paler with faint traces of barring (only noticeable at close range). Eye reddish-brown. Bill black. Legs greyish. **Juvenile** Slightly duller or tinged greenish. **VOICE** A loud, variable, emphatic note repeated in long series *tewp, tewp, tewp, tewp,...* or *peewk, peewk,...*; also transcribed as *pwee* and *tchiwup*. Calls include muffled *prree-pree*, nasal *maaaah* and sharp *tsik tsik*. [CD11:33] **HABITS** Singly or in pairs in thickets and dense scrub in wooded savanna. Tail slowly but constantly raised and lowered. **SIMILAR SPECIES** No other camaroptera has entirely grey plumage; if colour cannot be seen (e.g. in shade of dense vegetation), note tail length. **STATUS AND DISTRIBUTION** Not yet recorded in our region, but known to occur in Cabinda and in Congo-Kinshasa along Congo R., from Kinshasa to Mbandaka. **NOTE** Sometimes treated as race of *C. simplex* (Grey Wren Warbler). Close to *Camaroptera* and sometimes placed within that genus.

Genus *Macrosphenus*

Aberrant warblers with long straight bill, sharply hooked at tip, short wings and tail, and long, loose feathers on rump and flanks. Occur in dense undergrowth, tangles and vines of forest and forest edge. Song distinctive. Endemic to Africa.

YELLOW LONGBILL *Macrosphenus flavicans*   Plate 103 (8)
Nasique jaune

Other name F: Fauvette nasique jaune
L 13 cm (5"). **Adult *flavicans*** Head, upperparts and tail largely olive-green (darker, olive-grey on head; yellower on rump). Throat dirty white; rest of underparts olive-yellow, deeper on flanks. Eye yellow. Bill blackish, paler below. Legs grey. *M. f. hypochondriacus* more richly coloured below. **Juvenile** Head olive-green, uniform with upperparts. **VOICE** Song a series of single, clear whistles, descending in scale and slightly accelerating at end, *hee hu hu hu hu hu hu-hu-hu*, resembling that of Brown-chested Alethe. [CD11:55] **HABITS** In pairs or family parties in forest and near edges; mainly in dense tangles from mid-level to lower canopy. Often in mixed-species flocks.
SIMILAR SPECIES Juvenile Kemp's Longbill has lemon-yellow throat. Grey Longbill is much duller. Compare song with that of Brown-chested Alethe and Sabine's Puffback. **STATUS AND DISTRIBUTION** Uncommon to locally not uncommon forest resident, from SE Nigeria (east of Niger R.) to Cameroon, Bioko, Gabon and S Congo (nominate); also SE CAR (*hypochondriacus*).

KEMP'S LONGBILL *Macrosphenus kempi*   Plate 103 (6)
Nasique de Kemp

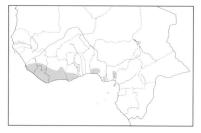

L 13 cm (5"). **Adult *kempi*** Distinctive, with orange-rufous flanks and undertail-coverts, and conspicuous yellow eye. Head, upperparts and tail dark grey-brown. Throat whitish; breast greyish-brown becoming grey in centre and on belly. Bill black with pale base. Legs dark grey. Eastern *flammeus* is richer coloured below. **Juvenile** Head, upperparts and tail olive-green; rump slightly yellower. Throat lemon-yellow; rest of underparts pale yellowish-green with olive wash to breast. Eye grey. Legs brownish. **VOICE** Song a distinctive, melodious *tee tuwe-tuwe-tuwe-tuwe tee*, with variations. Also a fast series of rather tuneless, scratchy, husky and nasal notes. Between songs, short *churrs*, sharp *tik-tik*s and various other short notes are often uttered. [CD11:56] **HABITS** Singly or in pairs in dense tangles within forest undergrowth and shrubbery at forest edges. **SIMILAR SPECIES** Compare juvenile with Yellow Longbill (whitish throat, yellow eye). **STATUS AND DISTRIBUTION** Uncommon forest resident, from Sierra Leone and SE Guinea to Ghana (nominate), and in W and SE Nigeria and SW Cameroon (Korup NP) (*flammeus*).

GREY LONGBILL *Macrosphenus concolor*   Plate 103 (7)
Nasique grise

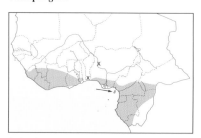

Other name F: Fauvette nasique grise
L 11.5 cm (4.5"). Small, short-tailed and long-billed forest species, lacking distinctive markings to its olive-green plumage. **Adult** Head, upperparts and tail dark olive-green; underparts paler, olive-grey. Feathers of lower back and flanks very long and occasionally puffed out. Eye greyish to dull orange. Bill dark with pale base. Legs pale pinkish. **Juvenile** Paler; underparts dirty white washed pale olive; breast tinged yellow. **VOICE** Song distinctive: a lively, sustained, extremely rapid warble; also slower versions of short, repeated phrases. Less frequent song consists of very rapid, repetitive imitations of other forest species (incl. Grey-throated Flycatcher, Blue-headed Crested Flycatcher, Rufous-vented Paradise Flycatcher, Gosling's Apalis, Pied Hornbill, Rufous Ant Thrush). [CD11:54] **HABITS** In pairs in forest and near edges, occurring at all levels and favouring vines and dense tangles. Often joins mixed-species flocks. **SIMILAR SPECIES** Fraser's Sunbird is similar in size and coloration, but has conspicuous eye-ring, dark legs and thinner, dark and slightly curved bill. Yellow Longbill distinguished by whitish throat contrasting with olive-yellow underparts, brighter eye, longer bill and entirely different song. **STATUS AND DISTRIBUTION** Common forest resident, from W Guinea to CAR and Congo; also Bioko.

Genus *Eremomela*

Small, active, insectivorous warblers, occurring in pairs or small parties and often joining mixed-species flocks. Range from arid *Acacia* country to lowland forest. Endemic to Africa.

YELLOW-BELLIED EREMOMELA *Eremomela icteropygialis*   Plate 101 (13)
Érémomèle à croupion jaune

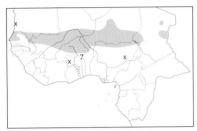

Other name F: Érémomèle gris-jaune
L 10 cm (4"). *E. i. alexanderi.* **Adult** Head and upperparts pale brownish-grey; faint supercilium, throat and breast greyish-white; rest of underparts lemon-yellow. Eye pale brown. Bill blackish above, pale horn below. Legs blackish. **Juvenile** Duller; wing-coverts and tertials tipped and edged buff. **VOICE** A short, clear song, similar to Northern Crombec's. [CD11:39] **HABITS** In pairs or small parties in arid *Acacia* scrub bordering desert. **SIMILAR SPECIES** Senegal Eremomela has yellowish-green upperparts and yellow on underparts extending onto breast. **STATUS AND DISTRIBUTION** Common to uncommon resident in arid belt just south of Sahara, from S Mauritania and N Senegal to Niger, N Nigeria and Chad. Also reported from N Ghana (rare; migrant?)

SALVADORI'S EREMOMELA *Eremomela salvadorii*   Plate 101 (14)
Érémomèle de Salvadori

L 10 cm (4"). **Adult** Similar to Yellow-bellied Eremomela (q.v.), but darker and more colourful, with olive-green upperparts, browner wings, whitish throat becoming pale grey on breast, and brighter yellow on rest of underparts. Eye dark brown or black. Bill blackish. Legs black. **Juvenile** No information, but presumably as juvenile Yellow-bellied. **VOICE** A short, clear song, similar to Yellow-bellied Eremomela's. [CD11:40] **HABITS** In pairs or small parties in wooded grassland. **SIMILAR SPECIES** Green-capped Eremomela, the only other eremomela sharing Salvadori's habitat and range, has head washed yellow, pale yellow eye, and lemon-yellow throat and breast. **STATUS AND DISTRIBUTION** Locally common resident, SE Gabon and Congo. **NOTE** Generally considered a race of Yellow-bellied Eremomela, but treated as distinct species by *BoA*.

SENEGAL EREMOMELA *Eremomela pusilla*   Plate 101 (11)
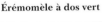
Érémomèle à dos vert

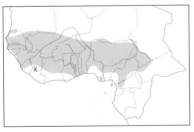

L 10 cm (4"). **Adult** Small with pale brownish-grey head and faint white supercilium; upperparts green, brighter, yellowish-green on rump; throat and upper breast white, rest of underparts bright lemon-yellow. Bill blackish above, pale below. Legs pale brownish. **Juvenile** Duller. **VOICE** A cheerful little chattering trill, constantly uttered when foraging. Dawn song a rhythmic, monotonous *whirp-whirp-whirp-...*, delivered from perch, prior to and during breeding season. Rarely heard territorial song a rapid harsh chattering. [CD11:44] **HABITS** In small parties, actively foraging and flitting between trees in wooded savanna. **SIMILAR SPECIES** Yellow-bellied Eremomela has uniformly grey head and upperparts, yellow on underparts restricted to belly, and shorter tail. Replaces Senegal Eremomela in arid north. **STATUS AND DISTRIBUTION** Common resident in broad savanna belt, from Senegambia to Sierra Leone east to W Chad and NW CAR. Rare, S Mauritania.

GREEN-BACKED EREMOMELA *Eremomela canescens*   Plate 101 (12)
Érémomèle grisonnante

L 10 cm (4"). *E. c. canescens.* **Adult** Similar to Senegal Eremomela (q.v.) but better marked and brighter, with well-defined, whiter supercilium, black eye-stripe and clear grey head sharply demarcated from yellowish-green upperparts. Bill black in most. **Juvenile** Duller. **VOICE** Song an unmusical but cheerful continuous chittering, less melodious and more complex than that of Senegal Eremomela (Chappuis 1979). Call a harsh, double note,

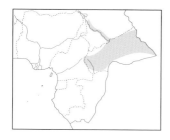

often repeated. [CD11:43] **HABITS** In small parties in open wooded savanna. Behaviour as Senegal Eremomela. **SIMILAR SPECIES** Senegal Eremomela (q.v. and above). Yellow-bellied Eremomela may overlap slightly but normally occurs in more arid country; is uniformly grey above and has yellow on underparts paler and restricted to belly. **STATUS AND DISTRIBUTION** Common resident in savanna belt in S Chad and CAR. Also E Cameroon, where intergradation with Senegal Eremomela reported along CAR border. **NOTE** Often treated as race of Senegal Eremomela.

GREEN-CAPPED EREMOMELA *Eremomela scotops* Plate 101 (9)
Érémomèle à calotte verte

Other name F: Érémomèle à tête verte
L 11 cm (*c.* 4.5"). *E. s. congensis.* Very yellow (slightly reminiscent of Yellow White-eye) with fairly open face. **Adult** *Head grey-green strongly washed yellow,* with narrow yellow line on forehead extending as faint supercilium, and dusky lores accentuating *pale yellow eye* (eye colour not always easy to see in field). Upperparts and tail grey-green; underparts lemon-yellow. Bill black. Legs brownish-pink. **Juvenile** Paler and duller. **VOICE** A twittering trill, uttered when foraging. Also harsh churring calls. Dawn song a rhythmic, monotonous *twurp-twurp-twurp-...* or *tlip-tlip-tlip-...* [CD11:41] **HABITS** In small, vocal parties restlessly foraging in wooded grassland. Also joins mixed-species flocks. **SIMILAR SPECIES** Salvadori's Eremomela, the only congener sharing its habitat and range, has whitish throat and breast, and dark eye. **STATUS AND DISTRIBUTION** Uncommon to locally common resident, SE Gabon and Congo.

RUFOUS-CROWNED EREMOMELA *Eremomela badiceps* Plate 101 (8)
Érémomèle à tête brune

L 11 cm (*c.* 4.5"). **Adult** *fantiensis* Forehead and crown bright rufous-chestnut bordered by blackish streak from lores to ear-coverts. Upperparts grey; wings and tail dusky-black. *Throat creamy-white bordered by black band;* rest of underparts pale grey, becoming creamy or yellowish in centre. Bill black. Legs reddish-brown. **Nominate** has whiter belly. **Juvenile** Lacks rufous crown and black gorget (or has only some traces of these); upperparts olive-grey; throat yellow; rest of underparts washed yellow. **VOICE** Soft little contact calls when foraging e.g. *ti-ti-tu ti-ti-tu.* Also characteristic short dry trills *trr trr trr trr ...,* similar in pitch to those of Black-capped Apalis (q.v.), but shorter. Mostly silent. [CD11:45] **HABITS** In pairs or small groups in canopy of lowland forest; sometimes descending to mid-level at edges. In Gabon also in small forest patches in savanna, venturing into surrounding woodland. Very active and often joining mixed-species flocks of small insectivores. **STATUS AND DISTRIBUTION** Common resident in forest zone, from Sierra Leone and SE Guinea to W Nigeria (*fantiensis*) east to SW CAR and Congo; also Bioko (nominate).

Genus *Sylvietta*

Very small with extremely short tails. Range from forest to arid scrub. Often in mixed-species flocks. Endemic to Africa.

NORTHERN CROMBEC *Sylvietta brachyura* Plate 102 (2)
Crombec sitelle

Other name F: Fauvette crombec
L 9 cm (3.5"). **Adult** *brachyura* Top of head and upperparts grey; underparts tawny (paler on throat and central belly); distinct buffish supercilium and dusky eye-stripe. *S. b. carnapi* has deeper coloured plumage, with slightly

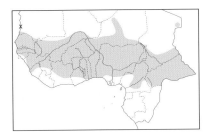

darker upperparts and almost chestnut underparts. **Juvenile** Wing-coverts tipped tawny. **VOICE** Song a short, sweet, clear warble, with variations. Call a sharp, double note. [CD11:50] **HABITS** Singly or in small parties in thorn scrub and wooded grassland. Actively inspects branches, occasionally hanging upside-down. Often joins mixed-species flocks of insectivores. **STATUS AND DISTRIBUTION** Common to uncommon resident in broad Sahel and savanna belt, from S Mauritania to Sierra Leone east to N Cameroon, Chad and N CAR (nominate), and from C Cameroon to S CAR (*carnapi*).

RED-CAPPED CROMBEC *Sylvietta ruficapilla*   Plate 102 (3)
Crombec à calotte rousse

Other name F: Crombec à joues rousses
L 10–12 cm (4.0–4.75"). *S. r. rufigenis.* Largest crombec. **Adult** Head greyish lightly washed rufous, becoming chestnut on ear-coverts; upperparts pale grey tinged olive. Lores and throat white mottled grey and bordered by chestnut crescent-shaped patch on upper breast (often forming collar). Rest of underparts pale grey tinged yellow; thighs rufous. Eye pale brown. Bill brownish-horn above, paler below. Legs pinkish-brown. **Juvenile** Flanks washed rufous. **VOICE** Song a clear little warble ending with 2 clear notes, with emphasis on last *trurrtritrut-trurrtritrut-twu-twee* or *richi-chichi-chichir.* Call *chik.* [CD11:46] **HABITS** In pairs or small parties foraging in small trees in wooded grassland. Joins mixed-species flocks. **STATUS AND DISTRIBUTION** Uncommon resident, SE Gabon and Congo.

GREEN CROMBEC *Sylvietta virens*   Plate 102 (1)
Crombec vert

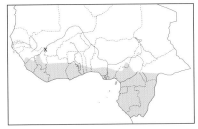

Other name F: Fauvette crombec verte
L 9 cm (3.5"). Very small and appearing tail-less. **Adult** *virens* Top of head dark brown with faint pale supercilium and pale brownish face. Upperparts dark olive-green. Throat and breast pale brown (tinged ginger on breast); lower underparts greyish becoming white on belly. Bill dark brownish above, paler below. Distinctive western race *flaviventris* more olive-green above; lower breast and upper belly yellow. Southern *tando* as nominate but upperparts more olive-green; throat and breast more rufous; belly whiter. **Juvenile** *virens* Paler than adult, with whitish throat and central belly tinged yellowish. Juvenile *flaviventris* has underparts mostly yellow. **VOICE** Song a short, sweet, rather high-pitched, clear whistle, descending in scale. Call a dry *prrt, prrt..* [CD11:48] **HABITS** In pairs or small family parties in second growth, dense shrub and tangles along tracks and clearings; mostly to mid-levels. Much more often heard than seen. **SIMILAR SPECIES** Lemon-bellied Crombec of ssp. *hardyi* distinguished from adult *S. v. flaviventris* by yellow extending from breast to undertail-coverts. In poor light difficult to distinguish from juvenile *flaviventris*, but Lemon-bellied more restricted to forest and usually frequents canopy; song distinctive. **STATUS AND DISTRIBUTION** Common resident throughout forest zone and adjacent wooded savanna, from Guinea to Liberia east to SW CAR and Congo. Also Gambia, S Senegal and S Mali. *S. v. flaviventris* west of Lower Niger R., Nigeria, *virens* to east. *S. v. tando* in Congo.

LEMON-BELLIED CROMBEC *Sylvietta denti*   Plate 102 (4)
Crombec à gorge tachetée

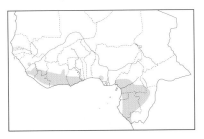

L 8 cm (*c.* 3"). **Adult** *denti* Tiny and inconspicuous, with olive-green upperparts and lemon-yellow underparts. Crown olive-grey; head-sides and throat greyish-white with some pale rufous on ear-coverts and lower throat. Breast washed olive. Eye reddish-brown. Bill short, black. Legs brownish. Western *hardyi* is deeper coloured, with breast to undertail-coverts brighter yellow. **Juvenile** Wing-coverts and tertials tipped greenish-white. **VOICE** Song a rapid series of identical, high-pitched notes (usually up to 7), *tswee-tswee-tswee-tswee-* ... easily unnoticed, or of 2 notes on different pitch *tsee-tsu-tsee-tsu-tsee-tsu-...* [CD11:49] **HABITS** Singly or in pairs within forest canopy, occasionally venturing lower at edges. Frequents mixed flocks of small insectivores. **SIMILAR SPECIES** Green Crombec of race *flaviventris* has white belly and buffish supercilium; juvenile with yellow underparts more difficult to distinguish, but note head and throat pattern, and coloration, habitat, height and voice. **STATUS AND DISTRIBUTION** Not uncommon forest resident, from Guinea and Sierra Leone to Ghana (*hardyi*), and from Nigeria to SW CAR and Congo (nominate).

Genus *Phylloscopus*

'Leaf warblers'. Very active, arboreal species with largely greenish upperparts, pale underparts, a pale supercilium and dusky eye-stripe.

WILLOW WARBLER *Phylloscopus trochilus*   Plate 101 (2)
Pouillot fitis

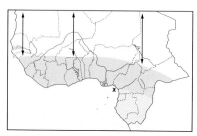

L 10.5–11.5 cm (4.5"). Small, slim and active. Upperparts olive-green or greyish-green; underparts pale yellowish on throat and breast, becoming whitish on belly. Yellowish supercilium bordered by indistinct, dusky eye-stripe. Eye dark. Bill appears mainly yellowish or pinkish; culmen brown. Legs pinkish or pale brown (exceptionally dark). Adults in fresh plumage (after moult Jul–Aug) and 1st-winters are yellower overall, with greener upperparts and evenly yellow underparts. Highly variable race **acredula** usually similar to **nominate** but paler overall, with browner upperparts and whiter underparts. **VOICE** Call a soft, plaintive *hooeet*, more disyllabic than Chiffchaff's. Song sweet, short and slightly melancholy, starting with a few faint notes, growing louder, then fading and ending in a short flourish *se-se-see-see-swee-swee-sweet-sweet-sweeut-sweetoo;* occasionally uttered in winter quarters, esp. before northward migration. [CD4:31] **HABITS** Frequents most wooded habitats, but avoids evergreen forest and arid north. Solitary or in small parties; sometimes in mixed-species flocks. Restless, very active, flicking wings and tail. Also flycatches. **SIMILAR SPECIES** Chiffchaff has less yellow in plumage, dark legs, shorter, less pointed wings, and darker looking bill. Wood Warbler is slightly larger, with brighter plumage and longer wings. Western Bonelli's Warbler is greyer and has a yellowish-green rump. **STATUS AND DISTRIBUTION** Common Palearctic passage migrant and winter visitor (Aug–Apr) throughout; also Cape Verde Is (rare), Bioko (vagrant), São Tomé and Príncipe (vagrant; on boat at sea between islands). Nominate from Cameroon west; *acredula* to east, also reported from Mauritania (rare).

CHIFFCHAFF *Phylloscopus collybita*   Plate 101 (4)
Pouillot véloce

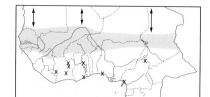

L 10–11 cm (4.5"). Resembles Willow Warbler, but is less yellow and has more uniform, darker plumage, dark legs, shorter primary projection (*c.* one third to half of tertial length), and rounder jizz. Upperparts dull olive-brown to olive-green; underparts dull buffish-white with variable amount of yellow. Eye dark. Bill blackish-brown, with pale base to lower mandible (restricted and often difficult to see). Races vary in plumage tones: **nominate** and **brehmii** mainly brownish-olive above and dull yellowish below; **abietinus** colder, greyer above, whiter, less yellow, below. Differences very slight and lone birds impossible to identify subspecifically with certainty in field. **First-winter** similar to adult. **VOICE** Call *hweet*, more monosyllabic than Willow Warbler's. Song of nominate and *abietinus* a diagnostic, rhythmic *chiff-chaff-chiff-chiff-chaff-chiff-chaff-...* introduced with low *krr-krr;* in winter quarters uttered mainly from Dec. *P. c. brehmii* has quite different, short rhythmic song ending in a short trill *tup-tup-tup-weet-weet-tsu-tchutututu.* [CD4:28, 30] **HABITS** Frequents various open wooded habitats, incl. *Acacia* scrub, savanna woodland, gardens and mangroves. Often pumps tail up and down when foraging. **SIMILAR SPECIES** Willow Warbler (q.v. and above); note also latter's better defined head markings, longer primary projection (about three quarters of, or equal to, tertial length). Western Bonelli's Warbler paler, with less distinct face pattern and contrasting yellowish-green rump. **STATUS AND DISTRIBUTION** Common Palearctic passage migrant and winter visitor (Sep–May) in dry savanna belt from Mauritania and Senegambia to C Guinea east to Chad. Vagrant/rare, Ivory Coast, Ghana, S Nigeria, S Cameroon and Cape Verde Is. Nominate and *abietinus* throughout; *brehmii* recorded from N Senegal. **NOTE** *P. c. brehmii* sometimes treated as separate species, Iberian Chiffchaff (Helbig *et al.* 1996). Nominate/*abietinus* then named Common Chiffchaff.

WOOD WARBLER *Phylloscopus sibilatrix*   Plate 101 (1)
Pouillot siffleur

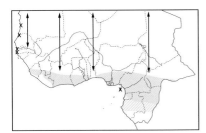

L 12 cm (4.75"). Relatively bright and well-marked arboreal warbler. Upperparts yellow-green; well-defined yellow supercilium; dusky eye-stripe; head-sides to throat and upper breast lemon-yellow; rest of underparts pure white. Dark tertials with contrasting yellowish fringes. Long wings (reaching beyond tail base) may produce short-tailed appearance. Eye dark. Bill appears mainly yellowish-pink; culmen blackish-brown. Legs yellowish-brown. Adults in fresh plumage (following moult Jun–Sep) and 1st-winters have less bright upperparts and paler yellow on underparts. **VOICE** Song a distinctive, accelerating series of *sip* notes ending in a shivering trill, *sip sip-sip-sipsipsip-sirrrrrrr*, often preceded by clear, piping *piu-piu-piu-piu*. Call a single *piu*, often repeated. Sings and calls irregularly in winter quarters, but increasingly prior to migration. [CD4:27] **HABITS** Winters in wooded savanna and forest, frequenting canopy at edges, often in mixed flocks of small insectivores. Active, but quieter than Willow Warbler, and does not flick wings and tail. **SIMILAR SPECIES** Willow Warbler (q.v.) is smaller, less attenuated. **STATUS AND DISTRIBUTION** Common to not uncommon Palearctic passage migrant and winter visitor (Aug–May). Winters in forest zone and adjacent savanna, from Guinea and Sierra Leone to CAR and Congo. Scarce/rare passage migrant in west (Mauritania to Burkina). Vagrant, Bioko.

WESTERN BONELLI'S WARBLER *Phylloscopus bonelli*   Plate 101 (3)
Pouillot de Bonelli

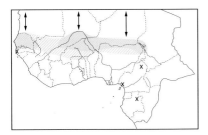

L 11.5 cm (4.5"). Pale. Upperparts pale greyish-brown with narrow yellowish-green edges to wing and tail feathers (*forming green panel on closed wing*, contrasting with dark-centred tertials), and *yellowish-green rump* (often obscured by wings). Underparts whitish with buffish suffusion. Dark eye obvious within bland face; faint pale supercilium. Bill appears mainly orangey-pink; culmen dark. Legs grey-brown. **VOICE** Call a disyllabic *hoo-eet*. Song (occasionally to frequently uttered in winter quarters) a short, shivering trill, reminiscent of finale of Wood Warbler's song, but slower, lower pitched and lacking acceleration. [CD4:26] **HABITS** Singly in dry open scrub and bushy savanna. Behaviour as Willow Warbler. Not shy. **SIMILAR SPECIES** Chiffchaff lacks bright edgings to wing feathers, has blackish bill, and often pumps tail up and down. Willow and Wood Warblers have yellow on breast and well-marked face pattern. **STATUS AND DISTRIBUTION** Common to not uncommon Palearctic winter visitor (Aug–Apr) in dry savanna belt, from Mauritania and Senegambia to N Cameroon and Chad. Vagrant, Togo, S Cameroon, NE Gabon and Cape Verde Is. **NOTE** Eastern Bonelli's Warbler *P. orientalis*, recently considered specifically distinct from *P. bonelli* (cf. *BWPC*), winters in Sudan and may reach the east of our region. Differs from *bonelli* in being greyer above and having paler yellow underwing-coverts.

BLACK-CAPPED WOODLAND WARBLER *Phylloscopus herberti*   Plate 101 (6)
Pouillot à tête noire

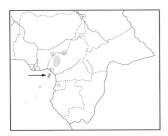

L *c.* 9 cm (3.5"). Well-marked *Phylloscopus* with very restricted range. **Adult** *camerunensis* Head to throat buff with *contrasting black crown, nape and eye-stripe*; upperparts golden-green; breast greyish becoming white on belly. **Nominate** has deeper buff on head and throat. Eye black. Bill black above, yellowish below. Legs brown. **Juvenile** Duller, with greenish supercilium and yellower underparts. **VOICE** Song a short, rapid, melodious whistle or warble, repeated at short intervals and usually delivered from open perch. Much faster, shorter and lower pitched than song of Uganda Woodland Warbler. [CD10:68] **HABITS** Occurs singly, in pairs or small parties, either mono-specifically or in association with sunbirds, in wet montane and submontane forest, from understorey to canopy (though usually high). Active, like congeners. **STATUS AND DISTRIBUTION** Common to not uncommon forest resident in highlands of SE Nigeria (Obudu and Mambilla Plateaux), W Cameroon (Mt Cameroon, Rumpi Hills, Mt Kupe, Mt Nlonako, Mt Manenguba) (*camerunensis*), and Bioko (nominate). Also Eq. Guinea (Mt Alen, above 800 m).

UGANDA WOODLAND WARBLER *Phylloscopus budongoensis*   Plate 101 (7)
Pouillot de l'Ouganda

L 10 cm (4"). Small with conspicuous whitish supercilium and black eye-stripe. **Adult** Head and upperparts plain olive-green; underparts greyish-white faintly tinged yellowish; undertail-coverts lemon. Yellow at wing bend. Bill black above, horn-coloured below. Legs pale brownish to dark grey. **Juvenile** Duller; breast tinged olive. **VOICE** Song a distinctive, short, and clear phrase of 4–8 sweet, high-pitched notes, *su-see-su-it* and *su-see-su-see-su-heeu*, with variations. [CD10:65] **HABITS** Singly or in pairs in interior of primary and secondary forest above 400 m, mainly foraging in canopy, also at mid-levels. Occasionally joins mixed-species flocks. **SIMILAR SPECIES** Green Hylia also has a conspicuous (though yellowish-green) supercilium, but is darker overall and noticeably larger; also has completely different voice. Palearctic *Phylloscopus* are paler below and lack bold face pattern. **STATUS AND DISTRIBUTION** Locally common mid-altitude forest resident, S Cameroon, Eq. Guinea, NE Gabon and NW Congo (Odzala NP).

ORIOLE WARBLER *Hypergerus atriceps*   Plate 102 (5)
Noircap loriot

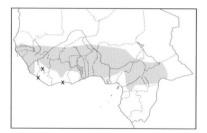

Other names E: Moho; F: Moho/Timalie à tête noire
L 20 cm (8"). Large and slender with unique jizz and plumage pattern. **Adult** Head, throat and upper breast black, feathers of crown and head-sides fringed silvery white (giving scaly appearance); upperparts and long, strongly graduated tail yellowish-olive; rest of underparts yellow. Slender, black bill. Eye reddish. Legs pale brownish. **Juvenile** Duller, with black of head and throat washed green; eye grey. **VOICE** Distinctive. Song a short, rapid and frequently repeated series of loud, melodious whistles *hu-hee-tee-teehu* and longer *hu-hee-tee-tu-hwu-hwuu* or *hu-heehee-heehu*, sometimes ending with descending *whee-whuu-whuu-whuu*, with variations. Female utters a *rikitikitikitik thuwthuwthuw* in duet. [CD11:36] **HABITS** Singly or in pairs in undergrowth of gallery forest with stands of oil palms and bamboo in wooded savanna; also on coast, in mangroves and dense shrubbery at forest edge and clearings. Often difficult to see, but emerges in the open to inspect palm fronds. **STATUS AND DISTRIBUTION** Locally common to rare resident, from Senegambia to Sierra Leone east to S Chad and CAR. **NOTE** Taxonomic position uncertain; sometimes considered a babbler on basis of morphology, but song unlike that of any babbler. Best treated as a warbler, pending further research (*BoA*).

Genus *Sylvia*

Palearctic scrub and woodland warblers. Sexes usually different. Females and immatures of some difficult to separate. Most have two characteristic vocalisations: a sharp *tak* and a harsh churr. Song a sweet pleasant warble.

BARRED WARBLER *Sylvia nisoria*   Plate 99 (2)
Fauvette épervière

L 15.5 cm (6"). *Robust*, grey or greyish-brown with relatively long tail. **Adult breeding** Upperparts greyish with whitish-tipped wing-coverts, tertials and uppertail-coverts; tail tipped and edged white (often evident as white corners in flight). Underparts white with crescentic barring (not always obvious in field). Eye pale to bright yellow. Bill rather heavy, dark brown with pale base to lower mandible. Legs strong, brownish-grey. Female often browner, with weaker barring on underparts and paler eyes. **Adult non-breeding** Following moult in Jun–Aug, upperparts browner, barring on underparts restricted to breast-sides and flanks. **First-winter** Upperparts brown to grey-brown with diagnostic *paler tips and fringes to wing-coverts, tertials and uppertail-coverts*; underparts buffish with faint barring restricted to flanks and undertail-coverts. Eye grey-brown, gradually paling and generally yellow in spring. **VOICE** Calls include a harsh *chak* and a rattling *chrrrrt-t-t-t-...* [CD10:61] **HABITS** In main wintering area frequents dry bush and woodland. Skulking, shy and rather slow. Frequently raises crown and flicks tail. Flight strong and usually heavy. **SIMILAR SPECIES** First-winter may recall Garden Warbler, but latter distinguished by smaller size, less heavy bill, shorter tail, uniform plumage (lacking markings on underparts and pale tips to wing feathers) and no white in tail. Compare also 1st-winter Orphean Warbler (neater, more uniform plumage; no pale tips to wing-coverts). **STATUS AND DISTRIBUTION** Palearctic vagrant, N Senegal (Feb, Apr), Gambia (Feb) and NE Nigeria (Oct). Also claimed from Sierra Leone. Winters in E Africa, principally S Sudan, Kenya, E Uganda and N Tanzania.

ORPHEAN WARBLER *Sylvia hortensis*
Fauvette orphée

Plate 99 (6)

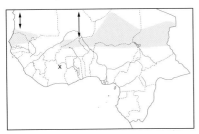

L 15 cm (6"). *S. h. hortensis*. Rather *large, stout* warbler with *black or dusky face*, white throat and *white outer tail feathers*. **Adult male** Blackish hood merges into greyish-brown upperparts. Throat white, rest of underparts whitish washed pinkish-buff on breast, flanks and undertail-coverts. Eye pale yellow to creamy-white, occasionally dark (when young?). Bill blackish with grey base to lower mandible. Legs greyish. **Adult female** Duller, browner with grey-brown crown, dusky-brown lores and ear-coverts, buffish-white underparts. **First-winter** As adult female, but paler. Crown brown as upperparts; outer tail feathers tinged buffish. Eye usually dark. **VOICE** Call a sharp *tak*. Alarm a rattling *trrrr*. [CD4:13] **HABITS** Singly in dry *Acacia* country and wooded grassland. Not shy, though often remains within cover, quietly gleaning insects in tops of trees and bushes. In flight may recall chat. **SIMILAR SPECIES** Adults distinctive, large size normally precluding confusion with other *Sylvia*, but 1st-winter may recall 1st-winter Barred Warbler (paler head-sides; pale tips to wing-coverts) or Garden Warbler (noticeably smaller; no white in tail; stubby bill). **STATUS AND DISTRIBUTION** Uncommon Palearctic winter visitor (Sep–May) to Sahel zone, in Mauritania, N Senegal, Gambia, Niger and Chad. A few records from Mali, Burkina Faso, N Ivory Coast and N Nigeria. Probably under-recorded.

GARDEN WARBLER *Sylvia borin*
Fauvette des jardins

Plate 99 (1)

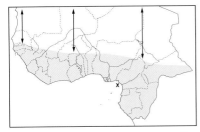

L 14 cm (5.5"). *S. b. borin*. *Featureless*, plain grey-brown and buff with gentle, open face, rather short, stubby bill, rounded crown and greyish legs. Very faint supercilium and eye-ring. Eye dark brown. Bill dark brownish with paler base to lower mandible. Lack of any distinctive features is in itself diagnostic. **VOICE** Calls a sharp *tek* and a low, harsh *chrrr*. Song, uttered from concealed perch, a sustained, melodious, even warble of largely mellow notes; in winter quarters often subdued. [CD4:14] **HABITS** In wooded savanna, bush, gardens, overgrown plantations and similar habitats. Unobtrusive and usually remaining hidden in foliage at mid-level. Often solitary but occasionally in small monospecific parties or in mixed-feeding flocks in savanna. **SIMILAR SPECIES** First-winter Barred Warbler is larger, more robust and has pale fringes or tips to wing feathers (forming indistinct wingbars), white corners to tail and variable faint barring on flanks and undertail-coverts. Compare also Olivaceous Warbler (longer bill, more elongated jizz) and 1st-winter Orphean (larger, heavier; white throat; white outer tail feathers). **STATUS AND DISTRIBUTION** Generally common Palearctic passage migrant and winter visitor (Sep–May) throughout. Vagrant, Bioko.

BLACKCAP *Sylvia atricapilla*
Fauvette à tête noire

Plate 99 (5)

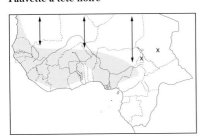

L 13 cm (5"). *S. a. atricapilla*. *Plain grey* with diagnostic *black or rufous cap*. **Adult male** Glossy black forehead and crown contrasting with grey head and upperparts; underparts paler grey becoming whitish on belly. Tail longish, dark grey (lacking white). Eye brown. Bill black. Legs dark greyish. **Adult female** Browner grey, with red-brown cap. **VOICE** Most frequently heard calls are a hard *tek* (like two pebbles knocked together; harder than Garden Warbler's) and *chrr*, both characteristic of genus. Song a rich warble starting with a hurried subdued jumble followed by clear whistles and ending in a loud flourish; recalls Garden Warbler's but typically shorter, louder, more varied and including higher pitched notes. In winter quarters mostly utters subdued, less varied version. [CD4:15] **HABITS** Solitary or in small flocks in savanna woodland, thickets, forest clearings and gardens; also mangroves (Gambia). Attracted to fruiting trees. **STATUS AND DISTRIBUTION** Generally scarce but locally common Palearctic passage migrant and winter visitor (Oct–May), from Mauritania to Liberia east to Cameroon. Vagrant, Chad. Common resident, Cape Verde Is (perhaps also Palearctic vagrant). **NOTE** Birds from Cape Verde Is usually separated as *S. a gularis*, but this race is poorly differentiated from nominate; wing usually shorter and bill longer (*BWP*).

COMMON WHITETHROAT *Sylvia communis*
Fauvette grisette

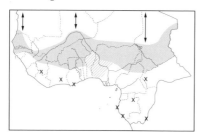

L 14 cm (5.5"). *S. c. communis.* Medium-sized, rather slim warbler with *rusty wing panel* and *white outer tail feathers.* **Adult male** Head grey tinged brown on crown and ear-coverts; upperparts grey-brown; wings dark brown, tertials and greater coverts broadly fringed rufous. Throat pure white; rest of underparts whitish with buffish-pink wash to breast. Eye pale brown, with narrow white eye-ring. Bill greyish-horn with pale base to lower mandible. Legs pale brown to pinkish. **Adult female** Head browner, underparts buffish, eye darker. **First-winter** Very much as adult female but duller; outer tail feathers fringed buffish-white. **VOICE** Calls include a sharp *tak*, a grating *charr* and a nasal *wed wed wed.* Song a brief, hurried, scratchy warble. [CD4:16] **HABITS** Usually solitary. Frequents thorn scrub, wooded savanna and gardens. **SIMILAR SPECIES** Spectacled and female/1st-winter Subalpine Warblers are noticeably smaller; Spectacled has slightly darker, more contrasting plumage, shorter tail and, in autumn, yellowish legs. **STATUS AND DISTRIBUTION** Common to scarce Palearctic winter visitor (Aug–May) in Sahel and dry savanna belt, from Mauritania and Senegambia to N Cameroon and Chad. Rare further south, in N Liberia, S Ivory Coast, S Ghana, S Nigeria, S Cameroon, CAR, Gabon, Congo.

LESSER WHITETHROAT *Sylvia curruca*

Plate 99 (3)

Fauvette babillarde

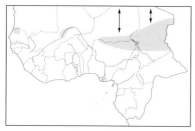

L 12.5–13.5 cm (*c.* 5.0–5.25"). *S. c. curruca.* Identified by greyish upperparts, whitish underparts and (usually) dark mask. **Adult** Head grey with darker lores and ear-coverts forming usually distinct mask (barely noticeable in some); upperparts grey-brown with darker wings; tail dark brown edged white. Throat and central belly white; breast and flanks tinged pinkish-buff. Eye grey to brown. Bill blackish with pale base to lower mandible. Legs dark grey. **First-winter** Very similar to adult but outer tail feathers buffish-white. **VOICE** Call a sharp, frequently uttered *tak* or *chek,* also *charr.* Song a brief muffled warble followed by a fast, far-carrying rattle *djedjedjedjedje.* [CD4:17] **HABITS** In *Acacia* woodland, thorny bushes in wadis and dry savanna. Skulking; often low. **SIMILAR SPECIES** Adult female Subalpine Warbler has buffish underparts and pinkish legs. First-winter Sardinian, Rüppell's and vagrant Ménétries's Warblers have pale brownish or red-brown legs. **STATUS AND DISTRIBUTION** Widespread Palearctic winter visitor (Sep–Apr), Mauritania and Senegambia (rare), Mali (uncommon), Niger (locally common), Nigeria (uncommon), Cameroon and Chad. Main wintering grounds from Chad east.

DESERT WARBLER *Sylvia nana*

Plate 100 (5)

Fauvette naine

L 11.5 cm (4.5"). *S. n. deserti.* Smallest and palest *Sylvia.* Easily identified by *sandy-buff upperparts* with rufous rump and uppertail-coverts, *white-edged, rufous-brown tail* and *yellowish eyes, bill and legs.* White eye-ring. Underparts white. **VOICE** Call a harsh *krrrr.* Song consists of short phrases usually starting with *krrrr* call and followed by a jaunty warble ending with a rising whistle. [CD4:18] **HABITS** Desert species occurring singly or in pairs and usually observed scurrying across sand before disappearing into low scrub. Not particularly shy. Tail held horizontally or slightly cocked. **STATUS AND DISTRIBUTION** Resident, Mauritania (rare), N Mali (uncommon and local) and N Niger (Aïr; rare?). Vagrant, Cape Verde Is. **NOTE** *S. n. deserti* considered specifically distinct from nominate *nana* (Middle East and C Asia; wintering in small numbers from NE Africa south to Somalia) on basis of morphology and vocalisations by Shirihai *et al.* (2001).

RÜPPELL'S WARBLER *Sylvia rueppelli*

Plate 100 (2)

Fauvette de Rüppell

L 14 cm (5.5"). **Adult male** Unmistakable. Crown, head-sides to upper breast black with *bold white submoustachial stripe.* Upperparts grey; underparts whitish with grey flanks. Wings black, tertials and greater coverts have broad,

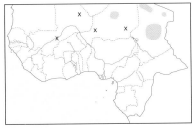

pale grey fringes. Tail black with white outer feathers. Eye red-brown; conspicuous red orbital ring. Bill blackish with paler base to lower mandible. Legs red-brown. **Adult female** Variable. Some as male but duller and browner, with mottled black crown and throat, less contrasting submoustachial stripe, grey-brown upperparts, brown-black wings with buffish-white fringes, and buffish-white underparts. Others have greyish head and whitish throat lacking dark mottling. **First-winter male** Black on head less uniform; upperparts browner. **First-winter female** Head grey; throat pale; no dark mottling; wings browner. **VOICE** Call a sparrow-like chatter; also a sharp *tak*. Song a series of short, fast, dry, chattering phrases. [CD10:62] **HABITS** Frequents low bushes in very arid country. Skulking but not shy. **SIMILAR SPECIES** Poorly marked females and 1st-winters could be confused with female/1st-winter Sardinian Warbler, but latter have browner upperparts and less conspicuous fringes to wing feathers. Lesser Whitethroat and much larger female/1st-winter Orphean Warbler lack red orbital ring and have black legs. **STATUS AND DISTRIBUTION** Not uncommon or common Palearctic winter visitor, Chad (Oct–Mar). Vagrant, Mali and Niger. Main wintering grounds from Chad to Sudan.

SARDINIAN WARBLER *Sylvia melanocephala*   Plate 100 (1)
Fauvette mélanocéphale

L 13.5 cm (5.25"). *S. m. melanocephala*. **Adult male** *Black hood contrasting with pure white throat.* Upperparts grey; *underparts whitish with greyish flanks.* Conspicuous red orbital ring. Tail black, graduated, edged and partially tipped white. Eye red-brown. Bill black with pale base to lower mandible. Legs pale brown. **Adult female** Hood dusky-grey; upperparts brown; flanks dull brown. **First-winter male/female** As respective adult but males have browner upperparts and females a brown hood; tail corners buffish; orbital ring brownish. **VOICE** Calls include a frequently uttered, harsh rattle *chretetikitikitik...*, fast, harsh *treet-treet* or *chreet* notes, and a hard *tsek*. Song an even, pleasing warble, incorporating harsh calls. [CD4:19] **HABITS** Mainly singly in bush country, wooded savanna, gardens. Skulking but not particularly shy; inquisitive. Frequently cocks tail. **SIMILAR SPECIES** Compare female/1st-winter with female/1st-winter Orphean Warbler (much larger, no red/brownish orbital ring, dark legs) and poorly marked females and 1st-winter Rüppell's Warbler (more grey-brown upperparts, distinct pale fringes to wing feathers). **STATUS AND DISTRIBUTION** Palearctic winter visitor (Sep–Apr), Mauritania (uncommon), Senegal (uncommon in north; reported once from south) and Niger (locally common in Ténéré area).

MÉNÉTRIES'S WARBLER *Sylvia mystacea*   Plate 100 (3)
Fauvette de Ménétries

L 13.5 cm (5.25"). **Adult male** Dull black hood merges into grey nape and upperparts. Wings dark brown with greyish fringes to tertials and greater coverts; tail black with white outer feathers. Underparts white tinged pink with pure white moustachial area. Prominent red-brown orbital ring. Eye red-brown. Bill brownish-horn with pale base to lower mandible. Legs brownish-pink. Races difficult to separate in field, *rubescens* overall paler than **nominate**; *turcmenica* as *rubescens* but breast pinker. **Adult female** Much paler and browner tinged. Entire upperparts sandy grey-brown; fringes to greater coverts sandy-brown. Underparts whitish tinged pale buff, esp. on flanks. Blackish, white-sided tail contrasts with rest of plumage. Narrow orbital ring pale yellowish to pinkish-brown. **First-winter** As female but outer tail feathers greyish-brown. Eye brown; orbital ring brownish. **VOICE** Calls include a sharp *tak*, a harsh *chrrr* and a sparrow-like chattering similar to that of Rüppell's Warbler. **HABITS** Frequents bushes in dry country. Skulking and restless, constantly flicking tail in all directions. **SIMILAR SPECIES** Compare female/1st-winter with very similar female/1st-winter Subalpine (paler tail; warmer buff on underparts) and Sardinian Warblers (darker; breast-sides and flanks dirty-brown). **STATUS AND DISTRIBUTION** Palearctic vagrant, N Nigeria (Kano, Apr; race unspecified, all three possible). Normal wintering range from Sudan to Somalia, Arabia and S Iran.

SUBALPINE WARBLER *Sylvia cantillans*   Plate 100 (7)
Fauvette passerinette

L 12 cm (4.75"). **Adult male** *cantillans* Easily identified by blue-grey upperparts, *rusty-orange throat, breast and flanks, distinct white moustachial stripe and red orbital ring.* Wings have pale buffish fringes to tertials and greater coverts; tail blackish edged and cornered white. Lower underparts become whitish washed pale rusty on central

belly and undertail-coverts. Eye pale brown. Bill blackish with pale base to lower mandible. Legs red-brown to yellowish-pink. *S. c. inornata* purer orange coloured (more orange-brown) below. In *albistriata* throat and breast brick-red (more chestnut-brown, less orange) and often flecked white; contrasting with pale rufous flanks and more extensive white on underparts; moustachial stripe slightly broader. *S. c. moltonii* pale brownish-pink below, lacking or with very little orange or chestnut tone (Gargallo 1994). **Adult female** Paler and duller. Upperparts pale grey-brown; underparts pinkish-buff to buffish-white on throat, breast and flanks; moustachial stripe more or less prominent. Orbital ring red to yellowish, usually surrounded by thin whitish eye-ring. **First-winter** Browner, less grey. Upperparts pale buffish-brown with pale sandy-brown fringes to tertials and greater coverts (often hardly contrasting with slightly darker, rather diffuse centres); tail edges buffish. Underparts buff on breast, flanks and vent, whitish on throat and central belly. Orbital ring whitish. **VOICE** Calls a hard but quiet *tek* (nominate and *inornata*), *trek-trek* (*albistriata*) or rattling *trrrt* (*moltonii*). Song (occasionally uttered in winter quarters) consists of regularly spaced phrases of a fast, clear and scratchy warble. [CD4:20] **HABITS** In dry Sahel country, esp. favouring *Acacia* scrub; also gardens and mangroves (Gambia and Sierra Leone). Skulking but not necessarily shy. Frequently cocks tail. **SIMILAR SPECIES** Female/1st-winter plumages sometimes difficult to separate from those of other *Sylvia*. Spectacled Warbler has rusty wing panel, very dark, contrasting tail, and slightly shorter wings. Common Whitethroat is noticeably larger. See also Tristram's Warbler. **STATUS AND DISTRIBUTION** Common to scarce Palearctic winter visitor (Sep–May) to Sahel belt, from Mauritania and Senegambia east to N Cameroon and Chad. Vagrant, Sierra Leone and Cape Verde Is. Nominate throughout; *inornata* from Senegal to W Niger; *albistriata* from E Mali eastwards. **NOTE** Race *moltonii*, breeding on Mediterranean islands, usually included in nominate (but see Gargallo 1994). Wintering range unknown.

SPECTACLED WARBLER *Sylvia conspicillata*   Plate 100 (4)
Fauvette à lunettes

L 12.5 cm (5"). Small and delicate *with striking rusty wing panel*. Resembles miniature Common Whitethroat. Narrow white eye-ring (broader above eye). Head rounded with rather steep forehead; wings relatively short. **Adult male *conspicillata*** Head grey with grey-black lores extending to below eye. Upperparts brownish; wings with broad orange-rufous fringes to tertials, secondaries and greater and median coverts, and contrasting black tertial centres. Throat pure white with centre of lower throat pale grey (white thus sometimes appears as broad malar stripe); rest of underparts greyish-pink, paler on belly. Tail blackish with white outer tail feathers. Eye red-brown. Bill dark with yellowish base to lower mandible. Legs reddish-brown to yellowish. Island race *orbitalis* somewhat more richly coloured. **Adult female** Browner. Head grey-brown; lores greyish; eye-ring less conspicuous. Underparts more buff. **First-winter** Resembles adult female, but upperparts more buffish-brown and contrasting less with uniformly pale buff underparts. Outer tail feathers buffish. Eye dark. **VOICE** Call a distinctive harsh rattle *trrrrrrr*; also *tek*. Song a typical, short, fast, high-pitched *Sylvia* chatter. [CD4:21] **HABITS** Mainly found singly or in pairs in dry scrub near coast, moving restlessly through and between low bushes, cocking tail and moving it sideways. Frequently hops on ground. **SIMILAR SPECIES** Common Whitethroat is distinctly larger and has narrower, less orangey (more chestnut) fringes to wing feathers, and longer wings; also moves tail very little. Tristram's Warbler has rufous on throat. Female/1st-winter Subalpine Warbler separated from female/1st-winter Spectacled by narrower, usually less rufous, tertial fringes, brown tail almost concolorous (not contrasting) with upperparts, longer wings and eye-ring of uniform width. **STATUS AND DISTRIBUTION** Nominate is Palearctic winter visitor (Aug–Apr), Mauritania (locally common near coast from Nouakchott to Senegal delta) and Senegambia (rare). Vagrant, N Niger. *S. c. orbitalis* is common resident, Cape Verde Is.

TRISTRAM'S WARBLER *Sylvia deserticola*   Plate 100 (6)
Fauvette de l'Atlas

Other name F: Fauvette du désert

L 12 cm (4.75"). **Adult male breeding** Head and upperparts slate-grey with *contrasting orange-rufous wing panel* formed by broad fringes to black-centred tertials, secondaries and greater coverts. Tail blackish with white edges. Underparts vinous-brown or terracotta, becoming whitish on central belly and undertail-coverts. Some whitish feathers on chin and throat may occasionally produce faint, ill-defined moustachial stripe. Eye yellow to brown with *narrow whitish eye-ring*. Bill dark with yellowish base to lower mandible. Legs yellowish-brown to red-brown.

Differences between races slight; **S. d. maroccana** slightly darker than **nominate**; dubious race **ticehursti** (perhaps an aberrant or hybrid between *S. deserticola* and *S. nana*) paler, more isabelline. **Adult female breeding** Generally paler, with browner, less grey upperparts and paler vinous underparts becoming cream-buff on throat and belly; chin white; eye-ring narrower, pale buff to off-white. **Adult male non-breeding** Similar to adult female breeding. **Adult female non-breeding** Paler, with more sandy-buff upperparts. **First-winter** Similar to respective adult non-breeding plumages. **VOICE** Call a frequently uttered, sharp *trk* or *trk-it*. [CD4:22] **HABITS** Restless and skulking, moving through scrub with cocked tail. **SIMILAR SPECIES** Compare Spectacled and Subalpine Warblers. Male Spectacled differs from male Tristram's in warmer upperparts and much paler underparts (with white throat and pinkish-tinged breast and flanks); male Subalpine has distinct moustachial stripe, red orbital ring and buffish (not orange-rufous) fringes to tertials and greater coverts. Female Spectacled very similar to 1st-winter and some female Tristram's, but upperparts usually more greyish-brown (less sandy-buff or buff-brown) and underparts paler. **STATUS AND DISTRIBUTION** Several unconfirmed records from Mauritania (Oct–Feb). This N African species is a partial migrant making only short-distance movements, some dispersing in post-breeding season south to edge of Sahara.

Genus *Hyliota*

Quite distinct, rather stocky, aberrant warblers with dark upperparts, whitish to orange underparts, and relatively short tails. Sexually dimorphic. Inhabit canopy of forest or savanna woodland; often in mixed-species flocks. Formerly regarded as flycatchers. Endemic to Africa.

YELLOW-BELLIED HYLIOTA *Hyliota flavigaster*   Plate 102 (8)

Hyliote à ventre jaune

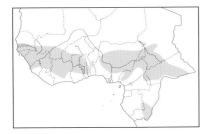

Other names F: Hyliota/Gobe-mouches à ventre jaune
L 12.5 cm (5"). **Adult male *flavigaster*** Upperparts glossy blue-black with large white wing patch; underparts pale peach. Tertials and innermost secondaries variably edged white (extending wing patch). Rather short tail with white-edged outer feathers. Bill black, base of lower mandible grey. Legs black. Southern ***barbozae*** has white patch extending further on wing, and breast usually deeper coloured. **Adult female** Duller, with dark grey-brown upperparts and paler underparts. **Juvenile** As adult female, but with some buffish mottling on upperparts and paler underparts. **VOICE** Usually silent. Call *twep twep*... Song a repeated *tiu-wheep tiu-wheep*... or higher-pitched *tseep hweet tseep-tseep hweet*... Also dry trills. [CD11:58] **HABITS** In pairs or small groups in savanna woodland, actively searching for insects among foliage of small trees, frequently upside-down. Often in mixed parties with eremomelas and Spotted Creeper. **SIMILAR SPECIES** In W Cameroon, compare extremely rare and local Southern Hyliota. **STATUS AND DISTRIBUTION** Locally not uncommon to scarce resident in savanna belt, from Senegambia to Sierra Leone east to Cameroon and CAR (nominate), and in SE Gabon and Congo (*barbozae*).

SOUTHERN HYLIOTA *Hyliota australis*   Plate 102 (7)

Hyliote australe

L 12.5 cm (5"). *H. a. slatini?* Very similar to Yellow-bellied Hyliota and almost indistinguishable in field. **Adult male** differs in having sooty-black (not glossy) upperparts and always lacks white edges to tertials and secondaries. **Adult female** has browner upperparts. **Juvenile** As adult female; feathers of upperparts and wings narrowly fringed and edged buffish; underparts paler. **VOICE** Twittering contact calls. Squeaky whistles, followed by a trilling warble. [CD11:59] **HABITS** As Yellow-bellied Hyliota. Some isolated populations north of main range occur in forest (habitat of Violet-backed Hyliota). **STATUS AND DISTRIBUTION** Puzzling isolated record from Cameroon of immature female, collected at Dikome Balue (Rumpi Hills) in secondary forest at 1300–1400 m (Serle 1965). Main range from Angola to Mozambique.

VIOLET-BACKED HYLIOTA *Hyliota violacea*   Plate 102 (6)

Hyliote à dos violet

L 12.5 cm (5"). **Adult male *violacea*** Head, upperparts and tail deep violet-blue with variable amount of white on inner greater coverts; underparts white with faint buffish wash on throat. Eye dark. Bill and legs black. Western ***nehrkorni*** has no white in wing. **Adult female** Similar but richer coloured, with throat and breast orange becoming dark buff on lower breast; nominate also has less white on wing. **Juvenile** Duller. **VOICE** Series of dry *tik-tik-tik-tik-*

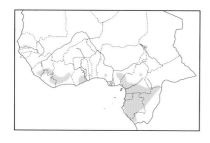

... on take-off. Song consists of short series of 4–5 sweet whistles *see-su-su-wit see-su-su-wit-wu* ... [CD11:60] **HABITS** Singly, in pairs or small family parties in canopy of lowland forest along edges and clearings. Often in mixed flocks of small insectivores. **STATUS AND DISTRIBUTION** Scarce or rare to locally not uncommon forest resident, from Guinea and Sierra Leone to Togo (*nehrkorni*) and from Nigeria to CAR and Congo (nominate). **NOTE** Some birds observed in S Cameroon (Campo, Mar 1999) had large white wing patch, similar in size to that of Southern Hyliota. Precise identity of these is unclear.

GREEN HYLIA *Hylia prasina* Plate 101 (5)
Hylia verte

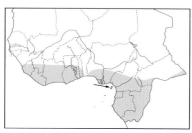

L 11.5 cm (4.5"). **Adult** *prasina* Head dark olive-green with conspicuous *pale yellowish-green supercilium* emphasised by blackish eye-stripe. Upperparts and tail dark olive-green; throat greenish-white becoming olivaceous-grey on rest of underparts. Eye dark brown. Bill short, black. Legs dark greyish. Island race *poensis* has whiter throat. **Juvenile** Greener overall; head concolorous with upperparts. Bill horn. Legs pale pinkish. **VOICE** One of the most typical forest sounds: a frequently uttered, loud, clear, double whistle *hee-hee*. Also a short dry rattle. [CD11:61] **HABITS** Singly or in pairs in lowland forest and relict forest patches. Forages at all levels, though mostly in mid-stratum, in manner recalling tit *Parus*. Often in mixed-species flocks. Betrays presence by characteristic, easily memorized, disyllabic whistle. **SIMILAR SPECIES** Scarce and local Uganda Woodland Warbler also has bold supercilium but is paler, esp. on underparts, and noticeably smaller; also has entirely different voice. **STATUS AND DISTRIBUTION** Common resident throughout forest zone, from Senegambia to Liberia east to CAR and Congo, also S Mali (*prasina*); Bioko (*poensis*). **NOTE** Relationships uncertain; variously treated as a warbler, tit, sunbird or weaver on basis of morphology. Provisionally placed in Sylviidae by most authorities (cf. *BoA*, Dowsett & Dowsett-Lemaire 1993).

SÃO TOMÉ SHORT-TAIL *Amaurocichla bocagei* Plate 145 (7)
Nasique de Bocage

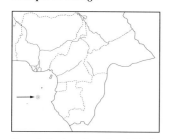

Other names E: Bocage's Longbill; F: Fauvette de Sao Tomé
L *c.* 11 cm (4.5"). Dark brown, short tailed, long billed and long legged. Affinities uncertain. Head, upperparts and tail dark chocolate-brown; upper throat white; rest of underparts pale rufous-brown becoming whitish on central belly. Bill straight and dark, with pale tip to lower mandible. Legs dark pinkish. **VOICE** Vocal. Call, frequently uttered by both sexes, a single, loud, very high-pitched, piercing *tsiiiii*, usually given from low branch or boulder; also a disyllabic *tsui-tsui* uttered in series. Song, given by male, a longer version of call (typically from branch *c.* 3 m from ground a few metres away from river; also uttered at night). [CD12:52] **HABITS** Little known. Found singly or in pairs within forest and on forested ridges with rocks and boulders, occurring on forest floor, fallen logs and low branches within 3 m of ground, and moss-covered boulders in undergrowth by rivers in primary forest, where forages rather like Grey Wagtail. Gait reminiscent of pipit *Anthus*. Flies reluctantly and for a few metres only; prefers to run; flight weak. Most active at dawn and dusk. **STATUS AND DISTRIBUTION** Endemic to São Tomé, where rare and confined to rainforest in south of island, esp. along R. São Miguel, Xufexufe, Quija, Io Grande and Ana Chaves. *TBW* category: THREATENED (Vulnerable).

FLYCATCHERS
Family MUSCICAPIDAE (World: 90 species; Africa: 37; WA: 25)

Large family of small to medium-sized insectivorous and arboreal birds, occurring in all wooded habitats, from forest to dry *Acacia* savanna. Sexes similar or not. Juveniles typically spotted. Most have broad, flattened bills with wide gape and rictal bristles. Catch insects in short sallying flight from perch, by pounding on prey on ground or by picking it from foliage in warbler-like manner. Mostly singly, in pairs or small family parties; some join mixed-species flocks. In W Africa 20 resident species and 5 Palearctic migrants, of which 2 are vagrants.

FRASER'S FOREST FLYCATCHER *Fraseria ocreata*   Plate 109 (13)
Gobemouche forestier

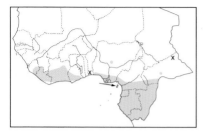

Other name E: Forest Flycatcher
L 18 cm (7"). Stout, bicoloured, rather shrike-like flycatcher. **Adult** *prosphora* Head, upperparts and tail sooty-grey. Underparts white with variable number of *dark slate-grey crescentic bars on lower throat, breast and flanks.* Bill and legs black. *F. o. kelsalli* slightly paler above; **nominate** has top of head blackish. **Juvenile** Head and upperparts sooty-brown sparsely spotted rufous-brown; underparts white washed rufous-brown on upper breast with irregular narrow bars. **VOICE** Vocal. Calls varied, mostly harsh and buzzing, but also a quavering *wruu hree hru*, a rapid *pink-pink-pink* and a whistled *weew*. Song melodious and varied, slightly quavering and interspersed by trills. Some short songs reminiscent of Velvet-mantled Drongo. Occasional imitations, e.g. of Black-and-white Flycatcher, Black-shouldered Puffback and West African Thrush. [CD11:69] **HABITS** Singly, in pairs or, most often, small parties (usually 3–7) in lowland and mid-elevation forest (up to 1600 m). Occurs at all levels, in forest proper, at edges and in clearings. Regularly joins mixed-species flocks. **SIMILAR SPECIES** White-browed Forest Flycatcher is smaller, quiet and unobtrusive, with white supraloral streak and less distinct crescents on underparts. **STATUS AND DISTRIBUTION** Uncommon to locally not uncommon resident in forest zone, from W Guinea and Sierra Leone (*kelsalli*) to Ghana (*prosphora*) and from S Benin and Nigeria (also Kagoro) to CAR, Congo and Bioko (nominate).

WHITE-BROWED FOREST FLYCATCHER *Fraseria cinerascens*   Plate 109 (12)
Gobemouche à sourcils blancs

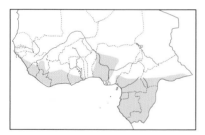

L *c.* 17 cm (*c.* 6.75"). **Adult** *cinerascens* Head, upperparts and tail dark sooty-grey; *white supraloral streak extending above eye usually conspicuous.* Underparts white with indistinct dark grey crescentic bars on throat and breast. Bill and legs black. Eastern *ruthae* has face (forehead, forecrown, lores and area below eye) black. **Juvenile** Head and upperparts dark brown densely spotted rufous-brown; underparts whitish with smudgy bars on breast and rufous-brown wash to breast and flanks. **VOICE** Song a series of drawn-out, very high-pitched *tsreee* notes, hard to detect. Call a high-pitched, drawn-out *tseee*. [CD11:70] **HABITS** Singly or in pairs in forest along rivers, lagoons and creeks. Usually low; often perches silently on bough above margin of forest stream or on stilt roots in swampy areas. Unobtrusive. **SIMILAR SPECIES** Fraser's Forest Flycatcher is larger, noisy, has more distinct crescents on underparts and lacks white supraloral streak. **STATUS AND DISTRIBUTION** Uncommon to locally not uncommon resident in forest zone, from S Senegal to Ghana (*cinerascens*) and from Nigeria to CAR and Congo (*ruthae*).

NIMBA FLYCATCHER *Melaenornis annamarulae*   Plate 108 (1)
Gobemouche du Libéria

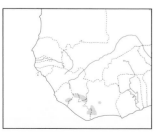

Other name F: Gobemouche noir du Nimba
L *c.* 19 cm (*c.* 7.5"). Large, robust and very dark. **Adult** Head, upperparts and tail blackish-plumbeous. Underparts slightly paler, slate-grey. Eye dark brown. Bill and legs black. **Juvenile** Unknown. **VOICE** Song consists of short, varied phrases of pleasant, melodious whistles. Calls include a thin, soft *wheep-wheep* and harsh, rather strident notes resembling some calls of drongos or Fraser's Forest Flycatcher. Usually silent. [CD11:67] **HABITS** Singly, in pairs or small parties in lowland rainforest. Mostly in canopy and upper mid-level. Also clearings. Perches upright like typical flycatcher, but also hops and runs along large branches like malimbe or cuckoo-shrike. Often inactive.
SIMILAR SPECIES Shining Drongo has similar size but wholly different jizz. **STATUS AND DISTRIBUTION** Rare to scarce and local forest resident in SE Sierra Leone (Gola Forest), SE Guinea, Liberia and W Ivory Coast. *TBW* category: THREATENED (Vulnerable).

NORTHERN BLACK FLYCATCHER *Melaenornis edolioides*   Plate 108 (2)
Gobemouche drongo

Other name E: Western Black Flycatcher
L *c.* 20 cm (*c.* 8"). **Adult male** Large and all black with long, slender tail. Eye dark; bill and legs black. **Nominate** has top of head, upperparts and tail glossed greenish-blue; eastern *lugubris* duller, more sooty-black. **Adult female** Duller; underparts washed slate-grey. **Juvenile** Duller and heavily spotted buffish-brown. **VOICE** Rather silent. Song a series of varied, mostly melodious phrases with some wheezy and scolding notes, delivered with frequent pauses. Also given by pair in duet. Call a drawn-out, high-pitched *tseeeu*; also a short *tsik*. Alarm a long, harsh churring note. [CD11:68] **HABITS** Singly or in pairs in various woodland types, incl. cultivated areas with trees. Hunts insects from perch, catching them in flight or on ground. Rather unobtrusive. **SIMILAR SPECIES** All-black plumage, upright posture and habitat may recall drongos, but these are conspicuous, vocal species with harsh voice and red eyes. Tail of similar-sized Square-tailed Drongo broader and slightly notched; that of slightly larger Fork-tailed Drongo strongly forked. Southern Black Flycatcher very similar but plumage glossed blue; ranges do not overlap. **STATUS AND DISTRIBUTION** Uncommon to not uncommon resident, from S Mauritania to Sierra Leone east to W Cameroon (nominate) and from E Cameroon to S Chad and CAR (*lugubris*).

SOUTHERN BLACK FLYCATCHER *Melaenornis pammelaina*   Plate 147 (9)
Gobemouche sud-africain

Other name F: Gobemouche noir austral
L 19–22 cm (7.5–8.75"). *M. p. tropicalis* (?). **Adult** Large, all-black flycatcher similar to Northern Black Flycatcher, but glossed violet above and steel-blue below. Long, slender tail squarer or slightly notched (not slightly rounded). **Juvenile** Duller, blackish-brown lacking gloss and heavily spotted buffish-brown. **VOICE** Rather silent. Song a series of various thin, high-pitched notes, in S Africa forming short, varied, quite melodious phrases, e.g. *tzi-tzi-teeu-teeu, wheeowheeowheeo, tzip-tseeu...* etc. [CD11:66] **HABITS** In wooded grassland. Behaviour as Northern Black Flycatcher. **SIMILAR SPECIES** Northern Black Flycatcher very similar but dull sooty-black; ranges do not overlap. For differences from drongos see Northern Black Flycatcher. **STATUS AND DISTRIBUTION** Rare and local resident, C and N Congo (ssp. undetermined, probably *tropicalis*).

PALE FLYCATCHER *Melaenornis pallidus*   Plate 108 (7)
Gobemouche pâle

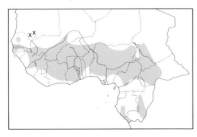

Other name E: Pallid Flycatcher
L 15–17 cm (6.0–6.75"). Medium-sized, rather nondescript savanna flycatcher. **Adult *pallidus*** Head, upperparts and tail earth-brown; underparts pale brown becoming white on throat and centre of belly; flanks pale rufous-brown. Flight feathers dark brown edged rufous-brown. Pale buff eye-ring. Bill blackish-brown with pale base to lower mandible. Legs blackish or brown. *M. p. modestus* very similar but slightly greyer below, esp. on breast and flanks. *M. p. murinus* greyer; flight feathers edged pale buff. **Juvenile** Head streaked buff; upperparts spotted buffish; underparts whitish irregularly streaked and mottled dark brown, esp. on breast and flanks. **VOICE** Mostly silent. Song a jumble of harsh churring notes. Calls include soft *churr* and thin *see-see*. [CD11:63] **HABITS** Singly or in pairs in woodland, wooded grassland, cultivation with trees, orchard bush and *Acacia* shrub. Perches on low branches and seizes prey on ground. Not shy. **STATUS AND DISTRIBUTION** Not uncommon resident throughout savanna zone (*pallidus* in north, *modestus* in south). Rare at limits of range (e.g. Gambia, Mali); vagrant, Mauritania. *M. p. murinus* Gabon and Congo. **NOTE** Often included in genus *Bradornis*.

SPOTTED FLYCATCHER *Muscicapa striata*   Plate 108 (3)
Gobemouche gris

L 13.5–14.5 cm (5.25–5.75"). Medium-sized, slender flycatcher with *faintly streaked crown and breast*, and *relatively*

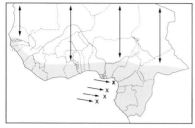

long wings. **Adult striata** Head, upperparts and tail greyish-brown; forehead and crown narrowly streaked dark brown; wing feathers with pale edges. Underparts dirty white diffusely streaked brownish on throat-sides, neck, breast and flanks. Long wings evident in flight; when perched, *wingtips extend beyond uppertail-coverts*. Bill black, base of lower mandible pinkish. Legs black. Races doubtfully separable in field. **M. s. balearica** paler and sandier above, slightly less streaked below; **tyrrhenica** warmer brown above, distinctly less streaked below. **First-winter** Greater coverts tipped buffish-white; tertials broadly edged pale buff. **VOICE** Mostly silent. Call a thin, scratchy *zreeht*.
Alarm *tik* or *tsee-tik*. [CD4:40] **HABITS** Singly, in all habitats with trees, incl. gardens; normally shuns tall forest. Characteristically perches upright on (often exposed) perch, making short sallies to catch flying insects. Hunts from ground to treetops. Frequently flicks wings and tail. Not shy. **SIMILAR SPECIES** Gambaga Flycatcher is slightly smaller and browner, with virtually invisible streaks and shorter wings. **STATUS AND DISTRIBUTION** Common to not uncommon Palearctic passage migrant and winter visitor (Sep–May) throughout. Vagrant, Cape Verde and Gulf of Guinea Is. Mostly nominate, but *balearica* (from Balearic Is) also recorded. Winter range and migration of *tyrrhenica* (from Corsica and Sardinia) unknown, but likely to occur in our region.

GAMBAGA FLYCATCHER *Muscicapa gambagae*   Plate 108 (4)
Gobemouche de Gambaga

L 12–13 cm (4.5–5.0"). Smaller, plainer and drabber version of Spotted Flycatcher, with distinctly shorter wings. **Adult** Head, upperparts and tail mouse-brown; crown largely unstreaked (some faint streaks may be visible at close range). Underparts dirty white washed brownish on breast (brown feather shafts can appear as very faint streaks). *Wingtips do not extend beyond uppertail-coverts*. Bill dark brownish-horn above, pale horn below; short. Legs brownish. **Juvenile** Head and upperparts spotted pale buff; underparts creamy-white with some dark brown streaks. **VOICE** Mostly silent. Call a sharp *tzik* (sharper than call of Spotted Flycatcher). [CD11:72]
HABITS Singly or in pairs, in dry savanna woodland. Perches upright on open branches, usually at mid-level. Flicks wings. **SIMILAR SPECIES** Spotted Flycatcher is larger and greyer brown, with crown and underparts noticeably streaked and bill and wings longer. African Dusky Flycatcher is smaller and slightly darker, greyer brown, with shorter tail and black bill with pale base to lower mandible; occurs in highlands. **STATUS AND DISTRIBUTION** Rare to scarce resident and partial migrant, patchily recorded in Mali, C Guinea, N Ivory Coast (Comoé NP), N Ghana, N Togo, Burkina Faso, Nigeria, Cameroon and Chad. Also Liberia (1 sight record, at foot of Mt Nimba). Possibly overlooked due to similarity with Spotted Flycatcher.

ASHY FLYCATCHER *Muscicapa caerulescens*   Plate 109 (5)
Gobemouche à lunettes

Other name E: Blue-grey Flycatcher
L 13 cm (5"). *Plain, pale grey flycatcher* of woodland and forest edge, with *white stripe from bill to eye and narrow white eye-ring emphasised by blackish lores*. **Adult brevicauda** Head and upperparts grey; flight feathers and tail dark brownish-grey. Underparts pale grey becoming whitish on throat and centre of belly. Indistinct, short, dark moustachial line. Tertials and secondaries fringed white. Bill blackish with pale base to lower mandible. Legs grey. Very similar western **nigrorum** slightly paler above, lightly washed brownish below. **Juvenile** Head and upperparts dark brown heavily spotted rusty-buff; underparts buffish-white heavily and irregularly spotted and barred brownish; flight and tail feathers edged and fringed rufous-brown. **VOICE** Mostly silent. Calls include high-pitched notes and a hoarse *zèèht*. Song a short, rhythmic series of a few similar notes ending with a more variable syllable, e.g. *tsip-tsip-tsip-tsip tsusu* or *tsip-tsip-tsip-tsip tsulupsip* etc. Also a more melodious, liquid phrase and a fast, sustained warble. [CD11:79] **HABITS** Singly or in pairs, frequenting forest edges, open gallery forest, and cultivated areas with large trees. At all levels, but rarely low. Catches insects in flight and by gleaning foliage, frequently changing perch. Sometimes in mixed-species flocks. Unobtrusive. **SIMILAR SPECIES** Cassin's Flycatcher is darker grey with contrasting blackish wings and tail, lacks white markings on face and is a forest species occurring near water. Dusky-blue Flycatcher much darker, bluish-slate, without white fringes to wing feathers. Other grey flycatchers frequent forest interior; thus easily

distinguished by habitat. **STATUS AND DISTRIBUTION** Uncommon resident, from Sierra Leone and SE Guinea to Togo (*nigrorum*) and from Benin and Nigeria to SW CAR and Congo (*brevicauda*).

SWAMP FLYCATCHER *Muscicapa aquatica*   Plate 109 (1)
Gobemouche des marais

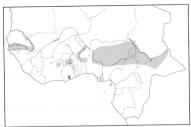

L 13 cm (5"). *M. a. aquatica. Drab flycatcher found near water.* **Adult** Head, upperparts and tail grey-brown. Underparts white washed brown on breast and flanks. Bill blackish with pale base to lower mandible. Legs blackish. **Juvenile** Head and upperparts spotted buff; underparts have some diffuse dark brown streaking on breast. **VOICE** Mostly silent. Song a mixture of harsh and soft high-pitched notes, followed by a little trill, with variations. Also duets. Call a sharp *tzitt*. [CD11:74] **HABITS** Singly or in pairs in trees, bushes or reeds bordering swamps, pools, lakes and rivers. Behaviour as congeners. **SIMILAR SPECIES** African Dusky Flycatcher is slightly smaller with shorter tail. Spotted Flycatcher larger with streaked breast. **STATUS AND DISTRIBUTION** Uncommon to locally common resident in savanna belt in Senegambia, Mali and NE Ivory Coast east to Cameroon, S Chad and N CAR.

CASSIN'S FLYCATCHER *Muscicapa cassini*   Plate 109 (2)
Gobemouche de Cassin

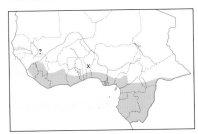

L 13 cm (5"). *Grey flycatcher of forest streams.* **Adult** Wholly ash-grey with blackish flight feathers and tail, and whitish throat and belly. Bill black. Legs blackish-brown. **Juvenile** Head and upperparts spotted buff; underparts dirty white with buffish wash to breast, belly and flanks, and diffuse dusky streaks, esp. on breast. **VOICE** Mostly silent. Calls high pitched and inconspicuous. Rarely heard song a varied, sustained series of high-pitched notes and trills. [CD11:76] **HABITS** Singly or in pairs by watercourses in forest and gallery forest. Perches on snags, rocks and branches over water. Not shy. Catches insects in flight. **SIMILAR SPECIES** Ashy Flycatcher is paler and has short, indistinct white supercilium and lower half of eye bordered by white crescent. **STATUS AND DISTRIBUTION** Common to not uncommon resident in forest belt and adjacent savanna, from Guinea and Sierra Leone to S CAR and Congo. Claims from Mali (Boucle du Baoulé) and Niger ('W' NP) are out of range and require confirmation.

OLIVACEOUS FLYCATCHER *Muscicapa olivascens*   Plate 108 (6)
Gobemouche olivâtre

L 14 cm (5.5"). Unobtrusive, drab forest flycatcher. **Adult** *olivascens* Head, upperparts and tail olivaceous-brown. Underparts whitish, washed olive-brown on breast and flanks. Lores greyish. Bill black above, pale horn below. Legs pale brownish. *M. o. nimbae* has upperparts more rufous-brown; underparts with more extensive and whiter area from belly to undertail-coverts; breast-band more pronounced. **Juvenile** Brown spotted rufous-brown; wing-coverts, tertials and innermost secondaries tipped and edged rufous-brown. Bill dark brown with base of lower mandible pale horn. **VOICE** Rather silent. Song a series of 6–10 high-pitched whistles, e.g. *tsee tsee tsee tsee tsu-tsui-tsui* (territorial) and a short, fast jumble of shrill, buzzy notes and trills. Calls include a thin *seeeee* and *wit-wit*. [CD11:75] **HABITS** Singly or in pairs in lowland rainforest. Perches quietly in open spaces, from mid-level to lower canopy, occasionally lower, from where it catches insects in flight or picks them from foliage. Frequently changes perch. Usually difficult to observe. **SIMILAR SPECIES** Other forest flycatchers are mainly grey, not brownish. **STATUS AND DISTRIBUTION** Rare or scarce to locally not uncommon forest resident, from Sierra Leone and SE Guinea to Ghana and from S Nigeria to SW CAR and Congo. *M. o. nimbae* only known with certainty from Mt Nimba.

AFRICAN DUSKY FLYCATCHER *Muscicapa adusta*   Plate 108 (5)
Gobemouche sombre

L 10–11 cm (4.0–4.25"). Small, unobtrusive, rather nondescript and dumpy flycatcher. **Adult** *obscura* Head,

upperparts and tail grey-brown. Underparts pale grey-brown with white throat and centre of belly, and indistinctly, coarsely striped breast and flanks. Narrow white supraloral streak and eye-ring; lores blackish. Bill blackish with paler base to lower mandible. Legs black. *M. a. pumila* darker brown above. **Juvenile** Head and upperparts spotted buff and rufous-brown; underparts mottled dusky-brown on breast. **VOICE** Mostly silent. Call a thin, high-pitched, drawn-out *seeeeu*; also a sharp *trt-trt-trtrt*. Song an unhurried series of short, sharp notes *tsr tsit tsr tsr tsitit* ... [CD11:73] **HABITS** Singly or in pairs in open forest, clearings and edges, open woodland, cultivated areas with scattered trees and strips of gallery forest. Perches at all heights on open branches. Catches insects on the wing. **SIMILAR SPECIES** Gambaga Flycatcher is slightly larger and browner, lacks eye-ring, has paler bill (dark brownish-horn above, pale horn below), and occurs in lowlands. Spotted Flycatcher even larger, with streaked crown and breast. **STATUS AND DISTRIBUTION** Locally not uncommon to rare resident. *M. a. obscura* in highlands of SE Nigeria (Obudu and Mambilla Plateaux), W and C Cameroon (Mt Cameroon, Rumpi Hills, Mt Manenguba, Bakossi Mts, Dschang, Bamenda Highlands, north to Tibati and Mt Genderu) and Bioko; *pumila* from N Cameroon (Adamawa Plateau) to NW CAR. **NOTE** Former subspecific treatment confusing, with 7 different names used; here all are subsumed within *obscura* except '*grotei*', which is included in *pumila*, following *BoA*.

LITTLE GREY FLYCATCHER *Muscicapa epulata*   Plate 109 (3)
Gobemouche cendré

L 9.5 cm (3.75"). Very small and unobtrusive. **Adult** Head and upperparts brownish-grey; flight feathers and rather short tail blackish-brown. Underparts largely grey indistinctly streaked white, becoming white on upper throat and belly. Bill black above, yellowish at base of lower mandible. Legs black. **Juvenile** Head and upperparts spotted buff; underparts scalloped dusky on throat and breast. **VOICE** Mostly silent. Call a very high-pitched *tsip*. Song a regular series of very high-pitched notes, e.g. *tsr tsr tsip tsr tsr tsip tsrsrsr tsip tsip* ... [CD11:77] **HABITS** Singly or in pairs at edges of lowland forest and second growth, mostly at mid-level. Catches insects in flight and by gleaning foliage. Frequently changes perch. **SIMILAR SPECIES** Yellow-footed Flycatcher has darker, bluish-slate plumage and yellow legs and feet. **STATUS AND DISTRIBUTION** Scarce forest resident, from Sierra Leone and SE Guinea to Togo, in Nigeria (Nindam FR), and from Cameroon to SW CAR and Congo.

YELLOW-FOOTED FLYCATCHER *Muscicapa sethsmithi*   Plate 109 (4)
Gobemouche à pattes jaunes

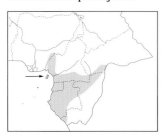

Other name F: Gobemouche de Seth-Smith
L 10.5 cm (4"). Small, compact forest flycatcher with *conspicuous yellow legs and feet* (diagnostic). **Adult** Wholly dark slate-grey, paler on underparts, with whitish throat, belly and undertail-coverts. Bill black above, yellow below. **Juvenile** Head and upperparts densely spotted rufous; underparts a dirty mixture of grey, rufous and white; legs and feet pale. **VOICE** Mostly silent. Call notes high pitched but rather far carrying, including *sii* and *sii-tzik sii-tzik-tzik* (alarm). Song a simple and high-pitched *tsrrrrwi-tsii tsrrrrwi-tsii* ... [CD11:78] **HABITS** In pairs within primary forest interior, frequenting lower and mid-strata; rarely higher. Catches insects on the wing from favoured perches. Very sedentary and often frequenting only small patch in territory. Not shy. **SIMILAR SPECIES** Little Grey Flycatcher has grey parts paler, white throat and belly ill defined, lower mandible yellow at base only, legs and feet black; also slightly smaller. Dusky-blue Flycatcher is larger and has black bill and legs. **STATUS AND DISTRIBUTION** Locally not uncommon to rare forest resident, from SE Nigeria (rare) and Cameroon (up to 1950 m) to SW CAR and Congo; also Bioko (scarce).

DUSKY-BLUE FLYCATCHER *Muscicapa comitata*   Plate 109 (7)
Gobemouche ardoisé

L 12 cm (4.75"). Dark, stocky flycatcher with *contrasting white throat*. **Adult male comitata** Head and upperparts dark bluish-slate; narrow white supraloral line extending to above eye, where merging with white eye-ring; flight feathers and tail black. Underparts blue-grey becoming whitish on throat and centre of belly. Bill broad (almost

triangular) and black. Legs blackish. Local race **camerunensis** has throat tinged fulvous. Western **aximensis** has face markings much less distinct (sometimes almost absent) and white of underparts more limited and variably sullied buff. **Adult female** As male, but very slightly paler and faintly washed brownish. **Juvenile** Similar to adult (unspotted); bill yellowish. **VOICE** Mostly silent. A dry *prrrrt.* Song short and unobtrusive, consisting of a few clear short notes. [CD11:80] **HABITS** In pairs in forest clearings and edges, farmland and plantations. Hunts insects on the wing from perch; usually low, occasionally higher than 10 m. May join mixed-species flocks. Breeds in old weaver nests. **SIMILAR SPECIES** Tessman's Flycatcher has distinctly more white on underparts, longer bill and slightly larger size. Cassin's Flycatcher is paler grey, esp. on underparts, and occurs exclusively near water. Ashy Flycatcher is even paler and has well-defined white face markings. Yellow-footed Flycatcher has similar plumage pattern but is noticeably smaller and has yellow lower mandible and legs. **STATUS AND DISTRIBUTION** Widespread and not uncommon to rare resident in forest zone, from Sierra Leone and SE Guinea to Togo and in Nigeria (*aximensis*), and from Cameroon to Congo and CAR (*comitata*). *M. c. camerunensis* restricted to Mt Cameroon.

TESSMANN'S FLYCATCHER *Muscicapa tessmanni*   Plate 109 (6)
Gobemouche de Tessmann

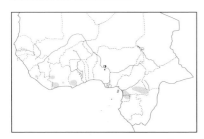

L *c.* 13 cm (5"). **Adult** Resembles Dusky-blue Flycatcher but slightly larger *with distinctly more white on underparts*: entire throat and belly pure white, slate-grey area restricted to breast and flanks. Narrow whitish supraloral streak. Bill blackish. Legs dark bluish-grey. **Juvenile** No information. An immature specimen described as having narrow rufous-buff tips to greater upperwing-coverts, tertials and undertail-coverts (*BoA*). **VOICE** Song melodious and thrush-like, with full, unhurried, whistled phrases, one motif being repeated a few times before switching to another. [CD11:81] **HABITS** Inadequately known. Occurs in lowland forest, mainly at mid-levels at edges and in small clearings; also second growth and abandoned tree plantations. **SIMILAR SPECIES** Dusky-blue Flycatcher (q.v. and above) has much darker underparts without obvious breast-band. Yellow-footed Flycatcher much smaller, with yellow legs and yellow lower mandible. **STATUS AND DISTRIBUTION** Rare and patchily distributed forest resident, recorded from SE Sierra Leone (Gola Forest), Liberia, Ivory Coast, SW Ghana, S Cameroon and Eq. Guinea. Two old records, Nigeria. *TBW* category: Data Deficient.

SOOTY FLYCATCHER *Muscicapa infuscata*   Plate 109 (10)
Gobemouche enfumé

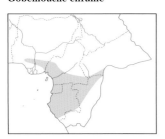

Other name F: Gobemouche fuligineux
L 13 cm (5"). *M. i. infuscata.* Dark forest flycatcher whose *long, pointed wings, slightly forked tail and flight action present martin-like appearance.* **Adult** Head, upperparts and tail dark brown. Underparts paler, rufous-brown, densely marked with indistinct brown streaks. Appears all dark at distance. Bill blackish. Legs brown. **Juvenile** Head and upperparts speckled pale buff; underparts a more distinctly streaked mix of dark brown, pale and rusty-buff. **VOICE** Mostly silent. Calls include a vigorous *teew* and a harsh *chep,* uttered singly or in fast series. Song inconspicuous and rarely heard, consisting of very short, simple but varied motifs. [CD11:82] **HABITS** As Ussher's Flycatcher. **SIMILAR SPECIES** May overlap with Ussher's Flycatcher in extreme west of range. Latter has similar jizz, but in favourable light darker underparts lacking streaks or mottling visible. **STATUS AND DISTRIBUTION** Uncommon to locally not uncommon forest resident, from S Nigeria (from Gambari east) to SW CAR and Congo. **NOTE** Often placed in genus *Artomyias*; then named *A. fuliginosa.*

USSHER'S FLYCATCHER *Muscicapa ussheri*   Plate 109 (9)
Gobemouche d'Ussher

L 13 cm (5"). **Adult** Wholly sooty-brown, paler on underparts. In flight *long and pointed wings, slightly forked tail and flight action give martin-like appearance.* Bill and legs blackish. **Juvenile** As adult, but wing and tail feathers narrowly tipped white; underparts with very small whitish specks. **VOICE** Mostly silent. Calls include *tssrip* and buzzy *dzip.* Rarely heard song a short, simple, pleasant whistled phrase, regularly repeated. [CD11:83] **HABITS** In pairs or

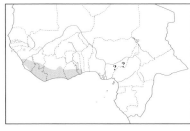

small parties. Perches high on dead branches along edges and clearings in lowland forest, from where it hawks insects in typical flycatcher manner, subsequently returning to perch. **SIMILAR SPECIES** May overlap with Sooty Flycatcher in extreme east of range. Latter has similar jizz, but in good light paler, warmer coloured underparts may be visible. **STATUS AND DISTRIBUTION** Locally not uncommon to rare forest resident, from Guinea and Sierra Leone to Togo; also S Nigeria (mouth of Benin R.; rare). Claimed from E Nigeria (Serti). **NOTE** This species and Sooty Flycatcher often placed in genus *Artomyias*, because of their long wings and short legs, which impart a distinctive jizz, quite unlike other *Muscicapa* flycatchers.

GREY-THROATED FLYCATCHER *Myioparus griseigularis*   Plate 109 (11)
Gobemouche à gorge grise

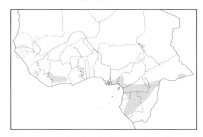

Other name E: Grey-throated Tit-Flycatcher
L 13 cm (5"). **Adult** All-grey flycatcher without distinctive markings. Upperparts slate-grey; flight feathers and longish tail blackish. Underparts grey becoming white on centre of belly and undertail-coverts. Bill black above, lower mandible pale horn (**nominate**) or black with trace of pale horn at base (**parelli**). Legs blackish. **Juvenile** Upper- and underparts densely spotted pale rufous. **VOICE** Song a thin, slightly quavering *truu-teee-thu-ee* with variations, somewhat reminiscent of Lead-coloured Flycatcher, and a whistled *hee-whuhee-heehee-hu*. [CD11:86] **HABITS** Occurs in mature lowland forest and edges, at all levels. Only occasionally catches insects on the wing, but actively searches foliage, using wings and cocked tail to dislodge prey. Joins mixed-species flocks. Easily passes unnoticed if vocalisations unknown. Sings in upright *Muscicapa* posture. **SIMILAR SPECIES** Lead-coloured Flycatcher may overlap at forest edge and has similar foraging behaviour, but has conspicuous white outer tail feathers. **STATUS AND DISTRIBUTION** Rare or scarce to locally uncommon forest resident, from E Sierra Leone (Gola Forest) to Ghana (*parelli*) and from SE Nigeria to CAR and Congo (*griseigularis*).

LEAD-COLOURED FLYCATCHER *Myioparus plumbeus*   Plate 109 (8)
Gobemouche mésange

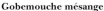

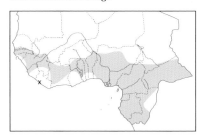

Other names E: Grey Tit-Flycatcher, Fan-tailed Flycatcher
L 14 cm (5.5"). *M. p. plumbeus*. Slender, warbler-like flycatcher with longish black tail and *conspicuous white outer tail feathers*. **Adult** Head and upperparts ashy-grey; underparts pale grey becoming white on centre of belly. White supraloral streak and eye-ring. Flight feathers blackish; tertials, secondaries and greater coverts narrowly edged white. Bill black with horn-coloured base. Legs blackish. **Juvenile** Head and upperparts spotted rusty-buff; tertials, secondaries and greater coverts edged rusty-buff; underparts mottled grey-buff and brown. **Immature** Grey parts of adult plumage tinged brownish. **VOICE** Song a soft, quavering *trruuu terree*, rising in pitch; unobtrusive, but quite distinctive once learnt. Variant: *truu tehee-hu-hee*. Call a high-pitched *heet*. [CD11:85] **HABITS** Singly or in pairs in wooded savanna, thickets, gallery forest and forest edges and clearings. Often cocks and slightly fans tail, displaying white outer feathers. Searches branches and gleans foliage for insects, frequently using fanned tail to disturb them. **SIMILAR SPECIES** Grey-throated Flycatcher has similar foraging behaviour but is more forest tied with more retiring habits and no white outer tail feathers. **STATUS AND DISTRIBUTION** Uncommon to rare resident in savanna (mainly) and forest zones, from Gambia and S Senegal to S Mali and Sierra Leone east to S Chad, CAR and Congo.

PIED FLYCATCHER *Ficedula hypoleuca*   Plate 108 (11)
Gobemouche noir

L 13 cm (5"). Rather compact, medium-sized flycatcher, mostly seen in non-breeding, female-like plumage. **Adult male breeding *hypoleuca*** Following moult (Jan–Feb) black to brownish above with conspicuous white, elongated wing patch (formed by white greater coverts and broad edges to tertials), small white patch on forehead (usually divided) and mainly white outer tail feathers. Underparts white. Very small white patch at base of inner primaries appears as downward-pointed line at border of primary-coverts (usually reaching 6th primary; sometimes absent).

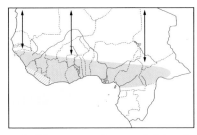

Indistinct greyish rump patch. N African race *speculigera* has larger white forehead patch and more white in wing, thus similar to Collared Flycatcher, but lacks white collar (although some possess half-collar like Semi-collared). Spanish *iberiae* intermediate; *tomensis* (E Siberia) dull grey-brown above. **Adult male non-breeding** After moult (Jun–Sep) brownish above often with paler forehead, pale brown eye-ring, and much narrower white wingbar and tertial edgings. Uppertail-coverts and tail blackish. Throat, breast and flanks buff, mottled darker on breast; belly and undertail-coverts white. **Adult female breeding** As male non-breeding, but uppertail-coverts and tail browner; underparts slightly paler. **Adult female non-breeding** As male non-breeding. **First-year** As male non-breeding, but median coverts often white tipped. **Second-year male** As adult, but primaries, primary-coverts, alula and usually some outermost greater coverts worn and paler brownish; white primary patch often absent. **VOICE** Mostly silent. Call a sharp *whit*. [CD4:37] **HABITS** Singly in wooded savanna and other rather open habitats with trees. Forages in typical flycatcher-manner, darting from perch, usually in middle or top of tree. Frequently changes perch and often feeds close to or on ground. Pumps tail up and down and flicks wings. Not shy but unobtrusive. **SIMILAR SPECIES** Birds in female-like plumage difficult, in some cases even impossible, to distinguish from similarly plumaged Collared and Semi-collared Flycatchers, but latter typically have larger white primary patch. Adult non-breeding males of latter two species differ from other adults and 1st-years in black rectrices, remiges, primary coverts, greater coverts and varying number of median coverts. First-years of all three often inseparable in field, as many have white tips to median coverts and small or no primary patch. Adult male breeding Semi-collared separated from *speculigera* by smaller forehead patch, white-tipped median coverts and large amount of white in tail. **STATUS AND DISTRIBUTION** Not uncommon to scarce Palearctic passage migrant and winter visitor (Aug–May) throughout, east to Cameroon and CAR; unknown further south, in Gabon/Congo. Vagrant, Cape Verde Is. All races presumably winter throughout; *speculigera* recorded from N Senegal and Ivory Coast.

COLLARED FLYCATCHER *Ficedula albicollis*   Plate 108 (9)
Gobemouche à collier

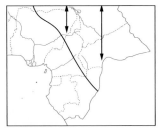

L 13 cm (5"). **Adult male breeding** Distinctive (after moult, starting Jan–Feb). Black head and upperparts separated by white collar; large white patch on forehead; wing patch formed by white greater coverts and broad white edges to tertials; relatively large white patch at base of primaries (almost rectangular; reaching 3rd primary); whitish rump; tail generally all black. Underparts white. **Adult male non-breeding** Following moult (Jun–Sep) very similar to Pied Flycatcher, but with larger primary patch and usually all-black tail. **Adult female breeding/non-breeding** Distinguished from female Pied by larger and broader primary patch; also slightly greyer brown above and rump usually paler. In breeding plumage neck-sides sometimes diffusely paler, extending as faint collar in a few. Some female Collared (esp. 2nd-years) extremely similar to Pied and impossible to separate. **Immature** See Pied Flycatcher. **VOICE** Mostly silent. Call a thin, sharp, far-carrying *seep*. [CD4:38] **HABITS** As Pied Flycatcher. **SIMILAR SPECIES** Pied and Semi-collared Flycatchers (q.v. and above). **STATUS AND DISTRIBUTION** Rare to scarce Palearctic passage migrant (Sep–Apr) recorded from Niger (Aïr), NE Nigeria, Chad, CAR and N Congo (Odzala NP). Inadequately known due to confusion in field with Pied Flycatcher, esp. *F. h. speculigera*. Claims from Ghana proved to be misidentified Pied (*BoA*); those from Mauritania, Senegal, Mali and CAR unsubstantiated. Winters south of equator (mainly Zambia, Zimbabwe, Malawi).

SEMI-COLLARED FLYCATCHER *Ficedula semitorquata*   Plate 108 (10)
Gobemouche à demi-collier

L 13 cm (5"). **Adult male breeding** Almost intermediate between Pied and Collared Flycatchers. Distinguished by white half-collar of variable extent, white-tipped median coverts forming narrow second upper wingbar (sometimes merging with greater coverts), and more white in tail. White patch on forehead small and usually divided (as in Pied); whitish rump and white patch at base of primaries closer to Collared. Many lack half-collar, 2nd-year males may also lack white tips to median coverts and have very small or no primary patch; all strongly resemble Pied, but differ in having larger amount of white in tail. **Adult male non-breeding** Following moult (Jun–Sep) as Collared, but usually distinguished by white-tipped median coverts. **Adult female/immature** Very similar to female/immature Collared, but usually differ in white-tipped median coverts (sometimes absent). **VOICE** Calls similar to Collared and Pied, a high *seep*, and a clicking *tsep* or *tek*. [CD11:71] **HABITS** As Pied and Collared. In known winter range recorded in wooded grassland, gallery forest and forest edge. **SIMILAR SPECIES** Birds in female-like plumage difficult, in some cases even impossible, to distinguish from similarly plumaged Collared and Pied Flycatchers, though these typically lack white tips to median coverts. Adult female Collared has less white in tail and usually larger primary patch; Pied has smaller (or no) primary patch. However, 1st-years of all three often have white tips to median coverts and small or no primary patch, rendering them inseparable in field. Adult male breeding Pied of ssp. *speculigera* distinguished by larger forehead patch, absence of white tips to median coverts and all-black tail. **STATUS AND DISTRIBUTION** Inadequately known, due to confusion with Collared and Pied Flycatchers. No certain records from our region. Main winter quarters in E Africa (S Sudan to Tanzania).

RED-BREASTED FLYCATCHER *Ficedula parva*   Plate 108 (8)
Gobemouche nain

L 11.5 cm (4.5"). *F. p. parva.* Small and active; *brown above with bold white sides to base of tail.* **Adult male** Head grey-brown; upperparts brown; tail blackish with diagnostic long white basal patches. Large-eyed appearance with narrow pale eye-ring. Throat and upper breast orange-red (usually paler and less extensive in autumn); rest of underparts whitish tinged creamy-buff on flanks. **Adult female** Head brown; throat and upper breast creamy-buff. **First-winter** As adult female, but with buffish tips and edges to greater coverts and tertials. Male acquires orange throat in 3rd or 4th calendar-year. **VOICE** Calls include a short, dry rattle *tzrrr*, a single *tzk*, a soft, plaintive *tulee* or *huwee*, and a weak *shrrr*. [CD4:39] **HABITS** Unobtrusive and usually keeping to cover. Hunts from perch but also forages like warbler. Frequently flicks wings and cocks and fans tail (showing distinctive pattern). **STATUS AND DISTRIBUTION** Palearctic vagrant, N Senegal (Nov). Normally winters in Asia.

MONARCHS
Family MONARCHIDAE (World: *c.* 100 species; Africa: 16–17; WA: 10–11)

A diverse family of small to medium-sized insectivorous and arboreal birds. Most African species have crests or incipient crests, and relatively long, graduated tails. Some male paradise flycatchers have extremely elongated median tail feathers when breeding. Most species sexually dimorphic. Juveniles unspotted. Occur in forest and wooded savanna. All are very active, gleaning most of their prey from branches in middle and lower storeys, often brushing foliage with fanned tail and more or less open wings to dislodge insects. Singly, in pairs or small family groups; often in mixed-species flocks.

CHESTNUT-CAPPED FLYCATCHER *Erythrocercus mccallii*   Plate 110 (6)
Érythrocerque à tête rousse

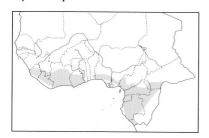

Other name F: Gobemouche à tête rousse
L 10 cm (4"). Unmistakable. *Very small, active forest flycatcher with rufous tail;* latter particularly conspicuous when bright light shines through feathers. **Adult** *nigeriae* Rufous forehead and crown contrast with olivaceous-brown nape and upperparts. Throat whitish; rest of underparts buffish-white. Eye pale yellow to reddish. Bill brownish-horn above, pale horn below. Legs pinkish-brown. **Nominate** has throat pale tawny. **Juvenile** Crown pale olivaceous-brown. **VOICE** Calls include high-pitched little notes, frequently uttered when foraging. Song a fast, sibilant, rather tuneless twittering. [CD12:20]
HABITS Usually in small mobile groups actively foraging in upper and middle strata of lowland forest, esp. near small gaps and at edges. Frequently fans graduated tail and droops half-open wings. Regularly joins mixed-species flocks. **STATUS AND DISTRIBUTION** Not uncommon forest resident, from W Guinea to Ghana, in SW Mali, and from S Benin to SW Nigeria (*nigeriae*); from SE Nigeria and Cameroon to SW CAR and Congo (nominate).

AFRICAN BLUE FLYCATCHER *Elminia longicauda*   Plate 110 (5)
Tchitrec bleu

Other name F: Gobemouche bleu
L 18 cm (7"). Unmistakable. **Adult** *longicauda* Pale, bright blue flycatcher with short crest and long, graduated tail. *Entire upperparts and tail cerulean-blue;* lores black; flight feathers blackish edged blue; underparts pale greyish-blue becoming whitish on belly. Bill and legs black. Eastern *teresita* paler, with whiter underparts. **Juvenile** Duller and faintly spotted buffish-grey on head and wing-coverts. Flight feathers and tail tipped white. Lower mandible

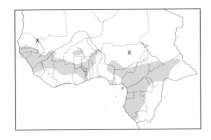

dark horn. **VOICE** Various little call notes, including *dzp* and *tsluip*. Song a jumble of sibilant notes. [CD12:21] **HABITS** Graceful, tame and active, foraging with fanned tail and half-open wings among foliage in various types of woodland, gallery forest, edges, farmbush with trees and gardens; also mangroves. Occasionally descends to ground and hunts insects in the air. Singly, in pairs or small family groups. Joins mixed-species flocks. **STATUS AND DISTRIBUTION** Locally not uncommon resident, from Senegambia to Sierra Leone east to Nigeria (nominate) and from Cameroon to S Chad, CAR and Congo (*teresita*). Scarce visitor during rains, S Mauritania.

DUSKY CRESTED FLYCATCHER *Elminia nigromitrata* Plate 110 (8)
Tchitrec à tête noire

Other name F: Gobemouche huppé à tête noire
L 11 cm (4.5"). **Adult** *colstoni* Small, all-dark, blue-slate bird with dull black crown ending in short, barely visible crest. Belly slightly paler. Flight feathers and tail brownish-black. Bill and legs bluish-black. **Nominate** Slightly darker and duller. **Juvenile** Wholly dull blackish-slate. **VOICE** Calls include a fast nasal sound, a soft *tsep* or *tsi-ep*, *ptee-diew-diew* and others. Song fast, highly varied and sustained, including many imitations (e.g. of Western Olive and Blue-throated Brown Sunbirds, Grey Longbill, Green Hylia, Emerald Cuckoo, Pied Hornbill, Forest Robin, malimbe sp., etc). [CD12:25] **HABITS** In pairs or small family groups in undergrowth (up to 2 m, occasionally to 4 m) of lowland rainforest. Favours most humid areas. Very active. Regularly joins mixed-species flocks. **SIMILAR SPECIES** Blue-headed Crested Flycatcher has obvious crest and glossy dark bluish upperparts contrasting with paler underparts; favours dense tangles of lianas, higher than Dusky Crested; voice different. White-bellied Crested Flycatcher is a montane species with white belly and duller plumage lacking blue cast. **STATUS AND DISTRIBUTION** Uncommon to locally not uncommon forest resident, patchily distributed from Sierra Leone and SE Guinea to Ghana and S Nigeria (*colstoni*), and from Cameroon to SW CAR and Congo, and in SE CAR (nominate). **NOTE** Formerly placed in *Trochocercus*.

WHITE-BELLIED CRESTED FLYCATCHER *Elminia albiventris* Plate 110 (7)
Tchitrec à ventre blanc

Other name F: Gobemouche huppé à ventre blanc
L 11 cm (4.5"). *E. a. albiventris.* **Adult** Small, all-dark, slate-grey bird with dull black head and short, barely visible crest, and dark grey breast becoming white on flanks and belly. Flight feathers and tail brownish-black. Bill and legs blackish. **Juvenile** Duller. **VOICE** Call a striking, sharp *pink* or *slip*. Song a soft, rather hesitant, melodious simple warble. [CD12:23] **HABITS** Restricted to montane forest (900–2500 m). Tame, inquisitive and very active, mainly foraging in undergrowth. Usually in pairs and often within mixed-species flocks. Frequently fans tail. **SIMILAR SPECIES** Dusky Crested Flycatcher is bluer and has darker underparts; also, like Blue-headed Crested Flycatcher, occurs at lower altitudes (but with overlap). **STATUS AND DISTRIBUTION** Not uncommon forest resident in highland areas of SE Nigeria (Obudu and Mambilla Plateaus), W Cameroon (Mt Cameroon, Rumpi Hills, Mt Kupe, Bakossi Mts, Mt Manenguba, Bamenda Highlands) and Bioko.

BLUE-HEADED CRESTED FLYCATCHER *Trochocercus nitens* Plate 110 (9)
Tchitrec noir

Other name F: Gobemouche huppé noir
L 15 cm (6"). **Adult male** *nitens* Wholly glossy blackish-blue with clearly delimited and contrasting grey underparts, from lower breast to undertail-coverts; short but obvious crest. Flight feathers brownish-black. Bill and legs black. Western **reichenowi** has darker grey underparts. **Adult female** Glossy blackish-blue restricted to crown; upperparts dark slate-grey with faint bluish tinge; underparts entirely grey. **Juvenile** As adult female but duller. **Immature male** As adult female, but head-sides, throat and upper breast finely speckled pale grey. **VOICE** Call a harsh 2-syllable *zwhee-zwheh*, very similar to call of paradise flycatchers. Song a very fast, hollow, far-carrying *hohohohohoho* often preceded by some harsh notes or a dry, rapid *tiktiktiktiktik*. [CD12:27] **HABITS** Singly or in pairs within dense lowland and mid-elevation forest (locally up to 1500 m), even in degraded forest patches. Favours dense

undergrowth and lianas at mid-level. Very active and restless, frequently fanning tail and spreading wings, using these as brush to disturb insects from thick vegetation. Very often in mixed-species flocks. Rather shy. **SIMILAR SPECIES** Dusky Crested Flycatcher lacks glossy plumage, obvious crest and clearly contrasting paler underparts; is also smaller and has entirely different voice. Where both species occur, Blue-headed Crested occupies stratum just above Dusky Crested. White-bellied Crested Flycatcher is restricted to montane forest. **STATUS AND DISTRIBUTION** Rare to not uncommon forest resident, from Guinea to Togo (*reichenowi*) and from Nigeria to SW CAR and Congo (nominate). **NOTE** *Trochocercus* may be an artificial genus, synonymous with *Terpsiphone* (Erard 1987).

Genus *Terpsiphone*

A characteristic genus; all species easily identified as paradise flycatchers. Some forms confusing, however: several species extremely variable and readily hybridise. Vocal and conspicuous. Vocalisations similar in all; typical call a rasping *zwhee-zwhèh*. More fieldwork required to better define species limits.

AFRICAN PARADISE FLYCATCHER *Terpsiphone viridis*   Plate 110 (1)
Tchitrec d'Afrique

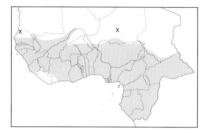

Other name F: Moucherolle de paradis
L *c.* 18 cm (7"); tail of male projecting up to 10–18 cm (4–7") beyond other rectrices. **Adult male *ferreti/speciosa*** Plumage extremely variable. All have glossy greenish blue-black head with distinct crest and much elongated central tail feathers. Bill, orbital ring and legs blue. Upperparts, wings and tail have varying amounts of rufous-chestnut or white (rarely glossy blue-black). Flight feathers blackish; a more or less distinct wingbar formed by white wing-coverts and white-edged secondaries. Underparts vary from blackish to greyish or white. Undertail-coverts grey, pale rufous or white. *T. v. ferreti* has greyer breast than *speciosa;* undertail-coverts grey to greyish-buff; tail (except central feathers) usually rufous. **Nominate** has upperparts deep rufous; white in wing; throat and breast blue-black; tail and undertail-coverts rufous. Migratory southern race *plumbeiceps* has duller and greyer head, rufous upperparts and tail, no white in wing and white undertail-coverts (sometimes washed pale rufous). **Adult female** Upperparts and tail always rufous; less or no white on wing; undertail-coverts greyish; tail shorter. **Juvenile** Resembles adult female but more brownish, esp. on head and on rather pale underparts. **Immature** As adult female but duller and with shorter crest. **VOICE** Call a rasping *zwhee-zwhèh*, typical of genus. Song variable, typically loud and cheerful, e.g. *twee-twee-twee-twee twee-twee-twee-twee.* [CD12:28] **HABITS** Singly, in pairs or small family parties in savanna woodland, gallery forest, second growth and farmland. Mostly at mid-level. Active, vocal and conspicuous. Joins mixed-species groups. **SIMILAR SPECIES** Rufous-vented Paradise Flycatcher has shorter crest, no white in wing and rufous undertail-coverts; male not variable. Juvenile very similar, but Rufous-vented somewhat more rufous. Bates's Paradise Flycatcher has blue-grey head, no crest, more orange-rufous upperparts and tail, which is usually shorter. **STATUS AND DISTRIBUTION** Widespread resident and intra-African migrant in savanna and forest zones, moving north in rainy season. North to S Mauritania, Mali (Tombouctou), Burkina Faso (at least to Ouagadougou), S Niger and S Chad. Usually common, but rare, absent or seasonal in certain parts of forest zone. *T. v. viridis* from Senegambia to Sierra Leone; *ferreti* from Mali and Ivory Coast to N Cameroon, Chad and N CAR; *speciosa* from S Cameroon to Congo. *T. v. plumbeiceps* is non-breeding visitor from south of the equator, reaching S Cameroon.

SÃO TOMÉ PARADISE FLYCATCHER *Terpsiphone atrochalybeia*   Plate 144 (5)
Tchitrec de Sao Tomé

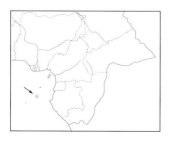

Other name F: Moucherolle de Sao Tomé
L *c.* 18 cm (7"); tail of male projecting 7.0–11.5 cm (up to 4.5") beyond other rectrices. **Adult male breeding** Distinctive. Wholly glossy blue-black with greatly elongated central tail feathers. **Adult male non-breeding** No tail streamers. **Adult female** Head greyish with dark crown; upperparts, edges of wing feathers and tail rufous; throat and upper breast grey merging into white of rest of underparts; undertail-coverts pale rufous-brown. **Juvenile** Similar to adult female. **VOICE** Calls and songs include those typical of genus. [CD12:31] **HABITS** Singly or in small groups in forest, forest edge, plantations with shade trees and patches of dry woodland in savanna. In

lower and mid-storeys. Joins mixed bird parties. Often tame and inquisitive. **STATUS AND DISTRIBUTION** Common resident, endemic to São Tomé, where the only paradise flycatcher.

RUFOUS-VENTED PARADISE FLYCATCHER *Terpsiphone rufocinerea*   Plate 110 (3)
Tchitrec du Congo

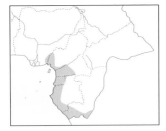

Other name F: Moucherolle du Congo

L *c.* 18 cm (7"); tail of male projecting 1–11 cm (up to 4.25") beyond other rectrices. **Adult male** Head glossy blue-black with very slight crest (barely visible in field). Upperparts and tail rufous. Underparts dark greyish-blue, paler on belly; *undertail-coverts rufous.* Central tail feathers elongated. Bill, orbital ring and legs blue. **Adult female** Underparts paler; tail shorter. **Juvenile** As adult female but duller. **VOICE** Calls and songs typical of genus, including harsh, rasping *zwhee-zwhèh* and cheerful *thululululululu* or *twee-twee-twee-twee.* [CD12:30] **HABITS** Singly, in pairs or small family groups in forest, old second growth, forest patches in savanna and adjacent woodland, and gardens. Behaviour as other paradise flycatchers. **SIMILAR SPECIES** Bates's Paradise Flycatcher has head blue-grey without crest, brighter, more orangey upperparts and usually no tail streamers. Male rufous morph and female African Paradise Flycatcher have white on wing and usually grey or white undertail-coverts (sometimes pale rufous); male also has longer tail streamers. Some without white in wing and with rufous undertail-coverts indistinguishable in field. Juveniles very similar, but African Paradise Flycatcher is more rufous-chestnut. **STATUS AND DISTRIBUTION** Common resident in forest zone, from SW Cameroon (and extreme SE Nigeria?) to W Gabon and S Congo. **NOTE** Appears to hybridise widely with *T. viridis* (conspecific?).

BATES'S PARADISE FLYCATCHER *Terpsiphone batesi*   Plate 110 (4)
Tchitrec de Bates

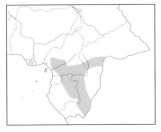

Other name F: Moucherolle de Bates

L *c.* 18 cm (7"); tail of male projecting 1.0–10.5 cm (up to 4") beyond other rectrices. **Adult male** *batesi* Similar to Rufous-vented Paradise Flycatcher, but *head blue-grey* (not blue-black) *without crest*; upperparts, tail and undertail-coverts brighter, *more orange-rufous* and central tail feathers less elongated (projecting 1–3 cm). *T. b. bannermani* has lower belly whiter and central tail feathers much longer (projecting 3.5–10.5 cm). **Adult female** Duller; tail shorter. **Juvenile** As adult female but duller and paler. **VOICE** Calls similar to those of other paradise flycatchers. Song a cheerful *twee-twee-twee-twee twee-twee-twee-twee* similar to that of Red-bellied Paradise Flycatcher, but typically higher pitched and slightly faster than more variable songs of African and Rufous-vented. [CD12:29] **HABITS** Occurs in lowland forest (up to 1300 m) and old second growth. Behaviour as other paradise flycatchers. **SIMILAR SPECIES** Rufous-vented Paradise Flycatcher (q.v.). Male rufous morph and female African Paradise Flycatcher have white on wing and usually grey or white undertail-coverts (sometimes pale rufous); male also has much longer tail streamers. Juveniles very similar, but African Paradise Flycatcher is more rufous-chestnut. **STATUS AND DISTRIBUTION** Rare to common resident in forest zone, from S Cameroon to SW CAR and Congo. Nominate throughout, except C Congo (Léfini), where *bannermani*. **NOTE** Often considered conspecific with Rufous-vented Paradise Flycatcher.

RED-BELLIED PARADISE FLYCATCHER *Terpsiphone rufiventer*   Plate 110 (2)
Tchitrec à ventre roux

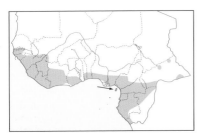

Other name F: Moucherolle à ventre roux

L *c.* 18 cm (7"); tail of male projecting up to 7 cm (2.75") beyond other rectrices. *The only paradise flycatcher with rufous underparts.* Plumage very variable; 9 races recognised in our region, with intermediates. All have head entirely blue-black; upperparts and tail rufous in most, but bluish-slate in 2. Bill and orbital ring bright cobalt-blue; legs slate-blue. **Adult male** *rufiventer* Small but distinct crest; wings largely black with long white panel (formed by white on coverts and edges of secondaries), and greatly elongated central tail feathers. *T. r. nigriceps* is slightly paler, more orange-rufous, with no crest and no white in wing, and much shorter tail (central feathers extending 2.5–7.0 cm). Following 3 races as *nigriceps*, with similar basic pattern and shape but differing in plumage tones: *fagani* has brownish-rufous upperparts and tail contrasting with bright rufous underparts and uppertail-coverts; *ignea* as *fagani* but with duller head; *mayombe* has blue-black head sharply defined and contrasting with orange-rufous body, and bluish-slate tail (thus very similar to Annobón Paradise Flycatcher, but slightly deeper rufous, esp. on back); *schubotzi* as *mayombe* but head greyer, less glossy. *T. r. neumanni* and *tricolor* distinguished from all preceding races by wholly bluish-slate upperparts and tail. Tail extensions: *fagani* 0.5–1.0 cm; *ignea* 0.5–2.5 cm; *schubotzi* 0.5–3.0 cm; *mayombe* 1.5–2.5 cm; *neumanni* 0.0–2.5 cm; *tricolor* 0.5–2.0 cm (*BoA*). **Adult female** As adult male but rather duller and lacking tail streamers. Nominate lacks white on wing. **Juvenile** As adult female, but duller, more brownish-rufous. **VOICE** Calls and songs typical of genus, including harsh, scolding *zwhee-zwhèh* and short, cheerful *tweedweedwee tweedweedwee*. [CD12:32] **HABITS** Singly, in pairs or small family groups in mid-storey of lowland and montane forest, clearings, edges, second growth, cocoa plantations, gallery forest and thickets in savanna. Often in mixed-species flocks. **SIMILAR SPECIES** Other paradise flycatchers of region have mostly dark bluish-grey, not rufous, underparts. **STATUS AND DISTRIBUTION** Common resident in forest and savanna zones. Approximate range of races: *rufiventer* Senegambia to W Guinea; *nigriceps* Guinea and Sierra Leone to SW Benin (migrant to Senegambia?); *fagani* Benin and SW Nigeria (west of Niger R.); *neumanni* SE Nigeria, Cameroon, Gabon and S Congo; *schubotzi* NE Congo and SW CAR; *mayombe* S Congo; *ignea* SE CAR; *tricolor* Bioko. **NOTES (1)** Local hybridisation with African Paradise Flycatcher reported from Cameroon and Gabon; with Bates's Paradise Flycatcher in Cameroon (*BoA*) **(2)** Some races behave as morphs in certain areas; e.g. in S Congo grey-backed and rufous-backed forms form mixed pairs (Dowsett-Lemaire & Dowsett 1991).

ANNOBÓN PARADISE FLYCATCHER *Terpsiphone (rufiventer) smithii*   Plate 143 (5)
Tchitrec d'Annobon

L *c.* 18 cm (7"); tail of male projecting 0.5–1.0 cm beyond other rectrices. **Adult male** Vividly coloured, similar in pattern to ssp. *nigriceps* of Red-bellied Paradise Flycatcher, but with blue-black head sharply defined and contrasting with rich orange-rufous body, and bluish-slate tail. **Adult female** Slightly duller and lacking tail extensions. **Juvenile** As female, but duller. **VOICE & HABITS** As other paradise flycatchers. Frequents cultivated areas, secondary forest and moist forest. **STATUS AND DISTRIBUTION** Common resident, endemic to Annobón. *TBW* category: THREATENED (Vulnerable). **NOTE** Usually considered a race of Red-bellied Paradise Flycatcher, but treated as separate species by *TBW*.

BATISES, WATTLE-EYES AND ALLIES
Family PLATYSTEIRIDAE (World: 28–30 species; Africa: 27–29; WA: 16)

A family of well-marked, mostly rather short-tailed birds. Sexually dimorphic. Catch insects in flight or by gleaning from leaves and twigs. Occur singly, in pairs or small family groups. Several species readily join mixed-species flocks. Endemic to Africa and Madagascar. Five genera occur in our region: *Megabyas* and *Bias* (both monotypic), *Batis* (7 species), *Dyaphorophyia* (5 species) and *Platysteira* (2 species). Relationships unclear, sometimes included within bush-shrikes, Malaconotidae (cf. Harris & Franklin 2000).

SHRIKE FLYCATCHER *Megabyas flammulatus*   Plate 111 (2)
Bias écorcheur

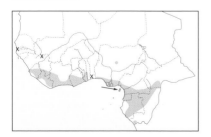

Other name F: Gobemouche écorcheur
L 15–16 cm (*c.* 6"). *M. f. flammulatus.* Distinctive, with shrike-like bill shape and male plumage. **Adult male** Head, upperparts and tail glossy blue-black; rump and underparts pure white. Eye brown to scarlet. Bill black. Legs reddish-brown. **Adult female** Head and upperparts earth-brown becoming rufous on lower back, rump and uppertail-coverts; tail blackish-brown edged rufous; underparts white with heavy, irregular earth-brown streaking; vent and undertail-coverts rufous. Flight feathers dark brown, tertials and secondaries fringed rufous-brown, coverts fringed dark rufous. **Juvenile** Head and upperparts dull earth-brown speckled buffish; underparts whitish with irregular dark brown streaking on throat and breast. **VOICE** Songs include a characteristic *chewee-cheweeet* or *chuah-chuwheesiu* and *chuchuwheechu*, a short, rather melodious trill (somewhat reminiscent of Senegal Eremomela but louder) and a fast *whee-whee-whee-*

chu, alternating with bill snapping. Calls include *tsu-tsu*, high-pitched, rather sharp whistles *whee-whee-whee* (contact call) and clicking *pit-pit-pit*. [CD12:1] **HABITS** In pairs or small family groups in lowland forest; also in farmland with scattered tall trees in forest zone and riparian woodland. Perches upright and characteristically moves tail sideways. Often in mixed-species flocks in canopy, occasionally lower at edges. **SIMILAR SPECIES** Male puffbacks *Dryoscopus* spp. resemble male, but have red (never brownish) eyes, and different shape (with longer tail), voice (often also quite noisy) and behaviour (with more horizontal posture). **STATUS AND DISTRIBUTION** Uncommon resident in forest zone, from Sierra Leone and SE Guinea to SW CAR and Congo; also C Nigeria (Jos Plateau) and Bioko. Vagrant, Gambia and SW Mali.

BLACK-AND-WHITE FLYCATCHER *Bias musicus*   Plate 111 (1)
Bias musicien

Other names E: Vanga Flycatcher; F: Gobemouche chanteur
L 16 cm (6.5"). *B. m. musicus.* Distinctive, highly vocal species with prominent spiky crest, broad rounded wings and short tail. **Adult male** Entire head to upper breast and flanks, upperparts and tail glossy black; remaining underparts white. White wing patch prominent in flight. Conspicuous yellow eye. Bill black. Legs yellow. **Adult female** Head dusky-brown; upperparts and tail rufous-chestnut; underparts white washed rufous on breast and flanks. Crest shorter than male's. **Juvenile** As adult female but duller and with some pale streaking on top of head and mantle, and small buffish spots on wing-coverts.
VOICE Song, uttered from perch or in flight, a loud, melodious, often repeated *wheet-tee-tee-tiuw-tiuw* with variations, e.g. *tee-tiu-tee-tiu* or *tiu-tititi-twEE-tu-tip-tiu-tititi-twEE-tu-...*; sometimes followed by a trill slightly reminiscent of Olive-bellied Sunbird. Calls include a harsh *tchèèèp* and *tk tk tk.* [CD12:2] **HABITS** Usually in pairs or small family parties. Frequents upper levels of edges and clearings in evergreen and gallery forest. Easily identified by characteristic jizz and flapping flight. **STATUS AND DISTRIBUTION** Widespread but generally uncommon resident in forest zone and gallery extensions, from Guinea and Sierra Leone to SW CAR and Congo. Vagrant, Gambia.

Genus *Dyaphorophyia*

Small, colourful, large-headed and very short-tailed birds with blue, purplish or green eye-wattles. Active forest species, often joining mixed-species flocks. Sometimes included in genus *Platysteira*.

CHESTNUT WATTLE-EYE *Dyaphorophyia castanea*   Plate 112 (4)
Pririt châtain

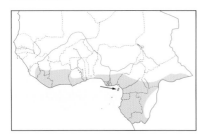

Other name F: Gobemouche caronculé châtain
L 10 cm (4"). **Adult male** *castanea* Head, upperparts and tail glossy blue-black; rump feathers long and white; underparts white with broad black breast-band. Eye-wattle purple-grey. Bill black. Legs purplish. Western *hormophora* has broad white neck collar. **Adult female** Head dark slate-grey separated from underparts by narrow white line from chin to below eye; upperparts chestnut with grey rump; tail blackish-brown; throat and breast chestnut; rest of underparts white. Flight feathers dark brown edged chestnut. Eye-wattles as male's. **Juvenile** Head dark brownish-grey; upperparts mainly chestnut (paler than adult female); underparts white with diffuse broad band of greyish and chestnut on upper breast. **Immature** Juvenile plumage progressively replaced by female-like dress, with pale rusty-grey throat and broad, rufous-grey breast-band indistinctly barred greyish-brown; throat and breast-band gradually becoming chestnut, top of head remaining brownish. Eye-wattles appear in 2nd year. **VOICE** Varied. Songs include rhythmic series of far-carrying *whop pEEEE* (in west), sharp *ptik-ptik-ptik-...* and *klonk-klonk-klonk-...* (in east), *ptik-kwonk* or *whep-pleenk*, high-pitched *hit-hit-hit-...* etc. Calls include a low *kwonk*, a soft *wop* and *wa*, and various nasal notes. Also frequent wing- and bill-snapping. [CD12:18] **HABITS** In pairs or small family groups in mid-strata of primary and secondary forest, edges, clearings, gallery forest and relict forest patches in savanna. Frequently joins mixed-species flocks. **SIMILAR SPECIES** Adult male White-spotted Wattle-eye similar to male *hormophora* but has short superciliary streak; female has black (not grey) crown. Usually occurs at higher levels. **STATUS AND DISTRIBUTION** Common forest resident, from Guinea and Sierra Leone to Benin (*hormophora*) and from Nigeria (where uncommon) to S CAR, Congo and Bioko (nominate).

WHITE-SPOTTED WATTLE-EYE *Dyaphorophyia tonsa*　　　　Plate 112 (3)
Pririt à taches blanches

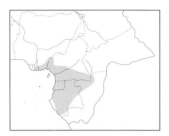

Other name F: Gobemouche caronculé à taches blanches
L 9.5 cm (3.75"). Similar to Chestnut Wattle-eye. **Adult male** Differs in having short white superciliary streak (partly hidden by eye-wattle); from eastern (nominate) race also by white neck collar. **Adult female** Distinguished by glossy black (not slate-grey) forehead and crown, partly hidden white superciliary streak and longer malar stripe. **Juvenile/immature** As corresponding plumages of Chestnut Wattle-eye. **VOICE** Songs include a series of whistled notes *hu-hee-hu-hu-hu hu-hee-hu ...*, rhythmic series of identical notes *hu-hu-hu-hu-... or ut ut ut ut ut ...* and *hee-hee-hee-hee-...*, and series of *ptok-ut*. Calls a soft *yup* and *ptek*. [CD12:19] **HABITS** In pairs or small family groups in upper levels of mature forest. Readily joins mixed-species flocks. **SIMILAR SPECIES** Chestnut Wattle-eye not only differs in plumage (see above), but is also more noisy and occurs at lower levels. **STATUS AND DISTRIBUTION** Rare to locally not uncommon forest resident, from SE Nigeria to SW CAR and Congo. Claims from Ivory Coast require confirmation.

RED-CHEEKED WATTLE-EYE *Dyaphorophyia blissetti*　　　　Plate 112 (6)
Pririt de Blissett

Other name F: Gobemouche caronculé à joues rouges
L 9 cm (3.5"). Very small. **Adult male** Head, upperparts, tail, throat and upper breast glossy blackish-green with broad triangular chestnut patch from below eye to sides of throat and neck; rest of underparts white. Conspicuous greenish-blue eye-wattle. Bill black. Legs greyish-purple. **Adult female** Greyer, less glossy; eye-wattle smaller. **Juvenile/immature** As adult female, but throat and upper breast tawny bordered chestnut. Band on breast gradually darkens and tawny throat gradually replaced by dark bottle-green feathers. Eye-wattle smaller. **VOICE** Song a far-carrying and difficult to locate series of identical high-pitched notes *hee-hee-hee-hee-...*; speed of delivery and length variable. Also wing- and bill-snapping and hoarse little noises (e.g. 'teeth gnashing' *k-k-sh-sh-sh* when excited). [CD12:15] **HABITS** Singly or in pairs in dense undergrowth of lowland forest (up to 700 m), gallery forest and relict patches in savanna. A frequent member of mixed-species flocks. **SIMILAR SPECIES** Black-necked Wattle-eye, which meets Red-cheeked near Mt Cameroon but there occurs at higher altitudes, has 'cheeks' glossy bottle-green and underparts pale yellow. **STATUS AND DISTRIBUTION** Uncommon to locally not uncommon forest resident, from Guinea to Togo and from Nigeria to extreme SW Cameroon.

BLACK-NECKED WATTLE-EYE *Dyaphorophyia chalybea*　　　　Plate 112 (5)
Pririt chalybée

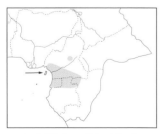

Other name F: Gobemouche caronculé à cou noir
L 9 cm (3.5"). Very small. **Adult male** As Red-cheeked Wattle-eye but lacks chestnut cheeks and has underparts washed pale yellow. Eye-wattle emerald-green. **Adult female** Duller; eye-wattle smaller. **Juvenile** Head and upperparts sooty brownish-grey; underparts white with broad tawny stripe on centre of throat bordered by chestnut. **Immature** As adult female, but throat and upper breast tawny bordered and mottled blackish-brown; eye-wattle small. **VOICE** Similar to Red-cheeked Wattle-eye, except advertising song, which consists of excited *ptiukteehee* or *ptiuk-ti-hi hihihu* followed by 4–6 descending sweet notes, reminiscent of Common Wattle-eye. Also long series of whistles when excited. Flight call *kwek-kwek-...* accelerating to *kwedekwedekwedek...* [CD12:17] **HABITS** Singly, in pairs or small family groups in lowland and montane forest (up to 1950 m). Frequents dense undergrowth of secondary forest and tangles, often near small clearings, in primary forest. Readily joins mixed-species flocks. **SIMILAR SPECIES** See Red-cheeked Wattle-eye; in Cameroon, where the two meet, Red-cheeked occurs at lower altitudes. **STATUS AND DISTRIBUTION** Locally not uncommon forest resident in Cameroon (from Mt Cameroon east, at 1050–1950 m), Eq. Guinea, Gabon and Bioko. **NOTE** Treated as race of Red-cheeked Wattle-eye in *BoA*, but usually considered a separate species. More fieldwork required.

YELLOW-BELLIED WATTLE-EYE *Dyaphorophyia concreta*   Plate 112 (7)
Pririt à ventre doré

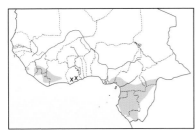

Other name F: Gobemouche caronculé à ventre doré
L 10 cm (4"). Brightly coloured, but elusive. **Adult male** *concreta*
Head, upperparts and tail dark slate-grey to olive-green; underparts
rich chestnut with golden-yellow upper throat. Yellowish supraloral
streak. *Conspicuous bright emerald-green wattle around eye.* Bill black.
Legs blue-grey. Eastern *graueri* has underparts deep orange-yellow
to bright yellow. **Adult female** *concreta* As male but underparts bright
yellow with throat and upper breast chestnut; eye-wattle smaller. In
graueri underparts are chestnut, becoming yellowish washed
chestnut on lower belly. **Juvenile** Head and upperparts greyish-olive
with some buffish-brown spots, esp. on head, and dirty yellowish supraloral streak; underparts dirty white washed
olive-grey on breast and tinged dull yellowish on throat and breast-sides. **Immature** Underparts dull yellow with
olive wash on breast and flanks, and some chestnut feathers on throat and upper breast; wing-coverts and tertials
tipped buffish-brown; small eye-wattle. **VOICE** Songs include short, rhythmic series of high-pitched, whistled
note(s) followed by lower note(s) *hee-hee-hwot-hwot* or *hee-hu hee-hu... whot* and *heet whot*; sometimes without *hwot*
note or preceded by buzzy notes. Also fast series of identical notes *hwrit-hwrit-hwrit-hwrit-...* Calls include various
hoarse and guttural notes. [CD12:14] **HABITS** Singly, in pairs or small family groups in understorey of lowland
and montane forest. Remains in shade of forest proper; not on edges. Usually shy and not easily seen. **STATUS
AND DISTRIBUTION** Rare to locally common forest resident, from Sierra Leone and SE Guinea to Ghana
(nominate) and from SE Nigeria to SW CAR and Congo (*graueri*). **NOTE** Former races '*harterti*' and '*kumbaensis*'
merged in *graueri*, following *BoA*.

Genus *Platysteira*

Distinguished by conspicuous scarlet eye-wattles. Behaviour similar to batises, but jizz different, due to longer tail
and larger size. Savanna and forest.

COMMON WATTLE-EYE *Platysteira cyanea*   Plate 112 (1)
Pririt à collier

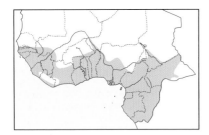

Other names E: Scarlet-spectacled Wattle-eye, Brown-throated
Wattle-eye; F: Gobemouche caronculé à collier
L 13 cm (5"). **Adult male** *cyanea* Head, upperparts and tail glossy
blue-black with conspicuous scarlet wattle above eye and long,
narrow white wingbar (formed by white tips to median coverts and
edges of tertials and inner secondaries); tail edged white. Underparts
white with glossy blue-black breast-band. Bill black. Legs blackish.
P. c. nyansae has narrow white frontal band extending above lores.
Adult female *cyanea* Similar to male but head dark slate-grey; throat
and upper breast dark chestnut bordered below with black; white
line from chin to below eye. Female *nyansae* has narrow white frontal band extending above lores (sometimes
also in nominate west to Nigeria). **Juvenile** Head and upperparts dull grey irregularly but densely spotted rusty-
buff; wingbar rusty-buff; underparts white with some buffish wash to throat-sides and breast. **Immature** Head and
upperparts dark olive-grey; wingbar rusty-buff; upper breast with greyish-chestnut mottling forming irregular
narrow band; throat and upper breast progressively more chestnut (as in adult female); eye-wattles small. **VOICE**
Song a diagnostic, far-carrying series of clear, melodious whistles *hee-hu-huu-ho hee-hu-ho* (2 parts, each descending
in pitch); also variations and shorter versions. Notes also given singly. Various harsh low calls, often included in
song; also wing- and bill-snapping. [CD12:12] **HABITS** Singly or in pairs in variety of habitats within savanna and
forest zones, incl. gardens, orchard bush, gallery forest, coastal thickets and mangrove; absent from arid north
and from forest proper. Tame. Gleans foliage for insects. **SIMILAR SPECIES** Batises lack scarlet eye-wattles and
are smaller. **STATUS AND DISTRIBUTION** Common resident throughout, esp. in savanna belt; absent from
arid north. *P.c. nyansae* in E CAR (Bangui and Fafa R. eastward); nominate in rest of range.

BANDED WATTLE-EYE *Platysteira laticincta*   Plate 112 (2)
Pririt du Bamenda

Other name F: Gobemouche caronculé à large bande
L 13 cm (5"). **Adult male** Head and upperparts glossy blue-black; scarlet wattle above eye. Underparts white with

broad glossy blue-black breast-band. Bill black. Legs purplish-black. **Adult female** Similar to male but throat to upper breast blue-black as upperparts; chin white. **Juvenile** Unknown; probably as juvenile of extralimital Black-throated Wattle-eye, which is similar to juvenile Common Wattle-eye but has very narrow, irregular wingbar. **Immature** Head and upperparts dark brownish with wing-coverts edged tawny; throat and upper breast brownish mixed with black on upper breast; rest of underparts white washed buffish-grey on flanks. **VOICE** Calls include a soft, rather grating note. [CD12:13] **HABITS** Singly or in pairs in montane forest above 1800 m. At all levels up to canopy, but usually in undergrowth. Snatches insects off foliage, in a hop or short flight, sometimes in mid-air. Joins mixed-species flocks. **SIMILAR SPECIES** Common Wattle-eye has distinct wingbar, white in adult, rusty-buff in juvenile. **STATUS AND DISTRIBUTION** Endemic to Bamenda Highlands, W Cameroon, where not uncommon in remaining forest patches. *TBW* category: THREATENED (Endangered). **NOTE** Often considered a race of Black-throated Wattle-eye *P. peltata* (cf. *BoA*), but here treated separately, following *TBW*.

Genus *Batis*

Small, active birds with distinctive jizz, created by relatively large head and short tail. In our region, breast-band typically black in male, chestnut in female. Some species hard to distinguish, esp. males. All have black masks, grey upperparts, blackish wings with long white bar from median coverts over edges of tertials, black tails with outer feathers edged and tipped white, and conspicuous yellow eyes. Bill, legs and feet black. Having lost the short-lived initial buffish-and-white speckles on head and upperparts, young are like adult females but have more or less pronounced rusty-buff tinge to head and upperparts, esp. noticeable on supercilium and wingbar; birds in this plumage cannot be sexed until some black appears on breast-band of male. Occur singly, in pairs or small family groups in forest, woodland and bush. Flight swift and bouncing, often on whirring wings. Snap bill loudly when catching insects.

CHINSPOT BATIS *Batis molitor* Plate 111 (6)
Pririt molitor

Other name F: Batis molitor

L 12 cm (4.75"). *B. m. pintoi*. **Adult male** Crown, nape and upperparts grey; black mask with conspicuous yellow eye bordered by *narrow* white supercilium. Wings black with long white bar. Underparts white with broad black breast-band. **Adult female** Similar but has chestnut breast-band and *large chestnut patch on centre of throat* (diagnostic). **Juvenile** Adult female-like pattern with buff-speckled head and upperparts, and variably indistinct throat patch and breast-band. Eye whitish. **Immature** As adult female but supercilium, nape, mantle and median coverts tinged rusty-buff. **VOICE** Song of male a vigorous, whistled, descending 2–3-note *hee-hu* or *hee-hee-hu* with variations and sometimes interspersed with rolling *kreew* notes; female duets with *wik* notes. Calls include *chik* (contact) and rapid *ch-ch-ch-ch* (alarm). Also snaps bill. [CD12:5] **HABITS** Frequents lower and middle strata of wooded savanna. Restless; flicking tail at each move. Joins mixed-species flocks. **SIMILAR SPECIES** Male Black-headed Batis with grey crown virtually indistinguishable (even in hand) from male Chinspot and only identifiable when in company of female; some distinguished by black top of head. Male of allopatric Grey-headed Batis smaller, with more distinct nape patch. **STATUS AND DISTRIBUTION** Locally scarce to rare resident, SE Gabon and Congo.

SENEGAL BATIS *Batis senegalensis* Plate 111 (3)
Pririt du Sénégal

Other names E: Senegal Puff-back Flycatcher; F: Batis du Sénégal

L 10.5 cm (4.25"). **Adult male** Crown dark slate-grey bordered by *broad white stripe from bill to nape*; black mask emphasising yellow eye; upperparts and tail dark slate-grey. Wings black with long white bar. Underparts white with black breast-band. **Adult female** Pattern as adult male but supercilium rusty-buff; upperparts and wingbar washed rusty-buff; breast-band pale chestnut (*palest breast-band of our region's batises*). **Juvenile** Similar to adult female, but head and wing-coverts spotted rusty-buff; breast mottled grey-brown. **Immature** As adult female but browner above; upperwing-coverts with some rusty-buff tips; white of outer tail washed pale buff. **VOICE** Song a harsh buzzing note preceded or occasionally followed by 1–2 short notes *whut-tzeet* and *tzit-tzit-zheet* or *zheet-tit*, with variations. Calls include *tek-tek-tek-...*, *tlup-tlup-tlup-...* and *peew*, and various buzzy and croaking notes. Also

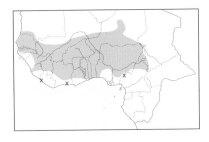

wing- and bill-snapping. [CD12:6] **HABITS** Frequents lower and middle strata of wooded savanna, farmland and dry *Acacia* woodland. Active and vocal. Often joins mixed-species flocks. **SIMILAR SPECIES** Senegal Batis is the only batis in most of its range, but meets Black-headed and Grey-headed in Cameroon; males of latter two have narrower and shorter supercilium; Black-headed also darker above, with blackish top of head. **STATUS AND DISTRIBUTION** Locally common to not uncommon resident in Sahel and savanna zones, from S Mauritania and Senegambia to Sierra Leone east to Niger, Nigeria and N Cameroon. Vagrant, Liberia.

GREY-HEADED BATIS *Batis orientalis*   Plate 111 (5)
Pririt à tête grise

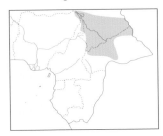

Other name F: Batis oriental

L 10 cm (4"). *B. o. chadensis.* Similar to allopatric Chinspot Batis but smaller. **Adult male** Differs from male Chinspot by *distinct white nape patch.* **Adult female** Lacks chestnut on throat and has darker chestnut breast-band; also has buffish wash to supercilium, nape patch, mantle and breast-sides. **Juvenile** Upperparts speckled white; breast-band blackish. **Immature** As adult female with rusty-buff supercilium and median coverts; mantle tinged brownish. **VOICE** Song an endless series of clear, whistled notes *hee hee-hu hee-hu hee-hu-hu-hu hee-hu-hu-hu-hu-...* Calls include a buzzy *dzek-dzek.* [CD12:7] **HABITS** Frequents lower and middle strata of savanna woodland and thorn scrub. Behaviour as congeners. **SIMILAR SPECIES** Chinspot Batis similar (q.v.) but range does not overlap. Black-headed Batis is slightly larger with usually darker top of head; breast-band slightly broader, in female also slightly darker chestnut. Senegal Batis has longer and broader supercilium. **STATUS AND DISTRIBUTION** Not uncommon to scarce or rare resident in eastern Sahel zone, from NE Nigeria (L. Chad area; scarce) and NE Cameroon to S Chad (also Ennedi, where rare) and CAR.

BLACK-HEADED BATIS *Batis minor*   Plate 111 (4)
Pririt à joues noires

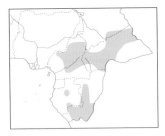

Other name F: Batis à joues noires

L *c.* 11.5 cm (*c.* 4.5"). *B. m. erlangeri.* Similar to Chinspot Batis. **Adult male** Top of head dark grey or black with slight gloss; birds with dark grey crowns indistinguishable from Chinspot. **Adult female** Lacks chestnut patch on throat and has darker chestnut breast-band. **Juvenile** Head and upperparts speckled buffish. Eye whitish. **Immature** As adult female but with rusty-buff supercilium and median coverts; mantle tinged brownish. **VOICE** Song a series of loud, penetrating, identical whistles *heet-heet-heet-heet-...,* uttered in flight with whirring wings and from perch; speed of delivery variable. Call a buzzing *dzip dzip...* [CD12:8] **HABITS** Frequents lower and middle strata of wooded grassland and orchard bush. Behaviour as congeners. **SIMILAR SPECIES** Chinspot Batis (q.v.); ranges overlap in Gabon and Congo. Grey-headed Batis very similar, but slightly smaller, with top of head paler grey; breast-band slightly narrower, in female also slightly paler chestnut. Senegal Batis is smaller, with broader and longer supercilium; female has breast-band distinctly paler chestnut. **STATUS AND DISTRIBUTION** Uncommon to locally common resident, from Cameroon to S Chad and CAR, and from S Gabon to Congo.

ANGOLA BATIS *Batis minulla*   Plate 111 (8)
Pririt de l'Angola

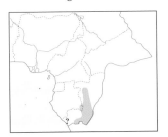

Other name F: Batis de l'Angola

L 10 cm (4"). Small batis with *broad breast-band* and *no supercilium.* Appears blacker than sympatric batises, with lack of supercilium giving rounder headed appearance. **Adult male** Top of head grey; distinct white nape patch; small white supraloral stripe and yellow eye emphasised by black mask; upperparts grey. Wings black with long white bar. Underparts white with broad black breast-band. **Adult female** Breast-band bright chestnut, broadening on flanks. **Juvenile/immature** Unknown. **VOICE** Song a series of weak, high-pitched, sucking *heep heep heep...* notes; speed of delivery variable. Calls include little buzzing notes. [CD12:9] **HABITS** Frequents lower and middle strata of relict forest patches, dense savanna woodland with thickets and adjacent orchard bush. Behaviour as congeners. **SIMILAR SPECIES** Chinspot and Black-headed Batises differ by noticeably larger size, white supercilium and narrower breast-band (which, in females, is also darker chestnut); usually also by preference for more open habitat. **STATUS AND DISTRIBUTION** Uncommon to locally common resident, SE Gabon and Congo.

VERREAUX'S BATIS *Batis minima*   Plate 111 (9)
Pririt de Verreaux

Other names E: Gabon Batis; F: Batis à tête grise
L 10 cm (4"). Small. *No supercilium.* **Adult male** Top of head blackish; distinct white nape patch; black mask with small, indistinct white supraloral spot; upperparts dark slate-grey. Wings black with long white bar. Underparts white with black breast-band. **Adult female** As male but *breast-band slate-grey* (unique among females in the region); crown dark grey. **Juvenile** Head and upperparts speckled whitish; breast-band grey. **Immature** As adult female but wing-coverts tinged rusty-buff and breast-band bordered above and below by rusty-buff. **VOICE** Song a series of high-pitched, identical whistles *heet-heet-heet-heet-...*; speed of delivery variable. Calls include various short buzzing and clucking notes. [CD12:10] **HABITS** In old second growth, mainly in canopy and mid-levels; occasionally also in primary forest, but never further than 500 m from edge. Behaviour as congeners. Often in mixed-species flocks. Unobtrusive. **SIMILAR SPECIES** Bioko Batis distinguished by more distinct supraloral spot, indistinct supercilium and slightly larger size; female Bioko has chestnut breast-band. **STATUS AND DISTRIBUTION** Scarce and local forest resident in S Cameroon, SW CAR, Eq. Guinea and Gabon. *TBW* category: Near Threatened.

BIOKO BATIS *Batis poensis*   Plates 111 (7) & 145 (12)
Pririt de Lawson

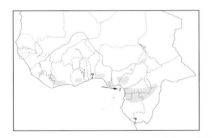

Other names E: Fernando Po Batis; F: Batis de Fernando Po
L *c.* 12 cm (*c.* 4.75"). **Adult male** *occulta* Similar to Verreaux's Batis but differs in black (not greyish-black) top of head, more distinct supraloral spot, narrow superciliary line and slightly larger size; upperparts grey. **Nominate** extremely similar; differences very slight, apparently not always consistent and not noticeable in field. Supraloral spot usually smaller, supercilium indistinct or absent, breast-band broader and white edges to tertials and outer tail feathers narrower. **Adult female** *occulta* Breast-band chestnut; rest of underparts pure white. In **nominate**, lower border of breast-band rather diffuse; flanks and upper belly slightly washed chestnut. **Juvenile** Normal batis pattern. **Immature** As adult female, but with rusty-buff tinge to wings, throat and breast. **VOICE** Songs include *tee-tee-tee-trrruuu tu-tee-tee-tee*, the second part often lacking; variation *trrruu-hu-hee-hee-hee*, often frequently repeated; series of shorter *trrru-hu trrru-hu trrru-hu*...; rhythmic *trrip-trrip-trrip-*...; and long series of very high-pitched, penetrating notes *heet-heet-heet-*... Calls include a dry *trrr.* [CD12:11] **HABITS** In canopy of evergreen forest, frequenting edges and clearings; also emergents in primary forest. Usually in mixed flocks of small insectivores. **SIMILAR SPECIES** Verreaux's Batis distinguished by very indistinct supraloral spot and slightly smaller size, but differences subtle and much caution needed. **STATUS AND DISTRIBUTION** Scarce or rare to locally not uncommon forest resident, from Sierra Leone to Ghana and from Nigeria to SW CAR, Gabon and N Congo (*occulta*). Claimed from Benin. The only batis on Bioko (nominate; below 800 m, locally up to 1100 m). **NOTE** Race *occulta* sometimes considered a separate species (West African Batis), but differences with *poensis* not clear-cut (*BoA*). More fieldwork required.

PICATHARTES
Family PICATHARTIDAE (World: 2 species; Africa: 2; WA: 2)

Unmistakable, strange-looking, slender, terrestrial birds with brightly coloured bare heads, long necks and tails, rather long, strong bills and long, strong legs. Sexes similar. Occur locally in closed rainforest and its extensions in savanna. Dependent on caves or overhanging rocks for breeding (hence alternative name 'rockfowl'). Nest is a bowl of mud plastered to rock face. Usually breed in small colonies. Sedentary, remaining close to nest-sites year-round. Endemic to W Africa.

YELLOW-HEADED PICATHARTES *Picathartes gymnocephalus* Plate 114 (9)
Picatharte de Guinée

Other names E: White-necked Picathartes, Yellow-headed Rockfowl; F: Picatharte à cou blanc, Picatharte chauve de Guinée
L *c.* 38 cm (15"). *Bare, bright yellow head* with contrasting dark eye and *large round black patch from hindcrown to ear-coverts*. Neck and underparts white. Upperparts slate-grey; wings and tail brown. Stout black bill. Legs bluish-grey. **Juvenile** Yellow on head slightly paler and mottled with dark spots; tail shorter. **VOICE** Mostly silent. [CD12:53] **HABITS** Singly, in pairs or small groups on ground of rainforest and gallery forest with caves and rocky outcrops. Fast and agile, progressing in long springing hops, without using wings. Seldom flies. Joins mixed-species flocks and attends ant-swarms. Secretive but not shy. **STATUS AND DISTRIBUTION** Scarce and very local resident in forest zone of SW and NE Guinea, Sierra Leone, Liberia, Ivory Coast and Ghana. *TBW* category: THREATENED (Vulnerable).

RED-HEADED PICATHARTES *Picathartes oreas* Plate 114 (10)
Picatharte du Cameroun

Other names E: Grey-necked Picathartes, Red-headed Rockfowl; F: Picatharte à cou gris
L *c.* 38 cm (15"). Bare tricoloured head: *bright cerulean blue from forecrown to nostril* (about middle of upper mandible), black on sides of head, *carmine hindcrown* separated from blue area by narrow black band. Upperparts and tail slate-grey with contrasting black primaries and primary-coverts. Neck and upper breast pale greyish. *Rest of underparts pale yellowish-buff.* Bill black. Eye dark. Legs greyish. **Juvenile** Bare skin of head less brightly coloured, grey-blue on forecrown, reddish-brown on hindcrown; tail shorter. **VOICE** Usually silent. Alarm, a muffled, drawn-out, hushing sound *kshhhhhhhhhh* uttered with body held erect and swollen neck. [CD12:54] **HABITS** As Yellow-headed Picathartes. **STATUS AND DISTRIBUTION** Scarce to locally not uncommon resident in forest zone, in SE Nigeria, Cameroon, Eq. Guinea and Gabon; also SW Bioko (at 250–900 m). Possibly also NW Congo. *TBW* category: THREATENED (Vulnerable).

BABBLERS
Family TIMALIIDAE (World: *c.* 260–280 species; Africa: 37; WA: 16)

Large and diverse family, represented in our region by rather small to medium-sized birds with largely sombre plumage. Legs and feet strong; bill stout. Sexes similar; no distinctive juvenile plumage. Sedentary. Many species gregarious; some are co-operative breeders. Parasitised by cuckoos.

Two main groups can be distinguished.

Illadopsises are secretive forest-dwellers of which most species forage in the undergrowth or on the ground. Plumage mainly brown above, paler below. Regularly in mixed-species flocks. Difficult to identify on plumage characters alone, but voice distinctive. Songs, uttered in duet or in groups, consist of pure, far-carrying whistles that are hard to locate. The genus *Illadopsis* was formerly placed in *Trichastoma* or *Malacocichla* by some authors.

Babblers (*Turdoides*) are sturdy, thrush-like birds that move in small, vocal groups in savanna and bush. Plumage mostly dull and sombre. Vocalisations loud and chattering. Forage on or close to the ground, tossing leaves aside with their strong bills. Parasitised by Levaillant's and Jacobin Cuckoos.

PALE-BREASTED ILLADOPSIS *Illadopsis rufipennis* Plate 113 (5)
Akalat à poitrine blanche

Other names E: White-breasted Akalat; F: Grive-akalat à poitrine blanche
L 15 cm (*c.* 6"). **Adult** *extrema* Similar to Brown Illadopsis, but head, upperparts and tail slightly darker, more dark russet-brown; underparts paler, with *pure white throat* (often puffed out), pale olive-brown breast, flanks and undertail-coverts, and white centre of belly. Bill black above, greyish-horn below. Legs bluish to brownish-grey or pale pinkish. **Nominate** has head-sides distinctly greyish. **Juvenile** Slightly more russet; lesser wing-coverts sometimes tipped tawny. **VOICE** Resembles that of Brown Illadopsis, but whistled notes higher pitched (thus generally impossible to imitate by human whistle). Two main song types distinguished. The first consists of a single, pure,

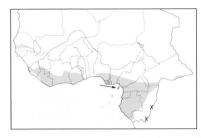

drawn-out whistle *wHEEE* repeated after a short pause and accompanied by rapid series of short clucking and hoarse notes (e.g. *chuk-kuk-kuk whiz-whiz-wheez-...*). The second comprises a short ascending series of 2–3 pure whistles, introduced and accompanied by short notes, e.g. *whit tew wHUUU wHEEE*, repeated constantly. Contact call a harsh *chuk*. Also a harsh *tzr*, sometimes lengthened to *tzrrzrzrzr* and *tiup-tiewtiewtiew-tiep-tiurrr-tirrr*. [CD12:36] **HABITS** Similar to Brown Illadopsis, but usually lower down and more often on forest floor, rarely higher than 5 m above ground. Particularly partial to clusters of dead leaves and decaying vegetation, which are examined systematically. **SIMILAR SPECIES** Brown Illadopsis (q.v. and above). Noticeably larger and more robust Rufous-winged Illadopsis has more white on underparts, greyish flanks and pale legs. Scaly-breasted Illadopsis (only recorded in CAR) has greyer head and underparts, scaly-looking breast and usually slightly paler legs. **STATUS AND DISTRIBUTION** Uncommon to locally common forest resident, from W Guinea and Sierra Leone to Ghana (*extrema*), and from SW Nigeria and Cameroon to CAR, Congo and Bioko (nominate).

BROWN ILLADOPSIS *Illadopsis fulvescens* — Plate 113 (2)
Akalat brun

Other names E: Brown Akalat; F: Grive-akalat brune
L *c.* 16 cm (6.25"). **Adult** *gularis* Head, upperparts and tail dark fulvous-brown appearing slightly darker on forehead and crown, and becoming dark greyish on head-sides; rump brighter. *Throat dirty white; rest of underparts pale fulvous-brown*. Bill black above, greyish-horn below. Legs bluish-grey. **Nominate** has forehead and crown slightly darker; in *ugandae* these are even darker and whitish throat is bordered by slate-grey submoustachial stripe. In *moloneyana* and *iboensis* underparts are darker fulvous-brown *lacking any white*. **Juvenile** Mainly dirty white below washed brownish on breast and flanks. **VOICE** Song a low, drawn-out, rather plaintive even whistle *wHEEEE* followed by a pause and another, similar whistle on lower pitch, *wHUUUU*, the whole interspersed with a few short *huit* and *tew* notes (only audible at close range) uttered by other group members. Also variations, with accompanying notes including a fast, hoarse *che-che-che-che*. Another song type (in east of range only?) superimposes somewhat twanging notes resembling *dict-a-phone*. Drawn-out whistles easily imitated by human. Various harsh and rattling contact calls. [CD12:35] **HABITS** In small groups within forest interior, thickets by roads in secondary forest and dense bush in clearings. Mostly perches low in vegetation (rarely on ground), but also higher (to lower canopy), esp. in dry season. Mainly active at dawn and dusk. When foraging, runs and climbs on branches, fallen logs and lianas, poking its bill into clusters of dead leaves. Readily joins mixed-species flocks. **SIMILAR SPECIES** Pale-breasted Illadopsis has pure white throat and whitish centre of belly (latter sometimes difficult to see); is also a shade more chestnut above and has brownish of underparts slightly paler (thus with greater contrast between upper- and underparts); whistles higher pitched. **STATUS AND DISTRIBUTION** Common resident in forest and forest-scrub mosaic, from S Senegal to Ghana, also SW Mali (*gularis*), from Cameroon to SW CAR and Congo (nominate), and in SE CAR (*ugandae*). Brown-throated birds, from S Ghana and Togo (*moloneyana*) to S Nigeria and W Cameroon, west of Mt Cameroon (*iboensis*).

SCALY-BREASTED ILLADOPSIS *Illadopsis albipectus* — Plate 113 (4)
Akalat à poitrine écaillée

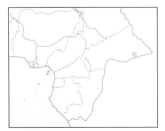

Other name F: Grive-akalat à poitrine écaillée
L 15 cm (*c.* 6"). **Adult** Rather similar to Pale-breasted Illadopsis, but *throat and breast pale grey, with breast feathers tipped olivaceous-brown giving faint scaly appearance* (lacking in some). Legs pinkish-grey to pale pink. **Juvenile** Brighter, with browner head and variable amount of rufous above eye. **VOICE** Similar to Blackcap Illadopsis. Far-carrying, rising series of 2–3 pure, drawn-out whistles, usually preceded by 1–2 short notes e.g. *wHU-HEEE* or *whit tiuk wHUU wHEE wHIII...*, endlessly repeated. [CD12:38] **HABITS** Usually singly or in pairs, examining clusters of dead leaves and decaying vegetation on forest floor. Rarely higher than a few metres above ground. **SIMILAR SPECIES** Pale-breasted Illadopsis has distinctly whiter underparts without scaly breast feathers and usually slightly darker legs. Brown Illadopsis is browner below and has darker legs. **STATUS AND DISTRIBUTION** Common resident, SE CAR.

BLACKCAP ILLADOPSIS *Illadopsis cleaveri* Plate 113 (3)
Akalat à tête noire

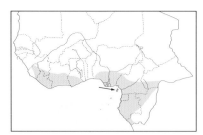

Other names E: Blackcap Akalat; F: Grive-akalat à tête noire
L 15 cm (*c.* 6"). **Adult** Distinguished from all other illadopsises by *dull black cap* (forehead to nape) separated from dusky head-sides by *whitish or pale greyish lores and supercilium.* Upperparts and tail dark russet-brown; underparts whitish washed grey on breast (forming diffuse band) and tinged brownish on breast-sides and flanks. Bill blackish, lower mandible paler. Legs pale pinkish. *I. c. johnsoni* darker than **nominate**, flanks paler; *batesi* has lores and supercilium darker greyish; *marchanti* and *poensis* have top of head olive-grey; *marchanti* also whiter below with less grey suffusion; *poensis* darker, with more dusky olivaceous-brown upperparts and darker tinged breast and flanks. **Juvenile** Similar to adult. **VOICE** Song consists of 1–2 resonant, far-carrying whistled notes, usually introduced by 1–2 short ones (only audible at close range), *ptk whit wHU wHEEE*, mainly uttered shortly after dawn and difficult to locate. Also a faster, rising series of 3 pure whistles introduced by 1–2 short notes, *whit wHU-wHEE-wHIII.* Both forms endlessly repeated. Contact calls include a dry *prrt prrt …* [CD12:37] **HABITS** Usually encountered singly or in pairs, occasionally in small groups, searching leaf litter and clusters of decaying vegetation on or just above forest floor. Rarely higher than 2 m above ground. Often in mixed-species flocks. **SIMILAR SPECIES** Brown-chested Alethe (Plate 92) rather similar in size and actions but distinguished principally by whiter supercilium and, in nominate race, brownish ear-coverts. **STATUS AND DISTRIBUTION** Not uncommon forest resident, from Sierra Leone and Liberia (*johnsoni*) to Ivory Coast (race?) and Ghana (nominate), in S Nigeria (east to Calabar, *marchanti*) and from SE Nigeria (Obudu Plateau) to CAR and Congo (*batesi*); also Bioko (*poensis*).

RUFOUS-WINGED ILLADOPSIS *Illadopsis rufescens* Plate 113 (6)
Akalat à ailes rousses

Other names E: Rufous-winged Akalat; F: Grive-akalat du Libéria
L 17 cm (6.75"). Relatively large and robust, with strong pale legs and feet. **Adult** Head and upperparts uniformly dark russet-brown; tail more chestnut. Underparts whitish with brownish wash on breast and olivaceous-grey flanks and undertail-coverts. Bill black, base of lower mandible pale horn. Legs pale whitish-pink. **Juvenile** Upperparts warmer coloured; legs blue-grey. **VOICE** Song a distinctive, far-carrying, rhythmic and fast *chk-chk-chk-HU-HU-HU* or shorter *chk-HU-HU.* [CD12:39] **HABITS** Mainly in pairs in lowland rainforest, on or near ground, occasionally higher. Often in mixed-species flocks. **SIMILAR SPECIES** Puvel's Illadopsis is slightly larger and more rufous, esp. on wings and tail. Other illadopsises are smaller, less thrush-like. Pale-breasted Illadopsis has less white on underparts; in Brown Illadopsis underparts are mainly pale fulvous-brown. **STATUS AND DISTRIBUTION** Uncommon to rare forest resident, recorded from S Senegal, Guinea-Bissau, Guinea, Sierra Leone, Liberia, Ivory Coast, Ghana and Togo. *TBW* category: Near Threatened.

PUVEL'S ILLADOPSIS *Illadopsis puveli* Plate 113 (7)
Akalat de Puvel

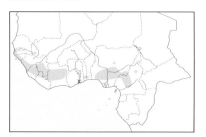

Other names E: Puvel's Akalat; F: Grive-akalat de Puvel
L 18 cm (*c.* 7"). Strong thrush-like build similar to Rufous-winged Illadopsis. **Adult** *puveli* Head, upperparts and tail russet-brown, esp. on wings and tail; lores whitish. Underparts white washed olivaceous-buff on breast, flanks and undertail-coverts. Bill blackish above, pale horn below. Legs whitish. Eastern *strenuipes* is more olivaceous-brown above. **Juvenile** Head and upperparts warmer brown. Legs bluish-white. **VOICE** Song far carrying, rhythmic and fast, resembling that of Rufous-winged Illadopsis in structure, but higher pitched and with slightly more varied notes and phrases, e.g. *chk-chk-whit HEE-HU-HUU* or *chk-whit-HE-HU-HU-HE-HU.* [CD12:40] **HABITS** Frequents forest edges, gallery forest and dense bush in forest–savanna mosaic. Mainly in pairs on or just above ground. **SIMILAR SPECIES** Rufous-winged Illadopsis has underparts washed browner on breast and greyer on flanks; also slightly smaller in size and darker russet on upperparts. Spotted Thrush Babbler has pure white underparts spotted brown on breast. **STATUS AND DISTRIBUTION** Uncommon resident in northern part of forest zone and forest outliers in savanna, from Gambia

and S Senegal to Ivory Coast and Ghana (nominate), and from Togo and S Benin (*strenuipes?*) to Nigeria and Cameroon (*strenuipes*). Also S Mali.

GREY-CHESTED ILLADOPSIS *Kakamega poliothorax* Plate 113 (8)
Akalat à poitrine grise

Other names E: Grey-chested Babbler; F: Grive-akalat à poitrine grise
L 17 cm (6.75") Relatively large and thrush-like with strong legs and distinctive coloration. **Adult** Head, upperparts and tail *rich chestnut-brown*; underparts *grey* paling to whitish on throat and centre of belly. Bill black above, pale horn below. Legs blue-grey. **Juvenile** Underparts darker, more olive-grey, with some chestnut or pale olive-tawny feathers on breast or centre of belly; throat has some olive-brown mottling. **VOICE** Song consists of short series of distinctive, loud, clear, melodious whistles with almost oriole-like quality, e.g. *tchlee tlu tluweeo* or *chee-wee-woo, wee-woo, wee-woo*, also a more rapid *trilutruleeo* and similar, frequently uttered sounds. Calls include a thrush-like chirp and a harsh, chattering alarm note. [CD12:33] **HABITS** Singly or in pairs on or near ground in mature montane forest. Joins other species and sometimes enters more open areas when following ant columns. Shy and unobtrusive. **SIMILAR SPECIES** African Hill Babbler also is dark chestnut above, but has black or dark slate-grey head, darker grey underparts, smaller size, and different voice and behaviour. Brown-chested Alethe has grey-brown face, white supercilium and paler underparts. **STATUS AND DISTRIBUTION** Generally uncommon to rare, but locally common forest resident in highland areas of SE Nigeria (Obudu Plateau, Gotel Mts), W Cameroon (Mt Cameroon, Rumpi Hills, Mt Kupe, Mt Nlonako, Bakossi Mts, Mt Manenguba, Bamenda Highlands) and Bioko.

AFRICAN HILL BABBLER *Pseudoalcippe abyssinica* Plate 113 (1)
Akalat à tête sombre

Other names E: Hill-Babbler; F: Alcippe à tête grise
L 14–15 cm (5.5–6.0"). **Adult** *atriceps* Head black; upperparts and tail dark chestnut-brown. Throat black slightly mottled grey; rest of underparts slate-grey, paler on belly and washed olive-rufous on flanks and undertail-coverts. Bill black above, grey below. Legs pale grey. Local races *monachus* and *claudei* paler with slate-grey head; in *claudei* grey of head extends further onto mantle. **Juvenile** *atriceps* Head brownish-black; underparts mainly olivaceous-brown. Juvenile *monachus* has top of head olive-brown, uniform with upper-parts. **VOICE** Characteristic, rich, clear, melodious warbling song, thrush-like and rather reminiscent of Blackcap or Garden Warbler. Both sexes sing year-round. Also a low, guttural chatter and a distinctive disyllabic call. [CD12:41] **HABITS** In dense undergrowth of montane forest (esp. in more open areas such as glades and clearings) and thickets of second growth. Generally in pairs within 2 m of (but not on) ground, gleaning insects from leaves and twigs, and taking seeds and berries; occasionally in canopy. Joins mixed-species flocks. **SIMILAR SPECIES** Grey-chested Illadopsis also is dark chestnut above, but has head concolorous with rest of upperparts (not black or slate-grey), paler grey underparts, larger size, different voice and different behaviour. **STATUS AND DISTRIBUTION** Locally common forest resident in highland areas. *P. a. atriceps* in E Nigeria (Gangirwal and Chappal Waddi, 1800–2300 m) and W Cameroon (Bamenda Highlands, above 2000 m); *monachus* endemic to Mt Cameroon (mainly at 1500–2100 m), *claudei* to Bioko (above 1100 m).

SPOTTED THRUSH BABBLER *Ptyrticus turdinus* Plate 114 (7)
Akalat à dos roux

Other name F: Grive-akalat à dos roux
L *c.* 20 cm (8"). Thrush-like, with strong, pale bill and legs. Resembles large illadopsis, esp. Puvel's, but even larger. **Adult** *harterti* Head, upperparts and tail *rufous*, brighter on rump; *underparts white with variable brown spotting on lower throat and breast* and brown patch on breast-sides. Lores whitish. Eye pale brown to red. Bill blackish above, greyish-white below. Legs whitish. **Nominate** has darker russet-brown plumage. **Juvenile** As adult but with dark eyes. Also reported to have smaller breast spotting with brown patches on sides not developed (Bannerman 1936), and to be rather more deep tawny, esp. on tail and outer webs of flight feathers (Mackworth-Praed & Grant 1973). **VOICE** Vocal. Song, mostly uttered at dawn and dusk, consists of loud, melodious, rather oriole-like

whistles, *tiow* or *kiuw*, often preceded by some subdued notes. Recalls song of Grey-chested Illadopsis, but is clearer and louder, and even more oriole-like. Territorial song uttered by pair in unison. Also low chuckling and rattling call notes. [CD12:42] **HABITS** In pairs or small parties within dense undergrowth of gallery forests, thickets and isolated moist forest patches. Elusive but not shy. **SIMILAR SPECIES** Puvel's Illadopsis has breast-sides and flanks washed pale brownish, lacks spots on breast and is smaller. **STATUS AND DISTRIBUTION** Uncommon to locally not uncommon resident, Cameroon (Adamawa Plateau) to W CAR (*harterti*) and in SE CAR (nominate).

BROWN BABBLER *Turdoides plebejus*   Plate 114 (4)
Cratérope brun

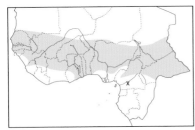

L 24 cm (9.5"). **Adult** *platycircus* Top of head, upperparts and tail *greyish-brown with contrasting pale buffish-brown rump* (conspicuous in flight); head-sides and underparts paler grey-brown. Feathers of head, throat and breast have darker centres giving scaly appearance. Plumage tone varies geographically. Eye bright orange or yellow. Bill black. Legs blackish. **Nominate** paler overall with browner head; *cinereus* darker. **Juvenile** Similar but eye brownish. **VOICE** Various harsh and scolding calls, incl. a fast *chuk-chuk-chuk*, a single, often repeated *kuk* and a clashing *kaaa*. Fast series of chuckling calls uttered simultaneously by several individuals result in a clamorous,

excited, raucous chattering. [CD12:43] **HABITS** In small, noisy groups of 4–12. Frequents various bushy habitats, incl. gallery forest, and shrubby edges of farmland and gardens. Forages on or near ground, usually in thick cover. Flies over short distances (next bush) with direct flight consisting of flapping wingbeats followed by a glide. Attracts attention by its frequent harsh chuckling and chattering calls. **SIMILAR SPECIES** In flight, sympatric Blackcap Babbler has uniformly brown upperparts (no paler rump). **STATUS AND DISTRIBUTION** Common to not uncommon resident in broad savanna zone. *T. p. platycircus* from Mauritania to Sierra Leone east to SW Nigeria; nominate from Niger and N Nigeria to S Chad and N CAR; *cinereus* from SE Nigeria to S CAR.

ARROW-MARKED BABBLER *Turdoides jardineii*   Plate 114 (5)
Cratérope fléché

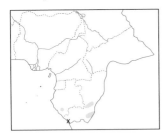

Other name F: Cratérope de Jardine
L 24 cm (9.5"). *T. j. hypostictus.* **Adult** Head and upperparts earth-brown with some scaly markings on head and mantle; tail dark brown. Throat and breast greyish with pointed whitish tips to feathers giving speckled appearance. Rest of underparts buffish-grey. Eye bright yellow with orange outer ring. Bill black. Legs dark grey. **Juvenile** Similar but eye dark brown. **VOICE & HABITS** Similar to Brown Babbler. [CD12:44] **STATUS AND DISTRIBUTION** Uncommon and local resident in S Gabon and Congo, where the only babbler.

BLACKCAP BABBLER *Turdoides reinwardtii*   Plate 114 (3)
Cratérope à tête noire

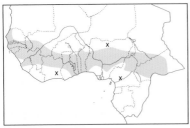

L 25 cm (10"). The only babbler with black head. **Adult** *reinwardtii* Upperparts and tail earth-brown. Throat buffish-white becoming pale greyish-brown on rest of underparts; feathers of throat and upper breast have darker centres giving scaly appearance. Under-parts darken from west to east, from nominate to *stictilaemus*; throat shades to greyish with dusky markings; black of head becomes less sharply defined, merging with upperparts on mantle. Eye creamy-white. Bill greenish-black; legs brown. **Juvenile** Underparts creamy-brown without 'scales'; eye dark. **VOICE** Various harsh and scolding calls, similar to those of Brown Babbler, but loud chorus of excited,

raucous chattering interspersed with less harsh, rather nasal *ko-kwee ko-kwee* ... [CD12:46] **HABITS** Similar to Brown Babbler. **SIMILAR SPECIES** Sympatric and more common Brown Babbler lacks black head and has contrasting paler rump in flight. **STATUS AND DISTRIBUTION** Uncommon to locally common resident in savanna zone, from Senegambia, Guinea-Bissau and Guinea to Mali (and Ivory Coast?) (nominate), from SE Mali and Ivory Coast to Chad and CAR (*stictilaemus*).

DUSKY BABBLER *Turdoides tenebrosus*

Plate 114 (6)

Cratérope ombré

Other name F: Cratérope fuligineux
L 24 cm (9.5"). **Adult** Wholly dark olive-brown with black lores and scaly-looking forehead, throat and upper breast. Eye yellowish-white. Bill black. Legs dark brown. **Juvenile** Slightly browner above; throat and upper breast greyish with fine streaking (not scaly); eye dark. **VOICE** An occasional hoarse *chow* and a nasal *what-kow*, repeated a number of times (Chapin 1953). **HABITS** In pairs or small groups in wooded habitats, frequenting dense undergrowth near water. Shy, secretive and much less vocal than other babblers. **SIMILAR SPECIES** Compare Brown Babbler. **STATUS AND DISTRIBUTION** Not recorded with certainty from our region. Claimed from NE CAR (Carroll 1988). Occurs at SW Sudan border, where uncommon to rare.

FULVOUS BABBLER *Turdoides fulvus*

Plate 114 (8)

Cratérope fauve

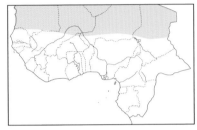

L 25 cm (10"). *T. f. buchanani.* **Adult** Unmistakable. Rather slender, entirely sandy-buff (paler below) with long, graduated tail. Some black streaks on head and nape. Eye colour variable, from white to brown and red. Bill brownish-black. Legs grey-green or brownish-yellow. **Juvenile** Upperparts more tawny; eye dark. **VOICE** Very varied. Commonest call a rather fast, slightly descending series of 6–9 whistled notes introduced by a similar, but more drawn-out *peeeew peew-peew-peew-peew-peew-peew*. Also a metallic trill *rirrrrrrrrrr* and short *pip* and clear *pee* notes. Alarm a sharp *pwit*. [CD4:41] **HABITS** Usually in small groups, in arid scrub country. Unobtrusive and much less vocal than other babblers, running or flying from bush to bush. Local movements with rains. **STATUS AND DISTRIBUTION** Locally not uncommon to rare in Sahel and desert zones, from Mauritania, N Senegal, Mali and N Burkina Faso to Niger and Chad. Recent southward extension of range due to drought. **NOTE** Perhaps better included in now defunct genus *Argya* given substantial vocal differences with other *Turdoides* babblers (Chappuis 2000).

CAPUCHIN BABBLER *Phyllanthus atripennis*

Plate 114 (2)

Phyllanthe capucin

Other name F: Cratérope capucin
L 24 cm (9.5"). **Adult** *atripennis* Distinctive, stout, very dark chestnut species, *appearing blackish in field*, with grey head, throat and upper breast, and conspicuous *pale greenish-yellow bill*. Lores black. Eye dark reddish-brown. Legs pale olive-green. *P. a. haynesi* has forehead and crown blackish and grey restricted to head-sides and upper throat; body plumage brighter, more reddish-chestnut. *P. a. bohndorffi* has throat chestnut. **Juvenile** As adult. **VOICE** Short chuckling notes and a loud, excited, raucous chattering similar to Brown Babbler. Also a drawn-out whistle *hu-wheew*. [CD12:50] **HABITS** Usually in small groups within dense scrub and undergrowth of secondary and gallery forest, forest edges and forest patches in savanna. Occasionally in mixed-species flocks. Rather shy. **STATUS AND DISTRIBUTION** Uncommon to not uncommon but local resident, from Senegambia to Ivory Coast (nominate), Ghana to Cameroon (*haynesi*) and in CAR (*bohndorffi*).

WHITE-THROATED MOUNTAIN BABBLER *Kupeornis gilberti*

Plate 114 (1)

Phyllanthe à gorge blanche

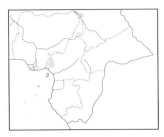

Other names E: Gilbert's Babbler; F: Timalie à gorge blanche
L 21 cm (8.25"). **Adult** Stout brown bird with striking white throat and upper breast extending to head-sides and neck. Forehead to nape dark chestnut; upperparts dark olive-brown; underparts slightly paler and more rufous. Flight feathers dark brown. Eye greyish-white. Bill dusky white, culmen brownish. Legs greenish-grey. **Juvenile** White area mixed with pale rusty-brown on face, neck-sides and upper breast. **VOICE** A harsh, explosive *chak*, usually uttered singly, but occasionally in short rapid series. Also a soft *kiorr*. Groups often give a harsh concerted chatter. [CD12:48] **HABITS** In noisy, active groups of up to 12 in canopy and mid-stratum of tall mature montane

rainforest; occasionally lower. Usually in mixed flocks with Grey-headed Greenbul, sometimes also other species. Often perches upside-down while searching for insects in moss, epiphytes and crevices in bark. **STATUS AND DISTRIBUTION** Locally not uncommon to common resident in primary montane forest of SE Nigeria (Obudu Plateau) and W Cameroon (Rumpi Hills, Mt Kupe, Mt Nlonako, Bakossi Mts, Foto near Dschang). *TBW* category: THREATENED (Endangered).

DOHRN'S THRUSH BABBLER *Horizorhinus dohrni*
Cratérope de Principé

Plate 145 (8)

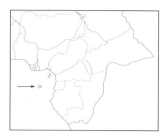

Other names E: Dohrn's Flycatcher; F: Gobemouche de Dohrn
L *c* 14 cm (5.5"). **Adult** Head, upperparts and tail uniform olivaceous-grey; underparts white washed pale yellow on belly, with olivaceous-grey breast-band and flanks. Bill dark brown, slightly paler below. Legs pale brownish. **Juvenile** Similar. **VOICE** Song, one of the most characteristic bird sounds on the island, melodious, vigorous and cheerful, starting with a drawn-out, descending note followed by a short one *tseeeeu-tu* then followed by fast, variable series of short notes, e.g. *tsitu-tsitu-tsitu-tutu- twitwitwitwi-tuwi-tuwi-tuwi tsitu-tsitu-tsitu.* Calls include rapid *tyentyentyentyen...* and long trill *trrrrrrirrrrrr...* [CD12:51] **HABITS** Singly, in pairs or small family parties in cocoa plantations, second growth and primary forest, mainly in open or dense understorey, mid-level and lower crowns of taller trees. Secretive but vocal. **STATUS AND DISTRIBUTION** Endemic to Príncipe, where common resident. **NOTE** Affinities uncertain; sometimes treated as a flycatcher (Muscicapidae).

TITS
Family PARIDAE (World: 55–57 species; Africa: 13–15; WA: 5)

Rather small, arboreal birds with short, relatively stout bills, almost square tails and strong legs and feet. Occur in various types of forest and woodland. Sexes similar. Juveniles similar to adults but duller. Active and agile, often hanging upside-down when searching for insects, nuts and seeds. Frequent and conspicuous members of mixed-species flocks, attracting attention by their rasping calls. Nest in holes.

DUSKY TIT *Parus funereus*
Mésange enfumée

Plate 115 (5)

Other name F: Mésange ardoisée
L 13–14 cm (5.0–5.5"). *P. f. funereus.* Distinctive. **Adult male** Wholly blackish with conspicuous red eye. Upperparts slightly glossed greenish. Bill black. Legs bluish-grey. **Adult female** Duller and greyer. **Juvenile** As adult male but duller; median and greater wing-coverts tipped white forming double wingbar. **VOICE** Varied. A clear, whistled *tsee-tu* and *ptsee-tu-tu* or *ptk-tsee-tu-tu* with variations; a series of drawn-out vibrant notes *teeeeeurrr, teeeeeurrr, whreeeEEp, teeeeeurrr ...;* also harsh, buzzing and churring notes. Some calls reminiscent of drongos or Fraser's Forest Flycatcher (Brosset & Erard 1986). [CD12:61] **HABITS** In pairs, small monospecific groups or mixed parties in canopy of high forest, often along clearings and edges. Noisy and active. **SIMILAR SPECIES** Compare juvenile Grey-crowned Negrofinch. Other black forest species (e.g. Shining and Square-tailed Drongos, Maxwell's Black Weaver) are larger and have different shape, behaviour and voice. **STATUS AND DISTRIBUTION** Rare to scarce rainforest resident, recorded from Sierra Leone (Gola Forest), SE Guinea, Liberia, SW and E Ivory Coast, S Ghana, Cameroon, SW CAR, NE Gabon and N Congo.

WHITE-BELLIED TIT *Parus albiventris*
Mésange à ventre blanc

Plate 115 (3)

Other name E: White-breasted Tit
L 14–15 cm (5.5–6.0"). *White underparts* diagnostic. **Adult male** Head, upperparts, throat and upper breast black faintly glossed blue. Wings black with white median coverts (forming shoulder patch), greater coverts fringed white, flight feathers edged white. Underparts from lower breast to undertail-coverts white. Tail black, edged and

tipped white. Eye dark brown. Bill black. Legs bluish-grey. **Adult female** Duller; black areas more sooty. **Juvenile** As adult female but slightly browner; white edges to flight feathers faintly washed yellow. **VOICE** A fast series of 2 sharp, high-pitched notes followed by 3–4 harsh, buzzing notes *tzii-tzii chèr-chèr-chèr*. [CD12:60] **HABITS** In pairs or family parties in savanna woodland and at edges and clearings of forests in highland forest–savanna mosaic. Restless and vocal, as other members of genus. **SIMILAR SPECIES** White-shouldered Black Tit has all-black underparts and no or very little white in tail. **STATUS AND DISTRIBUTION** Not uncommon to rare resident in highlands of SE Nigeria (Obudu Plateau; rare) and Cameroon (Bamenda Highlands and Adamawa Plateau; also reported from Yaoundé area).

RUFOUS-BELLIED TIT *Parus rufiventris*   Plate 115 (4)
Mésange à ventre cannelle

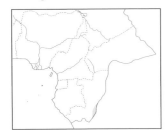

L 14–15 cm (6"). *P. r. rufiventris*. Distinctive colourful tit with *rufous belly*. **Adult** Head to throat black; upperparts and lower breast grey; rest of underparts cinnamon-rufous (diagnostic). Wing feathers black broadly edged and tipped white. Tail black, edged and tipped white. Conspicuous yellow eye. **Juvenile** Duller; black areas more sooty; upperparts tinged brown; underparts paler; edges to wing feathers washed yellowish; eye brown. **VOICE** A rasping *pzeet chrr chrr* and similar harsh notes. Also a fast *whee-tee-uw whee-tee-uw*. [CD12:64] **HABITS** In pairs or small parties in savanna woodland. Occasionally joins mixed-species flocks. Behaviour as other tits. **STATUS AND DISTRIBUTION** Rare resident, Congo (Brazzaville area).

WHITE-SHOULDERED BLACK TIT *Parus (leucomelas) guineensis*   Plate 115 (1)
Mésange gallonée

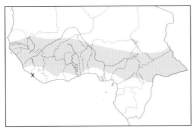

Other name F: Mésange à épaulettes blanches

L 14 cm (5.5"). Unmistakable and conspicuous. **Adult** Wholly glossy blue-black with white, elongated wing patch and *conspicuous yellow eye*. White on median and greater coverts form contrasting shoulder patch, extending on rest of wing as white edgings to secondaries. Bill and legs black. **Juvenile** Duller; white on wings washed yellow; eye grey or brown. **VOICE** Very varied. Calls harsh and buzzing, incl. *zeet-zeet* and *churr*, also *tsik tsik tsik*. Songs include short, often repeated, fast series of clear whistles *huwee-tEE-huwee-tee-huwèh* and *wheetu-tuleep wheetu-tuleep ...* or *ptiu-weet ptui-weet ptiu-weet ...*, and melodious series of various rolling notes (e.g. *tsrrrua tsrrrwèht*), with variations. [CD12:59] **HABITS** In pairs or small parties in open savanna woodland. Often joins mixed-species flocks. Restless and vocal. **STATUS AND DISTRIBUTION** Not uncommon to uncommon resident throughout savanna zone, from Senegambia to N Sierra Leone east to S Chad and CAR. Rare wet-season visitor, S Mauritania. **NOTE** Usually considered a race of *P. leucomelas* (cf. *BoA*), but here tentatively treated separately, following Harrap & Quinn (1996). Rationale for subspecific treatment presented by Dowsett & Dowsett-Lemaire (1993).

WHITE-WINGED BLACK TIT *Parus leucomelas*   Plate 115 (2)
Mésange à épaulettes

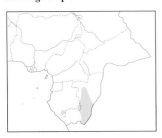

L 15 cm (6"). *P. l. insignis*. **Adult** Very similar to White-shouldered Black Tit but has *brown eye* and *outer tail feathers narrowly edged and tipped white*. Also slightly larger. **Juvenile** Duller; white on wings washed yellow; eye brown. **VOICE & HABITS** Similar to White-shouldered Black Tit (q.v.). **STATUS AND DISTRIBUTION** Not uncommon to uncommon resident in wooded grasslands of S Gabon and Congo. **NOTE** Usually considered conspecific with White-shouldered Black Tit (cf. *BoA*), but here tentatively treated separately, following Harrap & Quinn (1996). Rationale for subspecific treatment presented by Dowsett & Dowsett-Lemaire (1993).

PENDULINE TITS

Family REMIZIDAE (World: 13 species; Africa: 7; WA: 5)

Tiny, principally insectivorous and arboreal birds with short tails and short, sharply pointed bills. In African genus *Anthoscopus* ('kapok tits') sexes similar. Juveniles similar to adults. Occur from forest to arid *Acacia* zone. Active but unobtrusive; often in small parties; some species occasionally with other small insectivores. Nest a remarkable oval purse of felted plant or animal fibres, soft but very tight and strong, with closed tubular entrance near top.

The monotypic genus *Pholidornis* is provisionally placed here, following *BoA*, though its systematic position is a matter of debate.

FOREST PENDULINE TIT *Anthoscopus flavifrons*   Plate 115 (8)
Rémiz à front jaune

Other name E: Yellow-fronted Penduline Tit
L *c.* 9 cm (*c.* 3.5"). Tiny, inconspicuous canopy species. **Adult** *flavifrons* Forehead has narrow, deep golden band bordered by black lores (diagnostic, but usually very hard to see in field); crown and upperparts dark olive-green. Wings and tail blackish edged olive. Underparts olive-grey washed yellow, esp. on breast and belly. Eye blackish. Bill black above, paler below. Legs slate-grey. Western *waldroni* slightly yellower above and below. **Juvenile** Slightly duller, with buffish forehead. **VOICE** High-pitched, barely noticeable little calls. [CD12:58] **HABITS** In pairs or small groups (3–5) in lowland rainforest; usually in canopy of highest, flowering trees, rarely lower. Joins mixed flocks of small insectivores. Difficult to observe and easily overlooked due to small size, dull coloration and preference for canopy. **SIMILAR SPECIES** At great height, small female sunbirds may appear rather similar, but have longer, decurved (not short, sharply pointed) bills. **STATUS AND DISTRIBUTION** Rare to scarce (but probably overlooked) forest resident, from Liberia to Ghana (*waldroni*) and from Nigeria to Congo (nominate).

YELLOW PENDULINE TIT *Anthoscopus parvulus*   Plate 115 (10)
Rémiz à ventre jaune

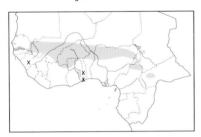

Other name E: West African Penduline Tit
L 7.5–8.0 cm (3"). **Adult** *senegalensis Tiny*, bright yellow bird. *Forehead and head-sides bright yellow*; crown and upperparts olive-yellow, becoming brighter on rump and uppertail-coverts. Wing feathers brown edged yellowish; greater coverts and tertials usually conspicuously fringed whitish (forming wingbar when fresh). Tail brown narrowly edged whitish. *Entire underparts bright yellow.* Small black dots on forehead visible at close range. Short, sharp-pointed bill dark horn, paler below. Eye blackish. Legs bluish-slate. **Nominate** slightly duller above, with more contrasting yellow forehead; *citrinus* slightly brighter overall; *aureus* duller overall, with forehead presenting little contrast with crown. **Juvenile** Duller. **VOICE** A fast *chipichipichipichipi...*, a thin, quiet *si sli-li-liii*, a high-pitched, slightly hoarse *bzee-bzee-bzee-bzee*, and a rhythmical buzzing *chura-chura-chura-...* or *duza-duza-duza-...*; also a short *ch, ch, ch* (Harrap & Quinn 1996). [CD12:55] **HABITS** In pairs or small parties in wooded savanna and dry *Acacia* scrub. Actively searches branches for insects, occasionally forming mixed flocks with Yellow White-eyes or other small insectivores. **SIMILAR SPECIES** Similar-sized Sennar Penduline Tit has duller coloured plumage with pale buffish underparts; also favours drier habitats. Yellow White-eye is larger and has obvious white eye-ring. **STATUS AND DISTRIBUTION** Local and uncommon to scarce or rare resident in Sahel and savanna zones. *A. p. senegalensis* almost throughout range, from S Mauritania and Senegambia east through Burkina Faso and N Ivory Coast to N Nigeria and L. Chad area; also W Guinea. Nominate in Chad; *citrinus* in CAR and (presumably this race) N Cameroon; *aureus* N Ghana.

SENNAR PENDULINE TIT *Anthoscopus punctifrons*   Plate 115 (9)
Rémiz du Soudan

Other name E: Sudan Penduline Tit
L 7.5–8.5 cm (3.0–3.25"). Tiny, inconspicuous pale bird. **Adult** Forehead yellowish finely speckled blackish (flecks only visible at close range); crown and upperparts yellowish-olive, slightly brighter on rump. Wings and tail dark

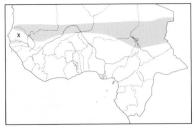

brown narrowly edged paler. Head-sides and throat off-white; rest of underparts pale pinkish-buff. Eye dark brown. Bill black above, paler below. Legs bluish-slate. **Juvenile** Slightly duller. **VOICE** Very high-pitched *tsii* notes, churring *whut-chrrrrrrrrrr* and harsh *tchuwhree-whree*, with variations. [CD12:57] **HABITS** In pairs or small groups in thorn scrub, esp. in wadis and near ponds and wells. Active and tame. Unobtrusive. **SIMILAR SPECIES** Similar-sized Yellow Penduline Tit has bright yellow underparts, variably distinct white wingbar and more southern distribution. **STATUS AND DISTRIBUTION** Uncommon to locally common resident in Sahel zone, recorded from Mauritania, N Senegal, Mali, Niger, NE Nigeria, N Cameroon and Chad.

GREY PENDULINE TIT *Anthoscopus caroli*   Plate 115 (6)
Rémiz de Carol

Other name E: African Penduline Tit
L 8.0–8.5 cm (3.0–3.25"). *A. c. ansorgei*. Tiny. **Adult** Forehead yellow; crown and upperparts yellowish-olive. Head-sides and underparts off-white. Wing and tail feathers dusky-brown edged olive; median and greater wing-coverts tipped yellowish-white forming double wingbar. Eye dark brown. Short, sharp-pointed bill bluish-grey with paler edges. Legs bluish-slate. **Juvenile** As adult. **VOICE** A thin squeaky *tseeep*, a fast, rhythmic *chisweep-chisweep-chisweep* and a rasping *chideZEE-chideZEE-chideZEE* or *chipchipzsee-chipchipzsee-chipchipzsee* with slight variations. [CD12:56] **HABITS** Singly, in pairs or small groups in wooded grassland. Actively searches branches of trees and bushes for small insects. Occasionally forms mixed flocks with other small insectivores. Unobtrusive. **SIMILAR SPECIES** Savanna-dwelling eremomelas are also small and active, but all have yellow on underparts. **STATUS AND DISTRIBUTION** Scarce resident, SE Gabon and Congo.

TIT-HYLIA *Pholidornis rushiae*   Plate 115 (7)
Mésangette rayée

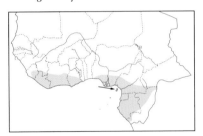

Other name F: Astrild-mésange
L 7.5 cm (3"). Tiny. **Adult** *ussheri* *Head, throat and breast pale greyish finely but densely streaked dusky-brown*; upperparts and tail blackish-brown *with contrasting yellow rump*. Wing feathers edged pale olive. Belly and undertail-coverts yellow faintly washed olive. Bill blackish, base of lower mandible yellow. Legs orange-yellow. **Nominate** has slightly coarser streaking; *denti* has rump and underparts brighter, more orange-yellow. Island race *bedfordi* more heavily streaked, with streaking broader, darker and extending onto flanks and belly; rump also streaked. **Juvenile** As adult but throat and breast almost unstreaked. **VOICE** Contact calls include little *tsik*, *ptu* or *ptiu* notes and a high-pitched *psee*. Alleged song, recorded in east of range, a fairly loud, fast, clear *tieetieetiee-chichichichi* or *puipuipui-tjitjitjitjitjitji*, sometimes preceded by 2 grating notes *ruirui* (Dowsett-Lemaire & Dowsett 1991); very similar to song of Yellow White-eye's forest race *stenocricotus*. [CD11:62] **HABITS** In small monospecific groups (3–8) in lowland forest, frequenting canopy of tall trees, including isolated ones in clearings, descending to top of lower trees in regrowth and along tracks and edges. Actively gleans leaves and twigs, sometimes hanging upside-down. Also joins mixed flocks of small insectivores. **SIMILAR SPECIES** Forest Penduline Tit is much duller and slightly larger, and restricted to high forest. **STATUS AND DISTRIBUTION** Not uncommon forest resident, from Sierra Leone and SE Guinea to Ghana (*ussheri*), from Nigeria to Congo (nominate), from SE Cameroon to CAR (*denti*) and on Bioko (*bedfordi*). Also Togo (1 record). Race of those in SW Nigeria (west of Niger R,) uncertain (*ussheri?*).

SPOTTED CREEPER
Family SALPORNITHIDAE (World: 1 species)

A small, arboreal bird with a long, decurved bill and short, strong legs. Sexes similar. Monotypic. Also occurs in India. Formerly often included in family Certhiidae (treecreepers).

SPOTTED CREEPER *Salpornis spilonotus*   Plate 115 (11)
Grimpereau tacheté

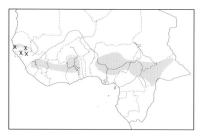

L 15 cm (6"). *S. s. emini*. Unmistakable. *Cryptically brown spotted white, with long decurved bill and short tail.* **Adult** Above dark brown, streaked whitish on crown, spotted whitish on upperparts; boldly barred whitish on wings and tail. Long whitish supercilium; dark eye-stripe. Underparts whitish variably spotted and barred brown. **Juvenile** Duller. **VOICE** Call a single, high-pitched *tseee*. Song a short series of clear, high-pitched notes *tsip-tsee-tsu-tuwee*, with slight variations. Also faster *tsitsutsitsutsitsitsu...* [CD12:66] **HABITS** Singly or in mixed-species flocks in wooded savanna. Clings to tree-trunks and branches, starting low down and working upwards in spirals, searching the bark. **STATUS AND DISTRIBUTION** Local and uncommon to rare resident in savanna belt, in Gambia, SE Senegal, Guinea-Bissau (rare), Guinea, N Sierra Leone, N Ivory Coast, N Ghana, Burkina Faso, Togo (rare), and from Nigeria to S Chad and N CAR. **NOTE** E and S African birds have much higher pitched, more wispy song *tseepy-tswee tseepy-tswee tswee-tswee-tswee...* [CD12:65]. Chappuis (2000) suggests they are best treated as a separate species *S. salvadori*.

SUNBIRDS
Family NECTARINIIDAE (World: *c.* 120 species; Africa: 81; WA: 36)

Large family of distinct, small to medium-sized passerines with long, slender, sharply pointed, decurved bills (shortest and least decurved in *Anthreptes*, *Deleornis* and *Hedydipna*; longer and more decurved in the other genera). Feed on nectar, insects and spiders. Sexes dissimilar in most species; males often have brightly coloured and iridescent plumage (which may, however, appear black in some lights); most females dull greenish or brownish. In some species, males have brightly coloured pectoral tufts (used in display, but otherwise mostly concealed under the wing) and/or elongated central tail feathers in breeding plumage. Some males moult into female-like eclipse plumage after breeding. Juveniles as adult females; some juvenile males moult directly into breeding plumage, others first assume intermediate ('immature') plumage. Young have bright orange, yellow or white skin at gape (black in adults), which initially can be seen in the field, but gradually becomes duller; in certain species it is retained for at least a year. Some females and juveniles are difficult to distinguish and are best identified by the presence of the male. Occur in a broad range of habitats, from arid scrub to wooded savanna, forest and gardens. Most are resident, but local or seasonal movements reported for several species. Usually in pairs, but may congregate at favourable food source. Active, restless and pugnacious, with rapid, dashing flight.

Having been initially considered to represent many genera, the Afrotropical Nectariniidae were more recently placed in just two, greatly expanded genera: *Anthreptes* (including *Anthreptes*, *Deleornis* and *Hedydipna*) and *Nectarinia* (the other genera). However, a review by Irwin (1999) led to the proposition, which has been followed here, to recognise 10 Afrotropical genera, 8 of which occur in our region.

WESTERN VIOLET-BACKED SUNBIRD *Anthreptes longuemarei*   Plate 116 (1)
Souimanga violet

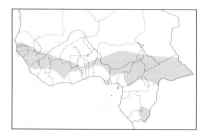

Other name E: Violet-backed Sunbird
L 13–14 cm (5.0–5.5"). Distinctive sunbird with unique plumage coloration. **Adult male *longuemarei*** Head, upperparts, tail and throat glossy violet, neatly contrasting with pure white underparts. Head-sides and flight feathers brownish-black. Pectoral tufts yellow. Bill rather short and straight; black above, dark horn below. Southern *angolensis* has breast washed pale buffish. **Adult female** Head, upperparts and tail brown with *white supercilium* and glossy violet on uppertail-coverts, tail and some wing-coverts. Underparts white with *lemon-yellow belly, flanks and undertail-coverts*. **Juvenile** Resembles adult female, but head and upperparts more olive; underparts entirely pale yellow. **VOICE** A harsh *chep* and *tit*, frequently repeated. Song a rapid, high-pitched jingle. [CD12:67] **HABITS** Usually singly or in pairs, occasionally in small parties, in wooded savanna and gallery forest. Frequently joins mixed-species flocks. Actively searches foliage and branches for insects, often hanging upside-down; sometimes captures prey in flight. **SIMILAR SPECIES** Adult female and juvenile may recall warbler, but underparts coloration and bill shape should prevent confusion. Violet-tailed Sunbird differs from corresponding plumages of Violet-backed as follows: adult male more bluish-green

(not purple) above and normally pale buffish (not white) below, pectoral tufts orange; adult female wholly glossy above, yellow on underparts brighter and more extensive; juvenile brighter yellow below. **STATUS AND DISTRIBUTION** Uncommon to rare resident in savanna belt, from Senegambia to Sierra Leone east to Chad and CAR (nominate), and in SE Gabon and S Congo (*angolensis*).

VIOLET-TAILED SUNBIRD *Anthreptes aurantium*　　　　　　Plate 116 (3)
Souimanga à queue violette

L 13–14 cm (5.0–5.5"). Distinctive sunbird of forested river banks. **Adult male** Head, upperparts, tail and throat a glossy mixture of deep violet-blue and green; underparts pale greyish-buff or dirty off-white. Pectoral tufts orange. Bill rather straight and short. **Adult female** Top of head and upperparts glossy green-blue; long white supercilium; throat and upper breast whitish becoming lemon-yellow on rest of underparts. **Juvenile** Head and upperparts olive-green (rapidly attaining some green gloss); supercilium and entire underparts lemon-yellow; tail dark with suggestion of gloss. **VOICE** A hard *tsip.* **HABITS** Restricted to forest and thickets along rivers and streams; also mangroves. Frequents lower and middle strata. Not attracted to flowering trees; takes some fruits. Usually in pairs or small family parties. Less restless than other sunbirds. **SIMILAR SPECIES** Violet-backed Sunbird differs from corresponding plumages of Violet-tailed as follows: adult male purple (not bluish-green) above and white (not pale buff-grey) below, pectoral tufts yellow; in adult female glossy areas limited to uppertail-coverts, tail and some wing-coverts, yellow on underparts less bright and more reduced; juvenile duller yellow below. **STATUS AND DISTRIBUTION** Uncommon and local forest resident in S Cameroon (north to Korup NP), S CAR, Gabon and Congo.

BROWN SUNBIRD *Anthreptes gabonicus*　　　　　　Plate 116 (2)
Souimanga brun

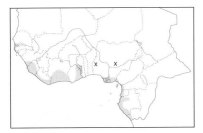

Other name E: Mouse-brown Sunbird
L 10 cm (4"). Distinctive. *Non-glossy, greyish-and-white species with short, almost straight bill, white lines above and below eye, and rather warbler-like appearance.* **Adult** Head and upperparts grey-brown with narrow white line over base of forehead extending below reddish-brown eye; white supercilium; wings greyish with white edge below bend. Underparts greyish-white. Tail brown; all feathers except central pair tipped white (only visible from below). **Juvenile** Head and upperparts olive; underparts washed lemon. **VOICE** A single, very high-pitched note, reminiscent of a small kingfisher, and little sibilant notes, such as *sipsipsipsip,* uttered in flight. Song described as *tser-tser-tsew-tsi-tsi-tsi-tsi-tsi-tsi-tseuuur* (Cheke 1999). [CD12:68] **HABITS** Singly or in pairs in mangroves and along forested river banks. Unobtrusive. **STATUS AND DISTRIBUTION** Uncommon to locally common resident, from S Senegambia to Liberia east to Ghana and from Nigeria to Congo; also SW Burkina Faso.

GREEN SUNBIRD *Anthreptes rectirostris*　　　　　　Plate 116 (6)
Souimanga à bec droit

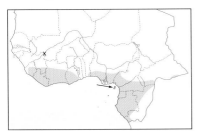

L 10 cm (4"). **Adult male** *rectirostris* ('Yellow-chinned Sunbird') Head and upperparts bright glossy green; wings and tail dusky-brown edged green. Throat yellow; bright glossy green breast-band narrowly bordered with orange below; lower breast pale grey; rest of underparts lemon-yellow washed olive on flanks. Bright yellow pectoral tufts. Bill short, rather stout and straight. Distinctive race *tephrolaemus* ('Grey-chinned Sunbird') has throat patch grey and smaller; lower underparts greyer. **Adult female** Wholly olive-green above, sometimes with some gloss; pale olive-yellow eye-ring and supercilium; underparts olive-yellow becoming lemon-yellow on centre of belly and undertail-coverts. Plumage and short bill give warbler-like appearance. **Juvenile** As adult female. **Immature male** Head-sides and upperparts have some glossy green feathers; lesser wing-coverts glossy green. **VOICE** A relatively loud *whseeet,* uttered singly or preceded by 2 short notes (then resembling French 'tout de suite', with accent on last word), a vigorous *peet* and a variable series of high-pitched notes, e.g. *seeu-seew-whu whu-see.* [CD12:69] **HABITS** In pairs and family parties in canopy beside forest tracks and clearings, gleaning foliage and branches for insects, spiders and small fruit. May move lower down along edges and in degraded habitat. Unobtrusive. **SIMILAR SPECIES** Male Collared Sunbird has throat and upper breast wholly glossy green; rest of underparts yellow, not grey. **STATUS AND DISTRIBUTION** Uncommon to locally common resident in forest zone, from Sierra Leone and SE Guinea to Ghana (nominate), and from Benin to CAR, Congo and Bioko (*tephrolaemus*). Also S Mali.

LITTLE GREEN SUNBIRD *Anthreptes seimundi*   Plate 116 (10)
Souimanga de Seimund

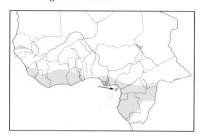

L 9.5 cm (3.75"). Very small, non-glossy species, resembling minia-ture Fraser's Sunbird. **Adult** *kruensis* Entire upperparts and short tail olive-green; underparts yellowish-green. Pale green eye-ring. Bill relatively short and straight, blackish above, pale pinkish-horn below. No pectoral tufts. **Nominate** brighter and larger; *traylori* intermediate. **Juvenile** Duller, more olivaceous. **VOICE** A soft *tsssip*. [CD12:71] **HABITS** Singly, in pairs or in small groups in canopy of lowland forest; lower down along clearings and edges. **SIMILAR SPECIES** Bates's Sunbird is very similar but has more decurved bill, slightly duller plumage with less contrast between upper- and underparts, black tail edged green and more upright stance. Adult female Olive-bellied and Tiny Sunbirds have blackish-blue tails and all-black bills. **STATUS AND DISTRIBUTION** Uncommon forest resident, from W Guinea and Sierra Leone to Togo (*kruensis*), and from Nigeria to S CAR and Congo (*traylori*); also Bioko (nominate). **NOTE** Sometimes placed in genus *Nectarinia*.

FRASER'S SUNBIRD *Deleornis fraseri*   Plate 116 (5)
Souimanga de Fraser

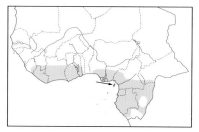

Other name E: Scarlet-tufted Sunbird
L 11–13 cm (4.5–5.0"). Medium-sized to fairly large, uncharacteristic, plain green forest sunbird with rather pale, pinkish, stout and straight bill. **Adult male** Head, upperparts and tail dark green; under-parts yellowish-green. Distinct pale green eye-ring. Pectoral tufts orange-red. Bill brownish above, pale horn below. Legs brownish or greyish-black. Size varies geographically: western *idius* smallest, island race *fraseri* largest, eastern *cameroonensis* intermediate. **Adult female** Similar but lacks pectoral tufts. **Juvenile** Similar to adult female, but slightly duller. **VOICE** Thin, high-pitched short notes, including *pseet*, *sriit* and *tswi* or *ssli*. [CD12:70] **HABITS** In pairs or small (family?) parties at lower and mid-levels (only occasionally higher) of lowland forest. Most often encountered in mixed-species flocks, with up to 10 in a group. Appears to be almost wholly insectivorous; rarely seen in flowering trees. Nest undescribed. **SIMILAR SPECIES** Grey Longbill may appear confusingly similar to observers unfamiliar with the species, but has greyer plumage, esp. on underparts (greyish-olive, not yellowish-green); bill similar in size but shape different, not gradually tapering to point; jizz and behaviour different. Conspicuous pale eye-ring may be reminiscent of Yellow White-eye if seen poorly. **STATUS AND DISTRIBUTION** Common resident in lowland forest, from W Guinea and Sierra Leone to Togo, and in S Mali (*idius*), and from Nigeria to SW CAR and Congo (*cameroonensis*); also Bioko (nominate). **NOTE** Formerly placed in genus *Anthreptes*.

REICHENBACH'S SUNBIRD *Anabathmis reichenbachii*   Plate 117 (1)
Souimanga de Reichenbach

L 13–14 cm (5.0–5.5"). **Adult male** Head, throat and upper breast dark glossy blue; upperparts olive-green, rump brighter, yellowish-olive. *Tail* blackish-brown, *graduated, all feathers except central pair with broad paler tips*. Underparts pale grey with *bright yellow lower belly, flanks and undertail-coverts*. Pectoral tufts yellow. **Adult female** Similar; pectoral tufts usually paler. **Juvenile** Wholly yellowish-green above; underparts a mix of yellow and grey with ventral area yellow as in adult; tail as adult. **VOICE** Call *chuwEE chuwEE*. Song a high-pitched jingle. [CD12:73] **HABITS** Singly or in pairs in gardens, edges of cultivation and forest, riverine thickets and swampy areas with bushes and palm trees. Never far from water. **SIMILAR SPECIES** Male Green-headed Sunbird has glossy parts blue-green and lacks yellow on lower underparts. **STATUS AND DISTRIBUTION** Uncommon to locally common resident, recorded in coastal areas from Liberia to Ghana, and from Nigeria to Congo. In east of range also inland along major rivers, reaching SW CAR. **NOTE** Formerly placed in genus *Nectarinia*.

PRÍNCIPE SUNBIRD *Anabathmis hartlaubii*   Plate 145 (9)
Souimanga de Hartlaub

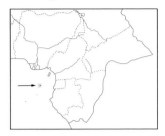

L 13–14 cm (5.0–5.5"). **Adult male** Head and upperparts dark olive-green. Throat and upper breast dark glossy violet-blue; lower breast dark olive-green paling to yellowish-olive. Tail graduated, blackish with blue gloss; all feathers, except central pair, tipped white (most obvious from below). **Adult female** As male, but throat and upper breast dark olive-green with paler tips to feathers giving scaled appearance. **Juvenile** As adult female, but throat and upper breast dark grey-brown. Young male similar to adult male but throat and upper breast dull sooty-grey. **VOICE** A clear *wheep* or *tuweet*, a fast chattering *chrrrp-ch-chep* or *chrtittt chrttt...* and various high-pitched notes. Song a fast rhythmic series of similar syllables, *t'k-tsit'k-tsit'k-tsit'k-tsit'k-tsit'k-* *...tsit* or *tsp-tsit'k-tsp-tsit'k-tsp-tsit'k-...* with slight variations. [CD12:74] **HABITS** Singly or in pairs, in primary and secondary forest, cocoa and coconut plantations, cultivated areas and gardens. **SIMILAR SPECIES** Western Olive Sunbird, the only other sunbird on Príncipe, is rather featureless, lacking any gloss, and has a more curved bill with orange base (bill in female Principe Sunbird relatively straight and wholly black). **STATUS AND DISTRIBUTION** Endemic to Príncipe, where common resident. **NOTE** Formerly placed in genus *Nectarinia*.

NEWTON'S SUNBIRD *Anabathmis newtonii*   Plate 145 (10)
Souimanga de Newton

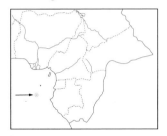

Other name E: Newton's Yellow-breasted Sunbird
L 10–11 cm (4"). Easily identified as it is the only sunbird on São Tomé, apart from Giant Sunbird. **Adult male** Head and upperparts dull brownish-olive. Throat and upper breast blue-green bordered with dark glossy violet-blue; rest of underparts yellow, brighter on lower breast. Tail black, graduated, all feathers except central pair broadly tipped white (obvious from below). Bill rather short and curved. **Adult female** Head and upperparts as male. Throat and upper breast dark olive-grey; rest of underparts pale yellow. **Juvenile** No information. Young male similar to adult female but slightly browner above; belly slightly paler. **VOICE** A frequently repeated *cheep* or *chit* and a harsh, high-pitched trill. Song a very rapid, high-pitched chattering jingle preceded by a drawn-out, high-pitched *whseeeew*. [CD12:75] **HABITS** Frequents forest, forest edge, cocoa and coffee plantations, gardens and areas of dry woodland in savanna. Usually in lower levels of forest, but occasionally in tall, flowering trees where many may congregate. Sometimes with São Tomé Speirops and São Tomé White-eyes. **STATUS AND DISTRIBUTION** Endemic to São Tomé, where common resident. **NOTE** Formerly placed in genus *Nectarinia*.

GIANT SUNBIRD *Dreptes thomensis*   Plate 145 (12)
Souimanga de Sao Tomé

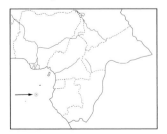

Other name E: São Tomé Sunbird
L male 20–23 cm (8–9"), female 18–19 cm (7.0–7.5"). Unmistakable. Very large, blackish-brown sunbird (appearing black in dim forest light). Largest family member. **Adult male** Head and upperparts sooty-black, feathers broadly fringed glossy blue. Tail long and graduated, blue-black, all feathers tipped white except central pair (visible from below). Underparts brownish-black, feathers narrowly tipped glossy blue; lower belly, flanks and undertail-coverts dirty yellow-green. **Adult female** Similar but smaller. **Juvenile** No information. **VOICE** A vigorous *cheep*, frequently uttered in short series *cheep-cheep-cheep* or faster, longer *chepchepchepchepchepchepchepchep...* [CD12:76] **HABITS** Restricted to primary forest and forest scrub, frequenting all levels though mainly in canopy. Vocal and active. **STATUS AND DISTRIBUTION** Rare to locally common forest resident, endemic to São Tomé, recorded from centre and south-west, in mid- to high-altitude forest (up to *c.* 2000 m) in Pico de São Tomé and Lagoa Amélia area, and in lowland forest along R. São Miguel, Xufexufe, Quija and Ana Chaves; also in lowland plantations in south-east. *TBW* category: THREATENED (Vulnerable). **NOTE** Formerly placed in genus *Nectarinia*.

GREEN-HEADED SUNBIRD *Cyanomitra verticalis*   Plate 117 (2)
Souimanga à tête verte

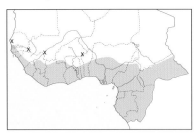

Other names E: Olive-backed Sunbird; F: Souimanga olive à tête bleue
L 13–14 cm (5.0–5.5"). **Adult male *verticalis*** Head, throat and upper breast glossy blue-green. Upperparts olive; rest of underparts dusky-grey. Flight and tail feathers dusky-brown edged olive. Pectoral tufts lemon-yellow. *C. v. bohndorffi* slightly darker above and below; *cyanocephala* more purplish on throat and upper breast, sootier on belly. **Adult female** As male but entire underparts pale grey; no pectoral tufts. **Juvenile/immature** Forehead and forecrown, throat and upper breast blackish-grey. Upperparts dull olive; rest of underparts greyish-olive, more yellowish on breast. **VOICE** A vigorous *chuwee* and *chi-ep*, a harsh *chee* and a high-pitched *chip* or *tsit*. Songs include a harsh, accelerating *chip-chip-chip-chichichichichichi* and a rapid jingle of high-pitched notes. [CD12:79] **HABITS** Singly, in pairs or small family parties in wooded savanna, gallery forest, thickets, rainforest, clearings, banana plantations and gardens. Mostly in canopy but also lower down. Joins mixed-species flocks. **SIMILAR SPECIES** Male Blue-throated Brown Sunbird is darker, has longer tail, less gloss on head and throat, and different voice. **STATUS AND DISTRIBUTION** Locally common resident in forest and wooded savanna zones throughout. Nominate from Senegambia to Cameroon; *bohndorffi* from Cameroon to CAR and interior of Gabon and Congo; *cyanocephala* from coastal Eq. Guinea to Congo. **NOTE** Formerly placed in genus *Nectarinia*.

BLUE-THROATED BROWN SUNBIRD *Cyanomitra cyanolaema*   Plate 117 (4)
Souimanga à gorge bleue

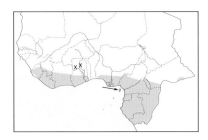

L 14–15 cm (5.5–6.0"). Fairly large, dark forest sunbird with relatively long tail. **Adult male *cyanolaema*** Forehead, forecrown and throat dark glossy blue. Head, upperparts and tail sooty-brown; underparts greyish-brown. Pectoral tufts lemon-yellow. *C. c. octaviae* similar; western *magnirostrata* has marginally longer and slightly less decurved bill. **Adult female** Head and upperparts dark olive; white supercilium and line below eye. Underparts dirty whitish or pale greyish mottled dusky on breast and washed yellowish-olive on flanks and undertail-coverts. Female *octaviae* usually paler above. **Juvenile** Grey-green above. Flight feathers and tail brown edged olive. Throat whitish with lemon wash; rest of underparts yellowish-grey. **Immature male** Sooty-brown above; throat sooty-grey; belly slightly washed olive. **Immature female** As adult female but duller, with slight olive wash. **VOICE** A characteristic short, high-pitched, descending trill *tsiiirrrrrrr* and a rapid, descending *tsitsitsitsup*. Also a short *tsk-tsk-tsk-...* [CD12:81] **HABITS** In pairs or small groups in canopy of forest and along edges and clearings. **SIMILAR SPECIES** Male Green-headed Sunbird has olive upperparts, shorter tail and more extensive glossy area. **STATUS AND DISTRIBUTION** Common forest resident. *C. c. magnirostrata* from Sierra Leone and SE Guinea to Ivory Coast and Ghana; *octaviae* from SW Nigeria to S CAR and Congo; nominate on Bioko. Rare, Togo (1 record). **NOTES (1)** Birds from Ghana usually considered *octaviae*, but gap in distribution suggests they are likely to be *magnirostrata* (*BoA*). **(2)** Formerly placed in genus *Nectarinia*.

CAMEROON SUNBIRD *Cyanomitra oritis*   Plate 117 (3)
Souimanga à tête bleue

Other name E: Cameroon Blue-headed Sunbird
L 12–13 cm (4.75–5.0"). Olive-green highland species with dark glossy bluish-purple head and throat. **Adult male *oritis*** Head, throat and upper breast dark glossy bluish-purple; nape and upperparts dark olive-green; rest of underparts dirty olivaceous-yellow. Flight and tail feathers dark brown edged olive-green. Pectoral tufts lemon-yellow. Bill long and decurved. *C. o. bansoensis* has head more greenish-blue and is rather brighter below; island race *poensis* slightly darker. **Adult female** As adult male but lacks pectoral tufts. **Juvenile/immature** Mainly dull olive with dusky head and throat. **VOICE** A soft *tik tik tik ...* or *chk chk chk...* uttered frequently and occasionally followed by high-pitched *seep*, a descending *tsi-tsi-tsi-tsup*, a series of disyllabic *pee-tsu-pee-tsu-pee-tsu-...* and a typical sunbird jingle of fast, high-pitched, metallic notes introduced by a few *chk* notes. Also a barely audible, sweet, sustained warble (Stuart & Jensen 1986). [CD12:82] **HABITS** Usually singly in undergrowth of mid-elevation and montane forest (down to 570 m on Mt Cameroon), esp. in interior and along streams; also at forest edge. Large

numbers may congregate at flowering trees. Unobtrusive. **SIMILAR SPECIES** Green-headed Sunbird may overlap at lower altitudes but usually occurs in higher strata; male has entire head (including nape) glossy blue-green, and grey (not olive-yellow) underparts. **STATUS AND DISTRIBUTION** Generally common forest resident in highland areas of SE Nigeria (Obudu and Mambilla Plateaux), W Cameroon (nominate: Mt Cameroon; *bansoensis*: Rumpi Hills, Mt Kupe, Bakossi Mts, Mt Manenguba, Dschang environs, Bamenda Highlands) and Bioko (*poensis*). **NOTE** Formerly placed in genus *Nectarinia*.

WESTERN OLIVE SUNBIRD *Cyanomitra obscura*   Plate 116 (7)
Souimanga olivâtre de l'Ouest

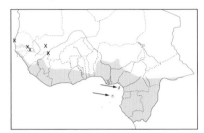

Other names E: Olive Sunbird; F: Souimanga olivâtre
L 13–15 cm (5–6"). Plain, olive-green, non-glossy sunbird with long, decurved bill. **Adult male** *guineensis* Head and upperparts olive-green; wings and tail dull blackish edged yellowish-olive. Underparts pale greyish-olive, paler on centre of belly. Pectoral tufts yellow. Bill black. Eastern *cephaelis* is slightly greener above and has pale base to lower mandible; island race *obscura* similar but larger. **Adult female** Similar but lacks pectoral tufts. **Juvenile** As adult female but underparts strongly washed yellow. **VOICE** Commonest call a vigorous, harsh *chip* or *cheep* (or *tsek*), frequently and rapidly repeated. Also a clear, loud *tieh-tieh-...* or *tuweet-tuweet-...* and *heet-heet-...* often uttered in long series. Song a series of clear, separate, piping notes, slightly accelerating and falling in pitch towards end, *TSIT, tsee tut tsiu tsu tu tsu tsu-tsu-tututu*. [CD12:84] **HABITS** Singly, in pairs or small family parties, frequenting mainly lower levels of forest interior, occasionally higher; also at edges and in thickets. Actively searches foliage, twigs and lianas for insects and spiders, often adopting acrobatic postures; occasionally takes nectar and fruit. Regularly joins mixed-species flocks. **SIMILAR SPECIES** Could be mistaken for unspecified female sunbird, but combination of plain olive plumage, long bill and relatively larger size should enable identification. Compare similar-sized Fraser's Sunbird, which has straighter bill. **STATUS AND DISTRIBUTION** Common resident in forest and gallery forest throughout. *C. o. guineensis* in west of range, from S Senegal to Togo, also W Senegal (Dakar area) and SW Mali; *cephaelis* from Togo to CAR and Congo; *obscura* on Bioko and Príncipe. One of the commonest forest species. **NOTE** Formerly included in Olive Sunbird *Nectarinia* (*Cyanomitra*) *olivacea*.

BUFF-THROATED SUNBIRD *Chalcomitra adelberti*   Plate 116 (12)
Souimanga à gorge rousse

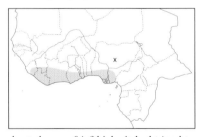

L 11.5–12.0 cm (4.5–4.75"). **Adult male** *adelberti* Distinctive. Forehead, forecrown and malar stripe glossy green; upperparts dark chocolate-brown with paler wing-coverts. *Underparts chestnut with striking straw-coloured throat* bordered by narrow blackish breast-band. Eastern *eboensis* more uniformly dark brown, with wing-coverts as rest of upperparts; underparts darker chestnut. **Adult female** Head and upperparts olive-brown with paler supercilium. Underparts off-yellowish washed pale brown and streaked olive-brown; throat whitish. **Juvenile** Dark greyish-olive above; paler below; dark bib; spotted breast. **VOICE** A vigorous, frequently repeated *che-pEEw*. Song consists of short phrases of 4–6 high-pitched *tsi* and *tsu* notes, e.g. *tsu-tsi tsi-tsu* and *tsu-tsi tsi-tsu tsitsu*. [CD12:85] **HABITS** Usually in pairs, frequenting flowering trees at forest edges and clearings, in plantations, gardens and gallery forest. **STATUS AND DISTRIBUTION** Not uncommon but generally local resident in forest zone, from Sierra Leone and SE Guinea to Togo (nominate) and from Benin to SE Nigeria, just reaching W Cameroon (*eboensis*). Local movements reported. (e.g. rare dry-season visitor, Nindam FR, C Nigeria). **NOTE** Formerly placed in genus *Nectarinia*.

GREEN-THROATED SUNBIRD *Chalcomitra rubescens*   Plate 117 (7)
Souimanga à gorge verte

L 12–13 cm (4.75–5.0"). **Adult male** *rubescens* Very dark brown (appearing black) with glossy green forehead, forecrown and throat bordered by narrow glossy violet band on crown and breast. No pectoral tufts. Island race *stangerii* similar. Distinctive race *crossensis* has entire underparts dark brown, lacking green on throat. **Adult female** *rubescens/crossensis* Head and upperparts brown with narrow pale yellowish supercilium. Underparts dirty yellowish, coarsely streaked and mottled dusky-brownish on throat, breast and flanks. Tail dark brown with paler tips. Female *stangerii* has heavier streaking below. **Juvenile** As adult female but upperparts tinged olive and underparts yellower, mottled dusky-brown on throat and breast (esp. densely on throat, which may appear wholly dusky, bordered by yellowish moustachial stripe). **Immature male** As adult female but with glossy green throat. **VOICE** A hard *tsik*.

Song includes a short, rapid phrase of frequently repeated single notes, e.g. *chep-chep-chep-...* sometimes accelerating into trill *chchchchchchch*, and a simple series of clear separate notes, e.g. *tseep tsup tsup* or *tsip tsu tsi tsi tsup* with variations, recalling song of Scarlet-chested Sunbird. [CD12:87] **HABITS** Singly or in pairs in forest clearings and edges, second growth, cultivation, and gardens with flowering trees and shrubs. Usually high. **SIMILAR SPECIES** Adult male Amethyst Sunbird distinguished by violet throat. Female quite similar to female Green-throated, but washed olive above and less heavily streaked below. Both sexes also have longer bill. **STATUS AND DISTRIBUTION** Common to not uncommon resident in forest zone and forest outliers in savanna, from SE Nigeria and Cameroon to S CAR and Congo (nominate); also Bioko (*stangerii*). *C. r. crossensis* restricted to small area in west of range, in extreme SE Nigeria and W Cameroon. **NOTE** Formerly placed in genus *Nectarinia*.

CARMELITE SUNBIRD *Chalcomitra fuliginosa* Plate 117 (5)
Souimanga carmélite

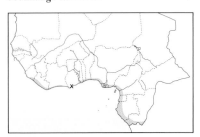

L 13–14 cm (5.0–5.5"). **Adult male** *aurea* Dark brown (looking black from distance) with dark glossy blue forehead and forecrown, and glossy violet throat and uppertail-coverts. Small violet patch on lesser wing-coverts. Pectoral tufts lemon-yellow. With wear head and upperparts become pale buffish-brown, even dirty whitish, contrasting strongly with dark brown wings, tail and underparts. **Nominate** darker above, with glossy area on forehead and throat more extensive. **Adult female** *Conspicuously pale.* Head and upperparts pale brownish; underparts even paler with brownish mottling on throat and faint brownish streaks on breast. Lores dark brown. Wings and tail brown. **Juvenile** Similar to adult female but head and upperparts tinged olive. Underparts dirty yellow mottled dusky-brown on throat and breast (esp. densely on throat, which may appear wholly dusky, bordered by yellowish moustachial stripe). **Immature male** Lores dark brown; throat glossy purple. **VOICE** A vigorous trill *srrrriiirrrr*. Song a variable series of short notes, e.g. *chep chi chew chi chup.* [CD12:86] **HABITS** Singly or in pairs in mangroves, dense bushes near water, open areas, gardens and plantations. **SIMILAR SPECIES** Male Copper Sunbird may also appear black at distance. Carmelite pairs easily identified by unusually pale female. **STATUS AND DISTRIBUTION** Locally common to rare resident in coastal areas, from Sierra Leone to Congo (*aurea*) and inland along Congo R. to Ngabé (nominate). **NOTE** Formerly placed in genus *Nectarinia*.

AMETHYST SUNBIRD *Chalcomitra amethystina* Plate 117 (6)
Souimanga améthyste

Other name E: Black Sunbird
L 13.5–14.0 cm (5.25–5.5"). *C. a. deminuta.* **Adult male** Sooty-black with glossy green forehead and crown, and glossy violet throat, lesser wing-coverts and uppertail-coverts. No pectoral tufts. **Adult female** Head and upperparts dull olive with faint pale yellowish supercilium. Underparts pale dirty yellowish indistinctly streaked dusky on breast and flanks. Tail blackish. **Juvenile** As adult female but entire underparts streaked dusky. **Immature male** As adult female but with glossy violet throat. **VOICE** A short, sharp *chut* or *tiu*, and *chak* or *chyek*, usually delivered in fast series. Alarm a stuttering *chchchchch* or *t-t-t-t-t.* Song a sustained, rapid series of high-pitched twittering notes, often interspersed with call notes and uttered for long periods from concealed perch. [CD12:88] **HABITS** Usually singly or in pairs, in wooded grassland, riparian bush, thicket and forest edge, and gardens. May gather in small parties, sometimes with other sunbirds, at favoured food source. Restless and aggressive. **SIMILAR SPECIES** Adult male Green-throated Sunbird distinguished from male Amethyst by glossy green (not violet) throat. Adult male Carmelite in fresh plumage is browner, esp. on upperparts, and has glossy blue (not green) forehead. Adult female and juvenile Green-throated Sunbird differ from corresponding plumages of Amethyst by shorter bills and browner heads, and upperparts. **STATUS AND DISTRIBUTION** Uncommon to locally common resident, SE Gabon and Congo. **NOTE** Formerly placed in genus *Nectarinia*.

SCARLET-CHESTED SUNBIRD *Chalcomitra senegalensis* Plate 117 (8)
Souimanga à poitrine rouge

Other name E: Scarlet-breasted Sunbird
L 14–15 cm (5.5–6.0"). **Adult male** *senegalensis* Unmistakable. Bulky, blackish sunbird with glossy green forehead,

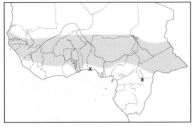

forecrown and upper throat, *brilliant red lower throat and breast* barred glossy violet (bars small and difficult to see in field), and long, decurved bill. Wings brownish-black and subject to bleaching, paling to milky brown. No pectoral tufts. Eastern *acik* has throat and breast paler red with less violet barring. **Adult female** Head and upperparts grey-brown without supercilium. Throat and upper breast mainly dusky-brown with yellowish tips to feathers; rest of underparts dull yellow broadly streaked brownish. Plumage becomes darker, duskier with wear. **Juvenile** As adult female but with sooty-grey throat; rest of underparts mottled and barred blackish. **Immature male** Glossy green feathers start appearing on upper throat, followed by red on lower throat and breast. **VOICE** A vigorous *chuw*, a harsh scolding *chep* or *chew* and (in flight) *tiup,* all usually uttered in rapid series. Song a short, simple series of clear separate notes *tip tiu tip tiu* or *tip tiu tip tip* with variations. [CD12:89] **HABITS** Singly or in pairs in various habitats within savanna zone, incl. woodland, riverine forest, gardens, etc. May gather in small feeding parties in dry season. **SIMILAR SPECIES** Female and juvenile Green-throated Sunbird are yellower, less heavily marked below, have distinct supercilium, and are less bulky, with slightly shorter bills. **STATUS AND DISTRIBUTION** Common to not uncommon resident in broad savanna belt. Nominate from S Mauritania to Sierra Leone east to Nigeria; *acik* from Cameroon (where intergrading with nominate) to Chad and CAR. Northward movement during rains reported. **NOTE** Formerly placed in genus *Nectarinia.*

COLLARED SUNBIRD *Hedydipna collaris*   Plate 116 (8)
Souimanga à collier

L 10 cm (4"). Easily identified, small, green-and-yellow sunbird. **Adult male** *subcollaris* Head, upperparts, throat and upper breast bright glossy green; rest of underparts bright yellow separated from throat by narrow purple breast-band. Wings blackish-brown edged dull yellowish-olive; tail blue-black edged glossy green. Pectoral tufts yellow. Bill short and only slightly decurved. Eastern *somereni* duller yellow below; island race *hypodilus* paler yellow and larger. **Adult female** As male, but entire underparts yellow (duller, greyish-yellow on throat). No pectoral tufts. **Juvenile** Dull olive above with small, short, pale yellowish-green lines above and below eye; throat greyish; rest of underparts pale lemon-yellow. Bill dark above, pinkish below. **Immature male** As adult female, but duller. **VOICE** A vigorous, frequently uttered *tsip* and *teew*, and a nasal *chee chee*. Song a series of high-pitched, separated notes, e.g. *swee swee swee swee swee see-see-see-see-see...* and *tsip-tsip-tsip-tsee-tsee tsee tsee tsee tsiup*, or *suweep-suweep-suweep-suweep-suweep-....*, with variations. [CD12:77] **HABITS** In pairs or small family groups in wide variety of habitats, incl. forest edge, gallery forest, thickets, woodland, farmbush and gardens. From lower levels to canopy. Regularly in mixed flocks of small insectivores. **SIMILAR SPECIES** Male Green Sunbird has yellow or grey throat and greyish (not yellow) lower underparts. Male Variable Sunbird has glossy violet throat and breast. Male Pygmy Sunbird without tail streamers differs by coppery green (not emerald-green) upperparts, purple rump, and absence of narrow purple breast-band. Juvenile Collared may recall Little Green Sunbird, but presence of parents and Little Green's occurrence in small, monospecific flocks (in west of range) should prevent confusion. **STATUS AND DISTRIBUTION** Common and widespread resident in forest and adjacent savanna zones. *H. c. subcollaris* in west of range, from Senegambia to Liberia east to Lower Niger R., also S Mali and Burkina Faso (rare); *somereni* east of Lower Niger R. to CAR and Congo; *hypodilus* Bioko. **NOTE** Formerly placed in genus *Anthreptes.*

PYGMY SUNBIRD *Hedydipna platura*   Plate 116 (4)
Souimanga pygmée

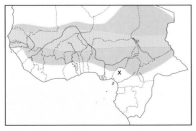

Other names E: Pigmy Long-tailed Sunbird; F: Petit Souimanga à longue queue
L 9–10 cm (3.5–4.0"); tail streamers of male projecting up to 7 cm (3"). **Adult male breeding** (dry season) Resembles Collared Sunbird with long tail streamers. Head, upperparts, throat and upper breast glossy coppery green; rump and uppertail-coverts dark glossy purple; wings and tail blackish; rest of underparts golden-yellow. Central tail feathers greatly elongated, slightly broader at tips. Bill short, slightly decurved. **Adult male non-breeding** (rainy season) Head and upperparts grey-brown with a few glossy green feathers; faint yellowish supercilium; some glossy green on wing-coverts. Underparts lemon-yellow with some blackish feathers

on throat sometimes forming small bib and partially glossed green. No tail streamers. **Adult female** As male non-breeding, but lacks glossy feathers. Becomes paler with wear. **Juvenile** As adult female, but duller. **Immature male** Some dark glossy feathers appear on head, upperparts and throat. **VOICE** Call *cheep* or *twee*. Song a fast, high-pitched jingle, including call notes and short trill. [CD12:78] **HABITS** Singly or in pairs when breeding, gregarious outside breeding season. In *Acacia* scrub, dry bush, wooded savanna and gardens. Readily joins mixed-species flocks. **SIMILAR SPECIES** Male Beautiful Sunbird, the only sympatric sunbird with tail streamers, has quite different coloration and longer, more decurved bill; retains tail streamers in non-breeding plumage (dry season). **STATUS AND DISTRIBUTION** Common in Sahel and savanna zones, from Mauritania and Guinea east to Niger (north to Aïr), Chad (north to Tibesti) and CAR. Intra-African migrant or locally dispersive, moving south to breed in dry season. Movements unclear: in some localities occurs throughout the year, in others only seasonally. **NOTE** Formerly placed in genus *Anthreptes*.

OLIVE-BELLIED SUNBIRD *Cinnyris chloropygius* Plate 117 (11)
Souimanga à ventre olive

L 10.5 cm (4"). **Adult male** *kempi* Head, upperparts and throat bright glossy green; breast bright red; rest of underparts dirty brownish-olive. Pectoral tufts yellow. Flight feathers and greater coverts brown. Tail black with dark blue gloss. **Nominate** has lower underparts slightly darker, less olivaceous. **Adult female** Head and upperparts dark olive; faint yellowish supercilium. Underparts dirty yellow with paler throat. Flight feathers and wing-coverts brown edged olive; tail blackish with dark blue gloss. **Juvenile** Similar to adult female but with greyish throat. **VOICE** Song consists of short *tsup* or *tsip* notes stuttering into a rapid, high-pitched, rising metallic jingle including many trills, descending at end. [CD13:1] **HABITS** In pairs or small parties, frequenting flowering trees and bushes in forest clearings, edges, gallery forest, thickets, second growth shrubbery, and gardens (even in city centres). Very active and not shy. **SIMILAR SPECIES** Male Tiny Sunbird has bluer upperparts and shorter, straighter bill. **STATUS AND DISTRIBUTION** Common resident in forest zone and moist savanna, from S Senegal to Liberia east to S Chad, CAR and Congo; also Gambia (rare) and Bioko. Western *kempi* meets eastern *chloropygius* in Nigeria. **NOTES (1)** The genus *Cinnyris* is treated here as masculine (*contra BoA*) following Cuvier's (1817) original treatment (J. Jobling & R.J. Dowsett *in litt.*). **(2)** Formerly placed in genus *Nectarinia*.

TINY SUNBIRD *Cinnyris minullus* Plate 117 (10)
Souimanga minule

L 9–10 cm (3.5–4.0"). **Adult male** Extremely similar to adult male Olive-bellied Sunbird but bill shorter and less decurved, lower mandible appearing almost straight except very tip (equally curved over most of its length in Olive-bellied). Glossy green upperparts slightly darker and *washed blue*, esp. on rump. Red breast has some narrow glossy blue bars but these virtually impossible to see in field. **Adult female** Differs from adult female Olive-bellied mainly by bill shape. Also slightly less yellow below, with greyish wash to throat and olive on lower underparts. **Juvenile** Similar to adult female but with greyish throat; belly slightly yellower; base of lower mandible pale. **VOICE** Calls include *tsi-tsi-tsi-tsup*. Song consists of short *tsup* or *tsip* notes stuttering into short, rapid, high-pitched metallic jingle; noticeably higher pitched and more simple than song of Olive-bellied Sunbird, and lacking trills. [CD13:2] **HABITS** Singly or in pairs in mature forest; locally also in farmbush and gardens. Usually in upper levels, lower at edges, in treefall gaps and clearings. **SIMILAR SPECIES** Olive-bellied Sunbird (q.v. and above) is more nervous, usually on the move, rarely singing for long time from same perch; also sings from lower in vegetation. **STATUS AND DISTRIBUTION** Scarce or rare to locally not uncommon forest resident, from Sierra Leone and SE Guinea to Ghana, in S Nigeria (rare; overlooked?), and from Cameroon to SW CAR and Congo; also Bioko. **NOTE** Formerly placed in genus *Nectarinia*.

NORTHERN DOUBLE-COLLARED SUNBIRD *Cinnyris reichenowi* Plate 118 (1)
Souimanga de Preuss

L 11.5 cm (4.5"). *C. r. preussi.* Resembles Olive-bellied Sunbird but differs in slightly larger size and montane habitat. **Adult male** Head, upperparts and throat bright glossy green; breast bright red separated from green throat by narrow violet band; rest of underparts dark brown; uppertail-coverts glossy violet. Pectoral tufts yellow. Flight feathers and greater coverts brown. Tail black with dark blue gloss. **Adult female** Head and upperparts

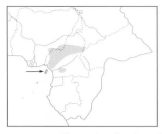

grey-green; underparts paler. No supercilium. Tail blackish. **Juvenile** Similar to adult female. **VOICE** Calls include a fast *chep-chep-chep-*... and a high-pitched *siip* (alarm). Song, uttered from perch or in flight, a vigorous, rapid, harsh and buzzy jingle, frequently repeated. [CD13:3] **HABITS** In small parties at all levels of open montane forest, clearings, edges, thicket, forested ravines, mature rainforest, secondary bush and scrubby grassland. In pairs when breeding (Oct–Mar). Frequently attracted, sometimes in large numbers, to flowering trees and shrubs. Very active. Males particularly pugnacious. Often in mixed-species flocks including other sunbirds. **SIMILAR SPECIES** Adult male Olive-bellied Sunbird has glossy green (not violet) uppertail-coverts, narrower red breast-band and much paler, more olive belly. Adult female Olive-bellied is paler and has supercilium. **STATUS AND DISTRIBUTION** Common resident in highlands of SE Nigeria (Obudu and Mambilla Plateaux), Cameroon (Mt Cameroon, Mt Kupe, Bakossi Mts, Mt Manenguba, Bamenda Highlands, Adamawa Plateau, hills in Yaoundé area) and extreme NW CAR; also Bioko. Recorded at lower altitudes (even sea level) in rainy season. **NOTE** Formerly named *Nectarinia preussi*. Not to be confused with extralimital Golden-winged Sunbird *Nectarinia reichenowi* (nor with Plain-backed Sunbird *Anthreptes reichenowi*).

BEAUTIFUL SUNBIRD *Cinnyris pulchellus* Plate 118 (8)
Souimanga à longue queue

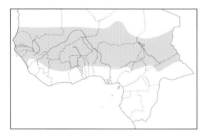

Other name E: Beautiful Long-tailed Sunbird
L 9–11 cm (3.5–4.5"); tail streamers of male projecting up to 6 cm (2.25"). *C. p. pulchellus.* **Adult male breeding** Easily identified by glossy green plumage, red breast patch bordered on sides by mix of yellow and glossy green, and narrow, elongated central tail feathers. Flight feathers and tail blackish. **Adult male non-breeding** (dry season) As adult female, but retains tail streamers and glossy green wing-coverts. **Adult female** Head and upperparts pale ashy-olive; underparts pale yellow with drabber throat; tail blackish tipped and edged white. **Juvenile** As adult female; young male has throat black. **VOICE** Call a hard, vigorous *chip* or *tut*, frequently repeated. Song a rapid series of similar short notes followed by a short, high-pitched jingle, e.g. *chiupchupchipchititit.* [CD13:8] **HABITS** Usually in pairs or small parties in wooded savanna, thorn scrub and gardens. **SIMILAR SPECIES** Only two other male sunbirds have tail streamers. Pygmy Sunbird is smaller and has quite different coloration and smaller bill. Congo Sunbird has all-red breast and dull black belly. Female Pygmy has underparts bright lemon-yellow. **STATUS AND DISTRIBUTION** Common or seasonally common throughout savanna and Sahel zones to desert edge (17°N), from S Mauritania to Sierra Leone east to Chad and CAR; in Niger also in Aïr. Largely resident but seasonal movements recorded (north with rains, south in dry season). **NOTE** Formerly placed in genus *Nectarinia*.

CONGO SUNBIRD *Cinnyris congensis* Plate 118 (2)
Souimanga du Congo

Other name E: Congo Black-bellied Sunbird
L 12–13 cm (4.75–5.0"); tail streamers of male projecting up to 7 cm (2.75") beyond other rectrices. **Adult male** Head, upperparts and throat bright glossy green, bluer on uppertail-coverts. Flight and tail feathers black; central tail feathers greatly elongated. Broad red breast-band separated from green throat by narrow blue band; rest of underparts black. No pectoral tufts. **Adult female** Head, upperparts, tail and throat grey-brown. Rest of underparts grey becoming whitish/yellowish-white on centre of belly and undertail-coverts. **Juvenile** No information. **VOICE** No information. **HABITS** Little known. Restricted to forested banks of large rivers. **STATUS AND DISTRIBUTION** Rare and local resident in Congo, where recorded along Ubangi and Léfini Rivers. May occur along other large rivers in the area. **NOTE** Formerly placed in genus *Nectarinia*.

PURPLE-BANDED SUNBIRD *Cinnyris bifasciatus* Plate 118 (4)
Souimanga bifascié

L 11–12 cm (4.5–4.75"). *C. b. bifasciatus.* **Adult male** Head, upperparts, throat and upper breast glossy green. Glossy blue-and-violet breast-band bordered below by dark purplish-red band. Rest of underparts, wings and tail black. No pectoral tufts. **Adult female** Head and upperparts olive-grey with faint pale supercilium; tail black

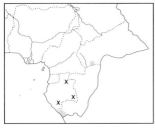

slightly glossed blue. Throat buffish-white or dusky; rest of underparts pale buffish-yellow narrowly and indistinctly streaked dusky on breast and flanks. **Juvenile** As adult female, but upperparts slightly more olive. **Immature male** As adult female but throat blackish mixed with glossy green and breast mottled blackish and olive-grey; wing-coverts possess some glossy feathers. **VOICE** A high-pitched *tsik-tsik-tsik* and buzzing *brrrzi*. Song a few short *tsup* or *tsip* notes stuttering into a variable twitter or trill. [CD13:11] **HABITS** Usually singly or in pairs, frequenting edges of forest and plantations, coastal scrub, second growth, thickets in savanna and gardens. May gather in numbers at favourable food source. Very active; aggressive towards other sunbirds.

SIMILAR SPECIES See Orange-tufted Sunbird. **STATUS AND DISTRIBUTION** Uncommon, SW CAR, Gabon and S Congo. Mainly resident but irregular local movements reported. **NOTE** Formerly placed in genus *Nectarinia*.

ORANGE-TUFTED SUNBIRD *Cinnyris bouvieri*   Plate 118 (3)
Souimanga de Bouvier

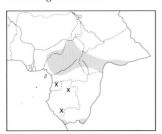

L 11.5–12.0 cm (4.5–4.75"). **Adult male** Head and upperparts glossy green *with glossy purple forehead*. Wings blackish-brown; tail blue-black. Throat and upper breast glossy green (with some glossy purple on malar area) separated from *broad, dark chestnut-red breast-band* by narrow glossy purple band; rest of underparts blackish-brown. Pectoral tufts orange-red and yellow. **Adult female** Head and upperparts greyish-brown with paler supercilium extending behind eye; tail blackish. Throat dusky; rest of underparts pale dirty yellowish indistinctly streaked and/or mottled dusky on breast. **Juvenile** Similar to adult female. Young male has darker throat. **VOICE** A low *cheep* and *chip-ip*; also *chew*. Song a few separate *tsik* notes followed by a fast, vigorous *cheepa-cheepa-cheepa-cheepa-cheep*, or *tsit-tsit-tsit-chewchewchew*, with shorter variations and also including hard churring *chrrrrrrr*. [CD13:9] **HABITS** Singly or in pairs at edges of montane forest, in forest–savanna mosaic, shrubby hillsides and occasionally around villages in lowland rainforest. **SIMILAR SPECIES** Adult male Purple-banded Sunbird lacks purple forehead and pectoral tufts. Adult female Purple-banded is slightly more olive above and paler below with narrower streaks. Palestine Sunbird has shorter bill; male has glossy violet-blue throat and upper breast; female has more greyish, plain breast. **STATUS AND DISTRIBUTION** Locally common to uncommon from SE Nigeria (Obudu and Mambilla Plateaux) and Cameroon (Mt Manenguba, Bamenda Highlands, Adamawa Plateau, etc.) to CAR. Also recorded from Eq. Guinea, NE and S Gabon and S Congo. Principally resident but irregular local movements reported. **NOTE** Formerly placed in genus *Nectarinia*.

PALESTINE SUNBIRD *Cinnyris oseus*   Plate 118 (5)
Souimanga de Palestine

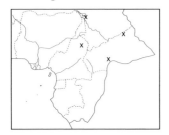

Other name E: Northern Orange-tufted Sunbird
L 10.0–11.5 cm (4.0–4.5"). *C. o. decorsei.* **Adult male breeding** Forehead and forecrown glossy violet-blue; rest of head and upperparts dark glossy green. Wings and tail blackish. Throat to upper breast glossy violet-blue; rest of underparts dull black. Pectoral tufts orange and yellow. Appears all black at distance. **Adult male non-breeding** Upperparts as adult female; underparts mainly dull black (mottled white when fresh) with some iridescent feathers; lower belly and undertail-coverts whitish. **Adult female** Head and upperparts ashy-brown tinged olive; underparts dusky-white tinged yellowish; breast slightly more greyish. **Juvenile** As adult female, but underparts yellower.
Immature male Dusky throat patch in some. **VOICE** A short *chip* or *chip-ip-ip-ip*; also a vigorous, sharp *chuWEEp*. Alarm a dry, hard *tek* or *tsik*, also given in stuttering series. Song a rapid series of hard, rather metallic *cheep* or *chwing* notes, often preceded by call note. [CD13:10] **HABITS** Singly or in pairs in open savannas with scattered shrubs and trees. **SIMILAR SPECIES** More widespread Orange-tufted Sunbird has longer bill; male has glossy green throat bordered by broad, dark chestnut-red band; female has yellower underparts. **STATUS AND DISTRIBUTION** Inadequately known. Locally common to rare intra-African migrant, CAR. Also E Cameroon (Tello and Meiganga area). Breeds in dry season; moves north with rains. **NOTE** Formerly placed in genus *Nectarinia*.

VARIABLE SUNBIRD *Cinnyris venustus*   Plate 117 (9)
Souimanga à ventre jaune

Other name E: Yellow-bellied Sunbird
L 10 cm (4"). **Adult male** *venustus* **breeding** Forehead and forecrown glossy violet; rest of head and upperparts

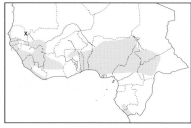

bright glossy green; uppertail-coverts glossy blue. Upper throat glossy violet, lower throat glossy green; breast glossy violet bordered below with dull black; rest of underparts lemon-yellow. Flight and tail feathers blackish. Pectoral tufts scarlet and yellow. Southern *falkensteini* has deeper yellow underparts. **Adult male non-breeding** As adult female but has glossy wing- and uppertail-coverts, and usually some glossy tips to body feathers. **Adult female** Head and upperparts grey-brown washed olive; underparts yellowish; tail black glossed green. **Juvenile** Similar to adult female. **Immature male** Head and upperparts olive-green; throat blackish; breast and upper belly mottled dusky; rest of underparts yellowish. **VOICE** A loud *chuw-chuw-chuw-...* and *chew-cheep*; also sharp *tik-tik-tik-... * Alarm *cheer-cheer-...* and *cheweep-cheweep-...* Song a series of call notes accelerating into a short, vigorous, cheerful jingle. [CD13:13] **HABITS** In pairs or small parties in various habitats, incl. wooded savanna, clearings, inselbergs, mangroves, coastal scrub, farmbush and gardens. **SIMILAR SPECIES** Male Collared Sunbird is superficially similar but has glossy areas less extensive and no violet on head and throat; underparts deeper yellow, bill less decurved. **STATUS AND DISTRIBUTION** Uncommon to locally common resident in savanna zones, from Senegambia to Liberia east to Cameroon and CAR (nominate), and in S Gabon and S Congo (*falkensteini*). Northward movement during rains reported in some areas. **NOTE** Formerly placed in genus *Nectarinia*.

JOHANNA'S SUNBIRD *Cinnyris johannae*   Plate 118 (10)
Souimanga de Johanna

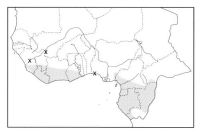

L 13–14 cm (5.0–5.5"). **Adult male *fasciatus*** Head, upperparts and upper throat glossy green. Wings and tail black. Lower throat and upper breast glossy purple; breast and belly red barred purple on lower breast; lower belly and undertail-coverts dark brown. Pectoral tufts yellow. **Nominate** has red on underparts darker and less bright. **Adult female** Easily distinguished from other female sunbirds. Head and upperparts olive-brown; throat and upper breast whitish; rest of underparts lemon; *entire underparts heavily streaked dark brown*. Tail blackish. **Juvenile** Similar to adult female but throat-sides dusky; streaks on underparts broader and greener. **VOICE** A characteristic, clear *tsik-peew* and a series of loud, separate notes *pee pee pee pseew pee pee...* or *tseew tseew tseew...* In flight a distinctive, slight *pit pit*. Song a fast jingle. [CD13:14] **HABITS** Usually in pairs in forest, principally frequenting canopy but descending to mid-levels by clearings and tracks; also in thickets and forest patches in forest–savanna mosaic. Occasionally in small groups. **SIMILAR SPECIES** Adult male Superb Sunbird is less stocky, with throat and upper breast dark glossy blue; rest of underparts dull dark red. **STATUS AND DISTRIBUTION** Rare to locally not uncommon forest resident, from Sierra Leone and SE Guinea to Ghana (*fasciatus*), and from SE Nigeria to SW CAR and Congo (nominate). Also S Mali and SW Nigeria (rare). **NOTE** Formerly placed in genus *Nectarinia*.

SUPERB SUNBIRD *Cinnyris superbus*   Plate 118 (9)
Souimanga superbe

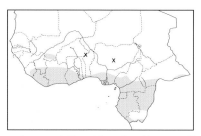

L 16 cm (6.25"). Largest sunbird, with very long bill. **Adult male *ashantiensis*** Head and upperparts glossy green with blue-green forehead and crown. Wings and tail dull black. Throat and upper breast dark glossy purple and blue; breast and belly non-glossy dark red; lower belly and undertail-coverts dark brown. **Nominate** slightly larger, with glossy cap more extensive and underparts rather paler; *nigeriae* paler and brighter than nominate. **Adult female** Head and upperparts olive with pale yellowish supercilium. Underparts olive-yellow with orange wash on undertail-coverts. Flight feathers and tail dark brown edged yellowish-olive. **Juvenile** As adult female, but underparts duller with undertail-coverts olive-yellow. **VOICE** A vigorous, typical sunbird *cheep*. Also clear, repeated *wheet wheet...* and *peeup peeup...* and similar notes. Song a short series of single notes, e.g. *chip chew chut*, with variations and occasionally followed by a short jingle. [CD13:16] **HABITS** Singly or in pairs, frequenting flowering trees along forest clearings and edges. Also sparingly in wooded grassland and gallery forest. **SIMILAR SPECIES** Male Johanna's Sunbird has head, throat and upper breast glossy green (concolorous with rest of upperparts); red on underparts brighter; tail shorter; shape more compact. **STATUS AND DISTRIBUTION** Not uncommon to uncommon resident in forest zone. *C. s. ashantiensis* from Sierra Leone and SE Guinea to Benin, also S Mali; *nigeriae* S Nigeria; nominate from Cameroon to CAR and Congo. **NOTE** Formerly placed in genus *Nectarinia*.

SPLENDID SUNBIRD *Cinnyris coccinigaster*

Plate 118 (7)

Souimanga éclatant

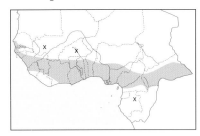

L 14 cm (5.5"). **Adult male** Head and throat glossy purple; upperparts glossy green with bluish uppertail-coverts; breast bright red mixed with glossy purple and blue; undertail-coverts glossy blue; rest of underparts and wings dull black; tail black fringed glossy green-blue. Pectoral tufts yellow. **Adult female** Head and upperparts brownish-olive with yellowish supercilium and slightly yellower rump. Underparts pale yellow streaked dusky on breast and flanks; throat paler. Flight feathers brown edged olive; tail blackish with slight green gloss. **Juvenile** As adult female, but has black throat and upper breast. Upperparts of young male slightly darker, less green, than in young female. **Immature male** As adult female but with glossy purple throat. **VOICE** Calls include a short *tsiup*, a harsh *chew-chew-...* (alarm), and *hueet* like *Phylloscopus* warbler. Song a series of 7–9 clear, separated notes *tip tiup tiup tiup...* with variations; recalling song of Scarlet-chested Sunbird but louder. [CD13:15] **HABITS** Singly or in pairs in savanna woodland, coastal thickets, shrubbery, large forest clearings and gardens. Not inside forest. **SIMILAR SPECIES** Female and juvenile Copper Sunbird smaller; tail longer, blue-black. **STATUS AND DISTRIBUTION** Common resident, from Senegambia to Liberia east to CAR. Seasonal migrant, NE Gabon. Northward movements in rainy season reported locally. **NOTE** Formerly placed in genus *Nectarinia*.

BATES'S SUNBIRD *Cinnyris batesi*

Plate 116 (9)

Souimanga de Bates

L 9.5 cm (3.75"). Very small, non-glossy species, resembling miniature Olive Sunbird. **Adult** Head and upperparts dark olive-green, paler around eye; tail black edged green. Throat greyish-olive becoming yellowish-olive on rest of underparts. Bill decurved, black with pale horn base to lower mandible. No pectoral tufts. In hand, white axillaries a useful feature. **Juvenile** Similar. **VOICE** A high-pitched *tsip*. Song a series of *tsip* notes interspersed with harsh *chep* calls and followed by subdued short trill. [CD13:17] **HABITS** Singly or in small groups in canopy of lowland forest. Strongly attracted to flowering trees. Joins mixed flocks of small insectivores. **SIMILAR SPECIES** Little Green Sunbird has straighter, paler bill, brighter coloured plumage and more horizontal posture. Female Green Sunbird *Anthreptes rectirostris* is similarly coloured, but has shorter, straighter bill and warbler-like appearance. **STATUS AND DISTRIBUTION** Rare to locally common forest resident, recorded from Liberia, Ivory Coast, Ghana, and from W Nigeria to SW CAR and Congo; also Bioko. **NOTE** Formerly placed in genus *Nectarinia*.

URSULA'S SUNBIRD *Cinnyris ursulae*

Plate 116 (11)

Souimanga d'Ursula

Other name E: Ursula's Mouse-coloured Sunbird

L 10 cm (4"). Very small, non-glossy sunbird of higher elevations. **Adult male** Forehead and crown grey; rest of upperparts olive-green. Head-sides and underparts greyish, paling towards belly and becoming yellowish-olive on lower belly and undertail-coverts. Pectoral tufts orange-red. Wing and tail feathers blackish edged yellowish-olive. Some crown feathers tipped glossy blue (visible at close range). Bill decurved, dark brown. **Adult female** Similar but lacks pectoral tufts. **Juvenile** As adult female, but paler, also lacks glossy blue feathers on crown. **VOICE** A soft, high-pitched *tsit-tsit*. Song a simple, rather vigorous, descending *tsee-see-see-see-see* and *tsee-tsee-tsee*. Also a fast, subdued jingle. [CD13:18] **HABITS** Singly, in pairs or in mixed parties with other sunbirds in mature mid-elevation and montane forest (above 950 m); also clearings. Very active, occurring at all levels. **SIMILAR SPECIES** Other small, dull-coloured sunbirds are more olive to yellowish-green on underparts. **STATUS AND DISTRIBUTION** Locally common forest resident, only known from some mountains in W Cameroon (Mt Cameroon, Rumpi Hills, Mt Kupe, Mt Nlonako, Bakossi Mts, Foto near Dschang) and from Bioko (1000–1200 m; rare). *TBW* category: Near Threatened. **NOTE** Formerly placed in genus *Nectarinia*.

COPPER SUNBIRD *Cinnyris cupreus*

Plate 118 (6)

Souimanga cuivré

L 12–13 cm (5"). *C. c. cupreus.* **Adult male breeding** Distinctive although appears all black at distance. Head,

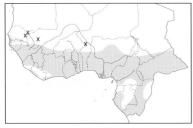

upperparts, throat and breast glossy coppery with purple reflections on back, rump, uppertail-coverts and lesser wing-coverts. Wings brownish; tail bluish-black. Belly to undertail-coverts dull black. No pectoral tufts. **Adult male non-breeding** (dry season) As adult female, but retaining glossy wing- and uppertail-coverts, and some iridescent feathers on body; often has dark bib. **Adult female** Head and upperparts olivaceous-green; underparts olive-yellow. Yellowish supercilium emphasised by dark lores. Flight feathers dark brown edged olive; tail blue-black. **Juvenile** Similar to adult female. Young male has dusky throat. **VOICE** A hoarse *chip-chip*, often repeated. Song a rapid series of *chip* notes followed by a short metallic jingle. Also a rattling alarm call. [CD13:19] **HABITS** Usually singly or in pairs, frequenting variety of habitats, incl. wooded grassland, coastal scrub, abandoned farmland, edges of gallery forest, and gardens. Not in forest. **SIMILAR SPECIES** Female and juvenile Splendid Sunbird are larger and have shorter, green-washed tail. Male Carmelite Sunbird may also look black at distance; Carmelite pairs easily identified by conspicuously pale female. **STATUS AND DISTRIBUTION** Common throughout savanna zone and in open habitats within forest zone. Mainly resident, but breeding migrant during rains in north of range. **NOTE** Formerly placed in genus *Nectarinia*.

WHITE-EYES
Family ZOSTEROPIDAE (World: *c.* 96 species; Africa: 11; WA: 7)

Small, arboreal, warbler-like species deriving their name from the distinct white eye-ring possessed by most species. Bill short and pointed. Occur in various types of forest and woodland. Sexes similar; juveniles similar to adults but duller and with narrower or no eye-rings. Active and restless. In pairs or small groups; also in mixed-species parties. Feed on insects, nectar and fruit; forage principally by foliage gleaning, but also catch insects on the wing. Two genera occur, *Zosterops* and the endemic *Speirops*. Only one species, *Z. senegalensis*, is widespread and has largely yellowish-olive and yellow plumage and a conspicuous eye-ring. In the five other species, two *Zosterops* and three *Speirops*, which are confined to the Gulf of Guinea islands and Mt Cameroon, the yellow pigment has been lost and the eye-rings are narrower and less conspicuous.

YELLOW WHITE-EYE *Zosterops senegalensis*   Plate 115 (12)
Zostérops jaune

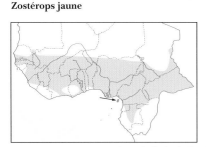

Other name F: Oiseau-lunettes jaune
L 10–11 cm (4.0–4.25"). Small and yellow with dark eye surrounded by *white ring* (diagnostic). **Adult** *senegalensis* Head and upperparts yellowish-olive with yellow forehead and black loral stripe. Entire underparts bright yellow. Flight and tail feathers brown edged yellowish-olive. Bill black. Forest forms *demeryi* and *stenocricotus* darker and greener. **Juvenile** As adult but darker above. **VOICE** Fairly loud, quavering notes *ti-trrruutrrruu-ti-trrruuti*, frequently uttered. Song, mainly given at sunrise, a series of burred notes, including calls. Distinct song of forest form *stenocricotus* a fast, vigorous phrase *huhihuhichewchwchwchwchew*. Calls include *chewp-chewp*, shorter *chip-chip* and *slip-slip*. Also a fast jumble of varied notes. [CD13:26–27] **HABITS** In pairs or small groups in various habitats, incl. thorn scrub, savanna woodland, lowland and montane forest, and gardens. Actively gleans foliage for insects and inspects flowers for nectar. Often joins other species, esp. sunbirds. **SIMILAR SPECIES** Similar-sized Senegal and Green-backed Eremomelas have grey head, white throat, lack white eye-ring and only occur in savanna. Yellow Penduline Tit is smaller and lacks obvious white eye-ring. **STATUS AND DISTRIBUTION** Common to uncommon but unevenly distributed resident. Nominate from S Mauritania and Senegambia to Chad and CAR; *demeryi* from Sierra Leone to Ivory Coast; *stenocricotus* from SE Nigeria to SW CAR, Congo and Bioko. **NOTE** *Z. s. stenocricotus* may be a separate species (*BoA*).

PRÍNCIPE WHITE-EYE *Zosterops ficedulinus*   Plate 145 (3)
Zostérops becfigue

Other names E: São Tomé White-eye; F: Zostérops de Principé
L 10.5 cm (4"). **Adult** *feae* Head and upperparts olive-green to dull greyish-olive, brighter on rump and uppertail-coverts; black loral stripe; narrow yellow supraloral line connecting on forehead; white eye-ring highlighted by

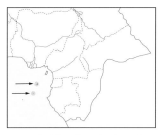

dark border. Underparts greyish-white tinged yellow on throat; breast and flanks tinged pale greyish-olive; white stripe on flanks. **Nominate** tinged brownish above; paler, with less grey wash below. **Juvenile** No information. **VOICE** Calls include a hard *trrrrr* or *prrrrip*, often repeated, and a sharp *pip pip pip...* Song a series of fairly loud, quavering notes *ptirrr ptirrr ptirrr pstirrrrr*; also a similar series of *ptirrr* notes interspersed with more melodious and varied syllables. **HABITS** In pairs or small groups of up to 20. On São Tomé in primary forest, forest regrowth, plantations and more open, degraded habitats with bushes and isolated trees, principally above 400 m. In closed forest frequents canopy; in open habitats much lower, down to 2 m above ground. Often in mixed parties with speirops, sunbirds, prinias and paradise flycatchers. On Príncipe usually recorded in forests of hilly interior; occasionally at lower altitudes. **STATUS AND DISTRIBUTION** Endemic to São Tomé (*feae*; uncommon) and Príncipe (nominate; rare). *TBW* category: THREATENED (Vulnerable).

ANNOBÓN WHITE-EYE *Zosterops griseovirescens*   Plate 145 (2)
Zostérops d'Annobon

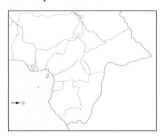

L 12 cm (4.75"). **Adult** Head and upperparts dull greyish-olive with brighter, more lemon-green rump; narrow white eye-ring emphasised by black lores and lemon supraloral stripe. Wings and tail dark brown edged olive. Underparts whitish tinged olive-yellow; breast and flanks washed fulvous. **Juvenile** No information. **VOICE** Constantly uttered chittering notes, similar to those of Príncipe White-eye; the dominant bird call on the island (Harrison 1990); in flight a rapid *plik-plik-plik*. Song relatively loud and melodious, described as 'an indefinite jumble of not unpleasing notes' (Fry 1961). **HABITS** In pairs and small parties wherever there are trees and bushes, in dry and moist forest, cultivation, etc. **STATUS AND DISTRIBUTION** Endemic to Annobón, where very common resident. *TBW* category: THREATENED (Vulnerable).

SÃO TOMÉ SPEIROPS *Speirops lugubris*   Plate 145 (4)
Speirops de Sao Tomé

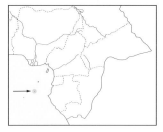

Other names E: Black-capped Speirops; F: Zostérops de Sao Tomé
L *c.* 14 cm (5.5"). *Dark greyish with black cap.* **Adult** Forehead and crown black; narrow white band on lores connecting on forehead; white eye-ring. Upperparts and tail greyish-olive, brighter on rump. Underparts grey with conspicuous white stripe from breast-side to flanks; throat flecked white; belly washed olive; undertail-coverts olive washed rufous; thighs white. Bill and legs pinkish. **Juvenile** No information. **VOICE** A soft, quavering trill *rrriirrr*, a short, hard *trrrr* or *rrrr* and a sharp *whseeeew*. **HABITS** In pairs or groups of up to 25, in all wooded habitats, incl. primary and secondary forest, coffee and cocoa plantations, clearings with bushes and small trees, and dry forest in savanna; at all altitudes and at all levels. Forages like other white-eyes, often in mixed parties with white-eyes, sunbirds, paradise flycatchers and weavers. Not shy. **STATUS AND DISTRIBUTION** Resident, São Tomé, where one of the commonest and most widespread endemics.

MOUNT CAMEROON SPEIROPS *Speirops melanocephalus*   Plate 115 (13)
Speirops du Cameroun

Other names E: Black-capped Speirops; F: Zostérops du Cameroun, Speirops à calotte noire
L 13 cm (5"). **Adult** As São Tomé Speirops but with *broader white band on forehead, very narrow eye-ring, white throat, and grey underparts washed fulvous on flanks.* **Juvenile** No information. **VOICE** Calls include a hard, rattling *trrrr* or *rrrr*. Also little calls, similar to those of Yellow White-eye (Serle 1954). Song a fast, vigorous, rich *tsip-tsip-twrr-twrr tsip-tsip-twrr-twrr-tsee-ti-tew*, with slight variations. Also a series of sweet notes, first rising, then falling in pitch, sometimes described as reminiscent of Blackcap (e.g. Stuart & Jensen 1986). [CD13:25] **HABITS** Singly, in pairs or small parties in canopy and mid-levels of more open forest and in clearings; common at upper forest–grassland boundary; higher up in thickets and patches of bush. Joins mixed flocks with White-bellied Crested Flycatchers, Yellow White-eyes and Northern Double-

collared Sunbirds. **SIMILAR SPECIES** Male Blackcap superficially similar but has dark bill and legs, and no white on forehead and throat. **STATUS AND DISTRIBUTION** Endemic to Mt Cameroon, where common resident at 1820–3000 m. *TBW* category: THREATENED (Vulnerable). **NOTE** Sometimes treated as race of São Tomé Speirops.

PRÍNCIPE SPEIROPS *Speirops leucophaeus*   Plate 145 (5)
Speirops de Príncipe

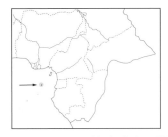

Other name F: Zostérops de Príncipe
L 13.5 cm (5.25"). Distinctive. **Adult** *Head white* washed grey on crown and nape; narrow white eye-ring; upperparts and tail greyish-brown. Throat white; rest of underparts soft ash-grey becoming white on centre of belly. Bill and legs greyish. **Juvenile** No information. **VOICE** A soft little trill *rrrrrrrr*, a short *tiup tup* and a fast, sibilant *whee-tsiu-tsiu-tseeu* and *tsee-tsitsiuu* or *tsiupti-ti-tiu*. [CD13:24] **HABITS** In pairs, small family parties or groups of up to 15 in forest regrowth, edges of primary forest and cocoa and coffee plantations, from lower mid-levels (at edges and in more open habitats) to canopy. **STATUS AND DISTRIBUTION** Endemic to Príncipe, where common resident but possibly declining. *TBW* category: Near Threatened.

FERNANDO PO SPEIROPS *Speirops brunneus*   Plate 145 (1)
Speirops de Fernando Po

Other name F: Zostérops de Fernando Po
L 13 cm (5"). **Adult** Entirely brown; forehead paler, crown darker; upperparts and tail tinged rusty. Throat greyish; rest of underparts slightly paler than upperparts. No eye-ring. Bill and legs blue-grey. **Juvenile** No information. **VOICE** A long trill *trrrrrrruuu* and a rapid *trik-trik-trik* (*TBW*). Also a soft, frequently uttered *peep* and *tweet* (*BoA*). **HABITS** In small groups of 4–5 in lichen forest and montane heathland. **STATUS AND DISTRIBUTION** Endemic to Bioko, where common resident on higher slopes of Pico Basilé, above 1900 m. *TBW* category: THREATENED (Vulnerable).

TRUE SHRIKES
Family LANIIDAE (World: 30–31 species; Africa: 19; WA: 13)

Small to medium-sized birds with strong, hooked bills and moderately long to very long tails. Plumage mainly combinations of black, grey, white and, in some, rufous. Bill and legs usually blackish. Sexes similar or female duller, but dissimilar in Red-backed Shrike. Juvenile typically finely barred above and below; bill paler. Capture prey by pouncing onto it from exposed perch. Food consists of insects, reptiles, young birds and small mammals. Some species impale prey on thorns. Flick, swing and fan tail when excited. Occur in a variety of habitats, including semi-desert, savanna, woodland and edges of cultivation.

COMMON FISCAL *Lanius collaris*   Plate 120 (1)
Pie-grièche fiscale

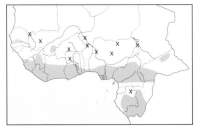

Other name E: Fiscal Shrike
L 21–23 cm (8.25–9.0"). Slender, black-and-white shrike with long, graduated tail. **Adult male** *smithii* Head and upperparts black slightly glossed bluish, with white scapulars forming elongated shoulder patch; wings black with white patch at base of primaries; rump grey. Tail black, all feathers except central pair broadly tipped white. Underparts white. Southern *capelli* duller black above, rump generally paler, outer pair of tail feathers all white. **Adult female** Sometimes with faint chestnut patch on flanks. **Juvenile** Above, rufous-brown densely barred blackish, incl. on largely whitish scapulars. Underparts whitish variably barred dusky. **VOICE** A harsh *chaaa-chaaa* (alarm). Northern *smithii* not vocal; rarely heard song is a subdued warble. Song of *capelli* includes various rather metallic motifs and imitations. [CD13:40] **HABITS**

Singly or in pairs, perching conspicuously. Inhabits a variety of open habitats, incl. farmland, roadsides and vicinity of habitations; neither in forest nor arid north. Bold and pugnacious. **SIMILAR SPECIES** Juvenile Sousa's Shrike is smaller and greyer than juvenile Common Fiscal, with narrower, wavy barring. **STATUS AND DISTRIBUTION** Common to uncommon resident, from W Guinea and Sierra Leone to CAR (*smithii*). Rare, S Mauritania, S Mali, Burkina Faso, W Niger, N Nigeria and N Cameroon. *L. c. capelli* in S Gabon and Congo (Brazzaville area; rare). Commonest and most widespread African shrike. **NOTE** Differences in vocalisations between northern and southern populations suggest they may be separate species (Harris & Franklin 2000).

SÃO TOMÉ FISCAL *Lanius newtoni* Plate 144 (5)
Pie-grièche de Sao Tomé

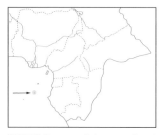

Other name E: Newton's Fiscal
L *c.* 23 cm (9"). Unmistakable. **Adult** Slender with long, narrow, graduated tail, resembling Common Fiscal but blacker above and tinged pale yellow below. Head and upperparts black with much less white on scapulars (forming irregular line); wings all black (without white patch); rump black (not grey). Tail black, tipped white on outer 3 feathers. **Juvenile** Above, brown with fine dark vermiculations; tail dark brown. Underparts tawny-yellowish. Bill brownish-horn. **Immature** Upperparts brown, slightly greyer on top of head; narrow buffish supercilium extending just behind eye; scapulars tawny forming distinct V; wing-coverts tipped buff. Bill black.
VOICE Song, mainly uttered at dawn and dusk, a series of *c.* 10 well-spaced, fluted, far-carrying notes *tiu tiu tiu tiu*..., and a more rapid, rhythmic series of slightly metallic notes *tsink-tsink-tsink-tsink-*... Vocalisations of young similar but more nasal, e.g. *tieu-tieu-tieu-*... or *kehe-kehe-kehe-*... Calls include a low squawk and a scolding churr. [CD13:41] **HABITS** Little known. Restricted to undisturbed primary forest, where probably frequents mid- and lower levels. Often perches on low vantage points, but also skulks in low bushes. Seen foraging on ground in closed-canopy forest, among boulders by stream and on flat ridge. Quiet and unobtrusive, but not necessarily shy. **STATUS AND DISTRIBUTION** Rare resident, endemic to São Tomé. Recorded in southern and central lowland rainforest, near R. São Miguel, Xufexufe, Quija and Io Grande, and on Formoso Pequeno, near Bombaím. The only shrike on the island. *TBW* category: THREATENED (Critical).

MACKINNON'S SHRIKE *Lanius mackinnoni* Plate 120 (4)
Pie-grièche de Mackinnon

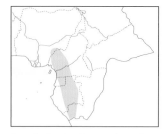

L 20 cm (8"). **Adult male** Top of head and upperparts grey with white scapulars forming elongated patch on shoulder; black mask bordered by white supercilium; *wings all black*. Tail feathers black tipped white, except all-black central pair. Underparts white. **Adult female** As male but with dark maroon patch on flanks. **Juvenile** Above, a mix of grey and brown with narrow dusky bars. Underparts white with variably distinct, narrow dusky markings on breast, flanks and undertail-coverts. **VOICE** Usually silent. Song varied, sustained and melodious, including whistling and scratchy syllables, and imitations. Both sexes sing. Calls include a musical *chik-erea* (Zimmerman *et al.* 1996) and a low drawn-out whistle (alarm). [CD13:36] **HABITS** Singly or in pairs in clearings and at edges of lowland and montane forest. Hunts from prominent perch; impales larger prey. Territorial and pugnacious. **SIMILAR SPECIES** All three other grey, white and black shrikes occurring in W Africa have white primary patches; Grey-backed Fiscal and Lesser Grey Shrike also separated by black forehead and lack of white on scapulars. See also distribution. **STATUS AND DISTRIBUTION** Locally not uncommon resident, from SE Nigeria (Obudu Plateau) through Cameroon, Eq. Guinea and Gabon to S Congo.

MASKED SHRIKE *Lanius nubicus* Plate 119 (6)
Pie-grièche masquée

L 17–18 cm (6.75–7.0"). **Adult male** Mainly black above and white below, with broad white forehead and supercilium, large white shoulder patch and white patch at base of primaries. Flanks have strong orange wash. Tail rather long and slender, black edged white. Moult starts prior to autumn migration and is completed in winter quarters; in fresh plumage, greater and median coverts and tertials have pale fringes. **Adult female** Slightly duller. **Juvenile** Head and upperparts brownish-grey closely barred blackish; narrow whitish forehead and faint supercilium accentuating dark ear-coverts; scapulars white with blackish fringes; greater and median coverts and tertials have white fringes and subterminal dark markings. Underparts off-white with dusky crescents on breast and flanks. **Immature** Gradually resembling adult female but has paler grey mantle and lacks orange wash on

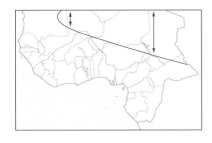

flanks. **VOICE** Mostly silent in winter quarters. Calls include a harsh *krret* and a scolding *krrrr.* [CD13:42] **HABITS** Singly in thorn scrub. Unlike other W Palearctic shrikes, usually rather skulking, quietly perching on lower branches or inside crown of trees rather than on top. **SIMILAR SPECIES** Immature Woodchat Shrike initially resembles immature Masked but has browner plumage, smaller primary patch, no pale forehead and stockier build. **STATUS AND DISTRIBUTION** Uncommon to rare Palearctic migrant to Sahel zone, recorded from Mauritania, Mali, Niger (rare), NE Nigeria (L. Chad area), N Cameroon, Chad and N CAR.

GREY-BACKED FISCAL *Lanius excubitoroides*   Plate 120 (5)
Pie-grièche à dos gris

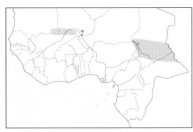

L 25 cm (10"). *L. e. excubitoroides.* Large, robust shrike. **Adult male** Crown and upperparts pale grey; *black mask extending to forehead* and on neck-sides; *scapulars* and wings *black* with white patch at base of primaries. Tail feathers black with narrow white tips and white basal half, except all-black central pair. Underparts white. **Adult female** As male, but some concealed dark maroon feathers on flanks. **Juvenile** Above, brownish narrowly barred dusky; mask smaller and not extending to forehead. Underparts white with faint buffish wash and some indistinct scaling on breast and flanks. **Immature** Gradually as adult but duller and with smaller mask; grey areas tinged brownish and finely barred dusky. **VOICE** A discordant, slightly hoarse and metallic chattering including liquid notes, e.g. *tcheh-tcheh-dleeo tchuleeoo chuh...*, with variations, often developing into excited chorus with entire group joining in. Alarm a harsh note. [CD13:39] **HABITS** Usually in small, noisy parties, reminiscent of babblers, in thorn scrub. Fly from tree to tree with seemingly weak, flapping flight. Often pumps and fans tail when perched. **SIMILAR SPECIES** Great Grey Shrike is much paler, with black mask not extending to forehead, white (not black) scapulars, and different tail pattern (all-white outer feathers). Smaller Lesser Grey Shrike also has black forehead, but lacks white border; tail shorter. **STATUS AND DISTRIBUTION** Rare resident in Sahel zone, from NE Nigeria (L. Chad area) through N Cameroon to Chad and N CAR. Also reported from Mauritania and Mali, where it appears to be absent during rains. Main range E Africa.

LESSER GREY SHRIKE *Lanius minor*   Plate 120 (3)
Pie-grièche à poitrine rose

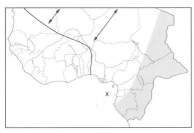

L 20–21 cm (8.0–8.25"). **Adult male** Resembles Great Grey Shrike but smaller, with shorter tail and *longer wings* (primary projection exceeds length of tertials), and *black mask extending to forehead*, not bordered by a white supercilium. Upperparts grey; wings black with white patch at base of primaries forming conspicuous bar in flight. Tail black edged white. Underparts white with variable pink wash on breast and flanks. **Adult female** Underparts less pink. **Immature** Duller and lacking black on forehead, underparts whiter. Remnants of barring on top of head, upperparts and flanks; wing-coverts and primaries narrowly tipped white. **VOICE** A harsh *chak.* [CD13:37] **HABITS** Mostly singly in various open habitats. **STATUS AND DISTRIBUTION** Rare Palearctic migrant/vagrant (Nov–Apr) recorded from Mauritania, Senegal, Mali, Niger, Nigeria, Cameroon, W Chad, Gabon, Congo and Príncipe; not uncommon in extreme E Chad. Mainly migrates through E Africa, wintering in SW Africa.

SOUTHERN GREY SHRIKE *Lanius meridionalis*   Plate 120 (2)
Pie-grièche méridionale

L 24 cm (9.5"). Large, grey, white and black shrike with long, graduated tail. **Adult *leucopygos*** Pale grey upperparts bordered along black wings by white scapular tips; black mask bordered by narrow white supercilium; white patch at base of primaries; tertials and secondaries tipped white; rump and uppertail-coverts white or greyish-white. Tail black, edged white. Underparts white washed pale buff. Saharan race *elegans* has rump and uppertail-coverts greyer and underparts pure white. *L. m. algeriensis* distinctly darker grey above; no supercilium; white patch at base of primaries smaller; underparts washed pale grey. **Juvenile** Upperparts sandy-grey; underparts white slightly washed buffish; mask brown-grey, smaller than adult Wings blackish-brown; greater and median

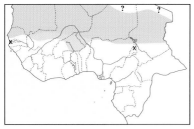

coverts and tertials broadly tipped sandy-buff. **Immature** As adult but greater coverts still tipped buffish. **VOICE** A dry *trr-trr* and a harsh *zzeh-zzeh*, both frequently repeated; also a subdued, varied warble, including imitations. Both sexes sing. [CD4:54, 13:38] **HABITS** Singly or in pairs in Sahel and thorn scrub at desert edge. Perches prominently atop low trees and bushes. Impales prey on thorns. **SIMILAR SPECIES** Grey-backed Fiscal is overall much darker with black forehead, black scapulars and no white tips to tertials and secondaries; also longer tail with white confined to tips and base. Smaller Lesser Grey Shrike has black forehead, no supercilium, longer wings, shorter tail and virtually no white on shoulders. **STATUS AND DISTRIBUTION** Not uncommon resident in arid (*elegans, algeriensis*) and semi-arid (*leucopygos*) zones. *L. m. elegans* breeds N Mauritania and NW Mali, possibly also NE Niger; *algeriensis* on Mauritanian coast; *leucopygos* from Mauritania through Mali, Niger and NE Nigeria (L. Chad area) to Chad. Local movements reported. *L. m. elegans* scarce dry-season visitor, N Senegal and N Burkina Faso; vagrant, Cape Verde Is; *leucopygos* vagrant (Nov–Apr), Gambia, N Ghana and (this race?) Annobón, also recorded from N Cameroon (status?). **NOTE** Often included within Great Grey Shrike *L. excubitor*.

ISABELLINE SHRIKE *Lanius isabellinus*   Plate 119 (4)
Pie-grièche isabelle

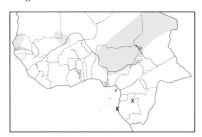

Other name E: Red-tailed Shrike
L 17–18 cm (6.75–7.0"). Rather plain, brownish shrike with rufous tail. **Adult male** *phoenicuroides* Crown rufous to greyish; black mask bordered by narrow white supercilium; *upperparts greyish-brown contrasting with rufous rump and tail.* Blackish flight feathers with white patch at base of primaries forming bar on outer wing in flight. Underparts white variably washed pale buffish. **Nominate** paler; crown and upperparts greyer; underparts creamy. **Adult female** Duller; usually with indistinct scaling on breast and flanks; white patch at base of primaries reduced or absent. **Immature** Resembles adult female but mask generally poorly developed; greater and median coverts and tertials fringed buffish with subterminal dark markings. **VOICE** A nasal *kihet* and a short *kzi-ek* or *tzea*. Subdued song, occasionally uttered in winter quarters, varied and sustained, including melodious whistles, chuckles, harsh notes and trills. [CD13:35] **HABITS** As Red-backed Shrike. **SIMILAR SPECIES** Adult female and immature Red-backed Shrike may be similar to immature Isabelline, but tail not as rufous and contrasting; upperparts of immature scaled (not plain). **STATUS AND DISTRIBUTION** Uncommon to scarce Palearctic migrant in Nigeria, Cameroon and Chad. A few records from Mauritania, N Senegal, Gambia, Mali and Gabon. Presumably *phoenicuroides*, but one specimen of nominate from N Nigeria. **NOTE** Specimen from N Nigeria originally identified as *L. i. speculigerus*; this race now demonstrated to be synonym of *L. i. isabellinus* (Pearson 2000).

RED-BACKED SHRIKE *Lanius collurio*   Plate 119 (2)
Pie-grièche écorcheur

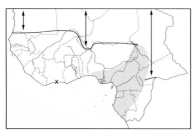

L 17–18 cm (6.75–7.0"). **Adult male** Crown, mantle and rump blue-grey; black mask; back and wing-coverts chestnut. Tail black, all but central pair with white at base and outermost pair entirely edged white. Underparts pinkish-white. **Adult female** Head and upperparts rufous-brown; buffish supercilium; usually no black mask. Tail dark brown edged white. Underparts buffish with dusky-brown crescents. **Immature** As adult female but upperparts with crescentic bars; greater and median coverts and tertials tipped buffish with subterminal dark markings; tail more rufous. **VOICE** Commonest call a harsh *chak chak*. Subdued song, occasionally uttered in winter quarters, includes imitations of African and European birds. [CD4:52] **HABITS** Singly in open habitats (savanna woodland, scrub and edges of farmland), where perches on small trees, bushes and wires. Flight direct, fast and undulating. Often fans, flicks and moves tail sideways when excited. **SIMILAR SPECIES** Adult male Emin's Shrike distinguished from adult male Red-backed by chestnut rump, rusty underparts and slightly smaller size. Adult female and immature Red-backed may be confused with corresponding plumages of the following. Isabelline Shrike typically has more rufous tail, lacking white sides, and contrasting with rest of plumage; white mark at base of primaries (sometimes absent); crescentic barring on underparts usually indistinct; upperparts plain in immature, with barring usually restricted to forehead, crown-sides, breast and flanks. Immature Woodchat Shrike has shoulders, base of primaries and rump whitish. Immature Masked Shrike is greyer, slimmer and has large white primary patch. **STATUS AND DISTRIBUTION** Rare Palearctic migrant/vagrant recorded in Mauritania, Mali, S Ivory Coast, Niger, Nigeria, Cameroon, Chad, Gabon, Congo, CAR and São Tomé. Main migration routes and winter quarters in E and S Africa. **NOTE** Considered monotypic, following *BWPC*. Variation between three commonly recognised races very slight. Birds in our region would be mostly (or all?) of nominate race; eastern *kobylini*, which may also occur, typically has chestnut on upperparts of adult male duller and often reduced, but is often inseparable from nominate.

EMIN'S SHRIKE *Lanius gubernator*　　　　　　　　　　　Plate 119 (3)
Pie-grièche à dos roux

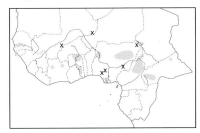

L 15–16 cm (6"). **Adult male** Resembles Red-backed Shrike but principally distinguished by *chestnut* (not grey) *rump* and *rusty* (not pinkish-white) *breast to undertail-coverts*, darker on flanks and becoming white on central belly; white patch at base of primaries (conspicuous in flight). Tail feathers blackish tipped and edged white. **Adult female** Duller and paler; mask less well defined; grey of crown and mantle extending over back to rufous-tinged rump; underparts whitish tinged buff. **Juvenile** Head and upperparts rufous-brown entirely barred blackish; underparts as adult and variably barred dusky on breast and flanks; outer tail feathers rusty. Young males have upperparts more rufous; young females greyer. **VOICE** Usually silent. Song consists of short, rapid series of a few simple, but varied, whistles, e.g. *tweet-u-wee-u-weet* and *trip-tu-trip-srtp*, interspersed by various low hoarse notes, e.g. *chweeeh*. **HABITS** Singly or in pairs in wooded savanna. Occasionally joins mixed-species flocks. **SIMILAR SPECIES** Male Red-backed Shrike (q.v. and above). Juvenile Emin's distinguished from other juvenile/immature shrikes by rufous on underparts. **STATUS AND DISTRIBUTION** Rare to locally uncommon resident in savanna zone, reported from Mali, NE Ivory Coast, N Ghana, Nigeria, Cameroon and CAR. Perhaps under-recorded through confusion with Red-backed Shrike.

SOUSA'S SHRIKE *Lanius souzae*　　　　　　　　　　　Plate 119 (1)
Pie-grièche de Sousa

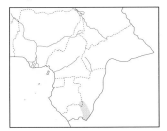

L 17–18 cm (6.75–7.0"). *L. s. souzae*. Resembles washed-out version of Red-backed Shrike with white V on upperparts and longish, narrow tail. **Adult male** Crown and mantle greyish becoming brownish on back; blackish mask bordered by whitish supercilium; scapulars white, forming elongated patch on shoulder; wings and rump rufous with fine blackish bars and vermiculations. Tail graduated, brownish with narrow blackish bars and white outer feathers. Underparts whitish. **Adult female** Flanks washed rufous. **Juvenile** Head and upperparts slightly paler, browner and entirely marked with dark narrow bars; underparts whitish with fine, indistinct wavy bars on breast. **VOICE** A drawn-out, muted whistle *peeeeht* (probably territorial call), a harsh *tzzer* (contact call) and *tzzjeht* or *tzzzik* (alarm). [CD13:34] **HABITS** Little known. Singly or in small (family?) groups in savanna woodland. Perches inconspicuously below canopy. **SIMILAR SPECIES** Red-backed Shrike lacks white V and has broader tail. Immature Red-backed distinguished from immature Sousa's by heavily scaled upper- and underparts. Juvenile Fiscal differs from juvenile Sousa's in much more rufous and more boldly barred head and upperparts; tail also much longer when fully grown. **STATUS AND DISTRIBUTION** Scarce and local resident, SE Gabon and Congo.

WOODCHAT SHRIKE *Lanius senator*　　　　　　　　　　　Plate 119 (5)
Pie-grièche à tête rousse

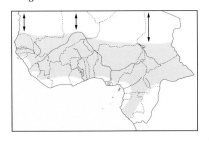

L 18–19 cm (7.0–7.5"). **Adult male** *senator* *Chestnut crown and nape* distinctive. Black mask; upperparts black with large white shoulder patch and white rump; white patch at base of primaries. Tail black fringed white. Underparts whitish. In fresh plumage, acquired in winter quarters, greater and median coverts and tertials fringed pale buff. Iberian and N African race *rutilans* slightly smaller. W Mediterranean *badius* has little or no white on primaries, less white on shoulders, narrower black forehead and stouter bill. **Adult female** Slightly duller; more white on forehead and lores. **Immature** Gradually as adult with variable amounts of juvenile plumage (such as buff wash and indistinct crescents on scapulars, breast and flanks). **VOICE** Mainly silent in winter quarters. Call a hard *chak chak chak*. [CD4:56] **HABITS** Singly in a various open habitats. Territorial. **SIMILAR SPECIES** Immature Masked Shrike initially similar to immature Woodchat but slimmer, with greyer plumage, larger white primary patch and paler forehead. Compare also immature Red-backed Shrike (q.v.). **STATUS AND DISTRIBU-TION** Widespread Palearctic migrant, from Mauritania to Liberia east to Chad and CAR. Common in most of W Africa, but rare in easternmost part, in Cameroon and CAR. Also recorded from Gabon. Nominate throughout; *rutilans* recorded from Senegal (probably more widespread); *badius* mainly south of nominate, at least from Ivory Coast to N Cameroon. **NOTE** Ssp. *rutilans* is often included within nominate.

YELLOW-BILLED SHRIKE *Corvinella corvina* Plate 120 (6)
Corvinelle à bec jaune

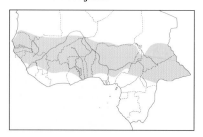

Other name E: Long-tailed Shrike

L 30–33 cm (12–13"), incl. tail of 18 cm (7"). Distinctive, brown shrike with *very long, graduated, often ragged tail* and *yellow bill*. **Adult male** *corvina* Top of head and upperparts earth-brown boldly streaked blackish-brown; crown and nape variably tinged rufous; blackish-brown mask bordered by narrow buffish supercilium; rounded, dark brown wings with rufous panel on primaries (conspicuous in flight). Underparts buff-white streaked blackish-brown on breast and flanks. Concealed cinnamon-rufous patch on flanks (visible when displaying or preening). Legs greenish. *C. c. togoensis* slightly more heavily streaked and less rufescent above. **Adult female** Concealed flank patch dark maroon. **Juvenile** Above, barred and mottled blackish-brown, appearing scalloped on head; below, whitish densely barred dusky-brown on breast and flanks. During moult from juvenile plumage flank patch originally pale cinnamon-rufous in both sexes. **VOICE** Various rasping, chirping and chattering calls, usually in series uttered simultaneously by several members of a group. Calls include imitations of other birds. [CD13:43] **HABITS** Usually in small, vocal parties, recalling babblers. Inhabit open savanna woodland and bush; absent from forest and adjoining moist savanna. Fly from tree to tree with weak and rarely sustained flight. Hunt from exposed perch. Breed cooperatively. **STATUS AND DISTRIBUTION** Locally common resident in broad savanna belt. Nominate from S Mauritania to Guinea-Bissau and N Guinea east to Niger and NW Nigeria; *togoensis* from W Guinea and Sierra Leone east to Chad and CAR. Some local movements reported (north with rains). **NOTE** *C. c. togoensis* sometimes included within extralimital *C. c. affinis*.

BUSH-SHRIKES
Family MALACONOTIDAE (World: 46 species; Africa: 46; WA: 27)

Small to medium-sized arboreal birds with stout, hooked bills, occurring in a variety of habitats, from semi-desert to rainforest. Sexes similar or not. Forage within foliage, preying on small vertebrates and invertebrates. Many species often encountered in pairs. Most are highly vocal, with loud, ringing calls. In W Africa, seven genera occur, all but one endemic to the Afrotropical region.

Typical bush-shrikes (*Malaconotus* and *Telophorus*, 10 species) are brightly coloured species of forest and savanna woodland, mostly with a grey head, green upperparts and yellow, orange or scarlet underparts. Despite their bright coloration the forest species are generally hard to observe, but betray their presence by melodious calls, and readily respond to playback or whistled imitations of these. Smaller species often placed in genus *Chlorophoneus*.

Tchagras (*Tchagra*, 2 species) are mainly brown above with rufous wings, and patterned head and tail. Sexes similar. Melodious calls and characteristic display flights attract attention. Sexually dimorphic *Antichromus* (Marsh Tchagra) is closely allied with *Tchagra* and often included in it.

Puffbacks (*Dryoscopus*, 4 species) are predominantly black and white in males and derive their name from the soft elongated feathers of lower back and rump, which are puffed out to produce a spectacular white ball in display. Occur in forest and gallery forest.

Boubous (*Laniarius*, 9 species) are either all black, or have black upperparts and white or red/yellow underparts. Skulk in heavy shrubbery and forest edges.

The **Brubru** (*Nilaus*) is the smallest shrike of the region and belongs to a distinctive, monotypic genus.

FIERY-BREASTED BUSH-SHRIKE *Malaconotus cruentus*   Plate 121 (9)
Gladiateur ensanglanté

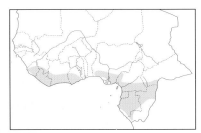

L 25 cm (10"). Large and colourful, with massive bill. **Adult** Head and mantle bluish-grey; lores and supraloral area greyish-white merging with white eye-ring; rest of upperparts green; tertials, secondaries and tail black broadly tipped yellow. Underparts vary from mainly scarlet to (rarely) all yellow, with intermediates; typically throat and breast scarlet becoming orange and bright yellow on lower underparts. Eye greyish. Bill black. **Juvenile** Head washed sandy-brown; underparts greenish-yellow. **Immature** Underparts soon turn orange, then scarlet. **VOICE** A clear, hollow, whistled *whoop* or *whoo-p*, recalling Many-coloured Bush-shrike's, but lacking its slightly melancholy quality and more rapidly repeated, forming short series of *c.* 3–12 similar notes. Depending on locality, call may be more monosyllabic *whoop*. Also various harsh calls. [CD13:71] **HABITS** Singly or in pairs in mature forest, second growth, forest edge and overgrown cultivation. Mainly forages at middle and upper levels, but also descends to understorey. Shy and unobtrusive, usually keeping to dense cover. **SIMILAR SPECIES** Lagden's Bush-shrike has darker, plain grey head, all-green tail only narrowly tipped yellow, black wing-coverts boldly tipped yellow and no scarlet on underparts. Also compare yellow-breasted form with rare Monteiro's Bush-shrike. **STATUS AND DISTRIBUTION** Generally uncommon to scarce, but locally not uncommon, forest resident, from Sierra Leone and SE Guinea to Togo and from SW Nigeria to SW CAR and Congo.

MONTEIRO'S BUSH-SHRIKE *Malaconotus monteiri*   Plate 121 (7)
Gladiateur de Monteiro

L *c.* 25 cm (*c.* 10"). *M. m. perspicillatus.* Very rare, enigmatic, large species with massive bill. **Adult** Similar to savanna-dwelling Grey-headed Bush-shrike but differs in having white of lores extending above and below eye, slightly darker grey head, and yellow underparts with slight or no orange wash on breast. Eye greyish. Bill black. **Juvenile** No information. **VOICE** Similar to Green-breasted Bush-shrike. **HABITS** Little known. Only recorded in upper levels of primary montane forest. **SIMILAR SPECIES** Grey-headed Bush-shrike (see above). Yellow-breasted form of Fiery-breasted Bush-shrike distinguished by black subterminal band on tail and boldly tipped black-and-yellow tertials. **STATUS AND DISTRIBUTION** Only 2 documented records in W Africa, both in Cameroon: a specimen collected on Mt Cameroon in 19th century and a sighting on Mt Kupe in 1992 (where also claimed in May 1997). Extralimitally on Angola escarpment (5 records; none since 1954). *TBW* category: Data Deficient. **NOTE** Birds in Cameroon suggested to be a morph of Green-breasted Bush-shrike (Williams 1998).

GREY-HEADED BUSH-SHRIKE *Malaconotus blanchoti*   Plate 121 (5)
Gladiateur de Blanchot

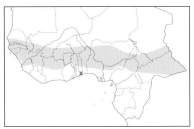

L 25 cm (10"). Large, with massive bill. **Adult** *blanchoti* Head and nape grey; lores white; upperparts and tail green; greater and median coverts, tertials, secondaries and tail narrowly tipped yellow. Underparts bright yellow washed orange on breast. Eye yellow to orange. Bill black. Eastern *catharoxanthus* typically lacks orange on underparts. **Immature** Head mottled brown; underparts paler. Eye paler. Bill greyish-horn. **VOICE** A far-carrying, drawn-out, single hollow whistle *whoooop* (easily imitated by human), occasionally rising at end (*whoooo-up*). Also various harsh and rasping notes, bill clicks and shorter whistles. [CD13:74] **HABITS** Singly or in pairs in upper and mid-levels of savanna woodland and gallery forest; occasionally lower. Shy and inconspicuous, but call betrays presence. May join mixed-species flocks. **SIMILAR SPECIES** Lagden's Bush-shrike is a scarce rainforest species with yellow tips to wing feathers emphasised by black. **STATUS AND DISTRIBUTION** Locally not uncommon to scarce resident in broad savanna belt, from Senegambia to Sierra Leone east to Cameroon (nominate) and from N Cameroon to Chad and CAR (*catharoxanthus*). Not in arid north nor rainforest.

LAGDEN'S BUSH-SHRIKE *Malaconotus lagdeni*   Plate 121 (8)
Gladiateur de Lagden

L 23–25 cm (9–10"). *M. l. lagdeni.* Large, with massive bill. **Adult** Head and nape grey; upperparts green; wing-

coverts, tertials and secondaries black boldly tipped yellow. Tail green narrowly tipped yellow. Underparts deep yellow washed orange on throat and breast. Eye greyish. Bill black. **Juvenile** Head and nape brownish-grey; wing feathers only narrowly tipped yellowish with faint blackish subterminal patch. Underparts greyish-white washed green on flanks and yellow on undertail-coverts. Bill horn or grey-brown. **Immature** Soon becomes more marked on wings and more yellow on underparts. Bill black. **VOICE** Various melodious hoots and whistles; most distinctive and typical are a far-carrying, slow *hoot, hoot-hoot* followed, after a pause, by two whistles *hweet-huuuu*. Also a harsh, grating *chrrrr*. [CD13:72] **HABITS** Singly or in pairs in mature forest. Inconspicuous; mainly in dense foliage and tangles of lianas at mid-level. **SIMILAR SPECIES** Grey-headed Bush-shrike has no black in wing and is a savanna species. Fiery-breasted Bush-shrike has paler head, black subterminal band in tail and usually scarlet on underparts. **STATUS AND DISTRIBUTION** Scarce forest resident in Sierra Leone, Liberia, Ivory Coast and SW Ghana. *TBW* category: Near Threatened.

GREEN-BREASTED BUSH-SHRIKE *Malaconotus gladiator*   Plate 121 (6)
Gladiateur à poitrine verte

L 25–28 cm (10–11”). Large and sombre coloured. **Adult** Head and nape dark grey; upperparts and tail plain dark olive-green; underparts uniformly yellowish-green. Eye greyish. Bill black. **Juvenile** No information. **VOICE** A far-carrying, drawn-out, mournful whistle, repeated up to 10 times; similar to that of Grey-headed Bush-shrike and easily imitated by a human. Also a harsh, grating call usually repeated with same timing as whistle, and a loud, unmusical, chattering alarm call. Agitated birds rattle with bill (Stuart & Jensen 1986). [CD13:73] **HABITS** Singly or in pairs in upper levels of montane forest (generally above 1250 m, up to 2300 m; down to 950 m on Mt Cameroon). **SIMILAR SPECIES** No other bush-shrike has similar dark underparts. **STATUS AND DISTRIBUTION** Scarce to locally not uncommon resident restricted to montane forest in SE Nigeria (Obudu Plateau) and W Cameroon (Mt Cameroon, Rumpi Hills, Mt Kupe, Mt Nlonako, Bakossi Mts, Mt Manenguba, Bamenda Highlands). *TBW* category: THREATENED (Vulnerable).

MOUNT KUPE BUSH-SHRIKE *Malaconotus kupeensis*   Plate 121 (4)
Gladiateur du Kupé

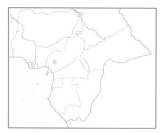

Other name F: Gladiateur du Mont Kupé
L 18–20 cm (7–8”). Unmistakable. **Adult** Black mask (extending from forehead to ear-coverts) bordered above by white line; crown and mantle grey; rest of upperparts and tail bright olive-green. Throat pure white, with or without round maroon patch in centre; rest of underparts grey becoming yellowish-green on lower belly and yellow on vent; short blackish line broadening in centre (forming patch), bordering white throat (‘necklace’), present in some. **Immature** Crown and mantle flecked olive-green; throat flecked pale yellow; rest of underparts flecked greenish-yellow; necklace reduced and indistinct or absent. **VOICE** Territorial song a short, loud, babbler-like introductory chatter *thek-thek, kh-kh-kh* followed by a series of 3–4 grating *tchraa* notes (up to *c.* 30 when excited). Rarely a short, ascending series of 3 clearly detached whistles sounding slightly out of tune. Noisily snaps wings like puffback *Dryoscopus* (Dowsett-Lemaire 1999). A soft, continuous insect-like grating also described (Bowden & Andrews 1994). [CD13:68] **HABITS** Singly or in pairs at mid-levels of primary montane forest. Also in mixed-species flocks. **STATUS AND DISTRIBUTION** Endemic to W Cameroon, where rare and only known from Mt Kupe (at 930–1450 m), nearby Bakossi Mts (at 1000–1250 m), and southern sector of Banyang Mbo Wildlife Sanctuary. *TBW* category: THREATENED (Critical). **NOTE** Presence of maroon throat patch and ‘necklace’ appears unrelated to sex (F. Dowsett-Lemaire pers. comm., *contra* BoA).

MANY-COLOURED BUSH-SHRIKE *Malaconotus multicolor*   Plate 121 (1)
Gladiateur multicolore

L 20 cm (8”). Beautiful but skulking; males occurring in three distinct forms with differently coloured underparts. **Adult male** *multicolor* Black mask (from forehead to ear-coverts) bordered by variable amount of white above (sometimes absent); crown and mantle grey; rest of upperparts green; flight feathers and tertials tipped yellow. Underparts deep yellow or scarlet becoming yellow on lower belly, or yellow with black throat and breast. Tail green with some black and broadly tipped yellow or reddish (according to morph). *M. m. batesi* has green in tail

replaced by black in all morphs. **Adult female** As male but head grey, without mask; tail green tipped yellow; underparts duller washed green on flanks. Never recorded with black throat. **Juvenile** As adult female but feathers of upper mantle green tipped yellow; wing-coverts tipped yellow; underparts barred dusky. **Immature** As adult female but has yellow-tipped wing-coverts. **VOICE** A single, melodious, resonant whistle *whoo-op*, often repeated and forming well-spaced out, slow series; distinctive but easily passing unnoticed. Also a double *whop-wheeu* or *whu-whee*. Female utters a drawn-out rasping note. [CD13:66] **HABITS** Singly or in pairs in canopy of high forest. Inconspicuous and very hard to observe, despite its bright coloration. **SIMILAR SPECIES** Similar-sized Sulphur-breasted Bush-shrike has yellow forehead and supercilium, and is a savanna species. Other forest-dwelling bush-shrikes with yellow or red underparts are much larger with massive bills. **STATUS AND DISTRIBUTION** Uncommon to scarce forest resident, from Sierra Leone and SE Guinea to Togo and from Nigeria to CAR and Congo; also S Mali. Races meet in Cameroon, nominate from Mt Cameroon west, *batesi* east.

BOCAGE'S BUSH-SHRIKE *Malaconotus bocagei* Plate 123 (4)
Gladiateur à front blanc

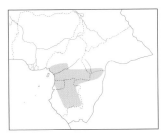

Other name E: Grey-green Bush-shrike

L 16.5 cm (6.5"). *M. b. bocagei*. Smallest and only rather dull-coloured bush-shrike. **Adult male** Head and upper mantle black with contrasting white forehead and supercilium (extending to neck); rest of upperparts plain grey. Tail blackish. Underparts white with buffish wash to breast and flanks. **Adult female** Tail washed grey. **Juvenile** Above, dull blackish-brown with buff or greenish-yellow tips to feathers producing strongly barred effect; buff supercilium; ear-coverts plain dark brown; flight feathers and tail dark brown tipped buff. Underparts whitish with indistinct dusky barring and buff or greenish-yellow wash to breast. **Immature** Head brownish-black speckled buff; upperparts and tail grey washed greenish; wing-coverts and tertials tipped buffish. Throat white; rest of underparts washed buffish with increasingly indistinct dusky barring. **VOICE** Varies geographically. Songs include rapid series of whistles *hu-hu-hu-hu-hu-huuu*, with variations, resembling Sulphur-breasted Bush-shrike's; clear, drawn-out, double notes *tliuu-theeee*; rhythmic *tliu-tliu-tliu*; fast *tiuptiuptiuptiuptiup*; series of disyllabic *kuli-kuli kuli-kuli* or *kukli-kukli kukli-kukli*, reminiscent of African Yellow Warbler; fast *klu, klihihihihihihihi...*, and others. May also duet, female answering with shorter whistles. Also harsh calls. [CD13:64] **HABITS** Singly or in pairs in overgrown forest clearings and edges, gallery forest and wooded savanna with dense bushes. Forages in middle and upper strata. Behaviour similar to Sulphur-breasted Bush-shrike. **SIMILAR SPECIES** None, but beware of confusing unseen calling birds with Sulphur-breasted Bush-shrike in more open, savanna habitats, where both species occur. **STATUS AND DISTRIBUTION** Uncommon resident in Cameroon (north to Matsari, Yoko area), Eq. Guinea, Gabon, Congo and SW CAR.

SULPHUR-BREASTED BUSH-SHRIKE *Malaconotus sulfureopectus* Plate 121 (3)
Gladiateur soufré

Other name E: Orange-breasted Bush-shrike

L 17–19 cm (6.75–7.5"). *M. s. sulfureopectus*. **Adult male** Forehead yellow, extending as narrow supercilium above black mask; fore-crown moss-green; hindcrown to mantle grey; rest of upperparts and tail moss-green; wing and tail feathers tipped lemon-yellow. Underparts bright yellow with strong orange wash to breast. Eye brown. **Adult female** Mask less black. **Juvenile** Entire head and mantle grey, feathers tipped greyish-buff with subterminal dark markings; rest of upperparts as adult but feathers tipped yellowish with dark subterminal bar, producing scaly effect. Underparts whitish washed yellow and barred dusky with some orange feathers on breast. **Immature** As adult but entire head grey; underparts duller. **VOICE** Series of far-carrying, clear whistles, varying in speed and motif, e.g. *hu-hu-hu-hweet* or *hu-wheet hu-wheet hu-wheet* and *wheet-wheet-huuu* or *huwhit-huhuu-whit*; also a shorter *hwut-heee*. Calls include *tzzzzrr*, used with bill-snapping in alarm and occasionally also by female in duet; also a single *puwheet*. [CD13:65] **HABITS** Singly or in pairs in savanna woodland and edges of gallery forest. Forages from canopy to mid-level, gleaning insects from leaves and branches. Rather skulking and inconspicuous, but loud melodious calls attract attention. **SIMILAR SPECIES** Grey-headed Bush-shrike has similar coloration and also occurs in savanna, but is much larger with massive bill; also lacks yellow forehead and black mask. Female Many-coloured Bush-shrike resembles immature Sulphur-breasted but has darker head-sides and deeper coloured underparts, and occurs in forest. **STATUS AND DISTRIBUTION** Not uncommon resident in broad savanna belt, from Senegambia to Sierra Leone east to Chad and CAR.

GORGEOUS BUSH-SHRIKE *Telophorus viridis*   Plate 121 (2)
Gladiateur vert

Other names E: Perrin's Bush-shrike, Four-coloured Bush-shrike
L 18–19 cm (7.0–7.5"). *T. v. viridis.* Distinctive and strikingly beautiful but skulking. **Adult male** Dark green above with short golden-yellow supercilium. Throat crimson encircled by black lores and black line from below eye to neck-sides and joining broad black breast-band, latter bordered by crimson band below; rest of underparts green with irregular chocolate-brown band on central belly; undertail-coverts dark reddish. Tail black. **Adult female** Duller, with paler red areas, narrower black breast-band and green wash to tail. **Juvenile** Olive-green above; all-yellow throat becoming paler, greenish-yellow on rest of underparts. **Immature** Upperparts as adult; tail washed dark green; throat and breast greenish-yellow with dark crescentic bars forming faint breast-band; rest of underparts yellowish-green. **VOICE** A characteristic, ringing *ko-ko-kwik ko-kwik.* [CD13:69] **HABITS** Singly or in pairs in undergrowth of dense thickets and shrubbery within wooded savanna and gallery forest. Difficult to observe, but call draws attention. **STATUS AND DISTRIBUTION** Locally not uncommon resident, SE Gabon and Congo. **NOTE (1)** Sometimes placed in genus *Malaconotus.* **(2)** Nominate race previously treated as separate species, Perrin's Bush-shrike *M. viridis*; S and E African forms known as Gorgeous or Four-coloured Bush-shrike *M. quadricolor.*

MARSH TCHAGRA *Antichromus minutus*   Plate 123 (7)
Tchagra des marais

Other names E: Blackcap Bush-shrike, Little Blackcap Tchagra
L 16–18 cm (6.25–7.0"). *A. m. minutus.* Rather stocky, short tailed and heavy billed. **Adult male** Resembles miniature coucal. Head jet-black; upperparts rufous-brown with black on mantle and scapulars forming variable V-shaped mark; wings deep chestnut. Tail black narrowly tipped buff. Throat whitish; rest of underparts rich buffish. Eye red. Bill black. **Adult female** As male but with broad white supercilium. **Juvenile** As adult female but duller; crown buffish or mottled tawny; supercilium buffish-white; bill horn coloured. **VOICE** Male song consists of a few melodious whistles, e.g. *ti-tweeu-tuwEEu tui-tiwee*, with variations, usually uttered during low, short display flight and preceded by wing-rattling. Female occasionally responds with nasal *cherrruu.* Calls include various hard churrs and clucks (e.g. *chrrr, chuk* or *tzik* and *klok*), often repeated; also subdued, muffled sounds. [CD13:50] **HABITS** Singly or in pairs in rank herbage along savanna streams, edges of forest, thickets and marshes, and in reedbeds. Usually skulks in dense vegetation, preferring to creep away than take flight when disturbed, though also perches on tall grass stems or atop bushes. Display a short rising flight on whirring wings, followed by song and ending in glide with fanned tail. **SIMILAR SPECIES** Other tchagras are larger and have more elongated shape emphasised by longer tails. **STATUS AND DISTRIBUTION** Uncommon and local resident, from Sierra Leone and SE Guinea to Togo and Nigeria to CAR and Congo. **NOTE** Often placed in genus *Tchagra.*

BROWN-CROWNED TCHAGRA *Tchagra australis*   Plate 123 (9)
Tchagra à tête brune

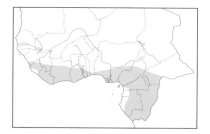

Other name E: Brown-headed Bush-shrike
L 18–19 cm (7.0–7.5"). **Adult *ussheri*** Differs from Black-crowned Tchagra in having *brown* crown bordered by narrow black line, and smaller size. Underparts pale greyish tinged olive on flanks. Bill black. Eastern *emini* slightly darker above and more olive on flanks. **Juvenile** Duller; bill dark horn. **VOICE** Male song a jaunty, liquid, descending series *tree-tree-treeu-treeuu-treeuu...* usually given in low display flight and preceded by wing-rattling. Female may respond with soft nasal *cheru-cheru.* Alarm *chuk-chuk-...* and *chrrr.* [CD13:51] **HABITS** Usually singly, low down at forest edges and in overgrown

clearings, thickets and scrub. Rather skulking, but attracts attention by display: a short rising flight on whirring wings, followed by a glide with fanned tail while singing. **SIMILAR SPECIES** Black-crowned Tchagra has all-black crown, different voice and is larger. **STATUS AND DISTRIBUTION** Not uncommon to uncommon resident, from W Guinea and Sierra Leone to CAR and Congo; also S Mali. Races meet in Nigeria, possibly separated by Lower Niger R., *ussheri* to west, *emini* east.

BLACK-CROWNED TCHAGRA *Tchagra senegala*   Plate 123 (8)
Tchagra à tête noire

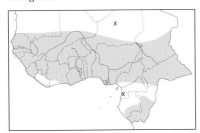

Other name E: Black-headed Bush-shrike
L 20–22 cm (8.0–8.75"). **Adult** *senegala Black crown* bordered by broad buff-white supercilium; black eye-stripe; mantle pale brownish becoming grey-brown on rump. Wings bright rufous (conspicuous in flight). Tail long and graduated, black boldly tipped white, except central feather pair, which is browner and faintly barred dusky. Underparts pale greyish. Bill black. *T. s. armena* slightly darker and overall more deeply coloured; *notha* paler; *remigialis* paler still, pale tawny-brown above, almost white below. **Juvenile** Duller, with brownish-black crown, buff-tipped tail and dark horn-coloured bill.

VOICE Male song distinctive and variable: a series of vigorous, far-carrying, melancholy whistles, given in display flight or from perch. Female may respond in duet with drawn-out *trrrrrrrrrr*. Also various explosive, growling whistles, e.g. *tok-tok-tok-tok* followed by rolling bubble (like melodious waterfall) and ending with *whu-heeuw*. [CD4:57] **HABITS** Usually singly, low down or on ground in thickets in savanna, and farmland; not in forest. Rather skulking and often only glimpsed flying low between bushes, spreading its tail on landing, thereby revealing distinctive black-and-white pattern before disappearing into cover. Spectacular display flight attracts attention: rises steeply on whirring wings, then starts singing while gliding down. **SIMILAR SPECIES** Brown-crowned Tchagra has brown crown, slightly smaller size and different voice. **STATUS AND DISTRIBUTION** Common to not uncommon resident throughout. Nominate from S Mauritania to Liberia east through Mali (north to Mopti) and Nigeria (north to *c.* 12°N) to S Chad and CAR; *notha* from N Mali (Tombouctou) to N Nigeria; *remigialis* C Chad (intergrading with *notha* in L. Chad area); *armena* from S Cameroon to S CAR, Gabon and Congo. **NOTE** *T. s. 'camerunensis'* and *'rufofusca'* included in *armena*, following *BoA*.

SABINE'S PUFFBACK *Dryoscopus sabini*   Plate 122 (4)
Cubla à gros bec

L 18–19 cm (7.0–7.5"). Forest puffback with relatively long, heavy bill. **Adult male** *sabini* Head, upperparts and tail black contrasting with pure white lower back, rump and underparts. Eye reddish-brown. Bill and legs blue-grey. *D. s. melanoleucus* similar. **Adult female** *sabini* Head grey; upperparts and tail tawny-brown; narrow pale tawny supraloral stripe to above eye; narrow white eye-ring. Underparts tawny-ochre becoming white on central belly. Female *melanoleucus* similar, but tail darker and browner. **Juvenile** Similar to adult female. **VOICE** Harsh calls; in flight *tok-tok-tok-...* Song a series of clear whistles, slowly descending *tsee tsu tsu tsu tsu tsu tsu ...* and variable in length; reminiscent of Yellow Longbill or Brown-chested Alethe. [CD13:49] **HABITS** Singly or in pairs in upper levels of lowland forest. Often in mixed-species flocks. **SIMILAR SPECIES** Male Black-shouldered Puffback distinguished by smaller bill and different habitat. Male Shrike Flycatcher is smaller, with shorter bill and tail, and has different voice and posture. Female Pink-footed Puffback has smaller bill, darker, more olivaceous upperparts and tail, and pinkish legs. **STATUS AND DISTRIBUTION** Locally not uncommon to rare forest resident, from Sierra Leone and SE Guinea to S Nigeria (nominate) and from Cameroon to SW CAR and Congo (*melanoleucus*).

PINK-FOOTED PUFFBACK *Dryoscopus angolensis*   Plate 122 (3)
Cubla à pieds roses

L 15–17 cm (6.0–6.75"). **Adult male** Head blue-black contrasting with dark grey upperparts and tail; lower back and rump paler. Underparts greyish-white. Eye dark brown. Legs and feet pinkish. **Adult female** Head and mantle blue-grey; upperparts and tail olivaceous-brown. Underparts tawny-orange becoming white on central belly. **Juvenile** As adult female but duller. **VOICE** A variety of harsh churring calls and rattles, and a series of clicking *tik-tik-tik-tik-tik-...* [CD13:48] **HABITS** In pairs in montane and submontane forest. Usually in upper and mid-strata of

mature forest; also in young, scrubby second growth and forest edge. Occasionally wanders to lower elevations. Often in mixed-species flocks. **SIMILAR SPECIES** Male Northern Puffback has wing feathers fringed greyish-white, entire mantle concolorous with head and orange-red eyes. Female Sabine's Puffback has heavier bill and tawny upperparts and tail. Female Northern has paler grey head (except in ssp. *malzacii*), pale buffish-brown fringes to wings and red eye. **STATUS AND DISTRIBUTION** Locally not uncommon to rare forest resident, in highlands of SE Nigeria (Obudu Plateau; rare), Cameroon (e.g. Mt Cameroon, Rumpi Hills, Mt Kupe, Mt Nlonako, Bakossi Mts, Bamenda Highlands, Adamawa Plateau) and Eq. Guinea (Mt Alen), and in S Congo.

BLACK-SHOULDERED PUFFBACK *Dryoscopus senegalensis* Plate 122 (2)
Cubla aux yeux rouges

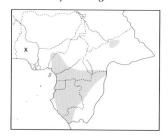

Other name E: Red-eyed Puffback
L 16–17 cm (6.5"). **Adult male** Head, upperparts and tail glossy black with white lower back and rump. Underparts pure white. Eye orange-red. Legs bluish-grey. **Adult female** As male but duller, with white supraloral stripe to halfway above eye, and grey rump. **Juvenile** As adult female but duller and buffish wash on wings; lower mandible horn coloured. **Immature** As adult but underparts and wings washed buffish. **VOICE** Male utters a loud, frequently repeated *KYow! KYow! KYow!* ...; also a harsh *kurrrWEERrr* and snapping sounds. Female responds with a rasping, weaver-like note. [CD13:47] **HABITS** In pairs or small family parties in upper levels of open second growth and forest clearings and edges (avoiding closed forest). Joins mixed-species flocks. **SIMILAR SPECIES** Male Northern Puffback has white fringes to wing feathers. Male Sabine's Puffback has more elongated shape emphasised by longer, heavier bill, and is a forest species. Male Pink-footed Puffback has greyish upperparts, greyish-white underparts and dark eye. **STATUS AND DISTRIBUTION** Resident from SE Nigeria (Obudu Plateau; rare) to CAR and Congo (locally not uncommon to common). Also SW Nigeria (1 record).

NORTHERN PUFFBACK *Dryoscopus gambensis* Plate 122 (1)
Cubla de Gambie

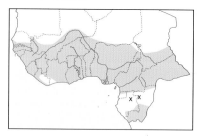

Other name E: Gambian Puff-back Shrike
L 18–19 cm (7.0–7.5"). The only W African puffback with, in both sexes, contrasting pale edges to wing feathers. **Adult male** *gambensis* Head, upperparts and tail glossy black with greyish-white scapulars, lower back and rump. Wings dark brown; flight feathers and wing-coverts fringed greyish-white. Underparts white with slight buffish-grey wash. Conspicuous orange-red eye. Legs bluish-grey. *D. g. malzacii* and *congicus* similar. **Adult female** *gambensis* Head grey; upperparts dull earth-brown; flight feathers and wing-coverts fringed buff; tail dark brown; underparts tawny-buff. Female *malzacii* distinctly darker and browner on head and upperparts; *congicus* richer tawny below. **Juvenile** As adult female but paler, more ashy-grey overall. Eye dark; bill horn. **VOICE** Varied. Typically a frequently repeated, loud, rasping *CHERP-CHERP-...* or *CHURK-CHURK-...* and harsh chattering, nasal and clicking sounds (such as *ptkew* and *ptkik*). [CD13:46] **HABITS** In pairs in canopy of savanna woodland and gallery forest edges; also forest clearings (avoiding closed forest) and mangroves. Forages by gleaning insects from leaves and branches. Attracts attention by its constant calling. **SIMILAR SPECIES** Male Black-shouldered Puffback is black and white, lacking grey tones and white fringes to wing feathers. Male Pink-footed Puffback has uniform wings; head and mantle contrasting neatly with rest of upperparts; eye dark. Female Pink-footed more colourful than female Northern, with head and mantle bluer grey, upperparts plain and more olivaceous, underparts more orange. Female Sabine's Puffback has much heavier bill, is overall tawnier and occurs in forest. **STATUS AND DISTRIBUTION** Not uncommon to common resident in savanna and forest zones throughout. Nominate replaced by *malzacii* in CAR and by *congicus* in S Gabon and Congo. Some evidence of local movements (Nigeria).

SOOTY BOUBOU *Laniarius leucorhynchus* Plate 122 (7)
Gonolek fuligineux

L 21.5 (8.5"). **Adult** Robust, all-black shrike. Faint dark blue gloss visible in certain lights. Bill and legs black.

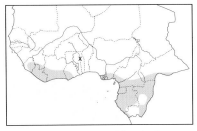

Immature Duller and with whitish bill. **VOICE** Male utters a far-carrying, clear, ringing *HOO-HOO*; also in long series of variable speed. Female answers with a drawn-out, plaintive whistle *hweeeew*. Other calls include repeated, hoarse *kchch-kchch-kchch* and *plt-plt-plt-plt* or *kl-kl-kl-kl*. [CD13:63] **HABITS** In pairs, skulking in tangles, rank vegetation, thickets and overgrown cultivation in large clearings and dense regrowth at forest edges. Secretive but vocal. Joins mixed-species flocks. **SIMILAR SPECIES** The only other all-black shrike of the region, Mountain Sooty Boubou, occurs in montane forest and is noticeably smaller. **STATUS AND DISTRIBUTION** Not uncommon to scarce resident in forest zone, from Sierra Leone and SE Guinea to Ghana and from S Nigeria to S CAR and Congo. Also Togo (1 record).

MOUNTAIN SOOTY BOUBOU *Laniarius poensis*   Plate 122 (5)
Gonolek de montagne

L 18 cm (7"). **Adult** *camerunensis* Uniformly black with slight dark blue gloss. Eye blackish to reddish-brown. Island race *poensis* slightly smaller. **Immature** Similar to adult but slightly duller. **VOICE** Calls very varied, including loud whistles, trills and rattles, squeals and harsh grating alarm notes. Most notable are duets of pair, *WHOO-EE* (male) followed by rasping *tchrerr* or *errgh* (female), with variations, e.g. *whooee tchreh* or *whee tcheh*; also *whishshsh*, rapid *huhuhuhuhuhu* and *kr-k-kr-k*. Many calls not separable from those of Yellow-breasted Boubou. [CD13:60–61] **HABITS** Usually in twos (not necessarily pairs) in dense undergrowth of montane forest edges and clearings, where difficult to observe. Generally commoner in lower montane zone than Yellow-breasted Boubou, and does not occur at highest altitudes. **SIMILAR SPECIES** Sooty Boubou is larger and inhabits lowland forest. Yellow-breasted Boubou often indistinguishable on call alone, but generally occurs at higher altitudes, though with broad overlap. **STATUS AND DISTRIBUTION** Locally common to scarce forest resident in highland areas of SE Nigeria (Obudu and Mambilla Plateaux, Gotel Mts), W Cameroon (Mt Cameroon, Rumpi Hills, Mt Kupe, Bakossi Mts, Mt Manenguba, Bamenda Highlands) and Bioko. *L. p. camerunensis* on mainland, nominate on Bioko. **NOTE** Sometimes considered conspecific with Fülleborn's (Black) Boubou *L. fuelleborni*.

LÜHDER'S BUSH-SHRIKE *Laniarius luehderi*   Plate 123 (5)
Gonolek de Lühder

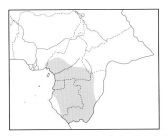

L 18–19 cm (7.0–7.5"). *L. l. luehderi*. **Adult** Above, black with *chestnut cap* and long white stripe on wing (formed by white tips to median and inner greater coverts and edges of innermost secondaries). Tail black. *Throat to upper belly cinnamon*; rest of underparts white. **Juvenile** Above, a mixture of olive, yellow and off-white barred dark brown; below, lemon-yellow strongly barred dusky; tail rufous. **Immature** Gradually loses dusky barring. Above, olivaceous; wingbar tinged yellowish; below, deep yellowish washed orange on breast. **VOICE** Varied. A guttural, quivering *krrrooh* or *keoooow*, repeated at regular intervals; occasionally uttered in duet by pair. Female also utters a dry *k-k-k-k-*... and a sharp *gkssss*. Also a rather melodious, hollow *hoo-up* similar to call of Many-coloured Bush-shrike. Other calls include *k-krrrr* and harsh and churring notes. [CD13:52] **HABITS** Singly or in pairs in overgrown forest clearings, second growth, dense scrub, thickets and abandoned cultivation with dense tangles. Difficult to observe, keeping to thick cover. **STATUS AND DISTRIBUTION** Uncommon to not uncommon resident in forest zone, from SE Nigeria (near Calabar; uncommon) to Congo.

TROPICAL BOUBOU *Laniarius aethiopicus*   Plate 122 (6)
Gonolek d'Abyssinie

Other names E: Bell Shrike; F: Gonolek à ventre blanc
L *c.* 23 cm (9"). *L. a. major.* **Adult** Head, upperparts and tail glossy blue-black with long white stripe on wing (formed by white on median and inner greater coverts and white edges to inner secondaries). Underparts white with slight and variable pinkish wash (often difficult to see in field). **Juvenile** Duller, with tawny tips to feathers of head and upperparts and dusky barring on underparts; bill dark horn. **VOICE** Duet consists of drawn-out, resonant and ringing *HOOO* whistles, uttered in short series of variable speed (male), answered by similar notes or grating *gkrzzz* (female), e.g. *hoo-hoo-hoo* or *hoo-hoo-HUloo*, fast *HUloo-HUloo-HUloo*, etc.; also *rrroh-gkzzz*. More variable than

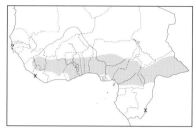

Turati's Boubou. Call a loud, explosive *KEK!*, often in long series. [CD13:53] **HABITS** Singly or in pairs at edges of gallery forest and in savanna woodland, thickets, bush and gardens. Forages on ground or by gleaning from branches and leaves. Parasitised by Black Cuckoo. **SIMILAR SPECIES** Turati's Boubou lacks white in wing; ranges known to overlap in N Sierra Leone. Swamp Boubou very similar but has shorter wing-stripe and generally whiter underparts; note also voice. **STATUS AND DISTRIBUTION** Not uncommon resident in savanna zone and vast cleared areas in forest zone, from NE Sierra Leone and C Guinea to Chad and CAR. Record from Senegal requires confirmation.

SWAMP BOUBOU *Laniarius bicolor*   Plate 122 (8)
Gonolek à ventre blanc

Other name F: Gonolek du Gabon
L *c.* 23 cm (9"). **Adult *bicolor*** As Tropical Boubou but edges of inner secondaries lack white (resulting in shorter wing-stripe) and underparts pure white. **L. b. guttatus** has white on 1–2 inner secondaries. **Juvenile** Similar to juvenile Tropical Boubou. **VOICE** Duet consists of 2–3 drawn-out, resonant and ringing *HOOO* or *HIOO* whistles (male), answered by hard rattling *K-K-K-KKKKK*, shorter *K-K* or rasping *gha gha gha* (female). Also a loud, explosive *KEK!* or *TUK!*, often in long series. [CD13:55] **HABITS** In pairs in savanna thickets, coastal scrub, forest regrowth and mangroves. Behaviour as Tropical Boubou. **STATUS AND DISTRIBUTION** Locally not uncommon resident in coastal areas from Gabon to Congo, and along lower Congo R. Also recorded in coastal Cameroon (rare). Nominate in north of range, intergrading with *guttatus* in Congo.

TURATI'S BOUBOU *Laniarius turatii*   Plate 122 (9)
Gonolek de Turati

Other name F: Gonolek de Verreaux
L *c.* 23 cm (9"). **Adult** As Tropical Boubou but *lacks any white on wing* and has pale pinkish-buff wash on underparts (often difficult to see in field); lower belly and undertail-coverts white. **Juvenile** See Tropical Boubou. **VOICE** Duet consists of series of resonant, ringing *HOO*'s (male) answered by a grating *gkrzzz* (female); similar to, but more stereotyped than Tropical Boubou. Also a loud, explosive *KEK!*, often in long series. [CD13:54] **HABITS** In pairs in wooded savanna, forest regrowth and forest edge. Behaviour as Tropical Boubou. **STATUS AND DISTRIBUTION** Common resident in Guinea-Bissau, W Guinea and W Sierra Leone. **NOTE** Sometimes treated as race of Tropical Boubou. Additional data on respective ranges and extent of any intergradation desirable.

YELLOW-CROWNED GONOLEK *Laniarius barbarus*   Plate 123 (2)
Gonolek de Barbarie

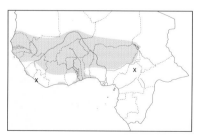

Other name E: Barbary Shrike
L *c.* 23 cm (9"). Beautiful bush-shrike with distinctive coloration. **Adult *barbarus*** Head, upperparts and tail jet-black but for golden-yellow crown. Underparts crimson; lower belly, thighs and undertail-coverts deep buff. **L. b. helenae** described as having deeper coloured, more orangey crown. **Juvenile** Feathers of upperparts blackish-brown tipped buff; underparts yellowish-buff narrowly barred dusky. **Immature** Barring on underparts gradually replaced by crimson; finally as adult but duller black above and with buff tips to feathers of underparts. **VOICE** Male song a clear, resonant and ringing double whistle, *WHEE-oo!*, likened to whip-lash, with variations, e.g. *whuoo-wHEE* and *oo-WHEE* or *whow-WHEE*; also a quivering *whiiiir*. Female usually responds with *kik-kik*. Various other, harsh calls. [CD13:56] **HABITS** Singly or in pairs, skulking in undergrowth of dense thickets, often near watercourses, in inland and coastal savanna, thorn scrub and mangroves. Occasionally perches in the open. **SIMILAR SPECIES** Black-headed Gonolek is all-black above, without yellow cap. **STATUS AND DISTRIBUTION** Locally not uncommon to common resident

throughout savanna zone, from S Mauritania to Guinea east to N Cameroon and W Chad, reaching coast in Dahomey Gap. Also coastal Liberia (rare) and Ivory Coast (locally common). *L. b. helenae* known only from coastal Sierra Leone.

BLACK-HEADED GONOLEK *Laniarius erythrogaster*　　　Plate 123 (1)
Gonolek à ventre rouge

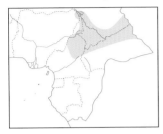

Other name F: Gonolek rouge et noir
L *c.* 23 cm (9"). **Adult** Head, upperparts and tail black. Underparts crimson; lower belly, thighs and undertail-coverts deep buff. **Juvenile** Feathers of upperparts blackish-brown tipped buff; underparts yellowish-buff narrowly barred black. **Immature** Barring on underparts gradually replaced by crimson; finally as adult but duller black above and with buff tips to feathers of underparts. **VOICE** Male song a clear, resonant and ringing whistle *tiu-WHEE-oo!* Female usually responds with harsh rasping. Call a hard *K-K-K-K.* [CD13:57] **HABITS** Singly or in pairs along grassy river banks. Shy and skulking, preferring to remain in dense cover but betraying its presence by loud calls. **SIMILAR SPECIES** Yellow-crowned Gonolek has yellow cap. **STATUS AND DISTRIBUTION** Common but local resident in NE Nigeria, Cameroon (from Adamawa Plateau north), Chad and CAR. **NOTE** Sometimes considered conspecific with S African Crimson-breasted Shrike *L. atrococcineus* (cf. Dowsett & Dowsett-Lemaire 1993).

YELLOW-BREASTED BOUBOU *Laniarius atroflavus*　　　Plate 123 (3)
Gonolek à ventre jaune

L 18–19 cm (7.0–7.5"). **Adult** Head, upperparts and tail jet-black. Underparts deep yellow; thighs and undertail-coverts buff. **Juvenile** No information. **Immature** Dull version of adult with feathers of crown and upperparts sparsely tipped yellowish-buff; wing-coverts and outer tail feathers tipped yellowish-buff. **VOICE** Most notable are duets of pair *WHEEw!* (male) followed by *chek* or harsh note (female), with variations, e.g. *WHEEo-WHEEo tcha; tchatcha rrrueh; kyoukyou tcha* and *WHEE tchou!* Also a variety of loud whistles (e.g. quivering *whi-i-i-i-ip*), swishing, rattling and harsh grating notes, most inseparable from those of Mountain Sooty Boubou. [CD13:59] **HABITS** Usually in pairs, within dense undergrowth of clearings, secondary scrub, small forest remnants and bamboo in highlands. Mainly forages in lower and mid-strata. Vocal. **SIMILAR SPECIES** None, but Mountain Sooty Boubou has many similar calls and, though generally occurring at lower altitudes, has wide altitudinal overlap. **STATUS AND DISTRIBUTION** Not uncommon but local resident. Endemic to montane forest in SE Nigeria (Obudu and Mambilla Plateaux, Chappal Hendu, Gashaka-Gumti NP) and W Cameroon (Mt Cameroon, Mt Manenguba, Bamenda Highlands).

BRUBRU *Nilaus afer*　　　Plate 123 (6)
Brubru africain

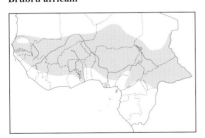

L 13–15 cm (5–6)". Small, active, pied shrike. **Adult male** *afer* Black crown bordered by white forehead and long, broad supercilium; upperparts boldly mottled black, white and buff; long buffish stripe on black wing. Tail black tipped and edged white. Underparts white with chestnut on breast-sides and flanks. *N. a. camerunensis* has underparts less clean white, more greyish. **Adult female** Duller, black areas brownish-grey, rufous on flanks paler and less extensive. **Juvenile** Above, dark brown speckled and mottled white; underparts whitish barred brown. **VOICE** Male gives a distinctive, far-carrying, drawn-out trill *brruuuu*, frequently repeated and often preceded by short *chuk* notes. Female may respond with a high-pitched *wheeeu*. Other calls include repeated *tu* and *peep* (contact) and *chK-chK-...* (alarm). [CD13:44] **HABITS** Singly or in pairs in savanna woodland. Actively forages at upper- and mid-levels, gleaning foliage and branches for insects, also hanging upside-down. Unobtrusive. Joins mixed-species flocks. Zigzag display flight on whirring wings. **STATUS AND DISTRIBUTION** Not uncommon in broad savanna belt, locally common in semi-arid zone. Nominate from S Mauritania to Sierra Leone east to Chad and Cameroon; *camerunensis* from Cameroon (Adamawa Plateau and highlands) to CAR. Probably largely resident, but north–south movements recorded, to 17°N (Mauritania, Mali) and 18°N (Niger) in rains, and south to Guinea savanna (even near coast, e.g. on Accra Plains, Ghana) in dry season.

HELMET-SHRIKES

Family PRIONOPIDAE (World: 8 species; Africa: 8; WA: 3)

Medium-sized, gregarious birds with rather stout, hooked bills, boldly patterned plumages and brightly coloured eye-wattles. Brush-like feathers on forehead produce 'helmeted' appearance. Sexes similar. Juveniles duller. Food mainly insects, also some fruit. In W Africa two species inhabit forest and one savanna. Conspicuous and vocal, constantly chattering and often snapping bill. Breed cooperatively.

WHITE HELMET-SHRIKE *Prionops plumatus*   Plate 120 (7)
Bagadais casqué

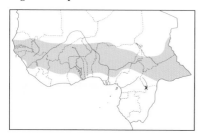

Other names E: Long-crested/White-crested Helmet-shrike
L 19–23 cm (7.5–9.0"). Distinctive pied species with white crest of variable length. **Adult** *plumatus* Head mainly white becoming pale grey on hindcrown and head-sides, with blackish crescent on ear-coverts. Crest bushy and rather short on forecrown, much longer and straighter on mid-crown. Upperparts black with long white stripe on wing and white patches in primaries; tail black edged and tipped white; underparts white. Conspicuous yellow wattle around yellow eye. Bill black. Legs orange. In flight, characteristic pattern with broad, rounded wings having white band above and below (formed by white patches on primaries). Eastern *concinnatus* has shorter and more curly crest; wing-stripe narrower. **Juvenile** Duller, without crest; head washed brownish or greyish; primary-coverts tipped and edged whitish. Eye brown; no eye-wattle. **VOICE** Various frequently uttered, hoarse growling sounds, e.g. *krrreew* and *kreepkrw*, often accompanied by bill-snapping. Alarm a high-pitched *tzzee-tzzee*. [CD14:2a] **HABITS** In parties in wooded savanna, mainly foraging on branches and trunks at mid- and lower levels. Sometimes joins mixed-species flocks. Restless and bold, moving from tree to tree in low, undulating flight. **STATUS AND DISTRIBUTION** Locally not uncommon in broad savanna belt, from S Mauritania to Sierra Leone east to Nigeria and N Cameroon (nominate), and from C Cameroon to Chad and CAR (*concinnatus*).

RED-BILLED HELMET-SHRIKE *Prionops caniceps*   Plate 120 (8)
Bagadais à bec rouge

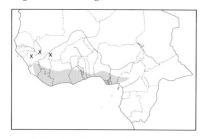

L *c.* 20 cm (8"). Unmistakable. **Adult** *caniceps* Top of head and lores have short, stiff, greyish-white feathers; rest of head, throat, upperparts and tail black; breast white becoming deep rufous-buff on rest of underparts. Conspicuous red bill and legs. Red orbital ring and yellow eye. In flight, *white patch in broad, rounded wings*. In **P. c. harterti** greyish-white of crown extends further onto head-sides; lower underparts paler. **Juvenile** Duller; crown washed brownish; ear-coverts whitish mottled grey; throat whitish or buff. Eye dark; no orbital ring. Bill blackish. **VOICE** Various indistinct hoarse, nasal whistles, a much heard combination being *chok rrrr* and a hoarse, muffled *kwèh-kwèh*; also a hard *chek*, a softer *tiuk* and frequent bill-snapping. Song a series of vigorous, ringing notes *kweeoo* or *whee-aw*. [CD14:3a] **HABITS** In pairs or small, noisy parties in mature forest; also in relict patches and gallery forest. Usually in mid- and upper levels; regularly joins mixed-species flocks. **SIMILAR SPECIES** In flight, rounded wings with white patch and harsh calls may recall male Black-and-white Flycatcher, but latter has shorter tail, all-black head and black bill. See also Rufous-bellied Helmet-shrike. **STATUS AND DISTRIBUTION** Not uncommon forest resident, from W Guinea and Sierra Leone to Togo (nominate) and from Benin to W Cameroon (*harterti*). **NOTE** Rufous-bellied Helmet-shrike (q.v.) is usually considered conspecific, but here treated separately following *BoA*.

RUFOUS-BELLIED HELMET-SHRIKE *Prionops rufiventris*   Plate 120 (9)
Bagadais à ventre roux

Other name E: Gabon Helmet-shrike
L *c.* 20 cm (8"). **Adult** *rufiventris* As Red-billed Helmet-shrike but greyish-white of crown tinged grey-blue (whiter on forehead) and extending onto head-sides, nape and upper throat; lower breast and belly orange-chestnut.

Eastern *mentalis* similar but colour of crown and lower underparts even deeper. **Juvenile** Similar to juvenile Red-billed Helmet-shrike, but underparts darker. **VOICE** Similar to Red-billed Helmet-shrike. Also a clear, melodious *huhu-huhu* and *kweekwee kweekwee*, and descending *chrrrrrrrr*. [CD14:3b] **HABITS & SIMILAR SPECIES** See Red-billed Helmet-shrike. **STATUS AND DISTRIBUTION** Not uncommon forest resident, from S Cameroon to SW CAR and Congo (nominate), and in SE CAR (*mentalis*). **NOTE** Usually considered conspecific with Red-billed Helmet-shrike.

ORIOLES
Family ORIOLIDAE (World: 29 species; Africa: 9; WA: 5)

Medium-sized, robust, arboreal birds of forest and woodland. Most occurring in Africa have males bright yellow and black, with black on head, wings and tail, and a strong, reddish bill. Females duller and/or streaked below. Juveniles mostly resemble dull females and have dusky bills. Adult plumage attained in 1–3 years. Feed on insects and fruit. Mostly in canopy, where often difficult to observe. Voice loud, fluty and melodious. Flight strong and undulating.

BLACK-WINGED ORIOLE *Oriolus nigripennis*   Plate 124 (3)
Loriot à ailes noires

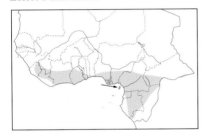

L *c.* 20 cm (8"). **Adult** Similar to Western Black-headed Oriole *but lacks small white patch on edge of wing* and has *central tail feathers black* (not greenish). Bill pinkish. Legs slate-grey. **Immature** Head to upper breast dull black flecked and streaked yellow, esp. on throat. Rest of underparts yellow variably streaked black on breast. Median and greater coverts tipped yellow. Bill blackish. **VOICE** Far-carrying, full, melodious whistles with characteristic oriole quality, resembling those of Western Black-headed Oriole but even more melodious, very liquid and slightly higher pitched, e.g. *whuteluw* or *pteeuw-ee-ooleo* and variations. Also short *oo-ik* or *kloo-ik* and *tiup*, sometimes followed by descending *kirrrrrrr*, clear *kpi-uw* and *hiu* or *hoo-whEE* (very similar to human whistle), often frequently repeated, and harsh *whrrrèèèr*. [CD13:30] **HABITS** Singly, in pairs or small family groups in canopy of lowland and montane forest (up to 2500 m). Where sympatric with Western Black-headed, Black-winged more in secondary forest, edges and forest outliers. **SIMILAR SPECIES** Adult Western Black-headed Oriole (q.v.). Immature Western has olive head. **STATUS AND DISTRIBUTION** Not uncommon to common resident in forest zone, from Sierra Leone and SE Guinea to S CAR and Congo. Also Bioko (rare).

WESTERN BLACK-HEADED ORIOLE *Oriolus brachyrhynchus*   Plate 124 (1)
Loriot à tête noire

Other names F: Loriot à tête noire occidental

L *c.* 21 cm (8.25"). **Adult** *brachyrhynchus* Head to upper breast black, separated from yellow-olive upperparts by narrow yellow collar; rest of underparts yellow. *Small white patch on edge of wing* formed by broad white tips to black primary-coverts. Flight feathers black narrowly edged greyish-white; tertials broadly edged olive-green. Tail feathers black broadly tipped yellow except *greenish central pair*. Bill brownish-pink. Legs slate-grey. Eastern *laetior* has broader yellow collar and yellower mantle. **Immature** Head dark olive; no yellow collar; throat and breast olive mottled yellow; white spot at edge of wing reduced.

Bill dusky, lower mandible paler. **VOICE** Various far-carrying, full, melodious whistles with characteristic oriole quality e.g. *uoo-dleeo, uoo-uoo, tioolioo, whoolioo, whee-whooliu, too-too-tuloo* etc. Notes usually more detached and rather lower pitched than those of Black-winged Oriole. Also a harsh *whit-chèèèw-chèèèw*. [CD13:29] **HABITS** Singly or in pairs in canopy of lowland forest (locally up to 1300 m); occasionally lower. Often joins mixed-species flocks. **SIMILAR SPECIES** Black-winged Oriole lacks small white patch on edge of wing and has black (not greenish) central tail feathers. Immature Black-winged has dull black head, throat black streaked yellow and blackish bill. **STATUS AND DISTRIBUTION** Not uncommon to common resident in forest zone, from W Guinea and Sierra Leone to Benin (nominate) and from Nigeria to CAR and Congo (*laetior*).

SÃO TOMÉ ORIOLE *Oriolus crassirostris*   Plate 143 (10)
Loriot de Sao Tomé

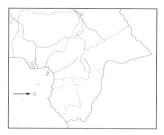

L 23–24 cm (9.0–9.5"). Distinctive oriole, with yellow pigment mostly lost. **Adult male** Head to upper breast black; upperparts greyish-olive separated from head by pale yellowish-white collar. Tail black edged and tipped yellow; central tail feathers greenish. Underparts whitish; undertail-coverts yellow. Bill red. **Adult female** Similar, but cheeks and throat appear dark greyish (due to whitish feather edgings). **Immature** Head more or less concolorous with upperparts; throat and upper breast streaked blackish. Bill dusky. **VOICE** Melodious fluting whistles, typically oriole-like in quality, e.g. *whuhoo tloo wleeiuw* and *whuloo-tloo* with variations. Also a drawn-out *heeeew* or *hoo-heeew* and a harsh *whrèèèh*. [CD13:33] **HABITS** Singly or in pairs in canopy and mid-levels of remote and undisturbed rainforest. **STATUS AND DISTRIBUTION** Common forest resident, endemic to São Tomé, where the only oriole. *TBW* category: THREATENED (Vulnerable).

EASTERN BLACK-HEADED ORIOLE *Oriolus larvatus*   Plate 124 (2)
Loriot masqué

Other name F: Loriot à tête noire oriental
L *c.* 21 cm (8.25"). *O. l. rolleti.* **Adult** Very similar to Western Black-headed Oriole but flight feathers, esp. secondaries, distinctly edged white; tertials edged olive-yellow. **Immature** Head to upper breast dull black flecked yellow; rest of underparts yellow narrowly streaked black on breast. Bill dusky. **VOICE** Short phrases of full, melodious whistles, incl. *wheeoo* and *whulEEoo* with variations. Also a short, clear *kulEEw*, often frequently repeated, and a harsh *whrreeaa* or *khwaarr*. [CD13:32] **HABITS** Singly or in pairs in open woodland. **SIMILAR SPECIES** Western Black-headed Oriole has most of flight feathers edged greyish; some primaries tipped with only a little white; visible part of tertials olive. Immature Western has olive head. Immature Black-winged Oriole similar but lacks wing patch and has black central tail feathers. **STATUS AND DISTRIBUTION** Claims from SW CAR (Haute Sangha and Lobaye) cannot be confirmed and probably result from misidentifications.

AFRICAN GOLDEN ORIOLE *Oriolus auratus*   Plate 124 (5)
Loriot doré

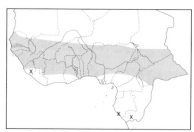

L *c.* 24 cm (9.5"). *O. a. auratus.* **Adult male** *Deep golden-yellow with black mask* (black of lores extending around and behind eye). Wing-coverts, secondaries and tertials black broadly edged and tipped golden-yellow; primaries and primary-coverts black with narrow yellow tips. Tail black tipped and edged yellow, except all-yellow outer feathers. Bill pink. Legs slate-grey. **Adult female** Duller. More olivaceous-yellow above; mask smaller and dusky; breast-sides variably but always very indistinctly streaked olive. **Immature** Yellowish-olive above, yellower on rump; no mask or just a trace of this. Throat to belly yellowish-white heavily streaked blackish; flanks and undertail-coverts golden-yellow. **VOICE** Various loud, melodious whistled phrases *wheetoliuw* and *tooleeoo* with longer variations; also a harsh mewing *whrèèèh* or *mwaaarr*. [CD13:28] **HABITS** In wooded savanna with large trees. Usually singly. Unobtrusive. **SIMILAR SPECIES** Eurasian Golden Oriole (both sexes and immature) lacks mask and yellow edges to wing feathers. **STATUS AND DISTRIBUTION** Locally common intra-African migrant in broad savanna belt, from Senegambia to Sierra Leone east to Chad and CAR. Rare, S Mauritania and Liberia. Also SW Gabon and S Congo (vagrant/rare). Southward movement in dry season; breeds during rains.

EURASIAN GOLDEN ORIOLE *Oriolus oriolus*   Plate 124 (4)
Loriot d'Europe

L *c.* 24 cm (9.5"). *O. o. oriolus.* **Adult male** *Golden-yellow with black lores, wings and tail.* Primary-coverts broadly tipped yellow, forming bright patch on wing-edge. Tail broadly tipped yellow; central pair narrowly so. Bill dark pink. Eye red. Legs slate-grey. **Adult female** Duller, yellow parts more olive, black more brownish; yellow tips to tail narrower; underparts narrowly streaked dusky. Streaks gradually disappear with age; older females (4 years

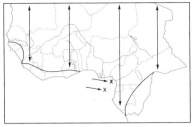

and older) may resemble males. **First-winter** Like adult female but greener, upperparts less yellowish; underparts whiter with extensive dark streaking and less yellow wash; primary-coverts have only narrow pale yellow fringe. Subsequent immature plumages often impossible to age and sex. **VOICE** A melodious, full, whistling *weehla-weeoo*; also a harsh note. Mostly silent in W Africa. [CD4:51] **HABITS** In wooded savanna. **SIMILAR SPECIES** Adult African Golden Oriole has richer yellow plumage, black lores extending back to form mask, and yellower wing. Immature has wing-coverts and tertials edged yellow. **STATUS AND DISTRIBUTION** Uncommon to rare Palearctic passage migrant and winter visitor (Sep–May/Jun) throughout. Vagrant, Bioko and Príncipe.

DRONGOS
Family DICRURIDAE (World: 23 species; Africa: 4; WA: 4)

Fairly small to medium-sized arboreal birds with black, usually glossy plumage and stout, slightly hooked bills. Eye red or orange in adults. Sexes similar. Juveniles less glossy, with shorter wings and tails. Mainly insectivorous. Hunt from perch, capturing insects on the wing like flycatchers. Occur in a variety of wooded habitats, from open savanna woodland and cultivation to rainforest. Conspicuous, bold, pugnacious and vocal. Voice consists of harsh, scolding notes interspersed with varied musical whistles. Main identification features are shape of tail, amount and distribution of gloss, and habitat.

SQUARE-TAILED DRONGO *Dicrurus ludwigii*   Plate 125 (4)
Drongo de Ludwig

L 19 cm (7.5"). *D. l. sharpei.* Smallest drongo. **Adult** All black slightly glossed purplish-blue with broad, *slightly notched, often almost square tail.* Inner webs of flight feathers blackish. Eye orange-red. **Juvenile** Duller. **VOICE** Varied. Call *whit whit.* Song consists of various series of short, explosive notes; also a characteristic, fast, melodious succession of call notes *whidididid* followed by a nasal, weaver-like *jeeeeezz;* a hard *rrrwee rrwee,* a rapid *chichichi* and *chi-rrrwee.* Also several rasping notes. [CD14:4b] **HABITS** Singly, in pairs or small family parties in second growth, forest edges, gallery forest and thickets in savanna. In Gabon also in primary forest, where mostly in canopy (Brosset & Erard 1986). Joins mixed-species flocks. **SIMILAR SPECIES** Shining Drongo is much glossier (but this may be hard to evaluate in shade of forest) and has longer, usually slightly more forked tail; voice different. **STATUS AND DISTRIBUTION** Not uncommon to scarce resident, from S Senegambia to N Liberia east to CAR and Congo.

SHINING DRONGO *Dicrurus atripennis*   Plate 125 (3)
Drongo de forêt

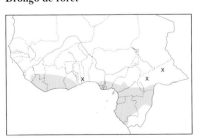

L *c.* 21 cm (8.5"). **Adult** All black with strong bluish-green gloss; tail notched or slightly forked (fork distinct in N Congo only). Eye dark red. **Juvenile** Duller. **VOICE** Varied. Short, rapid series of 4–6 explosive, ringing, discordant and harsh notes. Common phrase is fast *kwikwikwi-kwee-kwit* often preceded by a short rasping chatter. Also *whut-whut cheree* and *kzrr-kzrr twreet-twreet* and variations. Sometimes includes imitations, e.g. of Red-billed Helmet Shrike and Rufous Ant Thrush. [CD14:5a] **HABITS** Singly or in pairs, occurring exclusively in forest interior, where occupying mid-levels. Frequent and noisy member of mixed-species flocks. **SIMILAR SPECIES** Other drongos are less glossy (but beware: amount of gloss may be difficult to gauge in shade of forest). Square-tailed Drongo has even less forked and slightly shorter tail, and is smaller. Velvet-mantled Drongo has deeply forked tail, is larger and frequents upper levels. **STATUS AND DISTRIBUTION** Uncommon to common forest resident, from W Guinea and Sierra Leone to Togo and from S Nigeria to CAR and Congo.

FORK-TAILED DRONGO *Dicrurus adsimilis*   Plate 125 (1)
Drongo brillant

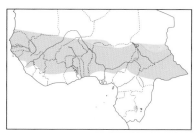

Other name E: Common Drongo
L 22.5–25.0 cm (9–10"). **Adult male** *divaricatus Wholly glossy blue-black* with broad, distinctly forked tail. *Inner webs of flight feathers ashy, giving pale, silvery appearance to wings in flight.* Eye red or brownish-red. Bill and legs black. In moult has double fork in tail. Southern *apivorus* similar, but has more deeply forked tail. **Adult female** Tail less deeply forked. **Juvenile** Dull black with feathers of upper- and underparts variably tipped greyish, forming crescentic bars. Eye brownish. **VOICE** Various vigorous, discordant, grating and twanging notes. Song a varied, loud medley of harsh, creaking, metallic notes and clear whistles. Sometimes includes imitations. [CD14:5b] **HABITS** Singly or in pairs in various types of woodland; not forest. Perches conspicuously and upright, hunting in short sallies, taking insects in flight or, occasionally, on ground, often returning to same lookout. Readily joins mixed-species flocks. Fearlessly mobs much larger birds, particularly raptors, owls and hornbills. **SIMILAR SPECIES** Northern Black Flycatcher occurs in similar habitat, but is a less conspicuous, silent bird with a more slender, rather square tail, slimmer bill and dark eye. **STATUS AND DISTRIBUTION** Resident almost throughout, outside forest. *D. a. divaricatus* from S Mauritania to Sierra Leone east to Chad and CAR (common); also Liberia (rare). *D. a. apivorus* SE Gabon and Congo (locally not uncommon to rare). **NOTE** *D. a. divaricatus* and *apivorus* often considered synonyms of nominate *adsimilis*.

VELVET-MANTLED DRONGO *Dicrurus modestus*   Plate 125 (2)
Drongo modeste

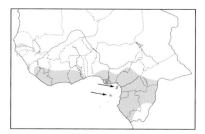

L 24–27 cm (9.5–10.5"). **Adult male** *coracinus* As Fork-tailed Drongo but has *unglossed velvet-black mantle* contrasting with rest of plumage, and more deeply forked 'fish-tail' with tip of outer feathers curved outwards. *Inner webs of flight feathers blackish.* Upper Guinea race *atactus* distinctly more glossed above. **Nominate** only slightly more glossy and usually larger. **Adult female** Tail less deeply forked. **Juvenile** *coracinus* Mainly dull black, almost entirely lacking greyish barring on upper- and underparts typical of juvenile Fork-tailed Drongo. Eye brownish. In **nominate**, barring more whitish and disappearing last from undertail-coverts, where initially quite distinct. **VOICE** As Fork-tailed Drongo. Songs of *coracinus* include e.g. vigorous *wheet-wheet-tiuw-rrweet wheet-wheet-tiuw-rrweet-tweet-tiuw* and *whut-tit-tew*, and variations. Commonest song of nominate a series of soft, melodious whistles *whee, hiu hiu hiuu*; also a rhythmic *tiu-wee-tiuh tiu-wee-tiuh tiu-wee-tiuh tiu-weeeh*. Calls include various harsh, disharmonic and nasal sounds. [CD14:6a] **HABITS** Singly or in pairs in forest along tracks, small clearings and edges, avoiding interior. Usually high, on bare branches below canopy. Behaviour much as Fork-tailed Drongo. **SIMILAR SPECIES** Fork-tailed Drongo (q.v. and above). Shining Drongo has more glossy plumage with only slightly forked tail and occurs at mid-levels within forest interior. **STATUS AND DISTRIBUTION** Common forest resident. *D. m. atactus* from Sierra Leone to Ghana, and possibly SW Nigeria; *coracinus* S Nigeria to Congo, also Bioko; *modestus* Príncipe. **NOTES (1)** Often treated as conspecific with Fork-tailed Drongo, *atactus* being considered intermediate between *coracinus* and *adsimilis/divaricatus*. When Velvet-mantled split from Fork-tailed, *atactus* often treated as race of latter. **(2)** Nominate sometimes considered a separate species, Príncipe Drongo. *TBW* category: Near Threatened.

CROWS
Family CORVIDAE (World: 112–117 species; Africa: 11; WA: 5)

Medium-sized to large birds (family includes largest of the passerines), with generally stout bills and strong legs and feet. Those of the Afrotropical region mostly belong to the genus *Corvus* (typical crows) and are either all black, including bill and legs, or black with some white, brown or grey. Sexes similar. Juveniles similar to adults. Omnivorous; foraging mainly on ground. Frequent various open habitats; avoid forest. Calls mostly loud and harsh. Flight strong.

BROWN-NECKED RAVEN *Corvus ruficollis*   Plate 125 (8)
Corbeau brun

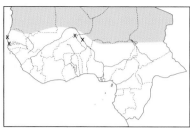

L 52–56 cm (20.5–22.0"). **Adult** Head, throat and breast brownish-black with crown, head-sides and nape browner (often difficult to see in field). Rest of plumage black with some bluish gloss. Tail wedge shaped with central feathers often slightly protruding. With wear head and body become browner, contrasting with blacker wings and tail. **Juvenile** Duller. **VOICE** A harsh *karr* or *aarg.* [CD4:65] **HABITS** Singly or in pairs in desert and semi-desert. At good food sources (e.g. rubbish dumps), sometimes in small flocks. **STATUS AND DISTRIBUTION** Resident in arid zone, from Mauritania (common) through Mali (not uncommon), N Burkina Faso, Niger and N Nigeria (uncommon) to Chad. Also N Senegal (scarce), Gambia (rare) and Cape Verde Is (common throughout).

PIED CROW *Corvus albus*   Plate 125 (6)
Corbeau pie

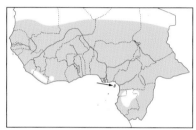

L 46–50 cm (18–20"). Unmistakable. **Adult** Large, robust, black-and-white crow. White of lower breast to upper belly extends as broad collar on base of hindneck. Black areas glossed blue and purple. **Juvenile** Duller; often with some dusky mottling within white. **VOICE** A harsh *kwaar,* a deep, guttural croak *kaarrh* or *kroh* and a hard *klok*; also various other sounds. [CD14:7a] **HABITS** Usually in pairs or small groups, but sometimes in large flocks. In all habitats except closed forest; typically near habitation and cultivation. Largely scavenges, foraging mainly on ground. Roosts communally. Flight powerful; often soars. Parasitised by Great Spotted Cuckoo. **STATUS AND DISTRIBUTION** Common throughout, except northernmost, very arid areas and rainforest. Mainly resident, but northward movements recorded during rains in north of range.

FAN-TAILED RAVEN *Corvus rhipidurus*   Plate 125 (7)
Corbeau à queue courte

L 47 cm (18.5"). Unmistakable. **Adult** All glossy blue-black with *very short tail.* At rest, primaries project well beyond tail. In flight, stumpy tail and broad wings impart distinctive, bat-like appearance. **Juvenile** Duller. **VOICE** A high-pitched *kraah-kraah,* a guttural *errrrow,* a hollow *wok!* and various other croaking, nasal and clucking sounds. [CD14:8a] **HABITS** Singly, in pairs or in flocks in desert, often near cliffs or crags. Mainly a scavenger, foraging on ground; often near habitation, in oases or at rubbish dumps. Frequently soars on thermals. **STATUS AND DISTRIBUTION** Scarce and local resident in Sahara, in N Mali, N Niger (Aïr) and Chad.

WESTERN JACKDAW *Corvus monedula*   Plate 147 (10)
Choucas des tours

L 33 cm (13"). Smallest typical crow recorded in the region. Black with grey sides and rear of head; whitish eye. Wingbeats faster and bill shorter than other crows. **VOICE** A diagnostic *kya* or *chak.* [CD4:62] **STATUS AND DISTRIBUTION** Palearctic vagrant (presumably *C. m. spermologus*), NW Mauritania (Nouadhibou: two, Jan–Oct 1985; five, Feb 1986).

PIAPIAC *Ptilostomus afer*   Plate 125 (5)
Piapiac africain

Other names E: Black Magpie; F: Piac-piac

L 35 cm (14"); tail up to 28 cm (11"). **Adult** Slender and black with very long, stiff, steeply graduated, brownish tail and drab brown primaries. Black areas have slight bluish gloss. In flight, greyish webs to flight feathers give ashy appearance to underwings. Eye purple to reddish. Bill black. **Juvenile/immature** Bill pinkish with black tip (remaining so for *c.* 1 year). **VOICE** Shrill, squeaking and scolding calls. [CD14:6b] **HABITS** Usually in small, noisy flocks (except when breeding). Inhabits wooded savanna, where partial to *Borassus* and other palms. Forages

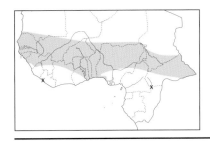

mainly on ground, running and hopping with great agility. Not shy; often common in or near villages. Frequently associates with domestic stock, occasionally perching on their backs and capturing insects disturbed by them. **SIMILAR SPECIES** Long-tailed Glossy Starling is much more glossy overall with supple, glossy purple-blue tail, slimmer bill and pale eye. Green Wood-hoopoe has white in wings and tail. **STATUS AND DISTRIBUTION** Locally common to uncommon resident in Sahel and savanna belts, from S Mauritania to Sierra Leone east through Mali and N Ivory Coast to N Cameroon, S Chad and CAR.

STARLINGS
Family STURNIDAE (World: *c.* 115 species; Africa: 50; WA: 20)

Small to medium-sized, mainly arboreal birds with strong, pointed bills and sturdy legs. Many are blackish, often with iridescent plumage. Sexes similar in many but markedly different in others. Juveniles a dull version of adult or different. Occur in various habitats, from dry country to rainforest. Omnivorous, most feeding on fruit and insects. Most forage in flocks, roost communally and nest in holes. Calls mostly harsh and grating but also including pleasing sounds.

In W Africa, seven genera are represented, two of which have more than one species.

Chestnut-winged starlings (*Onychognathus*, 3 species) are black with chestnut in flight feathers (diagnostic). Tail length and habitat important identification criteria. Sexually dimorphic; females with variable amount of grey on head. Juveniles like dull adult males.

Glossy starlings (*Lamprotornis*, 12 species) are mainly blackish strongly glossed blue, green and purple. Colour of brilliant reflections may change with light. Some species difficult to separate. Identification criteria include amount, distribution and colour of gloss (hard to see in bad light), colour of eye and habitat. Sexes similar. Juveniles much less glossy. Gregarious except when breeding. Endemic to Africa.

NARROW-TAILED STARLING *Poeoptera lugubris*   Plate 126 (1)
Rufipenne à queue étroite

Other name F: Étourneau à queue étroite
L 20–23 cm (8–9”), incl. tail of up to 12 cm (4.75”). *Slender with long, narrow, steeply graduated tail.* **Adult male** Glossy purple-black (appearing black in field) with dark brownish flight feathers. Edges of secondaries paler, becoming whitish-brown with wear. Eye yellow. **Adult female** Much greyer; upperparts have greenish-blue gloss; underparts dull grey. Basal part of inner webs of primaries chestnut, forming wing patch which is, at close range, visible in flight. **Juvenile** Similar to adult female; tail shorter. **VOICE** Usually silent. A clear, whistled *wheew* and *peeew*, in flight a medley of shrill, cheeping notes.

[CD14:9] **HABITS** Usually in small flocks in clearings and at edge of lowland forest. Frequents canopy of high trees. Flight fast and direct, in close formation. Associated with colonies of *Gymnobucco* barbets. **STATUS AND DISTRIBUTION** Not uncommon to scarce and rather local resident in forest zone, from Sierra Leone to Ghana, in Togo (rare), and from Nigeria to SW CAR and Congo. Also a single record from Bioko (1904).

WALLER'S STARLING *Onychognathus walleri*   Plate 126 (4)
Rufipenne de Waller

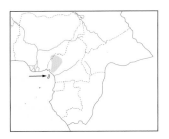

Other names E: Waller's Chestnut-winged Starling; F: Étourneau à bec court
L 23 cm (9”). *O. w. preussi.* Highland-forest starling, similar to Forest Chestnut-winged Starling but with shorter and only slightly graduated tail. **Adult male** Glossy purple-black with metallic-green reflections on head and throat. Dark chestnut on primaries forms conspicuous wing patch in flight. Eye red. **Adult female** Head and throat have some ash-grey streaking. Less easily visible than in Neumann's and Forest Chestnut-winged Starlings. **Juvenile** As adult male but duller. **VOICE** Clear, loud whistles *teeeuw-tee-wheew* (last syllable often dropped), with variations. Other calls include a melodious *preeti preeti*, clear *tewee tewee* and *wheet*, and a soft nasal note. Alarm a rather harsh *chrrra*.

[CD14:11] **HABITS** In small flocks of up to 15 in montane forest; in pairs when breeding. Frequents canopy of fruiting trees in mature forest, small forest patches and gallery extensions. **STATUS AND DISTRIBUTION** Common forest resident in highlands of SE Nigeria (Obudu Plateau), W Cameroon (Mt Cameroon, Rumpi Hills, Mt Kupe, Bakossi Mts, Bamenda Highlands) and Bioko.

FOREST CHESTNUT-WINGED STARLING *Onychognathus fulgidus*   Plate 126 (2)
Rufipenne de forêt

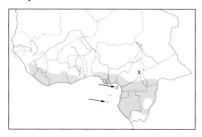

Other names E: Chestnut-winged Starling; F: Étourneau roupenne L 28–33 cm (11–13"). Blackish with long, strongly graduated tail and, in flight, conspicuous chestnut wing patch. **Adult male *hartlaubii*** Glossy purple-black with metallic-green reflections on head, throat and wings. Basal part of primaries dark chestnut (concealed when perched). Eye bright red. *O. f. intermedius* has reflections on head more bluish. **Nominate** similar but usually substantially larger. **Adult female** Head and throat streaked ash-grey. **Juvenile** As adult male but duller. **VOICE** Various resonant, rasping and melodious whistles. [CD14:13] **HABITS** In pairs or small flocks (usually 4–8, rarely up to 20) in canopy of mature forest, clearings and at edges. Restive, feeding actively on fruit, but rather inconspicuous and quiet. Joins glossy starlings in search of fruiting trees. **SIMILAR SPECIES** Neumann's Starling has larger and paler chestnut wing patch, broader tail, and occurs on crags. **STATUS AND DISTRIBUTION** Not uncommon resident in forest zone. *O. f. hartlaubii* from Sierra Leone and SE Guinea to Ghana, in Togo (rare), from Nigeria to S CAR, and Bioko; *intermedius* Gabon and Congo; nominate São Tomé.

NEUMANN'S STARLING *Onychognathus neumanni*   Plate 126 (3)
Rufipenne de Neumann

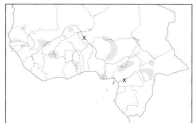

Other names E: Crag Chestnut-winged Starling; Red-winged Starling L 28–33 cm (11–13"). Large, with long graduated tail and black-tipped chestnut primaries forming conspicuous wing patch in flight. Associated with rocky habitats. **Adult male *neumanni*** Glossy purplish-black with more greenish gloss on wings and tail. Eye reddish. Western ***modicus*** has relatively shorter tail; eye described as brownish. **Adult female** Head, throat and upper breast ash-grey streaked blue-black. **Juvenile** Similar to adult male but duller. **VOICE** Various clear loud whistles, e.g. *twee-lee-uw*, *whutcheerrleeo* and *peeeo* or *wheo-peetoo* etc., melodious and oriole-like; uttered from perch and in flight.
[CD14:12] **HABITS** In pairs or flocks on rocky outcrops, inselbergs, crags and cliffs. Vocal, restless and conspicuous. Flight fast and direct. **SIMILAR SPECIES** Forest Chestnut-winged Starling occurs in forest, has smaller and darker wing patch, narrower tail and more metallic green on head. **STATUS AND DISTRIBUTION** Locally not uncommon to scarce resident, recorded from Mauritania, Senegal (esp. SE; scarce), Guinea, Mali, N Ivory Coast (scarce), SW Burkina Faso, Niger, Nigeria, Cameroon, Chad and CAR. *O. n. modicus* from Ivory Coast and W Mali (upper Niger R.) west; Nominate from E Mali east. **NOTE** Formerly included with E and S African forms in Red-winged Starling *O. morio* (Rufipenne morio/Étourneau roupenne d'Alexander) but considered specifically distinct by Feare & Craig (1998) and *BoA* on the basis of differences including bill shape and presence of feathers covering nostrils.

COPPER-TAILED GLOSSY STARLING *Lamprotornis cupreocauda*   Plate 126 (5)
Choucador à queue bronzée

Other name F: Merle métallique à dos bleu
L *c.* 20 cm (8"). Rather small and dark forest starling. **Adult** Head to upper breast mainly glossy purple with unglossed, velvety black forehead. Rest of plumage glossy blue-black; tail blackish slightly glossed bronzy. Eye yellow. **Juvenile** Duller; eye initially dark. **VOICE** Various harsh, squeaky notes. [CD14:15] **HABITS** In pairs or small groups in canopy of lowland rainforest. Occasionally joins mixed-species flocks. **SIMILAR SPECIES** Purple-headed Glossy Starling has mantle, back and lower underparts glossy metallic-green, and dark eye. **STATUS AND DISTRIBUTION** Not uncommon to locally common forest resident, from Sierra Leone and SE Guinea to Ghana. Rare, Togo. *TBW* category: Near Threatened. **NOTE** Placed in genus *Hylopsar* by Feare & Craig (1998) on basis of structure of melanin granules in feathers.

PURPLE-HEADED GLOSSY STARLING *Lamprotornis purpureiceps*   Plate 126 (6)
Choucador à tête pourprée

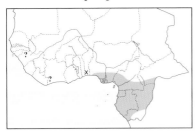

Other name F: Merle métallique à tête pourprée
L *c.* 20 cm (8"). **Adult** Head to upper breast glossy purple with unglossed, velvety purple crown. Rest of plumage glossy metallic blue-green with contrasting glossy blue wings. Tail blackish slightly glossed bronzy. *Eye brown.* **Juvenile** Much duller; underparts brownish-black with some gloss. **VOICE** Rather silent. Various chattering and whistling notes. In flight, a characteristic, metallic *pleep!* or *twink!* [CD14:16] **HABITS** Usually in small, wandering flocks in canopy of lowland forest, occasionally lower. In breeding season in pairs or singly. Associates with other frugivores (e.g. hornbills, starlings) at fruiting trees. Occasionally joins mixed flocks of insectivores. **SIMILAR SPECIES** Copper-tailed Glossy Starling has green areas replaced by glossy steel-blue, yellow eye (dark in juvenile) and occurs in Upper Guinea forest. **STATUS AND DISTRIBUTION** Uncommon to locally common forest resident, from S Benin to SW CAR and Congo. Claims from Guinea and W Ivory Coast require confirmation. **NOTE** Placed in genus *Hylopsar* by Feare & Craig (1998) on basis of structure of melanin granules in feathers.

PURPLE GLOSSY STARLING *Lamprotornis purpureus*   Plate 126 (9)
Choucador pourpré

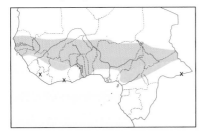

Other name F: Merle métallique pourpré
L *c.* 24 cm (9.5"). Large, with noticeably *longer bill and shorter tail* than other glossy starlings, producing distinctive jizz. **Adult** *purpureus Head, underparts and tail glossy purple* becoming metallic blue on nape and rump, *contrasting with iridescent blue-green upperparts.* Small velvety black spots on tips of wing-coverts and tertials. *Conspicuously large, yellow eye.* In flight, broad wings and short tail impart rather compact appearance. *L. p. amethystinus* described as slightly bluer, with somewhat longer tail. **Juvenile** Head and underparts dark sooty-brown with metallic blue feathers; upperparts and tail as adult but duller. Eye initially pale greenish, quickly becoming yellow. **VOICE** Various harsh, chattering and whistling notes. [CD14:17] **HABITS** In vocal flocks (of 10–100) in wooded savanna and thorn scrub. In pairs when breeding. Feeds on various fruits, particularly figs. Frequently walks on ground. **SIMILAR SPECIES** Other savanna starlings have purple on head and breast replaced by glossy metallic green. **STATUS AND DISTRIBUTION** Common to locally uncommon resident in savanna belt, from Senegambia and Guinea to Cameroon (nominate) east to Chad and CAR (*amethystinus*). Rare/vagrant, S Mauritania and Liberia.

CAPE GLOSSY STARLING *Lamprotornis nitens*   Plate 126 (10)
Choucador à épaulettes rouges

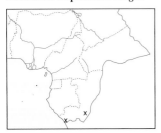

Other name F: Merle métallique à épaulettes rouge
L *c.* 23 cm (9"). *L. n. nitens.* Very uniformly coloured. **Adult** Glossy blue-green or greenish-blue with bronzy lesser coverts (forming small, narrow and rather inconspicuous shoulder patch), and purple-blue rump and tail. Velvety black spots on tips of median and greater coverts and tertials. Eye orange-yellow. **Juvenile** Duller; shoulder patch mainly blue; eye greyish. **VOICE** A pleasant, slurred, rolling *churweelee.* Song a sustained series of varied syllables, incorporating call. [CD14:18] **HABITS** As other glossy starlings. Inhabits savanna. **SIMILAR SPECIES** No other savanna starling occurs in its range. Purple-headed and Splendid Glossy Starlings are forest species that may frequent gallery forest and forest patches in cleared areas; both have more contrasting plumage; Purple-headed also smaller, with dark eye; Splendid much larger, with different voice. **STATUS AND DISTRIBUTION** A mainly S African species, reaching its northernmost limits in S Congo (2 records) and possibly Gabon.

BRONZE-TAILED GLOSSY STARLING *Lamprotornis chalcurus*   Plate 127 (6)
Choucador à queue violette

Other name F: Merle métallique à queue violette
L *c.* 21.5 cm (8.5"). **Adult** *chalcurus* Resembles Greater Blue-eared Starling but has *noticeably shorter, mainly purple*

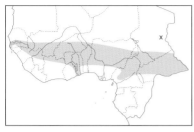

(not blue-green) *tail*, contrasting with iridescent blue-green upperparts. Ear-coverts and rump purple-blue. Tail may appear bronzy at some angles. *Eye orange-yellow to reddish-orange. L. c. emini* has rump more violet-blue; tail slightly longer. **Juvenile** Head and underparts dark sooty-brown with some iridescent green feathers; upperparts and tail as adult but duller; eye dark. **VOICE** A whining, drawn-out *weeaah*, an abrupt *plip!* and various other nasal, chattering and whistling notes. [CD14:19] **HABITS** As Lesser Blue-eared Starling and often together. In flight, wings make loud swishing sound. **SIMILAR SPECIES** Greater Blue-eared Starling (see above). Lesser Blue-eared Starling has blue-green tail and rump. Juvenile Greater Blue-eared differs from juvenile Bronze-tailed principally in longer tail; juvenile Lesser Blue-eared paler and browner below. **STATUS AND DISTRIBUTION** Locally common to uncommon resident in savanna belt, from Senegambia and Guinea through Mali and N Ivory Coast to N Cameroon (nominate) and from E Cameroon to S Chad and CAR (*emini*). Reaches coast in Ghana and W Nigeria.

GREATER BLUE-EARED STARLING *Lamprotornis chalybaeus*   Plate 127 (7)
Choucador à oreillons bleus

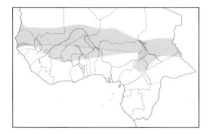

Other name F: Merle métallique à oreillons bleus
L 21–24 cm (8.25–9.5"). *L. c. chalybaeus*. **Adult male** Glossy metallic green with purple belly and more or less contrasting mask formed by black lores and bluish ear-coverts; rump bluish. Small velvety black spots at tips of wing-coverts (forming 2 rows) and tertials. Eye yellow (occasionally orange). **Adult female** Usually slightly smaller. **Juvenile** Head, upperparts and tail duller; underparts dark sooty-brown with some iridescent feathers; eye dark. **VOICE** A nasal, whining *wèèh-aa-ah* or *skwee-aar*. Song a varied series of whistling, mewing, chirping and clicking notes, including nasal call. [CD14:20] **HABITS** Usually in flocks in thorn scrub with trees to desert edge. Forages on ground and in fruiting trees. **SIMILAR SPECIES** Slightly smaller Lesser Blue-eared Starling (q.v.) is very similar and best distinguished by clear (not nasal) call. Juvenile noticeably paler below than juvenile Greater Blue-eared; presence of juvenile therefore aids identification. Lesser Blue-eared also has a more southerly range, although with some overlap and the two may occur together. Bronze-tailed Glossy Starling has similar size but distinctly shorter, largely purple tail and, typically, redder eye. **STATUS AND DISTRIBUTION** Common to not uncommon in semi-arid belt, from S Mauritania, Senegambia and Guinea east through Mali and N Ivory Coast (rare) to N Cameroon and Chad. Moves north with rains.

LESSER BLUE-EARED STARLING *Lamprotornis chloropterus*   Plate 127 (9)
Choucador de Swainson

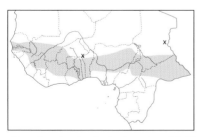

Other name F: Merle métallique de Swainson
L 19–20 cm (7.5–8.0"). *L. c. chloropterus*. **Adult** Strongly resembles Greater Blue-eared Starling but *ear-coverts blue-black, forming more contrasting and narrower mask*; tail relatively short; rump uniform with rest of upperparts; size smaller (noticeable when seen together). Eye yellow to yellow-orange. **Juvenile** Head and underparts *greyish fawn-brown* (almost 'café au lait') with some iridescent feathers; ear-coverts distinctly darker; upperparts and tail as adult but duller; eye initially dark. **VOICE** In flight, a clear, distinctive *wirree-wirree*, from perch various rather similar calls, incl. *wreet* and *cherwee*. Song consists of a rather simple, rhythmic series of 6–12 separated, chirruping and whistling syllables, such as *chirp chirp peelu chirp chrew whip* with variations. [CD14:21] **HABITS** In flocks in savanna woodland; typically more gregarious than Greater Blue-eared. Forages in fruiting trees and on ground. Occasionally associates with Bronze-tailed Glossy Starling. **SIMILAR SPECIES** Slightly larger Greater Blue-eared Starling is very similar and best distinguished by nasal, mewing call; juvenile darker, more blackish than juvenile Lesser Blue-eared; overall range also more northerly, though some overlap. Bronze-tailed Glossy Starling is also slightly larger and has shorter, purplish (not metallic green) tail; eye usually more orangey or reddish. **STATUS AND DISTRIBUTION** Common to not uncommon resident in savanna belt, from Senegambia and Guinea through Mali and N Ivory Coast to N Cameroon, Chad and CAR. Reaches coast in Ghana and Togo. Northward movements with rains reported.

PRÍNCIPE GLOSSY STARLING *Lamprotornis ornatus*
Choucador de Principé

Plate 143 (7)

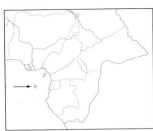

Other name F: Merle métallique de Principé
L *c.* 29 cm (11.5"). **Adult** Large, highly glossed starling with mainly bronzy plumage (at certain angles appearing almost black in field). Head and mantle glossy blue-green becoming bronze on back, scapulars and wings; rump and uppertail-coverts blue. Tail purple-bronze tipped glossy blue. *Underparts bronzy-green* with purplish-blue undertail-coverts. Eye whitish. **Juvenile** Duller, underparts greyish-brown; eye dark. **VOICE** Various nasal, twanging, wheezing and whistling notes. [CD14:23] **HABITS** In small, restless, noisy flocks in plantations, cultivation, forest regrowth and primary forest. Usually high in trees. In flight, wings make loud swishing sound. **SIMILAR SPECIES** Splendid Glossy Starling, the only other starling on Principe, has purplish (not bronzy-green) underparts that may appear two toned at certain angles (throat and breast bluish, rest of underparts purplish). **STATUS AND DISTRIBUTION** Endemic to Principe, where one of the commonest birds.

SPLENDID GLOSSY STARLING *Lamprotornis splendidus*
Choucador splendide

Plate 126 (7)

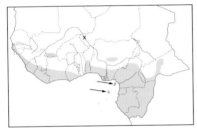

Other name F: Merle métallique à œil blanc
L 27–30 cm (10.5–12.0"). *Large, highly glossed, noisy forest starling.* **Adult** *splendidus* Head and upperparts iridescent green with metallic-blue and purple reflections; forehead and lores velvety black; small coppery patch on neck-sides. Wings have *black band on secondaries* and small subterminal black spots on coverts. Tail metallic-blue and purple with *broad blackish band* washed purple. Underparts glossy bluish-purple. *Eye creamy-white.* Western **chrysonotis** has greener crown and bluer throat; island race **lessoni** slightly larger. **Juvenile** Much duller; underparts blackish-brown with slight gloss; eye dark grey.
VOICE A variety of striking, harsh, piping and rasping calls with a nasal and metallic quality, e.g. *niyar-èh; spiyok!; kwank!*; an explosive *KYAH!* Makes considerable noise at roost. [CD14:24] **HABITS** In flocks within canopy of forest, gallery forest and forest patches in savanna. Large numbers may congregate at fruiting trees and roosts. In flight, wings produce loud swishing noise. **SIMILAR SPECIES** On Principe, compare Principe Glossy Starling. **STATUS AND DISTRIBUTION** Common to not uncommon in forest zone and, locally, outliers in savanna. *L. s. chrysonotis* from Togo west to Senegambia; nominate east to CAR and Congo. Also on Bioko (*lessoni*) and Principe (presumably nominate). Local movements reported.

LONG-TAILED GLOSSY STARLING *Lamprotornis caudatus*
Choucador à longue queue

Plate 127 (8)

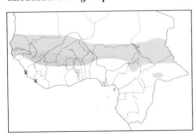

Other names E: Northern Long-tailed Starling; F: Merle métallique à longue queue
L *c.* 51 cm (21"), incl. tail of up to 33 cm (13"). Readily identified by *very long, strongly graduated, supple tail.* **Adult male** Head iridescent bronzy-green; rest of plumage glossy bluish-green with purplish rump, tail and belly. Black subterminal spots to wing-coverts and scapulars. Eye cream-white. **Adult female** Usually smaller, with shorter tail. **Juvenile** As adult but duller. **VOICE** Various shrill and harsh notes. [CD14:25] **HABITS** In small, lively and vocal flocks in open wooded savanna, farmland and thorn scrub. Often forages on ground. Flight slow and laborious on short rounded wings. **SIMILAR SPECIES** Piapiac also has long, but stiff (not supple) tail; plumage only slightly glossed; eye dark. **STATUS AND DISTRIBUTION** Not uncommon to common in semi-arid belt, from S Mauritania, Senegambia and Guinea to Chad and CAR. Moves north with rains.

EMERALD STARLING *Lamprotornis iris*
Choucador iris

Plate 127 (5)

Other name F: Merle métallique vert
L *c.* 20 cm (8"). **Adult** Small, distinctive starling with *bright glossy emerald-green plumage*, except glossy purple band from ear-coverts to neck-sides and glossy purple flanks and belly, with unglossed blackish centre. Eye dark brown.

Juvenile Duller; belly and flanks grey-brown with some purple feathers. VOICE Various squeaky notes. A drawn-out *wheeze* on take-off. HABITS Typically in small parties in wooded savanna where keeping to tops of tall, often dead, trees. Also forages on ground. Reported to feed more on insects, particularly ants, than other glossy starlings. SIMILAR SPECIES No other glossy starling has similar brilliant emerald plumage. STATUS AND DISTRIBUTION Scarce and local resident in narrow savanna belt of Guinea, Sierra Leone and Ivory Coast. Vagrant, SW Mali. *TBW* category: Data Deficient. NOTE Sometimes placed in genus *Coccycolius*.

CHESTNUT-BELLIED STARLING *Lamprotornis pulcher*   Plate 127 (4)
Choucador à ventre roux

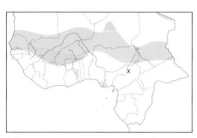

Other name F: Merle métallique à ventre roux
L 19 cm (7.5"). Rather small, squat, dingy-looking starling with distinctive *rufous-chestnut belly and undertail-coverts*. Adult Head dull greyish-brown; upperparts, throat and breast glossy green; tail short, glossy blue. Eye creamy-white. In flight, creamy inner webs of primaries produce *conspicuous pale wing patch*. Juvenile Duller, with ashy-brown throat and breast; bill pale. Eye dark. VOICE A vigorous *whirr*. Song a series of soft liquid notes. [CD14:27] HABITS In small flocks (except when breeding) in dry bushy country and farmland, often near villages. Feeds mainly on ground. SIMILAR SPECIES All congeners are much more glossy; none has similar green or chestnut colour in plumage. STATUS AND DISTRIBUTION Locally common to uncommon in semi-arid belt, from S Mauritania, Senegambia and Guinea through Mali and N Ivory Coast (rare) to Niger, N Cameroon and Chad. Moves north with rains. NOTE Previously placed in genus *Spreo*.

WHITE-COLLARED STARLING *Grafisia torquata*   Plate 126 (8)
Rufipenne à cou blanc

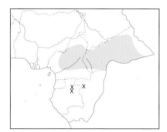

Other name F: Étourneau à poitrine blanche
L 21.5–23.0 cm (8.5–9.0"). Adult male Readily identified by wholly purple-black plumage with *broad white breast-band*. Eye orange-yellow. In flight, longish tail apparent. Adult female Head and upperparts greyish-brown slightly glossed purple. Underparts paler, dull greyish with paler tips to feathers producing faint scaly pattern (only visible at close range). Eye yellow. Juvenile As adult female but duller. Immature male As adult male but breast-band only indicated by white tips to purple-black feathers. VOICE Reminiscent of glossy starling *Lamprotornis*. Calls include chirruping notes and 3 short whistles. HABITS In small flocks in montane grasslands, forest–savanna mosaic and savannas just north of forest. Usually keeps to larger trees, actively feeding on figs and berries. Occasionally joins other starlings in search of fruit. STATUS AND DISTRIBUTION Uncommon to locally common resident (breeding visitor?), Cameroon (Bamenda Highlands, Adamawa Plateau) and CAR. Vagrant, Gabon and N Congo.

VIOLET-BACKED STARLING *Cinnyricinclus leucogaster*   Plate 127 (3)
Spréo améthyste

Other names E: Amethyst Starling, Plum-coloured Starling; F: Merle améthyste
L 16–18 cm (6.25–7.0"). Small, distinctive starling. Sexes markedly different. Adult male *leucogaster* Unmistakable. *Brilliant iridescent violet with contrasting pure white breast, belly and undertail-coverts*. Eye dark brown with conspicuous pale yellow rim. *C. l. verreauxi* has white base to outer web of outermost tail feather. Adult female Head, upperparts and tail brown with pale rufous-brown edges producing streaky and scale-like appearance to head and upperparts; *underparts white heavily streaked dark brown*. Juvenile Similar to adult female. Male gradually acquires adult plumage. VOICE A single, clear, high-pitched, drawn-out whistle *vfeeeee*. Also a soft nasal call, usually uttered on take-off. Song a short rapid series of varied whistling, twanging and clicking notes, some

reminiscent of drongos. [CD14:29] **HABITS** Usually in flocks, except when breeding; flocks often consist solely of females and young, or males. In various wooded habitats, from savanna woodland to forest clearings; avoids closed forest. Arboreal, mainly foraging in trees and often perching in treetops. Flight fast and direct. **STATUS AND DISTRIBUTION** Seasonally common almost throughout, except arid north, with irregular occurrence in some areas. Nominate throughout; *verreauxi* in Lower Congo area. Migratory; moves north with rains.

WATTLED STARLING *Creatophora cinerea* — Plate 127 (2)
Étourneau caronculé

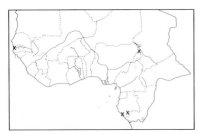

L 21.5 cm (8.5"). Pale, drab-grey starling with black wings and tail and *contrasting whitish rump* (conspicuous in flight). **Adult male non-breeding** Bare black malar stripe; greater coverts whitish; primary-coverts white. Bill pale pinkish. **Adult male breeding** Head loses feathers, exposing black skin and black wattles on forehead and throat, which contrast with bare bright yellow area on ear-coverts and hindcrown. **Adult female** As male non-breeding, but greater and primary-coverts blackish. Worn wings all brownish. **Juvenile** Similar to adult female but browner; bill dusky or brownish. **VOICE** Usually silent when not breeding. Song a rather soft jumble of high-pitched squeaky notes. Flocks often utter continuous creaking, gurgling, rasping and clicking chatter. [CD14:30] **HABITS** Highly nomadic and gregarious; attracted to locusts. Usually in dry grassy savanna; occasionally in more wooded habitat. **STATUS AND DISTRIBUTION** E and S African vagrant migrant. First recorded, Gambia (Dec 1975–Jan 1976). Since 1990 annually N Cameroon (Nov–Apr). Also S Gabon and S Congo (Jul, Oct).

COMMON STARLING *Sturnus vulgaris* — Plate 147 (11)
Étourneau sansonnet

Other name E: European Starling
L 19–22 cm (7.5–8.5"). *S. v. vulgaris.* **Adult non-breeding** Mainly glossy black, spotted buff above and white below. Bill long and pointed, brownish. In flight, characteristic silhouette with triangular wings and short, square tail. **VOICE** A grating *cheerr*; also various other harsh whistling sounds. [CD4:67] **HABITS** Gait upright and confident. Flight fast and direct; also glides. **STATUS AND DISTRIBUTION** Palearctic vagrant, N Mauritania (2 records, Nouadhibou area, Jan–Dec) and Cape Verde Is (2 records, Oct and Mar).

OXPECKERS
Family BUPHAGIDAE (World: 2 species; Africa: 2; WA: 1)

Medium-sized, mainly brown birds with moderately long, stiff tails, short legs and stout bills. Highly specialised, using their strong feet with sharp claws for clinging to herbivores, where they feed on parasites and wound tissue. Sexes similar. Juveniles duller with dusky bill. Usually forage in small flocks; roost communally and nest in holes. Calls mostly harsh. Formerly included within Sturnidae. Endemic to Africa.

YELLOW-BILLED OXPECKER *Buphagus africanus* — Plate 127 (1)
Piquebœuf à bec jaune

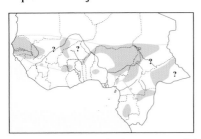

L 21–23 cm (8.25–9.0"). Slender, dark brown birds associated with large herbivores. **Adult *africanus*** Head to throat, upperparts and tail earth-brown becoming paler brownish-buff on rump and rest of underparts. *Bill bright yellow broadly tipped red.* Eye orange. In flight, *contrasting pale rump and uppertail-coverts*; characteristic silhouette with rather long and pointed wings and tail. ***B. a. langi*** slightly darker with greyer, less contrasting rump. **Juvenile** Bill dusky; eye brown. **VOICE** A hard rasping and hissing, metallic *krriz* or *pszrr* uttered in flight or when perched in tree; also a rattling *kzsriririri...* [CD14:31] **HABITS** Usually in small groups, in wooded savanna, grassland with trees, bush country and, locally, small openings within forest. Clambers on wild or domestic herbivores, esp. buffalo and cattle, searching for ticks (hence name 'tickbird') and other parasites, using stiff tail as prop. Perches in upright position on branches, with bill pointing skyward. **STATUS AND DISTRIBUTION**

Uncommon to not uncommon but local resident, mainly outside forest zone. Nominate from S Mauritania and Senegambia to Chad and CAR; *B. a. langi* Gabon and Congo. In certain areas where associates with cattle, numbers have decreased dramatically through improved veterinary care.

SPARROWS
Family PASSERIDAE (World: *c.* 35 species; Africa: 16–18; WA: 10–11)

Small, mostly brown and grey birds with short, conical bills. Sexes similar or markedly different. Juveniles similar to adult females. Bill of male typically changes from horn coloured to black in breeding season. Occur in a variety of mainly open habitats; also towns. Forage on or near ground, largely on seeds; also insects. Many species gregarious. Vocal, uttering variety of rather harsh chirps.

HOUSE SPARROW *Passer domesticus*   Plate 128 (1)
Moineau domestique

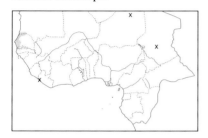

L 14–15 cm (5.5–6.0"). **Adult male *domesticus* breeding** *Grey crown* bordered by black lores and bright chestnut from eye to nape-sides; head-sides whitish. Mantle chestnut-brown streaked black; short wing-bar formed by white-tipped median coverts; *rump grey* (diagnostic; all indigenous *Passer* have rufous rumps). Underparts pale greyish with black throat and upper breast. Bill black. *P. d. tingitanus* similar; *indicus* has whiter cheeks and underparts; chestnut of upperparts deeper. **Adult male non-breeding** Upperparts duller; bib smaller, mixed with greyish. Bill horn coloured. **Adult female** Dull brown with buff supercilium (broader behind eye); mantle streaked black; 2 faint buff wingbars. Bill horn coloured. **Juvenile** Similar to adult female but duller; bill paler. **VOICE** A monotonous *chirrup* or *chilp*; a rattling *churrr-r-r-t-t* when alarmed or excited. Song a monotonous series of chirping notes, including call. [CD4:69, 14:33] **HABITS** Closely associated with man, occurring mainly in towns. Gregarious. Bold but wary. Flight fast and direct, with whirring wings. **SIMILAR SPECIES** Normally none within its West African range, but see Spanish Sparrow. **STATUS AND DISTRIBUTION** Introduced in Senegal in 1970s (*indicus*), where now common in towns; spread to Mauritania and Gambia. Also Liberia (Monrovia, Jul 1989 and Jun 1990). Introduced, Cape Verde Is (nominate; São Vicente, common). Claimed NE Niger (*tingitanus*, Dec 1970) and C Chad (race unknown).

SPANISH SPARROW *Passer hispaniolensis*   Plate 128 (3)
Moineau espagnol

L 15 cm (6"). *P. h. hispaniolensis*. **Adult male breeding** Resembles House Sparrow, but has *dark chestnut crown*, whiter head-sides, more heavily streaked upperparts, and *more black on underparts*, with *heavily streaked flanks*. Bill heavier, black. **Adult male non-breeding** Plumage duller, paler and less contrasting. Bill horn coloured. **Adult female** Largely indistinguishable from female House Sparrow. At close range, more streaked and contrasting upperparts, whiter tips to median and greater coverts, and faint dusky streaks on breast and flanks. **Juvenile** As juvenile House Sparrow. **VOICE** Similar to House Sparrow. [CD4:70] **HABITS** Closely associated with man, occurring mainly in towns, villages, cultivated areas and oases. Gregarious when not breeding. In Cape Verdes sometimes associates with Iago Sparrow. **STATUS AND DISTRIBUTION** Common resident, Cape Verde Is. Also recorded from Mauritania (Nouakchott; presumably accidentally introduced, none since 1983) and NE Chad (Fada; presumably genuine Palearctic vagrants).

TREE SPARROW *Passer montanus*   Plate 147 (12)
Moineau friquet

L 14 cm (5.5"). **Adult** Resembles adult male House Sparrow, but has chestnut crown, black ear-spot, neater black bib, partial white collar and buffish-brown rump. Sexes similar. [CD4:71] **STATUS AND DISTRIBUTION** Palearctic species, sighted in Conakry harbour, Guinea, Sep 1976 (Lambert 1978); presumably ship-assisted or escape, race unknown.

RUFOUS SPARROW *Passer motitensis*   Plate 128 (2)
Grand Moineau

Other name E: Great Sparrow
L 14 cm (5.5"). *P. m. cordofanicus*. **Adult male breeding** Forehead to nape grey bordered by rich chestnut behind

eye; supraloral streak and head-sides white; black eye-stripe. Upperparts rich chestnut streaked black on mantle. Underparts whitish with black bib. Bill black. **Adult male non-breeding** Bill horn coloured. **Adult female** Paler; bib dusky-grey. **Juvenile** Similar to adult but duller. **VOICE** Call and song consist of chirps, similar to House Sparrow's. [CD14:34] **HABITS** In dry *Acacia* scrub. In pairs when breeding; singly or in small flocks outside breeding season. The least social of the *Passer* sparrows. **SIMILAR SPECIES** Male House Sparrow has grey rump and larger bib. **STATUS AND DISTRIBUTION** Locally common resident, EC Chad. Vagrant, Mali?

IAGO SPARROW *Passer iagoensis*   Plate 143 (3)
Moineau du Cap-Vert

L 13 cm (5"). **Adult male breeding** Crown blackish becoming grey on nape and mantle; crown and nape bordered by rich chestnut behind eye; supraloral streak and head-sides white; black eye-stripe; scapulars chestnut streaked black; median coverts broadly tipped white forming wingbar. Underparts white washed grey, with narrow black bib. Bill black. **Adult male non-breeding** Bill dark horn, paler below. **Adult female** Dull brown with prominent pale creamy-buff supercilium (broader behind eye); mantle streaked black; median coverts tipped white, forming wingbar. Underparts white washed brownish. Bill dark horn, paler below. **Juvenile** Similar to adult female but duller. Young males distinguished at early age by cinnamon supercilium, browner upperparts and traces of black bib. **VOICE** Call and song consist of chirps, similar to House Sparrow's. [CD4:72] **HABITS** In various arid habitats, such as sparsely vegetated lava plains, cliffs and gorges; also at edges of cultivation, in oases, villages and towns (esp. where Spanish Sparrow absent). In flocks when not breeding, occasionally with Spanish Sparrow. Tame. **SIMILAR SPECIES** Adult female Spanish and House Sparrows mainly separated by less distinct, buff supercilium. **STATUS AND DISTRIBUTION** Common resident, endemic to Cape Verde Is (all islands).

DESERT SPARROW *Passer simplex*   Plate 128 (4)
Moineau blanc

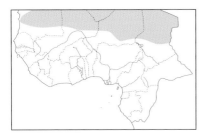

L 13.5 cm (5"). *P. s. simplex*. Very pale and rather small. **Adult male breeding** Unmistakable. Pale greyish above; pale buffish on rump and underparts; contrasting black mask, bib and bill. Black primary-coverts and black bases to greater coverts form distinctive pattern on wing. **Adult male non-breeding** Bill horn coloured. **Adult female** Pale sandy-buff (underparts paler) without face markings. Wing pattern much reduced. Bill pale horn. **Juvenile** Similar to adult female, but wing pattern duller; bill pale pinkish. **VOICE** Chirps similar to House Sparrow's. Also a soft melodious twitter, reminiscent of European Greenfinch. [CD4:73] **HABITS** In pairs or small groups at edge of desert in sandy areas with scattered trees or scrub, wadis, oases and human settlements. Feeds on ground. Rather shy. **SIMILAR SPECIES** Adult female may recall Bar-tailed Lark, female Sudan Golden Sparrow and female Trumpeter Finch, but unlikely to be misidentified as normally in company of male. **STATUS AND DISTRIBUTION** Uncommon and local resident in Sahara, from Mauritania (north of 17°N) and Mali (north of 16°N) to Niger (north of 15°N) and N Chad. Partially nomadic depending on effects of rains on availability of grass. **NOTE** Birds in Mauritania may belong to slightly paler and larger race *saharae* from NW Sahara.

NORTHERN GREY-HEADED SPARROW *Passer griseus*   Plate 128 (5)
Moineau gris

Other name E: Grey-headed Sparrow
L 14 cm (5.5"). **Adult** *griseus* **breeding** Head pale ash-grey; mantle plain grey-brown; rest of upperparts rufous-chestnut; tail brown. Underparts pale grey becoming white on throat and belly. Innermost median coverts tipped white, forming small wing patch (sometimes hidden or absent). Bill black. *P. g. laeneni* paler, *ugandae* darker overall. **Adult non-breeding** Bill dark horn. **Juvenile** As adult but with some dusky streaks on mantle, no white on wing-coverts and horn-coloured bill. **VOICE** A monotonous *chirp* or *cheerp*. Also a rattling *churrr-r-r-t-t* when alarmed or excited. Song a monotonous series of chirps, including call. [CD14:35] **HABITS** In small, vocal groups or pairs, in variety of habitats, to desert edge, though mainly in towns and villages; not closed forest. Feeds on ground or in vegetation. Bold but wary. **SIMILAR SPECIES** None in our region. Plain head distinguishes it from all other sparrows. **STATUS AND DISTRIBUTION** Common resident throughout; also Bioko. Nominate in most of range; *laeneni* from Mali and Burkina Faso to N Cameroon and Chad; *ugandae* S CAR, (part of?) Gabon and Congo.

SUDAN GOLDEN SPARROW *Passer luteus*
Moineau doré

Plate 128 (6)

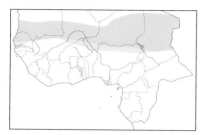

L 12 cm (4.75"). **Adult male breeding** Unmistakable. *Canary yellow with chestnut mantle, back and scapulars*, contrasting black bill and dark eye. Wings mainly blackish edged rufous; white tips to median and greater coverts form double wingbar (reduced or absent when worn). Tail dark greyish-brown, finely edged paler buff-brown. **Adult male non-breeding** Rump greyish washed yellow. Bill dark horn with pale base. **Adult female** Head and upperparts buffish-brown faintly streaked dusky on mantle; pale supercilium; *yellowish wash on head-sides and underparts*. Wings dull brown edged and tipped buffish. Bill pale horn. **Juvenile** As adult female but head greyer and underparts paler. **VOICE** A chirping *chilp* or *chirrup*. In flight also a fast *che-che-che-...* Song a monotonous series of chirps, including call. [CD14:36] **HABITS** Highly gregarious and nomadic, occurring in arid scrub, thornbush and dry savanna. Feeds on ground and in vegetation. Often in mixed flocks with other small seed-eaters, such as Red-billed Quelea. Breeds in large colonies. **SIMILAR SPECIES** Female and juvenile superficially resemble other small sparrows but distinguished by yellowish wash on head and underparts. **STATUS AND DISTRIBUTION** Common from Mauritania and N Senegal through N Mali, N Burkina Faso, Niger to Chad, wandering south to Gambia, N Nigeria and N Cameroon.

YELLOW-SPOTTED PETRONIA *Petronia pyrgita*
Moineau à point jaune

Plate 128 (11)

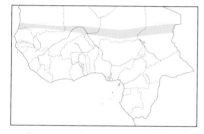

Other name F: Moineau soulcie à point jaune
L 15 cm (6"). *P. p. pallida.* **Adult male** Head, upperparts and tail plain grey-brown. Underparts creamy-white with lemon-yellow patch on throat (difficult to see in field). Bill brownish-horn above, paler below. **Adult female** Slightly smaller; usually with less distinct throat patch. **Juvenile** As adult but mantle usually indistinctly streaked dusky; no patch on throat. **VOICE** Sparrow-like notes, *chirp*, *chillip* and *wurli*. Usually silent. [CD14:37] **HABITS** In pairs or small groups, in thorn scrub, dry open woodland and edges of cultivation. Forages mainly in trees; also on ground. Associates with other small seed-eaters, such as bishops. **SIMILAR SPECIES** Bush Petronia is browner and smaller; female and juvenile further distinguished by well-defined supercilium and streaked mantle. **STATUS AND DISTRIBUTION** Scarce or uncommon and local resident in narrow dry belt, from S Mauritania and N Senegal to Mali (Gourma and Tamesna), Niger and Chad. Local movements outside breeding season reported. **NOTE (1)** Considered specifically distinct from extralimital *P. xanthocollis* (formerly *P. xanthosterna*) by most authors, following Hall & Moreau (1970). **(2)** Re-assigned to new genus *Gymnoris* by some authorities.

YELLOW-THROATED PETRONIA *Petronia superciliaris*
Moineau bridé

Plate 128 (8)

Other name F: Moineau soulcie austral
L 15 cm (6"). **Adult** Dusky earth-brown above with broad whitish supercilium emphasised by blackish-brown eye-stripe, dark-streaked mantle, and buffish tips to wing-coverts forming double pale wingbar. Throat whitish with lemon-yellow patch; rest of underparts washed buffish. Bill dark horn above, paler below. **Juvenile** As adult but upperparts warmer brown, supercilium brownish-buff, underparts paler without lemon patch on throat. **VOICE** Series of 3 identical sparrow-like chirps *chreep-chreep-chreep*; also a fast *chreechreechree* sometimes extended to *chreechreechreechreechree-chirp*. [CD14:38] **HABITS** In wooded grassland, bush and cultivation. Usually singly or in pairs and, outside breeding season, in small groups. Forages in trees, rarely on ground. **SIMILAR SPECIES** Streaky-headed Seedeater also has white supercilium but dark face and streaked crown; ranges not known to overlap. **STATUS AND DISTRIBUTION** Uncommon to locally common resident, SE Gabon and Congo.

BUSH PETRONIA *Petronia dentata* Plate 128 (7)
Petit Moineau

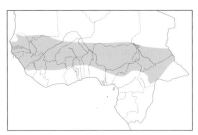

Other name F: Petit Moineau soulcie
L 13 cm (5"). **Adult male breeding** *dentata* Dull brown above with ash-grey crown and *chestnut stripe from eye around rear of ear-coverts.* Throat white with indistinct lemon-yellow patch (difficult to see in field; sometimes absent); rest of underparts pale greyish becoming whitish on belly. Bill black. **S. d. buchanani** paler below; intermediates known. **Adult male non-breeding** Bill dark brown with pale horn base to lower mandible. **Adult female** Similar to adult male but slightly paler; head brown with *buff supercilium*; mantle streaked dark brown. Bill horn. **Juvenile** As adult female but browner. **VOICE** Calls and song consist of sparrow-like chirps. [CD14:39] **HABITS** Singly, in pairs or small groups, in arid scrub and wooded grassland. Most often in trees; rarely on ground. Relatively tame. **SIMILAR SPECIES** Yellow-spotted Petronia is plainer and greyer, lacking distinct supercilium; also larger. Also compare other sparrows within range. **STATUS AND DISTRIBUTION** Uncommon to locally not uncommon resident in broad Sahel and savanna belts, from S Mauritania and Senegambia to Sierra Leone east to Chad and CAR (nominate). *S. d. buchanani* S Niger (Zinder to L. Chad). Dry-season breeder, wandering widely when not breeding.

WEAVERS
Family PLOCEIDAE (World: *c.* 113 species; Africa: 108; WA: 58)

Large, principally African family of small to medium-sized birds with generally strong, conical bills. Sexes more or less similar or female nondescript. Males often have a nondescript female-like non-breeding plumage. Juveniles similar to nondescript females. Found in many habitats, from rainforest to semi-arid scrub and cultivation. Most are mainly granivorous, but insects also taken; forest species predominantly insectivorous. Many species gregarious, feeding and roosting in flocks and nesting in colonies. Vocalisations typically consist of characteristic, drawn-out, wheezy, buzzy, chirping and chattering notes. Most sedentary; some migratory. The name refers to their elaborate, woven nests, typically consisting of a domed structure with downward-pointing entrance, suspended from tree or bush.

The family is usually subdivided into buffalo weavers (Bubalornithinae), scaly weavers (Sporopipinae), sparrow weavers (Plocepasserinae) and true weavers (Ploceinae).

The main genera within the true weavers are *Ploceus* (typical weavers), *Malimbus* (malimbes), *Quelea* (queleas) and *Euplectes* (bishops and widowbirds). Four other genera are monotypic: *Anaplectes*, *Brachycope*, *Anomalospiza* and *Amblyospiza*.

WHITE-BILLED BUFFALO WEAVER *Bubalornis albirostris* Plate 133 (1)
Alecto à bec blanc

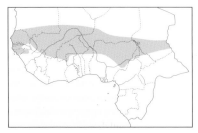

L 22–24 cm (8.75–9.5"). Readily identified by large size, black plumage, massive bill and gregarious habits. **Adult male breeding** All black with whitish, dusky-tipped bill. Feathers reveal white base when ruffled, esp. on back and flanks. Primaries narrowly edged white. **Adult female/adult male non-breeding** Similar, but bill black. **Juvenile** Dull dark brown; underparts white, densely mottled and streaked dark brown. **VOICE** Various harsh, croaking and high-pitched, squeaky sounds, resulting in cackling chatter. [CD14:40] **HABITS** In small flocks in dry savanna with *Acacia* and thorn scrub. Forages on ground. Huge communal stick nest with many chambers very conspicuous and centre of activities year-round. Noisy at nest. **STATUS AND DISTRIBUTION** Common resident in semi-arid belt, from S Mauritania to Guinea east to N Cameroon and Chad.

SPECKLE-FRONTED WEAVER *Sporopipes frontalis* Plate 128 (10)
Sporopipe quadrillé

Other name F: Moineau quadrillé
L 12 cm (4.5"). *S. f. frontalis.* Small; sparrow-like with conspicuous rufous nape. **Adult** Forehead, forecrown and

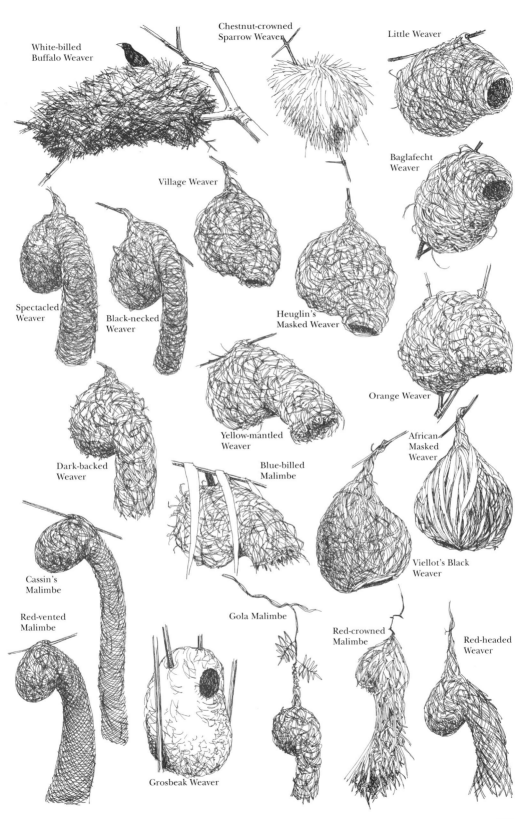

White-billed Buffalo Weaver

Chestnut-crowned Sparrow Weaver

Little Weaver

Baglafecht Weaver

Village Weaver

Spectacled Weaver

Black-necked Weaver

Heuglin's Masked Weaver

Orange Weaver

Dark-backed Weaver

Yellow-mantled Weaver

Blue-billed Malimbe

African Masked Weaver

Viellot's Black Weaver

Cassin's Malimbe

Red-vented Malimbe

Gola Malimbe

Red-crowned Malimbe

Red-headed Weaver

Grosbeak Weaver

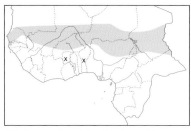

moustachial streak black speckled white; rest of crown, nape and neck-sides pale rufous; head-sides pale brownish-grey. Upperparts sandy-brown; wing and tail feathers dark brown fringed sandy-brown. Underparts white washed greyish on breast. Bill pale horn. **Juvenile** Rufous on crown and nape duller. **VOICE** A short, silvery trill *sriiii* or *tsrrrrk* and a series of mainly high-pitched, short notes *tsisit chep tsisisit che ...*; a rapid *tsip-tsip-tsip-tsip* when taking flight. [CD14:41] **HABITS** In small flocks in dry bush country; occasionally around villages. Forages on ground with characteristic hopping gait. **STATUS AND DISTRIBUTION** Uncommon to locally common in Sahel zone, from Mauritania and Senegambia to Chad and N CAR. **NOTE** Slightly paler desert populations often assigned to doubtful race *pallidior*.

CHESTNUT-CROWNED SPARROW WEAVER *Plocepasser superciliosus* Plate 128 (9)
Mahali à calotte marron

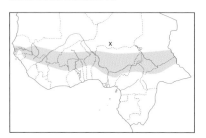

Other name F: Moineau-tisserin à calotte marron
L 17 cm (7"). Sparrow-like, with distinctive, well-marked head pattern. **Adult** Rufous-chestnut crown, pale rufous-brown cheeks, conspicuous white supercilium, white line below eye and black malar stripe. Upperparts dull brown slightly washed rufous; white tips to coverts form double wingbar. Throat white; rest of underparts dirty white. Bill horn. **Juvenile** Crown duller chestnut. **VOICE** A fast, metallic ticking trill *trrrrit* or *trrirr*, somewhat recalling *Euplectes*. Song a short, continuous phrase *witsweeweetsweeuseesweeuseeswee*. [CD14:42] **HABITS** In small groups in thorn scrub, wooded savanna and degraded savanna scrub. Arboreal. Not shy but inconspicuous and easily overlooked. **STATUS AND DISTRIBUTION** Uncommon to scarce resident in Sahel and savanna belts, from Senegambia through Mali, C Guinea and N Ivory Coast to Chad and N CAR. Local movements reported.

Genus *Ploceus*

Marked sexual dimorphism in many species, especially those in drier areas. Males in breeding dress mostly bright yellow; females dull and streaky (sparrow-like). In non-breeding dress (if any) males similar to females. All build strong, woven nests of varying shapes.

BAGLAFECHT WEAVER *Ploceus baglafecht*   Plate 130 (1)
Tisserin baglafecht

L 15 cm (6"). *P. b. neumanni.* **Adult male breeding** Distinctive *black mask* (from bill through eye to ear-coverts) contrasting with bright yellow forehead, forecrown and throat. Upperparts yellowish-green, feather centres dark brown imparting faintly streaked appearance; edges of flight feathers greenish-yellow. Tail dark brown washed yellowish-green. Underparts bright yellow, becoming white on belly. Bill black. Eye pale yellow. **Adult female breeding** Lacks mask; entire top of head yellowish-green as upperparts; lores dusky. **Adult male/female non-breeding** No mask, but area around eye dusky; more greyish-brown above (crown concolorous with upperparts); underparts whitish washed buff, esp. on throat and breast. Bill dusky. **Juvenile/immature** No mask; head dark yellowish-green. **VOICE** Shrill chattering and wheezy notes, characteristic of weavers. Main call a distinctive, loud, repeated *zwenk*. [CD14:44] **HABITS** Singly or in pairs, mainly at edge of montane forest (in west of range) and lowland forest (to east). **SIMILAR SPECIES** Bannerman's Weaver has brighter coloured plumage, mask extending to upper throat and entirely yellow underparts. **STATUS AND DISTRIBUTION** Scarce to uncommon resident, E Nigeria (Mambilla Plateau, Gashaka-Gumti NP) through Cameroon (Bamenda Highlands, Adamawa Plateau, forest edge in east) to W CAR.

BLACK-CHINNED WEAVER *Ploceus nigrimentum*   Plate 130 (4)
Tisserin à menton noir

L *c.* 17 cm (*c.* 6.75"). **Adult male** Black mask (from bill through eye and ear-coverts to throat) bordered by saffron-

yellow forehead, crown and lower throat; nape and rest of underparts golden-yellow. Upperparts black; rump and uppertail-coverts greenish-yellow. Wing feathers edged golden-yellow; tail yellow-green. Bill black. Eye greenish-yellow. **Adult female** Top of head and nape black. **Juvenile** As adult female, but black areas greenish-grey. **VOICE** Harsh calls. Song a rather short, distinctive *whit-pui-pui-trrrrr-pui*. **HABITS** In pairs or small groups in wooded grassland. Often perches atop trees. **STATUS AND DISTRIBUTION** Locally not uncommon resident, SE Gabon and C Congo (Bateke Plateau).

BANNERMAN'S WEAVER *Ploceus bannermani*   Plate 131 (10)
Tisserin de Bannerman

L 14 cm (5.5"). **Adult male** *Bright yellow* with *contrasting black mask* (from bill through eye and ear-coverts to upper throat) and *yellowish-green upperparts and tail*. Bill black. Eye yellow. **Adult female** Similar. **Juvenile** No information. **VOICE** A sharp *pritt* or *kit*. Song high pitched and with drawn-out wheeze, more pleasant than that of many other *Ploceus* weavers *psi-psisi-trrir-si-psiuuuu-tsisisi-irrrr*, with variations. [CD14:45] **HABITS** In pairs at edges and in clearings of montane forest, in secondary scrub and bushes near such forest, and in (even small) forest patches within montane grassland. Sometimes in small parties. **SIMILAR SPECIES** Male Baglafecht Weaver has mask not extending to upper throat; lower underparts white. **STATUS AND DISTRIBUTION** Locally not uncommon resident in highlands of SE Nigeria (Obudu and Mambilla Plateaux) and W Cameroon (Bakossi Mts, Mt Manenguba, Bamenda Highlands). *TBW* category: THREATENED (Vulnerable).

BATES'S WEAVER *Ploceus batesi*   Plate 131 (8)
Tisserin de Bates

L 14 cm (5.5"). **Adult male** *Head bright chestnut bordered by black throat and neck-sides*, and partially separated from bright yellowish-green upperparts by narrow yellow collar; tail olive-green. Rest of underparts bright yellow. Bill black. Eye dark brown. **Adult female** Chestnut on head replaced by black; throat yellow; underparts washed olive. **Juvenile** Resembles adult female, but head olive-green mixed with black. Bill pale horn. **VOICE** No information. Apparently usually silent. **HABITS** Inadequately known. Occurs in lowland forest. Climbs large branches and tree trunks in search of insects. Joins mixed-species flocks. **STATUS AND DISTRIBUTION** Rare forest resident, S Cameroon; also recorded further west (e.g. Limbe environs; Mt Kupe, at 900 m). *TBW* category: THREATENED (Endangered).

SLENDER-BILLED WEAVER *Ploceus pelzelni*   Plate 129 (2)
Tisserin de Pelzeln

Other name F: Tisserin nain
L 11 cm (4.5"). *P. p. monachus.* Smallest typical weaver. **Adult male** Forehead, forecrown, head-sides, throat and upper breast black; hindcrown, nape and neck-sides golden-yellow forming contrasting collar. Upperparts yellow-green, brighter on rump. Wing and tail feathers brownish edged greenish-yellow. Underparts golden-yellow. Bill black. Eye brown. **Adult female** No black on head. Top of head yellow-green, concolorous with upperparts; distinct yellow supercilium. Head-sides and underparts golden-yellow. Bill blackish. **Juvenile** Dull brownish-olive above; pale yellow supercilium; mantle streaked dusky; wing-coverts and tertials edged buff. Underparts pale yellowish. Bill horn. **Immature** As adult female but with horn-coloured bill. **VOICE** A jumble of subdued chattering and chirping notes, characteristic of weavers. [CD14:46] **HABITS** In small groups in mangroves and marshes, along coastal lagoons and river banks. Usually in small breeding colonies in trees overhanging water. Largely insectivorous, foraging like tit or warbler, actively examining trunks and branches, often hanging upside-down. Not shy. **SIMILAR SPECIES** Little Weaver in breeding plumage is similarly patterned but less deeply coloured, with variably streaked mantle; also has non-breeding

dress (q.v.); ranges do not normally overlap. Loango Weaver is duller, with olive-green upperparts and yellow areas washed brownish or buff; also slightly larger. **STATUS AND DISTRIBUTION** Locally not uncommon resident in coastal zone; also inland along some major rivers. Recorded in Sierra Leone, Ivory Coast, Ghana, Togo, Nigeria, Camëroon, Gabon and Congo (along Congo R.). Possibly also Liberia. **NOTE** Confusingly, Serle & Morel (1977), following Bannerman (1949), name this species Little Weaver, which is now generally used for *P. luteolus*.

LITTLE WEAVER *Ploceus luteolus* — Plate 129 (1)
Tisserin minule

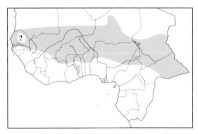

L 11.5 cm (4.5"). *P. l. luteolus*. Smallest savanna-dwelling typical weaver. **Adult male breeding** Forehead, forecrown, head-sides, throat and upper breast black; hindcrown and neck-sides bright yellow. Nape and upperparts greenish-yellow variably streaked dusky on mantle, becoming bright yellow on rump. Wings and tail brownish edged greenish-yellow. Underparts bright yellow. Bill black. Eye brown. **Adult male non-breeding** No black on head. Top of head yellow-green; mantle buff streaked dusky. Head-sides, throat, upper breast and flanks buff; lower breast and belly white; undertail-coverts lemon. Bill dusky-horn. **Adult female breeding** No black on head; top of head yellow-green, concolorous with upperparts; mantle streaked dusky. Head-sides, throat and breast pale yellow inclining to whitish on belly. Bill black. **Adult female non-breeding** Duller. Head-sides yellowish; underparts whitish washed buff on breast and flanks. Bill horn. **Juvenile** Similar to adult female non-breeding. **VOICE** Rather silent. Soft *tsssp* and harsh *chep* notes, and a jumble of chattering and chirping notes, typical of weavers. [CD14:47] **HABITS** In pairs or small family groups in dry wooded savanna; also farmland with large trees and gardens. Not usually gregarious, but may occasionally nest in small loose colonies. Not shy. **SIMILAR SPECIES** Slender-billed Weaver has richer coloured plumage, lacks non-breeding dress and is mainly confined to coasts. Adult female Slender-billed much brighter than female Little, with all-yellow underparts and unstreaked mantle. **STATUS AND DISTRIBUTION** Common to uncommon resident in northern savanna and Sahel zones, from Senegambia to Chad and CAR; scarce at limits of range, e.g. S Mauritania and N Ivory Coast. **NOTE** Confusingly, Serle & Morel (1977), following Bannerman (1949), name this species Slender-billed Weaver, which is now generally used for *P. pelzelni*.

LOANGO WEAVER *Ploceus subpersonatus* — Plate 129 (3)
Tisserin à bec grêle

Other names E: Loango Slender-billed Weaver; F: Tisserin de Cabinda
L 15 cm (6"). **Adult male** Forehead to mid-crown, head-sides and throat black (extending in point to upper breast); hindcrown, nape and head-sides to breast dirty orange forming broad collar. Upperparts olivaceous-green; rump paler, yellowish washed brown; tail greyish-olive. Rest of underparts yellow with dirty brown wash. Bill black. Eye brown. **Adult female** No black on head. Top of head olive-green, concolorous with upperparts, forehead yellower. Head-sides dull yellow; underparts dull yellow washed brownish. **Immature** As adult female but with darker dull olive forehead and paler, brownish bill. **VOICE** A soft jumble of characteristic weaver notes; also a more melodious, subdued song. [CD14:48] **HABITS** Little known. Occurs in beach scrub, landward mangrove edge, clumps of bushes and palm trees at edges of permanent pools, and savanna, within 2.5 km of coast. Skulking and shy. **SIMILAR SPECIES** Slender-billed Weaver is more brightly coloured and smaller. **STATUS AND DISTRIBUTION** Uncommon resident confined to narrow coastal strip from Gabon (Cap Lopez south) to Congo. *TBW* category: THREATENED (Vulnerable).

BLACK-NECKED WEAVER *Ploceus nigricollis* — Plate 130 (6)
Tisserin à cou noir

L 16–17 cm (6.25–6.5"). **Adult male** *brachypterus* Head golden-chestnut becoming bright yellowish-olive on hindcrown; black stripe through and around eye, forming small mask. Upperparts and tail *bright yellowish-olive*. Throat black bordered golden-chestnut; rest of underparts golden-yellow. Bill black. Eye creamy. Island race *po* similar but has slightly longer bill. Distinctive eastern *nigricollis* has entire head golden-chestnut, sharply delimited from *blackish-brown upperparts*; rump washed yellowish-olive; tail washed olivaceous. Eye brown. **Adult female** *brachypterus* Similar to adult male but top of head yellowish-olive, concolorous with upperparts; yellow supercilium; no black on throat. Adult female *nigricollis* also similar to adult male but top of head black; contrasting yellow

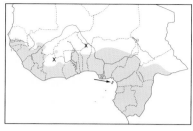

supercilium; no black on throat. **Juvenile** (both races) As adult female *brachypterus* but with horn-coloured bill. **Immature male** Similar to adult female. **VOICE** A harsh *chet-chet* and a twittering wheeze, characteristic of weavers. Usually rather silent. [CD14:49] **HABITS** In pairs within forest clearings and edges, gallery forest and wooded savanna. Not gregarious; often rather shy and skulking. Occasionally joins mixed-species flocks. Forages for insects in leafy trees. **SIMILAR SPECIES** Adult Spectacled Weaver (both sexes) separated from *brachypterus* by brighter, yellower forecrown and face, and more slender bill. Juveniles very similar, but Spectacled has more slender bill. **STATUS AND DISTRIBUTION** Common resident in forest and moist savanna zones throughout. *P. n. brachypterus* from Senegambia to Liberia east to W Cameroon; also S Mali, SW Burkina Faso and SW Niger. Nominate from E and S Cameroon to CAR and Congo. *P. n. po* Bioko. Both mainland forms may occur at same localities in Cameroon; intermediates found from W Cameroon to CAR border. **NOTE** Western races, lacking 'black neck', named 'Spectacled Weaver' by Bannerman (1949) and Serle & Morel (1977). Not to be confused with *P. ocularis*.

SPECTACLED WEAVER *Ploceus ocularis* — Plate 130 (5)
Tisserin à lunettes

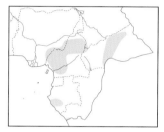

L 16–17 cm (6.25–6.5"). *P. o. crocatus*. **Adult male** Head to breast golden-chestnut becoming bright yellowish-olive on hindcrown and nape; black stripe through and around eye, forming small mask. Upperparts and tail bright yellowish-green. Throat black; rest of underparts golden-yellow. Bill black. Eye creamy. **Adult female** Similar, but lacks black on throat. **Juvenile** As adult female, but much duller. Top of head concolorous with upperparts; mask reduced to dusky mark. Bill horn. **VOICE** Distinctive. A fast, descending series of identical, resonant, piping notes *teeteeteeteeteetee*. Also *chirrrrdzrweew*, a clear *teet-teet-teet-teet* and a sharp, rasping *chit*. [CD14:50] **HABITS** Singly or in pairs in wooded grassland, scrub, forest edge, riverine bush and gallery forest. Forages at all levels, often creeping through dense vegetation and hanging upside-down in search of insects. Often joins mixed-species flocks. Rather shy and skulking, rarely emerging from cover. **SIMILAR SPECIES** Adult male Black-necked Weaver of race *brachypterus* distinguished by more chestnut head and more contrast between nape and mantle; adult female has entire top of head yellowish-olive. Juveniles very similar, but Black-necked has less slender bill. **STATUS AND DISTRIBUTION** Not uncommon to scarce resident from SE Nigeria (Mambilla and Obudu Plateaux, Oban Hills) and Cameroon to CAR, and in Gabon and S Congo.

BLACK-BILLED WEAVER *Ploceus melanogaster* — Plate 131 (9)
Tisserin à tête jaune

L 14 cm (5.5"). *P. m. melanogaster*. **Adult male** Black with golden-yellow head and narrow yellow band on upper breast, surrounding black throat; black eye-stripe. Bill black. Eye dark red. **Adult female** Lacks black on throat. **Juvenile** Sooty-black above with olive-tinged forehead and forecrown. Head-sides, throat and upper breast washed dull chestnut; rest of underparts dull olivaceous. Bill horn. **Immature** Gradually more like adult female but duller, with underparts becoming sooty-brown. Young males have some black feathers on throat. **VOICE** Song, uttered by male, consists of clear, ringing notes ending in a drawn-out rattling wheeze. Calls include a harsh *chet*. [CD14:51] **HABITS** In pairs or small groups in montane forest, favouring clearings and edges with dense undergrowth. Forages in dense vegetation, usually fairly low. Mainly insectivorous. Joins mixed-species flocks. **STATUS AND DISTRIBUTION** Locally not uncommon resident in montane forest, SE Nigeria (Obudu and Mambilla Plateaux), W Cameroon (Mt Cameroon, Rumpi Hills, Mt Kupe, Bakossi Mts, Mt Manenguba, Bamenda Highlands) and Bioko.

HOLUB'S GOLDEN WEAVER *Ploceus xanthops* — Plate 129 (9)
Tisserin safran

Other name F: Tisserin doré

L 17–18 cm (6.75–7.0"). Large, relatively unpatterned, mainly golden-yellow weaver with heavy black bill. **Adult male** Head and underparts golden-yellow washed orange on throat and upper breast; upperparts golden-green.

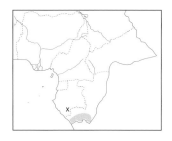

Eye yellow. **Adult female** Slightly duller. **Juvenile** Top and sides of head olive, concolorous with upperparts. Underparts yellow washed buffish on upper breast and flanks. Bill dusky-horn. **VOICE** A jumble of chattering and wheezy notes, typical of weavers. Also a harsh sparrow-like chirp. [CD14:53] **HABITS** Singly, in pairs or small groups in open savanna with rank vegetation and bushes, often near marshes and water. Not gregarious. **SIMILAR SPECIES** Juvenile Village Weaver differs from juvenile Holub's Golden in having white belly. **STATUS AND DISTRIBUTION** Rare or scarce and local resident, S Congo. Also recorded S Gabon.

ORANGE WEAVER *Ploceus aurantius* Plate 129 (7)
Tisserin orangé

L 15 cm (6"). *P. a. aurantius.* **Adult male** Brightly coloured with yellow-orange head and underparts. Upperparts and tail olive washed orange, becoming orange-yellow on rump. Wing feathers black broadly edged and fringed yellow. Short black eye-stripe. Bill black. Eye red to dark reddish-brown or pale greyish. **Adult female** Top of head, upperparts and tail olive; narrow olive-yellow supercilium. Underparts white with variable amount of pale yellow on head-sides, throat and upper breast. Bill brownish-horn. Combination of olive upperparts and largely white underparts unique among weavers in breeding plumage. **Juvenile** Similar to adult female. **VOICE** Characteristic weaver chattering at breeding colony. [CD14:54] **HABITS** In groups frequenting bushes and trees along coastal lagoons, mangroves, swamps and banks of large rivers. Breeding colonies typically near water. Appears to occur singly or in pairs, breeding in isolation, from S Cameroon east and south. **STATUS AND DISTRIBUTION** Locally not uncommon resident in coastal zone and inland along some major rivers, from Sierra Leone to Togo and from Nigeria to CAR and Congo.

PRÍNCIPE GOLDEN WEAVER *Ploceus princeps* Plate 144 (8)
Tisserin de Principé

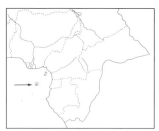

Other name F: Tisserin doré de Principé
L 18 cm (7"). The only weaver on Príncipe. **Adult male** Head strongly washed rufous-orange. Upperparts yellowish-green slightly mottled dusky; wing feathers blackish edged yellow; tail dark olive. Underparts bright golden-yellow. Bill blackish. Eye yellow. **Adult female** Differs in having top of head yellowish-green, concolorous with upperparts, and belly to undertail-coverts whitish. Bill horn. Eye yellow or dark. **Juvenile** As adult female but duller. **VOICE** A sharp *pzeep* and a hard, drawn-out wheeze, typical of weavers, preceded by a few call notes *pzeep-pzeep-pzzzzzzzzzrrrrrr.* [CD14:55] **HABITS** Usually in groups of up to 30; also singly. Occurs in all habitats with trees incl. secondary forest, plantations, gardens and villages. **STATUS AND DISTRIBUTION** Endemic to Príncipe, where one of the commonest birds.

LESSER MASKED WEAVER *Ploceus intermedius* Plate 129 (8)
Tisserin intermédiaire

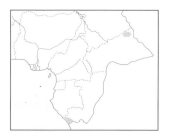

Other name F: Tisserin masqué austral
L 14 cm (5.5"). *P. i. intermedius.* **Adult male breeding** As African Masked Weaver but distinguished by *larger black mask, extending over mid-crown and in point to lower throat*; mantle indistinctly streaked dusky. Eye pale yellow. Legs bluish-grey. **Adult female/adult male non-breeding** No mask. Top of head and upperparts yellowish-green streaked dusky on mantle and becoming yellower on rump. Head-sides (incl. supercilium) and underparts yellow, becoming white on centre of belly. Bill horn. **Juvenile** As adult female but duller; underparts white washed buffish-yellow on breast. **VOICE** A jumble of chattering and wheezy notes, typical of weavers. [CD14:57] **HABITS** Usually in small colonies in vicinity of water. **SIMILAR SPECIES** Adult male breeding African Masked Weaver has smaller mask; all other plumages much browner on upperparts and pale buffish-yellow or buff on underparts; legs pinkish. **STATUS AND DISTRIBUTION** Rare or scarce and local resident, SW Congo.

AFRICAN MASKED WEAVER *Ploceus velatus*

Plate 129 (5) & 144 (4)

Tisserin à tête rousse

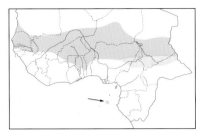

Other names E: Vitelline/Southern Masked Weaver; F: Tisserin masqué
L 14 cm (5.5"). **Adult male *vitellinus* breeding** Black mask (from narrow band on forehead over ear-coverts to upper throat); rest of head and underparts bright golden-yellow strongly tinged chestnut on crown and lower throat. Upperparts and tail yellow-olive, yellower on rump. Bill black. Eye reddish-orange. Legs pinkish. Island race *peixotoi* has black of throat extending in point to lower throat. **Adult female breeding/adult male non-breeding** No mask. Top of head and upperparts dull brownish-olive broadly streaked dusky on mantle and becoming greener on rump. Throat pale lemon-yellow, becoming buff on breast and white on belly. Bill pinkish-horn. **Adult female non-breeding** Similar, but buff on underparts paler. **Juvenile** Similar to adult female non-breeding. **VOICE** A jumble of chattering and wheezy notes, typical of weavers. Also a sharp *tzik* or *chek*. [CD14:58] **HABITS** In small colonies in dry woodland and open thorn scrub. Occasionally with Heuglin's Masked Weaver. **SIMILAR SPECIES** Breeding male Heuglin's Masked Weaver lacks chestnut wash on crown; black mask extends to lower throat. Adult male Lesser Masked has mask extending to mid-crown. Adult female Black-headed very similar to adult female African Masked, but distinguished by two-toned bill. **STATUS AND DISTRIBUTION** Common to not uncommon resident in Sahel and dry savanna zones, from S Mauritania and Senegambia to Chad and N CAR (*vitellinus*). Common resident, São Tomé (*peixotoi*). **NOTE** Sometimes treated as separate species, *P. vitellinus* (Vitelline Masked Weaver; Tisserin vitellin). Form *peixotoi* then remains race of *P. velatus* (Southern Masked Weaver).

HEUGLIN'S MASKED WEAVER *Ploceus heuglini*

Plate 129 (6)

Tisserin masqué

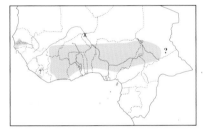

Other name F: Tisserin masqué de Heuglin
L 14 cm (5.5"). **Adult male breeding** Black mask extending in point to lower throat or upper breast but *not* onto forehead. *Top of head bright yellow without chestnut wash*; upperparts and tail yellowish-olive. Underparts bright yellow variably but slightly washed chestnut on breast. Bill black. Eye yellow. Legs brownish. **Adult male non-breeding** No mask. Top of head and upperparts olivaceous, indistinctly streaked dusky on mantle; yellow supercilium; head-sides and underparts yellow. Bill horn. **Adult female breeding** No mask. Top of head and upperparts yellowish-olive with dusky streaking on mantle. Underparts yellow washed buff on upper breast, and paling to white on belly. Bill brown to black. **Adult female non-breeding** Duller; bill horn. **Juvenile** Similar to adult female non-breeding. **VOICE** A jumble of chattering and wheezy notes, typical of weavers; more varied and including more melodious notes than most other *Ploceus*. [CD14:60] **HABITS** In pairs or small colonies in wooded savanna and farmland. Often breeds in small colonies close to raptors', wasps' and ants' nests. **SIMILAR SPECIES** Breeding male African Masked Weaver has mask extending in narrow band over forehead but restricted to upper throat; in Lesser Masked mask extends onto mid-crown; both also have chestnut wash to head and breast. Female African Masked virtually indistinguishable in field from non-breeding female Heuglin's, but eye reddish-orange and dusky streaking on mantle more distinct. Female Lesser Masked Weaver has more brightly coloured plumage than female non-breeding Heuglin's, and bluish-grey legs. **STATUS AND DISTRIBUTION** Uncommon to not uncommon resident in broad savanna belt (mainly south of African Masked Weaver but some overlap), from Senegambia through Mali and Ivory Coast (south to coast) to C and N Cameroon and Chad. Possibly also Liberia.

VIEILLOT'S BLACK WEAVER *Ploceus nigerrimus*

Plate 131 (2)

Tisserin noir

Other name F: Tisserin noir de Vieillot
L 17 cm (6.75"). Two distinct races. **Adult male *castaneofuscus*** Black with chestnut mantle, scapulars, back, rump, belly and undertail-coverts. Bill black. Eye bright yellow. **Nominate** is all black. **Adult female *castaneofuscus*** Top of head and upperparts dark brownish streaked dusky on mantle and becoming more rufous on rump. Wing-coverts and tertials broadly edged pale buffish. Underparts buffish-yellow, brighter on belly and washed pale rufous on breast-sides, flanks and undertail-coverts. Bill horn. Adult female **nominate** similar but rufous replaced by dark olivaceous. **Juvenile** Similar to adult female *castaneofuscus*. **VOICE** A shrill, sustained jumble of chattering notes,

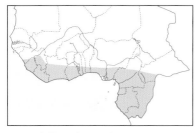

similar to that of other social weavers, esp. at breeding colony. [CD14:61-62] **HABITS** In noisy colonies, often in association with Village Weaver. In forest clearings, near villages, marshy habitats and edges of gallery forest. **SIMILAR SPECIES** Maxwell's Black Weaver distinguished from male nominate by paler eye, smaller size and, in nominate race, grey nuchal patch; it is also much less common and more restricted to high forest. Female Crowned and Cassin's Malimbes are also entirely black but have brown eyes. **STATUS AND DISTRIBUTION** Common resident in forest zone and southern savannas, from W Guinea to SE Nigeria (*castaneofuscus*) east to CAR, Gabon and Congo (nominate). Also Gambia (rare).

VILLAGE WEAVER *Ploceus cucullatus* Plate 130 (2)

Tisserin gendarme

Other name E: Spotted-backed Weaver, Black-headed Weaver
L 15.0–17.5 cm (6–7"). Large, conspicuous weaver, common in towns and villages. **Adult male** *cucullatus* **breeding** Head and throat black bordered by broad chestnut collar on nape and neck-sides. Upperparts golden-yellow with irregular black V bordering mantle; wing feathers blackish edged golden-yellow; tail olive. Black of throat extending in point to upper breast; rest of underparts yellow washed chestnut. Bill black and heavy. Eye red or orange. *P. c. bohndorffi* similar but with less black on crown and more chestnut on breast. *P. c. collaris* has all-black head, lacking chestnut collar (at most a narrow chestnut border); black mottling on upperparts denser and more even, not forming V; breast dark chestnut; rest of underparts yellow. **Adult female breeding** Top and sides of head yellow-olive; upperparts brownish-olive streaked dusky. Underparts bright lemon-yellow becoming white on central belly. Bill dark brown. **Adult male/female non-breeding** Top of head yellow-olive; head-sides olivaceous-yellow. Upperparts ashy-brown streaked dusky. Throat lemon-yellow; rest of underparts yellow, or white with buffish breast and flanks. Bill dark horn above, paler below. **Juvenile** Resembles adult female breeding, but upperparts darker, more olive-brown; underparts white with brownish wash on breast. Bill horn. **VOICE** Harsh chattering and twittering notes at colony. Song a jumble of rasping, chattering notes ending in a drawn-out rattling wheeze, *chit-chit chit-t-t-t-t shirrrzzzzwrrerr*. Alarm a sharp *zip*. In flight a short *chuk-chuk* or *chit-chit*. [CD14:63] **HABITS** In dense, noisy colonies, sometimes numbering 100s of pairs, in variety of habitats. Mainly near habitation; not in closed forest. Bold and aggressive. Breeding colony is centre of activities, with birds constantly coming and going, males building and repairing nests and trying to attract females by hanging below nest with fluttering wings. Parasitised by Didric Cuckoo. **SIMILAR SPECIES** Adult male Black-headed Weaver has much less or no chestnut in plumage, bright yellowish-olive upperparts and is smaller. Juvenile Holub's Golden Weaver differs from juvenile Village in all-yellow underparts. **STATUS AND DISTRIBUTION** Most widespread weaver, generally common resident throughout, though scarce in arid north; also on Bioko and São Tomé. Nominate in most of range; *bohndorffi* in most of Gabon and Congo; *collaris* in coastal Gabon and S Congo. Introduced, Cape Verde Is (recorded 1993).

GIANT WEAVER *Ploceus grandis* Plate 144 (10)
Tisserin géant

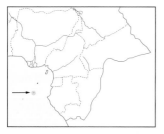

L 22 cm (8.75"). Unmistakable. Largest true weaver, with heaviest bill. **Adult male** Head black bordered by broad chestnut collar; upperparts yellowish-olive narrowly streaked dusky on mantle and back; rump and uppertail-coverts olive-yellow. Wings dusky-brown edged yellow; tail olive. Lower underparts golden-yellow washed chestnut on flanks. Bill very large, black. Eye yellow or pale brown. **Adult female** Top of head and upperparts greyish-olive streaked dusky. Head-sides to breast brownish-buff, paler on throat; belly white. Bill horn, paler below. **Juvenile** No information. **VOICE** A series of chattering notes ending in a typical weaver-like but rather short and tuneless wheeze. **HABITS** Singly or in small (family?) groups at forest edge, in secondary forest, plantations, scrub and patches of dry woodland in savanna. **STATUS AND DISTRIBUTION** Endemic to São Tomé, where common resident.

BLACK-HEADED WEAVER *Ploceus melanocephalus* Plate 130 (3)
Tisserin à tête noire

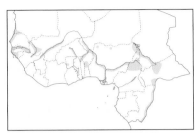

Other name E: Yellow-backed Weaver
L 14–15 cm (5.5–6.0"). **Adult male** *melanocephalus* breeding Head black, separated from yellowish-olive upperparts by golden-yellow collar. Rump golden-yellow; tail olive; wing feathers dusky-brown edged yellow. Black of head extending in point to upper breast; rest of underparts golden-yellow. Bill black. Eye creamy or brown. *P. m. capitalis* has breast washed chestnut. **Adult male non-breeding** Top and sides of head brownish-olive; creamy supercilium. Upperparts dull brownish streaked dusky on mantle; wing-coverts and tertials edged buff; flight feathers edged yellowish-green. Underparts whitish washed buffish on breast and flanks. Bill blackish above, pinkish-horn below. **Adult female** As male non-breeding but smaller. **Juvenile** As adult female. **VOICE** Discordant chattering and twittering; also a variety of calls, incl. hoarse, nasal *chè-èp, tshk* and *tseew*. [CD14:64] **HABITS** In scattered colonies near water within semi-arid savanna. Frequents high grass and bushes along river banks, margins of lakes, marshes and swamps. **SIMILAR SPECIES** Breeding male Village Weaver (both races) has much more chestnut on head and/or breast, black-and-yellow upperparts, and is larger. Adult female African Masked very similar to adult female Black-headed but bill wholly pinkish-horn (not two toned). **STATUS AND DISTRIBUTION** Locally common to not uncommon resident along rivers, lakes and wetlands, mainly in semi-arid belt. Nominate from S Mauritania, N Senegal and Gambia through Mali to Niger and Chad; *capitalis* from S Senegal, Guinea-Bissau and Guinea to N Ghana (rare), Togo (south to coast), Nigeria (south to Niger delta), Cameroon, S Chad and CAR. Also Congo (along Congo and Ubangi R.), where populations again lack chestnut wash on breast (*duboisi?*).

YELLOW-MANTLED WEAVER *Ploceus tricolor* Plate 131 (4)
Tisserin tricolore

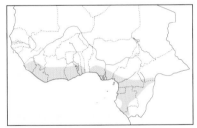

L 17 cm (6.75"). *P. t. tricolor.* **Adult male** Distinctive. *Head* (incl. throat) *and upperparts black with contrasting yellow collar on upper mantle. Underparts deep chestnut.* Bill black. Eye brown. **Adult female** Underparts usually slightly paler; some black throat feathers tipped chestnut. **Juvenile** Dull chestnut with dull black wings and tail. **VOICE** Usually silent. A sharp *cherrit* and, at nest, a typical weaver wheezing followed by a short whistled note. [CD14:65] **HABITS** Usually in pairs or small groups, in mature lowland forest, forest edge, clearings and second growth. Frequents upper and mid-strata. May cling to trunks like woodpecker when searching for insects. **SIMILAR SPECIES** Seen from below, adult male Vieillot's Black Weaver of race *castaneofuscus* may appear similar, but black of underparts extends to breast; also occurs in different habitat. **STATUS AND DISTRIBUTION** Not uncommon forest resident, from Sierra Leone and SE Guinea to S Cameroon and SW Congo. Rare, SW CAR. **NOTE** The two specimens from SW CAR, with dark brown underparts, are thought to belong to neighbouring race *interscapularis* (in which adult female is blackish-brown below) or to be intermediates (Germain & Cornet 1994).

MAXWELL'S BLACK WEAVER *Ploceus albinucha* Plate 131 (1)
Tisserin de Maxwell

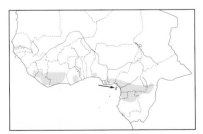

Other names E: White-naped Weaver; F: Tisserin noir de Maxwell
L 15 cm (6"). **Adult** *albinucha* All black with greyish-white bases to nape feathers showing through and forming indistinct patch. Bill black. Eye creamy or greyish-white. Eastern *holomelas* and island race *maxwelli* lack nuchal patch; latter also has less black, more blackish-brown underparts. **Juvenile** Dark sooty-grey; underparts paler, sometimes tinged yellowish-olive. **VOICE** A jumble of chattering and wheezy notes at colony, reminiscent of Village Weaver. [CD14:66] **HABITS** In small groups in canopy or mid-levels of mature lowland forest. Searches branches for insects, often hanging upside-down. Occasionally with malimbes. **SIMILAR SPECIES** Male nominate Vieillot's Black Weaver distinguished by bright yellow eye and larger size; it is also much more common and prefers edges and cultivation. Female Crowned and Cassin's Malimbes are also all black but have brown eyes. **STATUS AND DISTRIBUTION** Uncommon to not uncommon forest resident, from Sierra Leone and SE Guinea to Ghana (nominate) and from Nigeria and Cameroon to SW CAR, Gabon and N Congo (*holomelas*); also Bioko (*maxwelli*).

COMPACT WEAVER *Ploceus superciliosus*   Plate 129 (4)
Tisserin gros-bec

L 15 cm (6"). Distinctive stocky weaver with thick bill. **Adult male breeding** Black mask (from lores through eyes and ear-coverts to throat); forehead golden-chestnut becoming yellow on crown; nape and upperparts dark brown; mantle washed olive and indistinctly streaked dusky. Underparts yellow, becoming whitish or buffish on lower belly and undertail-coverts. Bill short and heavy, blackish-slate. Eye brown. **Adult female breeding** Top of head blackish. **Adult male/ female non-breeding** Dark earth-brown above; broad pale cinnamon supercilium contrasting with blackish top of head and eye-stripe; mantle streaked dusky. Underparts and neck-sides pale cinnamon. Bill horn or slate. **Juvenile** As adult non-breeding but supercilium, neck-sides and underparts washed pale yellow. Bill horn. **VOICE** Rapid, abrupt, crackling calls in flight. Also a harsh *cheee* and repeated *pink*. Song rather quiet and consisting of a series of hard, abrupt notes followed by a typical weaver sizzling *trrrip, trrp trrp trip trip, tru-tru-huu-tzzzzz*, with slight variations. [CD14:67] **HABITS** In small flocks or, when breeding, in pairs. Occurs in grasslands within forest and savanna zones, and in rank grass in moist areas and near rivers. **STATUS AND DISTRIBUTION** Patchily distributed, uncommon to locally not uncommon from S Senegal to Sierra Leone and N Liberia east to Cameroon, CAR and Congo. Subject to movements, apparently without seasonal pattern. **NOTE** Sometimes included in monotypic genus *Pachyphantes*.

DARK-BACKED WEAVER *Ploceus bicolor*   Plate 131 (6)
Tisserin bicolore

Other name E: Forest Weaver
L 15 cm (6"). *P. b. tephronotus.* **Adult** Distinctive. Head black; upperparts and tail greyish-black. Throat black mottled yellow; rest of underparts bright yellow. Bill blue-grey. Eye reddish-brown. **Juvenile** Slightly duller; underparts tinged olive. **VOICE** Song a short series of pleasant, melodious whistles ending in a drawn-out, hoarse, squeaky sizzling *who-he-who-he-hu kshshshrrr*, with variations. Duets, both sexes giving same song simultaneously. Calls include high-pitched *pink-pink* or *weet-weet* and *zzrree* or *wizzzz*; alarm a rapid *tsi-tsi-tsi-...* [CD14:68] **HABITS** Singly, in pairs or small parties, in canopy and mid-levels of montane and lowland forest edges and adjacent old plantations. Searches foliage and branches, often hanging upside-down. Joins mixed-species flocks. **STATUS AND DISTRIBUTION** Locally not uncommon forest resident, SE Nigeria (Obudu Plateau), Cameroon, Gabon and Congo; also Bioko (above 1000 m). **NOTE** Birds from lowland forest in S Cameroon and Gabon reportedly slightly darker above and paler below; sometimes distinguished as *P. b. analogus*.

BROWN-CAPPED WEAVER *Ploceus insignis*   Plate 131 (3)
Tisserin à cape brune

L 14 cm (5.5"). **Adult male** Top of head chestnut; head-sides and throat black; mantle to rump golden-yellow; rest of upperparts and tail black. Underparts golden-yellow. Bill slender, black. Eye reddish-brown. **Adult female** Head entirely black. **Juvenile** Similar to adult but duller; top and sides of head olive mottled black; no black on throat. Bill horn. Young males gradually acquire chestnut feathers on crown (initially often mixed with blackish); young females blackish crown feathers soon appear. **VOICE** Usually silent. A short *twit* and a wheezy sizzling, typical of weavers. [CD14:69] **HABITS** In pairs or small parties in canopy and mid-levels of montane and mid-elevation forest. Behaviour as Preuss's Golden-backed Weaver, typically searching large branches and tree trunks in manner of tit or nuthatch. **SIMILAR SPECIES** Adult male Preuss's Golden-backed Weaver differs from adult male Brown-capped in having golden-chestnut of crown gradually merging with yellow of nape, and chestnut wash to breast. Adult male Yellow-capped Weaver similar to Preuss's Golden-backed, but has black (not yellow) rump. Adult female Preuss's has black forehead and pale yellow-chestnut crown. Adult female Yellow-capped distinguished from female Brown-capped by black of head not extending to nape, and black rump. **STATUS AND DISTRIBUTION** Not uncommon forest resident in highland areas of SE Nigeria (Obudu and Mambilla Plateaux), W Cameroon (Mt Cameroon, Rumpi Hills, Mt Kupe (vagrant), Mt Manenguba, forest near Dschang, Bamenda Highlands) and Bioko. Also recorded from NE Gabon (Makokou) and SW Congo (Mbila, Du Chaillu Mts).

PREUSS'S GOLDEN-BACKED WEAVER *Ploceus preussi*
Tisserin de Preuss

Plate 131 (5)

Other name E: Golden-backed Weaver
L 14 cm (5.5"). **Adult male** Top of head golden-chestnut inclining to yellow on nape; head-sides and throat black. Upperparts and tail black with *broad, irregular golden-yellow stripe from mantle to uppertail-coverts.* Underparts golden-yellow with chestnut wash to breast. Bill black. Eye reddish-brown. Legs pinkish. **Adult female** Similar, but forehead and forecrown black. Hindcrown tinged pale chestnut; no chestnut wash on breast. **Juvenile** Similar to adult but duller and head predominantly yellow-olive; throat yellow. Bill horn. **VOICE** Usually silent. A harsh *chwep.* [CD14:70] **HABITS** Singly, in pairs or small parties in canopy of lowland forest. Climbs large branches and trunks of high trees in search of insects, often hanging upside-down. Joins mixed-species flocks. **SIMILAR SPECIES** Adult male Yellow-capped Weaver very similar to adult male Preuss's, but rump black (not yellow); also does not forage on bark. Adult male Brown-capped Weaver has dark chestnut crown sharply defined from yellow nape. Adult female Brown-capped and Yellow-capped distinguished from female Preuss's by all-black head. **STATUS AND DISTRIBUTION** Scarce to uncommon forest resident, from Sierra Leone and SE Guinea to Ghana and from Cameroon to SW CAR, Gabon and N Congo; probably reaching SE Nigeria (Obudu Plateau).

YELLOW-CAPPED WEAVER *Ploceus dorsomaculatus*
Tisserin à cape jaune

Plate 131 (7)

L 14 cm (5.5"). **Adult male** Top of head golden-chestnut; head-sides and throat black. Upperparts and tail black with irregular golden-yellow stripe on mantle and back. Underparts golden-yellow with slight chestnut wash on breast. Bill black. Eye reddish. **Adult female** Top of head black. **Juvenile** Similar to adult female but duller; head olivaceous; throat olive-yellow. Bill horn. **VOICE** Usually silent. **HABITS** Singly, in pairs or small parties in canopy of lowland forest. Searches foliage in search of insects; also flycatches in open canopy. Not known to forage on bark. Joins mixed-species flocks. **SIMILAR SPECIES** Adult male Preuss's Golden-backed Weaver has yellow rump and uppertail-coverts (at most mottled with some black); forages on bark. Brown-capped Weaver inhabits montane forest; adult male distinguished by dark chestnut crown sharply defined from yellow nape. Adult female Brown-capped distinguished from female Yellow-capped by black of head extending to nape; rump yellow. **STATUS AND DISTRIBUTION** Scarce forest resident, S Cameroon, Gabon and N Congo.

SÃO TOMÉ WEAVER *Ploceus sanctithomae*
Tisserin de Sao Tomé

Plate 144 (7)

L 14 cm (5.5"). **Adult male** Forehead, head-sides (from above eye) and underparts rusty-buff. Crown and centre of nape dark brown to blackish; upperparts dark olive-brown becoming rusty on rump. Greater and median coverts tipped whitish forming double wingbar; flight and tail feathers dark brownish edged olive; tertials edged whitish. Bill long and slender, dark horn. **Adult female** Duller, with brown crown, olive-brown upperparts and pale rusty underparts. **Juvenile** Crown olive, concolorous with upperparts; mantle vaguely streaked dark brown; face and underparts whitish washed buff. **VOICE** An abrupt *wik* or *pwith* often repeated in fast rhythmic series; also a clear, repeated *psink.* Song starts with a few clear call notes, followed by some high-pitched notes and a not unpleasant, accelerating chatter *pwink pwink psi-psi-psi-ch-ch-ch-chur-r-r-r-r;* initial call notes sometimes absent. [CD14:72] **HABITS** In pairs, family parties or small groups of up to 10, in all forested habitats, plantations, scrub and patches of dry woodland in savanna, and towns. Vocal, calling and singing frequently. Gleans insects from leaves. **STATUS AND DISTRIBUTION** Common resident, endemic to São Tomé.

Genus *Malimbus*

Distinctive black-and-red forest weavers. One also has yellow in its plumage and another is black and yellow, lacking red. Bill, legs and feet generally black. Juveniles are duller with horn-coloured bills variably tinged dusky. Calls harsh and rasping.

BLUE-BILLED MALIMBE *Malimbus nitens*　　　Plate 132 (1)
Malimbe à bec bleu

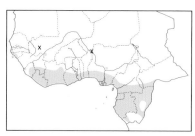

Other name E: Gray's Malimbe
L 17 cm (6.75"). **Adult male** All black with bright red patch from throat to upper breast. Bill bluish. **Adult female** Slightly duller black. **Juvenile** Red area duller, more diffuse, and extending to throat, head-sides and forehead. Bill horn. **VOICE** A harsh rasping *zheep* usually frequently repeated; various equally harsh but shorter calls, e.g. *zhep-zhep* and *zhwep*. Song a series of harsh *zheet* calls followed by some dissonant sounds and ending in a harsh, weaver-like sizzling. [CD14:73] **HABITS** Singly, in pairs or small family parties in dense humid undergrowth of forest interior, clearings and edges with watercourses or pools, and swamp forest; also high mangrove and even dense savanna woodland and thickets, provided trees overhanging water available. Frequently in mixed-species flocks and occasionally gathering in larger groups. Nests almost always above water; usually several (of which many are old) together. **SIMILAR SPECIES** Adult female Red-vented Malimbe also has red, though differently shaped, breast patch; distinguished by red undertail-coverts and black bill. **STATUS AND DISTRIBUTION** Common resident throughout forest zone and adjacent outliers in savanna, from S Senegal to Liberia east to SW CAR and Congo; also S Mali.

CRESTED MALIMBE *Malimbus malimbicus*　　　Plate 132 (2)
Malimbe huppé

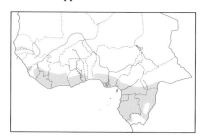

L 17 cm (6.75"). **Adult male** *malimbicus* Black with red crown, ear-coverts and breast. Short crest on hindcrown. Black around base of bill, on lores and around eyes. Bill and legs black. Upper Guinea race *nigrifrons* has slightly more black around base of bill and eyes, and somewhat shorter crest. **Adult female** Lacks crest; red on breast less extensive. **Juvenile** Red area duller, more rusty and extending to nape; throat blackish bordered by small rusty area on upper breast; no crest. Bill dirty horn. **VOICE** A harsh, rasping *zheet*, usually frequently repeated. Also various other harsh chirps occasionally interspersed with a few more musical whistles. [CD14:74] **HABITS** Singly, in pairs or small groups. Mainly mid-levels within lowland forest, at edges and in clearings. Joins mixed-species flocks. **SIMILAR SPECIES** Male Cassin's and Ibadan Malimbes have large black face patches and red areas extending to neck and lower on breast. **STATUS AND DISTRIBUTION** Not uncommon to uncommon forest resident, from W Guinea to Togo (*nigrifrons*) and from Nigeria to SW CAR and Congo (nominate).

CASSIN'S MALIMBE *Malimbus cassini*　　　Plate 132 (3)
Malimbe de Cassin

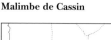

L 17 cm (6.75"). **Adult male** Black with bright red head, neck and breast, except black patch from lores and ear-coverts to upper throat. Bill black. Eye dark brown. **Adult female** Entirely black. **Juvenile** Black with pinkish-red area on head, throat and breast. Bill horn. **VOICE** Series of hard *tuk tuk tuk tuk...* or *tsip tsip tsep tsep...* followed by a drawn-out nasal sizzling wheeze *szszszzuiiiiiin.* [CD14:75] **HABITS** In small, vocal groups in canopy of humid, periodically flooded or dry mature lowland forest. Usually in mixed-species flocks. Dependent on palm trees for nesting. Remarkable nest of 'inverted sock' shape, with spout of up to 1 m. **SIMILAR SPECIES** Male Ibadan Malimbe not safely distinguishable from male Cassin's, though red on underparts usually more extensive and differently shaped at lower edge; identity only certain if pair observed while breeding, though ranges not known to overlap. Female Crowned Malimbe all black as female Cassin's; identified by presence of male. Maxwell's Black Weaver and male nominate Vieillot's Black Weaver also all black, but eye creamy-white or bright yellow. **STATUS AND DISTRIBUTION** Common forest resident, from S Cameroon to SW CAR, Gabon and Congo. **NOTE** Birds identified as male Cassin's Malimbe in Ghana (Grimes 1987) may be Ibadan.

IBADAN MALIMBE *Malimbus ibadanensis*　　　Plate 132 (4)
Malimbe d'Ibadan

L 17 cm (6.75"). **Adult male** Head and neck to breast and upper belly bright red, surrounding black facial patch

(from lores and ear-coverts to throat). Breast patch has black indentation in centre of lower edge. Rest of plumage black. Bill black. **Adult female** Red on underparts restricted to band of variable width from neck-sides over breast. **Juvenile** Head to breast dull reddish except blackish mask. Bill dusky-horn. **VOICE** Song a high-pitched mix of tinkling notes and drawn-out wheezes, resembling Red-headed Malimbe's (and unlike Red-vented's or Blue-billed's). A common phrase is *choop-ee-wurr* followed by a wheeze. Both sexes sing (Elgood 1958). **HABITS** In pairs and small parties at forest edge and in second growth, also gardens. Occasionally associates with Red-headed Malimbe; does not cling to trunks as does Red-headed. Nest of 'inverted sock' shape, with spout of *c.* 20–30 cm. **SIMILAR SPECIES** Male Cassin's Malimbe not safely distinguishable from male Ibadan as extent of red on underparts variable, though normally intermediate between male and female Ibadan; however, ranges do not overlap (but see Elgood 1992 for discussion of problem individuals in Ghana). Male Red-vented Malimbe has plumage pattern similar to female Ibadan but possesses red undertail-coverts. **STATUS AND DISTRIBUTION** Uncommon and local resident in small area of SW Nigeria (circumscribed by Ibadan, Ife, Iperu and Ilaro). Birds identified as Cassin's Malimbe at Tafo and Subri River FR, Ghana, may be Ibadan (only males observed). *TBW* category: THREATENED (Endangered).

RED-VENTED MALIMBE *Malimbus scutatus*   Plate 132 (5)
Malimbe à queue rouge

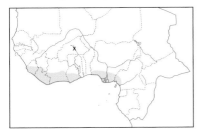

L 17 cm (6.75"). **Adult male** *scutatus* Head and neck to breast bright red, surrounding black facial patch (from lores and ear-coverts to throat). Rest of plumage black with *bright red undertail-coverts*. Bill black. In Lower Guinea race *scutopartitus* red extends lower on breast. **Adult female** *scutatus* As adult male, but head and neck black. In *scutopartitus* black of throat more extensive and extending, variably, through centre of red breast patch, separating it. **Juvenile** As adult female, but red areas paler, pinkish, and extending to throat, head-sides and forehead. Bill horn. **VOICE** A loud, very harsh rasping *zee-zee-zee-zee* and an abrupt, hard *tuk tuk-tuk tuktuktuk tuk ...,* often uttered by several together. [CD14:76] **HABITS** Canopy species occurring in small, noisy groups at forest edges and in clearings, second growth and coastal gallery forest. Nest, suspended from palm fronds (esp. raphia and oil-palms), a conspicuous, beautifully woven structure of 'inverted sock' shape with spout of up to 50 cm. Usually several nests together. **SIMILAR SPECIES** Male Cassin's and female Ibadan Malimbes have similar plumage pattern, but black (not red) undertail-coverts. **STATUS AND DISTRIBUTION** Not uncommon to common resident in forest zone, from Sierra Leone and Guinea to Togo (nominate) and from Benin to W Cameroon (*scutopartitus*).

RACHEL'S MALIMBE *Malimbus racheliae*   Plate 132 (6)
Malimbe de Rachel

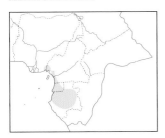

L 17 cm (6.75"). **Adult male** Similar to male Red-vented Malimbe but *lower part of breast patch and undertail-coverts bright orange-yellow*. Bill black. **Adult female** As adult male, but head and neck black. **Juvenile** Similar to adult female, but duller and with red of breast extending to throat. Some reddish around base of horn-coloured bill. **VOICE** A fairly harsh *zhep-zhep*, similar to Blue-billed Malimbe call, but less loud. Song short and ending in a harsh, weaver-like sizzling, reminiscent of Blue-billed Malimbe. **HABITS** In small family groups in canopy and mid-levels of lowland forest interior and edges, actively foraging among leaves. Often in mixed-species flocks. **STATUS AND DISTRIBUTION** Locally not uncommon to uncommon forest resident, from SE Nigeria and W Cameroon to Gabon.

GOLA MALIMBE *Malimbus ballmanni*   Plate 132 (8)
Malimbe de Ballmann

Other name F: Malimbe de Gola

L 17 cm (6.75"). The only black-and-yellow malimbe. **Adult male** Black with orange-yellow to cinnamon nape and bright yellow to orange-yellow breast and undertail-coverts. Breast patch has black indentation in centre of lower edge. Bill black. **Adult female** As adult male, but nape and neck black. **Juvenile** Both sexes have same plumage pattern as respective adults, but duller with yellow area extending to throat, forehead and crown. In juvenile

male yellow nape patch smaller and paler than in adult, often restricted to central nape. Bill horn. **VOICE** Male utters a series of discordant chattering sounds followed by a wheeze, *cheg chig cheg cheg-chega zzzzzzzzzzzzzzzzzzzz*, very similar to song of Village Weaver. Female utters similar series but lacking wheezing part, *cheg cheg chig chag chaaag cheg chiiig* (Gatter & Gardner 1993). **HABITS** Singly, in pairs or small family groups in canopy of lower storey (mainly at 8–22 m) within primary or old secondary lowland forest. Joins mixed-species flocks. **SIMILAR SPECIES** Yellow-mantled Weaver has yellow patch on upper mantle, but is easily distinguished by chestnut breast and belly. **STATUS AND DISTRIBUTION** Rare to locally common forest resident, from E Sierra Leone (Gola Forest) to W Liberia and from E Liberia to W Ivory Coast. *TBW* category: THREATENED (Endangered).

RED-HEADED MALIMBE *Malimbus rubricollis*   Plate 132 (7)
Malimbe à tête rouge

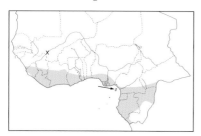

L 18 cm (7"). **Adult male** Bright red from forehead to nape and neck-sides; rest of plumage black. Bill black. In western *bartletti* head is deep crimson, in **nominate** orange-red; *nigeriae* intermediate; island race *rufovelatus* as nominate but with heavier bill. **Adult female** As adult male, but forehead and forecrown black. **Juvenile** As adult female but duller, with red area paler and reddish feathers showing through on head-sides, throat and breast. Bill dark horn. **VOICE** A harsh *zheet*. Song a few rather soft, variable, dissonant whistles ending in a short weaver-like sizzling, e.g. *whuduteew-whuduteew szszszrrrrr* or *whudutseetutseew szszszrrrrr* etc. [CD14:77] **HABITS** In pairs or small family groups in upper levels of lowland forest, along edges and clearings; also in small forest outliers within farmland. Regularly joins mixed-species flocks. Forages like woodpecker (unlike other malimbes), actively examining trunks and large branches, thereby adopting acrobatic, tit-like postures, often upside-down. **STATUS AND DISTRIBUTION** Not uncommon resident in forest zone and savanna outliers. *M. r. bartletti* from W Guinea and Sierra Leone to Togo; also S Mali (rare); *nigeriae* Benin to W Nigeria (west of Lower Niger R.); nominate E Nigeria to SW CAR and Congo; *rufovelatus* Bioko (where the only malimbe).

RED-BELLIED MALIMBE *Malimbus erythrogaster*   Plate 132 (9)
Malimbe à ventre rouge

L 17 cm (6.75"). Distinctive. **Adult male** *Head, neck and entire underparts bright red, except black facial patch* (from lores and ear-coverts to throat). Rest of plumage black. Bill black. **Adult female** As adult male, but with red throat. **Juvenile** As adult female, but head all red with slightly dusky mask; underparts dull pinkish-red becoming greyish-brown on belly. Bill horn. **VOICE** A short, dry *ptsik*, frequently uttered. Clicking notes in flight. [CD14:78] **HABITS** In small family groups in canopy of lowland forest edge. Often in mixed-species flocks. **STATUS AND DISTRIBUTION** Not uncommon to rare forest resident, in SE Nigeria (from Lower Niger R. east), Cameroon, CAR, Gabon and Congo.

RED-CROWNED MALIMBE *Malimbus coronatus*   Plate 132 (10)
Malimbe couronné

L 17 cm (6.75"). **Adult male** *All black except bright red crown*. Bill black. **Adult female** All black. **Juvenile** As adult male but crown patch rufous-brown. Bill horn. **VOICE** A characteristic, drawn-out, wheezy note; also various shorter calls. Song includes wheeze. [CD14:79] **HABITS** In small groups (3–7) in canopy of mature lowland forest. Readily joins mixed-species flocks. Unobtrusive and difficult to observe. **SIMILAR SPECIES** Female Cassin's Malimbe is all black and only identifiable by presence of male. Maxwell's Black Weaver and male nominate Vieillot's Black Weaver also all black, but eye creamy-white or bright yellow. **STATUS AND DISTRIBUTION** Not uncommon forest resident in S Cameroon, SW CAR, Eq. Guinea, Gabon and Congo.

RED-HEADED WEAVER *Anaplectes rubriceps*

Plate 133 (2)

Tisserin écarlate

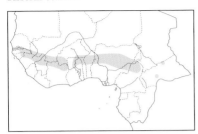

L 14–15 cm (5.5–6.0"). *A. r. leuconotus.* Distinctly coloured *with rather slender, bright red or orange-red bill.* **Adult male breeding** Head to throat and upper breast bright red surrounding black mask (incl. lores, ear-coverts and chin). Upperparts and tail grey-brown; mantle mottled red. Red edges to flight feathers form bright panel; tail also edged red. Rest of underparts white. Bill red or orange-red. **Adult female/adult male non-breeding** Differ in having head greyish-brown, concolorous with upperparts; underparts white washed greyish-buff on throat and breast; edges of flight and tail feathers usually paler red or even yellowish. Bill orangey or pinkish. **Juvenile** Similar to adult female but head to breast washed yellowish-buff; edges of flight and tail feathers orangey. Bill dusky. **VOICE** Usually silent. At nest a rapid sizzling, typical of weavers, but distinctly high pitched and squeaky *sizzi-sizzi-sizzi-sizzi tszrrrrr* etc. Call a sharp *tzik.* [CD14:80] **HABITS** Singly, in pairs or small family parties in open woodland, scrub and *Acacia* savanna. Forages among foliage of trees and bushes, often hanging upside-down. Joins mixed-species flocks. **STATUS AND DISTRIBUTION** Uncommon to rare and local resident, from Gambia and S Senegal to Guinea east to Cameroon and CAR. Also claimed from Gabon. **NOTE** Sometimes placed in genus *Malimbus.*

Genus *Quelea*

Small, gregarious seed-eaters of savanna and similar open habitats. Sexually dimorphic. Breeding males (rainy season) have distinctly coloured heads and bills; in non-breeding plumage (dry season) they resemble nondescript, sparrow-like females. Nest colonially.

RED-HEADED QUELEA *Quelea erythrops*

Plate 133 (7)

Travailleur à tête rouge

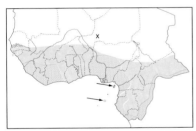

Other name E: Red-headed Dioch
L 11–12 cm (4.5–4.75"). **Adult male breeding** Easily identified by red head and throat contrasting with buff-brown, blackish-streaked upperparts. Flight and tail feathers dark brown narrowly edged yellow. Underparts pale buffish becoming white on central belly and undertail-coverts; some darker brown mottling on breast-sides. Bill blackish. Legs pinkish. **Adult female** Differs in lacking any red in plumage. Top of head brown; head-sides, incl. supercilium, washed pale yellow. Bill brownish-horn. **Adult male non-breeding** As adult female; rarely with red wash on head-sides and forehead.
Juvenile As adult female but with pale sandy edges to feathers of mantle, wing-coverts and tertials. **VOICE** Flocks utter a squeaky chattering, esp. at breeding colony. Song a slightly rising series of churring notes ending in a drawn-out wheeze. [CD14:82] **HABITS** In small or large flocks, sometimes with other small seed-eaters, in coastal thickets, forest clearings, moist grasslands and ricefields. **SIMILAR SPECIES** Adult female Yellow-crowned Bishop differs from female Red-headed Quelea by longer, more conspicuous, yellow supercilium. **STATUS AND DISTRIBUTION** Locally common to uncommon intra-African migrant and resident in moist savanna zone and clearings within forest zone, from Senegambia to Liberia east to Chad, CAR and Congo. Resident, São Tomé. Vagrant, Bioko. Movements may be obscured by similarity with other small weavers in non-breeding season.

RED-BILLED QUELEA *Quelea quelea*

Plate 133 (5)

Travailleur à bec rouge

Other name E: Black-faced Dioch
L 11–13 cm (4.5–5.0"). *Q. q. quelea.* **Adult male breeding** Distinguished by strong, conical *red* bill. Head pattern variable, most having black or dark brown mask of variable shape (usually from forehead over ear-coverts to throat). Mask sometimes lacking, replaced by white or buff-brown. Crown and nape pale to dark buff, often tinged pinkish. Upperparts streaked black and buff-brown; flight and tail feathers dark brown narrowly edged yellowish. Underparts buffish-white or white variably washed tawny or pinkish on breast and belly. Legs pink-orange. **Adult female breeding** Top of head brownish; long, pale buffish or off-white supercilium curving around dusky-grey ear-coverts; small off-white area below eye produces slightly spectacled appearance. Upperparts streaked

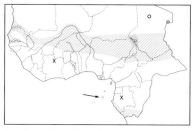

brownish; underparts whitish variably washed pale or dark buffish on breast and flanks. Bill red or gradually turning bright yellow. **Adult male/female non-breeding** As adult female breeding with red bill; legs paler pink. **Juvenile** As adult non-breeding; bill initially pale pink tipped dark grey. **VOICE** Flocks utter rasping and squeaky chattering, esp. at breeding colony. [CD14:83] **HABITS** Highly gregarious, sometimes in flocks of several million. Frequents grasslands, ricefields, open savanna and dry *Acacia* country. **SIMILAR SPECIES** The only other weaver with red bill and black mask is male Red-headed Weaver, which is easily identified by bright red head; mask smaller. Other female *Quelea* and *Euplectes* have dull horn-coloured bills at all seasons. **STATUS AND DISTRIBUTION** Locally common to uncommon resident and intra-African migrant in Sahel zone, from S Mauritania and Senegambia to Chad and N CAR, and in Gabon and S Congo. Esp. abundant in Senegal R. valley, interior delta of Niger R. (Mali) and L. Chad area. Itinerant breeder. Nomadic outside breeding season.

BOB-TAILED WEAVER *Brachycope anomala*   Plate 133 (6)
Travailleur à queue courte

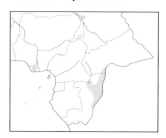

L 11 cm (4.5"). Small and compact with very short tail and relatively thick bill. Adults appear like young birds. **Adult male** Top of head and breast lemon-yellow; head-sides and throat blackish forming rather mottled mask. Upperparts brown heavily streaked darker. Belly to undertail-coverts buffish. Some have irregular, large blackish-brown spots on lower breast and flanks. Bill black. **Adult female** All brownish. Head and underparts buffish-brown; upperparts as adult male. **Juvenile** As adult female but paler; bill pinkish-brown. **VOICE** Call *chk-chk-chk-...* Song consists of a few call notes followed by a melodious, pleasant little trill, first rising in pitch, then descending; resembles combination of Village Weaver and Olive-bellied Sunbird songs. Juvenile utters a high-pitched, vigorous *tsee*, often repeated. [CD14:84] **HABITS** In pairs in villages and clearings along large rivers of Congo basin. **STATUS AND DISTRIBUTION** Locally common to uncommon resident along Congo R. and its larger tributaries, N Congo (Impfondo, Ouesso) and S CAR (Bangui). Vagrant, SE Cameroon.

Genus *Euplectes*

Bishops and widowbirds (also called widows or whydahs) occur in open grassland and at edges of marshes. Sexually dimorphic. Breeding males have conspicuous plumage and behaviour, and are easily identified. Breeding male bishops are black with bright yellow or red, and have short tails; in display they look like brilliantly coloured balls. Male widowbirds are largely black with long tails. In non-breeding season adult males resemble nondescript, sparrow-like females, but are often noticeably larger and, in some species, may retain coloured shoulder patches ('epaulettes'). Juveniles resemble adult females, but have buffy- or yellowish-toned plumage (retained for several months). Subadult males also similar to adult females; they do not acquire nuptial plumage in the first breeding season after hatching. Females and many males in non-breeding plumage are hard to identify and some are not safely separable in the field. Largely polygamous. Most breed during rains, forming mixed-species flocks in the off-season.

YELLOW-CROWNED BISHOP *Euplectes afer*   Plate 135 (5)
Euplecte vorabé

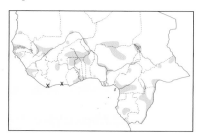

Other names E: Golden Bishop; F: Vorabé
L 11 cm (4.5"). *E. a. afer*. **Adult male breeding** Unmistakable. Bright yellow with large black mask (incl. head-sides and throat), mottled black collar on mantle, and black lower breast and belly. Wings and tail blackish-brown edged pale buff. Yellow breast-band, often tinged chestnut in centre. Bill black. **Adult female breeding** Head and upperparts boldly streaked brownish-buff and blackish; broad yellowish-white supercilium bordered by narrow dusky eye-stripe. Underparts white washed yellowish-buff and sparsely streaked dusky on breast and flanks. Bill horn. **Adult female/male non-breeding** As adult female breeding but more distinctly streaked on breast and flanks. **Juvenile** As adult female breeding but feathers of upperparts, wings and tail have broader and paler buff edges. **VOICE** Series of high-pitched chirping

and buzzing notes, e.g. *szit-szit-szit-...*, esp. in display flight. Call a sharp *tsip*. [CD14:85] **HABITS** Prefers moister habitats than red bishops: swampy areas, floodplains, rank vegetation near rivers, lakes and ponds, ricefields. Display of male consists of low, bouncing flight over territory with puffed-out feathers, uttering buzzing song. Outside breeding season often in mixed flocks of small granivores. **SIMILAR SPECIES** Adult female Red-headed Quelea differs from female Yellow-crowned Bishop by shorter, less conspicuous and less yellow supercilium. Non-breeding Northern Red Bishop slightly larger with less streaky breast. **STATUS AND DISTRIBUTION** Common but local resident throughout, from S Mauritania to Sierra Leone east to Chad, CAR and Congo. Rare, Liberia.

BLACK BISHOP *Euplectes gierowii* Plate 134 (3)
Euplecte de Gierow

L 15–16 cm (6.0–6.25"). *E. g. ansorgei*. **Adult male breeding** Black with bright orange hindcrown, nape, neck-sides and breast-band; mantle golden-yellow. Undertail-coverts fringed grey-buff. Bill black. **Adult female/adult male non-breeding/juvenile** Similar to corresponding plumages of Black-winged Red Bishop, but upperparts more broadly streaked black; entire underparts more washed yellowish-buff and more distinctly streaked on breast-sides. **VOICE** In display flight, male utters a rapid series of thin, silvery notes followed by clearer *tee-ee-ee-ee-eee* and sizzling *see-zee see-zee see-zhe see-zhe SEE-ZHEE*, accelerating as volume increases. Also a buzzing *zee-zee-zee-zee-zee*, often combined with a wheezy *hishaah, hishaah, SHAAAAAAAH, tse-tseet-tseet-tseet* (Zimmerman *et al.* 1996). **HABITS** Singly, in pairs or small groups, in high grass, rank herbage, overgrown farmland and marshy areas with tall grass and bushes. Joins mixed *Euplectes* flocks after breeding. **SIMILAR SPECIES** Black-winged Red Bishop (q.v. and above). **STATUS AND DISTRIBUTION** Scarce and local resident, Cameroon (Belo area, Bamenda Highlands and Meiganga area, Adamawa Plateau) and S CAR (Bangui).

BLACK-WINGED RED BISHOP *Euplectes hordeaceus* Plate 134 (4)
Euplecte monseigneur

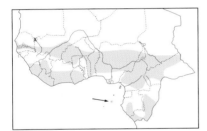

Other names E: Fire-crowned/Black-winged Bishop
L 13–14 cm (5.0–5.5"). *E. h. hordeaceus*. **Adult male breeding** Bright scarlet with contrasting black mask, breast, belly, wings and tail. Undertail-coverts buff. Bill black. **Adult female** Head and upperparts boldly streaked brownish-buff and blackish; yellowish supercilium. Underparts buff inclining to white on belly and faintly streaked brown on breast and flanks. Wings and tail dark brown. Bill horn. **Adult male non-breeding** As adult female but with broader, black streaks on upperparts; *wings black* (conspicuous in flight) with wing-coverts and tertials broadly edged buffish; tail black. **Juvenile** As adult female, but feathers of upperparts, wings and tail have broader buff edges, imparting more buffy appearance. **VOICE** A fast series of twittering, churring and buzzing notes, interspersed by occasional whistles. [CD14:86] **HABITS** Usually in pairs or small groups, in rank herbage within large forest clearings, savanna, edges of cultivation and fallow farmland. Male displays in slow, bouncing cruising flight with rapidly beating, rustling wings and puffed-out plumage, uttering wheezy song. Gathers in flocks with other weavers and bishops in off-season. **SIMILAR SPECIES** Adult male breeding Northern Red Bishop has black crown, brown wings and shorter tail hidden by long tail-coverts. Adult male non-breeding/adult female Northern Red distinguished from corresponding plumages of Black-winged Red principally by brown primaries and lack of yellowish wash to underparts; bill also smaller. Adult female Fan-tailed Widowbird very similar to female Black-winged Red but lesser coverts edged orange. Adult female Red-collared Widowbird also very similar but underparts unstreaked and less yellow, more fulvous. **STATUS AND DISTRIBUTION** Locally common resident, from Senegambia to Liberia east to Chad, CAR and Congo; also São Tomé. Rare, S Mauritania. Absent from arid north.

NORTHERN RED BISHOP *Euplectes franciscanus* Plate 134 (5)
Euplecte franciscain

L 11–12 cm (4.25–5.0"). **Adult male breeding** Bright scarlet with black crown and head-sides, and black lower breast and belly. Wings and tail brown. Scarlet upper- and undertail-coverts very long, covering short tail. Bill black. **Adult female/adult male non-breeding** Head and upperparts boldly streaked brownish-buff and black; buff supercilium. Underparts pale buff inclining to white on belly and faintly streaked brown on breast and flanks. Wings and tail dark brown edged and fringed buffish. Bill horn. **Juvenile** As adult female, but throat and breast washed yellow; feathers of upperparts, wings and tail have broader and paler buff edges. **VOICE** A fast

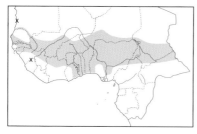

jumble of twittering, churring and wheezy notes, uttered in display flight and from perch. Call a sharp *chiz*. [CD14:88] **HABITS** Gregarious at all times, frequenting moist grassland, seasonal rank vegetation and reedbeds. Male displays from perch or in slow, bouncing cruising flight with rapidly beating, rustling wings and puffed-out plumage, uttering wheezy song. **SIMILAR SPECIES** Adult male breeding Black-winged Red Bishop has red crown, black wings, longer tail, shorter uppertail-coverts and buff undertail-coverts. Adult male non-breeding/adult female Black-winged Red distinguished from corresponding plumages of Northern Red by heavier bill and (in male) black flight feathers. Adult female Yellow-crowned very similar but more distinctly streaked on breast and flanks. **STATUS AND DISTRIBUTION** Common to not uncommon resident, from S Mauritania to Sierra Leone east to Chad and N CAR. Reaches coast from Ghana to W Nigeria.

GOLDEN-BACKED BISHOP *Euplectes aureus* — Plate 144 (9)
Euplecte doré

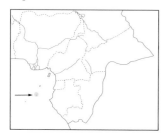

L 12 cm (5"). **Adult male breeding** Black with orange-yellow mantle, back and rump; uppertail-coverts ashy; wing and tail feathers blackish edged white and buff; lower belly and undertail-coverts white. Bill black. **Adult female/ adult male non-breeding** Head and upperparts *bright cinnamon-brown* (orange-buff) *heavily streaked black*. Face (incl. supercilium) and throat lemon; breast and flanks cinnamon-brown contrasting with pure white belly and vent; breast-sides streaked black. Bill horn. **Juvenile** As adult female but washed yellowish. **VOICE** A rasping *tzep* or *dzik*. Song, uttered in flight or from perch, a monotonous series of *dzik* notes. [CD14:89] **HABITS** In grassy areas near farmland. **SIMILAR SPECIES** Male non-breeding and female plumages are quite distinctive within genus. The only other *Euplectes* on São Tomé are Black-winged Red Bishop and White-winged Widowbird (q.v.). **STATUS AND DISTRIBUTION** Common resident, São Tomé.

YELLOW BISHOP *Euplectes capensis* — Plate 135 (4)
Euplecte à croupion jaune

Other name E: Yellow-rumped Widow
L male 14 cm (5.5"), female 11.0–12.5 cm (4.25–5.0"). *E. c. phoenicomerus.* The only *Euplectes* with yellow or yellowish rump in all plumages. **Adult male breeding** *Black with contrasting yellow shoulders and rump* (in fact, rump black but obscured by long yellow feathers of lower back). Wing feathers blackish-brown with pale edges. Bill black above, whitish below. **Adult female** Head and upperparts heavily streaked brownish-buff and blackish, with indistinct, narrow, pale supercilium, yellow-washed shoulders and *dull yellowish rump.* Underparts pale buffish-brown, streaked dark brown on breast and flanks. Bill dark horn above, paler below. **Adult male non-breeding** As adult female, but larger and *retaining yellow shoulders and rump.* **Juvenile** As adult female but lacks yellow on shoulders and rump. **VOICE** Thin *tseet* and *tsit.* In display flight loud wing rattling interspersed with short, fast series of high-pitched *tsip-tseep* or nasal *tzeep* notes. [CD14:90] **HABITS** Confined to grasslands in montane and hilly areas. Male displays from perch or in cruising flight with rustling wings and ruffled neck and rump plumes. **SIMILAR SPECIES** Female Yellow-mantled Widowbird has underparts much less streaked and lacks dull yellow rump. **STATUS AND DISTRIBUTION** Locally common resident in highlands of SE Nigeria (Obudu and Mambilla Plateaux), Cameroon (Mt Cameroon to Adamawa Plateau) and Bioko.

FAN-TAILED WIDOWBIRD *Euplectes axillaris* — Plate 134 (2)
Veuve à épaulettes orangées

Other name E: Red-shouldered Whydah
L male 17–18 cm (6.75–7.0"), female 13–14 cm (5.0–5.5"). **Adult male breeding** *bocagei Black with orange-yellow shoulders* (lesser wing-coverts) *above pale chestnut wing patch* (median and greater coverts). Tail rather short but broad, often fanned. Bill pale bluish. Western *batesi* has shoulders more reddish-orange. **Adult female** Head and upperparts buff-brown heavily streaked black; buff supercilium; *lesser coverts black edged orange* or *cinnamon*

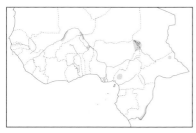

(diagnostic). Underparts whitish washed buff and faintly streaked brown on breast and flanks. Bill horn. **Adult male non-breeding** As adult female, but larger and *retaining diagnostic orangey shoulders* and *black flight feathers*. **Juvenile** As adult female. **VOICE** Series of weak *tseep* and similar notes, interspersed by rolling husky *twirrrlll*, uttered in display flight and from perch. [CD14:91] **HABITS** In seasonally flooded grasslands, swamps, marshes and edges of rivers and lakes with rank herbage. Male has slow flapping display flight ending with dive into grass. **SIMILAR SPECIES** Adult male breeding Marsh Widowbird has much longer tail and lacks pale chestnut wing patch. Adult male non-breeding/adult female/juvenile Marsh very similar to corresponding plumages of Fan-tailed, but have darker, more uniformly buff-brown and streaked underparts. Adult female Black-winged Red Bishop similar but lacks orange on shoulders. **STATUS AND DISTRIBUTION** Locally not uncommon to scarce resident. *E. a. batesi* along Niger R. from C to NE Mali and W Niger; *bocagei* in NE Nigeria, W Chad and N Cameroon (L. Chad area), and in W Cameroon (highlands), CAR and Congo (locally along Congo R.).

YELLOW-MANTLED WIDOWBIRD *Euplectes macrourus* Plate 135 (1)
Euplecte à dos d'or

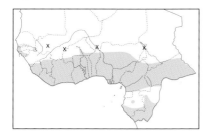

Other names E: Yellow-mantled Whydah; F: Veuve à dos d'or
L male breeding 19–21 cm (7.5–8.25"), female 13–14 cm (5.0–5.5").
E. m. macrourus. **Adult male breeding** Unmistakable. *Black with contrasting yellow mantle and shoulders, and long, graduated tail.* Bill blue-black. **Adult female** Head and upperparts streaked buff-brown and blackish; lesser coverts tipped yellow. Underparts pale buff, faintly streaked dusky on breast and flanks. Bill horn. Best identified through association with male. **Adult male non-breeding** As adult female, but larger and *retaining yellow shoulder patches*; flight feathers black. **Juvenile** As adult female but yellow-tinged underparts. **VOICE** Rhythmic series of thin, buzzing *zeet* and *tsweep* notes uttered in display flight and from perch. [CD14:92] **HABITS** In moist grassland, rank herbage on fallow farmland, marshy areas and large grassy clearings. Male displays in low cruising flight above territory with ruffled neck and rump plumes, uttering buzzing song. Outside breeding season in mixed flocks of bishops and widowbirds. **SIMILAR SPECIES** Adult female White-winged Widowbird very similar to female Yellow-mantled, but has white underwing-coverts; as these usually not seen, both species not safely distinguishable in field. **STATUS AND DISTRIBUTION** Common resident in savanna and forest zones, from S Senegambia to Liberia east to Chad, CAR and Congo. Rare, Mali. Absent from arid north. Local movements possibly occur where habitat dries out, but these obscured due to nondescript off-season plumage of males.

WHITE-WINGED WIDOWBIRD *Euplectes albonotatus* Plate 135 (3)
Euplecte à épaules blanches

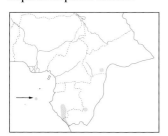

Other name E: White-winged Whydah
L male breeding up to 26 cm (*c.* 10"), non-breeding 14–15 cm; female 12–13 cm (4.75–5.0"). *E. a. asymmetrurus.* **Adult male breeding** Black with yellow shoulders, small but diagnostic white patch on closed wing (formed by white primary-coverts) and long, narrow tail. In flight, white bases to flight feathers produce conspicuous wingbar. Bill pale blue. **Adult female** Head and upperparts streaked buff-brown and black; lesser coverts tipped yellow or cinnamon. Head-sides (incl. supercilium) and throat washed pale yellow. Underparts whitish, washed buff and finely streaked brown on breast and flanks. Bill horn. **Adult male non-breeding** As adult female but larger and *retaining yellow shoulders and white in wings*. Flight feathers black. **Juvenile** As adult female, but head-sides, throat and breast more strongly washed yellowish-buff. **VOICE** Usually rather silent. A rapid, rhythmic series of identical buzzing or chirping notes; also a dry, rustling papery sound. Twittering contact calls. [CD14:93] **HABITS** In dry grasslands and areas with rank herbage, small trees and bushes. Usually in small flocks, often with other bishops and widowbirds. Male displays from perch with fanned tail or in rather slow cruising flight on rapid wingbeats, uttering buzzing song. **SIMILAR SPECIES** Adult female Yellow-mantled Widowbird may be distinguished from female White-winged by lack of yellowish wash to face, but this rather hard to see. **STATUS AND DISTRIBUTION** Uncommon and local resident, S Gabon and Congo (Bateke Plateau); also W CAR. Locally common, São Tomé.

RED-COLLARED WIDOWBIRD *Euplectes ardens*

Euplecte veuve-noire

Plate 134 (1)

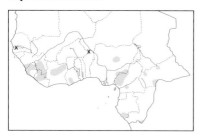

Other names E: Long-Tailed Black Whydah; F: Veuve noire
L male 14 cm (5.0–5.5"), tail up to *c.* 20 cm (10") when breeding;
female 12–13 cm (4.75–5.25"). *E. a. concolor.* **Adult male breeding**
Unmistakable. *All black with very long, supple, graduated tail.* Bill black.
In south of range (S Gabon, Congo) some have yellow, orange or
red crescent-shaped collar on lower throat. **Adult female** Head and
upperparts buff-brown heavily streaked black; wing feathers dusky
edged buff. Head-sides and throat washed yellow; rest of underparts
unstreaked fulvous washed yellow on breast and flanks and becoming
white on central belly. Bill horn. **Adult male non-breeding** As adult
female, but larger and has black wing feathers; undertail-coverts marked dusky. **Juvenile** As adult female but
usually has brighter yellow wash to supercilium and throat, and rather broader and paler tawny streaks above.
VOICE A rapid series of buzzes, chirps and ticks, uttered in display flight and from perch. [CD14:94] **HABITS**
Favours hillsides and plains with tall grass; also rank herbage at edges of farmland. Display of male consists of
slow, bouncing flight above territory with rapid wingbeats and depressed tail, uttering song; also short, low flight,
with intermittent drops into grass. Outside breeding season in flocks, often with other *Euplectes*, foraging mainly
on ground. **SIMILAR SPECIES** Adult female/juvenile Black-winged Red Bishop are more yellowish and faintly
streaked on throat and breast. **STATUS AND DISTRIBUTION** Patchily distributed, locally common resident,
from W Guinea to Togo and from C and SE Nigeria (Jos, Obudu and Mambilla Plateaux) to CAR and Congo.
Rare, Gambia (no recent records). Claimed from SW Niger.

MARSH WIDOWBIRD *Euplectes hartlaubi*

Euplecte des marais

Plate 135 (2)

Other names E: Marsh Whydah, Hartlaub's Marsh Widowbird
L male breeding 22 cm (8.75"), female 16 cm (6.25"). *E. h. humeralis.* **Adult
male breeding** Black with orange-yellow shoulders and long, broad black
tail. Strong, pale blue bill. **Adult female** Head and upperparts brown boldly
streaked black; lesser coverts edged dull yellow. Underparts buffish-brown,
streaked brown on breast and flanks. Bill horn. **Adult male non-breeding**
Like adult female, but larger and *retaining orange-yellow shoulder patches.*
Juvenile As adult female, but lesser coverts edged tawny and underparts browner. **VOICE** A short, abrupt, metallic chirping or loud *yek!* followed by a soft
high-pitched sizzling or buzzy trill, uttered in display. Male also gives a short
dry *krrrt.* [CD14:96] **HABITS** In moist grasslands, boggy marshes with sedges, swampy valleys with lush grass and
rank streamside vegetation, in both low- and highlands. Male displays in low cruising flight over territory with
ruffled neck plumes, often starting from particular vantage point; frequently flicks tail when perched. **SIMILAR
SPECIES** Breeding male Fan-tailed Widowbird has much shorter tail and pale chestnut patch below orange-
yellow shoulders; non-breeding male retains chestnut wing patch. Adult female/juvenile Fan-tailed have paler underparts with buffish-brown and streaking more confined to breast; belly whitish. **STATUS AND DISTRIBUTION**
Local and uncommon to scarce resident in E Nigeria (Mambilla Plateau), Cameroon, Gabon and Congo.

PARASITIC WEAVER *Anomalospiza imberbis*

Anomalospize parasite

Plate 133 (4)

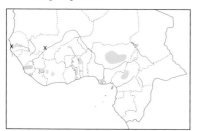

Other name F: Tisserin parasite
L 13 cm (5"). Small with short tail and diagnostic, *small, conical bill*
(unlike any *Ploceus*). **Adult male** Head and underparts bright yellow;
upperparts yellow-green streaked dark brown. Lower flanks streaked
dusky. Bill black. Plumage becomes brighter and yellower with wear.
Adult female Head and upperparts buff-brown boldly streaked black;
buff supercilium. Pale ear-coverts and lack of eye-stripe impart plain-
faced appearance. Underparts buffish becoming whitish on belly
and variably streaked dark brown on breast and flanks. Head-sides
to throat washed yellow in some. Bill dark horn above, paler below.
Juvenile As adult female, but more tawny or sandy-buff. **VOICE** Song a fast series of squeaky, sibilant notes *tslee-*
tslee-tslee-..., uttered from perch or in display flight. Flocks give soft chattering notes in flight. [CD14:97] **HABITS**

Singly or in pairs in open or lightly wooded grassland and cultivation. Gregarious in non-breeding season. Parasitises cisticolas, prinias and other small passerines. **SIMILAR SPECIES** Female Northern Red Bishop distinguished from female Parasitic Weaver by longer, less stubby bill. **STATUS AND DISTRIBUTION** Very local and patchily distributed, uncommon to rare resident, recorded from Sierra Leone, Guinea, Ivory Coast, Ghana, Togo, Nigeria and Cameroon. Single records from Gambia and Mali.

GROSBEAK WEAVER *Amblyospiza albifrons* Plate 133 (3)
Amblyospize à front blanc

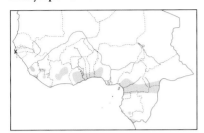

Other names E: Thick-billed Weaver, White-fronted Grosbeak; F: Grosbec à front blanc
L 18 cm (7"). Robust with massive bill. **Adult male** *capitalba* Head, mantle, throat and upper breast chestnut; *contrasting white patch on forehead and base of primaries.* Rest of upperparts and tail brownish-black. Lower breast and belly sooty-grey. Bill black. In flight, white primary patch forms conspicuous wingbar. Eastern *saturata* darker overall. **Adult female** Head and upperparts dark brown. *Underparts whitish heavily streaked brown.* Bill dark horn, paler below. **Juvenile** As adult female, but bill yellowish-horn. **VOICE** Usually silent. Song a vigorous, fast jumble of hard, rasping, chattering and churring notes. Chirping and twittering calls uttered in flight and at breeding colony. [CD14:98] **HABITS** Usually in small scattered colonies in rank grass, reedbeds and tall vegetation in swamps, marshy areas and along rivers; occasionally in tall herbage within dry areas. Forages in forest canopy. Male display consists of circular flights above territory and frequent flicking of wings and tail on perch (usually tall reed stem). **SIMILAR SPECIES** Adult female may recall similar-sized Thick-billed Seedeater, but latter has white forehead and double white wingbar and is a highland species with restricted range. **STATUS AND DISTRIBUTION** Locally not uncommon resident in forest zone and adjacent savanna, from Sierra Leone to SW Nigeria (*capitalba*) east to SW CAR, and in S Gabon and SW Congo (*saturata*). Also S Senegal (1 record). Local movements recorded.

ESTRILDID FINCHES
Family ESTRILDIDAE (World: *c.* 130 species; Africa: 69; WA: 49)

Large family of small and very small passerines with short, generally conical bills. Reach their greatest diversity in the Afrotropical region. Sexes similar or not. Juveniles a paler version of adults or quite different. No seasonal plumage variation. Occur from semi-desert and open savanna to forest. Predominantly granivorous, but *Nesocharis* and forest-dwelling *Nigrita* and *Parmoptila* largely or entirely insectivorous. Forest species usually in pairs or small parties; open country estrildids may gather in flocks, especially outside breeding season. Principally sedentary, but local movements may occur in drier parts of range. Songs weak, consisting of high-pitched notes. Some are parasitised by whydahs and indigobirds, nestlings of host and nest parasite having similar mouth markings and juvenile plumages.

RED-HEADED ANTPECKER *Parmoptila woodhousei* Plate 136 (4)
Parmoptile à gorge rousse

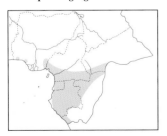

Other names E: Flower-pecker Weaver-Finch; F: Parmoptile à tête rouge
L 11 cm (4.25"). *P. w. woodhousei*. Small, warbler-like forest species with slim bill and rather short, rounded tail. **Adult male** Forehead has small red bars; head-sides and throat dull orange-rufous; crown, upperparts and tail dull brown. Underparts whitish densely speckled or mottled brown. **Adult female** Lacks forehead patch; underparts duller, less densely marked. **Juvenile** All dull brown, paler, more rufous on underparts. **VOICE** A soft, thin *tseeu* and *tsee*. [CD15:1] **HABITS** In pairs or small family groups in undergrowth of lowland forest; often along edges and small streams. Readily joins mixed-species flocks. Feeds largely on ants. Unobtrusive, but not shy. **STATUS AND DISTRIBUTION** Uncommon to scarce forest resident in S Nigeria, and from S Cameroon to SW CAR and Congo.

RED-FRONTED ANTPECKER *Parmoptila rubrifrons*　　　　　Plate 136 (2)
Parmoptile à front rouge

Other name E: Red-fronted Flower-pecker Weaver-Finch
L 10–11 cm (4.0–4.25"). *P. r. rubrifrons*. Small, warbler-like forest species with slim bill and rounded tail. **Adult male** Dark earth-brown above with bright red patch from forehead to mid-crown; underparts rufous-brown. Head-sides finely speckled white (hard to see in field). **Adult female** All dark earth-brown above; underparts pale cream densely spotted dark brown. **Juvenile** Earth-brown above; paler, more rufous below. **VOICE** A rather vigorous, hoarse *whseeeet*. **HABITS** In pairs or small family groups at lower levels (below 5 m) of lowland forest; occasionally at edges. Often joins mixed-species flocks. Feeds largely on ants. Unobtrusive and easily overlooked, but not shy.
STATUS AND DISTRIBUTION Inadequately known. Scarce to locally not uncommon forest resident, E Sierra Leone (Gola Forest), SE Guinea, Liberia, Ivory Coast (Taï NP, Yapo Forest) and Ghana. Also a record from extreme SE Mali (Misséni), in gallery forest.

GREY-CROWNED NEGROFINCH *Nigrita canicapilla*　　　　　Plate 137 (1)
Nigrette à calotte grise

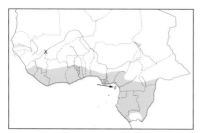

Other name F: Sénégali nègre
L 15 cm (6"). **Adult** *emiliae* Forehead, head-sides and underparts black; hindcrown and upperparts grey inclining to pale grey on rump; wings and tail black; wing-coverts and (in some) tertials have small, indistinct white tips. Bill black. Eye orange-red. In eastern (**nominate**) race wing-coverts and tertials are distinctly tipped with white spots; grey on head bordered by narrow white line on crown and along ear-coverts; rump inclining to whitish. **Juvenile** Wholly sooty-black; eye pale grey. **VOICE** Characteristic. Song a frequently repeated, short series of *c.* 2–7 distinctive, rather plaintive, whistles.
In western *emiliae* a stereotyped *thiuu hee tiuuu*; more varied in nominate, e.g. *theewhee theewhee thu-u-u-u* or *theehu theehu hu-u-u* with last note slightly quivering, faster *tiu-wheet-wheet-tiu wheet-wheet-tiu*, short *tiu-tiu tee*, longer *huhuhu heehee heehee* and other variations. [CD15:2] **HABITS** Singly, in pairs or small parties, in lowland and montane forest, edges, gallery forest and overgrown cultivation with tall trees. Mostly in canopy. Joins mixed-species flocks. Feeds mainly on insects and fruit. **SIMILAR SPECIES** Male Pale-fronted Negrofinch is noticeably smaller and has whitish- or buffish-tinged forehead. **STATUS AND DISTRIBUTION** Common to uncommon forest resident, from Sierra Leone and SE Guinea to Togo (*emiliae*) and from S Benin to CAR and Congo; also Bioko (nominate). A sight record from S Mali.

PALE-FRONTED NEGROFINCH *Nigrita luteifrons*　　　　　Plate 137 (4)
Nigrette à front jaune

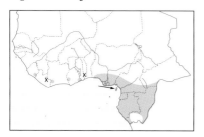

Other name F: Sénégali nègre à front jaune
L 11.5 cm (4.5"). **Adult male** *luteifrons* Forehead variably grey, pale buff or yellowish-buff; top of head and upperparts grey inclining to pale grey on rump; head-sides, underparts, flight feathers and tail black. Bill black. Eye red, pale grey or creamy-white. Legs brownish-pink. In island race *alexanderi* forehead and crown more extensively tinged yellowish-buff. **Adult female** All grey with small, ill-defined blackish mask; some whitish or yellowish-buff on forehead; black wings and tail. Eye pale. **Juvenile** Wholly dull grey; underparts slightly tinged brownish; eye pale. **VOICE** Song a simple, descending series of 4–6 whistles *wee-wee-wee-wee-wee-whuuuh*. [CD15:3] **HABITS** Singly, in pairs or small family parties at forest edges, in second growth and thickets in savanna. Mainly in canopy, but also lower. Joins mixed-species flocks. **SIMILAR SPECIES** Grey-headed Negrofinch is noticeably larger and has black forehead. **STATUS AND DISTRIBUTION** Uncommon to locally not uncommon resident in forest zone, from S Nigeria to SW CAR and Congo (*luteifrons*); also Bioko (*alexanderi*). Rare in Upper Guinea forest block in Sierra Leone, Liberia, SW Ivory Coast (Taï NP), SE Ghana and SW Togo.

CHESTNUT-BREASTED NEGROFINCH *Nigrita bicolor* Plate 137 (2)
Nigrette à ventre roux

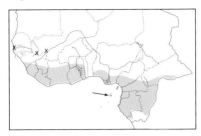

Other name F: Sénégali brun à ventre roux
L 11–12 cm (4.5"). **Adult male *bicolor*** Crown, upperparts and tail dark slate-grey; forehead, head-sides and underparts dark chestnut. Eastern ***brunnescens*** slightly duller, with more brown-grey upperparts. **Juvenile** Dark dull brown above; paler, warmer brown below. **VOICE** Song a series of cheerful whistles, e.g. *weet-weet-teeu weet-weet-teeu weet-weet-tiutiutiuhuu tiutiuhuu*, or rising *tuwee-tuwee-twee-twee* and *tititi-tiu-tiu-teeu tiu-tiu-teeu*. [CD15:4] **HABITS** Singly, in pairs or small family groups, frequenting all levels of lowland forest. Sometimes joins mixed-species flocks. **SIMILAR SPECIES** Juvenile Red-fronted Antpecker has slender bill and smaller size. **STATUS AND DISTRIBUTION** Locally common to uncommon resident in forest zone. Nominate in W Gambia (rare), SW Senegal, from W Guinea and Liberia east to Togo and (this race?) Benin, and in SW Mali; *brunnescens* from S Nigeria to S CAR and Congo, and on Príncipe.

WHITE-BREASTED NEGROFINCH *Nigrita fusconota* Plate 137 (3)
Nigrette à ventre blanc

Other name F: Sénégali brun à ventre blanc
L 10 cm (4"). Small, slender, warbler-like negrofinch with white underparts. **Adult *fusconota*** Head and nape glossy blue-black; upperparts brown; wings dark brown; uppertail-coverts and tail black. Throat pure white, rest of underparts off-white. Bill slender, black. Western ***uropygialis*** has rump noticeably paler, contrasting with back. **Juvenile** As adult but slightly duller. **VOICE** Variable. A short phrase of thin but vigorous, clear whistles *wheet-huw wheet heeew* and *hweet hwuut hweet* or *tsieet tsieet tsiuu tsiuu* and *tsip-tsip-rruuu tsee-uu*; also a melodious descending series, *hwitwitwitwitwitwit* and a trill, followed by some slower notes at end *trrrrrrrititititsiutsiutsiutsiu*. [CD15:5] **HABITS** Singly, in pairs or small parties in canopy of mature lowland forest, forest edges and gallery forest. Attracted to oil-palms, feeding on oily husk of palm nut. Joins mixed flocks of small insectivores. Unobtrusive, but vocalisations attract attention. **STATUS AND DISTRIBUTION** Uncommon to locally not uncommon forest resident. *N. f. uropygialis* from Sierra Leone and SE Guinea to S Ghana, in S Mali and W Nigeria (west of Lower Niger R.); nominate east to SW CAR, Congo and Bioko.

LITTLE OLIVEBACK *Nesocharis shelleyi* Plate 137 (6)
Dos-vert à tête noire

Other names E: Fernando Po/Shelley's Olive-back; F; Petit Sénégali vert
L 8.0–8.5 cm (3.0–3.25"). Very small and short tailed, with distinctive coloration. **Adult male *shelleyi*** Head black separated by grey nuchal collar from olive upperparts, latter becoming golden-yellow on rump. Tail black. Narrow white stripe at sides of black throat; breast yellowish-olive; rest of underparts grey. Bill rather stout, bluish with black cutting edges and culmen. *N. s. bansoensis* is somewhat darker above and usually slightly larger. **Adult female** White stripe on throat-sides reduced or lacking. Breast grey as rest of underparts. **Juvenile** As adult female but duller. **VOICE** An unobtrusive, continuous, sharp, very high-pitched twittering of *tsip* notes. [CD15:7] **HABITS** Singly, in pairs or small parties (of up to *c*. 10), in montane forest clearings and edges; also plantations and bushy savanna. Actively forages at all heights, often hanging upside-down, gleaning branches and leaves. Easily overlooked. **STATUS AND DISTRIBUTION** Locally common to uncommon forest resident above 1200 m, in SE Nigeria (Obudu and Mambilla Plateaux), W Cameroon (Mt Cameroon, Mt Kupe, Bakossi Mts, Mt Manenguba, Bamenda Highlands) and Bioko. *N. s. bansoensis* throughout, except Mt Cameroon and Bioko, where nominate. Records at 50–100 m on Bioko may indicate post-breeding altitudinal movements.

WHITE-CHEEKED OLIVEBACK *Nesocharis capistrata* Plate 137 (5)
Dos-vert à joues blanches

Other names E: Grey-headed Olive-back; F: Sénégali vert à joues blanches
L 12–13 cm (4.75–5.0"). **Adult** Thickset with characteristic head pattern: white face (white forehead extending

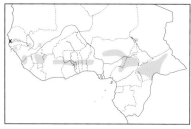

onto lores and area behind and below eye), grey crown and nape, and black throat extending as a line below ear-coverts. Upperparts and tail olive-green; underparts grey with conspicuous yellow flanks. Bill short and stubby, dark grey. **Juvenile** Duller, overall more dusky, lacking white face; bill whitish. **VOICE** Usually silent. Call a thin, high-pitched *tsip*. Song a fast, descending series of high-pitched notes *tsi-tsi-tsi-tsi-tsi* or slower, more drawn-out *tseee-tseee-tseee*. [CD15:8] **HABITS** Singly or in pairs in moist grassland and bush, forest edges and riparian woodland. Skulking and usually keeping low, though may search foliage at all heights in rather warbler-like manner. **STATUS AND DISTRIBUTION** Uncommon and local, patchily distributed resident, recorded from Gambia (rare), Guinea-Bissau, Guinea, Sierra Leone, S Mali, Burkina Faso, Ivory Coast, Ghana, Togo, Nigeria, Cameroon and CAR.

Genus *Pytilia*

Medium-sized estrildids with red rumps and tails, and rather slender bills. Frequently in trees and bushes, but feed on ground. Parasitised by paradise whydahs.

GREEN-WINGED PYTILIA *Pytilia melba*   Plate 136 (5)
Beaumarquet melba

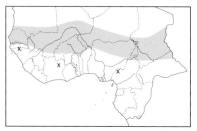

Other names E: Melba Finch; F: Beaumarquet
L 12–13 cm (4.75–5.0"). **Adult male breeding** *citerior* Forehead and forecrown to throat scarlet; hindcrown, nape and neck-sides grey; upperparts olive-green; uppertail-coverts and tail crimson. Upper breast golden-yellow; rest of underparts white with pale olive barring on lower breast and flanks. Bill red. **Nominate** has red face separated by grey area of head extending around eye and over lores; upper breast greenish. **Adult male non-breeding** Similar, but bill has dusky culmen. **Adult female** Red on head replaced by grey; upper breast white barred grey. Bill dull reddish with brownish culmen. **Juvenile** Much duller and browner than adult female; underparts buffish without bars, but sometimes with darker crescents, giving scaly appearance. Bill dark, becoming red with age. **VOICE** Usually silent. A drawn-out, thin and penetrating whistle *see-eh* or *heee*, often monotonously repeated and difficult to locate; also a short *wik* or *kip*. Song a variable mix of buzzing and croaking sounds interspersed with call notes. [CD15:9] **HABITS** Usually in pairs or small family parties in *Acacia* woodland and thorn scrub, to edge of desert; also in savanna woodland and edges of cultivation. Forages on ground. Unobtrusive and often shy. Parasitised by Sahel Paradise Whydah. **SIMILAR SPECIES** Orange-winged Pytilia has orange (not green) on wings and shorter tail; adult male also has grey upper breast. **STATUS AND DISTRIBUTION** Uncommon to locally not uncommon resident in semi-arid belt, from S Mauritania and Senegal to Chad and CAR, also Guinea-Bissau, Guinea and N Ivory Coast (*citerior*), and in SE Congo (nominate).

ORANGE-WINGED PYTILIA *Pytilia afra*   Plate 136 (8)
Beaumarquet à dos jaune

Other name E: Golden-backed Pytilia
L 11 cm (4.25"). **Adult male** Red face bordered by grey band from crown to upper breast; upperparts yellow-green; edges of wing feathers orange, forming bright panel on folded wing; rump and tail bright red. Breast to undertail-coverts olive finely barred whitish on lower breast, bars becoming broader on rest of underparts. Bill red. **Adult female** Duller with entirely grey head (no red face); barring on underparts starting on lower throat. **Juvenile** As adult female, but duller. **VOICE** A high-pitched, drawn-out *seee*, a short, abrupt *pit pit* or *pwit pwit* and a low *tiu*. [CD15:10] **HABITS** In pairs or small family parties in open wooded savanna. Perches atop trees and bushes; feeds on ground. **SIMILAR SPECIES** Male Green-winged Pytilia has yellow or greenish (not grey) breast-band; both sexes have all-green wings. **STATUS AND DISTRIBUTION** Locally not uncommon resident, SC Congo.

YELLOW-WINGED PYTILIA *Pytilia hypogrammica*   Plate 136 (7)
Beaumarquet à ailes jaunes

Other names E: Red-faced Pytilia, Golden-winged Pytilia
L 12.5–13.0 cm (5"). **Adult male** Grey with red face (incl. forehead, head-sides and throat), golden-yellow wings,

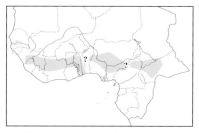

and crimson rump and tail. Belly to undertail-coverts barred whitish. Bill black. **Adult female** Paler, brown-grey, lacking red face and with lower breast to undertail-coverts pale grey-brown barred whitish. **Juvenile** Similar to adult female but paler, browner and duller, with less yellow in wing and less distinct barring on underparts. **VOICE** A frequently repeated *tsiee* or *tsee*. Usually silent. [CD15:11] **HABITS** Singly or in pairs in open savanna woodland, thickets and abandoned farmland. Occasionally with other waxbills or small weavers. Feeds on ground. Parasitised by Togo Paradise Whydah and probably also Exclamatory Paradise Whydah. **SIMILAR SPECIES** Red-winged Pytilia has red wings; male lacks red face. **STATUS AND DISTRIBUTION** Uncommon to locally not uncommon resident, from W Guinea and Sierra Leone to Cameroon (Adamawa Plateau and foothills of Bamenda Highlands), S Chad and CAR.

RED-WINGED PYTILIA *Pytilia phoenicoptera*   Plate 136 (10)
Beaumarquet aurore

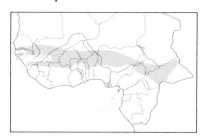

Other name E: Crimson-winged Pytilia
L 12.5–13.0 cm (5"). *P. p. phoenicoptera*. **Adult male** Grey with crimson-edged wing feathers forming bright panel on folded wing; rump and tail crimson. Lower breast to undertail-coverts delicately barred grey and white. Bill greyish-black. Eye scarlet. **Adult female** Paler, more brownish-grey. **Juvenile** Similar to adult female but browner and duller with fainter barring on underparts. **VOICE** A single *twink* or *klink*. Song a short, variable phrase, consisting of some hard and occasional whistles followed by a rolling, croaking sound; short version e.g. *chirp-chit che-chrrrreew*. [CD15:12] **HABITS** Singly or in pairs in various types of open wooded savanna, thickets and grassland; also edges of farmland with tall grass and bushes. Occasionally with other small granivores. Parasitised by Exclamatory Paradise Whydah. **SIMILAR SPECIES** None in its range. Perhaps superficially similar to Dybowski's Twinspot, when briefly glimpsed. **STATUS AND DISTRIBUTION** Uncommon to locally not uncommon resident, from Senegambia to Sierra Leone east to S Chad and CAR. **NOTE** Birds from N Cameroon east sometimes included in doubtfully distinct race *emini*, described as tending to have underwing-coverts and barring on underparts slightly darker.

RED-FACED CRIMSONWING *Cryptospiza reichenovii*   Plate 136 (6)
Sénégali de Reichenow

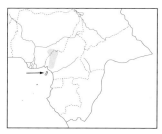

Other name F: Bengali de Reichenow
L 11–12 cm (4.25–4.75"). *C. r. reichenovii*. **Adult male** Head and underparts dull olive with contrasting bright red eye patch and deep red upperparts. Flight and tail feathers blackish. Red patch on rear flanks in some. Bill black. **Adult female** Eye patch yellowish-buff; upperparts duller. **Juvenile** Like adult female, but browner and duller; eye patch faint or absent. **VOICE** A high-pitched *tseet*. Song, only audible at very close range, includes sharp, insect-like syllables, such as drawn-out and slightly descending *tsrrrrrr* and short *tsrrp-tsrrp*. [CD15:13] **HABITS** In pairs or small parties in dense undergrowth and edges of montane forest. Occasionally joins flocks of other estrildids. Usually feeds on ground. Secretive and easily overlooked. **STATUS AND DISTRIBUTION** Not uncommon to uncommon resident in highlands of SE Nigeria (Obudu Plateau), W Cameroon (Mt Cameroon, Rumpi Hills, Mt Kupe, Bakossi Mts, Mt Manenguba, Bamenda Highlands) and Bioko.

BLACK-BELLIED SEEDCRACKER *Pyrenestes ostrinus*   Plate 137 (11)
Pyréneste ponceau

Other name F: Grosbec ponceau à ventre noir
L 15 cm (6"). ¶ Stout, with distinctive heavy, triangular bill. **Adult male** Entire head to breast and upper flanks, rump and tail bright red. Upperparts and rest of underparts blue-black. White, broken, orbital ring. Bill steel-blue; size variable. **Adult female** Head to throat and upper breast, rump and tail bright red; rest of plumage warm brown. **Juvenile** Wholly olive-brown with dull reddish rump and tail. **VOICE** Common call a soft, repeated *tak*; also a low metallic *peenk* and a hard *trrr*. Song a very fast, short, warbling phrase, abruptly ending with *swit*. [CD15:14] **HABITS** Usually in pairs or small parties, in marshes, ricefields, forest edges, clearings and gallery

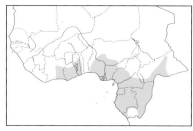

forest, usually near water. Shy and unobtrusive, foraging on or near ground, quickly diving into cover if disturbed. **SIMILAR SPECIES** Male Crimson Seedcracker is brown where Black-bellied is black; female and juvenile of both species very similar and inseparable in field. Male Red-headed Bluebill has black (not red) tail and less massive bill with red cutting edges; overlap very limited in our region. **STATUS AND DISTRIBUTION** Uncommon to locally common resident in forest zone, from E Ivory Coast (west at least to Dabou and Yapo Forest) to CAR and Congo. **NOTE** Three races described on basis of differences in bill and wing measurements (*frommi* large and larger billed than nominate; *rothschildi* small and smaller billed). However, all occur within same geographical area and bill sizes differ considerably, even at single locality.

CRIMSON SEEDCRACKER *Pyrenestes sanguineus*   Plate 137 (10)
Pyréneste gros-bec

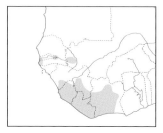

Other name F: Grosbec ponceau à ventre brun
L 13–14 cm (5.0–5.5"). Stout, with distinctive heavy, triangular bill and large, rounded head. **Adult male** *sanguineus* Entire head to breast and upper flanks, rump and tail bright red. Upperparts and rest of underparts warm earth-brown. White, broken, orbital ring. Bill steel-blue. *P. s. coccineus* has noticeably smaller bill. **Adult female** Red areas more restricted (not extending onto nape and flanks) and brown slightly paler, warm brown. In *coccineus* red less bright; nape and flanks washed red. **Juvenile** All brown with dull red rump and tail. **VOICE** A sharp *tsu-it* or *tsut*. Song a fast, short, pleasant warble, one version ending with drawn-out, sharp *tsee-ik*. [CD15:16] **HABITS** Singly, in pairs or small parties, frequenting marshes, ricefields, farmbush, rank vegetation at edges of forest and gallery forest, and suburban gardens. Shy and unobtrusive, usually foraging low near dense cover; occasionally with Western Bluebill. May fly high into trees when flushed. **SIMILAR SPECIES** Male Black-bellied Seedcracker is black where Crimson is brown; female and juvenile of both species very similar and inseparable in field. Western Bluebill appears superficially similar when glimpsed as it disappears in dense undergrowth, but separated by black (not red) tail and black top of head. **STATUS AND DISTRIBUTION** Uncommon to locally common resident in forest zone; also north along rivers in savanna zone, where scarce to rare. Nominate from Gambia and S Senegal (rare) to SW Mali, also Burkina Faso; *coccineus* from W Guinea and Sierra Leone to Ivory Coast.

GRANT'S BLUEBILL *Spermophaga poliogenys*   Plate 137 (7)
Sénégali à bec bleu

Other name F: Grosbec à front rouge
L 14 cm (5.5"). **Adult male** Forehead, crown and head-sides to throat, breast and flanks bright red. Rest of plumage jet-black with contrasting bright red rump and uppertail-coverts. Heavy bill red or pink with metallic blue at bases of upper and lower mandibles. **Adult female** Differs in having less intensely coloured plumage, entirely dark grey head, and dark grey underparts with dense, paired white spots and bars. **Juvenile** All dark slate-grey with dull reddish uppertail-coverts. Young females soon exhibit some ill-defined white spots on belly. **VOICE** Generally silent. Calls include a soft *tak, tak* (alarm) and *seeep* or *sreee*. **HABITS** Singly, in pairs or small parties in shaded understorey of rainforest. **SIMILAR SPECIES** Male Black-bellied Seedcracker distinguished from male by entirely red head, steel-blue triangular bill, and red (not black) tail. Western Bluebill has top of head black. Females of other bluebills differ from female in lacking entirely dark grey face. **STATUS AND DISTRIBUTION** Rare resident, N Congo.

WESTERN BLUEBILL *Spermophaga haematina*   Plate 137 (9)
Sénégali sanguin

Other names E: Blue-bill; F: Grosbec sanguin
L 15 cm (6"). Stout black-and-red estrildid with heavy, mainly blue, bill and black tail. **Adult male** *haematina* Glossy black with bright red throat, breast and flanks. Heavy bill metallic blue tipped red. Eastern *pustulata* has red of underparts extending onto lower head-sides and lores; uppertail-coverts red; red of bill extending along cutting edges. Intermediate *togoensis* as *pustulata* but has black head. **Adult female** Differs in having less intensely

coloured plumage; head-sides washed red, uppertail-coverts red, black of underparts densely spotted white. Female *pustulata* has head-sides brighter orange-red; *togoensis* similar. **Juvenile** All dark slate-grey with dull rusty uppertail-coverts. Young males have breast tinged reddish; young females start to exhibit white mottling on belly. **VOICE** A metallic *tswink-tswink-tswink-...*, a short *tsik* or *tsip-ip*, and a more drawn-out, high-pitched *tsee*; also a sharp *tak*, often uttered in alarm. Song variable and consisting of series of clear whistles, trills and/or rattles. Both sexes sing. [CD15:17] **HABITS** Singly or in pairs in forest, gallery forest, thickets, edges, clearings and overgrown cultivation. Low in undergrowth, occasionally mid-level. Sometimes in mixed-species flocks. Skulking and unobtrusive. **SIMILAR SPECIES** Male Black-bellied Seedcracker appears superficially similar when glimpsed diving into cover, but distinguished by red (not black) tail and entirely red head. Blue-billed Malimbe has slimmer bill, red not extending onto flanks, and is usually noisy and more easily observed. **STATUS AND DISTRIBUTION** Common resident in forest zone and gallery forest extensions throughout. Nominate from Gambia and S Senegal to Ghana, and S Mali; *togoensis* Togo to SW Nigeria; *pustulata* SE Nigeria (east of Lower Niger R.) to SW CAR and Congo.

RED-HEADED BLUEBILL *Spermophaga ruficapilla* Plate 137 (8)
Sénégali à tête rouge

Other name F: Grosbec à tête rouge
L 15 cm (6"). *S. r. ruficapilla.* **Adult male** Entire head to hindneck, throat, breast, flanks and uppertail-coverts bright red. Rest of plumage jet-black. Heavy bill metallic blue with red tip and cutting edges. **Adult female** Differs in having less intensely coloured plumage; black of underparts densely spotted white. **Juvenile** All dark slate-grey with dull reddish uppertail-coverts; young males have face and breast tinged reddish; young females may exhibit some white mottling on belly. **VOICE** A thin, squeaking *spit-spit-spit-spit...*, often barely audible, and a brief *skwee* or *speek* (Zimmerman *et al.* 1996). **HABITS** As Western Bluebill. **SIMILAR SPECIES** Male Grant's Bluebill distinguished from male Red-headed by black of upperparts extending to nape and blue on bill restricted to bases of upper and lower mandibles. Western Bluebill of race *pustulata* has top of head black. Male Black-bellied Seedcracker has red (not black) tail and all-dark, triangular bill. Females of other bluebills differ from female Red-headed in having top or entire head dark grey. **STATUS AND DISTRIBUTION** Common resident, SE CAR.

GREEN TWINSPOT *Mandingoa nitidula* Plate 136 (9)
Sénégali vert

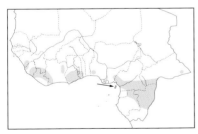

Other names E: Green-backed Twinspot; F: Sénégali vert tacheté
L 10–11 cm (4.0–4.5"). Diagnostic combination of green above and white spots on black below. **Adult male** *schlegeli* Head, upperparts and tail bright green with variable golden tinge, becoming yellowish or rusty-orange on rump. Face bright red. Throat and upper breast green tinged deep yellow or orange; lower breast to belly black densely spotted white; undertail-coverts pale olive. Bill black with red tip and cutting edges. Legs pinkish. Island race *virginiae* has upperparts tinged orange; yellowish-orange on rump extending onto back; bill red with black base. **Adult female** Paler and duller; face pale yellowish-buff; rump not or only slightly tinged yellowish. **Juvenile** Dull greyish-olive above; face pale buff; drab olive-grey below. **VOICE** A sharp, unobtrusive *tsk*, a drawn-out *tseeet* and a short high-pitched trill *tsrrriii*. Song variable, including series of high-pitched whistles and trills, interspersed by call notes. [CD15:19] **HABITS** Singly, in pairs or small groups in dense undergrowth at forest edges and along tracks; sometimes higher up. Unobtrusive and shy, occasionally feeding in the open, but rapidly retreating into nearby cover if disturbed. **SIMILAR SPECIES** Male Green-winged Pytilia has more red on face; Orange-winged Pytilia has orange in wing; both have red rumps and longer, red tails, and lack white spots on black below. **STATUS AND DISTRIBUTION** Uncommon to rare resident in forest zone, from W Guinea and Sierra Leone to Togo and from Nigeria to SW CAR and Congo, and in SE CAR (*schlegeli*); also Bioko (*virginiae*). **NOTE** Twinspots take their name from 'paired' white underpart spots (one each side of feather shaft).

BROWN TWINSPOT *Clytospiza monteiri*   Plate 136 (1)
Sénégali brun

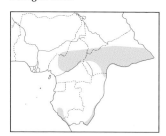

L 13 cm (5"). Grey headed and dark above, with red rump. *Rich brown, white-spotted underparts* diagnostic. **Adult male** Head (incl. throat) grey; upperparts brown; lower rump and uppertail-coverts crimson; tail blackish. Red median throat stripe broadening towards breast; rest of underparts reddish-chestnut densely spotted white. Bill black. Eye red. Legs brownish-pink. **Adult female** Similar but with white throat stripe. **Juvenile** Resembles adult, but much duller; underparts orange-brown without spots; no throat stripe; eye brown. **VOICE** A sharp *chk* or *chuk*, frequently repeated and occasionally as rattling *chukchukchkrrrrk*, and a harsh, nasal *chèèp*. Song, given by male in display, a fast, variable series of twitters and chirps. [CD15:20] **HABITS** Singly, in pairs or small parties, at edges of forest, gallery forest, and adjacent bush, grassland and cultivated areas with thick cover. **STATUS AND DISTRIBUTION** Uncommon to locally not uncommon resident, from SE Nigeria (Mambilla Plateau and nearby lowlands) and Cameroon (from edge of forest zone to Adamawa Plateau) to Chad and CAR, and in S Gabon and Congo.

DYBOWSKI'S TWINSPOT *Euschistospiza dybowskii*   Plate 136 (3)
Sénégali à ventre noir

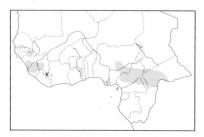

L 12 cm (4.75"). Distinctive, dark, little-known finch. **Adult male** Head to upper mantle, throat and upper breast slate-grey; lower mantle to uppertail-coverts crimson; wings and tail blackish. Lower breast greyish-black; belly, flanks and undertail-coverts black with white spots on flanks. Bill black. Eye reddish-brown. Legs dark grey to black. **Adult female** Generally paler grey on head and breast; rest of underparts greyer. **Juvenile** Much duller than adult; upperparts washed rusty; underparts wholly slate-grey without spots. **VOICE** Calls include *chip-chip-...* or *whist-whist-..* (probably alarm); also *see, tsit-tsit-* and *tswink-tswink-...* Song a varied medley of trills, melodious whistles and bubbling notes, usually uttered from cover or in display. Female has similar, but softer, song. [CD15:21] **HABITS** In pairs or small parties in tall grass and bushes in wooded savanna and at forest edges. Forms mixed feeding flocks with Black-bellied and African Firefinches. **STATUS AND DISTRIBUTION** Uncommon to scarce and local resident, recorded from SE Senegal (rare), Guinea, Sierra Leone, W Ivory Coast, Nigeria, Cameroon, S Chad and CAR. Presumably also Liberia.

Genus *Lagonosticta*

Small, ground-feeding waxbills with much crimson or wine-red in male plumage and usually some small white spots on breast-sides. Parasitised by indigobirds.

BAR-BREASTED FIREFINCH *Lagonosticta rufopicta*   Plate 139 (1)
Amarante pointé

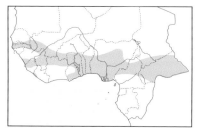

L 11 cm (4.25"). *L. r. rufopicta*. **Adult male** Face pinkish-red; crown, nape and upperparts earth-brown; rump and base of tail crimson; rest of tail blackish. Underparts pinkish-red becoming greyish on lower belly and dirty cream on undertail-coverts. Breast-sides and upper flanks have small white spots (in good view crescent or bar shape noticeable). Bill pinkish-red with dusky culmen. **Adult female** Some have slightly paler face and breast. **Juvenile** Dull grey-brown washed pinkish-red on face and breast; rump and tail-base dull crimson. Bill dusky. **VOICE** A hard, high-pitched *chip* or *pik* (alarm). Song a fast jumble of short, mostly clear notes. [CD15:22] **HABITS** In pairs or small parties in various open areas, incl. grassy forest clearings, rank vegetation near water, open wooded savanna, farmland, coastal thickets and suburban gardens. Feeds on ground, often in company of other firefinches and waxbills. Parasitised by Wilson's Indigobird. **SIMILAR SPECIES** Red-billed Firefinch also has red bill, but male is brighter red, with all-red head and distinct yellow orbital ring. **STATUS AND DISTRIBUTION** Not uncommon resident, from Senegambia through S Mali to Chad and CAR, and in Guinea, Sierra Leone and coastal W Liberia.

RED-BILLED FIREFINCH *Lagonosticta senegala*   Plate 139 (2)
Amarante du Sénégal

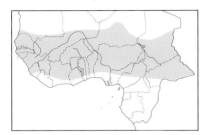

Other names E: Senegal Firefinch; F: Amarante commun
L 10 cm (4"). Commonest and most widespread firefinch, associated with habitation. Male readily identified by *entirely red head* and *yellowish orbital ring.* Both sexes have red bill. **Adult male *senegala*** Head bright pinkish-red; upperparts brown washed reddish; rump and tail-base crimson; rest of tail blackish. Underparts pinkish-red becoming buff-brown on lower belly and buffish on undertail-coverts. Breast-sides variably marked with small white spots (sometimes absent). Bill pinkish-red with dusky culmen. ***L. s. rhodopsis*** is slightly paler, with brown and buff parts faintly tinged yellowish.
Adult female Head and upperparts mainly buffish-brown; underparts paler, more yellowish-buff. Red patch in front of eye; narrow whitish orbital ring. Breast usually more extensively speckled white; rump and tail as male.
Juvenile Similar to adult female, but no red patch in front of eye, no white spots on breast, and black bill. **VOICE** A soft, high-pitched *dwee* or *pfweet*; alarm an abrupt *chep* or *trrp*. Songs include a short *tsi-tu-tuuuuit* or a rising series of fluting notes, e.g. *tu-tweet-tweet-...* (sometimes abbreviated to *tu-tweet*), occasionally introduced by *tzep*. [CD15:24] **HABITS** In pairs or small groups in towns and villages, farmland, scrub and open savanna. Often with other firefinches or waxbills. Tame. Parasitised by Village Indigobird. **SIMILAR SPECIES** Bar-breasted Firefinch, the only other firefinch with red bill, is darker above, with top of head brown and no yellowish orbital ring. **STATUS AND DISTRIBUTION** Common resident in broad savanna belt, from SW Mauritania to Sierra Leone east to Chad and CAR. Rare, N Liberia. *L. s. rhodopsis* in north of range.

BLACK-BELLIED FIREFINCH *Lagonosticta rara*   Plate 139 (6)
Amarante à ventre noir

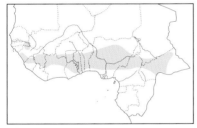

L 10 cm (4"). **Adult male *forbesi*** Head and mantle deep wine-red; wings brown; rump and uppertail-coverts bright crimson; tail mainly black. Underparts deep wine-red; *central lower breast to undertail-coverts jet-black.* No small white spots on breast-sides. Bill two toned: bluish-black with reddish or whitish base to lower mandible. **Nominate** less brightly coloured. **Adult female** Head greyish with red patch in front of eye; upperparts dull dark brown, variably washed wine-red and becoming wine-red on rump and uppertail-coverts; tail blackish-brown. Throat pale greyish-cream; breast and upper flanks dull pinkish-carmine; rest of underparts and bill as male. **Juvenile** Dull buff-brown above; paler below; rump and uppertail-coverts dull crimson. Bill blackish, base of lower mandible pinkish. **VOICE** A short *pwit, pwit,* a hard *chep* and a nasal *cheeay* and *chew*; also a short, pleasant trill. Song *tiu-tiu-tiu-tiu-tiu* and *tsiuw-tsiuw-tsiuw.* [CD15:25] **HABITS** In pairs or small parties, in grassy savanna, edges of gallery forest and abandoned farmland. Often with other firefinches. Parasitised by Cameroon Indigobird. **SIMILAR SPECIES** Male Blue-billed Firefinch has grey-brown (not red) mantle and back, small white spots on breast-sides and black restricted to lower belly. Female Blue-billed similar to male but paler. Male Rock Firefinch has greyish (not red) top of head, white spots on breast and black restricted to lower belly. Female Rock also has greyish head, but throat concolorous with rest of underparts (not buff) and black on lower underparts less extensive. **STATUS AND DISTRIBUTION** Not uncommon resident in savanna zone. *L. r. forbesi* in SE Senegal (rare) and W Guinea, Sierra Leone and N Liberia (rare) to Nigeria; nominate from Cameroon to Chad and CAR.

BLUE-BILLED FIREFINCH *Lagonosticta rubricata*   Plate 139 (4)
Amarante flambé

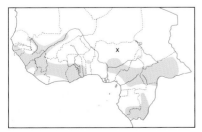

Other names E: African Firefinch; F: Amarante foncé
L 10.0–11.5 cm (4.0–4.5"). *Blue bill* and *black undertail-coverts.* **Adult male *polionota*** Crown, nape and upperparts grey-brown; rump and tail-base deep carmine-red; rest of tail blackish. Face and underparts deep red becoming greyish on central belly and black on vent and undertail-coverts. Breast-sides and upper flanks have small white spots. Bill deep bluish. ***L. r. ugandae*** has top of head and nape variably washed pinkish-red, and slightly paler upperparts. ***L. r. congica*** has more earth-brown upperparts. Number of specks on breast in both races variable, sometimes absent. ***L. r. landanae*** similar to *congica,* but face brighter red and bill bluish above (sometimes tinged pinkish), and pink to reddish below with dark tip.

Adult female Similar but paler overall, with top of head washed pinkish and central belly buff-brown. Female *landanae* has top of head washed pinkish; bill as male. **Juvenile** Dull brown to buff-brown above; paler, more buffish below; rump and uppertail-coverts deep red. Bill blackish, paler below. Juvenile *landanae* has forehead reddish and head-sides washed pinkish; bill dark with pinkish base. **VOICE** A high-pitched twittering *trrrrrrr-t* or *rititititit*, a loud *chew-chew-chew* or *chwee-chwee-chwee* and *feeeu-feeeu*, and a rapid series *wink-wink-wink* or *whitwhitwhitwhit...* Alarm a dry, rattling *prttt* and *prrititit.* [CD15:26] **HABITS** In pairs or small parties in savanna bush, woodland, thickets, gallery forest and forest edge. Occasionally with other firefinches. Parasitised by Variable, Cameroon and Jos Plateau Indigobirds. **SIMILAR SPECIES** Male Black-bellied Firefinch has red (not grey-brown) mantle and back, no white spots on breast, and black extending to lower breast. Female Black-bellied has greyish head; black also much more extensive. Male Rock Firefinch has red mantle and two-toned bill. Female Rock is darker, more reddish above and richer red below. **STATUS AND DISTRIBUTION** Not uncommon resident in forest–savanna mosaic and savanna zone. *L. r. polionota* from SW Senegal (rare) to Liberia east to Nigeria; *ugandae* from Cameroon and Chad to CAR and N Congo; *congica* from Gabon to SW and C Congo. *L. r. landanae* reported in SW Congo (rare). **NOTE** *L. r. landanae* sometimes treated as separate species, Pale-billed or Landana Firefinch (Amarante de Landana), but considered a race of *rubricata* by Payne (1982) and Dowsett & Dowsett-Lemaire (1993).

REICHENOW'S FIREFINCH *Lagonosticta* (*rhodopareia*) *umbrinodorsalis* Plate 139 (7)
Amarante de Reichenow

Other names E: Chad Firefinch; F: Amarante de Jameson
L 10–11 cm (4.0–4.5"). Resembles Blue-billed Firefinch but has greyish top of head. **Adult male** Crown and nape grey; upperparts brown; rump and tail-base deep carmine-red; rest of tail blackish. Face and underparts bright pinkish-red with central belly to undertail-coverts black. Breast-sides and upper flanks with small white spots. Bill deep bluish. **Adult female** Similar but red areas slightly paler. **Juvenile** Very similar to juvenile Blue-billed Firefinch. Warm brown above; paler, more buffish below; rump and uppertail-coverts dull pinkish-red. Bill blackish, paler below. **VOICE** Similar to Blue-billed Firefinch. A soft *tsit, tsit* or faster *ti-ti-ti-ti*, a high-pitched twittering *rititititit* and a soft *chew*. Alarm a dry *pitpitpit.* Song a series of identical notes, uttered at variable speed, e.g. *tsee-tsee-tsee-tsee* and *wit-wit-wit-wit* or *tsip-tsip-tsip-tsip* and variations. [CD15:27] **HABITS** In pairs on rocky hillsides with grassy areas, bushes, bare granitic slabs and scattered clumps of large trees. Wary. **SIMILAR SPECIES** Black-bellied Firefinch (q.v.), the only other firefinch occurring in Reichenow's restricted W African range, has quite different plumage and frequents level, shrubby areas. Also compare Rock Firefinch (q.v.). **STATUS AND DISTRIBUTION** Locally not uncommon in very limited, isolated area of extreme SW Chad; also N Cameroon (rare). **NOTES (1)** Variously treated as separate species or as race of Jameson's Firefinch *L. rhodopareia.* **(2)** *L. r. bruneli* is junior synonym of *umbrinodorsalis* (Payne & Louette 1983).

ROCK FIREFINCH *Lagonosticta sanguinodorsalis*   Plate 139 (3)
Amarante des rochers

L *c.* 10 cm (4"). **Adult male** Combination of *grey crown*, *bright red mantle* and two-toned bill, black and blue-grey, distinctive. Wings brown; rump and uppertail-coverts deep red; tail mainly black. Head-sides and underparts deep red with small white spots on upper flanks; undertail-coverts black. Bill black with bluish-grey base to lower mandible. **Adult female** Head greyish with red patch in front of eye. *Mantle and back bright reddish-brown* (distinctive). Wings brown; rump and uppertail-coverts deep red; tail black. Throat, breast and upper belly pinkish-red with small white spots on upper flanks; lower belly deep red; undertail-coverts black. Bill as male. **Juvenile** Similar to adult female but head more greyish-brown; no red patch in front of eye; mantle and back slightly less red. Underparts pale greyish-brown to reddish-brown, paler on throat; no white spots. Bill as adult but with pale grey gape flanges. **VOICE** A characteristic rapid descending trill *treeeee.* Also a short *chew*, a rising *chwee*, drawn-out whistles *feeew* and *feeee-eeeee*, and a low, repeated *too.* Alarm a dry *pitpitpit.* [CD15:28] **HABITS** In pairs in wooded grassland and thickets with rocky outcrops and on wooded, grassy inselbergs. Parasitised by Jos Plateau Indigobird. **SIMILAR SPECIES** Male Black-bellied Firefinch has red (not grey) top of head, no white spots on breast and black extending onto lower breast. Female Black-bellied also has greyish head, but has buff throat and black on lower underparts much more extensive. Male Blue-billed Firefinch has brownish upperparts. Female Blue-billed is paler and less reddish above, and paler red below. **STATUS AND DISTRIBUTION** Not uncommon resident, NC and NE Nigeria (Jos Plateau and area to north; Mandara Mts). **NOTE** Recently described species; account based on Payne (1998).

KULIKORO FIREFINCH *Lagonosticta virata*   Plate 139 (5)
Amarante de Kulikoro

Other name E: Mali Firefinch
L 10–11 cm (4.0–4.5"). **Adult male** Similar to Blue-billed Firefinch but slightly duller and usually has lower forehead reddish and breast-sides with fewer white spots. Bill deep bluish with black tip (often all black above); narrow orbital ring pale yellow. **Adult female** Very similar but usually slightly paler. **Juvenile** Pale grey-brown to buff-brown above; paler, more buffish below; rump and uppertail-coverts dull red. **VOICE** A low-pitched, nasal, slightly harsh *chew*, alarm a rattling trill *chrrrrrrr*. Song a series of drawn-out, plaintive whistles *tseeeeeu*; also a trill. [CD15:29] **HABITS** In rocky areas with bushes and grass. Behaviour as other firefinches. **SIMILAR SPECIES** Blue-billed Firefinch (q.v.); range partially overlaps. **STATUS AND DISTRIBUTION** Uncommon and local resident, Mali (Upper Niger R. from Mopti to Bamako; Mandingue Mts at border with Guinea); also SE Senegal (1 record). **NOTE** Taxonomic status debatable. Here tentatively treated as separate species, but also variously considered a race of either Jameson's Firefinch *L. rhodopareia* or Blue-billed Firefinch.

BLACK-FACED FIREFINCH *Lagonosticta larvata*   Plate 139 (8)
Amarante masqué

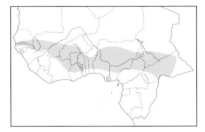

L 11.5 cm (4.5"). **Adult male** *vinacea* Distinctive firefinch with *black face*. Top of head soft grey, washed pinkish on nape; head-sides to upper throat black. Upperparts pinkish-mauve; rump and central tail deep red; outer tail blackish. Underparts bright pink with some small white spots on lower breast-sides and upper flanks; belly to undertail-coverts black. Bill black, occasionally with pink on upper mandible. *L. l. togoensis* has top of head and upperparts grey; underparts paler grey with variable pinkish wash (sometimes absent). *L. l. nigricollis* similar to *togoensis* but slightly darker on head and underparts. **Adult female** *vinacea* Head pale grey-brown; upperparts as adult male. Throat buff tinged greyish; rest of underparts paler than adult male, belly grey-brown becoming dark grey on vent and undertail-coverts. *L. l. togoensis* has upperparts grey-brown; underparts soft grey-brown becoming buff-brown on belly and undertail-coverts. *L. l. nigricollis* similar to *togoensis* but somewhat darker, with slightly greyer upperparts. **Juvenile** Entirely dull brown with dull red rump and uppertail-coverts. **VOICE** A high-pitched, piercing *seesee* and a hard *twit-it-it* (alarm). Song a variable series of short notes ending in a clear, rising double note, e.g. *pwit twet twet-te-twet twiweet*, with variations; also a fast *tweetweetweetwee*. [CD15:30] **HABITS** Usually in pairs or small groups in wooded grassland, rank herbage near water and moist thickets. Associates with other firefinches and waxbills. Parasitised by Baka Indigobird. **SIMILAR SPECIES** Adult females of other firefinches differ from female Black-faced by dark undertail-coverts and/or red bills. **STATUS AND DISTRIBUTION** Uncommon to locally not uncommon resident in savanna belt, from Senegambia, Guinea-Bissau and Guinea to S Burkina Faso (*vinacea*), N Ivory Coast to N Cameroon and S Chad (*togoensis*) and E Cameroon to CAR (*nigricollis*).

Genus *Estrilda*

Small, active and agile estrildids, which feed mainly on grass seeds taken from growing plants. The sealing wax-like red bills of some species provide the genus name. Parasitised by Pin-tailed Whydah.

LAVENDER WAXBILL *Estrilda caerulescens*   Plate 138 (5)
Astrild queue-de-vinaigre

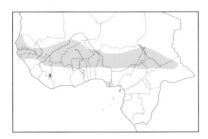

Other names F: Astrild gris-bleu, Queue-de-vinaigre
L 10 cm (4"). Unmistakable. *Wholly soft grey* with contrasting *red rump and tail*. **Adult male** Head and body mainly plain, slightly bluish-grey; lores to eye black; lower back, rump, tail and undertail-coverts crimson; flight feathers dark grey. Lower belly blackish with some white speckles on lower flanks (hard to see in field). Bill blackish, reddish-brown at base. **Adult female** Belly usually less blackish. Probably best separated on call. **Juvenile** Paler and duller; lower flanks lack speckles. **VOICE** A clear, vigorous, high-pitched *tsee-tsee* and a descending *tseeu*; occasionally a dry *tik-tik tik tik-tik...* and a

drawn-out *skweeep* (alarm). Female has di- or trisyllabic *tseeeht-tseeet* or *tseeeht-tseeeht-tseeet* contact, answered by male with *seeyou* (Goodwin 1982). Song a series of high-pitched, thin notes *see-see-swee-see*. [CD15:31] **HABITS** In pairs or small groups in various habitats in savanna zone, incl. grassland with scattered trees and bushes, edges of thickets and cultivation, roadsides and gardens. Often with other waxbills. Not shy. **STATUS AND DISTRIBUTION** Locally common to scarce resident in savanna belt, from Senegambia to Guinea east to Chad and N CAR.

GREY WAXBILL *Estrilda perreini* Plate 138 (7)
Astrild à queue noire

Other name E: Black-tailed Waxbill
L 11 cm (4.25"). *E. p. perreini.* Resembles Lavender Waxbill (no overlap), but has brighter red rump and longer, *black tail.* **Adult male** Head and body mainly plain grey; lores to eye black; rump and uppertail-coverts bright crimson. Tail black, slightly graduated. Lower belly to undertail-coverts blackish (no white speckles on lower flanks). Bill blue-grey tipped black. **Adult female** Vent usually greyer. **Juvenile** Duller. **VOICE** A high-pitched *psee* or *pseeu* and a more drawn-out *pseeeee*, occasionally followed by a few short notes. [CD15:32] **HABITS** Singly, in pairs or small groups in wooded grassland, edges of gallery forest and forest regrowth. **STATUS AND DISTRIBUTION** Uncommon to scarce resident, S Gabon and Congo.

FAWN-BREASTED WAXBILL *Estrilda paludicola* Plate 138 (2)
Astrild à poitrine fauve

L 11.5 cm (4.5"). Pale, dull-coloured waxbill with red bill and red rump. **Adult male** *ruthae* Head pale ashy-grey; upperparts tawny-brown, very finely barred darker (only visible at close range). Rump and uppertail-coverts bright red; tail black. Underparts whitish. Bill bright red. **Nominate** is slightly darker above and below: underparts washed pale creamy-buff, darker on flanks and belly; central lower belly washed pinkish when fresh; undertail-coverts whitish. **Adult female** Similar. **Nominate** has no or only slight trace of pinkish wash on belly. **Juvenile** Similar to adult but duller and with black bill. **Nominate** lacks pink on belly. **VOICE** A nasal *chyeek* and a shorter *chyep*; alarm *chyee-krr*. Song a harsh *tek tek tek teketree teketree* (Goodwin 1982). [CD15:35] **HABITS** In pairs or small groups in moist grassland, dambos and grassy clearings. Often with other waxbills. **SIMILAR SPECIES** Orange-cheeked Waxbill rather similar but immediately distinguished by orange face. **STATUS AND DISTRIBUTION** Locally common resident, S Gabon and Congo (*ruthae*) and E CAR (nominate).

ANAMBRA WAXBILL *Estrilda poliopareia* Plate 138 (3)
Astrild du Niger

L 11–12 cm (4.5"). Very local, dull-coloured waxbill with red bill and red rump. **Adult male** Head pale, plain brown faintly washed grey; upperparts buffish-brown, very finely barred darker (only visible at close range). Rump and uppertail-coverts bright red; tail blackish-brown. Underparts pale yellowish-buff, darker on flanks and belly, sometimes with reddish wash extending to breast-sides. Bill bright red. Eye pale. **Adult female** Some slightly paler and duller, with orange-tinged rump. **Juvenile** No information. Probably similar to adult but duller, with black bill. **VOICE** Reported to give a variety of nondescript waxbill-type calls (Collar & Stuart 1985). **HABITS** Inadequately known. Occurs in tall grass and herbage along rivers and on lagoon sand banks. **SIMILAR SPECIES** The only other waxbill with red bill and red rump is adult Orange-cheeked Waxbill, which is easily distinguished by orange face. **STATUS AND DISTRIBUTION** Very local resident, endemic to S Nigeria (recorded from Lower Niger R. to extreme southwest, though mainly in Onitsha area). Status inadequately known; formerly considered locally not uncommon, but no recent sightings. *TBW* category: THREATENED (Vulnerable). **NOTE** Often treated as race of Fawn-breasted Waxbill.

ORANGE-CHEEKED WAXBILL *Estrilda melpoda* Plate 138 (1)
Astrild à joues oranges

Other name F: Joues-orange
L 10 cm (4"). *Orange face* diagnostic in all plumages. **Adult** *melpoda* Head pale grey with bright orange face.

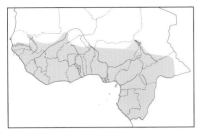

Upperparts brown; rump and uppertail-coverts bright red or reddish-orange; tail blackish. Underparts pale greyish washed yellow on belly. Bill bright red. Plumage tones variable throughout range, but birds recognised as *tschadensis* consistently have deep reddish-orange face and paler brown upperparts. **Juvenile** Generally buffier, with pale orange face and paler rump; bill black. **VOICE** A pleasant short trill *tsiririririt* and a soft, high-pitched *tsee-tsee* or *tsit-tsit*, frequently repeated. [CD15:36] **HABITS** In pairs or small flocks in rank grass in savanna, large forest clearings and swampy areas, farmbush, and gardens. Associates with other waxbills. **STATUS AND DISTRIBUTION** Common resident throughout, north to Senegambia and Mali (15°N) to SW Niger, L. Chad and W CAR; also SC Mauritania. Nominate in most of range; *tschadensis* in N Cameroon (N Adamawa) and W Chad (L. Chad area).

BLACK-RUMPED WAXBILL *Estrilda troglodytes*   Plate 138 (8)
Astrild cendré

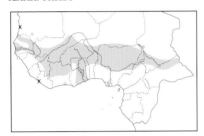

Other name F: Bec-de-corail cendré
L 10 cm (4"). Red mask, *black rump* and *white undertail-coverts*. **Adult male** Top of head grey; broad red eye-stripe; head-sides whitish or pale buff; upperparts grey-brown, very finely barred darker (only visible at close range). Rump, uppertail-coverts and tail black. Underparts pale buffish washed pink. Bill bright red. **Adult female** Similar, but no or only slight trace of pinkish wash on underparts. **Juvenile** Similar to adult but duller, lacking red mask and with darkish eye-stripe and black bill. **VOICE** A vigorous, drawn-out *cheeeu* and a shorter *chup*, frequently repeated; in flight, flocks utter a constant *chiup-chiup-chiup*. Song of male a disyllabic *che-cheer* or *chu-weee*; that of female a shorter *pwitch pwitch* or *pwitch cheee*. [CD15:37] **HABITS** In pairs and small (sometimes large) flocks in dry grassy savanna, scrub, edges of thickets and farmbush. Flicks tail from side to side. **SIMILAR SPECIES** Common Waxbill has brown rump and black undertail-coverts. **STATUS AND DISTRIBUTION** Locally common resident, from S Mauritania and Senegambia to Guinea east to Chad and N CAR.

COMMON WAXBILL *Estrilda astrild*   Plate 138 (9)
Astrild ondulé

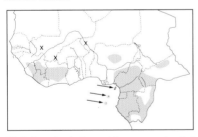

Other name F: Astrild bec-de-corail
L 11 cm (4.25"). Red mask, *brown rump* and *black undertail-coverts*. **Adult male** *occidentalis* Top of head grey-brown; broad red eye-stripe; head-sides whitish; upperparts, incl. rump and long uppertail-coverts, brown, finely but distinctly barred darker; tail blackish. Underparts pale buffish finely barred darker from breast-sides to flanks and variably but lightly washed pink on breast and belly. Bill bright red. Plumage tones variable throughout range. *E. a. kempi* and *jagoensis* similar to *occidentalis*; *rubriventris* richer coloured and heavily tinged pink or pinkish-red above and below. **Adult female** Slightly paler below, with no or only slight trace of pinkish wash. Ssp. *rubriventris* much less pink than male. **Juvenile** Similar to adult but duller, plainer and more buff-brown, with paler and smaller red eye-stripe and blackish bill. **VOICE** A sharp, abrupt *pit* or *chik*, a nasal *pcher-pcher* and a soft *chip*. Flocks utter a buzzy twittering. Song a variable fast series of harsh buzzy notes, e.g. *cheche-puwirz*. [CD15:38] **HABITS** Usually in small flocks in rank grass, scrub, edges of thickets and cultivation, and gardens. Flocks larger in non-breeding season. Flicks tail from side to side. Tame. **SIMILAR SPECIES** Black-rumped Waxbill has black rump and white undertail-coverts. **STATUS AND DISTRIBUTION** Locally common resident, from W Guinea, Sierra Leone and Liberia (*kempi*) to S Mali and Ghana, and from Nigeria to CAR, Congo and Bioko (*occidentalis*). *E. a. rubriventris* in coastal Gabon and SW Congo. Introduced (from Angola) to Cape Verde Is, São Tomé and Príncipe (*jagoensis*).

BLACK-CROWNED WAXBILL *Estrilda nonnula*   Plate 138 (4)
Astrild nonnette

Other names F: Astrild/Sénégali à cape noire
L 11 cm (4.25"). Contrastingly patterned. **Adult male** *nonnula* Top of head jet-black sharply contrasting with white cheeks; upperparts grey, finely barred darker (only visible at close range); lower back, rump and uppertail-coverts

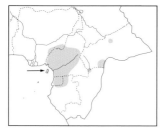

bright crimson; tail black. Underparts white washed pale greyish on breast, flanks and undertail-coverts, and tinged reddish on lower flanks. Bill black with red sides to upper mandible and red base to lower mandible. Local races *elizae* and *eisentrauti* have underparts more silky-grey. **Adult female** Upperparts have slight brownish tinge, flanks and bill less red. **Juvenile** Similar to adult but duller; upperparts greyish-brown without bars; underparts pale buffish-brown; bill black. **VOICE** A thin, high-pitched *tsree* and *sreeu*; also a short, sharp *psit*. Song a short series of similar, high-pitched notes. [CD15:39] **HABITS** In pairs or small flocks in grassland, forest edge and clearings, thickets in savanna, and gardens. Tame. **SIMILAR SPECIES** Adult Black-headed Waxbill is darker, with grey underparts and black vent and undertail-coverts; red on rump deeper and brighter; red on flanks more extensive. Juvenile Black-headed has grey underparts. **STATUS AND DISTRIBUTION** Locally common resident, SE Nigeria (Obudu and Mambilla Plateaux), Cameroon, CAR, Eq. Guinea and Gabon (nominate), and Bioko (*elizae*, above 1200 m). *E. n. eisentrauti* restricted to Mt Cameroon.

BLACK-HEADED WAXBILL *Estrilda atricapilla*   Plate 138 (6)
Astrild à tête noire

Other name F: Sénégali à tête noire

L 10 cm (4"). *E. a. atricapilla*. Similar to Black-crowned Waxbill but darker, esp. on underparts. **Adult male** Top of head jet-black; cheeks white washed grey; upperparts dark grey, finely barred darker (only visible at close range); rump and uppertail-coverts bright red; tail black. Underparts smoke-grey *becoming black on belly and undertail-coverts*; flanks reddish. Bill black with red base to lower mandible. **Adult female** Slightly duller; upperparts with slight brownish tinge; flanks with less red. **Juvenile** Similar to adult female but duller; upperparts dark greyish-brown; underparts tinged dark brown; bill black. **VOICE** A thin, high-pitched *psee* and *tsree*, and a short, sharp *psit*. [CD15:40] **HABITS** Usually in small flocks in grassy clearings and roadsides, and low forest regrowth. Tame. **SIMILAR SPECIES** Adult Black-crowned Waxbill, which is more of a highland species, is paler, more contrasting, with mainly white underparts and less red on flanks. Juvenile Black-crowned distinguished from juvenile Black-headed by paler, brownish-buff underparts. **STATUS AND DISTRIBUTION** Locally common resident, from S Cameroon to N and SW Congo.

Genus *Uraeginthus*

Cordon-bleus. Tame, with pale brown upperparts and conspicuously pale blue underparts and tails, the latter rather long and graduated.

SOUTHERN CORDON-BLEU *Uraeginthus angolensis*   Plate 138 (10)
Cordonbleu de l'Angola

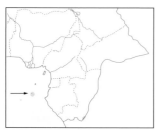

Other name E: Blue Waxbill

L 12–13 cm (4.75–5.0"). *U. a. angolensis*. **Adult male** As Red-cheeked Cordon-bleu, but lacks crimson patch on ear-coverts. Upperparts slightly colder, more greyish-brown; tail slightly shorter. Bill greyish. **Adult female** Slightly duller; blue on underparts less extensive. **Juvenile** Similar to adult female, but even duller and paler; blue on underparts confined to throat and breast; bill dark slate. **VOICE** A thin, high-pitched *tseeup-tseeup* and a short *tsip*, often repeated and similar to calls of Red-cheeked Cordon-bleu. Alarm a rattling *chchchrrrt*. Song a variable series of high-pitched sibilant notes interspersed by dry rattles and drawn-out harsh sounds. [CD15:41] **HABITS** In pairs or small flocks at edges of cultivation, in gardens, villages and towns, and wooded grassland. Often with firefinches and waxbills. Tame. **SIMILAR SPECIES** Male Red-cheeked Cordon-bleu has crimson cheek patch; ranges do not overlap. **STATUS AND DISTRIBUTION** Uncommon resident, Congo (Brazzaville area). Locally common, São Tomé (where probably introduced).

RED-CHEEKED CORDON-BLEU *Uraeginthus bengalus*   Plate 138 (11)
Cordonbleu à joues rouges

Other name F: Cordon-bleu

L 13 cm (5"). *U. b. bengalus*. Unmistakable. **Adult male** Top of head and upperparts pale brown; rump and tail

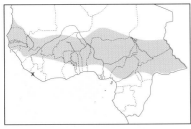

pale cerulean-blue. Head-sides to breast and flanks pale cerulean-blue; *bright crimson patch on ear-coverts*; rest of underparts pale buffish-brown. Bill pinkish. **Adult female** Lacks crimson cheek-patch and is slightly duller. **Juvenile** Similar to adult female, but duller and paler; blue on underparts more confined to throat. Bill dark slate. **VOICE** A thin, high-pitched *sweep-sweep* or *tseep-tseep*. Alarm a rattling *chchchrrrt*. Song a thin, high-pitched *tsee-tsee-tsu-tsuu*, with variations, frequently repeated. [CD15:42] **HABITS** In pairs or small flocks in savanna grassland, edges of cultivation, gardens and villages. Often with firefinches and waxbills. Tame. **SIMILAR SPECIES** Male Southern Cordon-bleu lacks crimson patch; ranges do not overlap. **STATUS AND DISTRIBUTION** Common resident in broad savanna belt, from S Mauritania to Guinea east to Chad and CAR.

ZEBRA WAXBILL *Amandava subflava*   Plate 138 (13)
Bengali zébré

Other names E: Goldbreast; F: Ventre-orange
L 9–10 cm (3.75–4.0"). Very small and active with, in male, diagnostic bright orange breast. **Adult male** *subflava* Top of head and upperparts olive-brown; bright red supercilium; rump and uppertail-coverts bright scarlet. Tail short, blackish. Throat yellow; rest of underparts orange, usually paler on central belly, barred olivaceous and yellow on breast-sides and flanks. Bill bright red. Southern *clarkei* mainly yellow below, with orange restricted to variable patch on breast and to undertail-coverts. **Adult female** Duller; no red supercilium; under-parts pale yellow washed dusky on breast; barring on flanks less distinct; undertail-coverts orange. **Juvenile** Much duller; brownish above; no red supercilium; rump washed orange; underparts dull buff without barring; bill blackish. **VOICE** A high-pitched, squeaky *cheep* or *sweep*; in flight a fast *trip-trp-trp-trp* and a soft metallic twittering. [CD15:43] **HABITS** Usually in small flocks (larger when not breeding) in various grassy habitats, often near water. Tame. **STATUS AND DISTRIBUTION** Locally common to scarce resident, from Senegambia to Liberia east to S Chad and CAR (nominate), and in Gabon and S Congo (*clarkei*).

Genus *Ortygospiza*

Quailfinches. Very small, short-tailed estrildids occurring on ground in grasslands, often near water.

LOCUST FINCH *Ortygospiza locustella*   Plate 138 (15)
Astrild-caille à gorge rouge

L 9–10 cm (3.5–4.0"). *O. l. uelensis*. Very small, short tailed and dumpy with reddish or orange wings. **Adult male** Unmistakable. Mainly blackish with head-sides to breast bright scarlet, wings bright reddish, rump-sides and uppertail-coverts red. Tail blackish. Bill red (breeding) or with black upper mandible. Eye yellowish. Resembles black-and-red ball in flight. Red on wings fades with wear. **Adult female** Head and upperparts greyish-brown; wings orangey; underparts buff-white boldly barred black on flanks. Bill as male. **Juvenile** Similar to adult female but browner; wing feathers edged brown; breast dusky; bill dark. **VOICE** Mainly silent. An abrupt *chup-chup-...* or *chilp-chilp-...* in flight. [CD15:44] **HABITS** In pairs or family groups in moist grassland; in larger flocks when not breeding. Entirely terrestrial; never perches. **STATUS AND DISTRIBUTION** Uncommon and local resident in SE Cameroon, SE Gabon and Congo. Local movements reported.

AFRICAN QUAILFINCH *Ortygospiza atricollis*   Plate 138 (12)
Astrild-caille à lunettes

Other name F: Astrild-caille
L 9.5–10.0 cm (3.75–4.0"). Small, dumpy and unobtrusive; usually first noticed when flushed. **Adult male** *atricollis* Head grey-brown with black face and throat; upperparts dark earth-brown. *Tail short*, blackish with *white tips to outer feathers* (conspicuous in flight). Chin white; breast finely and flanks broadly barred black and white; belly rusty-brown, fading to orange-buff on undertail-coverts. Bill red (breeding) or reddish with dusky culmen. *O. a. ansorgei* darker with larger black mask (extending to ear-coverts and upper breast). **Adult female** Duller; head

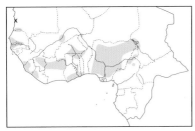

greyer and lacking black mask; underparts paler. Bill dusky above, dull reddish below. **Juvenile** Similar to adult female but even duller and paler, with barring on breast and flanks faint; bill dusky with pinkish base. **VOICE** A hard, repeated metallic *trrink-trrink* or *krreep-krreep*, uttered in flight; distinctive but easily passing unnoticed. Song, given from ground, a rapid series of *klik-klak-kloik-kluk* notes, often repeated. [CD15:45] **HABITS** In small flocks of a few to 50, in open grassy habitats, often near water. Entirely terrestrial; never perches. Shy, well camouflaged and generally hard to observe. When flushed, flies with short jerky movements, usually dropping into nearby cover. **SIMILAR SPECIES** Black-chinned Quailfinch (q.v.) very similar but ranges do not overlap. **STATUS AND DISTRIBUTION** Patchily distributed, locally common resident, from S Mauritania (rare) and Senegambia east to Cameroon and Chad (nominate); *ansorgei* in coastal Guinea-Bissau to Togo.

BLACK-CHINNED QUAILFINCH *Ortygospiza gabonensis*   Plate 138 (14)
Astrild-caille à gorge noire

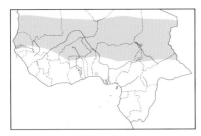

L 9.5–10.0 cm (3.75–4.0"). *O. g. gabonensis.* **Adult male** Similar to African Quailfinch (no overlap), but brown head and upperparts have dark feather centres producing slightly streaked appearance. Chin black. Barring on breast more conspicuous; belly paler, more orange-buff, becoming buff on lower belly and undertail-coverts. Bill bright red. **Adult female** Differences from male as female African Quailfinch. Barring on underparts quite pronounced in some. Bill colour variable; like African Quailfinch or more reddish. **Juvenile** Similar to juvenile African Quailfinch. **VOICE & HABITS** Similar to African Quailfinch. [CD15:46] **STATUS AND DISTRIBUTION** Uncommon to common and local resident in Eq. Guinea, Gabon and Congo.

AFRICAN SILVERBILL *Lonchura cantans*   Plate 139 (10)
Capucin bec-d'argent

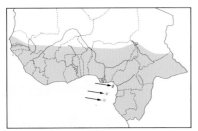

Other names E: Warbling Silverbill; F: Spermète bec-d'argent
L 10 cm (4"). *L. c. cantans.* Small, slender, pale estrildid with black tail. **Adult** Head and upperparts pale sandy-brown; inner greater coverts, tertials and back very finely barred darker (only visible at close range); primaries, rump and tail black. Tail pointed and graduated. Throat and breast sandy-buff, becoming white on rest of underparts. Bill blue-grey. Eye dark with narrow blue orbital ring. **Juvenile** Lacks fine barring; tail edged brown. **VOICE** A short *tsik*, frequently repeated. In flight, flocks give a twittering *chip-chi-chi-chi-chip...* Song a soft, fast series of gentle rising and descending muffled sounds. [CD15:47] **HABITS** Usually in small flocks in dry *Acacia* savanna. Sometimes near habitation. Tame. **SIMILAR SPECIES** Unlikely to be confused. Adult Black-rumped and Common Waxbills darker, with red bills and red masks; juveniles have dark bills. **STATUS AND DISTRIBUTION** Uncommon or scarce to locally common resident in broad savanna belt, from S Mauritania and Senegambia to Chad and CAR. **NOTE** Sometimes considered conspecific with white-rumped Indian Silverbill *L. malabarica.*

BRONZE MANNIKIN *Lonchura cucullata*   Plate 139 (11)
Capucin nonnette

Other name F: Spermète-nonnette
L 9 cm (3.5"). *L. c. cucullata.* Very small, dumpy, grey-brown and white. **Adult** Head to throat and central breast blackish; upperparts brown; rump and uppertail-coverts barred black and white; tail short, blackish. Metallic bronze-green shoulder patch. Underparts white barred brownish on lower breast-sides, flanks and undertail-coverts. Bill black above, blue-grey below. **Juvenile** Entirely dull buff-brown, paler below. Bill dark. **VOICE** A rather soft, rolling *rreep* or *treep*, frequently uttered in flight and sounding like wheezy twittering *chrreep-reep-chrrepchreepchrewprreep...* when coming from flock. [CD15:48] **HABITS** Usually in small flocks in farmbush, scrub, rank grass and gardens. Frequently flicks tail sideways; in alarm also flicks wings. Often with other mannikins and waxbills. Very tame. **SIMILAR SPECIES**

Magpie Mannikin is distinctly larger and lacks barring on rump. Black-and-white Mannikin lacks brown in plumage and has heavier, all-blue bill. Juveniles of the three mannikins similar and most easily identified by company of parents. Juvenile Black-and-white more contrasting: darker above and paler below. Juvenile Magpie noticeably larger, also paler below. **STATUS AND DISTRIBUTION** Common resident throughout, from Senegambia to Liberia east to Chad, CAR and Congo; also Bioko, São Tomé, Príncipe and Annobón.

BLACK-AND-WHITE MANNIKIN *Lonchura bicolor* Plate 139 (13)
Capucin bicolore

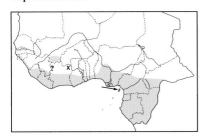

Other name F: Spermète à bec bleu

L 9.5–10.5 cm (3.75–4.0"). **Adult** *bicolor* Head, upperparts, tail, throat and breast glossy black. Rest of underparts pure white barred black on flanks. Bill heavy, pale grey-blue. ***L. b. poensis*** has some fine white barring on wings, rump and uppertail-coverts. **Juvenile** Dull earth-brown above, blackish on primaries and tail; pale buff-brown below. Bill dark. **VOICE** A clear, soft *kip* or *tsik* and a more drawn-out *tseew.* [CD15:49] **HABITS** In rank grass and bushes within forest clearings, along edges, near water and in cultivation. Usually in small groups; often with Bronze Mannikins. **SIMILAR SPECIES** Compare Bronze and Magpie Mannikins. **STATUS AND DISTRIBUTION** Common resident in forest zone, from Guinea-Bissau to W Cameroon, and in S Mali (nominate) east to CAR, Congo and Bioko (*poensis*). Many in Cameroon are intergrades.

MAGPIE MANNIKIN *Lonchura fringilloides* Plate 139 (12)
Capucin pie

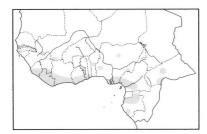

Other name F: Spermète-pie

L 11.5–12.0 cm (4.5–4.75"). *L. f. fringilloides.* Resembles Bronze Mannikin but noticeably larger and *has heavier, weaver-like bill.* **Adult** Head to throat glossy blue-black; upperparts dark brown with some buff edgings and streaks to feathers of mantle, scapulars and back; rump and tail black. Underparts pure white with *black patch on breast-sides* and black-and-chestnut bars on flanks; undertail-coverts creamy. Bill dusky above, blue-grey below. **Juvenile** Dusky-brown above; rump and tail blackish; pale buff-brown below; flanks more peach coloured. Bill dark. **VOICE** A fairly loud *teeoo,* frequently repeated; alarm a thin *cheep* or short *tsek.* Song a simple series of similar notes *teeo-teeo-twee teew-tee twee-twee-twee...* [CD15:50] **HABITS** In rank grass and bushes within forest clearings, along edges, and in cultivation, esp. ricefields. Partial to indigenous bamboo seeds. Usually in small groups; often with Bronze and Black-and-white Mannikins, but always in smaller numbers. **SIMILAR SPECIES** Both adult and juvenile Bronze and Black-and-white Mannikins are distinctly smaller with less heavy bills. Adult Bronze also has paler head and barred rump. Adult Black-and-white lacks brown in plumage and has all-blue bill. **STATUS AND DISTRIBUTION** Patchily distributed, locally uncommon to rare resident in forest zone and forest–savanna mosaic, from Senegambia to CAR and Gabon. Vagrant, NW Congo.

CUT-THROAT *Amadina fasciata* Plate 139 (9)
Amadine cou-coupé

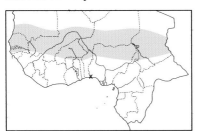

Other names E: Cut-throat Finch; F: Cou-coupé

L 11–12 cm (4.5–4.75"). *A. f. fasciata.* Sandy coloured and scaly looking. **Adult male** Distinctive red band across throat extending to ear-coverts (hence name). Top of head and upperparts pale sandy-brown densely barred black; face whitish; flight feathers dark grey-brown edged and tipped buffish. Tail blackish edged and tipped white. Underparts buff becoming whitish on lower belly; irregular chestnut patch on lower breast and upper belly; lower breast, flanks and undertail-coverts with black crescents or chevrons. Bill horn coloured. **Adult female** Less prominently barred, without red throat band; face barred; no chestnut on belly. **Juvenile** Similar to adult female, but mantle usually unbarred and underparts poorly marked. In juvenile male red throat soon appears but is obscured by black-and-white feather tips. **VOICE** A sparrow-like, abrupt *chilp;* also a nasal *dzèèp.* In flight, a thin *eee-eee-...* Song a fast medley of churring, ticking, rattling, chirping and clear whistling notes. [CD15:51] **HABITS** In pairs or small flocks, often with other estrildids outside breeding season. In dry woodland, thorn scrub and edges of cultivation. **STATUS AND**

DISTRIBUTION Locally common resident in Sahel zone, from S Mauritania and Senegambia to N Cameroon and Chad. Dry-season visitor in south of range.

INDIGOBIRDS AND WHYDAHS
Family VIDUIDAE (World: 18 species; Africa: 18; WA: 12)

Small, seed-eating African finches with short, conical bills. Sexually highly dimorphic in breeding plumage. Females, non-breeding males and juveniles are sparrow-like and often specifically indistinguishable. Occur mainly in savanna. Forage on ground. Breed mainly during rains and gather in flocks in off-season. Mainly sedentary, but local movements may occur in drier parts of range. All are polygamous, mostly species-specific brood parasites, laying eggs in nests of estrildids. Nestlings do not evict the host's offspring. Breeding males indulge in conspicuous aerial displays, but their songs are weak and consist of high-pitched notes, most mimicking those of their hosts.

Two groups can be distinguished.

Indigobirds are confusingly similar and often only distinguishable in the field if song of foster species is known. Males in breeding plumage are black and short tailed, differing only in glossy tone to plumage, wing colour, bill and leg colour, and, most importantly, songs. Latter, uttered from certain perches, include non-mimetic chattering and clear whistles mimicking hosts. Parasitise mainly firefinches *Lagonosticta* spp. The number of recognised species has increased as more has become known of their biology: up to ten species are now recognised, of which eight occur in our region. Status and distribution of several still imperfectly known; presence of host an indication of species possibly involved.

Whydahs. Males in breeding plumage are black with yellow and chestnut or white, and have four greatly elongated central tail feathers. Paradise whydahs are very similar but differ mainly in length, width and shape of tail; nape and breast colour may also differ but these features are rather variable and do not constitute reliable field marks. Field identification is tricky because apparent width of tail may vary with angle of view and longest tail streamers may not have attained full length or may have been lost. As with indigobirds, taxonomy is debated (with up to five species now recognised, of which three occur in our area), and status and distribution are inadequately known. Parasitise waxbills *Estrilda* and pytilias *Pytilia*.

Taxonomy and species accounts principally based on studies of Payne (1982, 1985, 1991, 1996).

VILLAGE INDIGOBIRD *Vidua chalybeata*   Plate 140 (6)
Combassou du Sénégal

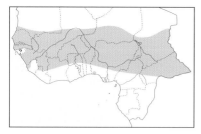

L 11.5 cm (4.5"). **Adult male breeding** The only indigobird in the region with *orange to pinkish-red legs*. Plumage glossy green (western *chalybeata*) grading to blue (eastern *neumanni*) with black flight feathers narrowly edged whitish. Bill white or pinkish-white. **Adult male non-breeding/adult female** Top of head has broad buff median crown-stripe bordered by bold dark brown lateral stripes; buff supercilium; upperparts buff-brown streaked dark brown; underparts pale buffish-brown becoming whitish on throat and belly. Bill white; *legs orange*. **Juvenile** Similar to adult female but crown plain dusky-brown, edges of wing feathers and upperparts slightly more rufous, head-sides and breast slightly browner. **VOICE** Mimics Red-billed Firefinch. Mimicry song, uttered from conspicuous perch or in display flight, consists of alarm call *chik* followed by a series of simple, clear whistled notes *pea-pea-pea*. Flight call a hoarse *cheh cheh cheh* ... [CD15:52] **HABITS** Generally tame, occurring near human habitation and in farmland. Male displays with hovering and bouncing flight before female. Parasitises Red-billed Firefinch. **SIMILAR SPECIES** All other male indigobirds in breeding dress have pale purplish legs. Female and non-breeding male Pin-tailed Whydah distinguished from corresponding Village Indigobird by pinkish bills and dark legs. Female and non-breeding male paradise whydahs have either dark bills and legs (Sahel P.W.) or pale orange bills and pinkish legs (Exclamatory P.W.). **STATUS AND DISTRIBUTION** Not uncommon resident in arid and semi-arid zones, from S Mauritania and Senegambia to Sierra Leone and Mali (nominate) and from Mali and N Ivory Coast to N Cameroon and Chad (*neumanni*).

JAMBANDU INDIGOBIRD *Vidua raricola*   Not illustrated
Combassou de Jambandu

Other name E: Goldbreast Indigobird

L 11.5 cm (4.5"). **Adult male breeding** Glossy blue to green with brown flight feathers. Bill white; legs pale purplish.

Adult male non-breeding/adult female/juvenile As Village Indigobird (q.v.) but with pale purplish legs. **VOICE & HABITS** Mimics and parasitises Zebra Waxbill. Mimicry song includes contact calls *chit chit* and *chirp*, social call *ink churr churr*, a series of twittering notes, flight call *trip* and juvenile begging calls. [CD15:54] **SIMILAR SPECIES** All other indigobirds. See also Village Indigobird. **STATUS AND DISTRIBUTION** Uncommon resident, recorded from Sierra Leone, Ghana, Nigeria and Cameroon.

BAKA INDIGOBIRD *Vidua larvaticola* Not illustrated
Combassou de Baka

Other name E: Black-faced Firefinch Indigobird
L 11.5 cm (4.5"). **Adult male breeding** Glossy blue to green-blue with pale brown flight feathers. Bill white; legs pale purplish. **Adult male non-breeding/adult female/juvenile** As Jambandu Indigobird (q.v.). **VOICE & HABITS** Mimics and parasitises Black-faced Firefinch. Mimicry song includes short series of slow, slurred whistles *whuu-ee whuu-ee* and *whee-uu whee-uu*, sharp alarm *dwit-it-it* and shrill contact call *seesee*. [CD15:55] **SIMILAR SPECIES** All other indigobirds. See also Village Indigobird. **STATUS AND DISTRIBUTION** Probably throughout, from Senegambia and Guinea to Chad and CAR. Recorded with certainty from Gambia, Nigeria and Cameroon; probably also Guinea, Mali, Ivory Coast, Ghana, Chad and CAR.

VARIABLE INDIGOBIRD *Vidua funerea* Not illustrated
Combassou noir

Other names E: Dusky Indigobird; F: Combassou variable
L 11.5 cm (4.5"). *V. f. nigerrima.* **Adult male breeding** Dull purplish-blue with dark brown flight feathers narrowly edged whitish. Bill white; legs pale purplish. **Adult male non-breeding/adult female/juvenile** As Jambandu Indigobird (q.v.). **VOICE & HABITS** Mimics and parasitises Blue-billed Firefinch. Mimicry song includes fast *pit-pit-pit-...* (alarm), slurred whistles and series of whistled *too-too* notes. [CD15:56] **SIMILAR SPECIES** All other indigobirds. See also Village Indigobird. **STATUS AND DISTRIBUTION** Uncommon resident, Nigeria and Cameroon.

CAMEROON INDIGOBIRD *Vidua camerunensis* Plate 140 (5)
Combassou du Cameroun

L 11.5 cm (4.5"). **Adult male breeding** Glossy blue with brown flight feathers narrowly edged whitish. Bill white; legs pale purplish. **Adult male non-breeding/adult female/juvenile** As Jambandu Indigobird (q.v.). **VOICE & HABITS** Mimics and parasitises Blue-billed Firefinch, Black-bellied Firefinch, Brown Twinspot and Dybowski's Twinspot; each male mimics only one host species. Mimicry songs include (1) Blue-billed Firefinch: fast *pit-pit-pit-...* (alarm), slurred whistles and series of whistled *too-too* notes; (2) Black-bellied Firefinch: plaintive low, quickly rising and slowly descending *peeuuh* (contact), sharp *chek* (alarm) and series of 4 or more low whistled notes *tew-tew-tew-tew*, (3) Brown Twinspot: *vay vay* (contact), sharp *tek tek tek* (alarm) and various whistles and trills; (4) Dybowski's Twinspot: *kek* and *churr* (contact), *zet* (alarm) and complex song including canary-like whistled trill *rrrrrrr* and distinctive buzzy whistle *vweee*. [CD15:60] **SIMILAR SPECIES** All other indigobirds. Jos Plateau Indigobird also mimics Blue-billed Firefinch but has glossy green plumage. See also Village Indigobird. **STATUS AND DISTRIBUTION** Resident, probably throughout from Senegal to Cameroon. Recorded with certainty from Sierra Leone, Ghana, Togo, Nigeria and Cameroon; probably also Senegal. **NOTE** Although 'camerunensis' was considered a *nomen dubium* by Payne (1982, 1985) because of doubt to which species the type specimen belonged, the name was subsequently used, by the same researcher, for indigobirds that were associated with above-mentioned hosts (Payne & Payne 1994, Payne 1996). In view of the multiple host species, the status of *V. camerunensis* is considered uncertain.

JOS PLATEAU INDIGOBIRD *Vidua maryae* Not illustrated
Combassou du Plateau de Jos

L 11.5 cm (4.5"). **Adult male breeding** Glossy green with pale brown flight feathers. Bill white; legs pale purplish. **Adult male non-breeding/adult female/juvenile** As Jambandu Indigobird (q.v.). **VOICE & HABITS** Mimics and parasitises Rock Firefinch. Mimicry song includes *pit-pit-pit-...* (alarm), series of whistled *too-too* notes and a descending trill. [CD15:57] **SIMILAR SPECIES** All other indigobirds. Some Cameroon Indigobirds, suspected to occur in SE Nigeria, mimic Blue-billed Firefinch and are glossy blue. See also Village Indigobird. **STATUS AND DISTRIBUTION** Known only from C and N Nigeria (Panshanu Pass, Taboru, Kagoro).

QUAILFINCH INDIGOBIRD *Vidua nigeriae* Not illustrated
Combassou du Nigéria

L 11.5 cm (4.5"). **Adult male breeding** Dull green (not distinctive in field; less bright than Jambandu Indigobird), with brown flight feathers narrowly edged whitish. Bill white; legs pale purplish. **Adult male non-breeding/adult**

female/juvenile As Jambandu Indigobird (q.v.). **VOICE & HABITS** Mimics and parasitises African Quailfinch. Mimicry song includes metallic *tink* and *trillink* and rapid series of *klik-klak-kloik-kluk* notes. [CD15:59] **SIMILAR SPECIES** All other indigobirds. See also Village Indigobird. **STATUS AND DISTRIBUTION** Known from Gambia, Nigeria and Cameroon.

WILSON'S INDIGOBIRD *Vidua wilsoni*   Not illustrated
Combassou de Wilson

Other name E: Pale-winged Indigobird
L 11.5 cm (4.5"). **Adult male breeding** Glossy purplish with dark brown flight feathers narrowly edged whitish. Bill white; legs purplish. **Adult male non-breeding/adult female/juvenile** As Jambandu Indigobird (q.v.). **VOICE & HABITS** Mimics and parasitises Bar-breasted Firefinch. Mimicry song includes a sharp *pik* or *tzet* (alarm) and rapid series of harsh metallic and low nasal notes. [CD15:58] **SIMILAR SPECIES** All other indigobirds. See also Village Indigobird. **STATUS AND DISTRIBUTION** Uncommon resident, recorded from Guinea-Bissau, Guinea, Ghana, Togo, Nigeria, Cameroon and CAR; probably also Senegambia.

PIN-TAILED WHYDAH *Vidua macroura*   Plate 140 (4)
Veuve dominicaine

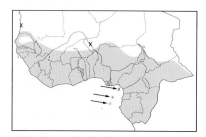

L 12.5 cm (5"); tail streamers up to *c.* 20 cm (8"). **Adult male breeding** Unmistakable. Black and white with 4 greatly elongated, ribbon-like central tail feathers and red bill. Black cap, upperparts and tail contrast with pure white head-sides (extending as collar on nape), rump, uppertail-coverts and underparts. Lesser and median coverts white forming wing patch. Legs blackish. **Adult female breeding** Broad tawny median crown-stripe bordered by bold blackish lateral stripes; supercilium and head-sides buff; black stripe behind eye and around ear-coverts. Upperparts buff-brown streaked dark brown. Underparts pale buffish-brown becoming whitish on throat and belly, and distinctly streaked on breast-sides. Wing and tail feathers blackish edged pale buffish. Bill mainly dark brownish. Legs dark. **Adult male/female non-breeding** Similar to adult female breeding but has pinkish bill. **Juvenile** Top of head, upperparts and tail plain pale brown; underparts buffish becoming buff-white on throat and central belly. Bill dusky. **VOICE** Call *tsip-tsip*. Song, uttered in display flight and at rest, a vigorous jerky series of high-pitched variations on the call note *tsip tsweep tswup tsweeu tswip* ... [CD15:61] **HABITS** Tame and gregarious, occurring in variety of open habitats, incl. savanna, grass- and farmland, and gardens. Male in breeding season very conspicuous, singing in low hovering and bouncing display flight over territory. Usually one adult male with some females; congregate in larger flocks in off-season. Parasitises mainly waxbills (incl. Orange-cheeked, Black-rumped, Common, Black-crowned); also Bronze Mannikin. **SIMILAR SPECIES** Female and non-breeding male indigobirds distinguished by white bills and pale or, in one species, orange legs. Paradise whydahs in corresponding plumages have less extensive dark markings on head-sides and only little and faint streaking on breast-sides; also either dark bill and legs (Sahel P.W.), or pale orange bill and pinkish legs (Exclamatory P.W.). **STATUS AND DISTRIBUTION** Common resident in suitable habitat throughout, incl. Bioko, São Tomé and Príncipe, avoiding desert and closed forest.

SAHEL PARADISE WHYDAH *Vidua orientalis*   Plate 140 (1)
Veuve à collier d'or

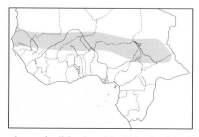

Other name F: Veuve de paradis
L 12.5 cm (5.5"); tail streamers up to *c.* 24 cm (9.5"). **Adult male breeding** Head, throat and upperparts black; broad nuchal collar chestnut (*aucupum*) or more golden-buff (**nominate**); breast chestnut; rest of underparts creamy-buff. *Tail black with greatly elongated feathers, central pair broad and relatively short with long bare shaft, next pair broad and very long with broad tips.* Bill and legs black. **Adult male non-breeding/adult female** Broad creamy-white median crown-stripe bordered by bold blackish lateral stripes; creamy-white supercilium; dark brown eye-stripe. Upperparts buff-brown streaked blackish; wing and tail feathers blackish edged buffish. Underparts buffish-brown becoming whitish on throat and central belly, with a few faint streaks on breast-sides. No tail streamers. Bill and legs dark. **Juvenile** Plain earth-brown above; pale buffish-brown below becoming white on lower breast and belly. Bill dark horn. **VOICE** Relatively silent. Call a sharp *chip*. Mimicry song includes vocalisations of Green-winged Pytilia. **HABITS** In small flocks, in open savanna woodland and adjacent cultivation. Displaying male reaches considerable height, circles with flapping wings, then dives to perch. Often perches conspicuously on high treetop. Parasitises Green-winged Pytilia.

[CD15:62] **SIMILAR SPECIES** Compare longer and more slender tailed Exclamatory and Togo Paradise Whydahs. Female and non-breeding male Pin-tailed Whydah distinguished from corresponding paradise whydahs by more extensive dark markings on head-sides and more distinct streaking on breast-sides. Female and non-breeding male indigobirds distinguished by white bills and pale or, in one species, orange legs. **STATUS AND DISTRIBUTION** Rather uncommon resident in broad savanna belt; recorded from S Mauritania, Senegambia, Guinea, Mali, Burkina Faso, Niger and Nigeria (*aucupum*) and from N Cameroon to Chad (nominate).

EXCLAMATORY PARADISE WHYDAH *Vidua interjecta*   Plate 140 (2)
Veuve nigérienne

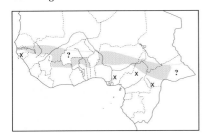

Other names E: Uelle Paradise Whydah; F: Veuve d'Uelle
L 12.5 cm (5.5"); tail streamers >27 cm (>10.5"). **Adult male breeding** As Sahel Paradise Whydah but with longer tail streamers; these also slightly broader than in Togo P.W. (width 30–40 mm). Reported to differ from latter in darker nape and more extensive chestnut on underparts, but these variable. **Adult male non-breeding** Similar to Sahel P.W. but with *pale orange bill* and *pinkish-grey legs*. **Adult female** Similar to male non-breeding but with *pinkish legs*. **Juvenile** Unknown. **VOICE** Mimicry song includes *pik* calls and *to-wit-to-wit* song of Red-winged Pytilia. [CD15:63] **HABITS** Savanna woodlands. Parasitises Red-winged and (probably) Yellow-winged Pytilias. **SIMILAR SPECIES** Compare Togo and Sahel Paradise Whydahs. **STATUS AND DISTRIBUTION** Common to uncommon resident, recorded from Gambia, Guinea, Mali, N Ivory Coast, Ghana, Togo, Nigeria, Cameroon, Chad and CAR. Distribution more southern than Sahel Paradise Whydah.

TOGO PARADISE WHYDAH *Vidua togoensis*   Plate 140 (3)
Veuve du Togo

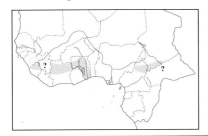

L 12.5 cm (5.5"); tail streamers >27 cm (>10.5"). **Adult male breeding** The longest tailed paradise whydah with relatively narrow tail streamers (width <30mm). **Adult male non-breeding/adult female/juvenile** Similar to other paradise whydahs; details unknown. **VOICE & HABITS** Mimics and parasitises Yellow-winged Pytilia. **SIMILAR SPECIES** Exclamatory and Sahel Paradise Whydahs (q.v.) have broader and shorter tails. **STATUS AND DISTRIBUTION** Uncommon resident, recorded from Sierra Leone, N Liberia (rare), Mali, N Ivory Coast, Ghana, Togo, N Cameroon and Chad.

TRUE FINCHES
Family FRINGILLIDAE (World: *c.* 130 species; Africa: 41; WA: 14)

Small seed-eaters with stout, conical bills. Most species sexually dimorphic. Juveniles similar to adult females, but often more streaked. Widespread and occurring in a variety of habitats, from forest to farmland and desert. Arboreal or terrestrial. Several have remarkable, melodious songs.

COMMON CHAFFINCH *Fringilla coelebs*   Plate 141 (8)
Pinson des arbres

L 15 cm (6"). *F. c. africana.* In all plumages, white shoulder, wingbar and outer tail feathers conspicuous, both when perched and in bouncing, undulating flight. **Adult male** Head bluish-grey; mantle, back and rump olive-green. Wings black with white on coverts forming 2 broad bars. Tail slightly forked, mainly black with white outer feathers. Underparts pale pink. Bill blue-grey, turning horn in non-breeding plumage. **Adult female** Mainly grey-brown, paler on underparts; wing and tail pattern as male. Bill brownish. **VOICE** Calls include a distinctive, clear *pink*; in flight a low *yup*. **STATUS AND DISTRIBUTION** Palearctic vagrant, N Mauritania (male, Nouadhibou, Oct 1988).

BRAMBLING *Fringilla montifringilla*   Plate 141 (9)
Pinson du Nord

L 15.5 cm (6"). All plumages distinguished by *orangey breast and shoulder*, and *white rump* (conspicuous in flight). **Adult male non-breeding** Head grey-brown with blackish face and dark centres to crown and nape feathers;

mantle feathers black broadly fringed orange-brown. Wings black with orangey shoulders and whitish to orangey wingbar. Tail slightly forked, black. Underparts orangey, becoming white on central belly and vent; flanks spotted black. Bill yellowish tipped black. **Adult male breeding** More colourful and contrasting, with black head and mantle (acquired through wear). Bill blue-black. **Adult female** As male non-breeding but duller, with broad buff-grey supercilium and without blackish face. **VOICE** Most characteristic call a drawn-out nasal *kèèhp*; in flight *yup*. [CD4:82] **STATUS AND DISTRIBUTION** Palearctic vagrant, N Mauritania (2 records of males, Nouadhibou, Jan and Apr 1988) and 60 km offshore S Senegal (male, Nov 1976).

BLACK-FACED CANARY *Serinus capistratus*   Plate 141 (1)
Serin à masque noir

L 11.5 cm (4.5"). *S. c. capistratus.* Small and brightly coloured with, in male, distinctive *black face.* **Adult male** Black face (formed by narrow band on base of forehead, above eye to cheeks and chin) bordered above by bright yellow upper forehead and supercilium; crown and upperparts pale olive-green finely streaked blackish; rump and uppertail-coverts yellower and unstreaked. Wing and tail feathers blackish edged greenish-yellow; yellowish tips to median and greater coverts form double wingbar. Underparts bright yellow slightly washed olive on flanks. Bill stubby, pale horn. Legs brownish. **Adult female** Similar but lacking black face and slightly more olive-green; head-sides olivaceous; throat, breast and flanks finely streaked dusky-olive. **Juvenile** Similar to adult female but paler or duller, and more heavily streaked. **VOICE** Various *chissik* notes. Song a jumble of whistles, trills and twittering phrases. [CD15:67] **HABITS** In pairs or small flocks at edges of forest, thickets and cultivation, and in gardens, usually near water. Occasionally in mixed flocks with *Ploceus* weavers. **STATUS AND DISTRIBUTION** Uncommon to locally common resident, Gabon and Congo.

WHITE-RUMPED SEEDEATER *Serinus leucopygius*   Plate 141 (2)
Serin à croupion blanc

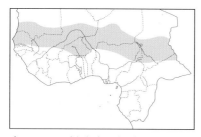

Other names E: Grey Canary, White-rumped Serin; F: Chanteur d'Afrique
L 10.0–11.5 cm (4.0–4.5"). Small and drab coloured with contrasting *white rump* (conspicuous in flight). **Adult *riggenbachi*** Head and upper-parts pale grey-brown streaked dark brownish. Wing and tail feathers dark brown edged paler; off-white tips to greater coverts form narrow wingbar. Underparts whitish washed and spotted grey-brown on breast. Bill pale horn. Legs pinkish. *S. l. pallens* slightly paler and greyer. **Juvenile** Paler or warmer brown, and more heavily streaked. **VOICE** Call *twee-eet.* Song, delivered from perch, a varied series of clear, sweet whistled and twittering phrases, resembling (but superior to) Yellow-fronted Canary's. [CD15:69] **HABITS** In small flocks in *Acacia* scrub, farmland, grassland near water and gardens. Tame. **SIMILAR SPECIES** Combination of dull grey-brown plumage and white rump distinctive, eliminating all other seedeaters within range. **STATUS AND DISTRIBUTION** Locally common to scarce resident in Sahel and northern savanna zones, from Mauritania and Senegambia to Chad and CAR. Rare, Togo. *S. l. riggenbachi* throughout; *pallens* in N Niger (Aïr)

BLACK-THROATED SEEDEATER *Serinus atrogularis*   Plate 141 (4)
Serin à gorge noire

Other names E: Yellow-rumped Seedeater, Black-throated Serin
L 11–12 cm (4.5"). *S. a. lwenarum.* Small and drab coloured with contrasting *bright yellow rump* (conspicuous in flight). **Adult** Head and upperparts grey-brown streaked dark brownish; indistinct pale buff supercilium. Wing and tail feathers dark brown edged buffish; buff tips to greater and median coverts form narrow double wingbar. Throat variably mottled blackish; rest of underparts washed pale buff-brown. Bill brownish-horn. Legs pinkish-brown. **Juvenile** Warmer buff-brown and more heavily streaked. **VOICE** Call a clear rising *tweee*; in flight a double *chirrup.* Song, often uttered from treetop, a rapid, sustained jumble of trills and whistles, reminiscent of Yellow-fronted Canary. [CD15:70] **HABITS** In pairs or small flocks in open woodland and cultivation. Forages mostly on ground. **SIMILAR SPECIES** Combination of dull grey-brown plumage and bright yellow rump distinctive, eliminating all other seedeaters within range. **STATUS AND DISTRIBUTION** Locally not uncommon resident, S Gabon and Congo.

YELLOW-FRONTED CANARY *Serinus mozambicus*   Plate 141 (5)
Serin du Mozambique

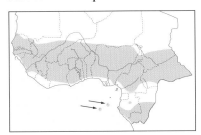

Other name E: Yellow-eyed Canary
L 11–13 cm (4.5–5.0"). Small with distinctive face pattern, bright yellow underparts and yellow rump (conspicuous in flight). **Adult male caniceps** *Yellow forehead and supercilium* contrasting with grey crown and broad *dark eye-stripe*; cheeks yellow contrasting with *dark moustachial stripe*; nape and neck-sides greyish. Upperparts greyish-green lightly streaked dusky; wings dusky edged yellowish-green; yellow tips to median and greater coverts form indistinct narrow wingbars; tail blackish tipped yellowish-buff. Underparts yellow washed olive on breast-sides. Bill horn. Legs dark. Geographical variation difficult to assess in field: *punctigula* has olive-green crown and upperparts; *barbatus* similar but brighter overall; *tando* similar but less bright; poorly defined *santhome* has crown more grey-brown. **Adult female** Slightly duller. In *caniceps* lower throat usually has dark spots, forming necklace (more distinct in *punctigula*). **Juvenile** As adult female but duller; yellow areas paler; some streaks or spots on breast-sides. **VOICE** A single or double *tseeu* or *tsssp* and *tuwu-tsilip* (or *tuwu-tswirri*) or *tuwu-tuwee*. Song, delivered from treetop, a fairly simple but melodious, rapid jumble of twittering and whistled phrases. [CD15:71] **HABITS** In pairs or small flocks in open woodland, edges of cultivation, gardens and near villages. Forages mostly on ground. Tame. A popular cage bird. **STATUS AND DISTRIBUTION** Locally common resident throughout, except arid north; also São Tomé and Annobón. *S. m. caniceps* in most of range; *punctigula* C and S Cameroon; *barbatus* CAR; *tando* Gabon and Congo; *santhome* São Tomé.

STREAKY-HEADED SEEDEATER *Serinus gularis*   Plate 141 (6)
Serin gris

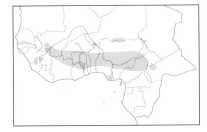

L 15 cm (6"). Mainly brown with distinctive head pattern. **Adult canicapillus** Blackish face bordered by bold white supercilium and white throat; crown streaked dark brown and white. Upperparts indistinctly streaked; rump plain. Underparts buffish becoming white on central belly. Bill heavy, horn coloured. Races differ slightly in coloration: *montanorum* darker and warmer brown; *elgonensis* slightly greyer. **Juvenile** Duller and more streaky; breast and flanks streaked dark brown. **VOICE** Calls include a nasal *shewee-uee*, a short *chip*, a high-pitched *tseeee* and *chiririt*. Song a clear, rapid jumble of varied notes typical of canary, occasionally uttered in display flight. [CD15:74] **HABITS** Usually singly, in pairs or small flocks in open woodland, degraded savanna, edges of cultivation and scrub. Quiet and unobtrusive. **STATUS AND DISTRIBUTION** Locally not uncommon to scarce resident, in Guinea and Sierra Leone, and from S Mali and N Ivory Coast to S Niger and N Cameroon (*canicapillus*), C Cameroon highlands (*montanorum*) and NW CAR (*elgonensis*).

THICK-BILLED SEEDEATER *Serinus burtoni*   Plate 141 (7)
Serin de Burton

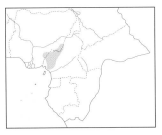

L 18 cm (7"). *S. b. burtoni*. Distinctive; robust and dark brown with very heavy bill. **Adult** Head and upperparts earth-brown streaked dusky, with white forehead and blackish face; buffish-white tips to median and greater coverts form double wingbar. Underparts paler brown becoming buffish on belly and undertail-coverts; flanks streaked dark brown. Bill dark horn above, paler below. **Juvenile** Paler brown; forehead to upper throat whitish; bill blackish. **VOICE** Mostly silent. A high-pitched *srip-sreep*. Song a soft jumble of tinkling notes, typical of canary. [CD15:77] **HABITS** Singly, in pairs or small parties at edges of montane forest and, at higher elevations, in patches of scrub and small trees. Mainly in dense undergrowth, but may perch conspicuously near tops of bushes or small trees. Unobtrusive and easily overlooked. **SIMILAR SPECIES** Adult female Thick-billed Weaver has similar size but lacks white on forehead and wings; also is a more widespread lowland species. **STATUS AND DISTRIBUTION** Scarce or uncommon to locally common resident in highlands of SE Nigeria (Obudu and Mambilla Plateaux) and W Cameroon (Mt Cameroon, Mt Manenguba, Bamenda Highlands).

PRÍNCIPE SEEDEATER *Serinus rufobrunneus*   Plate 144 (11)
Serin roux

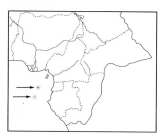

Other name F: Serin de Principé
L 11.0–12.5 cm (4.5–5.0"). **Adult** *thomensis* All grey-brown, slightly warmer on underparts, with blackish-streaked crown and paler throat. Bill dark horn above, paler below. **Nominate** all rufous-brown with blackish streaks of crown extending onto mantle; paler throat with some dark marks; tail shorter. **S. r. fradei** as nominate but slightly more rufous. **Juvenile** Duller. **VOICE** A hard, tuneless whistle *whsiiii*. Song a fast, melodious jumble of typical canary-like twitters and trills. [CD15:78] **HABITS** Usually in pairs, foraging high up along edges of forest and plantations. On São Tomé in variety of habitats, incl. city centres and dry woodland, up to 900 m; on Príncipe less often in towns. **STATUS AND DISTRIBUTION** Resident, endemic to São Tomé (*thomensis*; very common), Príncipe (nominate; uncommon to scarce) and Caroco (south of Príncipe; *fradei*).

SÃO TOMÉ GROSBEAK *Neospiza concolor*   Plate 144 (12)
Néospize de Sao Tomé

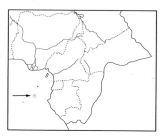

Other name F: Grosbec de Sao Tomé
L 19–20 cm (7.5–8.0"). **Adult** All dark chestnut with large head and massive, dark horn-coloured bill. Short blackish streaks on crown and ear-coverts visible at close range. Sexes presumed alike. **Juvenile** Unknown. **VOICE** A brief series of 4–5 short, thin whistles given in a canary-like fashion (Sargeant *et al.* 1992). **HABITS** Restricted to dense lowland primary rainforest, where it feeds in closed canopy of tall trees. Biology unknown. **STATUS AND DISTRIBUTION** Endemic to São Tomé, where extremely rare resident, confined to rainforest in south of island. Rediscovered in 1991, after 100 years, in small forest clearing above R. Xufexufe. *TBW* category: THREATENED (Critical)

ORIOLE FINCH *Linurgus olivaceus*   Plate 141 (3)
Linurge loriot

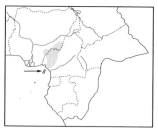

Other name F: Pinson-loriot
L 13 cm (5"). *L. o. olivaceus.* **Adult male** Head and throat black separated from bright yellowish-green upperparts by golden-yellow collar; greater coverts broadly tipped yellow; flight feathers black, with secondaries and tertials broadly fringed yellow; tail edged and tipped yellow. Underparts yellow; upper breast tinged orange. Bill bright orange-yellow. **Adult female** Mainly olive-green with greyish head, becoming more yellowish on belly; wing pattern as male but edges of secondaries and tertials greener. Bill yellowish. **Juvenile** As adult female but duller; underparts paler and faintly streaked on breast and flanks. Bill dusky. **VOICE** Mostly silent. A wheezy, high-pitched *tssit* or *twee*. Song a sustained, high-pitched twittering and soft churring followed by a melodious whistle. [CD15:79] **HABITS** Singly, in pairs or small parties in montane forest, esp. at edges; also in scrub-filled ravines above tree line and in bushes, abandoned farmland, bamboo and *Eucalyptus* plantations. From undergrowth to treetops. **SIMILAR SPECIES** Some adult male weavers may superficially resemble male Oriole Finch, but latter always separated by orange bill. Forest orioles are much larger, have black in tail and long, slender bill. **STATUS AND DISTRIBUTION** Not uncommon resident in highlands of SE Nigeria (Obudu and Mambilla Plateaux), W Cameroon (Mt Cameroon, Rumpi Hills, Mt Manenguba, Bamenda Highlands, Tchabal Mbabo) and Bioko.

EUROPEAN GREENFINCH *Carduelis chloris*   Plate 141 (10)
Verdier d'Europe

L 15 cm (6"). Robust with stout bill and distinctive, bright yellow patches on primaries and sides of relatively short, notched tail (conspicuous in flight). **Adult male** Mainly olive-green; plumage brightening with wear. Bill pinkish-horn. **Adult female** Duller, more grey-brown. **VOICE** A soft *dsooeet*; in flight a characteristic *djururrup*. Song includes a distinctive, drawn-out wheeze *dzwèèèèèèèh*. [CD4:76] **HABITS** On ground or perched in trees and bushes. Feeds on variety of seeds. **STATUS AND DISTRIBUTION** Palearctic vagrant, coastal Mauritania (probably N African and S European *aurantiiventris*): Nouadhibou (Oct–May); also Nouakchott (May–Jun; introduced or escaped?).

COMMON LINNET *Carduelis cannabina*
Plate 141 (11)

Linotte mélodieuse

L 14 cm (5.5"). Brownish with white edges to primaries and tail (fairly conspicuous in flight). **Adult male** Head brownish-grey with some dark streaking on crown; upperparts chestnut with some darker streaks; narrow white edges to blackish primaries form panel on folded wing. Tail slightly forked, feathers black edged white. Underparts pale buff-brown becoming whitish in centre and softly streaked brown on breast-sides and flanks. Plumage becomes more colourful and contrasting with wear; by Apr head more uniformly grey with crimson forehead; breast crimson; upperparts without streaking. **Adult female** Resembles adult male but lacks crimson and is more heavily streaked on upper- and underparts. **VOICE** Call a dry stuttering *knutnutnutnut*, usually in flight. [CD4:79] **HABITS** On ground or perched in bushes. Flight undulating. **STATUS AND DISTRIBUTION** Palearctic vagrant (probably N African and European nominate), Mauritania (mainly Dec–Jan; also Mar, Jun) and N Senegal (one, Mar 1971). **NOTE** Sometimes placed in genus *Acanthis*.

TRUMPETER FINCH *Bucanetes githagineus*
Plate 141 (12)

Roselin githagine

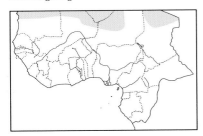

Other name F: Bouvreuil githagine

L 12.5 cm (5"). *B. g. zedlitzi*. Small, inconspicuous, dumpy desert finch with short, stubby bill. **Adult male breeding** Head grey; rest of plumage buff-grey with distinct pink wash to face, wings, rump and esp. underparts; tail brownish edged pink. Bill coral-red; legs orange-pink. **Adult male non-breeding/adult female** Duller and rather featureless. Pale sandy-grey with slightly paler rump; throat, tips of greater coverts and base of tail-sides washed pink. Bill pale horn, tinged yellow or pink. **Juvenile** As adult female but lacks pink; edges of wing feathers pale buffish-brown. **VOICE** Call a short *tip* or *chee-chup*; in flight a soft, nasal *veechp*. Song a distinctive drawn-out nasal wheeze, often described as recalling a toy trumpet. [CD4:85] **HABITS** Usually in pairs or, outside breeding season, in small flocks in rocky desert, usually near water. Forages on ground. Tame but unobtrusive. **STATUS AND DISTRIBUTION** Not uncommon but local resident, Mauritania, N Mali, Niger (esp. Aïr) and N Chad. Local movements. Vagrant, Cape Verde Is. **NOTE** Sometimes placed in genus *Rhodopechys*.

BUNTINGS

Family EMBERIZIDAE (World: *c.* 290 species; Africa: 15; WA: 8)

Small, mainly seed-eating passerines with stout, conical bills. Most sexually dimorphic. Juveniles similar to adult females but duller. Predominantly terrestrial. Flight typically undulating. Widespread and occurring in variety of open habitats. Usually singly or in pairs, but some gregarious outside breeding season. Vocalisations usually rather simple.

ORTOLAN BUNTING *Emberiza hortulana*
Plate 142 (7)

Bruant ortolan

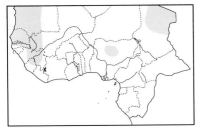

L 16.5 cm (6.5"). In all plumages distinguished by *sulphur-yellow moustachial stripe, throat and eye-ring*, and *pink bill*. **Adult male** Head to upper breast olive-grey, contrasting with orange-rufous under-parts. Upperparts rufous-brown boldly streaked black. Outer tail feathers with large amount of white. Legs pink. With wear head becomes greyer. **Adult female** Duller; head more olive-brown and finely streaked on crown; breast distinctly streaked dark. **First-year** As adult female but duller and more distinctly and extensively streaked, with narrow dark streaks on flanks. **VOICE** Calls, mostly in flight, include a short *twit* and a clear metallic *seee*. Song a short series of ringing notes followed by 2–3 lower, more melancholy units *dzee-dzee-dzee-dzee truuh-truuh*. [CD4:91] **HABITS** Winter visitors frequent open uplands. Singly or in small parties, often with Cinnamon-breasted Rock Bunting. Forages mainly on ground, but also in trees. **SIMILAR SPECIES** Adult female Cretzschmar's Bunting differs from female Ortolan in bluish-grey tinge to head and pale rusty moustachial stripe and throat. First-year Cretzschmar's very similar to 1st-year Ortolan, but more rusty. House Bunting lacks characteristic head pattern and white in tail, and has two-toned bill. **STATUS AND DISTRIBUTION** Rare or uncommon and local Palearctic passage migrant and winter visitor, recorded from Mauritania, Senegambia, Mali, Sierra Leone, Guinea (Fouta Djalon, Mt Nimba), Liberia (Mt Nimba), Nigeria (Zaria, Jos Plateau) and Chad.

CRETZSCHMAR'S BUNTING *Emberiza caesia*   Plate 142 (5)
Bruant cendrillard

L 15.5 cm (6"). Plumage pattern as Ortolan Bunting but more rusty. **Adult male** *Head to upper breast bluish-grey* with creamy eye-ring and *rusty moustachial stripe and throat.* Underparts deep rusty-brown. Upperparts rufous-brown boldly streaked black; rump plain; white on outer tail feathers. Bill horn-grey (non-breeding) or pink (breeding). Legs pink. **Adult female** Much duller with paler moustachial stripe and throat; crown and breast finely streaked dark. **First-year** As adult female but duller and more distinctly and extensively streaked, with narrow dark streaks on flanks. **VOICE** Calls include a harsh *zee* or *chip*. Song resembles Ortolan's but shorter and harsher. **HABITS** Frequents dry open habitats. Mainly terrestrial. **SIMILAR SPECIES** Adult female Ortolan Bunting separated from female Cretzschmar's by olive-grey tinge to head and pale yellow moustachial stripe and throat. First-year Ortolan very similar to 1st-year Cretzschmar's, but more olive-toned. **STATUS AND DISTRIBUTION** Palearctic vagrant, Chad (Abéché, Jan 1963). Wintering grounds further east, from Sudan to Eritrea.

HOUSE BUNTING *Emberiza striolata*   Plate 142 (6)
Bruant striolé

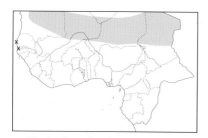

L 14 cm (5.5"). **Adult male** *sahari* Head and breast grey streaked black with dusky eye-stripe and moustachial stripe. Rest of plumage rusty-brown, slightly paler on underparts. Tail dark brown edged rufous (lacking white). Bill grey above, orange-yellow below. Legs pinkish. Colour saturation and distinctness of streaking variable. Birds in Ennedi (Chad) more prominently streaked in all plumages. **Adult female** Duller and paler; head slightly more grey-brown; streaking more indistinct, only obvious on crown. **Juvenile** Even paler than adult female, with browner head. Bill dusky-horn. **VOICE** A nasal *chzwee* and a thin *chik*. Song, uttered from perch or ground, a rapid, simple and rather monotonous series of similar syllables *werdji-werdi-wedi-werdji* or *cherwi-cherwi-chwrii wi-cherr*. [CD4:92] **HABITS** In pairs and small groups in rocky wadis, cultivation and villages in desert and Sahel. Not shy. In off-season sometimes with Cinnamon-breasted Rock Bunting. **SIMILAR SPECIES** Cinnamon-breasted Rock Bunting has prominent head pattern and less rufous in wings. **STATUS AND DISTRIBUTION** Uncommon to locally common resident in arid zone, Mauritania, Mali, Niger and Chad. Vagrant, Senegambia. Some local movements may occur. **NOTE** Given variation in plumage tone, validity of race *sanghae*, described from Dogon Plateau (Mali), which has reportedly darker and more chestnut body, is doubtful (cf. Lamarche 1981).

EURASIAN ROCK BUNTING *Emberiza cia*   Plate 147 (13)
Bruant fou

L 16 cm (6.25"). **Adult male** Head and breast grey with distinct pattern of black eye-stripe and moustachial stripe joining around ear-coverts, and lateral crown-stripe. Upper- and underparts rufous-brown boldly streaked black on mantle; white tips to black median coverts form white wingbar. Bill *grey.* Legs pinkish. **Adult female** Much duller. **Juvenile** Dull brown and streaky. Distinguished by dark stripe through eye and around ear-coverts; underparts streaked dusky. **VOICE** A sharp *tsee* and a short *tup*; in flight *chelut* and twittering *cheedeedee*. Song varied, fast and high pitched, including short clear trills and twitters. [CD4:93] **HABITS** In dry rocky habitats. Unobtrusive and terrestrial. Associates with other buntings in off-season. **SIMILAR SPECIES** House Bunting lacks prominent head stripes and has plain mantle and bicoloured bill; juvenile has unstreaked underparts. **STATUS AND DISTRIBUTION** One mist-netted in Chad (Abéché, Jan 1963) claimed as juvenile of this Palearctic species (Salvan 1969).

CINNAMON-BREASTED ROCK BUNTING *Emberiza tahapisi*   Plate 142 (4)
Bruant cannelle

L 14 cm (5"). Rufous-brown with distinctly striped head. **Adult male** *goslingi* Head boldly striped black and white: median crown-stripe, supercilium, stripe above ear-coverts and submoustachial stripe white, alternating with black lateral crown-stripe, eye-stripe and moustachial stripe. Upperparts rufous-brown boldly streaked black; secondaries rufous (forming bright panel); tail blackish-brown, without white. Throat grey; rest of underparts

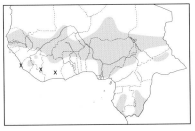

plain cinnamon-brown. Bill dark above, yellowish below. Legs pinkish. **Nominate** has black throat and less rufous wings. **Adult female** Duller; head has buffish-white and blackish-brown stripes. **Juvenile** Similar to adult female but duller with less contrasting head pattern; throat paler with some dusky streaking. Bill grey. **VOICE** A nasal *wee-eh* or *perwee-er*; alarm a thin, drawn-out *sweeee*. Song short and rapid, with emphasis on final note, e.g. *chrr-trr-erl-CHEEP* or *chrr-terr-erl-CHIrlp*, frequently repeated. [CD15:80] **HABITS** In pairs or small parties in open, mostly dry, savanna with rocky outcrops and eroded areas. Forages on ground. Unobtrusive but not shy. **STATUS AND DISTRIBUTION** Patchily distributed, locally common resident and migrant, recorded from most countries in the region. *E. t. goslingi* throughout north of range; nominate in Gabon and Congo. Southward movements in dry season recorded.

AFRICAN GOLDEN-BREASTED BUNTING *Emberiza flaviventris*   Plate 142 (1)
Bruant à poitrine dorée

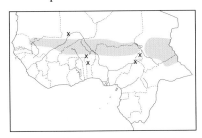

L 15.5 cm (6"). Bold black-and-white head pattern and golden-yellow underparts. **Adult male** *flavigaster* Broad white median crown-stripe, supercilium and stripe from bill to ear-coverts alternating with black lateral crown-stripe, eye-stripe and moustachial stripe. Mantle rufous separated from head by paler collar and contrasting with *pale grey rump and uppertail-coverts; median coverts and tips to greater coverts white, forming double wingbar.* Tail blackish-brown, 3 outer feathers broadly tipped white. Underparts yellow washed orangey on breast and becoming whitish on belly and undertail-coverts. Bill dark horn above, paler below. Legs pinkish. **Nominate** has deeper coloured plumage with chestnut mantle and richer yellow underparts. **Adult female** Dark head markings browner; often some yellow in median crown-stripe and supercilium; breast slightly paler. **Juvenile** Duller, with head pattern brown and buff; feathers of upperparts broadly edged buffish. **VOICE** A subdued *chruteeu*; in flight a soft *chup* or *tsip*. Song, usually uttered from rather concealed perch, a rapid succession of variable series of 4–10 similar syllables, e.g. *tsee-ep tsee-ep tsee-ep...*, melodious *tueet-tueet-tueet...*, faster *tsewtsewtsewtsew...*, drawn-out sibilant *hseeuw-hseeuw-hseeuw-...*etc. [CD15:81] **HABITS** Singly or in pairs in *Acacia* steppe and dry open savanna; nominate in various types of open woodland. Forages mainly on ground. **SIMILAR SPECIES** Brown-rumped Bunting has similar pattern but lacks white wingbars and has brown (not grey) rump; also a whiter face. Cabanis's Bunting is darker above with all-dark cheeks. **STATUS AND DISTRIBUTION** Not uncommon to rare and local resident in Sahel zone, from S Mauritania and Senegal to Chad and CAR (*flavigaster*); also S Congo (nominate).

BROWN-RUMPED BUNTING *Emberiza affinis*   Plate 142 (2)
Bruant à ventre jaune

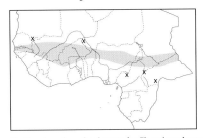

L 14 cm (5.5"). Resembles African Golden-breasted Bunting but principally separated by brown (not grey) rump and absence of white wingbars. **Adult male** *nigeriae* breeding Head white with black lateral crown-stripe and stripe from bill below ear-coverts joining post-ocular stripe. Upperparts chestnut, indistinctly streaked on mantle, and separated from head by greyish collar; wings lack white. Almost entire underparts yellow; undertail-coverts white. Bill dusky above, paler below. Legs greyish or brownish. *E. a. vulpecula* has darker, more chestnut upperparts. **Adult male non-breeding** Mantle paler. **Adult female** As adult male but head markings browner. **Juvenile** Duller; head washed rusty-buff; stripes brown. **VOICE** A liquid melodious *pidrewdrlwi*; in flight a short *chip*. Song, uttered from perch, a rather short, slightly harsh warble *rwidji-drrrwidji-drr-djirree*, with variations. [CD15:83] **HABITS** Singly or in pairs in open savanna and farmland. Forages on ground. Nest undescribed. **SIMILAR SPECIES** In addition to above-mentioned differences, African Golden-breasted Bunting has bolder black-and-white head pattern, white lower underparts, stronger orangey wash on breast, slightly larger size and different, monotonous song. **STATUS AND DISTRIBUTION** Uncommon to locally not uncommon resident in savanna zone, recorded from Senegambia and Guinea, through S Mali and N Ivory Coast to S Chad and W CAR. Vagrant, S Mauritania. *E. a. nigeriae* throughout most of range, east to L. Chad area and N Cameroon; *vulpecula* in Cameroon (Adamawa Plateau), extreme S Chad and W CAR. Northward movement with rains recorded. **NOTE** Often treated under specific name *E. forbesi* (for discussion, cf. Dowsett & Dowsett-Lemaire 1993).

CABANIS'S BUNTING *Emberiza cabanisi*
Bruant de Cabanis

Plate 142 (3)

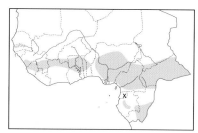

L 16.5 cm (6.5"). **Adult male** *cabanisi* Head blackish with long white supercilium; upperparts blackish indistinctly streaked grey; median and greater coverts tipped white, forming double wingbar. Tail black tipped and edged white (conspicuous in flight). Throat white; underparts yellow; undertail-coverts white. Bill dark horn above, pale below. Legs pinkish-brown. **E. c. cognominata** has narrow grey median crown-stripe; upperparts grey and brown distinctly streaked black. **Adult female** Head and upperparts browner; supercilium and stripe bordering head-sides tinged buffish. **Juvenile** Head and upperparts dark brown; supercilium and tips of wing-coverts rufous-buff; underparts paler yellow with rusty-brown wash to breast. **VOICE** A soft *tureee*; in flight also *tsip*; alarm a drawn-out, high-pitched *seeee*. Song, uttered from conspicuous perch in tree, consists of rapid and variable series of 3–6 similar syllables, similar to those of African Golden-breasted Bunting e.g. high-pitched, drawn-out *heest-heest-heest*, sweet *cheedewu- cheedewu- cheedewu-...*, clear *tsuwee-tsuwee-tsuwee-...* and faster *chewichewichewichew*. [CD15:82] **HABITS** Singly or in pairs in wooded savanna. Forages mainly on ground. **SIMILAR SPECIES** The only other buntings in the region with yellow underparts, Brown-rumped and African Golden-breasted, have white on ear-coverts and brown upperparts. **STATUS AND DISTRIBUTION** Uncommon to locally not uncommon resident in savanna zone, from W Guinea to Sierra Leone, and from S Mali (rare) and N Ivory Coast to Chad, CAR and Congo. Nominate throughout north of range; *cognominata* in S Gabon and Congo.

CORN BUNTING *Miliaria calandra*
Bruant proyer

Plate 142 (8)

L 18 cm (7"). Bulky, brown, streaked bunting with heavy bill and no white in tail. **Adult** Head and upperparts buff-brown or grey-brown heavily streaked blackish-brown. Wing feathers blackish-brown edged and fringed buff-brown. Underparts pale buffish heavily streaked on throat-sides, breast (esp. centre) and flanks. Bill horn. Legs pinkish. Plumage becomes more contrasting and colder with wear. **VOICE** Most characteristic call a dry, hard *kwit* or *tuk* and *kwitit*, uttered on take-off and in flight. [CD4:96] **HABITS** In our region mostly singly. In open habitat along coast. **STATUS AND DISTRIBUTION** Palearctic vagrant, coastal N Mauritania (singles, Oct, Jan and Feb) and N Senegal (three, Feb 1974). **NOTE** Some times placed in genus *Emberiza* (cf. Sibley 1996).

RECENT ADDITIONS

NORTHERN FULMAR *Fulmarus glacialis*
Fulmar boréal

Not illustrated

L 43–52 cm (17.0–20.5"); WS 101–117 cm (40–46"). A rather stocky seabird with relatively large head and short neck and tail. Even at great distance easily distinguised from gulls by flight on stiff, straight, relatively narrower and shorter wings. Typical form has head and underparts white. Upperparts pale grey with diffuse white patch at base of primaries; tail paler grey. Wings lack white trailing edge. Small dark spot in front of eye. Bill short and stubby, with prominent tubed nostrils; greyish and yellowish. **HABITS** Flight consists of rapid wingbeats followed by long periods of gliding low over waves. Floats high when swimming; take-off requires pattering run along surface. Often follows ships. Takes fish offal and carrion. **SIMILAR SPECIES** Shearwaters have more slender appearance, with longer, narrower wings, narrower tail and thinner bill, and are darker above, incl. on head. Large gulls have different jizz , with buoyant flight on long, narrow, rather pointed, black-tipped wings; upperwings have white trailing edge. **STATUS AND DISTRIBUTION** Palearctic vagrant. Single record, the first for the Afrotropics: one freshly dead off Dakar, Senegal, early Sep 1999 (ringed as nestling in Ireland; Clark *et al.* 2001).

LESSER SAND PLOVER *Charadrius mongolus*
Pluvier de Mongolie

Not illustrated

L 18–20 cm (7–8"). *C. m. pamirensis*. In all plumages very similar to Greater Sand Plover (p. 432) but slightly smaller, with shorter, blunter-tipped bill, more rounded head (with steeper forehead) and shorter legs (esp. tibiae). Legs dark grey to dull blackish. Feet do not project beyond tail in flight. **VOICE** A clear, hard *krip* or *tirrip*. **HABITS** In W Africa singly, on intertidal mudflats. **SIMILAR SPECIES** Greater Sand Plover (see above). Common Ringed Plover appears more compact, with relatively shorter legs and bill. Kentish Plover has similar structure (with relatively large head and without noticeably long tibiae), but is smaller and differs, in non-breeding plumage, by smaller breast patches and white nuchal collar. **STATUS AND DISTRIBUTION** Palearctic vagrant, claimed from Gabon (Aug/Sep). Winters mainly on E and S African coasts (Aug-early May).

REFERENCES

The references are in four parts: (1) general and regional references, (2) country references, (3) family and species references and (4) acoustic reference. Titles in (1) are not repeated in (2) and (3).

1. GENERAL AND REGIONAL REFERENCES

Bannerman, D. A. (1930–1951) *The Birds of Tropical West Africa*, Vols. 1–8. Crown Agents, London.

Bannerman, D. A. (1953) *The Birds of West and Equatorial Africa*. 2 vols. Oliver & Boyd, Edinburgh & London.

Bates, G. L. (1930) *Handbook of the Birds of West Africa*. John Bale, Sons & Danielsson, London.

Beaman, M. (1994) *Palearctic Birds: a Checklist of the Birds of Europe, North Africa and Asia north of the foothills of the Himalayas*. Harrier Publications, Stonyhurst.

Beaman, M. & Madge, S. (1998) *The Handbook of Bird Identification for Europe and the Western Palearctic*. Christopher Helm, London.

Benson, C. W., Brooke, R. K., Dowsett, R. J. & Irwin, M. P. S. (1971) *The Birds of Zambia*. Collins, London.

BirdLife International (2000) *Threatened Birds of the World*. Lynx Edicions, Barcelona & BirdLife International, Cambridge.

Bouet, G. (1955–1961) *Oiseaux de l'Afrique Tropicale*, Vols. 1–2. Office de la Recherche Scientifique et Technique Outre-Mer, Paris.

Brown, L. H., Urban, E. K. & Newman, K. (1982) *The Birds of Africa*, Vol. 1. Academic Press, London.

Chapin, J. P. (1932–1954) The birds of the Belgian Congo. *Bull. Am. Mus. Nat. Hist.* 65, 75, 75A, 75B.

Chappuis, C. (1974–1985) Illustration sonore de problèmes bioacoustiques posés par les oiseaux de la zone éthiopienne. *Alauda* 42: 197–222; 42: 467–500; 43: 427–474; 46: 327–355; 47: 195–212; 49: 35–58; 53: 115–136.

Collar, N. J., Crosby, M. J. & Stattersfield, A. J. (1994) *Birds to Watch 2: the World List of Threatened Birds*. BirdLife International, Cambridge.

Collar, N. J. & Stuart, S. N. (1985) *Threatened Birds of Africa and Related Islands: the ICBP/IUCN Red Data Book*. International Council for Bird Preservation & International Union for Conservation of Nature and Natural Resources, Cambridge.

Commission Internationale des Noms Français des Oiseaux (1993) *Noms Français des Oiseaux du Monde*. Éditions MultiMondes, Sainte-Foy & Chabaud, Bayonne.

Cramp, S. (ed.) (1985–1992) *The Birds of the Western Palearctic*, Vols. 4–6. Oxford University Press, Oxford.

Cramp, S. & Perrins, C. M. (eds.) (1993–1994) *The Birds of the Western Palearctic*, Vols. 7–9. Oxford University Press, Oxford.

Cramp, S. & Simmons, K. E. L. (eds.) (1977–1983) *The Birds of the Western Palearctic*, Vols. 1–3. Oxford University Press, Oxford.

Curry-Lindahl, K. (1981) *Bird Migration in Africa*. 2 vols. Academic Press, London.

Dekeyser, P. L. & Derivot, J. H. (1966–1968) *Les Oiseaux de l'Ouest Africain*. Fasc. 1–3. Initiations et Études africaines No. 19. Inst. fon. Afr. noire, Dakar.

Dodman, T. & Taylor, V. (1996) *African Waterfowl Census 1996*. Wetlands International, Wageningen.

Dowsett, R. J. & Forbes-Watson, A. D. (1993) *Checklist of Birds of the Afrotropical and Malagasy Regions. Volume 1: Species limits and distribution*. Tauraco Press, Liège.

Dowsett, R. J. & Dowsett-Lemaire, F. (eds.) (1993) *A Contribution to the Distribution and Taxonomy of Afrotropical and Malagasy Birds*. Tauraco Research Report No. 5. Tauraco Press, Liège.

Dowsett, R. J., Fry, C. H. & Dowsett-Lemaire, F. (eds.) (1997) *A Bibliography of Afrotropical Birds, 1971–1990*. Tauraco Research Report No. 7. Tauraco Press, Liège.

Fry, C. H., Keith, S. & Urban, E. K. (eds.) (1988, 2000) *The Birds of Africa*, Vols 3 & 6. Academic Press, London.

Ginn, P. J., McIlleron, W. G. & Milstein, P. le S. (1989) *The Complete Book of Southern African Birds*. Struik Winchester, Cape Town.

Hall, B. P. & Moreau, R. E. (1970) *An Atlas of Speciation in African Passerine Birds*. British Museum (Natural History), London.

Harris, A., Tucker, L. & Vinicombe, K. (1989) *The Macmillan Field Guide to Bird Identification*. Macmillan, London.

Harris, A., Shirihai, H. & Christie, D. A. (1996) *The Macmillan Birder's Guide to European and Middle Eastern Birds*. Macmillan, London.

del Hoyo, J., Elliot, A. & Sargatal, J. (eds.) (1992–2001) *Handbook of the Birds of the World*, Vols. 1–6. Lynx Edicions, Barcelona.

Inskipp, T., Lindsey, N. & Duckworth, W. (1996) *An Annotated Checklist of the Birds of the Oriental Region*. Oriental Bird Club, Sandy.

Jonsson, L. (1992) *Birds of Europe with North Africa and the Middle East*. Christopher Helm, London.

Keith, S., Urban, E. K. & Fry, C.H. (eds.) (1992) *The Birds of Africa*, Vol. 4. Academic Press, London.

Lewington, I., Alström, P. & Colston, P. (1991) *A Field Guide to the Rare Birds of Britain and Europe*. HarperCollins, London.

Lippens, L. & Wille, H. (1976) *Les Oiseaux du Zaïre*. Lannoo, Tielt.

Mackworth-Praed, C. W. & Grant, C. H. B. (1970–1973) *Birds of West Central and Western Africa*. 2 vols. Longmans, London.

Maclean, G. L. (1985) *Roberts' Birds of Southern Africa* (5th edn.). John Voelcker Bird Book Fund, Cape Town.

Mullarney, K., Svensson, L., Zetterström, D. & Grant, P. J. (1999) *Collins Bird Guide*. HarperCollins, London.

Newman, K. (1983) *Newman's Birds of Southern Africa*. Macmillan South Africa, Johannesburg.

Perennou, C. (1991) *Les Recensements Internationaux d'Oiseaux d'Eau en Afrique Tropicale*. IWRB Special Publication No. 15. The International Waterfowl and Wetlands Research Bureau, Slimbridge.

Perennou, C. (1992) *African Waterfowl Census 1992*. The International Waterfowl and Wetlands Research Bureau, Slimbridge.

Prigogine, A. (1971–1984) *Les Oiseaux de l'Itombwe et de son Hinterland*, Vols. 1–3. Koninklijk Museum voor Midden-Afrika, Tervuren.

Schouteden, H. (1963) La faune ornithologique des districts du Bas-Uele et du Haut-Uele. *Doc. zool. Mus. roy. Afr. centr.* 4: 1–241.

Scott, D. A. & Rose, P. M. (1996) *Atlas of Anatidae Populations in Africa and Western Eurasia*. Wetlands International Publication No. 41. Wetlands International, Wageningen.

Serle, W., Morel, G. & Hartwig, W. (1977) *A Field Guide to the Birds of West Africa*. Collins, London.

Shirihai, H., Gargallo, G. & Helbig, A. J. (2001) *Sylvia Warblers: Identification, Taxonomy and Phylogeny of the Genus Sylvia*. Christopher Helm, London.

Sibley, C. G. & Monroe, B. L. (1990) *Distribution and Taxonomy of Birds of the World*. Yale University Press, New Haven & London.

Sibley, C. G. & Monroe, B. L. (1993) *A Supplement to Distribution and Taxonomy of Birds of the World*. Yale University Press, New Haven & London.

Sibley, C. G. (1996) *Birds of the World*. Version 2.0. Thayer Birding Software, Cincinnati.

Sinclair, I., Hockey, P. & Tarboton, W. (1993) *Sasol Birds of Southern Africa*. Struik, Cape Town.

Snow, D.W. (1978) *An Atlas of Speciation in African Non-Passerine Birds*. British Museum (Natural History), London.

Snow, D. & Perrins, C. (1998) *The Birds of the Western Palearctic. Concise Edition*. 2 vols. Oxford University Press, Oxford.

Stattersfield, A. J., Crosby, M. J., Long, A. J. & Wege, D. C. (1998) *Endemic Bird Areas of the World: priorities for biodiversity conservation*. BirdLife International, Cambridge.

Svensson, L. (1992) *Identification Guide to European Passerines* (4th edn.). Privately published, Stockholm.

Thiollay, J.-M. (1977) Distribution saisonnière des rapaces diurnes en Afrique occidentale. *Oiseau & R.F.O.* 47: 253–294.

Urban, E. K., Fry, C. H. & Keith, S. (eds.) (1986, 1997) *The Birds of Africa*, Vols. 2 & 5. Academic Press, London.

Zimmerman, D. A., Turner, D. A. & Pearson, D.J. (1996) *Birds of Kenya and Northern Tanzania*. Christopher Helm, London.

2. COUNTRY REFERENCES

Benin

Anciaux, M.-R. (1996) Aperçu de l'avifaune dans différents milieux de l'intérieur des terres du Sud-Bénin. Plateau d'Allada et sud de la dépression de la Lama. *Cahiers d'Éthologie* 16: 79–98.

Anciaux, M.-R. (2000) Approche de la phénologie de la migration des migrateurs intra-africains de l'intérieur des terres du Sud-Bénin (plateau d'Allada et sud de la dépression de la Lama). 1. Les non-Passériformes et les non-Coraciiformes. *Alauda* 68: 311–320.

Berlioz, J. (1956) Étude d'une collection d'oiseaux du Dahomey. *Bull. Mus. Nat. Hist. Nat.* 3: 261–264.

Brunel, J. (1958) Observations sur les oiseaux du Bas-Dahomey. *Oiseau & R.F.O.* 28: 1–38.

Cheke, R. A. (1996) Historical records of birds from the Republic of Benin. *Malimbus* 18: 58–59.

Claffey, P. M. (1995) Notes on the avifauna of the Bétérou area, Borgou Province, Republic of Benin. *Malimbus* 17: 63–84.

Claffey, P. M. (1997a) The status of Pel's Fishing Owl *Scotopelia peli* in the Togo–Bénin Gap. *Bull. Afr. Bird Club* 4: 135–136.

Claffey, P. M. (1997b) Western Red-footed Falcon *Falco vespertinus*, a new addition to the Republic of Benin list. *Malimbus* 19: 95–96.

Claffey, P. M. (1998a) The status of Black Cuckoo *Cuculus clamosus* and Red-throated Cuckoo *C. solitarius* in Benin. *Malimbus* 20: 56–57.

Claffey, P. M. (1998b) Blackcap *Sylvia atricapilla*, new to Benin. *Malimbus* 20: 57–58.

Claffey, P. (1998c) New breeding records of Verreaux's Eagle Owl *Bubo lacteus* in Bénin, West Africa. *Bull. Afr. Bird Club* 5: 127.

Claffey, P. (1999a) Occurrence of Red-billed Dwarf Hornbill *Tockus camurus* in the Republic of Bénin. *Bull. Afr. Bird Club* 6: 107–108.

Claffey, P. M. (1999b) Crowned Eagle *Stephanoaetus coronatus* and White-breasted Negro-Finch *Nigrita fusconota*, new to the Benin list. *Malimbus* 21: 51–53.

Claffey, P. M. (1999c) Sharp decline in the population of Pin-tailed Whydah *Vidua macroura* in Benin. *Malimbus* 21: 53–54.

Douaud, J. (1955) Les oiseaux du Dahomey et du Niger. Notes de voyage. *Oiseau & R.F.O.* 25: 295–307.

Green, A. A. & Sayer, J. A. (1979) The birds of Pendjari and Arli National Parks (Benin and Upper Volta). *Malimbus* 1: 14–28.

Holyoak, D. T. & Seddon, M. B. (1990) Distributional notes on the birds of Benin. *Malimbus* 11: 128–134.

Thonnerieux, Y. (1985) Notes complémentaires sur l'avifaune des Parcs Nationaux d'Arli (Burkina) et de la Pendjari (Benin). *Malimbus* 7: 137–139.

Walsh, J. F. (1991) On the occurrence of the Black Stork *Ciconia nigra* in West Africa. *Bull. Brit. Orn. Club* 111: 209–215.

Waltert, M. & Mühlenberg, M. (1999) Notes on the avifauna of the Noyau Central, Forêt Classée de la Lama, Republic of Benin. *Malimbus* 21: 82–92.

Burkina Faso

Balança, G. & de Visscher, M.-N. (1993) Nouvelles données de distribution pour deux espèces d'oiseaux au Burkina Faso. *Malimbus* 15: 89–90.

Balança, G. & de Visscher, M.-N. (1996) Observations sur la reproduction et les déplacements du Rollier d'Abyssinie *Coracias abyssinica*, du Rollier varié *C. naevia* et du Rolle africain *Eurystomus glaucurus*, au nord du Burkina Faso. *Malimbus* 18: 44–57.

Balança, G. & de Visscher, M.-N. (1997) Composition et évolution saisonnière d'un peuplement d'oiseaux au nord du Burkina Faso (nord-Yatenga). *Malimbus* 19: 68–94.

Fishpool, L., Oueda, G. & Compaoré, P. (2000) Kordofan Bush Lark *Mirafra cordofanica* and Desert Lark *Ammomanes deserti*, additions to the avifauna of Burkina Faso. *Malimbus* 22: 49–54.

Green, A. A. & Sayer, J. A. (1979) The birds of Pendjari and Arli National Parks (Benin and Upper Volta). *Malimbus* 1: 14–28.

Holyoak, D. T. & Seddon, M. B. (1989) Distributional notes on the birds of Burkina Faso. *Bull. Brit. Orn. Club* 109: 205–216.

Mauvais, G. (1998) Recensement des espèces d'oiseaux du Bois de Boulogne à Ouagadougou (Burkina Faso) en saison des pluies. *Alauda* 66: 324–328.

Portier, B. (in press) Red-necked Nightjar *Caprimulgus ruficollis*, new to Burkina Faso. *Bull. Afr. Bird Club*

Thonnerieux, Y. (1985) Notes complémentaires sur l'avifaune des Parcs Nationaux d'Arli (Burkina) et de la Pendjari (Benin). *Malimbus* 7: 137–139.

Thonnerieux, Y. (1988) Commentaires sur la distribution de quelques migrateurs paléarctiques au Burkina Faso. *Gerfaut* 78: 317–362.

Thonnerieux, Y. (1988) État des connaissances sur la reproduction de l'avifaune du Burkina Faso (ex Haute-Volta). *Oiseau & R.F.O.* 58: 120–146.

Thonnerieux, Y., Walsh, J. F. & Bortoli, L. (1989) L'avifaune de la ville de Ouagadougou et ses environs (Burkina Faso). *Malimbus* 11: 7–40. [Errata 1990, 12: 58–59.]

Weesie, P. D. M. (1996) Les oiseaux d'eau du Sahel burkinabé: peuplement d'hiver, capacité de charge des sites. *Alauda* 64: 307–332.

Weesie, P. D. M. & Belemsogbo, U. (1997) Les rapaces diurnes du ranch de gibier de Nazinga (Burkina Faso): liste commentée, analyse du peuplement et cadre biogéographique. *Alauda* 65: 263–278.

Cameroon

Bobo, S. K., Anye, D. N., Njabo, K. Y. & Nayuoh, L. (2000) First records of Tufted Duck *Aythya fuligula* in Cameroon. *Malimbus* 22: 91–92.

Bowden, C. G. R. & Andrews, S. M. (1994) Mount Kupe and its birds. *Bull. Afr. Bird Club* 1: 13–16.

Clark, W. S. (1999) Ayres' Hawk-Eagle *Hieraaetus* (*dubius*) *ayresi* sightings in Cameroon. *Bull. Afr. Bird Club* 6: 115–116.

Demey, R. & Njabo, K. Y. (2001) A new sight record in Cameroon of the distinctive race *crossensis* of Green-throated Sunbird *Nectarinia rubescens*. *Malimbus* 23: 66–67.

Dowsett, R. J. & Dowsett-Lemaire, F. (2000) New species and amendments to the avifauna of Cameroon. *Bull. Brit. Orn. Club* 120: 179–185.

Dowsett-Lemaire, F. & Dowsett, R.J. (2000) Birds of the Lobéké Faunal Reserve, Cameroon, and its regional importance for conservation. *Bird Conserv. Internatn.* 10: 67–87.

Erard, C. & Colston, P. R. (1988) *Batis minima* (Verreaux) new for Cameroon. *Bull. Brit. Orn. Club* 108: 182–184.

Fossé, A. (1997) L'Aigrette intermédiaire *Egretta intermedia* au Cameroon. *Malimbus* 19: 38.

Fotso, R. C. (1990) Notes sur les oiseaux d'eau de la région de Yaoundé. *Malimbus* 12: 25–30.

Germain, M., Dragesco, J., Roux, F. & Garcin, H (1973) Contribution à l'ornithologie du sud-Cameroun. *Oiseau & R.F.O.* 43: 119–182, 212–259.

Girard, O. & Thal, J. (1996) Quelques observations ornithologiques dans la région de Garoua, Cameroun. *Malimbus* 18: 142–148.

Green, A. A. (1996) More bird records from Rio del Rey estuary, Cameroon. *Malimbus* 18: 112–121.

Green, A. A. & Rodewald, P. G. (1996) New bird records from Korup National Park and environs, Cameroon. *Malimbus* 18: 122–133.

Heaton, A. M. (1979) Birds of the Parc National de la Bénoué. *Malimbus* 1: 146–147.

Holyoak, D. T. & Seddon, M. B. (1990) Notes on some birds of western Cameroon. *Malimbus* 11: 123–127.

Hopkins, M. (1998) Buff-throated Sunbird *Nectarinia adelberti* and Fire-bellied Woodpecker *Dendropicos pyrrhogaster* in Cameroon. *Malimbus* 20: 124–125.

Languy, M. & Lambin, X. (2001) First record of Spotted Sandpiper *Actitis macularia* for Cameroon and Central Africa. *Bull. Afr. Bird Club* 8: 50.

Louette, M. (1981) The birds of Cameroon. An annotated check-list. *Verhandel. Kon. Acad. Wetensch. Lett. Schone Kunst. Belg., Kl. Wetensch.* 163: 1–295.

Louette, M. & Prevost, J. (1987) Passereaux collectés par J. Prévost au Cameroun. *Malimbus* 9: 83–96.

Mahé, E. (1988) Contribution à la liste des oiseaux du Parc National de la Bénoué Nord, Cameroun. *Malimbus* 10: 218–221.

Mahé, E. (1988) Sur la nidification du Pélican blanc *Pelecanus onocrotalus* à Mogodé (Nord-Cameroun). *Alauda* 56: 180–181.

Manners, G. R., Burtch, P., Bowden, C. G. R., Bowden, E. M. & Williams, E. (1993) Purple Gallinule *Porphyrio porphyrio,* further sightings in Cameroon. *Malimbus* 14: 59.

Martinez, I., Elliott, V. A. & Field, G. D. (1996) Yellow-billed Egret *Egretta intermedia* on the coast of Cameroon. *Malimbus* 18: 58.

McNiven, D. (1994) Finding fishing owls in southern Cameroon. *Bull. Afr. Bird Club* 1: 75.

Quantrill, B & Quantrill, R. (1995) First record of Little Gull *Larus minutus* in Cameroon. *Malimbus* 17: 103.

Quantrill, B & Quantrill, R. (1998) The birds of the Parcours Vita, Yaoundé, Cameroon. *Malimbus* 20: 1–14.

Robertson, I. (1992) New information on birds in Cameroon. *Bull. Brit. Orn. Club* 112: 36–42.

Robertson, I. (1993a) Unusual records from Cameroon. *Malimbus* 14: 62–63.

Robertson, I. (1993b) Horus Swift *Apus horus*, new to Cameroon. *Malimbus* 14: 63–64.

Rodewald, P. G. & Bowden, C. G. R. (1995) First record of Kemp's Longbill *Macrosphenus kempi* in Cameroon. *Bull. Brit. Orn. Club* 115: 66–68.

Rodewald, P. G., Dejaifve, P.-A. & Green, A. (1994) The birds of Korup National Park and Korup Project Area, Southwest Province, Cameroon. *Bird Conserv. Internatn.* 4: 1–68.

Sala, A. (1991) La Talève poule-sultane *Porphyrio porphyrio madagascariensis* à Yaoundé, Cameroun. *Malimbus* 13: 78.

Scholte, P., de Kort, S., & van Weerd, M. (1999) The birds of the Waza–Logone area, North Province, Cameroon. *Malimbus* 21: 16–50.

Scholte, P. & Dowsett, R. J. (2000) Birds of Waza new to Cameroon: corrigenda and addenda. *Malimbus* 22: 29–31.

Serle, W. (1950) A contribution to the ornithology of the British Cameroons. *Ibis* 92: 343–376, 602–638.

Serle, W. (1954) A second contribution to the ornithology of the British Cameroons. *Ibis* 96: 47–80.

Serle, W. (1965) A third contribution to the ornithology of the British Cameroons. *Ibis* 107: 60–94, 230–246.

Smith, T. B. & McNiven, D. (1993) Preliminary survey of the avifauna of Mt Tchabal Mbabo, west-central Cameroon. *Bird Conserv. Internatn.* 3: 13–19.

Sørensen, U. G., Bech, J. & Krabbe, E. (1996) New and unusual records of birds in Cameroon. *Bull. Brit. Orn. Club* 116: 145–155.

Stuart, S.N. (1986) Records of other species of bird from western Cameroon. Pp. 106–129 in Stuart, S.N. (ed.) *Conservation of Cameroon Montane Forests.* International Council for Bird Preservation, Cambridge.

Swarth, C.W. (1987) First sight record of the Black-headed Gull for Cameroon, West Africa. *Malimbus* 9: 127–128.

Taylor, P. B. (1981) Bates's Weaver *Ploceus batesi* near Victoria, and other observations from western Cameroon. *Malimbus* 3: 49–59.

Thomas, J. (1991) Birds of the Korup National Park, Cameroon. *Malimbus* 13: 11–23. [Erratum 1992, 14: 26.]

Thomas, J. (1995) Birds of the Rio del Rey estuary, Cameroon. *Malimbus* 17: 7–18.

Van Beirs, M. (1997) Black-billed Barbet *Lybius guifsobalito,* new to Cameroon and W Africa. *Malimbus* 19: 32.

Van Beirs, M. (1999) First record of Scaly-fronted Warbler *Spiloptila clamans* in Cameroon. *Malimbus* 21: 110.

Van Ouwerkerk, R. (1992) Grey Phalarope *Phalaropus fulicarius* in northern Cameroon. *Malimbus* 14: 25.

Williams, E. (1991) Black-necked Grebe *Podiceps nigricollis,* new to Cameroon. *Malimbus* 13: 40.

Williams, E. (1992) Red-shouldered Cuckoo-Shrike *Campephaga phoenicea* on Mt Oku, Cameroon. *Malimbus* 14: 15.

Williams, E. (1995) Recent records of White-naped Pigeon *Columba albinucha* from Cameroon. *Malimbus* 17: 104–106.

Wilson, J. D. (1989) Range extensions of some bird species of Cameroon. *Bull. Brit. Orn. Club* 109: 110–115.

Cape Verde Islands

Dufourny, H. (1999) White-tailed Tropicbird in Cape Verde Islands in February 1999. *Dutch Birding* 21: 254–255.

Geniez, P. & Lopez-Jurado, L.-F. (1998) Nouvelles observations ornithologiques aux Îles du Cap Vert. *Alauda* 66: 307–311.

Hazevoet, C. J. (1995) *The Birds of the Cape Verde Islands.* Check-list No. 13. British Ornithologists' Union, Tring.

Hazevoet, C. J. (1996) A record of the Blue-cheeked Bee-eater *Merops persicus* from the Cape Verde Islands and status of the species in West Africa. *Bull. Brit. Orn. Club* 116: 50–52.

Hazevoet, C. J. (1997) Notes on distribution, conservation, and taxonomy of birds from the Cape Verde Islands, including records of six species new to the archipelago. *Bull. zool. Mus. Univ. Amsterdam* 15: 89–100.

Hazevoet, C. J. (1998) Third annual report on birds from the Cape Verde Islands, including records of seven taxa new to the archipelago. *Bull. zool. Mus. Univ. Amsterdam* 17: 19–32.

Hazevoet, C. J. (1999a) Notes on birds from the Cape Verde Islands in the collection of the Centro de Zoologia, Lisbon, with comments on taxonomy and distribution. *Bull. Brit. Orn. Club* 119: 25–31.

Hazevoet, C. J. (1999b) Fourth report on birds from the Cape Verde Islands, including notes on conservation and records of 11 taxa new to the archipelago. *Bull. zool. Mus. Univ. Amsterdam* 17: 19–32.

Hazevoet, C. J., Fischer, S. & Deloison, G. (1996) Ornithological news from the Cape Verde Islands in 1995, including records of species new to the archipelago. *Bull. zool. Mus. Univ. Amsterdam* 15: 21–27.

Hazevoet, C. J., Monteiro, L. R. & Ratcliffe, N. (1999) Rediscovery of the Cape Verde Cane Warbler *Acrocephalus brevipennis* on São Nicolau in February 1998. *Bull. Brit. Orn. Club* 119: 68–71.

Hille, S, & Thiollay, J.-M. (2000) The imminent extinction of the Kites *Milvus milvus fasciicauda* and *Milvus m. milvus* on the Cape Verde Islands. *Bird Conserv. Internatn.* 10: 361–369.

Palacios, C.-J. & Barone, R. (2001) Le Héron cendré *Ardea cinerea*, nouvelle espèce nidificatrice aux Îles du Cap Vert. *Alauda* 69: 18.

Ratcliffe, N., Monteiro, L. R. & Hazevoet, C. J. (1999) Status of Raso Lark *Alauda razae* with notes on threats and foraging behaviour. *Bird Conserv. Internatn.* 9: 43–46.

Central African Republic

Berlioz, J. (1934) Étude d'une collection d'oiseaux de l'Oubangui-Chari. *Bull. Mus. Natl. Hist. Nat.* 6: 228–234.

Berlioz, J. (1935) Étude d'une collection d'oiseaux de l'Afrique Équatoriale Française. *Bull. Mus. Natl. Hist. Nat.* 7: 349–353.

Berlioz, J. (1939) Étude d'une nouvelle collection d'oiseaux de l'Oubangui-Chari (A.E.F.). *Bull. Mus. Natl. Hist. Nat.* 11: 526–530.

Blancou, L. (1933) Contribution à l'étude des oiseaux de l'Oubangui-Chari (Bassins de la Ouaka et de la Kandjia). *Oiseau et R.F.O.* 3: 8–58, 299–336.

Blancou, L. (1938–1939) Contribution à l'étude des oiseaux de l'Oubangui-Chari occidental (Bassin supérieur de l'Ouham). *Oiseau et R.F.O.* 8: 405–430, 642–649; 9: 58–88, 255–277, 410–485.

Bouet, G. (1944) Révision des collections d'oiseaux recueillis au Congo et dans l'Oubangui par la mission J. Dybowski (avril 1891–mai 1892). *Oiseau & R.F.O.* 14: 44–88.

Bretagnolle, F. (1993) An annotated checklist of birds of north-eastern Central African Republic. *Malimbus* 15: 6–16.

Carroll, R. W. (1988) Birds of the Central African Republic. *Malimbus* 10: 177–200.

Dowsett, R. J., Christy, P. & Germain, M. (1999) Additions and corrections to the avifauna of Central African Republic. *Malimbus* 21: 1–15.

Dowsett, R. J. (1997) Birds of interest from the Sangba area, adjacent to the Bamingui-Bangoran National Park, Central African Republic. *Malimbus* 19: 102–103.

Friedmann, H. (1978) Results of the Lathrop Central African Republic Expedition 1976, ornithology. *Contrib. Sci.* 287: 1–22.

Fry, C. H. & Carroll, R. W. (1987) Range extension and probable breeding records of the Brown Nightjar (*Caprimulgus binotatus* Bonaparte) in Central African Republic. *Malimbus* 9: 125–127.

Germain, M. (1992) Sur quelques données erronées concernant l'avifaune de la Lobaye, République Centrafricaine. *Malimbus* 14: 1–6.

Germain, M. & Cornet, J. P. (1994) Oiseaux nouveaux pour la République Centrafricaine ou dont les notifications de ce pays sont peu nombreuses. *Malimbus* 16: 30–51. [Corrigendum 1994, 16: 125.]

Green, A. A. (1983) The birds of Bamingui-Bangoran National Park, Central African Republic. *Malimbus* 5: 17–30.

Green, A. A. (1984) Additional bird records from Bamingui-Bangoran National Park, Central African Republic. *Malimbus* 6: 70–72.

Green, A. A. (1990) Corrections to the list of birds of Bamingui-Bangoran National Park, Central African Republic. *Malimbus* 12: 53–54.

Green, A. A. & Carroll, R. W. (1991) The avifauna of Dzanga-Ndoki National Park and Dzanga-Sangha Rainforest Reserve, Central African Republic. *Malimbus* 13: 49–66.

Jehl, H. (1974) Quelques migrateurs paléarctiques en République Centrafricaine. *Alauda* 42: 397–406.

Jehl, H. (1976) Les oiseaux de l'Île Kembé (R.C.A.). *Alauda* 44: 153–167.

Quantrill, R. (1995) Red-tailed Ant-Thrush *Neocossyphus rufus* in Central African Republic. *Malimbus* 17: 103–104.

de Schauensee, R. M. (1949) Results of the Carpenter African Expedition 1947–1948. *Notulae Naturae* 219: 1–16.

Chad

Berlioz, J. (1938) Étude d'une collection d'oiseaux du Tchad. *Bull. Mus.* 10: 252–259.

de Boer, W. F. & Legoupil, F. (1993) Observations sur la présence et l'abondance des oiseaux au Tchad. *Malimbus* 15: 17–23.

Dowsett, R. J. (1971) Quelques observations sur les oiseaux du Tchad. *Oiseau & R.F.O.* 41: 83–85

Friedmann, H. (1966) The Machris expedition to Tchad, Africa. Birds. *Contrib. Sci.* 59: 3–27.

Guichard, K. M. (1955) The birds of Fezzan and Tibesti. *Ibis* 97: 393–424.

Louchart, A. (1999a) Observation du Moineau bridé *Petronia superciliaris* à N'Djamena (Tchad). *Alauda* 67: 72–73.

Louchart, A. (1999b) Invalidité de la note 3321 (*Alauda* 67(1): 72–73): "Observation du Moineau bridé *Petronia superciliaris* à N'Djamena (Tchad)". *Alauda* 67: 356.

Newby, J. E. (1979–80) The birds of the Ouadi Rimé–Ouadi Achim Faunal Reserve, a contribution to the study of the Chadian avifauna. *Malimbus* 1: 90–109; 2: 29–50.

Salvan, J. (1967–1969) Contribution à l'étude des oiseaux du Tchad. *Oiseau & R.F.O.* 37: 255–284; 38: 53–85, 127–150, 249–273; 39: 38–69. [Erratum 1971, 41: 181.]

Simon, P. (1965) Synthèse de l'avifaune du massif montagneux du Tibesti. *Gerfaut* 55: 26–71.

Thiollay, J.-M. (1975) Les rapaces des parcs nationaux du Tchad méridional. *Oiseau & R.F.O.* 45: 27–40.

Vielliard, J. (1971–72) Données biogéographiques sur l'avifaune d'Afrique centrale. *Alauda* 39: 227–248; 40: 63–92.

Congo

Bulens, P. & Dowsett, R.J. (2001) Little-known African bird: observations on Loango Slender-billed Weaver *Ploceus subpersonatus* in Congo-Brazzaville. *Bull. Afr. Bird Club* 8: 57-58.

Dowsett, R. J. & Dowsett-Lemaire, F. (1989a) Liste préliminaire des oiseaux du Congo. Pp. 29–51 in Dowsett, R. J. (ed.) *Enquête Faunistique dans la Forêt du Mayombe et Check-liste des Oiseaux et des Mammifères du Congo.* Tauraco Research Report No. 2. Tauraco Press, Liège.

Dowsett, R. J. & Dowsett-Lemaire, F. (1989b) Avifaune du Congo: additions and corrections. Pp. 17–19 in Dowsett, R. J. (ed.) *Enquête Faunistique dans la Forêt du Mayombe et Check-liste des Oiseaux et des Mammifères du Congo.* Tauraco Research Report No.2. Tauraco Press, Liège.

Dowsett, R. J. & Dowsett-Lemaire, F. (1989c) Liste commentée des oiseaux de la forêt du Mayombe (Congo). Pp. 5–16 in Dowsett, R. J. (ed.) *Enquête Faunistique dans la Forêt du Mayombe et Check-liste des Oiseaux et des Mammifères du Congo.* Tauraco Research Report No. 2. Tauraco Press, Liège.

Dowsett-Lemaire, F. (1997a) The avifauna of Odzala National Park, northern Congo. Pp. 15–48 in Dowsett, R. J. & Dowsett-Lemaire, F. (eds.) *Flore et Faune du Parc National d'Odzala, Congo.* Tauraco Research Report No. 6. Tauraco Press, Liège.

Dowsett-Lemaire, F. (1997b) The avifauna of Nouabalé-Ndoki National Park, northern Congo. Pp. 111–124 in Dowsett, R. J. & Dowsett-Lemaire, F. (eds.) *Flore et Faune du Parc National d'Odzala, Congo.* Tauraco Research Report No. 6. Tauraco Press, Liège.

Dowsett-Lemaire, F. (1997c) The avifauna of the Léfini Reserve, Téké Plateau (Congo). Pp. 125–134 in Dowsett, R. J. & Dowsett-Lemaire, F. (eds.) *Flore et Faune du Parc National d'Odzala, Congo.* Tauraco Research Report No. 6. Tauraco Press, Liège.

Dowsett-Lemaire, F. & Dowsett, R. J. (1991) The avifauna of the Kouilou basin in Congo. Pp. 189–239 in Dowsett, R. J. & Dowsett-Lemaire, F. (eds.) *Flore et Faune du Bassin du Kouilou (Congo) et leur Exploitation.* Tauraco Research Report No. 4. Tauraco Press, Liège.

Dowsett-Lemaire, F. & Dowsett, R. J. (1996) Découverte de *Phylloscopus budongoensis* et autres espèces à caractère montagnard dans les forêts d'Odzala (Cuvette congolaise). *Alauda* 64: 364–367.

Dowsett-Lemaire, F. & Dowsett, R. J. (1998) Further additions to and deletions from the avifauna of Congo-Brazzaville. *Malimbus* 20: 15–32.

Dowsett-Lemaire, F., Dowsett, R. J. & Bulens, P. (1993) Additions and corrections to the avifauna of Congo. *Malimbus* 15: 68–80.

Dowsett, R. J. & Simpson, R. D. H. (1991) The status of seabirds off the coast of Congo. Pp. 241–250 in Dowsett, R. J. & Dowsett-Lemaire, F. (eds.) *Flore et Faune du Bassin du Kouilou (Congo) et leur Exploitation.* Tauraco Research Report No. 4. Tauraco Press, Liège.

Ikonga, J.M. & Bockandza, P. (2001) Première observation de la Cigogne blanche *Ciconia ciconia* au Congo-Brazzaville. *Bull. Afr. Bird Club* 8: 61.

Maisels, F. & Cruickshank, A. (2000) New breeding records of African River Martin *Pseudochelidon eurystomina* and Rosy Bee-eater *Merops malimbicus* in Conkouati Reserve, Republic of Congo. *Bull. Afr. Bird Club* 7: 48–49.

Rand, A. L., Friedmann, H. & Traylor, M. A. (1959) Birds from Gabon and Moyen Congo. *Fieldiana Zool.* 41: 221–411.

Equatorial Guinea – including Bioko (Fernando Po) and Annobón (Pagalú)

Basilio, A. (1963) *Aves de la Isla de Fernando Poo.* Editorial Coculsa, Madrid.

Butynski, T. M. & Koster, S. H. (1989) Grey-necked Picathartes *Picathartes oreas* found on Bioko Island (Fernando Po). *Tauraco* 1: 186–189.

Butynski, T. M., Schaaf, C. D & Hearn, G.W. (1997) The Grey-necked Picathartes *Picathartes oreas* on Bioko Island, Equatorial Guinea. *Ostrich* 69: 90–93.

Dowsett-Lemaire, F. & Dowsett, R. J. (1999) Birds of the Parque Nacional de Monte Alen, mainland Equatorial Guinea, with an updating of the country's list. *Alauda* 67: 179–188.

Eisentraut, M. (1972) Die Wirbeltierfauna von Fernando Poo und Westkamerun. *Bonn. zool. Monogr.* 3: 1–400.

Fry, C. H. (1961) Notes on the birds of Annobon and other islands in the Gulf of Guinea. *Ibis* 103a: 267–276.

Harrison, M. J. S. (1990) A recent survey of the birds of Pagalu (Annobon). *Malimbus* 11: 135–143.

Moore, A. (2000) Comment on species rejected from and added to the avifauna of Bioko Island (Equatorial Guinea). *Malimbus* 22: 31–33.

Pérez del Val, J. (1996) *Las Aves de Bioko, Guinea Ecuatorial: Guía de Campo.* Edilesa, León.

Pérez del Val, J. (2000) Reply to Moore. *Malimbus* 22: 33–34.

Pérez del Val, J. (2001) A survey of birds of Annobón Island, Equatorial Guinea: preliminary report. *Bull. Afr. Bird Club* 8: 54.

Pérez del Val, J., Castroviejo, J. & Purroy, F. J. (1997) Species rejected from and added to the avifauna of Bioko island (Equatorial Guinea). *Malimbus* 19: 19–31.

Gabon

Alexander-Marrack, P. (1992) Nearctic vagrant waders in the Cap Lopez area, Gabon. *Malimbus* 14: 7–10.

Berlioz, J. (1954) Étude d'une collection d'oiseaux du Gabon. *Bull. Mus. Nat. Hist. Nat. Paris* 27: 185–192.

Berlioz, J. (1959) Étude d'une nouvelle collection d'oiseaux du Gabon. *Bull. Mus. Nat. Hist. Nat. Paris* 31: 395–400.

Brosset, A. & Erard, C. (1977) New faunistic records from Gabon. *Bull. Brit. Orn. Club* 97: 125–132.

Brosset, A. & Erard, C. (1986) *Les Oiseaux des Régions forestières du Nord-Est du Gabon*, Vol. 1. Société Nationale de Protection de la Nature, Paris.

Christy, P. (1982a) L'Aigrette ardoisée *Egretta ardesiaca* au Gabon. *Oiseau & R.F.O.* 52: 91–92.

Christy, P. (1982b) Notes sur les migrateurs paléarctiques observés sur le littoral gabonais. *Oiseau & R.F.O.* 52: 251–258.

Christy, P. (1983) La Mouette rieuse *Larus ridibundus* au Gabon. *Oiseau & R.F.O.* 53: 293.

Christy, P. (1984a) L'Hirondelle des rochers d'Angola (*Petrochelidon rufigula*) au Gabon. *Oiseau & R.F.O.* 54: 362–363.

Christy, P. (1984b) L'Engoulevent terne *Caprimulgus inornatus* au Gabon. *Oiseau & R.F.O.* 54: 364–365.

Christy, P. (1990) New records of palaearctic migrants in Gabon. *Malimbus* 11: 117–122.

Christy, P. & Clarke, W. (1994) *Guide des Oiseaux de la Réserve de la Lopé*. ÉCOFAC, Libreville.

Erard, C. (1989) *Phylloscopus* (=*Seicercus*) *laurae* au Gabon? *Oiseau & R.F.O.* 59: 86–88.

Erard, C. & Chappuis, C. (1991) Sur le *Phylloscopus* (=*Seicercus*) mentionné au Gabon et sur son appartenance probable à une espèce non encore décrite. *Oiseau & R.F.O.* 61: 155–161.

Greth, A. (1996) Concentration saisonnière d'oiseaux aquatiques au lac Kivoro, sud-ouest du Gabon. *Malimbus* 18: 149–151.

Macaulay, L. & Sinclair, J. C. (1999) Perrin's Bush-Shrike *Telophorus viridis*, new to Gabon. *Malimbus* 21: 110–111.

Malbrant, R. & Maclatchy, A. (1949) *Faune de l'Équateur Africain Français*, Vol. 1. Lechevalier, Paris.

Malbrant, R. & Maclatchy, R. (1958) A propos de l'occurrence de deux oiseaux d'Afrique australe au Gabon: le Manchot du Cap, *Spheniscus demersus* Linn, et la Grue couronnée, *Balearica regulorum* Bennett. *Oiseau & R.F.O.* 28: 84–86.

Rand, A. L., Friedmann, H. & Traylor, M. A. (1959) Birds from Gabon and Moyen Congo. *Fieldiana Zool.* 41: 221–411.

Robertson, I. S. (1995) Black-throated Apalis *Apalis jacksoni*, a new bird for Gabon. *Malimbus* 17: 28.

Sargeant, D. (1993) A birders guide to Gabon. Privately published.

Schepers, F. J. & Marteijn, E. C. L. (eds.) (1992) *Coastal Waterbirds in Gabon, Winter 1992*. WIWO Report No. 41. Foundation Working Group International Wader and Waterfowl Research, Zeist.

The Gambia

Baillon, F. & Dubois, P. J. (1992) Nearctic gull species in Senegal and The Gambia. *Dutch Birding* 14: 49–50.

Barlow, C., Wacher, T. & Disley, T. (1997) *A Field Guide to Birds of The Gambia and Senegal*. Pica Press, Mountfield.

Barlow, C. R. & Gale, G. (1999) Information obtained from nine road-killed Red-necked Nightjar *Caprimulgus ruficollis* in The Gambia, in winters 1990–1997. *Bull. Afr. Bird Club* 6: 48–51.

Barnett, L.K. & Emms, C. (2001) New species and breeding records for The Gambia. *Bull. Afr. Bird Club* 8: 44-45.

Barnett, L.K., Emms, C. & Camara, A. (2001) The birds of Bijol Island, Tanji River (Karinti) Bird Reserve, The Gambia. *Bull. Afr. Bird Club* 8: 39-43,

Borrow, N. (1997) Red-crested Bustard *Eupodotis ruficrista* and Adamawa Turtle Dove *Streptopelia hypopyrrha*, new to The Gambia, and sightings of Great Snipe *Gallinago media*. *Malimbus* 19: 36–38.

Brown, R. G. B. (1979) Seabirds of the Senegal upwelling and adjacent waters. *Ibis* 121: 283–292.

Geeson, J. & Geeson, J. (1990) First Red Kite record for The Gambia. *Malimbus* 11: 144.

Gore, M. E. J. (1990) *Birds of The Gambia* (2nd edn.). Check-list No. 3. British Ornithologists' Union, Tring.

Jones, R. M. (1991) The status of larks in The Gambia, including first records of Sun Lark *Galerida modesta*. *Malimbus* 13: 67–73.

Jones, R. M. (1992) European Crag Martin *Ptyonoprogne rupestris* in the Gambia. *Malimbus* 14: 21–23.

King, M. (2000) Noteworthy records from Ginal Island, The Gambia. *Malimbus* 22: 77–85.

Kirtland, C. A. E. & Rogers, E. P. (1997) First record and nesting of White-crested Tiger-Heron *Tigriornis leucolophus* in The Gambia. *Bull. Afr. Bird Club* 4: 105–106.

Morel, G. J. & Morel, M.-Y. (1990) *Les Oiseaux de Sénégambie*. ORSTOM, Paris.

Payne, R. B., Payne, L. L. & Barlow, C. R. (1997) Observation of Savile's Bustard *Eupodotis savilei* in The Gambia. *Malimbus* 19: 97–99.

Payne, R. B., Barlow, C. R. & Wacher, T. (2000) Adamawa Turtle Dove *Streptopelia hypopyrrha* in The Gambia, with comparison of its calls in The Gambia and Nigeria. *Malimbus* 22: 37–40.

Ranner, A., Tebb, G. & Craig, M. (2000) First record of Little Crake *Porzana parva* in The Gambia. *Bull. Afr. Bird Club* 7: 51–52.

Robel, D. (1998) An observation of Ayres' Hawk-Eagle *Hieraaetus dubius* in The Gambia. *Bull Afr. Bird Club* 5: 128–129.

Rumsey, S.J.R. (2001) Recoveries of birds ringed in Sénégal and The Gambia during recent years. *Bull. Afr. Bird Club* 8: 46-47.

Wacher, T. (1993) Some new observations of forest birds in The Gambia. *Malimbus* 15: 24–37.

Ghana

Dutson, G. & Branscombe, J. (1990) *Rainforest Birds in South-west Ghana*. Study Report No. 46. International Council for Bird Preservation, Cambridge.

Grimes, L. G. (1987) *The Birds of Ghana*. Check-list No. 9. British Ornithologists' Union, London.

Hedenstrom, A., Bensch, S., Hasselquist, D. & Ottosson, U. (1990) Observations of Palaearctic migrants rare to Ghana. *Bull. Brit. Orn. Club* 110: 194–197.

Helsens, T. (1996) New information on birds in Ghana, April 1991 to October 1993. *Malimbus* 18: 1–9.

Ntiamoa-Baidu, Y., Asamoah, S. A., Owusu, E. H. & Owusu-Boateng, K. (2000) Avifauna of two upland evergreen forest reserves, the Atewa range and Tano Offin, Ghana. *Ostrich* 71: 277–281.

Ntiamoa-Baidu, Y., Owusu, E. H., Asamoah, S. & Owusu-Boateng, K. (2000) Distribution and abundance of forest birds in Ghana. *Ostrich* 71: 262–268.

Taylor, I. R. & Macdonald, M. A. (1979) A population of *Anthus similis* on the Togo range in eastern Ghana. *Bull. Brit. Orn. Club* 99: 29–30.

Van Gastel, A. J. G. & Van Gastel, E. R. (1999) Cream-coloured Courser *Cursorius cursor*, new for Ghana. *Malimbus* 21: 54–55.

Walsh, J. F., Sowah, S. A. & Yamagata, Y. (1987) Newly discovered colonies of the northern Carmine Bee-eater *Merops nubicus* in Ghana and Togo. *Malimbus* 9: 129–130.

Guinea

Altenburg, W. & Van der Kamp, J. (1989) *Étude Ornithologique Préliminaire de la Zone Côtière du Nord-Est de la Guinée.* Study Report No. 30. International Council for Bird Preservation, Cambridge

Altenburg, W. & Van der Kamp, J. (1991) *Ornithological Importance of Coastal Wetlands in Guinea.* Study Report No. 47. International Council for Bird Preservation, Cambridge.

Brosset, A. (1984) Oiseaux migrateurs européens hivernant dans la partie guinéenne du Mont Nimba. *Alauda* 52: 81–101.

de Bournonville, D. (1967) Notes d'ornithologie guinéenne. *Gerfaut* 57: 145–158.

Demey, R. (1995) Notes on the birds of the coastal and Kindia areas, Guinea. *Malimbus* 17: 85–99.

Halleux, D. (1994) Annotated bird list of Macenta Prefecture, Guinea. *Malimbus* 16: 10–29.

Hayman, P. V., Prangley, M., Barnett, A. & Diawara, D. (1995) The birds of the Kounounkan Massif, Guinea. *Malimbus* 17: 53–62.

Lambert, K. (1978) Feldsperling, *Passer montanus*, und Bergfink, *Fringilla montifringilla*, in den Tropen. *Beitr. Vogelkd.* 24: 103.

Mead, C. J. & Clark, J. A. (1987) Report on bird-ringing for 1987 [error,=1986]. *Ringing & Migr.* 8: 135–200.

Morel, G. J. & Morel, M.-Y. (1988) Liste des oiseaux de Guinée. *Malimbus* 10: 143–176.

Nikolaus, G. (2000) The birds of the Parc National du Haut Niger, Guinea. *Malimbus* 22: 1–22.

Richards, D. K. (1982) The birds of Conakry and Kakulima, Democratic Republic of Guinée. *Malimbus* 4: 93–103.

Trolliet, B. & Fouquet, M. (2001) Observation de Bruants ortolans *Emberiza hortulana* hivernant en Moyenne-Guinée. *Alauda* 69: 327-328.

Walsh, J. F. (1985) Extension of known range of the African Black Duck *Anas sparsa* in West Africa. *Bull. Brit. Orn. Club* 105: 117.

Walsh, J. F. (1987) Records of birds seen in north-eastern Guinea in 1984–1985. *Malimbus* 9: 105–122.

Wilson, R. (1990) Annotated bird lists for the Forêts Classées de Diécké and Ziama and their immediate environs. Unpublished report commissioned by IUCN.

Guinea-Bissau

Altenburg, W. & Van der Kamp, J. (1989) Utilization of mangroves by birds in Guinea-Bissau. *Ardea* 77: 57–74.

Catry, P. & Mendes, L. (1998) Red-throated Pipit *Anthus cervinus* and Icterine Warbler *Hippolais icterinus*, new to Guinea-Bissau. *Malimbus* 20: 123–124.

Collar, N. J. (1998) Wattled Cranes in Guinea-Bissau. *Bull. Brit. Orn. Club* 118: 57–58.

Frade, F. & Bacelar, A. (1955) Catalogo das Aves da Guiné Portuguesa. I. Non Passeres. *Anais Estud. Zool. (Lisboa)* 10, IV (2): 1–194.

Hazevoet, C. J. (1996) Birds observed in Guinea-Bissau, January 1986, with a review of current ornithological knowledge of the country. *Malimbus* 18: 10–24.

Hazevoet, C. J. (1997) On a record of the Wattled Crane *Bugeranus carunculatus* from Guinea-Bissau. *Bull. Brit. Orn. Club* 117: 56–59.

Rodwell, S. P. (1996) Notes on the distribution and abundance of birds observed in Guinea-Bissau, 21 February to 3 April 1992. *Malimbus* 18: 25–43.

Zwarts, L. (1988) Numbers and distribution of coastal waders in Guinea-Bissau. *Ardea* 76: 42–55.

Ivory Coast

Balchin, C. S. (1988) Recent observations of birds from Ivory Coast. *Malimbus* 10: 201–206.

Balchin, C. S. (1990) Further observations of birds from Ivory Coast. *Malimbus* 12: 52–53.

Cable, T. T. (1994) First record of Three-banded Plover *Charadrius tricollaris* in Ivory Coast. *Malimbus* 16: 57–58.

Cheke, R. A. (1987) Sooty Shearwater—new to Ivory Coast. *Malimbus* 9: 134.

Cheke, R. A. (1993) Manx Shearwater *Puffinus puffinus*—new to Ivory Coast. *Malimbus* 15: 92.

Cheke, R. A. & Fishpool, L. D. C. (1992) British Storm-Petrels *Hydrobates pelagicus* off Côte d'Ivoire. *Malimbus* 14: 24–25.

Demey, R. (1986) Two new species for Ivory Coast. *Malimbus* 8: 44.

Demey, R. & Fishpool, L. D. C. (1991) Additions and annotations to the avifauna of Côte d'Ivoire. *Malimbus* 12: 61–86.

Demey, R. & Fishpool, L. D. C. (1994) The birds of Yapo Forest, Ivory Coast. *Malimbus* 16: 100–122.

Eccles, S. D. (1985a) Reichenbach's Sunbird *Nectarinia reichenbachii* new to Ivory Coast. *Malimbus* 7: 140.

Eccles, S. D. (1985b) Oriole Babbler *Hypergerus atriceps* near coast in Ivory Coast. *Malimbus* 7: 140.

Eccles, S. D. (1985c) Western Reef Heron *Egretta gularis* inland in Ivory Coast and Nigeria. *Malimbus* 7: 140.

Falk, K. H. & Salewski, V. (1999) First records of Golden-tailed Woodpecker *Campethera abingoni* in Ivory Coast. *Bull. Afr. Bird Club* 6: 101–102.

Fishpool, L. D. C. & Demey, R. (1991) The occurrence of both species of 'Lesser Golden Plover' and of Nearctic Scolopacids in Côte d'Ivoire. *Malimbus* 13: 3–10.

Fry, C. H. (1985) Plain Nightjar at sea off Ivory Coast. *Malimbus* 7: 129.

Gartshore, M. E. (1989) *An Avifaunal Survey of Tai National Park, Ivory Coast.* Study Report No. 39. International Council for Bird Preservation, Cambridge.

Gartshore, M. E., Taylor, P. D. & Francis, I. S. (1995) *Forest Birds in Côte d'Ivoire. A survey of Tai National Park and other forests and forestry plantations, 1989–1991.* Study Report No. 58. BirdLife International, Cambridge.

Holyoak, D. T. & Seddon, M. B. (1990) Notes on some birds of the Ivory Coast. *Malimbus* 11: 146–148.

Kühn, I. & Späth, J. (1991) A new record of the Lesser Jacana *Microparra capensis* in northern Côte d'Ivoire, with notes on habitat. *Malimbus* 12: 91–93.

Salewski, V. (1997) Notes on some bird species from Comoé National Park, Ivory Coast. *Malimbus* 19: 61–67.

Salewski, V. (1998) Brown-throated Sand Martin *Riparia paludicola*, new for Ivory Coast. *Malimbus* 20: 127–128.

Salewski, V. (1998) A record of an immature Ovambo Sparrowhawk *Accipiter ovampensis* from Ivory Coast. *Bull. Afr. Bird Club* 5: 120–121.

Salewski, V. (2000) The birds of Comoé National Park, Ivory Coast. *Malimbus* 22: 55–76.

Salewski, V., Bobek, M., Peske, L. & Pojer, F. (2000) Status of the Black Stork *Ciconia nigra* in Ivory Coast. *Malimbus* 22: 92–93.

Salewski, V. & Göken, F. (1999) A southern record of Cinnamon-breasted Rock Bunting *Emberiza tahapisi* in Lamto, Ivory Coast. *Malimbus* 21: 121–122.

Salewski, V. & Korb, J. (1998) New bird records from Comoé National Park, Ivory Coast. *Malimbus* 20: 54–55.

Salewski, V. & Schmidt, S. (2000) New breeding records of the Grey-headed Kingfisher *Halcyon leucocephala* in Côte d'Ivoire. *Bull. Afr. Bird Club* 7: 67–68.

Thiollay, J. M. (1985) The birds of Ivory Coast. *Malimbus* 7: 1–59.

Thonnerieux, Y. (1987) Présence du Martinet pâle (*Apus pallidus*) entre autres migrateurs paléarctiques sur le versant ivoirien du Mont Nimba. *Malimbus* 9: 56–57.

Walsh, J. F. (1986) Notes on the birds of Ivory Coast. *Malimbus* 8: 89–93.

Walsh, J. F. (1991) On the occurrence of the Black Stork *Ciconia nigra* in West Africa. *Bull. Brit. Orn. Club* 111: 209–215.

Waltert, M., Yaokokore-Beibro, K.H., Mühlenberg, M. & Waitkuwait, W. E. (1999) Preliminary check-list of the birds of the Bossematié area, Ivory Coast. *Malimbus* 21: 93–109.

Williams, E. (1997) Unusual records of Palaearctic warblers in Ivory Coast. *Malimbus* 19: 33–34.

Liberia

Colston, P. R. & Curry-Lindahl, K. (1986) *The Birds of Mount Nimba, Liberia.* British Museum (Natural History), London.

Dickerman, R. W., Cane, W. P., Carter, M. F., Chapman, A. & Schmitt, C. G. (1994) Report on three collections of birds from Liberia. *Bull. Brit. Orn. Club* 114: 267–274.

Gatter, W. (1997) *Birds of Liberia.* Pica Press, Mountfield.

Mali

Balança, G. & de Visscher, M. N. (1993) Notes sur les oiseaux observés sur le Plateau Dogon au Mali. *Malimbus* 14: 52–58.

Cheke, R. A. & Howe, M. A. (1990) White-rumped Swift *Apus caffer* new to Mali. *Malimbus* 12: 54.

Cheke, R. A. (1995) A historical breeding record in Mali and description of the young of the Grasshopper Buzzard *Butastur rufipennis*. *Malimbus* 17: 106–107.

Curry, P. J. & Sayer, J. A. (1979) The inundation zone of the Niger as an environment for Palaearctic migrants. *Ibis* 121: 20–40.

De Bie, S. & Morgan, N. (1989) Les oiseaux de la Réserve de la Biosphère 'Boucle du Baoulé', Mali. *Malimbus* 11: 41–60.

Duhart, F. & Descamps, M. (1963) Notes sur l'avifaune du delta central Nigérien et régions voisinantes. *Oiseau & R.F.O.* 33 (no. spécial): 1–106.

Goar, J.-L. & Rutkowski, T. (2000) Reproduction de l'Aigle royal *Aquila chrysaetos* au Mali. *Alauda* 68: 327–328.

Lamarche, B. (1980–1981) Liste commentée des oiseaux du Mali. *Malimbus* 2: 121–158; 3: 73–102.

Salewski, V. (1998) Yellow-breasted Apalis *Apalis flavida*: a new bird for Mali. *Bull. Afr. Bird Club* 5: 59.

Skinner, J. (1987) Complément d'information sur les Tourterelles des bois dans la zone d'inondation du Niger au Mali. *Malimbus* 9: 133–134.

Skinner, J., Wallace, J. P., Altenburg, W. & Fourana, B. (1987) The status of the heron colonies in the inner Niger delta, Mali. *Malimbus* 9: 65–82.

Spierenburg, P. (1998) Migration of swifts over Bougouni, southern Mali. *Malimbus* 20: 69–74.

Spierenburg, P. (2000) Nouvelles observations de six espèces d'oiseaux au Mali. *Malimbus* 22: 23–28.

Tréca, B. (1993) Quelques données sur les reprises de bagues au Mali. *Malimbus* 14: 37–43.

Mauritania

Balança, G. (1996) Notes sur la nidification de quatre espèces d'oiseaux en Mauritanie. *Malimbus* 18: 151–153.

Beaman, M. (1990) Identification of Black Noddy and validity of December 1984 record in Mauritania. *Dutch Birding* 12: 245–248.

Bengtsson, K. (1997) Some interesting bird observations in Mauritania and Senegal. *Malimbus* 19: 96–97.

Farnsworth, S. J. (1994) Corn Bunting *Emberiza calandra* in Mauritania and West Africa. *Malimbus* 16: 124–125.

Farnsworth, S. J. (1995) Bird observations in Kaédi and Foum Gleïta, southern Mauritania. *Malimbus* 17: 2–6.

Isenmann, P. (1991) Breeding of Tawny Pipit in southern Mauritania. *Bull. Brit. Orn. Club* 111: 172–173.

Lamarche, B. (1988) Liste commentée des oiseaux de Mauritanie. *Études Sahariennes et Ouest-Africaines* 1(4): 1–162. Privately published, Nouakchott & Paris.

Meininger, P. L., Duiven, P., Marteijn, E. C. L. & Van Spanje, T. M. (1990) Notable bird observations from Mauritania. *Malimbus* 12: 19–24.

Pineau, O., Kayser, Y., Sall, M., Gueye, A. & Hafner, H. (2001) The Kelp Gull at Banc d'Arguin – a new Western Palearctic bird. *Birding World* 14: 110-111.

Robel, D. (1999) Zur Verbreitung einiger Vogelarten in Mauretanien. *Orn. Jber. Mus. Heineanum* 17: 117–122.

Triplet, P. & Yésou, P. (1993) Observation de *Sphenoeacus* (=*Melocichla*) *mentalis* en Mauritanie. *Oiseau & R.F.O.* 63: 222.

Williams, R. S. R. & Jacoby, M. C. (1996) Semipalmated Sandpiper *Calidris pusilla* in the Banc d'Arguin National Park, Mauritania: a new species for Africa. *Bull. Afr. Bird Club* 3: 132–133.

Yésou, P. & Triplet, P. (1996) Un site d'importance internationale pour l'avifaune risque de disparaître: le lac d'Aleg (République Islamique de Mauritanie). *Alauda* 64: 455–457.

Niger

Brouwer, J. (1992) Road-kills of three nightjar species near Niamey, Niger. *Malimbus* 14: 16–18.

Brouwer, J. & Mullié, W. C. (1992) Range extensions of two nightjar species in Niger, with a note on prey. *Malimbus* 14: 11–14.

Debout, G., Meister, P. & Ventelon, M. (2000) Notes complémentaires sur l'avifaune du Niger. *Malimbus* 22: 87–88.

Demey, R., Dowsett, R. J. & Fishpool, L. D. C. (in press) Comments on Black-throated Coucal *Centropus leucogaster* claimed from Niger. *Malimbus*

Giraudoux, P., Degauquier, R., Jones, P. J., Weigel, J. & Isenmann, P. (1988) Avifaune du Niger: état des connaissances en 1986. *Malimbus* 10: 1–140.

Holyoak, D. T. & Seddon, M. B. (1991) Notes sur la répartition des oiseaux du Niger. *Alauda* 59: 55–57, 116–120.

Koster, S. H. & Grettenberger, J. F. (1983) A preliminary survey of birds in Park W, Niger. *Malimbus* 5: 62–72.

Newby, J., Grettenberger, J. & Watkins, J. (1987) The birds of the northern Aïr, Niger. *Malimbus* 9: 4–16.

Sauvage, A. (1993) Notes complémentaires sur l'avifaune du Niger. *Malimbus* 14: 44–47.

Sharland, R. E. (1989) Birds of Niger. *Malimbus* 11: 99.

Shull, B., Grettenberger, J. F. & Newby, J. (1986) Recent observations of birds in W National Park (Niger). *Malimbus* 8: 23–24.

Walsh, J. F. (1991) On the occurrence of the Black Stork *Ciconia nigra* in West Africa. *Bull. Brit. Orn. Club* 111: 209–215.

Nigeria

Debski, I. (1995) Mallard *Anas platyrhynchos* in Nigeria. *Malimbus* 17: 31–32.

Dowsett, R. J. & Moore, A. (1997) Swamp warblers *Acrocephalus gracilirostris* and *A. rufescens* at Lake Chad, Nigeria. *Bull. Brit. Orn. Club* 117: 48–51.

Elgood, J. H., Heigham, J. B., Moore, A. M., Nason, A. N., Sharland, R. E. & Skinner, N. J. (1994) *The Birds of Nigeria* (2nd edn.). Check-list No. 4. British Ornithologists' Union, Tring.

Fry, C. H. (1981) Black-necked Grebe new to Nigeria and West Africa. *Malimbus* 3: 54.

Fry, C. H. (1986) First Yellow-billed Duck record for Nigeria. *Malimbus* 8: 43.

Gretton, A. (1991) *The Ecology and Conservation of the Slender-billed Curlew (Numenius tenuirostris).* Monograph No. 6. International Council for Bird Preservation, Cambridge.

Hopkins, M. T. E., Demey, R. & Barker, J. C. (1999) First documented records of Green-throated Sunbird *Nectarinia rubescens* for Nigeria, with a discussion of the distinctive race *crossensis. Malimbus* 21: 57–60.

Jones, P. J. (1984) The status of the Pygmy Kingfisher *Ceyx picta* in north-eastern Nigeria. *Malimbus* 6: 11–14.

Künzel, T. & Künzel, S. (1999) First Nigerian record of Red-fronted Parrot *Poicephalus gulielmi*, and other notable records from SE Nigeria. *Malimbus* 21: 111–113.

Loske, K.-H. (1996) Ein wichtiger Schlafplatz europäischer Rauchschwalben *Hirundo rustica* in Nigeria und seine Bedrohung. *Limicola* 10: 42–48.

Manu, S. A. & Demey, R. (1997) First record of Brown-backed Scrub Robin *Cercotrichas hartlaubii* in Nigeria. *Malimbus* 19: 103–104.

Manu, S. A. & Demey, R. (1999) First records of Xavier's Greenbul *Phyllastrephus xavieri* in Nigeria. *Malimbus* 21: 55–57.

Nicolai, J (1976) Beobachtungen an eigenen paläarktischen Wintergästen in Ost-Nigeria. *Vogelwarte* 28: 274–278.

Ottoson, U., Hjort, C. & Hall, P. (2001) The Lake Chad Bird Migration Project: Malamfatori revisited. *Bull. Afr. Bird Club* 8: 121-125.

Payne, R. B., Payne, L. L. & Hopkins, M. T. E. (1997) Grey-headed Broadbill *Smithornis sharpei*, new to Nigeria. *Malimbus* 19: 100–101.
Sharland, R. E. (1997) Ringing recoveries between Nigeria and eastern Europe. *Malimbus* 19: 103.
Wilkinson, R. & Aidley, D. J. (1982) The status of Savi's Warbler in Nigeria. *Malimbus* 4: 49.
Wilkinson, R. & Beecroft, R. (1988) *Kagoro Forest Conservation Study.* Study Report No. 28. International Council for Bird Preservation, Cambridge.
Wilkinson, R., Beecroft, R. Ezealor, A. U. & Sharland, R. E. (1989) Brown-chested Wattled Plover breeding in Nigeria. *Malimbus* 11: 95.

São Tomé & Príncipe

Atkinson, P., Peet, N. & Alexander, J. (1991) The status and conservation of the endemic bird species of São Tomé and Príncipe, West Africa. *Bird Conserv. Internatn.* 1: 255–282.
Christy, P. & Clarke, W. V. (1998) *Guide des Oiseaux de São Tomé et Príncipe.* ÉCOFAC, São Tomé.
Jones, P. J. & Tye, A. (1988) *A Survey of the Avifauna of São Tomé and Príncipe.* Study Report No. 24. International Council for Bird Preservation, Cambridge.
de Naurois, R. (1994) *Les Oiseaux des Îles du Golfe de Guinée (São Tomé, Prince et Annobon).* Instituto de Investigaçao Científica Tropical, Lisboa.
Sargeant, D. E. (1994) Recent ornithological observations from São Tomé and Príncipe Islands. *Bull. Afr. Bird Club* 1: 96–102.
Sargeant, D. E., Gullick, T., Turner, D. A. & Sinclair, J. C. (1992) The rediscovery of the São Tomé Grosbeak *Neospiza concolor* in south-western São Tomé. *Bird Conserv. Internatn.* 2: 157–159.
Schollaert, V. & Willem, G. (2001) A new site for Newton's Fiscal *Lanius newtonii. Bull. Afr. Bird Club* 8: 21-22.

Senegal

Bailleul, P. & Le Gal, P.-Y. (1991) Observation du Grèbe à cou noir au Sénégal. *Malimbus* 13: 40–41.
Baillon, F. (1991) Une nouvelle espèce pour le Sénégal: le Goéland à bec cerclé *Larus delawarensis. Alauda* 59: 113.
Baillon, F. (1992) *Streptopelia cf. hypopyrrha*, nouvelle espèce de tourterelle pour le Sénégal. *Oiseau & R.F.O.* 62: 320–334.
Baillon, F. & Dubois, P. J. (1992) Nearctic gull species in Senegal and The Gambia. *Dutch Birding* 14: 49–50.
Barlow, C., Wacher, T. & Disley, T. (1997) *A Field Guide to Birds of The Gambia and Senegal.* Pica Press, Mountfield.
Bengtsson, K. (1995) More observations of Audouin's Gulls *Larus audouinii* in Senegal. *Malimbus* 17: 102.
Bengtsson, K. (1997) Some interesting bird observations in Mauritania and Senegal. *Malimbus* 19: 96–97.
Brown, R. G. B. (1979) Seabirds of the Senegal upwelling and adjacent waters. *Ibis* 121: 283–292.
Clark, J.A., Wernham, C.V., Balmer, D.E., Adams, S.Y., Griffin, B.M., Blackburn, J.R., Anning, D. & Milne, L.J. (2001) Bird Ringing in Britain and Ireland in 1999. *Ringing & Migr.* 20: 239-288.
Condamin, M. (1987) Le Pluvier de Leschenault (*Charadrius leschenaultii*), espèce nouvelle pour le Sénégal. *Malimbus* 9: 131–133.
Delaporte, P. & Dubois, P. J. (1990) Premier recensement de laridés sur les côtes sénégambiennes. *Alauda* 58: 163–172.
Devisse, R. (1992) Première observation au Sénégal du Martinet marbré *Tachymarptis aequatorialis. Malimbus* 14: 16.
Dubois, P.J. (1990) Audouin's Gulls in Senegal. *Dutch Birding* 12: 25–26.
Ericsson, S. (1989) Notes on birds observed in Gambia and Senegal in November 1984. *Malimbus* 11: 88–94.
Gruwier, C., Claerbout, S. & Portier, B. (2001) Bergeronnette citrine *Motacilla citreola* à Dakar: première mention pour le Sénégal. *Bull. Afr. Bird Club* 8: 139-140.
Guillou, J.-J. (1987) Nidification de *Thalassornis leuconotus* au Sénégal. *Alauda* 55: 149.
King, M. (2000) Breeding of Swallow-tailed Kite in Senegal. *Malimbus* 22: 90–91.
Lambert, K. (1978) Feldsperling, *Passer montanus*, und Bergfink, *Fringilla montifringilla*, in den Tropen. *Beitr. Vogelkd.* 24: 103.
La Rédaction (1988) [Rejection of *Larus belcheri* in Senegal]. *Alauda* 56: 68.
Mackrill, E. J. (1989) Audouin's Gulls in Senegal in January 1989. *Dutch Birding* 11: 122–123.
Marr, T., Newell, D. & Porter, R. (1998) Seabirds off Senegal, West Africa. *Bull. Afr. Bird Club* 5: 22–28.
Marr, T. & Porter, R. (1992) Spring seabird passage off Senegal. *Birding World* 5: 391–394.
Meeth, P. (1981) Baird's Sandpiper in Senegal in December 1965. *Dutch Birding* 3: 51.
Morel, G. J. & Morel, M.-Y. (1989) Une héronnière mixte sur le lac de Guier (Sénégal) avec référence spéciale à *Ixobrychus m. minutus* et *Platalea leucorodia. Oiseau & R.F.O.* 59: 290–295.
Morel, G. J. & Morel, M.-Y. (1990) *Les Oiseaux de Sénégambie.* ORSTOM, Paris.
Payne, R. B. (1997) The Mali Firefinch *Lagonosticta virata* in Senegal. *Malimbus* 19: 39–41.
Printemps, T., Rouillon, Y. & Morel, G. J. (1999) Observation de la Bernache cravant *Branta bernicla* au Sénégal. *Malimbus* 21: 114–115.
Riddiford, N. (1990) Collared Flycatcher *Ficedula albicollis* in Senegal. *Malimbus* 11: 149–150.
Rodwell, S. P., Sauvage, A., Rumsey, S. J. R. & Bräunlich, A. (1996) An annotated check-list of birds occurring at the Parc National des Oiseaux du Djoudj in Senegal, 1984–1994. *Malimbus* 18: 74–111.
Rouchouse, C (1985) Sédentarisation de *Monticola solitarius* au Cap de Naze, Sénégal. *Malimbus* 7: 91–94.

Rumsey, S. (1992) More news from Senegal. *B.T.O. News* 183: 13.

Rumsey, S.J.R. (2001) Recoveries of birds ringed in Sénégal and The Gambia during recent years. *Bull. Afr. Bird Club* 8: 46-47.

Sauvage, A. & Rodwell, S. P. (1998) Notable observations of birds in Senegal (excluding Parc National des Oiseaux du Djoudj), 1984–1994. *Malimbus* 20: 75–122.

Sauvage, A., Rumsey, S. & Rodwell, S. (1998) Recurrence of Palaearctic birds in the lower Senegal river valley. *Malimbus* 20: 33–53.

Schricke, V., Triplet, P. & Leray, G. (2001) La Foulque macroule *Fulica atra*, une nouvelle espèce nicheuse au Sénégal. *Alauda* 69: 328.

Tréca, B. & Sakho, M. (1995) Confirmation de la présence du Martinet alpin *Apus melba* au Sénégal. *Malimbus* 17: 100–101.

Tréca, B. (1996) Fréquentation de bassins de lagunage par le Grèbe castagneux *Tachybaptus ruficollis* et le Grèbe à cou noir *Podiceps nigricollis* au Sénégal. *Alauda* 64: 421–428.

Tréca, B. (1997) Les Chevaliers combattants *Philomachus pugnax* L. dans le nord du Sénégal. Composition des populations par sexe et classes d'âge. *Alauda* 65: 161–166.

Triplet, P., Schricke, V. & Tréca, B. (1995) L'exploitation de la basse vallée du Sénégal par les anatidés paléarctiques. Une actualisation des données. *Alauda* 63: 15–24.

Triplet, P. & Yésou, P. (1999) La Spatule blanche *Platalea leucorodia* hivernant dans le delta du Fleuve Sénégal. *Malimbus* 21: 77–81.

Trolliet, B., Fouquet, M., Triplet, P., Girard, O. & Yésou, P. (1995) A propos de l'hivernage de la barge à queue noire *Limosa limosa* dans le delta du Sénégal. *Alauda* 63: 246–247.

Yésou, P. & Triplet, P. (1995) The Common Gull *Larus canus* in Senegal. *Malimbus* 17: 26–27.

Yésou, P. & Triplet, P. (1995) La Mouette atricille *Larus atricilla* au Sénégal. *Alauda* 63: 335.

Sierra Leone

Allport, G., Ausden, M., Hayman, P. V., Robertson, P. & Wood, P. (1989) *The Conservation of the Birds of Gola Forest, Sierra Leone*. Study Report No. 38. International Council for Bird Preservation, Cambridge.

Bannerman, D. A. (1921) A systematic list of the birds of Sierra Leone. *Ibis* (11)3: 283–302.

Field, G. D. (1973a) Ortolan and Blue Rock Thrush in Sierra Leone. *Bull. Brit. Orn. Club* 93: 81–82.

Field, G. D. (1973b) Subalpine and Grasshopper Warblers in Sierra Leone. *Bull. Brit. Orn. Club* 93: 101–103.

Field, G. D. (1974a) *Birds of the Freetown Peninsula*. Fourah Bay College Bookshop, Freetown.

Field, G. D. (1974b) Nearctic waders in Sierra Leone. Lesser Golden Plover and Buff-breasted Sandpiper. *Bull. Brit. Orn. Club* 94: 76–78.

Field, G. D. (1974c) The distribution and behaviour of *Apalis* warblers in Sierra Leone. *Ostrich* 45: 258–260.

Field, G. D. (1975) The Yellow-billed Egret *Egretta intermedia* in Sierra Leone. *Bull. Niger. Orn. Soc.* 11: 53–55.

Field, G. D. (1978) Status of Ciconiiformes in Sierra Leone. *Bull. Niger. Orn. Soc.* 14: 42–46.

Field, G. D. (1979) The *Laniarius* bushshrikes in Sierra Leone. *Bull. Brit. Orn. Club* 99: 42–44.

Field, G. D. & Owen, D. F. (1969) Little Gull in Sierra Leone. *Bull. Brit. Orn. Club* 89: 94.

Happel, R. E. (1985) Birds of Outamba area, Northwest Sierra Leone. *Malimbus* 7: 101–102.

Harding, D. P. & Harding, R. S. O. (1982) A preliminary checklist of the birds of the Kilimi Area of Northwest Sierra Leone. *Malimbus* 4: 64–68.

Harkrider, J. R. (1993) Garden and farm-bush birds of Njala, Sierra Leone. *Malimbus* 15: 38–46.

Harrop, J. H. (1961) African Serpent Eagle in Sierra Leone. *Bull. Brit. Orn. Club* 81: 52.

Iles, D. (1993) White-backed Night Heron *Nycticorax leuconotus* in Sierra Leone. *Malimbus* 15: 47–48.

Serle, W. (1948–1949) Notes on the birds of Sierra Leone. *Ostrich* 19: 129–141, 187–199; 20: 70–85, 114–126.

Tye, A. (1985) Preuss's Cliff Swallow *Hirundo preussi* breeding in Sierra Leone. *Malimbus* 7: 95–96.

Togo

Cheke, R. A. (1986) The supposed occurrence of the White-necked Picathartes *Picathartes gymnocephalus* in Togo. *Bull. Brit. Orn. Club* 106: 152.

Cheke, R. A. & Walsh, J. F. (1996) *The Birds of Togo*. Check-list No. 14, British Ornithologists' Union, Tring.

Demey, R. (1998) A wet season record of Cut-throat Finch *Amadina fasciata* from Togo. *Malimbus* 20: 125–126.

3. FAMILY AND SPECIES REFERENCES

Alexander-Marrack, P. (1994) Notes on a breeding colony of the African River Martin *Pseudochelidon eurystomina* in Gabon. *Malimbus* 16: 1–9.

Allen, D. G. (1997) Field identification of African *Accipiter* species and similar-looking hawks. *Bull. Afr. Bird Club* 4: 74–82.

Allport, G. (1991) The status and conservation of threatened birds in the Upper Guinea forest. *Bird Conserv. Internatn.* 1: 53–74.

Allport, G. A., Ausden, M. J., Fishpool, L. D. C., Hayman, P. V., Robertson, P. & Wood, P. (1996) Identification of Illadopsises *Illadopsis* spp. in the Upper Guinea forest. *Bull. Afr. Bird Club* 3: 26–30.

Allport, G. A. & Fanshawe, J. R. (1994) Is the Thick-billed Cuckoo *Pachycoccyx audeberti* a forest dependent species in West Africa? *Malimbus* 16: 52–53.

Alström, P. (1989) Identification of marsh terns in juvenile and winter plumages. *Brit. Birds* 82: 296–319.

Alström, P. (1990) Calls of American and Pacific Golden Plovers. *Brit. Birds* 83: 70–72.

Andrews, S. M. (1994) Rediscovery of the Monteiro's Bush-shrike *Malaconotus monteiri* in Cameroon. *Bull. Afr. Bird Club* 1: 26–27.

Atkinson, P. W., Koroma, A. P., Ranft, R., Rowe, S. G. & Wilkinson, R. (1994) The status, identification and vocalisations of African fishing owls with particular reference to the Rufous Fishing Owl *Scotopelia ussheri*. *Bull. Afr. Bird Club* 1: 67–72.

Barthel, P. H. (1993) Bemerkungen zur bestimmung von Nachtigall *Luscinia megarhynchos* und Sprosser *L. luscinia*. *Limicola* 7: 57–76.

Becker, P. (1995) Identification of Water Rail and *Porzana* crakes in Europe. *Dutch Birding* 17: 181–211.

Beintema, A. J. & Drost, N. (1986) Migration of the Black-tailed Godwit. *Gerfaut* 76: 37–62.

Bennun, L. A. (1985) Notes on the behaviour and plumage dimorphism in Lagden's Bush Shrike *Malaconotus lagdeni*. *Scopus* 9: 111–114.

Beresford, P. & Cracraft, J. (1999) Speciation in African forest robins (*Stiphrornis*): species limits, phylogenetic relationships, and molecular biogeography. *Am. Mus. Novitates* 3270: 1–21.

Berthold, P. (1988) The biology of the genus *Sylvia*—a model and a challenge for Afro–European co-operation. *Tauraco* 1: 3–28.

Bird, D. (1994) The field characters of distant Great and Cory's Shearwaters. *Birding World* 7: 279–282.

Bourne, W. R. P. (1983) The Soft-plumaged Petrel, the Gon-gon and the Freira, *Pterodroma mollis*, *P. feae* and *P. madeira*. *Bull. Brit. Orn. Club* 103: 52–58.

Bourne, W. R. P., Mackrill & E. J., Paterson, A. M. & Yésou, P. (1988) The Yelkouan Shearwater *Puffinus* (*puffinus?*) *yelkouan*. *Brit. Birds* 81: 306–319.

Bowden, C. G. R., Hayman, P. V., Martins, R. P., Robertson, P. A., Mudd, S. H. & Woodcock, M. W. (1995) The *Melignomon* honeyguides: a review of recent range extensions and some remarks on their identification, with a description of the song of Zenker's Honeyguide. *Bull. Afr. Bird Club* 2: 32–38.

Bradshaw, C. (1993) Separating juvenile Little and Baillon's Crakes in the field. *Brit. Birds* 86: 303–311.

Bretagnolle, V. (1995) Systematics of the Soft-plumaged Petrel *Pterodroma mollis* (Procellariidae): new insight from the study of vocalizations. *Ibis* 137: 207–218.

Brosset, A. (1978) Social organization and nest-building in the forest weaver birds of the genus *Malimbus* (Ploceinae). *Ibis* 120: 27–37.

Brouwer, J. & Mullié, W.C. (2000) The Barbary Falcon *Falco pelegrinoides* in the Sahel. *Alauda* 68: 158–161.

Brown, L. (1972) *African Birds of Prey* (2nd edn.). Collins, London.

Brown, L. H. (1974) The races of the European Snake Eagle *Circaetus gallicus*. *Bull. Brit. Orn. Club* 94: 126–128.

Brown, L. & Amadon, D. (1968) *Eagles, Hawks and Falcons of the World*. 2 vols. Hamlyn, Middlesex.

Brunel, J., Chappuis, C. & Erard, C. (1980) Data on *Lagonosticta rhodopareia bruneli*. *Bull. Brit. Orn. Club* 100: 164–170.

Burton, J. A. (ed.) (1984) *Owls of the World: Their Evolution, Structure and Ecology* (New edn.). Peter Lowe.

Chandler, R. J. (1989) *The Macmillan Field Guide to North Atlantic Shorebirds*. Macmillan, London.

Chantler, P. & Driessens, G. (1995) *Swifts: A Guide to the Swifts and Treeswifts of the World*. Pica Press, Mountfield.

Chappuis, C. (1976) Note sur le genre *Bathmocercus* Reichenow, 2me partie: Discrimination acoustique de *B. rufus* et *B. cerviniventris*. *Rev. Zool. Afr.* 90: 1028–1031.

Chappuis, C. (1980) Study and analysis of certain vocalizations as an aid in classifying African Sylviidae. *Proc. IV Pan-Afr. Orn. Congr.* 57–63.

Chappuis, C. & Erard, C. (1973) A new race of Pectoral-patch Cisticola from Cameroon. *Bull. Brit. Orn. Club* 93: 143–144.

Chappuis, C. & Erard, C. (1991) A new cisticola from west-central Africa. *Bull. Brit. Orn. Club* 111: 59–70.

Chappuis, C. & Erard, C. (1993) Species limits in the genus *Bleda* Bonaparte, 1857 (Aves, Pycnonotidae). *Z. zool. Syst. Evolut.-forsch.* 31: 280–299.

Chappuis, C., Erard, C. & Morel, G. J. (1979) Données comparatives sur la morphologie et les vocalisations des diverses formes d'*Eupodotis ruficrista* (Smith). *Malimbus* 1: 74–89.

Chappuis, C., Erard, C. & Morel, G. J. (1989) Type specimens of *Prinia subflava* (Gmelin) and *Prinia fluviatilis* Chappuis. *Bull. Brit. Orn. Club* 109: 108–110.

Chappuis, C., Erard, C. & Morel, G. J. (1992) Morphology, habitat, vocalisations and distribution of the River Prinia *Prinia fluviatilis* Chappuis. *Proc. VII Pan-Afr. Orn. Congr.* 481–488.

Cheke, R. A. (1999) Vocalisations of the Mouse-brown Sunbird *Anthreptes gabonicus*. *Malimbus* 21: 51.

Christie, D. A., Shirihai, H. & Harris, A. (1996) Field identification of Little and Baillon's Crakes. *Brit. Birds* 89: 54–59.

Clancey, P. A. (1984) On the so-called Mountain Pipit of the Afrotropics. *Durban Mus. Novitates* 13: 189–194.

Clancey, P. A. (1984) The Long-billed Pipit *Anthus similis* in equatorial West Africa. *Durban Mus. Novitates* 14: 226–227.

Clancey, P. A. (1985) Subspeciation in *Anthus brachyurus* Sundevall 1850. *Bull. Brit. Orn. Club* 105: 133–135.

Clancey, P. A. (1985) Species limits in the Long-billed Pipits of the southern Afrotropics. *Ostrich* 56: 157–169.

Clancey, P. A. (1986) On the status of *Anthus richardi bannermani* Bates, 1930. *Durban Mus. Novitates* 14: 19–23.

Clancey, P. A. (1987) Long-billed Pipit systematics. *Ostrich* 58: 45–46.

Clancey, P. A. (1990) A review of the indigenous pipits (genus *Anthus* Bechstein: Motacillidae) of the Afrotropics. *Durban Mus. Novitates* 15: 42–72.

Clark, W. S. (1992) The taxonomy of Steppe and Tawny Eagles, with criteria for separation of museum specimens and live eagles. *Bull. Brit. Orn. Club* 112: 150–157.

Clark, W. S. (1996) Ageing Steppe Eagles. *Birding World* 9: 268–274.

Clark, W. S. (1997) Identification of perched Montagu's and Pallid Harriers. *Birding World* 10: 267–269.

Clark, W. S. (1999) Plumage differences and taxonomic status of three similar *Circaetus* snake-eagles. *Bull. Brit. Orn. Club* 119: 56–59.

Clark, W. S. (1999) *A Field Guide to the Raptors of Europe, The Middle East and North Africa.* Oxford University Press, Oxford.

Clark, W. S. (2000) Field identification of Beaudouin's Snake Eagle *Circaetus (gallicus) beaudouini. Bull. Afr. Bird Club* 7: 13–17.

Clark, W. S. & Shirihai, H. (1995) Identification of Barbary Falcon. *Birding World* 8: 336–343.

Cleere, N. (1995) The identification, taxonomy and distribution of the Mountain Nightjar *Caprimulgus poliocephalus/* Fiery-necked Nightjar *C. pectoralis* complex. *Bull. Afr. Bird Club* 2: 86–97.

Cleere, N. & Nurney, D. (1998) *Nightjars: A Guide to Nightjars and Related Nightbirds.* Pica Press, Mountfield.

Clement, P. (1987) Field identification of West Palearctic wheatears. *Brit. Birds* 80: 135–157, 187–238.

Clement, P. (1995) Identification pitfalls and assessment problems. 17. Woodchat Shrike. *Brit. Birds* 88: 291–295.

Clement, P. (1999) The African *Zoothera* thrushes—identification, distribution and some problems with classification. *Bull. Afr. Bird Club* 6: 17–24.

Clement, P., Harris, A. & Davis, J. (1993) *Finches and Sparrows: An Identification Guide.* Christopher Helm, London.

Colebrook-Robjent, J. F. R. & Griffith, J. E. (1996) Forbes's Plover *Charadrius forbesi* breeding in Central Africa. *Bull. Brit. Orn. Club* 116: 244–246.

Colston, P. R. & Morel, G. J. (1984) A new subspecies of the African Reed Warbler *Acrocephalus baeticatus* from Senegal. *Bull. Brit. Orn. Club* 104: 3–5.

Colston, P. R. & Morel, G. J. (1985) A new subspecies of the Rufous Swamp Warbler *Acrocephalus rufescens* from Senegal. *Malimbus* 7: 61–62.

Corso, A. (1997) Variability of identification characters of Isabelline Wheatear. *Dutch Birding* 19: 153–165.

Corso, A. (2000) Identification of European Lanner. *Birding World* 13: 200–213.

Craig, A. J. F. K. (1992) The identification of *Euplectes* species in non-breeding plumage. *Bull. Brit. Orn. Club* 112: 102–108.

Craig, A. J. F. K. (1993) Geographical variation and taxonomy of the genus *Euplectes* (Aves, Ploceidae). Parts 1 & 2. *J. Afr. Zool.* 107: 83–96,139–151.

Craig, A. J. F. K. & Hartley, A. H. (1985) The arrangement and structure of feather melanin granules as a taxonomic character in African starlings (Sturnidae). *Auk* 102: 629–630.

Craig, A. J. F. K. & Villet, M. H. (1998) Sexual dimorphism and tail-length in widowbirds and bishopbirds (Ploceidae: *Euplectes* spp.): a reassessment. *Ibis* 140: 137–143.

Crook, J. H. (1960) Nest form and construction in certain West African weaver-birds. *Ibis* 102: 1–25.

Crook, J. H. (1963) A comparative analysis of nest structure in the weaver birds (Ploceinae). *Ibis* 105: 238–262.

Crowe, T. M., Harley, E. H., Jakutowicz, M. B., Komen, J. & Crowe, A. A. (1992) Phylogenetic, taxonomic and biogeographical implications of genetic, morphological and behavioral variation in francolins (Phasianidae: *Francolinus*). *Auk* 109: 24–42.

Cruickshank, A. J., Gautier, J-P. & Chappuis, C. (1993) Vocal mimicry in wild African Grey Parrots *Psittacus erithacus*. *Ibis* 135: 293–299.

Curtis, W. F., Lassey, P. A. & Wallace, D. I. M. (1985) Identifying the smaller shearwaters. *Brit. Birds* 78: 123–138.

Dean, W. R. J. (1989) A review of the genera *Calandrella, Spizocorys* & *Eremalauda* (Alaudidae). *Bull. Brit. Orn. Club* 109: 95–100.

Delacour, J. & Edmond-Blanc, F. (1933) Monographie des veuves (révision des genres *Euplectes* and *Vidua*). *Oiseau* 3: 519–562.

Demey, R. & Fishpool, L. D. C. (1998) On the existence of a melanistic morph of the Long-tailed Hawk *Urotriorchis macrourus*. *Bull. Brit. Orn. Club* 118: 105–108.

Dickerman, R. W. (1989) Notes on the Malachite Kingfisher *Corythornis (Alcedo) cristata*. *Bull. Brit. Orn. Club* 109: 158–159.

Dickerman, R. W. (1994) Notes on birds from Africa with descriptions of three new subspecies. *Bull. Brit. Orn. Club* 114: 274–278.

Dickerman, R. W., Cane, W. P., Carter, M. F., Chapman, A. & Schmitt, C. G. (1994) Report on three collections of birds from Liberia. *Bull. Brit. Orn. Club* 114: 267–274.

Dowsett, R. J. (1971) The Lesser Grey Shrike in Africa. *Ostrich* 42: 259–270.

Dowsett, R. J. (1972) Geographical variation in *Pseudhirundo griseopyga*. *Bull. Brit. Orn. Club* 92: 97–100.

Dowsett, R. J. (1980) The migration of coastal waders from the Palaearctic across Africa. *Gerfaut* 70: 3–35.

Dowsett, R. J. (1989) The nomenclature of some African barbets of the genus *Tricholaema*. *Bull. Brit. Orn. Club* 109: 180–181.

Dowsett, R. J. & Dowsett-Lemaire, F. (1986) Long-billed Pipit systematics. *Ostrich* 57: 115.

Dowsett, R. J., Olson, S. L., Roy, M. S. & Dowsett-Lemaire, F. (1999) Systematic status of the Black-collared Bulbul *Neolestes torquatus*. *Ibis* 141: 22–28.

Dowsett, R.J. & Stjernstedt, R. (1979) The *Bradypterus cinnamomeus–mariae* complex in Central Africa. *Bull. Brit. Orn. Club* 99: 86–94.

Dowsett-Lemaire, F. (1990) Eco-ethology, distribution and status of Nyungwe Forest birds (Rwanda). Pp. 31–85 in

Dowsett, R. J. (ed.) *Enquête Faunistique et Floristique dans la Forêt de Nyungwe, Rwanda.* Tauraco Research Report No. 3. Tauraco Press, Liège.

Dowsett-Lemaire, F. (1992) On the vocal behaviour and habitat of the Maned Owl *Jubula lettii* in south-western Congo. *Bull. Brit. Orn. Club* 112: 213–218.

Dowsett-Lemaire, F. (1994) The song of the Seychelles Warbler *Acrocephalus seychellensis* and its African relatives. *Ibis* 136: 489–491.

Dowsett-Lemaire, F. (1996a) A comment on the voice and status of Vermiculated Fishing-Owl *Scotopelia bouvieri* and a correction to Dowsett-Lemaire (1992) on the Maned Owl *Jubula lettii. Bull. Afr. Bird Club* 3: 134–135.

Dowsett-Lemaire, F. (1996b) Observations of two *Cuculus* species fed by forest hosts in the Congo. *Malimbus* 18: 153–154.

Dowsett-Lemaire, F. (1997) Avian frugivore assemblages at three small-fruited tree species in the forests of northern Congo. *Ostrich* 67: 94–95.

Dowsett-Lemaire, F. (1999a) First observations on the territorial song and display of the Kupe Bush Shrike *Malaconotus kupeensis. Malimbus* 21: 115–117.

Dowsett-Lemaire, F. (1999b) Hybridization in paradise flycatchers (*Terpsiphone rufiventer, T. batesi* and *T. viridis*) in Odzala National Park, Northern Congo. *Ostrich* 70: 123–126.

Dowsett-Lemaire, F. & Dowsett, R. J. (1987a) European and African Reed Warblers, *Acrocephalus scirpaceus* and *A. baeticatus*: vocal and other evidence for a single species. *Bull. Brit. Orn. Club* 107: 74–85.

Dowsett-Lemaire, F. & Dowsett, R. J. (1987b) European Reed and Marsh Warblers in Africa: migration patterns, moult and habitat. *Ostrich* 58: 65–85.

Dowsett-Lemaire, F. & Dowsett, R. J. (1988) Vocalisations of the green turacos (*Tauraco* species) and their systematic status. *Tauraco* 1: 64–71.

Dowsett-Lemaire, F. & Dowsett, R. J. (1989) Zoogeography and taxonomic relationships of forest birds of the Cameroon Afromontane region. Pp. 48–56 in Dowsett, R. J. (ed.) *A Preliminary Natural History Survey of Mambilla Plateau and some Lowland Forests of Eastern Nigeria.* Tauraco Research Report No. 1. Tauraco Press, Ely.

Dowsett-Lemaire, F. & Dowsett, R. J. (1998) Vocal and other pecularities of Brown Nightjar *Caprimulgus binotatus. Bull. Afr. Bird Club* 5: 35–38.

Dubois, P. J. (1991) Identification forum: Royal, Lesser Crested and Elegant Terns. *Birding World* 4: 120–123.

Dubois, P. J. & Yésou, P. (1995) Identification of Western Reef Egrets and dark Little Egrets. *Brit. Birds* 88: 307–319.

Earlé, R. A. (1987) Notes on *Hirundo fuliginosa* and its status as a "cliff swallow". *Bull. Brit. Orn. Club* 107: 59–63.

Elgood, J. H. (1958) A new species of *Malimbus. Ibis* 100: 621–624.

Elgood, J. H. (1975) *Malimbus ibadanensis*: a fresh statement of biology and status. *Bull. Brit. Orn. Club* 95: 78–80.

Elgood, J. H. (1988) Rediscovery of *Malimbus ibadanensis* Elgood, 1958. *Bull. Brit. Orn. Club* 108: 184–185.

Elgood, J. H. (1982) The case for the retention of *Anaplectes* as a separate genus. *Bull. Brit. Orn. Club* 102: 70–75.

Elgood, J. H. (1992) The range of *Malimbus ibadanensis. Bull. Brit. Orn. Club* 112: 205–207.

Enticott, J. W. (1991) Identification of Soft-plumaged Petrel. *Brit. Birds* 84: 245–264.

Enticott, J. & Tipling, D. (1997) *A Photographic Handbook of the Seabirds of the World.* New Holland, London.

Erard, C. (1987–1990) Écologie et comportement des gobe-mouches (Aves: Muscicapinae, Platysteirinae, Monarchinae) du Nord-Est du Gabon. Vols. 1 & 2. *Mém. Mus. natn. Hist. nat. Série A, Zoologie* 138: 1–256; 146: 1–234.

Erard, C. (1989) *Phylloscopus (=Seicercus) laurae* au Gabon? *Oiseau & R.F.O.* 59: 86–88.

Erard, C. (1991) Variation géographique de *Bleda canicapilla* (Hartlaub) 1854 (Aves, Pycnonotidae). Description d'une sous-espèce nouvelle en Sénégambie. *Oiseau & R.F.O.* 61: 66–67.

Erard, C. (1992) *Bleda canicapilla morelorum,* émendation du nom subspécifique d'un bulbul récemment décrit du Sénégal. *Oiseau & R.F.O.* 62: 288.

Erard, C. & Chappuis, C. (1991) Sur le *Phylloscopus (=Seicercus)* mentionné au Gabon et sur son appartenance probable à une espèce non encore décrite. *Oiseau & R.F.O.* 61: 155–161.

Erard, C., Guillou, J. & Maynaud, N. (1986) Le Héron blanc du Banc d'Arguin *Ardea monicae.* Ses affinités morphologiques, son histoire. *Alauda* 54: 161–169.

Erard, C. & Morel, G. J. (1994) La sous-espèce du Cochevis modeste *Galerida modesta* en Sénégambie. *Malimbus* 16: 56–57.

Erard, C. & Roche, J. (1977) Un nouveau *Lagonosticta* du Tchad méridional. *Oiseau & R.F.O.* 47: 335–343.

Erard, C. & Roux, F. (1983) La Chevêchette du Cap *Glaucidium capense* dans l'ouest Africain. Description d'une race géographique nouvelle. *Oiseau & R.F.O.* 53: 97–104.

Erasmus, R. P. B. (1992) Notes on the call of the Grass Owl *Tyto capensis. Ostrich* 63: 184–185.

Feare, C. & Craig, A. (1998) *Starlings and Mynas.* Christopher Helm, London.

Field, G. D. (1979) A new species of *Malimbus* sighted in Sierra Leone and a review of the genus. *Malimbus* 1: 2–13.

Fishpool, L. D. C. (1999) Little-known African bird: Baumann's Greenbul *Phyllastrephus baumanni. Bull. Afr. Bird Club* 6: 137.

Fishpool, L. D. C. (2000) A review of the status, distribution and habitat of Baumann's Greenbul *Phyllastrephus baumanni. Bull. Brit. Orn. Club* 120: 213–229.

Fishpool, L. D. C. (in prep.) A reassessment of the systematic position of *Nicator* (Hartlaub and Finsch 1870)—fused toes resolve confused affinities?

Fishpool, L. D. C., Demey, R., Allport, G. & Hayman, P. V. (1994) Notes on the field identification of the bulbuls (Pycnonotidae) of Upper Guinea. Parts 1 & 2. *Bull. Afr. Bird Club* 1: 32–38, 90–95.

Forshaw, J. M. (1989) *The Parrots of the World* (3rd edn.). Landsdowne Editions, Melbourne.

Forsman, D. (1999) *The Raptors of Europe and The Middle East. A Handbook of Field Identification.* T. & A. D. Poyser, London.

Fotso, R. C. (1993) Contribution à l'étude de la biologie du Picatharte chauve du Cameroun *Picathartes oreas. Proc. VIII Pan-Afr. Orn. Congr.* 431–437.

Friedmann, H. (1948) *The Parasitic Cuckoos of Africa.* Monograph 1. Washington Academy of Sciences, Washington D.C.

Friedmann, H. (1960) *The Parasitic Weaverbirds.* Bulletin 223. Smithsonian Institution, US National Museum, Washington D.C.

Fry, C. H. (1982) Spanish Black Kites in West Africa. *Malimbus* 4: 48.

Fry, C. H. (1984) *The Bee-eaters.* T. & A. D. Poyser, Calton.

Fry, C. H. (1988) Skull, songs and systematics of African nightjars. *Proc. VI Pan-Afr. Orn. Congr.* 105–132.

Fry, C. H., Fry, K. & Harris, A. (1992) *Kingfishers, Bee-eaters & Rollers.* Christopher Helm, London.

Fry, C. H., Keith, S & Urban, E. K. (1985) Evolutionary expositions from *The Birds of Africa: Halcyon* song phylogeny; cuckoo host partitioning; systematics of *Aplopelia* and *Bostrychia. Proc. Intern. Symp. Afr. Vert. (Bonn 1984):* 163–180.

Fry, C. H. & Naurois, R. de (1984) *Corythornis* systematics and character release in the Gulf of Guinea islands. *Proc. V Pan-Afr. Orn. Congr.* 47–62

Furness, R. W. (1987) *The Skuas.* T. & A. D. Poyser, Calton.

Gantlett, S. (1995) Identification forum: field separation of Fea's, Zino's and Soft-plumaged Petrels. *Birding World* 8: 256–260.

Gantlett, S. & Harrap, S. (1992) Identification forum: South Polar Skua. *Birding World* 5: 256–270.

Gantlett, S. J. M. & Harris, A. (1987) Identification of large terns. *Brit. Birds* 80: 257–276.

Gargallo, G. (1994) On the taxonomy of the western Mediterranean island populations of Subalpine Warbler *Sylvia cantillans. Bull. Brit. Orn. Club* 114: 31–36.

Gatter, W. (1985) Ein neuer Bulbul aus Westafrika (Aves, Pycnonotidae). *J. Orn.* 126: 155–161.

Gatter, W. (1993) The status of the Black Swift *Apus barbatus* in western West Africa. *Malimbus* 15: 90–91.

Gatter, W. & Gardner, R (1993) The biology of the Gola Malimbe *Malimbus ballmanni* Wolters 1974. *Bird Conserv. Internatn.* 3: 87–103.

Gatter, W., Peal, A., Steiner, C. & Weick, F. (1988) Die unbekannten Jugendkleider des seltenen Weissbrüstperlhuhns (*Agelastes meleagrides* Bonaparte, 1850). *Okol. Voel.* 10: 105–111.

Gensbøl, B. (1984) *Collins Guide to the Birds of Prey of Britain and Europe, North Africa and the Middle East.* Collins, London.

Golley, M. & Stoddart, A. (1991) Identification of American and Pacific Golden Plovers. *Birding World* 4: 195–204.

Goodwin, D. (1982) *Estrildid Finches of the World.* British Museum (Natural History), London & Oxford University Press, Oxford.

Goodwin, D. (1983) *Pigeons and Doves of the World.* British Museum (Natural History), London.

Goodwin, D. (1986) *Crows of the World* (2nd edn.). British Museum (Natural History), London.

Grant, P. J. (1986) *Gulls: A Guide to Identification* (2nd edn.). T. & A. D. Poyser, Calton.

Grimes, L. G. (1974) Duetting in *Hypergerus atriceps* and its taxonomic relationship to *Eminia lepida. Bull. Brit. Orn. Club* 93: 89–96.

Grimes, L. G. (1975) The dawn song of the Grey-backed Eremomela *Eremomela pusilla. Bull. Brit. Orn. Club* 95: 92–93.

Grimes, L. G. (1976a) The vocalizations of the Green Longtail *Urolais epichlora. Bull. Brit. Orn. Club* 96: 99–101.

Grimes, L. G. (1976b) The duets of *Laniarius atroflavus, Cisticola discolor* and *Bradypterus barratti. Bull. Brit. Orn. Club* 96: 113–120.

Grimes, L. G. (1979) Sexual dimorphism in the Yellow-billed Shrike *Corvinella corvina* and other African shrikes (subfamily Laniidae). *Bull. Brit. Orn. Club* 99: 33–36.

Guillou, J. J. & Pages, J. J. (1987) Le Flamant nain *Phoeniconaias minor* pénètre à l'intérieur des terres de l'Afrique de l'Ouest. *Alauda* 55: 233–234.

Hall, B. P. (1963) The francolins, a study in speciation. *Bull. Brit. Mus. Nat. Hist. Zool.* 10(2): 105–204.

Hall, B. P., Moreau, R. E. & Galbraith, I. C. J. (1966) Polymorphism and parallelism in the African bush-shrikes of the genus *Malaconotus* (including *Chlorophoneus*). *Ibis* 108: 161–182.

Hancock, J. & Kushlan, J. (1984) *The Herons Handbook.* Croom Helm, London & Sydney.

Hancock, J. A., Kushlan, J. A. & Kahl, M. P. (1992) *Storks, Ibises and Spoonbills of the World.* Academic Press, London.

Harrap, S. (1996) The vocalisations of African black tits (*Parus niger* complex). *Bull. Afr. Bird Club* 3: 99–104.

Harrap, S. & Quinn, D. (1996) *Tits, Nuthatches & Treecreepers.* Christopher Helm, London.

Harris, T. & Arnott, G. (1988) *Shrikes of Southern Africa.* Struik Winchester, Cape Town.

Harris, T. & Franklin, K. (2000) *Shrikes & Bush-shrikes.* Christopher Helm, London.

Harrison, P. (1983) *Seabirds: An Identification Guide.* Croom Helm, Beckenham.

Harrison, P. (1987) *Seabirds of the World: A Photographic Guide.* Christopher Helm, London.

Hayman, P., Marchant, J. & Prater, T. (1986) *Shorebirds: An Identification Guide to the Waders of the World.* Croom Helm, London & Sydney.

Helbig, A. J. (1991) Identification of juvenile Verreaux's Eagle. *Brit. Birds* 84: 287–289.

Helbig, A. J., Martens, J., Siebold, I. Henning, F. Schottler, B. & Wink, M. (1996) Phylogeny and species limits in the Palaearctic chiffchaff *Phylloscopus collybita* complex: mitochondrial genetic differentiation and bioacoustic evidence. *Ibis* 138: 650–666.

Herroelen, P. (1987) Bill and leg colour in Forbes' Plover. *Malimbus* 9: 57–58.

Herroelen, P., Louette, M. & Adams, M. (1999) A reassessment of the subspecies in the owl *Glaucidium tephronotum*, with notes on its biology. *Bull. Brit. Orn. Club* 119: 151–162.

Hume, R. A. (1993) Common, Arctic and Roseate Terns: an identification review. *Brit. Birds* 86: 210–217.

Hume, R. & Boyer, T. (1991) *Owls of the World*. Dragon's World, Limpsfield.

Imber, M. J. (1985) Origins, phylogeny and taxonomy of the gadfly petrels *Pterodroma* spp. *Ibis* 127: 197–229.

Irwin, M. P. S. (1987) What are the affinities of the Black-capped Apalis *Apalis nigriceps*? *Malimbus* 9: 130–131.

Irwin, M. P. S. (1999) The genus *Nectarinia* and the evolution and diversification of sunbirds: an Afrotropical perspective. *Honeyguide* 45: 45–58.

Jiguet, F. (1997) Identification of South Polar Skua: the Brown Skua pitfall. *Birding World* 10: 306–310.

Jiguet, F. (2000a) The two giant petrels. *Birding World* 13: 108–115.

Jiguet, F. (2000b) Identification and ageing of Black-browed Albatross at sea. *Brit. Birds* 93: 263–276.

Johnsgard, P. A. (1983) *Cranes of the World*. Croom Helm, London & Canberra.

Johnsgard, P. A. (1988) *The Quails, Partridges, and Francolins of the World*. Oxford University Press, Oxford.

Johnsgard, P. A. (1991) *Bustards, Hemipodes, and Sandgrouse: Birds of Dry Places*. Oxford University Press, Oxford.

Johnson, D. F. & Horner, R. F. (1986) Identifying widows, bishops and queleas in female plumage. *Bokmakierie* 38: 13–17.

Johnson, D. N. (1989) The feeding habits of some Afrotropical forest bulbuls. *Ostrich Suppl.* 14: 49–56.

Juniper, T. & Parr, M. (1998) *Parrots. A Guide to the Parrots of the World*. Pica Press, Mountfield.

Kemp, A. C. (1986) The Gabar Goshawk: taxonomy, ecology and further research. *Gabar* 1: 4–6.

Kemp, A. (1995) *The Hornbills*. Oxford University Press, Oxford.

Kemp, A. & Calburn, S. (1987) *The Owls of Southern Africa*. Struik Winchester, Cape Town.

Kemp, A. & Kemp, M. (1998) *Sasol Birds of Prey of Africa and its Islands*. New Holland, London.

King, J. (1996) Identification of nightingales. *Birding World* 9: 179–189.

King, J. & Shirihai, H. (1996) Identification and ageing of Audouin's Gull. *Birding World* 9: 52–61.

Kirkham, I. R. & Nisbet, I. C. T. (1987) Feeding techniques and field identification of Arctic, Common and Roseate Terns. *Brit. Birds* 80: 41–47.

König, C., Weick, F. & Becking, J.-H. (1999) *Owls. A Guide to the Owls of the World*. Pica Press, Mountfield.

Lambert, K. (1967) Beobachtungen zum Zug und Winterquartier der Schwalbenmöwe (*Xema sabini*) im östlichen Atlantik. *Vogelwarte* 24: 99–106.

Lambert, K. (1969) Sommerbeobachtungen einjäriger Swalbenmöwen (*Xema sabini*) im tropischen Atlantik. *J. Orn.* 110: 219.

Lawson, W. J. (1984) The West African mainland forest dwelling population of *Batis*; a new species. *Bull. Brit. Orn. Club* 104: 144–146.

Lawson, W. J. (1986) Speciation in the forest-dwelling populations of the avian genus *Batis*. *Durban Mus. Novitates* 13 (21): 285–304.

Lawson, W. J. (1987) Systematics and evolution in the savanna species of the genus *Batis* (Aves) in Africa. *Bonn. Zool. Beitr.* 38: 19–45.

Lontkowski, J. (1995) Die Unterscheidung von Korn- *Circus cyaneus*, Wiesen- *C. pygargus* und Steppenweihe *C. macrourus*. *Limicola* 9: 233–275.

Louette, M. (1976) Notes on the genus *Bathmocercus* Reichenow (Aves, Sylviidae). Part 1. The different plumages of *B. cerviniventris* and its relationship to *B. rufus*. *Rev. Zool. Afr.* 90: 1021–1027.

Louette, M. (1980) The populations of *Nectarinia preussi* in the Cameroon montane area. *Proc. IV Pan-Afr. Orn. Congr.* 9–16.

Louette, M. (1982) Allopatric species of birds approaching in western Cameroon: the *Nectarinia adelberti, N. rubescens* example. *Bonn. Zool. Beitr.* 33: 303–312.

Louette, M. (1990) A new species of nightjar from Zaïre. *Ibis* 132: 349–353.

Louette, M. (1992) The identification of forest *Accipiters* in central Africa. *Bull. Brit. Orn. Club* 112: 50–53.

Louette, M. (2000) Evolutionary exposition from plumage pattern in African *Accipiter*. *Ostrich* 71: 45–50.

Louette, M. & Benson, C. W. (1982) Swamp-dwelling weavers of the *Ploceus velatus/vitellinus* complex, with the description of a new species. *Bull. Brit. Orn. Club* 102: 24–31.

Louette, M. & Herroelen, P. (1994) A revised key for *Cercococcyx* cuckoos, taxonomic status of *C. montanus patulus* and its occurrence in Zaïre. *Bull. Brit. Orn. Club* 114: 144–149.

Lynes, H. (1925) On the birds of North and Central Darfur, with notes on the West-Central Kordofan and North Nuba Provinces of British Sudan. Part V. *Ibis* (12) 1: 541–590.

Lynes, H. (1930) Review of the genus *Cisticola*. *Ibis* (12) 6, Suppl.: 1–673.

Madge, S. & Burn, H. (1988) *Wildfowl: An Identification Guide to the Ducks, Geese and Swans of the World*. Christopher Helm, London.

Madge, S. & Burn, H. (1994) *Crows and Jays: A Guide to the Crows, Jays and Magpies of the World*. Christopher Helm, London.

Malling Olsen, K. (1989) Head pattern of Brown Noddy. *Dutch Birding* 11: 126–127.

Malling Olsen, K. & Larsson, H. (1994) *Terns of Europe and North America*. Christopher Helm, London.

Malling Olsen, K. & Larsson, H. (1997) *Skuas and Jaegers. A Guide to the Skuas and Jaegers of the World*. Pica Press, Mountfield.

Mann, C. F., Burton, P. J. K. & Lennerstedt, I. (1978) A re-appraisal of the systematic position of *Trichastoma poliothorax* (Timaliinae, Muscicapidae). *Bull. Brit. Orn. Club* 98: 131–140.

Meyburg, B. U., Meyburg, C. & Barbraud, J.-C. (1998) Migration strategies of an adult Short-toed Eagle *Circaetus gallicus* tracked by satellite. *Alauda* 66: 39–48.

Meyburg, B.-U., Eichacker, X., Meyburg, C. & Paillat, P. (1995) Migration of an adult Spotted Eagle tracked by satellite. *Brit. Birds* 88: 357–361.

Mild, K. (1993) Die Bestimmung der europäischen schwarzweißen Fliegenschnapper *Ficedula*. *Limicola* 7: 222–276.

Mild, K. (1994) Field identification of Pied, Collared and Semi-collared Flycatchers. Part 3: first-winters and non-breeding adults. *Birding World* 7: 325–334.

Moreau, R. E. (1957) Variation in the western Zosteropidae (Aves). *Bull. Br. Mus. nat. Hist. Zool.* 4(7): 309–433.

Moreau, R. E. (1958) The *Malimbus* spp. as an evolutionary problem. *Rev. Zool. Bot. Afr.* 57: 243–255.

Moreau, R. E. (1960) Conspectus and classification of the Ploceine weavers. *Ibis* 102: 298-321, 443–471.

Morel, G. J. & Chappuis, C. (1992) Past and future taxonomic research in West Africa. *Bull. Brit. Orn. Club* 112A: 217–224.

Morel, G. J. & Morel, M.-Y. (1979) La Tourterelle des bois dans l'extrême Ouest-Africain. *Malimbus* 1: 66–67.

Morel, G. J. & Morel, M.-Y. (1992) Habitat use by Palaearctic migrant passerine birds in West Africa. *Ibis* 134 Supplement 1: 83–88.

Morel, M.-Y. (1987) La Tourterelle des bois dans l'Ouest Africain: mouvements migratoires et régime alimentaire. *Malimbus* 9: 23–42.

Morel, M.-Y. (1988) Successful establishment of the House Sparrow *Passer domesticus* in Senegambia. *Proc. VI Pan-Afr. Orn. Congr.* 159–162.

Morlion, M. L. (1980) Pterylosis as a secondary criterion in the taxonomy of the African Ploceidae and Estrildidae. *Proc. IV Pan-Afr. Orn. Congr.* 27–41.

Mudd, H. & Martins, R. (1996) Possible display behaviour of White-necked Picathartes. *Bull. Brit. Orn. Club* 116: 15–17.

Mullarney, K. (1988) Identification of Roseate Tern in juvenile plumage. *Dutch Birding* 10: 109–120.

Mullarney, K. (1988) Identification of adult Roseate Tern. *Dutch Birding* 10: 136–137.

Mullarney, K. (1988) Identification of a Roseate x Common Tern hybrid. *Dutch Birding* 10: 133–135.

Mullié, W. C., Brouwer, J. & Albert, C. (1992) Gregarious behaviour of African Swallow-tailed Kite *Chelictinia riocourii* in response to high grasshopper densities near Ouallam, western Niger. *Malimbus* 14: 19–21.

Mundy, P., Butchart, D., Ledger, J. & Piper, S. (1992) *The Vultures of Africa*. Academic Press, London.

de Naurois, R. (1975) The Grey Heron of the Banc d'Arguin (Mauritania) *Ardea cinerea monicae*. *Bull. Brit. Orn. Club* 95: 135–140.

de Naurois, R. & Wolters, H. E. (1975) The affinities of the São Tomé Weaver *Textor grandis* (Gray, 1844). *Bull. Brit. Orn. Club* 95: 122–126.

Ndao, B. (1989) Au Sénégal, un Gonolek de Barbarie (*Laniarius barbarus*) à dessous jaune apparié à un sujet normal. *Malimbus* 11: 97–98.

Ndao, B. (1999a) Observations de Gonolek de Barbarie *Laniarius barbarus* à dessous jaune au Sénégal. *Bull. Afr. Bird Club* 6: 60.

Ndao, B. (1999b) Le Pique-boeufs à bec jaune *Buphagus africanus* buveur occasionnel de lait de vache. *Bull. Afr. Bird Club* 6: 59.

Newell, D., Porter, R. & Marr, T. (1997) South Polar Skua—an overlooked bird in the eastern Atlantic. *Birding World* 10: 229–235.

Newton, I. (1995) Relationship between breeding and wintering ranges in Palaearctic–African migrants. *Ibis* 137: 241–249.

Nicolai, J. (1982) Comportement, voix et parenté de l'Amarante du Mali (*Lagonosticta virata*). *Malimbus* 4: 9–14.

Ogilvie, M. & Ogilvie, C. (1986) *Flamingos*. Alan Sutton, Gloucester.

Oreel, G. J. (1980) On field identification of Short-toed Larks. *Dutch Birding* 2: 115.

Osborne, P., Collar, N. & Goriup, P. (1984) *Bustards*. Dubai Wildlife Research Centre, Dubai.

Page, D. (1999) Identification of Bonelli's Warblers. *Brit. Birds* 92: 524–531.

Park, P. O. (1975) Mixed nesting colonies of *Quelea quelea* and *Quelea erythrops* in the Lake Chad basin. *Bull. Nigerian Orn. Soc.* 11 (40): 74–76.

Parmenter, T. & Byers, C (1991) *A Guide to the Warblers of the Western Palaearctic*. Bruce Coleman, Uxbridge.

Parrott, J. (1979) Kaffir Rail (*Rallus caerulescens*) in West Africa. *Malimbus* 1: 145–146.

Payne, R. B. (1970) Temporal patterns of duetting in the Barbary Shrike *Laniarius barbarus*. *Ibis* 112: 106–108.

Payne, R. B. (1982) Species limits in the indigobirds (Ploceidae, Vidua) of West Africa: mouth mimicry, song mimicry, and the description of new species. *Misc. Publ. Mus. Zool. Univ. Michigan* 162: 1–96.

Payne, R. B. (1985) The species of parasitic finches in West Africa. *Malimbus* 7: 103–113.

Payne, R. B. (1991) Female and first-year male plumages of paradise whydahs *Vidua interjecta*. *Bull. Brit. Orn. Club* 111: 95–100.

Payne, R. B. (1996) Field identification of the Indigobirds. *Bull. Afr. Bird Club* 3: 14–25.

Payne, R. B. (1997) Field identification of the brood-parasitic whydahs *Vidua* and Cuckoo Finch *Anamalospiza imberbis*. *Bull. Afr. Bird Club* 4: 18–28.

Payne, R. B. (1998) A new species of firefinch *Lagonosticta* from northern Nigeria and its association with the Jos Plateau Indigobird *Vidua maryae*. *Ibis* 140: 368–381.

Payne, R. B. & Louette, M. (1983) What is *Lagonosticta umbrinodorsalis* Reichenow, 1910? *Mitt. Zool. Mus. Berlin* 59, *Ann. Orn.* 7: 157–161.

Payne, R. B. & Payne, L. L. (1994) Song mimicry and species associations of west African indigobirds *Vidua* with Quail-finch *Ortygospiza atricollis*, Goldbreast *Amandava subflava* and Brown Twinspot *Clytospiza monteiri*. *Ibis* 136: 291–304.

Pearson, D. J. (2000) The races of the Isabelline Shrike *Lanius isabellinus* and their nomenclature. *Bull. Brit. Orn. Club* 120: 22–27.

Porter, R., Newell, D. Marr, T. & Jolliffe, R. (1997) Identification of Cape Verde Shearwater. *Birding World* 10: 222–228.

Porter, R. F., Willis, I., Christensen, S. & Nielsen, B. P. (1981) *Flight Identification of European Raptors* (3rd edn.). T. & A. D. Poyser, Calton.

Prigogine, A. (1974) Note sur deux gladiateurs (*Malaconotus*). *Gerfaut* 74: 75–81.

Prigogine, A. (1981) A new species of *Malimbus* from Sierra Leone? *Malimbus* 3: 55.

Prigogine, A. (1986) Le plumage immature du Gladiateur de Lagden, *Malaconotus lagdeni*, et du Gladiateur ensanglanté, *Malaconotus cruentus*. *Gerfaut* 76: 255–261.

Ratcliffe, N., Monteiro, L. R. & Hazevoet, C. J. (1999) Status of Raso Lark *Alauda razae* with notes on threats and foraging behaviour. *Bird Conserv. Internatn* 9: 43–46.

Restall, R. (1996) *Munias and Mannikins*. Pica Press, Mountfield.

Riddiford, N. (1991) A field character for identification of Collared Flycatcher in female and non-breeding plumages. *Brit. Birds* 84: 19–23.

Round, P. D. & Walsh, T. A. (1981) The field identification of Dunn's Lark. *Sandgrouse* 3: 78–83.

Rowan, M. K. (1983) *The Doves, Parrots, Louries and Cuckoos of Southern Africa*. David Philip, Cape Town & Johannesburg.

Salewski, V. (1997) The immature plumage of Sun Lark *Galerida modesta*. *Bull. Afr. Bird Club* 4: 136.

Salewski, V. & Grafe, T. U. (1999) New tape recordings of three West African birds. *Malimbus* 21: 117–121.

Serle, W. (1952) Colour variation in *Malaconotus cruentus* (Lesson). *Bull. Brit. Orn. Club* 72: 27–28.

Serle, W. (1963) A new race of sunbird from West Africa. *Bull. Brit. Orn. Club* 83: 118–119.

Sharland, R. E. (1981) Black-headed and Grey-headed Gulls in West Africa. *Malimbus* 3: 54.

Shirihai, H. (1994) Separation of Tawny Eagle from Steppe Eagle in Israel. *Brit. Birds* 87: 396.

Shirihai, H. (1994) Field identification of Dunn's, Bar-tailed Desert and Desert Larks. *Dutch Birding* 16: 1–9.

Shirihai, H. (1988) Iris colour of *Sylvia* warblers. *Brit. Birds* 81: 325–328.

Shirihai, H., Christie, D. A. & Harris, A. (1996) Identification of *Hippolais* warblers. *Brit. Birds* 89: 114–138.

Shirihai, H., Harris, A. & Cottridge, D. (1991) Identification of Spectacled Warbler. *Brit. Birds* 84: 423–430.

Shirihai, H., Roselaar, C. S., Helbig, A. J., Barthel, P. H. & van Loon, A. J. (1995) Identification and taxonomy of large *Acrocephalus* warblers. *Dutch Birding* 17: 229–239.

Short, L. (1982) *Woodpeckers of the World*. Delaware Museum of Natural History, Monograph 4. Greenville, Delaware.

Short, L. L. (1982) On the status of *Lybius* (*minor*) *macclounii*. *Bull. Brit. Orn. Club* 102: 142–148.

Short, L. L. & Horne, J. F. M. (1983) A review of duetting, sociality and speciation in some African barbets (Capitonidae). *Condor* 85: 323–332.

Short, L. L. & Horne, J. F. M. (1985) Social behavior and systematics of African barbets (Aves: Capitonidae). *Proc. Intern. Symp. Afr. Vert. (Bonn 1984)*: 255–278.

Simms, E. (1985) *British Warblers*. Collins, London.

Simms, E. (1992) *British Larks, Pipits and Wagtails*. HarperCollins, London.

Skead, C. J. (1967) *The Sunbirds of Southern Africa*. Balkema, Cape Town.

Skilleter, M. (1995) Winter site fidelity of Redstart *Phoenicurus phoenicurus* in Nigeria. *Malimbus* 17: 101–102.

Smith, T. B. (1987) Bill size polymorphism and intraspecific niche utilization in an African finch. *Nature* 329: 717–719.

Smith, T. B. (1990) Resource use by bill morphs of an African finch: evidence for intraspecific competition. *Ecology* 71: 1246–1257.

Smith, T. B. (1990) Natural selection on bill characters in the two bill morphs of the African finch *Pyrenestes ostrinus*. *Evolution* 44: 832–842.

Smith, T. B. (1990) Comparative breeding biology of the two bill morphs of the Black-bellied Seedcracker. *Auk* 107: 153–160.

Steyn, P. (1973) *Eagle Days. A Study of African Eagles at the Nest*. Sable Publishers, Sandton.

Steyn, P. (1982) *Birds of Prey of Southern Africa. Their Identification and Life Histories*. David Philip, Cape Town & Johannesburg.

Steyn, P. (1984) *A Delight of Owls. African Owls Observed*. David Philip, Cape Town.

Stuart, S. N. & Gartshore, M. E. (1986) The Red-capped Robin-chat *Cossypha natalensis* in West Africa. *Malimbus* 8: 73–76.

Stuart, S. N. & Jensen, F. P. (1986) The status and ecology of montane forest bird species in western Cameroon. Pp. 38–105 in Stuart, S. N. (ed.) *Conservation of Cameroon Montane Forests*. International Council for Bird Preservation, Cambridge.

Summers-Smith, J. D. (1988) *The Sparrows*. T. & A. D. Poyser, Calton.

Svensson, L. (1988) Field identification of black-headed Yellow Wagtails. *Brit. Birds* 81: 77–78.

Taylor, B. & Van Perlo, B. (1998) *Rails. A Guide to the Rails, Crakes, Gallinules and Coots of the World*. Pica Press, Mountfield.

Thiollay, J.-M. (1974) Nidification du Martinet pâle *Apus pallidus* et du Martinet alpin *Apus melba* en Afrique occidentale. *Alauda* 47: 223–225

Thompson, H. S. (1993) Status of the White-necked Picathartes—another reason for the conservation of the Peninsula Forest, Sierra Leone. *Oryx* 27: 155–158.

Thompson, H. S. & Fotso, R. (1995) Rockfowl. The genus *Picathartes*. *Bull. Afr. Bird Club* 2: 25–28.

Thompson, H. S. & Fotso, R. (2000) Conservation of two threatened species: *Picathartes*. *Ostrich* 71: 154–156.

Tove, M. (1994) *Pterodroma* identification. *Birding World* 7: 286–287.

Tove, M. (2001) Verification of suspected field identification differences in Fea's and Zino's Petrels. *Birding World* 14: 283-288.

Tréca, B. & Erard, C. (2000) A new subspecies of the Red-billed Hornbill, *Tockus erythrorhynchus*, from West Africa. *Ostrich* 71: 363–366.

Turner, A. (1989) *A Handbook to the Swallows and Martins of the World*. Christopher Helm, London.

Tye, A. (1988) Vocalizations and territorial behaviour by wheatears *Oenanthe* spp. in winter quarters. *Proc. VI Pan-Afr. Orn. Congr.* 297–305.

Tye, A. (1991) A new subspecies of Forest Scrub-Robin *Cercotrichas leucosticta* from West Africa. *Malimbus* 13: 74–77.

Tye, A. & Macaulay, L. R. (1993) The races of Olive Sunbird *Nectarinia olivacea* on the Gulf of Guinea islands. *Malimbus* 14: 65–66.

Urban, E. K. (1988) Status of cranes in Africa. *Proc. VI Pan-Afr. Orn. Congr.* 315–329.

Ullman, M. (1991) Distinguishing Demoiselle Crane from Crane. *Dutch Birding* 13: 90–93.

Van de Kam, J., Ens, B., Piersma, T. & Zwarts, L. (1999) *Ecologische Atlas van de Nederlandse Wadvogels*. Schuyt & Co., Haarlem.

Van den Berg, M. & Oreel, G. J. (1985) Field identification and status of black-headed Yellow Wagtails in Western Europe. *Brit. Birds* 78: 176–183.

Van den Bossche, W. & Jadoul, G. (1998) Ageing of White Stork and Black Stork. *Birding World* 11: 195–199.

Van den Elzen, R. (1985) Systematics and evolution of African canaries and seedeaters (Aves: Carduelidae). *Proc. Intern. Symp. Afr. Vert. (Bonn 1984):* 435–451.

Vaurie, C. (1949) A revision of the bird family Dicruridae. *Bull. Am. Mus. Nat. Hist.* 93: 199–342.

Vernon, C. J. & Dean, W. R. J. (1975) On the systematic position of *Pholidornis rushiae*. *Bull. Brit. Orn. Club* 95: 20.

Veron, G. & Winney, B. J. (2000) Phylogenetic relationships within the turacos (Musophagidae). *Ibis* 142: 446–456.

Voisin, C. (1983) Les Ardéidés du fleuve Sénégal. *Oiseau & R.F.O.* 53: 335–369.

Walsh, J. F. (1977) Nesting of the Jabiru Stork *Ephippiorrhynchus senegalensis* in West Africa. *Bull. Brit. Orn. Club* 97: 136.

Walsh, J. F. (1987) Inland records of Western Reef Heron *Egretta gularis*. *Malimbus* 9: 58.

Walsh, J. F. (1991) On the occurrence of the Black Stork *Ciconia nigra* in West Africa. *Bull. Brit. Orn. Club* 111: 209–215.

Waltert, M. & Faber, K. (2000) Olive-bellied Sunbird *Nectarinia chloropygia* host to Cassin's Honeybird *Prodotiscus insignis*. *Malimbus* 22: 86.

Ward, P. (1966) Distribution, systematics, and polymorphism of the African weaver-bird *Quelea quelea*. *Ibis* 108: 34–40.

Wilkinson, R. (1984) Variation of eye-colour of Blue-eared Glossy Starlings. *Malimbus* 6: 2–4.

Williams, E. (1998) Green-breasted Bush-shrike *Malaconotus gladiator* and its relationship with Monteiro's Bush-shrike *M. monteiri*. *Bull. Afr. Bird Club* 5: 101–104.

Wilson, E. O. (1986) West African thrushes as safari ant followers. *Gerfaut* 76: 95–108.

Wilson, R. T. (1993) Distribution and ecology of the African Corvidae. *Proc. VIII Pan-Afr. Orn. Congr.* 371–378.

Winkler, H., Christie, D. A. & Nurney, D. (1995) *Woodpeckers. A Guide to the Woodpeckers, Piculets and Wrynecks of the World*. Pica Press, Mountfield.

Wolters, H. E. (1974) Ein neuer *Malimbus* (Ploceidae, Aves) von der Elfenbeinküste. *Bonn. Zool. Beitr.* 25: 290–291.

Wolters, H. E. (1977) Die Gattungen der Nectariniidae (Aves, Passeriformes). *Bonn. Zool. Beitr.* 28: 82–101.

Wood, B. (1975) Observations on the Adamawa Turtle Dove. *Bull. Brit. Orn. Club* 95: 68–73.

Yésou, P. & Paterson, A. M. (1999) Puffin yelkouan et Puffin des Baléares: une ou deux espèces? *Ornithos* 6: 20–31.

Yésou, P., Paterson, A. M., Mackrill, E. J. & Bourne, W. R. P. (1990) Plumage variation and identification of the 'Yelkouan Shearwater'. *Brit. Birds* 83: 299–319.

Zino, F., Oliveira, P., King, S., Buckle, A., Biscoito, M., Neves, H.C. & Vasconcelos, A. (2001) Conservation of Zino's petrel *Pterodroma madeira* in the archipelago of Madeira. *Oryx* 35: 128-136.

Zonfrillo, B. (1994) The soft-plumaged petrel group. *Birding World* 7: 71–72.

4. ACOUSTIC REFERENCE

Chappuis, C. (2000) *African Bird Sounds: Birds of North, West and Central Africa and Neighbouring Atlantic Islands*. 15 CDs. Société d'Études Ornithologiques de France, Paris.

SCIENTIFIC INDEX

Scientific names are presented as follows:
(1) Family name, in CAPITALS (e.g. DICRURIDAE)
(2) Generic name, in **bold** (e.g. *Dicrurus*)
(3) Specific name followed by generic name (e.g. *adsimilis, Dicrurus*)
(4) Subspecific name followed by both specific and generic names (e.g. *apivorus, Dicrurus adsimilis*)
Alternative scientific names and other names mentioned in the species accounts are shown within parentheses.
Numbers in roman type refer to the page of the relevant main species account; those in **bold** refer to the colour
plate numbers.

FRENCH INDEX

Les chiffres en caractères romains renvoient à la page du texte relatif à l'espèce; ceux en caractères **gras** aux numéros des planches en couleurs. Les synonymes et noms d'autres espèces mentionnés dans les textes relatifs aux espèces principales apparaissent entre parenthèses.

ENGLISH INDEX

Numbers in roman type refer to the page of the relevant main species account; those in **bold** refer to the colour plate numbers. Alternative names and other names mentioned in the species accounts are shown within parentheses.

DATE DUE

DEMCO 38-296

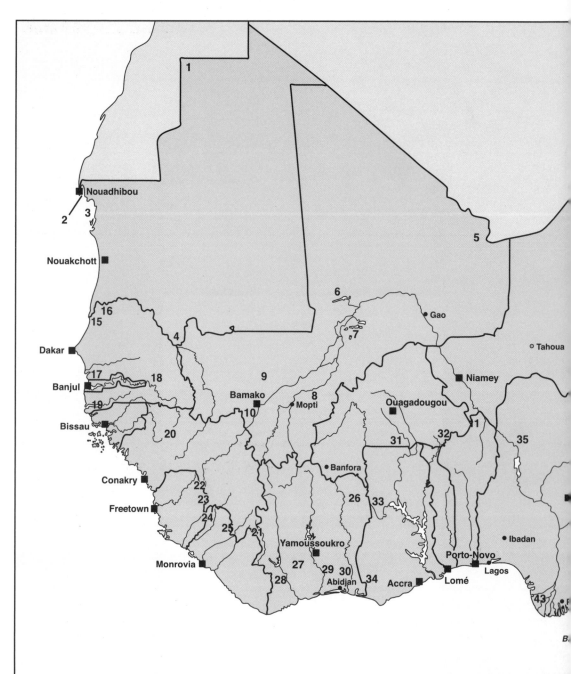

Nouadhibou

2 3

Nouakchott

16

15

Dakar

17   18

Banjul

19

Bissau

20

Conakry

22

23

Freetown

24

25   21

Monrovia

28

1

5

6

Gao

7

9

Bamako

10

8

Mopti

Ouagadougou

31

32

11

35

Banfora

26   33

Ibadan

Yamoussoukro

27   29 30   34

Abidjan

Accra

Porto-Novo

Lomé

Lagos

43

Niamey

Tahoua

PRÍNCIPE

SÃO TOMÉ

ANNOBÓN